SPECTROMETRIC IDENTIFICATION OF ORGANIC COMPOUNDS

SPECTROMETRIC IDENTIFICATION OF ORGANIC COMPOUNDS

FOURTH EDITION

Robert M. Silverstein
SUNY College of Environmental Science and Forestry

G. Clayton Bassler
Los Altos, California

Terence C. Morrill
Rochester Institute of Technology

JOHN WILEY & SONS
New York • Chichester • Brisbane • Toronto • Singapore

Library of Congress Cataloging in Publication Data

Silverstein, Robert Milton, 1916-
Spectrometric identification of organic compounds.

Includes bibliographical references and index.
1. Spectrum analysis, 2. Chemistry, Organic.
I. Bassler, G. Clayton, joint author, II. Morrill, Terence C., joint author. III. Title.
QD272.S6S55 1980 547.3'0858 80-20548
ISBN 0-471-02990-4

Printed in the United States of America

10 9 8 7

PREFACE

The major impetus for the preparation of the fourth edition was the rapid development of ^{13}C NMR spectrometry, which was treated in a few paragraphs in the third edition but which now constitutes a full chapter. Many of the "problem sets" now consist of 5 spectra – the ^{13}C spectrum having been added to the mass, infrared, proton NMR, and UV spectra of the earlier editions.

Other chapters have been revised, resulting in some expansion and, we hope, in some clarification. The revision has also prompted reorganization of reference material.

We include one paragraph from the preface to the third edition:

"In the first edition (1963) we referred to "methods now available and used in research but *not systematically taught* in universities. . . ." The concept behind the first edition was that organic structures could be elucidated from the complementary information available from mass, infrared, NMR, and UV spectrometry. Since that time, several excellent books have appeared, organized similarly to this text, and brief treatments of the four areas of spectrometry are a part of almost every elementary organic textbook. In the first and second editions we stated: "In one form or another, such material [i.e., spectrometric interpretation] should soon become part of the training of every organic chemist." We are now constrained to change "should soon" to "has." One of us (T.C.M.) has a unique perspective. He can look back over the past decade to his first contact with spectrometry in a graduate course at San Jose State College in 1962. The instructors were R.M. Silverstein and G.C. Bassler, whose lecture notes were the basis for the first edition of this text."

Several of the revisions are results of stimulating and enjoyable discussions with our colleagues, Dr. Robert R. LaLonde, Dr. Conrad Schuerch, and Dr. Alan Harvey (SUNY College of Environmental Science and Forestry), and Dr. Craig Van Antwerp (Rochester Institute of Technology). The comments of several reviewers were most helpful.

For those whose primary concerns are teaching and research, the writing or revising of a textbook is largely an extracurricular activity, and the burden of compensating for such preoccupation devolves on the author's wife. Our wives have our deepest appreciation.

R.M. Silverstein
T.C. Morrill

PREFACE TO FIRST EDITION

During the past several years, we have been engaged in isolating small amounts of organic compounds from complex mixtures and identifying these compounds spectrometrically.

At the suggestion of Dr. A. J. Castro of San Jose State College, we developed a one unit course entitled "Spectrometric Identification of Organic Compounds," and presented it to a class of graduate students and industrial chemists during the 1962 spring semester. This book has evolved largely from the material gathered for the course and bears the same title as the course.

We should first like to acknowledge the financial support we received from two sources: The Perkin-Elmer Corporation and Stanford Research Institute.

A large debt of gratitude is owed to our colleagues at Stanford Research Institute. We have taken advantage of the generosity of too many of them to list them individually, but we should like to thank Dr. S. A. Fuqua, in particular, for many helpful discussions of NMR spectrometry. We wish to acknowledge also the cooperation at the management level, of Dr. C. M. Himel, chairman of the Organic Research Department, and Dr. D. M. Coulson, chairman of the Analytical Research Department.

Varian Associates contributed the time and talents of its NMR Applications Laboratory. We are indebted to Mr. N. S. Bhacca, Mr. L. F. Johnson, and Dr. J. N. Shoolery for the NMR spectra and for their generous help with points of interpretation.

The invitation to teach at San Jose State College was extended by Dr. Bert M. Morris, head of the Department of Chemistry, who kindly arranged the administrative details.

The bulk of the manuscript was read by DR. R. H. Eastman of the Stanford University whose comments were most helpful and are deeply appreciated.

Finally, we want to thank our wives. As a test of a wife's patience, there are few things to compare with an author in the throes of composition. Our wives not only endured, they also encouraged, assisted, and inspired.

R. M. Silverstein
G. C. Bassler

Menlo Park, California
April 1963

CONTENTS

SPECTROMETRIC IDENTIFICATION OF ORGANIC COMPOUNDS

one

introduction

Our purpose in writing this book is to teach the organic chemist how to identify organic compounds from the complementary information afforded by four spectra: mass, infrared, nuclear magnetic resonance, and ultraviolet. Essentially, the molecule in question is subjected to four energy probes, and the molecule's responses are recorded as spectra.

The small amounts of pure compounds that can be isolated from complex mixtures by chromatography present a challenge to the chemist concerned with identification and structure elucidation of organic compounds. These techniques have two characteristics: they are rapid, and they are most effective in milligram and microgram quantities. There is neither enough time nor enough material to accommodate the classical manipulations involving sodium fusion, boiling point, refractive index, solubility tests, functional group tests, derivative preparation, mixture melting point, combustion analysis, molecular weight, and degradation with similar manipulations of the degradation products.

Our goal in this book is a rather modest level of sophistication and expertise in each of the four areas of spectrometry. Even this level will permit solution of a gratifying number of identification problems with *no history and no other chemical or physical data.* Of course, in practice other information is usually available: the sample source, details of isolation, a synthesis sequence, or information on analogous material. Often complex molecules can be identified because partial structures are known, and specific questions can be formulated; the process is more confirmation than identification. In practice, however, difficulties arise in physical handling of minute amounts of compound: trapping, elution from adsorbents, solvent removal, prevention of contamination, and decomposition of unstable compounds. Water, air, stopcock greases, solvent impurities, and plasticizers have frustrated many investigations. The quality of spectra obtained in practice is usually inferior to that presented here.

For pedagogical reasons, we deal only with pure organic compounds. *Pure* in this context is a relative term, and all we can say is: the purer, the better. A good criterion of purity (for a sufficiently volatile compound) is gas chromatographic homogeneity on both polar and nonpolar substrate capillary columns. Various forms of liquid-phase chromatography (adsorption and liquid-liquid columns, paper, thin layer) are applicable to many compounds. The spectra presented in this book were obtained on purified samples.

In many cases, identification can be made on a fraction of a milligram, or even on several micrograms, of sample. Identification on the milligram scale is routine. Of course, not all molecules yield so easily. Chemical manipulations may be necessary but the information obtained from the four spectra will permit intelligent selection of chemical treatment, and the energy probe methodology can be applied to the resulting products.

There are limitations to the methodology we espouse. A mass spectrum is dependent on a degree of volatility and thermal stability. However, mass spectra have been obtained on many high molecular weight compounds—steroids, terpenoids, peptides, polysaccharides, and alkaloids—by obtaining positive ions by special techniques. Solubility is a limiting factor in nuclear magnetic resonance spectrometry. However, the availability of many deuterated solvents and the development of microsample tubes and Fourier transform NMR make it possible to obtain good spectra from many kinds of compounds.

The problem of cost of necessary instrumentation is raised and answered by pointing to the amazing evolution of commercial instruments. The time saved, the smaller samples required, and the information made available far outweigh the cost. Identifications are frequently made on the basis of several hours of a student's, technician's, or analyst's time. Under a classical regime, much larger samples and several days or even weeks of a skilled analyst's time would probably be necessary. Infrared and ultraviolet spectrometers have been developed beyond the stage of reliable instruments in the hands of a trained technician. They are now cheap, rugged, and simple enough to be used as bench tools by the organic chemist. Certain types of simple nuclear magnetic resonance spectrometers are now available to even the more modest institutions; these are essentially as easy to run as infrared instruments. More sophisticated magnetic resonance spectrometers, especially those with computer facets, require skilled technicians.

Almost from its inception, the utility of nuclear magnetic resonance spectrometry to the organic chemist has been evident. Mass spectrometry, however, has had a somewhat different history. Developed by the physicist and utilized extensively by the petroleum chemist, it has been ignored until recently by the organic chemist concerned with identification and structure determination. Mass spectrometers have become available to a wide range of institutions.

We spend very little time on instrumentation, per se, for three reasons: it is not requisite to our goal; we are not qualified; and excellent treatises on the subject are available. The five chapters on spectrometry are designed to give the analyst an appreciation of the potentialities of each technique as applied to the identification of organic compounds. The rest of the book consists of selected spectra. These mass, infrared, nuclear magnetic resonance (^{1}H and ^{13}C), and ultraviolet spectra are presented as sets, each set representing a compound. These are translated, as exercises, into the chemical structure (Chapter 7), identified by a Beilstein reference, or presented without identification (Chapter 8). Aside from practical applications, the considerations involved in translating spectra into organic compounds lead to an appreciation of modern concepts of structural organic chemistry.

If we have been judicious in our selection of spectra, they should serve as useful references for teachers and for chemists in industry. In one form or another such material has become part of the training of every organic chemist.

two

mass spectrometry

I. INTRODUCTION

A mass spectrometer bombards the substance under investigation with an electron beam and quantitatively records the result as a spectrum of positive ion fragments. This record is a mass spectrum. Separation of the positive ion fragments is on the basis of mass (strictly, mass/charge, but the majority of ions are singly charged). How this is accomplished will be sketched in sufficient detail to impart some appreciation for the potentialities of mass spectrometry as applied to compound identification to the organic chemist.

A number of good texts are available for the broad field of mass spectrometry;[1-36] these include general[1-15, 34-36] and specialized[30-33] monographs. Journals include *Organic Mass Spectrometry,* the *Journal of Mass Spectrometry and Ion Physics,* and *Biomedical Mass Spectrometry.* Compilations of data and spectra are available;[16-29] references include Appendix I (organized by compound class) of Hamming and Foster,[7] the *Atlas of Mass Spectral Data*[28], the CRC collection,[22] and the NIH-CIS collection.[27] Chapter 7 of Middleditch[11] contains an extensive listing of other compilations. Indexes to mass spectral information include the *ASTM* index,[16, 17] and the chemical compound index of the journal *Org. Mass Spec.*

Mass spectrometers are still characterized by high cost and the need for skilled technicians for operation and maintenance. Mass spectral data, however, are routinely reported in professional journals. In addition, mass spectral discussions are commonplace in organic chemistry textbooks and problem manuals for structure determination.

II. INSTRUMENTATION

In this section we describe the electron impact method for obtaining mass spectra.

Mass spectrometers for structure elucidation can be classified according to the method of separating the charged particles:

A. Magnetic Field Deflection (Direction Focusing)
   1. Magnetic field only (unit resolution)
   2. Double focusing (electrostatic field before magnetic field, high resolution)
B. Time of Flight
C. Quadrupole

The minimum instrumental requirement for the organic chemist is the ability to record the molecular weight of the compound under examination to the nearest whole number. Thus, the recording should show a peak at, say, mass 400, which is distinguishable from a peak at mass 399 or at mass 401. In order to select possible molecular formulas by measuring isotope peak intensities (see Section IV), adjacent peaks must be quite cleanly separated. Arbitrarily, a valley between two adjacent peaks should not be more than about 10% of the height of the larger peak. This latter degree of resolution is termed "unit" resolution and can be obtained up to about mass 500 on several single-focusing (magnetic-focusing) instruments.

To determine the resolution of an instrument, consider two adjacent peaks of approximately equal intensity. These

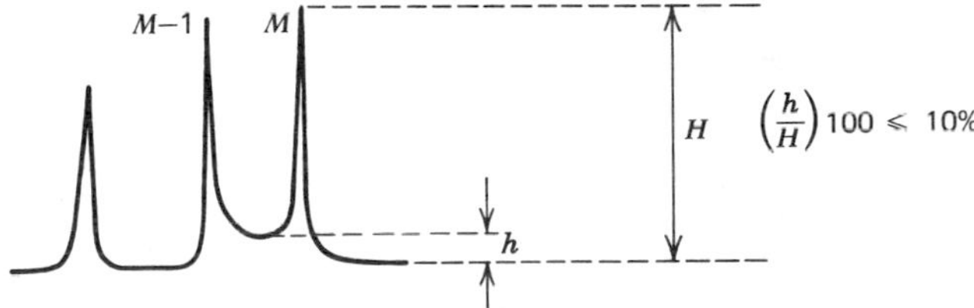

peaks should be chosen so that the height of the valley between the peaks is about 10% of the intensity of the peaks. The resolution (R) is

$$R = \frac{M}{\Delta M}$$

where M is the higher mass number of the two peaks, and ΔM is the difference between the two mass numbers.

There are two important categories of magnetic-deflection mass spectrometers: low resolution and high resolution. Low-resolution instruments can be defined arbitrarily as the instruments that separate unit masses up to m/e 2000 [$R = 2000/(2000-1999) = 2000$]. Unit mass (or low-resolution) spectra are obtained from these instruments. An instrument is generally considered high resolution if it can separate two ions differing in mass by at least one part in ten thousand to fifteen thousand (R = 10,000–15,000). An instrument with 10,000 resolution can separate an ion of mass 500.00 from one of mass 499.95 ($R = 500/0.05 = 10{,}000$). This important class of mass spectrometers can measure the mass of an ion with sufficient accuracy to determine its atomic composition. High-resolution mass spectrometry will be discussed only briefly, since the instruments are not available in most laboratories.

A schematic diagram of a typical 180° single-focusing mass spectrometer is shown in Figure 1. There are five component parts in this low-resolution system.

1. SAMPLE HANDLING SYSTEM. This consists of a device for introducing the sample, a micromanometer for determining the amount of sample introduced, a device (molecular leak) for metering the sample to the ionization chamber, and a pumping system. Introduction of gases is usually a simple matter of transfer from a gas bulb into the metering volume, thence to the ionizing chamber. Liquids are introduced by touching a micropipette to a sintered glass disc or an orifice under mercury or gallium, or simply by hypodermic needle injection through a serum cap. A bulb containing the sample may be pumped out under dry ice, then warmed to vaporize the sample into the inlet system. Heated inlet systems are used for less volatile liquids and for solids. Insertion of the sample directly into the ionization chamber further reduces the limitations imposed by lack of volatility and of thermal stability.

This direct-insertion technique decreases the degree of volatility necessary for analysis. Since ionization must take place in the vapor state, some degree of volatility and thermal stability is still required. Reproducible breakdown patterns have been obtained on high molecular weight terpenoids, steroids, polysaccharides, peptides, and alkaloids. Even with these special techniques, a compound must be stable at a temperature at which its vapor pressure is of the order of 10^{-7} to 10^{-6} Torr. For routine work, using a molecular leak, a vapor pressure of about 10^{-1} to 10^{-3} Torr is desired. Sample sizes for liquids and solids range from several milligrams to less than a microgram, depending on the method of introduction and the detector.

2. IONIZATION AND ACCELERATING CHAMBERS. The gas stream from the molecular leak enters the ionization chamber (operated at a pressure of about 10^{-6} to

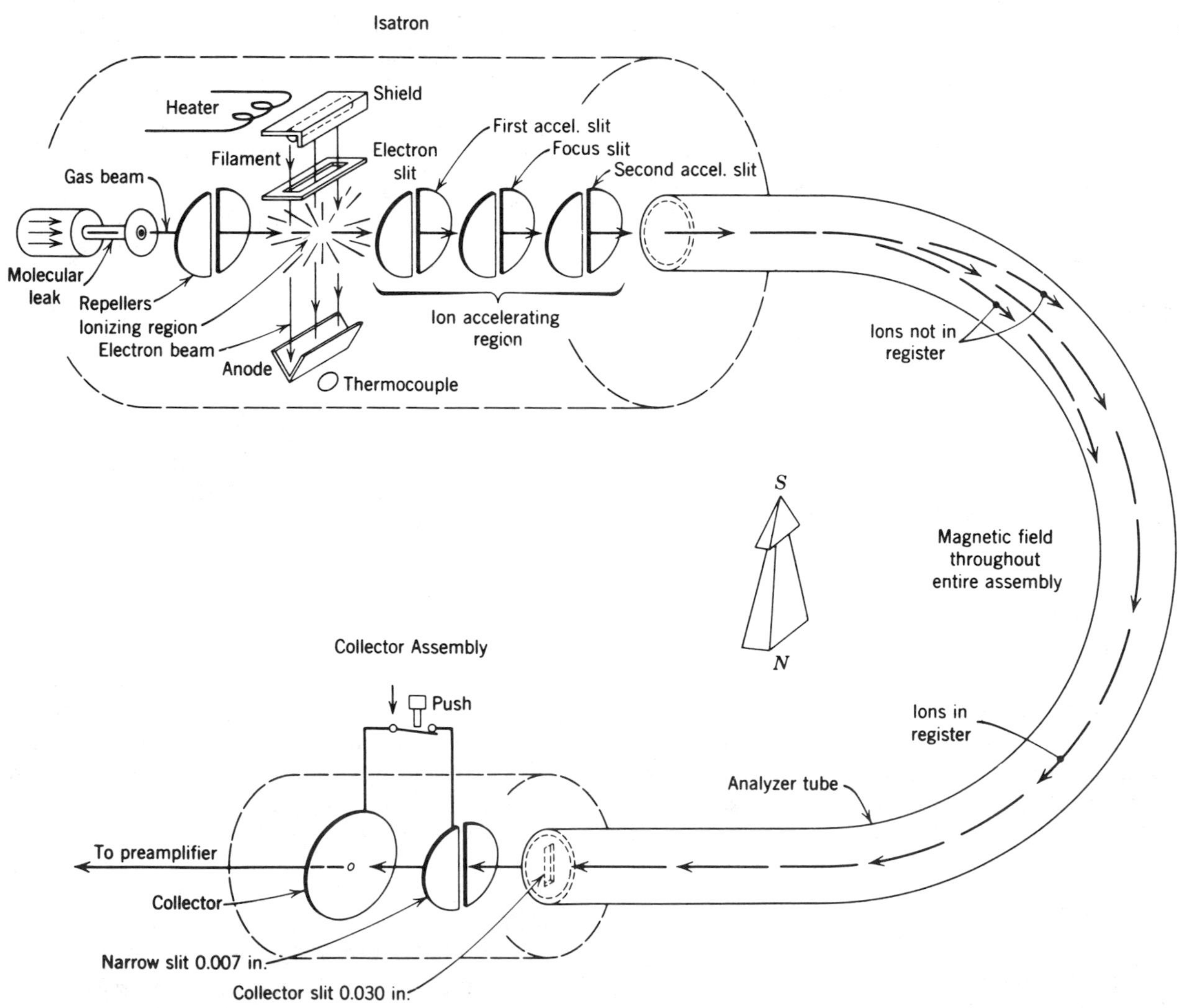

Figure 1. *Schematic diagram of CEC model 21-103 Mass Spectrometer, a single-focusing, 180° sector mass analyzer. The magnetic field is perpendicular to the page.*

10^{-5} Torr) in which it is bombarded at right angles by an electron beam emitted from a hot filament. Positive ions produced by interaction with the electron beam are forced through the first accelerating slit by a small electrostatic field between the repellers and the first accelerating slit. A strong electrostatic field between the first and second accelerating slits accelerates the ions to their final velocities. Additional focusing of the ion beam is provided between the accelerating slits. To obtain a spectrum, either the magnetic field applied to the analyzer tube (Figure 1) or the accelerating voltage between the first and second ion slits is varied. Thus, the ions are successively focused at the collector slit as a function of mass (strictly, mass/charge). In most instruments, a scan from mass 12 to mass 500 may be performed in 1 to 4 minutes. However, scan speeds of a few seconds have been used to obtain spectra of gas chromatography fractions (see below).

3. ANALYZER TUBE AND MAGNET. The analyzer tube is an evacuated (10^{-7} to 10^{-8} Torr), curved (in Figure 1, a 180° curve), metal tube through which the ion beam passes from ion source to collector. The magnetic pole pieces (electromagnets are usually used for the larger instruments) are mounted perpendicular to the plane of the diagram (Figure 1). The main requirement is a uniform, stable magnetic field.

4. ION COLLECTOR AND AMPLIFIER. A typical ion collector consists of one or more collimating slits and a Faraday cylinder; the ion beam impinges axially into the collector, and the signal is amplified by an electron multiplier.

5. RECORDER. A widely used recorder employs five separate galvanometers that record simultaneously. Figure 2a presents a spectrum traced by a five-element galvanometer system at sensitivity levels decreasing from top to bottom in the ratios of 1:3:10:30:100. Peak heights from the base line are read on the most sensitive trace remaining on scale and are multiplied by the appropriate sensitivity factor. Peak heights are proportional to the number of ions of each mass. The recorder trace can be presented as a table or as a graph (Figures 2*b* and 2*c*).

Throughout this book, and in most published work, a peak is reported as a value of mass divided by charge (m/e). Since we are usually dealing with singly charged ions (i.e., $e = 1$), such a value is the mass (m) of the ion corresponding to that peak. That multiple ionization does occur is indicated by the peaks at half-mass units (e.g., m/e 43.5) in the tracing; these represent odd-numbered masses that carry a double charge. A broad, weak, "metastable" peak (see below) can be seen between m/e 90 and 91 (Figure 2a).*

Assignment of mass to the peaks of the recorder tracing can be a problem at the high mass end of the scan. The common practice is to start at the low mass end of the scan, which can be accurately set, and count the peaks to the last recorded peak. This is often feasible because the most sensitive galvanometer will record a slight ion current at each mass unit. Sometimes the peaks at the high end of the spectrum may be widely spaced, and the background trace indistinct; in this case, a calibration compound may be added to the sample. Some instruments are equipped with automatic mass markers, but often these are not reliable where they are most needed—at the high mass end of the spectrum. A mass digitizer that prints out the mass number and relative intensities is a valuable auxiliary piece of equipment. However, a slight maladjustment may result in loss or gain of a full mass unit in a scan; this can be disastrous. It is well to check the digitizer printout against the galvanometer trace.

A mass analyzer using four electric poles (a "quadrupole") and no magnetic field has been developed (Figure 3). Ions entering from the top travel with constant velocity in the direction parallel to the poles (z direction), but acquire oscillations in the x and y directions. This is accomplished by application of both a dc voltage and a radio frequency voltage to the poles. There is a "stable oscillation" that allows an ion to pass from one end of the quadrupole to the other without striking the poles; this oscillation is dependent on the mass-to-charge ratio of an ion. Therefore, ions of only a single m/e value will traverse the entire length of the analyzer. All other ions will have unstable oscillations and will strike the poles. Mass scanning is carried out by varying each of the dc and radio frequencies while keeping their ratios constant. Resolutions as high as $R = 10{,}000$ have been achieved with this type of analyzer.

The introduction of an electrostatic field ahead of the magnetic field (double focusing) permits high resolution so that the mass of a particle can be obtained to three or four decimal places. Figure 4 shows an example of such a double-focusing instrument. A positive ion in an electric field experiences a force in the direction of the field; the path of an ion moving through the field is thus curved. In a radial electric field (always perpendicular to the direction of flight of the ions) the radius of curvature, r_e, of the ion path is dependent on the energy of the ion and the strength of the electric field. The electric field is an energy analyzer, instead of a mass analyzer, and serves to limit the energy spread of the ion beam before it enters the magnetic field. The energy spread is one of the major factors in decreasing the resolution of the simple magnetic deflection analyzers. Proper choice of the angle of deflection in the electric field also results in directional focusing of an ion beam. The Nier-Johnson double-focusing mass analyzer incorporates 90° electrostatic and 90° magnetic analyzers (Figure 4). The multimass ion beam is uniform in energy as it leaves the electrostatic analyzer, and the magnetic field produces the desired mass dispersion while focusing each unimass ion beam to the same point. Another common double-focusing arrangement, the Mattauch-Herzog mass analyzer,

*The symbol z has been recommended as a substitute for e.

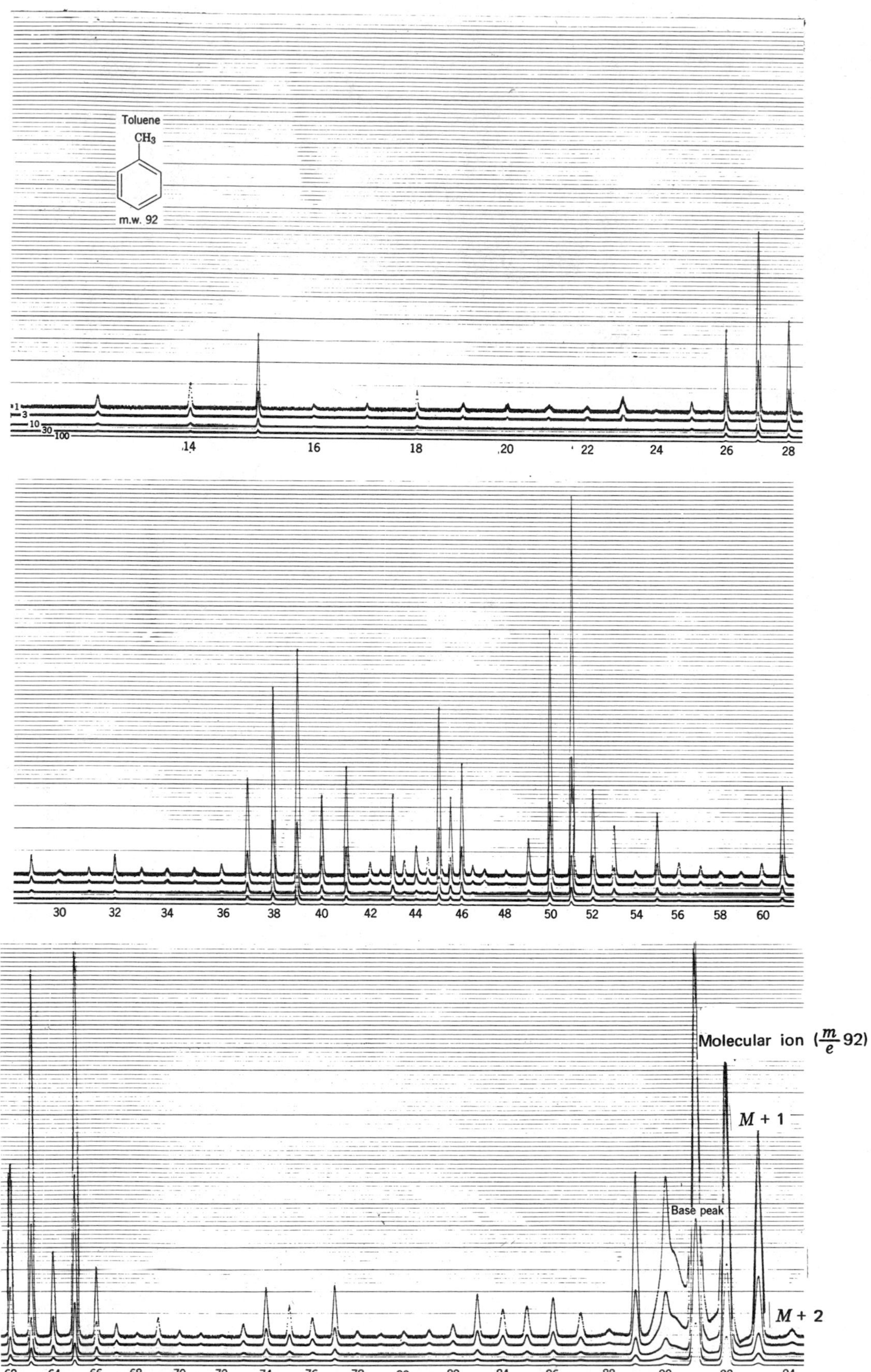

Figure 2a. *Mass Spectrum traced by a five-element galvanometer. Note the metastable peak at* $m^* = (91)^2/(92) = ca. 90$. *The relative peak intensities decrease from top to bottom trace, by 1:3:10:30:100.*

has electric and magnetic sectors of 90° and 31° 50′ respectively, and the ions are focused at different points in the detector.

Resolutions on the order of 40,000 are routinely obtainable with the high-resolution commercial instruments available (either geometry) when relatively monoenergetic ions are produced; both geometries have been used extensively for accurate measurement of the masses of ions in organic structure determinations. However, the high cost and complexity make this instrument available to relatively few laboratories.

Toluene
CH_3
m.w. 92

m/e	% of Base Peak	Isotope Abundances m/e	% of M
38	4.4	92 (M)	100
39	5.3	93 ($M + 1$)	7.23
45	3.9	94 ($M + 2$)	0.29
50	6.3		
51	9.1		
62	4.1		
63	8.6		
65	11		
91	100 (Base)		
92	68 (Molecular Ion Peak)		
93	4.9 ($M + 1$)		
94	0.21 ($M + 2$)		

Figure 2b. *Tabular presentation of Figure 2a.*

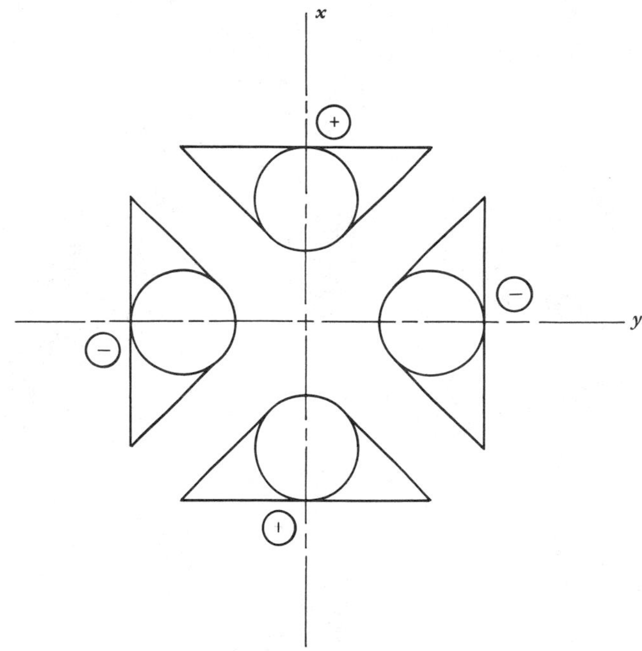

Figure 3. *Quadrupole arrangement.*

Several firms offer a gas chromatographic instrument coupled to a mass spectrometer through an interface that enriches the concentration of the sample in the carrier gas by taking advantage of the higher diffusivity of the carrier gas, (Figure 5). Scan times are rapid enough so that several mass spectra can be obtained during the elution of a single peak from the gas chromatographic unit. The enrichment device decreases the pressure of gas emitted from the gas chromatograph to a pressure suitable for the spectrometer detector.

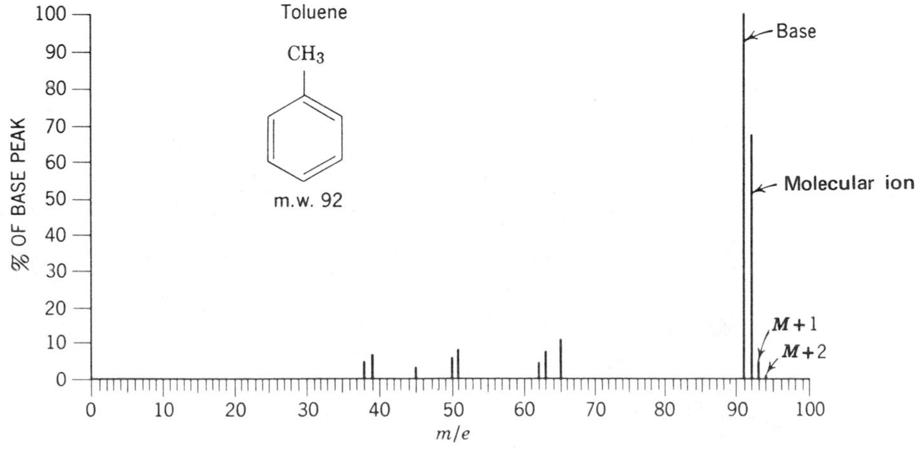

Figure 2c. *Graphical presentation of Figure 2a.*

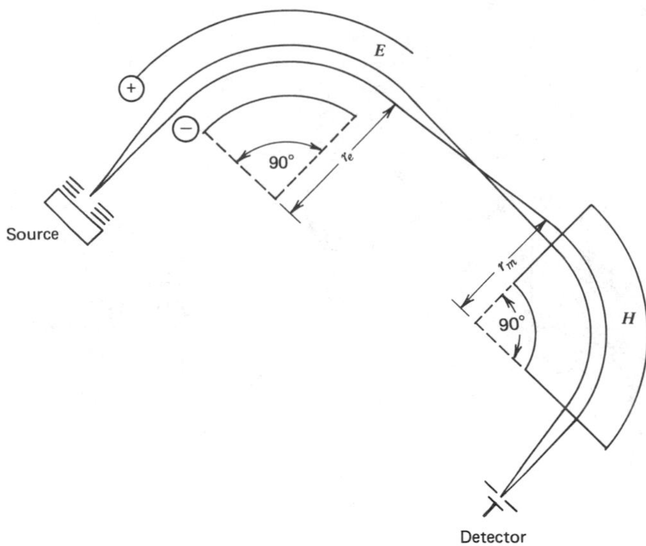

Figure 4. *Nier-Johnson double-focusing mass analyzer. E = electric field. H = magnetic field.*

III. THE MASS SPECTRUM

Mass spectra are routinely obtained at an electron beam energy of 70 electron volts. The simplest event that occurs is the removal of a single electron from the molecule in the gas phase by an electron of the electron beam to form a molecular ion, which is a radical cation ($M^{\dagger}$). For example, methanol forms a molecular ion:

$$CH_3OH + e \rightarrow CH_3OH^{\dagger} (m/e\ 32) + 2e$$

When the charge can be localized on one particular atom, the charge is shown on that atom:

$$CH_3\overset{\cdot +}{\underset{\cdot\cdot}{O}}H$$

The single dot represents the odd electron. Many of these molecular ions disintegrate in 10^{-10} to 10^{-3} second to give, in the simplest case, a positively charged fragment and a radical. A number of fragment ions are thus formed, and each of these can cleave to yield smaller fragments. Again, illustrating with methanol:

$$CH_3OH^{\dagger} \rightarrow CH_2OH^+ (m/e\ 31) + H\cdot$$

$$CH_3OH^{\dagger} \rightarrow CH_3^+ (m/e\ 15) + \cdot OH$$

$$CH_2OH^+ \rightarrow CHO^+ (m/e\ 29) + H_2$$

If some of the molecular (parent) ions remain intact long enough (about 10^{-6} seconds) to reach the detector, we see a molecular ion peak. It is important to recognize the molecular ion peak because this gives the molecular weight of the compound. With unit resolution, this molecular weight is the molecular weight to the nearest whole number, and not merely the approximation obtained by all other molecular weight determinations familiar to the organic chemist.

A mass spectrum is a presentation of the masses of the positively charged fragments (including the molecular ion) versus their relative concentrations. The most intense peak in the spectrum, called the base peak, is assigned a value of 100%, and the intensities (height × sensitivity factor) of the other peaks, including the molecular ion peak, are reported as percentages of the base peak. Of course, the molecular ion peak may sometimes be the base peak. In Figure 2a, the molecular ion peak is *m/e* 92, and the base peak is *m/e* 91.

A tabular or graphic presentation of a spectrum may be used. A graph has the advantage of presenting patterns that, with experience, can be quickly recognized. However, a graph must be drawn so that there is no difficulty in distinguishing mass units. Mistaking a peak at, say, *m/e* 79 for *m/e* 80 can result in total confusion. Unfortunately, many of the graphs presented in the journals are illegible. Computer programs are available for printing out good line graphs. The grid that we use permits ready recognition of individual mass units as well as patterns. Except for isotope peaks and molecular ion peaks, peaks of less than 3%

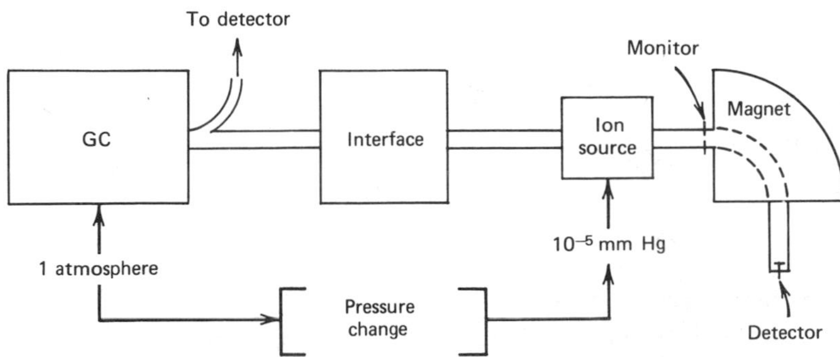

Figure 5. *Gas Chromatograph-mass spectrometer combination.*

intensity are not shown unless they have special significance.

The molecular ion peak is usually the peak of highest mass number except for the isotope peaks. These isotope peaks are present because a certain number of molecules contain heavier isotopes than the common isotopes. We shall show how the intensities of the isotope peaks relative to the parent peak can lead to the determination of a molecular formula. In a separate table accompanying the mass spectral graphs in the problems in Chapters 7 and 8, the molecular ion peak is given an intensity of 100%, and the isotope peak intensities are given relative to the molecular ion peak intensity (Figure 2*b*).

If an ion (m_1) fragments after acceleration but before entering the magnetic field, it will have been accelerated as mass m_1 but dispersed in the magnetic field as m_2. The resulting ion current will be recorded as a low-intensity, broad peak at apparent mass m^*. The numerical value of m^* is given by

$$m^* = \frac{(m_2)^2}{m_1}$$

The peak caused by the ion current corresponding to mass m^* is called a *metastable peak*. Measurement of the mass of the metastable peak affords information that m_2 is derived directly from m_1 by loss of a neutral fragment. An example of a metastable peak is the broad peak at about m/e 90 in Figure 2*a*. The m_1 peak is the molecular ion (m/e 92) and the m_2 peak is at m/e 91. Although this technique is useful in interpretations of spectra, it will not be used in this book. Its use is discussed in References 1, 4, and 5.

IV. DETERMINATION OF A MOLECULAR FORMULA

A unique molecular formula (or fragment formula) can often be derived from a sufficiently accurate mass measurement alone (high-resolution mass spectrometry). This is possible because the atomic masses are not integers (see Table I). For example, we can distinguish at a nominal mass of 28 among CO, N_2, CH_2N, and C_2H_4:

^{12}C	12.0000	$^{14}N_2$	28.0062	^{12}C	12.0000	$^{12}C_2$	24.0000
^{16}O	15.9949			$^{1}H_2$	2.0156	$^{1}H_4$	4.0312
				^{14}N	14.0031		
	27.9949				28.0187		28.0312

Thus, the mass observed for the molecular ion of CO is the sum of the exact masses of the most abundant isotope of carbon and of oxygen. This differs from a molecular weight of CO based on atomic weights that are the average of weights of all natural isotopes of an element (e.g., C = 12.01, O = 15.999). It is a tedious task to find a molecular formula by arithmetic trial and error from the output of a high-resolution mass spectrometer. Tables (Appendix A), algorithms, and computer programs have been assembled for this purpose.[13,18,21,23,24] We will use unit mass resolution results and intensities of isotope peaks to arrive at possible molecular formulas, since most laboratories have limited access to high-resolution spectrometers. Determination of a possible molecular formula from the isotope peak intensities is limited to the cases in which the molecular ion peak is relatively intense so that the isotope peaks are large enough to be measured accurately. In a high-resolution spectrum, the position of even a very weak molecular ion peak can be accurately measured.

Table I. Exact Masses of Isotopes

Element	Atomic Weight	Nuclide	Mass
Hydrogen	1.00797	^{1}H	1.00783
		D (^{2}H)	2.01410
Carbon	12.01115	^{12}C	12.00000 (std)
		^{13}C	13.00336
Nitrogen	14.0067	^{14}N	14.0031
		^{15}N	15.0001
Oxygen	15.9994	^{16}O	15.9949
		^{17}O	16.9991
		^{18}O	17.9992
Fluorine	18.9984	^{19}F	18.9984
Silicon	28.086	^{28}Si	27.9769
		^{29}Si	28.9765
		^{30}Si	29.9738
Phosphorus	30.974	^{31}P	30.9738
Sulfur	32.064	^{32}S	31.9721
		^{33}S	32.9715
		^{34}S	33.9679
Chlorine	35.453	^{35}Cl	34.9689
		^{37}Cl	36.9659
Bromine	79.909	^{79}Br	78.9183
		^{81}Br	80.9163
Iodine	126.904	^{127}I	126.9045

Table II lists the principal stable isotopes of the common elements and their relative abundance calculated on the basis of 100 molecules containing the most common isotope. Note that this presentation differs from many isotope abundance tables in which the sum of all the isotopes of an element adds up to 100%.

Suppose that a compound contains 1 carbon atom. Then for every 100 molecules containing a ^{12}C atom, about 1.08 "molecules" contain a ^{13}C atom, and these molecules will produce an $M + 1$ peak of about 1.08% the intensity of the molecular ion peak; the ^{2}H atoms present will make an additional very small contribution to the $M + 1$ peak. If a compound contains one sulfur atom, the $M + 2$ peak will be about 4.4% of the parent peak. The $M + 1$ and $M + 2$ peaks are so designated in Figure 2.

Selection of likely molecular formulas appropriate to particular mass and isotope abundance measurements is greatly facilitated by the table constructed by Beynon.* This table, which had later been extended[18] to mass 500, can be used for high resolution mass spectrometry. Beyond about mass 250, use of the isotopes to determine a molecular formula loses its effectiveness. A reduced, modified version of Beynon's table is presented as Appendix A. Its use will become more evident as we work through the spectra in this book. In practice, the measured isotope peaks are usually slightly higher than the calculated contributions because of incomplete resolution, bimolecular collisions (see below), or a contribution from the coincident peak of an impurity. The table is limited to compounds containing C, H, O, and N. The presence of S, Cl, or Br is usually readily apparent because of a large isotope contribution to $M + 2$. We shall see that the number of sulfur, chlorine, and bromine atoms can be determined. Iodine, fluorine, and phosphorus are monoisotopic. Their presence can usually be deduced from a suspiciously small $M + 1$ peak relative to the molecular weight, from the fragmentation pattern, from the other spectra with which we are concerned, or from the history of the compound.

If only C, H, N, O, F, P, I are present, the approximate expected % $(M + 1)$ and % $(M + 2)$ data can be calculated by use of the following formulas:

$$\%(M+1) = 100\left\{\frac{[M+1]}{[M]}\right\}$$

$$\simeq 1.1 \times \text{number of C atoms} + 0.36 \times \text{number of N atoms}^{\dagger}$$

$$\%(M+2) = 100\left\{\frac{[M+2]}{[M]}\right\}$$

$$\simeq \frac{(1.1 \times \text{number of C atoms})^2}{200} + 0.20 \times \text{number of O atoms}^{\dagger}$$

These equations are useful for cases in which one has a preconceived notion about the molecular formula for the compound of interest.

It is difficult to overemphasize the importance of locating the molecular ion peak. It will be stressed again that this gives an exact numerical molecular weight. Even in cases in which the molecular ion peak is very small (and therefore an accurate determination of $M + 1$ and $M + 2$ is impossible), only a little extra information can often lead to identification. This information may be available from the source and history of the sample, from the fragmentation pattern, and from other spectra. Let us work through the selection of a molecular formula from the isotope abundance data obtained on an organic compound. We are given the following information:

m/e	%
150 (M)	100
151 ($M + 1$)	10.2
152 ($M + 2$)	0.88

The molecular ion peak is mass 150; thus we have the molecular weight. The $M + 2$ peak obviously does not allow for the presence of sulfur or halogen atoms. We look in Appendix A under mass 150. Our $M + 1$ peak is 10.2% of the parent peak. We list the formulas whose calculated isotope contribution to the $M + 1$ peak falls—to be arbitrary—between 9.0 and 11.0; we also list the calculated $M + 2$ values:

Formula	$M + 1$	$M + 2$
$C_7H_{10}N_4$	9.25	0.38
$C_8H_8NO_2$	9.23	0.78
$C_8H_{10}N_2O$	9.61	0.61
$C_8H_{12}N_3$	9.98	0.45
$C_9H_{10}O_2$	9.96	0.84
$C_9H_{12}NO$	10.34	0.68
$C_9H_{14}N_2$	10.71	0.52

On the basis of the "nitrogen rule" (see Section V), we immediately eliminate three of these formulas because they contain an odd number of nitrogen atoms. Our $M + 2$ peak is 0.88% of the parent. This best fits $C_9H_{10}O_2$. However, $C_8H_{10}N_2O$ cannot be ruled out without additional evidence.

Table II. Relative Isotope Abundances of Common Elements

Elements						
Carbon	^{12}C 100		^{13}C	1.08		
Hydrogen	^{1}H 100		^{2}H	0.016		
Nitrogen	^{14}N 100		^{15}N	0.38		
Oxygen	^{16}O 100		^{17}O	0.04	^{18}O	0.20
Fluorine	^{19}F 100					
Silicon	^{28}Si 100		^{29}Si	5.10	^{30}Si	3.35
Phosphorus	^{31}P 100					
Sulfur	^{32}S 100		^{33}S	0.78	^{34}S	4.40
Chlorine	^{35}Cl 100		^{37}Cl	32.5		
Bromine	^{79}Br 100		^{81}Br	98.0		
Iodine	^{127}I 100					

*Beynon's first table is based on O = 16.000000. His expanded table is based on the currently accepted standard, C = 12.000000.

†[M + 1] is the intensity of the M + 1 peak, and [M] is the intensity of the M peak.

Note that mass 150 is the sum of the masses of the common isotopes of these molecular formulas; the isotope masses used are whole numbers (12 for carbon, 14 for nitrogen, etc.).

When elements other than C, H, O, and N are present, their kind and number must be determined (see the discussion under the appropriate chemical class) and their mass subtracted from the molecular weight. The composition of the remainder of the molecule is then determined from Appendix A.

Quite often the organic chemist must identify a by-product or an unexpected main product from a reaction. Under these circumstances, even rather complex molecules can be handled expeditiously. Since the compound has a history, an intelligent guess can be made as to what elements are present. The following example may be instructive. In the course of polymerizing the fluorinated silicon-containing monomer, a crystalline sublimate was

monomer

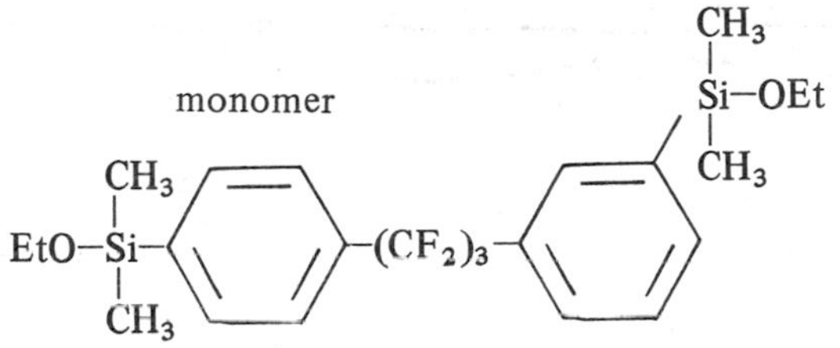

obtained as a by-product. The parent peak of the mass spectrum of the sublimate was m/e 434, and the base peak, m/e 419 (i.e., $M - 15$). The $M + 1$ peak was 31.2% of the molecular ion peak, and the $M + 2$ peak, 10.4%. On the assumption that the compound contained six fluorine and two silicon atoms, the molecular formula $C_{19}H_{20}F_6OSi_2$ was written. The intensities of the $M + 1$ and $M + 2$ peaks for this formula were calculated as follows:

Contributor	$M + 1$	$M + 2$	Source
$C_{19}H_{20}$ (mass 248)	20.85	2.06	Appendix A*
O	0.04	0.20	Table II
F_6	–	–	Table II
Si_2	10.20†	6.70†	Table II
Sum:	31.09	8.96	

*The C, H, and O content could have been obtained directly from Beynon's extended tables.[4] An alternative method for obtaining C, H, and O contributions is the use of the equations given earlier, thus:

$$\%(M + 1) = (1.1)(19) + (0.36)(0) = 20.9\% \text{ of parent}$$

$$\%(M + 2) = [(1.1)(19)]^2/200 + 0.20(1) = 2.38\% \text{ of parent}$$

†2 × (contribution to $M + 1$ or $M + 2$ per Si atom).

The good agreement between the calculated and the determined values provided strong support for the molecular formula written. This information combined with the fragmentation pattern and the infrared and NMR spectra led to the following structure.

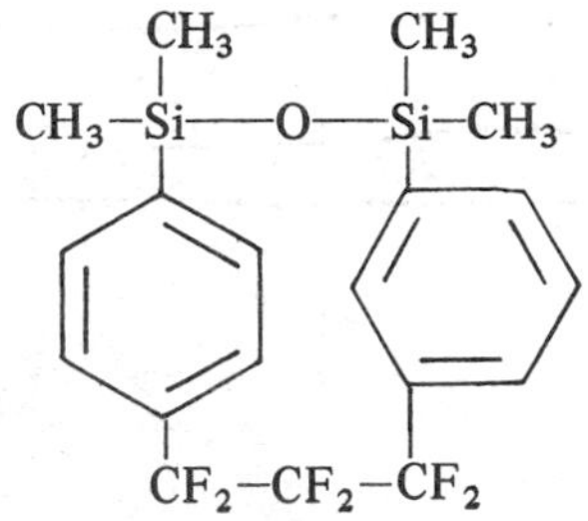

sublimate

The foregoing is a good example of both the possibilities and limitations of spectrometric identification without any chemical manipulation on the compound. It is doubtful that even an experienced mass spectrometrist would have derived a molecular formula without some indication that silicon and fluorine were present. Confirmation is certainly easier than diagnosis.

A reasonable amount of caution must be used in developing a molecular formula from the relative intensities of the M, $M + 1$, and $M + 2$ peaks. Clearly, if these peaks are small, the error associated with their relative measurements will be large. Several spectra should be obtained to determine the reproducibility of these measurements on a specific instrument. As we shall see in Chapter 5, a ^{13}C NMR spectrum is a powerful aid in establishing the molecular formula.

V. RECOGNITION OF THE MOLECULAR ION PEAK

There are two situations in which identification of the molecular ion peak may be difficult.

1. The molecular ion does not appear or is very weak. The obvious remedy in most cases is to run the spectrum at maximum sensitivity (and accept the resulting loss in resolution) and to use a larger sample. (Sometimes a large sample exaggerates the $M + 1$ peak. See below.) Still the molecular ion may not be evident, and other sources of information may be useful. The type of compound may be known, and the molecular mass may be deduced from the breakdown pattern. For example, alcohols usually give a very weak parent molecular ion peak but often show a pronounced peak resulting from loss of water ($M - 18$) A combustion analysis together with consideration of the fragmentation pattern may help us to arrive at the parent mass. Preparation of a suitable derivative is another device.
2. The molecular ion is present but is one of several peaks which may be as prominent or even more prominent. In this situation, the first question is that of purity. If the compound can be assumed to be pure, the usual problem is to distinguish the molecular ion peak from a more

prominent $M - 1$ peak. One good test is to reduce the energy of the bombarding electron beam to near the appearance potential. This will reduce the intensities of all peaks, but will increase the intensity of the molecular ion relative to other peaks, including fragmentation peaks (but not molecular ion peaks) of impurities. Another test frequently used is to increase the size of the sample, or increase the time the sample spends in the ionization chamber by decreasing the ion repeller voltage. In either case, the net effect is to increase the opportunity for bimolecular collisions to occur in the ion chamber. The most common result of a bimolecular collision of a molecular ion containing a heteroatom (O, N, or S) is a contribution to the $M + 1$ peak (i.e., the net effect is the transfer of a hydrogen atom from a neutral molecule to the molecular ion).

$$RCH_2-\overset{+}{\ddot{O}}-CH_2R + RCH_2-O-CH_2R \rightarrow$$
$$(M^{\dot{+}})$$
$$RCH_2-\underset{H}{\overset{+}{\ddot{O}}}-CH_2R + R\dot{C}H-\ddot{O}-CH_2R$$
$$(M+1)^+$$

Thus, an increase in peak size relative to other peaks, as sample size is increased or the repeller voltage is decreased, designates that peak as the $M + 1$ peak and affords an indirect identification of the molecular ion. Of course, the dependence of the $M + 1$ peak on sample size must be kept in mind when this peak is used to establish a molecular formula of a compound containing a heteroatom.

Many peaks can be ruled out as possible molecular ions simply on grounds of reasonable structure requirements. The "nitrogen rule" is often helpful in this regard. It states that a molecule of even-numbered molecular weight must contain no nitrogen or an even number of nitrogen atoms; an odd-numbered molecular weight requires an odd number of nitrogen atoms. This rule holds for all compounds containing carbon, hydrogen, oxygen, nitrogen, sulfur, and the halogens, as well as many of the less usual atoms such as phosphorus, boron, silicon, arsenic, and the alkaline earths. A useful corollary states that fragmentation at a single bond gives an odd-numbered ion fragment from an even-numbered molecular ion, and an even-numbered ion fragment from an odd-numbered molecular ion. For this corollary to hold, the ion fragment must contain all of the nitrogen (if any) of the molecular ion. Consideration of the breakdown pattern coupled with other information will also assist in identifying molecular ions. It should be kept in mind that Appendix A contains fragment formulas as well as molecular formulas. Some of the formulas may be regarded as trivial when used in attempts to solve practical problems.

The presence of appreciable amounts of impurities that give rise to prominent peaks near the molecular ion can be troublesome. Here again, the expediency of reducing the energy of the electron beam will cause a relative increase in the intensity of the molecular ion peak (and also of the molecular ion peak of an impurity). The fragmentation pattern will often furnish clues.

Another useful technique for detection of impurities is microeffusiometry. A fixed volume of sample is allowed to flow through a molecular leak, and the logarithms of the intensities of the peaks in question are plotted as a function of time. All peaks belonging to the same molecule will give lines of the same slope. Those due to other components of different volatility will give lines of different slopes. In essence, a fractionation is effected.

The intensity of the molecular ion peak depends on the stability of the molecular ion. The most stable molecular ions are those of purely aromatic systems. If substituents that have favorable modes of cleavage are present, the molecular ion peak will be less intense, and the fragment peaks relatively more intense. In general, the following group of compounds will, in order of decreasing ability, give prominent molecular ion peaks: aromatic compounds > conjugated alkenes > cyclic compounds > organic sulfides > short, normal alkanes > mercaptans. Recognizable molecular ions are usually produced for these compound in order of decreasing ability: ketones > amines > esters > ethers > carboxylic acids ~ aldehydes ~ amides ~ halides. The molecular ion is frequently not detectable in aliphatic alcohols, nitrites, nitrates, nitro compounds, nitriles, and in highly branched compounds.

The presence of an $M - 15$ peak (loss of CH_3) or an $M - 18$ peak (loss of H_2O) or an $M - 31$ peak (loss of OCH_3 from methyl esters), etc., is taken as confirmation of a molecular ion peak. An $M - 1$ peak is common, and occasionally an $M - 2$ peak (loss of H_2 by either fragmentation or thermolysis), or even a rare $M - 3$ peak (from alcohols) is reasonable. Peaks in the range of $M - 3$ to $M - 14$, however, indicate that contaminants may be present or that the presumed molecular ion peak is actually a fragment ion peak. Losses of fragments of masses 19 to 25 are also unlikely (except for loss of $F = 19$ or $HF = 20$ from fluorinated compounds). Loss of 16 (O), 17 (OH), or 18 (H_2O) are likely only if an oxygen atom is in the molecule.

In addition to the electron impact (EI) method described above, several promising techniques have been developed for obtaining and locating the molecular ion peak for compounds that give very weak or nonexistant molecular ions.

In chemical ionization (CI) the sample is introduced at about 1 Torr pressure with a carrier gas, such as methane. The methane is ionized by electron impact to the primary ions: $CH_4^{\dot{+}} + CH_3^+$, etc. These react with the excess methane to give secondary ions:

$$CH_4^{\dot{+}} + CH_4 \rightarrow CH_5^+ \text{ and } CH_3\cdot$$
$$CH_3^+ + CH_4 \rightarrow C_2H_5^+ \text{ and } H_2$$
$$CH_4 + C_2H_5^+ \rightarrow C_3H_5^+ + 2H_2$$

The secondary ions react with the sample (RH):

$$CH_5^+ + RH \rightarrow RH_2^+ \text{ and } CH_4$$

$$C_2H_5^+ + RH \rightarrow RH_2^+ \text{ and } C_2H_4$$

These $M + 1$ ions (*quasi-molecular ions*) are often prominent. They can fragment (loss of H_2) to give prominent $M - 1$ ions. Since the $M + 1$ ions that are chemically produced do not have the great excess of energy associated with ionization by electron impact, they undergo less fragmentation. For example, the EI spectrum of 3,4-dimethoxyacetophenone shows, in addition to the molecular ion at *m/e* 180, 49 fragment peaks in the range of *m/e* 40-167; these include the base peak at *m/e* 165 and prominent peaks at *m/e* 137 and *m/e* 77. The CI spectrum shows the quasi-molecular ion (MH^+, *m/e* 181) as the base peak (100%), and virtually the only other peaks, each of just a few % intensity, are the molecular ion peak, *m/e* 180, and *m/e* 209 and 221 peaks due to reaction with carbocations.

$Ar(C{=}O)CH_3 + CH_5^+ \rightarrow [Ar(C{=}O)CH_3 + H]^+$ *m/e* 181, base peak

$[Ar(C{=}O)CH_3 + H]^+ \rightarrow H\cdot + [Ar(C{=}O)CH_3]^{+\cdot}$ *m/e* 180, molecular ion

$Ar(C{=}O)CH_3 + C_2H_5^+ \rightarrow [Ar(C{=}O)CH_3 + C_2H_5]^+$ *m/e* 209

$Ar(C{=}O)CH_3 + C_3H_5^+ \rightarrow [Ar(C{=}O)CH_3 + C_3H_5]^+$ *m/e* 221

Ar = 3,4-dimethoxyphenyl

Field ionization (FI) is a technique that, like CI and EI techniques, requires a reasonable concentration of the sample in the gas phase. In FI spectrometry, gaseous sample molecules approach a sharp metal surface that has a high positive potential (10^8 volts/cm), which causes an election to "tunnel" from the sample molecule to the metal surface. This process occurs without exciting the sample molecules, and, since the molecular ion is quickly accelerated, the time scale for analysis is much shorter than for EI spectrometry. Thus, there is less fragmentation with a more intense molecular ion peak.

Molecular ions can be obtained from samples of very low volatility by the field desorption (FD) technique. Here the solid sample is placed directly on the field ion emitter. The ion source is operated only slightly above room temperature, and the ions produced thus have little or no internal energy. Since the internal energy necessary for fragmentation is not present, predominantly molecular ions are formed. These ions are desorbed from the solid by a strong electric field. Related techniques include plasma desorption (PD) mass spectrometry, in which ^{252}Cf fission fragments provide short energy pulses that vaporize the solid samples. The vaporized sample particle quickly forms molecular or quasi-molecular ions, and these ions are immediately accelerated. Ionization takes place quickly and little energy is converted to vibrational motion. Most of the ion's energy is used for translational motion. Thus, molecular ions or quasi-molecular ions of substantial intensity are produced. Similar procedures in which the short pulse of energy is provided by a laser have been developed.

Applications and comparisons of EI, CI, FI, and FD techniques are illustrated in the portion on natural products. (See Figures 17-19 and related text.)

VI. USE OF THE MOLECULAR FORMULA

If an organic chemist had to choose a single item of information above all others that are usually available to him from spectra or from chemical manipulations, he would certainly choose the molecular formula.

In addition to the kinds and numbers of atoms, the molecular formula gives the index of hydrogen deficiency. The index of hydrogen deficiency is the number of *pairs* of hydrogen atoms that must be removed from the "saturated" formula (e.g., C_nH_{2n+2} for alkanes) to produce the molecular formula of the compound of interest. The index of hydrogen deficiency has also been called the number of "sites (or degrees) of unsaturation"; this description is unsatisfactory since hydrogen deficiency can be due to cyclic structure features as well as to multiple bonds. The index is thus the sum of the number of rings, the number of double bonds, and twice the number of triple bonds.

The index of hydrogen deficiency can be calculated for compounds containing carbon, hydrogen, nitrogen, halogen, oxygen, and sulfur from the formula

$$\text{Index} = \text{carbons} - \frac{\text{hydrogens}}{2} - \frac{\text{halogens}}{2} + \frac{\text{nitrogens}}{2} + 1$$

Thus, the compound C_7H_7NO has an index of $7 - 3.5 + 0.5 + 1 = 5$. Note that divalent atoms (oxygen and sulfur) are not counted in the formula.

For the generalized molecular formula $\alpha_I \beta_{II} \gamma_{III} \delta_{IV}$, the index = $IV - I/2 + III/2 + 1$, where

α is H,D or halogen (i.e., any monovalent atom)
β is O,S or any other bivalent atom
γ is N,P or any other trivalent atom.
δ is C,Si or any other tetravalent atom

I,II,III, and IV designate the numbers of the mono-, di-, tri-, and tetravalent atoms, respectively.

For simple molecular formulas, we can arrive at the index by comparison of the formula of interest with the molecular formula of the corresponding saturated compound. Compare C_6H_6 and C_6H_{14}; the index is 4 for the former and 0 for the latter.

Polar structures must be used for compounds containing an atom in a higher valence state, such as sulfur or phosphorus. Thus, if we treat sulfur in dimethyl sulfoxide formally as a divalent atom, the calculated index, zero, is compatible with the structure $CH_3-\overset{+}{S}(-\ddot{O}:^-)-CH_3$. We must use only formulas with filled valence shells; that is, the Lewis octet rule must be obeyed.

Similarly, if we treat the nitrogen in nitromethane as a trivalent atom, the index is 1, which is compatible with

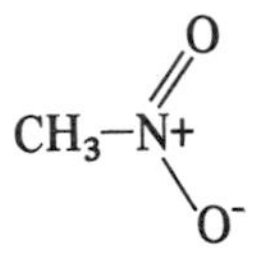

$$CH_3-N^+(=O)-O^-$$

. If we treat phosphorus in triphenylphosphine oxide as trivalent, the index is 12, which fits $(C_6H_5)_3P^+-O^-$. As an example, consider the molecular formula $C_{13}H_9N_2O_4BrS$. The index of hydrogen deficiency would be $13 - 10/2 + 2/2 + 1 = 10$ and a consistent structure would be:

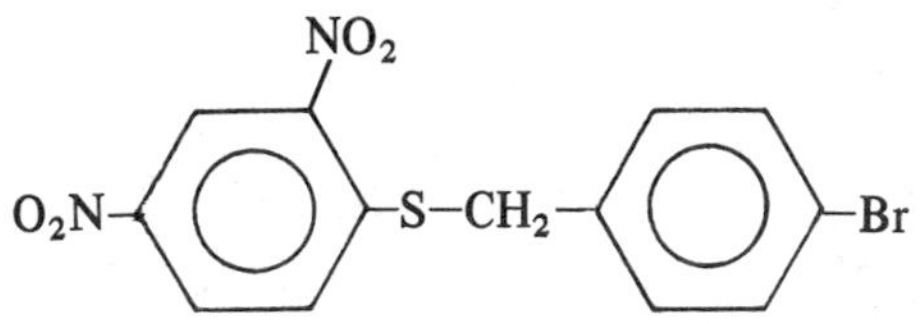

$$O_2N-C_6H_3(NO_2)-S-CH_2-C_6H_4-Br$$

(Index of hydrogen deficiency = 4 per benzene ring and 1 per NO_2 group.)

The formula above for the index can be applied to fragment ions as well as to the molecular ion. When it is applied to even-electron (all electrons paired) ions, the result is always an odd multiple of 0.5. As an example, consider $C_7H_5O^+$ with an index of 5.5. A reasonable structure is

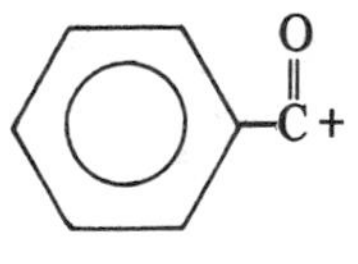

$$C_6H_5-\overset{+}{C}=O$$

since 5½ pairs of hydrogens would be necessary to obtain the corresponding saturated formula $C_7H_{16}O(C_nH_{2n+2}O)$. Odd-electron fragment ions will always give integer values of the index.

Terpenes often present a choice between a double bond and a ring structure. This question can readily be resolved on a microgram scale by catalytically hydrogenating the compound and rerunning the mass spectrum. If no other easily reducible groups are present, the increase in the mass of the molecular peak is a measure of the number of double bonds; and other "unsaturated sites" must be rings.

Such simple considerations give the chemist very ready information about structure. As another example, a compound containing a single oxygen atom may quickly be determined to be an ether or a carbonyl compound simply by counting "unsaturated sites."

VII. FRAGMENTATION

As a first impression, fragmenting a molecule with a huge excess of energy would seem a brute-force approach to molecular structure. The rationalizations used to correlate spectral patterns with structure, however, can only be described as elegant. The insight of such pioneers as McLafferty, Beynon, Stenhagen, Ryhage, and Meyerson have led to a number of rational mechanisms for fragmentation. These have been masterfully summarized and elaborated by Biemann.

Generally, the tendency has been to represent the molecular ion with a delocalized charge. Djerassi's approach has been to localize the positive charge on either a π bond (except in conjugated systems), or on a heteroatom. Whether or not this concept is totally rigorous, it is at the least a *pedagogic tour de force*. We shall use such locally charged molecular ions in this book.

A single-barbed fishhook (⇀) designates the shift of a single electron. Cleavage of a bond requires the movement of two electrons. However, to prevent clutter, only one of a pair of fishhooks will be drawn. This practice can be illustrated for cleavage of a C–C bond next to a heteroatom:

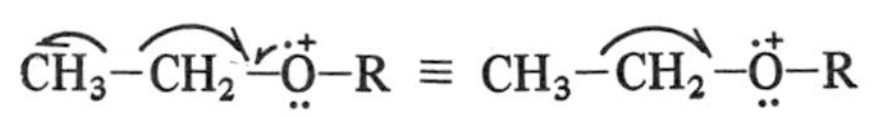

$$CH_3-CH_2-\overset{+\cdot}{\ddot{O}}-R \equiv CH_3-CH_2-\overset{+\cdot}{\ddot{O}}-R$$

The probability of cleavage of a particular bond is related to the bond strength, to the possibility of low-energy transitions, and to the stability of the fragments both charged and uncharged formed in the fragmentation process. Our knowledge of pyrolytic cleavages can be used, to some extent, to predict likely modes of cleavage of the molecular ion. Because of the extremely low vapor pressure in the mass spectrometer, there are very few fragment collisions; we are dealing largely with unimolecular decompositions. This assumption, backed by a file of reference spectra, is the basis for the vast amount of information available from the fragmentation pattern of a molecule. Whereas conventional organic chemistry deals with reactions initiated by chemical reagents, by thermal energy, or by light, mass spectrometry is concerned with the consequences suffered by an organic molecule struck by an ionizing electronic beam at a vapor pressure of about 10^{-5} mm Hg.

A number of general rules for predicting prominent peaks in electron impact spectra can be written and rationalized by using standard concepts of physical organic chemistry:

1. The relative height of the molecular ion peak is greatest for the straight-chain compound and decreases as the degree of branching increases. (See Rule 3.)
2. The relative height of the molecular ion peak usually decreases with increasing molecular weight in a homologous series. Fatty esters appear to be an exception.
3. Cleavage is favored at alkyl substituted carbons; the more substituted, the more likely is cleavage. This is a consequence of the increased stability of a tertiary carbocation over a secondary, which in turn is more stable than a primary.

$$[\mathrm{R{-}\overset{|}{\underset{|}{C}}}]^{+\cdot} \rightarrow \mathrm{R}^{\cdot} + {}^{+}\overset{|}{\underset{|}{\mathrm{C}}}{-}$$

Cation stability order:

$$\mathrm{CH_3^+ < R'CH_2^+ < R_2'CH^+ < R_3'C^+}$$

Generally, the largest substituent at a branch is eliminated most readily as a radical, presumably because a long-chain radical can achieve some stability by delocalization of the lone electron.

4. Double bonds, cyclic structures, and especially aromatic (or heteroaromatic) rings stabilize the molecular ion, and thus increase the probability of its appearance.
5. Double bonds favor allylic cleavage and give the resonance-stabilized allylic carbonium ion.

$$\mathrm{CH_2^{+\cdot}{:}CH{-}CH_2{-}R \xrightarrow{-R^\cdot} \overset{+}{C}H_2{-}CH{=}CH_2 \leftrightarrow CH_2{=}CH{-}\overset{+}{C}H_2}$$

6. Saturated rings tend to lose side chains at the α-bond. This is merely a special case of branching (Rule 3). The positive charge tends to stay with the ring fragment. Unsaturated rings can undergo a *retro*-Diels-Alder reaction:

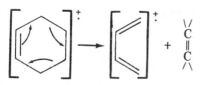

$$[\text{cyclohexene}]^{+\cdot} \rightarrow [\text{butadiene}]^{+\cdot} + \mathrm{{>}C{=}C{<}}$$

7. In alkyl-substituted aromatic compounds, cleavage is very probable at the bond beta to the ring, giving the resonance-stabilized benzyl ion or, more likely, the tropylium ion:

$$[\mathrm{C_6H_5{-}CH_2{-}R}]^{+\cdot} \xrightarrow{-R^\cdot} \mathrm{C_6H_5CH_2^+} \leftrightarrow \ldots$$

$$[\mathrm{C_6H_5CH_2R}]^{+\cdot} \rightarrow [\mathrm{C_7H_8}]^{+} \xrightarrow{\text{1,2-H SHIFT}} \mathrm{C_7H_7^+}\ (\text{tropylium})$$

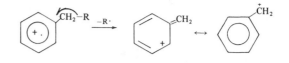

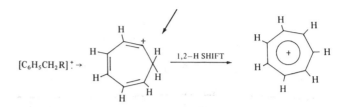

8. C–C bonds next to a heteroatom are frequently cleaved, leaving the charge on the fragment containing the heteroatom whose nonbonding electrons provide resonance stabilization.

$$\mathrm{CH_3{-}CH_2{-}\overset{+\cdot}{\ddot{Y}}{-}R \xrightarrow{-CH_3^\cdot} CH_2{=}\overset{+}{Y}{-}R \leftrightarrow {}^{+}CH_2{-}\ddot{Y}{-}R}$$

Y = O, N, or S

$$\mathrm{R{-}\underset{\underset{+}{\cdot O:}}{\overset{}{C}}{-}CH_2R' \xrightarrow{-R^\cdot} C{-}CH_2R'\ (C{\equiv}\overset{+}{O}{:}) \leftrightarrow \overset{+}{C}{-}CH_2R'\ (C{=}\ddot{O}{:})}$$

9. Cleavage is often associated with elimination of small stable neutral molecules, such as carbon monoxide, olefins, water, ammonia, hydrogen sulfide, hydrogen cyanide, mercaptans, ketene, or alcohols.

It should be kept in mind that the fragmentation rules above apply to EI mass spectrometry. Since other ionizing (CI, etc.) techniques often produce molecular ions with much lower energy or quasi-molecular ions with very different fragmentation patterns, different rules govern the fragmentation of these molecular ions.

VIII. REARRANGEMENTS

Rearrangement ions are fragments whose origin cannot be described by simple cleavage of bonds in the molecular ion, but are a result of intramolecular atomic rearrangement during fragmentation. Rearrangements involving migration of hydrogen atoms in molecules that contain a heteroatom are especially common. One important example is the so-called McLafferty rearrangement:

$$\text{:}\overset{+\cdot}{O}\text{ (H) }CR_2 \quad \xrightarrow[Y = H, R, OH, OR, NR_2]{-R_2C{=}CH_2} \quad Y{-}C(\overset{+\cdot}{O}H){-}CH_2 \longleftrightarrow$$

$$Y{-}C(=\overset{+}{O}H){-}\dot{C}H_2 \longleftrightarrow Y{-}\overset{+}{C}(OH){-}\dot{C}H_2$$

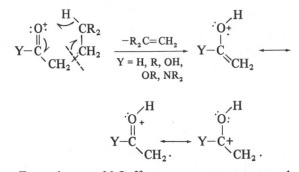

To undergo a McLafferty rearrangement, a molecule must possess: an appropriately located heteroatom (e.g., O), a π-system (usually a double bond) and an abstractable hydrogen γ to the C=O system.

Such rearrangements often account for prominent characteristic peaks and are consequently very useful for our purpose. They can frequently be rationalized on the basis of low-energy transitions and increased stability of the products. Rearrangements resulting in elimination of a stable neutral molecule are common (e.g., the olefin product in the McLafferty rearrangement) and will be encountered in the discussion of mass spectra of chemical classes.

Rearrangement peaks can be recognized by considering the mass (m/e) number for fragment ions and for their corresponding molecular ions. A simple (no rearrangement) cleavage of an even-numbered molecular ion gives an odd-numbered fragment ion and simple cleavage of an odd-numbered molecular ion gives an even-numbered fragment. Observation of a fragment ion mass different by 1 unit from that expected for a fragment resulting from simple cleavage (e.g., an even-numbered fragment mass from an even-numbered molecular ion mass) indicates rearrangement of hydrogen has accompanied fragmentation. Rearrangement peaks may be recognized by considering the corollary to the "nitrogen rule" (Section V). Thus, an even-numbered peak derived from an even-numbered molecular ion is a result of two cleavages, which may involve a rearrangement.

"Random" rearrangements of hydrocarbons were noted by the early mass spectrometrists in the petroleum industry. For example:

$$\left[CH_3{-}\underset{CH_3}{\overset{CH_3}{C}}{-}CH_3\right]^{+\cdot} \rightarrow [C_2H_5]^+$$

These rearrangements defy straightforward explanations based on ground-state theory.

IX. DERIVATIVES

If a compound has low volatility or if the parent mass cannot be determined, it may be possible to prepare a suitable derivative. The derivative selected should provide enhanced volatility, a predictable mode of cleavage, a simplified fragmentation pattern, or an increased stability of the parent ion.

Compounds containing several polar groups may have very low volatility (e.g., sugars, peptides, and dibasic carboxylic acids). Acetylation of hydroxyl and amino groups and methylation of free acids are obvious and effective choices to increase volatility and give characteristic peaks. Perhaps less immediately obvious is the use of trimethylsilyl derivatives of hydroxyl, amino, sulfhydryl, and carboxylic acid groups. Trimethylsilyl derivatives of sugars and of amino acids are volatile enough to pass through gas chromatographic columns. The molecular ion peak of trimethylsilyl derivatives may not always be present, but the $M - 15$ peak due to cleavage of one of the Si–CH_3 bonds is always prominent.

Reduction of ketones to hydrocarbons has been used to elucidate the carbon skeletón of the ketone molecule. Polypeptides have been reduced with $LiAlH_4$ to give polyamino alcohols that are volatile with predictable fragmentation patterns. Methylation and trifluoroacylation of tri- and tetrapeptides have been used to obtain mass spectra.

X. MASS SPECTRA OF SOME CHEMICAL CLASSES

Mass spectra of a number of chemical classes are briefly described in this section in terms of the most useful generalizations for identification. For more details, the references cited (in particular, the thorough treatment by Budzikiewicz, Djerassi, and Williams)[1-3] should be consulted. The references are selective rather than comprehensive. A table of frequently encountered fragment ions is given in Appendix B. A table of fragments (uncharged) that are commonly eliminated is presented in Appendix C. More exhaustive listings of common fragment ions have been compiled in the books by Hamming and Foster and by McLafferty. The cleavage patterns described here in Section X are for EI spectra, unless stated otherwise.

HYDROCARBONS

Saturated Hydrocarbons

Most of the work in mass spectrometry has been done on hydrocarbons of interest to the petroleum industry. Rules 1 to 3 (p. 15) apply quite generally; rearrangement peaks, though common, are not usually intense (random rearrangements), and numerous reference spectra are available.

The molecular ion peak (*M*) of a straight-chain, saturated hydrocarbon is always present, though of low intensity for long-chain compounds. The fragmentation pattern is characterized by clusters of peaks, and the corresponding peaks of each cluster are 14 (CH_2) mass units apart. The largest peak in each cluster represents a C_nH_{2n+1} fragment; this is accompanied by C_nH_{2n} and C_nH_{2n-1} fragments. The most abundant fragments are at C_3 and C_4, and the fragment abundances decrease in a smooth curve down to M–C_2H_5; the M–CH_3 peak is characteristically very weak or missing. Compounds containing more than 8 carbon atoms show fairly similar spectra; identification then depends on the molecular ion peak.

Spectra of branched saturated hydrocarbons are grossly similar to those of straight-chain compounds, but the smooth curve of decreasing intensities is broken by preferred fragmentaiton at each branch. The smooth curve for the *n*-alkane in Figure 6a is in contrast to the discontinuity at C_{10} for the branched alkane (Figure 6b). This discontinuity indicates that the longest branch of 5-methylpentadecane has 10 carbons.

In Figure 6b, the peaks at *m/e* 169 and 85 represent cleavage on either side of the branch with charge retention on the substituted carbon atom. Subtraction of the molecular weight from the sum of these fragments accounts for the fragment –CH–CH_3. Again we appreciate the absence of a C_{11} unit, which cannot form by a single cleavage. Finally, the presence of a distinct $M - 15$ peak also indicates a methyl branch. The fragment resulting from cleavage at a branch tends to lose a single hydrogen atom so that the resulting C_nH_{2n} peak is prominent and sometimes more intense than the corresponding C_nH_{2n+1} peak. Random rearrangements are common, and the use of reference compounds for final identification is good practice.

A saturated ring in a hydrocarbon increases the relative intensity of the molecular ion peak, and favors cleavage at the bond connecting the ring to the rest of the molecule (Rule 6, p. 15). Fragmentation of the ring is usually characterized by loss of two carbon atoms as C_2H_4 (28) and C_2H_5 (29). This tendency to lose even numbered fragments, such as C_2H_4, gives a spectrum that contains a greater proportion of even-numbered mass ions than the spectrum of an acyclic hydrocarbon. As in branched hydrocarbons, C–C cleavage is accompanied by loss of a hvdrogen atom. The characteristic peaks are therefore in the C_nH_{2n-1} and C_nH_{2n-2} series.

The mass spectrum of cyclohexane (Figure 6c) shows a much more intense molecular ion than those of acyclic compounds, since fragmentation requires the cleavage of two carbon-carbon bonds. This spectrum has its base peak

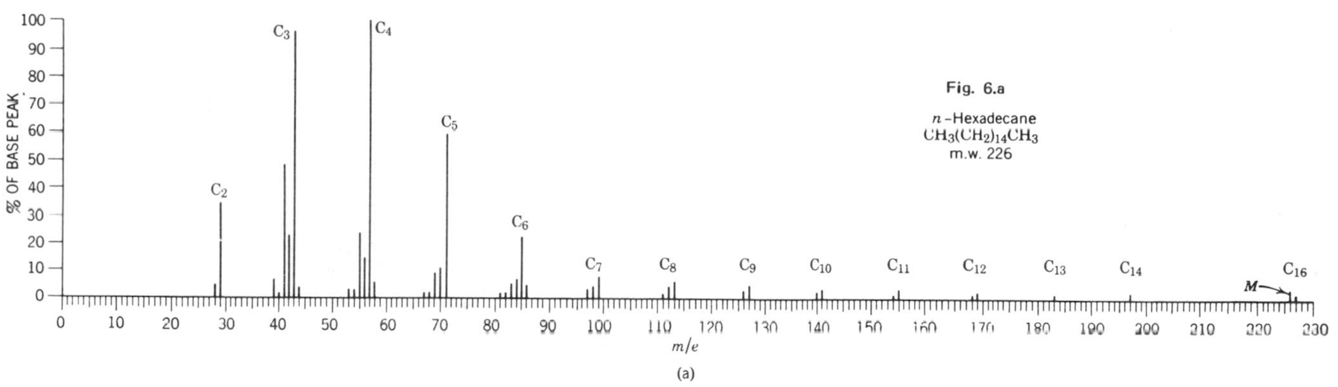

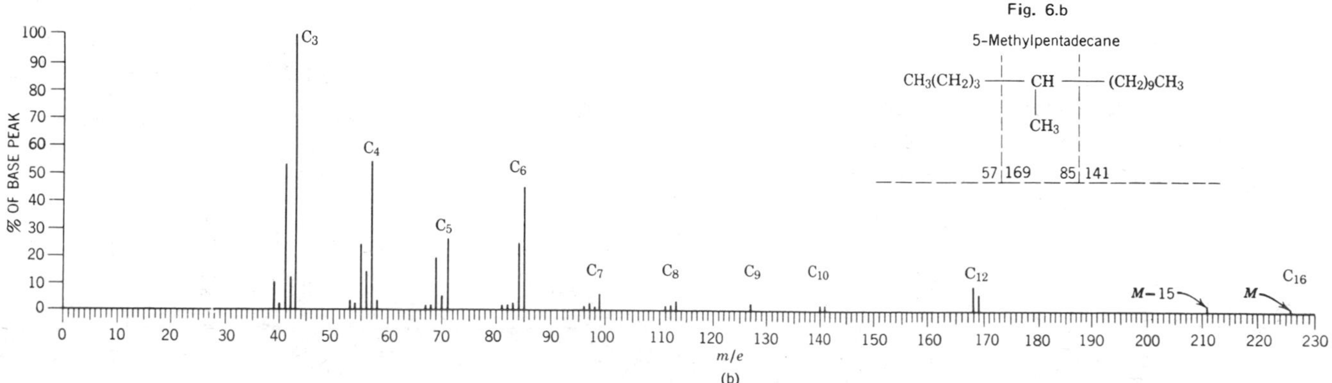

Figure 6 a and b. *Isomeric C_{16} hydrocarbons.*

at m/e 56 (due to loss of C_2H_4) and a large peak at m/e 41, which is a fragment in the C_nH_{2n-1} series with n = 3.

Olefins

The molecular ion peak of olefins, especially polyolefins, is usually distinct. Location of the double bond in acyclic olefins is difficult because of its facile migration in the fragments. In cyclic (especally polycyclic) olefins, location of the double bond is frequently evident as a result of a strong tendency for allylic cleavage without much double bond migration (Rule 5, p. 15). Conjugation with a carbonyl group also fixes the position of the double bond. As with saturated hydrocarbons, acyclic olefins are characterized by clusters of peaks at intervals of 14 units. In these clusters the C_nH_{2n-1} and C_nH_{2n} peaks are more intense than the C_nH_{2n+1} peaks.

The mass spectrum of β-myrcene, a terpene, is shown in Figure 7. The peaks at m/e 41, 55, and 69 correspond to the formula C_nH_{2n-1} with n = 3, 4, and 5, respectively. Formation of the m/e 41 peak must involve rearrangement. The peaks at m/e 67 and 69 are the fragments from cleavage of a bi-allylic bond.

m/e 69

m/e 67

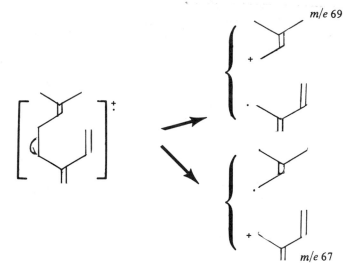

The peak at m/e 93 may be rationalized as a structure of formula $C_7H_9^+$ formed by isomerization (resulting in increased conjugation), followed by allylic cleavage.

m/e 93

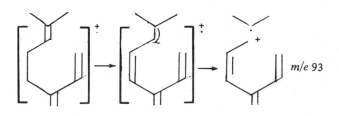

Cyclic olefins usually show a distinct molecular ion peak. A unique mode of cleavage is a type of homolytic *retro*-Diels-Alder reaction as shown by limonene.

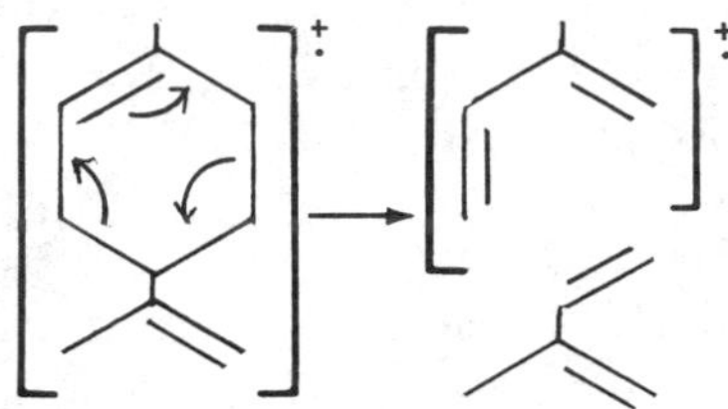

Aromatic and Aralkyl Hydrocarbons

An aromatic ring in a molecule stabilizes the molecular ion peak (Rule 4, p. 15), which is usually sufficiently large that accurate measurements can be made on the $M+1$ and $M+2$ peaks.

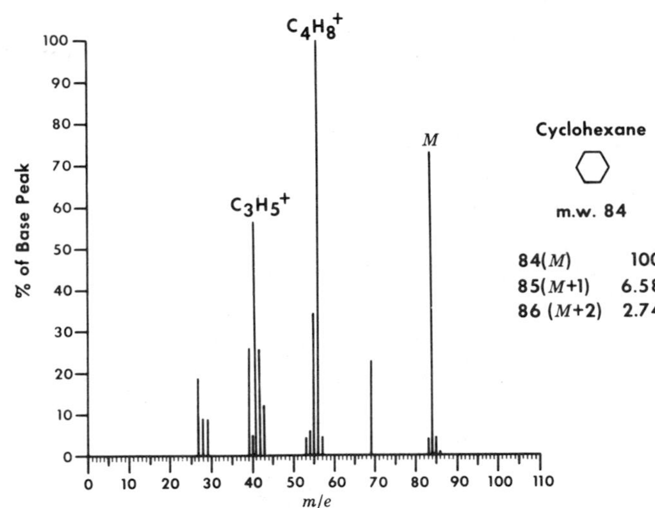

Figure 6c. *Cyclohexane.*

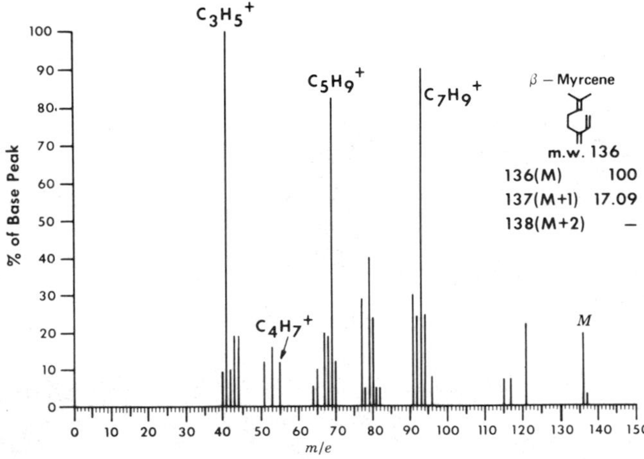

Figure 7. *β-Myrcene.*

Figure 8 is the mass spectrum of naphthalene. The molecular ion peak is also the base peak, and the largest fragment peak, *m/e* 51, is only 12.5% as intense as the molecular ion peak.

A prominent peak (often the base peak) at *m/e* 91 ($C_6H_5CH_2^+$) is indicative of an alkyl substituted benzene ring. Branching at the α-carbon leads to masses higher than 91 by increments of 14, the largest substituent being eliminated most readily (Rule 3, p. 15). The mere presence of a peak at mass 91, however, does not preclude branching at the α-carbon because this highly stabilized fragment may result from rearrangements. A distinct and sometimes prominent $M - 1$ peak results from similar benzylic cleavage of a C—H bond.

It has been shown that, in most cases, the ion of mass 91 is a tropylium rather than a benzylic cation. This explains the ready loss of a methyl group from xylenes although toluene does not easily lose a methyl group. The incipient molecular ion rearranges to the parent tropylium radical ion, which then cleaves to the simple tropylium ion ($C_7H_7^+$):

$$\text{xylene}^{+\cdot}\ (\text{CH}_3,\ \text{CH}_3) \rightarrow \text{methyltropylium radical ion}\ (\text{H},\ \text{H},\ \text{CH}_3) \xrightarrow{-\text{CH}_3\cdot} \text{tropylium}^+\quad m/e\ 91$$

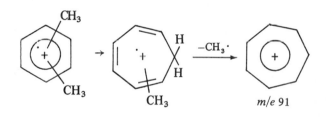

The frequently observed peak at *m/e* 65 results from elimination of a neutral acetylene molecule from the tropylium ion:

$$\text{tropylium}^+ \xrightarrow{-\text{HC}\equiv\text{CH}} C_5H_5^+\quad m/e\ 65$$

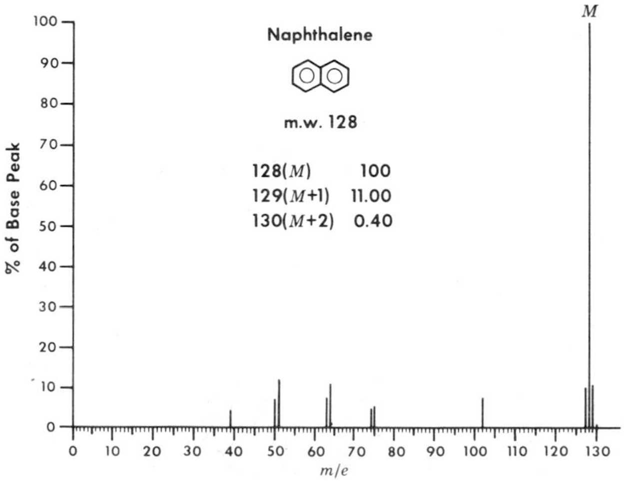

Figure 8. *Naphthalene.*

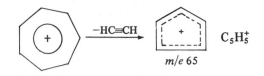

Hydrogen migration with elimination of a neutral olefin molecule accounts for the peak at *m/e* 92 observed when the alkyl group is longer than C_2.

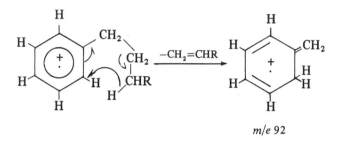

$$C_6H_5\text{—}CH_2\text{—}CH_2\text{—}CHR\text{—}H^{+\cdot} \xrightarrow{-\text{CH}_2=\text{CHR}} C_7H_8^{+\cdot}\ (\text{CH}_2,\ \text{H},\ \text{H})\quad m/e\ 92$$

A characteristic cluster of ions due to α-cleavage and hydrogen migration in monalkylbenzenes appears at *m/e* 77 ($C_6H_5^+$), 78 ($C_6H_6^+$), and 79 ($C_6H_7^+$).

Alkylated polyphenyls and alkylated polycyclic aromatic hydrocarbons exhibit the same β-cleavage as alkylbenzene compounds.

HYDROXY COMPOUNDS

Alcohols

The molecular ion peak of a primary or secondary alcohol is quite small and for a tertiary alcohol is undetectable. The molecular ion of 1-pentanol is extremely weak compared with its near homologs. The expedients mentioned above may be used to obtain the molecular weight.

Cleavage of the C—C bond next to the oxygen atom is of general occurrence (Rule 8, p. 15). Thus, primary alcohols show a prominent peak due to $CH_2{=}\overset{+}{\ddot{O}}H$ (*m/e* 31). Secondary and tertiary alcohols cleave analogously to give a prominent peak due to $\text{R(H)C}{=}\overset{+}{\ddot{O}}H$ (*m/e* 45, 59, 73, etc.), and $\text{RR'C}{=}\overset{+}{\ddot{O}}H$ (*m/e* 59, 73, 87, etc.), respectively. The largest substituent is expelled most readily (Rule 3).

$$\text{RR'R''C—}\dot{\ddot{O}}^{+}\text{—H} \xrightarrow{-\text{R}''\cdot} \text{RR'C}{=}\overset{+}{\ddot{O}}\text{H} \longleftrightarrow \text{RR'}\overset{+}{\text{C}}\text{—}\ddot{O}\text{H}$$

(R″ > R′ or R) = alkyl

When R and/or R′ = H, an $M - 1$ peak can usually be seen.

Primary alcohols, in addition to the principal C—C cleavage next to the oxygen atom, show a homologous series of peaks of progressively decreasing intensity resulting from cleavage at C—C bonds successively removed from the oxygen atom In long-chain ($>C_6$) alcohols, the fragmentation becomes dominated by the hydrocarbon pattern; in fact, the spectrum resembles that of the corresponding olefin.

The spectrum in the vicinity of the very weak or missing molecular ion peak of a primary alcohol is sometimes complicated by weak $M - 2$ (R—CH=$\ddot{O}$) and $M - 3$ (R—C≡$\overset{+}{O}$) peaks.

A distinct and sometimes prominent peak can usually be found at $M - 18$ from loss of water. This peak is most noticeable in spectra of primary alcohols. This elimination by electron impact has been rationalized as follows:

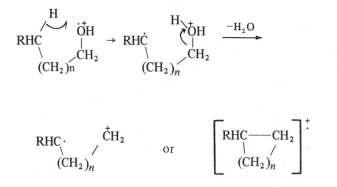

$$\text{RHC(H)}-(CH_2)_n-CH_2-\overset{\cdot +}{O}H \rightarrow R\dot{H}C-(CH_2)_n-CH_2-\overset{+}{O}H(H) \xrightarrow{-H_2O}$$

$$R\dot{H}C-(CH_2)_n-\overset{+}{C}H_2 \quad \text{or} \quad \left[\text{RHC—CH}_2,\ (CH_2)_n \text{ ring}\right]^{\dot{+}}$$

This pathway is consistent with the loss of the OH and γ-hydrogen ($n = 1$) or δ-hydrogen ($n = 2$); the ring structure is not proven by the observations and is merely one possible structure for the product radical cation. The $M - 18$ peak is frequently exaggerated by thermal decomposition of higher alcohols on hot inlet surfaces. Elimination of water, together with elimination of an olefin from primary alcohols, accounts for the presence of a peak at $M - (\text{olefin} + H_2O)$, that is, a peak at $M - 46$, $M - 74$, $M - 102, \ldots$

$$\text{RCH(H)}-CH_2-CH_2-CH_2-\overset{\cdot +}{O}H \xrightarrow[-H_2O]{-CH_2=CH_2} [CH_2=CHR]^{\dot{+}}$$

$M-(\text{olefin} + H_2O)$

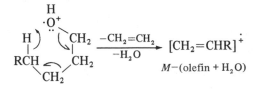

The olefinic ion then decomposes by successive eliminations of ethylene.

Alcohols containing branched methyl groups (e.g., terpene alcohols) frequently show a fairly strong peak at $M - 33$ resulting from loss of CH_3 and H_2O.

Cyclic alcohols undergo fragmentation by complicated pathways; for example, cyclohexanol ($M = m/e$ 100) forms $C_6H_{11}O^+$ by simple loss of the α-hydrogen, loses H_2O to form $C_6H_{10}^{\dot{+}}$ (which appears to have more than one possible bridged bicyclic structure), and forms $C_3H_5O^+$ (m/e 57) by a complex ring cleavage pathway.

A peak at m/e 31 (see above) is quite diagnostic for a primary alcohol provided it is more intense than peaks at m/e 45, 59, 73. . . . However, the first-formed ion of a secondary alcohol can decompose further to give a moderately intense m/e 31 ion.

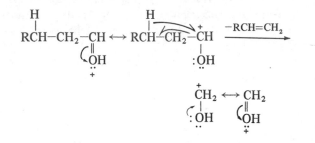

$$\text{RCH(H)}-CH_2-CH=\overset{+}{O}H \longleftrightarrow \text{RCH(H)}-CH_2-\overset{+}{C}H-\ddot{O}H \xrightarrow{-RCH=CH_2}$$

$$\overset{+}{C}H_2-\ddot{O}H \longleftrightarrow CH_2=\overset{+}{O}H$$

Figure 9 gives the characteristic spectra of isomeric primary, secondary, and tertiary C_5 alcohols.

Benzyl alcohols and their substituted homologs and analogs constitute a distinct class. Generally the parent peak is strong. A moderate benzylic peak (M – OH) is present as expected from cleavage beta to the ring. A complicated sequence leads to prominent $M - 1$, $M - 2$, and $M - 3$ peaks. Benzyl alcohol itself fragments to give sequentially the $M - 1$ ion, the C_6H_7 ion by loss of CO, and the C_6H_5 ion by loss of H_2:

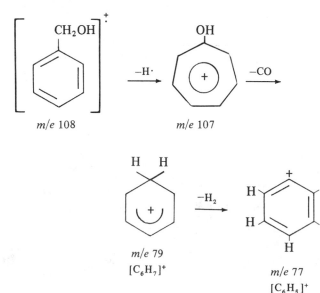

$$[C_6H_5CH_2OH]^{\dot{+}} \xrightarrow{-H\cdot} C_7H_6OH^+ \xrightarrow{-CO}$$

m/e 108 $\quad$ m/e 107

$$C_6H_7^+ \xrightarrow{-H_2} C_6H_5^+$$

m/e 79 $[C_6H_7]^+$ $\quad$ m/e 77 $[C_6H_5]^+$

Loss of H_2O to give a distinct $M - 18$ peak is a common feature, especially pronounced and mechanistically straightforward in some ortho-substituted benzyl alcohols.

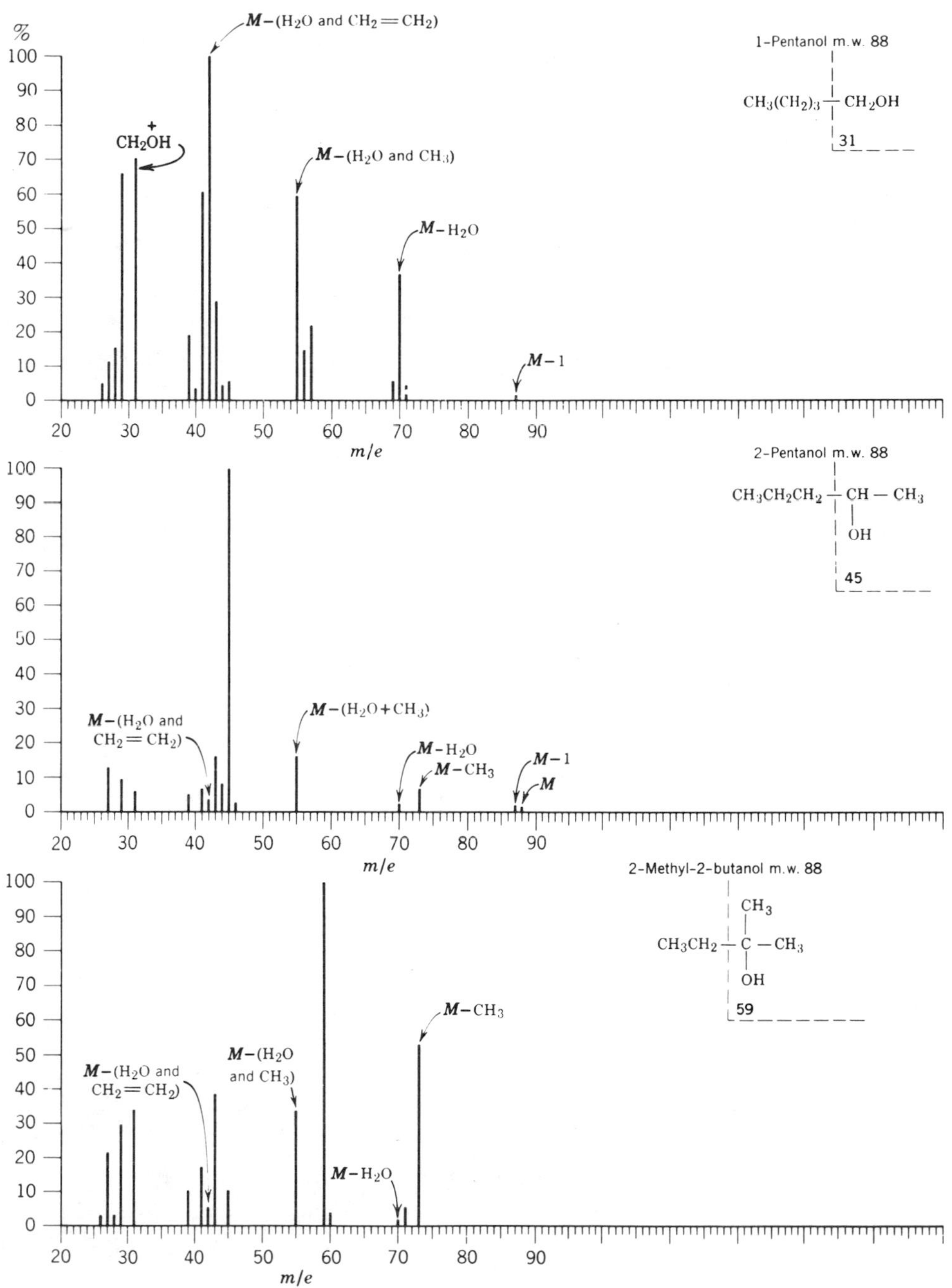

Figure 9. *Isomeric pentanols.*

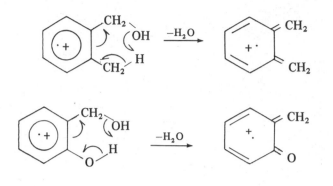

$-H_2O$

$-H_2O$

The aromatic cluster at *m/e* 77, 78, and 79 resulting from complex degradation is prominent here also.

Phenols

A conspicuous molecular ion peak facilitates identification of phenols. In phenol itself, the molecular ion peak is the base peak, and the $M-1$ peak is small. In cresols, the $M-1$ peak is larger than the parent peak as a result of a facile benzylic C–H cleavage. A rearrangement peak at *m/e* 77 and peaks resulting from loss of CO ($M-28$) and CHO ($M-29$) are usually found in phenols:

$-CO$ $-H\cdot$

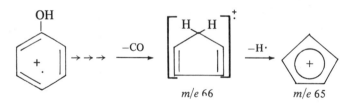

m/e 66   *m/e* 65

The mass spectrum of a typical phenol is shown in Figure 10. This spectrum shows that a methyl group is lost much more readily than an α-hydrogen.

$-CH_3\cdot$ *m/e* 107 (100%)

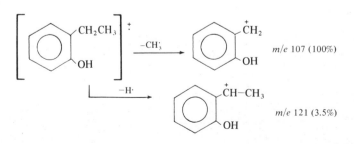

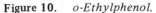

$-H\cdot$ *m/e* 121 (3.5%)

ETHERS

Aliphatic Ethers

The molecular ion peak (two mass units larger than that of an analogous hydrocarbon) is small, but larger sample size usually will make the molecular ion peak or the $M+1$ peak obvious (H· transfer during ion-molecule collision, see part 2, Section V).

The presence of an oxygen atom can be deduced from strong peaks at *m/e* 31, 45, 59, 73, ... These peaks represent the RO^+ and $ROCH_2^+$ fragments.

Fragmentation occurs in two principal ways:

a. Cleavage of the C–C bond next to the oxygen atom (α-β bond, Rule 8, Section VII)

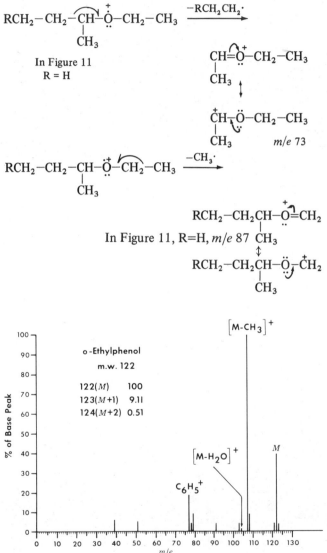

$RCH_2-CH_2-CH(CH_3)-\overset{+}{O}-CH_2-CH_3 \xrightarrow{-RCH_2CH_2\cdot} CH(CH_3)=\overset{+}{O}-CH_2-CH_3 \leftrightarrow \overset{+}{C}H(CH_3)-O-CH_2-CH_3$ *m/e* 73

In Figure 11
R = H

$RCH_2-CH_2-CH(CH_3)-\overset{+}{O}-CH_2-CH_3 \xrightarrow{-CH_3\cdot} RCH_2-CH_2CH(CH_3)-\overset{+}{O}=CH_2 \leftrightarrow RCH_2-CH_2CH(CH_3)-O-\overset{+}{C}H_2$

In Figure 11, R=H, *m/e* 87

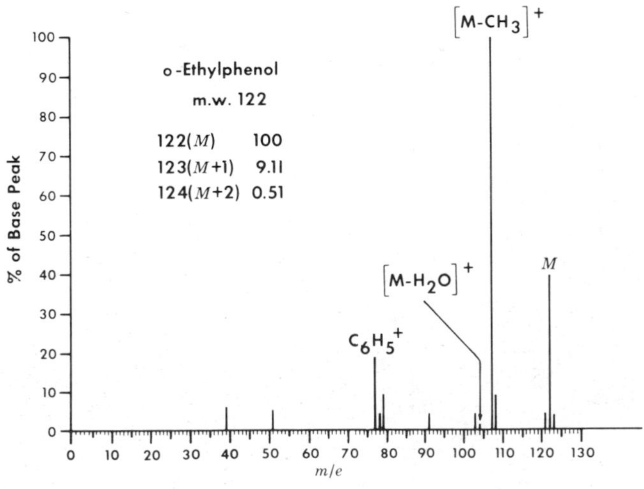

Figure 10. *o-Ethylphenol.*

One or the other of these oxygen-containing ions may account for the base peak. In the case shown, the first cleavage (i.e., at the branch positions to lose the larger fragment) is preferred. However, the first-formed fragment decomposes further by the following process, often to give the base peak (Figure 11); the decomposition is important when the α–C is substituted.

$$\underset{\underset{CH_3\ H-CH_2}{|\quad\ \ |}}{CH=\overset{+}{\ddot{O}}-CH_2} \xrightarrow{-CH_2=CH_2} \underset{\underset{CH_3}{|}}{CH=\overset{+}{\ddot{O}}H} \longleftrightarrow \underset{\underset{CH_3}{|}}{\overset{+}{C}H-\ddot{O}H}$$

m/e 45

b. C–O bond cleavage with the charge remaining on the alkyl fragment.

$$R-\overset{+}{\ddot{O}}-R' \xrightarrow{-\cdot\ddot{O}R'} R^+$$

$$R-\overset{+}{\ddot{O}}-R' \xrightarrow{-R\ddot{O}\cdot} R'^+$$

As expected, the spectrum of long-chain ethers becomes dominated by the hydrocarbon pattern.

Acetals are a special class of ethers. Their mass spectra are characterized by an extremely weak molecular ion peak, by the prominent peaks at *M* minus R, and *M* minus OR, and a weak peak at *M* minus H. Each of these cleavages is mediated by an oxygen atom and thus facile. As usual elimination of the largest group is preferred. As with aliphatic ethers, the first-formed oxygen-containing fragments can decompose further with hydrogen migration and olefin elimination.

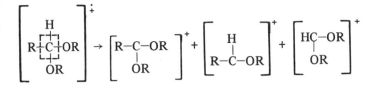

$$\left[\underset{\underset{OR}{|}}{\overset{\overset{H}{|}}{R-C-OR}}\right]^{+\cdot} \rightarrow \left[\underset{\underset{OR}{|}}{R-C-OR}\right]^{+} + \left[\overset{\overset{H}{|}}{R-C-OR}\right]^{+} + \left[\underset{\underset{OR}{|}}{HC-OR}\right]^{+}$$

Ketals behave similarly.

Aromatic Ethers

The molecular ion peak of aromatic ethers is prominent. Primary cleavage occurs at the bond beta to the ring, and the first-formed ion can decompose further. Thus anisole, m.w. 108, gives ions of *m/e* 93 and 65.

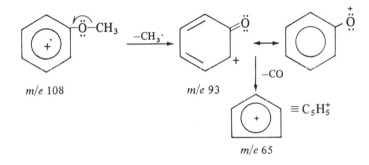

$$C_6H_5\ddot{O}CH_3^{+\cdot}\ (m/e\ 108) \xrightarrow{-CH_3\cdot} C_6H_5O^+\ (m/e\ 93) \xrightarrow{-CO} C_5H_5^+\ (m/e\ 65)$$

m/e 108 *m/e* 93 $\equiv C_5H_5^+$ *m/e* 65

The characteristic aromatic peaks at *m/e* 78 and 77 may arise from anisole as follows.

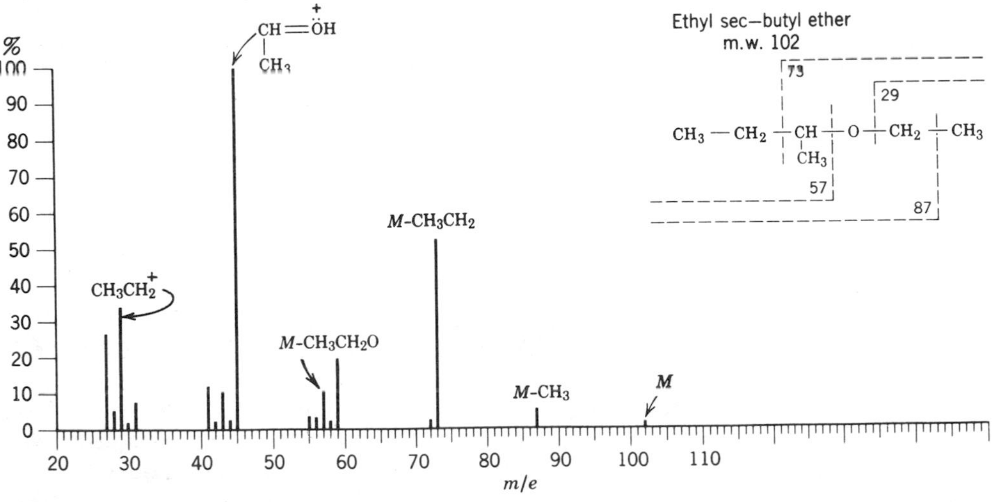

Figure 11. *Ethyl sec-butyl ether.*

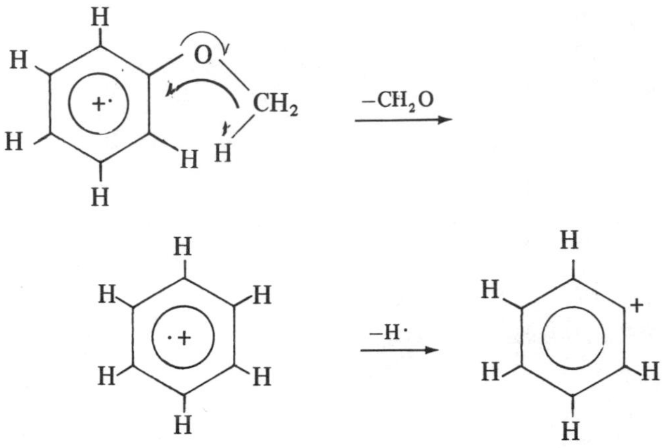

$$\xrightarrow{-CH_2O}$$

m/e 78 $\xrightarrow{-H\cdot}$ *m/e* 77

When the alkyl portion of an aromatic alkyl ether is C_2 or larger, cleavage beta to the ring is accompanied by hydrogen migration as noted above for alkylbenzenes. Clearly, cleavage is mediated by the ring rather than by the oxygen atom; C−C cleavage next to the oxygen atom is insignificant.

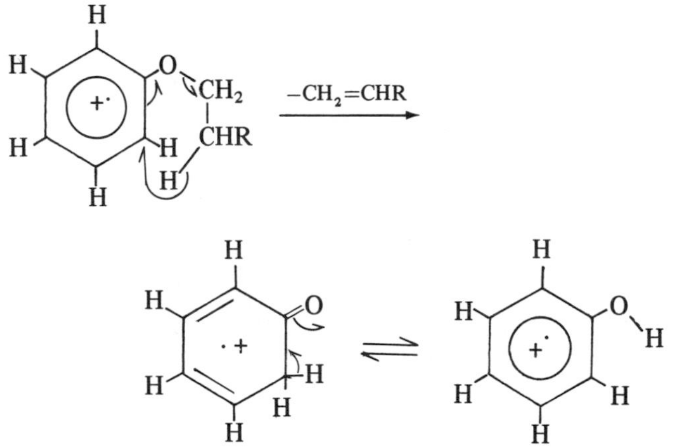

$$\xrightarrow{-CH_2{=}CHR}$$

m/e 94

Diphenyl ethers show peaks at $M - H$, $M - CO$, and $M - CHO$ by complex rearrangements.

KETONES

Aliphatic Ketones

The molecular ion peak of ketones is usually quite pronounced. Major fragmentation peaks of aliphatic ketones result from cleavage at the C−C bonds adjacent to the oxygen atom, the charge remaining with the oxygenated fragment. Thus, as with alcohols and ethers, cleavage is again at the C−C bond next to the oxygen atom.

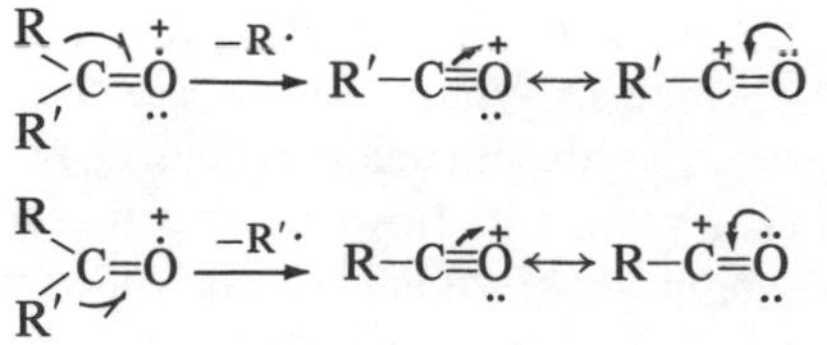

$$RR'C{=}\dot{O}^{+} \xrightarrow{-R\cdot} R'{-}C{\equiv}O^{+} \longleftrightarrow R'{-}\overset{+}{C}{=}O$$

$$RR'C{=}\dot{O}^{+} \xrightarrow{-R'\cdot} R{-}C{\equiv}O^{+} \longleftrightarrow R{-}\overset{+}{C}{=}O$$

This cleavage gives rise to a peak at *m/e* 43 or 57 or 71.... The base peak very often results from loss of the larger alkyl group.

When one of the alkyl chains attached to the C=O group is C_3 or longer, cleavage of the C−C bond once removed ($\alpha\beta$ bond) from the C=O group occurs with hydrogen migration to give a major peak (McLafferty rearrangement):

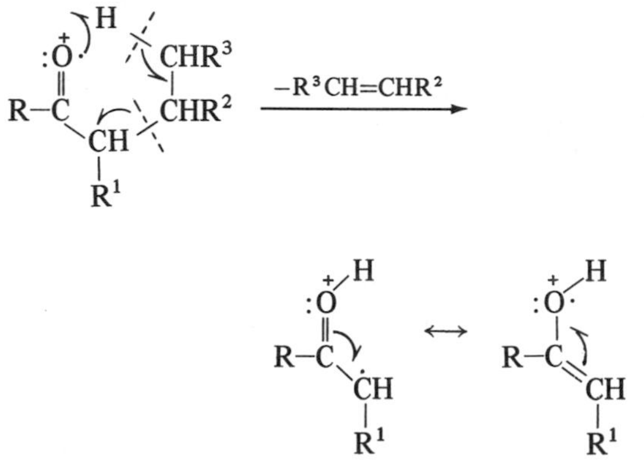

$$\xrightarrow{-R^3CH{=}CHR^2}$$

Simple cleavage of the $\alpha\beta$ bond, which does not occur to any extent, would give an ion of low stability with two adjacent positive centers $R{-}\overset{\delta+}{C}(=\overset{\delta-}{O}){-}\overset{+}{C}H_2$. When R is C_3 or longer, the first-formed ion can cleave again with hydrogen migration:

$$\xrightarrow{-CH_2{=}CH_2}$$

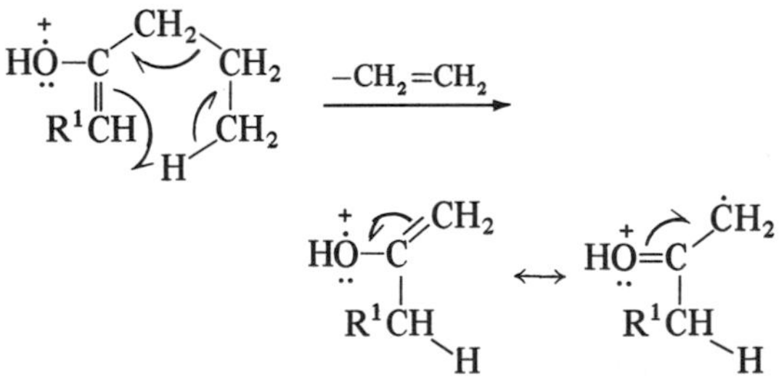

Note that in long-chain ketones the hydrocarbon peaks are indistinguishable (without the aid of high-resolution techniques) from the acyl peaks, since the mass of the CO unit (C=O, 28) is the same as two methylene units. The multiple cleavage modes in ketones sometimes make difficult the determination of the carbon chain configuration. Reduction of the carbonyl group to a methylene group yields the corresponding hydrocarbon whose fragmentation pattern leads to the carbon skeleton.

Cyclic Ketones

The molecular ion peak in cyclic ketones is prominent. As with aliphatic ketones, the primary cleavage of cyclic ketones is adjacent to the C=O group, but the ion thus formed must undergo further cleavage in order to produce a fragment. The base peak in the spectrum of cyclopentanone and of cyclohexanone is *m/e* 55. The mechanism is similar in both cases: hydrogen shift to convert a primary radical to a conjugated secondary radical followed by formation of the resonance-stable ion, *m/e* 55.

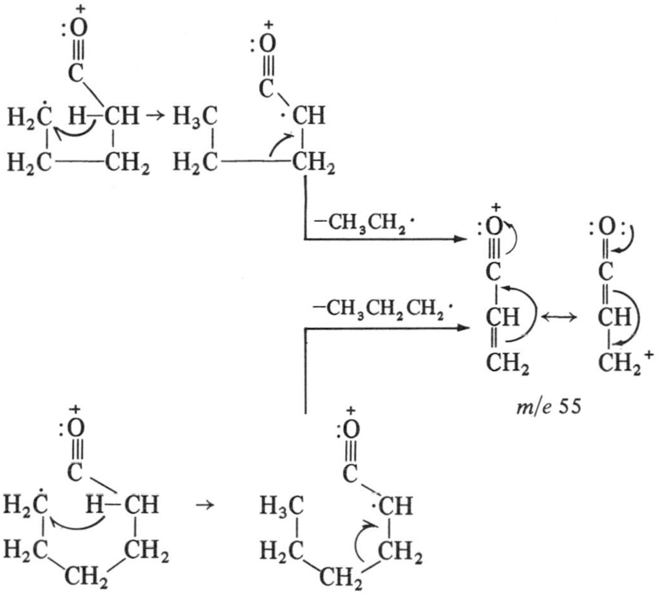

−CH_3CH_2·

−$CH_3CH_2CH_2$·

m/e 55

The other distinctive peaks at *m/e* 83 and 42 in the spectrum of cyclohexanone have been rationalized as follows.

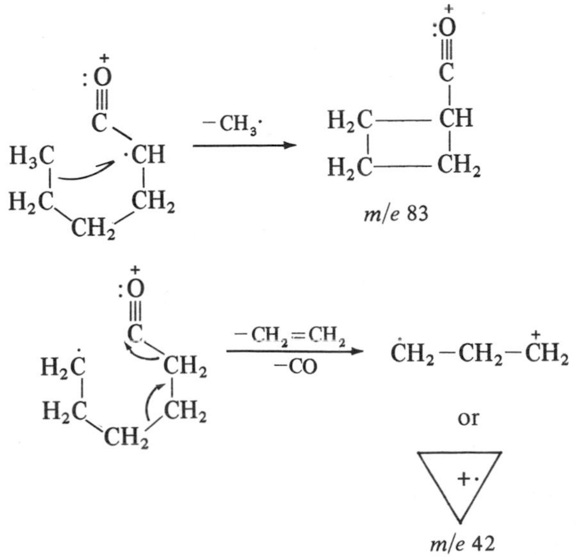

−CH_3·

m/e 83

−CH_2=CH_2
−CO

$\dot{C}H_2$−CH_2−$\overset{+}{C}H_2$

or

m/e 42

Aromatic Ketones

The molecular ion peak of aromatic ketones is prominent. Cleavage of aryl alkyl ketones occurs at the bond beta to the ring, leaving a characteristic $ArC\equiv\overset{+}{O}$: fragment which usually accounts for the base peak. Loss of CO from this fragment gives the "aryl" ion (*m/e* 77 in the case of acetophenone). Cleavage of the bond adjacent to the ring to form a $RC\equiv\overset{+}{O}$ fragment is less important though somewhat enhanced by electron-withdrawing groups (and diminished by electron-donating groups) in the *para* position of the Ar group.

When the alkyl chain is C_3 or longer, cleavage of the C−C bond once-removed from the C=O group occurs with hydrogen migration. This is the same cleavage noted for aliphatic ketones that proceeds through a cyclic transition state and results in elimination of an olefin and formation of a stable ion.

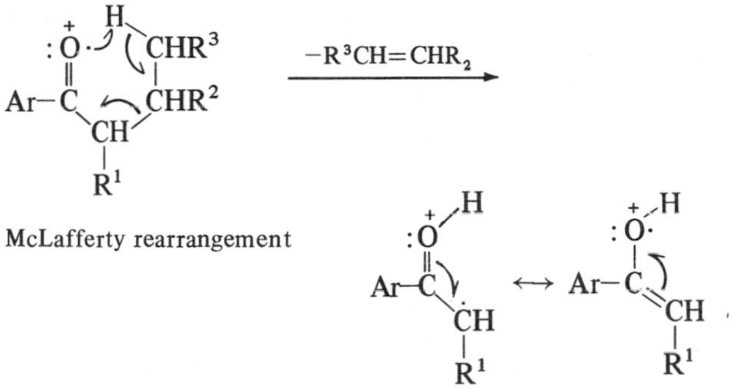

−R^3CH=CHR_2

McLafferty rearrangement

The mass spectrum of an unsymmetrical diaryl ketone, *p*-chlorobenzophenone, is displayed in Figure 12. The molecular ion peak (*m/e* 216) is prominent and the intensity (33.99%) of the $M + 2$ peak (relative to the molecular ion peak) demonstrates that chlorine is in the structure (see the discussion of Table III and Figure 18 below).

Since the intensity of the *m/e* 141 peak is about ⅓ the intensity of the *m/e* 139 peak, these peaks correspond to fragments containing 1 chlorine each. The same can be said about the fragments producing the *m/e* 111 and 113 peaks.

The major peaks in Figure 12 arise as follows.

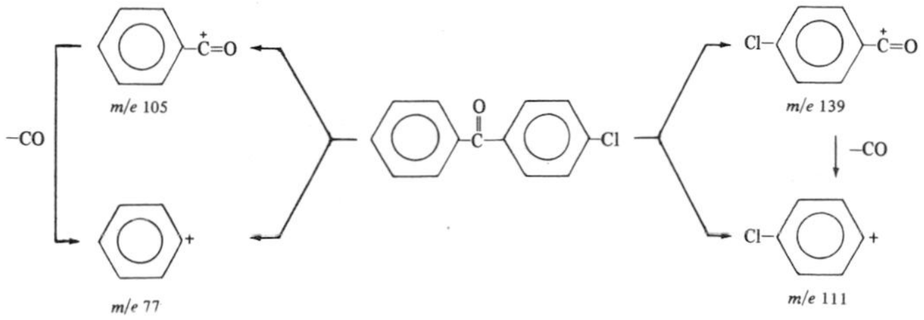

m/e 105 · *m/e* 139 · −CO · −CO · *m/e* 77 · *m/e* 111

The observed peak intensities make it clear that formation of a $ArCO^+$ fragment occurs more readily than does Ar^+ formation. In addition, for either of these two types of fragments, the structure bearing the chlorine is less favored than that bearing no chlorine because the electron-withdrawal effect of the chlorine destabilizes a positively charged fragment.

ALDEHYDES

Aliphatic Aldehydes

The molecular ion peak of aliphatic aldehydes is usually discernible. Cleavage of the C—H and C—C bonds next to the oxygen atom results in an $M-1$ peak and in an $M-R$ peak (m/e 29, CHO^+). The $M-1$ peak is a good diagnostic peak even for long-chain aldehydes, but the m/e 29 peak present in C_4 and higher aldehydes is due to the hydrocarbon $C_2H_5^+$ ion.

In the C_4 and higher aldehydes, McLafferty cleavage of the $\alpha\beta$ C—C bond occurs to give a major peak at m/e 44, 58, or 72, . . . , depending on the α-substituents. This is the resonance-stabilized ion formed through the cyclic transition state as shown above for aliphatic ketones (R = H):

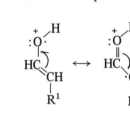

$$\overset{+}{:\!\dot{O}}\text{—H},\ \text{HC}{=}\text{CH—R}^1 \longleftrightarrow \overset{+}{:\!O}\text{(—H)}{=}\text{HC—}\dot{\text{C}}\text{H—R}^1$$

In straight-chain aldehydes, the other unique, diagnostic peaks are at $M-18$ (loss of water), $M-28$ (loss of ethylene), $M-43$ (loss of $CH_2{=}CH{-}O\cdot$), and $M-44$ (loss of $CH_2{=}CH{-}OH$). The rearrangements leading to these peaks have been rationalized (see Reference 2). As the chain lengthens, the hydrocarbon pattern (m/e 29, 43, 57, 71, . . .) becomes dominant. These features are evident in the spectrum of nonanal (Figure 13).

Aromatic Aldehydes

Aromatic aldehydes are characterized by a large molecular ion peak and by an $M-1$ peak ($Ar{-}C{\equiv}\overset{+}{O}$:) that is always large and may be larger than the molecular ion peak. The $M-1$ ion $C_6H_5C{\equiv}\ddot{O}^+$ eliminates CO to give the phenyl ion (m/e 77), which in turn eliminates HC≡CH to give the $C_4H_3^+$ ion (m/e 51).

CARBOXYLIC ACIDS

Aliphatic Acids

The molecular ion peak of a straight-chain monocarboxylic acid is weak but usually discernible. The most characteristic

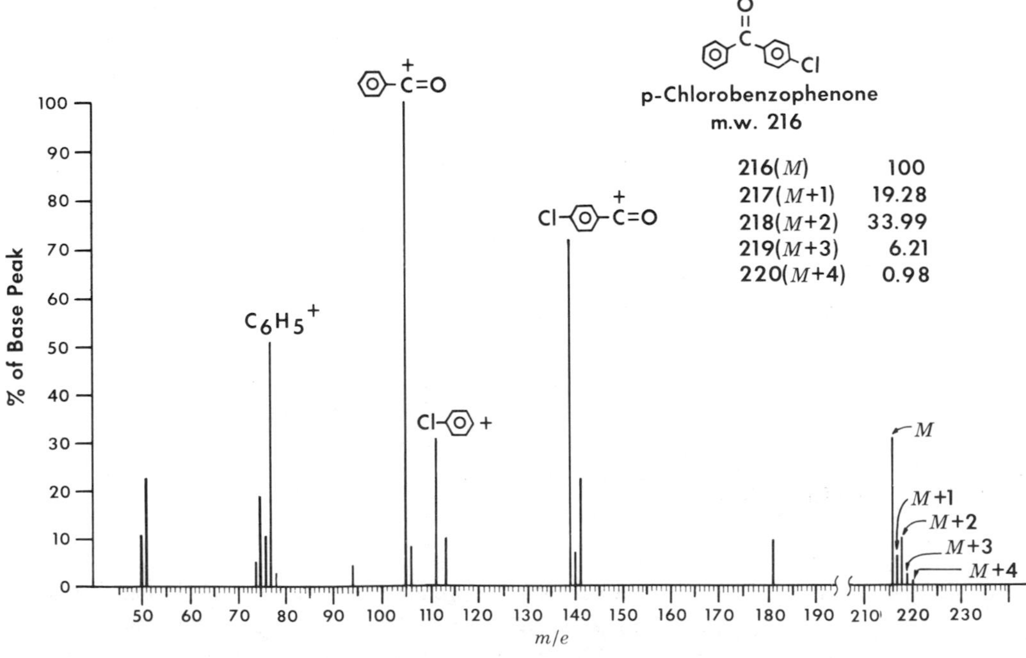

Figure 12. *p-Chlorobenzophenone.*

(sometimes the base) peak is *m/e* 60 due to the McLafferty rearrangement. Branching at the α-carbon enhances this cleavage.

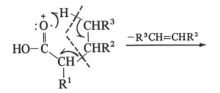

$$\mathrm{HO{-}C({=}\overset{+}{O}{\cdot}\cdots H)\text{-}CH(R^1)\text{-}CHR^2\text{-}CHR^3} \xrightarrow{-\mathrm{R^3CH{=}CHR^2}}$$

McLafferty rearrangement

$$\mathrm{HO{-}C({=}\overset{+}{O}H){-}\dot{C}H{-}R^1} \longleftrightarrow \mathrm{HO{-}C(\overset{+}{O}{\cdot}H){=}CH{-}R^1}$$

In short-chain acids, peaks at $M - \mathrm{OH}$ and $M - \mathrm{COOH}$ are prominent; these represent cleavage of bonds next to C=O. In long-chain acids, the spectrum consists of two series of peaks resulting from cleavage at each C—C bond with retention of charge either on the oxygen-containing fragment (*m/e* 45, 59, 73, 87, . . .) or on the alkyl fragment (*m/e* 29, 43, 57, 71, 85, . . .). As previously discussed, the hydrocarbon pattern also shows peaks at *m/e* 27, 28; 41, 42; 55, 56; 69, 70; . . . . In summary, besides the McLafferty rearrangement peak, the spectrum of a long-chain acid resembles the series of "hydrocarbon" clusters at interval of 14 mass units. In each cluster, however, is a prominent peak at $C_nH_{2n-1}O_2$. Hexanoic acid (m.w. 116) for example, cleaves as follows:

$CH_3(CH_2)_4CO^+$ (small) 99

$CH_3(CH_2)_4^+$ 71

$CH_3(CH_2)_3^+$ 57

$CH_3(CH_2)_2^+$ 43

$CH_3CH_2^+$ 29

$$\mathrm{CH_3{-}CH_2{\vdots}CH_2{\vdots}CH_2{\vdots}CH_2{\vdots}C({=}O){\vdots}OH}$$

45 CO_2H^+

59 (small) $CH_2CO_2H^+$

73 $(CH_2)_2CO_2H^+$

87 $(CH_2)_3CO_2H^+$

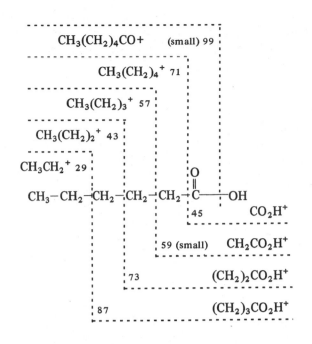

Dibasic acids are usually converted to esters to increase volatility.

Aromatic Acids

The molecular ion peak of aromatic acids is large. The other prominent peaks are formed by loss of OH($M - 17$) and of COOH ($M - 45$). Loss of H_2O ($M - 18$) is noted if a hydrogen-bearing ortho group is available. This is one example of the general "ortho effect" noted when the

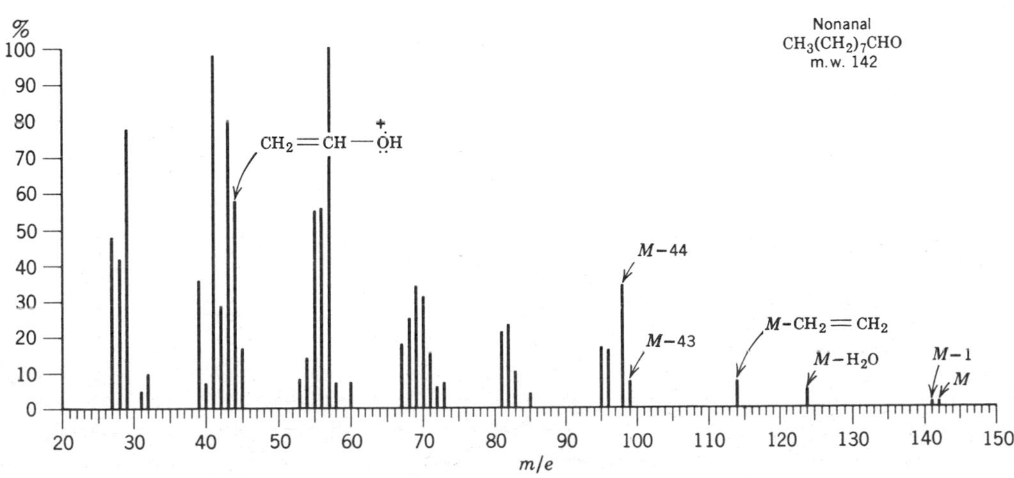

Figure 13. *Nonanal.*

substituents can be in a 6-membered transition state to facilitate loss of a neutral molecule of H_2O, ROH, or NH_3.

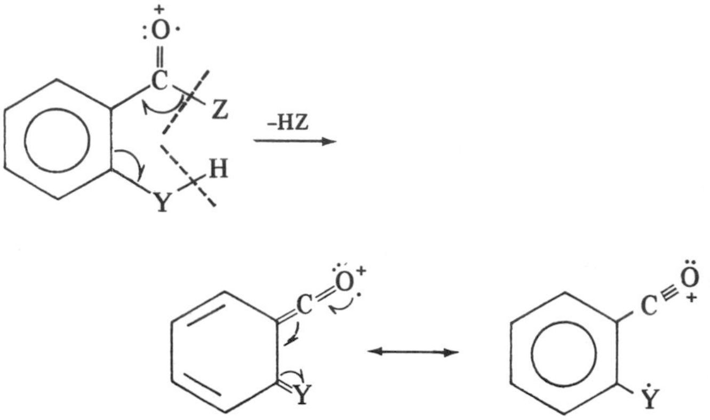

$Z = OH, OR, NH_2$
$Y = CH_2, O, NH$

CARBOXYLIC ESTERS

Aliphatic Esters

The molecular ion peak of a methyl ester of a straight-chain aliphatic acid is usually distinct. Even waxes usually show a discernible molecular ion peak. The molecular ion peak is weak in the range *m/e* 130 to about 200, but becomes somewhat more intense beyond this range. The most characteristic peak is due to the familiar McLafferty rearrangement and cleavage one bond removed from the C=O group. Thus a methyl ester of an aliphatic acid unbranched at the α-carbon gives a strong peak at *m/e* 74, which, in fact, is the base peak in straight-chain methyl esters from C_6 to C_{26}. The alcohol moiety and/or the α-substituent can often be deduced by the location of the peak resulting from this cleavage.

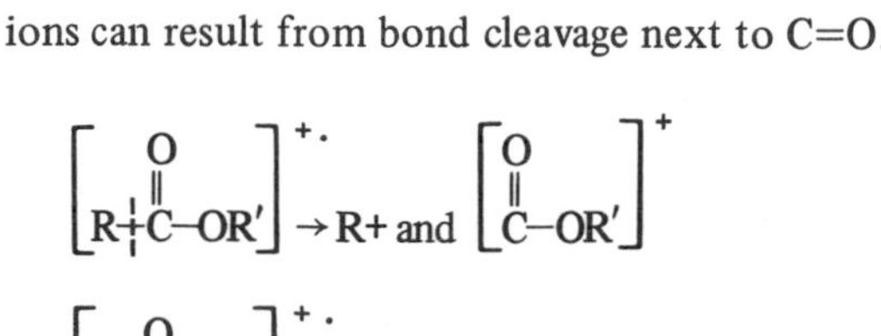

$-R^3CH=CHR^2$

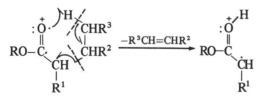

McLafferty rearrangement

Four ions can result from bond cleavage next to C=O.

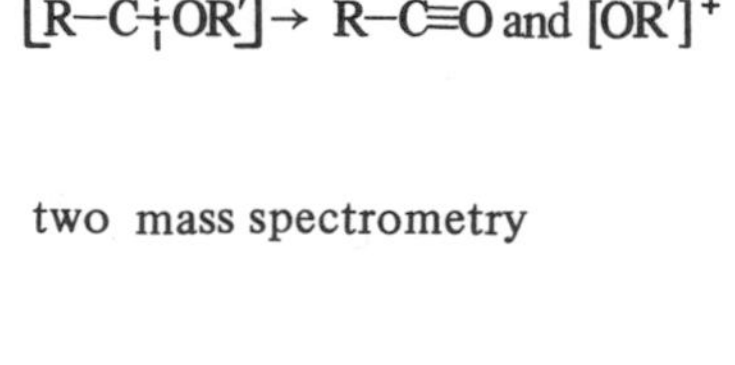

$$\left[R-\overset{O}{\overset{\|}{C}}-OR'\right]^{+\cdot} \rightarrow R^+ \text{ and } \left[\overset{O}{\overset{\|}{C}}-OR'\right]^+$$

$$\left[R-\overset{O}{\overset{\|}{C}}-OR'\right]^{+\cdot} \rightarrow R-C\equiv\overset{+}{O} \text{ and } [OR']^+$$

The ion R^+ is prominent in the short chain esters, but diminishes rapidly with increasing chain length and is barely perceptible in methyl hexanoate. The ion $R-C\equiv\overset{+}{O}$ gives an excellent diagnostic peak for esters. In methyl esters it occurs at $M-31$. It is the base peak in methyl acetate, and is still 4% of the base peak in the C_{26} methyl ester. The ions $[OR']^+$ and $\left[\overset{O}{\overset{\|}{C}}OR'\right]^+$ are usually of little importance. The latter is discernible when $R' = CH_3$ (see *m/e* 59 peak of Figure 14).

Consider, first, esters in which the acid portion is the predominant portion of the molecule. The fragmentation pattern for methyl esters of straight-chain acids can be described in the same terms used for the pattern of the free acid. Cleavage at each C–C bond gives an alkyl ion (*m/e* 29, 43, 57, . . .) and an oxygen-containing ion, $C_nH_{2n-1}O_2^+$ (59, 73, 87, . . .). Thus, there are hydrocarbon clusters at intervals of 14 mass units; in each cluster is a prominent peak at $C_nH_{2n-1}O_2$. The peak (*m/e* 87) representing the ion $[CH_2CH_2COOCH_3]^+$ is always more intense than its homologs, but the reason is not immediately obvious. However, it seems clear that the $C_nH_{2n-1}O_2$ ions do not at all arise from simple cleavage.

The spectrum of methyl octanoate is presented as Figure 14. This spectrum illustrates one difficulty previously mentioned (Section V, part 2) in using the $M+1$ peak to arrive at a molecular formula. The measured value for the $M+1$ peak is 12.9%. The calculated value (Appendix A) is 10.0%. The measured value is high due to an ion-molecule reaction because a relatively large sample was used in order to see the weak molecular ion peak. The utility of the 5-galvanometer recorder is implied by this spectrum; thus, the $M+1$ peak can be measured accurately even though its intensity is only 0.11% of that of the base peak. The accuracy of the $M+2$ peak measurement is marginal.

Now let us consider esters in which the alcohol portion is the predominant portion of the molecule. Esters of fatty alcohols (except methyl esters) eliminate a molecule of acid in the same manner that alcohols eliminate water. A scheme similar to that described earlier for alcohols, involving a single hydrogen transfer to the alcohol oxygen of the ester, can be written. An alternative mechanism involves a hydride transfer to the carbonyl oxygen (McLafferty rearrangement):

$-CH_3COOH$

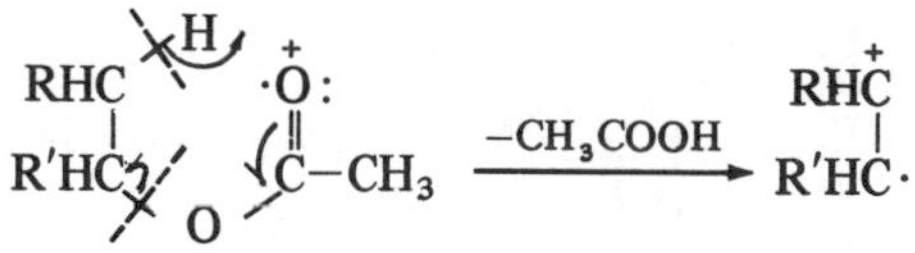

The preceding loss of acetic acid is so facile in steroidal acetates that they frequently show no detectable molecular ion peak. Steroidal systems also seem unusual in that they often display significant molecular ions as alcohols, even when the corresponding acetates do not.

Esters of long-chain alcohols show a diagnostic peak at *m/e* 61, 75, or 89, . . . from elimination of the alkyl moiety and transfer of *two* hydrogen atoms to the fragment containing the oxygen atoms.

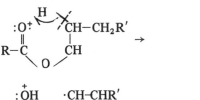

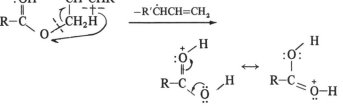

$$R-C(=O^{+\cdot})-O-CH(-CH-CH_2R')\ (\text{H transfer}) \rightarrow$$

$$R-C(=\overset{+}{O}H)-O\cdots\cdot CH-CHR'(CH_2H) \xrightarrow{-R'\dot{C}HCH=CH_2}$$

$$R-C(=\overset{+}{O}-H)-\ddot{O}-H \longleftrightarrow R-C(-\ddot{O}-H)=\overset{+}{O}-H$$

Esters of dibasic acids $RO\overset{\displaystyle O}{\overset{\|}{C}}(CH_2)_n\overset{\displaystyle O}{\overset{\|}{C}}OR$, in general, give recognizable molecular ion peaks. Intense peaks are found at $(RO\overset{\displaystyle O}{\overset{\|}{C}}(CH_2)_n\overset{\displaystyle O}{\overset{\|}{C}})^+$ and at $(RO\overset{\displaystyle O}{\overset{\|}{C}}(CH_2)_n)^+$.

Benzyl and Phenyl Esters

Benzyl acetate (also furfuryl acetate and other similar acetates) and phenyl acetate eliminate the neutral molecule ketene; frequently this gives rise to the base peak.

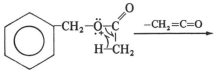

$$C_6H_5-CH_2-\overset{+\cdot}{\ddot{O}}-C(=O)-CH_2-H \xrightarrow{-CH_2=C=O} [C_6H_5-CH_2OH]^{+\cdot}$$

m/e 108

Of course, the *m/e* 43 peak ($CH_3C\equiv\overset{+}{O}$) and *m/e* 91 ($C_7H_7^+$) peaks are prominent for benzyl acetate.

Esters of Aromatic Acids

The molecular ion peak of methyl esters of aromatic acids is prominent. As the size of the alcohol moiety increases, the intensity of the molecular ion peak decreases rapidly to practically zero at C_5. The base peak results from elimination of ·OR, and elimination of ·COOR accounts for another prominent peak. In methyl esters, these peaks are at $M - 31$, and $M - 59$, respectively.

As the alkyl moiety increases in length, three modes of cleavage become important: (1) McLafferty rearrangement, (2) rearrangement of two hydrogen atoms with elimination of an allylic radical, and (3) retention of the positive charge by the alkyl group.

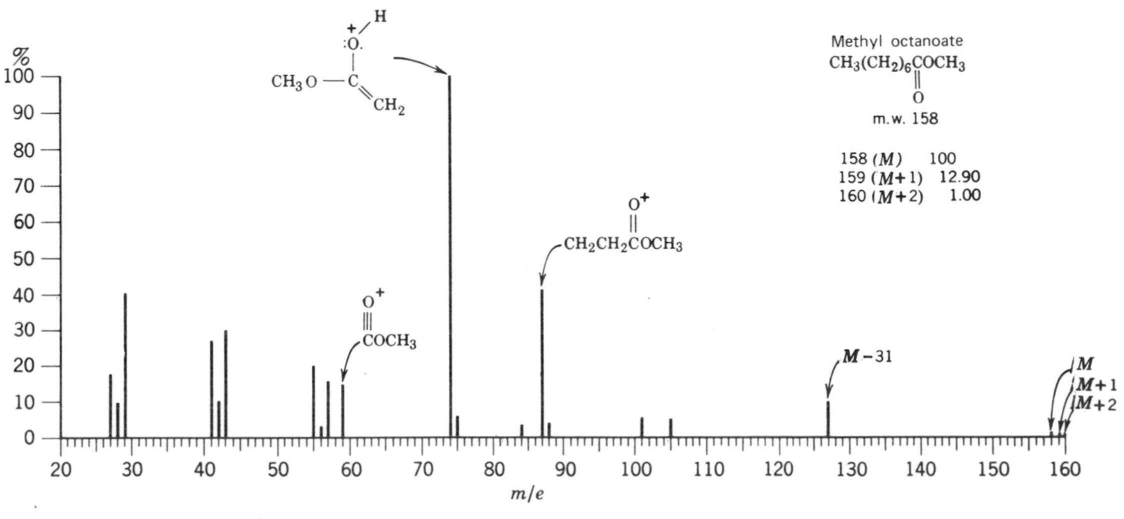

Figure 14. *Methyl octanoate.*

(1) [McLafferty rearrangement scheme: $-CH_2{=}CHR$]

(2) [scheme: $-R'\dot{C}HCH{=}CH_2$]

(3) [scheme: $-Ar-C(=O)-\ddot{O}\cdot$ → R+]

Appropriately, ortho-substituted benzoates eliminate ROH through the general "ortho" effect described above under aromatic acids. Thus, the base peak in the spectrum of methyl salicylate is *m/e* 120; this ion eliminates carbon monoxide to give a strong peak at *m/e* 92.

A strong characteristic peak at mass 149 is found in the spectra of all esters of phthalic acid, starting with the diethyl ester. This peak is not significant in the dimethyl or methyl ethyl ester of phthalic acid, nor in esters of isophthalic or terephthalic acids, all of which give the expected peaks at $M - R$, $M - 2R$, $M - COOR$, and $M - 2COOR$. Since long-chain phthalate esters are widely used as plasticizers and in oil diffusion pumps (often in the diffusion pump of the mass spectrometer inlet), a strong peak at *m/e* 149 may indicate contamination. The *m/e* 149 fragment is probably formed by two ester cleavages involving the shift of two hydrogen atoms and then another hydrogen atom, followed by elimination of H_2O:

[scheme]

m/e 149

LACTONES

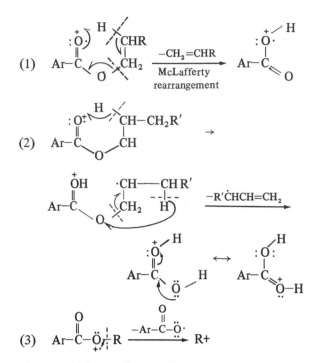

The molecular ion peak of 5-membered ring lactones is distinct but is weaker when an alkyl substituent is present at C_4. Facile cleavage of the side chain at C_4 (Rules 3 and 8, p. 15) gives a strong peak at M minus alkyl.

The base peak (*m/e* 56) of γ-valerolactone and the same strong peak of butyrolactone probably arise as follows.

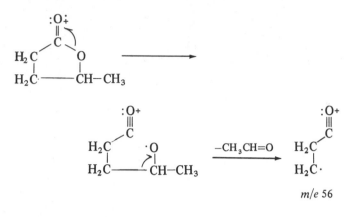

[scheme: $-CH_3CH{=}O$]

m/e 56

Labeling experiments indicate that some of the *m/e* 56 peak in γ-valerolactone arises from the $C_4H_8^+$ ion. The other intense peaks in γ-valerolactone are at *m/e* 27 ($C_2H_3^+$), 28 ($C_2H_4^+$), 29 ($C_2H_5^+$), 41 ($C_3H_5^+$), and 43 ($C_3H_7^+$), and 85 ($C_4H_5O_2^+$, loss of the methyl group). In butyrolactone, there are strong peaks at *m/e* 27, 28, 29, 41, and 42 ($C_3H_6^+$).

AMINES

Aliphatic Amines

The molecular ion peak of an aliphatic monoamine is an odd number, but it is usually quite weak and, in long-chain or highly branched amines, undetectable. The base peak frequently results from C—C cleavage next ($\alpha\beta$) to the nitrogen atom (Rule 8, p. 15); for primary amines unbranched at the α-carbon, this is *m/e* 30 ($CH_2NH_2^+$). This cleavage accounts for the base peak in all primary amines and secondary and tertiary amines that are not branched at the α-carbon. Loss of the largest branch from the α-C atom is preferred.

[scheme: $-\cdot R^2$]

($R^2 > R^1$ or R) = alkyl

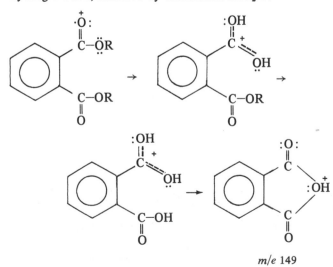

When R and/or R^1 = H, an $M - 1$ peak is usually visible. This is the same type of cleavage noted above for alcohols. The effect is more pronounced in amines because of the better resonance stabilization of the ion fragment by the less electronegative N atom compared with the O atom.

Primary straight-chain amines show a homologous series of peaks of progressively decreasing intensity (the cleavage at the ϵ bond is slightly more important than at the neighboring bonds) at *m/e* 30, 44, 58, . . . resulting from cleavage at C–C bonds successively removed from the nitrogen atom with retention of the charge on the N-containing fragment. These peaks are accompanied by the hydrocarbon pattern of C_nH_{2n+1}, C_nH_{2n}, and C_nH_{2n-1} ions. Thus, we note characteristic clusters at intervals of 14 mass units, each cluster containing a peak due to a $C_nH_{2n+2}N$ ion. Because of the very facile cleavage to form the base peak, the fragmentation pattern in the high mass region becomes extremely weak.

Cyclic fragments apparently occur during the fragmentation of longer chain amines:

$$R{-}CH_2{-}(CH_2)_n{-}\overset{+\cdot}{N}H_2 \rightarrow R\cdot + \overset{}{CH_2}{-}(CH_2)_n{-}\overset{+}{N}H_2 \text{ (ring)}$$

$n = 3,4$ *m/e* 72,86

A peak at *m/e* 30 is good though not conclusive evidence for a straight-chain primary amine. Further decomposition of the first-formed ion from a secondary or tertiary amine leads to a peak at *m/e* 30, 44, 58, 72, ... This is a process similar to that described for aliphatic alcohols and ethers above, and similarly is enhanced by branching at one of the α-carbon atoms:

$$RCH_2{-}\underset{R''}{CH}{-}\overset{+\cdot}{N}H{-}CH_2CH_2R' \xrightarrow{-RCH_2\cdot} \underset{R''}{CH}{=}\overset{+}{N}H{-}CH_2{-}H{-}CHR' \xrightarrow{-CH_2=CHR'} \underset{R''}{CH}{=}\overset{+}{N}H_2$$

$R'' = CH_3$, *m/e* 44, more intense
$R'' = H$, *m/e* 30, less intense

Cleavage of amino acid esters occurs at both C–C bonds (*a, b,* below) next to the nitrogen atom, loss of the carbalkoxy group being preferred (*a*). The aliphatic amine fragment decomposes further to give a peak at *m/e* 30.

$$\underset{\overset{\cdot+}{N}H_2}{CH}{-}COOR' \xleftarrow{b} RCH_2CH_2 \overset{b}{\vdots} \underset{\cdot\overset{+}{N}H_2}{CH} \overset{a}{\vdots} COOR' \xrightarrow{a} RCH_2CH_2CH{=}\overset{+}{N}H_2 \xrightarrow{-RCH=CH_2} CH_2{=}\overset{+}{N}H_2$$

m/e 30

Cyclic Amines

In contrast to acyclic amines, the molecular ion peaks of cyclic amines are usually intense; e.g., the molecular ion peak of pyrrolidine is strong. Primary cleavage at the bonds next to the N atom leads either to loss of an α-H atom to give a strong $M-1$ peak or to opening of the ring; the latter event is followed by elimination of ethylene to give $\cdot CH_2\overset{+}{N}H{=}CH_2$ (*m/e* 43, base peak), thence by loss of a hydrogen atom to give $CH_2{=}\overset{+}{N}{=}CH_2$ (*m/e* 42). *N*-Methyl pyrrolidine also gives a $C_2H_4N^+$ (*m/e* 42) peak, apparently by more than one pathway.

Piperidine likewise shows a strong molecular ion and $M-1$ (base) peak. Ring opening followed by several available sequences leads to characteristic peaks at *m/e* 70, 57. 56, 44, 43, 42, 30, 29, and 28. Substituents are cleaved from the ring (Rule 6, p. 15).

Anilines

The molecular ion peak (odd number) of an aromatic monoamine is intense. Loss of one of the amino H atoms of aniline gives a moderately intense $M-1$ peak; loss of a neutral molecule of HCN followed by loss of a hydrogen atom gives prominent peaks at *m/e* 66 and 65, respectively.

It was noted above that cleavage of alkyl aryl ethers occurs with rearrangement involving cleavage of the ArO–R bond; that is, cleavage was controlled by the ring rather than by the oxygen atom. In the case of alkylarylamines, cleavage of the C–C bond next to the nitrogen atom is dominant; i.e., the heteroatom controls cleavage:

$$C_6H_5{-}\overset{+\cdot}{N}H{-}CH_2{-}CH_2R \xrightarrow{-CH_2R\cdot} C_6H_5{-}\overset{+}{N}H{=}CH_2 \longleftrightarrow C_6H_5{-}\ddot{N}H{-}\overset{+}{C}H_2$$

m/e 106

ALIPHATIC AMIDES

The molecular ion peak of straight-chain monoamides is usually discernible. The dominant modes of cleavage depend on the length of the acyl moiety, and on the lengths and number of the alkyl groups attached to the nitrogen atom.

The base peak in all straight-chain primary amides higher than propionamide results from the familiar McLafferty rearrangement:

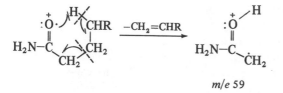

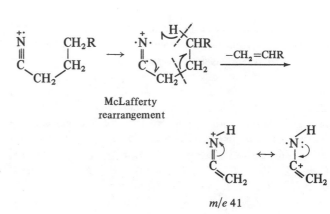

m/e 59

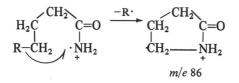

Branching at the α-carbon (CH_3, etc.) gives a homologous peak at m/e 73 or 87, . . .

Primary amides give a strong peak at m/e 44 from cleavage of the R–$CONH_2$ bond: ($\overset{+}{O}\equiv C-\ddot{N}H_2 \longleftrightarrow \ddot{O}=C=\overset{+}{N}H_2$); this is the base peak in C_1-C_3 primary amides and in isobutyramide. A moderate peak at m/e 86 results from γδ C–C cleavage, possibly accompanied by cyclization:

m/e 86

Secondary and tertiary amides with an available hydrogen on the γ-carbon of the acyl moiety and methyl groups on the N atom show the dominant peak resulting from the McLafferty rearrangement. When the N-alkyl groups are C_2 or longer and the acyl moiety is shorter than C_3, another mode of cleavage predominates. This is cleavage of the N-alkyl group beta to the N atom, and cleavage of the carbonyl C–N bond with migration of an α-H atom of the acyl moiety:

m/e 30

ALIPHATIC NITRILES

The molecular ion peaks of aliphatic nitriles (except for acetonitrile and propionitrile) are weak or absent, but the $M+1$ peak can usually be located by its behavior on increasing inlet pressure or decreasing repeller voltage (Section V, part 2). A weak but diagnostically useful $M-1$ peak is formed by loss of an α-hydrogen to form the stable ion: $R\dot{C}H-C\equiv N^{+} \longleftrightarrow RCH=C=\dot{N}^{+}$

The base peak of straight-chain nitriles between C_4 and C_9 is m/e 41. This peak is the ion resulting from hydrogen rearrangement in a 6-membered transition state:

McLafferty rearrangement

m/e 41

However, this peak lacks diagnostic value because of the presence of the C_3H_5 (m/e 41) for all molecules containing a hydrocarbon chain.

A peak at m/e 97 is characteristic and intense (sometimes the base peak) in straight-chain nitriles C_8 and higher. The following mechanism has been depicted:

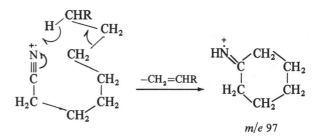

m/e 97

Simple cleavage at each C–C bond (except the one next to the N atom) gives a characteristic series of homologous peaks of even mass number down the entire length of the chain (m/e 40, 54, 68, 82, . . .) due to the $(CH_2)_nC\equiv N^+$ ions. Accompanying these peaks are the usual peaks of the hydrocarbon pattern.

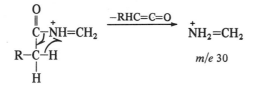

NITRO COMPOUNDS

Aliphatic Nitro Compounds

The molecular ion peak (odd number) of an aliphatic mononitro compound is weak or absent (except in the lower homologs). The main peaks are attributable to the hydrocarbon fragments up to $M-NO_2$. Presence of a nitro group is indicated by an appreciable peak at m/e 30 (NO^+) and a smaller peak at mass 46 (NO_2^+).

Aromatic Nitro Compounds

The molecular ion peak of aromatic nitro compounds (odd number for one N atom) is strong. Prominent peaks result from elimination of an NO_2 radical ($M-46$, the base peak in nitrobenzene), and of a neutral NO molecule with rearrangement to form the phenoxy cation ($M - 30$); both are good diagnostic peaks. Loss of HC≡CH from the $M - 46$ ion accounts for a strong peak at $M - 72$; loss of CO from the $M - 30$ ion gives a peak at $M - 58$. A diagnostic peak at m/e 30 results from the NO^+ ion.

The isomeric *o*-, *m*-, and *p*-nitroanilines each give a strong molecular ion (even number). They all give prominent peaks resulting from two sequences:

$$m/e\ 138\ (M) \xrightarrow{-NO_2} m/e\ 92 \xrightarrow{-HCN} m/e\ 65$$
$$m/e\ 138\ (M) \xrightarrow{-NO} m/e\ 108 \xrightarrow{-CO} m/e\ 80$$

Aside from differences in intensities, the three isomers give very similar spectra. The *m*- and *p*-compounds give a small peak at m/e 122 from loss of an O atom, whereas the *o*-compound eliminates $\cdot\ddot{O}H$ as follows to give a small peak at m/e 121.

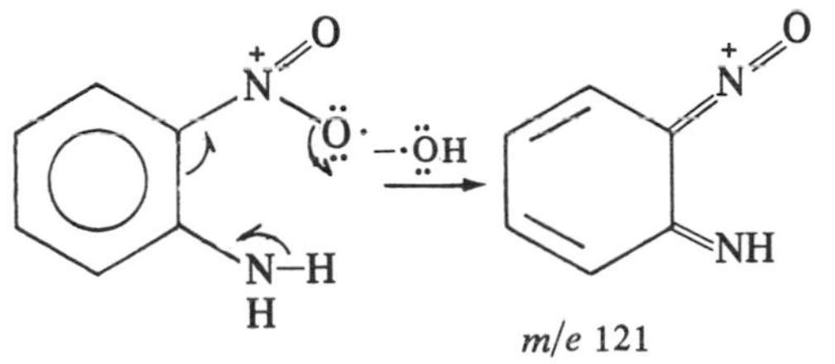

m/e 121

ALIPHATIC NITRITES

The molecular ion peak (odd number) of aliphatic nitrites (1 N present) is weak or absent. The peak at m/e 30 (NO^+) is always large and is often the base peak. There is a large peak at m/e 60 ($CH_2{=}\overset{+}{O}NO$) in all nitrites unbranched at the α-carbon; this represents cleavage of the C–C bond next to the ONO group. An α-branch can be identified by a peak at m/e 74, 88, or 102, . . . Absence of a large peak at m/e 46 permits differentiation from nitro compounds. Hydrocarbon peaks are prominent, and their distribution and intensities describe the arrangement of the carbon chain.

ALIPHATIC NITRATES

The molecular ion peak (odd number) of aliphatic nitrates (1 N present) is weak or absent. A prominent (frequently the base) peak is formed by cleavage of the C–C bond next to the ONO_2 group with loss of the heaviest alkyl group attached to the α-carbon.

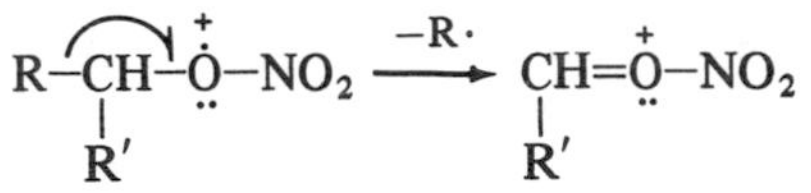

$$R{-}\underset{\substack{|\\R'}}{CH}{-}\overset{+}{\ddot{O}}{-}NO_2 \xrightarrow{-R\cdot} \underset{\substack{|\\R'}}{CH}{=}\overset{+}{\ddot{O}}{-}NO_2$$

$$R > R'$$

The NO_2^+ peak at m/e 46 is also prominent. As in the case of aliphatic nitrites, the hydrocarbon fragment ions are distinct.

SULFUR COMPOUNDS

The contribution (4.4%, see Table II) of the ^{34}S isotope to the $M + 2$ peak, and often to a [fragment + 2] peak, affords ready recognition of sulfur-containing compounds. A homologous series of sulfur-containing fragments is four mass units higher than the hydrocarbon fragment series. The number of sulfur atoms can be determined from the size of the contribution of the ^{34}S isotope to the $M + 2$ peak. The mass of the sulfur atom(s) present is subtracted from the molecular weight. The formula of the rest of the molecule is now determined from the $M + 1$ peak after subtracting the contribution of the ^{33}S isotope. The large correction applied to the $M + 2$ peak makes this peak unreliable for use in Appendix A. In di-isopentyl disulfide, for example, the molecular weight is 206, and the molecule contains two sulfur atoms. The formula for the rest of the molecule is therefore found under mass 142, that is, 206 minus (2 × 32). The corrected intensity for the $M + 1$ peak is 12.5 minus (2 × 0.78) which gives 10.9.

Aliphatic Mercaptans

The molecular ion peak of aliphatic mercaptans, except for higher tertiary mercaptans, is usually strong enough so that the $M + 2$ peak can be accurately measured. In general, the cleavage modes resemble those of alcohols. Cleavage of the C–C bond (αβ bond) next to the SH group gives the characteristic ion $CH_2{=}\overset{+}{S}H \longleftrightarrow \overset{+}{C}H_2{-}\ddot{S}H$ (m/e 47). Sulfur is poorer than nitrogen, but better than oxygen, at stabilizing such a fragment. Cleavage at the βγ bond gives a peak at m/e 61 of about one-half the intensity of the m/e 47 peak. Cleavage at the γδ bond gives a small peak at m/e 75, and cleavage at the δε bond gives a peak at m/e 89 that is more intense than the peak at m/e 75; presumably the m/e 89 ion is stabilized by cyclization:

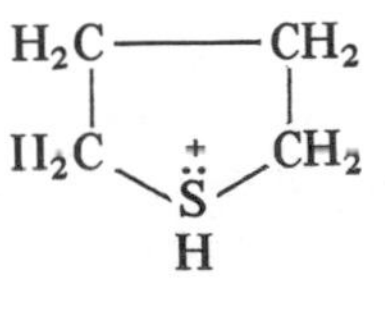

Again analogous to alcohols, primary mercaptans split out H_2S to give a strong $M - 34$ peak, the resulting ion then eliminating ethylene; thus the homologous series $M - H_2S - (CH_2{=}CH_2)_n$ arises:

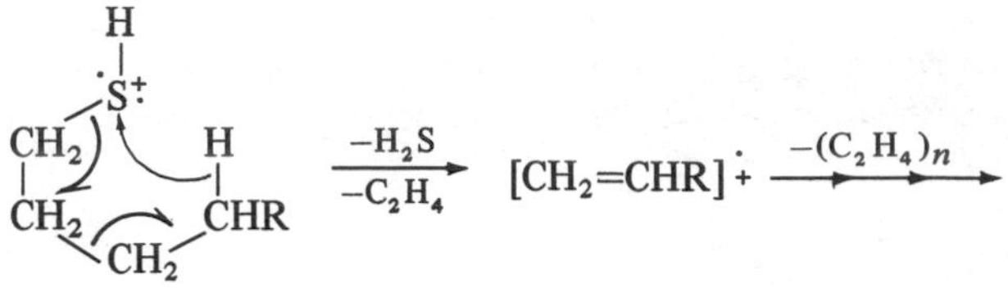

$$[CH_2{=}CHR]^{+\cdot} \xrightarrow{-(C_2H_4)_n}$$

Secondary and tertiary mercaptans cleave at the α-C with loss of the largest group to give a prominent peak $M{-}CH_3$, $M{-}C_2H_5$, $M{-}C_3H_7$, ... However, a peak at m/e 47 may also appear as a rearrangement peak of secondary and tertiary mercaptans. A peak at $M - 33$ (loss of HS) is usually present for secondary mercaptans.

In long-chain mercaptans, the hydrocarbon pattern is superimposed on the mercaptan pattern. As for alcohols, the alkenyl peaks (i.e., m/e 41, 55, 69, ...) are as large or larger than the alkyl peaks (m/e 43, 57, 71, ...).

Aliphatic Sulfides

The molecular ion peak of aliphatic sulfides is usually intense enough so that the $M + 2$ peak can be accurately measured. The cleavage modes generally resemble those of ethers. Cleavage of one or the other of the αβ C–C bonds occurs, with loss of the largest group being favored. These first-formed ions decompose further with hydrogen transfer and elimination of an olefin. The steps for aliphatic ethers also occur for sulfides; the end result is the ion $RCH{=}\overset{+}{S}H$ (see p. 35 for an example.)

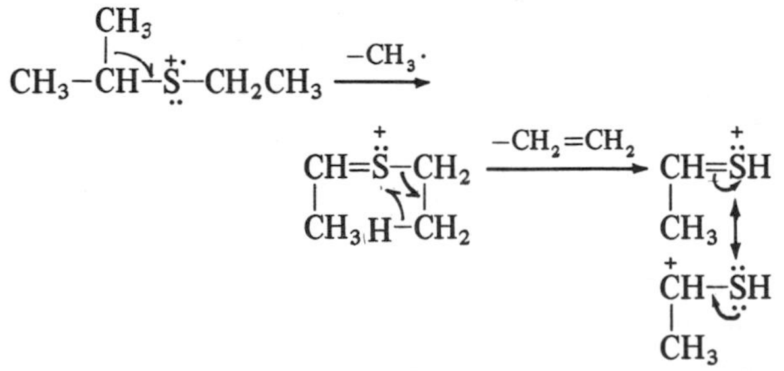

$$CH_3{-}\overset{CH_3}{CH}{-}\overset{+\cdot}{S}{-}CH_2CH_3 \xrightarrow{-CH_3\cdot} CH{=}\overset{+}{S}{-}CH_2 \xrightarrow{-CH_2=CH_2} CH{=}\overset{+}{S}H \leftrightarrow \overset{+}{C}H{-}SH$$

m/e 61

For a sulfide unbranched at either α-C, this ion is $CH_2{=}\overset{+}{S}H$ (m/e 47), and its intensity may lead to confusion with the same ion derived from a mercaptan. However, the absence of $M - H_2S$ or $M - SH$ peaks in sulfide spectra makes the distinction.

A moderate to strong peak at m/e 61 is present (see alkyl sulfide cleavage above) in the spectrum of all except tertiary sulfides. When an α-methyl substituient is present, m/e 61 is the ion $CH_3CH{=}\overset{+}{S}H$ resulting from the double cleavage described above. Methyl primary sulfides cleave at the αβ bond to give the m/e 61 ion, $CH_3{-}\overset{+}{S}{=}CH_2$.

However, a strong m/e 61 peak in the spectrum of a completely straight-chain sulfide calls for a different explanation. The following rationalization is offered

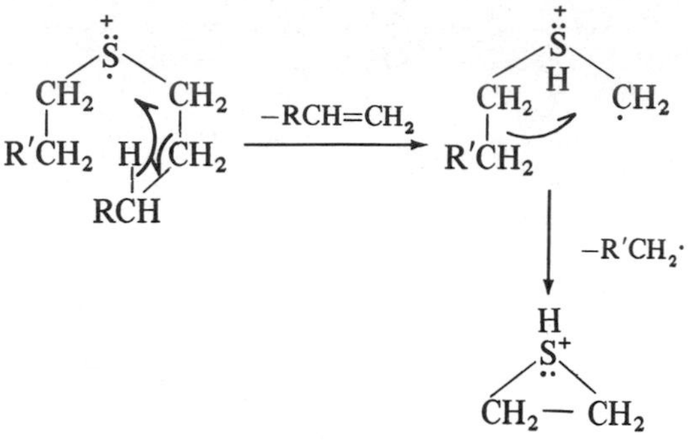

$$\xrightarrow{-RCH=CH_2} \quad \xrightarrow{-R'CH_2\cdot}$$

m/e 61

Sulfides give a characteristic ion by cleavage of the C–S bond with retention of charge on sulfur. The resulting $R\overset{+}{S}$ ion gives a peak at m/e $32 + CH_3$, $32 + C_2H_5$, $32 + C_3H_7$, ... The ion of m/e 103 seems especially favored possibly because of formation of a rearranged cyclic ion:

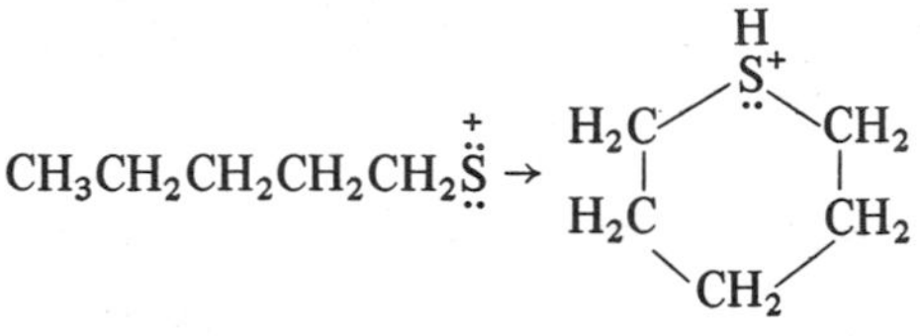

$$CH_3CH_2CH_2CH_2CH_2\overset{+}{S} \rightarrow$$

m/e 103

These features are illustrated by the spectrum of di-*n*-pentyl sulfide (Figure 15).

As with long-chain ethers, the hydrocarbon pattern may dominate the spectrum of long-chain sulfides; the C_nH_{2n} peaks seem especially prominent. In branched-chain sulfides, cleavage at the branch may reduce the relative intensity of the characteristic sulfide peaks.

Aliphatic Disulfides

The molecular ion peak, at least up to C_{10} disulfides, is strong (Chapter 8. Compound No. 8-29).

A major peak results from cleavage of one of the C–S bonds with retention of the charge on the alkyl fragment. Another major peak results from the same cleavage with shift of a hydrogen atom to form the RSSH fragment which retains the charge. Other peaks apparently result from cleavage between the sulfur atoms without rearrangement, and with migration of one or two hydrogen atoms to give, respectively, $R\overset{+}{S}$, $R\overset{+}{S} - 1$, and $R\overset{+}{S} - 2$.

HALOGEN COMPOUNDS

A compound that contains 1 chlorine atom will have an $M+2$ peak approximately one-third the intensity of the molecular ion peak because of the presence of a molecular ion containing the ^{37}Cl isotope (see Table II). A compound that contains one bromine atom will have an $M+2$ peak almost equal in intensity to the molecular ion because of the presence of a molecular ion containing the ^{81}Br isotope. A compound that contains two chlorines, or two bromines, or one chlorine and one bromine, will show a distinct $M+4$ peak, in addition to the $M+2$ peak, because of the presence of a molecular ion containing 2 atoms of the heavy isotope. In general, the number of chlorine and/or bromine atoms in a molecule can be ascertained by the number of alternate peaks beyond the molecular ion peak. Thus, 3 Cl atoms in a molecule will give peaks at $M+2$, $M+4$, and $M+6$; in polychloro compounds, the peak of highest mass may be so weak as to escape notice.

The relative abundances of the peaks (molecular ion, $M+2$, $M+4$, and so forth,) have been calculated by Beynon for compounds containing chlorine and bromine (atoms other than chlorine and bromine were ignored). A portion of these results is presented here, somewhat modified, as Table III. We can now tell what combination of chlorine and bromine atoms is present. It should be noted that Table III presents the isotope contributions in terms of percent of the molecular ion peak.

As required by Table III, the $M+2$ peak in the spectrum of *p*-chlorobenzophenone (Figure 12) is about one-third the intensity of the molecular ion peak (*m/e* 218). As mentioned earlier, the chlorine-containing fragments (*m/e* 141 and 113) show [fragment + 2] peaks of the proper intensity.

Table III. Intensities of Isotope Peaks (Relative to the Molecular Ion) for Combinations of Bromine and Chlorine

Halogen Present	% $M+2$	% $M+4$	% $M+6$	% $M+8$	% $M+10$	% $M+12$
Br	97.9					
Br_2	195.0	95.5				
Br_3	293.0	286.0	93.4			
Cl	32.6					
Cl_2	65.3	10.6				
Cl_3	97.8	31.9	3.47			
Cl_4	131.0	63.9	14.0	1.15		
Cl_5	163.0	106.0	34.7	5.66	0.37	
Cl_6	196.0	161.0	69.4	17.0	2.23	0.11
BrCl	130.0	31.9				
Br_2Cl	228.0	159.0	31.2			
Cl_2Br	163.0	74.4	10.4			

The $M+1$ peak is still useful for arriving at a molecular formula by use of Appendix A, after we subtract the masses of the appropriate number of chlorine and bromine atoms.

Unfortunately, the application of isotope contributions, though generally useful for aromatic halogen compounds, is limited by the weak molecular ion peak of many aliphatic halogen compounds of more than about six carbon atoms for a straight chain, or fewer for a branched chain. However, the halogen-containing fragments are recognizable by the ratio of the [fragment + 2] peaks to fragment peaks in monochlorides or monobromides. In polychloro- or bromo-

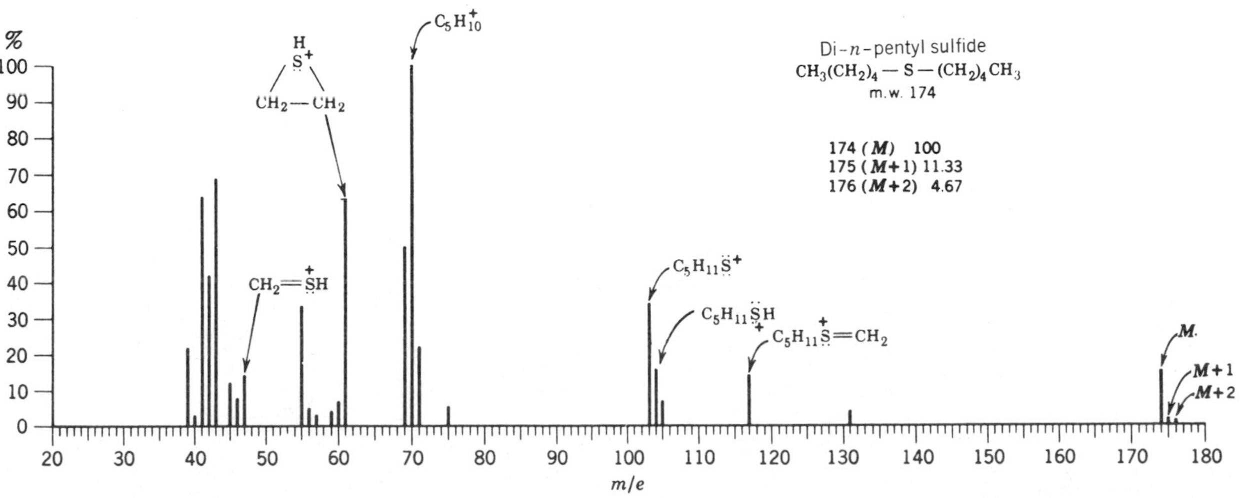

Figure 15. *Di-n-pentyl sulfide.*

compounds, these [fragment + isotope] peaks form a distinctive series of multiplets (Figure 16). Coincidence of fragment ion with one of the isotope fragments, with another disruption of the characteristic ratios, must always be kept in mind.

Neither fluorine nor iodine has a heavier isotope.

Aliphatic Chlorides

The molecular ion peak is detectable only in the lower monochlorides. Fragmentation of the molecular ion is mediated by the chlorine atom but to a much lesser degree than is the case in oxygen-, nitrogen-, or sulfur-containing compounds. Thus, cleavage of a straight-chain monochloride at the C–C bond adjacent to the chlorine atom accounts for a small peak at *m/e* 49 (and, of course, the isotope peak at *m/e* 51).

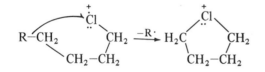

$$\mathrm{R{-}CH_2{-}\overset{+}{\ddot{C}l}{:}} \xrightarrow{-\mathrm{R}\cdot} \mathrm{CH_2{=}\overset{+}{\ddot{C}l}{:}} \longleftrightarrow \mathrm{\overset{+}{C}H_2{-}\ddot{C}l{:}}$$

Cleavage of the C–Cl bond leads to a small Cl^+ peak and to a R^+ peak which is prominent in the lower chlorides, but quite small when the chain is longer than about C_5.

Straight-chain chlorides longer than C_6 give $C_3H_6\overset{+}{Cl}$, $C_4H_8\overset{+}{Cl}$, and $C_5H_{10}\overset{+}{Cl}$ ions. Of these the $C_4H_8\overset{+}{Cl}$ ion forms the most intense (sometimes the base) peak; a five-membered cyclic structure may explain its stability.

$$\mathrm{R{-}CH_2CH_2CH_2CH_2\overset{+}{\ddot{C}l}} \xrightarrow{-\mathrm{R}\cdot} \mathrm{H_2C{-}\overset{+}{\ddot{C}l}{-}CH_2 \ (ring: \ H_2C,\ CH_2{-}CH_2,\ CH_2)}$$

Loss of HCl occurs, possibly by 1,3-elimination, to give a peak (weak or moderate) at $M - 36$.

In general, the spectrum of an aliphatic monochloride is dominated by the hydrocarbon pattern to a greater extent than that of a corresponding alcohol, amine, or mercaptan.

Aliphatic Bromides

The remarks under aliphatic chlorides apply quite generally to the corresponding bromides.

Aliphatic Iodides

Aliphatic iodides give the strongest molecular ion peak of the aliphatic halides. Since iodine is monoisotopic, there is no distinctive isotope peak. The presence of an iodine atom can sometimes be deduced from isotope peaks that are suspiciously low in relation to the molecular weight, and from several distinctive peaks; in polyiodo compounds, the large interval between major peaks is characteristic.

Iodides cleave much as do chlorides and bromides, but the $C_4H_8\overset{+}{I}$ ion is not as evident as the corresponding chloride and bromide ions.

Aliphatic Fluorides

Aliphatic fluorides give the weakest molecular ion peak of the aliphatic halides. Fluorine is monoisotopic, and its detection in polyfluoro compounds depends on suspiciously small isotopic peaks relative to the molecular ion, on the intervals between peaks, and on characteristic peaks. Of

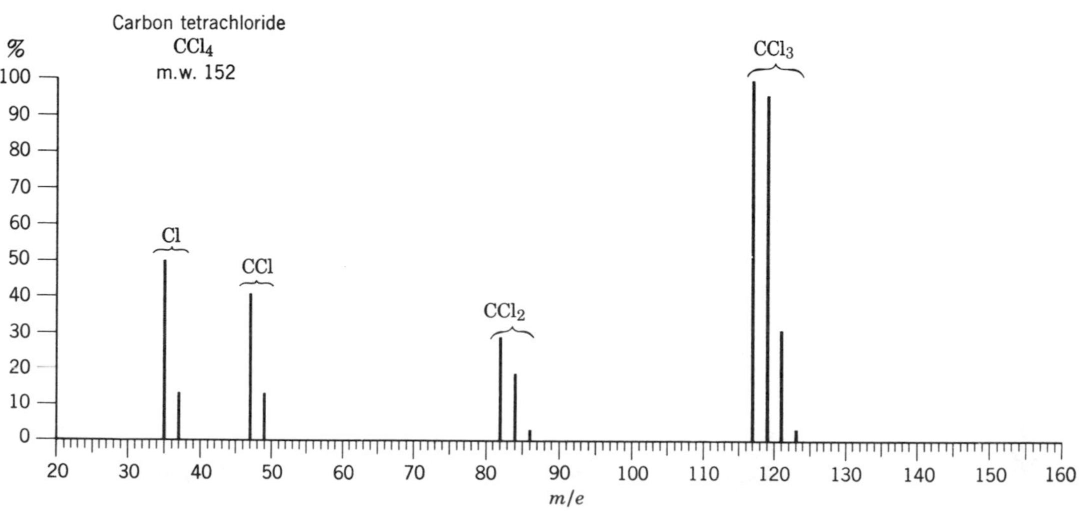

Figure 16. *Carbon tetrachloride.*

these, the most characteristic is *m/e* 69 due to the ion CF_3^+, which is the base peak in all perfluorocarbons. Prominent peaks are noted at *m/e* 119, 169, 219. . . ; these are increments of CF_2. The stable ions $C_3F_5^+$ and $C_4F_7^+$ give large peaks at *m/e* 131 and 181. The $M-F$ peak is frequently visible in perfluorinated compounds. In monofluorides, cleavage of the $\alpha\beta$ C–C bond is less important than in the other monohalides, but cleavage of a C–H bond on the α–C is more important. This reversal is a consequence of the high electronegativity of the F atom, and is rationalized by placing the positive charge on the α-carbon. The secondary carbonium ion thus depicted by a loss of a hydrogen atom is more stable than the primary carbonium ion resulting from loss of an alkyl radical.

$$[R{-}CH_2{-}F]^{+\cdot} \xrightarrow{-H\cdot} R{-}\overset{+}{C}H{-}F$$
$$[R{-}CH_2{-}F]^{+\cdot} \xrightarrow{-R\cdot} \overset{+}{C}H_2{-}F$$

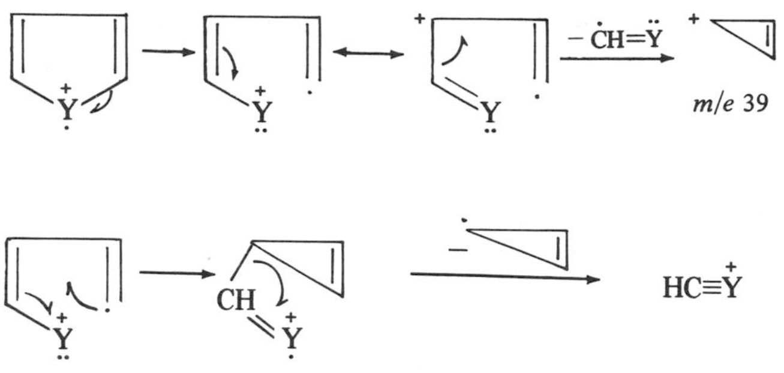

m/e 39

Y = O, S, NH

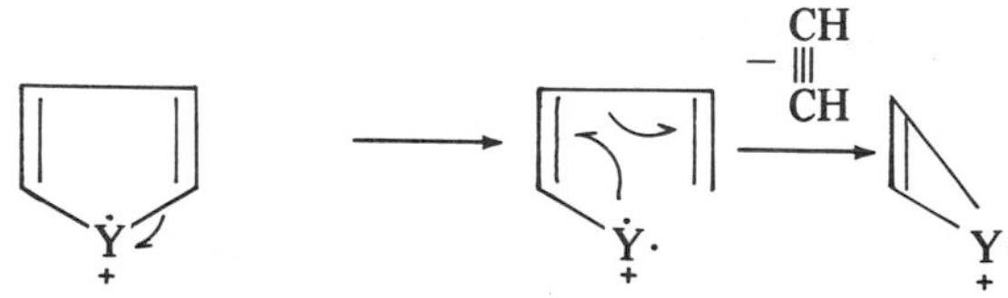

Y = S, NH

Thus furan exhibits two principal peaks: $C_3H_3^+$ (*m/e* 39) and $HC{\equiv}\overset{+}{O}$ (*m/e* 29). For thiophene, there are three: $C_3H_3^+$ (*m/e* 39), $HC{\equiv}\overset{+}{S}$ (*m/e* 45), and $C_2H_2\overset{+\cdot}{S}$ (*m/e* 58). And for pyrrole, there are three: $C_3H_3^+$ (*m/e* 39), $HC{\equiv}\overset{+\cdot}{N}H$ (*m/e* 28) and $C_2H_2\overset{+\cdot}{N}H$ (*m/e* 41). Pyrrole also eliminates a neutral molecule of HCN to give an intense peak at *m/e* 40. The base peak of 2,5-dimethylfuran is *m/e* 43 ($CH_3C{\equiv}\overset{+}{O}$).

Cleavage of the β C–C bond in alkylpyridines depends on the position of the ring substitution, being more pronounced when the alkyl group is in the 3-position. An alkyl group of more than 3 carbon atoms in the 2 position can undergo migration of a hydrogen atom to the ring nitrogen.

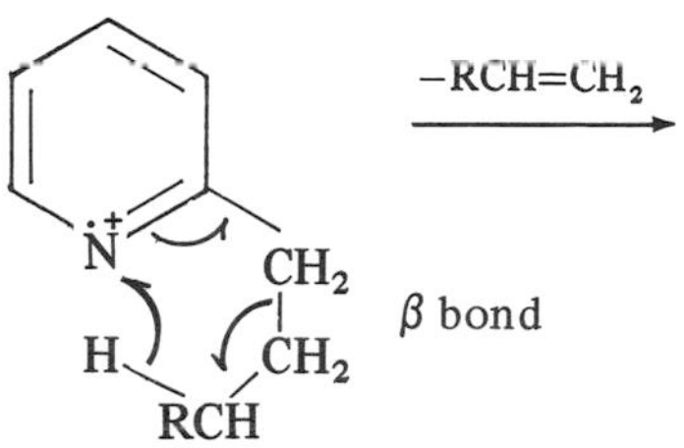

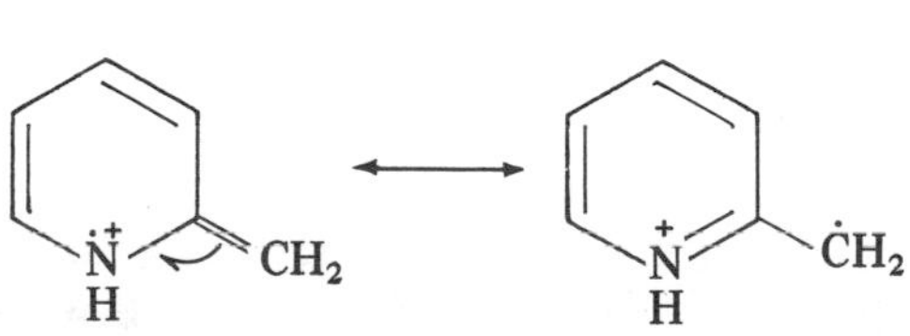

Benzyl Halides

The molecular ion peak of benzyl halides is usually detectable. The benzyl (or tropylium) ion from loss of the halide (Section VII, Rule 8) is favored even over β-bond cleavage of an alkyl substituent. A substituted phenyl ion (α-bond cleavage) is prominent when the ring is polysubstituted.

Aromatic Halides

The molecular ion peak of an aryl halide is readily apparent. The $M-X$ peak is large for all compounds in which X is attached directly to the ring.

HETEROAROMATIC COMPOUNDS

The molecular ion peak of heteroaromatics and alkylated heteroaromatics is intense. Cleavage of the bond beta to the ring, as in alkylbenzenes, is the general rule; in pyridine, the position of substitution determines the ease of cleavage of the beta bond (see below).

Localizing the charge of the molecular ion on the heteroatom, rather than in the ring π structure, provides a satisfactory rationale for the observed mode of cleavage. The present treatment follows that used by Djerassi.[2]

The five-membered ring heteroaromatics (furan, thiophene, and pyrrole) show very similar ring cleavage patterns. The first step in each case is cleavage of the carbon-heteroatom bond.

–RCH=CH$_2$

β bond

A similar cleavage is found in pyrazines since all ring substituents are necessarily ortho to one of the nitrogen atoms.

NATURAL PRODUCTS

Amino Acids

Detection of the molecular ion peaks of amino acids can be difficult. If we examine the mass spectra of amino acids, as well as of steroids and triglycerides, by a variety of ionization techniques, we can appreciate their relative merits.

The EI spectra of amino acids (Figure 17a) or their esters give weak or nonexistent molecular ion peaks, but CI and FI (Figure 17b, c) give either molecular or quasi-molecular ion peaks. The weak molecular ions in the EI spectra arise since amino acids easily lose their carboxyl group and the esters easily lose their carboalkoxyl group upon electron impact.

$$R{-}CH(C(=O){-}O{-}H)\overset{+\cdot}{N}H_2 \xrightarrow{-[CO_2H]^{\cdot}} R{-}CH{=}\overset{+}{N}H_2 \qquad \text{EI}$$

$$M - 45$$

$$R{-}CH(C(=O)OR')\overset{+\cdot}{N}H_2 \xrightarrow{-[CO_2R']^{\cdot}} R{-}CH{=}\overset{+}{N}H_2 \qquad \text{EI}$$

$$M - [44 + R']$$

The FI and FD fragmentation patterns for leucine are similar. An ion-molecule reaction produces an MH^+ (m/e 132) ion, which readily loses a carboxyl group to form the

$$(CH_3)_2CHCH_2CH(CO_2H)\overset{+}{N}H_3 \xrightarrow{-[CO_2H]} (CH_3)_2CHCH_2\dot{C}H\overset{+}{N}H_3$$

$$m/e\ 87$$

$$\downarrow -H\cdot$$

$$(CH_3)_2CHCH_2CH{=}\overset{+}{N}H_2$$

$$m/e\ 86$$

m/e 87 ion, which in turn loses a hydrogen atom to form the m/e 86 ion.

STEROIDS

Polyhydroxy steroids (e.g., the tetrol for which spectra are shown in Figure 18) give EI spectra that often show weak or nonexistent molecular ion peaks. For this tetrol facile

(a)

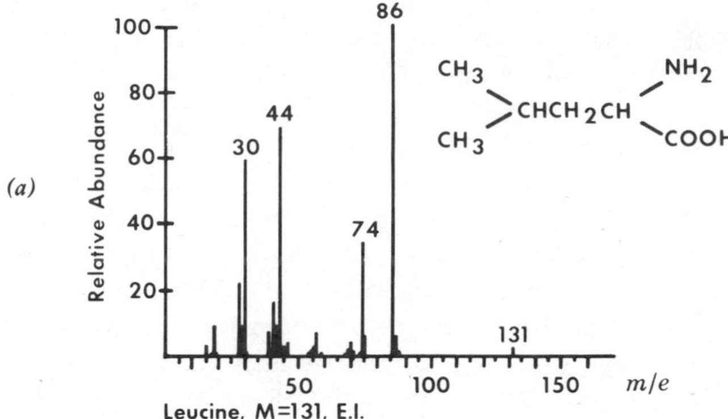

(b)

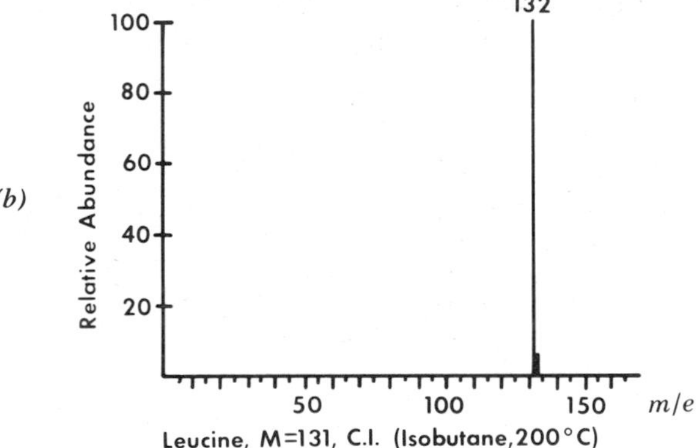

(c)

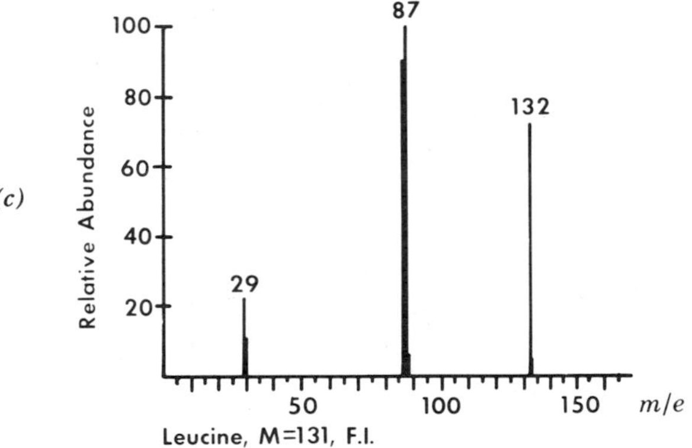

(d)

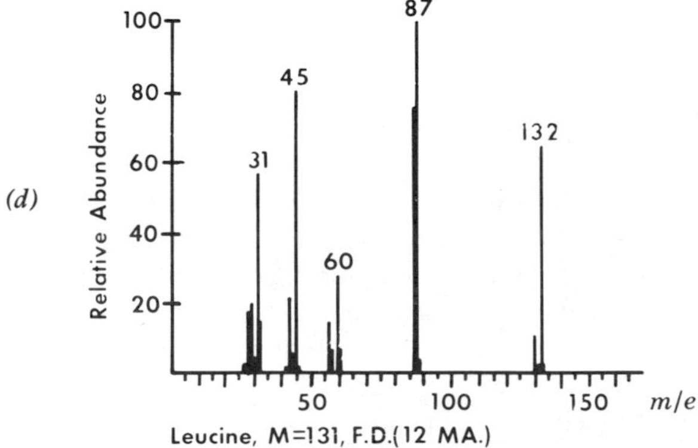

Figure 17. *Mass spectra of leucine. (a) Electron impact (EI). (b) Chemical ionization (CI). (c) Field ionization (FI. (d) Field desorption (FD). See reference a, Table IV.*

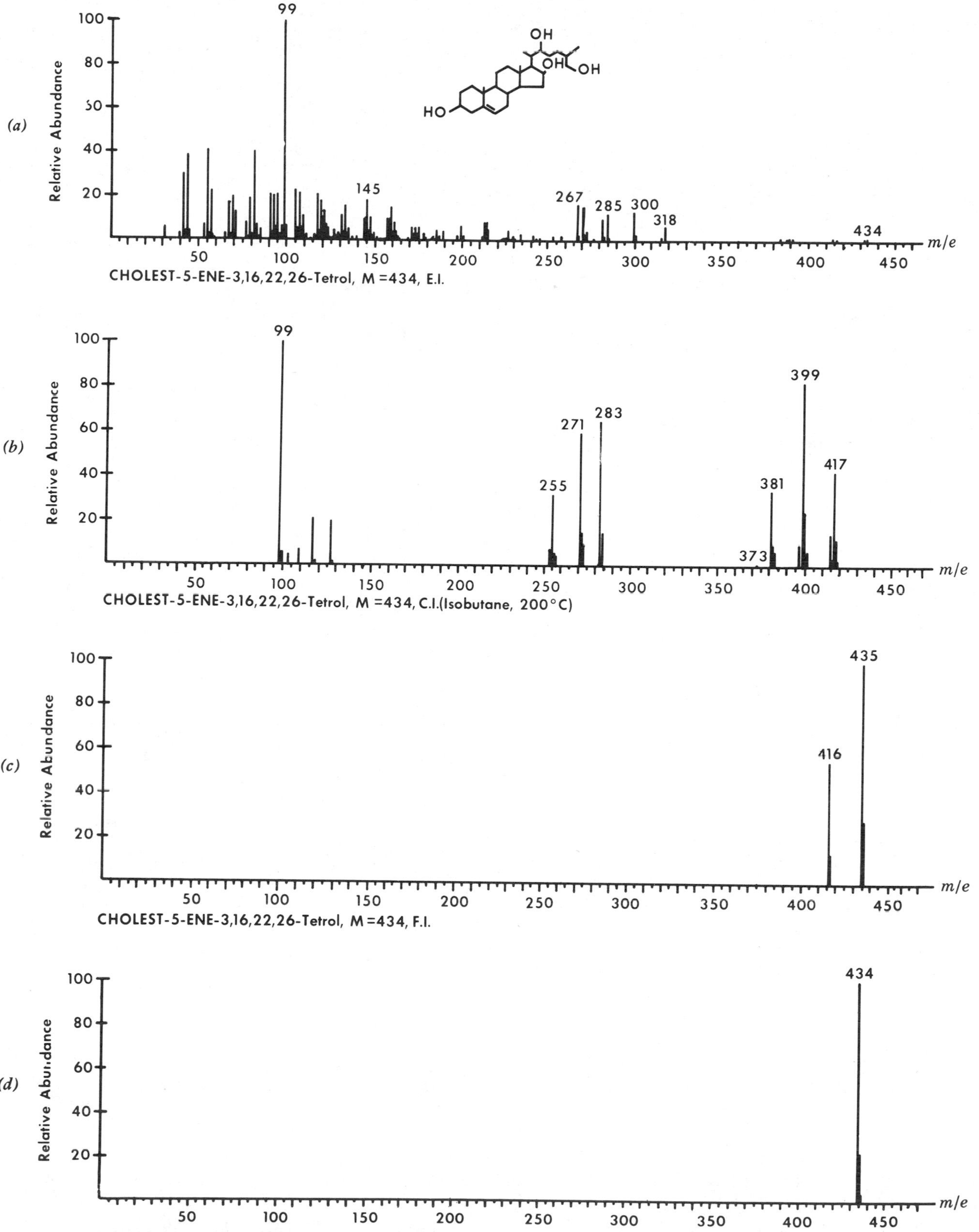

Figure 18. *Mass spectra of cholest-5-ene-3,16,22,26-tetrol. (a) Electron Impact (EI) (b) Chemical Ionization (CI). (c) Field Ionization (FI). (d) Field Desorption (FD). See reference a, Table IV.*

dehydration is not entirely thermally induced since the

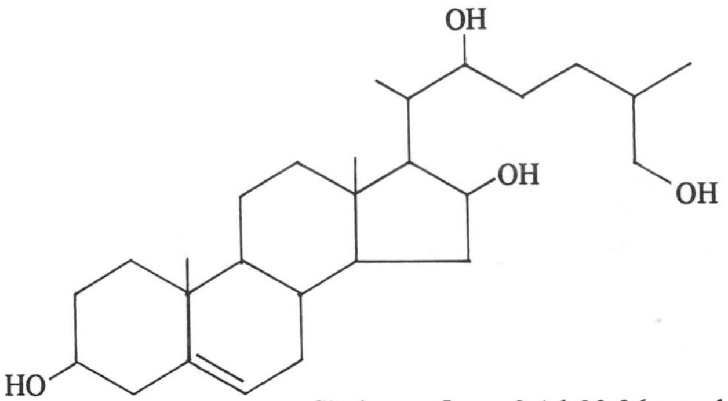

Cholesten-5-ene-3,16,22,26-tetrol

FI spectrum (Figure 18c) shows an intense M + H peak. CI (Figure 18b) is not useful since the protonated molecular ion also readily dehydrates. Dehydration is, however, a secondary path in FI (Figure 18c) and is not observed at all in the FD spectrum (Figure 18d) which shows only a molecular ion peak.

Generalizations about steroids are not easily made since the following steroidal diol dione behaves quite differently. Here the FD spectrum shows the $M - H_2O$ peak as the

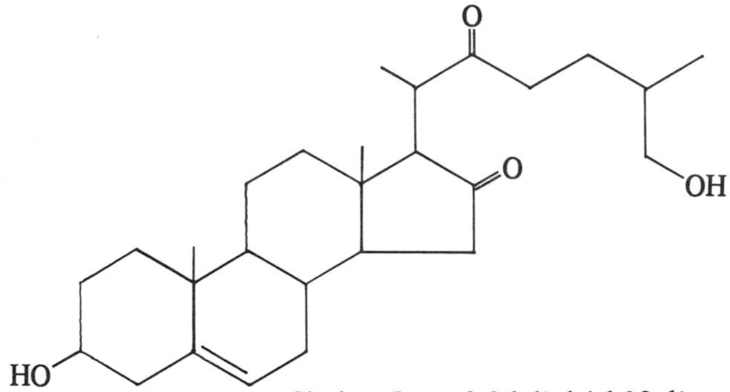

Cholest-5-ene-3,26-diol-16,22-dione

base peak, accompanied by a reasonably intense molecular ion peak. The FI spectrum is very similar, whereas neither the CI nor EI spectrum shows molecular ion peaks.

Triglycerides

Naturally occuring triglycerides, which have high molecular weights and low volatility, give weak or nonexistent molecular ion (or MH^+) peaks by EI, CI, and FI techniques. These triester samples require high source temperatures, and this may lead to pyrolytic elimination of the elements of carboxylic acid.

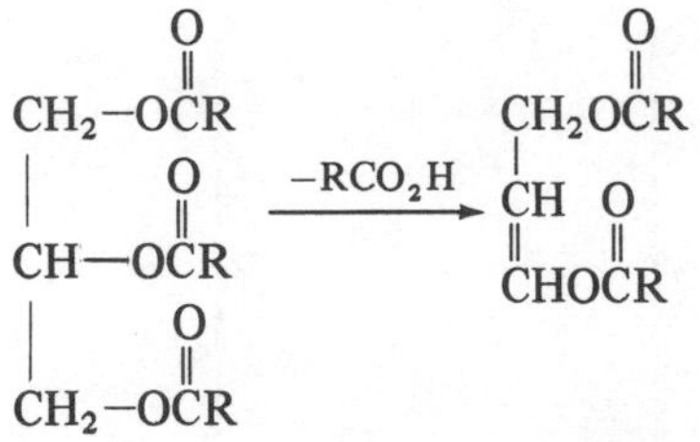

$$\begin{array}{l} CH_2{-}OC(=O)R \\ | \\ CH{-}OC(=O)R \\ | \\ CH_2{-}OC(=O)R \end{array} \xrightarrow{-RCO_2H} \begin{array}{l} CH_2OC(=O)R \\ | \\ CH \\ \| \\ CHOC(=O)R \end{array}$$

The EI spectrum of trilaurin (Figure 19a) shows a modest molecular ion peak (*m/e* 638) but a much more intense $M - RCO_2$ peak (*m/e* 439). In the CI and FI spectra (Figure 19), the molecular ion peaks (or MH^+ peaks) are not visible, and the *m/e* 439 peak is the base peak. The FD spectrum provides an MH^+ peak (*m/e* 639), which is the base peak.

In summary, it is clear that no simple instructions can be set forth for choosing the best method of producing molecular ions. EI analyses are useful for those instances in which the samples are sufficiently thermally stable and have reasonable vapor pressure. EI instrumentation is also widely available as are a substantial number of published spectra, and data compilations are available for comparison. Table IV provides guidelines for selecting an appropriate ionization technique. CI, FI and FD techniques can be valuable alternatives when EI is not successful.

MISCELLANEOUS CLASSES

The following classes of organic compounds are discussed in Biemann's book (B)[1], or in Djerassi's books (D[2], DI[3], DII[3]).

Alkaloids	D, Chapter 5; DI; DII, Chapter 17; B, p. 305
Amino Acids & Peptides	DII, Chapter 26; B, Chapter 7
Antibiotics	DII, p. 172
Carbohydrates	DII, Chapter 27
Cyanides & Isothiocyanates	D, Chapter 11
Estrogens	DII, Chapter 19
Glycerides	B, p. 255
Sapogenins	DII, Chapter 22
Silicones	B, p. 172
Steroids	B, Chapter 9; DII, Chapters 17, 20, 21, 22
Terpenes	B, p. 334; DII, Chapters 23, 24
Thioketals	D, Chapter 7
Tropone & Tropolones	D, Chapter 18

REFERENCES

General Monographs

1. Biemann, K. *Mass Spectrometry, Applications to Organic Chemistry.* New York: McGraw-Hill, 1962. (Reference B above.)
2. Budzikiewicz, H., Djerassi, C., Williams, D. H. *Mass Spectrometry of Organic Compounds.* San Francisco: Holden-Day,1967. This is the most complete general reference written for the organic chemist interested in employing mass spectrometry in structure determination.
3. Budzikiewicz, H., Djerassi, C., Williams, D. H. *Structure Elucidation of Natural Products by Mass Spectrometry.* San Francisco: Holden-Day, 1964, Vol. I, Vol. II.
4. Beynon, J. H. *Mass Spectrometry and Its Application to Organic Chemistry.* Amsterdam: Elsevier, 1960.

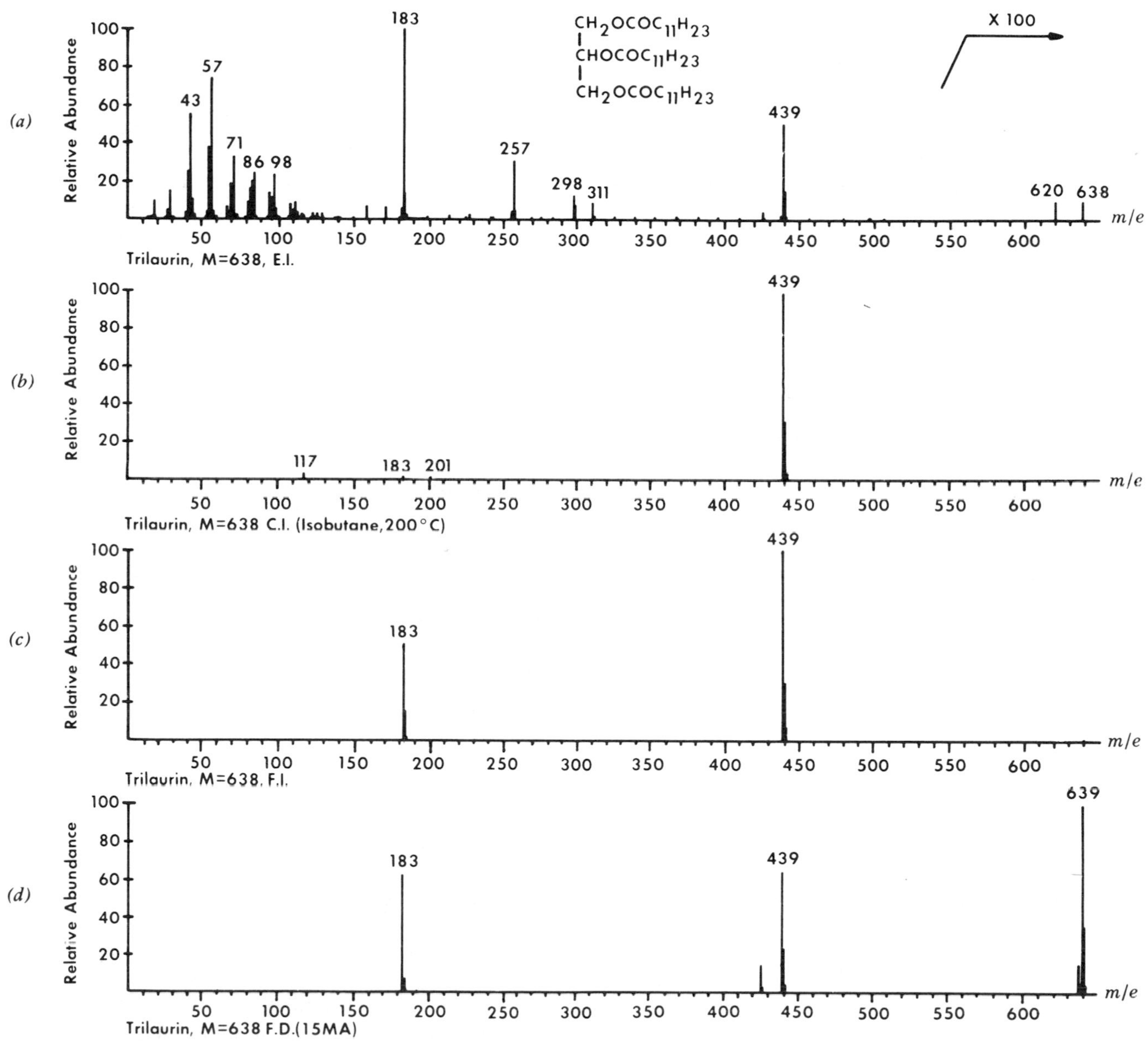

Figure 19. *Mass spectra of trilaurin: (a) Electron impact (EI). (b) Chemical ionization (CI). (c) Field ionization (FI). (d) Field desorption (FD).* Note: *that the molecular ion peak (*m/e *638) would be 1% the intensity shown if it were at the same sensitivity as the peaks from* m/e *0 to 450. See Reference a, Table IV.*

Table IV. Molecular Ion (or MH^+) Peak and Fragmentation Characteristics for Four Compounds not Easily Analyzed by EI[a]

			EI			CI		FI		FD	
Compound	Class	Volatility	Stability to Heat	M^+ Intensity	Fragmentation	MH^+ Intensity	Fragmentation	MH^+ Intensity	Fragmentation	M^+ Intensity	Fragmentation
leucine	amino acid	modest	acceptable	very weak	extensive	base peak	none	intense	moderate	intense	extensive
tetrol[b]	steroid	very low	poor	very weak	extensive	none	extensive	base	minor	base	none
diol dione[c]	steroid	very low	poor	none	extensive	none	extensive	base	minor	base	none
trilaurin	triglycerides	very low	high	weak	extensive	none	limited[d]	none	modest	MH^+ base	substantial

[a]Taken from H.M. Fales, et al., *Anal. Chem.*, **1975**, *47*, 207.
[b]For structure, see discussion in text, pp. 38-41.
[c]Cholest-5-ene-3, 26-diol-16, 22-dione.
[d]$M - RCO_2H$ only.

5. Beynon, J.H., Saunders, R.A., Williams, A.E. *The Mass Spectra of Organic Molecules.* New York: American Elsevier, 1968.
6. Frigerio, A. *Essential Aspects of Mass Spectrometry.* New York: Spectrum Publishers, 1974.
7. Hamming, M. and Foster, N. *Interpretation of Mass Spectra of Organic Compounds.* New York: Academic Press, 1972.
8. Johnstone, R. A. W. *Mass Spectrometry for Organic Chemists.* Cambridge: Cambridge University Press, 1972.
9. McLafferty, F.W. *Interpretation of Mass Spectra.* 2nd ed. New York: W. A. Benjamin, 1973.
10. McLafferty, F.W. (ed.) *Mass Spectrometry of Organic Ions.* New York: Academic Press, 1963.
11. Middeditch, B. S. (ed.) *Practical Mass Spectrometry.* New York: Plenum, 1979.
12. Milne, G. W. A. *Mass Spectrometry: Techniques and Applications.* New York: Wiley-Interscience, 1971.
13. Shrader, S. R. *Introduction to Mass Spectrometry.* Boston: Allyn and Bacon, 1971.
14. Williams, D. H. (senior reporter). *Mass Spectrometry, A Specialist Periodical Report.* Vols. I-V. London: Chemical Society, June 1968-June 1979.
15. Williams, D.H. and Howe, I. *Principles of Organic Mass Spectrometry.* New York: McGraw-Hill, 1972.
15a. Gross, M. L. (ed.) *High Performance Mass Spectrometry: Chemical Applications.* Washington: American Chemical Society, 1978.
15b. Millard, B. J. (ed.) *Quantitative Mass Spectrometry.* London: Heyden & Son. 1978.

Data and Spectral Compilations

16. ASTM (1963). "Index of Mass Spectral Data," American Society for Testing and Materials. STP-356, 244 pp.
17. ASTM (1969). "Index of Mass Spectral Data," American Society for Testing and Materials, AMD 11, 632 pp.
18. Beynon, J. H., and Williams, A. E. *Mass and Abundance Tables for Use in Mass Spectrometry.* Amsterdam: Elsevier, 1963.
19. Beynon, J.H., Saunders, R.A., Williams, A.E. *Table of Meta-Stable Transitions.* New York: Elsevier, 1965.
20. *Catalog of Selected Mass Spectral Data* (1947 to date), American Petroleum Institute Research Project 44 and Thermodynamics Research Center (formerly MCA Research Project). College Station, Texas. Texas A & M University. Dr. Bruno Zwolinski, Director.
21. Cornu, A. and Massot, R. *Compilation of Mass Spectral Data.* London: Heyden, 1966. First and second supplements issued.
22. *Handbook of Spectroscopy.* Vol II. Cleveland: CRC Press Inc., 1974, pp. 317-330. EI data for 15 compounds in each of 16 classes of organic compounds.
23. Lederberg, J. *Computation of Molecular Formulas for Mass Spectrometry.* San Francisco: Holden-Day, 1964.
24. Lederberg, J. *Tables and an Algorithm for Calculating Functional Groups of Organic Molecules in High Resolution Mass Spectrometry.* NASA Sci. Techn. Aerosp. Rep., N64-21426 (1964).
25. McLafferty, F.W. *Mass Spectral Correlations.* Advances in Chemistry Series No. 40. Washington, D. C.: American Chemical Society, 1963.
26. McLafferty, F. W. and Penzelik, J. *Index and Bibliography of Mass Spectrometry* 1963-1965. New York: Interscience-Wiley, 1967.
27. Heller, S.R., Milne, G.W. *EPA/NIH* Mass Spectral Search System (MSSS), A Division of CIS, Washington: U.S. Government Printing Office. An interactive computer searching system containing the spectra of over 32,000 compounds. These can be searched on the basis of peak intensities as well as by Biemann and probability matching techniques.
28. Stenhagen, E., Abrahamsson, S., McLafferty, F.

(eds.), *Atlas of Mass Spectral Data.* New York: Interscience-Wiley, 1969. Vol. 1: Molecular Weight: 16.313 to 142.0089; Vol. 2: Molecular Weight: 142.0185 to 213.2456; Vol. 3: Molecular Weight: 213.8629 to 702.7981. The 3 volumes have complete EI data for ca. 6000 compounds.

29. Stenhagen, E., Abrahamsson, S., McLafferty, F. (eds.) *Registry of Mass Spectral Data.* New York: Wiley-Interscience, 1974. Vol. I: MW 16.0313 to 187.1572, Formula CH_4 to $C_{10}H_{21}NO_2$; Vol. II: MW 187.1572 to 270.1256, Formula $C_{10}H_{21}NO_2$ to $C_{17}H_{18}O_3$; Vol. III: MW 270.2671 to 402.3305, Formula $C_{16}H_{34}N_2O$ to $C_{25}H_{13}O_3B$; Vol. IV MW 40.3305 to 1519.806, Formula $C_{25}H_3O_3B$ to $(C_9H_{18}D_3O_3Al)_n$. The 4 volumes contain bar graphs for 18,806 compounds. Vol. IV also contains the index for all 4 volumes.

Special Monographs

30. McFadden, W. H. *Techniques of Combined Gas Chromatography/Mass Spectrometry: Applications in Organic Analysis,* New York: Wiley-Interscience, 1973.
31. Porter, Q. N. and Baldas, J. *Mass Spectrometry of Heterocyclic Compounds* in General Heterocycle Chemistry Series, A. Weissberger and E. C. Taylor (eds.) New York: Wiley-Interscience, 1971.
32. Safe, S. and Hutzinger, O. *Mass Spectrometry of Pesticides and Pollutants.* Cleveland: CRC Press, 1973.
33. Waller, G. R. (ed.) *Biochemical Applications of Mass Spectrometry.* New York: Wiley-Interscience, 1972; Waller, G. R., Dermer, O. C., (eds.) *Biochemical Applications of Mass Spectrometry,* First Supplementary Volume, New York: Wiley-Interscience, 1980.

Chapters from More General Texts

34. Drago, R. S. *Physical Methods in Chemistry.* Philadelphia: W. B. Saunders, 1977, ch. 16.
35. Lambert, J. B., Shurvel, H. F., Verbit, L. Cooks, R. G., Stout, G. H. *Organic Structural Analysis.* New York: Mac Millan, 1976, part 4, pp. 405-517.
36. Pasto, D. J. and Johnson, C. R. *Organic Chemistry.* Englewood Cliffs, N. J.: Prentice Hall, 1979, ch. 7.

PROBLEMS

2.1 A compound gives rise to the following intensities for peaks in the region of the molecular ion (M):

m/e	*intensity*
110 (M)	100
111 ($M+1$)	6.96
112 ($M+2$)	0.60

Deduce the molecular formula for this compound.

2.2 The exact mass of a compound determined by high-resolution mass spectrometry is 70.0419. What is the molecular formula of the compound?

2.3 The low-resolution mass spectrum of 2-butenal shows a peak at m/e 69 that is 28.9% as intense as the base peak. Propose at least one fragmentation route to account for this peak, and explain why this fragment would be reasonably stable.

2.4 The low-resolution mass spectrum of 3-butyn-2-ol shows the base peak at m/e 55. Explain why the fragment giving rise to this peak would be very stable.

2.5 Consider the mass spectra on p. 44 (low resolution) of 2 isomers of molecular formula $C_{10}H_{14}$. Determine their structures and explain the major spectral features for each.

2.6 The mass spectrum of *o*-nitrotoluene shows a substantial peak at m/e 120. Similar analysis of α,α,α-trideutero-o-nitrotoluene does not give a peak at m/e 120 but rather at m/e 122. Explain.

2.7 Determine the structure for the mass spectrum shown on p. 45.

Problem 2.5

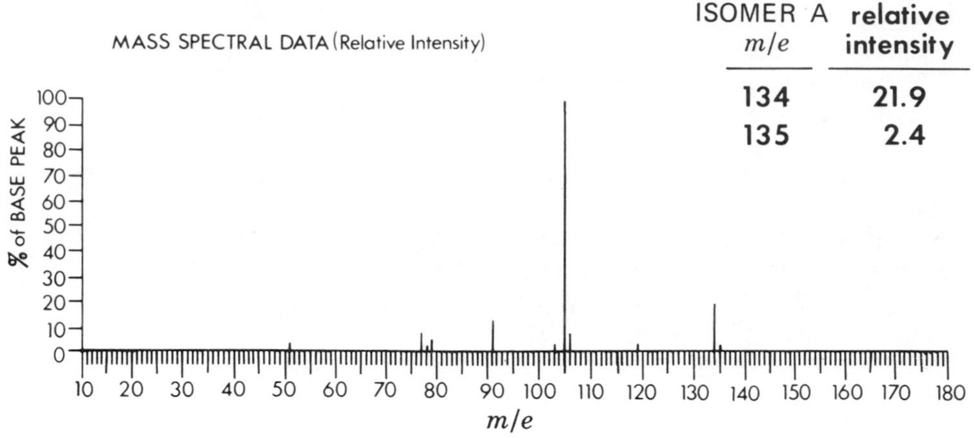

ISOMER A

m/e	relative intensity
134	21.9
135	2.4

Problem 2.5

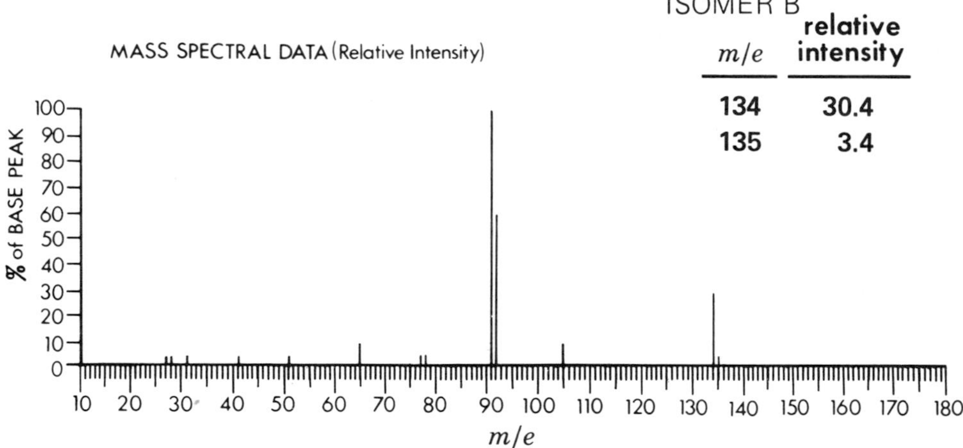

ISOMER B

m/e	relative intensity
134	30.4
135	3.4

Problem 2.7

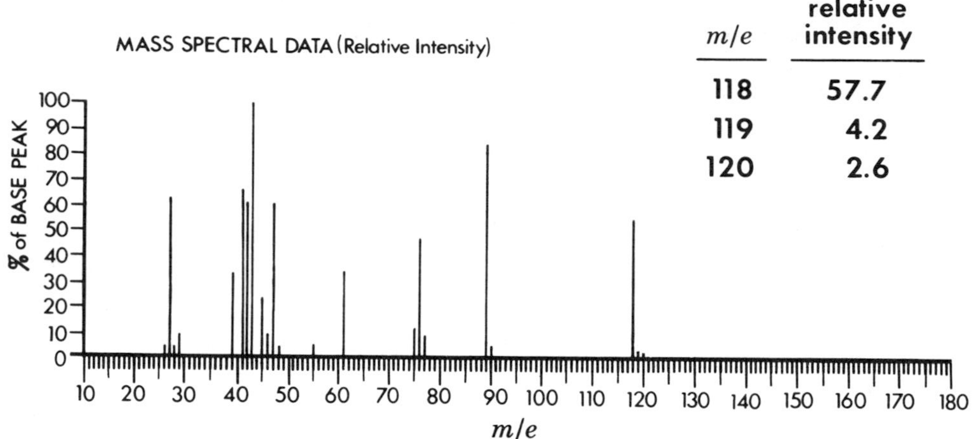

m/e	relative intensity
118	57.7
119	4.2
120	2.6

appendix a masses and isotopic abundance ratios for various combinations of carbon, hydrogen, nitrogen and oxygen*

	M+1	M+2	FM
12			
C	1.08	0.00	12.0000
13			
CH	1.10	0.00	13.0078
14			
N	0.38	0.00	14.0031
CH_2	1.11	0.00	14.0157
15			
HN	0.40	0.00	15.0109
CH_3	1.13	0.00	15.0235
16			
O	0.04	0.20	15.9949
H_2N	0.41	0.00	16.0187
CH_4	1.14	0.00	16.0313
17			
HO	0.06	0.20	17.0027*
H_3N	0.43	0.00	17.0266
18			
H_2O	0.07	0.20	18.0106
24			
C_2	2.16	0.01	24.0000
25			
C_2H	2.18	0.01	25.0078
26			
CN	1.46	0.00	26.0031
C_2H_2	2.19	0.01	26.0157
27			
CHN	1.48	0.00	27.0109
C_2H_3	2.21	0.01	27.0235
28			
N_2	0.76	0.00	28.0062
CO	1.12	0.20	27.9949
CH_2N	1.49	0.00	28.0187
C_2H_4	2.23	0.01	28.0313
29			
HN_2	0.78	0.00	29.0140
CHO	1.14	0.20	29.0027
CH_3N	1.51	0.00	29.0266
C_2H_5	2.24	0.01	29.0391
30			
NO	0.42	0.20	29.9980
H_2N_2	0.79	0.00	30.0218
CH_2O	1.15	0.20	30.0106
CH_4N	1.53	0.01	30.0344
C_2H_6	2.26	0.01	30.0470
31			
HNO	0.44	0.20	31.0058
H_3N_2	0.81	0.00	31.0297
CH_3O	1.17	0.20	31.0184
CH_5N	1.54	0.01	31.0422
32			
O_2	0.08	0.40	31.9898
H_2NO	0.45	0.20	32.0136
H_4N_2	0.83	0.00	32.0375
CH_4O	1.18	0.20	32.0262
33			
HO_2	0.09	0.40	32.9976
H_3NO	0.47	0.20	33.0215
34			
H_2O_2	0.11	0.40	34.0054
36			
C_3	3.24	0.04	36.0000
37			
C_3H	3.26	0.04	37.0078
38			
C_2N	2.54	0.02	38.0031
C_3H_2	3.27	0.04	38.0157
39			
C_2HN	2.56	0.02	39.0109
C_3H_3	3.29	0.04	39.0235
40			
CN_2	1.84	0.01	40.0062
C_2O	2.20	0.21	39.9949
C_2H_2N	2.57	0.02	40.0187
C_3H_4	3.31	0.04	40.0313
41			
CHN_2	1.86	0.01	41.0140
C_2HO	2.22	0.21	41.0027
C_2H_3N	2.59	0.02	41.0266
C_3H_5	3.32	0.04	41.0391
42			
N_3	1.14	0.00	42.0093
CNO	1.50	0.21	41.9980
CH_2N_2	1.88	0.01	42.0218
C_2H_2O	2.23	0.21	42.0106
C_2H_4N	2.61	0.02	42.0344
C_3H_6	3.34	0.04	42.0470
43			
HN_3	1.16	0.00	43.0171
CHNO	1.52	0.21	43.0058
CH_3N_2	1.89	0.01	43.0297
C_2H_3O	2.25	0.21	43.0184
C_2H_5N	2.62	0.02	43.0422
C_3H_7	3.35	0.04	43.0548
44			
N_2O	0.80	0.20	44.0011
H_2N_3	1.18	0.00	44.0249
CO_2	1.16	0.40	43.9898
CH_2NO	1.53	0.21	44.0136
CH_4N_2	1.91	0.01	44.0375
C_2H_4O	2.26	0.21	44.0262
C_2H_6N	2.64	0.02	44.0501
C_3H_8	3.37	0.04	44.0626
45			
HN_2O	0.82	0.20	45.0089
H_3N_3	1.19	0.00	45.0328
CHO_2	1.17	0.40	44.9976
CH_3NO	1.55	0.21	45.0215
CH_5N_2	1.92	0.01	45.0453
C_2H_5O	2.28	0.21	45.0340
C_2H_7N	2.65	0.02	45.0579
46			
NO_2	0.46	0.40	45.9929
H_2N_2O	0.83	0.20	46.0167
H_4N_3	1.21	0.01	46.0406
CH_2O_2	1.19	0.40	46.0054
CH_4NO	1.57	0.21	46.0293
CH_6N_2	1.94	0.01	46.0532
C_2H_6O	2.30	0.22	46.0419
47			
HNO_2	0.48	0.40	47.0007
H_3N_2O	0.85	0.20	47.0246
H_5N_3	1.22	0.01	47.0484
CH_3O_2	1.21	0.40	47.0133
CH_5NO	1.58	0.21	47.0371
48			
O_3	0.12	0.60	47.9847
H_2NO_2	0.49	0.40	48.0085
H_4N_2O	0.87	0.20	48.0324
CH_4O_2	1.22	0.40	48.0211
C_4	4.32	0.07	48.0000

The M+1 and M+2 data were taken, with permission, from J.H. Beynon, *Mass Spectrometry and its Application to Organic Chemistry*, Elsevier, Amsterdam, 1960. These data form a self-consistent set and can be used regardless of the mass standard.

The columns headed FM contain the formula masses based on the exact mass of the most abundant isotope of each element; these masses are based on the most abundant isotope of carbon having a mass of 12.0000.

	M+1	M+2	FM
49			
HO_3	0.13	0.60	48.9925
H_3NO_2	0.51	0.40	49.0164
C_4H	4.34	0.07	49.0078
50			
H_2O_3	0.15	0.60	50.0003
C_3N	3.62	0.05	50.0031
C_4H_2	4.35	0.07	50.0157
51			
C_3HN	3.64	0.05	51.0109
C_4H_3	4.37	0.07	51.0235
52			
C_2N_2	2.92	0.03	52.0062
C_3O	3.28	0.24	51.9949
C_3H_2N	3.66	0.05	52.0187
C_4H_4	4.39	0.07	52.0313
53			
C_2HN_2	2.94	0.03	53.0140
C_3HO	3.30	0.24	53.0027
C_3H_3N	3.67	0.05	53.0266
C_4H_5	4.40	0.07	53.0391
54			
CN_3	2.22	0.02	54.0093
C_2NO	2.58	0.22	53.9980
$C_2H_2N_2$	2.96	0.03	54.0218
C_3H_2O	3.31	0.24	54.0106
C_3H_4N	3.69	0.05	54.0344
C_4H_6	4.42	0.07	54.0470
55			
CHN_3	2.24	0.02	55.0171
C_2HNO	2.60	0.22	55.0058
$C_2H_3N_2$	2.97	0.03	55.0297
C_3H_3O	3.33	0.24	55.0184
C_3H_5N	3.70	0.05	55.0422
C_4H_7	4.43	0.07	55.0548
56			
N_4	1.53	0.01	56.0124
CN_2O	1.88	0.21	56.0011
CH_2N_3	2.26	0.02	56.0249
C_2O_2	2.24	0.41	55.9898
C_2H_2NO	2.61	0.22	56.0136
$C_2H_4N_2$	2.99	0.03	56.0375
C_3H_4O	3.34	0.24	56.0262
C_3H_6N	3.72	0.05	56.0501
C_4H_8	4.45	0.08	56.0626
57			
HN_4	1.54	0.01	57.0202
CHN_2O	1.90	0.21	57.0089
CH_3N_3	2.27	0.02	57.0328
C_2HO_2	2.26	0.41	56.9976
C_2H_3NO	2.63	0.22	57.0215
$C_2H_5N_2$	3.00	0.03	57.0453
C_3H_5O	3.36	0.24	57.0340
C_3H_7N	3.74	0.05	57.0579
C_4H_9	4.47	0.08	57.0705
58			
N_3O	1.18	0.21	58.0042
H_2N_4	1.56	0.01	58.0280
CNO_2	1.54	0.41	57.9929
CH_2N_2O	1.91	0.21	58.0167
CH_4N_3	2.29	0.02	58.0406
$C_2H_2O_2$	2.27	0.42	58.0054
C_2H_4NO	2.65	0.22	58.0293
$C_2H_6N_2$	3.02	0.03	58.0532
C_3H_6O	3.38	0.24	58.0419
C_3H_8N	3.75	0.05	58.0657
C_4H_{10}	4.48	0.08	58.0783
59			
HN_3O	1.20	0.21	59.0120
H_3N_4	1.57	0.01	59.0359
$CHNO_2$	1.56	0.41	59.0007
CH_3N_2O	1.93	0.21	59.0246
CH_5N_3	2.30	0.02	59.0484
$C_2H_3O_2$	2.29	0.42	59.0133
C_2H_5NO	2.66	0.22	59.0371
$C_2H_7N_2$	3.04	0.03	59.0610
C_3H_7O	3.39	0.24	59.0497
C_3H_9N	3.77	0.05	59.0736
60			
N_2O_2	0.84	0.40	59.9960
H_2N_3O	1.22	0.21	60.0198
H_4N_4	1.59	0.01	60.0437
CO_3	1.20	0.60	59.9847
CH_2NO_2	1.57	0.41	60.0085
CH_4N_2O	1.95	0.21	60.0324
CH_6N_3	2.32	0.02	60.0563
$C_2H_4O_2$	2.30	0.42	60.0211
C_2H_6NO	2.68	0.22	60.0449
$C_2H_8N_2$	3.05	0.03	60.0688
C_3H_8O	3.41	0.24	60.0575
C_5	5.40	0.12	60.0000
61			
HN_2O_2	0.86	0.40	61.0038
H_3N_3O	1.23	0.21	61.0277
H_5N_4	1.61	0.01	61.0515
CHO_3	1.21	0.60	60.9925
CH_3NO_2	1.59	0.41	61.0164
CH_5N_2O	1.96	0.21	61.0402
CH_7N_3	2.34	0.02	61.0641
$C_2H_5O_2$	2.32	0.42	61.0289
C_2H_7NO	2.69	0.22	61.0528
C_5H	5.42	0.12	61.0078
62			
NO_3	0.50	0.60	61.9878
$H_2N_2O_2$	0.87	0.40	62.0116
H_4N_3O	1.25	0.21	62.0355
H_6N_4	1.62	0.01	62.0594
CH_2O_3	1.23	0.60	62.0003
CH_4NO_2	1.60	0.41	62.0242
CH_6N_2O	1.98	0.21	62.0480
$C_2H_6O_2$	2.34	0.42	62.0368
C_4N	4.70	0.09	62.0031
C_5H_2	5.44	0.12	62.0157
63			
HNO_3	0.51	0.60	62.9956
$H_3N_2O_2$	0.89	0.40	63.0195
H_5N_3O	1.26	0.21	63.0433
CH_3O_3	1.25	0.60	63.0082
CH_5NO_2	1.62	0.41	63.0320
C_4HN	4.72	0.09	63.0109
C_5H_3	5.45	0.12	63.0235
64			
O_4	0.16	0.80	63.9796
H_2NO_3	0.53	0.60	64.0034
$H_4N_2O_2$	0.91	0.40	64.0273
CH_4O_3	1.26	0.60	64.0160
C_3N_2	4.00	0.06	64.0062
C_4O	4.36	0.27	63.9949
C_4H_2N	4.74	0.09	64.0187
C_5H_4	5.47	0.12	64.0313
65			
HO_4	0.17	0.80	64.9874
H_3NO_3	0.55	0.60	65.0113
C_3HN_2	4.02	0.06	65.0140
C_4HO	4.38	0.27	65.0027
C_4H_3N	4.75	0.09	65.0266
C_5H_5	5.48	0.12	65.0391
66			
H_2O_4	0.19	0.80	65.9953
C_2N_3	3.31	0.04	66.0093
C_3NO	3.66	0.25	65.9980
$C_3H_2N_2$	4.04	0.06	66.0218
C_4H_2O	4.39	0.27	66.0106
C_4H_4N	4.77	0.09	66.0344
C_5H_6	5.50	0.12	66.0470
67			
C_2HN_3	3.32	0.04	67.0171
C_3HNO	3.68	0.25	67.0058
$C_3H_3N_2$	4.05	0.06	67.0297
C_4H_3O	4.41	0.27	67.0184
C_4H_5N	4.78	0.09	67.0422
C_5H_7	5.52	0.12	67.0548
68			
CN_4	2.61	0.03	68.0124
C_2N_2O	2.96	0.23	68.0011
$C_2H_2N_3$	3.34	0.04	68.0249
C_3O_2	3.32	0.44	67.9898
C_3H_2NO	3.69	0.25	68.0136
$C_3H_4N_2$	4.07	0.06	68.0375
C_4H_4O	4.43	0.28	68.0262
C_4H_6N	4.80	0.09	68.0501
C_5H_8	5.53	0.12	68.0626
69			
CHN_4	2.62	0.03	69.0202
C_2HN_2O	2.98	0.23	69.0089
$C_2H_3N_3$	3.35	0.04	69.0328

	M+1	M+2	FM
C_3HO_2	3.34	0.44	68.9976
C_3H_3NO	3.71	0.25	69.0215
$C_3H_5N_2$	4.08	0.06	69.0453
C_4H_5O	4.44	0.28	69.0340
C_4H_7N	4.82	0.09	69.0579
C_5H_9	5.55	0.12	69.0705
70			
CN_3O	2.26	0.22	70.0042
CH_2N_4	2.64	0.03	70.0280
C_2NO_2	2.62	0.42	69.9929
$C_2H_2N_2O$	3.00	0.23	70.0167
$C_2H_4N_3$	3.37	0.04	70.0406
$C_3H_2O_2$	3.35	0.44	70.0054
C_3H_4NO	3.73	0.25	70.0293
$C_3H_6N_2$	4.10	0.07	70.0532
C_4H_6O	4.46	0.28	70.0419
C_4H_8N	4.83	0.09	70.0657
C_5H_{10}	5.56	0.13	70.0783
71			
CHN_3O	2.28	0.22	71.0120
CH_3N_4	2.65	0.03	71.0359
C_2HNO_2	2.64	0.42	71.0007
$C_2H_3N_2O$	3.01	0.23	71.0246
$C_2H_5N_3$	3.39	0.04	71.0484
$C_3H_3O_2$	3.37	0.44	71.0133
C_3H_5NO	3.74	0.25	71.0371
$C_3H_7N_2$	4.12	0.07	71.0610
C_4H_7O	4.47	0.28	71.0497
C_4H_9N	4.85	0.09	71.0736
C_5H_{11}	5.58	0.13	71.0861
72			
N_4O	1.56	0.21	72.0073
CN_2O_2	1.92	0.41	71.9960
CH_2N_3O	2.30	0.22	72.0198
CH_4N_4	2.67	0.03	72.0437
C_2O_3	2.28	0.62	71.9847
$C_2H_2NO_2$	2.65	0.42	72.0085
$C_2H_4N_2O$	3.03	0.23	72.0324
$C_2H_6N_3$	3.40	0.04	72.0563
$C_3H_4O_2$	3.38	0.44	72.0211
C_3H_6NO	3.76	0.25	72.0449
$C_3H_8N_2$	4.13	0.07	72.0688
C_4H_8O	4.49	0.28	72.0575
$C_4H_{10}N$	4.86	0.09	72.0814
C_5H_{12}	5.60	0.13	72.0939
C_6	6.48	0.18	72.0000
73			
HN_4O	1.58	0.21	73.0151
CHN_2O_2	1.94	0.41	73.0038
CH_3N_3O	2.31	0.22	73.0277
CH_5N_4	2.69	0.03	73.0515
C_2HO_3	2.29	0.62	72.9925
$C_2H_3NO_2$	2.67	0.42	73.0164
$C_2H_5N_2O$	3.04	0.23	73.0402
$C_2H_7N_3$	3.42	0.04	73.0641
$C_3H_5O_2$	3.40	0.44	73.0289
C_3H_7NO	3.77	0.25	73.0528
$C_3H_9N_2$	4.15	0.07	73.0767
C_4H_9O	4.51	0.28	73.0653
$C_4H_{11}N$	4.88	0.09	73.0892
C_6H	6.50	0.18	73.0078
74			
N_3O_2	1.22	0.41	73.9991
H_2N_4O	1.60	0.21	74.0229
CNO_3	1.58	0.61	73.9878
$CH_2N_2O_2$	1.95	0.41	74.0116
CH_4N_3O	2.33	0.22	74.0355
CH_6N_4	2.70	0.03	74.0594
$C_2H_2O_3$	2.31	0.62	74.0003
$C_2H_4NO_2$	2.68	0.42	74.0242
$C_2H_6N_2O$	3.06	0.23	74.0480
$C_2H_8N_3$	3.43	0.05	74.0719
$C_3H_6O_2$	3.42	0.44	74.0368
C_3H_8NO	3.79	0.25	74.0606
$C_3H_{10}N_2$	4.16	0.07	74.0845
$C_4H_{10}O$	4.52	0.28	74.0732
C_5N	5.78	0.14	74.0031
C_6H_2	6.52	0.18	74.0157
75			
HN_3O_2	1.24	0.41	75.0069
H_3N_4O	1.61	0.21	75.0308
$CHNO_3$	1.60	0.61	74.9956
$CH_3N_2O_2$	1.97	0.41	75.0195
CH_5N_3O	2.34	0.22	75.0433
CH_7N_4	2.72	0.03	75.0672
$C_2H_3O_3$	2.33	0.62	75.0082
$C_2H_5NO_2$	2.70	0.43	75.0320
$C_2H_7N_2O$	3.08	0.23	75.0559
$C_2H_9N_3$	3.45	0.05	75.0798
$C_3H_7O_2$	3.43	0.44	75.0446
C_3H_9NO	3.81	0.25	75.0684
C_5HN	5.80	0.14	75.0109
C_6H_3	6.53	0.18	75.0235
76			
N_2O_3	0.88	0.60	75.9909
$H_2N_3O_2$	1.25	0.41	76.0147
H_4N_4O	1.63	0.21	76.0386
CO_4	1.24	0.80	75.9796
CH_2NO_3	1.61	0.61	76.0034
$CH_4N_2O_2$	1.99	0.41	76.0273
CH_6N_3O	2.36	0.22	76.0511
CH_8N_4	2.73	0.03	76.0750
$C_2H_4O_3$	2.34	0.62	76.0160
$C_2H_6NO_2$	2.72	0.43	76.0399
$C_2H_8N_2O$	3.09	0.24	76.0637
$C_3H_8O_2$	3.45	0.44	76.0524
C_4N_2	5.09	0.10	76.0062
C_5O	5.44	0.32	75.9949
C_5H_2N	5.82	0.14	76.0187
C_6H_4	6.55	0.18	76.0313
77			
HN_2O_3	0.90	0.60	76.9987
$H_3N_3O_2$	1.27	0.41	77.0226
H_5N_4O	1.64	0.21	77.0464
CHO_4	1.25	0.80	76.9874
CH_3NO_3	1.63	0.61	77.0113
$CH_5N_2O_2$	2.00	0.41	77.0351
CH_7N_3O	2.38	0.22	77.0590
$C_2H_5O_3$	2.36	0.62	77.0238
$C_2H_7NO_2$	2.73	0.43	77.0477
C_4HN_2	5.10	0.11	77.0140
C_5HO	5.46	0.32	77.0027
C_5H_3N	5.83	0.14	77.0266
C_6H_5	6.56	0.18	77.0391
78			
NO_4	0.54	0.80	77.9827
$H_2N_2O_3$	0.91	0.60	78.0065
$H_4N_3O_2$	1.29	0.41	78.0304
H_6N_4O	1.66	0.21	78.0542
CH_2O_4	1.27	0.80	77.9953
CH_4NO_3	1.64	0.61	78.0191
$CH_6N_2O_2$	2.02	0.41	78.0429
$C_2H_6O_3$	2.37	0.62	78.0317
C_3N_3	4.39	0.08	78.0093
C_4NO	4.74	0.29	77.9980
$C_4H_2N_2$	5.12	0.11	78.0218
C_5H_2O	5.47	0.32	78.0106
C_5H_4N	5.85	0.14	78.0344
C_6H_6	6.58	0.18	78.0470
79			
HNO_4	0.55	0.80	78.9905
$H_3N_2O_3$	0.93	0.60	79.0144
$H_5N_3O_2$	1.30	0.41	79.0382
CH_3O_4	1.28	0.00	70.0031
CH_5NO_3	1.66	0.61	79.0269
C_3HN_3	4.40	0.08	79.0171
C_4HNO	4.76	0.29	79.0058
$C_4H_3N_2$	5.13	0.11	79.0297
C_5H_3O	5.49	0.32	79.0184
C_5H_5N	5.86	0.14	79.0422
C_6H_7	6.60	0.18	79.0548
80			
H_2NO_4	0.57	0.80	79.9983
$H_4N_2O_3$	0.94	0.60	80.0222
CH_4O_4	1.30	0.80	80.0109
C_2N_4	3.69	0.05	80.0124
C_3N_2O	4.04	0.26	80.0011
$C_3H_2N_3$	4.42	0.08	80.0249
C_4O_2	4.40	0.47	79.9898
C_4H_2NO	4.77	0.29	80.0136
$C_4H_4N_2$	5.15	0.11	80.0375
C_5H_4O	5.51	0.32	80.0262
C_5H_6N	5.88	0.14	80.0501

	M+1	M+2	FM
C_6H_8	6.61	0.18	80.0626
81			
H_3NO_4	0.59	0.80	81.0062
C_2HN_4	3.70	0.05	81.0202
C_3HN_2O	4.06	0.26	81.0089
$C_3H_3N_3$	4.43	0.08	81.0328
C_4HO_2	4.42	0.48	80.9976
C_4H_3NO	4.79	0.29	81.0215
$C_4H_5N_2$	5.17	0.11	81.0453
C_5H_5O	5.52	0.32	81.0340
C_5H_7N	5.90	0.14	81.0579
C_6H_9	6.63	0.18	81.0705
82			
C_2N_3O	3.34	0.24	82.0042
$C_2H_2N_4$	3.72	0.05	82.0280
C_3NO_2	3.70	0.45	81.9929
$C_3H_2N_2O$	4.08	0.26	82.0167
$C_3H_4N_3$	4.45	0.08	82.0406
$C_4H_2O_2$	4.43	0.48	82.0054
C_4H_4NO	4.81	0.29	82.0293
$C_4H_6N_2$	5.18	0.11	82.0532
C_5H_6O	5.54	0.32	82.0419
C_5H_8N	5.91	0.14	82.0657
C_6H_{10}	6.64	0.19	82.0783
83			
C_2HN_3O	3.36	0.24	83.0120
$C_2H_3N_4$	3.74	0.06	83.0359
C_3HNO_2	3.72	0.45	83.0007
$C_3H_3N_2O$	4.09	0.27	83.0246
$C_3H_5N_3$	4.47	0.08	83.0484
$C_4H_3O_2$	4.45	0.48	83.0133
C_4H_5NO	4.82	0.29	83.0371
$C_4H_7N_2$	5.20	0.11	83.0610
C_5H_7O	5.55	0.33	83.0497
C_5H_9N	5.93	0.15	83.0736
C_6H_{11}	6.66	0.19	83.0861
84			
CN_4O	2.65	0.23	84.0073
$C_2N_2O_2$	3.00	0.43	83.9960
$C_2H_2N_3O$	3.38	0.24	84.0198
$C_2H_4N_4$	3.75	0.06	84.0437
C_3O_3	3.36	0.64	83.9847
$C_3H_2NO_2$	3.73	0.45	84.0085
$C_3H_4N_2O$	4.11	0.27	84.0324
$C_3H_6N_3$	4.48	0.08	84.0563
$C_4H_4O_2$	4.46	0.48	84.0211
C_4H_6NO	4.84	0.29	84.0449
$C_4H_8N_2$	5.21	0.11	84.0688
C_5H_8O	5.57	0.33	84.0575
$C_5H_{10}N$	5.94	0.15	84.0814
C_6H_{12}	6.68	0.19	84.0939
C_7	7.56	0.25	84.0000
85			
CHN_4O	2.66	0.23	85.0151
$C_2HN_2O_2$	3.02	0.43	85.0038
$C_2H_3N_3O$	3.39	0.24	85.0277
$C_2H_5N_4$	3.77	0.06	85.0515
C_3HO_3	3.38	0.64	84.9925
$C_3H_3NO_2$	3.75	0.45	85.0164
$C_3H_5N_2O$	4.12	0.27	85.0402
$C_3H_7N_3$	4.50	0.08	85.0641
$C_4H_5O_2$	4.48	0.48	85.0289
C_4H_7NO	4.85	0.29	85.0528
$C_4H_9N_2$	5.23	0.11	85.0767
C_5H_9O	5.59	0.33	85.0653
$C_5H_{11}N$	5.96	0.15	85.0892
C_6H_{13}	6.69	0.19	85.1018
C_7H	7.58	0.25	85.0078
86			
CN_3O_2	2.30	0.42	85.9991
CH_2N_4O	2.68	0.23	86.0229
C_2NO_3	2.66	0.62	85.9878
$C_2H_2N_2O_2$	3.03	0.43	86.0116
$C_2H_4N_3O$	3.41	0.24	86.0355
$C_2H_6N_4$	3.78	0.06	86.0594
$C_3H_2O_3$	3.39	0.64	86.0003
$C_3H_4NO_2$	3.77	0.45	86.0242
$C_3H_6N_2O$	4.14	0.27	86.0480
$C_3H_8N_3$	4.51	0.08	86.0719
$C_4H_6O_2$	4.50	0.48	86.0368
C_4H_8NO	4.87	0.30	86.0606
$C_4H_{10}N_2$	5.25	0.11	86.0845
$C_5H_{10}O$	5.60	0.33	86.0732
$C_5H_{12}N$	5.98	0.15	86.0970
C_6H_{14}	6.71	0.19	86.1096
C_6N	6.87	0.20	86.0031
C_7H_2	7.60	0.25	86.0157
87			
CHN_3O_2	2.32	0.42	87.0069
CH_3N_4O	2.69	0.23	87.0308
C_2HNO_3	2.68	0.62	86.9956
$C_2H_3N_2O_2$	3.05	0.43	87.0195
$C_2H_5N_3O$	3.42	0.25	87.0433
$C_2H_7N_4$	3.80	0.06	87.0672
$C_3H_3O_3$	3.41	0.64	87.0082
$C_3H_5NO_2$	3.78	0.45	87.0320
$C_3H_7N_2O$	4.16	0.27	87.0559
$C_3H_9N_3$	4.53	0.08	87.0798
$C_4H_7O_2$	4.51	0.48	87.0446
C_4H_9NO	4.89	0.30	87.0684
$C_4H_{11}N_2$	5.26	0.11	87.0923
$C_5H_{11}O$	5.62	0.33	87.0810
$C_5H_{13}N$	5.99	0.15	87.1049
C_6HN	6.88	0.20	87.0109
C_7H_3	7.61	0.25	87.0235
88			
N_4O_2	1.60	0.41	88.0022
CN_2O_3	1.96	0.61	87.9909
$CH_2N_3O_2$	2.34	0.42	88.0147
CH_4N_4O	2.71	0.23	88.0386
C_2O_4	2.32	0.82	87.9796
$C_2H_2NO_3$	2.69	0.63	88.0034
$C_2H_4N_2O_2$	3.07	0.43	88.0273
$C_2H_6N_3O$	3.44	0.25	88.0511
$C_2H_8N_4$	3.82	0.06	88.0750
$C_3H_4O_3$	3.42	0.64	88.0160
$C_3H_6NO_2$	3.80	0.45	88.0399
$C_3H_8N_2O$	4.17	0.27	88.0637
$C_3H_{10}N_3$	4.55	0.08	88.0876
$C_4H_8O_2$	4.53	0.48	88.0524
$C_4H_{10}NO$	4.90	0.30	88.0763
$C_4H_{12}N_2$	5.28	0.11	88.1001
$C_5H_{12}O$	5.63	0.33	88.0888
C_5N_2	6.17	0.16	88.0062
C_6O	6.52	0.38	87.9949
C_6H_2N	6.90	0.20	88.0187
C_7H_4	7.63	0.25	88.0313
89			
HN_4O_2	1.62	0.41	89.0100
CHN_2O_3	1.98	0.61	88.9987
$CH_3N_3O_2$	2.35	0.42	89.0226
CH_5N_4O	2.73	0.23	89.0464
C_2HO_4	2.33	0.82	88.9874
$C_2H_3NO_3$	2.71	0.63	89.0113
$C_2H_5N_2O_2$	3.08	0.44	89.0351
$C_2H_7N_3O$	3.46	0.25	89.0590
$C_2H_9N_4$	3.83	0.06	89.0829
$C_3H_5O_3$	3.44	0.64	89.0238
$C_3H_7NO_2$	3.81	0.46	89.0477
$C_3H_9N_2O$	4.19	0.27	89.0715
$C_3H_{11}N_3$	4.56	0.08	89.0954
$C_4H_9O_2$	4.54	0.48	89.0603
$C_4H_{11}NO$	4.92	0.30	89.0841
C_5HN_2	6.18	0.16	89.0140
C_6HO	6.54	0.38	89.0027
C_6H_3N	6.91	0.20	89.0266
C_7H_5	7.64	0.25	89.0391
90			
N_3O_3	1.26	0.61	89.9940
$H_2N_4O_2$	1.64	0.41	90.0178
CNO_4	1.62	0.81	89.9827
$CH_2N_2O_3$	1.99	0.61	90.0065
$CH_4N_3O_2$	2.37	0.42	90.0304
CH_6N_4O	2.74	0.23	90.0542
$C_2H_2O_4$	2.35	0.82	89.9953
$C_2H_4NO_3$	2.72	0.63	90.0191
$C_2H_6N_2O_2$	3.10	0.44	90.0429
$C_2H_8N_3O$	3.47	0.25	90.0668
$C_2H_{10}N_4$	3.85	0.06	90.0907
$C_3H_6O_3$	3.46	0.64	90.0317
$C_3H_8NO_2$	3.83	0.46	90.0555
$C_3H_{10}N_2O$	4.20	0.27	90.0794
$C_4H_{10}O_2$	4.56	0.48	90.0681
C_4N_3	5.47	0.12	90.0093
C_5NO	5.82	0.34	89.9980
$C_5H_2N_2$	6.20	0.16	90.0218

	M+1	M+2	FM
C_6H_2O	6.55	0.38	90.0106
C_6H_4N	6.93	0.20	90.0344
C_7H_6	7.66	0.25	90.0470
91			
HN_3O_3	1.28	0.61	91.0018
$H_3N_4O_2$	1.65	0.41	91.0257
$CHNO_4$	1.63	0.81	90.9905
$CH_3N_2O_3$	2.01	0.61	91.0144
$CH_5N_3O_2$	2.38	0.42	91.0382
CH_7N_4O	2.76	0.23	91.0621
$C_2H_3O_4$	2.37	0.82	91.0031
$C_2H_5NO_3$	2.74	0.63	91.0269
$C_2H_7N_2O_2$	3.11	0.44	91.0508
$C_2H_9N_3O$	3.49	0.25	91.0746
$C_3H_7O_3$	3.47	0.64	91.0395
$C_3H_9NO_2$	3.85	0.46	91.0634
C_4HN_3	5.48	0.12	91.0171
C_5HNO	5.84	0.34	91.0058
$C_5H_3N_2$	6.21	0.16	91.0297
C_6H_3O	6.57	0.38	91.0184
C_6H_5N	6.95	0.21	91.0422
C_7H_7	7.68	0.25	91.0548
92			
N_2O_4	0.92	0.80	91.9858
$H_2N_3O_3$	1.29	0.61	92.0096
$H_4N_4O_2$	1.67	0.41	92.0335
CH_2NO_4	1.65	0.81	91.9983
$CH_4N_2O_3$	2.02	0.61	92.0222
$CH_6N_3O_2$	2.40	0.42	92.0460
CH_8N_4O	2.77	0.23	92.0699
$C_2H_4O_4$	2.38	0.82	92.0109
$C_2H_6NO_3$	2.76	0.63	92.0348
$C_2H_8N_2O_2$	3.13	0.44	92.0586
$C_3H_8O_3$	3.49	0.64	92.0473
C_3N_4	4.77	0.09	92.0124
C_4N_2O	5.12	0.31	92.0011
$C_4H_2N_3$	5.50	0.13	92.0249
C_5O_2	5.48	0.52	91.9898
C_5H_2NO	5.86	0.34	92.0136
$C_5H_4N_2$	6.23	0.16	92.0375
C_6H_4O	6.59	0.38	92.0262
C_6H_6N	6.96	0.21	92.0501
C_7H_8	7.69	0.25	92.0626
93			
HN_2O_4	0.94	0.80	92.9936
$H_3N_3O_3$	1.31	0.61	93.0175
$H_5N_4O_2$	1.68	0.41	93.0413
CH_3NO_4	1.67	0.81	93.0062
$CH_5N_2O_3$	2.04	0.61	93.0300
$CH_7N_3O_2$	2.42	0.42	93.0539
$C_2H_5O_4$	2.40	0.82	93.0187
$C_2H_7NO_3$	2.77	0.63	93.0426
C_3HN_4	4.78	0.09	93.0202
C_4HN_2O	5.14	0.31	93.0089

	M+1	M+2	FM
$C_4H_3N_3$	5.51	0.13	93.0328
C_5HO_2	5.50	0.52	92.9976
C_5H_3NO	5.87	0.34	93.0215
$C_5H_5N_2$	6.25	0.16	93.0453
C_6H_5O	6.60	0.38	93.0340
C_6H_7N	6.98	0.21	93.0579
C_7H_9	7.71	0.26	93.0705
94			
$H_2N_2O_4$	0.95	0.80	94.0014
$H_4N_3O_3$	1.33	0.61	94.0253
$H_6N_4O_2$	1.70	0.41	94.0491
CH_4NO_4	1.68	0.81	94.0140
$CH_6N_2O_3$	2.06	0.62	94.0379
$C_2H_6O_4$	2.41	0.82	94.0266
C_3N_3O	4.43	0.28	94.0042
$C_3H_2N_4$	4.80	0.09	94.0280
C_4NO_2	4.78	0.49	93.9929
$C_4H_2N_2O$	5.16	0.31	94.0167
$C_4H_4N_3$	5.53	0.13	94.0406
$C_5H_2O_2$	5.51	0.52	94.0054
C_5H_4NO	5.89	0.34	94.0293
$C_5H_6N_2$	6.26	0.17	94.0532
C_6H_6O	6.62	0.38	94.0419
C_6H_8N	6.99	0.21	94.0657
C_7H_{10}	7.72	0.26	94.0783
95			
$H_3N_2O_4$	0.97	0.81	95.0093
$H_5N_3O_3$	1.34	0.61	95.0331
CH_5NO_4	1.70	0.81	95.0218
C_3HN_3O	4.44	0.28	95.0120
$C_3H_3N_4$	4.82	0.10	95.0359
C_4HNO_2	4.80	0.49	95.0007
$C_4H_3N_2O$	5.17	0.31	95.0246
$C_4H_5N_3$	5.55	0.13	95.0484
$C_5H_3O_2$	5.53	0.52	95.0133
C_5H_5NO	5.90	0.34	95.0371
$C_5H_7N_2$	6.28	0.17	95.0610
C_6H_7O	6.63	0.39	95.0497
C_6H_9N	7.01	0.21	95.0736
C_7H_{11}	7.74	0.26	95.0861
96			
$H_4N_2O_4$	0.98	0.81	96.0171
C_2N_4O	3.73	0.26	96.0073
$C_3N_2O_2$	4.08	0.47	95.9960
$C_3H_2N_3O$	4.46	0.28	96.0198
$C_3H_4N_4$	4.83	0.10	96.0437
C_4O_3	4.44	0.68	95.9847
$C_4H_2NO_2$	4.81	0.49	96.0085
$C_4H_4N_2O$	5.19	0.31	96.0324
$C_4H_6N_3$	5.56	0.13	96.0563
$C_5H_4O_2$	5.55	0.53	96.0211
C_5H_6NO	5.92	0.35	96.0449
$C_5H_8N_2$	6.29	0.17	96.0688
C_6H_8O	6.65	0.39	96.0575

	M+1	M+2	FM
$C_6H_{10}N$	7.03	0.21	96.0814
C_7H_{12}	7.76	0.26	96.0939
C_8	8.64	0.33	96.0000
97			
C_2HN_4O	3.74	0.26	97.0151
$C_3HN_2O_2$	4.10	0.47	97.0038
$C_3H_3N_3O$	4.47	0.28	97.0277
$C_3H_5N_4$	4.85	0.10	97.0515
C_4HO_3	4.46	0.68	96.9925
$C_4H_3NO_2$	4.83	0.49	97.0164
$C_4H_5N_2O$	5.20	0.31	97.0402
$C_4H_7N_3$	5.58	0.13	97.0641
$C_5H_5O_2$	5.56	0.53	97.0289
C_5H_7NO	5.94	0.35	97.0528
$C_5H_9N_2$	6.31	0.17	97.0767
C_6H_9O	6.67	0.39	97.0653
$C_6H_{11}N$	7.04	0.21	97.0892
C_7H_{13}	7.77	0.26	97.1018
C_8H	8.66	0.33	97.0078
98			
$C_2N_3O_2$	3.38	0.44	97.9991
$C_2H_2N_4O$	3.76	0.26	98.0229
C_3NO_3	3.74	0.65	97.9878
$C_3H_2N_2O_2$	4.11	0.47	98.0116
$C_3H_4N_3O$	4.49	0.28	98.0355
$C_3H_6N_4$	4.86	0.10	98.0594
$C_4H_2O_3$	4.47	0.68	98.0003
$C_4H_4NO_2$	4.85	0.49	98.0242
$C_4H_6N_2O$	5.22	0.31	98.0480
$C_4H_8N_3$	5.59	0.13	98.0719
$C_5H_6O_2$	5.58	0.53	98.0368
C_5H_8NO	5.95	0.35	98.0606
$C_5H_{10}N_2$	6.33	0.17	98.0845
$C_6H_{10}O$	6.68	0.39	98.0732
$C_6H_{12}N$	7.06	0.21	98.0970
C_7H_{14}	7.79	0.26	98.1096
C_7N	7.95	0.27	98.0031
C_8H_2	8.68	0.33	98.0157
99			
$C_2HN_3O_2$	3.40	0.44	99.0069
$C_2H_3N_4O$	3.77	0.26	99.0308
C_3HNO_3	3.76	0.65	98.9956
$C_3H_3N_2O_2$	4.13	0.47	99.0195
$C_3H_5N_3O$	4.51	0.28	99.0433
$C_3H_7N_4$	4.88	0.10	99.0672
$C_4H_3O_3$	4.49	0.68	99.0082
$C_4H_5NO_2$	4.86	0.50	99.0320
$C_4H_7N_2O$	5.24	0.31	99.0559
$C_4H_9N_3$	5.61	0.13	99.0798
$C_5H_7O_2$	5.59	0.53	99.0446
C_5H_9NO	5.97	0.35	99.0684
$C_5H_{11}N_2$	6.34	0.17	99.0923
$C_6H_{11}O$	6.70	0.39	99.0810
$C_6H_{13}N$	7.07	0.21	99.1049

	M+1	M+2	FM
C_7H_{15}	7.80	0.26	99.1174
C_7HN	7.96	0.28	99.0109
C_8H_3	8.69	0.33	99.0235
100			
CN_4O_2	2.68	0.43	100.0022
$C_2N_2O_3$	3.04	0.63	99.9909
$C_2H_2N_3O_2$	3.42	0.45	100.0147
$C_2H_4N_4O$	3.79	0.26	100.0386
C_3O_4	3.40	0.84	99.9796
$C_3H_2NO_3$	3.77	0.65	100.0034
$C_3H_4N_2O_2$	4.15	0.47	100.0273
$C_3H_6N_3O$	4.52	0.28	100.0511
$C_3H_8N_4$	4.90	0.10	100.0750
$C_4H_4O_3$	4.50	0.68	100.0160
$C_4H_6NO_2$	4.88	0.50	100.0399
$C_4H_8N_2O$	5.25	0.31	100.0637
$C_4H_{10}N_3$	5.63	0.13	100.0876
$C_5H_8O_2$	5.61	0.53	100.0524
$C_5H_{10}NO$	5.98	0.35	100.0763
$C_5H_{12}N_2$	6.36	0.17	100.1001
$C_6H_{12}O$	6.71	0.39	100.0888
$C_6H_{14}N$	7.09	0.22	100.1127
C_6N_2	7.25	0.23	100.0062
C_7H_{16}	7.82	0.26	100.1253
C_7O	7.60	0.45	99.9949
C_7H_2N	7.98	0.28	100.0187
C_8H_4	8.71	0.33	100.0313
101			
CHN_4O_2	2.70	0.43	101.0100
$C_2HN_2O_3$	3.06	0.64	100.9987
$C_2H_3N_3O_2$	3.43	0.45	101.0226
$C_2H_5N_4O$	3.81	0.26	101.0464
C_3HO_4	3.41	0.84	100.9874
$C_3H_3NO_3$	3.79	0.65	101.0113
$C_3H_5N_2O_2$	4.16	0.47	101.0351
$C_3H_7N_3O$	4.54	0.28	101.0590
$C_3H_9N_4$	4.91	0.10	101.0829
$C_4H_5O_3$	4.52	0.68	101.0238
$C_4H_7NO_2$	4.89	0.50	101.0477
$C_4H_9N_2O$	5.27	0.31	101.0715
$C_4H_{11}N_3$	5.64	0.13	101.0954
$C_5H_9O_2$	5.63	0.53	101.0603
$C_5H_{11}NO$	6.00	0.35	101.0841
$C_5H_{13}N_2$	6.37	0.17	101.1080
$C_6H_{13}O$	6.73	0.39	101.0967
$C_6H_{15}N$	7.11	0.22	101.1205
C_6HN_2	7.26	0.23	101.0140
C_7HO	7.62	0.45	101.0027
C_7H_3N	7.99	0.28	101.0266
C_8H_5	8.72	0.33	101.0391
102			
CN_3O_3	2.34	0.62	101.9940
$CH_2N_4O_2$	2.72	0.43	102.0178
C_2NO_4	2.70	0.83	101.9827

	M+1	M+2	FM
$C_2H_2N_2O_3$	3.07	0.64	102.0065
$C_2H_4N_3O_2$	3.45	0.45	102.0304
$C_2H_6N_4O$	3.82	0.26	102.0542
$C_3H_2O_4$	3.43	0.84	101.9953
$C_3H_4NO_3$	3.80	0.66	102.0191
$C_3H_6N_2O_2$	4.18	0.47	102.0429
$C_3H_8N_3O$	4.55	0.28	102.0668
$C_3H_{10}N_4$	4.93	0.10	102.0907
$C_4H_6O_3$	4.54	0.68	102.0317
$C_4H_8NO_2$	4.91	0.50	102.0555
$C_4H_{10}N_2O$	5.28	0.32	102.0794
$C_4H_{12}N_3$	5.66	0.13	102.1032
$C_5H_{10}O_2$	5.64	0.53	102.0681
$C_5H_{12}NO$	6.02	0.35	102.0919
$C_5H_{14}N_2$	6.39	0.17	102.1158
C_5N_3	6.55	0.18	102.0093
$C_6H_{14}O$	6.75	0.39	102.1045
C_6NO	6.90	0.40	101.9980
$C_6H_2N_2$	7.28	0.23	102.0218
C_7H_2O	7.64	0.45	102.0106
C_7H_4N	8.01	0.28	102.0344
C_8H_6	8.74	0.34	102.0470
103			
CHN_3O_3	2.36	0.62	103.0018
$CH_3N_4O_2$	2.73	0.43	103.0257
C_2HNO_4	2.72	0.83	102.9905
$C_2H_3N_2O_3$	3.09	0.64	103.0144
$C_2H_5N_3O_2$	3.46	0.45	103.0382
$C_2H_7N_4O$	3.84	0.26	103.0621
$C_3H_3O_4$	3.45	0.84	103.0031
$C_3H_5NO_3$	3.82	0.66	103.0269
$C_3H_7N_2O_2$	4.19	0.47	103.0508
$C_3H_9N_3O$	4.57	0.29	103.0746
$C_3H_{11}N_4$	4.94	0.10	103.0985
$C_4H_7O_3$	4.55	0.68	103.0395
$C_4H_9NO_2$	4.93	0.50	103.0634
$C_4H_{11}N_2O$	5.30	0.32	103.0872
$C_4H_{13}N_3$	5.67	0.14	103.1111
$C_5H_{11}O_2$	5.66	0.53	103.0759
$C_5H_{13}NO$	6.03	0.35	103.0998
C_5HN_3	6.56	0.18	103.0171
C_6HNO	6.92	0.40	103.0058
$C_6H_3N_2$	7.29	0.23	103.0297
C_7H_3O	7.65	0.45	103.0184
C_7H_5N	8.03	0.28	103.0422
C_8H_7	8.76	0.34	103.0548
104			
CN_2O_4	2.00	0.81	103.9858
$CH_2N_3O_3$	2.37	0.62	104.0096
$CH_4N_4O_2$	2.75	0.43	104.0335
$C_2H_2NO_4$	2.73	0.83	103.9983
$C_2H_4N_2O_3$	3.11	0.64	104.0222
$C_2H_6N_3O_2$	3.48	0.45	104.0460

	M+1	M+2	FM
$C_2H_8N_4O$	3.85	0.26	104.0699
$C_3H_4O_4$	3.46	0.84	104.0109
$C_3H_6NO_3$	3.84	0.66	104.0348
$C_3H_8N_2O_2$	4.21	0.47	104.0586
$C_3H_{10}N_3O$	4.59	0.29	104.0825
$C_3H_{12}N_4$	4.96	0.10	104.1063
$C_4H_8O_3$	4.57	0.68	104.0473
$C_4H_{10}NO_2$	4.94	0.50	104.0712
$C_4H_{12}N_2O$	5.32	0.32	104.0950
C_4N_4	5.85	0.14	104.0124
$C_5H_{12}O_2$	5.67	0.53	104.0837
C_5N_2O	6.20	0.36	104.0011
$C_5H_2N_3$	6.58	0.19	104.0249
C_6O_2	6.56	0.58	103.9898
C_6H_2NO	6.94	0.41	104.0136
$C_6H_4N_2$	7.31	0.23	104.0375
C_7H_4O	7.67	0.45	104.0262
C_7H_6N	8.04	0.28	104.0501
C_8H_8	8.77	0.34	104.0626
105			
CHN_2O_4	2.02	0.81	104.9936
$CH_3N_3O_3$	2.39	0.62	105.0175
$CH_5N_4O_2$	2.76	0.43	105.0413
$C_2H_3NO_4$	2.75	0.83	105.0062
$C_2H_5N_2O_3$	3.12	0.64	105.0300
$C_2H_7N_3O_2$	3.50	0.45	105.0539
$C_2H_9N_4O$	3.87	0.26	105.0777
$C_3H_5O_4$	3.48	0.84	105.0187
$C_3H_7NO_3$	3.85	0.66	105.0426
$C_3H_9N_2O_2$	4.23	0.47	105.0664
$C_3H_{11}N_3O$	4.60	0.29	105.0903
$C_4H_9O_3$	4.58	0.68	105.0552
$C_4H_{11}NO_2$	4.96	0.50	105.0790
C_4HN_4	5.86	0.15	105.0202
C_5HN_2O	6.22	0.36	105.0089
$C_5H_3N_3$	6.60	0.19	105.0328
C_6HO_2	6.58	0.58	104.9976
C_6H_3NO	6.95	0.41	105.0215
$C_6H_5N_2$	7.33	0.23	105.0453
C_7H_5O	7.68	0.45	105.0340
C_7H_7N	8.06	0.28	105.0579
C_8H_9	8.79	0.34	105.0705
106			
$CH_2N_2O_4$	2.03	0.82	106.0014
$CH_4N_3O_3$	2.41	0.62	106.0253
$CH_6N_4O_2$	2.78	0.43	106.0491
$C_2H_4NO_4$	2.76	0.83	106.0140
$C_2H_6N_2O_3$	3.14	0.64	106.0379
$C_2H_8N_3O_2$	3.51	0.45	106.0617
$C_2H_{10}N_4O$	3.89	0.26	106.0856
$C_3H_6O_4$	3.49	0.85	106.0266
$C_3H_8NO_3$	3.87	0.66	106.0504
$C_3H_{10}N_2O_2$	4.24	0.47	106.0743
$C_4H_{10}O_3$	4.60	0.68	106.0630

	M+1	M+2	FM
C_4N_3O	5.51	0.33	106.0042
$C_4H_2N_4$	5.88	0.15	106.0280
C_5NO_2	5.86	0.54	105.9929
$C_5H_2N_2O$	6.24	0.36	106.0167
$C_5H_4N_3$	6.61	0.19	106.0406
$C_6H_2O_2$	6.59	0.58	106.0054
C_6H_4NO	6.97	0.41	106.0293
$C_6H_6N_2$	7.34	0.23	106.0532
C_7H_6O	7.70	0.46	106.0419
C_7H_8N	8.07	0.28	106.0657
C_8H_{10}	8.80	0.34	106.0783
107			
$CH_3N_2O_4$	2.05	0.82	107.0093
$CH_5N_3O_3$	2.42	0.62	107.0331
$CH_7N_4O_2$	2.80	0.43	107.0570
$C_2H_5NO_4$	2.78	0.83	107.0218
$C_2H_7N_2O_3$	3.15	0.64	107.0457
$C_2H_9N_3O_2$	3.53	0.45	107.0695
$C_3H_7O_4$	3.51	0.85	107.0344
$C_3H_9NO_3$	3.88	0.66	107.0583
C_4HN_3O	5.52	0.33	107.0120
$C_4H_3N_4$	5.90	0.15	107.0359
C_5HNO_2	5.88	0.54	107.0007
$C_5H_3N_2O$	6.25	0.37	107.0246
$C_5H_5N_3$	6.63	0.19	107.0484
$C_6H_3O_2$	6.61	0.58	107.0133
C_6H_5NO	6.98	0.41	107.0371
$C_6H_7N_2$	7.36	0.23	107.0610
C_7H_7O	7.72	0.46	107.0497
C_7H_9N	8.09	0.29	107.0736
C_8H_{11}	8.82	0.34	107.0861
108			
$CH_4N_2O_4$	2.06	0.82	108.0171
$CH_6N_3O_3$	2.44	0.62	108.0410
$CH_8N_4O_2$	2.81	0.43	108.0648
$C_2H_6NO_4$	2.80	0.83	108.0297
$C_2H_8N_2O_3$	3.17	0.64	108.0535
$C_3H_8O_4$	3.53	0.85	108.0422
C_3N_4O	4.81	0.30	108.0073
$C_4N_2O_2$	5.16	0.51	107.9960
$C_4H_2N_3O$	5.54	0.33	108.0198
$C_4H_4N_4$	5.91	0.15	108.0437
C_5O_3	5.52	0.72	107.9847
$C_5H_2NO_2$	5.89	0.54	108.0085
$C_5H_4N_2O$	6.27	0.37	108.0324
$C_5H_6N_3$	6.64	0.19	108.0563
$C_6H_4O_2$	6.63	0.59	108.0211
C_6H_6NO	7.00	0.41	108.0449
$C_6H_8N_2$	7.37	0.24	108.0688
C_7H_8O	7.73	0.46	108.0575
$C_7H_{10}N$	8.11	0.29	108.0814
C_8H_{12}	8.84	0.34	108.0939
C_9	9.73	0.42	108.0000
109			
$CH_5N_2O_4$	2.08	0.82	109.0249
$CH_7N_3O_3$	2.45	0.62	109.0488
$C_2H_7NO_4$	2.81	0.83	109.0375
C_3HN_4O	4.82	0.30	109.0151
$C_4HN_2O_2$	5.18	0.51	109.0038
$C_4H_3N_3O$	5.55	0.33	109.0277
$C_4H_5N_4$	5.93	0.15	109.0515
C_5HO_3	5.54	0.73	108.9925
$C_5H_3NO_2$	5.91	0.55	109.0164
$C_5H_5N_2O$	6.29	0.37	109.0402
$C_5H_7N_3$	6.66	0.19	109.0641
$C_6H_5O_2$	6.64	0.59	109.0289
C_6H_7NO	7.02	0.41	109.0528
$C_6H_9N_2$	7.39	0.24	109.0767
C_7H_9O	7.75	0.46	109.0653
$C_7H_{11}N$	8.12	0.29	109.0892
C_8H_{13}	8.85	0.35	109.1018
C_9H	9.74	0.42	109.0078
110			
$CH_6N_2O_4$	2.10	0.82	110.0328
$C_3N_3O_2$	4.46	0.48	109.9991
$C_3H_2N_4O$	4.84	0.30	110.0229
C_4NO_3	4.82	0.69	109.9878
$C_4H_2N_2O_2$	5.20	0.51	110.0116
$C_4H_4N_3O$	5.57	0.33	110.0355
$C_4H_6N_4$	5.94	0.15	110.0594
$C_5H_2O_3$	5.55	0.73	110.0003
$C_5H_4NO_2$	5.93	0.55	110.0242
$C_5H_6N_2O$	6.30	0.37	110.0480
$C_5H_8N_3$	6.68	0.19	110.0719
$C_6H_6O_2$	6.66	0.59	110.0368
C_6H_8NO	7.03	0.41	110.0606
$C_6H_{10}N_2$	7.41	0.24	110.0845
$C_7H_{10}O$	7.76	0.46	110.0732
$C_7H_{12}N$	8.14	0.29	110.0970
C_8H_{14}	8.87	0.35	110.1096
C_8N	9.03	0.36	110.0031
C_9H_2	9.76	0.42	110.0157
111			
$C_3HN_3O_2$	4.48	0.48	111.0069
$C_3H_3N_4O$	4.85	0.30	111.0308
C_4HNO_3	4.84	0.69	110.9956
$C_4H_3N_2O_2$	5.21	0.51	111.0195
$C_4H_5N_3O$	5.59	0.33	111.0433
$C_4H_7N_4$	5.96	0.15	111.0672
$C_5H_3O_3$	5.57	0.73	111.0082
$C_5H_5NO_2$	5.94	0.55	111.0320
$C_5H_7N_2O$	6.32	0.37	111.0559
$C_5H_9N_3$	6.69	0.19	111.0798
$C_6H_7O_2$	6.67	0.59	111.0446
C_6H_9NO	7.05	0.41	111.0684
$C_6H_{11}N_2$	7.42	0.24	111.0923
$C_7H_{11}O$	7.78	0.46	111.0810
$C_7H_{13}N$	8.15	0.29	111.1049
C_8H_{15}	8.88	0.35	111.1174
C_8HN	9.04	0.36	111.0109
C_9H_3	9.77	0.43	111.0235
112			
$C_2N_4O_2$	3.77	0.46	112.0022
$C_3N_2O_3$	4.12	0.67	111.9909
$C_3H_2N_3O_2$	4.50	0.48	112.0147
$C_3H_4N_4O$	4.87	0.30	112.0386
C_4O_4	4.48	0.88	111.9796
$C_4H_2NO_3$	4.85	0.70	112.0034
$C_4H_4N_2O_2$	5.23	0.51	112.0273
$C_4H_6N_3O$	5.60	0.33	112.0511
$C_4H_8N_4$	5.98	0.15	112.0750
$C_5H_4O_3$	5.58	0.73	112.0160
$C_5H_6NO_2$	5.96	0.55	112.0399
$C_5H_8N_2O$	6.33	0.37	112.0637
$C_5H_{10}N_3$	6.71	0.19	112.0876
$C_6H_8O_2$	6.69	0.59	112.0524
$C_6H_{10}NO$	7.06	0.41	112.0763
$C_6H_{12}N_2$	7.44	0.24	112.1001
$C_7H_{12}O$	7.80	0.46	112.0888
$C_7H_{14}N$	8.17	0.29	112.1127
C_7N_2	8.33	0.30	112.0062
C_8H_{16}	8.90	0.35	112.1253
C_8O	8.68	0.53	111.9949
C_8H_2N	9.06	0.36	112.0187
C_9H_4	9.79	0.43	112.0313
113			
$C_2HN_4O_2$	3.78	0.46	113.0100
$C_3HN_2O_3$	4.14	0.67	112.9987
$C_3H_3N_3O_2$	4.51	0.48	113.0226
$C_3H_5N_4O$	4.89	0.30	113.0464
C_4HO_4	4.49	0.88	112.9874
$C_4H_3NO_3$	4.87	0.70	113.0113
$C_4H_5N_2O_2$	5.24	0.51	113.0351
$C_4H_7N_3O$	5.62	0.33	113.0590
$C_4H_9N_4$	5.99	0.15	113.0829
$C_5H_5O_3$	5.60	0.73	113.0238
$C_5H_7NO_2$	5.97	0.55	113.0477
$C_5H_9N_2O$	6.35	0.37	113.0715
$C_5H_{11}N_3$	6.72	0.19	113.0954
$C_6H_9O_2$	6.71	0.59	113.0603
$C_6H_{11}NO$	7.08	0.42	113.0841
$C_6H_{13}N_2$	7.45	0.24	113.1080
$C_7H_{13}O$	7.81	0.46	113.0967
$C_7H_{15}N$	8.19	0.29	113.1205
C_7HN_2	8.34	0.31	113.0140
C_8H_{17}	8.92	0.35	113.1331
C_8HO	8.70	0.53	113.0027
C_8H_3N	9.07	0.36	113.0266
C_9H_5	9.81	0.43	113.0391
114			
$C_2N_3O_3$	3.42	0.85	113.9940

	M+1	M+2	FM
$C_2H_2N_4O_2$	3.80	0.46	114.0178
C_3NO_4	3.78	0.86	113.9827
$C_3H_2N_2O_3$	4.15	0.67	114.0065
$C_3H_4N_3O_2$	4.53	0.48	114.0304
$C_3H_6N_4O$	4.90	0.30	114.0542
$C_4H_2O_4$	4.51	0.88	113.9953
$C_4H_4NO_3$	4.89	0.70	114.0191
$C_4H_6N_2O_2$	5.26	0.51	114.0429
$C_4H_8N_3O$	5.63	0.33	114.0668
$C_4H_{10}N_4$	6.01	0.15	114.0907
$C_5H_6O_3$	5.62	0.73	114.0317
$C_5H_8NO_2$	5.99	0.55	114.0555
$C_5H_{10}N_2O$	6.37	0.37	114.0794
$C_5H_{12}N_3$	6.74	0.20	114.1032
$C_6H_{10}O_2$	6.72	0.59	114.0681
$C_6H_{12}NO$	7.10	0.42	114.0919
$C_6H_{14}N_2$	7.47	0.24	114.1158
C_6N_3	7.63	0.25	114.0093
$C_7H_{14}O$	7.83	0.47	114.1045
$C_7H_{16}N$	8.20	0.29	114.1284
C_7NO	7.98	0.48	113.9980
$C_7H_2N_2$	8.36	0.31	114.0218
C_8H_{18}	8.93	0.35	114.1409
C_8H_2O	8.72	0.53	114.0106
C_8H_4N	9.09	0.37	114.0344
C_9H_6	9.82	0.43	114.0470
115			
$C_2HN_3O_3$	3.44	0.65	115.0018
$C_2H_3N_4O_2$	3.81	0.46	115.0257
C_3HNO_4	3.80	0.86	114.9905
$C_3H_3N_2O_3$	4.17	0.67	115.0144
$C_3H_5N_3O_2$	4.54	0.48	115.0382
$C_3H_7N_4O$	4.92	0.30	115.0621
$C_4H_3O_4$	4.53	0.88	115.0031
$C_4H_5NO_3$	4.90	0.70	115.0269
$C_4H_7N_2O_2$	5.28	0.52	115.0508
$C_4H_9N_3O$	5.65	0.33	115.0746
$C_4H_{11}N_4$	6.02	0.16	115.0985
$C_5H_7O_3$	5.63	0.73	115.0395
$C_5H_9NO_2$	6.01	0.55	115.0634
$C_5H_{11}N_2O$	6.38	0.37	115.0872
$C_5H_{13}N_3$	6.76	0.20	115.1111
$C_6H_{11}O_2$	6.74	0.59	115.0759
$C_6H_{13}NO$	7.11	0.42	115.0998
$C_6H_{15}N_2$	7.49	0.24	115.1236
C_6HN_3	7.64	0.25	115.0171
$C_7H_{15}O$	7.84	0.47	115.1123
$C_7H_{17}N$	8.22	0.30	115.1362
C_7HNO	8.00	0.48	115.0058
$C_7H_3N_2$	8.38	0.31	115.0297
C_8H_3O	8.73	0.53	115.0184
C_8H_5N	9.11	0.37	115.0422
C_9H_7	9.84	0.43	115.0548
116			
$C_2N_2O_4$	3.08	0.84	115.9858
$C_2H_2N_3O_3$	3.45	0.65	116.0096
$C_2H_4N_4O_2$	3.83	0.46	116.0335
$C_3H_2NO_4$	3.81	0.86	115.9983
$C_3H_4N_2O_3$	4.19	0.67	116.0222
$C_3H_6N_3O_2$	4.56	0.49	116.0460
$C_3H_8N_4O$	4.93	0.30	116.0699
$C_4H_4O_4$	4.54	0.88	116.0109
$C_4H_6NO_3$	4.92	0.70	116.0348
$C_4H_8N_2O_2$	5.29	0.52	116.0586
$C_4H_{10}N_3O$	5.67	0.34	116.0825
$C_4H_{12}N_4$	6.04	0.16	116.1063
$C_5H_8O_3$	5.65	0.73	116.0473
$C_5H_{10}NO_2$	6.02	0.55	116.0712
$C_5H_{12}N_2O$	6.40	0.37	116.0950
$C_5H_{14}N_3$	6.77	0.20	116.1189
C_5N_4	6.93	0.21	116.0124
$C_6H_{12}O_2$	6.75	0.59	116.0837
$C_6H_{14}NO$	7.13	0.42	116.1076
$C_6H_{16}N_2$	7.50	0.24	116.1315
C_6N_2O	7.29	0.43	116.0011
$C_6H_2N_3$	7.66	0.26	116.0249
$C_7H_{16}O$	7.86	0.47	116.1202
C_7O_2	7.64	0.65	115.9898
C_7H_2NO	8.02	0.48	116.0136
$C_7H_4N_2$	8.39	0.31	116.0375
C_8H_4O	8.75	0.54	116.0262
C_8H_6N	9.12	0.37	116.0501
C_9H_8	9.85	0.43	116.0626
117			
$C_2HN_2O_4$	3.10	0.84	116.9936
$C_2H_3N_3O_3$	3.47	0.65	117.0175
$C_2H_5N_4O_2$	3.85	0.46	117.0413
$C_3H_3NO_4$	3.83	0.86	117.0062
$C_3H_5N_2O_3$	4.20	0.67	117.0300
$C_3H_7N_3O_2$	4.58	0.49	117.0539
$C_3H_9N_4O$	4.95	0.30	117.0777
$C_4H_5O_4$	4.56	0.88	117.0187
$C_4H_7NO_3$	4.93	0.70	117.0426
$C_4H_9N_2O_2$	5.31	0.52	117.0664
$C_4H_{11}N_3O$	5.68	0.34	117.0903
$C_4H_{13}N_4$	6.06	0.16	117.1142
$C_5H_9O_3$	5.66	0.73	117.0552
$C_5H_{11}NO_2$	6.04	0.55	117.0790
$C_5H_{13}N_2O$	6.41	0.38	117.1029
$C_5H_{15}N_3$	6.79	0.20	117.1267
C_5HN_4	6.94	0.21	117.0202
$C_6H_{13}O_2$	6.77	0.60	117.0916
$C_6H_{15}NO$	7.14	0.42	117.1154
C_6HN_2O	7.30	0.43	117.0089
$C_6H_3N_3$	7.68	0.26	117.0328
C_7HO_2	7.66	0.65	116.9976
C_7H_3NO	8.03	0.48	117.0215
$C_7H_5N_2$	8.41	0.31	117.0453
C_8H_5O	8.76	0.54	117.0340
C_8H_7N	9.14	0.37	117.0579
C_9H_9	9.87	0.43	117.0705
118			
$C_2H_2N_2O_4$	3.11	0.84	118.0014
$C_2H_4N_3O_3$	3.49	0.65	118.0253
$C_2H_6N_4O_2$	3.86	0.46	118.0491
$C_3H_4NO_4$	3.84	0.86	118.0140
$C_3H_6N_2O_3$	4.22	0.67	118.0379
$C_3H_8N_3O_2$	4.59	0.49	118.0617
$C_3H_{10}N_4O$	4.97	0.30	118.0856
$C_4H_6O_4$	4.57	0.88	118.0266
$C_4H_8NO_3$	4.95	0.70	118.0504
$C_4H_{10}N_2O_2$	5.32	0.52	118.0743
$C_4H_{12}N_3O$	5.70	0.34	118.0981
$C_4H_{14}N_4$	6.07	0.16	118.1220
$C_5H_{10}O_3$	5.68	0.73	118.0630
$C_5H_{12}NO_2$	6.05	0.55	118.0868
$C_5H_{14}N_2O$	6.43	0.38	118.1107
C_5N_3O	6.59	0.39	118.0042
$C_5H_2N_4$	6.96	0.21	118.0280
$C_6H_{14}O_2$	6.79	0.60	118.0994
C_6NO_2	6.94	0.61	117.9929
$C_6H_2N_2O$	7.32	0.43	118.0167
$C_6H_4N_3$	7.69	0.26	118.0406
$C_7H_2O_2$	7.67	0.65	118.0054
C_7H_4NO	8.05	0.48	118.0293
$C_7H_6N_2$	8.42	0.31	118.0532
C_8H_6O	8.78	0.54	118.0419
C_8H_8N	9.15	0.37	118.0657
C_9H_{10}	9.89	0.44	118.0783
119			
$C_2H_3N_2O_4$	3.13	0.84	119.0093
$C_2H_5N_3O_3$	3.50	0.65	119.0331
$C_2H_7N_4O_2$	3.88	0.46	119.0570
$C_3H_5NO_4$	3.86	0.86	119.0218
$C_3H_7N_2O_3$	4.23	0.67	119.0457
$C_3H_9N_3O_2$	4.61	0.49	119.0695
$C_3H_{11}N_4O$	4.98	0.30	119.0934
$C_4H_7O_4$	4.59	0.88	119.0344
$C_4H_9NO_3$	4.97	0.70	119.0583
$C_4H_{11}N_2O_2$	5.34	0.52	119.0821
$C_4H_{13}N_3O$	5.71	0.34	119.1060
$C_5H_{11}O_3$	5.70	0.73	119.0708
$C_5H_{13}NO_2$	6.07	0.56	119.0947
C_5HN_3O	6.60	0.39	119.0120
$C_5H_3N_4$	6.98	0.21	119.0359
C_6HNO_2	6.96	0.61	119.0007
$C_6H_3N_2O$	7.33	0.43	119.0246
$C_6H_5N_3$	7.71	0.26	119.0484
$C_7H_3O_2$	7.69	0.66	119.0133
C_7H_5NO	8.06	0.48	119.0371
$C_7H_7N_2$	8.44	0.31	119.0610

	M+1	M+2	FM
C_8H_7O	8.80	0.54	119.0497
C_8H_9N	9.17	0.37	119.0736
C_9H_{11}	9.90	0.44	119.0861
120			
$C_2H_4N_2O_4$	3.14	0.84	120.0171
$C_2H_6N_3O_3$	3.52	0.65	120.0410
$C_2H_8N_4O_2$	3.89	0.46	120.0648
$C_3H_6NO_4$	3.88	0.86	120.0297
$C_3H_8N_2O_3$	4.25	0.67	120.0535
$C_3H_{10}N_3O_2$	4.62	0.49	120.0774
$C_3H_{12}N_4O$	5.00	0.30	120.1012
$C_4H_8O_4$	4.61	0.88	120.0422
$C_4H_{10}NO_3$	4.98	0.70	120.0661
$C_4H_{12}N_2O_2$	5.36	0.52	120.0899
C_4N_4O	5.89	0.35	120.0073
$C_5H_{12}O_3$	5.71	0.74	120.0786
$C_5N_2O_2$	6.24	0.57	119.9960
$C_5H_2N_3O$	6.62	0.39	120.0198
$C_5H_4N_4$	6.99	0.21	120.0437
C_6O_3	6.60	0.78	119.9847
$C_6H_2NO_2$	6.98	0.61	120.0085
$C_6H_4N_2O$	7.35	0.43	120.0324
$C_6H_6N_3$	7.72	0.26	120.0563
$C_7H_4O_2$	7.71	0.66	120.0211
C_7H_6NO	8.08	0.49	120.0449
$C_7H_8N_2$	8.46	0.32	120.0688
C_8H_8O	8.81	0.54	120.0575
$C_8H_{10}N$	9.19	0.37	120.0814
C_9H_{12}	9.92	0.44	120.0939
C_{10}	10.81	0.53	120.0000
121			
$C_2H_5N_2O_4$	3.16	0.84	121.0249
$C_2H_7N_3O_3$	3.53	0.65	121.0488
$C_2H_9N_4O_2$	3.91	0.46	121.0726
$C_3H_7NO_4$	3.89	0.86	121.0375
$C_3H_9N_2O_3$	4.27	0.67	121.0614
$C_3H_{11}N_3O_2$	4.64	0.49	121.0852
$C_4H_9O_4$	4.62	0.89	121.0501
$C_4H_{11}NO_3$	5.00	0.70	121.0739
C_4HN_4O	5.90	0.35	121.0151
$C_5HN_2O_2$	6.26	0.57	121.0038
$C_5H_3N_3O$	6.63	0.39	121.0277
$C_5H_5N_4$	7.01	0.21	121.0515
C_6HO_3	6.62	0.79	120.9925
$C_6H_3NO_2$	6.99	0.61	121.0164
$C_6H_5N_2O$	7.37	0.44	121.0402
$C_6H_7N_3$	7.74	0.26	121.0641
$C_7H_5O_2$	7.72	0.66	121.0289
C_7H_7NO	8.10	0.49	121.0528
$C_7H_9N_2$	8.47	0.32	121.0767
C_8H_9O	8.83	0.54	121.0653
$C_8H_{11}N$	9.20	0.38	121.0892
C_9H_{13}	9.93	0.44	121.1018
$C_{10}H$	10.82	0.53	121.0078
122			
$C_2H_6N_2O_4$	3.18	0.84	122.0328
$C_2H_8N_3O_3$	3.55	0.65	122.0566
$C_2H_{10}N_4O_2$	3.93	0.46	122.0805
$C_3H_8NO_4$	3.91	0.86	122.0453
$C_3H_{10}N_2O_3$	4.28	0.67	122.0692
$C_4H_{10}O_4$	4.64	0.89	122.0579
$C_4N_3O_2$	5.54	0.53	121.9991
$C_4H_2N_4O$	5.92	0.35	122.0229
C_5NO_3	5.90	0.75	121.9878
$C_5H_2N_2O_2$	6.28	0.57	122.0116
$C_5H_4N_3O$	6.65	0.39	122.0355
$C_5H_6N_4$	7.02	0.21	122.0594
$C_6H_2O_3$	6.63	0.79	122.0003
$C_6H_4NO_2$	7.01	0.61	122.0242
$C_6H_6N_2O$	7.38	0.44	122.0480
$C_6H_8N_3$	7.76	0.26	122.0719
$C_7H_6O_2$	7.74	0.66	122.0368
C_7H_8NO	8.11	0.49	122.0606
$C_7H_{10}N_2$	8.49	0.32	122.0845
$C_8H_{10}O$	8.84	0.54	122.0732
$C_8H_{12}N$	9.22	0.38	122.0970
C_9H_{14}	9.95	0.44	122.1096
C_9N	10.11	0.46	122.0031
$C_{10}H_2$	10.84	0.53	122.0157
123			
$C_2H_7N_2O_4$	3.19	0.84	123.0406
$C_2H_9N_3O_3$	3.57	0.65	123.0644
$C_3H_9NO_4$	3.92	0.86	123.0532
$C_4HN_3O_2$	5.56	0.53	123.0069
$C_4H_3N_4O$	5.94	0.35	123.0308
C_5HNO_3	5.92	0.75	122.9956
$C_5H_3N_2O_2$	6.29	0.57	123.0195
$C_5H_5N_3O$	6.67	0.39	123.0433
$C_5H_7N_4$	7.04	0.22	123.0672
$C_6H_3O_3$	6.65	0.79	123.0082
$C_6H_5NO_2$	7.02	0.61	123.0320
$C_6H_7N_2O$	7.40	0.44	123.0559
$C_6H_9N_3$	7.77	0.26	123.0798
$C_7H_7O_2$	7.75	0.66	123.0446
C_7H_9NO	8.13	0.49	123.0684
$C_7H_{11}N_2$	8.50	0.32	123.0923
$C_8H_{11}O$	8.86	0.55	123.0810
$C_8H_{13}N$	9.23	0.38	123.1049
C_9H_{15}	9.97	0.44	123.1174
C_9HN	10.12	0.46	123.0109
$C_{10}H_3$	10.85	0.53	123.0235
124			
$C_2H_8N_2O_4$	3.21	0.84	124.0484
$C_3N_4O_2$	4.85	0.50	124.0022
$C_4N_2O_3$	5.20	0.71	123.9909
$C_4H_2N_3O_2$	5.58	0.53	124.0147
$C_4H_4N_4O$	5.95	0.35	124.0386
C_5O_4	5.56	0.93	123.9796
$C_5H_2NO_3$	5.93	0.75	124.0034
$C_5H_4N_2O_2$	6.31	0.57	124.0273
$C_5H_6N_3O$	6.68	0.39	124.0511
$C_5H_8N_4$	7.06	0.22	124.0750
$C_6H_4O_3$	6.66	0.79	124.0160
$C_6H_6NO_2$	7.04	0.61	124.0399
$C_6H_8N_2O$	7.41	0.44	124.0637
$C_6H_{10}N_3$	7.79	0.27	124.0876
$C_7H_8O_2$	7.77	0.66	124.0524
$C_7H_{10}NO$	8.14	0.49	124.0763
$C_7H_{12}N_2$	8.52	0.32	124.1001
$C_8H_{12}O$	8.88	0.55	124.0888
$C_8H_{14}N$	9.25	0.38	124.1127
C_8N_2	9.41	0.39	124.0062
C_9H_{16}	9.98	0.45	124.1253
C_9O	9.76	0.62	123.9949
C_9H_2N	10.14	0.46	124.0187
$C_{10}H_4$	10.87	0.53	124.0313
125			
$C_3HN_4O_2$	4.86	0.50	125.0100
$C_4HN_2O_3$	5.22	0.71	124.9987
$C_4H_3N_3O_2$	5.59	0.53	125.0226
$C_4H_5N_4O$	5.97	0.35	125.0464
C_5HO_4	5.58	0.93	124.9874
$C_5H_3NO_3$	5.95	0.75	125.0113
$C_5H_5N_2O_2$	6.32	0.57	125.0351
$C_5H_7N_3O$	6.70	0.39	125.0590
$C_5H_9N_4$	7.07	0.22	125.0829
$C_6H_5O_3$	6.68	0.79	125.0238
$C_6H_7NO_2$	7.06	0.61	125.0477
$C_6H_9N_2O$	7.43	0.44	125.0715
$C_6H_{11}N_3$	7.80	0.27	125.0954
$C_7H_9O_2$	7.79	0.66	125.0603
$C_7H_{11}NO$	8.16	0.49	125.0841
$C_7H_{13}N_2$	8.54	0.32	125.1080
$C_8H_{13}O$	8.89	0.55	125.0967
$C_8H_{15}N$	9.27	0.38	125.1205
C_8HN_2	9.42	0.40	125.0140
C_9H_{17}	10.00	0.45	125.1331
C_9HO	9.78	0.63	125.0027
C_9H_3N	10.15	0.46	125.0266
$C_{10}H_5$	10.89	0.53	125.0391
126			
$C_3N_3O_3$	4.50	0.68	125.9940
$C_3H_2N_4O_2$	4.88	0.50	126.0178
C_4NO_4	4.86	0.90	125.9827
$C_4H_2N_2O_3$	5.23	0.71	126.0065
$C_4H_4N_3O_2$	5.61	0.53	126.0304
$C_4H_6N_4O$	5.98	0.35	126.0542
$C_5H_2O_4$	5.59	0.93	125.9953
$C_5H_4NO_3$	5.97	0.75	126.0191
$C_5H_6N_2O_2$	6.34	0.57	126.0429
$C_5H_8N_3O$	6.71	0.39	126.0668
$C_5H_{10}N_4$	7.09	0.22	126.0907

	M+1	M+2	FM
$C_6H_6O_3$	6.70	0.79	126.0317
$C_6H_8NO_2$	7.07	0.62	126.0555
$C_6H_{10}N_2O$	7.45	0.44	126.0794
$C_6H_{12}N_3$	7.82	0.27	126.1032
$C_7H_{10}O_2$	7.80	0.66	126.0681
$C_7H_{12}NO$	8.18	0.49	126.0919
$C_7H_{14}N_2$	8.55	0.32	126.1158
C_7N_3	8.71	0.34	126.0093
$C_8H_{14}O$	8.91	0.55	126.1045
$C_8H_{16}N$	9.28	0.38	126.1284
C_8NO	9.07	0.56	125.9980
$C_8H_2N_2$	9.44	0.40	126.0218
C_9H_{18}	10.01	0.45	126.1409
C_9H_2O	9.80	0.63	126.0106
C_9H_4N	10.17	0.46	126.0344
$C_{10}H_6$	10.90	0.54	126.0470
127			
$C_3HN_3O_3$	4.52	0.68	127.0018
$C_3H_3N_4O_2$	4.89	0.50	127.0257
C_4HNO_4	4.88	0.90	126.9905
$C_4H_3N_2O_3$	5.25	0.71	127.0144
$C_4H_5N_3O_2$	5.62	0.53	127.0382
$C_4H_7N_4O$	6.00	0.35	127.0621
$C_5H_3O_4$	5.61	0.93	127.0031
$C_5H_5NO_3$	5.98	0.75	127.0269
$C_5H_7N_2O_2$	6.36	0.57	127.0508
$C_5H_9N_3O$	6.73	0.40	127.0746
$C_5H_{11}N_4$	7.10	0.22	127.0985
$C_6H_7O_3$	6.71	0.79	127.0395
$C_6H_9NO_2$	7.09	0.62	127.0634
$C_6H_{11}N_2O$	7.46	0.44	127.0872
$C_6H_{13}N_3$	7.84	0.27	127.1111
$C_7H_{11}O_2$	7.82	0.67	127.0759
$C_7H_{13}NO$	8.19	0.49	127.0998
$C_7H_{15}N_2$	8.57	0.32	127.1236
C_7HN_3	8.72	0.34	127.0171
$C_8H_{15}O$	8.92	0.55	127.1123
$C_8H_{17}N$	9.30	0.38	127.1362
C_8HNO	9.08	0.57	127.0058
$C_8H_3N_2$	9.46	0.40	127.0297
C_9H_{19}	10.03	0.45	127.1488
C_9H_3O	9.81	0.63	127.0184
C_9H_5N	10.19	0.47	127.0422
$C_{10}H_7$	10.92	0.54	127.0548
128			
$C_3N_2O_4$	4.16	0.87	127.9858
$C_3H_2N_3O_3$	4.54	0.68	128.0096
$C_3H_4N_4O_2$	4.91	0.50	128.0335
$C_4H_2NO_4$	4.89	0.90	127.9983
$C_4H_4N_2O_3$	5.27	0.72	128.0222
$C_4H_6N_3O_2$	5.64	0.53	128.0460
$C_4H_8N_4O$	6.02	0.36	128.0699
$C_5H_4O_4$	5.62	0.93	128.0109
$C_5H_6NO_3$	6.00	0.75	128.0348
$C_5H_8N_2O_2$	6.37	0.57	128.0586
$C_5H_{10}N_3O$	6.75	0.40	128.0825
$C_5H_{12}N_4$	7.12	0.22	128.1063
$C_6H_8O_3$	6.73	0.79	128.0473
$C_6H_{10}NO_2$	7.10	0.62	128.0712
$C_6H_{12}N_2O$	7.48	0.44	128.0950
$C_6H_{14}N_3$	7.85	0.27	128.1189
C_6N_4	8.01	0.28	128.0124
$C_7H_{12}O_2$	7.83	0.67	128.0837
$C_7H_{14}NO$	8.21	0.50	128.1076
$C_7H_{16}N_2$	8.58	0.33	128.1315
C_7N_2O	8.37	0.51	128.0011
$C_7H_2N_3$	8.74	0.34	128.0249
$C_8H_{16}O$	8.94	0.55	128.1202
C_8O_2	8.72	0.73	127.9898
$C_8H_{18}N$	9.31	0.39	128.1440
C_8H_2NO	9.10	0.57	128.0136
$C_8H_4N_2$	9.47	0.40	128.0375
C_9H_{20}	10.05	0.45	128.1566
C_9H_4O	9.83	0.63	128.0262
C_9H_6N	10.20	0.47	128.0501
$C_{10}H_8$	10.93	0.54	128.0626
129			
$C_3HN_2O_4$	4.18	0.87	128.9936
$C_3H_3N_3O_3$	4.55	0.69	129.0175
$C_3H_5N_4O_2$	4.93	0.50	129.0413
$C_4H_3NO_4$	4.91	0.90	129.0062
$C_4H_5N_2O_3$	5.28	0.72	129.0300
$C_4H_7N_3O_2$	5.66	0.54	129.0539
$C_4H_9N_4O$	6.03	0.36	129.0777
$C_5H_5O_4$	5.64	0.93	129.0187
$C_5H_7NO_3$	6.01	0.75	129.0426
$C_5H_9N_2O_2$	6.39	0.57	129.0664
$C_5H_{11}N_3O$	6.76	0.40	129.0903
$C_5H_{13}N_4$	7.14	0.22	129.1142
$C_6H_9O_3$	6.74	0.79	129.0552
$C_6H_{11}NO_2$	7.12	0.62	129.0790
$C_6H_{13}N_2O$	7.49	0.44	129.1029
$C_6H_{15}N_3$	7.87	0.27	129.1267
C_6HN_4	8.03	0.28	129.0202
$C_7H_{13}O_2$	7.85	0.67	129.0916
$C_7H_{15}NO$	8.22	0.50	129.1154
$C_7H_{17}N_2$	8.60	0.33	129.1393
C_7HN_2O	8.38	0.51	129.0089
$C_7H_3N_3$	8.76	0.34	129.0328
$C_8H_{17}O$	8.96	0.55	129.1280
C_8HO_2	8.74	0.74	128.9976
$C_8H_{19}N$	9.33	0.39	129.1519
C_8H_3NO	9.11	0.57	129.0215
$C_8H_5N_2$	9.49	0.40	129.0453
C_9H_5O	9.84	0.63	129.0340
C_9H_7N	10.22	0.47	129.0579
$C_{10}H_9$	10.95	0.54	129.0705
130			
$C_3H_2N_2O_4$	4.19	0.87	130.0014
$C_3H_4N_3O_3$	4.57	0.69	130.0253
$C_3H_6N_4O_2$	4.94	0.50	130.0491
$C_4H_4NO_4$	4.92	0.90	130.0140
$C_4H_6N_2O_3$	5.30	0.72	130.0379
$C_4H_8N_3O_2$	5.67	0.54	130.0617
$C_4H_{10}N_4O$	6.05	0.36	130.0856
$C_5H_6O_4$	5.66	0.93	130.0266
$C_5H_8NO_3$	6.03	0.75	130.0504
$C_5H_{10}N_2O_2$	6.40	0.58	130.0743
$C_5H_{12}N_3O$	6.78	0.40	130.0981
$C_5H_{14}N_4$	7.15	0.22	130.1220
$C_6H_{10}O_3$	6.76	0.79	130.0630
$C_6H_{12}NO_2$	7.14	0.62	130.0868
$C_6H_{14}N_2O$	7.51	0.45	130.1107
$C_6H_{16}N_3$	7.88	0.27	130.1346
C_6N_3O	7.67	0.46	130.0042
$C_6H_2N_4$	8.04	0.29	130.0280
$C_7H_{14}O_2$	7.87	0.67	130.0994
$C_7H_{16}NO$	8.24	0.50	130.1233
C_7NO_2	8.02	0.68	129.9929
$C_7H_{18}N_2$	8.62	0.33	130.1471
$C_7H_2N_2O$	8.40	0.51	130.0167
$C_7H_4N_3$	8.77	0.34	130.0406
$C_8H_{18}O$	8.97	0.56	130.1358
$C_8H_2O_2$	8.75	0.74	130.0054
C_8H_4NO	9.13	0.57	130.0293
$C_8H_6N_2$	9.50	0.40	130.0532
C_9H_6O	9.86	0.63	130.0419
C_9H_8N	10.23	0.47	130.0657
$C_{10}H_{10}$	10.97	0.54	130.0783
131			
$C_3H_3N_2O_4$	4.21	0.87	131.0093
$C_3H_5N_3O_3$	4.58	0.69	131.0331
$C_3H_7N_4O_2$	4.96	0.50	131.0570
$C_4H_5NO_4$	4.94	0.90	131.0218
$C_4H_7N_2O_3$	5.31	0.72	131.0457
$C_4H_9N_3O_2$	5.69	0.54	131.0695
$C_4H_{11}N_4O$	6.06	0.36	131.0934
$C_5H_7O_4$	5.67	0.93	131.0344
$C_5H_9NO_3$	6.05	0.75	131.0583
$C_5H_{11}N_2O_2$	6.42	0.58	131.0821
$C_5H_{13}N_3O$	6.79	0.40	131.1060
$C_5H_{15}N_4$	7.17	0.22	131.1298
$C_6H_{11}O_3$	6.78	0.80	131.0708
$C_6H_{13}NO_2$	7.15	0.62	131.0947
$C_6H_{15}N_2O$	7.53	0.45	131.1185
$C_6H_{17}N_3$	7.90	0.27	131.1424
C_6HN_3O	7.68	0.46	131.0120
$C_6H_3N_4$	8.06	0.29	131.0359
$C_7H_{15}O_2$	7.88	0.67	131.1072
$C_7H_{17}NO$	8.26	0.50	131.1311
C_7HNO_2	8.04	0.68	131.0007

	M+1	M+2	FM
$C_7H_3N_2O$	8.41	0.51	131.0246
$C_7H_5N_3$	8.79	0.34	131.0484
$C_8H_3O_2$	8.77	0.74	131.0133
C_8H_5NO	9.15	0.57	131.0371
$C_8H_7N_2$	9.52	0.40	131.0610
C_9H_7O	9.88	0.64	131.0497
C_9H_9N	10.25	0.47	131.0736
$C_{10}H_{11}$	10.98	0.54	131.0861
132			
$C_3H_4N_2O_4$	4.23	0.87	132.0171
$C_3H_6N_3O_3$	4.60	0.69	132.0410
$C_3H_8N_4O_2$	4.97	0.50	132.0648
$C_4H_6NO_4$	4.96	0.90	132.0297
$C_4H_8N_2O_3$	5.33	0.72	132.0535
$C_4H_{10}N_3O_2$	5.70	0.54	132.0774
$C_4H_{12}N_4O$	6.08	0.36	132.1012
$C_5H_8O_4$	5.69	0.93	132.0422
$C_5H_{10}NO_3$	6.06	0.76	132.0661
$C_5H_{12}N_2O_2$	6.44	0.58	132.0899
$C_5H_{14}N_3O$	6.81	0.40	132.1138
$C_5H_{16}N_4$	7.18	0.23	132.1377
C_5N_4O	6.97	0.41	132.0073
$C_6H_{12}O_3$	6.79	0.80	132.0786
$C_6H_{14}NO_2$	7.17	0.62	132.1025
$C_6H_{16}N_2O$	7.54	0.45	132.1264
$C_6N_2O_2$	7.32	0.63	131.9960
$C_6H_2N_3O$	7.70	0.46	132.0198
$C_6H_4N_4$	8.07	0.29	132.0437
$C_7H_{16}O_2$	7.90	0.67	132.1151
C_7O_3	7.68	0.86	131.9847
$C_7H_2NO_2$	8.06	0.68	132.0085
$C_7H_4N_2O$	8.43	0.51	132.0324
$C_7H_6N_3$	8.80	0.34	132.0563
$C_8H_4O_2$	8.79	0.74	132.0211
C_8H_6NO	9.16	0.57	132.0449
$C_8H_8N_2$	9.54	0.41	132.0688
C_9H_8O	9.89	0.64	132.0575
$C_9H_{10}N$	10.27	0.47	132.0814
$C_{10}H_{12}$	11.00	0.55	132.0939
C_{11}	11.89	0.64	132.0000
133			
$C_3H_5N_2O_4$	4.24	0.87	133.0249
$C_3H_7N_3O_3$	4.62	0.69	133.0488
$C_3H_9N_4O_2$	4.99	0.50	133.0726
$C_4H_7NO_4$	4.97	0.90	133.0375
$C_4H_9N_2O_3$	5.35	0.72	133.0614
$C_4H_{11}N_3O_2$	5.72	0.54	133.0852
$C_4H_{13}N_4O$	6.10	0.36	133.1091
$C_5H_9O_4$	5.70	0.94	133.0501
$C_5H_{11}NO_3$	6.08	0.76	133.0739
$C_5H_{13}N_2O_2$	6.45	0.58	133.0978
$C_5H_{15}N_3O$	6.83	0.40	133.1216
C_5HN_4O	6.98	0.41	133.0151
$C_6H_{13}O_3$	6.81	0.80	133.0865
$C_6H_{15}NO_2$	7.18	0.62	133.1103
$C_6HN_2O_2$	7.34	0.63	133.0038
$C_6H_3N_3O$	7.72	0.46	133.0277
$C_6H_5N_4$	8.09	0.29	133.0515
C_7HO_3	7.70	0.86	132.9925
$C_7H_3NO_2$	8.07	0.69	133.0164
$C_7H_5N_2O$	8.45	0.51	133.0402
$C_7H_7N_3$	8.82	0.35	133.0641
$C_8H_5O_2$	8.80	0.74	133.0289
C_8H_7NO	9.18	0.57	133.0528
$C_8H_9N_2$	9.55	0.41	133.0767
C_9H_9O	9.91	0.64	133.0653
$C_9H_{11}N$	10.28	0.48	133.0892
$C_{10}H_{13}$	11.01	0.55	133.1018
$C_{11}H$	11.90	0.64	133.0078
134			
$C_3H_6N_2O_4$	4.26	0.87	134.0328
$C_3H_8N_3O_3$	4.63	0.69	134.0566
$C_3H_{10}N_4O_2$	5.01	0.51	134.0805
$C_4H_8NO_4$	4.99	0.90	134.0453
$C_4H_{10}N_2O_3$	5.36	0.72	134.0692
$C_4H_{12}N_3O_2$	5.74	0.54	134.0930
$C_4H_{14}N_4O$	6.11	0.36	134.1169
$C_5H_{10}O_4$	5.72	0.94	134.0579
$C_5H_{12}NO_3$	6.09	0.76	134.0817
$C_5H_{14}N_2O_2$	6.47	0.58	134.1056
$C_5N_3O_2$	6.63	0.59	133.9991
$C_5H_2N_4O$	7.00	0.41	134.0229
$C_6H_{14}O_3$	6.82	0.80	134.0943
C_6NO_3	6.98	0.81	133.9878
$C_6H_2N_2O_2$	7.36	0.64	134.0116
$C_6H_4N_3O$	7.73	0.46	134.0355
$C_6H_6N_4$	8.11	0.29	134.0594
$C_7H_2O_3$	7.71	0.86	134.0003
$C_7H_4NO_2$	8.09	0.69	134.0242
$C_7H_6N_2O$	8.46	0.52	134.0480
$C_7H_8N_3$	8.84	0.35	134.0719
$C_8H_6O_2$	8.82	0.74	134.0368
C_8H_8NO	9.19	0.58	134.0606
$C_8H_{10}N_2$	9.57	0.41	134.0845
$C_9H_{10}O$	9.92	0.64	134.0732
$C_9H_{12}N$	10.30	0.48	134.0970
$C_{10}H_{14}$	11.03	0.55	134.1096
$C_{10}N$	11.19	0.57	134.0031
$C_{11}H_2$	11.92	0.65	134.0157
135			
$C_3H_7N_2O_4$	4.27	0.87	135.0406
$C_3H_9N_3O_3$	4.65	0.69	135.0644
$C_3H_{11}N_4O_2$	5.02	0.51	135.0883
$C_4H_9NO_4$	5.00	0.90	135.0532
$C_4H_{11}N_2O_3$	5.38	0.72	135.0770
$C_4H_{13}N_3O_2$	5.75	0.54	135.1009
$C_5H_{11}O_4$	5.74	0.94	135.0657
$C_5H_{13}NO_3$	6.11	0.76	135.0896
$C_5HN_3O_2$	6.64	0.59	135.0069
$C_5H_3N_4O$	7.02	0.41	135.0308
C_6HNO_3	7.00	0.81	134.9956
$C_6H_3N_2O_2$	7.37	0.64	135.0195
$C_6H_5N_3O$	7.75	0.46	135.0433
$C_6H_7N_4$	8.12	0.29	135.0672
$C_7H_3O_3$	7.73	0.86	135.0082
$C_7H_5NO_2$	8.10	0.69	135.0320
$C_7H_7N_2O$	8.48	0.52	135.0559
$C_7H_9N_3$	8.85	0.35	135.0798
$C_8H_7O_2$	8.84	0.74	135.0446
C_8H_9NO	9.21	0.58	135.0684
$C_8H_{11}N_2$	9.58	0.41	135.0923
$C_9H_{11}O$	9.94	0.64	135.0810
$C_9H_{13}N$	10.31	0.48	135.1049
$C_{10}H_{15}$	11.05	0.55	135.1174
$C_{10}HN$	11.20	0.57	135.0109
$C_{11}H_3$	11.93	0.65	135.0235
136			
$C_3H_8N_2O_4$	4.29	0.87	136.0484
$C_3H_{10}N_3O_3$	4.66	0.69	136.0723
$C_3H_{12}N_4O_2$	5.04	0.51	136.0961
$C_4H_{10}NO_4$	5.02	0.90	136.0610
$C_4H_{12}N_2O_3$	5.39	0.72	136.0848
$C_4N_4O_2$	5.93	0.55	136.0022
$C_5H_{12}O_4$	5.75	0.94	136.0735
$C_5N_2O_3$	6.28	0.77	135.9909
$C_5H_2N_3O_2$	6.66	0.59	136.0147
$C_5H_4N_4O$	7.03	0.42	136.0386
C_6O_4	6.64	0.99	135.9796
$C_6H_2NO_3$	7.01	0.81	136.0034
$C_6H_4N_2O_2$	7.39	0.64	136.0273
$C_6H_6N_3O$	7.76	0.46	136.0511
$C_6H_8N_4$	8.14	0.29	136.0750
$C_7H_4O_3$	7.75	0.86	136.0160
$C_7H_6NO_2$	8.12	0.69	136.0399
$C_7H_8N_2O$	8.49	0.52	136.0637
$C_7H_{10}N_3$	8.87	0.35	136.0876
$C_8H_8O_2$	8.85	0.75	136.0524
$C_8H_{10}NO$	9.23	0.58	136.0763
$C_8H_{12}N_2$	9.60	0.41	136.1001
$C_9H_{12}O$	9.96	0.64	136.0888
$C_9H_{14}N$	10.33	0.48	136.1127
C_9N_2	10.49	0.50	136.0062
$C_{10}H_{16}$	11.06	0.55	136.1253
$C_{10}O$	10.85	0.73	135.9949
$C_{10}H_2N$	11.22	0.57	136.0187
$C_{11}H_4$	11.95	0.65	136.0313
137			
$C_3H_9N_2O_4$	4.31	0.88	137.0563
$C_3H_{11}N_3O_3$	4.68	0.69	137.0801
$C_4H_{11}NO_4$	5.04	0.90	137.0688
$C_4HN_4O_2$	5.94	0.55	137.0100
$C_5HN_2O_3$	6.30	0.77	136.9987

Formula	M+1	M+2	FM
$C_5H_3N_3O_2$	6.67	0.59	137.0226
$C_5H_5N_4O$	7.05	0.42	137.0464
C_6HO_4	6.66	0.99	136.9874
$C_6H_3NO_3$	7.03	0.81	137.0113
$C_6H_5N_2O_2$	7.40	0.64	137.0351
$C_6H_7N_3O$	7.78	0.47	137.0590
$C_6H_9N_4$	8.15	0.29	137.0829
$C_7H_5O_3$	7.76	0.86	137.0238
$C_7H_7NO_2$	8.14	0.69	137.0477
$C_7H_9N_2O$	8.51	0.52	137.0715
$C_7H_{11}N_3$	8.88	0.35	137.0954
$C_8H_9O_2$	8.87	0.75	137.0603
$C_8H_{11}NO$	9.24	0.58	137.0841
$C_8H_{13}N_2$	9.62	0.41	137.1080
$C_9H_{13}O$	9.97	0.65	137.0967
$C_9H_{15}N$	10.35	0.48	137.1205
C_9HN_2	10.50	0.50	137.0140
$C_{10}H_{17}$	11.08	0.56	137.1331
$C_{10}HO$	10.86	0.73	137.0027
$C_{10}H_3N$	11.24	0.57	137.0266
$C_{11}H_5$	11.97	0.65	137.0391
138			
$C_3H_{10}N_2O_4$	4.32	0.88	138.0641
$C_4N_3O_3$	5.58	0.73	137.9940
$C_4H_2N_4O_2$	5.96	0.55	138.0178
C_5NO_4	5.94	0.95	137.9827
$C_5H_2N_2O_3$	6.32	0.77	138.0065
$C_5H_4N_3O_2$	6.69	0.59	138.0304
$C_5H_6N_4O$	7.06	0.42	138.0542
$C_6H_2O_4$	6.67	0.99	137.9953
$C_6H_4NO_3$	7.05	0.81	138.0191
$C_6H_6N_2O_2$	7.42	0.64	138.0429
$C_6H_8N_3O$	7.80	0.47	138.0668
$C_6H_{10}N_4$	8.17	0.30	138.0907
$C_7H_6O_3$	7.78	0.86	138.0317
$C_7H_8NO_2$	8.15	0.69	138.0555
$C_7H_{10}N_2O$	8.53	0.52	138.0794
$C_7H_{12}N_3$	8.90	0.35	138.1032
$C_8H_{10}O_2$	8.88	0.75	138.0681
$C_8H_{12}NO$	9.26	0.58	138.0919
$C_8H_{14}N_2$	9.63	0.42	138.1158
C_8N_3	9.79	0.43	138.0093
$C_9H_{14}O$	9.99	0.65	138.1045
$C_9H_{16}N$	10.36	0.48	138.1284
C_9NO	10.15	0.66	137.9980
$C_9H_2N_2$	10.52	0.50	138.0218
$C_{10}H_{18}$	11.09	0.56	138.1409
$C_{10}H_2O$	10.88	0.73	138.0106
$C_{10}H_4N$	11.25	0.57	138.0344
$C_{11}H_6$	11.98	0.65	138.0470
139			
$C_4HN_3O_3$	5.60	0.73	139.0018
$C_4H_3N_4O_2$	5.97	0.55	139.0257
C_5HNO_4	5.96	0.95	138.9905
$C_5H_3N_2O_3$	6.33	0.77	139.0144
$C_5H_5N_3O_2$	6.71	0.59	139.0382
$C_5H_7N_4O$	7.08	0.42	139.0621
$C_6H_3O_4$	6.69	0.99	139.0031
$C_6H_5NO_3$	7.06	0.82	139.0269
$C_6H_7N_2O_2$	7.44	0.64	139.0508
$C_6H_9N_3O$	7.81	0.47	139.0746
$C_6H_{11}N_4$	8.19	0.30	139.0985
$C_7H_7O_3$	7.79	0.86	139.0395
$C_7H_9NO_2$	8.17	0.69	139.0634
$C_7H_{11}N_2O$	8.54	0.52	139.0872
$C_7H_{13}N_3$	8.92	0.35	139.1111
$C_8H_{11}O_2$	8.90	0.75	139.0759
$C_8H_{13}NO$	9.27	0.58	139.0998
$C_8H_{15}N_2$	9.65	0.42	139.1236
C_8HN_3	9.81	0.43	139.0171
$C_9H_{15}O$	10.00	0.65	139.1123
$C_9H_{17}N$	10.38	0.49	139.1362
C_9HNO	10.16	0.66	139.0058
$C_9H_3N_2$	10.54	0.50	139.0297
$C_{10}H_{19}$	11.11	0.56	139.1488
$C_{10}H_3O$	10.89	0.74	139.0184
$C_{10}H_5N$	11.27	0.58	139.0422
$C_{11}H_7$	12.00	0.66	139.0548
140			
$C_4N_2O_4$	5.24	0.91	139.9858
$C_4H_2N_3O_3$	5.62	0.73	140.0096
$C_4H_4N_4O_2$	5.99	0.55	140.0335
$C_5H_2NO_4$	5.97	0.95	139.9983
$C_5H_4N_2O_3$	6.35	0.77	140.0222
$C_5H_6N_3O_2$	6.72	0.60	140.0460
$C_5H_8N_4O$	7.10	0.42	140.0699
$C_6H_4O_4$	6.70	0.99	140.0109
$C_6H_6NO_3$	7.08	0.82	140.0348
$C_6H_8N_2O_2$	7.45	0.64	140.0586
$C_6H_{10}N_3O$	7.83	0.47	140.0825
$C_6H_{12}N_4$	8.20	0.30	140.1063
$C_7H_8O_3$	7.81	0.87	140.0473
$C_7H_{10}NO_2$	8.18	0.69	140.0712
$C_7H_{12}N_2O$	8.56	0.52	140.0950
$C_7H_{14}N_3$	8.93	0.36	140.1189
C_7N_4	9.09	0.37	140.0124
$C_8H_{12}O_2$	8.92	0.75	140.0837
$C_8H_{14}NO$	9.29	0.58	140.1076
$C_8H_{16}N_2$	9.66	0.42	140.1315
C_8N_2O	9.45	0.60	140.0011
$C_8H_2N_3$	9.82	0.43	140.0249
$C_9H_{16}O$	10.02	0.65	140.1202
C_9O_2	9.80	0.83	139.9898
$C_9H_{18}N$	10.39	0.49	140.1440
C_9H_2NO	10.18	0.67	140.0136
$C_9H_4N_2$	10.55	0.50	140.0375
$C_{10}H_{20}$	11.13	0.56	140.1566
$C_{10}H_4O$	10.91	0.74	140.0262
$C_{10}H_6N$	11.28	0.58	140.0501
$C_{11}H_8$	12.01	0.66	140.0626
141			
$C_4HN_2O_4$	5.26	0.92	140.9936
$C_4H_3N_3O_3$	5.63	0.73	141.0175
$C_4H_5N_4O_2$	6.01	0.56	141.0413
$C_5H_3NO_4$	5.99	0.95	141.0062
$C_5H_5N_2O_3$	6.36	0.77	141.0300
$C_5H_7N_3O_2$	6.74	0.60	141.0539
$C_5H_9N_4O$	7.11	0.42	141.0777
$C_6H_5O_4$	6.72	0.99	141.0187
$C_6H_7NO_3$	7.09	0.82	141.0426
$C_6H_9N_2O_2$	7.47	0.64	141.0664
$C_6H_{11}N_3O$	7.84	0.47	141.0903
$C_6H_{13}N_4$	8.22	0.30	141.1142
$C_7H_9O_3$	7.83	0.87	141.0552
$C_7H_{11}NO_2$	8.20	0.70	141.0790
$C_7H_{13}N_2O$	8.57	0.53	141.1029
$C_7H_{15}N_3$	8.95	0.36	141.1267
C_7HN_4	9.11	0.37	141.0202
$C_8H_{13}O_2$	8.93	0.75	141.0916
$C_8H_{15}NO$	9.31	0.59	141.1154
$C_8H_{17}N_2$	9.68	0.42	141.1393
C_8HN_2O	9.46	0.60	141.0089
$C_8H_3N_3$	9.84	0.43	141.0328
$C_9H_{17}O$	10.04	0.65	141.1280
C_9HO_2	9.82	0.83	140.9976
$C_9H_{19}N$	10.41	0.49	141.1519
C_9H_3NO	10.19	0.67	141.0215
$C_9H_5N_2$	10.57	0.50	141.0453
$C_{10}H_{21}$	11.14	0.56	141.1644
$C_{10}H_5O$	10.93	0.74	141.0340
$C_{10}H_7N$	11.30	0.58	141.0579
$C_{11}H_9$	12.03	0.66	141.0705
142			
$C_4H_2N_2O_4$	5.27	0.92	142.0014
$C_4H_4N_3O_3$	5.65	0.74	142.0253
$C_4H_6N_4O_2$	6.02	0.56	142.0491
$C_5H_4NO_4$	6.00	0.95	142.0140
$C_5H_6N_2O_3$	6.38	0.77	142.0379
$C_5H_8N_3O_2$	6.75	0.60	142.0617
$C_5H_{10}N_4O$	7.13	0.42	142.0856
$C_6H_6O_4$	6.74	0.99	142.0266
$C_6H_8NO_3$	7.11	0.82	142.0504
$C_6H_{10}N_2O_2$	7.48	0.64	142.0743
$C_6H_{12}N_3O$	7.86	0.47	142.0981
$C_6H_{14}N_4$	8.23	0.30	142.1220
$C_7H_{10}O_3$	7.84	0.87	142.0630
$C_7H_{12}NO_2$	8.22	0.70	142.0868
$C_7H_{14}N_2O$	8.59	0.53	142.1107
$C_7H_{16}N_3$	8.96	0.36	142.1346
C_7N_3O	8.75	0.54	142.0042
$C_7H_2N_4$	9.12	0.37	142.0280

	M+1	M+2	FM
$C_8H_{14}O_2$	8.95	0.75	142.0994
$C_8H_{16}NO$	9.32	0.59	142.1233
C_8NO_2	9.10	0.77	141.9929
$C_8H_{18}N_2$	9.70	0.42	142.1471
$C_8H_2N_2O$	9.48	0.60	142.0167
$C_8H_4N_3$	9.85	0.44	142.0406
$C_9H_{18}O$	10.05	0.65	142.1358
$C_9H_2O_2$	9.84	0.83	142.0054
$C_9H_{20}N$	10.43	0.49	142.1597
C_9H_4NO	10.21	0.67	142.0293
$C_9H_6N_2$	10.58	0.51	142.0532
$C_{10}H_{22}$	11.16	0.56	142.1722
$C_{10}H_6O$	10.94	0.74	142.0419
$C_{10}H_8N$	11.32	0.58	142.0657
$C_{11}H_{10}$	12.05	0.66	142.0783
143			
$C_4H_3N_2O_4$	5.29	0.92	143.0093
$C_4H_5N_3O_3$	5.66	0.74	143.0331
$C_4H_7N_4O_2$	6.04	0.56	143.0570
$C_5H_5NO_4$	6.02	0.95	143.0218
$C_5H_7N_2O_3$	6.40	0.78	143.0457
$C_5H_9N_3O_2$	6.77	0.60	143.0695
$C_5H_{11}N_4O$	7.14	0.42	143.0934
$C_6H_7O_4$	6.75	0.99	143.0344
$C_6H_9NO_3$	7.13	0.82	143.0583
$C_6H_{11}N_2O_2$	7.50	0.65	143.0821
$C_6H_{13}N_3O$	7.88	0.47	143.1060
$C_6H_{15}N_4$	8.25	0.30	143.1298
$C_7H_{11}O_3$	7.86	0.87	143.0708
$C_7H_{13}NO_2$	8.23	0.70	143.0947
$C_7H_{15}N_2O$	8.61	0.53	143.1185
$C_7H_{17}N_3$	8.98	0.36	143.1424
C_7HN_3O	8.76	0.54	143.0120
$C_7H_3N_4$	9.14	0.37	143.0359
$C_8H_{15}O_2$	8.96	0.76	143.1072
$C_8H_{17}NO$	9.34	0.59	143.1311
C_8HNO_2	9.12	0.77	143.0007
$C_8H_{19}N_2$	9.71	0.42	143.1549
$C_8H_3N_2O$	9.49	0.60	143.0246
$C_8H_5N_3$	9.87	0.44	143.0484
$C_9H_{19}O$	10.07	0.65	143.1436
$C_9H_3O_2$	9.85	0.83	143.0133
$C_9H_{21}N$	10.44	0.49	143.1675
C_9H_5NO	10.23	0.67	143.0371
$C_9H_7N_2$	10.60	0.51	143.0610
$C_{10}H_7O$	10.96	0.74	143.0497
$C_{10}H_9N$	11.33	0.58	143.0736
$C_{11}H_{11}$	12.00	0.66	143.0861
144			
$C_4H_4N_2O_4$	5.31	0.92	144.0171
$C_4H_6N_3O_3$	5.68	0.74	144.0410
$C_4H_8N_4O_2$	6.05	0.56	144.0648
$C_5H_6NO_4$	6.04	0.95	144.0297
$C_5H_8N_2O_3$	6.41	0.78	144.0535
$C_5H_{10}N_3O_2$	6.79	0.60	144.0774
$C_5H_{12}N_4O$	7.16	0.42	144.1012
$C_6H_8O_4$	6.77	1.00	144.0422
$C_6H_{10}NO_3$	7.14	0.82	144.0661
$C_6H_{12}N_2O_2$	7.52	0.65	144.0899
$C_6H_{14}N_3O$	7.89	0.47	144.1138
$C_6H_{16}N_4$	8.27	0.30	144.1377
C_6N_4O	8.05	0.49	144.0073
$C_7H_{12}O_3$	7.87	0.87	144.0786
$C_7H_{14}NO_2$	8.25	0.70	144.1025
$C_7H_{16}N_2O$	8.62	0.53	144.1264
$C_7N_2O_2$	8.41	0.71	143.9960
$C_7H_{18}N_3$	9.00	0.36	144.1502
$C_7H_2N_3O$	8.78	0.54	144.0198
$C_7H_4N_4$	9.15	0.38	144.0437
$C_8H_{16}O_2$	8.98	0.76	144.1151
C_8O_3	8.76	0.94	143.9847
$C_8H_{18}NO$	9.35	0.59	144.1389
$C_8H_2NO_2$	9.14	0.77	144.0085
$C_8H_{20}N_2$	9.73	0.42	144.1628
$C_8H_4N_2O$	9.51	0.60	144.0324
$C_8H_6N_3$	9.89	0.44	144.0563
$C_9H_{20}O$	10.08	0.66	144.1515
$C_9H_4O_2$	9.87	0.84	144.0211
C_9H_6NO	10.24	0.67	144.0449
$C_9H_8N_2$	10.62	0.51	144.0688
$C_{10}H_8O$	10.97	0.74	144.0575
$C_{10}H_{10}N$	11.35	0.58	144.0814
$C_{11}H_{12}$	12.08	0.67	144.0939
C_{12}	12.97	0.77	144.0000
145			
$C_4H_5N_2O_4$	5.32	0.92	145.0249
$C_4H_7N_3O_3$	5.70	0.74	145.0488
$C_4H_9N_4O_2$	6.07	0.56	145.0726
$C_5H_7NO_4$	6.05	0.96	145.0375
$C_5H_9N_2O_3$	6.43	0.78	145.0614
$C_5H_{11}N_3O_2$	6.80	0.60	145.0852
$C_5H_{13}N_4O$	7.18	0.43	145.1091
$C_6H_9O_4$	6.78	1.00	145.0501
$C_6H_{11}NO_3$	7.16	0.82	145.0739
$C_6H_{13}N_2O_2$	7.53	0.65	145.0978
$C_6H_{15}N_3O$	7.91	0.48	145.1216
$C_6H_{17}N_4$	8.28	0.30	145.1455
C_6HN_4O	8.06	0.49	145.0151
$C_7H_{13}O_3$	7.89	0.87	145.0865
$C_7H_{15}NO_2$	8.26	0.70	145.1103
$C_7H_{17}N_2O$	8.64	0.53	145.1342
$C_7HN_2O_2$	8.42	0.71	145.0038
$C_7H_{19}N_3$	9.01	0.36	145.1580
$C_7H_3N_3O$	8.80	0.54	145.0277
$C_7H_5N_4$	9.17	0.38	145.0515
$C_8H_{17}O_2$	9.00	0.76	145.1229
C_8HO_3	8.78	0.94	144.9925
$C_8H_{19}NO$	9.37	0.59	145.1467
$C_8H_3NO_2$	9.15	0.77	145.0164
$C_8H_5N_2O$	9.53	0.61	145.0402
$C_8H_7N_3$	9.90	0.44	145.0641
$C_9H_5O_2$	9.88	0.84	145.0289
C_9H_7NO	10.26	0.67	145.0528
$C_9H_9N_2$	10.63	0.51	145.0767
$C_{10}H_9O$	10.99	0.75	145.0653
$C_{10}H_{11}N$	11.36	0.59	145.0892
$C_{11}H_{13}$	12.09	0.67	145.1018
$C_{12}H$	12.98	0.77	145.0078
146			
$C_4H_6N_2O_4$	5.34	0.92	146.0328
$C_4H_8N_3O_3$	5.71	0.74	146.0566
$C_4H_{10}N_4O_2$	6.09	0.56	146.0805
$C_5H_8NO_4$	6.07	0.96	146.0453
$C_5H_{10}N_2O_3$	6.44	0.78	146.0692
$C_5H_{12}N_3O_2$	6.82	0.60	146.0930
$C_5H_{14}N_4O$	7.19	0.43	146.1169
$C_6H_{10}O_4$	6.80	1.00	146.0579
$C_6H_{12}NO_3$	7.17	0.82	146.0817
$C_6H_{14}N_2O_2$	7.55	0.65	146.1056
$C_6H_{16}N_3O$	7.92	0.48	146.1295
$C_6N_3O_2$	7.71	0.66	145.9991
$C_6H_{18}N_4$	8.30	0.31	146.1533
$C_6H_2N_4O$	8.08	0.49	146.0229
$C_7H_{14}O_3$	7.91	0.87	146.0943
$C_7H_{16}NO_2$	8.28	0.70	146.1182
C_7NO_3	8.06	0.88	145.9878
$C_7H_{18}N_2O$	8.65	0.53	146.1420
$C_7H_2N_2O_2$	8.44	0.71	146.0116
$C_7H_4N_3O$	8.81	0.55	146.0355
$C_7H_6N_4$	9.19	0.38	146.0594
$C_8H_{18}O_2$	9.01	0.76	146.1307
$C_8H_2O_3$	8.79	0.94	146.0003
$C_8H_4NO_2$	9.17	0.77	146.0242
$C_8H_6N_2O$	9.54	0.61	146.0480
$C_8H_8N_3$	9.92	0.44	146.0719
$C_9H_6O_2$	9.90	0.84	146.0368
C_9H_8NO	10.27	0.67	146.0606
$C_9H_{10}N_2$	10.65	0.51	146.0845
$C_{10}H_{10}O$	11.01	0.75	146.0732
$C_{10}H_{12}N$	11.38	0.59	146.0970
$C_{11}H_{14}$	12.11	0.67	146.1096
$C_{11}N$	12.27	0.69	146.0031
$C_{12}H_2$	13.00	0.77	146.0157
147			
$C_4H_7N_2O_4$	5.35	0.92	147.0406
$C_4H_9N_3O_3$	5.73	0.74	147.0644
$C_4H_{11}N_4O_2$	6.10	0.56	147.0883
$C_5H_9NO_4$	6.08	0.96	147.0532
$C_5H_{11}N_2O_3$	6.46	0.78	147.0770
$C_5H_{13}N_3O_2$	6.83	0.60	147.1009
$C_5H_{15}N_4O$	7.21	0.43	147.1247

	M+1	M+2	FM
$C_6H_{11}O_4$	6.82	1.00	147.0657
$C_6H_{13}NO_3$	7.19	0.82	147.0896
$C_6H_{15}N_2O_2$	7.56	0.65	147.1134
$C_6H_{17}N_3O$	7.94	0.48	147.1373
$C_6HN_3O_2$	7.72	0.66	147.0069
$C_6H_3N_4O$	8.10	0.49	147.0308
$C_7H_{15}O_3$	7.92	0.87	147.1021
$C_7H_{17}NO_2$	8.30	0.70	147.1260
C_7HNO_3	8.08	0.89	146.9956
$C_7H_3N_2O_2$	8.45	0.72	147.0195
$C_7H_5N_3O$	8.83	0.55	147.0433
$C_7H_7N_4$	9.20	0.38	147.0672
$C_8H_3O_3$	8.81	0.94	147.0082
$C_8H_5NO_2$	9.18	0.78	147.0320
$C_8H_7N_2O$	9.56	0.61	147.0559
$C_8H_9N_3$	9.93	0.44	147.0798
$C_9H_7O_2$	9.92	0.84	147.0446
C_9H_9NO	10.29	0.68	147.0684
$C_9H_{11}N_2$	10.66	0.51	147.0923
$C_{10}H_{11}O$	11.02	0.75	147.0810
$C_{10}H_{13}N$	11.40	0.59	147.1049
$C_{11}H_{15}$	12.13	0.67	147.1174
$C_{11}HN$	12.28	0.69	147.0109
$C_{12}H_3$	13.02	0.78	147.0235
148			
$C_4H_8N_2O_4$	5.37	0.92	148.0484
$C_4H_{10}N_3O_3$	5.74	0.74	148.0723
$C_4H_{12}N_4O_2$	6.12	0.56	148.0961
$C_5H_{10}NO_4$	6.10	0.96	148.0610
$C_5H_{12}N_2O_3$	6.48	0.78	148.0848
$C_5H_{14}N_3O_2$	6.85	0.60	148.1087
$C_5H_{16}N_4O$	7.22	0.43	148.1325
$C_5N_4O_2$	7.01	0.61	148.0022
$C_6H_{12}O_4$	6.83	1.00	148.0735
$C_6H_{14}NO_3$	7.21	0.83	148.0974
$C_6H_{16}N_2O_2$	7.58	0.65	148.1213
$C_6N_2O_3$	7.36	0.84	147.9909
$C_6H_2N_3O_2$	7.74	0.66	148.0147
$C_6H_4N_4O$	8.11	0.49	148.0386
$C_7H_{16}O_3$	7.94	0.88	148.1100
C_7O_4	7.72	1.06	147.0706
$C_7H_2NO_3$	8.09	0.89	148.0034
$C_7H_4N_2O_2$	8.47	0.72	148.0273
$C_7H_6N_3O$	8.84	0.55	148.0511
$C_7H_8N_4$	9.22	0.38	148.0750
$C_8H_4O_3$	8.83	0.94	148.0160
$C_8H_6NO_2$	9.20	0.78	148.0399
$C_8H_8N_2O$	9.57	0.61	148.0637
$C_8H_{10}N_3$	9.95	0.45	148.0876
$C_9H_8O_2$	9.93	0.84	148.0524
$C_9H_{10}NO$	10.31	0.68	148.0763
$C_9H_{12}N_2$	10.68	0.52	148.1001
$C_{10}H_{12}O$	11.04	0.75	148.0888

	M+1	M+2	FM
$C_{10}H_{14}N$	11.41	0.59	148.1127
$C_{10}N_2$	11.57	0.61	148.0062
$C_{11}H_{16}$	12.14	0.67	148.1253
$C_{11}O$	11.93	0.85	147.9949
$C_{11}H_2N$	12.30	0.69	148.0187
$C_{12}H_4$	13.03	0.78	148.0313
149			
$C_4H_9N_2O_4$	5.39	0.92	149.0563
$C_4H_{11}N_3O_3$	5.76	0.74	149.0801
$C_4H_{13}N_4O_2$	6.13	0.56	149.1040
$C_5H_{11}NO_4$	6.12	0.96	149.0688
$C_5H_{13}N_2O_3$	6.49	0.78	149.0927
$C_5H_{15}N_3O_2$	6.87	0.61	149.1165
$C_5HN_4O_2$	7.02	0.62	149.0100
$C_6H_{13}O_4$	6.85	1.00	149.0814
$C_6H_{15}NO_3$	7.22	0.83	149.1052
$C_6HN_2O_3$	7.38	0.84	148.9987
$C_6H_3N_3O_2$	7.75	0.66	149.0226
$C_6H_5N_4O$	8.13	0.49	149.0464
C_7HO_4	7.74	1.06	148.9874
$C_7H_3NO_3$	8.11	0.89	149.0113
$C_7H_5N_2O_2$	8.49	0.72	149.0351
$C_7H_7N_3O$	8.86	0.55	149.0590
$C_7H_9N_4$	9.23	0.38	149.0829
$C_8H_5O_3$	8.84	0.95	149.0238
$C_8H_7NO_2$	9.22	0.78	149.0477
$C_8H_9N_2O$	9.59	0.61	149.0715
$C_8H_{11}N_3$	9.97	0.45	149.0954
$C_9H_9O_2$	9.95	0.84	149.0603
$C_9H_{11}NO$	10.32	0.68	149.0841
$C_9H_{13}N_2$	10.70	0.52	149.1080
$C_{10}H_{13}O$	11.05	0.75	149.0967
$C_{10}H_{15}N$	11.43	0.59	149.1205
$C_{10}HN_2$	11.58	0.61	149.0140
$C_{11}H_{17}$	12.16	0.67	149.1331
$C_{11}HO$	11.94	0.85	149.0027
$C_{11}H_3N$	12.32	0.69	149.0266
$C_{12}H_5$	13.05	0.78	149.0391
150			
$C_4H_{10}N_2O_4$	5.40	0.92	150.0641
$C_4H_{12}N_3O_3$	5.78	0.74	150.0879
$C_4H_{14}N_4O_2$	6.15	0.56	150.1118
$C_5H_{12}NO_4$	6.13	0.96	150.0766
$C_5H_{14}N_2O_3$	6.51	0.78	150.1005
$C_5N_3O_3$	6.66	0.79	149.9940
$C_5H_2N_4O_2$	7.04	0.62	150.0178
$C_6H_{14}O_4$	6.86	1.00	150.0892
C_6NO_4	7.02	1.01	149.9827
$C_6H_2N_2O_3$	7.40	0.84	150.0065
$C_6H_4N_3O_2$	7.77	0.67	150.0304
$C_6H_6N_4O$	8.14	0.49	150.0542
$C_7H_2O_4$	7.75	1.06	149.9953
$C_7H_4NO_3$	8.13	0.89	150.0191
$C_7H_6N_2O_2$	8.50	0.72	150.0429

	M+1	M+2	FM
$C_7H_8N_3O$	8.88	0.55	150.0668
$C_7H_{10}N_4$	9.25	0.38	150.0907
$C_8H_6O_3$	8.86	0.95	150.0317
$C_8H_8NO_2$	9.23	0.78	150.0555
$C_8H_{10}N_2O$	9.61	0.61	150.0794
$C_8H_{12}N_3$	9.98	0.45	150.1032
$C_9H_{10}O_2$	9.96	0.84	150.0681
$C_9H_{12}NO$	10.34	0.68	150.0919
$C_9H_{14}N_2$	10.71	0.52	150.1158
C_9N_3	10.87	0.54	150.0093
$C_{10}H_{14}O$	11.07	0.75	150.1045
$C_{10}H_{16}N$	11.44	0.60	150.1284
$C_{10}NO$	11.23	0.77	149.9980
$C_{10}H_2N_2$	11.60	0.61	150.0218
$C_{11}H_{18}$	12.17	0.68	150.1409
$C_{11}H_2O$	11.96	0.85	150.0106
$C_{11}H_4N$	12.33	0.70	150.0344
$C_{12}H_6$	13.06	0.78	150.0470
151			
$C_4H_{11}N_2O_4$	5.42	0.92	151.0719
$C_4H_{13}N_3O_3$	5.79	0.74	151.0958
$C_5H_{13}NO_4$	6.15	0.96	151.0845
$C_5HN_3O_3$	6.68	0.79	151.0018
$C_5H_3N_4O_2$	7.05	0.62	151.0257
C_6HNO_4	7.04	1.01	150.9905
$C_6H_3N_2O_3$	7.41	0.84	151.0144
$C_6H_5N_3O_2$	7.79	0.67	151.0382
$C_6H_7N_4O$	8.16	0.50	151.0621
$C_7H_3O_4$	7.77	1.06	151.0031
$C_7H_5NO_3$	8.14	0.89	151.0269
$C_7H_7N_2O_2$	8.52	0.72	151.0508
$C_7H_9N_3O$	8.89	0.55	151.0746
$C_7H_{11}N_4$	9.27	0.39	151.0985
$C_8H_7O_3$	8.87	0.95	151.0395
$C_8H_9NO_2$	9.25	0.78	151.0634
$C_8H_{11}N_2O$	9.62	0.62	151.0872
$C_8H_{13}N_3$	10.00	0.45	151.1111
$C_9H_{11}O_2$	9.98	0.85	151.0759
$C_9H_{13}NO$	10.35	0.68	151.0998
$C_9H_{15}N_2$	10.73	0.52	151.1236
C_9HN_3	10.89	0.54	151.0171
$C_{10}H_{15}O$	11.09	0.76	151.1123
$C_{10}H_{17}N$	11.46	0.60	151.1362
$C_{10}HNO$	11.24	0.77	151.0058
$C_{10}H_3N_2$	11.62	0.61	151.0297
$C_{11}H_{19}$	12.19	0.68	151.1488
$C_{11}H_3O$	11.97	0.85	151.0184
$C_{11}H_5N$	12.35	0.70	151.0422
$C_{12}H_7$	13.08	0.79	151.0548
152			
$C_4H_{12}N_2O_4$	5.43	0.92	152.0797
$C_5N_2O_4$	6.32	0.97	151.9858
$C_5H_2N_3O_3$	6.70	0.79	152.0096
$C_5H_4N_4O_2$	7.07	0.62	152.0335

	M+1	M+2	FM
$C_6H_2NO_4$	7.05	1.01	151.9983
$C_6H_4N_2O_3$	7.43	0.84	152.0222
$C_6H_6N_3O_2$	7.80	0.67	152.0460
$C_6H_8N_4O$	8.18	0.50	152.0699
$C_7H_4O_4$	7.78	1.06	152.0109
$C_7H_6NO_3$	8.16	0.89	152.0348
$C_7H_8N_2O_2$	8.53	0.72	152.0586
$C_7H_{10}N_3O$	8.91	0.55	152.0825
$C_7H_{12}N_4$	9.28	0.39	152.1063
$C_8H_8O_3$	8.89	0.95	152.0473
$C_8H_{10}NO_2$	9.26	0.78	152.0712
$C_8H_{12}N_2O$	9.64	0.62	152.0950
$C_8H_{14}N_3$	10.01	0.45	152.1189
C_8N_4	10.17	0.47	152.0124
$C_9H_{12}O_2$	10.00	0.85	152.0837
$C_9H_{14}NO$	10.37	0.68	152.1076
$C_9H_{16}N_2$	10.74	0.52	152.1315
C_9N_2O	10.53	0.70	152.0011
$C_9H_2N_3$	10.90	0.54	152.0249
$C_{10}H_{16}O$	11.10	0.76	152.1202
$C_{10}O_2$	10.88	0.93	151.9898
$C_{10}H_{18}N$	11.48	0.60	152.1440
$C_{10}H_2NO$	11.26	0.78	152.0136
$C_{10}H_4N_2$	11.63	0.62	152.0375
$C_{11}H_{20}$	12.21	0.68	152.1566
$C_{11}H_4O$	11.99	0.86	152.0262
$C_{11}H_6N$	12.36	0.70	152.0501
$C_{12}H_8$	13.10	0.79	152.0626
153			
$C_5HN_2O_4$	6.34	0.97	152.9936
$C_5H_3N_3O_3$	6.71	0.80	153.0175
$C_5H_5N_4O_2$	7.09	0.62	153.0413
$C_6H_3NO_4$	7.07	1.02	153.0062
$C_6H_5N_2O_3$	7.44	0.84	153.0300
$C_6H_7N_3O_2$	7.82	0.67	153.0539
$C_6H_9N_4O$	8.19	0.50	153.0777
$C_7H_5O_4$	7.80	1.07	153.0187
$C_7H_7NO_3$	8.17	0.89	153.0426
$C_7H_9N_2O_2$	8.55	0.72	153.0664
$C_7H_{11}N_3O$	8.92	0.56	153.0903
$C_7H_{13}N_4$	9.30	0.39	153.1142
$C_8H_9O_3$	8.91	0.95	153.0552
$C_8H_{11}NO_2$	9.28	0.78	153.0790
$C_8H_{13}N_2O$	9.65	0.62	153.1029
$C_8H_{15}N_3$	10.03	0.45	153.1267
C_8HN_4	10.19	0.47	153.0202
$C_9H_{13}O_2$	10.01	0.85	153.0916
$C_9H_{15}NO$	10.39	0.69	153.1154
$C_9H_{17}N_2$	10.76	0.52	153.1393
C_9HN_2O	10.54	0.70	153.0089
$C_9H_3N_3$	10.92	0.54	153.0328
$C_{10}H_{17}O$	11.12	0.76	153.1280
$C_{10}HO_2$	10.90	0.94	152.9976
$C_{10}H_{19}N$	11.49	0.60	153.1519
$C_{10}H_3NO$	11.27	0.78	153.0215
$C_{10}H_5N_2$	11.65	0.62	153.0453
$C_{11}H_{21}$	12.22	0.68	153.1644
$C_{11}H_5O$	12.01	0.86	153.0340
$C_{11}H_7N$	12.38	0.70	153.0579
$C_{12}H_9$	13.11	0.79	153.0705
154			
$C_5H_2N_2O_4$	6.35	0.97	154.0014
$C_5H_4N_3O_3$	6.73	0.80	154.0253
$C_5H_6N_4O_2$	7.10	0.62	154.0491
$C_6H_4NO_4$	7.09	1.02	154.0140
$C_6H_6N_2O_3$	7.46	0.84	154.0379
$C_6H_8N_3O_2$	7.83	0.67	154.0617
$C_6H_{10}N_4O$	8.21	0.50	154.0856
$C_7H_6O_4$	7.82	1.07	154.0266
$C_7H_8NO_3$	8.19	0.90	154.0504
$C_7H_{10}N_2O_2$	8.57	0.73	154.0743
$C_7H_{12}N_3O$	8.94	0.56	154.0981
$C_7H_{14}N_4$	9.31	0.39	154.1220
$C_8H_{10}O_3$	8.92	0.95	154.0630
$C_8H_{12}NO_2$	9.30	0.79	154.0868
$C_8H_{14}N_2O$	9.67	0.62	154.1107
$C_8H_{16}N_3$	10.05	0.46	154.1346
C_8N_3O	9.83	0.63	154.0042
$C_8H_2N_4$	10.20	0.47	154.0280
$C_9H_{14}O_2$	10.03	0.85	154.0994
$C_9H_{16}NO$	10.40	0.69	154.1233
C_9NO_2	10.19	0.87	153.9929
$C_9H_{18}N_2$	10.78	0.53	154.1471
$C_9H_2N_2O$	10.56	0.70	154.0167
$C_9H_4N_3$	10.93	0.54	154.0406
$C_{10}H_{18}O$	11.13	0.76	154.1358
$C_{10}H_2O_2$	10.92	0.94	154.0054
$C_{10}H_{20}N$	11.51	0.60	154 1597
$C_{10}H_4NO$	11.29	0.78	154.0293
$C_{10}H_6N_2$	11.66	0.62	154.0532
$C_{11}H_{22}$	12.24	0.68	154.1722
$C_{11}H_6O$	12.02	0.86	154.0419
$C_{11}H_8N$	12.40	0.70	154.0657
$C_{12}H_{10}$	13.13	0.79	154.0783
155			
$C_5H_3N_2O_4$	6.37	0.97	155.0093
$C_5H_5N_3O_3$	6.74	0.80	155.0331
$C_5H_7N_4O_2$	7.12	0.62	155.0570
$C_6H_5NO_4$	7.10	1.02	155.0218
$C_6H_7N_2O_3$	7.48	0.84	155.0457
$C_6H_9N_3O_2$	7.85	0.67	155.0695
$C_6H_{11}N_4O$	8.22	0.50	155.0934
$C_7H_7O_4$	7.83	1.07	155.0344
$C_7H_9NO_3$	8.21	0.90	155.0583
$C_7H_{11}N_2O_2$	8.58	0.73	155.0821
$C_7H_{13}N_3O$	8.96	0.56	155.1060
$C_7H_{15}N_4$	9.33	0.39	155.1298
$C_8H_{11}O_3$	8.94	0.95	155.0708
$C_8H_{13}NO_2$	9.31	0.79	155.0947
$C_8H_{15}N_2O$	9.69	0.62	155.1185
$C_8H_{17}N_3$	10.06	0.46	155.1424
C_8HN_3O	9.84	0.64	155.0120
$C_8H_3N_4$	10.22	0.47	155.0359
$C_9H_{15}O_2$	10.04	0.85	155.1072
$C_9H_{17}NO$	10.42	0.69	155.1311
C_9HNO_2	10.20	0.87	155.0007
$C_9H_{19}N_2$	10.79	0.53	155.1549
$C_9H_3N_2O$	10.58	0.71	155.0246
$C_9H_5N_3$	10.95	0.54	155.0484
$C_{10}H_{19}O$	11.15	0.76	155.1436
$C_{10}H_3O_2$	10.93	0.94	155.0133
$C_{10}H_{21}N$	11.52	0.60	155.1675
$C_{10}H_5NO$	11.31	0.78	155.0371
$C_{10}H_7N_2$	11.68	0.62	155.0610
$C_{11}H_{23}$	12.25	0.69	155.1801
$C_{11}H_7O$	12.04	0.86	155.0497
$C_{11}H_9N$	12.41	0.71	155.0736
$C_{12}H_{11}$	13.14	0.79	155.0861
156			
$C_5H_4N_2O_4$	6.39	0.98	156.0171
$C_5H_6N_3O_3$	6.76	0.80	156.0410
$C_5H_8N_4O_2$	7.14	0.62	156.0648
$C_6H_6NO_4$	7.12	1.02	156.0297
$C_6H_8N_2O_3$	7.49	0.85	156.0535
$C_6H_{10}N_3O_2$	7.87	0.67	156.0774
$C_6H_{12}N_4O$	8.24	0.50	156.1012
$C_7H_8O_4$	7.85	1.07	156.0422
$C_7H_{10}NO_3$	8.22	0.90	156.0661
$C_7H_{12}N_2O_2$	8.60	0.73	156.0899
$C_7H_{14}N_3O$	8.97	0.56	156.1138
$C_7H_{16}N_4$	9.35	0.39	156.1377
C_7N_4O	9.13	0.57	156.0073
$C_8H_{12}O_3$	8.95	0.96	156.0786
$C_8H_{14}NO_2$	9.33	0.79	156.1025
$C_8H_{16}N_2O$	9.70	0.62	156.1264
$C_8N_2O_2$	9.49	0.80	155.9960
$C_8H_{18}N_3$	10.08	0.46	156.1502
$C_8H_2N_3O$	9.86	0.64	156.0198
$C_8H_4N_4$	10.23	0.47	156.0437
$C_9H_{16}O_2$	10.06	0.85	156.1151
C_9O_3	9.84	1.03	155.9847
$C_9H_{18}NO$	10.43	0.69	156.1389
$C_9H_2NO_2$	10.22	0.87	156.0085
$C_9H_{20}N_2$	10.81	0.53	156.1628
$C_9H_4N_2O$	10.59	0.71	156.0324
$C_9H_6N_3$	10.97	0.55	156.0563
$C_{10}H_{20}O$	11.17	0.77	156.1515
$C_{10}H_4O_2$	10.95	0.94	156.0211
$C_{10}H_{22}N$	11.54	0.61	156.1753
$C_{10}H_6NO$	11.32	0.78	156.0449

	M+1	M+2	FM
$C_{10}H_{8}N_{2}$	11.70	0.62	150.0008
$C_{11}H_{24}$	12.27	0.69	156.1879
$C_{11}H_{8}O$	12.05	0.86	156.0575
$C_{11}H_{10}N$	12.43	0.71	156.0814
$C_{12}H_{12}$	13.16	0.80	156.0939
C_{13}	14.05	0.91	156.0000
157			
$C_{5}H_{5}N_{2}O_{4}$	6.40	0.98	157.0249
$C_{5}H_{7}N_{3}O_{3}$	6.78	0.80	157.0488
$C_{5}H_{9}N_{4}O_{2}$	7.15	0.62	157.0726
$C_{6}H_{7}NO_{4}$	7.13	1.02	157.0375
$C_{6}H_{9}N_{2}O_{3}$	7.51	0.85	157.0614
$C_{6}H_{11}N_{3}O_{2}$	7.88	0.67	157.0852
$C_{6}H_{13}N_{4}O$	8.26	0.50	157.1091
$C_{7}H_{9}O_{4}$	7.86	1.07	157.0501
$C_{7}H_{11}NO_{3}$	8.24	0.90	157.0739
$C_{7}H_{13}N_{2}O_{2}$	8.61	0.73	157.0978
$C_{7}H_{15}N_{3}O$	8.99	0.56	157.1216
$C_{7}H_{17}N_{4}$	9.36	0.39	157.1455
$C_{7}HN_{4}O$	9.15	0.57	157.0151
$C_{8}H_{13}O_{3}$	8.97	0.96	157.0865
$C_{8}H_{15}NO_{2}$	9.34	0.79	157.1103
$C_{8}H_{17}N_{2}O$	9.72	0.62	157.1342
$C_{8}HN_{2}O_{2}$	9.50	0.80	157.0038
$C_{8}H_{19}N_{3}$	10.09	0.46	157.1580
$C_{8}H_{3}N_{3}O$	9.88	0.64	157.0277
$C_{8}H_{5}N_{4}$	10.25	0.48	157.0515
$C_{9}H_{17}O_{2}$	10.08	0.86	157.1229
$C_{9}HO_{3}$	9.86	1.03	156.9925
$C_{9}H_{19}NO$	10.45	0.69	157.1467
$C_{9}H_{3}NO_{2}$	10.23	0.87	157.0164
$C_{9}H_{21}N_{2}$	10.82	0.53	157.1706
$C_{9}H_{5}N_{2}O$	10.61	0.71	157.0402
$C_{9}H_{7}N_{3}$	10.98	0.55	157.0641
$C_{10}H_{21}O$	11.18	0.77	157.1593
$C_{10}H_{5}O_{2}$	10.96	0.94	157.0289
$C_{10}H_{23}N$	11.56	0.61	157.1832
$C_{10}H_{7}NO$	11.34	0.78	157.0528
$C_{10}H_{9}N_{2}$	11.71	0.63	157.0767
$C_{11}H_{9}O$	12.07	0.86	157.0653
$C_{11}H_{11}N$	12.44	0.71	157.0802
$C_{12}H_{13}$	13.18	0.80	157.1018
$C_{13}H$	14.06	0.91	157.0078
158			
$C_{5}H_{6}N_{2}O_{4}$	6.42	0.98	158.0328
$C_{5}H_{8}N_{3}O_{3}$	6.79	0.80	158.0566
$C_{5}H_{10}N_{4}O_{2}$	7.17	0.63	158.0805
$C_{6}H_{8}NO_{4}$	7.15	1.02	158.0453
$C_{6}H_{10}N_{2}O_{3}$	7.52	0.85	158.0692
$C_{6}H_{12}N_{3}O_{2}$	7.90	0.68	158.0930
$C_{6}H_{14}N_{4}O$	8.27	0.50	158.1169
$C_{7}H_{10}O_{4}$	7.88	1.07	158.0579
$C_{7}H_{12}NO_{3}$	8.25	0.90	158.0817
$C_{7}H_{14}N_{2}O_{2}$	8.63	0.73	158.1056
$C_{7}H_{16}N_{3}O$	0.00	0.50	150.1295
$C_{7}N_{3}O_{2}$	8.79	0.74	157.9991
$C_{7}H_{18}N_{4}$	9.38	0.40	158.1533
$C_{7}H_{2}N_{4}O$	9.16	0.58	158.0229
$C_{8}H_{14}O_{3}$	8.99	0.96	158.0943
$C_{8}H_{16}NO_{2}$	9.36	0.79	158.1182
$C_{8}NO_{3}$	9.14	0.97	157.9878
$C_{8}H_{18}N_{2}O$	9.73	0.63	158.1420
$C_{8}H_{2}N_{2}O_{2}$	9.52	0.81	158.0116
$C_{8}H_{20}N_{3}$	10.11	0.46	158.1659
$C_{8}H_{4}N_{3}O$	9.89	0.64	158.0355
$C_{8}H_{6}N_{4}$	10.27	0.48	158.0594
$C_{9}H_{18}O_{2}$	10.09	0.86	158.1307
$C_{9}H_{2}O_{3}$	9.87	1.04	158.0003
$C_{9}H_{20}NO$	10.47	0.69	158.1546
$C_{9}H_{4}NO_{2}$	10.25	0.87	158.0242
$C_{9}H_{22}N_{2}$	10.84	0.53	158.1784
$C_{9}H_{6}N_{2}O$	10.62	0.71	158.0480
$C_{9}H_{8}N_{3}$	11.00	0.55	158.0719
$C_{10}H_{22}O$	11.20	0.77	158.1671
$C_{10}H_{6}O_{2}$	10.98	0.95	158.0368
$C_{10}H_{8}NO$	11.35	0.79	158.0606
$C_{10}H_{10}N_{2}$	11.73	0.63	158.0845
$C_{11}H_{10}O$	12.09	0.87	158.0732
$C_{11}H_{12}N$	12.46	0.71	158.0970
$C_{12}H_{14}$	13.19	0.80	158.1096
$C_{12}N$	13.35	0.82	158.0031
$C_{13}H_{2}$	14.08	0.92	158.0157
159			
$C_{5}H_{7}N_{2}O_{4}$	6.43	0.98	159.0406
$C_{5}H_{9}N_{3}O_{3}$	6.81	0.80	159.0644
$C_{5}H_{11}N_{4}O_{2}$	7.18	0.63	159.0883
$C_{6}H_{9}NO_{4}$	7.17	1.02	159.0532
$C_{6}H_{11}N_{2}O_{3}$	7.54	0.85	159.0770
$C_{6}H_{13}N_{3}O_{2}$	7.91	0.68	159.1009
$C_{6}H_{15}N_{4}O$	8.29	0.51	159.1247
$C_{7}H_{11}O_{4}$	7.90	1.07	159.0657
$C_{7}H_{13}NO_{3}$	8.27	0.90	159.0896
$C_{7}H_{15}N_{2}O_{2}$	8.65	0.73	159.1134
$C_{7}H_{17}N_{3}O$	9.02	0.56	159.1373
$C_{7}HN_{3}O_{2}$	8.80	0.75	159.0069
$C_{7}H_{19}N_{4}$	9.39	0.40	159.1611
$C_{7}H_{3}N_{4}O$	9.18	0.58	159.0308
$C_{8}H_{15}O_{3}$	9.00	0.96	159.1021
$C_{8}H_{17}NO_{2}$	9.38	0.79	159.1260
$C_{8}HNO_{3}$	9.16	0.97	158.9956
$C_{8}H_{19}N_{2}O$	9.75	0.63	159.1498
$C_{8}H_{3}N_{2}O_{2}$	9.53	0.81	159.0195
$C_{8}H_{21}N_{3}$	10.13	0.46	159.1737
$C_{8}H_{5}N_{3}O$	9.91	0.64	159.0433
$C_{8}H_{7}N_{4}$	10.28	0.48	159.0672
$C_{9}H_{19}O_{2}$	10.11	0.86	159.1385
$C_{9}H_{3}O_{3}$	9.89	1.04	159.0082
$C_{9}H_{21}NO$	10.40	0.70	159.1024
$C_{9}H_{5}NO_{2}$	10.27	0.87	159.0320
$C_{9}H_{7}N_{2}O$	10.64	0.71	159.0559
$C_{9}H_{9}N_{3}$	11.01	0.55	159.0798
$C_{10}H_{7}O_{2}$	11.00	0.95	159.0446
$C_{10}H_{9}NO$	11.37	0.79	159.0684
$C_{10}H_{11}N_{2}$	11.74	0.63	159.0923
$C_{11}H_{11}O$	12.10	0.87	159.0810
$C_{11}H_{13}N$	12.48	0.71	159.1049
$C_{12}H_{15}$	13.21	0.80	159.1174
$C_{12}HN$	13.36	0.82	159.0109
$C_{13}H_{3}$	14.10	0.92	159.0235
160			
$C_{5}H_{8}N_{2}O_{4}$	6.45	0.98	160.0484
$C_{5}H_{10}N_{3}O_{3}$	6.82	0.80	160.0723
$C_{5}H_{12}N_{4}O_{2}$	7.20	0.63	160.0961
$C_{6}H_{10}NO_{4}$	7.18	1.02	160.0610
$C_{6}H_{12}N_{2}O_{3}$	7.56	0.85	160.0848
$C_{6}H_{14}N_{3}O_{2}$	7.93	0.68	160.1087
$C_{6}H_{16}N_{4}O$	8.30	0.51	160.1325
$C_{6}N_{4}O_{2}$	8.09	0.69	160.0022
$C_{7}H_{12}O_{4}$	7.91	1.07	160.0735
$C_{7}H_{14}NO_{3}$	8.29	0.90	160.0974
$C_{7}H_{16}N_{2}O_{2}$	8.66	0.73	160.1213
$C_{7}N_{2}O_{3}$	8.44	0.92	159.9909
$C_{7}H_{18}N_{3}O$	9.04	0.57	160.1451
$C_{7}H_{2}N_{3}O_{2}$	8.82	0.75	160.0147
$C_{7}H_{20}N_{4}$	9.41	0.40	160.1690
$C_{7}H_{4}N_{4}O$	9.19	0.58	160.0386
$C_{8}H_{16}O_{3}$	9.02	0.96	160.1100
$C_{8}O_{4}$	8.80	1.14	159.9796
$C_{8}H_{18}NO_{2}$	9.39	0.79	160.1338
$C_{8}H_{2}NO_{3}$	9.18	0.97	160.0034
$C_{8}H_{20}N_{2}O$	9.77	0.63	160.1577
$C_{8}H_{4}N_{2}O_{2}$	9.55	0.81	160.0273
$C_{8}H_{6}N_{3}O$	9.92	0.64	160.0511
$C_{8}H_{8}N_{4}$	10.30	0.48	160.0750
$C_{9}H_{20}O_{2}$	10.12	0.86	160.1464
$C_{9}H_{4}O_{3}$	9.91	1.04	160.0160
$C_{9}H_{6}NO_{2}$	10.28	0.88	160.0399
$C_{9}H_{8}N_{2}O$	10.66	0.71	160.0637
$C_{9}H_{10}N_{3}$	11.03	0.55	160.0876
$C_{10}H_{8}O_{2}$	11.01	0.95	160.0524
$C_{10}H_{10}NO$	11.39	0.79	160.0763
$C_{10}H_{12}N_{2}$	11.76	0.63	160.1001
$C_{11}H_{12}O$	12.12	0.87	160.0888
$C_{11}H_{14}N$	12.49	0.72	160.1127
$C_{11}N_{2}$	12.65	0.73	160.0062
$C_{12}H_{16}$	13.22	0.80	160.1253
$C_{12}O$	13.01	0.98	159.9949
$C_{12}H_{2}N$	13.38	0.82	160.0187
$C_{13}H_{4}$	14.11	0.92	160.0313
161			
$C_{5}H_{9}N_{2}O_{4}$	6.47	0.98	161.0563

	M+1	M+2	FM
$C_5H_{11}N_3O_3$	6.84	0.80	161.0801
$C_5H_{13}N_4O_2$	7.22	0.63	161.1040
$C_6H_{11}NO_4$	7.20	1.03	161.0688
$C_6H_{13}N_2O_3$	7.57	0.85	161.0927
$C_6H_{15}N_3O_2$	7.95	0.68	161.1165
$C_6H_{17}N_4O$	8.32	0.51	161.1404
$C_6HN_4O_2$	8.10	0.69	161.0100
$C_7H_{13}O_4$	7.93	1.08	161.0814
$C_7H_{15}NO_3$	8.30	0.90	161.1052
$C_7H_{17}N_2O_2$	8.68	0.74	161.1291
$C_7HN_2O_3$	8.46	0.92	160.9987
$C_7H_{19}N_3O$	9.05	0.57	161.1529
$C_7H_3N_3O_2$	8.83	0.75	161.0226
$C_7H_5N_4O$	9.21	0.58	161.0464
$C_8H_{17}O_3$	9.03	0.96	161.1178
C_8HO_4	8.82	1.14	160.9874
$C_8H_{19}NO_2$	9.41	0.80	161.1416
$C_8H_3NO_3$	9.19	0.98	161.0113
$C_8H_5N_2O_2$	9.57	0.81	161.0351
$C_8H_7N_3O$	9.94	0.65	161.0590
$C_8H_9N_4$	10.31	0.48	161.0829
$C_9H_5O_3$	9.92	1.04	161.0238
$C_9H_7NO_2$	10.30	0.88	161.0477
$C_9H_9N_2O$	10.67	0.72	161.0715
$C_9H_{11}N_3$	11.05	0.56	161.0954
$C_{10}H_9O_2$	11.03	0.95	161.0603
$C_{10}H_{11}NO$	11.40	0.79	161.0841
$C_{10}H_{13}N_2$	11.78	0.63	161.1080
$C_{11}H_{13}O$	12.13	0.87	161.0967
$C_{11}H_{15}N$	12.51	0.72	161.1205
$C_{11}HN_2$	12.67	0.74	161.0140
$C_{12}H_{17}$	13.24	0.81	161.1331
$C_{12}HO$	13.02	0.98	161.0027
$C_{12}H_3N$	13.40	0.83	161.0266
$C_{13}H_5$	14.13	0.92	161.0391
162			
$C_5H_{10}N_2O_4$	6.48	0.98	162.0641
$C_5H_{12}N_3O_3$	6.86	0.81	162.0879
$C_5H_{14}N_4O_2$	7.23	0.63	162.1118
$C_6H_{12}NO_4$	7.21	1.03	162.0766
$C_6H_{14}N_2O_3$	7.59	0.85	162.1005
$C_6H_{16}N_3O_2$	7.96	0.68	162.1244
$C_6N_3O_3$	7.75	0.86	161.9940
$C_6H_{18}N_4O$	8.34	0.51	162.1482
$C_6H_2N_4O_2$	8.12	0.69	162.0178
$C_7H_{14}O_4$	7.94	1.08	162.0892
$C_7H_{16}NO_3$	8.32	0.91	162.1131
C_7NO_4	8.10	1.09	161.9827
$C_7H_{18}N_2O_2$	8.69	0.74	162.1369
$C_7H_2N_2O_3$	8.48	0.92	162.0065
$C_7H_4N_3O_2$	8.85	0.75	162.0304
$C_7H_6N_4O$	9.23	0.58	162.0542
$C_8H_{18}O_3$	9.05	0.96	162.1256

	M+1	M+2	FM
$C_8H_2O_4$	8.83	1.15	161.9953
$C_8H_4NO_3$	9.21	0.98	162.0191
$C_8H_6N_2O_2$	9.58	0.81	162.0429
$C_8H_8N_3O$	9.96	0.65	162.0668
$C_8H_{10}N_4$	10.33	0.48	162.0907
$C_9H_6O_3$	9.94	1.04	162.0317
$C_9H_8NO_2$	10.31	0.88	162.0555
$C_9H_{10}N_2O$	10.69	0.72	162.0794
$C_9H_{12}N_3$	11.06	0.56	162.1032
$C_{10}H_{10}O_2$	11.04	0.95	162.0681
$C_{10}H_{12}NO$	11.42	0.79	162.0919
$C_{10}H_{14}N_2$	11.79	0.64	162.1158
$C_{10}N_3$	11.95	0.65	162.0093
$C_{11}H_{14}O$	12.15	0.87	162.1045
$C_{11}H_{16}N$	12.52	0.72	162.1284
$C_{11}NO$	12.31	0.89	161.9980
$C_{11}H_2N_2$	12.68	0.74	162.0218
$C_{12}H_{18}$	13.26	0.81	162.1409
$C_{12}H_2O$	13.04	0.98	162.0106
$C_{12}H_4N$	13.41	0.83	162.0344
$C_{13}H_6$	14.14	0.92	162.0470
163			
$C_5H_{11}N_2O_4$	6.50	0.98	163.0719
$C_5H_{13}N_3O_3$	6.87	0.81	163.0958
$C_5H_{15}N_4O_2$	7.25	0.63	163.1196
$C_6H_{13}NO_4$	7.23	1.03	163.0845
$C_6H_{15}N_2O_3$	7.60	0.85	163.1083
$C_6H_{17}N_3O_2$	7.98	0.68	163.1322
$C_6HN_3O_3$	7.76	0.87	163.0018
$C_6H_3N_4O_2$	8.14	0.69	163.0257
$C_7H_{15}O_4$	7.96	1.08	163.0970
$C_7H_{17}NO_3$	8.33	0.91	163.1209
C_7HNO_4	8.12	1.09	162.9905
$C_7H_3N_2O_3$	8.49	0.92	163.0144
$C_7H_5N_3O_2$	8.87	0.75	163.0382
$C_7H_7N_4O$	9.24	0.58	163.0621
$C_8H_3O_4$	8.85	1.15	163.0031
$C_8H_5NO_3$	9.22	0.98	163.0269
$C_8H_7N_2O_2$	9.60	0.81	163.0508
$C_8H_9N_3O$	9.97	0.65	163.0746
$C_8H_{11}N_4$	10.35	0.49	163.0985
$C_9H_7O_3$	9.95	1.04	163.0395
$C_9H_9NO_2$	10.33	0.88	163.0634
$C_9H_{11}N_2O$	10.70	0.72	163.0872
$C_9H_{13}N_3$	11.08	0.56	163.1111
$C_{10}H_{11}O_2$	11.06	0.95	163.0759
$C_{10}H_{13}NO$	11.43	0.80	163.0998
$C_{10}H_{15}N_2$	11.81	0.64	163.1236
$C_{10}HN_3$	11.97	0.66	163.0171
$C_{11}H_{15}O$	12.17	0.88	163.1123
$C_{11}H_{17}N$	12.54	0.72	163.1362
$C_{11}HNO$	12.32	0.89	163.0058
$C_{11}H_3N_2$	12.70	0.74	163.0297

	M+1	M+2	FM
$C_{12}H_{19}$	13.27	0.81	163.1488
$C_{12}H_3O$	13.05	0.98	163.0184
$C_{12}H_5N$	13.43	0.83	163.0422
$C_{13}H_7$	14.16	0.93	163.0548
164			
$C_5H_{12}N_2O_4$	6.51	0.98	164.0797
$C_5H_{14}N_3O_3$	6.89	0.81	164.1036
$C_5H_{16}N_4O_2$	7.26	0.63	164.1275
$C_6H_{14}NO_4$	7.25	1.03	164.0923
$C_6H_{16}N_2O_3$	7.62	0.86	164.1162
$C_6N_2O_4$	7.40	1.04	163.9858
$C_6H_2N_3O_3$	7.78	0.87	164.0096
$C_6H_4N_4O_2$	8.15	0.70	164.0335
$C_7H_{16}O_4$	7.98	1.08	164.1049
$C_7H_2NO_4$	8.13	1.09	163.9983
$C_7H_4N_2O_3$	8.51	0.92	164.0222
$C_7H_6N_3O_2$	8.88	0.75	164.0460
$C_7H_8N_4O$	9.26	0.59	164.0699
$C_8H_4O_4$	8.87	1.15	164.0109
$C_8H_6NO_3$	9.24	0.98	164.0348
$C_8H_8N_2O_2$	9.61	0.81	164.0586
$C_8H_{10}N_3O$	9.99	0.65	164.0825
$C_8H_{12}N_4$	10.36	0.49	164.1063
$C_9H_8O_3$	9.97	1.05	164.0473
$C_9H_{10}NO_2$	10.35	0.88	164.0712
$C_9H_{12}N_2O$	10.72	0.72	164.0950
$C_9H_{14}N_3$	11.09	0.56	164.1189
C_9N_4	11.25	0.58	164.0124
$C_{10}H_{12}O_2$	11.08	0.96	164.0837
$C_{10}H_{14}NO$	11.45	0.80	164.1076
$C_{10}H_{16}N_2$	11.82	0.64	164.1315
$C_{10}N_2O$	11.61	0.81	164.0011
$C_{10}H_2N_3$	11.98	0.66	164.0249
$C_{11}H_{16}O$	12.18	0.88	164.1202
$C_{11}O_2$	11.96	1.05	163.9898
$C_{11}H_{18}N$	12.56	0.72	164.1440
$C_{11}H_2NO$	12.34	0.90	164.0136
$C_{11}H_4N_2$	12.71	0.74	164.0375
$C_{12}H_{20}$	13.29	0.81	164.1566
$C_{12}H_4O$	13.07	0.98	164.0262
$C_{12}H_6N$	13.44	0.83	164.0501
$C_{13}H_8$	14.18	0.93	164.0626
165			
$C_5H_{13}N_2O_4$	6.53	0.98	165.0876
$C_5H_{15}N_3O_3$	6.90	0.81	165.1114
$C_6H_{15}NO_4$	7.26	1.03	165.1001
$C_6HN_2O_4$	7.42	1.04	164.9936
$C_6H_3N_3O_3$	7.79	0.87	165.0175
$C_6H_5N_4O_2$	8.17	0.70	165.0413
$C_7H_3NO_4$	8.15	1.09	165.0062
$C_7H_5N_2O_3$	8.52	0.92	165.0300
$C_7H_7N_3O_2$	8.90	0.75	165.0539
$C_7H_9N_4O$	9.27	0.59	165.0777
$C_8H_5O_4$	8.88	1.15	165.0187

	M+1	M+2	FM
$C_8H_7NO_3$	9.26	0.98	165.0426
$C_8H_9N_2O_2$	9.63	0.82	165.0664
$C_8H_{11}N_3O$	10.00	0.65	165.0903
$C_8H_{13}N_4$	10.38	0.49	165.1142
$C_9H_9O_3$	9.99	1.05	165.0552
$C_9H_{11}NO_2$	10.36	0.88	165.0790
$C_9H_{13}N_2O$	10.74	0.72	165.1029
$C_9H_{15}N_3$	11.11	0.56	165.1267
C_9HN_4	11.27	0.58	165.0202
$C_{10}H_{13}O_2$	11.09	0.96	165.0916
$C_{10}H_{15}NO$	11.47	0.80	165.1154
$C_{10}H_{17}N_2$	11.84	0.64	165.1393
$C_{10}HN_2O$	11.62	0.82	165.0089
$C_{10}H_3N_3$	12.00	0.66	165.0328
$C_{11}H_{17}O$	12.20	0.88	165.1280
$C_{11}HO_2$	11.98	1.05	164.9976
$C_{11}H_{19}N$	12.57	0.73	165.1519
$C_{11}H_3NO$	12.36	0.90	165.0215
$C_{11}H_5N_2$	12.73	0.74	165.0453
$C_{12}H_{21}$	13.30	0.81	165.1644
$C_{12}H_5O$	13.09	0.99	165.0340
$C_{12}H_7N$	13.46	0.84	165.0579
$C_{13}H_9$	14.19	0.93	165.0705
166			
$C_5H_{14}N_2O_4$	6.55	0.99	166.0954
$C_6H_2N_2O_4$	7.43	1.04	166.0014
$C_6H_4N_3O_3$	7.81	0.87	166.0253
$C_6H_6N_4O_2$	8.18	0.70	166.0491
$C_7H_4NO_4$	8.17	1.09	166.0140
$C_7H_6N_2O_3$	8.54	0.92	166.0379
$C_7H_8N_3O_2$	8.91	0.76	166.0617
$C_7H_{10}N_4O$	9.29	0.59	166.0856
$C_8H_6O_4$	8.90	1.15	166.0266
$C_8H_8NO_3$	9.27	0.98	166.0504
$C_8H_{10}N_2O_2$	9.65	0.82	166.0743
$C_8H_{12}N_3O$	10.02	0.65	166.0981
$C_8H_{14}N_4$	10.39	0.49	166.1220
$C_9H_{10}O_3$	10.00	1.05	166.0630
$C_9H_{12}NO_2$	10.38	0.89	166.0868
$C_9H_{14}N_2O$	10.75	0.72	166.1107
$C_9H_{16}N_3$	11.13	0.56	166.1346
C_9N_3O	10.91	0.74	166.0042
$C_9H_2N_4$	11.28	0.58	166.0280
$C_{10}H_{14}O_2$	11.11	0.96	166.0994
$C_{10}H_{16}NO$	11.48	0.80	166.1233
$C_{10}NO_2$	11.27	0.98	165.9929
$C_{10}H_{18}N_2$	11.86	0.64	166.1471
$C_{10}H_2N_2O$	11.64	0.82	166.0167
$C_{10}H_4N_3$	12.01	0.66	166.0406
$C_{11}H_{18}O$	12.21	0.88	166.1358
$C_{11}H_2O_2$	12.00	1.06	166.0054
$C_{11}H_{20}N$	12.59	0.73	166.1597
$C_{11}H_4NO$	12.37	0.90	166.0293
$C_{11}H_6N_2$	12.75	0.75	166.0532
$C_{12}H_{22}$	13.32	0.82	166.1722
$C_{12}H_6O$	13.10	0.99	166.0419
$C_{12}H_8N$	13.48	0.84	166.0657
$C_{13}H_{10}$	14.21	0.93	166.0783
167			
$C_6H_3N_2O_4$	7.45	1.04	167.0093
$C_6H_5N_3O_3$	7.83	0.87	167.0331
$C_6H_7N_4O_2$	8.20	0.70	167.0570
$C_7H_5NO_4$	8.18	1.10	167.0218
$C_7H_7N_2O_3$	8.56	0.93	167.0457
$C_7H_9N_3O_2$	8.93	0.76	167.0695
$C_7H_{11}N_4O$	9.31	0.59	167.0934
$C_8H_7O_4$	8.91	1.15	167.0344
$C_8H_9NO_3$	9.29	0.99	167.0583
$C_8H_{11}N_2O_2$	9.66	0.82	167.0821
$C_8H_{13}N_3O$	10.04	0.66	167.1060
$C_8H_{15}N_4$	10.41	0.49	167.1298
$C_9H_{11}O_3$	10.02	1.05	167.0708
$C_9H_{13}NO_2$	10.39	0.89	167.0947
$C_9H_{15}N_2O$	10.77	0.73	167.1185
$C_9H_{17}N_3$	11.14	0.57	167.1424
C_9HN_3O	10.92	0.74	167.0120
$C_9H_3N_4$	11.30	0.58	167.0359
$C_{10}H_{15}O_2$	11.12	0.96	167.1072
$C_{10}H_{17}NO$	11.50	0.80	167.1311
$C_{10}HNO_2$	11.28	0.98	167.0007
$C_{10}H_{19}N_2$	11.87	0.64	167.1549
$C_{10}H_3N_2O$	11.66	0.82	167.0246
$C_{10}H_5N_3$	12.03	0.66	167.0484
$C_{11}H_{19}O$	12.23	0.88	167.1436
$C_{11}H_3O_2$	12.01	1.06	167.0133
$C_{11}H_{21}N$	12.60	0.73	167.1675
$C_{11}H_5NO$	12.39	0.90	167.0371
$C_{11}H_7N_2$	12.76	0.75	167.0610
$C_{12}H_{23}$	13.34	0.82	167.1801
$C_{12}H_7O$	13.12	0.99	167.0497
$C_{12}H_9N$	13.49	0.84	167.0736
$C_{13}H_{11}$	14.22	0.94	167.0861
168			
$C_6H_4N_2O_4$	7.47	1.04	168.0171
$C_6H_6N_3O_3$	7.84	0.87	168.0410
$C_6H_8N_4O_2$	8.22	0.70	168.0648
$C_7H_6NO_4$	8.20	1.10	168.0297
$C_7H_8N_2O_3$	8.57	0.93	168.0535
$C_7H_{10}N_3O_2$	8.95	0.76	168.0774
$C_7H_{12}N_4O$	9.32	0.59	168.1012
$C_8H_8O_4$	8.93	1.15	168.0422
$C_8H_{10}NO_3$	9.30	0.99	168.0661
$C_8H_{12}N_2O_2$	9.68	0.82	168.0899
$C_8H_{14}N_3O$	10.05	0.66	168.1138
$C_8H_{16}N_4$	10.43	0.49	168.1377
C_8N_4O	10.21	0.67	168.0073
$C_9H_{12}O_3$	10.03	1.05	168.0786
$C_9H_{14}NO_2$	10.41	0.89	168.1025
$C_9H_{16}N_2O$	10.78	0.73	168.1264
$C_9N_2O_2$	10.57	0.91	167.9960
$C_9H_{18}N_3$	11.16	0.57	168.1502
$C_9H_2N_3O$	10.94	0.74	168.0198
$C_9H_4N_4$	11.32	0.58	168.0437
$C_{10}H_{16}O_2$	11.14	0.96	168.1151
$C_{10}O_3$	10.92	1.14	167.9847
$C_{10}H_{18}NO$	11.51	0.80	168.1389
$C_{10}H_2NO_2$	11.30	0.98	168.0085
$C_{10}H_{20}N_2$	11.89	0.65	168.1628
$C_{10}H_4N_2O$	11.67	0.82	168.0324
$C_{10}H_6N_3$	12.05	0.67	168.0563
$C_{11}H_{20}O$	12.25	0.89	168.1515
$C_{11}H_4O_2$	12.03	1.06	168.0211
$C_{11}H_{22}N$	12.62	0.73	168.1753
$C_{11}H_6NO$	12.40	0.90	168.0449
$C_{11}H_8N_2$	12.78	0.75	168.0688
$C_{12}H_{24}$	13.35	0.82	168.1879
$C_{12}H_8O$	13.13	0.99	168.0575
$C_{12}H_{10}N$	13.51	0.84	168.0814
$C_{13}H_{12}$	14.24	0.94	168.0939
C_{14}	15.13	1.06	168.0000
169			
$C_6H_5N_2O_4$	7.48	1.05	169.0249
$C_6H_7N_3O_3$	7.86	0.87	169.0488
$C_6H_9N_4O_2$	8.23	0.70	169.0726
$C_7H_7NO_4$	8.21	1.10	169.0375
$C_7H_9N_2O_3$	8.59	0.93	169.0614
$C_7H_{11}N_3O_2$	8.96	0.76	169.0852
$C_7H_{13}N_4O$	9.34	0.59	169.1091
$C_8H_9O_4$	8.95	1.16	169.0501
$C_8H_{11}NO_3$	9.32	0.99	169.0739
$C_8H_{13}N_2O_2$	9.69	0.82	169.0978
$C_8H_{15}N_3O$	10.07	0.66	169.1216
$C_8H_{17}N_4$	10.44	0.50	169.1455
C_8HN_4O	10.23	0.67	169.0151
$C_9H_{13}O_3$	10.05	1.05	169.0865
$C_9H_{15}NO_2$	10.43	0.89	169.1103
$C_9H_{17}N_2O$	10.80	0.73	169.1342
$C_9HN_2O_2$	10.58	0.91	169.0038
$C_9H_{19}N_3$	11.17	0.57	169.1580
$C_9H_3N_3O$	10.96	0.75	169.0277
$C_9H_5N_4$	11.33	0.59	169.0515
$C_{10}H_{17}O_2$	11.16	0.96	169.1229
$C_{10}HO_3$	10.94	1.14	168.9925
$C_{10}H_{19}NO$	11.53	0.81	169.1467
$C_{10}H_3NO_2$	11.31	0.98	169.0164
$C_{10}H_{21}N_2$	11.90	0.65	169.1706
$C_{10}H_5N_2O$	11.69	0.82	169.0402
$C_{10}H_7N_3$	12.06	0.67	169.0641
$C_{11}H_{21}O$	12.26	0.89	169.1593
$C_{11}H_5O_2$	12.04	1.06	169.0289

	M+1	M+2	FM
$C_{11}H_{23}N$	12.64	0.73	169.1832
$C_{11}H_7NO$	12.42	0.91	169.0528
$C_{11}H_9N_2$	12.79	0.75	169.0767
$C_{12}H_{25}$	13.37	0.82	169.1957
$C_{12}H_9O$	13.15	1.00	169.0653
$C_{12}H_{11}N$	13.52	0.84	169.0892
$C_{13}H_{13}$	14.26	0.94	169.1018
$C_{14}H$	15.14	1.07	169.0078
170			
$C_6H_6N_2O_4$	7.50	1.05	170.0328
$C_6H_8N_3O_3$	7.87	0.87	170.0566
$C_6H_{10}N_4O_2$	8.25	0.70	170.0805
$C_7H_8NO_4$	8.23	1.10	170.0453
$C_7H_{10}N_2O_3$	8.60	0.93	170.0692
$C_7H_{12}N_3O_2$	8.98	0.76	170.0930
$C_7H_{14}N_4O$	9.35	0.59	170.1169
$C_8H_{10}O_4$	8.96	1.16	170.0579
$C_8H_{12}NO_3$	9.34	0.99	170.0817
$C_8H_{14}N_2O_2$	9.71	0.82	170.1056
$C_8H_{16}N_3O$	10.08	0.66	170.1295
$C_8N_3O_2$	9.87	0.84	169.9991
$C_8H_{18}N_4$	10.46	0.50	170.1533
$C_8H_2N_4O$	10.24	0.68	170.0229
$C_9H_{14}O_3$	10.07	1.06	170.0943
$C_9H_{16}NO_2$	10.44	0.89	170.1182
C_9NO_3	10.22	1.07	169.9878
$C_9H_{18}N_2O$	10.82	0.73	170.1420
$C_9H_2N_2O_2$	10.60	0.91	170.0116
$C_9H_{20}N_3$	11.19	0.57	170.1659
$C_9H_4N_3O$	10.97	0.75	170.0355
$C_9H_6N_4$	11.35	0.59	170.0594
$C_{10}H_{18}O_2$	11.17	0.97	170.1307
$C_{10}H_2O_3$	10.96	1.14	170.0003
$C_{10}H_{20}NO$	11.55	0.81	170.1546
$C_{10}H_4NO_2$	11.33	0.98	170.0242
$C_{10}H_{22}N_2$	11.92	0.65	170.1784
$C_{10}H_6N_2O$	11.70	0.83	170.0480
$C_{10}H_8N_3$	12.08	0.67	170.0719
$C_{11}H_{22}O$	12.28	0.89	170.1671
$C_{11}H_6O_2$	12.06	1.06	170.0368
$C_{11}H_{24}N$	12.65	0.74	170.1910
$C_{11}H_8NO$	12.44	0.91	170.0606
$C_{11}H_{10}N_2$	12.81	0.75	170.0845
$C_{12}H_{26}$	13.38	0.83	170.2036
$C_{12}H_{10}O$	13.17	1.00	170.0732
$C_{12}H_{12}N$	13.54	0.85	170.0970
$C_{13}H_{14}$	14.27	0.94	170.1096
$C_{13}N$	14.43	0.96	170.0031
$C_{14}H_2$	15.16	1.07	170.0157
171			
$C_6H_7N_2O_4$	7.51	1.05	171.0406
$C_6H_9N_3O_3$	7.89	0.88	171.0644
$C_6H_{11}N_4O_2$	8.26	0.70	171.0883
$C_7H_9NO_4$	8.25	1.10	171.0532
$C_7H_{11}N_2O_3$	8.62	0.93	171.0770
$C_7H_{13}N_3O_2$	8.99	0.76	171.1009
$C_7H_{15}N_4O$	9.37	0.60	171.1247
$C_8H_{11}O_4$	8.98	1.16	171.0657
$C_8H_{13}NO_3$	9.35	0.99	171.0896
$C_8H_{15}N_2O_2$	9.73	0.83	171.1134
$C_8H_{17}N_3O$	10.10	0.66	171.1373
$C_8HN_3O_2$	9.88	0.84	171.0069
$C_8H_{19}N_4$	10.47	0.50	171.1611
$C_8H_3N_4O$	10.26	0.68	171.0308
$C_9H_{15}O_3$	10.08	1.06	171.1021
$C_9H_{17}NO_2$	10.46	0.89	171.1260
C_9HNO_3	10.24	1.07	170.9956
$C_9H_{19}N_2O$	10.83	0.73	171.1498
$C_9H_3N_2O_2$	10.61	0.91	171.0195
$C_9H_{21}N_3$	11.21	0.57	171.1737
$C_9H_5N_3O$	10.99	0.75	171.0433
$C_9H_7N_4$	11.36	0.59	171.0672
$C_{10}H_{19}O_2$	11.19	0.97	171.1385
$C_{10}H_3O_3$	10.97	1.14	171.0082
$C_{10}H_{21}NO$	11.56	0.81	171.1624
$C_{10}H_5NO_2$	11.35	0.99	171.0320
$C_{10}H_{23}N_2$	11.94	0.65	171.1863
$C_{10}H_7N_2O$	11.72	0.83	171.0559
$C_{10}H_9N_3$	12.09	0.67	171.0798
$C_{11}H_{23}O$	12.29	0.89	171.1750
$C_{11}H_7O_2$	12.08	1.07	171.0446
$C_{11}H_{25}N$	12.67	0.74	171.1988
$C_{11}H_9NO$	12.45	0.91	171.0684
$C_{11}H_{11}N_2$	12.83	0.76	171.0923
$C_{12}H_{11}O$	13.18	1.00	171.0810
$C_{12}H_{13}N$	13.56	0.85	171.1049
$C_{13}H_{15}$	14.29	0.94	171.1174
$C_{13}HN$	14.45	0.97	171.0109
$C_{14}H_3$	15.18	1.07	171.0235
172			
$C_6H_8N_2O_4$	7.53	1.05	172.0484
$C_6H_{10}N_3O_3$	7.91	0.88	172.0723
$C_6H_{12}N_4O_2$	8.28	0.71	172.0961
$C_7H_{10}NO_4$	8.26	1.10	172.0610
$C_7H_{12}N_2O_3$	8.64	0.93	172.0848
$C_7H_{14}N_3O_2$	9.01	0.76	172.1087
$C_7H_{16}N_4O$	9.39	0.60	172.1325
$C_7N_4O_2$	9.17	0.78	172.0022
$C_8H_{12}O_4$	8.99	1.16	172.0735
$C_8H_{14}NO_3$	9.37	0.99	172.0974
$C_8H_{16}N_2O_2$	9.74	0.83	172.1213
$C_8N_2O_3$	9.52	1.01	171.9909
$C_8H_{18}N_3O$	10.12	0.66	172.1451
$C_8H_2N_3O_2$	9.90	0.84	172.0147
$C_8H_{20}N_4$	10.49	0.50	172.1690
$C_8H_4N_4O$	10.27	0.68	172.0386
$C_9H_{16}O_3$	10.10	1.06	172.1100
C_9O_4	9.88	1.24	171.9796
$C_9H_{18}NO_2$	10.47	0.90	172.1338
$C_9H_2NO_3$	10.26	1.07	172.0034
$C_9H_{20}N_2O$	10.85	0.73	172.1577
$C_9H_4N_2O_2$	10.63	0.91	172.0273
$C_9H_{22}N_3$	11.22	0.57	172.1815
$C_9H_6N_3O$	11.00	0.75	172.0511
$C_9H_8N_4$	11.38	0.59	172.0750
$C_{10}H_{20}O_2$	11.20	0.97	172.1464
$C_{10}H_4O_3$	10.99	1.15	172.0160
$C_{10}H_{22}NO$	11.58	0.81	172.1702
$C_{10}H_6NO_2$	11.36	0.99	172.0399
$C_{10}H_{24}N_2$	11.95	0.65	172.1941
$C_{10}H_8N_2O$	11.74	0.83	172.0637
$C_{10}H_{10}N_3$	12.11	0.67	172.0876
$C_{11}H_{24}O$	12.31	0.89	172.1828
$C_{11}H_8O_2$	12.09	1.07	172.0524
$C_{11}H_{10}NO$	12.47	0.91	172.0763
$C_{11}H_{12}N_2$	12.84	0.76	172.1001
$C_{12}H_{12}O$	13.20	1.00	172.0888
$C_{12}H_{14}N$	13.57	0.85	172.1127
$C_{12}N_2$	13.73	0.87	172.0062
$C_{13}H_{16}$	14.30	0.95	172.1253
$C_{13}O$	14.09	1.12	171.9949
$C_{13}H_2N$	14.46	0.97	172.0187
$C_{14}H_4$	15.19	1.07	172.0313
173			
$C_6H_9N_2O_4$	7.55	1.05	173.0563
$C_6H_{11}N_3O_3$	7.92	0.88	173.0801
$C_6H_{13}N_4O_2$	8.30	0.71	173.1040
$C_7H_{11}NO_4$	8.28	1.10	173.0688
$C_7H_{13}N_2O_3$	8.65	0.93	173.0927
$C_7H_{15}N_3O_2$	9.03	0.77	173.1165
$C_7H_{17}N_4O$	9.40	0.60	173.1404
$C_7HN_4O_2$	9.18	0.78	173.0100
$C_8H_{13}O_4$	9.01	1.16	173.0814
$C_8H_{15}NO_3$	9.38	0.99	173.1052
$C_8H_{17}N_2O_2$	9.76	0.83	173.1291
$C_8HN_2O_3$	9.54	1.01	172.9987
$C_8H_{19}N_3O$	10.13	0.66	173.1529
$C_8H_3N_3O_2$	9.92	0.84	173.0226
$C_8H_{21}N_4$	10.51	0.50	173.1768
$C_8H_5N_4O$	10.29	0.68	173.0464
$C_9H_{17}O_3$	10.11	1.06	173.1178
C_9HO_4	9.90	1.24	172.9874
$C_9H_{19}NO_2$	10.49	0.90	173.1416
$C_9H_3NO_3$	10.27	1.08	173.0113
$C_9H_{21}N_2O$	10.86	0.74	173.1655
$C_9H_5N_2O_2$	10.65	0.91	173.0351
$C_9H_{23}N_3$	11.24	0.58	173.1894
$C_9H_7N_3O$	11.02	0.75	173.0590
$C_9H_9N_4$	11.40	0.59	173.0829

	M+1	M+2	FM
$C_{10}H_{21}O_2$	11.22	0.97	173.1542
$C_{10}H_5O_3$	11.00	1.15	173.0238
$C_{10}H_{23}NO$	11.59	0.81	173.1781
$C_{10}H_7NO_2$	11.38	0.99	173.0477
$C_{10}H_9N_2O$	11.75	0.83	173.0715
$C_{10}H_{11}N_3$	12.13	0.67	173.0954
$C_{11}H_9O_2$	12.11	1.07	173.0603
$C_{11}H_{11}NO$	12.48	0.91	173.0841
$C_{11}H_{13}N_2$	12.86	0.76	173.1080
$C_{12}H_{13}O$	13.21	1.00	173.0967
$C_{12}H_{15}N$	13.59	0.85	173.1205
$C_{12}HN_2$	13.75	0.87	173.0140
$C_{13}H_{17}$	14.32	0.95	173.1331
$C_{13}HO$	14.10	1.12	173.0027
$C_{13}H_3N$	14.48	0.97	173.0266
$C_{14}H_5$	15.21	1.07	173.0391
174			
$C_6H_{10}N_2O_4$	7.56	1.05	174.0641
$C_6H_{12}N_3O_3$	7.94	0.88	174.0879
$C_6H_{14}N_4O_2$	8.31	0.71	174.1118
$C_7H_{12}NO_4$	8.29	1.10	174.0766
$C_7H_{14}N_2O_3$	8.67	0.93	174.1005
$C_7H_{16}N_3O_2$	9.04	0.77	174.1244
$C_7N_3O_3$	8.83	0.95	173.9940
$C_7H_{18}N_4O$	9.42	0.60	174.1482
$C_7H_2N_4O_2$	9.20	0.78	174.0178
$C_8H_{14}O_4$	9.03	1.16	174.0892
$C_8H_{16}NO_3$	9.40	1.00	174.1131
C_8NO_4	9.18	1.18	173.9827
$C_8H_{18}N_2O_2$	9.77	0.83	174.1369
$C_8H_2N_2O_3$	9.56	1.01	174.0065
$C_8H_{20}N_3O$	10.15	0.67	174.1608
$C_8H_4N_3O_2$	9.93	0.85	174.0304
$C_8H_{22}N_4$	10.52	0.50	174.1846
$C_8H_6N_4O$	10.31	0.68	174.0542
$C_9H_{18}O_3$	10.13	1.06	174.1256
$C_9H_2O_4$	9.91	1.24	173.9953
$C_9H_{20}NO_2$	10.51	0.90	174.1495
$C_9H_4NO_3$	10.29	1.08	174.0191
$C_9H_{22}N_2O$	10.88	0.74	174.1733
$C_9H_6N_2O_2$	10.66	0.92	174.0429
$C_9H_8N_3O$	11.04	0.75	174.0668
$C_9H_{10}N_4$	11.41	0.60	174.0907
$C_{10}H_{22}O_2$	11.24	0.97	174.1620
$C_{10}H_6O_3$	11.02	1.15	174.0317
$C_{10}H_8NO_2$	11.39	0.99	174.0555
$C_{10}H_{10}N_2O$	11.77	0.83	174.0794
$C_{10}H_{12}N_3$	12.14	0.68	174.1032
$C_{11}H_{10}O_2$	12.12	1.07	174.0681
$C_{11}H_{12}NO$	12.50	0.92	174.0919
$C_{11}H_{14}N_2$	12.87	0.76	174.1158
$C_{11}N_3$	13.03	0.78	174.0093
$C_{12}H_{14}O$	13.23	1.01	174.1045

	M+1	M+2	FM
$C_{12}H_{16}N$	13.60	0.85	174.1284
$C_{12}NO$	13.39	1.03	173.9980
$C_{12}H_2N_2$	13.76	0.88	174.0218
$C_{13}H_{18}$	14.34	0.95	174.1409
$C_{13}H_2O$	14.12	1.12	174.0106
$C_{13}H_4N$	14.49	0.97	174.0344
$C_{14}H_6$	15.22	1.08	174.0470
175			
$C_6H_{11}N_2O_4$	7.58	1.05	175.0719
$C_6H_{13}N_3O_3$	7.95	0.88	175.0958
$C_6H_{15}N_4O_2$	8.33	0.71	175.1196
$C_7H_{13}NO_4$	8.31	1.11	175.0845
$C_7H_{15}N_2O_3$	8.68	0.94	175.1083
$C_7H_{17}N_3O_2$	9.06	0.77	175.1322
$C_7HN_3O_3$	8.84	0.95	175.0018
$C_7H_{19}N_4O$	9.43	0.60	175.1560
$C_7H_3N_4O_2$	9.22	0.78	175.0257
$C_8H_{15}O_4$	9.04	1.16	175.0970
$C_8H_{17}NO_3$	9.42	1.00	175.1209
C_8HNO_4	9.20	1.18	174.9905
$C_8H_{19}N_2O_2$	9.79	0.83	175.1447
$C_8H_3N_2O_3$	9.57	1.01	175.0144
$C_8H_{21}N_3O$	10.16	0.67	175.1686
$C_8H_5N_3O_2$	9.95	0.85	175.0382
$C_8H_7N_4O$	10.32	0.68	175.0621
$C_9H_{19}O_3$	10.15	1.06	175.1334
$C_9H_3O_4$	9.93	1.24	175.0031
$C_9H_{21}NO_2$	10.52	0.90	175.1573
$C_9H_5NO_3$	10.30	1.08	175.0269
$C_9H_7N_2O_2$	10.68	0.92	175.0508
$C_9H_9N_3O$	11.05	0.76	175.0746
$C_9H_{11}N_4$	11.43	0.60	175.0985
$C_{10}H_7O_3$	11.04	1.15	175.0395
$C_{10}H_9NO_2$	11.41	0.99	175.0634
$C_{10}H_{11}N_2O$	11.78	0.83	175.0872
$C_{10}H_{13}N_3$	12.16	0.68	175.1111
$C_{11}H_{11}O_2$	12.14	1.07	175.0759
$C_{11}H_{13}NO$	12.52	0.92	175.0998
$C_{11}H_{15}N_2$	12.89	0.77	175.1236
$C_{11}HN_3$	13.05	0.78	175.0171
$C_{12}H_{15}O$	13.25	1.01	175.1123
$C_{12}H_{17}N$	13.62	0.86	175.1362
$C_{12}HNO$	13.40	1.03	175.0058
$C_{12}H_3N_2$	13.78	0.88	175.0297
$C_{13}H_{19}$	14.35	0.95	175.1488
$C_{13}H_3O$	14.13	1.12	175.0184
$C_{13}H_5N$	14.51	0.98	175.0422
$C_{14}H_7$	15.24	1.08	175.0548
176			
$C_6H_{12}N_2O_4$	7.59	1.05	176.0797
$C_6H_{14}N_3O_3$	7.97	0.88	176.1036
$C_6H_{16}N_4O_2$	8.34	0.71	176.1275
$C_7H_{14}NO_4$	8.33	1.11	176.0923
$C_7H_{16}N_2O_3$	8.70	0.94	176.1162

	M+1	M+2	FM
$C_7N_2O_4$	8.48	1.12	175.9858
$C_7H_{18}N_3O_2$	9.07	0.77	176.1400
$C_7H_2N_3O_3$	8.86	0.95	176.0096
$C_7H_{20}N_4O$	9.45	0.60	176.1639
$C_7H_4N_4O_2$	9.23	0.78	176.0335
$C_8H_{16}O_4$	9.06	1.17	176.1049
$C_8H_{18}NO_3$	9.43	1.00	176.1287
$C_8H_2NO_4$	9.21	1.18	175.9983
$C_8H_{20}N_2O_2$	9.81	0.83	176.1526
$C_8H_4N_2O_3$	9.59	1.01	176.0222
$C_8H_6N_3O_2$	9.96	0.85	176.0460
$C_8H_8N_4O$	10.34	0.69	176.0699
$C_9H_{20}O_3$	10.16	1.07	176.1413
$C_9H_4O_4$	9.95	1.24	176.0109
$C_9H_6NO_3$	10.32	1.08	176.0348
$C_9H_8N_2O_2$	10.69	0.92	176.0586
$C_9H_{10}N_3O$	11.07	0.76	176.0825
$C_9H_{12}N_4$	11.44	0.60	176.1063
$C_{10}H_8O_3$	11.05	1.15	176.0473
$C_{10}H_{10}NO_2$	11.43	0.99	176.0712
$C_{10}H_{12}N_2O$	11.80	0.84	176.0950
$C_{10}H_{14}N_3$	12.17	0.68	176.1189
$C_{10}N_4$	12.33	0.70	176.0124
$C_{11}H_{12}O_2$	12.16	1.08	176.0837
$C_{11}H_{14}NO$	12.53	0.92	176.1076
$C_{11}H_{16}N_2$	12.91	0.77	176.1315
$C_{11}N_2O$	12.69	0.94	176.0011
$C_{11}H_2N_3$	13.06	0.79	176.0249
$C_{12}H_{16}O$	13.26	1.01	176.1202
$C_{12}O_2$	13.05	1.18	175.9898
$C_{12}H_{18}N$	13.64	0.86	176.1440
$C_{12}H_2NO$	13.42	1.03	176.0136
$C_{12}H_4N_2$	13.79	0.88	176.0375
$C_{13}H_{20}$	14.37	0.96	176.1566
$C_{13}H_4O$	14.15	1.13	176.0262
$C_{13}H_6N$	14.53	0.98	176.0501
$C_{14}H_8$	15.26	1.08	176.0626
177			
$C_6H_{13}N_2O_4$	7.61	1.06	177.0876
$C_6H_{15}N_3O_3$	7.99	0.88	177.1114
$C_6H_{17}N_4O_2$	8.36	0.71	177.1353
$C_7H_{15}NO_4$	8.34	1.11	177.1001
$C_7H_{17}N_2O_3$	8.72	0.94	177.1240
$C_7HN_2O_4$	8.50	1.12	176.9936
$C_7H_{19}N_3O_2$	9.09	0.77	177.1478
$C_7H_3N_3O_3$	8.87	0.95	177.0175
$C_7H_5N_4O_2$	9.25	0.78	177.0413
$C_8H_{17}O_4$	9.07	1.17	177.1127
$C_8H_{19}NO_3$	9.45	1.00	177.1365
$C_8H_3NO_4$	9.23	1.18	177.0062
$C_8H_5N_2O_3$	9.60	1.01	177.0300
$C_8H_7N_3O_2$	9.98	0.85	177.0539
$C_8H_9N_4O$	10.35	0.69	177.0777

	M+1	M+2	FM
$C_9H_5O_4$	9.96	1.25	177.0187
$C_9H_7NO_3$	10.34	1.08	177.0426
$C_9H_9N_2O_2$	10.71	0.92	177.0664
$C_9H_{11}N_3O$	11.08	0.76	177.0903
$C_9H_{13}N_4$	11.46	0.60	177.1142
$C_{10}H_9O_3$	11.07	1.16	177.0552
$C_{10}H_{11}NO_2$	11.44	1.00	177.0790
$C_{10}H_{13}N_2O$	11.82	0.84	177.1029
$C_{10}H_{15}N_3$	12.19	0.68	177.1267
$C_{10}HN_4$	12.35	0.70	177.0202
$C_{11}H_{13}O_2$	12.17	1.08	177.0916
$C_{11}H_{15}NO$	12.55	0.92	177.1154
$C_{11}H_{17}N_2$	12.92	0.77	177.1393
$C_{11}HN_2O$	12.70	0.94	177.0089
$C_{11}H_3N_3$	13.08	0.79	177.0328
$C_{12}H_{17}O$	13.28	1.01	177.1280
$C_{12}HO_2$	13.06	1.18	176.9976
$C_{12}H_{19}N$	13.65	0.86	177.1519
$C_{12}H_3NO$	13.44	1.03	177.0215
$C_{12}H_5N_2$	13.81	0.88	177.0453
$C_{13}H_{21}$	14.38	0.96	177.1644
$C_{13}H_5O$	14.17	1.13	177.0340
$C_{13}H_7N$	14.54	0.98	177.0579
$C_{14}H_9$	15.27	1.08	177.0705
178			
$C_6H_{14}N_2O_4$	7.63	1.06	178.0954
$C_6H_{16}N_3O_3$	8.00	0.88	178.1193
$C_6H_{18}N_4O_2$	8.38	0.71	178.1431
$C_7H_{16}NO_4$	8.36	1.11	178.1080
$C_7H_{18}N_2O_3$	8.73	0.94	178.1318
$C_7H_2N_2O_4$	8.52	1.12	178.0014
$C_7H_4N_3O_3$	8.89	0.95	178.0253
$C_7H_6N_4O_2$	9.26	0.79	178.0491
$C_8H_{18}O_4$	9.09	1.17	178.1205
$C_8H_4NO_4$	9.25	1.18	178.0140
$C_8H_6N_2O_3$	9.62	1.02	178.0379
$C_8H_8N_3O_2$	10.00	0.85	178.0617
$C_8H_{10}N_4O$	10.37	0.69	178.0856
$C_9H_6O_4$	9.98	1.25	178.0266
$C_9H_8NO_3$	10.35	1.08	178.0504
$C_9H_{10}N_2O_2$	10.73	0.92	178.0743
$C_9H_{12}N_3O$	11.10	0.76	178.0981
$C_9H_{14}N_4$	11.48	0.60	178.1220
$C_{10}H_{10}O_3$	11.08	1.16	178.0630
$C_{10}H_{12}NO_2$	11.46	1.00	178.0868
$C_{10}H_{14}N_2O$	11.83	0.84	178.1107
$C_{10}H_{16}N_3$	12.21	0.68	178.1346
$C_{10}N_3O$	11.99	0.86	178.0042
$C_{10}H_2N_4$	12.36	0.70	178.0280
$C_{11}H_{14}O_2$	12.19	1.08	178.0994
$C_{11}H_{16}NO$	12.56	0.92	178.1233
$C_{11}NO_2$	12.35	1.10	177.9929
$C_{11}H_{18}N_2$	12.94	0.77	178.1471
$C_{11}H_2N_2O$	12.72	0.94	178.0167
$C_{11}H_4N_3$	13.09	0.79	178.0406
$C_{12}H_{18}O$	13.29	1.01	178.1358
$C_{12}H_2O_2$	13.08	1.19	178.0054
$C_{12}H_{20}N$	13.67	0.86	178.1597
$C_{12}H_4NO$	13.45	1.03	178.0293
$C_{12}H_6N_2$	13.83	0.88	178.0532
$C_{13}H_{22}$	14.40	0.96	178.1722
$C_{13}H_6O$	14.18	1.13	178.0419
$C_{13}H_8N$	14.56	0.98	178.0657
$C_{14}H_{10}$	15.29	1.09	178.0783
179			
$C_6H_{15}N_2O_4$	7.64	1.06	179.1032
$C_6H_{17}N_3O_3$	8.02	0.89	179.1271
$C_7H_{17}NO_4$	8.37	1.11	179.1158
$C_7H_3N_2O_4$	8.53	1.12	179.0093
$C_7H_5N_3O_3$	8.91	0.95	179.0331
$C_7H_7N_4O_2$	9.28	0.79	179.0570
$C_8H_5NO_4$	9.26	1.18	179.0218
$C_8H_7N_2O_3$	9.64	1.02	179.0457
$C_8H_9N_3O_2$	10.01	0.85	179.0695
$C_8H_{11}N_4O$	10.39	0.69	179.0934
$C_9H_7O_4$	9.99	1.25	179.0344
$C_9H_9NO_3$	10.37	1.09	179.0583
$C_9H_{11}N_2O_2$	10.74	0.92	179.0821
$C_9H_{13}N_3O$	11.12	0.76	179.1060
$C_9H_{15}N_4$	11.49	0.60	179.1298
$C_{10}H_{11}O_3$	11.10	1.16	179.0708
$C_{10}H_{13}NO_2$	11.47	1.00	179.0947
$C_{10}H_{15}N_2O$	11.85	0.84	179.1185
$C_{10}H_{17}N_3$	12.22	0.69	179.1424
$C_{10}HN_3O$	12.01	0.86	179.0120
$C_{10}H_3N_4$	12.38	0.71	179.0359
$C_{11}H_{15}O_2$	12.20	1.08	179.1072
$C_{11}H_{17}NO$	12.58	0.93	179.1311
$C_{11}HNO_2$	12.36	1.10	179.0007
$C_{11}H_{19}N_2$	12.95	0.77	179.1549
$C_{11}H_3N_2O$	12.74	0.95	179.0246
$C_{11}H_5N_3$	13.11	0.79	179.0484
$C_{12}H_{19}O$	13.31	1.02	179.1436
$C_{12}H_3O_2$	13.09	1.19	179.0133
$C_{12}H_{21}N$	13.68	0.87	179.1675
$C_{12}H_5NO$	13.47	1.04	179.0371
$C_{12}H_7N_2$	13.84	0.89	179.0610
$C_{13}H_{23}$	14.42	0.96	179.1801
$C_{13}H_7O$	14.20	1.13	179.0497
$C_{13}H_9N$	14.57	0.99	179.0736
$C_{14}H_{11}$	15.30	1.09	179.0861
180			
$C_6H_{16}N_2O_4$	7.66	1.06	180.1111
$C_7H_4N_2O_4$	8.55	1.12	180.0171
$C_7H_6N_3O_3$	8.92	0.96	180.0410
$C_7H_8N_4O_2$	9.30	0.79	180.0648
$C_8H_6NO_4$	9.28	1.18	180.0297
$C_8H_8N_2O_3$	9.65	1.02	180.0535
$C_8H_{10}N_3O_2$	10.03	0.85	180.0774
$C_8H_{12}N_4O$	10.40	0.69	180.1012
$C_9H_8O_4$	10.01	1.25	180.0422
$C_9H_{10}NO_3$	10.38	1.09	180.0661
$C_9H_{12}N_2O_2$	10.76	0.93	180.0899
$C_9H_{14}N_3O$	11.13	0.77	180.1138
$C_9H_{16}N_4$	11.51	0.61	180.1377
C_9N_4O	11.29	0.78	180.0073
$C_{10}H_{12}O_3$	11.12	1.16	180.0786
$C_{10}H_{14}NO_2$	11.49	1.00	180.1025
$C_{10}H_{16}N_2O$	11.86	0.84	180.1264
$C_{10}N_2O_2$	11.65	1.02	179.9960
$C_{10}H_{18}N_3$	12.24	0.69	180.1502
$C_{10}H_2N_3O$	12.02	0.86	180.0198
$C_{10}H_4N_4$	12.40	0.71	180.0437
$C_{11}H_{16}O_2$	12.22	1.08	180.1151
$C_{11}O_3$	12.00	1.26	179.9847
$C_{11}H_{18}NO$	12.60	0.93	180.1389
$C_{11}H_2NO_2$	12.38	1.10	180.0085
$C_{11}H_{20}N_2$	12.97	0.78	180.1628
$C_{11}H_4N_2O$	12.75	0.95	180.0324
$C_{11}H_6N_3$	13.13	0.80	180.0563
$C_{12}H_{20}O$	13.33	1.02	180.1515
$C_{12}H_4O_2$	13.11	1.19	180.0211
$C_{12}H_{22}N$	13.70	0.87	180.1753
$C_{12}H_6NO$	13.48	1.04	180.0449
$C_{12}H_8N_2$	13.86	0.89	180.0688
$C_{13}H_{24}$	14.43	0.97	180.1879
$C_{13}H_8O$	14.21	1.13	180.0575
$C_{13}H_{10}N$	14.59	0.99	180.0814
$C_{14}H_{12}$	15.32	1.09	180.0939
C_{15}	16.21	1.23	180.0000
181			
$C_7H_5N_2O_4$	8.56	1.13	181.0249
$C_7H_7N_3O_3$	8.94	0.96	181.0488
$C_7H_9N_4O_2$	9.31	0.79	181.0726
$C_8H_7NO_4$	9.29	1.19	181.0375
$C_8H_9N_2O_3$	9.67	1.02	181.0614
$C_8H_{11}N_3O_2$	10.04	0.86	181.0852
$C_8H_{13}N_4O$	10.42	0.69	181.1091
$C_9H_9O_4$	10.03	1.25	181.0501
$C_9H_{11}NO_3$	10.40	1.09	181.0739
$C_9H_{13}N_2O_2$	10.77	0.93	181.0978
$C_9H_{15}N_3O$	11.15	0.77	181.1216
$C_9H_{17}N_4$	11.52	0.61	181.1455
C_9HN_4O	11.31	0.78	181.0151
$C_{10}H_{13}O_3$	11.13	1.16	181.0865
$C_{10}H_{15}NO_2$	11.51	1.00	181.1103
$C_{10}H_{17}N_2O$	11.88	0.85	181.1342
$C_{10}HN_2O_2$	11.66	1.02	181.0038
$C_{10}H_{19}N_3$	12.25	0.69	181.1580
$C_{10}H_3N_3O$	12.04	0.86	181.0277

	M+1	M+2	FM
$C_{10}H_5N_4$	12.41	0.71	181.0515
$C_{11}H_{17}O_2$	12.24	1.09	181.1229
$C_{11}HO_3$	12.02	1.26	180.9925
$C_{11}H_{19}NO$	12.61	0.93	181.1467
$C_{11}H_3NO_2$	12.39	1.10	181.0164
$C_{11}H_{21}N_2$	12.99	0.78	181.1706
$C_{11}H_5N_2O$	12.77	0.95	181.0402
$C_{11}H_7N_3$	13.14	0.80	181.0641
$C_{12}H_{21}O$	13.34	1.02	181.1593
$C_{12}H_5O_2$	13.13	1.19	181.0289
$C_{12}H_{23}N$	13.72	0.87	181.1832
$C_{12}H_7NO$	13.50	1.04	181.0528
$C_{12}H_9N_2$	13.87	0.89	181.0767
$C_{13}H_{25}$	14.45	0.97	181.1957
$C_{13}H_9O$	14.23	1.14	181.0653
$C_{13}H_{11}N$	14.61	0.99	181.0892
$C_{14}H_{13}$	15.34	1.09	181.1018
$C_{15}H$	16.22	1.23	181.0078
182			
$C_7H_6N_2O_4$	8.58	1.13	182.0328
$C_7H_8N_3O_3$	8.95	0.96	182.0566
$C_7H_{10}N_4O_2$	9.33	0.79	182.0805
$C_8H_8NO_4$	9.31	1.19	182.0453
$C_8H_{10}N_2O_3$	9.69	1.02	182.0692
$C_8H_{12}N_3O_2$	10.06	0.86	182.0930
$C_8H_{14}N_4O$	10.43	0.70	182.1169
$C_9H_{10}O_4$	10.04	1.25	182.0579
$C_9H_{12}NO_3$	10.42	1.09	182.0817
$C_9H_{14}N_2O_2$	10.79	0.93	182.1056
$C_9H_{16}N_3O$	11.16	0.77	182.1295
$C_9N_3O_2$	10.95	0.95	181.9991
$C_9H_{18}N_4$	11.54	0.61	182.1533
$C_9H_2N_4O$	11.32	0.79	182.0229
$C_{10}H_{14}O_3$	11.15	1.16	182.0943
$C_{10}H_{16}NO_2$	11.52	1.01	182.1182
$C_{10}NO_3$	11.30	1.18	181.9878
$C_{10}H_{18}N_2O$	11.90	0.85	182.1420
$C_{10}H_2N_2O_2$	11.68	1.02	182.0116
$C_{10}H_{20}N_3$	12.27	0.69	182.1659
$C_{10}H_4N_3O$	12.05	0.87	182.0355
$C_{10}H_6N_4$	12.43	0.71	182.0594
$C_{11}H_{18}O_2$	12.25	1.09	182.1307
$C_{11}H_2O_3$	12.04	1.26	182.0003
$C_{11}H_{20}NO$	12.63	0.93	182.1546
$C_{11}H_4NO_2$	12.41	1.11	182.0242
$C_{11}H_{22}N_2$	13.00	0.78	182.1784
$C_{11}H_6N_2O$	12.78	0.95	182.0480
$C_{11}H_8N_3$	13.16	0.80	182.0719
$C_{12}H_{22}O$	13.36	1.02	182.1671
$C_{12}H_6O_2$	13.14	1.19	182.0368
$C_{12}H_{24}N$	13.73	0.87	182.1910
$C_{12}H_8NO$	13.52	1.04	182.0606
$C_{12}H_{10}N_2$	13.89	0.89	182.0845

	M+1	M+2	FM
$C_{13}H_{26}$	14.46	0.97	182.2036
$C_{13}H_{10}O$	14.25	1.14	182.0732
$C_{13}H_{12}N$	14.62	0.99	182.0970
$C_{14}H_{14}$	15.35	1.10	182.1096
$C_{14}N$	15.51	1.12	182.0031
$C_{15}H_2$	16.24	1.23	182.0157
183			
$C_7H_7N_2O_4$	8.60	1.13	183.0406
$C_7H_9N_3O_3$	8.97	0.96	183.0644
$C_7H_{11}N_4O_2$	9.34	0.79	183.0883
$C_8H_9NO_4$	9.33	1.19	183.0532
$C_8H_{11}N_2O_3$	9.70	1.02	183.0770
$C_8H_{13}N_3O_2$	10.08	0.86	183.1009
$C_8H_{15}N_4O$	10.45	0.70	183.1247
$C_9H_{11}O_4$	10.06	1.26	183.0657
$C_9H_{13}NO_3$	10.43	1.09	183.0896
$C_9H_{15}N_2O_2$	10.81	0.93	183.1134
$C_9H_{17}N_3O$	11.18	0.77	183.1373
$C_9HN_3O_2$	10.96	0.95	183.0069
$C_9H_{19}N_4$	11.56	0.61	183.1611
$C_9H_3N_4O$	11.34	0.79	183.0308
$C_{10}H_{15}O_3$	11.16	1.17	183.1021
$C_{10}H_{17}NO_2$	11.54	1.01	183.1260
$C_{10}HNO_3$	11.32	1.18	182.9956
$C_{10}H_{19}N_2O$	11.91	0.85	183.1498
$C_{10}H_3N_2O_2$	11.70	1.02	183.0195
$C_{10}H_{21}N_3$	12.29	0.69	183.1737
$C_{10}H_5N_3O$	12.07	0.87	183.0433
$C_{10}H_7N_4$	12.44	0.71	183.0672
$C_{11}H_{19}O_2$	12.27	1.09	183.1385
$C_{11}H_3O_3$	12.05	1.26	183.0082
$C_{11}H_{21}NO$	12.64	0.93	183.1624
$C_{11}H_5NO_2$	12.43	1.11	183.0320
$C_{11}H_{23}N_2$	13.02	0.78	183.1863
$C_{11}H_7N_2O$	12.80	0.95	183.0559
$C_{11}H_9N_3$	13.17	0.80	183.0798
$C_{12}H_{23}O$	13.37	1.02	183.1750
$C_{12}H_7O_2$	13.16	1.20	183.0446
$C_{12}H_{25}N$	13.75	0.87	183.1988
$C_{12}H_9NO$	13.53	1.05	183.0684
$C_{12}H_{11}N_2$	13.91	0.90	183.0923
$C_{13}H_{27}$	14.48	0.97	183.2114
$C_{13}H_{11}O$	14.26	1.14	183.0810
$C_{13}H_{13}N$	14.64	0.99	183.1049
$C_{14}H_{15}$	15.37	1.10	183.1174
$C_{14}HN$	15.53	1.12	183.0109
$C_{15}H_3$	16.26	1.23	183.0235
184			
$C_7H_8N_2O_4$	8.61	1.13	184.0484
$C_7H_{10}N_3O_3$	8.99	0.96	184.0723
$C_7H_{12}N_4O_2$	9.36	0.80	184.0961
$C_8H_{10}NO_4$	9.34	1.19	184.0610
$C_8H_{12}N_2O_3$	9.72	1.03	184.0848

	M+1	M+2	FM
$C_8H_{14}N_3O_2$	10.09	0.86	184.1087
$C_8H_{16}N_4O$	10.47	0.70	184.1325
$C_8N_4O_2$	10.25	0.88	184.0022
$C_9H_{12}O_4$	10.07	1.26	184.0735
$C_9H_{14}NO_3$	10.45	1.09	184.0974
$C_9H_{16}N_2O_2$	10.82	0.93	184.1213
$C_9N_2O_3$	10.61	1.11	183.9909
$C_9H_{18}N_3O$	11.20	0.77	184.1451
$C_9H_2N_3O_2$	10.98	0.95	184.0147
$C_9H_{20}N_4$	11.57	0.61	184.1690
$C_9H_4N_4O$	11.35	0.79	184.0386
$C_{10}H_{16}O_3$	11.18	1.17	184.1100
$C_{10}O_4$	10.96	1.34	183.9796
$C_{10}H_{18}NO_2$	11.55	1.01	184.1338
$C_{10}H_2NO_3$	11.34	1.18	184.0034
$C_{10}H_{20}N_2O$	11.93	0.85	184.1577
$C_{10}H_4N_2O_2$	11.71	1.03	184.0273
$C_{10}H_{22}N_3$	12.30	0.70	184.1815
$C_{10}H_6N_3O$	12.09	0.87	184.0511
$C_{10}H_8N_4$	12.46	0.71	184.0750
$C_{11}H_{20}O_2$	12.28	1.09	184.1464
$C_{11}H_4O_3$	12.07	1.27	184.0160
$C_{11}H_{22}NO$	12.66	0.94	184.1702
$C_{11}H_6NO_2$	12.44	1.11	184.0399
$C_{11}H_{24}N_2$	13.03	0.78	184.1941
$C_{11}H_8N_2O$	12.82	0.96	184.0637
$C_{11}H_{10}N_3$	13.19	0.80	184.0876
$C_{12}H_{24}O$	13.39	1.03	184.1828
$C_{12}H_8O_2$	13.17	1.20	184.0524
$C_{12}H_{26}N$	13.76	0.88	184.2067
$C_{12}H_{10}NO$	13.55	1.05	184.0763
$C_{12}H_{12}N_2$	13.92	0.90	184.1001
$C_{13}H_{28}$	14.50	0.97	184.2192
$C_{13}H_{12}O$	14.28	1.14	184.0888
$C_{13}H_{14}N$	14.65	1.00	184.1127
$C_{13}N_2$	14.81	1.02	184.0062
$C_{14}H_{16}$	15.38	1.10	184.1253
$C_{14}O$	15.17	1.27	183.9949
$C_{14}H_2N$	15.54	1.13	184.0187
$C_{15}H_4$	16.27	1.24	184.0313
185			
$C_7H_9N_2O_4$	8.63	1.13	185.0563
$C_7H_{11}N_3O_3$	9.00	0.96	185.0801
$C_7H_{13}N_4O_2$	9.38	0.80	185.1040
$C_8H_{11}NO_4$	9.36	1.19	185.0688
$C_8H_{13}N_2O_3$	9.73	1.03	185.0927
$C_8H_{15}N_3O_2$	10.11	0.86	185.1165
$C_8H_{17}N_4O$	10.48	0.70	185.1404
$C_8HN_4O_2$	10.26	0.88	185.0100
$C_9H_{13}O_4$	10.09	1.26	185.0814
$C_9H_{15}NO_3$	10.46	1.10	185.1052
$C_9H_{17}N_2O_2$	10.84	0.93	185.1291
$C_9HN_2O_3$	10.62	1.11	184.9987

	M+1	M+2	FM
$C_9H_{19}N_3O$	11.21	0.77	185.1529
$C_9H_3N_3O_2$	11.00	0.95	185.0226
$C_9H_{21}N_4$	11.59	0.62	185.1768
$C_9H_5N_4O$	11.37	0.79	185.0464
$C_{10}H_{17}O_3$	11.20	1.17	185.1178
$C_{10}HO_4$	10.98	1.35	184.9874
$C_{10}H_{19}NO_2$	11.57	1.01	185.1416
$C_{10}H_3NO_3$	11.35	1.19	185.0113
$C_{10}H_{21}N_2O$	11.94	0.85	185.1655
$C_{10}H_5N_2O_2$	11.73	1.03	185.0351
$C_{10}H_{23}N_3$	12.32	0.70	185.1894
$C_{10}H_7N_3O$	12.10	0.87	185.0590
$C_{10}H_9N_4$	12.48	0.72	185.0829
$C_{11}H_{21}O_2$	12.30	1.09	185.1542
$C_{11}H_5O_3$	12.08	1.27	185.0238
$C_{11}H_{23}NO$	12.68	0.94	185.1781
$C_{11}H_7NO_2$	12.46	1.11	185.0477
$C_{11}H_{25}N_2$	13.05	0.79	185.2019
$C_{11}H_9N_2O$	12.83	0.96	185.0715
$C_{11}H_{11}N_3$	13.21	0.81	185.0954
$C_{12}H_{25}O$	13.41	1.03	185.1906
$C_{12}H_9O_2$	13.19	1.20	185.0603
$C_{12}H_{27}N$	13.78	0.88	185.2145
$C_{12}H_{11}NO$	13.56	1.05	185.0841
$C_{12}H_{13}N_2$	13.94	0.90	185.1080
$C_{13}H_{13}O$	14.29	1.15	185.0967
$C_{13}H_{15}N$	14.67	1.00	185.1205
$C_{13}HN_2$	14.83	1.02	185.0140
$C_{14}H_{17}$	15.40	1.10	185.1331
$C_{14}HO$	15.18	1.27	185.0027
$C_{14}H_3N$	15.56	1.13	185.0266
$C_{15}H_5$	16.29	1.24	185.0391
186			
$C_7H_{10}N_2O_4$	8.64	1.13	186.0641
$C_7H_{12}N_3O_3$	9.02	0.96	186.0879
$C_7H_{14}N_4O_2$	9.39	0.80	186.1118
$C_8H_{12}NO_4$	9.37	1.19	186.0766
$C_8H_{14}N_2O_3$	9.75	1.03	186.1005
$C_8H_{16}N_3O_2$	10.12	0.86	186.1244
$C_8N_3O_3$	9.91	1.04	185.9940
$C_8H_{18}N_4O$	10.50	0.70	186.1482
$C_8H_2N_4O_2$	10.28	0.88	186.0178
$C_9H_{14}O_4$	10.11	1.26	186.0892
$C_9H_{16}NO_3$	10.48	1.10	186.1131
C_9NO_4	10.26	1.28	185.9827
$C_9H_{18}N_2O_2$	10.85	0.94	186.1369
$C_9H_2N_2O_3$	10.64	1.11	186.0065
$C_9H_{20}N_3O$	11.23	0.78	186.1608
$C_9H_4N_3O_2$	11.01	0.95	186.0304
$C_9H_{22}N_4$	11.60	0.62	186.1846
$C_9H_6N_4O$	11.39	0.79	186.0542
$C_{10}H_{18}O_3$	11.21	1.17	186.1256
$C_{10}H_2O_4$	10.99	1.35	185.9953

	M+1	M+2	FM
$C_{10}H_{20}NO_2$	11.59	1.01	186.1495
$C_{10}H_4NO_3$	11.37	1.19	186.0191
$C_{10}H_{22}N_2O$	11.96	0.86	186.1733
$C_{10}H_6N_2O_2$	11.74	1.03	186.0429
$C_{10}H_{24}N_3$	12.33	0.70	186.1972
$C_{10}H_8N_3O$	12.12	0.87	186.0668
$C_{10}H_{10}N_4$	12.49	0.72	186.0907
$C_{11}H_{22}O_2$	12.32	1.10	186.1620
$C_{11}H_6O_3$	12.10	1.27	186.0317
$C_{11}H_{24}NO$	12.69	0.94	186.1859
$C_{11}H_8NO_2$	12.47	1.11	186.0555
$C_{11}H_{26}N_2$	13.07	0.79	186.2098
$C_{11}H_{10}N_2O$	12.85	0.96	186.0794
$C_{11}H_{12}N_3$	13.22	0.81	186.1032
$C_{12}H_{26}O$	13.42	1.03	186.1985
$C_{12}H_{10}O_2$	13.21	1.20	186.0681
$C_{12}H_{12}NO$	13.58	1.05	186.0919
$C_{12}H_{14}N_2$	13.95	0.90	186.1158
$C_{12}N_3$	14.11	0.92	186.0093
$C_{13}H_{14}O$	14.31	1.15	186.1045
$C_{13}H_{16}N$	14.69	1.00	186.1284
$C_{13}NO$	14.47	1.17	185.9980
$C_{13}H_2N_2$	14.84	1.02	186.0218
$C_{14}H_{18}$	15.42	1.11	186.1409
$C_{14}H_2O$	15.20	1.27	186.0106
$C_{14}H_4N$	15.57	1.13	186.0344
$C_{15}H_6$	16.30	1.24	186.0470
187			
$C_7H_{11}N_2O_4$	8.66	1.13	187.0719
$C_7H_{13}N_3O_3$	9.03	0.97	187.0958
$C_7H_{15}N_4O_2$	9.41	0.80	187.1196
$C_8H_{13}NO_4$	9.39	1.20	187.0845
$C_8H_{15}N_2O_3$	9.77	1.03	187.1083
$C_8H_{17}N_3O_2$	10.14	0.87	187.1322
$C_8HN_3O_3$	9.92	1.04	187.0018
$C_8H_{19}N_4O$	10.51	0.70	187.1560
$C_8H_3N_4O_2$	10.30	0.88	187.0257
$C_9H_{15}O_4$	10.12	1.26	187.0970
$C_9H_{17}NO_3$	10.50	1.10	187.1209
C_9HNO_4	10.28	1.28	186.9905
$C_9H_{19}N_2O_2$	10.87	0.94	187.1447
$C_9H_3N_2O_3$	10.65	1.11	187.0144
$C_9H_{21}N_3O$	11.24	0.78	187.1686
$C_9H_5N_3O_2$	11.03	0.95	187.0382
$C_9H_{23}N_4$	11.62	0.62	187.1925
$C_9H_7N_4O$	11.40	0.80	187.0621
$C_{10}H_{19}O_3$	11.23	1.17	187.1334
$C_{10}H_3O_4$	11.01	1.35	187.0031
$C_{10}H_{21}NO_2$	11.60	1.01	187.1573
$C_{10}H_5NO_3$	11.38	1.19	187.0269
$C_{10}H_{23}N_2O$	11.98	0.86	187.1811
$C_{10}H_7N_2O_2$	11.76	1.03	187.0508
$C_{10}H_{25}N_3$	12.35	0.70	187.2050

	M+1	M+2	FM
$C_{10}H_9N_3O$	12.13	0.88	187.0746
$C_{10}H_{11}N_4$	12.51	0.72	187.0985
$C_{11}H_{23}O_2$	12.33	1.10	187.1699
$C_{11}H_7O_3$	12.12	1.27	187.0395
$C_{11}H_{25}NO$	12.71	0.94	187.1937
$C_{11}H_9NO_2$	12.49	1.12	187.0634
$C_{11}H_{11}N_2O$	12.86	0.96	187.0872
$C_{11}H_{13}N_3$	13.24	0.81	187.1111
$C_{12}H_{11}O_2$	13.22	1.20	187.0759
$C_{12}H_{13}NO$	13.60	1.05	187.0998
$C_{12}H_{15}N_2$	13.97	0.90	187.1236
$C_{12}HN_3$	14.13	0.93	187.0171
$C_{13}H_{15}O$	14.33	1.15	187.1123
$C_{13}H_{17}N$	14.70	1.00	187.1362
$C_{13}HNO$	14.48	1.17	187.0058
$C_{13}H_3N_2$	14.86	1.03	187.0297
$C_{14}H_{19}$	15.43	1.11	187.1488
$C_{14}H_3O$	15.22	1.28	187.0184
$C_{14}H_5N$	15.59	1.13	187.0422
$C_{15}H_7$	16.32	1.24	187.0548
188			
$C_7H_{12}N_2O_4$	8.68	1.14	188.0797
$C_7H_{14}N_3O_3$	9.05	0.97	188.1036
$C_7H_{16}N_4O_2$	9.42	0.80	188.1275
$C_8H_{14}NO_4$	9.41	1.20	188.0923
$C_8H_{16}N_2O_3$	9.78	1.03	188.1162
$C_8N_2O_4$	9.56	1.21	187.9858
$C_8H_{18}N_3O_2$	10.16	0.87	188.1400
$C_8H_2N_3O_3$	9.94	1.05	188.0096
$C_8H_{20}N_4O$	10.53	0.71	188.1639
$C_8H_4N_4O_2$	10.31	0.88	188.0335
$C_9H_{16}O_4$	10.14	1.26	188.1049
$C_9H_{18}NO_3$	10.51	1.10	188.1287
$C_9H_2NO_4$	10.30	1.28	187.9983
$C_9H_{20}N_2O_2$	10.89	0.94	188.1526
$C_9H_4N_2O_3$	10.67	1.12	188.0222
$C_9H_{22}N_3O$	11.26	0.78	188.1764
$C_9H_6N_3O_2$	11.04	0.96	188.0460
$C_9H_{24}N_4$	11.64	0.62	188.2003
$C_9H_8N_4O$	11.42	0.80	188.0699
$C_{10}H_{20}O_3$	11.24	1.18	188.1413
$C_{10}H_4O_4$	11.03	1.35	188.0109
$C_{10}H_{22}NO_2$	11.62	1.02	188.1651
$C_{10}H_6NO_3$	11.40	1.19	188.0348
$C_{10}H_{24}N_2O$	11.99	0.86	188.1890
$C_{10}H_8N_2O_2$	11.78	1.03	188.0586
$C_{10}H_{10}N_3O$	12.15	0.88	188.0825
$C_{10}H_{12}N_4$	12.52	0.72	188.1063
$C_{11}H_{24}O_2$	12.35	1.10	188.1777
$C_{11}H_8O_3$	12.13	1.27	188.0473
$C_{11}H_{10}NO_2$	12.51	1.12	188.0712
$C_{11}H_{12}N_2O$	12.88	0.96	188.0950
$C_{11}H_{14}N_3$	13.25	0.81	188.1189

	M+1	M+2	FM
$C_{11}N_4$	13.41	0.83	188.0124
$C_{12}H_{12}O_2$	13.24	1.21	188.0837
$C_{12}H_{14}NO$	13.61	1.06	188.1076
$C_{12}H_{16}N_2$	13.99	0.91	188.1315
$C_{12}N_2O$	13.77	1.08	188.0011
$C_{12}H_2N_3$	14.14	0.93	188.0249
$C_{13}H_{16}O$	14.34	1.15	188.1202
$C_{13}O_2$	14.13	1.32	187.9898
$C_{13}H_{18}N$	14.72	1.01	188.1440
$C_{13}H_2NO$	14.50	1.18	188.0136
$C_{13}H_4N_2$	14.87	1.03	188.0375
$C_{14}H_{20}$	15.45	1.11	188.1566
$C_{14}H_4O$	15.23	1.28	188.0262
$C_{14}H_6N$	15.61	1.14	188.0501
$C_{15}H_8$	16.34	1.25	188.0626
189			
$C_7H_{13}N_2O_4$	8.69	1.14	189.0876
$C_7H_{15}N_3O_3$	9.07	0.97	189.1114
$C_7H_{17}N_4O_2$	9.44	0.80	189.1353
$C_8H_{15}NO_4$	9.42	1.20	189.1001
$C_8H_{17}N_2O_3$	9.80	1.03	189.1240
$C_8HN_2O_4$	9.58	1.21	188.9936
$C_8H_{19}N_3O_2$	10.17	0.87	189.1478
$C_8H_3N_3O_3$	9.95	1.05	189.0175
$C_8H_{21}N_4O$	10.55	0.71	189.1717
$C_8H_5N_4O_2$	10.33	0.88	189.0413
$C_9H_{17}O_4$	10.15	1.26	189.1127
$C_9H_{19}NO_3$	10.53	1.10	189.1365
$C_9H_3NO_4$	10.31	1.28	189.0062
$C_9H_{21}N_2O_2$	10.90	0.94	189.1604
$C_9H_5N_2O_3$	10.69	1.12	189.0300
$C_9H_{23}N_3O$	11.28	0.78	189.1842
$C_9H_7N_3O_2$	11.06	0.96	189.0539
$C_9H_9N_4O$	11.43	0.80	189.0777
$C_{10}H_{21}O_3$	11.26	1.18	189.1491
$C_{10}H_5O_4$	11.04	1.35	189.0187
$C_{10}H_{23}NO_2$	11.63	1.02	189.1730
$C_{10}H_7NO_3$	11.42	1.19	189.0426
$C_{10}H_9N_2O_2$	11.79	1.04	189.0664
$C_{10}H_{11}N_3O$	12.17	0.88	189.0903
$C_{10}H_{13}N_4$	12.54	0.72	189.1142
$C_{11}H_9O_3$	12.15	1.28	189.0552
$C_{11}H_{11}NO_2$	12.52	1.12	189.0790
$C_{11}H_{13}N_2O$	12.90	0.97	189.1029
$C_{11}H_{15}N_3$	13.27	0.81	189.1267
$C_{11}HN_4$	13.43	0.83	189.0202
$C_{12}H_{13}O_2$	13.25	1.21	189.0916
$C_{12}H_{15}NO$	13.63	1.06	189.1154
$C_{12}H_{17}N_2$	14.00	0.91	189.1393
$C_{12}HN_2O$	13.79	1.08	189.0089
$C_{12}H_3N_3$	14.16	0.93	189.0328
$C_{13}H_{17}O$	14.36	1.16	189.1280
$C_{13}HO_2$	14.14	1.33	188.9976

	M+1	M+2	FM
$C_{13}H_{19}N$	14.73	1.01	189.1519
$C_{13}H_3NO$	14.52	1.18	189.0215
$C_{13}H_5N_2$	14.89	1.03	189.0453
$C_{14}H_{21}$	15.46	1.11	189.1644
$C_{14}H_5O$	15.25	1.28	189.0340
$C_{14}H_7N$	15.62	1.14	189.0579
$C_{15}H_9$	16.35	1.25	189.0705
190			
$C_7H_{14}N_2O_4$	8.71	1.14	190.0954
$C_7H_{16}N_3O_3$	9.08	0.97	190.1193
$C_7H_{18}N_4O_2$	9.46	0.80	190.1431
$C_8H_{16}NO_4$	9.44	1.20	190.1080
$C_8H_{18}N_2O_3$	9.81	1.03	190.1318
$C_8H_2N_2O_4$	9.60	1.21	190.0014
$C_8H_{20}N_3O_2$	10.19	0.87	190.1557
$C_8H_4N_3O_3$	9.97	1.05	190.0253
$C_8H_{22}N_4O$	10.56	0.71	190.1795
$C_8H_6N_4O_2$	10.34	0.89	190.0491
$C_9H_{18}O_4$	10.17	1.27	190.1205
$C_9H_{20}NO_3$	10.54	1.10	190.1444
$C_9H_4NO_4$	10.33	1.28	190.0140
$C_9H_{22}N_2O_2$	10.92	0.94	190.1682
$C_9H_6N_2O_3$	10.70	1.12	190.0379
$C_9H_8N_3O_2$	11.08	0.96	190.0617
$C_9H_{10}N_4O$	11.45	0.80	190.0856
$C_{10}H_{22}O_3$	11.28	1.18	190.1569
$C_{10}H_6O_4$	11.06	1.35	190.0266
$C_{10}H_8NO_3$	11.43	1.20	190.0504
$C_{10}H_{10}N_2O_2$	11.81	1.04	190.0743
$C_{10}H_{12}N_3O$	12.18	0.88	190.0981
$C_{10}H_{14}N_4$	12.56	0.73	190.1220
$C_{11}H_{10}O_3$	12.16	1.28	190.0630
$C_{11}H_{12}NO_2$	12.54	1.12	190.0868
$C_{11}H_{14}N_2O$	12.91	0.97	190.1107
$C_{11}H_{16}N_3$	13.29	0.82	190.1346
$C_{11}N_3O$	13.07	0.99	190.0042
$C_{11}H_2N_4$	13.44	0.84	190.0280
$C_{12}H_{14}O_2$	13.27	1.21	190.0994
$C_{12}H_{16}NO$	13.64	1.06	190.1233
$C_{12}NO_2$	13.43	1.23	189.9929
$C_{12}H_{18}N_2$	14.02	0.91	190.1471
$C_{12}H_2N_2O$	13.80	1.08	190.0167
$C_{12}H_4N_3$	14.18	0.93	190.0406
$C_{13}H_{18}O$	14.37	1.16	190.1358
$C_{13}H_2O_2$	14.16	1.33	190.0054
$C_{13}H_{20}N$	14.75	1.01	190.1597
$C_{13}H_4NO$	14.53	1.18	190.0293
$C_{13}H_6N_2$	14.91	1.03	190.0532
$C_{14}H_{22}$	15.48	1.12	190.1722
$C_{14}H_6O$	15.26	1.28	190.0419
$C_{14}H_8N$	15.64	1.14	190.0657
$C_{15}H_{10}$	16.37	1.25	190.0783
191			
$C_7H_{15}N_2O_4$	8.72	1.14	191.1032

	M+1	M+2	FM
$C_7H_{17}N_3O_3$	9.10	0.97	191.1271
$C_7H_{19}N_4O_2$	9.47	0.81	191.1509
$C_8H_{17}NO_4$	9.45	1.20	191.1158
$C_8H_{19}N_2O_3$	9.83	1.04	191.1396
$C_8H_3N_2O_4$	9.61	1.22	191.0093
$C_8H_{21}N_3O_2$	10.20	0.87	191.1635
$C_8H_5N_3O_3$	9.99	1.05	191.0331
$C_8H_7N_4O_2$	10.36	0.89	191.0570
$C_9H_{19}O_4$	10.19	1.27	191.1284
$C_9H_{21}NO_3$	10.56	1.11	191.1522
$C_9H_5NO_4$	10.34	1.28	191.0218
$C_9H_7N_2O_3$	10.72	1.12	191.0457
$C_9H_9N_3O_2$	11.09	0.96	191.0695
$C_9H_{11}N_4O$	11.47	0.80	191.0934
$C_{10}H_7O_4$	11.07	1.36	191.0344
$C_{10}H_9NO_3$	11.45	1.20	191.0583
$C_{10}H_{11}N_2O_2$	11.82	1.04	191.0821
$C_{10}H_{13}N_3O$	12.20	0.88	191.1060
$C_{10}H_{15}N_4$	12.57	0.73	191.1298
$C_{11}H_{11}O_3$	12.18	1.28	191.0708
$C_{11}H_{13}NO_2$	12.55	1.12	191.0947
$C_{11}H_{15}N_2O$	12.93	0.97	191.1185
$C_{11}H_{17}N_3$	13.30	0.82	191.1424
$C_{11}HN_3O$	13.09	0.99	191.0120
$C_{11}H_3N_4$	13.46	0.84	191.0359
$C_{12}H_{15}O_2$	13.29	1.21	191.1072
$C_{12}H_{17}NO$	13.66	1.06	191.1311
$C_{12}HNO_2$	13.44	1.23	191.0007
$C_{12}H_{19}N_2$	14.03	0.91	191.1549
$C_{12}H_3N_2O$	13.82	1.08	191.0246
$C_{12}H_5N_3$	14.19	0.93	191.0484
$C_{13}H_{19}O$	14.39	1.16	191.1436
$C_{13}H_3O_2$	14.17	1.33	191.0133
$C_{13}H_{21}N$	14.77	1.01	191.1675
$C_{13}H_5NO$	14.55	1.18	191.0371
$C_{13}H_7N_2$	14.92	1.04	191.0610
$C_{14}H_{23}$	15.50	1.12	191.1801
$C_{14}H_7O$	15.28	1.29	191.0497
$C_{14}H_9N$	15.65	1.14	191.0736
$C_{15}H_{11}$	16.39	1.25	191.0861
192			
$C_7H_{16}N_2O_4$	8.74	1.14	192.1111
$C_7H_{18}N_3O_3$	9.11	0.97	192.1349
$C_7H_{20}N_4O_2$	9.49	0.81	192.1588
$C_8H_{18}NO_4$	9.47	1.20	192.1236
$C_8H_{20}N_2O_3$	9.85	1.04	192.1475
$C_8H_4N_2O_4$	9.63	1.22	192.0171
$C_8H_6N_3O_3$	10.00	1.05	192.0410
$C_8H_8N_4O_2$	10.38	0.89	192.0648
$C_9H_{20}O_4$	10.20	1.27	192.1362
$C_9H_6NO_4$	10.36	1.29	192.0297
$C_9H_8N_2O_3$	10.73	1.12	192.0535
$C_9H_{10}N_3O_2$	11.11	0.96	192.0774

	M+1	M+2	FM
$C_9H_{12}N_4O$	11.48	0.80	192.1012
$C_{10}H_8O_4$	11.09	1.36	192.0422
$C_{10}H_{10}NO_3$	11.46	1.20	192.0661
$C_{10}H_{12}N_2O_2$	11.84	1.04	192.0899
$C_{10}H_{14}N_3O$	12.21	0.89	192.1138
$C_{10}H_{16}N_4$	12.59	0.73	192.1377
$C_{10}N_4O$	12.37	0.90	192.0073
$C_{11}H_{12}O_3$	12.20	1.28	192.0786
$C_{11}H_{14}NO_2$	12.57	1.13	192.1025
$C_{11}H_{16}N_2O$	12.94	0.97	192.1264
$C_{11}N_2O_2$	12.73	1.15	191.9960
$C_{11}H_{18}N_3$	13.32	0.82	192.1502
$C_{11}H_2N_3O$	13.10	0.99	192.0198
$C_{11}H_4N_4$	13.48	0.84	192.0437
$C_{12}H_{16}O_2$	13.30	1.22	192.1151
$C_{12}O_3$	13.08	1.39	191.9847
$C_{12}H_{18}NO$	13.68	1.06	192.1389
$C_{12}H_2NO_2$	13.46	1.24	192.0085
$C_{12}H_{20}N_2$	14.05	0.92	192.1628
$C_{12}H_4N_2O$	13.83	1.09	192.0324
$C_{12}H_6N_3$	14.21	0.94	192.0563
$C_{13}H_{20}O$	14.41	1.16	192.1515
$C_{13}H_4O_2$	14.19	1.33	192.0211
$C_{13}H_{22}N$	14.78	1.02	192.1753
$C_{13}H_6NO$	14.56	1.18	192.0449
$C_{13}H_8N_2$	14.94	1.04	192.0688
$C_{14}H_{24}$	15.51	1.12	192.1879
$C_{14}H_8O$	15.30	1.29	192.0575
$C_{14}H_{10}N$	15.67	1.15	192.0814
$C_{15}H_{12}$	16.40	1.26	192.0939
C_{16}	17.29	1.40	192.0000
193			
$C_7H_{17}N_2O_4$	8.76	1.14	193.1189
$C_7H_{19}N_3O_3$	9.13	0.98	193.1427
$C_8H_{19}NO_4$	9.49	1.20	193.1315
$C_8H_5N_2O_4$	9.64	1.22	193.0249
$C_8H_7N_3O_3$	10.02	1.05	193.0488
$C_8H_9N_4O_2$	10.39	0.89	193.0726
$C_9H_7NO_4$	10.38	1.29	193.0375
$C_9H_9N_2O_3$	10.75	1.13	193.0614
$C_9H_{11}N_3O_2$	11.12	0.96	193.0852
$C_9H_{13}N_4O$	11.50	0.81	193.1091
$C_{10}H_9O_4$	11.11	1.36	193.0501
$C_{10}H_{11}NO_3$	11.48	1.20	193.0739
$C_{10}H_{13}N_2O_2$	11.86	1.04	193.0978
$C_{10}H_{15}N_3O$	12.23	0.89	193.1216
$C_{10}H_{17}N_4$	12.60	0.73	193.1455
$C_{10}HN_4O$	12.39	0.91	193.0151
$C_{11}H_{13}O_3$	12.21	1.28	193.0865
$C_{11}H_{15}NO_2$	12.59	1.13	193.1103
$C_{11}H_{17}N_2O$	12.96	0.97	193.1342
$C_{11}HN_2O_2$	12.74	1.15	193.0038
$C_{11}H_{19}N_3$	13.34	0.82	193.1580

	M+1	M+2	FM
$C_{11}H_3N_3O$	13.12	0.99	193.0277
$C_{11}H_5N_4$	13.49	0.84	193.0515
$C_{12}H_{17}O_2$	13.32	1.22	193.1229
$C_{12}HO_3$	13.10	1.39	192.9925
$C_{12}H_{19}NO$	13.69	1.07	193.1467
$C_{12}H_3NO_2$	13.47	1.24	193.0164
$C_{12}H_{21}N_2$	14.07	0.92	193.1706
$C_{12}H_5N_2O$	13.85	1.09	193.0402
$C_{12}H_7N_3$	14.22	0.94	193.0641
$C_{13}H_{21}O$	14.42	1.16	193.1593
$C_{13}H_5O_2$	14.21	1.33	193.0289
$C_{13}H_{23}N$	14.80	1.02	193.1832
$C_{13}H_7NO$	14.58	1.19	193.0528
$C_{13}H_9N_2$	14.95	1.04	193.0767
$C_{14}H_{25}$	15.53	1.12	193.1957
$C_{14}H_9O$	15.31	1.29	193.0653
$C_{14}H_{11}N$	15.69	1.15	193.0892
$C_{15}H_{13}$	16.42	1.26	193.1018
$C_{16}H$	17.31	1.40	193.0078
194			
$C_7H_{18}N_2O_4$	8.77	1.14	194.1267
$C_8H_6N_2O_4$	9.66	1.22	194.0328
$C_8H_8N_3O_3$	10.03	1.06	194.0566
$C_8H_{10}N_4O_2$	10.41	0.89	194.0805
$C_9H_8NO_4$	10.39	1.29	194.0453
$C_9H_{10}N_2O_3$	10.77	1.13	194.0692
$C_9H_{12}N_3O_2$	11.14	0.97	194.0930
$C_9H_{14}N_4O$	11.51	0.81	194.1169
$C_{10}H_{10}O_4$	11.12	1.36	194.0579
$C_{10}H_{12}NO_3$	11.50	1.20	194.0817
$C_{10}H_{14}N_2O_2$	11.87	1.05	194.1056
$C_{10}H_{16}N_3O$	12.25	0.89	194.1295
$C_{10}N_3O_2$	12.03	1.06	193.9991
$C_{10}H_{18}N_4$	12.62	0.73	194.1533
$C_{10}H_2N_4O$	12.40	0.91	194.0229
$C_{11}H_{14}O_3$	12.23	1.28	194.0943
$C_{11}H_{16}NO_2$	12.60	1.13	194.1182
$C_{11}NO_3$	12.39	1.30	193.9878
$C_{11}H_{18}N_2O$	12.98	0.98	194.1420
$C_{11}H_2N_2O_2$	12.76	1.15	194.0116
$C_{11}H_{20}N_3$	13.35	0.82	194.1659
$C_{11}H_4N_3O$	13.13	1.00	194.0355
$C_{11}H_6N_4$	13.51	0.85	194.0594
$C_{12}H_{18}O_2$	13.33	1.22	194.1307
$C_{12}H_2O_3$	13.12	1.39	194.0003
$C_{12}H_{20}NO$	13.71	1.07	194.1546
$C_{12}H_4NO_2$	13.49	1.24	194.0242
$C_{12}H_{22}N_2$	14.08	0.92	194.1784
$C_{12}H_6N_2O$	13.87	1.09	194.0480
$C_{12}H_8N_3$	14.24	0.94	194.0719
$C_{13}H_{22}O$	14.44	1.17	194.1671
$C_{13}H_6O_2$	14.22	1.34	194.0368
$C_{13}H_{24}N$	14.81	1.02	194.1910

	M+1	M+2	FM
$C_{13}H_8NO$	14.60	1.19	194.0606
$C_{13}H_{10}N_2$	14.97	1.04	194.0845
$C_{14}H_{26}$	15.54	1.13	194.2036
$C_{14}H_{10}O$	15.33	1.29	194.0732
$C_{14}H_{12}N$	15.70	1.15	194.0970
$C_{15}H_{14}$	16.43	1.26	194.1096
$C_{15}N$	16.59	1.29	194.0031
$C_{16}H_2$	17.32	1.41	194.0157
195			
$C_8H_7N_2O_4$	9.68	1.22	195.0406
$C_8H_9N_3O_3$	10.05	1.06	195.0644
$C_8H_{11}N_4O_2$	10.42	0.89	195.0883
$C_9H_9NO_4$	10.41	1.29	195.0532
$C_9H_{11}N_2O_3$	10.78	1.13	195.0770
$C_9H_{13}N_3O_2$	11.16	0.97	195.1009
$C_9H_{15}N_4O$	11.53	0.81	195.1247
$C_{10}H_{11}O_4$	11.14	1.36	195.0657
$C_{10}H_{13}NO_3$	11.51	1.21	195.0896
$C_{10}H_{15}N_2O_2$	11.89	1.05	195.1134
$C_{10}H_{17}N_3O$	12.26	0.89	195.1373
$C_{10}HN_3O_2$	12.04	1.07	195.0069
$C_{10}H_{19}N_4$	12.64	0.74	195.1611
$C_{10}H_3N_4O$	12.42	0.91	195.0308
$C_{11}H_{15}O_3$	12.24	1.29	195.1021
$C_{11}H_{17}NO_2$	12.62	1.13	195.1260
$C_{11}HNO_3$	12.40	1.31	194.9956
$C_{11}H_{19}N_2O$	12.99	0.98	195.1498
$C_{11}H_3N_2O_2$	12.78	1.15	195.0195
$C_{11}H_{21}N_3$	13.37	0.83	195.1737
$C_{11}H_5N_3O$	13.15	1.00	195.0433
$C_{11}H_7N_4$	13.52	0.85	195.0672
$C_{12}H_{19}O_2$	13.35	1.22	195.1385
$C_{12}H_3O_3$	13.13	1.39	195.0082
$C_{12}H_{21}NO$	13.72	1.07	195.1624
$C_{12}H_5NO_2$	13.51	1.24	195.0320
$C_{12}H_{23}N_2$	14.10	0.92	195.1863
$C_{12}H_7N_2O$	13.88	1.09	195.0559
$C_{12}H_9N_3$	14.26	0.94	195.0798
$C_{13}H_{23}O$	14.45	1.17	195.1750
$C_{13}H_7O_2$	14.24	1.34	195.0446
$C_{13}H_{25}N$	14.83	1.02	195.1988
$C_{13}H_9NO$	14.61	1.19	195.0684
$C_{13}H_{11}N_2$	14.99	1.05	195.0923
$C_{14}H_{27}$	15.56	1.13	195.2114
$C_{14}H_{11}O$	15.34	1.30	195.0810
$C_{14}H_{13}N$	15.72	1.15	195.1049
$C_{15}H_{15}$	16.45	1.27	195.1174
$C_{15}HN$	16.61	1.29	195.0109
$C_{16}H_3$	17.34	1.41	195.0235
196			
$C_8H_8N_2O_4$	9.69	1.22	196.0484
$C_8H_{10}N_3O_3$	10.07	1.06	196.0723
$C_8H_{12}N_4O_2$	10.44	0.90	196.0961

	M+1	M+2	FM
$C_9H_{10}NO_4$	10.42	1.29	196.0610
$C_9H_{12}N_2O_3$	10.80	1.13	196.0848
$C_9H_{14}N_3O_2$	11.17	0.97	196.1087
$C_9H_{16}N_4O$	11.55	0.81	196.1325
$C_9N_4O_2$	11.33	0.99	196.0022
$C_{10}H_{12}O_4$	11.15	1.37	196.0735
$C_{10}H_{14}NO_3$	11.53	1.21	196.0974
$C_{10}H_{16}N_2O_2$	11.90	1.05	196.1213
$C_{10}N_2O_3$	11.69	1.22	195.9909
$C_{10}H_{18}N_3O$	12.28	0.89	196.1451
$C_{10}H_2N_3O_2$	12.06	1.07	196.0147
$C_{10}H_{20}N_4$	12.65	0.74	196.1690
$C_{10}H_4N_4O$	12.43	0.91	196.0386
$C_{11}H_{16}O_3$	12.26	1.29	196.1100
$C_{11}O_4$	12.04	1.46	195.9796
$C_{11}H_{18}NO_2$	12.63	1.13	196.1338
$C_{11}H_2NO_3$	12.42	1.31	196.0034
$C_{11}H_{20}N_2O$	13.01	0.98	196.1577
$C_{11}H_4N_2O_2$	12.79	1.15	196.0273
$C_{11}H_{22}N_3$	13.38	0.83	196.1815
$C_{11}H_6N_3O$	13.17	1.00	196.0511
$C_{11}H_8N_4$	13.54	0.85	196.0750
$C_{12}H_{20}O_2$	13.37	1.22	196.1464
$C_{12}H_4O_3$	13.15	1.40	196.0160
$C_{12}H_{22}NO$	13.74	1.07	196.1702
$C_{12}H_6NO_2$	13.52	1.24	196.0399
$C_{12}H_{24}N_2$	14.11	0.92	196.1941
$C_{12}H_8N_2O$	13.90	1.09	196.0637
$C_{12}H_{10}N_3$	14.27	0.95	196.0876
$C_{13}H_{24}O$	14.47	1.17	196.1828
$C_{13}H_8O_2$	14.25	1.34	196.0524
$C_{13}H_{26}N$	14.85	1.03	196.2067
$C_{13}H_{10}NO$	14.63	1.19	196.0763
$C_{13}H_{12}N_2$	15.00	1.05	196.1001
$C_{14}H_{28}$	15.58	1.13	196.2192
$C_{14}H_{12}O$	15.36	1.30	196.0888
$C_{14}H_{14}N$	15.73	1.16	196.1127
$C_{14}N_2$	15.89	1.18	196.0062
$C_{15}H_{16}$	16.47	1.27	196.1253
$C_{15}O$	16.25	1.43	195.9949
$C_{15}H_2N$	16.62	1.29	196.0187
$C_{16}H_4$	17.35	1.41	196.0313
197			
$C_8H_9N_2O_4$	9.71	1.22	197.0563
$C_8H_{11}N_3O_3$	10.08	1.06	197.0801
$C_8H_{13}N_4O_2$	10.46	0.90	197.1040
$C_9H_{11}NO_4$	10.44	1.29	197.0688
$C_9H_{13}N_2O_3$	10.81	1.13	197.0927
$C_9H_{15}N_3O_2$	11.19	0.97	197.1165
$C_9H_{17}N_4O$	11.56	0.81	197.1404
$C_9HN_4O_2$	11.35	0.99	197.0100
$C_{10}H_{13}O_4$	11.17	1.37	197.0814
$C_{10}H_{15}NO_3$	11.54	1.21	197.1052
$C_{10}H_{17}N_2O_2$	11.92	1.05	197.1291
$C_{10}HN_2O_3$	11.70	1.23	196.9987
$C_{10}H_{19}N_3O$	12.29	0.90	197.1529
$C_{10}H_3N_3O_2$	12.08	1.07	197.0226
$C_{10}H_{21}N_4$	12.67	0.74	197.1768
$C_{10}H_5N_4O$	12.45	0.91	197.0464
$C_{11}H_{17}O_3$	12.28	1.29	197.1178
$C_{11}HO_4$	12.06	1.46	196.9874
$C_{11}H_{19}NO_2$	12.65	1.14	197.1416
$C_{11}H_3NO_3$	12.43	1.31	197.0113
$C_{11}H_{21}N_2O$	13.02	0.98	197.1655
$C_{11}H_5N_2O_2$	12.81	1.16	197.0351
$C_{11}H_{23}N_3$	13.40	0.83	197.1894
$C_{11}H_7N_3O$	13.18	1.00	197.0590
$C_{11}H_9N_4$	13.56	0.85	197.0829
$C_{12}H_{21}O_2$	13.38	1.23	197.1542
$C_{12}H_5O_3$	13.16	1.40	197.0238
$C_{12}H_{23}NO$	13.76	1.08	197.1781
$C_{12}H_7NO_2$	13.54	1.25	197.0477
$C_{12}H_{25}N_2$	14.13	0.93	197.2019
$C_{12}H_9N_2O$	13.91	1.10	197.0715
$C_{12}H_{11}N_3$	14.29	0.95	197.0954
$C_{13}H_{25}O$	14.49	1.17	197.1906
$C_{13}H_9O_2$	14.27	1.34	197.0603
$C_{13}H_{27}N$	14.86	1.03	197.2145
$C_{13}H_{11}NO$	14.64	1.20	197.0841
$C_{13}H_{13}N_2$	15.02	1.05	197.1080
$C_{14}H_{29}$	15.59	1.13	197.2270
$C_{14}H_{13}O$	15.38	1.30	197.0967
$C_{14}H_{15}N$	15.75	1.16	197.1205
$C_{14}HN_2$	15.91	1.18	197.0140
$C_{15}H_{17}$	16.48	1.27	197.1331
$C_{15}HO$	16.26	1.44	197.0027
$C_{15}H_3N$	16.64	1.30	197.0266
$C_{16}H_5$	17.37	1.42	197.0391
198			
$C_8H_{10}N_2O_4$	9.72	1.23	198.0641
$C_8H_{12}N_3O_3$	10.10	1.06	198.0879
$C_8H_{14}N_4O_2$	10.47	0.90	198.1118
$C_9H_{12}NO_4$	10.46	1.30	198.0766
$C_9H_{14}N_2O_3$	10.83	1.13	198.1005
$C_9H_{16}N_3O_2$	11.20	0.97	198.1244
$C_9N_3O_3$	10.99	1.15	197.9940
$C_9H_{18}N_4O$	11.58	0.82	198.1482
$C_9H_2N_4O_2$	11.36	0.99	198.0178
$C_{10}H_{14}O_4$	11.19	1.37	198.0892
$C_{10}H_{16}NO_3$	11.56	1.21	198.1131
$C_{10}NO_4$	11.34	1.39	197.9827
$C_{10}H_{18}N_2O_2$	11.94	1.05	198.1369
$C_{10}H_2N_2O_3$	11.72	1.23	198.0065
$C_{10}H_{20}N_3O$	12.31	0.90	198.1608
$C_{10}H_4N_3O_2$	12.09	1.07	198.0304
$C_{10}H_{22}N_4$	12.68	0.74	198.1846
$C_{10}H_6N_4O$	12.47	0.92	198.0542
$C_{11}H_{18}O_3$	12.29	1.29	198.1256
$C_{11}H_2O_4$	12.07	1.47	197.9953
$C_{11}H_{20}NO_2$	12.67	1.14	198.1495
$C_{11}H_4NO_3$	12.45	1.31	198.0191
$C_{11}H_{22}N_2O$	13.04	0.99	198.1733
$C_{11}H_6N_2O_2$	12.82	1.16	198.0429
$C_{11}H_{24}N_3$	13.42	0.83	198.1972
$C_{11}H_8N_3O$	13.20	1.01	198.0668
$C_{11}H_{10}N_4$	13.57	0.85	198.0907
$C_{12}H_{22}O_2$	13.40	1.23	198.1620
$C_{12}H_6O_3$	13.18	1.40	198.0317
$C_{12}H_{24}NO$	13.77	1.08	198.1859
$C_{12}H_8NO_2$	13.55	1.25	198.0555
$C_{12}H_{26}N_2$	14.15	0.93	198.2098
$C_{12}H_{10}N_2O$	13.93	1.10	198.0794
$C_{12}H_{12}N_3$	14.30	0.95	198.1032
$C_{13}H_{26}O$	14.50	1.18	198.1985
$C_{13}H_{10}O_2$	14.29	1.35	198.0681
$C_{13}H_{28}N$	14.88	1.03	198.2223
$C_{13}H_{12}NO$	14.66	1.20	198.0919
$C_{13}H_{14}N_2$	15.03	1.05	198.1158
$C_{13}N_3$	15.19	1.08	198.0093
$C_{14}H_{30}$	15.61	1.14	198.2349
$C_{14}H_{14}O$	15.39	1.30	198.1045
$C_{14}H_{16}N$	15.77	1.16	198.1284
$C_{14}NO$	15.55	1.33	197.9980
$C_{14}H_2N_2$	15.92	1.18	198.0218
$C_{15}H_{18}$	16.50	1.27	198.1409
$C_{15}H_2O$	16.28	1.44	198.0106
$C_{15}H_4N$	16.65	1.30	198.0344
$C_{16}H_6$	17.39	1.42	198.0470
199			
$C_8H_{11}N_2O_4$	9.74	1.23	199.0719
$C_8H_{13}N_3O_3$	10.11	1.06	199.0958
$C_8H_{15}N_4O_2$	10.49	0.90	199.1196
$C_9H_{13}NO_4$	10.47	1.30	199.0845
$C_9H_{15}N_2O_3$	10.85	1.14	199.1083
$C_9H_{17}N_3O_2$	11.22	0.98	199.1322
$C_9HN_3O_3$	11.00	1.15	199.0018
$C_9H_{19}N_4O$	11.59	0.82	199.1560
$C_9H_3N_4O_2$	11.38	0.99	199.0257
$C_{10}H_{15}O_4$	11.20	1.37	199.0970
$C_{10}H_{17}NO_3$	11.58	1.21	199.1209
$C_{10}HNO_4$	11.36	1.39	198.9905
$C_{10}H_{19}N_2O_2$	11.95	1.06	199.1447
$C_{10}H_3N_2O_3$	11.73	1.23	199.0144
$C_{10}H_{21}N_3O$	12.33	0.90	199.1686
$C_{10}H_5N_3O_2$	12.11	1.07	199.0382
$C_{10}H_{23}N_4$	12.70	0.75	199.1925
$C_{10}H_7N_4O$	12.48	0.92	199.0621
$C_{11}H_{19}O_3$	12.31	1.29	199.1334
$C_{11}H_3O_4$	12.09	1.47	199.0031

	M+1	M+2	FM
$C_{11}H_{21}NO_2$	12.68	1.14	199.1573
$C_{11}H_5NO_3$	12.47	1.31	199.0269
$C_{11}H_{23}N_2O$	13.06	0.99	199.1811
$C_{11}H_7N_2O_2$	12.84	1.16	199.0508
$C_{11}H_{25}N_3$	13.43	0.84	199.2050
$C_{11}H_9N_3O$	13.21	1.01	199.0746
$C_{11}H_{11}N_4$	13.59	0.86	199.0985
$C_{12}H_{23}O_2$	13.41	1.23	199.1699
$C_{12}H_7O_3$	13.20	1.40	199.0395
$C_{12}H_{25}NO$	13.79	1.08	199.1937
$C_{12}H_9NO_2$	13.57	1.25	199.0634
$C_{12}H_{27}N_2$	14.16	0.93	199.2176
$C_{12}H_{11}N_2O$	13.95	1.10	199.0872
$C_{12}H_{13}N_3$	14.32	0.95	199.1111
$C_{13}H_{27}O$	14.52	1.18	199.2063
$C_{13}H_{11}O_2$	14.30	1.35	199.0759
$C_{13}H_{29}N$	14.89	1.03	199.2301
$C_{13}H_{13}NO$	14.68	1.20	199.0998
$C_{13}H_{15}N_2$	15.05	1.06	199.1236
$C_{13}HN_3$	15.21	1.08	199.0171
$C_{14}H_{15}O$	15.41	1.31	199.1123
$C_{14}H_{17}N$	15.78	1.16	199.1362
$C_{14}HNO$	15.56	1.33	199.0058
$C_{14}H_3N_2$	15.94	1.19	199.0297
$C_{15}H_{19}$	16.51	1.28	199.1488
$C_{15}H_3O$	16.30	1.44	199.0184
$C_{15}H_5N$	16.67	1.30	199.0422
$C_{16}H_7$	17.40	1.42	199.0548
200			
$C_8H_{12}N_2O_4$	9.76	1.23	200.0797
$C_8H_{14}N_3O_3$	10.13	1.07	200.1036
$C_8H_{16}N_4O_2$	10.50	0.90	200.1275
$C_9H_{14}NO_4$	10.49	1.30	200.0923
$C_9H_{16}N_2O_3$	10.86	1.14	200.1162
$C_9N_2O_4$	10.64	1.31	199.9858
$C_9H_{18}N_3O_2$	11.24	0.98	200.1400
$C_9H_2N_3O_3$	11.02	1.15	200.0096
$C_9H_{20}N_4O$	11.61	0.82	200.1639
$C_9H_4N_4O_2$	11.39	0.99	200.0335
$C_{10}H_{16}O_4$	11.22	1.37	200.1049
$C_{10}H_{18}NO_3$	11.59	1.21	200.1287
$C_{10}H_2NO_4$	11.38	1.39	199.9983
$C_{10}H_{20}N_2O_2$	11.97	1.06	200.1526
$C_{10}H_4N_2O_3$	11.75	1.23	200.0222
$C_{10}H_{22}N_3O$	12.34	0.90	200.1764
$C_{10}H_6N_3O_2$	12.12	1.08	200.0460
$C_{10}H_{24}N_4$	12.72	0.75	200.2003
$C_{10}H_8N_4O$	12.50	0.92	200.0699
$C_{11}H_{20}O_3$	12.32	1.30	200.1413
$C_{11}H_4O_4$	12.11	1.47	200.0109
$C_{11}H_{22}NO_2$	12.70	1.14	200.1651
$C_{11}H_6NO_3$	12.48	1.32	200.0348
$C_{11}H_{24}N_2O$	13.07	0.99	200.1890

	M+1	M+2	FM
$C_{11}H_8N_2O_2$	12.86	1.16	200.0586
$C_{11}H_{26}N_3$	13.45	0.84	200.2129
$C_{11}H_{10}N_3O$	13.23	1.01	200.0825
$C_{11}H_{12}N_4$	13.60	0.86	200.1063
$C_{12}H_{24}O_2$	13.43	1.23	200.1777
$C_{12}H_8O_3$	13.21	1.40	200.0473
$C_{12}H_{26}NO$	13.80	1.08	200.2015
$C_{12}H_{10}NO_2$	13.59	1.25	200.0712
$C_{12}H_{28}N_2$	14.18	0.93	200.2254
$C_{12}H_{12}N_2O$	13.96	1.10	200.0950
$C_{12}H_{14}N_3$	14.34	0.96	200.1189
$C_{12}N_4$	14.49	0.98	200.0124
$C_{13}H_{28}O$	14.53	1.18	200.2141
$C_{13}H_{12}O_2$	14.32	1.35	200.0837
$C_{13}H_{14}NO$	14.69	1.20	200.1076
$C_{13}H_{16}N_2$	15.07	1.06	200.1315
$C_{13}N_2O$	14.85	1.23	200.0011
$C_{13}H_2N_3$	15.22	1.08	200.0249
$C_{14}H_{16}O$	15.42	1.31	200.1202
$C_{14}O_2$	15.21	1.48	199.9898
$C_{14}H_{18}N$	15.80	1.17	200.1440
$C_{14}H_2NO$	15.58	1.33	200.0136
$C_{14}H_4N_2$	15.96	1.19	200.0375
$C_{15}H_{20}$	16.53	1.28	200.1566
$C_{15}H_4O$	16.31	1.44	200.0262
$C_{15}H_6N$	16.69	1.30	200.0501
$C_{16}H_8$	17.42	1.42	200.0626
201			
$C_8H_{13}N_2O_4$	9.77	1.23	201.0876
$C_8H_{15}N_3O_3$	10.15	1.07	201.1114
$C_8H_{17}N_4O_2$	10.52	0.90	201.1353
$C_9H_{15}NO_4$	10.50	1.30	201.1001
$C_9H_{17}N_2O_3$	10.88	1.14	201.1240
$C_9HN_2O_4$	10.66	1.32	200.9936
$C_9H_{19}N_3O_2$	11.25	0.98	201.1478
$C_9H_3N_3O_3$	11.04	1.16	201.0175
$C_9H_{21}N_4O$	11.63	0.82	201.1717
$C_9H_5N_4O_2$	11.41	1.00	201.0413
$C_{10}H_{17}O_4$	11.23	1.37	201.1127
$C_{10}H_{19}NO_3$	11.61	1.22	201.1365
$C_{10}H_3NO_4$	11.39	1.39	201.0062
$C_{10}H_{21}N_2O_2$	11.98	1.06	201.1604
$C_{10}H_5N_2O_3$	11.77	1.23	201.0300
$C_{10}H_{23}N_3O$	12.36	0.90	201.1842
$C_{10}H_7N_3O_2$	12.14	1.08	201.0539
$C_{10}H_{25}N_4$	12.73	0.75	201.2081
$C_{10}H_9N_4O$	12.51	0.92	201.0777
$C_{11}H_{21}O_3$	12.34	1.30	201.1491
$C_{11}H_5O_4$	12.12	1.47	201.0187
$C_{11}H_{23}NO_2$	12.71	1.14	201.1730
$C_{11}H_7NO_3$	12.50	1.32	201.0426
$C_{11}H_{25}N_2O$	13.09	0.99	201.1968
$C_{11}H_9N_2O_2$	12.87	1.16	201.0664

	M+1	M+2	FM
$C_{11}H_{27}N_3$	13.46	0.84	201.2207
$C_{11}H_{11}N_3O$	13.25	1.01	201.0903
$C_{11}H_{13}N_4$	13.62	0.86	201.1142
$C_{12}H_{25}O_2$	13.45	1.23	201.1855
$C_{12}H_9O_3$	13.23	1.41	201.0552
$C_{12}H_{27}NO$	13.82	1.08	201.2094
$C_{12}H_{11}NO_2$	13.60	1.26	201.0790
$C_{12}H_{13}N_2O$	13.98	1.11	201.1029
$C_{12}H_{15}N_3$	14.35	0.96	201.1267
$C_{12}HN_4$	14.51	0.98	201.0202
$C_{13}H_{13}O_2$	14.33	1.35	201.0916
$C_{13}H_{15}NO$	14.71	1.21	201.1154
$C_{13}H_{17}N_2$	15.08	1.06	201.1393
$C_{13}HN_2O$	14.87	1.23	201.0089
$C_{13}H_3N_3$	15.24	1.08	201.0328
$C_{14}H_{17}O$	15.44	1.31	201.1280
$C_{14}HO_2$	15.22	1.48	200.9976
$C_{14}H_{19}N$	15.81	1.17	201.1519
$C_{14}H_3NO$	15.60	1.33	201.0215
$C_{14}H_5N_2$	15.97	1.19	201.0453
$C_{15}H_{21}$	16.55	1.28	201.1644
$C_{15}H_5O$	16.33	1.45	201.0340
$C_{15}H_7N$	16.70	1.31	201.0579
$C_{16}H_9$	17.43	1.43	201.0705
202			
$C_8H_{14}N_2O_4$	9.79	1.23	202.0954
$C_8H_{16}N_3O_3$	10.16	1.07	202.1193
$C_8H_{18}N_4O_2$	10.54	0.91	202.1431
$C_9H_{16}NO_4$	10.52	1.30	202.1080
$C_9H_{18}N_2O_3$	10.89	1.14	202.1318
$C_9H_2N_2O_4$	10.68	1.32	202.0014
$C_9H_{20}N_3O_2$	11.27	0.98	202.1557
$C_9H_4N_3O_3$	11.05	1.16	202.0253
$C_9H_{22}N_4O$	11.64	0.82	202.1795
$C_9H_6N_4O_2$	11.43	1.00	202.0491
$C_{10}H_{18}O_4$	11.25	1.38	202.1205
$C_{10}H_{20}NO_3$	11.62	1.22	202.1444
$C_{10}H_4NO_4$	11.41	1.39	202.0140
$C_{10}H_{22}N_2O_2$	12.00	1.06	202.1682
$C_{10}H_6N_2O_3$	11.78	1.24	202.0379
$C_{10}H_{24}N_3O$	12.37	0.91	202.1921
$C_{10}H_8N_3O_2$	12.16	1.08	202.0617
$C_{10}H_{26}N_4$	12.75	0.75	202.2160
$C_{10}H_{10}N_4O$	12.53	0.92	202.0856
$C_{11}H_{22}O_3$	12.35	1.30	202.1569
$C_{11}H_6O_4$	12.14	1.47	202.0266
$C_{11}H_{24}NO_2$	12.73	1.15	202.1808
$C_{11}H_8NO_3$	12.51	1.32	202.0504
$C_{11}H_{26}N_2O$	13.10	0.99	202.2046
$C_{11}H_{10}N_2O_2$	12.89	1.17	202.0743
$C_{11}H_{12}N_3O$	13.26	1.01	202.0981
$C_{11}H_{14}N_4$	13.64	0.86	202.1220
$C_{12}H_{26}O_2$	13.46	1.24	202.1934
$C_{12}H_{10}O_3$	13.24	1.41	202.0630

	M+1	M+2	FM
$C_{12}H_{12}NO_2$	13.62	1.26	202.0868
$C_{12}H_{14}N_2O$	13.99	1.11	202.1107
$C_{12}H_{16}N_3$	14.37	0.96	202.1346
$C_{12}N_3O$	14.15	1.13	202.0042
$C_{12}H_2N_4$	14.53	0.98	202.0280
$C_{13}H_{14}O_2$	14.35	1.35	202.0994
$C_{13}H_{16}NO$	14.72	1.21	202.1233
$C_{13}NO_2$	14.51	1.38	201.9929
$C_{13}H_{18}N_2$	15.10	1.06	202.1471
$C_{13}H_2N_2O$	14.88	1.23	202.0167
$C_{13}H_4N_3$	15.26	1.09	202.0406
$C_{14}H_{18}O$	15.46	1.31	202.1358
$C_{14}H_2O_2$	15.24	1.48	202.0054
$C_{14}H_{20}N$	15.83	1.17	202.1597
$C_{14}H_4NO$	15.61	1.34	202.0293
$C_{14}H_6N_2$	15.99	1.19	202.0532
$C_{15}H_{22}$	16.56	1.28	202.1722
$C_{15}H_6O$	16.34	1.45	202.0419
$C_{15}H_8N$	16.72	1.31	202.0657
$C_{16}H_{10}$	17.45	1.43	202.0783
203			
$C_8H_{15}N_2O_4$	9.80	1.23	203.1032
$C_8H_{17}N_3O_3$	10.18	1.07	203.1271
$C_8H_{19}N_4O_2$	10.55	0.91	203.1509
$C_9H_{17}NO_4$	10.54	1.30	203.1158
$C_9H_{19}N_2O_3$	10.91	1.14	203.1396
$C_9H_3N_2O_4$	10.69	1.32	203.0093
$C_9H_{21}N_3O_2$	11.28	0.98	203.1635
$C_9H_5N_3O_3$	11.07	1.16	203.0331
$C_9H_{23}N_4O$	11.66	0.82	203.1873
$C_9H_7N_4O_2$	11.44	1.00	203.0570
$C_{10}H_{19}O_4$	11.27	1.38	203.1284
$C_{10}H_{21}NO_3$	11.64	1.22	203.1522
$C_{10}H_5NO_4$	11.42	1.40	203.0218
$C_{10}H_{23}N_2O_2$	12.02	1.06	203.1761
$C_{10}H_7N_2O_3$	11.80	1.24	203.0457
$C_{10}H_{25}N_3O$	12.39	0.91	203.1999
$C_{10}H_9N_3O_2$	12.17	1.08	203.0695
$C_{10}H_{11}N_4O$	12.55	0.93	203.0934
$C_{11}H_{23}O_3$	12.37	1.30	203.1648
$C_{11}H_7O_4$	12.15	1.48	203.0344
$C_{11}H_{25}NO_2$	12.75	1.15	203.1886
$C_{11}H_9NO_3$	12.53	1.32	203.0583
$C_{11}H_{11}N_2O_2$	12.90	1.17	203.0821
$C_{11}H_{13}N_3O$	13.28	1.02	203.1060
$C_{11}H_{15}N_4$	13.65	0.86	203.1298
$C_{12}H_{11}O_3$	13.26	1.41	203.0708
$C_{12}H_{13}NO_2$	13.63	1.26	203.0947
$C_{12}H_{15}N_2O$	14.01	1.11	203.1185
$C_{12}H_{17}N_3$	14.38	0.96	203.1424
$C_{12}HN_3O$	14.17	1.13	203.0120
$C_{12}H_3N_4$	14.54	0.98	203.0359
$C_{13}H_{15}O_2$	14.37	1.36	203.1072

	M+1	M+2	FM
$C_{13}H_{17}NO$	14.74	1.21	203.1311
$C_{13}HNO_2$	14.52	1.38	203.0007
$C_{13}H_{19}N_2$	15.11	1.06	203.1549
$C_{13}H_3N_2O$	14.90	1.23	203.0246
$C_{13}H_5N_3$	15.27	1.09	203.0484
$C_{14}H_{19}O$	15.47	1.32	203.1436
$C_{14}H_3O_2$	15.25	1.48	203.0133
$C_{14}H_{21}N$	15.85	1.17	203.1675
$C_{14}H_5NO$	15.63	1.34	203.0371
$C_{14}H_7N_2$	16.00	1.20	203.0610
$C_{15}H_{23}$	16.58	1.29	203.1801
$C_{15}H_7O$	16.36	1.45	203.0497
$C_{15}H_9N$	16.73	1.31	203.0736
$C_{16}H_{11}$	17.47	1.43	203.0861
204			
$C_8H_{16}N_2O_4$	9.82	1.24	204.1111
$C_8H_{18}N_3O_3$	10.19	1.07	204.1349
$C_8H_{20}N_4O_2$	10.57	0.91	204.1588
$C_9H_{18}NO_4$	10.55	1.31	204.1236
$C_9H_{20}N_2O_3$	10.93	1.14	204.1475
$C_9H_4N_2O_4$	10.71	1.32	204.0171
$C_9H_{22}N_3O_2$	11.30	0.98	204.1713
$C_9H_6N_3O_3$	11.08	1.16	204.0410
$C_9H_{24}N_4O$	11.67	0.83	204.1952
$C_9H_8N_4O_2$	11.46	1.00	204.0648
$C_{10}H_{20}O_4$	11.28	1.38	204.1362
$C_{10}H_{22}NO_3$	11.66	1.22	204.1600
$C_{10}H_6NO_4$	11.44	1.40	204.0297
$C_{10}H_{24}N_2O_2$	12.03	1.06	204.1839
$C_{10}H_8N_2O_3$	11.81	1.24	204.0535
$C_{10}H_{10}N_3O_2$	12.19	1.08	204.0774
$C_{10}H_{12}N_4O$	12.56	0.93	204.1012
$C_{11}H_{24}O_3$	12.39	1.30	204.1726
$C_{11}H_8O_4$	12.17	1.48	204.0422
$C_{11}H_{10}NO_3$	12.55	1.32	204.0661
$C_{11}H_{12}N_2O_2$	12.92	1.17	204.0899
$C_{11}H_{14}N_3O$	13.29	1.02	204.1138
$C_{11}H_{16}N_4$	13.67	0.87	204.1377
$C_{11}N_4O$	13.45	1.04	204.0073
$C_{12}H_{12}O_3$	13.28	1.41	204.0786
$C_{12}H_{14}NO_2$	13.65	1.26	204.1025
$C_{12}H_{18}N_2O$	14.03	1.11	204.1264
$C_{12}N_2O_2$	13.81	1.28	203.9960
$C_{12}H_{18}N_3$	14.40	0.96	204.1502
$C_{12}H_2N_3O$	14.18	1.13	204.0198
$C_{12}H_4N_4$	14.56	0.99	204.0437
$C_{13}H_{16}O_2$	14.38	1.36	204.1151
$C_{13}O_3$	14.17	1.53	203.9847
$C_{13}H_{18}NO$	14.76	1.21	204.1389
$C_{13}H_2NO_2$	14.54	1.38	204.0085
$C_{13}H_{20}N_2$	15.13	1.07	204.1628
$C_{13}H_4N_2O$	14.91	1.24	204.0324
$C_{13}H_6N_3$	15.29	1.09	204.0563
$C_{14}H_{20}O$	15.49	1.32	204.1515

	M+1	M+2	FM
$C_{14}H_4O_2$	15.27	1.49	204.0211
$C_{14}H_{22}N$	15.86	1.18	204.1753
$C_{14}H_6NO$	15.64	1.34	204.0449
$C_{14}H_8N_2$	16.02	1.20	204.0688
$C_{15}H_{24}$	16.59	1.29	204.1879
$C_{15}H_8O$	16.38	1.45	204.0575
$C_{15}H_{10}N$	16.75	1.31	204.0814
$C_{16}H_{12}$	17.48	1.43	204.0939
C_{17}	18.37	1.59	204.0000
205			
$C_8H_{17}N_2O_4$	9.84	1.24	205.1189
$C_8H_{19}N_3O_3$	10.21	1.07	205.1427
$C_8H_{21}N_4O_2$	10.58	0.91	205.1666
$C_9H_{19}NO_4$	10.57	1.31	205.1315
$C_9H_{21}N_2O_3$	10.94	1.15	205.1553
$C_9H_5N_2O_4$	10.72	1.32	205.0249
$C_9H_{23}N_3O_2$	11.32	0.99	205.1791
$C_9H_7N_3O_3$	11.10	1.16	205.0488
$C_9H_9N_4O_2$	11.47	1.00	205.0726
$C_{10}H_{21}O_4$	11.30	1.38	205.1440
$C_{10}H_{23}NO_3$	11.67	1.22	205.1679
$C_{10}H_7NO_4$	11.46	1.40	205.0375
$C_{10}H_9N_2O_3$	11.83	1.24	205.0614
$C_{10}H_{11}N_3O_2$	12.20	1.09	205.0852
$C_{10}H_{13}N_4O$	12.58	0.93	205.1091
$C_{11}H_9O_4$	12.19	1.48	205.0501
$C_{11}H_{11}NO_3$	12.56	1.33	205.0739
$C_{11}H_{13}N_2O_2$	12.94	1.17	205.0978
$C_{11}H_{15}N_3O$	13.31	1.02	205.1216
$C_{11}H_{17}N_4$	13.68	0.87	205.1455
$C_{11}HN_4O$	13.47	1.04	205.0151
$C_{12}H_{13}O_3$	13.29	1.41	205.0865
$C_{12}H_{15}NO_2$	13.67	1.26	205.1103
$C_{12}H_{17}N_2O$	14.04	1.11	205.1342
$C_{12}HN_2O_2$	13.82	1.28	205.0038
$C_{12}H_{19}N_3$	14.42	0.97	205.1580
$C_{12}H_3N_3O$	14.20	1.14	205.0277
$C_{12}H_5N_4$	14.57	0.99	205.0515
$C_{13}H_{17}O_2$	14.40	1.36	205.1229
$C_{13}HO_3$	14.18	1.53	204.9925
$C_{13}H_{19}NO$	14.77	1.21	205.1467
$C_{13}H_3NO_2$	14.56	1.38	205.0164
$C_{13}H_{21}N_2$	15.15	1.07	205.1706
$C_{13}H_5N_2O$	14.93	1.24	205.0402
$C_{13}H_7N_3$	15.30	1.09	205.0641
$C_{14}H_{21}O$	15.50	1.32	205.1593
$C_{14}H_5O_2$	15.29	1.49	205.0289
$C_{14}H_{23}N$	15.88	1.18	205.1832
$C_{14}H_7NO$	15.66	1.34	205.0528
$C_{14}H_9N_2$	16.04	1.20	205.0767
$C_{15}H_{25}$	16.61	1.29	205.1957
$C_{15}H_9O$	16.39	1.46	205.0653
$C_{15}H_{11}N$	16.77	1.32	205.0892
$C_{16}H_{13}$	17.50	1.44	205.1018

	M+1	M+2	FM
$C_{17}H$	18.39	1.59	205.0078
206			
$C_8H_{18}N_2O_4$	9.85	1.24	206.1267
$C_8H_{20}N_3O_3$	10.23	1.08	206.1506
$C_8H_{22}N_4O_2$	10.60	0.91	206.1744
$C_9H_{20}NO_4$	10.58	1.31	206.1393
$C_9H_{22}N_2O_3$	10.96	1.15	206.1631
$C_9H_6N_2O_4$	10.74	1.32	206.0328
$C_9H_8N_3O_3$	11.12	1.16	206.0566
$C_9H_{10}N_4O_2$	11.49	1.01	206.0805
$C_{10}H_{22}O_4$	11.31	1.38	206.1518
$C_{10}H_8NO_4$	11.47	1.40	206.0453
$C_{10}H_{10}N_2O_3$	11.85	1.24	206.0692
$C_{10}H_{12}N_3O_2$	12.22	1.09	206.0930
$C_{10}H_{14}N_4O$	12.59	0.93	206.1169
$C_{11}H_{10}O_4$	12.20	1.48	206.0579
$C_{11}H_{12}NO_3$	12.58	1.33	206.0817
$C_{11}H_{14}N_2O_2$	12.95	1.17	206.1056
$C_{11}H_{16}N_3O$	13.33	1.02	206.1295
$C_{11}N_3O_2$	13.11	1.19	205.9991
$C_{11}H_{18}N_4$	13.70	0.87	206.1533
$C_{11}H_2N_4O$	13.48	1.04	206.0229
$C_{12}H_{14}O_3$	13.31	1.42	206.0943
$C_{12}H_{16}NO_2$	13.68	1.27	206.1182
$C_{12}NO_3$	13.47	1.44	205.9878
$C_{12}H_{18}N_2O$	14.06	1.12	206.1420
$C_{12}H_2N_2O_2$	13.84	1.29	206.0116
$C_{12}H_{20}N_3$	14.43	0.97	206.1659
$C_{12}H_4N_3O$	14.21	1.14	206.0355
$C_{12}H_6N_4$	14.59	0.99	206.0594
$C_{13}H_{18}O_2$	14.41	1.36	206.1307
$C_{13}H_2O_3$	14.20	1.53	206.0003
$C_{13}H_{20}NO$	14.79	1.22	206.1546
$C_{13}H_4NO_2$	14.57	1.39	206.0242
$C_{13}H_{22}N_2$	15.16	1.07	206.1784
$C_{13}H_6N_2O$	14.95	1.24	206.0480
$C_{13}H_8N_3$	15.32	1.10	206.0719
$C_{14}H_{22}O$	15.52	1.32	206.1671
$C_{14}H_6O_2$	15.30	1.49	206.0368
$C_{14}H_{24}N$	15.89	1.18	206.1910
$C_{14}H_8NO$	15.68	1.35	206.0606
$C_{14}H_{10}N_2$	16.05	1.21	206.0845
$C_{15}H_{26}$	16.63	1.29	206.2036
$C_{15}H_{10}O$	16.41	1.46	206.0732
$C_{15}H_{12}N$	16.78	1.32	206.0970
$C_{16}H_{14}$	17.51	1.44	206.1096
$C_{16}N$	17.67	1.47	206.0031
$C_{17}H_2$	18.40	1.59	206.0157
207			
$C_8H_{19}N_2O_4$	9.87	1.24	207.1345
$C_8H_{21}N_3O_3$	10.24	1.08	207.1584
$C_9H_{21}NO_4$	10.60	1.31	207.1471
$C_9H_7N_2O_4$	10.76	1.33	207.0406
$C_9H_9N_3O_3$	11.13	1.17	207.0644
$C_9H_{11}N_4O_2$	11.51	1.01	207.0883
$C_{10}H_9NO_4$	11.49	1.40	207.0532
$C_{10}H_{11}N_2O_3$	11.86	1.25	207.0770
$C_{10}H_{13}N_3O_2$	12.24	1.09	207.1009
$C_{10}H_{15}N_4O$	12.61	0.93	207.1247
$C_{11}H_{11}O_4$	12.22	1.48	207.0657
$C_{11}H_{13}NO_3$	12.59	1.33	207.0896
$C_{11}H_{15}N_2O_2$	12.97	1.18	207.1134
$C_{11}H_{17}N_3O$	13.34	1.02	207.1373
$C_{11}HN_3O_2$	13.13	1.20	207.0069
$C_{11}H_{19}N_4$	13.72	0.87	207.1611
$C_{11}H_3N_4O$	13.50	1.04	207.0308
$C_{12}H_{15}O_3$	13.32	1.42	207.1021
$C_{12}H_{17}NO_2$	13.70	1.27	207.1260
$C_{12}HNO_3$	13.48	1.44	206.9956
$C_{12}H_{19}N_2O$	14.07	1.12	207.1498
$C_{12}H_3N_2O_2$	13.86	1.29	207.0195
$C_{12}H_{21}N_3$	14.45	0.97	207.1737
$C_{12}H_5N_3O$	14.23	1.14	207.0433
$C_{12}H_7N_4$	14.61	0.99	207.0672
$C_{13}H_{19}O_2$	14.43	1.37	207.1385
$C_{13}H_3O_3$	14.21	1.54	207.0082
$C_{13}H_{21}NO$	14.80	1.22	207.1624
$C_{13}H_5NO_2$	14.59	1.39	207.0320
$C_{13}H_{23}N_2$	15.18	1.07	207.1863
$C_{13}H_7N_2O$	14.96	1.24	207.0559
$C_{13}H_9N_3$	15.34	1.10	207.0798
$C_{14}H_{23}O$	15.54	1.33	207.1750
$C_{14}H_7O_2$	15.32	1.49	207.0446
$C_{14}H_{25}N$	15.91	1.18	207.1988
$C_{14}H_9NO$	15.69	1.35	207.0684
$C_{14}H_{11}N_2$	16.07	1.21	207.0923
$C_{15}H_{27}$	16.64	1.30	207.2114
$C_{15}H_{11}O$	16.42	1.46	207.0810
$C_{15}H_{13}N$	16.80	1.32	207.1049
$C_{16}H_{15}$	17.53	1.44	207.1174
$C_{16}HN$	17.69	1.47	207.0109
$C_{17}H_3$	18.42	1.60	207.0235
208			
$C_8H_{20}N_2O_4$	9.88	1.24	208.1424
$C_9H_8N_2O_4$	10.77	1.33	208.0484
$C_9H_{10}N_3O_3$	11.15	1.17	208.0723
$C_9H_{12}N_4O_2$	11.52	1.01	208.0961
$C_{10}H_{10}NO_4$	11.50	1.40	208.0610
$C_{10}H_{12}N_2O_3$	11.88	1.25	208.0848
$C_{10}H_{14}N_3O_2$	12.25	1.09	208.1087
$C_{10}H_{16}N_4O$	12.63	0.94	208.1325
$C_{10}N_4O_2$	12.41	1.11	208.0022
$C_{11}H_{12}O_4$	12.24	1.49	208.0735
$C_{11}H_{14}NO_3$	12.61	1.33	208.0974
$C_{11}H_{16}N_2O_2$	12.98	1.18	208.1213
$C_{11}N_2O_3$	12.77	1.35	207.9909
$C_{11}H_{18}N_3O$	13.36	1.03	208.1451
$C_{11}H_2N_3O_2$	13.14	1.20	208.0147
$C_{11}H_{20}N_4$	13.73	0.88	208.1690
$C_{11}H_4N_4O$	13.52	1.05	208.0386
$C_{12}H_{16}O_3$	13.34	1.42	208.1100
$C_{12}O_4$	13.12	1.59	207.9796
$C_{12}H_{18}NO_2$	13.71	1.27	208.1338
$C_{12}H_2NO_3$	13.50	1.44	208.0034
$C_{12}H_{20}N_2O$	14.09	1.12	208.1577
$C_{12}H_4N_2O_2$	13.87	1.29	208.0273
$C_{12}H_{22}N_3$	14.46	0.97	208.1815
$C_{12}H_6N_3O$	14.25	1.14	208.0511
$C_{12}H_8N_4$	14.62	1.00	208.0750
$C_{13}H_{20}O_2$	14.45	1.37	208.1464
$C_{13}H_4O_3$	14.23	1.54	208.0160
$C_{13}H_{22}NO$	14.82	1.22	208.1702
$C_{13}H_6NO_2$	14.60	1.39	208.0399
$C_{13}H_{24}N_2$	15.19	1.08	208.1941
$C_{13}H_8N_2O$	14.98	1.24	208.0637
$C_{13}H_{10}N_3$	15.35	1.10	208.0876
$C_{14}H_{24}O$	15.55	1.33	208.1828
$C_{14}H_8O_2$	15.33	1.49	208.0524
$C_{14}H_{26}N$	15.93	1.19	208.2067
$C_{14}H_{10}NO$	15.71	1.35	208.0763
$C_{14}H_{12}N_2$	16.08	1.21	208.1001
$C_{15}H_{28}$	16.66	1.30	208.2192
$C_{15}H_{12}O$	16.44	1.46	208.0888
$C_{15}H_{14}N$	16.81	1.33	208.1127
$C_{15}N_2$	16.97	1.35	208.0062
$C_{16}H_{16}$	17.55	1.45	208.1253
$C_{16}O$	17.33	1.61	207.9949
$C_{16}H_2N$	17.70	1.47	208.0187
$C_{17}H_4$	18.43	1.60	208.0313
209			
$C_9H_9N_2O_4$	10.79	1.33	209.0563
$C_9H_{11}N_3O_3$	11.16	1.17	209.0801
$C_9H_{13}N_4O_2$	11.54	1.01	209.1040
$C_{10}H_{11}NO_4$	11.52	1.41	209.0688
$C_{10}H_{13}N_2O_3$	11.89	1.25	209.0927
$C_{10}H_{15}N_3O_2$	12.27	1.09	209.1165
$C_{10}H_{17}N_4O$	12.64	0.94	209.1404
$C_{10}HN_4O_2$	12.43	1.11	209.0100
$C_{11}H_{13}O_4$	12.25	1.49	209.0814
$C_{11}H_{15}NO_3$	12.63	1.33	209.1052
$C_{11}H_{17}N_2O_2$	13.00	1.18	209.1291
$C_{11}HN_2O_3$	12.78	1.35	208.9987
$C_{11}H_{19}N_3O$	13.37	1.03	209.1529
$C_{11}H_3N_3O_2$	13.16	1.20	209.0226
$C_{11}H_{21}N_4$	13.75	0.88	209.1768
$C_{11}H_5N_4O$	13.53	1.05	209.0464
$C_{12}H_{17}O_3$	13.36	1.42	209.1176
$C_{12}HO_4$	13.14	1.60	208.9874
$C_{12}H_{19}NO_2$	13.73	1.27	209.1416

	M+1	M+2	FM
$C_{12}H_{3}NO_{3}$	13.51	1.44	209.0113
$C_{12}H_{21}N_{2}O$	14.11	1.12	209.1655
$C_{12}H_{5}N_{2}O_{2}$	13.89	1.29	209.0351
$C_{12}H_{23}N_{3}$	14.48	0.98	209.1894
$C_{12}H_{7}N_{3}O$	14.26	1.15	209.0590
$C_{12}H_{9}N_{4}$	14.64	1.00	209.0829
$C_{13}H_{21}O_{2}$	14.46	1.37	209.1542
$C_{13}H_{5}O_{3}$	14.25	1.54	209.0238
$C_{13}H_{23}NO$	14.84	1.22	209.1781
$C_{13}H_{7}NO_{2}$	14.62	1.39	209.0477
$C_{13}H_{25}N_{2}$	15.21	1.08	209.2019
$C_{13}H_{9}N_{2}O$	14.99	1.25	209.0715
$C_{13}H_{11}N_{3}$	15.37	1.10	209.0954
$C_{14}H_{25}O$	15.57	1.33	209.1906
$C_{14}H_{9}O_{2}$	15.35	1.50	209.0603
$C_{14}H_{27}N$	15.94	1.19	209.2145
$C_{14}H_{11}NO$	15.72	1.35	209.0841
$C_{14}H_{13}N_{2}$	16.10	1.21	209.1080
$C_{15}H_{29}$	16.67	1.30	209.2270
$C_{15}H_{13}O$	16.46	1.47	209.0967
$C_{15}H_{15}N$	16.83	1.33	209.1205
$C_{15}HN_{2}$	16.99	1.35	209.0140
$C_{16}H_{17}$	17.56	1.45	209.1331
$C_{16}HO$	17.34	1.61	209.0027
$C_{16}H_{3}N$	17.72	1.48	209.0266
$C_{17}H_{5}$	18.45	1.60	209.0391
210			
$C_{9}H_{10}N_{2}O_{4}$	10.80	1.33	210.0641
$C_{9}H_{12}N_{3}O_{3}$	11.18	1.17	210.0879
$C_{9}H_{14}N_{4}O_{2}$	11.55	1.01	210.1118
$C_{10}H_{12}NO_{4}$	11.54	1.41	210.0766
$C_{10}H_{14}N_{2}O_{3}$	11.91	1.25	210.1005
$C_{10}H_{16}N_{3}O_{2}$	12.28	1.09	210.1244
$C_{10}N_{3}O_{3}$	12.07	1.27	209.9940
$C_{10}H_{18}N_{4}O$	12.66	0.94	210.1482
$C_{10}H_{2}N_{4}O_{2}$	12.44	1.11	210.0178
$C_{11}H_{14}O_{4}$	12.27	1.49	210.0892
$C_{11}H_{16}NO_{3}$	12.64	1.34	210.1131
$C_{11}NO_{4}$	12.42	1.51	209.9827
$C_{11}H_{18}N_{2}O_{2}$	13.02	1.18	210.1369
$C_{11}H_{2}N_{2}O_{3}$	12.80	1.35	210.0065
$C_{11}H_{20}N_{3}O$	13.39	1.03	210.1608
$C_{11}H_{4}N_{3}O_{2}$	13.17	1.20	210.0304
$C_{11}H_{22}N_{4}$	13.76	0.88	210.1846
$C_{11}H_{6}N_{4}O$	13.55	1.05	210.0542
$C_{12}H_{18}O_{3}$	13.37	1.43	210.1256
$C_{12}H_{2}O_{4}$	13.16	1.60	209.9953
$C_{12}H_{20}NO_{2}$	13.75	1.28	210.1495
$C_{12}H_{4}NO_{3}$	13.53	1.45	210.0191
$C_{12}H_{22}N_{2}O$	14.12	1.13	210.1733
$C_{12}H_{6}N_{2}O_{2}$	13.90	1.30	210.0429
$C_{12}H_{24}N_{3}$	14.50	0.98	210.1972
$C_{12}H_{8}N_{3}O$	14.28	1.15	210.0668
$C_{12}H_{10}N_{4}$	14.65	1.00	210.0907
$C_{13}H_{22}O_{2}$	14.48	1.37	210.1620
$C_{13}H_{6}O_{3}$	14.26	1.54	210.0317
$C_{13}H_{24}NO$	14.85	1.23	210.1859
$C_{13}H_{8}NO_{2}$	14.64	1.40	210.0555
$C_{13}H_{26}N_{2}$	15.23	1.08	210.2098
$C_{13}H_{10}N_{2}O$	15.01	1.25	210.0794
$C_{13}H_{12}N_{3}$	15.38	1.11	210.1032
$C_{14}H_{26}O$	15.58	1.33	210.1985
$C_{14}H_{10}O_{2}$	15.37	1.50	210.0681
$C_{14}H_{28}N$	15.96	1.19	210.2223
$C_{14}H_{12}NO$	15.74	1.36	210.0919
$C_{14}H_{14}N_{2}$	16.12	1.22	210.1158
$C_{14}N_{3}$	16.27	1.24	210.0093
$C_{15}H_{30}$	16.69	1.31	210.2349
$C_{15}H_{14}O$	16.47	1.47	210.1045
$C_{15}H_{16}N$	16.85	1.33	210.1284
$C_{15}NO$	16.63	1.49	209.9980
$C_{15}H_{2}N_{2}$	17.00	1.36	210.0218
$C_{16}H_{18}$	17.58	1.45	210.1409
$C_{16}H_{2}O$	17.36	1.61	210.0106
$C_{16}H_{4}N$	17.74	1.48	210.0344
$C_{17}H_{6}$	18.47	1.61	210.0470
211			
$C_{9}H_{11}N_{2}O_{4}$	10.82	1.33	211.0719
$C_{9}H_{13}N_{3}O_{3}$	11.20	1.17	211.0958
$C_{9}H_{15}N_{4}O_{2}$	11.57	1.01	211.1196
$C_{10}H_{13}NO_{4}$	11.55	1.41	211.0845
$C_{10}H_{15}N_{2}O_{3}$	11.93	1.25	211.1083
$C_{10}H_{17}N_{3}O_{2}$	12.30	1.10	211.1322
$C_{10}HN_{3}O_{3}$	12.08	1.27	211.0018
$C_{10}H_{19}N_{4}O$	12.67	0.94	211.1560
$C_{10}H_{3}N_{4}O_{2}$	12.46	1.12	211.0257
$C_{11}H_{15}O_{4}$	12.28	1.49	211.0970
$C_{11}H_{17}NO_{3}$	12.66	1.34	211.1209
$C_{11}HNO_{4}$	12.44	1.51	210.9905
$C_{11}H_{19}N_{2}O_{2}$	13.03	1.18	211.1447
$C_{11}H_{3}N_{2}O_{3}$	12.81	1.36	211.0144
$C_{11}H_{21}N_{3}O$	13.41	1.03	211.1686
$C_{11}H_{5}N_{3}O_{2}$	13.19	1.20	211.0382
$C_{11}H_{23}N_{4}$	13.78	0.88	211.1925
$C_{11}H_{7}N_{4}O$	13.56	1.05	211.0621
$C_{12}H_{19}O_{3}$	13.39	1.43	211.1334
$C_{12}H_{3}O_{4}$	13.17	1.60	211.0031
$C_{12}H_{21}NO_{2}$	13.76	1.28	211.1573
$C_{12}H_{5}NO_{3}$	13.55	1.45	211.0269
$C_{12}H_{23}N_{2}O$	14.14	1.13	211.1811
$C_{12}H_{7}N_{2}O_{2}$	13.92	1.30	211.0508
$C_{12}H_{25}N_{3}$	14.51	0.98	211.2050
$C_{12}H_{9}N_{3}O$	14.29	1.15	211.0746
$C_{12}H_{11}N_{4}$	14.67	1.00	211.0985
$C_{13}H_{23}O_{2}$	14.49	1.38	211.1699
$C_{13}H_{7}O_{3}$	14.28	1.54	211.0395
$C_{13}H_{25}NO$	14.87	1.23	211.1937
$C_{13}H_{9}NO_{2}$	14.65	1.40	211.0634
$C_{13}H_{27}N_{2}$	15.24	1.08	211.2176
$C_{13}H_{11}N_{2}O$	15.03	1.25	211.0872
$C_{13}H_{13}N_{3}$	15.40	1.11	211.1111
$C_{14}H_{27}O$	15.60	1.34	211.2063
$C_{14}H_{11}O_{2}$	15.38	1.50	211.0759
$C_{14}H_{29}N$	15.97	1.19	211.2301
$C_{14}H_{13}NO$	15.76	1.36	211.0998
$C_{14}H_{15}N_{2}$	16.13	1.22	211.1236
$C_{14}HN_{3}$	16.29	1.24	211.0171
$C_{15}H_{31}$	16.71	1.31	211.2427
$C_{15}H_{15}O$	16.49	1.47	211.1123
$C_{15}H_{17}N$	16.86	1.33	211.1362
$C_{15}HNO$	16.65	1.50	211.0058
$C_{15}H_{3}N_{2}$	17.02	1.36	211.0297
$C_{16}H_{19}$	17.59	1.45	211.1488
$C_{16}H_{3}O$	17.38	1.62	211.0184
$C_{16}H_{5}N$	17.75	1.48	211.0422
$C_{17}H_{7}$	18.48	1.61	211.0548
212			
$C_{9}H_{12}N_{2}O_{4}$	10.84	1.34	212.0797
$C_{9}H_{14}N_{3}O_{3}$	11.21	1.18	212.1036
$C_{9}H_{16}N_{4}O_{2}$	11.59	1.02	212.1275
$C_{10}H_{14}NO_{4}$	11.57	1.41	212.0923
$C_{10}H_{16}N_{2}O_{3}$	11.94	1.25	212.1162
$C_{10}N_{2}O_{4}$	11.73	1.43	211.9858
$C_{10}H_{18}N_{3}O_{2}$	12.32	1.10	212.1400
$C_{10}H_{2}N_{3}O_{3}$	12.10	1.27	212.0096
$C_{10}H_{20}N_{4}O$	12.69	0.94	212.1639
$C_{10}H_{4}N_{4}O_{2}$	12.47	1.12	212.0335
$C_{11}H_{16}O_{4}$	12.30	1.49	212.1049
$C_{11}H_{18}NO_{3}$	12.67	1.34	212.1287
$C_{11}H_{2}NO_{4}$	12.46	1.51	211.9983
$C_{11}H_{20}N_{2}O_{2}$	13.05	1.19	212.1526
$C_{11}H_{4}N_{2}O_{3}$	12.83	1.36	212.0222
$C_{11}H_{22}N_{3}O$	13.42	1.03	212.1764
$C_{11}H_{6}N_{3}O_{2}$	13.21	1.21	212.0460
$C_{11}H_{24}N_{4}$	13.80	0.88	212.2003
$C_{11}H_{8}N_{4}O$	13.58	1.06	212.0699
$C_{12}H_{20}O_{3}$	13.40	1.43	212.1413
$C_{12}H_{4}O_{4}$	13.19	1.60	212.0109
$C_{12}H_{22}NO_{2}$	13.78	1.28	212.1651
$C_{12}H_{6}NO_{3}$	13.56	1.45	212.0348
$C_{12}H_{24}N_{2}O$	14.15	1.13	212.1890
$C_{12}H_{8}N_{2}O_{2}$	13.94	1.30	212.0586
$C_{12}H_{26}N_{3}$	14.53	0.98	212.2129
$C_{12}H_{10}N_{3}O$	14.31	1.15	212.0825
$C_{12}H_{12}N_{4}$	14.69	1.01	212.1063
$C_{13}H_{24}O_{2}$	14.51	1.38	212.1777
$C_{13}H_{8}O_{3}$	14.29	1.55	212.0473
$C_{13}H_{26}NO$	14.88	1.23	212.2015
$C_{13}H_{10}NO_{2}$	14.67	1.40	212.0712

	M+1	M+2	FM
$C_{13}H_{28}N_2$	15.26	1.09	212.2254
$C_{13}H_{12}N_2O$	15.04	1.25	212.0950
$C_{13}H_{14}N_3$	15.42	1.11	212.1189
$C_{13}N_4$	15.57	1.13	212.0124
$C_{14}H_{28}O$	15.62	1.34	212.2141
$C_{14}H_{12}O_2$	15.40	1.50	212.0837
$C_{14}H_{30}N$	15.99	1.20	212.2380
$C_{14}H_{14}NO$	15.77	1.36	212.1076
$C_{14}H_{16}N_2$	16.15	1.22	212.1315
$C_{14}N_2O$	15.93	1.39	212.0011
$C_{14}H_2N_3$	16.30	1.25	212.0249
$C_{15}H_{32}$	16.72	1.31	212.2505
$C_{15}H_{16}O$	16.50	1.47	212.1202
$C_{15}O_2$	16.29	1.64	211.9898
$C_{15}H_{18}N$	16.88	1.34	212.1440
$C_{15}H_2NO$	16.66	1.50	212.0136
$C_{15}H_4N_2$	17.04	1.36	212.0375
$C_{16}H_{20}$	17.61	1.46	212.1566
$C_{16}H_4O$	17.39	1.62	212.0262
$C_{16}H_6N$	17.77	1.48	212.0501
$C_{17}H_8$	18.50	1.61	212.0626
213			
$C_9H_{13}N_2O_4$	10.85	1.34	213.0876
$C_9H_{15}N_3O_3$	11.23	1.18	213.1114
$C_9H_{17}N_4O_2$	11.60	1.02	213.1353
$C_{10}H_{15}NO_4$	11.58	1.41	213.1001
$C_{10}H_{17}N_2O_3$	11.96	1.26	213.1240
$C_{10}HN_2O_4$	11.74	1.43	212.9936
$C_{10}H_{19}N_3O_2$	12.33	1.10	213.1478
$C_{10}H_3N_3O_3$	12.12	1.27	213.0175
$C_{10}H_{21}N_4O$	12.71	0.95	213.1717
$C_{10}H_5N_4O_2$	12.49	1.12	213.0413
$C_{11}H_{17}O_4$	12.32	1.50	213.1127
$C_{11}H_{19}NO_3$	12.69	1.34	213.1365
$C_{11}H_3NO_4$	12.47	1.51	213.0062
$C_{11}H_{21}N_2O_2$	13.06	1.19	213.1604
$C_{11}H_5N_2O_3$	12.85	1.36	213.0300
$C_{11}H_{23}N_3O$	13.44	1.04	213.1842
$C_{11}H_7N_3O_2$	13.22	1.21	213.0539
$C_{11}H_{25}N_4$	13.81	0.89	213.2081
$C_{11}H_9N_4O$	13.60	1.06	213.0777
$C_{12}H_{21}O_3$	13.42	1.43	213.1491
$C_{12}H_5O_4$	13.20	1.60	213.0187
$C_{12}H_{23}NO_2$	13.79	1.28	213.1730
$C_{12}H_7NO_3$	13.58	1.45	213.0426
$C_{12}H_{25}N_2O$	14.17	1.13	213.1968
$C_{12}H_9N_2O_2$	13.95	1.30	213.0664
$C_{12}H_{27}N_3$	14.54	0.99	213.2207
$C_{12}H_{11}N_3O$	14.33	1.15	213.0903
$C_{12}H_{13}N_4$	14.70	1.01	213.1142
$C_{13}H_{25}O_2$	14.53	1.38	213.1855
$C_{13}H_9O_3$	14.31	1.55	213.0552
$C_{13}H_{27}NO$	14.90	1.23	213.2094

	M+1	M+2	FM
$C_{13}H_{11}NO_2$	14.68	1.40	213.0790
$C_{13}H_{29}N_2$	15.27	1.09	213.2332
$C_{13}H_{13}N_2O$	15.06	1.26	213.1029
$C_{13}H_{15}N_3$	15.43	1.11	213.1267
$C_{13}HN_4$	15.59	1.14	213.0202
$C_{14}H_{29}O$	15.63	1.34	213.2219
$C_{14}H_{13}O_2$	15.41	1.51	213.0916
$C_{14}H_{31}N$	16.01	1.20	213.2458
$C_{14}H_{15}NO$	15.79	1.36	213.1154
$C_{14}H_{17}N_2$	16.16	1.22	213.1393
$C_{14}HN_2O$	15.95	1.39	213.0089
$C_{14}H_3N_3$	16.32	1.25	213.0328
$C_{15}H_{17}O$	16.52	1.48	213.1280
$C_{15}HO_2$	16.30	1.64	212.9976
$C_{15}H_{19}N$	16.89	1.34	213.1519
$C_{15}H_3NO$	16.68	1.50	213.0215
$C_{15}H_5N_2$	17.05	1.36	213.0453
$C_{16}H_{21}$	17.63	1.46	213.1644
$C_{16}H_5O$	17.41	1.62	213.0340
$C_{16}H_7N$	17.78	1.49	213.0579
$C_{17}H_9$	18.51	1.61	213.0705
214			
$C_9H_{14}N_2O_4$	10.87	1.34	214.0954
$C_9H_{16}N_3O_3$	11.24	1.18	214.1193
$C_9H_{18}N_4O_2$	11.62	1.02	214.1431
$C_{10}H_{16}NO_4$	11.60	1.42	214.1080
$C_{10}H_{18}N_2O_3$	11.97	1.26	214.1318
$C_{10}H_2N_2O_4$	11.76	1.43	214.0014
$C_{10}H_{20}N_3O_2$	12.35	1.10	214.1557
$C_{10}H_4N_3O_3$	12.13	1.28	214.0253
$C_{10}H_{22}N_4O$	12.72	0.95	214.1795
$C_{10}H_6N_4O_2$	12.51	1.12	214.0491
$C_{11}H_{18}O_4$	12.33	1.50	214.1205
$C_{11}H_{20}NO_3$	12.71	1.34	214.1444
$C_{11}H_4NO_4$	12.49	1.52	214.0140
$C_{11}H_{22}N_2O_2$	13.08	1.19	214.1682
$C_{11}H_6N_2O_3$	12.86	1.36	214.0379
$C_{11}H_{24}N_3O$	13.45	1.04	214.1921
$C_{11}H_8N_3O_2$	13.24	1.21	214.0617
$C_{11}H_{26}N_4$	13.83	0.89	214.2160
$C_{11}H_{10}N_4O$	13.61	1.06	214.0856
$C_{12}H_{22}O_3$	13.44	1.43	214.1569
$C_{12}H_6O_4$	13.22	1.61	214.0266
$C_{12}H_{24}NO_2$	13.81	1.28	214.1808
$C_{12}H_8NO_3$	13.59	1.45	214.0504
$C_{12}H_{26}N_2O$	14.19	1.14	214.2046
$C_{12}H_{10}N_2O_2$	13.97	1.30	214.0743
$C_{12}H_{28}N_3$	14.56	0.99	214.2285
$C_{12}H_{12}N_3O$	14.34	1.16	214.0981
$C_{12}H_{14}N_4$	14.72	1.01	214.1220
$C_{13}H_{26}O_2$	14.54	1.38	214.1934
$C_{13}H_{10}O_3$	14.33	1.55	214.0630
$C_{13}H_{28}NO$	14.92	1.24	214.2172

	M+1	M+2	FM
$C_{13}H_{12}NO_2$	14.70	1.40	214.0868
$C_{13}H_{30}N_2$	15.29	1.09	214.2411
$C_{13}H_{14}N_2O$	15.07	1.26	214.1107
$C_{13}H_{16}N_3$	15.45	1.12	214.1346
$C_{13}N_3O$	15.23	1.28	214.0042
$C_{13}H_2N_4$	15.61	1.14	214.0280
$C_{14}H_{30}O$	15.65	1.34	214.2298
$C_{14}H_{14}O_2$	15.43	1.51	214.0994
$C_{14}H_{16}NO$	15.80	1.37	214.1233
$C_{14}NO_2$	15.59	1.53	213.9929
$C_{14}H_{18}N_2$	16.18	1.23	214.1471
$C_{14}H_2N_2O$	15.96	1.39	214.0167
$C_{14}H_4N_3$	16.34	1.25	214.0406
$C_{15}H_{18}O$	16.54	1.48	214.1358
$C_{15}H_2O_2$	16.32	1.64	214.0054
$C_{15}H_{20}N$	16.91	1.34	214.1597
$C_{15}H_4NO$	16.69	1.51	214.0293
$C_{15}H_6N_2$	17.07	1.37	214.0532
$C_{16}H_{22}$	17.64	1.46	214.1722
$C_{16}H_6O$	17.42	1.63	214.0419
$C_{16}H_8N$	17.80	1.49	214.0657
$C_{17}H_{10}$	18.53	1.62	214.0783
215			
$C_9H_{15}N_2O_4$	10.88	1.34	215.1032
$C_9H_{17}N_3O_3$	11.26	1.18	215.1271
$C_9H_{19}N_4O_2$	11.63	1.02	215.1509
$C_{10}H_{17}NO_4$	11.62	1.42	215.1158
$C_{10}H_{19}N_2O_3$	11.99	1.26	215.1396
$C_{10}H_3N_2O_4$	11.77	1.44	215.0093
$C_{10}H_{21}N_3O_2$	12.36	1.10	215.1635
$C_{10}H_5N_3O_3$	12.15	1.28	215.0331
$C_{10}H_{23}N_4O$	12.74	0.95	215.1873
$C_{10}H_7N_4O_2$	12.52	1.12	215.0570
$C_{11}H_{19}O_4$	12.35	1.50	215.1284
$C_{11}H_{21}NO_3$	12.72	1.35	215.1522
$C_{11}H_5NO_4$	12.50	1.52	215.0218
$C_{11}H_{23}N_2O_2$	13.10	1.19	215.1761
$C_{11}H_7N_2O_3$	12.88	1.37	215.0457
$C_{11}H_{25}N_3O$	13.47	1.04	215.1999
$C_{11}H_9N_3O_2$	13.25	1.21	215.0695
$C_{11}H_{27}N_4$	13.84	0.89	215.2238
$C_{11}H_{11}N_4O$	13.63	1.06	215.0934
$C_{12}H_{23}O_3$	13.45	1.44	215.1648
$C_{12}H_7O_4$	13.24	1.61	215.0344
$C_{12}H_{25}NO_2$	13.83	1.29	215.1886
$C_{12}H_9NO_3$	13.61	1.46	215.0583
$C_{12}H_{27}N_2O$	14.20	1.14	215.2125
$C_{12}H_{11}N_2O_2$	13.98	1.31	215.0821
$C_{12}H_{29}N_3$	14.58	0.99	215.2363
$C_{12}H_{13}N_3O$	14.36	1.16	215.1060
$C_{12}H_{15}N_4$	14.73	1.01	215.1298
$C_{13}H_{27}O_2$	14.56	1.38	215.2012
$C_{13}H_{11}O_3$	14.34	1.55	215.0708
$C_{13}H_{29}NO$	14.93	1.24	215.2250

	M+1	M+2	FM
$C_{13}H_{13}NO_2$	14.72	1.41	215.0947
$C_{13}H_{15}N_2O$	15.09	1.26	215.1185
$C_{13}H_{17}N_3$	15.46	1.12	215.1424
$C_{13}HN_3O$	15.25	1.28	215.0120
$C_{13}H_3N_4$	15.62	1.14	215.0359
$C_{14}H_{15}O_2$	15.45	1.51	215.1072
$C_{14}H_{17}NO$	15.82	1.37	215.1311
$C_{14}HNO_2$	15.60	1.54	215.0007
$C_{14}H_{19}N_2$	16.20	1.23	215.1549
$C_{14}H_3N_2O$	15.98	1.39	215.0246
$C_{14}H_5N_3$	16.35	1.25	215.0484
$C_{15}H_{19}O$	16.55	1.48	215.1436
$C_{15}H_3O_2$	16.34	1.65	215.0133
$C_{15}H_{21}N$	16.93	1.34	215.1675
$C_{15}H_5NO$	16.71	1.51	215.0371
$C_{15}H_7N_2$	17.08	1.37	215.0610
$C_{16}H_{23}$	17.66	1.47	215.1801
$C_{16}H_7O$	17.44	1.63	215.0497
$C_{16}H_9N$	17.82	1.49	215.0736
$C_{17}H_{11}$	18.55	1.62	215.0861
216			
$C_9H_{16}N_2O_4$	10.90	1.34	216.1111
$C_9H_{18}N_3O_3$	11.28	1.18	216.1349
$C_9H_{20}N_4O_2$	11.65	1.02	216.1588
$C_{10}H_{18}NO_4$	11.63	1.42	216.1236
$C_{10}H_{20}N_2O_3$	12.01	1.26	216.1475
$C_{10}H_4N_2O_4$	11.79	1.44	216.0171
$C_{10}H_{22}N_3O_2$	12.38	1.11	216.1713
$C_{10}H_6N_3O_3$	12.16	1.28	216.0410
$C_{10}H_{24}N_4O$	12.75	0.95	216.1952
$C_{10}H_8N_4O_2$	12.54	1.13	216.0648
$C_{11}H_{20}O_4$	12.36	1.50	216.1362
$C_{11}H_{22}NO_3$	12.74	1.35	216.1600
$C_{11}H_6NO_4$	12.52	1.52	216.0297
$C_{11}H_{24}N_2O_2$	13.11	1.19	216.1839
$C_{11}H_8N_2O_3$	12.89	1.37	216.0535
$C_{11}H_{26}N_3O$	13.49	1.04	216.2077
$C_{11}H_{10}N_3O_2$	13.27	1.21	216.0774
$C_{11}H_{28}N_4$	13.86	0.89	216.2316
$C_{11}H_{12}N_4O$	13.64	1.06	216.1012
$C_{12}H_{24}O_3$	13.47	1.44	216.1726
$C_{12}H_8O_4$	13.25	1.61	216.0422
$C_{12}H_{26}NO_2$	13.84	1.29	216.1965
$C_{12}H_{10}NO_3$	13.63	1.46	216.0661
$C_{12}H_{28}N_2O$	14.22	1.14	216.2203
$C_{12}H_{12}N_2O_2$	14.00	1.31	216.0899
$C_{12}H_{14}N_3O$	14.37	1.16	216.1138
$C_{12}H_{16}N_4$	14.75	1.01	216.1377
$C_{12}N_4O$	14.53	1.18	216.0073
$C_{13}H_{28}O_2$	14.57	1.39	216.2090
$C_{13}H_{12}O_3$	14.36	1.56	216.0786
$C_{13}H_{14}NO_2$	14.73	1.41	216.1025
$C_{13}H_{16}N_2O$	15.11	1.26	216.1264

	M+1	M+2	FM
$C_{13}N_2O_2$	14.89	1.43	215.9960
$C_{13}H_{18}N_3$	15.48	1.12	216.1502
$C_{13}H_2N_3O$	15.26	1.29	216.0198
$C_{13}H_4N_4$	15.64	1.14	216.0437
$C_{14}H_{16}O_2$	15.46	1.51	216.1151
$C_{14}O_3$	15.25	1.68	215.9847
$C_{14}H_{18}NO$	15.84	1.37	216.1389
$C_{14}H_2NO_2$	15.62	1.54	216.0085
$C_{14}H_{20}N_2$	16.21	1.23	216.1628
$C_{14}H_4N_2O$	15.99	1.40	216.0324
$C_{14}H_6N_3$	16.37	1.26	216.0563
$C_{15}H_{20}O$	16.57	1.49	216.1515
$C_{15}H_4O_2$	16.35	1.65	216.0211
$C_{15}H_{22}N$	16.94	1.35	216.1753
$C_{15}H_6NO$	16.73	1.51	216.0449
$C_{15}H_8N_2$	17.10	1.37	216.0688
$C_{16}H_{24}$	17.67	1.47	216.1879
$C_{16}H_8O$	17.46	1.63	216.0575
$C_{16}H_{10}N$	17.83	1.50	216.0814
$C_{17}H_{12}$	18.56	1.62	216.0939
C_{18}	19.45	1.79	216.0000
217			
$C_9H_{17}N_2O_4$	10.92	1.34	217.1189
$C_9H_{19}N_3O_3$	11.29	1.18	217.1427
$C_9H_{21}N_4O_2$	11.67	1.03	217.1666
$C_{10}H_{19}NO_4$	11.65	1.42	217.1315
$C_{10}H_{21}N_2O_3$	12.02	1.26	217.1553
$C_{10}H_5N_2O_4$	11.81	1.44	217.0249
$C_{10}H_{23}N_3O_2$	12.40	1.11	217.1791
$C_{10}H_7N_3O_3$	12.18	1.28	217.0488
$C_{10}H_{25}N_4O$	12.77	0.95	217.2030
$C_{10}H_9N_4O_2$	12.55	1.13	217.0726
$C_{11}H_{21}O_4$	12.38	1.50	217.1440
$C_{11}H_{23}NO_3$	12.75	1.35	217.1679
$C_{11}H_7NO_4$	12.54	1.52	217.0375
$C_{11}H_{25}N_2O_2$	13.13	1.20	217.1917
$C_{11}H_9N_2O_3$	12.91	1.37	217.0614
$C_{11}H_{27}N_3O$	13.50	1.05	217.2156
$C_{11}H_{11}N_3O_2$	13.29	1.22	217.0852
$C_{11}H_{13}N_4O$	13.66	1.07	217.1091
$C_{12}H_{25}O_3$	13.48	1.44	217.1804
$C_{12}H_9O_4$	13.27	1.61	217.0501
$C_{12}H_{27}NO_2$	13.86	1.29	217.2043
$C_{12}H_{11}NO_3$	13.64	1.46	217.0739
$C_{12}H_{13}N_2O_2$	14.02	1.31	217.0978
$C_{12}H_{15}N_3O$	14.39	1.16	217.1216
$C_{12}H_{17}N_4$	14.77	1.02	217.1455
$C_{12}HN_4O$	14.55	1.19	217.0151
$C_{13}H_{13}O_3$	14.37	1.56	217.0865
$C_{13}H_{15}NO_2$	14.75	1.41	217.1103
$C_{13}H_{17}N_2O$	15.12	1.27	217.1342
$C_{13}HN_2O_2$	14.90	1.43	217.0038
$C_{13}H_{19}N_3$	15.50	1.12	217.1580

	M+1	M+2	FM
$C_{13}H_3N_3O$	15.28	1.29	217.0277
$C_{13}H_5N_4$	15.65	1.15	217.0515
$C_{14}H_{17}O_2$	15.48	1.52	217.1229
$C_{14}HO_3$	15.26	1.68	216.9925
$C_{14}H_{19}NO$	15.85	1.37	217.1467
$C_{14}H_3NO_2$	15.64	1.54	217.0164
$C_{14}H_{21}N_2$	16.23	1.23	217.1706
$C_{14}H_5N_2O$	16.01	1.40	217.0402
$C_{14}H_7N_3$	16.38	1.26	217.0641
$C_{15}H_{21}O$	16.58	1.49	217.1593
$C_{15}H_5O_2$	16.37	1.65	217.0289
$C_{15}H_{23}N$	16.96	1.35	217.1832
$C_{15}H_7NO$	16.74	1.51	217.0528
$C_{15}H_9N_2$	17.12	1.38	217.0767
$C_{16}H_{25}$	17.69	1.47	217.1957
$C_{16}H_9O$	17.47	1.63	217.0653
$C_{16}H_{11}N$	17.85	1.50	217.0892
$C_{17}H_{13}$	18.58	1.63	217.1018
$C_{18}H$	19.47	1.79	217.0078
218			
$C_9H_{18}N_2O_4$	10.93	1.35	218.1267
$C_9H_{20}N_3O_3$	11.31	1.19	218.1506
$C_9H_{22}N_4O_2$	11.68	1.03	218.1744
$C_{10}H_{20}NO_4$	11.66	1.42	218.1393
$C_{10}H_{22}N_2O_3$	12.04	1.27	218.1631
$C_{10}H_6N_2O_4$	11.82	1.44	218.0328
$C_{10}H_{24}N_3O_2$	12.41	1.11	218.1870
$C_{10}H_8N_3O_3$	12.20	1.28	218.0566
$C_{10}H_{26}N_4O$	12.79	0.96	218.2108
$C_{10}H_{10}N_4O_2$	12.57	1.13	218.0805
$C_{11}H_{22}O_4$	12.40	1.51	218.1518
$C_{11}H_{24}NO_3$	12.77	1.35	218.1757
$C_{11}H_8NO_4$	12.55	1.52	218.0453
$C_{11}H_{26}N_2O_2$	13.14	1.20	218.1996
$C_{11}H_{10}N_2O_3$	12.93	1.37	218.0692
$C_{11}H_{12}N_3O_2$	13.30	1.22	218.0930
$C_{11}H_{14}N_4O$	13.68	1.07	218.1169
$C_{12}H_{26}O_3$	13.50	1.44	218.1883
$C_{12}H_{10}O_4$	13.28	1.61	218.0579
$C_{12}H_{12}NO_3$	13.66	1.46	218.0817
$C_{12}H_{14}N_2O_2$	14.03	1.31	218.1050
$C_{12}H_{16}N_3O$	14.41	1.17	218.1295
$C_{12}N_3O_2$	14.19	1.34	217.9991
$C_{12}H_{18}N_4$	14.78	1.02	218.1533
$C_{12}H_2N_4O$	14.56	1.19	218.0229
$C_{13}H_{14}O_3$	14.39	1.56	218.0943
$C_{13}H_{16}NO_2$	14.76	1.41	218.1102
$C_{13}NO_3$	14.55	1.58	217.9878
$C_{13}H_{18}N_2O$	15.14	1.27	218.1420
$C_{13}H_2N_2O_2$	14.92	1.44	218.0116
$C_{13}H_{20}N_3$	15.51	1.13	218.1659
$C_{13}H_4N_3O$	15.30	1.29	218.0355
$C_{13}H_6N_4$	15.67	1.15	218.0594
$C_{14}H_{18}O_2$	15.49	1.52	218.1307

	M+1	M+2	FM
$C_{14}H_2O_3$	15.28	1.69	218.0003
$C_{14}H_{20}NO$	15.87	1.38	218.1546
$C_{14}H_4NO_2$	15.65	1.54	218.0242
$C_{14}H_{22}N_2$	16.24	1.24	218.1784
$C_{14}H_6N_2O$	16.03	1.40	218.0480
$C_{14}H_8N_3$	16.40	1.26	218.0719
$C_{15}H_{22}O$	16.60	1.49	218.1671
$C_{15}H_6O_2$	16.38	1.66	218.0368
$C_{15}H_{24}N$	16.97	1.35	218.1910
$C_{15}H_8NO$	16.76	1.52	218.0606
$C_{15}H_{10}N_2$	17.13	1.38	218.0845
$C_{16}H_{26}$	17.71	1.47	218.2036
$C_{16}H_{10}O$	17.49	1.64	218.0732
$C_{16}H_{12}N$	17.86	1.50	218.0970
$C_{17}H_{14}$	18.59	1.63	218.1096
$C_{17}N$	18.75	1.66	218.0031
$C_{18}H_2$	19.48	1.79	218.0157
219			
$C_9H_{19}N_2O_4$	10.95	1.35	219.1345
$C_9H_{21}N_3O_3$	11.32	1.19	219.1584
$C_9H_{23}N_4O_2$	11.70	1.03	219.1822
$C_{10}H_{21}NO_4$	11.68	1.42	219.1471
$C_{10}H_{23}N_2O_3$	12.05	1.27	219.1710
$C_{10}H_7N_2O_4$	11.84	1.44	219.0406
$C_{10}H_{25}N_3O_2$	12.43	1.11	219.1948
$C_{10}H_9N_3O_3$	12.21	1.29	219.0644
$C_{10}H_{11}N_4O_2$	12.59	1.13	219.0883
$C_{11}H_{23}O_4$	12.41	1.51	219.1597
$C_{11}H_{25}NO_3$	12.79	1.35	219.1835
$C_{11}H_9NO_4$	12.57	1.53	219.0532
$C_{11}H_{11}N_2O_3$	12.94	1.37	219.0770
$C_{11}H_{13}N_3O_2$	13.32	1.22	219.1009
$C_{11}H_{15}N_4O$	13.69	1.07	219.1247
$C_{12}H_{11}O_4$	13.30	1.62	219.0657
$C_{12}H_{13}NO_3$	13.67	1.47	219.0896
$C_{12}H_{15}N_2O_2$	14.05	1.32	219.1134
$C_{12}H_{17}N_3O$	14.42	1.17	219.1373
$C_{12}HN_3O_2$	14.21	1.34	219.0069
$C_{12}H_{19}N_4$	14.80	1.02	219.1611
$C_{12}H_3N_4O$	14.58	1.19	219.0308
$C_{13}H_{15}O_3$	14.41	1.56	219.1021
$C_{13}H_{17}NO_2$	14.78	1.42	219.1260
$C_{13}HNO_3$	14.56	1.59	218.9956
$C_{13}H_{19}N_2O$	15.15	1.27	219.1498
$C_{13}H_3N_2O_2$	14.94	1.44	219.0195
$C_{13}H_{21}N_3$	15.53	1.13	219.1737
$C_{13}H_5N_3O$	15.31	1.29	219.0433
$C_{13}H_7N_4$	15.69	1.15	219.0672
$C_{14}H_{19}O_2$	15.51	1.52	219.1385
$C_{14}H_3O_3$	15.29	1.69	219.0082
$C_{14}H_{21}NO$	15.89	1.38	219.1624
$C_{14}H_5NO_2$	15.67	1.55	219.0320
$C_{14}H_{23}N_2$	16.26	1.24	219.1863
$C_{14}H_7N_2O$	16.04	1.40	219.0559
$C_{14}H_9N_3$	16.42	1.26	219.0798
$C_{15}H_{23}O$	16.62	1.49	219.1750
$C_{15}H_7O_2$	16.40	1.66	219.0446
$C_{15}H_{25}N$	16.99	1.36	219.1988
$C_{15}H_9NO$	16.77	1.52	219.0684
$C_{15}H_{11}N_2$	17.15	1.38	219.0923
$C_{16}H_{27}$	17.72	1.48	219.2114
$C_{16}H_{11}O$	17.50	1.64	219.0810
$C_{16}H_{13}N$	17.88	1.50	219.1049
$C_{17}H_{15}$	18.61	1.63	219.1174
$C_{17}HN$	18.77	1.66	219.0109
$C_{18}H_3$	19.50	1.80	219.0235
220			
$C_9H_{20}N_2O_4$	10.96	1.35	220.1424
$C_9H_{22}N_3O_3$	11.34	1.19	220.1662
$C_9H_{24}N_4O_2$	11.71	1.03	220.1901
$C_{10}H_{22}NO_4$	11.70	1.43	220.1549
$C_{10}H_{24}N_2O_3$	12.07	1.27	220.1788
$C_{10}H_8N_2O_4$	11.85	1.44	220.0484
$C_{10}H_{10}N_3O_3$	12.23	1.29	220.0723
$C_{10}H_{12}N_4O_2$	12.60	1.13	220.0961
$C_{11}H_{24}O_4$	12.43	1.51	220.1675
$C_{11}H_{10}NO_4$	12.58	1.53	220.0610
$C_{11}H_{12}N_2O_3$	12.96	1.38	220.0848
$C_{11}H_{14}N_3O_2$	13.33	1.22	220.1087
$C_{11}H_{16}N_4O$	13.71	1.07	220.1325
$C_{11}N_4O_2$	13.49	1.24	220.0022
$C_{12}H_{12}O_4$	13.32	1.62	220.0735
$C_{12}H_{14}NO_3$	13.69	1.47	220.0974
$C_{12}H_{16}N_2O_2$	14.06	1.32	220.1213
$C_{12}N_2O_3$	13.85	1.49	219.9909
$C_{12}H_{18}N_3O$	14.44	1.17	220.1451
$C_{12}H_2N_3O_2$	14.22	1.34	220.0147
$C_{12}H_{20}N_4$	14.81	1.02	220.1690
$C_{12}H_4N_4O$	14.60	1.19	220.0386
$C_{13}H_{16}O_3$	14.42	1.57	220.1100
$C_{13}O_4$	14.20	1.73	219.9796
$C_{13}H_{18}NO_2$	14.80	1.42	220.1338
$C_{13}H_2NO_3$	14.58	1.59	220.0034
$C_{13}H_{20}N_2O$	15.17	1.27	220.1577
$C_{13}H_4N_2O_2$	14.95	1.44	220.0273
$C_{13}H_{22}N_3$	15.54	1.13	220.1815
$C_{13}H_6N_3O$	15.33	1.30	220.0511
$C_{13}H_8N_4$	15.70	1.15	220.0750
$C_{14}H_{20}O_2$	15.53	1.52	220.1464
$C_{14}H_4O_3$	15.31	1.69	220.0160
$C_{14}H_{22}NO$	15.90	1.38	220.1702
$C_{14}H_6NO_2$	15.68	1.55	220.0399
$C_{14}H_{24}N_2$	16.28	1.24	220.1941
$C_{14}H_8N_2O$	16.06	1.41	220.0637
$C_{14}H_{10}N_3$	16.43	1.27	220.0876
$C_{15}H_{24}O$	16.63	1.50	220.1828
$C_{15}H_8O_2$	16.42	1.66	220.0524
$C_{15}H_{26}N$	17.01	1.36	220.2067
$C_{15}H_{10}NO$	16.79	1.52	220.0763
$C_{15}H_{12}N_2$	17.16	1.38	220.1001
$C_{16}H_{28}$	17.74	1.48	220.2192
$C_{16}H_{12}O$	17.52	1.64	220.0888
$C_{16}H_{14}N$	17.90	1.51	220.1127
$C_{16}N_2$	18.05	1.53	220.0062
$C_{17}H_{16}$	18.63	1.64	220.1253
$C_{17}O$	18.41	1.80	219.9949
$C_{17}H_2N$	18.78	1.66	220.0187
$C_{18}H_4$	19.51	1.80	220.0313
221			
$C_9H_{21}N_2O_4$	10.98	1.35	221.1502
$C_9H_{23}N_3O_3$	11.36	1.19	221.1741
$C_{10}H_{23}NO_4$	11.71	1.43	221.1628
$C_{10}H_9N_2O_4$	11.87	1.45	221.0563
$C_{10}H_{11}N_3O_3$	12.24	1.29	221.0801
$C_{10}H_{13}N_4O_2$	12.62	1.14	221.1040
$C_{11}H_{11}NO_4$	12.60	1.53	221.0688
$C_{11}H_{13}N_2O_3$	12.97	1.38	221.0927
$C_{11}H_{15}N_3O_2$	13.35	1.23	221.1165
$C_{11}H_{17}N_4O$	13.72	1.07	221.1404
$C_{11}HN_4O_2$	13.51	1.25	221.0100
$C_{12}H_{13}O_4$	13.33	1.62	221.0814
$C_{12}H_{15}NO_3$	13.71	1.47	221.1052
$C_{12}H_{17}N_2O_2$	14.08	1.32	221.1291
$C_{12}HN_2O_3$	13.86	1.49	220.9987
$C_{12}H_{19}N_3O$	14.45	1.17	221.1529
$C_{12}H_3N_3O_2$	14.24	1.34	221.0226
$C_{12}H_{21}N_4$	14.83	1.03	221.1768
$C_{12}H_5N_4O$	14.61	1.20	221.0464
$C_{13}H_{17}O_3$	14.44	1.57	221.1178
$C_{13}HO_4$	14.22	1.74	220.9874
$C_{13}H_{19}NO_2$	14.81	1.42	221.1416
$C_{13}H_3NO_3$	14.59	1.59	221.0113
$C_{13}H_{21}N_2O$	15.19	1.28	221.1655
$C_{13}H_5N_2O_2$	14.97	1.44	221.0351
$C_{13}H_{23}N_3$	15.56	1.13	221.1894
$C_{13}H_7N_3O$	15.34	1.30	221.0590
$C_{13}H_9N_4$	15.72	1.16	221.0829
$C_{14}H_{21}O_2$	15.54	1.53	221.1542
$C_{14}H_5O_3$	15.33	1.69	221.0238
$C_{14}H_{23}NO$	15.92	1.38	221.1781
$C_{14}H_7NO_2$	15.70	1.55	221.0477
$C_{14}H_{25}N_2$	16.29	1.24	221.2019
$C_{14}H_9N_2O$	16.07	1.41	221.0715
$C_{14}H_{11}N_3$	16.45	1.27	221.0954
$C_{15}H_{25}O$	16.65	1.50	221.1906
$C_{15}H_9O_2$	16.43	1.66	221.0603
$C_{15}H_{27}N$	17.02	1.36	221.2145
$C_{15}H_{11}NO$	16.81	1.52	221.0841
$C_{15}H_{13}N_2$	17.18	1.39	221.1080

Formula	M+1	M+2	FM
$C_{16}H_{29}$	17.75	1.48	221.2270
$C_{16}H_{13}O$	17.54	1.64	221.0967
$C_{16}H_{15}N$	17.91	1.51	221.1205
$C_{16}HN_2$	18.07	1.54	221.0140
$C_{17}H_{17}$	18.64	1.64	221.1331
$C_{17}HO$	18.43	1.80	221.0027
$C_{17}H_3N$	18.80	1.67	221.0266
$C_{18}H_5$	19.53	1.80	221.0391
222			
$C_9H_{22}N_2O_4$	11.00	1.35	222.1580
$C_{10}H_{10}N_2O_4$	11.89	1.45	222.0641
$C_{10}H_{12}N_3O_3$	12.26	1.29	222.0879
$C_{10}H_{14}N_4O_2$	12.63	1.14	222.1118
$C_{11}H_{12}NO_4$	12.62	1.53	222.0766
$C_{11}H_{14}N_2O_3$	12.99	1.38	222.1005
$C_{11}H_{16}N_3O_2$	13.37	1.23	222.1244
$C_{11}N_3O_3$	13.15	1.40	221.9940
$C_{11}H_{18}N_4O$	13.74	1.08	222.1482
$C_{11}H_2N_4O_2$	13.52	1.25	222.0178
$C_{12}H_{14}O_4$	13.35	1.62	222.0892
$C_{12}H_{16}NO_3$	13.72	1.47	222.1131
$C_{12}NO_4$	13.51	1.64	221.9827
$C_{12}H_{18}N_2O_2$	14.10	1.32	222.1369
$C_{12}H_2N_2O_3$	13.88	1.49	222.0065
$C_{12}H_{20}N_3O$	14.47	1.18	222.1608
$C_{12}H_4N_3O_2$	14.25	1.34	222.0304
$C_{12}H_{22}N_4$	14.85	1.03	222.1846
$C_{12}H_6N_4O$	14.63	1.20	222.0542
$C_{13}H_{18}O_3$	14.45	1.57	222.1256
$C_{13}H_2O_4$	14.24	1.74	221.9953
$C_{13}H_{20}NO_2$	14.83	1.42	222.1495
$C_{13}H_4NO_3$	14.61	1.59	222.0191
$C_{13}H_{22}N_2O$	15.20	1.28	222.1733
$C_{13}H_6N_2O_2$	14.98	1.45	222.0429
$C_{13}H_{24}N_3$	15.58	1.13	222.1972
$C_{13}H_8N_3O$	15.36	1.30	222.0668
$C_{13}H_{10}N_4$	15.73	1.16	222.0907
$C_{14}H_{22}O_2$	15.56	1.53	222.1620
$C_{14}H_6O_3$	15.34	1.70	222.0317
$C_{14}H_{24}NO$	15.93	1.39	222.1859
$C_{14}H_8NO_2$	15.72	1.55	222.0555
$C_{14}H_{26}N_2$	16.31	1.25	222.2098
$C_{14}H_{10}N_2O$	16.09	1.41	222.0794
$C_{14}H_{12}N_3$	16.46	1.27	222.1032
$C_{15}H_{26}O$	16.66	1.50	222.1985
$C_{15}H_{10}O_2$	16.45	1.67	222.0681
$C_{15}H_{28}N$	17.04	1.36	222.2223
$C_{15}H_{12}NO$	16.82	1.53	222.0919
$C_{15}H_{14}N_2$	17.20	1.39	222.1158
$C_{15}N_3$	17.35	1.42	222.0093
$C_{16}H_{30}$	17.77	1.49	222.2349
$C_{16}H_{14}O$	17.55	1.65	222.1045
$C_{16}H_{16}N$	17.93	1.51	222.1284
$C_{16}NO$	17.71	1.67	221.9980
$C_{16}H_2N_2$	18.08	1.54	222.0218
$C_{17}H_{18}$	18.66	1.64	222.1409
$C_{17}H_2O$	18.44	1.80	222.0106
$C_{17}H_4N$	18.82	1.67	222.0344
$C_{18}H_6$	19.55	1.81	222.0470
223			
$C_{10}H_{11}N_2O_4$	11.90	1.45	223.0719
$C_{10}H_{13}N_3O_3$	12.28	1.29	223.0958
$C_{10}H_{15}N_4O_2$	12.65	1.14	223.1196
$C_{11}H_{13}NO_4$	12.63	1.53	223.0845
$C_{11}H_{15}N_2O_3$	13.01	1.38	223.1083
$C_{11}H_{17}N_3O_2$	13.38	1.23	223.1322
$C_{11}HN_3O_3$	13.16	1.40	223.0018
$C_{11}H_{19}N_4O$	13.76	1.08	223.1560
$C_{11}H_3N_4O_2$	13.54	1.25	223.0257
$C_{12}H_{15}O_4$	13.36	1.62	223.0970
$C_{12}H_{17}NO_3$	13.74	1.47	223.1209
$C_{12}HNO_4$	13.52	1.65	222.9905
$C_{12}H_{19}N_2O_2$	14.11	1.33	223.1447
$C_{12}H_3N_2O_3$	13.90	1.50	223.0144
$C_{12}H_{21}N_3O$	14.49	1.18	223.1686
$C_{12}H_5N_3O_2$	14.27	1.35	223.0382
$C_{12}H_{23}N_4$	14.86	1.03	223.1925
$C_{12}H_7N_4O$	14.64	1.20	223.0621
$C_{13}H_{19}O_3$	14.47	1.57	223.1334
$C_{13}H_3O_4$	14.25	1.74	223.0031
$C_{13}H_{21}NO_2$	14.84	1.43	223.1573
$C_{13}H_5NO_3$	14.63	1.59	223.0269
$C_{13}H_{23}N_2O$	15.22	1.28	223.1811
$C_{13}H_7N_2O_2$	15.00	1.45	223.0508
$C_{13}H_{25}N_3$	15.59	1.14	223.2050
$C_{13}H_9N_3O$	15.38	1.30	223.0746
$C_{13}H_{11}N_4$	15.75	1.16	223.0985
$C_{14}H_{23}O_2$	15.57	1.53	223.1699
$C_{14}H_7O_3$	15.36	1.70	223.0395
$C_{14}H_{25}NO$	15.95	1.39	223.1937
$C_{14}H_9NO_2$	15.73	1.56	223.0634
$C_{14}H_{27}N_2$	16.32	1.25	223.2176
$C_{14}H_{11}N_2O$	16.11	1.41	223.0872
$C_{14}H_{13}N_3$	16.48	1.27	223.1111
$C_{15}H_{27}O$	16.68	1.50	223.2063
$C_{15}H_{11}O_2$	16.46	1.67	223.0759
$C_{15}H_{29}N$	17.05	1.37	223.2301
$C_{15}H_{13}NO$	16.84	1.53	223.0998
$C_{15}H_{15}N_2$	17.21	1.39	223.1236
$C_{15}HN_3$	17.37	1.42	223.0171
$C_{16}H_{31}$	17.79	1.49	223.2427
$C_{16}H_{15}O$	17.57	1.65	223.1123
$C_{16}H_{17}N$	17.94	1.52	223.1362
$C_{16}HNO$	17.73	1.68	223.0058
$C_{16}H_3N_2$	18.10	1.54	223.0297
$C_{17}H_{19}$	18.67	1.64	223.1488
$C_{17}H_3O$	18.46	1.80	223.0184
$C_{17}H_5N$	18.83	1.67	223.0422
$C_{18}H_7$	19.56	1.81	223.0548
224			
$C_{10}H_{12}N_2O_4$	11.92	1.45	224.0797
$C_{10}H_{14}N_3O_3$	12.29	1.30	224.1036
$C_{10}H_{16}N_4O_2$	12.67	1.14	224.1275
$C_{11}H_{14}NO_4$	12.65	1.54	224.0923
$C_{11}H_{16}N_2O_3$	13.02	1.38	224.1162
$C_{11}N_2O_4$	12.81	1.56	223.9858
$C_{11}H_{18}N_3O_2$	13.40	1.23	224.1400
$C_{11}H_2N_3O_3$	13.18	1.40	224.0096
$C_{11}H_{20}N_4O$	13.77	1.08	224.1639
$C_{11}H_4N_4O_2$	13.55	1.25	224.0335
$C_{12}H_{16}O_4$	13.38	1.63	224.1049
$C_{12}H_{18}NO_3$	13.75	1.48	224.1287
$C_{12}H_2NO_4$	13.54	1.65	223.9983
$C_{12}H_{20}N_2O_2$	14.13	1.33	224.1526
$C_{12}H_4N_2O_3$	13.91	1.50	224.0222
$C_{12}H_{22}N_3O$	14.50	1.18	224.1764
$C_{12}H_6N_3O_2$	14.29	1.35	224.0460
$C_{12}H_{24}N_4$	14.88	1.03	224.2003
$C_{12}H_8N_4O$	14.66	1.20	224.0699
$C_{13}H_{20}O_3$	14.49	1.57	224.1413
$C_{13}H_4O_4$	14.27	1.74	224.0109
$C_{13}H_{22}NO_2$	14.86	1.43	224.1651
$C_{13}H_6NO_3$	14.64	1.60	224.0348
$C_{13}H_{24}N_2O$	15.23	1.28	224.1890
$C_{13}H_8N_2O_2$	15.02	1.45	224.0586
$C_{13}H_{26}N_3$	15.61	1.14	224.2129
$C_{13}H_{10}N_3O$	15.39	1.31	224.0825
$C_{13}H_{12}N_4$	15.77	1.16	224.1063
$C_{14}H_{24}O_2$	15.59	1.53	224.1777
$C_{14}H_8O_3$	15.37	1.70	224.0473
$C_{14}H_{26}NO$	15.97	1.39	224.2015
$C_{14}H_{10}NO_2$	15.75	1.56	224.0712
$C_{14}H_{28}N_2$	16.34	1.25	224.2254
$C_{14}H_{12}N_2O$	16.12	1.42	224.0950
$C_{14}H_{14}N_3$	16.50	1.28	224.1189
$C_{14}N_4$	16.65	1.30	224.0124
$C_{15}H_{28}O$	16.70	1.51	224.2141
$C_{15}H_{12}O_2$	16.48	1.67	224.0837
$C_{16}H_{30}N$	17.07	1.37	224.2380
$C_{15}H_{14}NO$	16.85	1.53	224.1076
$C_{15}H_{16}N_2$	17.23	1.39	224.1315
$C_{15}N_2O$	17.01	1.56	224.0011
$C_{15}H_2N_3$	17.39	1.42	224.0249
$C_{16}H_{32}$	17.80	1.49	224.2505
$C_{16}H_{16}O$	17.58	1.65	224.1202
$C_{16}O_2$	17.37	1.82	223.9898
$C_{16}H_{18}N$	17.96	1.52	224.1440
$C_{16}H_2NO$	17.74	1.68	224.0136
$C_{16}H_4N_2$	18.12	1.55	224.0375
$C_{17}H_{20}$	18.69	1.65	224.1566
$C_{17}H_4O$	18.47	1.81	224.0262

	M+1	M+2	FM
$C_{17}H_6N$	18.85	1.68	224.0501
$C_{18}H_8$	19.58	1.81	224.0626
225			
$C_{10}H_{13}N_2O_4$	11.93	1.45	225.0876
$C_{10}H_{15}N_3O_3$	12.31	1.30	225.1114
$C_{10}H_{17}N_4O_2$	12.68	1.14	225.1353
$C_{11}H_{15}NO_4$	12.66	1.54	225.1001
$C_{11}H_{17}N_2O_3$	13.04	1.39	225.1240
$C_{11}HN_2O_4$	12.82	1.56	224.9936
$C_{11}H_{19}N_3O_2$	13.41	1.23	225.1478
$C_{11}H_3N_3O_3$	13.20	1.41	225.0175
$C_{11}H_{21}N_4O$	13.79	1.08	225.1717
$C_{11}H_5N_4O_2$	13.57	1.25	225.0413
$C_{12}H_{17}O_4$	13.40	1.63	225.1127
$C_{12}H_{19}NO_3$	13.77	1.48	225.1365
$C_{12}H_3NO_4$	13.55	1.65	225.0062
$C_{12}H_{21}N_2O_2$	14.14	1.33	225.1604
$C_{12}H_5N_2O_3$	13.93	1.50	225.0300
$C_{12}H_{23}N_3O$	14.52	1.18	225.1842
$C_{12}H_7N_3O_2$	14.30	1.35	225.0539
$C_{12}H_{25}N_4$	14.89	1.04	225.2081
$C_{12}H_9N_4O$	14.68	1.20	225.0777
$C_{13}H_{21}O_3$	14.50	1.58	225.1491
$C_{13}H_5O_4$	14.28	1.75	225.0187
$C_{13}H_{23}NO_2$	14.88	1.43	225.1730
$C_{13}H_7NO_3$	14.66	1.60	225.0426
$C_{13}H_{25}N_2O$	15.25	1.29	225.1968
$C_{13}H_9N_2O_2$	15.03	1.45	225.0664
$C_{13}H_{27}N_3$	15.62	1.14	225.2207
$C_{13}H_{11}N_3O$	15.41	1.31	225.0903
$C_{13}H_{13}N_4$	15.78	1.17	225.1142
$C_{14}H_{25}O_2$	15.61	1.54	225.1855
$C_{14}H_9O_3$	15.39	1.70	225.0552
$C_{14}H_{27}NO$	15.98	1.39	225.2094
$C_{14}H_{11}NO_2$	15.76	1.56	225.0790
$C_{14}H_{29}N_2$	16.36	1.25	225.2332
$C_{14}H_{13}N_2O$	16.14	1.42	225.1029
$C_{14}H_{15}N_3$	16.51	1.28	225.1267
$C_{14}HN_4$	16.67	1.30	225.0202
$C_{15}H_{29}O$	16.71	1.51	225.2219
$C_{15}H_{13}O_2$	16.50	1.67	225.0916
$C_{15}H_{31}N$	17.09	1.37	225.2458
$C_{15}H_{15}NO$	16.87	1.54	225.1154
$C_{15}H_{17}N_2$	17.24	1.40	225.1393
$C_{15}HN_2O$	17.03	1.56	225.0089
$C_{15}H_3N_3$	17.40	1.42	225.0328
$C_{16}H_{33}$	17.82	1.49	225.2584
$C_{16}H_{17}O$	17.60	1.66	225.1280
$C_{16}HO_2$	17.38	1.82	224.9976
$C_{16}H_{19}N$	17.98	1.52	225.1519
$C_{16}H_3NO$	17.76	1.68	225.0215
$C_{16}H_5N_2$	18.13	1.55	225.0453
$C_{17}H_{21}$	18.71	1.65	225.1644

	M+1	M+2	FM
$C_{17}H_5O$	18.49	1.81	225.0340
$C_{17}H_7N$	18.86	1.68	225.0579
$C_{18}H_9$	19.59	1.81	225.0705
226			
$C_{10}H_{14}N_2O_4$	11.95	1.46	226.0954
$C_{10}H_{16}N_3O_3$	12.32	1.30	226.1193
$C_{10}H_{18}N_4O_2$	12.70	1.15	226.1431
$C_{11}H_{16}NO_4$	12.68	1.54	226.1080
$C_{11}H_{18}N_2O_3$	13.05	1.39	226.1318
$C_{11}H_2N_2O_4$	12.84	1.56	226.0014
$C_{11}H_{20}N_3O_2$	13.43	1.24	226.1557
$C_{11}H_4N_3O_3$	13.21	1.41	226.0253
$C_{11}H_{22}N_4O$	13.80	1.09	226.1795
$C_{11}H_6N_4O_2$	13.59	1.26	226.0491
$C_{12}H_{18}O_4$	13.41	1.63	226.1205
$C_{12}H_{20}NO_3$	13.79	1.48	226.1444
$C_{12}H_4NO_4$	13.57	1.65	226.0140
$C_{12}H_{22}N_2O_2$	14.16	1.33	226.1682
$C_{12}H_6N_2O_3$	13.94	1.50	226.0379
$C_{12}H_{24}N_3O$	14.53	1.18	226.1921
$C_{12}H_8N_3O_2$	14.32	1.35	226.0617
$C_{12}H_{26}N_4$	14.91	1.04	226.2160
$C_{12}H_{10}N_4O$	14.69	1.21	226.0856
$C_{13}H_{22}O_3$	14.52	1.58	226.1569
$C_{13}H_6O_4$	14.30	1.75	226.0266
$C_{13}H_{24}NO_2$	14.89	1.43	226.1808
$C_{13}H_8NO_3$	14.67	1.60	226.0504
$C_{13}H_{26}N_2O$	15.27	1.29	226.2046
$C_{13}H_{10}N_2O_2$	15.05	1.46	226.0743
$C_{13}H_{28}N_3$	15.64	1.14	226.2285
$C_{13}H_{12}N_3O$	15.42	1.31	226.0981
$C_{13}H_{14}N_4$	15.80	1.17	226.1220
$C_{14}H_{26}O_2$	15.62	1.54	226.1934
$C_{14}H_{10}O_3$	15.41	1.71	226.0630
$C_{14}H_{28}NO$	16.00	1.40	226.2172
$C_{14}H_{12}NO_2$	15.78	1.56	226.0868
$C_{14}H_{30}N_2$	16.37	1.26	226.2411
$C_{14}H_{14}N_2O$	16.15	1.42	226.1107
$C_{14}H_{16}N_3$	16.53	1.28	226.1346
$C_{14}N_3O$	16.31	1.45	226.0042
$C_{14}H_2N_4$	16.69	1.31	226.0280
$C_{15}H_{30}O$	16.73	1.51	226.2298
$C_{15}H_{14}O_2$	16.51	1.68	226.0994
$C_{15}H_{32}N$	17.10	1.37	226.2536
$C_{15}H_{16}NO$	16.89	1.54	226.1233
$C_{15}NO_2$	16.67	1.70	225.9929
$C_{15}H_{18}N_2$	17.26	1.40	226.1471
$C_{15}H_2N_2O$	17.04	1.56	226.0167
$C_{15}H_4N_3$	17.42	1.43	226.0406
$C_{16}H_{34}$	17.83	1.50	226.2662
$C_{16}H_{18}O$	17.62	1.66	226.1358
$C_{16}H_2O_2$	17.40	1.82	226.0054
$C_{16}H_{20}N$	17.99	1.52	226.1597

	M+1	M+2	FM
$C_{16}H_4NO$	17.77	1.69	226.0293
$C_{16}H_6N_2$	18.15	1.55	226.0532
$C_{17}H_{22}$	18.72	1.65	226.1722
$C_{17}H_6O$	18.51	1.81	226.0419
$C_{17}H_8N$	18.88	1.68	226.0657
$C_{18}H_{10}$	19.61	1.82	226.0783
227			
$C_{10}H_{15}N_2O_4$	11.97	1.46	227.1032
$C_{10}H_{17}N_3O_3$	12.34	1.30	227.1271
$C_{10}H_{19}N_4O_2$	12.71	1.15	227.1509
$C_{11}H_{17}NO_4$	12.70	1.54	227.1158
$C_{11}H_{19}N_2O_3$	13.07	1.39	227.1396
$C_{11}H_3N_2O_4$	12.85	1.56	227.0093
$C_{11}H_{21}N_3O_2$	13.45	1.24	227.1635
$C_{11}H_5N_3O_3$	13.23	1.41	227.0331
$C_{11}H_{23}N_4O$	13.82	1.09	227.1873
$C_{11}H_7N_4O_2$	13.60	1.26	227.0570
$C_{12}H_{19}O_4$	13.43	1.63	227.1284
$C_{12}H_{21}NO_3$	13.80	1.48	227.1522
$C_{12}H_5NO_4$	13.59	1.65	227.0218
$C_{12}H_{23}N_2O_2$	14.18	1.33	227.1761
$C_{12}H_7N_2O_3$	13.96	1.50	227.0457
$C_{12}H_{25}N_3O$	14.55	1.19	227.1999
$C_{12}H_9N_3O_2$	14.33	1.36	227.0695
$C_{12}H_{27}N_4$	14.93	1.04	227.2238
$C_{12}H_{11}N_4O$	14.71	1.21	227.0934
$C_{13}H_{23}O_3$	14.53	1.58	227.1648
$C_{13}H_7O_4$	14.32	1.75	227.0344
$C_{13}H_{25}NO_2$	14.91	1.44	227.1886
$C_{13}H_9NO_3$	14.69	1.60	227.0583
$C_{13}H_{27}N_2O$	15.28	1.29	227.2125
$C_{13}H_{11}N_2O_2$	15.06	1.46	227.0821
$C_{13}H_{29}N_3$	15.66	1.15	227.2363
$C_{13}H_{13}N_3O$	15.44	1.31	227.1060
$C_{13}H_{15}N_4$	15.81	1.17	227.1298
$C_{14}H_{27}O_2$	15.64	1.54	227.2012
$C_{14}H_{11}O_3$	15.42	1.71	227.0708
$C_{14}H_{29}NO$	16.01	1.40	227.2250
$C_{14}H_{13}NO_2$	15.80	1.57	227.0947
$C_{14}H_{31}N_2$	16.39	1.26	227.2489
$C_{14}H_{15}N_2O$	16.17	1.42	227.1185
$C_{14}H_{17}N_3$	16.54	1.28	227.1424
$C_{14}HN_3O$	16.33	1.45	227.0120
$C_{14}H_3N_4$	16.70	1.31	227.0359
$C_{15}H_{31}O$	16.74	1.51	227.2376
$C_{15}H_{15}O_2$	16.53	1.68	227.1072
$C_{15}H_{33}N$	17.12	1.38	227.2615
$C_{15}H_{17}NO$	16.90	1.54	227.1311
$C_{15}HNO_2$	16.68	1.70	227.0007
$C_{15}H_{19}N_2$	17.28	1.40	227.1549
$C_{15}H_3N_2O$	17.06	1.57	227.0246
$C_{15}H_5N_3$	17.43	1.43	227.0484
$C_{16}H_{19}O$	17.63	1.66	227.1436

	M+1	M+2	FM		M+1	M+2	FM		M+1	M+2	FM
$C_{16}H_3O_2$	17.42	1.82	227.0133	$C_{15}H_{20}N_2$	17.29	1.41	228.1628	$C_{15}HO_3$	16.34	1.85	228.9925
$C_{16}H_{21}N$	18.01	1.53	227.1675	$C_{15}H_4N_2O$	17.07	1.57	228.0324	$C_{15}H_{19}NO$	16.93	1.55	229.1467
$C_{16}H_5NO$	17.79	1.69	227.0371	$C_{15}H_6N_3$	17.45	1.43	228.0563	$C_{15}H_3NO_2$	16.72	1.71	229.0164
$C_{16}H_7N_2$	18.16	1.55	227.0610	$C_{16}H_{20}O$	17.65	1.66	228.1515	$C_{15}H_{21}N_2$	17.31	1.41	229.1706
$C_{17}H_{23}$	18.74	1.66	227.1801	$C_{16}H_4O_2$	17.43	1.83	228.0211	$C_{15}H_5N_2O$	17.09	1.57	229.0402
$C_{17}H_7O$	18.52	1.82	227.0497	$C_{16}H_{22}N$	18.02	1.53	228.1753	$C_{15}H_7N_3$	17.47	1.44	229.0641
$C_{17}H_9N$	18.90	1.69	227.0736	$C_{16}H_6NO$	17.81	1.69	228.0449	$C_{16}H_{21}O$	17.66	1.67	229.1593
$C_{18}H_{11}$	19.63	1.82	227.0861	$C_{16}H_8N_2$	18.18	1.56	228.0688	$C_{16}H_5O_2$	17.45	1.83	229.0289
228				$C_{17}H_{24}$	18.75	1.66	228.1879	$C_{16}H_{23}N$	18.04	1.53	229.1832
$C_{10}H_{16}N_2O_4$	11.98	1.46	228.1111	$C_{17}H_8O$	18.54	1.82	228.0575	$C_{16}H_7NO$	17.82	1.69	229.0528
$C_{10}H_{18}N_3O_3$	12.36	1.30	228.1349	$C_{17}H_{10}N$	18.91	1.69	228.0814	$C_{16}H_9N_2$	18.20	1.56	229.0767
$C_{10}H_{20}N_4O_2$	12.73	1.15	228.1588	$C_{18}H_{12}$	19.64	1.82	228.0939	$C_{17}H_{25}$	18.77	1.66	229.1957
$C_{11}H_{18}NO_4$	12.71	1.54	228.1236	C_{19}	20.53	2.00	228.0000	$C_{17}H_9O$	18.55	1.82	229.0653
$C_{11}H_{20}N_2O_3$	13.09	1.39	228.1475	229				$C_{17}H_{11}N$	18.93	1.69	229.0892
$C_{11}H_4N_2O_4$	12.87	1.56	228.0171	$C_{10}H_{17}N_2O_4$	12.00	1.46	229.1189	$C_{18}H_{13}$	19.66	1.83	229.1018
$C_{11}H_{22}N_3O_2$	13.46	1.24	228.1713	$C_{10}H_{19}N_3O_3$	12.37	1.31	229.1427	$C_{19}H$	20.55	2.00	229.0078
$C_{11}H_6N_3O_3$	13.24	1.41	228.0410	$C_{10}H_{21}N_4O_2$	12.75	1.15	229.1666	230			
$C_{11}H_{24}N_4O$	13.84	1.09	228.1952	$C_{11}H_{19}NO_4$	12.73	1.55	229.1315	$C_{10}H_{18}N_2O_4$	12.01	1.46	230.1267
$C_{11}H_8N_4O_2$	13.62	1.26	228.0648	$C_{11}H_{21}N_2O_3$	13.10	1.39	229.1553	$C_{10}H_{20}N_3O_3$	12.39	1.31	230.1506
$C_{12}H_{20}O_4$	13.44	1.64	228.1362	$C_{11}H_5N_2O_4$	12.89	1.57	229.0249	$C_{10}H_{22}N_4O_2$	12.76	1.15	230.1744
$C_{12}H_{22}NO_3$	13.82	1.49	228.1600	$C_{11}H_{23}N_3O_2$	13.48	1.24	229.1791	$C_{11}H_{20}NO_4$	12.74	1.55	230.1393
$C_{12}H_6NO_4$	13.60	1.66	228.0297	$C_{11}H_7N_3O_3$	13.26	1.41	229.0488	$C_{11}H_{22}N_2O_3$	13.12	1.40	230.1631
$C_{12}H_{24}N_2O_2$	14.19	1.34	228.1839	$C_{11}H_{25}N_4O$	13.85	1.09	229.2030	$C_{11}H_6N_2O_4$	12.90	1.57	230.0328
$C_{12}H_8N_2O_3$	13.98	1.51	228.0535	$C_{11}H_9N_4O_2$	13.63	1.26	229.0726	$C_{11}H_{24}N_3O_2$	13.49	1.24	230.1870
$C_{12}H_{26}N_3O$	14.57	1.19	228.2077	$C_{12}H_{21}O_4$	13.46	1.64	229.1440	$C_{11}H_8N_3O_3$	13.28	1.42	230.0566
$C_{12}H_{10}N_3O_2$	14.35	1.36	228.0774	$C_{12}H_{23}NO_3$	13.83	1.49	229.1679	$C_{11}H_{26}N_4O$	13.87	1.09	230.2108
$C_{12}H_{28}N_4$	14.94	1.04	228.2316	$C_{12}H_7NO_4$	13.62	1.66	229.0375	$C_{11}H_{10}N_4O_2$	13.65	1.27	230.0805
$C_{12}H_{12}N_4O$	14.72	1.21	228.1012	$C_{12}H_{25}N_2O_2$	14.21	1.34	229.1917	$C_{12}H_{22}O_4$	13.48	1.64	230.1518
$C_{13}H_{24}O_3$	14.55	1.58	228.1726	$C_{12}H_9N_2O_3$	13.99	1.51	229.0614	$C_{12}H_{24}NO_3$	13.85	1.49	230.1757
$C_{13}H_8O_4$	14.33	1.75	228.0422	$C_{12}H_{27}N_3O$	14.58	1.19	229.2156	$C_{12}H_8NO_4$	13.63	1.66	230.0453
$C_{13}H_{26}NO_2$	14.92	1.44	228.1965	$C_{12}H_{11}N_3O_2$	14.37	1.36	229.0852	$C_{12}H_{26}N_2O_2$	14.22	1.34	230.1996
$C_{13}H_{10}NO_3$	14.71	1.61	228.0661	$C_{12}H_{29}N_4$	14.96	1.05	229.2394	$C_{12}H_{10}N_2O_3$	14.01	1.51	230.0692
$C_{13}H_{28}N_2O$	15.30	1.29	228.2203	$C_{12}H_{13}N_4O$	14.74	1.21	229.1091	$C_{12}H_{28}N_3O$	14.60	1.19	230.2234
$C_{13}H_{12}N_2O_2$	15.08	1.46	228.0899	$C_{13}H_{25}O_3$	14.57	1.59	229.1804	$C_{12}H_{12}N_3O_2$	14.38	1.36	230.0930
$C_{13}H_{30}N_3$	15.67	1.15	228.2442	$C_{13}H_9O_4$	14.35	1.76	229.0501	$C_{12}H_{30}N_4$	14.97	1.05	230.2473
$C_{13}H_{14}N_3O$	15.46	1.32	228.1138	$C_{13}H_{27}NO_2$	14.94	1.44	229.2043	$C_{12}H_{14}N_4O$	14.76	1.22	230.1169
$C_{13}H_{16}N_4$	15.83	1.17	228.1377	$C_{13}H_{11}NO_3$	14.72	1.61	229.0739	$C_{13}H_{26}O_3$	14.50	1.59	230.1883
$C_{13}N_4O$	15.61	1.34	228.0073	$C_{13}H_{29}N_2O$	15.31	1.30	229.2281	$C_{13}H_{10}O_4$	14.36	1.76	230.0579
$C_{14}H_{28}O_2$	15.65	1.54	228.2090	$C_{13}H_{13}N_2O_2$	15.10	1.46	229.0978	$C_{13}H_{28}NO_2$	14.96	1.44	230.2121
$C_{14}H_{12}O_3$	15.44	1.71	228.0786	$C_{13}H_{31}N_3$	15.69	1.15	229.2520	$C_{13}H_{12}NO_3$	14.74	1.61	230.0817
$C_{14}H_{30}NO$	16.03	1.40	228.2329	$C_{13}H_{15}N_3O$	15.47	1.32	229.1216	$C_{13}H_{30}N_2O$	15.[illegible]	1.[illegible]	2[illegible]0.2[illegible]
$C_{14}H_{14}NO_2$	15.81	1.57	228.1025	$C_{13}H_{17}N_4$	15.85	1.18	229.1455	$C_{13}H_{14}N_2O_2$	15.11	1.47	230.1056
$C_{14}H_{32}N_2$	16.40	1.26	228.2567	$C_{13}HN_4O$	15.63	1.34	229.0151	$C_{13}H_{16}N_3O$	15.49	1.32	230.1295
$C_{14}H_{16}N_2O$	16.19	1.43	228.1264	$C_{14}H_{29}O_2$	15.67	1.55	229.2168	$C_{13}N_3O_2$	15.27	1.49	229.9991
$C_{14}N_2O_2$	15.97	1.59	227.9960	$C_{14}H_{13}O_3$	15.45	1.71	229.0865	$C_{13}H_{18}N_4$	15.86	1.18	230.1533
$C_{14}H_{18}N_3$	16.56	1.29	228.1502	$C_{14}H_{31}NO$	16.05	1.41	229.2407	$C_{13}H_2N_4O$	15.64	1.35	230.0229
$C_{14}H_2N_3O$	16.34	1.46	228.01[illegible]	$C_{14}H_{15}NO_2$	15.83	1.57	229.1103	$C_{14}H_{30}O_2$	15.69	1.55	230.2247
$C_{14}H_4N_4$	16.72	1.31	228.0437	$C_{14}H_{17}N_2O$	16.20	1.43	229.1342	$C_{14}H_{14}O_3$	15.47	1.72	230.0943
$C_{15}H_{32}O$	16.76	1.52	228.2454	$C_{14}HN_2O_2$	15.99	1.60	229.0038	$C_{14}H_{16}NO_2$	15.84	1.57	230.1182
$C_{15}H_{16}O_2$	16.54	1.68	228.1151	$C_{14}H_{19}N_3$	16.58	1.29	229.1580	$C_{14}NO_3$	15.63	1.74	229.9878
$C_{15}O_3$	16.33	1.85	227.9847	$C_{14}H_3N_3O$	16.36	1.45	229.0277	$C_{14}H_{18}N_2O$	16.22	1.43	230.1420
$C_{15}H_{18}NO$	16.92	1.54	228.1389	$C_{14}H_5N_4$	16.73	1.32	229.0515	$C_{14}H_2N_2O_2$	16.00	1.60	230.0116
$C_{15}H_2NO_2$	16.70	1.71	228.0085	$C_{15}H_{17}O_2$	16.56	1.68	229.1229	$C_{14}H_{20}N_3$	16.59	1.29	230.1659

	M+1	M+2	FM
$C_{14}H_{4}N_{3}O$	16.38	1.46	230.0355
$C_{14}H_{6}N_{4}$	16.75	1.32	230.0594
$C_{15}H_{18}O_{2}$	16.58	1.69	230.1307
$C_{15}H_{2}O_{3}$	16.36	1.85	230.0003
$C_{15}H_{20}NO$	16.95	1.55	230.1546
$C_{15}H_{4}NO_{2}$	16.73	1.71	230.0242
$C_{15}H_{22}N_{2}$	17.32	1.41	230.1784
$C_{15}H_{6}N_{2}O$	17.11	1.57	230.0480
$C_{15}H_{8}N_{3}$	17.48	1.44	230.0719
$C_{16}H_{22}O$	17.68	1.67	230.1671
$C_{16}H_{6}O_{2}$	17.46	1.83	230.0368
$C_{16}H_{24}N$	18.06	1.54	230.1910
$C_{16}H_{8}NO$	17.84	1.70	230.0606
$C_{16}H_{10}N_{2}$	18.21	1.56	230.0845
$C_{17}H_{26}$	18.79	1.67	230.2036
$C_{17}H_{10}O$	18.57	1.83	230.0732
$C_{17}H_{12}N$	18.94	1.69	230.0970
$C_{18}H_{14}$	19.67	1.83	230.1096
$C_{18}N$	19.83	1.86	230.0031
$C_{19}H_{2}$	20.56	2.00	230.0157
231			
$C_{10}H_{19}N_{2}O_{4}$	12.03	1.47	231.1345
$C_{10}H_{21}N_{3}O_{3}$	12.40	1.31	231.1584
$C_{10}H_{23}N_{4}O_{2}$	12.78	1.16	231.1822
$C_{11}H_{21}NO_{4}$	12.76	1.55	231.1471
$C_{11}H_{23}N_{2}O_{3}$	13.13	1.40	231.1710
$C_{11}H_{7}N_{2}O_{4}$	12.92	1.57	231.0406
$C_{11}H_{25}N_{3}O_{2}$	13.51	1.25	231.1948
$C_{11}H_{9}N_{3}O_{3}$	13.29	1.42	231.0644
$C_{11}H_{27}N_{4}O$	13.88	1.10	231.2187
$C_{11}H_{11}N_{4}O_{2}$	13.67	1.27	231.0883
$C_{12}H_{23}O_{4}$	13.49	1.64	231.1597
$C_{12}H_{25}NO_{3}$	13.87	1.49	231.1835
$C_{12}H_{9}NO_{4}$	13.65	1.66	231.0532
$C_{12}H_{27}N_{2}O_{2}$	14.24	1.34	231.2074
$C_{12}H_{11}N_{2}O_{3}$	14.02	1.51	231.0770
$C_{12}H_{29}N_{3}O$	14.61	1.20	231.2312
$C_{12}H_{13}N_{3}O_{2}$	14.40	1.37	231.1009
$C_{12}H_{15}N_{4}O$	14.77	1.22	231.1247
$C_{13}H_{27}O_{3}$	14.60	1.59	231.1961
$C_{13}H_{11}O_{4}$	14.38	1.76	231.0657
$C_{13}H_{29}NO_{2}$	14.97	1.45	231.2199
$C_{13}H_{13}NO_{3}$	14.75	1.61	231.0896
$C_{13}H_{15}N_{2}O_{2}$	15.13	1.47	231.1134
$C_{13}H_{17}N_{3}O$	15.50	1.32	231.1373
$C_{13}HN_{3}O_{2}$	15.29	1.49	231.0069
$C_{13}H_{19}N_{4}$	15.88	1.18	231.1611
$C_{13}H_{3}N_{4}O$	15.66	1.35	231.0308
$C_{14}H_{15}O_{3}$	15.49	1.72	231.1021
$C_{14}H_{17}NO_{2}$	15.86	1.58	231.1260
$C_{14}HNO_{3}$	15.64	1.74	230.9956
$C_{14}H_{19}N_{2}O$	16.23	1.44	231.1498
$C_{14}H_{3}N_{2}O_{2}$	16.02	1.60	231.0195

	M+1	M+2	FM
$C_{14}H_{21}N_{3}$	16.61	1.30	231.1737
$C_{14}H_{5}N_{3}O$	16.39	1.46	231.0433
$C_{14}H_{7}N_{4}$	16.77	1.32	231.0672
$C_{15}H_{19}O_{2}$	16.59	1.69	231.1385
$C_{15}H_{3}O_{3}$	16.37	1.85	231.0082
$C_{15}H_{21}NO$	16.97	1.55	231.1624
$C_{15}H_{5}NO_{2}$	16.75	1.72	231.0320
$C_{15}H_{23}N_{2}$	17.34	1.41	231.1863
$C_{15}H_{7}N_{2}O$	17.12	1.58	231.0559
$C_{15}H_{9}N_{3}$	17.50	1.44	231.0798
$C_{16}H_{23}O$	17.70	1.67	231.1750
$C_{16}H_{7}O_{2}$	17.48	1.84	231.0446
$C_{16}H_{25}N$	18.07	1.54	231.1988
$C_{16}H_{9}NO$	17.85	1.70	231.0684
$C_{16}H_{11}N_{2}$	18.23	1.57	231.0923
$C_{17}H_{27}$	18.80	1.67	231.2114
$C_{17}H_{11}O$	18.59	1.83	231.0810
$C_{17}H_{13}N$	18.96	1.70	231.1049
$C_{18}H_{15}$	19.69	1.83	231.1174
$C_{18}HN$	19.85	1.86	231.0109
$C_{19}H_{3}$	20.58	2.01	231.0235
232			
$C_{10}H_{20}N_{2}O_{4}$	12.05	1.47	232.1424
$C_{10}H_{22}N_{3}O_{3}$	12.42	1.31	232.1662
$C_{10}H_{24}N_{4}O_{2}$	12.79	1.16	232.1901
$C_{11}H_{22}NO_{4}$	12.78	1.55	232.1549
$C_{11}H_{24}N_{2}O_{3}$	13.15	1.40	232.1788
$C_{11}H_{8}N_{2}O_{4}$	12.93	1.57	232.0484
$C_{11}H_{26}N_{3}O_{2}$	13.53	1.25	232.2026
$C_{11}H_{10}N_{3}O_{3}$	13.31	1.42	232.0723
$C_{11}H_{28}N_{4}O$	13.90	1.10	232.2265
$C_{11}H_{12}N_{4}O_{2}$	13.68	1.27	232.0961
$C_{12}H_{24}O_{4}$	13.51	1.64	232.1675
$C_{12}H_{26}NO_{3}$	13.88	1.49	232.1914
$C_{12}H_{10}NO_{4}$	13.67	1.66	232.0610
$C_{12}H_{28}N_{2}O_{2}$	14.26	1.35	232.2152
$C_{12}H_{12}N_{2}O_{3}$	14.04	1.52	232.0848
$C_{12}H_{14}N_{3}O_{2}$	14.41	1.37	232.1087
$C_{12}H_{16}N_{4}O$	14.79	1.22	232.1325
$C_{12}N_{4}O_{2}$	14.57	1.39	232.0022
$C_{13}H_{28}O_{3}$	14.61	1.59	232.2039
$C_{13}H_{12}O_{4}$	14.40	1.76	232.0735
$C_{13}H_{14}NO_{3}$	14.77	1.62	232.0974
$C_{13}H_{16}N_{2}O_{2}$	15.14	1.47	232.1213
$C_{13}N_{2}O_{3}$	14.93	1.64	231.9909
$C_{13}H_{18}N_{3}O$	15.52	1.33	232.1451
$C_{13}H_{2}N_{3}O_{2}$	15.30	1.49	232.0147
$C_{13}H_{20}N_{4}$	15.89	1.18	232.1690
$C_{13}H_{4}N_{4}O$	15.68	1.35	232.0386
$C_{14}H_{16}O_{3}$	15.50	1.72	232.1100
$C_{14}O_{4}$	15.28	1.89	231.9796
$C_{14}H_{18}NO_{2}$	15.88	1.58	232.1338
$C_{14}H_{2}NO_{3}$	15.66	1.74	232.0034

	M+1	M+2	FM
$C_{14}H_{20}N_{2}O$	16.25	1.44	232.1577
$C_{14}H_{4}N_{2}O_{2}$	16.03	1.60	232.0273
$C_{14}H_{22}N_{3}$	16.62	1.30	232.1815
$C_{14}H_{6}N_{3}O$	16.41	1.46	232.0511
$C_{14}H_{8}N_{4}$	16.78	1.32	232.0750
$C_{15}H_{20}O_{2}$	16.61	1.69	232.1464
$C_{15}H_{4}O_{3}$	16.39	1.86	232.0160
$C_{15}H_{22}NO$	16.98	1.55	232.1702
$C_{15}H_{6}NO_{2}$	16.76	1.72	232.0399
$C_{15}H_{24}N_{2}$	17.36	1.42	232.1941
$C_{15}H_{8}N_{2}O$	17.14	1.58	232.0637
$C_{15}H_{10}N_{3}$	17.51	1.44	232.0876
$C_{16}H_{24}O$	17.71	1.68	232.1828
$C_{16}H_{8}O_{2}$	17.50	1.84	232.0524
$C_{16}H_{26}N$	18.09	1.54	232.2067
$C_{16}H_{10}NO$	17.87	1.70	232.0763
$C_{16}H_{12}N_{2}$	18.24	1.57	232.1001
$C_{17}H_{28}$	18.82	1.67	232.2192
$C_{17}H_{12}O$	18.60	1.83	232.0888
$C_{17}H_{14}N$	18.98	1.70	232.1127
$C_{17}N_{2}$	19.13	1.73	232.0062
$C_{18}H_{16}$	19.71	1.84	232.1253
$C_{18}O$	19.49	1.99	231.9949
$C_{18}H_{2}N$	19.86	1.87	232.0187
$C_{19}H_{4}$	20.60	2.01	232.0313
233			
$C_{10}H_{21}N_{2}O_{4}$	12.06	1.47	233.1502
$C_{10}H_{23}N_{3}O_{3}$	12.44	1.31	233.1741
$C_{10}H_{25}N_{4}O_{2}$	12.81	1.16	233.1979
$C_{11}H_{23}NO_{4}$	12.79	1.56	233.1628
$C_{11}H_{25}N_{2}O_{3}$	13.17	1.40	233.1866
$C_{11}H_{9}N_{2}O_{4}$	12.95	1.57	233.0563
$C_{11}H_{27}N_{3}O_{2}$	13.54	1.25	233.2105
$C_{11}H_{11}N_{3}O_{3}$	13.32	1.42	233.0801
$C_{11}H_{13}N_{4}O_{2}$	13.70	1.27	233.1040
$C_{12}H_{25}O_{4}$	13.52	1.65	233.1753
$C_{12}H_{27}NO_{3}$	13.90	1.50	233.1992
$C_{12}H_{11}NO_{4}$	13.68	1.67	233.0688
$C_{12}H_{13}N_{2}O_{3}$	14.06	1.52	233.0927
$C_{12}H_{15}N_{3}O_{2}$	14.43	1.37	233.1165
$C_{12}H_{17}N_{4}O$	14.80	1.22	233.1404
$C_{12}HN_{4}O_{2}$	14.59	1.39	233.0100
$C_{13}H_{13}O_{4}$	14.41	1.76	233.0814
$C_{13}H_{15}NO_{3}$	14.79	1.62	233.1052
$C_{13}H_{17}N_{2}O_{2}$	15.16	1.47	233.1291
$C_{13}HN_{2}O_{3}$	14.94	1.64	232.9987
$C_{13}H_{19}N_{3}O$	15.54	1.33	233.1529
$C_{13}H_{3}N_{3}O_{2}$	15.32	1.50	233.0226
$C_{13}H_{21}N_{4}$	15.91	1.19	233.1768
$C_{13}H_{5}N_{4}O$	15.69	1.35	233.0464
$C_{14}H_{17}O_{3}$	15.52	1.72	233.1178
$C_{14}HO_{4}$	15.30	1.89	232.9874
$C_{14}H_{19}NO_{2}$	15.89	1.58	233.1416
$C_{14}H_{3}NO_{3}$	15.68	1.75	233.0113

	M+1	M+2	FM
$C_{14}H_{21}N_2O$	16.27	1.44	233.1655
$C_{14}H_5N_2O_2$	16.05	1.61	233.0351
$C_{14}H_{23}N_3$	16.64	1.30	233.1894
$C_{14}H_7N_3O$	16.42	1.47	233.0590
$C_{14}H_9N_4$	16.80	1.33	233.0829
$C_{15}H_{21}O_2$	16.62	1.70	233.1542
$C_{15}H_5O_3$	16.41	1.86	233.0238
$C_{15}H_{23}NO$	17.00	1.56	233.1781
$C_{15}H_7NO_2$	16.78	1.72	233.0477
$C_{15}H_{25}N_2$	17.37	1.42	233.2019
$C_{15}H_9N_2O$	17.16	1.58	233.0715
$C_{15}H_{11}N_3$	17.53	1.45	233.0954
$C_{16}H_{25}O$	17.73	1.68	233.1906
$C_{16}H_9O_2$	17.51	1.84	233.0603
$C_{16}H_{27}N$	18.10	1.54	233.2145
$C_{16}H_{11}NO$	17.89	1.71	233.0841
$C_{16}H_{13}N_2$	18.26	1.57	233.1080
$C_{17}H_{29}$	18.83	1.67	233.2270
$C_{17}H_{13}O$	18.62	1.83	233.0967
$C_{17}H_{15}N$	18.99	1.70	233.1205
$C_{17}HN_2$	19.15	1.73	233.0140
$C_{18}H_{17}$	19.72	1.84	233.1331
$C_{18}HO$	19.51	2.00	233.0027
$C_{18}H_3N$	19.88	1.87	233.0266
$C_{19}H_5$	20.61	2.01	233.0391
234			
$C_{10}H_{22}N_2O_4$	12.08	1.47	234.1580
$C_{10}H_{24}N_3O_3$	12.45	1.32	234.1819
$C_{10}H_{26}N_4O_2$	12.83	1.16	234.2057
$C_{11}H_{24}NO_4$	12.81	1.56	234.1706
$C_{11}H_{26}N_2O_3$	13.18	1.40	234.1945
$C_{11}H_{10}N_2O_4$	12.97	1.58	234.0641
$C_{11}H_{12}N_3O_3$	13.34	1.42	234.0879
$C_{11}H_{14}N_4O_2$	13.71	1.27	234.1118
$C_{12}H_{26}O_4$	13.54	1.65	234.1832
$C_{12}H_{12}NO_4$	13.70	1.67	234.0766
$C_{12}H_{14}N_2O_3$	14.07	1.52	234.1005
$C_{12}H_{16}N_3O_2$	14.45	1.37	234.1244
$C_{12}N_3O_3$	14.23	1.54	233.9940
$C_{12}H_{18}N_4O$	14.82	1.23	234.1482
$C_{12}H_2N_4O_2$	14.60	1.39	234.0178
$C_{13}H_{14}O_4$	14.43	1.77	234.0892
$C_{13}H_{16}NO_3$	14.80	1.62	234.1131
$C_{13}NO_4$	14.59	1.79	233.9827
$C_{13}H_{18}N_2O_2$	15.18	1.48	234.1369
$C_{13}H_2N_2O_3$	14.96	1.64	234.0065
$C_{13}H_{20}N_3O$	15.55	1.33	234.1608
$C_{13}H_4N_3O_2$	15.33	1.50	234.0304
$C_{13}H_{22}N_4$	15.93	1.19	234.1846
$C_{13}H_6N_4O$	15.71	1.36	234.0542
$C_{14}H_{18}O_3$	15.53	1.73	234.1256
$C_{14}H_2O_4$	15.32	1.89	233.9953
$C_{14}H_{20}NO_2$	15.91	1.58	234.1495
$C_{14}H_4NO_3$	15.69	1.75	234.0191
$C_{14}H_{22}N_2O$	16.28	1.44	234.1733
$C_{14}H_6N_2O_2$	16.07	1.61	234.0429
$C_{14}H_{24}N_3$	16.66	1.30	234.1972
$C_{14}H_8N_3O$	16.44	1.47	234.0668
$C_{14}H_{10}N_4$	16.81	1.33	234.0907
$C_{15}H_{22}O_2$	16.64	1.70	234.1620
$C_{15}H_6O_3$	16.42	1.86	234.0317
$C_{15}H_{24}NO$	17.01	1.56	234.1859
$C_{15}H_8NO_2$	16.80	1.72	234.0555
$C_{15}H_{26}N_2$	17.39	1.42	234.2098
$C_{15}H_{10}N_2O$	17.17	1.59	234.0794
$C_{15}H_{12}N_3$	17.55	1.45	234.1032
$C_{16}H_{26}O$	17.74	1.68	234.1985
$C_{16}H_{10}O_2$	17.53	1.84	234.0681
$C_{16}H_{28}N$	18.12	1.55	234.2223
$C_{16}H_{12}NO$	17.90	1.71	234.0919
$C_{16}H_{14}N_2$	18.28	1.58	234.1158
$C_{16}N_3$	18.43	1.60	234.0093
$C_{17}H_{30}$	18.85	1.68	234.2349
$C_{17}H_{14}O$	18.63	1.84	234.1045
$C_{17}H_{16}N$	19.01	1.71	234.1284
$C_{17}NO$	18.79	1.87	233.9980
$C_{17}H_2N_2$	19.17	1.74	234.0218
$C_{18}H_{18}$	19.74	1.84	234.1409
$C_{18}H_2O$	19.52	2.00	234.0106
$C_{18}H_4N$	19.90	1.87	234.0344
$C_{19}H_6$	20.63	2.02	234.0470
235			
$C_{10}H_{23}N_2O_4$	12.09	1.47	235.1659
$C_{10}H_{25}N_3O_3$	12.47	1.32	235.1897
$C_{11}H_{25}NO_4$	12.82	1.56	235.1784
$C_{11}H_{11}N_2O_4$	12.98	1.58	235.0719
$C_{11}H_{13}N_3O_3$	13.36	1.43	235.0958
$C_{11}H_{15}N_4O_2$	13.73	1.28	235.1196
$C_{12}H_{13}NO_4$	13.71	1.67	235.0845
$C_{12}H_{15}N_2O_3$	14.09	1.52	235.1083
$C_{12}H_{17}N_3O_2$	14.46	1.37	235.1322
$C_{12}HN_3O_3$	14.24	1.54	235.0018
$C_{12}H_{19}N_4O$	14.84	1.23	235.1560
$C_{12}H_3N_4O_2$	14.62	1.40	235.0257
$C_{13}H_{15}O_4$	14.44	1.77	235.0970
$C_{13}H_{17}NO_3$	14.82	1.62	235.1209
$C_{13}HNO_4$	14.60	1.79	234.9905
$C_{13}H_{19}N_2O_2$	15.19	1.48	235.1447
$C_{13}H_3N_2O_3$	14.98	1.65	235.0144
$C_{13}H_{21}N_3O$	15.57	1.33	235.1686
$C_{13}H_5N_3O_2$	15.35	1.50	235.0382
$C_{13}H_{23}N_4$	15.94	1.19	235.1925
$C_{13}H_7N_4O$	15.72	1.36	235.0621
$C_{14}H_{19}O_3$	15.55	1.73	235.1334
$C_{14}H_3O_4$	15.33	1.90	235.0031
$C_{14}H_{21}NO_2$	15.92	1.59	235.1573
$C_{14}H_5NO_3$	15.71	1.75	235.0269
$C_{14}H_{23}N_2O$	16.30	1.45	235.1811
$C_{14}H_7N_2O_2$	16.08	1.61	235.0508
$C_{14}H_{25}N_3$	16.67	1.31	235.2050
$C_{14}H_9N_3O$	16.46	1.47	235.0746
$C_{14}H_{11}N_4$	16.83	1.33	235.0985
$C_{15}H_{23}O_2$	16.66	1.70	235.1699
$C_{15}H_7O_3$	16.44	1.86	235.0395
$C_{15}H_{25}NO$	17.03	1.56	235.1937
$C_{15}H_9NO_2$	16.81	1.73	235.0634
$C_{15}H_{27}N_2$	17.40	1.43	235.2176
$C_{15}H_{11}N_2O$	17.19	1.59	235.0872
$C_{15}H_{13}N_3$	17.56	1.45	235.1111
$C_{16}H_{27}O$	17.76	1.68	235.2063
$C_{16}H_{11}O_2$	17.54	1.85	235.0759
$C_{16}H_{29}N$	18.14	1.55	235.2301
$C_{16}H_{13}NO$	17.92	1.71	235.0998
$C_{16}H_{15}N_2$	18.29	1.58	235.1236
$C_{16}HN_3$	18.45	1.61	235.0171
$C_{17}H_{31}$	18.87	1.68	235.2427
$C_{17}H_{15}O$	18.65	1.84	235.1123
$C_{17}H_{17}N$	19.02	1.71	235.1362
$C_{17}HNO$	18.81	1.87	235.0058
$C_{17}H_3N_2$	19.18	1.74	235.0297
$C_{18}H_{19}$	19.75	1.85	235.1488
$C_{18}H_3O$	19.54	2.00	235.0184
$C_{18}H_5N$	19.91	1.88	235.0422
$C_{19}H_7$	20.64	2.02	235.0548
236			
$C_{10}H_{24}N_2O_4$	12.11	1.48	236.1737
$C_{11}H_{12}N_2O_4$	13.00	1.58	236.0797
$C_{11}H_{14}N_3O_3$	13.37	1.43	236.1036
$C_{11}H_{16}N_4O_2$	13.75	1.28	236.1275
$C_{12}H_{14}NO_4$	13.73	1.67	236.0923
$C_{12}H_{16}N_2O_3$	14.10	1.52	236.1162
$C_{12}N_2O_4$	13.89	1.69	235.9858
$C_{12}H_{18}N_3O_2$	14.48	1.38	236.1400
$C_{12}H_2N_3O_3$	14.26	1.55	236.0090
$C_{12}H_{20}N_4O$	14.85	1.23	236.1639
$C_{12}H_4N_4O_2$	14.64	1.40	236.0335
$C_{13}H_{16}O_4$	14.46	1.77	236.1049
$C_{13}H_{18}NO_3$	14.83	1.63	236.1207
$C_{13}H_2NO_4$	14.62	1.79	235.9983
$C_{13}H_{20}N_2O_2$	15.21	1.48	236.1526
$C_{13}H_4N_2O_3$	14.99	1.65	236.0222
$C_{13}H_{22}N_3O$	15.58	1.34	236.1764
$C_{13}H_6N_3O_2$	15.37	1.50	236.0460
$C_{13}H_{24}N_4$	15.96	1.19	236.2003
$C_{13}H_8N_4O$	15.74	1.36	236.0699
$C_{14}H_{20}O_3$	15.57	1.73	236.1413
$C_{14}H_4O_4$	15.35	1.90	236.0109
$C_{14}H_{22}NO_2$	15.94	1.59	236.1651
$C_{14}H_6NO_3$	15.72	1.76	236.0348
$C_{14}H_{24}N_2O$	16.31	1.45	236.1890

	M+1	M+2	FM
$C_{14}H_8N_2O_2$	16.10	1.61	236.0586
$C_{14}H_{26}N_3$	16.69	1.31	236.2129
$C_{14}H_{10}N_3O$	16.47	1.47	236.0825
$C_{14}H_{12}N_4$	16.85	1.33	236.1063
$C_{15}H_{24}O_2$	16.67	1.70	236.1777
$C_{15}H_8O_3$	16.45	1.87	236.0473
$C_{15}H_{26}NO$	17.05	1.56	236.2015
$C_{15}H_{10}NO_2$	16.83	1.73	236.0712
$C_{15}H_{28}N_2$	17.42	1.43	236.2254
$C_{15}H_{12}N_2O$	17.20	1.59	236.0950
$C_{15}H_{14}N_3$	17.58	1.46	236.1189
$C_{15}N_4$	17.73	1.48	236.0124
$C_{16}H_{28}O$	17.78	1.69	236.2141
$C_{16}H_{12}O_2$	17.56	1.85	236.0837
$C_{16}H_{30}N$	18.15	1.55	236.2380
$C_{16}H_{14}NO$	17.93	1.71	236.1076
$C_{16}H_{16}N_2$	18.31	1.58	236.1315
$C_{16}N_2O$	18.09	1.74	236.0011
$C_{16}H_2N_3$	18.47	1.61	236.0249
$C_{17}H_{32}$	18.88	1.68	236.2505
$C_{17}H_{16}O$	18.67	1.84	236.1202
$C_{17}O_2$	18.45	2.00	235.9898
$C_{17}H_{18}N$	19.04	1.71	236.1440
$C_{17}H_2NO$	18.82	1.87	236.0136
$C_{17}H_4N_2$	19.20	1.74	236.0375
$C_{18}H_{20}$	19.77	1.85	236.1566
$C_{18}H_4O$	19.55	2.01	236.0262
$C_{18}H_6N$	19.93	1.88	236.0501
$C_{19}H_8$	20.66	2.02	236.0626
237			
$C_{11}H_{13}N_2O_4$	13.01	1.58	237.0876
$C_{11}H_{15}N_3O_3$	13.39	1.43	237.1114
$C_{11}H_{17}N_4O_2$	13.76	1.28	237.1353
$C_{12}H_{15}NO_4$	13.75	1.68	237.1001
$C_{12}H_{17}N_2O_3$	14.12	1.53	237.1240
$C_{12}HN_2O_4$	13.90	1.70	236.9936
$C_{12}H_{19}N_3O_2$	14.49	1.38	237.1478
$C_{12}H_3N_3O_3$	14.28	1.55	237.0175
$C_{12}H_{21}N_4O$	14.87	1.23	237.1717
$C_{12}H_5N_4O_2$	14.65	1.40	237.0413
$C_{13}H_{17}O_4$	14.48	1.77	237.1127
$C_{13}H_{19}NO_3$	14.85	1.63	237.1365
$C_{13}H_3NO_4$	14.63	1.80	237.0062
$C_{13}H_{21}N_2O_2$	15.22	1.48	237.1604
$C_{13}H_5N_2O_3$	15.01	1.65	237.0300
$C_{13}H_{23}N_3O$	15.60	1.34	237.1842
$C_{13}H_7N_3O_2$	15.38	1.51	237.0539
$C_{13}H_{25}N_4$	15.97	1.20	237.2081
$C_{13}H_9N_4O$	15.76	1.36	237.0777
$C_{14}H_{21}O_3$	15.58	1.73	237.1491
$C_{14}H_5O_4$	15.36	1.90	237.0187
$C_{14}H_{23}NO_2$	15.96	1.59	237.1730
$C_{14}H_7NO_3$	15.74	1.76	237.0426

	M+1	M+2	FM
$C_{14}H_{25}N_2O$	16.33	1.45	237.1968
$C_{14}H_9N_2O_2$	16.11	1.62	237.0664
$C_{14}H_{27}N_3$	16.70	1.31	237.2207
$C_{14}H_{11}N_3O$	16.49	1.48	237.0903
$C_{14}H_{13}N_4$	16.86	1.34	237.1142
$C_{15}H_{25}O_2$	16.69	1.71	237.1855
$C_{15}H_9O_3$	16.47	1.87	237.0552
$C_{15}H_{27}NO$	17.06	1.57	237.2094
$C_{15}H_{11}NO_2$	16.84	1.73	237.0790
$C_{15}H_{29}N_2$	17.44	1.43	237.2332
$C_{15}H_{13}N_2O$	17.22	1.59	237.1029
$C_{15}H_{15}N_3$	17.59	1.46	237.1267
$C_{15}HN_4$	17.75	1.48	237.0202
$C_{16}H_{29}O$	17.79	1.69	237.2219
$C_{16}H_{13}O_2$	17.58	1.85	237.0916
$C_{16}H_{31}N$	18.17	1.56	237.2458
$C_{16}H_{15}NO$	17.95	1.72	237.1154
$C_{16}H_{17}N_2$	18.32	1.58	237.1393
$C_{16}HN_2O$	18.11	1.75	237.0089
$C_{16}H_3N_3$	18.48	1.61	237.0328
$C_{17}H_{33}$	18.90	1.69	237.2584
$C_{17}H_{17}O$	18.68	1.85	237.1280
$C_{17}HO_2$	18.46	2.01	236.9976
$C_{17}H_{19}N$	19.06	1.72	237.1519
$C_{17}H_3NO$	18.84	1.87	237.0215
$C_{17}H_5N_2$	19.21	1.75	237.0453
$C_{18}H_{21}$	19.79	1.85	237.1644
$C_{18}H_5O$	19.57	2.01	237.0340
$C_{18}H_7N$	19.94	1.88	237.0579
$C_{19}H_9$	20.68	2.03	237.0705
238			
$C_{11}H_{14}N_2O_4$	13.03	1.59	238.0954
$C_{11}H_{16}N_3O_3$	13.40	1.43	238.1193
$C_{11}H_{18}N_4O_2$	13.78	1.28	238.1431
$C_{12}H_{16}NO_4$	13.76	1.68	238.1080
$C_{12}H_{18}N_2O_3$	14.14	1.53	238.1318
$C_{12}H_2N_2O_4$	13.92	1.70	238.0014
$C_{12}H_{20}N_3O_2$	14.51	1.38	238.1557
$C_{12}H_4N_3O_3$	14.29	1.55	238.0253
$C_{12}H_{22}N_4O$	14.88	1.24	238.1795
$C_{12}H_6N_4O_2$	14.67	1.40	238.0491
$C_{13}H_{18}O_4$	14.49	1.78	238.1205
$C_{13}H_{20}NO_3$	14.87	1.63	238.1444
$C_{13}H_4NO_4$	14.65	1.80	238.0140
$C_{13}H_{22}N_2O_2$	15.24	1.48	238.1682
$C_{13}H_6N_2O_3$	15.02	1.65	238.0379
$C_{13}H_{24}N_3O$	15.62	1.34	238.1921
$C_{13}H_8N_3O_2$	15.40	1.51	238.0617
$C_{13}H_{26}N_4$	15.99	1.20	238.2160
$C_{13}H_{10}N_4O$	15.77	1.37	238.0856
$C_{14}H_{22}O_3$	15.60	1.74	238.1569
$C_{14}H_6O_4$	15.38	1.90	238.0266
$C_{14}H_{24}NO_2$	15.97	1.59	238.1808
$C_{14}H_8NO_3$	15.76	1.76	238.0504

	M+1	M+2	FM
$C_{14}H_{26}N_2O$	16.35	1.45	238.2046
$C_{14}H_{10}N_2O_2$	16.13	1.62	238.0743
$C_{14}H_{28}N_3$	16.72	1.31	238.2285
$C_{14}H_{12}N_3O$	16.50	1.48	238.0981
$C_{14}H_{14}N_4$	16.88	1.34	238.1220
$C_{15}H_{26}O_2$	16.70	1.71	238.1934
$C_{15}H_{10}O_3$	16.49	1.87	238.0630
$C_{15}H_{28}NO$	17.08	1.57	238.2172
$C_{15}H_{12}NO_2$	16.86	1.73	238.0868
$C_{15}H_{30}N_2$	17.45	1.43	238.2411
$C_{15}H_{14}N_2O$	17.24	1.60	238.1107
$C_{15}H_{16}N_3$	17.61	1.46	238.1346
$C_{15}N_3O$	17.39	1.62	238.0042
$C_{15}H_2N_4$	17.77	1.49	238.0280
$C_{16}H_{30}O$	17.81	1.69	238.2298
$C_{16}H_{14}O_2$	17.59	1.85	238.0994
$C_{16}H_{32}N$	18.18	1.56	238.2536
$C_{16}H_{16}NO$	17.97	1.72	238.1233
$C_{16}NO_2$	17.75	1.88	237.9929
$C_{16}H_{18}N_2$	18.34	1.59	238.1471
$C_{16}H_2N_2O$	18.12	1.75	238.0167
$C_{16}H_4N_3$	18.50	1.62	238.0406
$C_{17}H_{34}$	18.91	1.69	238.2662
$C_{17}H_{18}O$	18.70	1.85	238.1358
$C_{17}H_2O_2$	18.48	2.01	238.0054
$C_{17}H_{20}N$	19.07	1.72	238.1597
$C_{17}H_4NO$	18.85	1.88	238.0293
$C_{17}H_6N_2$	19.23	1.75	238.0532
$C_{18}H_{22}$	19.80	1.86	238.1722
$C_{18}H_6O$	19.59	2.01	238.0419
$C_{18}H_8N$	19.96	1.89	238.0657
$C_{19}H_{10}$	20.69	2.03	238.0783
239			
$C_{11}H_{15}N_2O_4$	13.05	1.59	239.1032
$C_{11}H_{17}N_3O_3$	13.42	1.44	239.1271
$C_{11}H_{19}N_4O_2$	13.79	1.29	239.1509
$C_{12}H_{17}NO_4$	13.78	1.68	239.1158
$C_{12}H_{19}N_2O_3$	14.15	1.53	239.1396
$C_{12}H_3N_2O_4$	13.93	1.70	239.0093
$C_{12}H_{21}N_3O_2$	14.53	1.38	239.1635
$C_{12}H_5N_3O_3$	14.31	1.55	239.0331
$C_{12}H_{23}N_4O$	14.90	1.24	239.1873
$C_{12}H_7N_4O_2$	14.68	1.41	239.0570
$C_{13}H_{19}O_4$	14.51	1.78	239.1284
$C_{13}H_{21}NO_3$	14.88	1.63	239.1522
$C_{13}H_5NO_4$	14.67	1.80	239.0218
$C_{13}H_{23}N_2O_2$	15.26	1.49	239.1761
$C_{13}H_7N_2O_3$	15.04	1.66	239.0457
$C_{13}H_{25}N_3O$	15.63	1.34	239.1999
$C_{13}H_9N_3O_2$	15.41	1.51	239.0695
$C_{13}H_{27}N_4$	16.01	1.20	239.2238
$C_{13}H_{11}N_4O$	15.79	1.37	239.0934
$C_{14}H_{23}O_3$	15.61	1.74	239.1648

	M+1	M+2	FM		M+1	M+2	FM		M+1	M+2	FM
$C_{14}H_7O_4$	15.40	1.91	239.0344	$C_{13}H_{28}N_4$	16.02	1.20	240.2316	$C_{13}H_{21}O_4$	14.54	1.78	241.1440
$C_{14}H_{25}NO_2$	15.99	1.60	239.1886	$C_{13}H_{12}N_4O$	15.80	1.37	240.1012	$C_{13}H_{23}NO_3$	14.91	1.64	241.1679
$C_{14}H_9NO_3$	15.77	1.76	239.0583	$C_{14}H_{24}O_3$	15.63	1.74	240.1726	$C_{13}H_7NO_4$	14.70	1.81	241.0375
$C_{14}H_{27}N_2O$	16.36	1.46	239.2125	$C_{14}H_8O_4$	15.41	1.91	240.0422	$C_{13}H_{25}N_2O_2$	15.29	1.49	241.1917
$C_{14}H_{11}N_2O_2$	16.15	1.62	239.0821	$C_{14}H_{26}NO_2$	16.00	1.60	240.1965	$C_{13}H_9N_2O_3$	15.07	1.66	241.0614
$C_{14}H_{29}N_3$	16.74	1.32	239.2363	$C_{14}H_{10}NO_3$	15.79	1.77	240.0661	$C_{13}H_{27}N_3O$	15.66	1.35	241.2156
$C_{14}H_{13}N_3O$	16.52	1.48	239.1060	$C_{14}H_{28}N_2O$	16.38	1.46	240.2203	$C_{13}H_{11}N_3O_2$	15.45	1.52	241.0852
$C_{14}H_{15}N_4$	16.89	1.34	239.1298	$C_{14}H_{12}N_2O_2$	16.16	1.62	240.0899	$C_{13}H_{29}N_4$	16.04	1.21	241.2394
$C_{15}H_{27}O_2$	16.72	1.71	239.2012	$C_{14}H_{30}N_3$	16.75	1.32	240.2442	$C_{13}H_{13}N_4O$	15.82	1.37	241.1091
$C_{15}H_{11}O_3$	16.50	1.88	239.0708	$C_{14}H_{14}N_3O$	16.54	1.48	240.1138	$C_{14}H_{25}O_3$	15.65	1.74	241.1804
$C_{15}H_{29}NO$	17.09	1.57	239.2250	$C_{14}H_{16}N_4$	16.91	1.35	240.1377	$C_{14}H_9O_4$	15.43	1.91	241.0501
$C_{15}H_{13}NO_2$	16.88	1.74	239.0947	$C_{14}N_4O$	16.69	1.51	240.0073	$C_{14}H_{27}NO_2$	16.02	1.60	241.2043
$C_{15}H_{31}N_2$	17.47	1.44	239.2489	$C_{15}H_{28}O_2$	16.74	1.71	240.2090	$C_{14}H_{11}NO_3$	15.80	1.77	241.0739
$C_{15}H_{15}N_2O$	17.25	1.60	239.1185	$C_{15}H_{12}O_3$	16.52	1.88	240.0786	$C_{14}H_{29}N_2O$	16.39	1.46	241.2281
$C_{15}H_{17}N_3$	17.63	1.46	239.1424	$C_{15}H_{30}NO$	17.11	1.58	240.2329	$C_{14}H_{13}N_2O_2$	16.18	1.63	241.0978
$C_{15}HN_3O$	17.41	1.63	239.0120	$C_{15}H_{14}NO_2$	16.89	1.74	240.1025	$C_{14}H_{31}N_3$	16.77	1.32	241.2520
$C_{15}H_3N_4$	17.78	1.49	239.0359	$C_{15}H_{32}N_2$	17.48	1.44	240.2567	$C_{14}H_{15}N_3O$	16.55	1.49	241.1216
$C_{16}H_{31}O$	17.82	1.70	239.2376	$C_{15}H_{16}N_2O$	17.27	1.60	240.1264	$C_{14}H_{17}N_4$	16.93	1.35	241.1455
$C_{16}H_{15}O_2$	17.61	1.86	239.1072	$C_{15}N_2O_2$	17.05	1.77	239.9960	$C_{14}HN_4O$	16.71	1.51	241.0151
$C_{16}H_{33}N$	18.20	1.56	239.2615	$C_{15}H_{18}N_3$	17.64	1.47	240.1502	$C_{15}H_{29}O_2$	16.75	1.72	241.2168
$C_{16}H_{17}NO$	17.98	1.72	239.1311	$C_{15}H_2N_3O$	17.42	1.63	240.0198	$C_{15}H_{13}O_3$	16.53	1.88	241.0865
$C_{16}HNO_2$	17.77	1.88	239.0007	$C_{15}H_4N_4$	17.80	1.49	240.0437	$C_{15}H_{31}NO$	17.13	1.58	241.2407
$C_{16}H_{19}N_2$	18.36	1.59	239.1549	$C_{16}H_{32}O$	17.84	1.70	240.2454	$C_{15}H_{15}NO_2$	16.91	1.74	241.1103
$C_{16}H_3N_2O$	18.14	1.75	239.0246	$C_{16}H_{16}O_2$	17.62	1.86	240.1151	$C_{15}H_{33}N_2$	17.50	1.44	241.2646
$C_{16}H_5N_3$	18.51	1.62	239.0484	$C_{16}O_3$	17.41	2.02	239.9847	$C_{15}H_{17}N_2O$	17.28	1.60	241.1342
$C_{17}H_{35}$	18.93	1.69	239.2740	$C_{16}H_{34}N$	18.22	1.56	240.2693	$C_{15}HN_2O_2$	17.07	1.77	241.0030
$C_{17}H_{19}O$	18.71	1.85	239.1436	$C_{16}H_{18}NO$	18.00	1.73	240.1389	$C_{15}H_{19}N_3$	17.66	1.47	241.1580
$C_{17}H_3O_2$	18.50	2.01	239.0133	$C_{16}H_2NO_2$	17.78	1.89	240.0085	$C_{15}H_3N_3O$	17.44	1.63	241.0277
$C_{17}H_{21}N$	19.09	1.72	239.1675	$C_{16}H_{20}N_2$	18.37	1.59	240.1628	$C_{15}H_5N_4$	17.81	1.50	241.0515
$C_{17}H_5NO$	18.87	1.88	239.0371	$C_{16}H_4N_2O$	18.16	1.75	240.0324	$C_{16}H_{33}O$	17.86	1.70	241.2533
$C_{17}H_7N_2$	19.25	1.75	239.0610	$C_{16}H_6N_3$	18.53	1.62	240.0563	$C_{16}H_{17}O_2$	17.64	1.86	241.1229
$C_{18}H_{23}$	19.82	1.86	239.1801	$C_{17}H_{36}$	18.95	1.70	240.2819	$C_{16}HO_3$	17.42	2.03	240.9925
$C_{18}H_7O$	19.60	2.02	239.0497	$C_{17}H_{20}O$	18.73	1.86	240.1515	$C_{16}H_{35}N$	18.23	1.57	241.2771
$C_{18}H_9N$	19.98	1.89	239.0736	$C_{17}H_4O_2$	18.51	2.02	240.0211	$C_{16}H_{19}NO$	18.01	1.73	241.1467
$C_{19}H_{11}$	20.71	2.03	239.0861	$C_{17}H_{22}N$	19.10	1.72	240.1753	$C_{16}H_3NO_2$	17.80	1.89	241.0164
				$C_{17}H_6NO$	18.89	1.88	240.0449	$C_{16}H_{21}N_2$	18.39	1.60	241.1706
240				$C_{17}H_8N_2$	19.26	1.75	240.0688	$C_{16}H_5N_2O$	18.17	1.76	241.0402
$C_{11}H_{16}N_2O_4$	13.06	1.59	240.1111	$C_{18}H_{24}$	19.83	1.86	240.1879	$C_{16}H_7N_3$	18.55	1.62	241.0641
$C_{11}H_{18}N_3O_3$	13.44	1.44	240.1349	$C_{18}H_8O$	19.62	2.02	240.0575	$C_{17}H_{21}O$	18.75	1.86	241.1593
$C_{11}H_{20}N_4O_2$	13.81	1.29	240.1588	$C_{18}H_{10}N$	19.99	1.89	240.0814	$C_{17}H_5O_2$	18.53	2.02	241.0289
$C_{12}H_{18}NO_4$	13.79	1.68	240.1236	$C_{19}H_{12}$	20.72	2.04	240.0939	$C_{17}H_{23}N$	19.12	1.73	241.1832
$C_{12}H_{20}N_2O_3$	14.17	1.53	240.1475	C_{20}	21.61	2.22	240.0000	$C_{17}H_7NO$	18.90	1.89	241.0528
$C_{12}H_4N_2O_4$	13.95	1.70	240.0171					$C_{17}H_9N_2$	19.28	1.76	241.0767
$C_{12}H_{22}N_3O_2$	14.54	1.39	240.1713	241				$C_{18}H_{25}$	19.85	1.87	241.1957
$C_{12}H_6N_3O_3$	14.32	1.56	240.0410	$C_{11}H_{17}N_2O_4$	13.08	1.59	241.1189	$C_{18}H_9O$	19.63	2.02	241.0653
$C_{12}H_{24}N_4O$	14.92	1.24	240.1952	$C_{11}H_{19}N_3O_3$	13.45	1.44	241.1427	$C_{18}H_{11}N$	20.01	1.90	241.0892
$C_{12}H_8N_4O_2$	14.70	1.41	240.0648	$C_{11}H_{21}N_4O_2$	13.83	1.29	241.1666	$C_{19}H_{13}$	20.74	2.04	241.1018
$C_{13}H_{20}O_4$	14.52	1.78	240.1362	$C_{12}H_{19}NO_4$	13.81	1.68	241.1315	$C_{20}H$	21.63	2.22	241.0078
$C_{13}H_{22}NO_3$	14.90	1.63	240.1600	$C_{12}H_{21}N_2O_3$	14.18	1.54	241.1553				
$C_{13}H_6NO_4$	14.68	1.80	240.0297	$C_{12}H_5N_2O_4$	13.97	1.71	241.0249	242			
$C_{13}H_{24}N_2O_2$	15.27	1.49	240.1839	$C_{12}H_{23}N_3O_2$	14.56	1.39	241.1791	$C_{11}H_{18}N_2O_4$	13.09	1.59	242.1267
$C_{13}H_8N_2O_3$	15.06	1.66	240.0535	$C_{12}H_7N_3O_3$	14.34	1.56	241.0488	$C_{11}H_{20}N_3O_3$	13.47	1.44	242.1506
$C_{13}H_{26}N_3O$	15.65	1.35	240.2077	$C_{12}H_{25}N_4O$	14.93	1.24	241.2030	$C_{11}H_{22}N_4O_2$	13.84	1.29	242.1744
$C_{13}H_{10}N_3O_2$	15.43	1.51	240.0774	$C_{12}H_9N_4O_2$	14.72	1.41	241.0726	$C_{12}H_{20}NO_4$	13.83	1.69	242.1393

	M+1	M+2	FM
$C_{12}H_{22}N_2O_3$	14.20	1.54	242.1631
$C_{12}H_6N_2O_4$	13.98	1.71	242.0328
$C_{12}H_{24}N_3O_2$	14.57	1.39	242.1870
$C_{12}H_8N_3O_3$	14.36	1.56	242.0566
$C_{12}H_{26}N_4O$	14.95	1.24	242.2108
$C_{12}H_{10}N_4O_2$	14.73	1.41	242.0805
$C_{13}H_{22}O_4$	14.56	1.79	242.1518
$C_{13}H_{24}NO_3$	14.93	1.64	242.1757
$C_{13}H_8NO_4$	14.71	1.81	242.0453
$C_{13}H_{26}N_2O_2$	15.30	1.49	242.1996
$C_{13}H_{10}N_2O_3$	15.09	1.66	242.0692
$C_{13}H_{28}N_3O$	15.68	1.35	242.2234
$C_{13}H_{12}N_3O_2$	15.46	1.52	242.0930
$C_{13}H_{30}N_4$	16.05	1.21	242.2473
$C_{13}H_{14}N_4O$	15.84	1.38	242.1169
$C_{14}H_{26}O_3$	15.66	1.75	242.1883
$C_{14}H_{10}O_4$	15.44	1.91	242.0579
$C_{14}H_{28}NO_2$	16.04	1.60	242.2121
$C_{14}H_{12}NO_3$	15.82	1.77	242.0817
$C_{14}H_{30}N_2O$	16.41	1.46	242.2360
$C_{14}H_{14}N_2O_2$	16.19	1.63	242.1056
$C_{14}H_{32}N_3$	16.78	1.32	242.2598
$C_{14}H_{16}N_3O$	16.57	1.49	242.1295
$C_{14}N_3O_2$	16.35	1.65	241.9991
$C_{14}H_{18}N_4$	16.94	1.35	242.1533
$C_{14}H_2N_4O$	16.73	1.51	242.0229
$C_{15}H_{30}O_2$	16.77	1.72	242.2247
$C_{15}H_{14}O_3$	16.55	1.88	242.0943
$C_{15}H_{32}NO$	17.14	1.58	242.2485
$C_{15}H_{16}NO_2$	16.92	1.74	242.1182
$C_{15}NO_3$	16.71	1.91	241.9878
$C_{15}H_{34}N_2$	17.52	1.44	242.2724
$C_{15}H_{18}N_2O$	17.30	1.61	242.1420
$C_{15}H_2N_2O_2$	17.08	1.77	242.0116
$C_{15}H_{20}N_3$	17.67	1.47	242.1659
$C_{15}H_4N_3O$	17.46	1.63	242.0355
$C_{15}H_6N_4$	17.83	1.50	242.0594
$C_{16}H_{34}O$	17.87	1.70	242.2611
$C_{16}H_{18}O_2$	17.66	1.87	242.1307
$C_{16}H_2O_3$	17.44	2.03	242.0003
$C_{16}H_{20}NO$	18.03	1.73	242.1546
$C_{16}H_4NO_2$	17.81	1.89	242.0242
$C_{16}H_{22}N_2$	18.40	1.60	242.1784
$C_{16}H_6N_2O$	18.19	1.76	242.0480
$C_{16}H_8N_3$	18.56	1.63	242.0719
$C_{17}H_{22}O$	18.76	1.86	242.1671
$C_{17}H_6O_2$	18.54	2.02	242.0368
$C_{17}H_{24}N$	19.14	1.73	242.1910
$C_{17}H_8NO$	18.92	1.89	242.0606
$C_{17}H_{10}N_2$	19.29	1.76	242.0845
$C_{18}H_{26}$	19.87	1.87	242.2036
$C_{18}H_{10}O$	19.65	2.03	242.0732
$C_{18}H_{12}N$	20.02	1.90	242.0970
$C_{19}H_{14}$	20.76	2.04	242.1096
$C_{19}N$	20.91	2.08	242.0031
$C_{20}H_2$	21.64	2.23	242.0157
243			
$C_{11}H_{19}N_2O_4$	13.11	1.60	243.1345
$C_{11}H_{21}N_3O_3$	13.48	1.44	243.1584
$C_{11}H_{23}N_4O_2$	13.86	1.29	243.1822
$C_{12}H_{21}NO_4$	13.84	1.69	243.1471
$C_{12}H_{23}N_2O_3$	14.22	1.54	243.1710
$C_{12}H_7N_2O_4$	14.00	1.71	243.0406
$C_{12}H_{25}N_3O_2$	14.59	1.39	243.1948
$C_{12}H_9N_3O_3$	14.37	1.56	243.0644
$C_{12}H_{27}N_4O$	14.96	1.25	243.2187
$C_{12}H_{11}N_4O_2$	14.75	1.42	243.0883
$C_{13}H_{23}O_4$	14.57	1.79	243.1597
$C_{13}H_{25}NO_3$	14.95	1.64	243.1835
$C_{13}H_9NO_4$	14.73	1.81	243.0532
$C_{13}H_{27}N_2O_2$	15.32	1.50	243.2074
$C_{13}H_{11}N_2O_3$	15.10	1.66	243.0770
$C_{13}H_{29}N_3O$	15.70	1.35	243.2312
$C_{13}H_{13}N_3O_2$	15.48	1.52	243.1009
$C_{13}H_{31}N_4$	16.07	1.21	243.2551
$C_{13}H_{15}N_4O$	15.85	1.38	243.1247
$C_{14}H_{27}O_3$	15.68	1.75	243.1961
$C_{14}H_{11}O_4$	15.46	1.92	243.0657
$C_{14}H_{29}NO_2$	16.05	1.61	243.2199
$C_{14}H_{13}NO_3$	15.84	1.77	243.0896
$C_{14}H_{31}N_2O$	16.43	1.47	243.2438
$C_{14}H_{15}N_2O_2$	16.21	1.63	243.1134
$C_{14}H_{33}N_3$	16.80	1.33	243.2677
$C_{14}H_{17}N_3O$	16.58	1.49	243.1373
$C_{14}HN_3O_2$	16.37	1.66	243.0069
$C_{14}H_{19}N_4$	16.96	1.35	243.1611
$C_{14}H_3N_4O$	16.74	1.52	243.0308
$C_{15}H_{31}O_2$	16.78	1.72	243.2325
$C_{15}H_{15}O_3$	16.57	1.89	243.1021
$C_{15}H_{33}NO$	17.16	1.58	243.2564
$C_{15}H_{17}NO_2$	16.94	1.75	243.1260
$C_{15}HNO_3$	16.72	1.91	242.9956
$C_{15}H_{19}N_2O$	17.32	1.61	243.1498
$C_{15}H_3N_2O_2$	17.10	1.77	243.0195
$C_{15}H_{21}N_3$	17.69	1.47	243.1737
$C_{15}H_5N_3O$	17.47	1.64	243.0433
$C_{15}H_7N_4$	17.85	1.50	243.0672
$C_{16}H_{19}O_2$	17.67	1.87	243.1385
$C_{16}H_3O_3$	17.45	2.03	243.0082
$C_{16}H_{21}NO$	18.05	1.73	243.1624
$C_{16}H_5NO_2$	17.83	1.90	243.0320
$C_{16}H_{23}N_2$	18.42	1.60	243.1863
$C_{16}H_7N_2O$	18.20	1.76	243.0559
$C_{16}H_9N_3$	18.58	1.63	243.0798
$C_{17}H_{23}O$	18.78	1.86	243.1750
$C_{17}H_7O_2$	18.56	2.02	243.0446
$C_{17}H_{25}N$	19.15	1.73	243.1988
$C_{17}H_9NO$	18.93	1.89	243.0684
$C_{17}H_{11}N_2$	19.31	1.76	243.0923
$C_{18}H_{27}$	19.88	1.87	243.2114
$C_{18}H_{11}O$	19.67	2.03	243.0810
$C_{18}H_{13}N$	20.04	1.90	243.1049
$C_{19}H_{15}$	20.77	2.05	243.1174
$C_{19}HN$	20.93	2.08	243.0109
$C_{20}H_3$	21.66	2.23	243.0235
244			
$C_{11}H_{20}N_2O_4$	13.13	1.60	244.1424
$C_{11}H_{22}N_3O_3$	13.50	1.45	244.1662
$C_{11}H_{24}N_4O_2$	13.87	1.30	244.1901
$C_{12}H_{22}NO_4$	13.86	1.69	244.1549
$C_{12}H_{24}N_2O_3$	14.23	1.54	244.1788
$C_{12}H_8N_2O_4$	14.01	1.71	244.0484
$C_{12}H_{26}N_3O_2$	14.61	1.40	244.2026
$C_{12}H_{10}N_3O_3$	14.39	1.56	244.0723
$C_{12}H_{28}N_4O$	14.98	1.25	244.2265
$C_{12}H_{12}N_4O_2$	14.76	1.42	244.0961
$C_{13}H_{24}O_4$	14.59	1.79	244.1675
$C_{13}H_{26}NO_3$	14.96	1.64	244.1914
$C_{13}H_{10}NO_4$	14.75	1.81	244.0610
$C_{13}H_{28}N_2O_2$	15.34	1.50	244.2152
$C_{13}H_{12}N_2O_3$	15.12	1.67	244.0848
$C_{13}H_{30}N_3O$	15.71	1.36	244.2391
$C_{13}H_{14}N_3O_2$	15.49	1.52	244.1087
$C_{13}H_{32}N_4$	16.09	1.21	244.2629
$C_{13}H_{16}N_4O$	15.87	1.38	244.1325
$C_{13}N_4O_2$	15.65	1.55	244.0022
$C_{14}H_{28}O_3$	15.69	1.75	244.2039
$C_{14}H_{12}O_4$	15.48	1.92	244.0735
$C_{14}H_{30}NO_2$	16.07	1.61	244.2278
$C_{14}H_{14}NO_3$	15.85	1.78	244.0974
$C_{14}H_{32}N_2O$	16.44	1.47	244.2516
$C_{14}H_{16}N_2O_2$	16.23	1.63	244.1213
$C_{14}N_2O_3$	16.01	1.80	243.9909
$C_{14}H_{18}N_3O$	16.60	1.49	244.1451
$C_{14}H_2N_3O_2$	16.38	1.66	244.0147
$C_{14}H_{20}N_4$	16.97	1.36	244.1690
$C_{14}H_4N_4O$	16.76	1.52	244.0386
$C_{15}H_{32}O_2$	16.80	1.72	244.2403
$C_{15}H_{16}O_3$	16.58	1.89	244.1100
$C_{15}O_4$	16.37	2.05	243.9796
$C_{15}H_{18}NO_2$	16.96	1.75	244.1338
$C_{15}H_2NO_3$	16.74	1.91	244.0034
$C_{15}H_{20}N_2O$	17.33	1.61	244.1577
$C_{15}H_4N_2O_2$	17.11	1.78	244.0273
$C_{15}H_{22}N_3$	17.71	1.48	244.1815
$C_{15}H_6N_3O$	17.49	1.64	244.0511
$C_{15}H_8N_4$	17.86	1.50	244.0750
$C_{16}H_{20}O_2$	17.69	1.87	244.1464
$C_{16}H_4O_3$	17.47	2.03	244.0160

	M+1	M+2	FM
$C_{16}H_{22}NO$	18.06	1.74	244.1702
$C_{16}H_{6}NO_{2}$	17.85	1.90	244.0399
$C_{16}H_{24}N_{2}$	18.44	1.60	244.1941
$C_{16}H_{8}N_{2}O$	18.22	1.77	244.0637
$C_{16}H_{10}N_{3}$	18.59	1.63	244.0876
$C_{17}H_{24}O$	18.79	1.87	244.1828
$C_{17}H_{8}O_{2}$	18.58	2.03	244.0524
$C_{17}H_{26}N$	19.17	1.74	244.2067
$C_{17}H_{10}NO$	18.95	1.90	244.0763
$C_{17}H_{12}N_{2}$	19.33	1.77	244.1001
$C_{18}H_{28}$	19.90	1.87	244.2192
$C_{18}H_{12}O$	19.68	2.03	244.0888
$C_{18}H_{14}N$	20.06	1.91	244.1127
$C_{18}N_{2}$	20.21	1.94	244.0062
$C_{19}H_{16}$	20.79	2.05	244.1253
$C_{19}O$	20.57	2.21	243.9949
$C_{19}H_{2}N$	20.94	2.08	244.0187
$C_{20}H_{4}$	21.68	2.23	244.0313
245			
$C_{11}H_{21}N_{2}O_{4}$	13.14	1.60	245.1502
$C_{11}H_{23}N_{3}O_{3}$	13.52	1.45	245.1741
$C_{11}H_{25}N_{4}O_{2}$	13.89	1.30	245.1979
$C_{12}H_{23}NO_{4}$	13.87	1.69	245.1628
$C_{12}H_{25}N_{2}O_{3}$	14.25	1.54	245.1866
$C_{12}H_{9}N_{2}O_{4}$	14.03	1.71	245.0563
$C_{12}H_{27}N_{3}O_{2}$	14.62	1.40	245.2105
$C_{12}H_{11}N_{3}O_{3}$	14.40	1.57	245.0801
$C_{12}H_{29}N_{4}O$	15.00	1.25	245.2343
$C_{12}H_{13}N_{4}O_{2}$	14.78	1.42	245.1040
$C_{13}H_{25}O_{4}$	14.00	1.79	245.1755
$C_{13}H_{27}NO_{3}$	14.98	1.65	245.1992
$C_{13}H_{11}NO_{4}$	14.76	1.81	245.0688
$C_{13}H_{29}N_{2}O_{2}$	15.35	1.50	245.2230
$C_{13}H_{13}N_{2}O_{3}$	15.14	1.67	245.0927
$C_{13}H_{31}N_{3}O$	15.73	1.36	245.2469
$C_{13}H_{15}N_{3}O_{2}$	15.51	1.53	245.1165
$C_{13}H_{17}N_{4}O$	15.88	1.38	245.1404
$C_{13}HN_{4}O_{2}$	15.67	1.55	245.0100
$C_{14}H_{29}O_{3}$	15.71	1.75	245.2117
$C_{14}H_{13}O_{4}$	15.49	1.92	245.0814
$C_{14}H_{31}NO_{2}$	16.08	1.61	245.2356
$C_{14}H_{15}NO_{3}$	15.87	1.78	245.1052
$C_{14}H_{17}N_{2}O_{2}$	16.24	1.64	245.1291
$C_{14}HN_{2}O_{3}$	16.02	1.80	244.9987
$C_{14}H_{19}N_{3}O$	16.62	1.50	245.1529
$C_{14}H_{3}N_{3}O_{2}$	16.40	1.66	245.0226
$C_{14}H_{21}N_{4}$	16.99	1.36	245.1768
$C_{14}H_{5}N_{4}O$	16.77	1.52	245.0464
$C_{15}H_{17}O_{3}$	16.60	1.89	245.1178
$C_{15}HO_{4}$	16.38	2.06	244.9874
$C_{15}H_{19}NO_{2}$	16.97	1.75	245.1416
$C_{15}H_{3}NO_{3}$	16.76	1.92	245.0113
$C_{15}H_{21}N_{2}O$	17.35	1.62	245.1655

	M+1	M+2	FM
$C_{15}H_{5}N_{2}O_{2}$	17.13	1.78	245.0351
$C_{15}H_{23}N_{3}$	17.72	1.48	245.1894
$C_{15}H_{7}N_{3}O$	17.50	1.64	245.0590
$C_{15}H_{9}N_{4}$	17.88	1.51	245.0829
$C_{16}H_{21}O_{2}$	17.70	1.87	245.1542
$C_{16}H_{5}O_{3}$	17.49	2.04	245.0238
$C_{16}H_{23}NO$	18.08	1.74	245.1781
$C_{16}H_{7}NO_{2}$	17.86	1.90	245.0477
$C_{16}H_{25}N_{2}$	18.45	1.61	245.2019
$C_{16}H_{9}N_{2}O$	18.24	1.77	245.0715
$C_{16}H_{11}N_{3}$	18.61	1.64	245.0954
$C_{17}H_{25}O$	18.81	1.87	245.1906
$C_{17}H_{9}O_{2}$	18.59	2.03	245.0603
$C_{17}H_{27}N$	19.18	1.74	245.2145
$C_{17}H_{11}NO$	18.97	1.90	245.0841
$C_{17}H_{13}N_{2}$	19.34	1.77	245.1080
$C_{18}H_{29}$	19.91	1.88	245.2270
$C_{18}H_{13}O$	19.70	2.04	245.0967
$C_{18}H_{15}N$	20.07	1.91	245.1205
$C_{18}HN_{2}$	20.23	1.94	245.0140
$C_{19}H_{17}$	20.80	2.05	245.1331
$C_{19}HO$	20.59	2.21	245.0027
$C_{19}H_{3}N$	20.96	2.09	245.0266
$C_{20}H_{5}$	21.69	2.24	245.0391
246			
$C_{11}H_{22}N_{2}O_{4}$	13.16	1.60	246.1580
$C_{11}H_{24}N_{3}O_{3}$	13.53	1.45	246.1819
$C_{11}H_{26}N_{4}O_{2}$	13.91	1.30	246.2057
$C_{12}H_{24}NO_{4}$	13.89	1.70	246.1706
$C_{12}H_{26}N_{2}O_{3}$	14.20	1.55	246.1945
$C_{12}H_{10}N_{2}O_{4}$	14.05	1.72	246.0641
$C_{12}H_{28}N_{3}O_{2}$	14.64	1.40	246.2183
$C_{12}H_{12}N_{3}O_{3}$	14.42	1.57	246.0879
$C_{12}H_{30}N_{4}O$	15.01	1.25	246.2422
$C_{12}H_{14}N_{4}O_{2}$	14.80	1.42	246.1118
$C_{13}H_{26}O_{4}$	14.62	1.79	246.1832
$C_{13}H_{28}NO_{3}$	14.99	1.65	246.2070
$C_{13}H_{12}NO_{4}$	14.78	1.82	246.0766
$C_{13}H_{30}N_{2}O_{2}$	15.37	1.50	246.2309
$C_{13}H_{14}N_{2}O_{3}$	15.15	1.67	246.1005
$C_{13}H_{16}N_{3}O_{2}$	15.53	1.53	246.1244
$C_{13}N_{3}O_{3}$	15.31	1.70	245.9940
$C_{13}H_{18}N_{4}O$	15.90	1.39	246.1482
$C_{13}H_{2}N_{4}O_{2}$	15.68	1.55	246.0178
$C_{14}H_{30}O_{3}$	15.73	1.76	246.2196
$C_{14}H_{14}O_{4}$	15.51	1.92	246.0892
$C_{14}H_{16}NO_{3}$	15.88	1.78	246.1131
$C_{14}NO_{4}$	15.67	1.95	245.9827
$C_{14}H_{18}N_{2}O_{2}$	16.26	1.64	246.1369
$C_{14}H_{2}N_{2}O_{3}$	16.04	1.80	246.0065
$C_{14}H_{20}N_{3}O$	16.63	1.50	246.1608
$C_{14}H_{4}N_{3}O_{2}$	16.41	1.66	246.0304
$C_{14}H_{22}N_{4}$	17.01	1.36	246.1846

	M+1	M+2	FM
$C_{14}H_{6}N_{4}O$	16.79	1.53	246.0542
$C_{15}H_{18}O_{3}$	16.61	1.89	246.1256
$C_{15}H_{2}O_{4}$	16.40	2.06	245.9953
$C_{15}H_{20}NO_{2}$	16.99	1.76	246.1495
$C_{15}H_{4}NO_{3}$	16.77	1.92	246.0191
$C_{15}H_{22}N_{2}O$	17.36	1.62	246.1733
$C_{15}H_{6}N_{2}O_{2}$	17.15	1.78	246.0429
$C_{15}H_{24}N_{3}$	17.74	1.48	246.1972
$C_{15}H_{8}N_{3}O$	17.52	1.65	246.0668
$C_{15}H_{10}N_{4}$	17.89	1.51	246.0907
$C_{16}H_{22}O_{2}$	17.72	1.88	246.1620
$C_{16}H_{6}O_{3}$	17.50	2.04	246.0317
$C_{16}H_{24}NO$	18.09	1.74	246.1859
$C_{16}H_{8}NO_{2}$	17.88	1.90	246.0555
$C_{16}H_{26}N_{2}$	18.47	1.61	246.2098
$C_{16}H_{10}N_{2}O$	18.25	1.77	246.0794
$C_{16}H_{12}N_{3}$	18.63	1.64	246.1032
$C_{17}H_{26}O$	18.83	1.87	246.1985
$C_{17}H_{10}O_{2}$	18.61	2.03	246.0681
$C_{17}H_{28}N$	19.20	1.74	246.2223
$C_{17}H_{12}NO$	18.98	1.90	246.0919
$C_{17}H_{14}N_{2}$	19.36	1.77	246.1158
$C_{17}N_{3}$	19.51	1.80	246.0093
$C_{18}H_{30}$	19.93	1.88	246.2349
$C_{18}H_{14}O$	19.71	2.04	246.1045
$C_{18}H_{16}N$	20.09	1.91	246.1284
$C_{18}NO$	19.87	2.07	245.9980
$C_{18}H_{2}N_{2}$	20.25	1.94	246.0218
$C_{19}H_{18}$	20.82	2.06	246.1409
$C_{19}H_{2}O$	20.60	2.21	246.0106
$C_{19}H_{4}N$	20.98	2.09	246.0344
$C_{20}H_{6}$	21.71	2.24	246.0470
247			
$C_{11}H_{23}N_{2}O_{4}$	13.17	1.60	247.1659
$C_{11}H_{25}N_{3}O_{3}$	13.55	1.45	247.1897
$C_{11}H_{27}N_{4}O_{2}$	13.92	1.30	247.2136
$C_{12}H_{25}NO_{4}$	13.91	1.70	247.1784
$C_{12}H_{27}N_{2}O_{3}$	14.28	1.55	247.2023
$C_{12}H_{11}N_{2}O_{4}$	14.06	1.72	247.0719
$C_{12}H_{29}N_{3}O_{2}$	14.65	1.40	247.2261
$C_{12}H_{13}N_{3}O_{3}$	14.44	1.57	247.0958
$C_{12}H_{15}N_{4}O_{2}$	14.81	1.42	247.1196
$C_{13}H_{27}O_{4}$	14.64	1.80	247.1910
$C_{13}H_{29}NO_{3}$	15.01	1.65	247.2148
$C_{13}H_{13}NO_{4}$	14.79	1.82	247.0845
$C_{13}H_{15}N_{2}O_{3}$	15.17	1.67	247.1083
$C_{13}H_{17}N_{3}O_{2}$	15.54	1.53	247.1322
$C_{13}HN_{3}O_{3}$	15.33	1.70	247.0018
$C_{13}H_{19}N_{4}O$	15.92	1.39	247.1560
$C_{13}H_{3}N_{4}O_{2}$	15.70	1.55	247.0257
$C_{14}H_{15}O_{4}$	15.52	1.93	247.0970
$C_{14}H_{17}NO_{3}$	15.90	1.78	247.1209
$C_{14}HNO_{4}$	15.68	1.95	246.9905
$C_{14}H_{19}N_{2}O_{2}$	16.27	1.64	247.1447

	M+1	M+2	FM
$C_{14}H_3N_2O_3$	16.06	1.81	247.0144
$C_{14}H_{21}N_3O$	16.65	1.50	247.1686
$C_{14}H_5N_3O_2$	16.43	1.67	247.0382
$C_{14}H_{23}N_4$	17.02	1.36	247.1925
$C_{14}H_7N_4O$	16.81	1.53	247.0621
$C_{15}H_{19}O_3$	16.63	1.90	247.1334
$C_{15}H_3O_4$	16.41	2.06	247.0031
$C_{15}H_{21}NO_2$	17.00	1.76	247.1573
$C_{15}H_5NO_3$	16.79	1.92	247.0269
$C_{15}H_{23}N_2O$	17.38	1.62	247.1811
$C_{15}H_7N_2O_2$	17.16	1.78	247.0508
$C_{15}H_{25}N_3$	17.75	1.49	247.2050
$C_{15}H_9N_3O$	17.54	1.65	247.0746
$C_{15}H_{11}N_4$	17.91	1.51	247.0985
$C_{16}H_{23}O_2$	17.74	1.88	247.1699
$C_{16}H_7O_3$	17.52	2.04	247.0395
$C_{16}H_{25}NO$	18.11	1.75	247.1937
$C_{16}H_9NO_2$	17.89	1.91	247.0634
$C_{16}H_{27}N_2$	18.48	1.61	247.2176
$C_{16}H_{11}N_2O$	18.27	1.77	247.0872
$C_{16}H_{13}N_3$	18.64	1.64	247.1111
$C_{17}H_{27}O$	18.84	1.88	247.2063
$C_{17}H_{11}O_2$	18.62	2.04	247.0759
$C_{17}H_{29}N$	19.22	1.75	247.2301
$C_{17}H_{13}NO$	19.00	1.91	247.0998
$C_{17}H_{15}N_2$	19.37	1.78	247.1236
$C_{17}HN_3$	19.53	1.81	247.0171
$C_{18}H_{31}$	19.95	1.88	247.2427
$C_{18}H_{15}O$	19.73	2.04	247.1123
$C_{18}H_{17}N$	20.10	1.92	247.1362
$C_{18}HNO$	19.89	2.07	247.0058
$C_{18}H_3N_2$	20.26	1.95	247.0297
$C_{19}H_{19}$	20.84	2.06	247.1488
$C_{19}H_3O$	20.62	2.22	247.0184
$C_{19}H_5N$	20.99	2.09	247.0422
$C_{20}H_7$	21.72	2.24	247.0548
248			
$C_{11}H_{24}N_2O_4$	13.19	1.61	248.1737
$C_{11}H_{26}N_3O_3$	13.56	1.45	248.1976
$C_{11}H_{28}N_4O_2$	13.94	1.31	248.2214
$C_{12}H_{26}NO_4$	13.92	1.70	248.1863
$C_{12}H_{28}N_2O_3$	14.30	1.55	248.2101
$C_{12}H_{12}N_2O_4$	14.08	1.72	248.0797
$C_{12}H_{14}N_3O_3$	14.45	1.57	248.1036
$C_{12}H_{16}N_4O_2$	14.83	1.43	248.1275
$C_{13}H_{28}O_4$	14.65	1.80	248.1988
$C_{13}H_{14}NO_4$	14.81	1.82	248.0923
$C_{13}H_{16}N_2O_3$	15.18	1.68	248.1162
$C_{13}N_2O_4$	14.97	1.84	247.9858
$C_{13}H_{18}N_3O_2$	15.56	1.53	248.1400
$C_{13}H_2N_3O_3$	15.34	1.70	248.0096
$C_{13}H_{20}N_4O$	15.93	1.39	248.1639
$C_{13}H_4N_4O_2$	15.72	1.56	248.0335
$C_{14}H_{16}O_4$	15.54	1.93	248.1049
$C_{14}H_{18}NO_3$	15.92	1.79	248.1287
$C_{14}H_2NO_4$	15.70	1.95	247.9983
$C_{14}H_{20}N_2O_2$	16.29	1.64	248.1526
$C_{14}H_4N_2O_3$	16.07	1.81	248.0222
$C_{14}H_{22}N_3O$	16.66	1.50	248.1764
$C_{14}H_6N_3O_2$	16.45	1.67	248.0460
$C_{14}H_{24}N_4$	17.04	1.37	248.2003
$C_{14}H_8N_4O$	16.82	1.53	248.0699
$C_{15}H_{20}O_3$	16.65	1.90	248.1413
$C_{15}H_4O_4$	16.43	2.06	248.0109
$C_{15}H_{22}NO_2$	17.02	1.76	248.1651
$C_{15}H_6NO_3$	16.80	1.92	248.0348
$C_{15}H_{24}N_2O$	17.40	1.62	248.1890
$C_{15}H_8N_2O_2$	17.18	1.79	248.0586
$C_{15}H_{26}N_3$	17.77	1.49	248.2129
$C_{15}H_{10}N_3O$	17.55	1.65	248.0825
$C_{15}H_{12}N_4$	17.93	1.52	248.1063
$C_{16}H_{24}O_2$	17.75	1.88	248.1777
$C_{16}H_8O_3$	17.53	2.05	248.0473
$C_{16}H_{26}NO$	18.13	1.75	248.2015
$C_{16}H_{10}NO_2$	17.91	1.91	248.0712
$C_{16}H_{28}N_2$	18.50	1.62	248.2254
$C_{16}H_{12}N_2O$	18.28	1.78	248.0950
$C_{16}H_{14}N_3$	18.66	1.64	248.1189
$C_{16}N_4$	18.82	1.67	248.0124
$C_{17}H_{28}O$	18.86	1.88	248.2141
$C_{17}H_{12}O_2$	18.64	2.04	248.0837
$C_{17}H_{30}N$	19.23	1.75	248.2380
$C_{17}H_{14}NO$	19.01	1.91	248.1076
$C_{17}H_{16}N_2$	19.39	1.78	248.1315
$C_{17}N_2O$	19.17	1.94	248.0011
$C_{17}H_2N_3$	19.55	1.81	248.0249
$C_{18}H_{32}$	19.96	1.89	248.2505
$C_{18}H_{16}O$	19.75	2.04	248.1202
$C_{18}O_2$	19.53	2.20	247.9898
$C_{18}H_{18}N$	20.12	1.92	248.1440
$C_{18}H_2NO$	19.90	2.08	248.0136
$C_{18}H_4N_2$	20.28	1.95	248.0375
$C_{19}H_{20}$	20.85	2.06	248.1566
$C_{19}H_4O$	20.63	2.22	248.0262
$C_{19}H_6N$	21.01	2.10	248.0501
$C_{20}H_8$	21.74	2.25	248.0626
249			
$C_{11}H_{25}N_2O_4$	13.21	1.61	249.1815
$C_{11}H_{27}N_3O_3$	13.58	1.46	249.2054
$C_{12}H_{27}NO_4$	13.94	1.70	249.1941
$C_{12}H_{13}N_2O_4$	14.09	1.72	249.0876
$C_{12}H_{15}N_3O_3$	14.47	1.58	249.1114
$C_{12}H_{17}N_4O_2$	14.84	1.43	249.1353
$C_{13}H_{15}NO_4$	14.83	1.82	249.1001
$C_{13}H_{17}N_2O_3$	15.20	1.68	249.1240
$C_{13}HN_2O_4$	14.98	1.85	248.9936
$C_{13}H_{19}N_3O_2$	15.57	1.54	249.1478
$C_{13}H_3N_3O_3$	15.36	1.70	249.0175
$C_{13}H_{21}N_4O$	15.95	1.39	249.1717
$C_{13}H_5N_4O_2$	15.73	1.56	249.0413
$C_{14}H_{17}O_4$	15.56	1.93	249.1127
$C_{14}H_{19}NO_3$	15.93	1.79	249.1365
$C_{14}H_3NO_4$	15.71	1.95	249.0062
$C_{14}H_{21}N_2O_2$	16.31	1.65	249.1604
$C_{14}H_5N_2O_3$	16.09	1.81	249.0300
$C_{14}H_{23}N_3O$	16.68	1.51	249.1842
$C_{14}H_7N_3O_2$	16.46	1.67	249.0539
$C_{14}H_{25}N_4$	17.05	1.37	249.2081
$C_{14}H_9N_4O$	16.84	1.53	249.0777
$C_{15}H_{21}O_3$	16.66	1.90	249.1491
$C_{15}H_5O_4$	16.45	2.07	249.0187
$C_{15}H_{23}NO_2$	17.04	1.76	249.1730
$C_{15}H_7NO_3$	16.82	1.93	249.0426
$C_{15}H_{25}N_2O$	17.41	1.63	249.1968
$C_{15}H_9N_2O_2$	17.19	1.79	249.0664
$C_{15}H_{27}N_3$	17.79	1.49	249.2207
$C_{15}H_{11}N_3O$	17.57	1.65	249.0903
$C_{15}H_{13}N_4$	17.94	1.52	249.1142
$C_{16}H_{25}O_2$	17.77	1.89	249.1855
$C_{16}H_9O_3$	17.55	2.05	249.0552
$C_{16}H_{27}NO$	18.14	1.75	249.2094
$C_{16}H_{11}NO_2$	17.93	1.91	249.0790
$C_{16}H_{29}N_2$	18.52	1.62	249.2332
$C_{16}H_{13}N_2O$	18.30	1.78	249.1029
$C_{16}H_{15}N_3$	18.67	1.65	249.1267
$C_{16}HN_4$	18.83	1.68	249.0202
$C_{17}H_{29}O$	18.87	1.88	249.2219
$C_{17}H_{13}O_2$	18.66	2.04	249.0916
$C_{17}H_{31}N$	19.25	1.75	249.2458
$C_{17}H_{15}NO$	19.03	1.91	249.1154
$C_{17}H_{17}N_2$	19.41	1.78	249.1393
$C_{17}HN_2O$	19.19	1.94	249.0089
$C_{17}H_3N_3$	19.56	1.81	249.0328
$C_{18}H_{33}$	19.98	1.89	249.2584
$C_{18}H_{17}O$	19.76	2.05	249.1280
$C_{18}HO_2$	19.54	2.21	248.9976
$C_{18}H_{19}N$	20.14	1.92	249.1519
$C_{18}H_3NO$	19.92	2.08	249.0215
$C_{18}H_5N_2$	20.29	1.95	249.0453
$C_{19}H_{21}$	20.87	2.07	249.1644
$C_{19}H_5O$	20.65	2.22	249.0340
$C_{19}H_7N$	21.02	2.10	249.0579
$C_{20}H_9$	21.76	2.25	249.0705
250			
$C_{11}H_{26}N_2O_4$	13.22	1.61	250.1894
$C_{12}H_{14}N_2O_4$	14.11	1.73	250.0954
$C_{12}H_{16}N_3O_3$	14.48	1.58	250.1193
$C_{12}H_{18}N_4O_2$	14.86	1.43	250.1431
$C_{13}H_{16}NO_4$	14.84	1.83	250.1080
$C_{13}H_{18}N_2O_3$	15.22	1.68	250.1318

	M+1	M+2	FM
$C_{13}H_{2}N_{2}O_{4}$	15.00	1.85	250.0014
$C_{13}H_{20}N_{3}O_{2}$	15.59	1.54	250.1557
$C_{13}H_{4}N_{3}O_{3}$	15.37	1.71	250.0253
$C_{13}H_{22}N_{4}O$	15.96	1.40	250.1795
$C_{13}H_{6}N_{4}O_{2}$	15.75	1.56	250.0491
$C_{14}H_{18}O_{4}$	15.57	1.93	250.1205
$C_{14}H_{20}NO_{3}$	15.95	1.79	250.1444
$C_{14}H_{4}NO_{4}$	15.73	1.96	250.0140
$C_{14}H_{22}N_{2}O_{2}$	16.32	1.65	250.1682
$C_{14}H_{6}N_{2}O_{3}$	16.10	1.82	250.0379
$C_{14}H_{24}N_{3}O$	16.70	1.51	250.1921
$C_{14}H_{8}N_{3}O_{2}$	16.48	1.67	250.0617
$C_{14}H_{26}N_{4}$	17.07	1.37	250.2160
$C_{14}H_{10}N_{4}O$	16.85	1.54	250.0856
$C_{15}H_{22}O_{3}$	16.68	1.90	250.1569
$C_{15}H_{6}O_{4}$	16.46	2.07	250.0266
$C_{15}H_{24}NO_{2}$	17.05	1.77	250.1808
$C_{15}H_{8}NO_{3}$	16.84	1.93	250.0504
$C_{15}H_{26}N_{2}O$	17.43	1.63	250.2046
$C_{15}H_{10}N_{2}O_{2}$	17.21	1.79	250.0743
$C_{15}H_{28}N_{3}$	17.80	1.49	250.2285
$C_{15}H_{12}N_{3}O$	17.58	1.66	250.0981
$C_{15}H_{14}N_{4}$	17.96	1.52	250.1220
$C_{16}H_{26}O_{2}$	17.78	1.89	250.1934
$C_{16}H_{10}O_{3}$	17.57	2.05	250.0630
$C_{16}H_{28}NO$	18.16	1.75	250.2172
$C_{16}H_{12}NO_{2}$	17.94	1.92	250.0868
$C_{16}H_{30}N_{2}$	18.53	1.62	250.2411
$C_{16}H_{14}N_{2}O$	18.32	1.78	250.1107
$C_{16}H_{16}N_{3}$	18.69	1.65	250.1346
$C_{16}N_{3}O$	18.47	1.81	250.0042
$C_{16}H_{2}N_{4}$	18.85	1.68	250.0280
$C_{17}H_{30}O$	18.89	1.89	250.2298
$C_{17}H_{14}O_{2}$	18.67	2.04	250.0994
$C_{17}H_{32}N$	19.26	1.76	250.2536
$C_{17}H_{16}NO$	19.05	1.91	250.1233
$C_{17}NO_{2}$	18.83	2.07	249.9929
$C_{17}H_{18}N_{2}$	19.42	1.79	250.1471
$C_{17}H_{2}N_{2}O$	19.20	1.94	250.0167
$C_{17}H_{4}N_{3}$	19.58	1.82	250.0406
$C_{18}H_{34}$	19.99	1.89	250.2662
$C_{18}H_{18}O$	19.78	2.05	250.1358
$C_{18}H_{2}O_{2}$	19.56	2.21	250.0054
$C_{18}H_{20}N$	20.15	1.92	250.1597
$C_{18}H_{4}NO$	19.94	2.08	250.0293
$C_{18}H_{6}N_{2}$	20.31	1.96	250.0532
$C_{19}H_{22}$	20.88	2.07	250.1722
$C_{19}H_{6}O$	20.67	2.23	250.0419
$C_{19}H_{8}N$	21.04	2.10	250.0657
$C_{20}H_{10}$	21.77	2.25	250.0783

appendix b

COMMON FRAGMENT IONS

Not all members of homologous and isomeric series are given. The list is meant to be suggestive rather than exhaustive. Appendix II of Hamming and Foster,[7] Table A-7 of McLafferty's Interpretative book,[9] and the high-resolution ion data of McLafferty are recommended as supplements.

m/e	Ions*
14	CH_2
15	CH_3
16	O
17	OH
18	H_2O, NH_4
19	F, H_3O
26	$C{\equiv}N$
27	C_2H_3
28	C_2H_4, CO, N_2 (air), CH=NH
29	C_2H_5, CHO
30	CH_2NH_2, NO
31	CH_2OH, OCH_3
32	O_2 (Air)
33	SH, CH_2F
34	H_2S
35	Cl
36	HCl
39	C_3H_3
40	$CH_2C{=}N$, Ar(Air)
41	C_3H_5, $CH_2C{=}N + H$, C_2H_2NH
42	C_3H_6
43	C_3H_7, $CH_3C{=}O$, C_2H_5N
44	$CH_2\overset{H}{\overset{\vert}{C}}{=}O + H$, CH_3CHNH_2, CO_2, $NH_2C{=}O$, $(CH_3)_2N$
45	$\overset{CH_3}{\overset{\vert}{C}}HOH$, CH_2CH_2OH, CH_2OCH_3, $\overset{O}{\overset{\Vert}{C}}{-}OH$, $CH_3CH{-}O + H$
46	NO_2
47	CH_2SH, CH_3S
48	$CH_3S + H$
49	CH_2Cl
51	CHF_2
53	C_4H_5
54	$CH_2CH_2C{\equiv}N$
55	C_4H_7, $CH_2{=}CHC{=}O$
56	C_4H_8
57	C_4H_9, $C_2H_5C{=}O$
58	$CH_3{-}C({=}O)(CH_2) + H$, $C_2H_5CHNH_2$, $(CH_3)_2NCH_2$, $C_2H_5NHCH_2$, C_2H_2S
59	$(CH_3)_2COH$, $CH_2OC_2H_5$, $\overset{O}{\overset{\Vert}{C}}{-}OCH_3$, $NH_2\underset{CH_2}{\underset{\vert}{C}}{=}O + H$, CH_3OCHCH_3, CH_3CHCH_2OH
60	$CH_2C({=}O)(OH) + H$, CH_2ONO
61	$CH_3\overset{O}{\overset{\Vert}{C}}{-}O + 2H$, CH_2CH_2SH, CH_2SCH_3
65	$\equiv C_5H_5$
66	$\equiv C_5H_6$
67	C_5H_7
68	$CH_2CH_2CH_2C{\equiv}N$
69	C_5H_9, CF_3, $CH_3CH{=}CHC{=}O$, $CH_2{=}C(CH_3)C{=}O$
70	C_5H_{10}
71	C_5H_{11}, $C_3H_7C{=}O$
72	$C_2H_5C({=}O)(CH_2) + H$, $C_3H_7CHNH_2$, $(CH_3)_2N{=}C{=}O$, $C_2H_5NHCHCH_3$, and isomers
73	Homologs of 59
74	$CH_2{-}\overset{O}{\overset{\Vert}{C}}{-}OCH_3 + H$
75	$\overset{O}{\overset{\Vert}{C}}{-}OC_2H_5 + 2H$, $CH_2SC_2H_5$, $(CH_3)_2CSH$, $(CH_3O)_2CH$
77	C_6H_5
78	$C_6H_5 + H$
79	$C_6H_5 + 2H$, Br

*Ions indicated as a fragment + nH (n = 1, 2, 3, ...) are ions that arise via rearrangement involving hydrogen transfer.

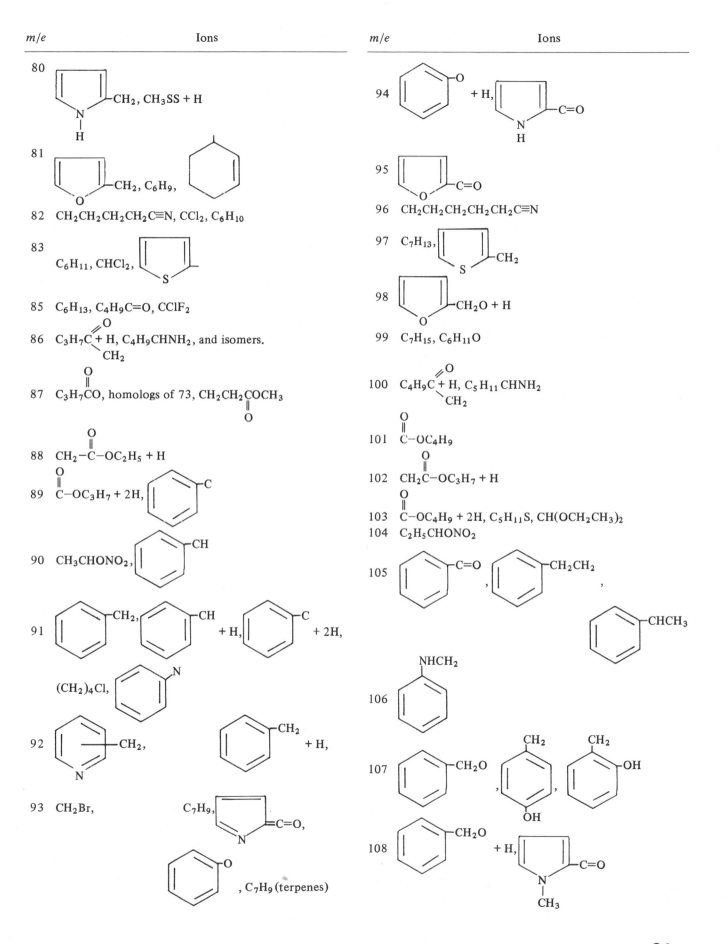

m/e	Ions
80	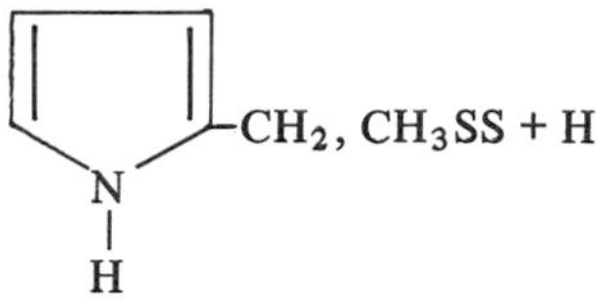 —CH_2, $CH_3SS + H$ (N, H)
81	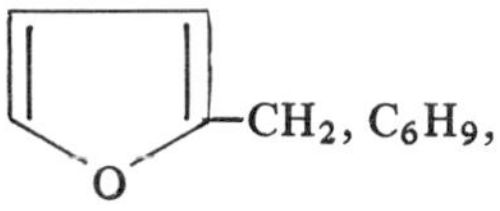 —CH_2, C_6H_9, 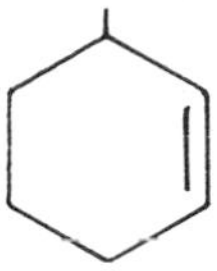 (O)
82	$CH_2CH_2CH_2CH_2C{\equiv}N$, CCl_2, C_6H_{10}
83	C_6H_{11}, $CHCl_2$, 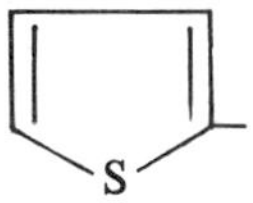 (S)
85	C_6H_{13}, $C_4H_9C{=}O$, $CClF_2$
86	$C_3H_7C(=O)(CH_2) + H$, $C_4H_9CHNH_2$, and isomers.
87	$C_3H_7\overset{O}{\overset{\Vert}{C}}O$, homologs of 73, $CH_2CH_2\underset{O}{\underset{\Vert}{C}}OCH_3$
88	$CH_2{-}\overset{O}{\overset{\Vert}{C}}{-}OC_2H_5 + H$
89	$\overset{O}{\overset{\Vert}{C}}{-}OC_3H_7 + 2H$, —C
90	CH_3CHONO_2, 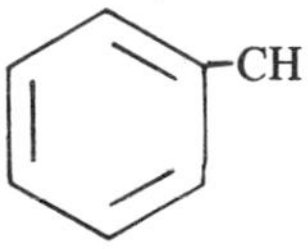 —CH
91	—CH_2, —CH + H, —C + 2H,
	$(CH_2)_4Cl$, N
92	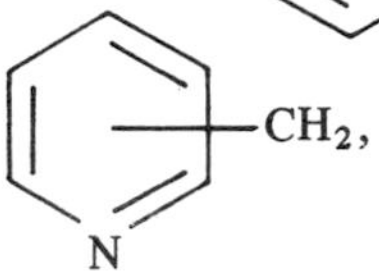 —CH_2, 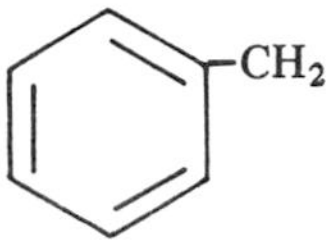 (N) —CH_2 + H,
93	CH_2Br, C_7H_9, 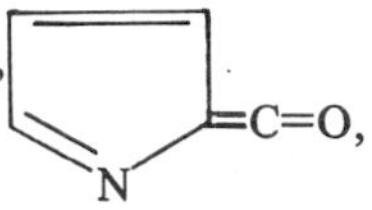 (N) —C=O, 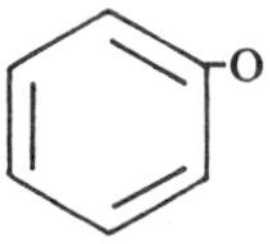 —O, C_7H_9 (terpenes)
94	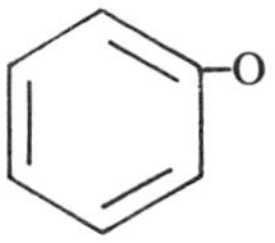 —O + H, 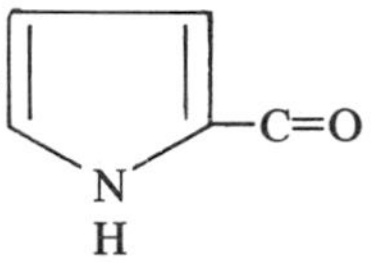 (N, H) —C=O
95	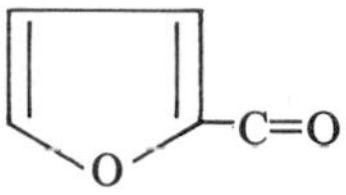 (O) —C=O
96	$CH_2CH_2CH_2CH_2CH_2C{\equiv}N$
97	C_7H_{13}, 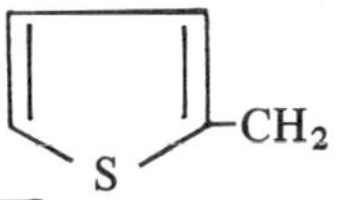 (S) —CH_2
98	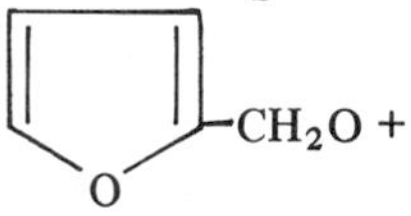 (O) —$CH_2O + H$
99	C_7H_{15}, $C_6H_{11}O$
100	$C_4H_9C(=O)(CH_2) + H$, $C_5H_{11}CHNH_2$
101	$\overset{O}{\overset{\Vert}{C}}{-}OC_4H_9$
102	$CH_2\overset{O}{\overset{\Vert}{C}}{-}OC_3H_7 + H$
103	$\overset{O}{\overset{\Vert}{C}}{-}OC_4H_9 + 2H$, $C_5H_{11}S$, $CH(OCH_2CH_3)_2$
104	$C_2H_5CHONO_2$
105	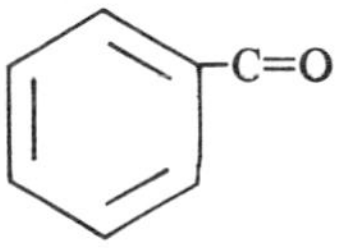 —C=O, 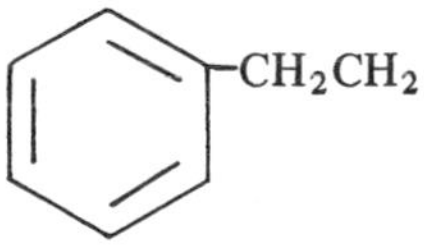 —CH_2CH_2, 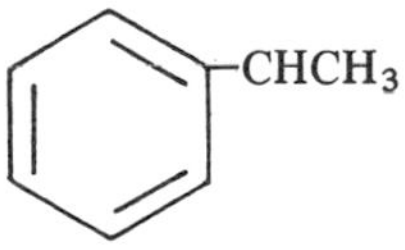 —$CHCH_3$
106	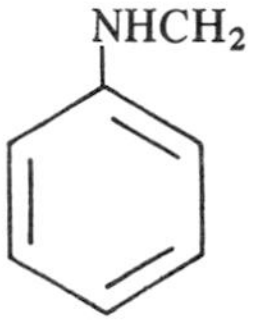 $NHCH_2$
107	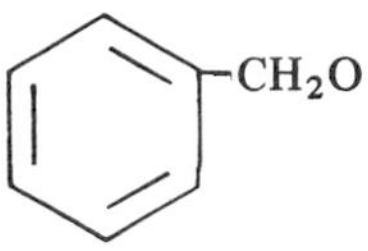 —CH_2O, 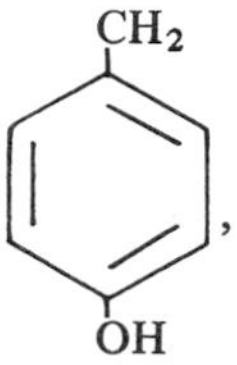 CH_2 (OH), 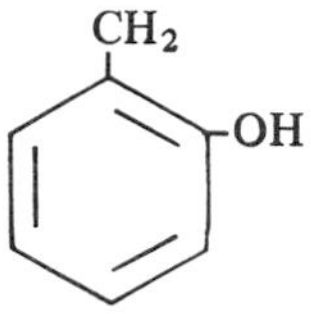 CH_2 —OH
108	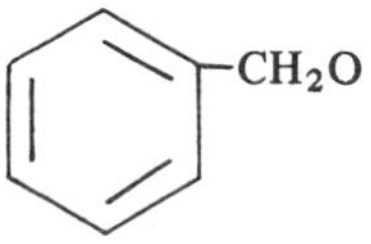 —CH_2O + H, (N, CH_3) —C=O

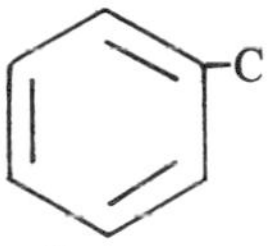

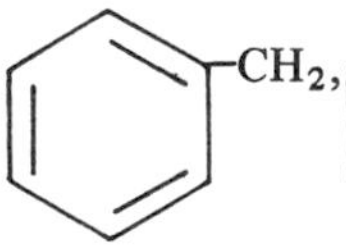

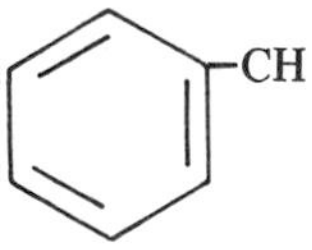

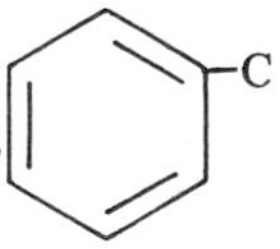

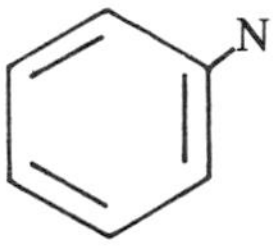

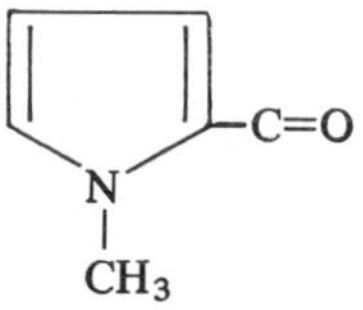

m/e	Ions
109	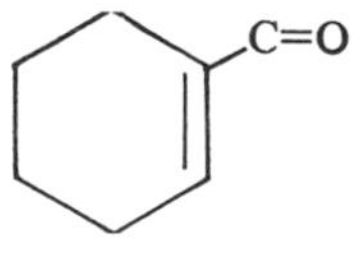 C=O
111	C=O, S
119	CF_3CF_2, 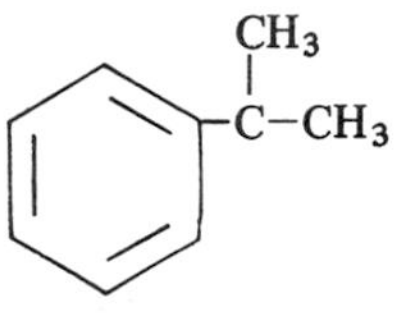 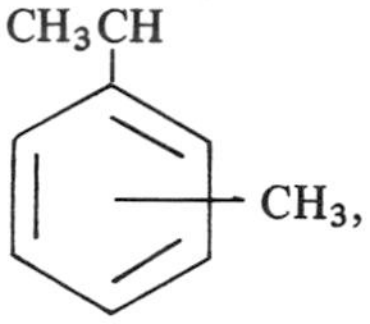 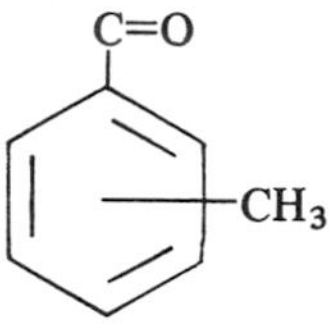 CH₃–C–CH₃ (with CH_3), CH_3CH ... CH_3, C=O ... CH_3
120	C=O, O
121	C=O ... OH, OCH_3 ... CH_2, N=O ... NH
	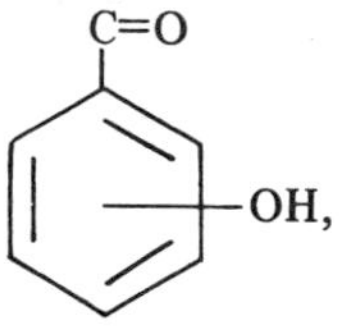 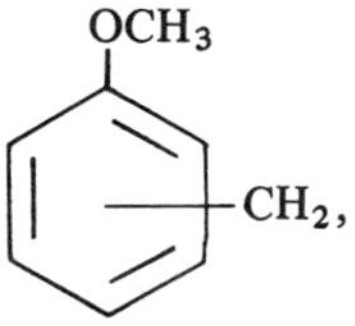 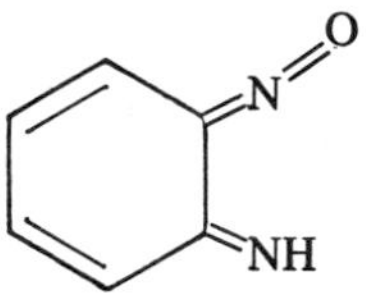 C_9H_{13} (terpenes)
123	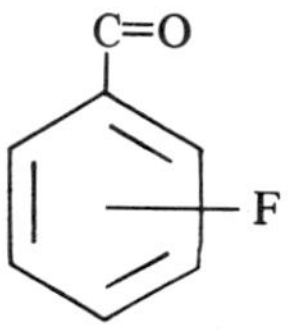 C=O ... F
125	S → O
127	I
131	C_3F_5, 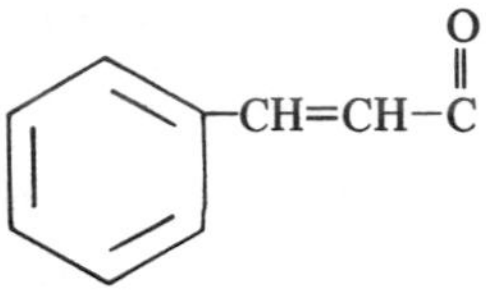 CH=CH–C=O
135	$(CH_2)_4Br$
138	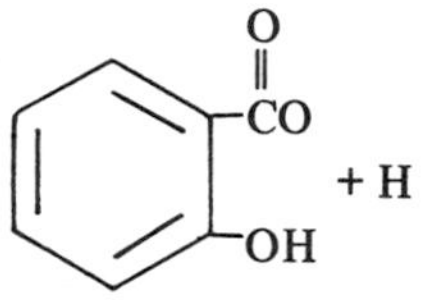 CO (C=O), OH, + H
139	C=O ... Cl
149	C=O, O + H, C=O
154	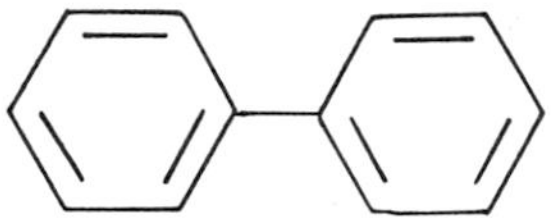

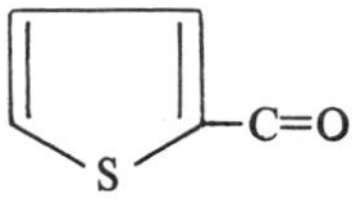

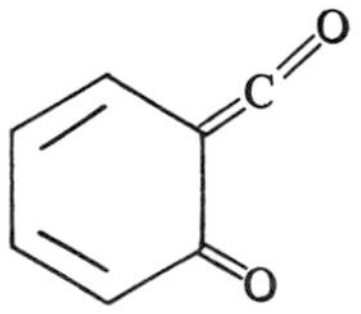

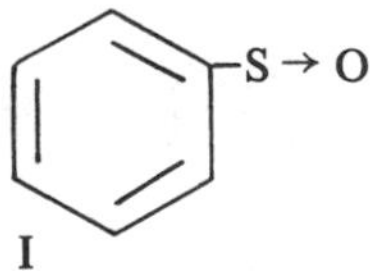

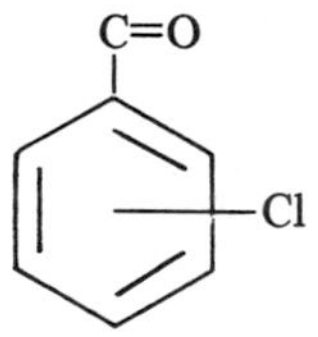

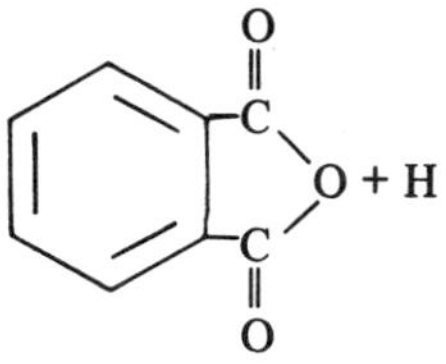

appendix c

COMMON FRAGMENTS LOST

This list is suggestive rather than comprehensive. It should be used in conjunction with Appendix B. Table 5-19 of Hamming and Foster[7] and Table A-5 of McLafferty[9] are recommended as supplements. All of these are lost as neutral species.

Molecular Ion Minus	Fragment Lost
1	H·
15	CH_3·
17	HO·
18	H_2O
19	F·
20	HF
26	CH≡CH, ·C≡N
27	CH_2=CH·, HC≡N
28	CH_2=CH_2, CO, (HCN + H)
29	CH_3CH_2·, ·CHO
30	NH_2CH_2·, CH_2O, NO
31	·OCH_3, ·CH_2OH, CH_3NH_2
32	CH_3OH, S
33	HS·, (·CH_3 and H_2O)
34	H_2S
35	Cl·
36	HCl, $2H_2O$
37	H_2Cl (or HCl + H)
38	C_3H_2·, C_2N, F_2
39	C_3H_3, HC_2N
40	CH_3C≡CH
41	CH_2=$CHCH_2$·
42	CH_2=$CHCH_3$, CH_2=C=O, CH_2–CH_2 (ring, with CH_2), NCO, $NCNH_2$
43	C_3H_7·, $CH_3C(=O)$·, CH_2=CH–O·, [CH_3· and CH_2=CH_2], HCNO
44	CH_2=CHOH, CO_2, N_2O, $CONH_2$, $NHCH_2CH_3$
45	CH_3CHOH, CH_3CH_2O·, CO_2H, $CH_3CH_2NH_2$
46	[H_2O and CH_2=CH_2], CH_3CH_2OH, ·NO_2
47	CH_3S·
48	CH_3SH, SO, O_3
49	·CH_2Cl
51	·CHF_2
52	C_4H_4, C_2N_2
53	C_4H_5
54	CH_2=CH–CH=CH_2
55	CH_2=$CHCHCH_3$
56	CH_2=$CHCH_2CH_3$, CH_3CH=$CHCH_3$, 2CO
57	C_4H_9·
58	·NCS, (NO + CO), CH_3COCH_3
59	$CH_3OC(=O)$·, $CH_3C(=O)NH_2$, (three-membered ring with H–S·)
60	C_3H_7OH
61	CH_3CH_2S·, (three-membered ring with H–S·)
62	[H_2S and CH_2=CH_2]
63	·CH_2CH_2Cl
64	C_5H_4, S_2, SO_2
68	CH_2=$C(CH_3)$–CH=CH_2
69	CF_3·, C_5H_9·
71	C_5H_{11}·
73	$CH_3CH_2OC(=O)$·
74	C_4H_9OH
75	C_6H_3
76	C_6H_4, CS_2
77	C_6H_5, CS_2H
78	C_6H_6, CS_2H_2, C_5H_4N
79	Br·, C_5H_5N
80	HBr
85	·$CClF_2$
100	CF_2=CF_2
119	CF_3–CF_2·
122	C_6H_5COOH
127	I·
128	HI

three

infrared spectrometry

The importance of infrared spectrometry as a tool of the practicing organic chemist is readily apparent from the number of books devoted wholly or in part to discussions of applications of infrared spectrometry (see References[1-31] at the end of the chapter.)

The text by Colthup, Daly, and Wiberley[3] is a thorough up-to-date discourse on infrared and Raman spectroscopy covering theory, practice, and structure correlations. Bellamy's books[8,9] include extensive discussions of infrared correlations from the literature and many tables of data. The book by Conley[1] is an introduction to theory, spectral determination, and spectral interpretation at an elementary level. There are many compilations of spectra as well as indices to spectral collections and to the literature.[7-25] Among the more commonly used compilations are those by Sadtler[25] and by Aldrich.[7]

I. INTRODUCTION

Infrared radiation refers broadly to that part of the electromagnetic spectrum between the visible and microwave regions. Of greatest practical use to the organic chemist is the limited portion between 4000 cm^{-1} and 666 cm^{-1} (2.5-15.0 μm). Recently there has been some interest in the near infrared region, 14,290-4000 cm^{-1} (0.7-2.5 μm) and the far infrared region, 700-200 cm^{-1} (14.3-50 μm).

From the brief theoretical discussion that follows, it is clear that even a very simple molecule can give an extremely complex spectrum. The organic chemist takes advantage of this complexity when he matches the spectrum of an unknown compound against that of an authentic sample. A peak-by-peak correlation is excellent evidence for identity. It is unlikely that any two compounds, except enantiomers, give the same infrared spectrum.

Although the infrared spectrum is characteristic of the entire molecule, it is true that certain groups of atoms give rise to bands at or near the same frequency regardless of the structure of the rest of the molecule. It is the persistence of these characteristic bands that permits the chemist to obtain useful structural information by simple inspection and reference to generalized charts of characteristic group frequencies. We shall rely heavily on these characteristic group frequencies.

Since we are not solely dependent on infrared spectra for identification, a detailed analysis of the spectrum will not be required. Following our general plan, we shall present only sufficient theory to accomplish our purpose: utilization of infrared spectra in conjunction with other spectral data to determine molecular structure.

Because most academic and industrial laboratories make infrared spectrometers available as bench tools for the organic chemist, we shall describe instrumentation and sample preparation in somewhat more detail than is given in the other chapters.

II. THEORY

Infrared radiation of frequencies less than about 100 cm^{-1} (wavelengths longer than 100 μm) is absorbed and converted by an organic molecule into energy of molecular rotation. This absorption is quantized; thus a molecular rotation spectrum consists of discrete lines.

Infrared radiation in the range from about 10,000-100 cm^{-1} (1-100 μm) is absorbed and converted by an organic molecule into energy of molecular vibration. This absorption is also quantized, but vibrational spectra appear as bands rather than as lines because a single vibrational energy change is accompanied by a number of rotational energy changes. It is with these vibrational-rotational bands, particularly those occurring between 4000 cm^{-1} and 666 cm^{-1} (2.5-15 μm), that we shall be concerned. The frequency or wavelength of absorption depends on the relative masses of the atoms, the force constants of the bonds, and the geometry of the atoms.

Band positions in infrared spectra are presented either as wavenumbers or wavelengths. The wavenumber unit (cm^{-1}, reciprocal centimeters) is used most often since it is directly proportional to the energy of the vibration and since most modern instruments are linear in the cm^{-1} scale. Wavelength is reported in micrometers (μm, 10^{-6} meters) although the older literature refers to this quantity as the micron (μ). Wavenumbers are reciprocally related to wavelength as follows:

$$\text{cm}^{-1} = \frac{1}{\mu\text{m}} \times 10^4$$

Specific absorptions and absorption ranges will be given in cm^{-1}, with the corresponding μm data adjoining (usually in parentheses).

Notice also that the wavenumbers ($\bar{\nu}$) are often called "frequencies"; this is not rigorously correct, since wavenumbers are $1/\lambda$ and frequency (ν) is c/λ. Such "frequency" for ν terminology is quite common and is probably not a serious error if the missing speed-of-light term (c) is remembered. Spectra linear in μm *and* in cm^{-1} are used in this book. A spectrum linear in wavenumbers has a very different appearance from one linear in wavelength, as shown later (Figure 7, p. 105).

Band intensities are expressed either as transmittance (T) or absorbance (A). Transmittance is the ratio of the radiant power transmitted by a sample to the radiant power incident on the sample. Absorbance is the logarithm, to the base 10, of the reciprocal of the transmittance; $A = \log_{10} (1/T)$.

There are two types of molecular vibrations: stretching and bending. A stretching vibration is a rhythmical movement along the bond axis such that the interatomic distance is increasing or decreasing. A bending vibration may consist of a change in bond angle between bonds with a common atom or the movement of a group of atoms with respect to the remainder of the molecule without movement of the atoms in the group with respect to one another. For example, twisting, rocking, and torsional vibrations involve a change in bond angles with reference to a set of coordinates arbitrarily set up within the molecule.

Only those vibrations that result in a rhythmical change in the dipole moment of the molecule are observed in the infrared. The alternating electric field, produced by the changing charge distribution accompanying a vibration, couples the molecular vibration with the oscillating electric field of the electromagnetic radiation.

A molecule has as many degrees of freedom as the total degrees of freedom of its individual atoms. Each atom has 3 degrees of freedom corresponding to the Cartesian coordinates (x, y, z) necessary to describe its position relative to other atoms in the molecule. A molecule of n atoms therefore has $3n$ degrees of freedom. For nonlinear molecules, 3 degrees of freedom describe rotation and 3 describe translation; the remaining $3n - 6$ degrees of freedom are vibrational degrees of freedom or fundamental vibrations. Linear molecules have $3n - 5$ vibrational degrees of freedom, for only two degrees of freedom are required to describe rotation.

Fundamental vibrations involve no change in the center of gravity of the molecule.

The three fundamental vibrations of the nonlinear, triatomic water molecule can be depicted as follows:

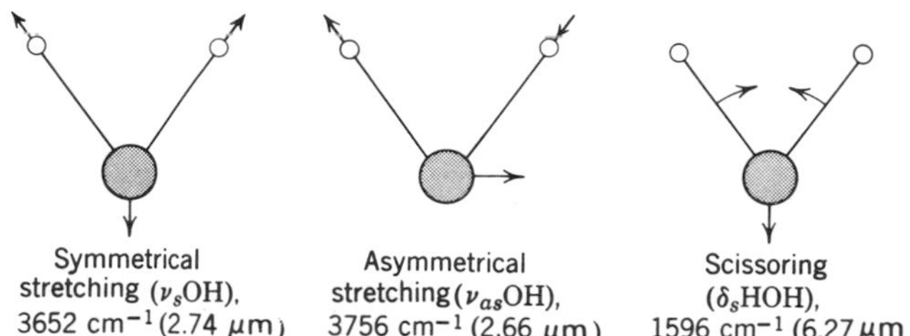

Note the very close spacing of the interacting ("coupled") asymmetric and symmetric stretching above compared with the far removed scissoring mode. This will be shown to be useful later in classification of absorptions and application to structure determination.

The CO_2 molecule is linear and contains three atoms; therefore it has four fundamental vibrations ((3 X 3) – 5):

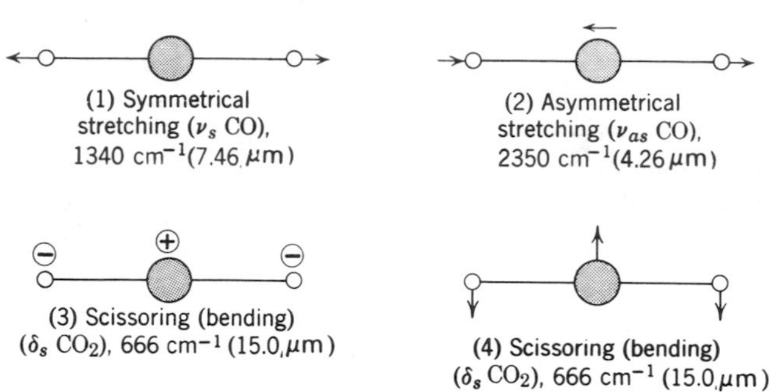

⊕ and ⊖ indicate movement perpendicular to the plane of the page.

The symmetrical stretching vibration (1) is inactive in the infrared since it produces no change in the dipole moment of the molecule. The bending vibrations (3) and (4) are equivalent, and are the resolved components of bending motion oriented at any angle to the internuclear axis; they have the same frequency and are said to be doubly degenerate.

The various stretching and bending modes for an AX_2 group appearing as a portion of a molecule, for example, the CH_2 group in a hydrocarbon molecule, are shown in Figure 1. The $3n - 6$ rule does not apply since the CH_2 represents only a portion of a molecule.

The theoretical number of fundamental vibrations (absorption frequencies) will seldom be observed because overtones (multiples of a given frequency) and combination tones (sum of two other vibrations) increase the number of bands, whereas other phenomena reduce the number of bands. The following will reduce the theoretical number of bands:

1. Fundamental frequencies that fall outside of the 2.5-15 μm region.
2. Fundamental bands that are too weak to be observed.
3. Fundamental vibrations that are so close that they coalesce.
4. The occurrence of a degenerate band from several absorptions of the same frequency in highly symmetrical molecules.
5. The failure of certain fundamental vibrations to appear in the infrared because of lack of required change in dipole character of the molecule.

Assignments for stretching frequencies can be approximated by the application of Hooke's law. In the application

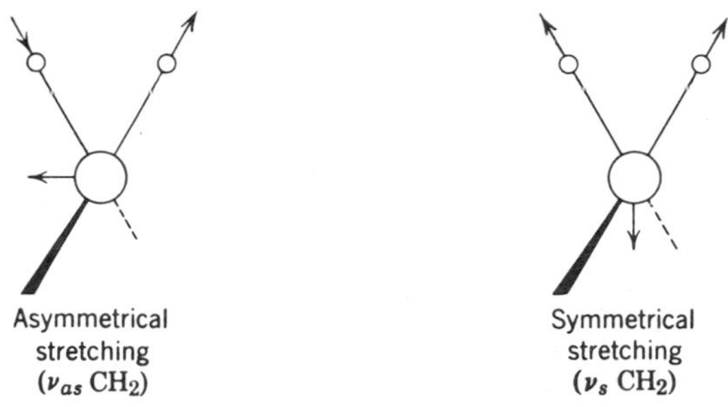

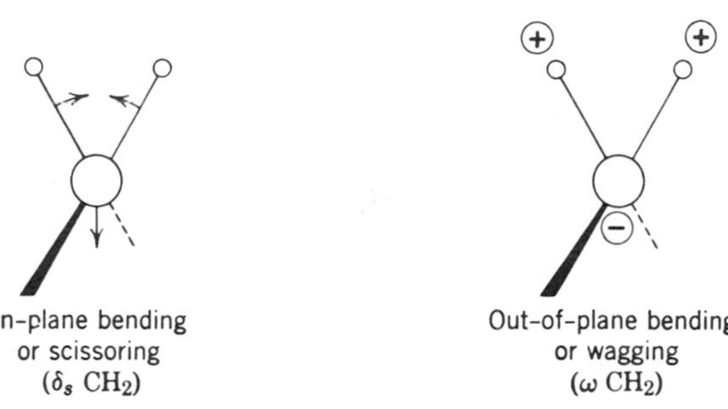

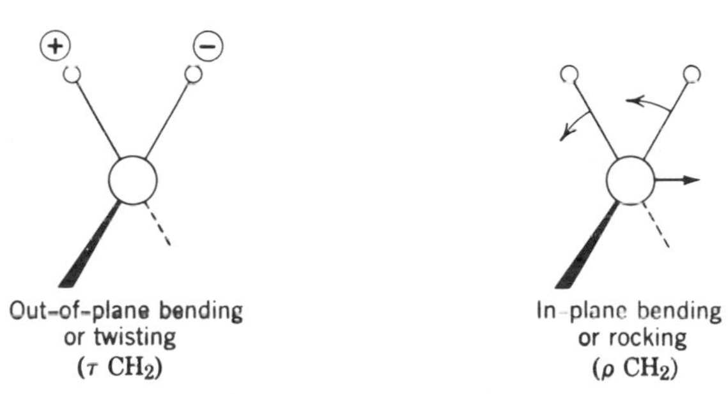

Figure 1. *Vibrational modes for a CH_2 group. (⊕ and ⊖ indicate movement perpendicular to the plane of the page.)*

of the law, 2 atoms and their connecting bond are treated as a simple harmonic oscillator composed of 2 masses joined by a spring. The following equation, derived from Hooke's law, states the relationship between frequency of oscillation, atomic masses, and the force constant of the bond.

$$\bar{\nu} = \frac{1}{2\pi c}\left(\frac{f}{\frac{MxMy}{Mx + My}}\right)^{1/2}$$

where $\bar{\nu}$ = the vibrational frequency (cm^{-1})
c = velocity of light (cm/sec)
f = force constant of bond (dynes/cm)
Mx and My = mass (g) of atom x and atom y, respectively.

The value of f is approximately 5×10^5 dynes per cm for single bonds and approximately 2 and 3 times this value for double bonds and triple bonds, respectively.

Application of the formula to C–H stretching using 19.8×10^{-24} g and 1.64×10^{-24} g as mass values for C and H, respectively, places the frequency of the C–H bond vibration at 3040 cm^{-1} (3.30 μm). Actually, C–H stretching vibrations, associated with methyl and methylene groups, are generally observed in the region between 2960-2850 cm^{-1} (3.38-3.51 μm). The calculation is not precise because effects arising from the environment of the C–H within a molecule have been ignored. The frequency of infrared absorption is commonly used to calculate the force constants of bonds.

The shift in absorption frequency following deuteration is often employed in the assignment of C–H stretching frequencies. The above equation can be used to estimate the change in stretching frequency as the result of deuteration. The term $MxMy/(Mx + My)$ will be equal to $M_C M_H/(M_C + M_H)$ for the C–H compound. Since $M_C >> M_H$, this term is approximately equal to $M_C M_H/M_C$ or to M_H. Thus, for the C–D compound the term is equal to M_D, and the frequency by Hooke's law application is inversely proportional to the square root of the mass of the isotope of hydrogen, and the ratio of the C–H to C–D stretching frequencies should equal $\sqrt{2}$. If the ratio of the frequencies, following deuteration, is much less than $\sqrt{2}$, we can assume that the vibration is not simply a C–H stretching vibration, but rather a mixed vibration involving interaction (coupling) with another vibration.

Calculations place the stretching frequencies of the following bonds in the general absorption regions indicated.

C–C, C–O, C–N	1300-800 cm^{-1} (7.7-12.5 μm)
C=C, C=O, C=N, N=O	1900-1500 cm^{-1} (5.3- 6.7 μm)
C≡C, C≡N	2300-2000 cm^{-1} (4.4- 5.0 μm)
C–H, O–H, N–H	3800-2700 cm^{-1} (2.6- 3.7 μm)

To approximate the vibrational frequencies of bond stretching by Hooke's law, the relative contributions of bond strengths and atomic masses must be considered. For example, a superficial comparison of the C–H group with the F–H group, on the basis of atomic masses, might lead to the conclusion that the stretching frequency of the F–H bond should occur at a lower frequency than that for the C–H bond. However, the increase in the force constant from left to right across the first two rows of the periodic table has a greater effect than the mass increase. Thus, the F–H group absorbs at a higher frequency (4138 cm^{-1}, 2.42 μm) than the C–H group (3040 cm^{-1}, 3.30 μm).

In general, functional groups that have a strong dipole give rise to strong absorptions in the infrared.

COUPLED INTERACTIONS

When two bond oscillators share a common atom, they seldom behave as individual oscillators unless the individual oscillation frequencies are widely different. This is because there is mechanical coupling interaction between the oscillators. For example, the carbon dioxide molecule, which consists of two C=O bonds with a common carbon atom, has two fundamental stretching vibrations: an asymmetrical and a symmetrical stretching mode. The symmetrical stretching mode consists of an in-phase stretching or contracting of the C to O bonds, and absorption occurs at a wavelength longer than that observed for the carbonyl group in an aliphatic ketone. The symmetrical stretching mode produces no change in the dipole moment of the molecule and is therefore "inactive" in the infrared, but is easily observed in the Raman spectrum* near 1340 cm^{-1} (7.46 μm). In the asymmetrical stretching mode, the two C to O bonds stretch out of phase; one C=O bond stretches as the other contracts. The asymmetrical stretching mode, since it produces a change in the dipole moment, is infrared active; the absorption (2350 cm^{-1}, 4.26 μm) is at a shorter wavelength (higher frequency) than observed for a carbonyl group in aliphatic ketones.

This difference in carbonyl absorption frequencies displayed by the carbon dioxide molecule results from strong mechanical coupling or interaction. In contrast, 2 ketonic carbonyl groups separated by 1 or more carbon atoms show normal carbonyl absorption near 1715 cm^{-1} (4.83 μm) because appreciable coupling is apparently prevented by the intervening carbon atom(s).

Coupling accounts for the two N–H stretching bands in the 3497-3077 cm^{-1} (2.86-3.25 μm) region in primary amine and primary amide spectra, for the two C=O stretching bands in the 1818-1720 cm^{-1} (5.50-5.81 μm) region in carboxylic anhydride and imide spectra, and for the two C–H stretching bands in the 3000-2760 cm^{-1} (3.33-3.62 μm) region for both methylene and methyl groups.

Useful characteristic group frequency bands often involve coupled vibrations. The spectra of alcohols have a strong band in the region between 1212 to 1000 cm^{-1} (8.25-10.00 μm), which is usually designated as the "C–O stretching band." In the spectrum of methanol this band is at 1034 cm^{-1} (9.67 μm); in the spectrum of ethanol it occurs at 1053 cm^{-1} (9.50 μm). Branching and unsaturation produce absorption characteristic of these structures (see alcohols). It is evident that we are not dealing with an isolated C–O stretching vibration, but rather a coupled asymmetric vibration involving C–C–O stretching.

Vibrations resulting from bond angle changes frequently couple in a manner similar to stretching vibrations. Thus, the ring C–H out-of-plane bending frequencies of aromatic molecules depend on the number of adjacent hydrogen atoms on the ring; coupling between the hydrogen atoms is affected by the bending of the C–C bond in the ring to which the hydrogen atoms are attached.

Interaction arising from coupling of stretching and bending vibrations is illustrated by the absorption of secondary acyclic amides. Secondary acyclic amides, which exist predominantly in the *trans* conformation, show strong absorption in the 1563-1515 cm^{-1} (6.40-6.60 μm) region; this absorption involves coupling of the N–H bending and C–N stretching vibrations.

The requirements for effective coupling interaction may be summarized as follows:

1. The vibrations must be of the same symmetry species if interaction is to occur.
2. Strong coupling between stretching vibrations requires a common atom between the groups.
3. Interaction is greatest when the coupled groups absorb, individually, near the same frequency.
4. Coupling between bending and stretching vibrations can occur if the stretching bond forms one side of the changing angle.
5. A common bond is required for coupling of bending vibrations.
6. Coupling is negligible when groups are separated by one or more carbon atoms and the vibrations are mutually perpendicular.

HYDROGEN BONDING. Hydrogen bonding can occur in any system containing a proton donor group (X–H) and a proton acceptor (Y) if the *s* orbital of the proton can effectively overlap the *p* or *π* orbital of the acceptor group. Atoms X and Y are electronegative with Y possessing lone pair electrons. The common proton donor groups in organic molecules are carboxyl, hydroxyl, amine, or amide groups. Common proton acceptor atoms are oxygen, nitrogen, and the halogens. Unsaturated groups, such as the ethylenic linkage, can also act as proton acceptors.

The strength of the hydrogen bond is at a maximum when the proton donor group and the axis of the lone pair orbital are collinear. The strength of the bond is inversely proportional to the distance between X and Y.

Hydrogen bonding alters the force constant of both groups; thus, the frequencies of both stretching and bending vibrations are altered. The X–H stretching bands move to lower frequencies (longer wavelengths) usually with increased intensity and band widening. The stretching frequency of the acceptor group, for example, C=O, is also reduced but to a lesser degree than the proton donor

*Band intensity in Raman spectra depends upon bond polarizability rather than molecular dipole changes.

group. The H—X bending vibration usually shifts to a shorter wavelength when bonding occurs; this shift is less pronounced than that of the stretching frequency.

*Inter*molecular hydrogen bonding involves association of two or more molecules of the same or different compounds. Intermolecular bonding may result in dimer molecules (as observed for carboxylic acids) or in polymer molecules, which exist in neat samples or concentrated solutions of monohydroxy alcohols. *Intra*molecular hydrogen bonds are formed when the proton donor and acceptor are present in a single molecule under spatial conditions that allow the required overlap of orbitals, for example, the formation of a 5- or 6-membered ring. The extent of both inter- and intra-molecular bonding is temperature dependent. The effect of concentration on intermolecular and intramolecular hydrogen bonding is markedly different. The bands that result from intermolecular bonding generally disappear at low concentrations (less than about 0.01 *M* in nonpolar solvents). Intramolecular hydrogen bonding is an internal effect and persists at very low concentrations.

The change in frequency between "free" OH absorption and bonded OH absorption is a measure of the strength of the hydrogen bond. Ring strain, molecular geometry, and the relative acidity and basicity of the proton donor and acceptor groups affect the strength of bonding. Intramolecular bonding involving the same bonding groups is stronger when a 6-membered ring is formed than when a smaller ring results from bonding. Hydrogen bonding is strongest when the bonded structure is stabilized by resonance.

The effects of hydrogen bonding on the stretching frequencies of hydroxyl and carbonyl groups are summarized in Table I.

An important aspect of hydrogen bonding involves interaction between functional groups of solvent and solute. If the solute is polar, then it is important to note the solvent used and the solute concentration.

FERMI RESONANCE. As we have seen in our discussion of interaction, coupling of two fundamental vibrational modes will produce two new modes of vibration, with frequencies higher and lower than that observed when interaction is absent. Interaction can also occur between fundamental vibrations and overtones or combination-tone vibrations. Such interaction is known as Fermi resonance. One example of Fermi resonance is afforded by the absorption pattern of carbon dioxide. In our discussion of interaction, we indicated that the symmetrical stretching band of CO_2 appears in the Raman spectrum near 1340 cm^{-1} (7.46 μm). Actually two bands are observed; one at 1286 cm^{-1} (7.78 μm), one at 1388 cm^{-1} (7.20 μm). The splitting results from coupling between the fundamental C=O stretching vibration, near 1340 cm^{-1} (7.46 μm), and the first overtone of the bending vibration. The fundamental bending vibration occurs near 666 cm^{-1} (15.00 μm), the first overtone near 1334 cm^{-1} (7.55 μm).

Fermi resonance is a common phenomenon in infrared and Raman spectra. It requires that the vibrational levels be of the same symmetry species and that the interacting groups be located in the molecule so that mechanical coupling is appreciable.

An example of Fermi resonance in an organic structure is the "doublet" appearance of the C=O stretch of cyclopentanone under sufficient resolution conditions. Figure 2 shows the appearance of the spectrum of cyclopentanone under the usual conditions. With adequate resolution (Fig-

Table I. Stretching Frequencies in Hydrogen Bonding

	Intermolecular Bonding			Intramolecular Bonding		
X—H · · · Y Strength	Frequency Reduction (cm^{-1})[a] νOH	νC=O[b]	Compound Class	Frequency Reduction (cm^{-1})[a] νOH	νC=O[b]	Compound Class
Weak	300	15	Alcohols, phenols, and intermolecular hydroxyl to carbonyl bonding.	< 100	10	1,2-diols; α- and most β-hydroxy ketones; *o*-chloro and *o*-alkoxy phenols
Medium				100-300	50	1,3-diols; some β-hydroxy ketones; β-hydroxy amino compounds; β-hydroxy nitro compounds
Strong	> 500	50	RCOOH dimers	> 300	100	*o*-hydroxy aryl ketones; *o*-hydroxy aryl acids; *o*-hydroxy aryl esters; β-diketones; tropolones.

[a]Frequency shift relative to "free" stretching frequencies.
[b]Carbonyl stretching only where applicable.

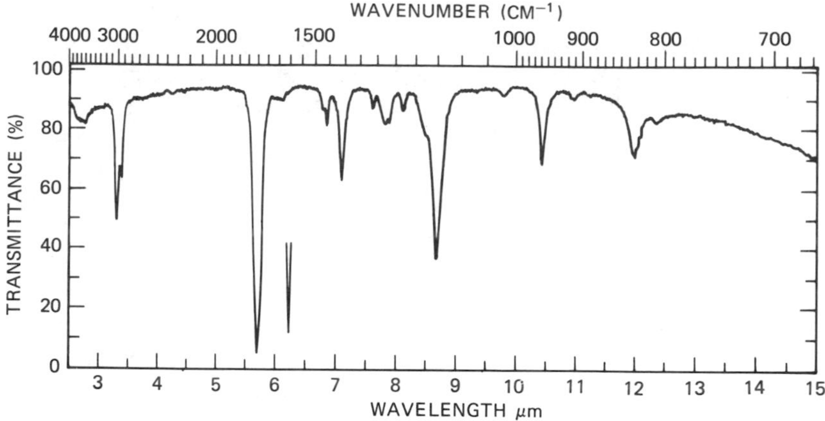

Figure 2. *Cyclopentanone, liquid film.*

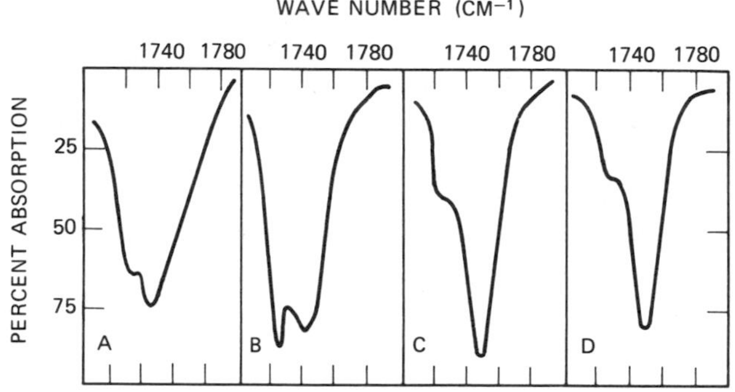

Figure 3. *Infrared spectrum of cyclopentanone in various media. A. Carbon tetrachloride solution (0.15 molar). B. Carbon disulfide solution (0.023 molar). C. Chloroform solution (0.025 molar). D. Liquid state (thin films). (Computed spectral slit width 2 cm^{-1}.)*

ure 3), Fermi resonance with an overtone or combination band of an α-methylene group shows two absorptions in the carbonyl stretch region.

III. INSTRUMENTATION

The modern double-beam infrared spectrophotometer consists of five principal sections: source (radiation), sampling area, photometer, grating (monochromator), and detector (thermocouple). A diagram of the optical system of a double-beam infrared spectrophotometer is shown in Figure 4.

RADIATION SOURCE. Infrared radiation is produced by electrically heating a source, often a Nernst filament or

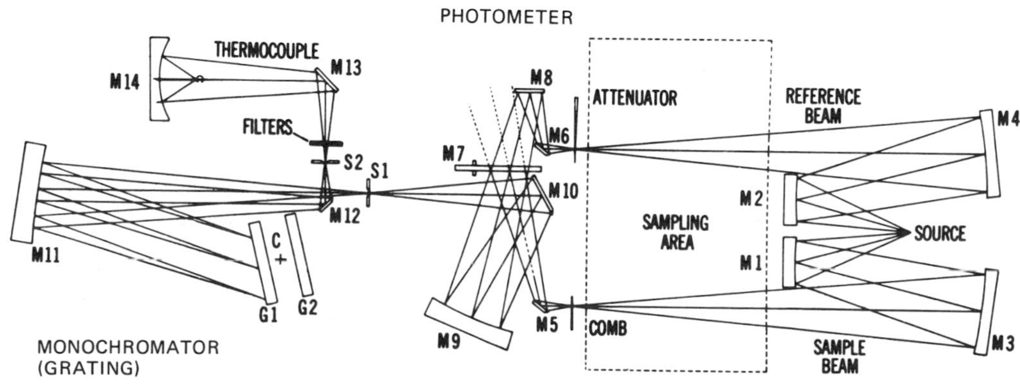

Figure 4. *Optical system of double-beam infrared spectrophotometer.*

a Globar, to 1000-1800°C. The Nernst filament is fabricated from a binder and oxides of zirconium, thorium, and cerium. The Globar is a small rod of silicon carbide. The image of the source must be wider than the maximum width of the slits (see Reference 1). The maximum radiation for the Globar occurs in the 5500-5000 cm^{-1} (1.8-2.0 μm) region and drops off by a factor of about 600 as the 600 cm^{-1} (16.7 μm) region is approached. The Nernst filament furnishes maximum radiation energy at about 7100 cm^{-1} (1.4 μm) and drops by a factor of about 1000 as the lower frequency region is approached.

The radiation from the source is divided into two beams by mirrors $M1$ and $M2$. The two beams, reference beam and sample beam, are focused in to the sample area by mirrors $M3$ and $M4$.

SAMPLING AREA. Reference and sample beams enter the sampling area and pass through the reference cell and sampling cell, respectively. Opaque shutters, mounted on the source housing, permit blocking of either beam independently. The sampling area of a precision spectrophotometer accommodates a wide variety of sampling accessories varying from gas cells of 40 m effective path to microcells.

PHOTOMETER. The reference beam passes through the attenuator (see below) and is reflected by mirrors $M6$ and $M8$ to the rotating sector mirror $M7$, which alternately reflects the reference beam out of the optical system and transmits the beam to mirror $M9$. The reference beam is now an intermittent beam with a "frequency" of from 8 to 13 cycles per second depending on the particular instrument. This beam is focused by mirror $M10$ on the slit $S1$. The sample beam passes through the comb (see below) and is reflected by mirror $M5$ to the rotating sector mirror $M7$, which alternately transmits the beam out of the optical system and reflects it to mirror $M9$, thence to mirror $M10$ and slit $S1$.

At any given moment, the beam focused on slit $S1$ is either the reference beam, which was transmitted by the rotating sector mirror $M7$, or the sample beam, which was reflected by $M7$. In other words, the reference beam and the sample beam have been combined into a single beam of alternating segments; this establishes a switching frequency at the detector equal to the speed of rotation of $M7$

When the beams are of equal intensity, the instrument is at an optical null. The comb in the sample beam permits balancing the beams. The recording pen is then at 100% transmittance when no sample is present.

The attenuator is driven in and out of the reference beam in response to the signal created at the detector by the sample beam. Thus, when the sample beam is absorbed by the sample, the attenuator is driven into the reference beam until its intensity matches that of the sample beam.

MONOCHROMATOR. The combined beam passes through the monochromator entrance slit $S1$ to the mirror $M11$, which reflects it to the diffraction grating $G1$. This beam is dispersed into various frequencies; this beam is reflected back to mirror $M11$, subsequently to mirror $M12$. $M12$ focuses the beam on exit slit $S2$. The width of the frequency range focused on slit $S2$ is determined by the width of entrance slit $S1$ and the dispersing power of the grating. The frequency band of one dispersed beam focused on slit $S2$ at any moment is determined by the angle of grating $G1$ at that moment. Rotation of grating $G1$ produces a scan of frequency bands at exit slit $S2$ and thus at the detector. Filters are automatically inserted into the radiation path at the exit slit in a sequence during scanning to eliminate all unwanted radiation; this unwanted radiation corresponds to harmonic multiples of the frequency to be measured.

Maximum resolution is obtained by using gratings only in the range of greatest dispersing effectiveness. Therefore, often in a high-resolution instrument, two or more gratings are used ($G1$ and $G2$ of Figure 4).

The narrower the slit width, the greater is the resolution. Here again, some compromise is necessary because of the decreased energy output of the source at lower frequencies (longer wavelengths). On most instruments, the slit width is programmed to increase as the emitted source energy decreases, so that constant reference beam energy enters the monochromator.

DETECTOR (THERMOCOUPLE). After leaving the exit slit of the monochromator, the beam is reflected by a flat mirror $M13$ to an ellipsoidal mirror $M14$. The foci of the ellipsoidal mirror are the exit slit $S2$ and the detector.

The detector is a device that measures radiant energy by means of its heating effect. Two common types of detectors are the thermocouple and bolometer. In the thermocouple detector, the radiant energy heats one of two bimetallic junctions, and an emf is produced between the two junctions proportional to the degree of heating. The bolometer changes its resistance upon heating. It serves as one arm of a bridge so that a change in temperature will cause an unbalanced signal across the circuit. The unbalanced signal can be amplified and recorded or used to activate a servomechanism to re-establish a balance.

Since the detector sees alternately the reference and the sample beam at a switching frequency determined by the rotation of the sector mirror, any change in the intensity of the radiation due to absorption is detected as an off-null signal.

The amplified off-null signal of the detector is used to position the optical attenuator so that the radiation from the reference and sample beams are kept at equal intensity. The amount of attenuation required is a direct measure of the absorption by the sample. The movement of the attenuator is recorded by the recording chart pen.

In recent years Fourier Transform instruments have become available. FT IR allows analysis of very small samples.

IV. SAMPLE HANDLING

Infrared spectra may be obtained for gases, liquids, or solids.

1. The spectra of gases or low-boiling liquids may be obtained by expansion of the sample into an evacuated cell. Cells equipped with freeze-out tips are used for sample concentration and cell evacuation prior to expansion of the sample into the cell. Gas cells are available in lengths of a few centimeters to 40 meters. The sampling area of a standard infrared spectrophotometer will not accommodate cells much longer than 10 cm; long paths are achieved by multiple reflection optics.

   The vapor phase technique is limited because of the relatively large percentage of compounds that do not have sufficiently high vapor pressures to produce a useful absorption spectrum. However, the usefulness of the technique can be extended by heating the cell.

2. Liquids may be examined neat or in solution. Neat liquids are examined between salt plates usually without a spacer. Pressing a liquid sample between flat plates produces a film of 0.01 mm or less in thickness, the plates being held together by capillarity. Samples of 1-10 mg are required. Thick samples of neat liquids usually absorb too strongly to produce a satisfactory spectrum. Volatile liquids are examined in sealed cells with very thin spacers. Silver chloride or KRS-5 plates may be used for samples that dissolve sodium chloride plates.

   Solutions are handled in cells of 0.1-1 mm thickness. Volumes of 0.1-1 ml of 0.05-10% solutions are required for readily available cells (Figure 5). A compensating cell, containing pure solvent, is placed in the reference beam. The spectrum thus obtained is that of the solute except in those regions in which the solvent absorbs strongly. For example, thick samples of carbon tetrachloride absorb strongly near 800 cm^{-1} (12.50 μm); compensation for this band is ineffective since strong absorption prevents any radiation from reaching the detector.

   The solvent selected must be dry and reasonably transparent in the region of interest. When the entire spectrum is of interest, several solvents must be used. A common pair of solvents is carbon tetrachloride and carbon disulfide. Carbon tetrachloride is relatively free of absorption at frequencies above 1333 cm^{-1} (shorter than 7.50 μm) whereas carbon disulfide shows little absorption below 1333 cm^{-1} (longer than 7.50 μm). Solvent and solute combinations that react must be avoided. For example, carbon disulfide cannot be used as a solvent for primary or secondary amines. Amino alcohols react slowly with carbon disulfide and carbon tetrachloride.

   When only very small samples are available, ultramicro cavity cells are used in conjunction with a beam condenser. The smallest commercially available cell has a path length of approximately 0.05 mm and a capacity of about 0.8 microliter.* A spectrum can thus be obtained on a few micrograms of sample in solution. When volatility permits, the solute can be recovered for examination by other spectrometric techniques.

   The absorption patterns of selected solvents and mulling oils are presented in Appendix A.

3. Solids are usually examined as a mull, a pressed disc, or as a deposited glassy film.

   Mulls are prepared by thoroughly grinding 2-5 mg of a solid in a smooth agate mortar. Grinding is continued after the addition of 1 or 2 drops of the mulling oil. The suspended particles must be less than 2 μm to avoid excessive scattering of radiation. The mull is examined as a thin film between flat salt plates. Nujol (a high-boiling petroleum oil) is commonly used as a mulling agent. When hydrocarbon bands interfere with the spectrum, Fluorolube (a completely halogenated polymer containing F and Cl) or hexachlorobutadiene may be used. The use of both Nujol and Fluorolube mulls makes possible a scan, essentially free of interfering

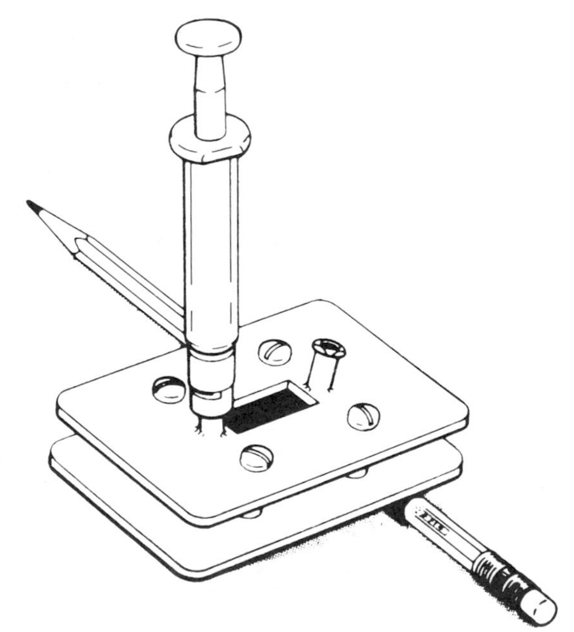

Figure 5. *Correct way to fill a sealed cell.*

*Barnes Engineering Co., Instrument Division, Stamford, Conn.

bands, over the 4000 cm^{-1}-250 cm^{-1} (2.5-40.0 μm) region.

The pressed-disc technique depends upon the fact that dry, powdered potassium bromide (or other alkali metal halides) can be pressed under pressure *in vacuo* to form transparent discs. The sample (0.5-1.0 mg) is intimately mixed with approximately 100 mg of dry, powdered potassium bromide. Mixing can be effected by thorough grinding in a smooth agate mortar or, more efficiently, with a small vibrating ball mill, or by lyophilization. The mixture is pressed with special dies under a pressure of 10,000-15,000 pounds per square inch into a transparent disc. The quality of the spectrum depends on the intimacy of mixing and the reduction of the suspended particles to 2 μm or less. Micro-discs, 0.5 to 1.5 mm in diameter, can be used with a beam condenser. The micro-disc technique permits examination of samples as small as 1 microgram. Bands near 3448 cm^{-1} and 1639 cm^{-1} (2.9 and 6.1 μm), due to moisture, frequently appear in spectra obtained by the pressed disc technique (see H_2O spectrum in Appendix B).

The use of KBr discs or pellets has often been avoided because of the demanding task of making good pellets. Such KBr techniques can be less formidable through the Mini-Press (Figure 6) that affords a simple procedure; the KBr-sample mixture is placed in the nut portion of the assembly with one bolt in place. The second bolt is introduced, and pressure is applied by tightening the bolts. Removal of the bolts leaves a pellet in the nut that now serves as a cell.

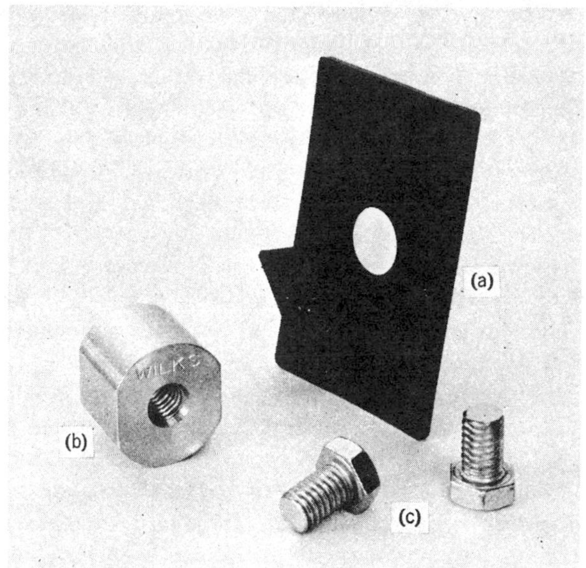

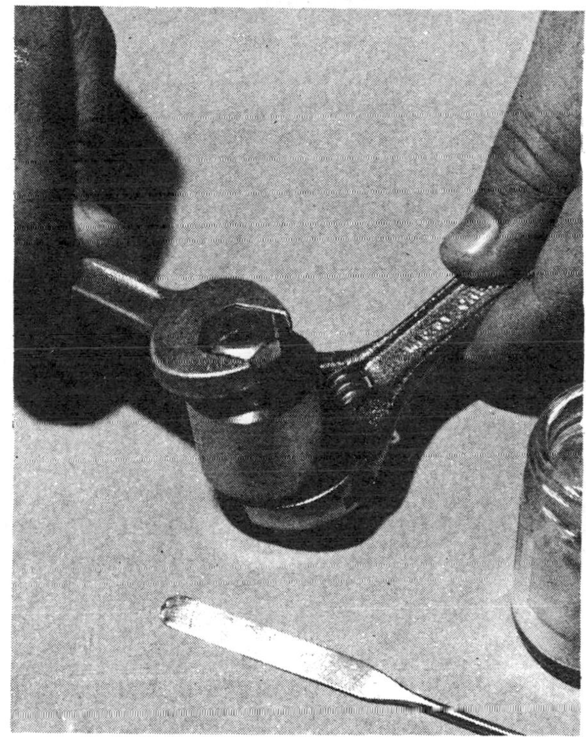

Figure 6. *Mini-press and operation. (a) Cell holder. (b) Nut = cell. (c) Bolts.*

Deposited films are useful only when the material can be deposited from solution or cooled from a melt as microcrystals or as a glassy film. Crystalline films generally lead to excessive light scattering. Specific crystal orientation may lead to spectra differing from those observed for randomly oriented particles such as exist in a mull or halide disc. The deposited film technique is particularly useful for obtaining spectra of resins and plastics. Care must be taken to free the sample of solvent by vacuum treatment or gentle heating.

A technique known as attenuated total reflection or internal reflection spectroscopy is now available for obtaining qualitative spectra of solids regardless of thickness. The technique depends on the fact that a beam of light that is internally reflected from the surface of a transmitting medium passes a short distance beyond the reflecting boundary and returns to the transmitting medium as a part of the process of reflection. If a material (i.e., the sample) of lower refractive index than the transmitting medium is brought in contact with the reflecting surface, the light passes through the material to the depth of a few μm, producing an absorption spectrum. An extension of the technique provides for multiple internal reflections along the surface of the sample. The multiple internal reflection technique results in spectra with intensities comparable to transmission spectra.

In general, a dilute solution in a nonpolar solvent furnishes the best (i.e., least distorted) spectrum. Nonpolar compounds give essentially the same spectra in the condensed phase (i.e., neat liquid, a mull, a KBr disc, or a film) as they give in nonpolar solvents. Polar compounds, however, often show hydrogen bonding effects in the condensed phase. Unfortunately, polar compounds are frequently insoluble in nonpolar solvents, and the spectrum must be obtained either in a condensed phase or in a polar solvent; the latter introduces the possibility of solute-solvent hydrogen bonding.

Reasonable care must be taken in handling salt cells and plates. Moisture-free samples should be used. Fingers should not come in contact with the optical surfaces. Care should be taken to prevent contamination with silicones, which are hard to remove and have strong absorption patterns.

V. INTERPRETATION OF SPECTRA

There are no rigid rules for interpreting an infrared spectrum. Certain requirements, however, must be met before an attempt is made to interpret a spectrum.

1. The spectrum must be adequately resolved and of adequate intensity.
2. The spectrum should be that of a reasonably pure compound.
3. The spectrophotometer should be calibrated so that the bands are observed at their proper frequencies or wavelengths. Proper calibration can be made with reliable standards, such as polystyrene film.
4. The method of sample handling must be specified. If a solvent is employed, the solvent, concentration, and the cell thickness should be indicated.

A precise treatment of the vibrations of a complex molecule is not feasible; thus, the infrared spectrum must be interpreted from empirical comparison of spectra, and extrapolation of studies of simpler molecules. Many questions arising in the interpretation of an infrared spectrum can be answered by data obtained from the mass, ultraviolet, and NMR spectra.

Infrared absorption of organic molecules is summarized in the chart of characteristic group frequencies in Appendix C. Many of the group frequencies vary over a wide range because the bands arise from complex interacting vibrations within the molecule. Absorption bands may, however, represent predominantly a single vibrational mode. Certain absorption bands, for example, those arising from the C—H, O—H, and C=O stretching modes, remain within fairly narrow regions of the spectrum. Important details of structure may be revealed by the exact position of an absorption band within these narrow regions. Shifts in absorption position and changes in band contours, accompanying changes in molecular environment, may also suggest important structural details.

The two important areas for a preliminary examination of a spectrum are the region 4000-1300 cm^{-1} (2.5-7.7 μm) and the 909-650 cm^{-1} (11.0-15.4 μm) region. The high frequency portion of the spectrum is called the functional group region. The characteristic stretching frequencies for important functional groups such as OH, NH, and C=O occur in this portion of the spectrum. The absence of absorption in the assigned ranges for the various functional groups can usually be used as evidence for the absence of such groups in the molecule. Care must be exercised, however, in such interpretations since certain structural characteristics may cause a band to become extremely broad so that it may go unnoticed. For example, intramolecular hydrogen bonding in the enolic form of acetylacetone results in a broad OH band, which may be overlooked. The absence of absorption in the 1850-1540 cm^{-1} (5.40-6.50 μm) region excludes a structure containing a carbonyl group. Weak bands in the high frequency region, resulting from the fundamental absorption of functional groups, such as S—H and C≡C, are extremely valuable in the determination of structure. Such weak bands would be of little value in the more complicated regions of the spectrum. Overtones and combination tones of lower frequency bands frequently appear in the high frequency region of the spectrum. Overtone- and combination-tone bands are characteristically weak except when Fermi resonance occurs. Strong skeletal bands for aromatics and heteroaromatics fall in the 1600-1300 cm^{-1} (6.25-7.70 μm) region of the spectrum.

The lack of strong absorption bands in the 909-650 cm^{-1} (11.0-15.4 μm) region generally indicates a nonaromatic structure. Aromatic and heteroaromatic compounds display strong out-of-plane C—H bending and ring bending absorption bands in this region that can frequently be correlated with the substitution pattern. Broad, moderately intense absorption in the low frequency region suggests the presence of carboxylic acid dimers, amines, or amides, all of which show out-of-plane bending in this region. If the region is extended to 1,000 cm^{-1} (10.0 μm) absorption bands characteristic of olefinic structures are included.

The intermediate portion of the spectrum, 1300-909 cm^{-1} (7.7-11.0 μm), is usually referred to as the "fingerprint" region. The absorption pattern in this region is frequently complex, with the bands originating in interacting vibrational modes. This portion of the spectrum is extremely valuable when examined in reference to the other regions. For example, if alcoholic or phenolic O—H stretching absorption appears in the high frequency region of the spectrum, the position of the C—C—O absorption band in the 1260-1000 cm^{-1} (7.93-10.00 μm) region frequently makes it possible to assign the O—H absorption to alcohols and phenols with highly specific structures. Absorption in this intermediate region is probably unique for every molecular species.

Any conclusions reached after examination of a particular band should be confirmed where possible by examination of other portions of the spectrum. For example, the assignment of a carbonyl band to an aldehyde should be confirmed by the appearance of a band or a pair of bands in the 2900-2695 cm^{-1} (3.45-3.71 μm) region of the spectrum, arising from the C—H stretching vibrations of the aldehyde group. Similarly, the assignment of a carbonyl band to an ester should be confirmed by observation of a strong band in the C—O stretching region, 1300-1100 cm^{-1} (7.6-9.1 μm).

Finally, in a "fingerprint" comparison of spectra, or any other situation in which the *shapes* of peaks are important, we should anticipate the substantial change in the general appearance of the spectrum in changing from a spectrum that is linear in wavenumber to one that is linear in wavelength (Figure 7).

VI. CHARACTERISTIC GROUP FREQUENCIES OF ORGANIC MOLECULES

A table of characteristic group frequencies is presented as Appendix C. The ranges presented for group frequencies

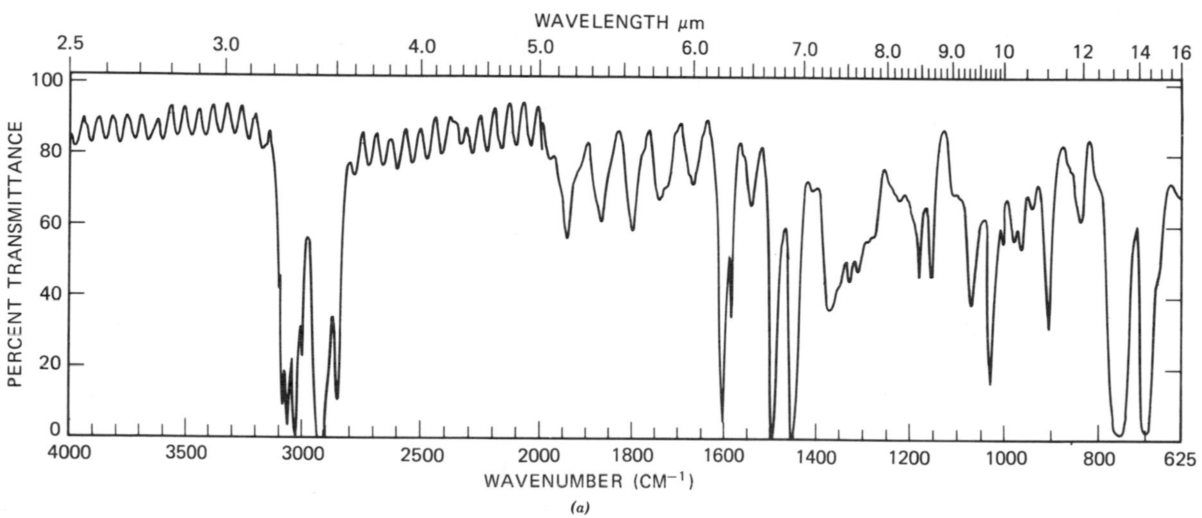

(a)

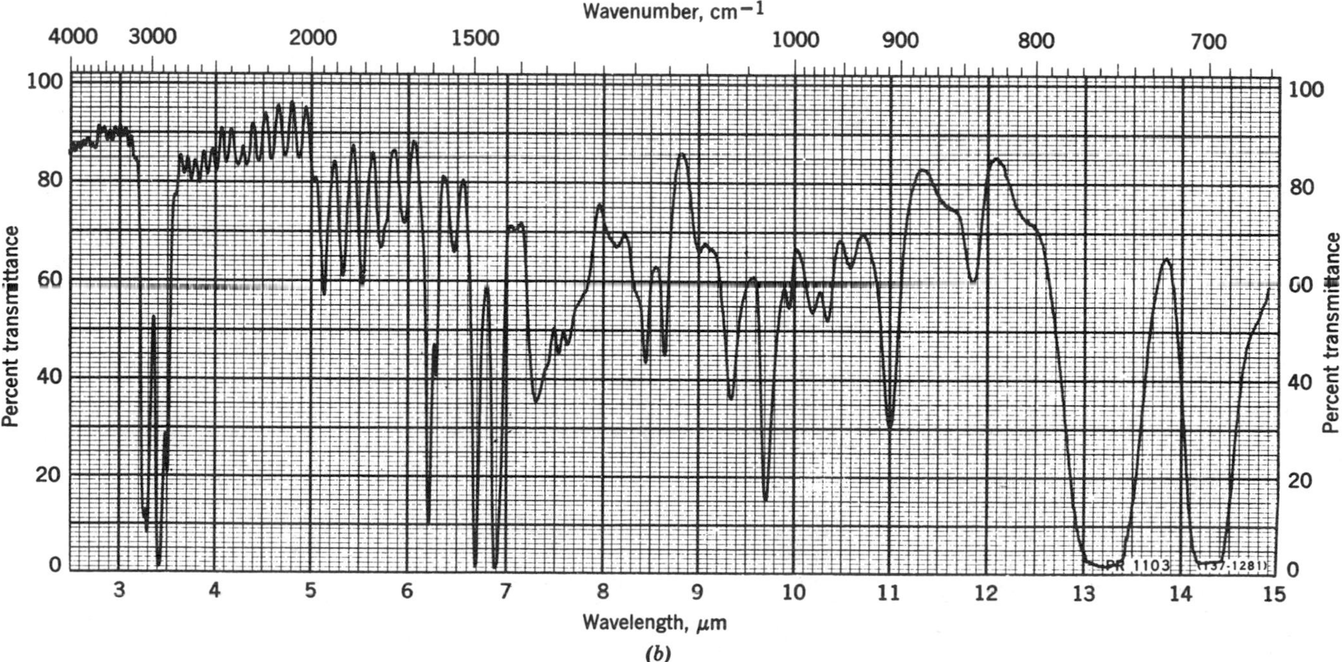

(b)

Figure 7. *Polystyrene, same sample for both a and b. a, Linear in wavenumber (cm^{-1}); b, Linear in wavelength (μm).*

have been assigned following the examination of many compounds in which the groups occur. Although the ranges are quite well defined, the precise frequency or wavelength at which a specific group absorbs is dependent on its environment within the molecule and on its physical state.

This section of the chapter is concerned with a comprehensive look at these characteristic group frequencies and their relationship to molecular structure. As a major type or class of molecule or functional group is introduced in the succeeding sections, an example of an infrared spectrum with the important peak assignments will be given. Spectra of common laboratory substances, representing many of the chemical classes listed below, are shown in Appendix B. Characteristic group absorptions are found in Appendix C.

References that serve to supplement this section have been cited.[1-3,7-25]

NORMAL ALKANES (PARAFFINS)

The spectra of normal paraffins can be interpreted in terms of 4 vibrations, namely the stretching and bending of C—H and C—C bonds. Detailed analysis of the spectra of the lower members of the alkane series has made detailed assignments of the spectral positions of specific vibrational modes possible.

Not all of the possible absorption frequencies of the paraffin molecule are of equal value in the assignment of structure. The C—C bending vibrations occur at very low frequencies (below 500 cm^{-1}, longer than 20 μm) and therefore do not appear in our spectra. The bands assigned to C—C stretching vibrations are weak and appear in the broad region of 1200-800 cm^{-1} (8.3-12.5 μm); they are generally of little value for identification.

The most characteristic vibrations are those arising from C—H stretching and bending. Of these vibrations, those arising from methylene twisting and wagging are usually of limited diagnostic value because of their weakness and instability. This instability is a result of strong coupling to the remainder of the molecule.

The vibrational modes of alkanes are common to many organic molecules. Although the positions of C—H stretching and bending frequencies of methyl and methylene groups remain nearly constant in hydrocarbons, the attachment of CH_3 or CH_2 to atoms other than carbon, or to a carbonyl group or aromatic ring, may result in appreciable shifts of the C—H stretching and bending frequencies. These shifts are summarized in Tables I and II of Appendix D.

The spectrum of dodecane, Figure 8, is that of a typical straight-chain hydrocarbon.

C—H Stretching Vibrations

Absorption arising from C—H stretching in the alkanes occurs in the general region of 3,000-2,840 cm^{-1} (3.3-3.5 μm). The positions of the C—H stretching vibrations are among the most stable in the spectrum. When a spectrum is obtained with an instrument using a sodium chloride prism, the bands in this region are frequently unresolved (as in Figure 7b). Resolution (Figure 7a) is achieved through the use of a fluoride prism or a grating instrument.

METHYL GROUPS. An examination of a large number of saturated hydrocarbons containing methyl groups

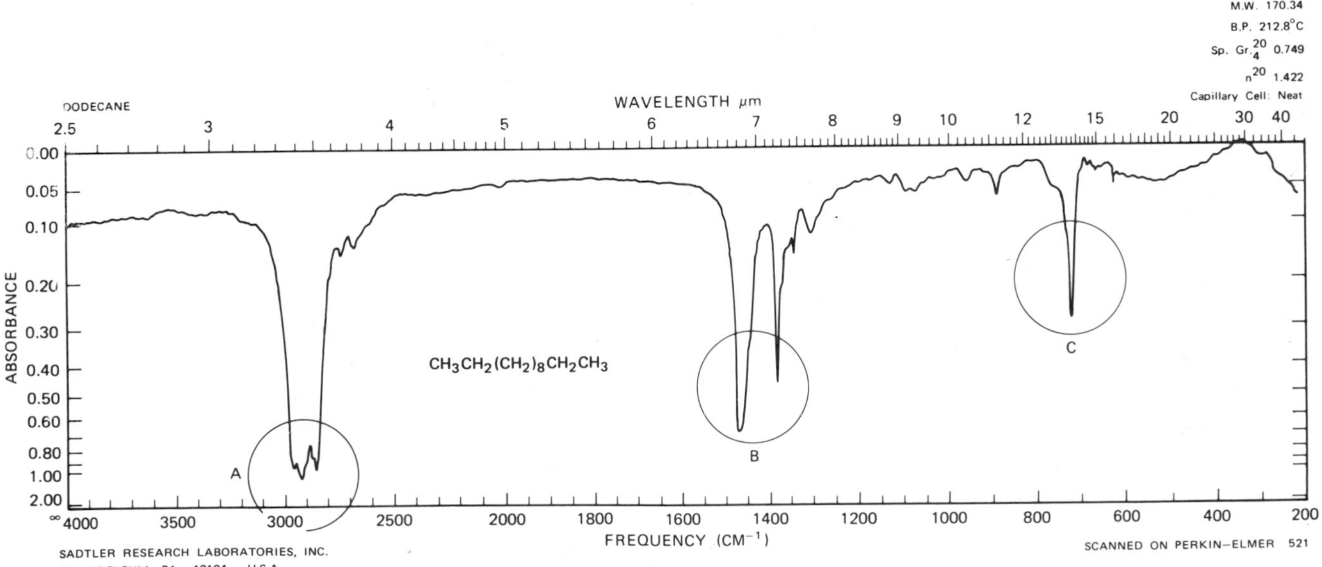

Figure 8. **A.** C—H *stretch: 2962 cm^{-1} (3.38 μm)* ν_{as} CH_3, *2872 cm^{-1} (3.48 μm)* ν_s CH_3, *2926 cm^{-1} (3.43 μm)* ν_{as} CH_2, *2853 cm^{-1} (3.51 μm)* ν_s CH_2. **B.** C—H *bend: 1465 cm^{-1} (6.83 μm)* δ_s CH_2, *1450 cm^{-1} (6.90 μm)* δ_{as} CH_3, *1375 cm^{-1} (7.28 μm)* δ_s CH_3. **C.** CH_2 *rock: 722 cm^{-1} (13.9 μm)* ρ CH_2.

showed, in all cases, 2 distinct bands occurring at 2962 cm^{-1} (3.38 μm) and at 2872 cm^{-1} (3.48 μm). The first of these results from the asymmetrical stretching mode in which 2 C—H bonds of the methyl group are extending while the third one is contracting ($\nu_{as}CH_3$). The second arises from symmetrical stretching (ν_sCH_3) in which all 3 of the C—H bonds extend and contract in phase. The presence of several methyl groups in a molecule results in strong absorption at these positions.

METHYLENE GROUPS. The asymmetrical stretching ($\nu_{as}CH_2$) and symmetrical stretching (ν_sCH_2) occur, respectively, near 2926 cm^{-1} (3.43 μm) and 2853 cm^{-1} 3.51 μm). The positions of these bands do not vary more than ± 10 cm^{-1} in the aliphatic and nonstrained cyclic hydrocarbons. The frequency of methylene stretching is increased when the methylene group is part of a strained ring.

C—H Bending Vibrations

METHYL GROUPS. Two bending vibrations can occur within a methyl group. The first of these, the symmetrical bending vibration, involves the in-phase bending of the C—H bonds (I). The second, the asymmetrical bending vibration, involves out-of-phase bending of the C—H bonds (II).

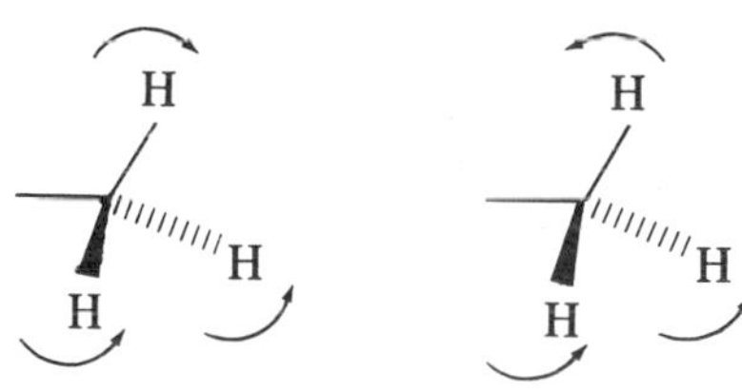

In I, the C—H bonds are moving like the closing petals of a flower; in II, one petal opens as two petals close.

The symmetrical bending vibration (δ_sCH_3) occurs near 1375 cm^{-1} (7.28 μm), the asymmetrical bending vibration ($\delta_{as}CH_3$) near 1450 cm^{-1} (6.90 μm).

The asymmetrical vibration generally overlaps the scissoring vibration of the methylene groups (see below). Two distinct bands are observed, however, in compounds such as diethyl ketone, in which the methylene scissoring band has been shifted to a lower frequency, 1439-1399 cm^{-1} (6.95-7.15 μm) and increased in intensity because of its proximity to the carbonyl group.

The absorption band near 1375 cm^{-1} (7.28 μm), arising from the symmetrical bending of the methyl C—H bonds, is very stable in position when the methyl group is attached to another carbon atom. The intensity of this band is greater for each methyl group in the compound than that for the asymmetrical methyl bending vibration or the methylene scissoring vibration.

METHYLENE GROUPS. The bending vibrations of the C—H bonds in the methylene group have been shown schematically in Figure 1. The four bending vibrations are referred to as *scissoring, rocking, wagging,* and *twisting.*

The scissoring band (δ_sCH_2) in the spectra of hydrocarbons occurs at a nearly constant position near 1465 cm^{-1} (6.83 μm).

The band resulting from the methylene rocking vibration (ρCH_2), in which all of the methylene groups rock in phase, appears near 720 cm^{-1} (13.9 μm) for straight-chain alkanes of seven or more carbons. This band may appear as a doublet in the spectra of solid samples. In the lower members of the *n*-alkane series, the band appears at somewhat higher frequencies.

Absorption of hydrocarbons, due to methylene twisting and wagging vibrations, is observed in the 1350-1150 cm^{-1} 7.4-8.7 μm) region. These bands are generally appreciably weaker than those resulting from methylene scissoring. A series of bands in this region, arising from the methylene group, is characteristic of the spectra of solid samples of long-chain acids, amides, and esters.

BRANCHED-CHAIN ALKANES

In general, the changes brought about in the spectrum of a hydrocarbon by branching result from changes in skeletal stretching vibrations and methyl bending vibrations; these occur below 1500 cm^{-1} (longer than 6.7 μm). The spectrum of Figure 9 is that of a typical branched alkane.

C—H Stretching Vibrations

TERTIARY C—H GROUPS. Absorption resulting from this vibrational mode is very weak and is usually lost in other aliphatic C—H absorption. Absorption in hydrocarbons occurs near 2890 cm^{-1} (3.46 μm).

C—H Bending Vibrations

GEM-DIMETHYL GROUPS. Configurations in which 2 methyl groups are attached to the same carbon atom exhibit distinctive absorption in the C—H bending region. The isopropyl group shows a strong doublet, with peaks of almost equal intensity at 1385-1380 cm^{-1} (7.22-7.25 μm) and at 1370-1365 cm^{-1} (7.30-7.33 μm). The tertiary butyl group gives rise to 2 C—H bending bands, 1 in the 1395-1385 cm^{-1} (7.17-7.22 μm) region and 1 near 1370 cm^{-1} (7.30 μm). In the *t*-butyl doublet, the long wavelength band is more intense. When the gem-dimethyl group occurs at an internal position, a doublet is observed in essentially the same region where absorption occurs for the isopropyl and *t*-butyl groups. Doublets are observed for gem-dimethyl groups because of interaction between the in-phase and out-of-phase symmetrical CH_3

bending of the 2 methyl groups attached to a common carbon atom.

Weak bands result from methyl rocking vibrations in isopropyl and *t*-butyl groups. These vibrations are sensitive to mass and interaction with skeletal stretching modes and are generally less reliable than the C—H bending vibrations. The following assignments have been made: isopropyl group, 922-919 cm^{-1} (10.85-10.88 μm), and *t*-butyl group, 932-926 cm^{-1} (10.73-10.80 μm).

CYCLIC ALKANES

C—H Stretching Vibrations

The methylene stretching vibrations of unstrained cyclic polymethylene structures are much the same as those observed for acyclic paraffins. Increasing ring strain moves the C—H stretching bands progressively to high frequencies. The ring CH_2 and CH groups in a monoalkylcyclopropane ring absorb in the region of 3100-2990 cm^{-1} (3.23-3.34 μm).

C—H Bending Vibrations

Cyclization decreases the frequency of the CH_2 scissoring vibration. Cyclohexane absorbs at 1452 cm^{-1} (6.89 μm), whereas *n*-hexane absorbs at 1468 cm^{-1} (6.81 μm). Cyclopentane absorbs at 1455 cm^{-1} (6.87 μm), cyclopropane at 1442 cm^{-1} (6.94 μm). This shift frequently makes it possible to observe distinct bands for methylene and methyl absorption in this region. Spectra of other saturated hydrocarbons appear in Appendix B: hexane-no. 1, nujol-no. 2 and cyclohexane-no. 3.

OLEFINIC HYDROCARBONS (ALKENES)

Olefinic structure introduces several new modes of vibration into a hydrocarbon molecule: a C=C stretching vibration, C—H stretching vibrations in which the carbon atom is present in the olefinic linkage, and in-plane and out-of-plane bending of the olefinic C—H bond. The spectrum of Figure 10 is that of a typical terminal olefin.

C=C Stretching Vibrations

UNCONJUGATED LINEAR OLEFINS. The C=C stretching mode of unconjugated olefins usually shows moderate to weak absorption at 1667-1640 cm^{-1} (6.00-6.10 μm). Monosubstituted olefins, that is, vinyl groups, absorb near 1640 cm^{-1} (6.10 μm) with moderate intensity. Disubstituted *trans*-olefins, tri-, and tetra-alkyl substituted olefins absorb at or near 1670 cm^{-1} (5.99 μm); disubstituted *cis*-olefins and vinylidene olefins absorb near 1650 cm^{-1} (6.06 μm).

The absorption of symmetrical disubstituted *trans*-olefins or tetrasubstituted olefins may be extremely weak or absent. *Cis*-olefins, which lack the symmetry of the *trans* structure, absorb more strongly than *trans*-olefins. Internal double bonds generally absorb more weakly than terminal double bonds because of pseudosymmetry.

Abnormally high absorption frequency is observed for $-CH{=}CF_2$ and $-CF{=}CF_2$ groups. The former absorbs

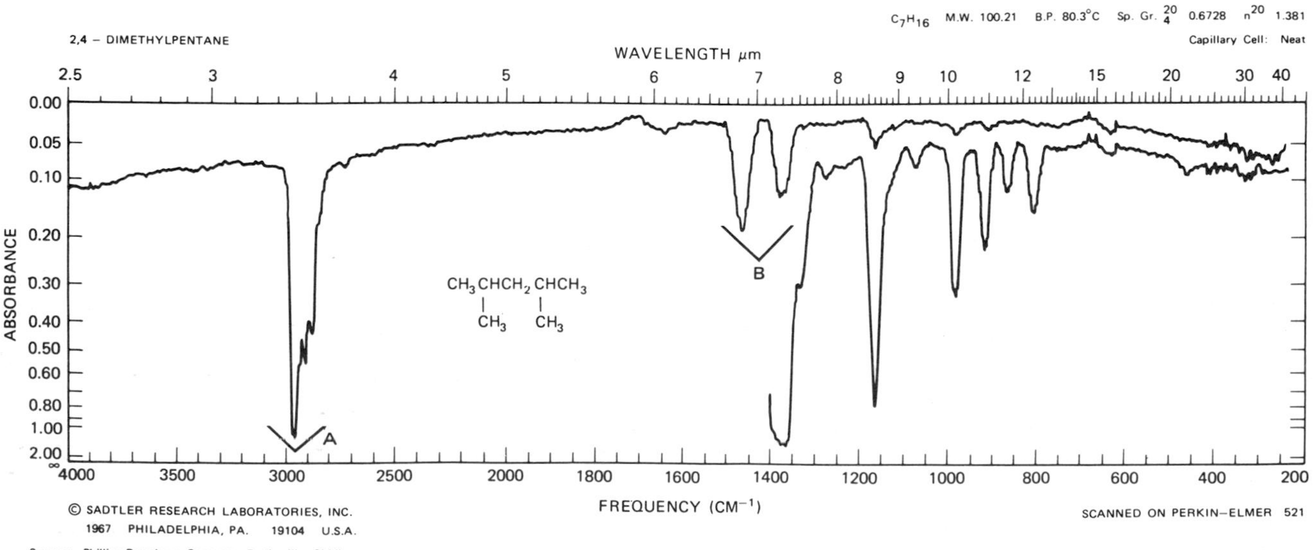

Figure 9. **A.** C–H *stretch (see Figure 8).* **B.** C–H *bend (see Figure 8). There is an unresolved gem-dimethyl band from 1385-1395* cm^{-1}. *Compare the weak methylene rocking band(s) (800-1000* cm^{-1}*) to that for Figure 8.*

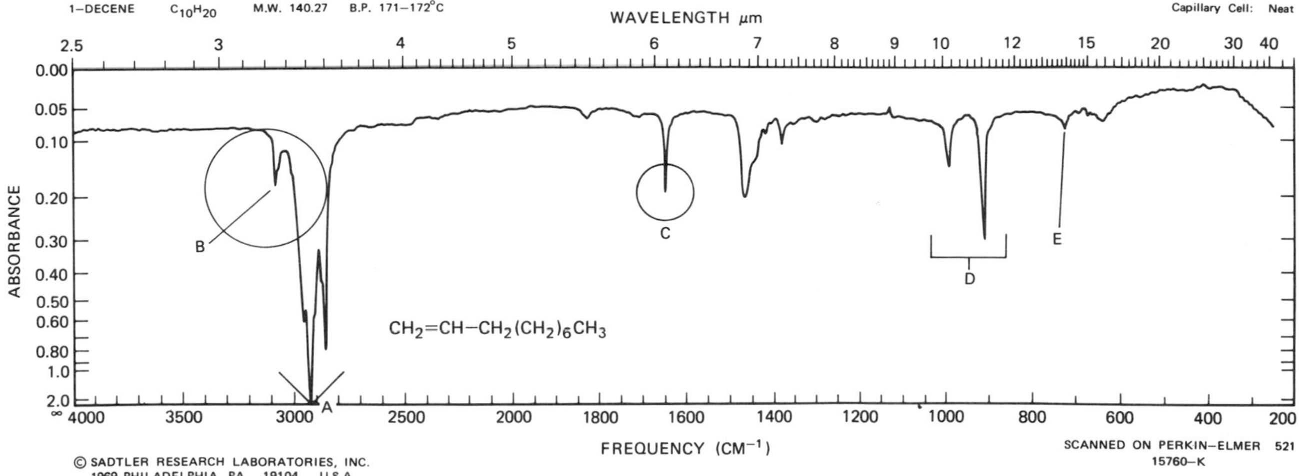

Figure 10. A. C–H *stretch (see Figure 8). Note olefinic* C–H *stretch (B) at 3049 cm⁻¹ (3.28 μm).* **C.** C=C *stretch, 1645 cm⁻¹ (6.08 μm), see Table III of Appendix D.* **D.** *Out-of-plane* C–H *bend: 986 cm⁻¹ (10.14 μm), (olefinic) 907 cm⁻¹ (11.03 μm).* **E.** *(methylene rock: 720 cm⁻¹ (13.88 μm).*

near 1754 cm⁻¹ (5.70 μm), the latter near 1786 cm⁻¹ (5.60 μm). In contrast, the absorption frequency is reduced by the attachment of chlorine, bromine, or iodine.

CYCLOOLEFINS. Absorption of the internal double bond in the unstrained cyclohexene system is essentially the same as that of a *cis* isomer in an acyclic system. The C=C stretch vibration is coupled with the C–C stretching of the adjacent bonds. As the angle α C–C–C (angle α at central C) becomes smaller the interaction becomes less until it is at a minimum at 90° in cyclobutene (1566 cm⁻¹, 6.39 μm). In the cyclopropene structure, interaction again becomes appreciable, and the absorption frequency increases (1641 cm⁻¹, 6.09 μm).

The substitution of alkyl groups for an α-hydrogen atom in strained ring systems serves to increase the frequency of C=C absorption. Cyclobutene absorbs at 1566 cm⁻¹ (6.39 μm), 1-methylcyclobutene at 1641 cm⁻¹ (6.09 μm).

The absorption frequency of external exocyclic-olefinic bonds increases with decreasing ring size. Methylenecyclohexane absorbs at 1650 cm⁻¹ (6.06 μm), methylenecyclopropane at 1781 cm⁻¹ (5.62 μm).

CONJUGATED SYSTEMS. The olefinic bond stretching vibrations in conjugated dienes without a center of symmetry interact to produce two C=C stretching bands. The spectrum of an unsymmetrical conjugated diene, such as 1,3-pentadiene, shows absorption near 1650 cm⁻¹ (6.06 μm) and near 1600 cm⁻¹ (6.25 μm). The symmetrical molecule 1,3-butadiene shows only one band near 1600 cm⁻¹ (6.25 μm), resulting from asymmetric stretching; the symmetrical stretching band is inactive in the infrared. The infrared spectrum of isoprene (Figure 11) illustrates many of these features.

Conjugation of an olefinic double bond with an aromatic ring produces enhanced olefinic absorption near 1625 cm⁻¹ (6.15 μm).

The absorption frequency of the olefinic bond in conjugation with a carbonyl group is lowered by about 30 cm⁻¹ (ca 0.11 μm); the intensity of absorption is increased. In *s-cis* structures, the olefinic absorption may be as intense as that of the carbonyl group. *s-Trans* structures absorb more weakly than *s-cis* structures.

CUMULATED OLEFINS. A cumulated double bond system, as occurs in the allenes (>C=C=CH_2), absorbs near 2000-1900 cm⁻¹ (5.00-5.26 μm). The absorption results from asymmetric C=C=C stretching. The absorption may be considered an extreme case of exocyclic C=C absorption.

Olefinic C–H Stretching Vibrations

In general, any C–H stretching bands above 3000 cm⁻¹ (below 3.33 μm) result from aromatic, heteroaromatic, acetylenic, or olefinic C–H stretching. Also found in the same region are the C–H stretching in small rings, such as cyclopropane, and the C–H in halogenated alkyl groups. The frequency and intensity of olefinic C–H stretching absorption are influenced by the pattern of substitution. With proper resolution, multiple bands are observed for structures in which stretching interaction may occur. For example, the vinyl group produces three closely spaced C–H stretching bands. Two of these result from symmetrical and asymmetrical stretching of the terminal C–H groups, and the third from the stretching of the remaining single C–H.

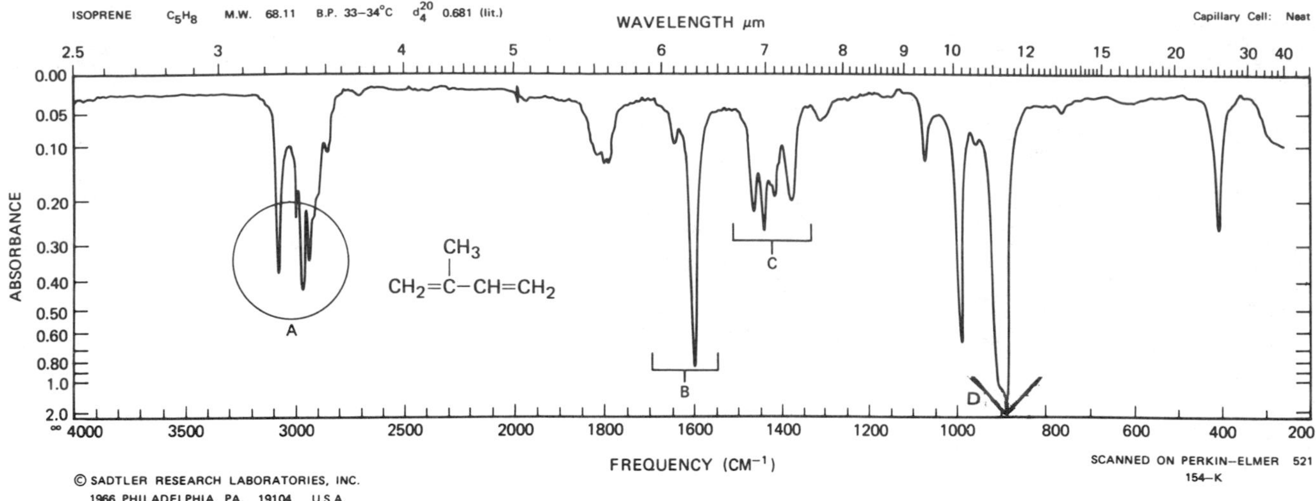

Figure 11. **A.** C–H *stretch:* =C–H *3078 cm^{-1} (3.25 μm).* **B.** *Coupled* C=C–C=C *stretch: symmetric 1640 cm^{-1} (6.10 μm) (weak), asymmetric 1598 cm^{-1} (6.26 μm) (strong).* **C.** C–H *bend (saturated, olefinic in-plane).* **D.** C–H *out-of-plane bend: 980 cm^{-1} (10.20 μm), 880 cm^{-1} (11.36 μm) (see vinyl, Table III, Appendix D).*

Olefinic C–H Bending Vibrations

Olefinic C–H bonds can undergo bending either in the same plane as the C=C bond or perpendicular to it; the bending vibrations can be either in-phase or out-of-phase with respect to each other.

Assignments have been made for a few of the more prominent and reliable in-plane bending vibrations. The vinyl group absorbs near 1416 cm^{-1} (7.06 μm) due to a scissoring vibration of the terminal methylene. The C–H rocking vibration of a *cis*-disubstituted olefin occurs in the same general region.

The most characteristic vibrational modes of olefins are the out-of-plane C–H bending vibrations between 1000 and 650 cm^{-1} (10.0 and 15.4 μm). These bands are usually the strongest in the spectra of olefins. The most reliable bands are those of the vinyl group, the vinylidene group, and the *trans* disubstituted olefins. Olefinic absorption is summarized in Tables III to V of Appendix D.

In allene structures, strong absorption is observed near 850 cm^{-1} (11.76 μm), arising from =CH_2 wagging. The first overtone of this band may also be seen. Some spectra showing olefinic features are shown in Appendix B: trichloroethylene-no. 12 and tetrachloroethylene-no. 13.

ACETYLENIC HYDROCARBONS (ALKYNES)

The two stretching vibrations in acetylenic molecules involve C≡C and C–H stretching. Absorption due to C–H bending is characteristic of acetylene and monosubstituted alkynes. The spectrum of Figure 12 is that of a typical terminal alkyne.

C≡C Stretching Vibrations

The weak C≡C stretching band of acetylenic molecules occurs in the region of 2260-2100 cm^{-1} (4.43-4.76 μm). Because of symmetry, no C≡C band is observed in the infrared for acetylene and symmetrically substituted acetylenes. In the infrared spectra of monosubstituted acetylenes, the band appears at 2140-2100 cm^{-1} (4.67-4.76 μm). Disubstituted acetylenes, in which the substituents are different, absorb near 2260-2190 cm^{-1} (4.43-4.57 μm). When the substituents are similar in mass, or produce similar inductive and resonance effects, the band may be so weak as to be unobserved in the infrared spectrum. For reasons of symmetry, a terminal C≡C produces a stronger band than an internal C≡C (pseudosymmetry). The intensity of the C≡C stretching band is increased by conjugation with a carbonyl group.

C–H Stretching Vibrations

The C–H stretching band of monosubstituted acetylenes occurs in the general region of 3333-3267 cm^{-1} (3.00-3.06 μm). This is a strong band and is narrower than the hydrogen-bonded OH and NH bands occurring in the same region.

C–H Bending Vibrations

The C–H bending vibration of acetylene or monosubstituted acetylenes leads to strong, broad absorption in the 700-610 cm^{-1} (14.29-16.39 μm) region. The first overtone

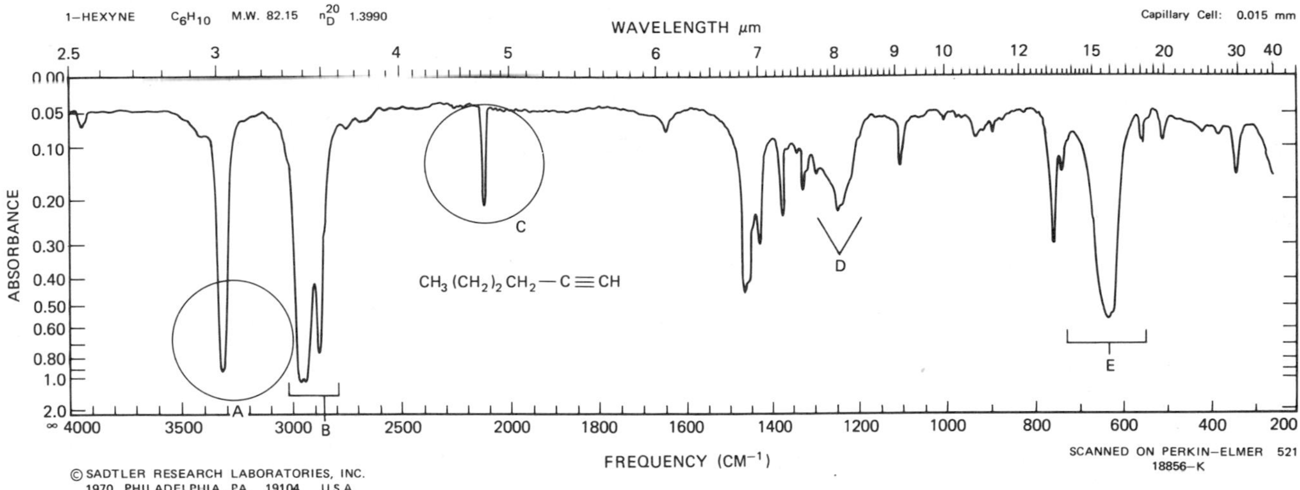

Figure 12. **A.** ≡C—H *stretch, 3268 cm*$^{-1}$ *(3.06 μm).* **B.** *Normal* C—H *stretch (see Figure 8), 2857-2941 cm*$^{-1}$ *(3.4-3.5 μm).* **C.** C≡C *stretch, 2110 cm*$^{-1}$ *(4.74 μm).* **D.** ≡C—H *bend overtone, 1247 cm*$^{-1}$ *(8.02 μm).* **E.** ≡C—H *bend fundamental, 630 cm*$^{-1}$ *(15.87 μm).*

of the C—H bending vibration appears as a weak, broad band in the 1370-1220 cm^{-1} (7.30-8.20 μm) region.

MONONUCLEAR AROMATIC HYDROCARBONS

The most prominent and most informative bands in the spectra of aromatic compounds occur in the low-frequency range between 900-675 cm^{-1} (11.11 and 14.82 μm). These strong absorption bands result from the out-of-plane bending of the ring C—H bonds. In-plane bending bands appear in the 1300-1000 cm^{-1} (7.70-10.00 μm) region. Skeletal vibrations, involving carbon to carbon stretching within the ring, absorb in the 1600-1585 cm^{-1} (6.25-6.31 μm) and in the 1500-1400 cm^{-1} (6.67-7.14 μm) regions. The skeletal bands frequently appear as doublets, depending upon the nature of the ring substituents.

Aromatic C—H stretching bands occur between 3100 and 3000 cm^{-1} (3.23-3.33 μm).

Weak combination and overtone bands appear in the 2000-1650 cm^{-1} (5.00-6.06 μm) region. The pattern of the overtone bands is characteristic of the substitution pattern of the ring. Because they are weak, the overtone and combination bands are most readily observed in spectra obtained from thick samples. The spectrum of Figure 13 is that of a typical aromatic (benzenoid) compound.

Out-of-Plane C—H Bending Vibrations

The in-phase, out-of-plane bending of a ring hydrogen atom is strongly coupled to adjacent hydrogen atoms. The position of absorption of the out-of-plane bending bands is, therefore, characteristic of the number of adjacent hydrogen atoms on the ring. The bands are frequently intense, and may be used for the quantitative determination of the relative concentrations of isomers in mixtures.

Assignments for C—H out-of-plane bending bands in the spectra of substituted benzenes appear in the chart of characteristic group frequencies (Appendix C). These assignments are usually reliable for alky-substituted benzenes, but caution must be observed in the interpretation of spectra when polar groups are attached directly to the ring, for example, in nitrobenzenes, aromatic acids, and esters or amides of aromatic acids.

The absorption band that frequently appears in the spectra of substituted benzenes near 710-675 cm^{-1} (14.08-14.81 μm) is attributed to out-of-plane ring bending. Some spectra showing typical aromatic absorption are shown in Appendix B: benzene-no.4, indene-no.8, diethyl phthalate-no.21, and *m*-xylene-no.6.

POLYNUCLEAR AROMATIC COMPOUNDS

Polynuclear aromatic compounds, like the mononuclear aromatics, show characteristic absorption in 3 regions of the spectrum.

The aromatic C—H stretching and the skeletal vibrations absorb in the same regions as observed for the mononuclear aromatics. The most characteristic absorption of polynuclear aromatics results from C—H out-of-plane bending in the 900-675 cm^{-1} (11.11-14.81 μm) region. These bands can be correlated with the number of adjacent hydrogen atoms on the rings. Most β-substituted naphthalenes, for example, show three absorption bands due to out-of-plane

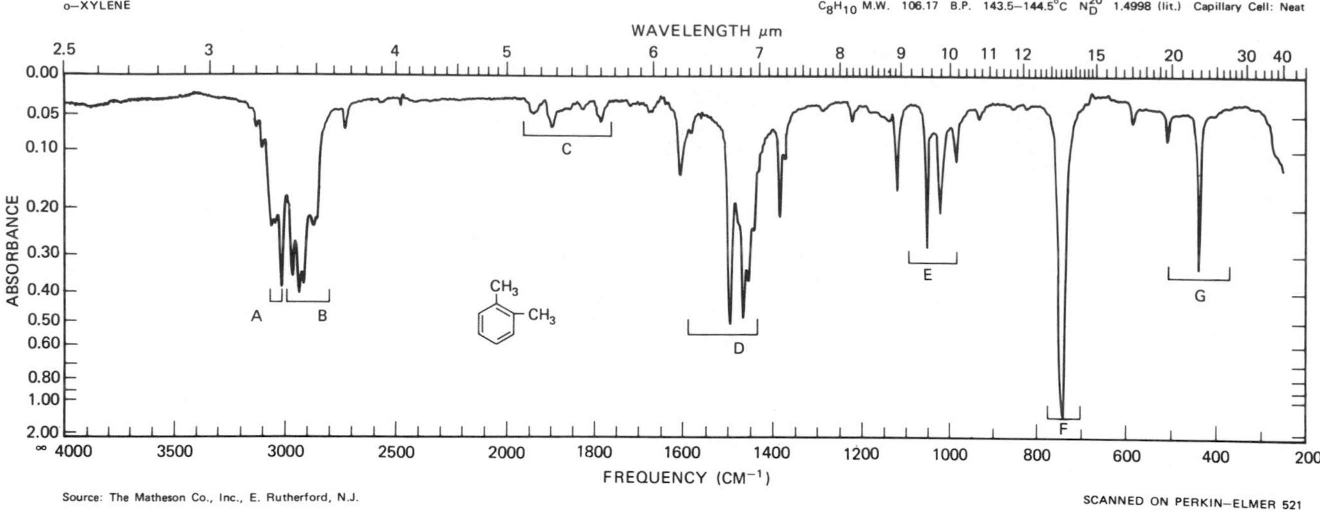

Figure 13. **A.** *Aromatic* C—H *stretch, 3008 cm*$^{-1}$ *(3.32 μm).* **B.** *Methyl* C—H *stretch, 2965, 2938, 2918, 2875 cm*$^{-1}$ *(3.37, 3.40, 3.43, 3.48 μm) (see Figure 8).* **C.** *Overtone or combination bands, 2000-1667 cm*$^{-1}$ *(5.0-6.0 μm) (see Figure 14.)* **D.** C⩵C *ring stretch, 1605, 1495, 1466 cm*$^{-1}$ *(6.23, 6.69, 6.82 μm).* **E.** *In-plane* C—H *bend, 1052, 1022 cm*$^{-1}$ *(9.51, 9.78 μm).* **F.** *Out-of-plane* ⩵C—H *bend, 742 cm*$^{-1}$ *(13.48 μm).* **G.** *Out-of-plane ring* C⩵C *bend, 438 cm*$^{-1}$ *(22.83 μm).*

C—H bending; these correspond to an isolated hydrogen atom and 2 adjacent hydrogen atoms on one ring and 4 adjacent hydrogen atoms on the other ring.

C—H Out-of-Plane Bending Vibrations of a β-Substituted Naphthalene

Isolated hydrogen	862-835 cm^{-1} (11.60-11.98 μm)
2 Adjacent hydrogens	835-805 cm^{-1} (11.98-12.42 μm)
4 Adjacent hydrogens	760-735 cm^{-1} (13.16-13.61 μm)

In the spectra of α-substituted naphthalenes the bands for the isolated hydrogen and the two adjacent hydrogens of β-naphthalenes are replaced by a band for three adjacent hydrogens. This band is near 810-785 cm^{-1} (12.34-12.74 μm).

Additional bands may appear due to ring bending vibrations. The position of absorption bands for more highly substituted naphthalenes and other polynuclear aromatics are summarized by Colthup[3] and by Conley.[1]

ALCOHOLS AND PHENOLS

The characteristic bands observed in the spectra of alcohols and phenols result from O—H stretching and C—O stretching. These vibrations are sensitive to hydrogen bonding. The C—O stretching and O—H bending modes are not independent vibrational modes because they couple with the vibrations of adjacent groups. Some typical spectra of alcohols and a phenol are shown in Figures 14 to 16.

O—H Stretching Vibrations

The unbonded or "free" hydroxyl group of alcohols and phenols absorbs strongly in the 3650-3584 cm^{-1} (2.74-2.79 μm) region. Sharp, "free" hydroxyl bands are observed only in the vapor phase or in very dilute solution in nonpolar solvents. Intermolecular hydrogen bonding increases as the concentration of the solution increases, and additional bands start to appear at lower frequencies, 3550-3200 cm^{-1} (2.82-3.13 μm), at the expense of the "free" hydroxyl band. This effect is illustrated in Figure 17 in which the absorption bands in the O—H stretching region are shown for two different concentrations of cyclohexylcarbinol in carbon tetrachloride. For comparisons of this type, the path length of the cell must be altered with changing concentration, so that the same number of absorbing molecules will be present in the infrared beam at each concentration. The band at 3623 cm^{-1} (2.76 μm) results from the monomer, whereas the broad absorption near 3333 cm^{-1} (3.00 μm) arises from "polymeric" structures.

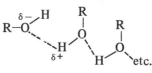

R—O(δ−)—H ··· H(δ+)—O(R) ··· H—O(R) etc.

Strong intramolecular hydrogen bonding occurs in *o*-hydroxyacetophenone. The resulting absorption at 3077 cm^{-1} (3.25 μm) is broad, shallow, and independent of concentration (Figure 18).

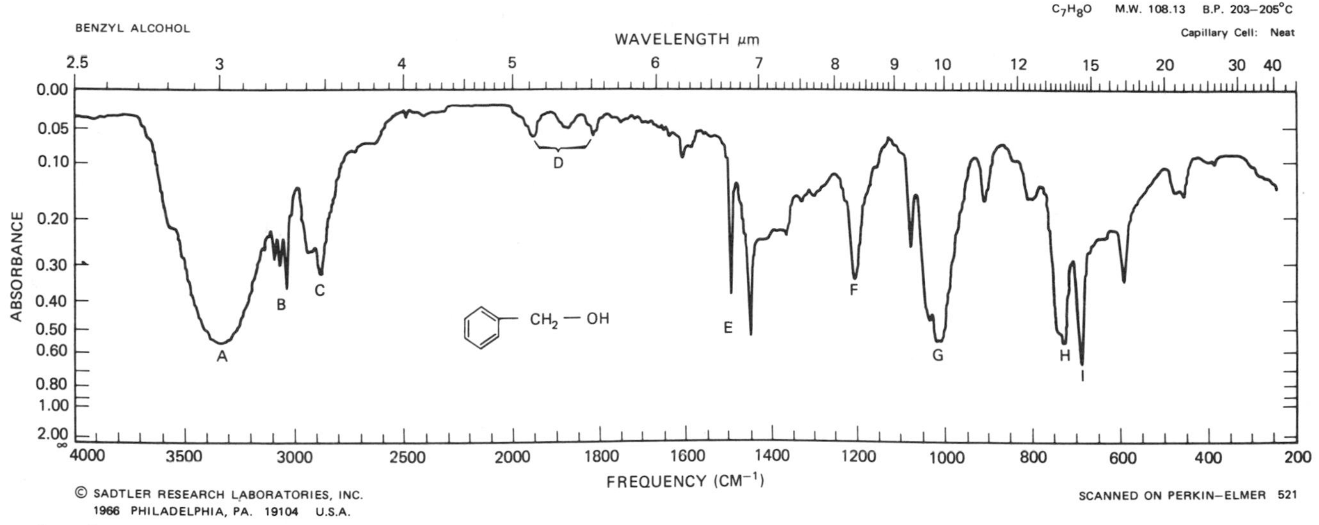

Figure 14. **A.** O—H *stretch: intermolecular hydrogen bonded, 3300* cm^{-1} *(3.03 μm).* **B.** C—H *stretch: aromatic 3100-3000* cm^{-1} *(3.22-3.33 μm).* **C.** C—H *stretch: methylene, 2980-2840* cm^{-1} *(3.36-3.52 μm).* **D.** *Overtone or combination bands, 2000-1667* cm^{-1} *(5.0-6.0 μm).* **E.** C⋯C *ring stretch, 1497, 1453* cm^{-1} *(6.68, 6.88 μm), overlapped by* CH_2 *scissoring, ca. 1471* cm^{-1} *(ca. 6.8 μm).* **F.** O—H *bend, possibly augmented by* C—H *in plane bend, 1208* cm^{-1} *(8.28 μm).* **G.** C—O *stretch, primary alcohol (see Table II) 1017* cm^{-1} *(9.83 μm).* **H.** *Out-of-plane aromatic* C—H *bend, 735* cm^{-1} *(13.60 μm).* **I.** *Ring* C⋯C *bend, 697* cm^{-1} *(14.35 μm).*

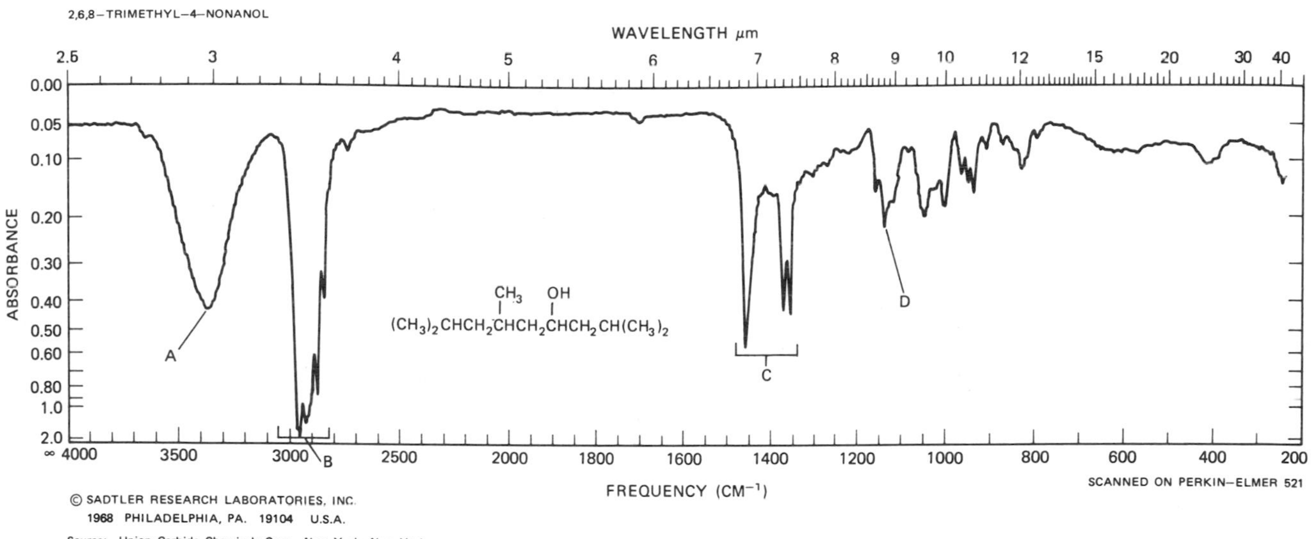

Figure 15. **A.** O—H *stretch, intermolecular hydrogen bonding 3355* cm^{-1} *(2.98 μm).* **B.** C—H *stretch (see Figure 8), 3000-2800* cm^{-1} *(3.33-3.57 μm).* **C.** C—H *bend (see Figure 8). Note the pair of bands for the gem di-methyl groups at 1373 and 1355* cm^{-1} *(7.28 and 7.39 μm).* **D.** C—O *stretch 1138* cm^{-1}*, (8.79 μm). This appears to be weak only because the* C—H *bands are very intense due to the large number of* C—H *bonds in the molecule.*

In contrast, *p*-hydroxyacetophenone

$$HO-C_6H_4-\overset{O}{\overset{\|}{C}}-CH_3$$

shows a sharp "free" hydroxyl peak at 3600 cm^{-1} (2.78 μm) in dilute CCl_4 solution as well as a broad strong intermolecular peak at 3100 cm^{-1} (3.20 μm) in the spectrum of a neat sample.

In structures, such as 2,6-di-*t*-butylphenol, in which steric hindrance prevents hydrogen bonding, no bonded hydroxyl band is observed, not even in spectra of neat samples.

C O Stretching Vibrations

The C—O stretching vibrations in alcohols and phenols produce a strong band in the 1260-1000 cm^{-1} (7.93-

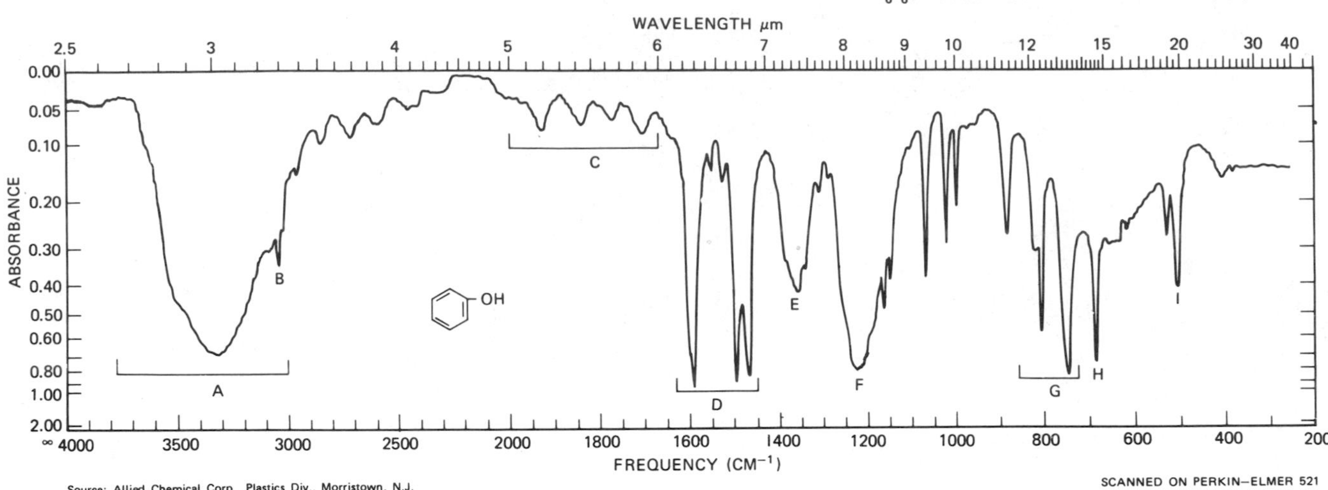

Figure 16. **A.** *Broad, intermolecular hydrogen bonded,* O–H *stretch, 3333 cm*$^{-1}$ *(3.0 μm).* **B.** *Aromatic* C–H *stretch, 3045 cm*$^{-1}$ *(3.28 μm).* **C.** *Overtone or combination bands (see Figure 14), 2000-1667 cm*$^{-1}$ *(5.0-6.0 μm).* **D.** C$\doteq$C *ring stretch, 1580, 1495, 1468 cm*$^{-1}$ *(6.33, 6.69, 6.81 μm).* **E.** *In plane* O–H *bend, 1359 cm*$^{-1}$ *(7.36 μm).* **F.** C–O *stretch, 1223 cm*$^{-1}$ *(8.18 μm).* **G.** *Out-of-plane* C–H *bend, 805, 745 cm*$^{-1}$ *(12.40, 13.43 μm).* **H.** *Out-of-plane ring* C$\doteq$C *bend, 685 cm*$^{-1}$ *(14.60 μm).* **I.** *(Broad) hydrogen bonded, out-of-plane* O–H *bend, ca. 650 cm*$^{-1}$ *(ca. 15.0 μm).*

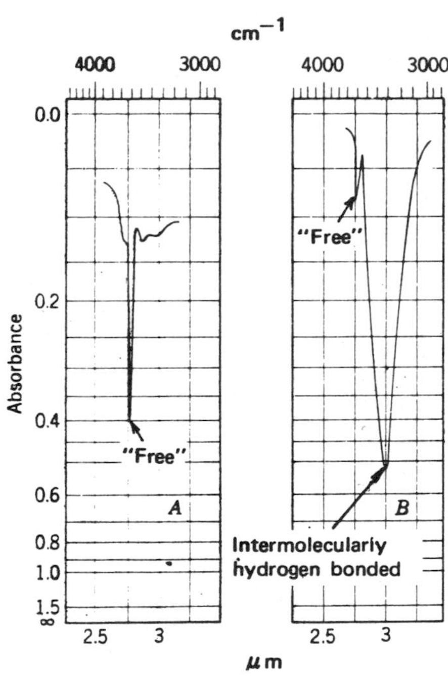

Figure 17. *Infrared spectrum of the* O–H *stretching region of cyclohexylcarbinol in* CCl_4. **A.** *0.03 M (0.406 mm cell);* **B.** *1.00 M (0.014 mm cell).*

10.00 μm) region of the spectrum. The C–O stretching mode is coupled with the adjacent C–C stretching vibration; thus in primary alcohols the vibration might better be described as an asymmetric C–C–O stretching vibration. The vibrational mode is further complicated by branching and α,β-unsaturation. These effects are summarized as follows for a series of secondary alcohols (neat samples):

Methylethylcarbinol	1105 cm^{-1} (9.05 μm)
Methylisopropylcarbinol	1091 cm^{-1} (9.17 μm)
Methylphenylcarbinol	1073 cm^{-1} (9.32 μm)
Methylvinylcarbinol	1058 cm^{-1} (9.45 μm)
Diphenylcarbinol	1014 cm^{-1} (9.86 μm)

The absorption ranges of the various types of alcohols appear in Table II, p. 115. These values are for neat samples of the alcohols.

Mulls, pellets or melts of phenols absorb at 1390-1330 cm^{-1} (7.20-7.52 μm) and 1260-1180 cm^{-1} (7.93-8.48 μm). These bands apparently result from interaction between O–H bending and C–O stretching. The long wavelength band is the stronger and both bands appear at longer wavelengths in spectra observed in solution. The spectrum of Figure 16 was determined on a melt, to show a high degree of association.

O–H Bending Vibrations

The O–H in-plane bending vibration occurs in the general region of 1420-1330 cm^{-1} (7.04-7.52 μm). In primary and secondary alcohols, the O–H in-plane bending couples with the C–H wagging vibrations to produce 2 bands; the

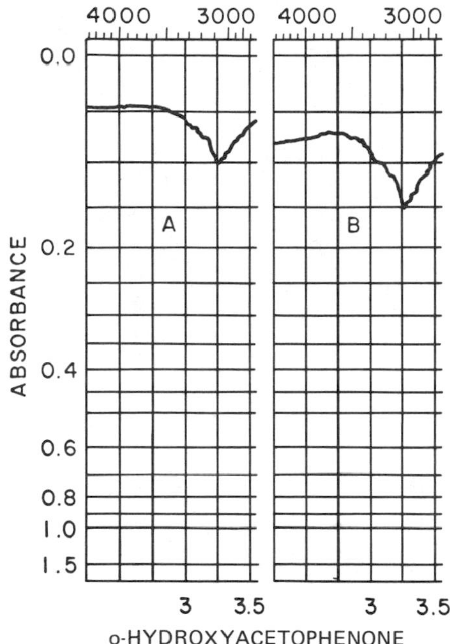

Figure 18. *A portion of the infrared spectra of o-hydroxyacetophenone.* **A.** *0.03 M, cell thickness: 0.41 mm.* **B.** *1.0 M, cell thickness: 0.015 mm.*

first near 1420 cm^{-1} (7.04 μm), the second near 1330 cm^{-1} (7.52 μm). These bands are of little diagnostic value. Tertiary alcohols, in which no coupling can occur, show a single band in this region, the position depending on the degree of hydrogen bonding.

The spectra of alcohols and phenols determined in the liquid state, show a broad absorption band in the 769-650 cm^{-1} (13.00-15.40 μm) region because of out-of-plane bending of the bonded O–H group. Some spectra showing typical alcoholic absorptions are shown in Appendix B: ethyl alcohol-no.16 and methanol-no.15.

ETHERS, EPOXIDES, AND PEROXIDES

C–O Stretching Vibrations

The characteristic response of ethers in the infrared is associated with the stretching vibration of the C–O–C system. Since the vibrational characteristics of this system would not be expected to differ greatly from the C–C–C system, it is not surprising to find the response to C–O–C stretching in the same general region. However, since vibrations involving oxygen atoms result in greater dipole moment changes than those involving carbon atoms, more intense infrared bands are observed for ethers. The C–O–C stretching bands of ethers, as is the case with the C–O stretching bands of alcohols, involve coupling with other vibrations within the molecule. The spectrum of Figure 19 is that of a typical aryl alkyl ether. In addition, the spectra of ethyl ether-no.22 and *p*-dioxane (a cyclic diether, no.23) are shown in Appendix B.

In the spectra of aliphatic ethers, the most characteristic absorption is a strong band in the 1150-1085 cm^{-1} (8.70-9.23 μm) region because of asymmetrical C–O–C stretching; this band usually occurs near 1125 cm^{-1} (8.89 μm). The symmetrical stretching band is usually weak and is more readily observed in the Raman spectrum.

The C–O–C group in a 6-membered ring absorbs at the same frequency as in an acyclic ether. As the ring becomes smaller, the asymmetrical C–O–C stretching vibration moves progressively to lower wavenumbers (longer wavelengths), whereas the symmetrical C–O–C stretching vibration (ring breathing frequency) moves to higher wavenumbers.

Table II. Alcoholic C–O Absorption

Alcohol Type	Absorption Range
(1) Saturated tertiary (2) Secondary, highly symmetrical	1205-1124 cm^{-1} (8.30-8.90 μm)
(1) Saturated secondary (2) α-Unsaturated or cyclic tert.	1124-1087 cm^{-1} (8.90-9.20 μm)
(1) Secondary, α-unsaturated (2) Secondary, alicyclic 5- or 6-membered ring (3) Saturated primary	1085-1050 cm^{-1} (9.22-9.52 μm)
(1) Tertiary, highly α-unsaturated (2) Secondary, di-α-unsaturated (3) Secondary, α-unsaturated and α-branched (4) Secondary, alicyclic 7- or 8-membered ring (5) Primary, α-unsaturated and/or α-branched	$<$1050 cm^{-1} ($>$9.52 μm)

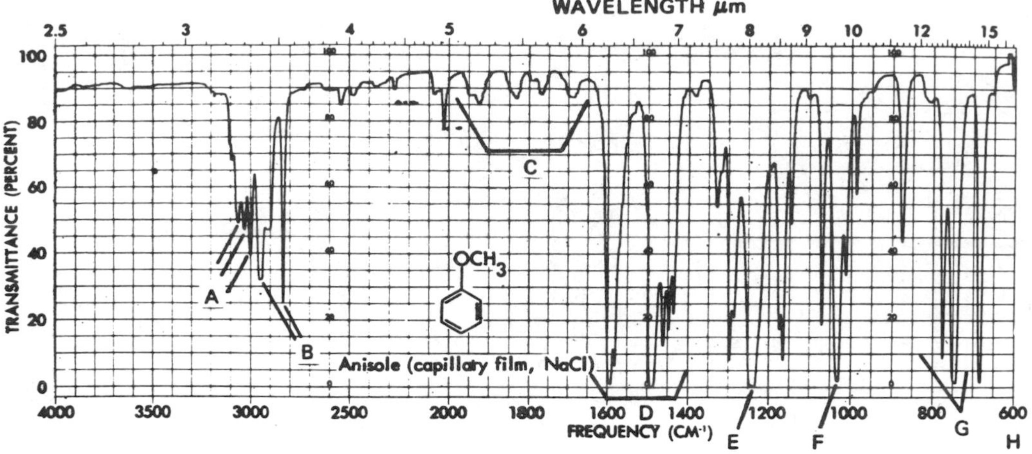

Anisole. (Source: Daniel J. Pasto and Carl R. Johnson, *Organic Structure Determination,* copyright 1969, p. 372. Reprinted by permission of Prentice-Hall, Inc., Englewood Cliffs, N.J.)

Figure 19. **A.** *Aromatic* C—H *stretch, 3060, 3030, 3000 cm*$^{-1}$ *(3.27, 3.30, 3.33 µm).* **B.** *Methyl* C—H *stretch, 2950, 2835 cm*$^{-1}$ (3.39, 3.53 *µm).* **C.** *Overtone or combination region, 2000-1650 cm*$^{-1}$ (5.0-6.0 *µm), see Figure 13).* **D.** C$\overset{...}{-}$C *ring stretch, 1590, 1480 cm*$^{-1}$ *(6.29, 6.76 µm).* **E.** *Asymmetric* C—O—C *stretch, 1245 cm*$^{-1}$ *(8.03 µm).* **F.** *Symmetric* C—O—C *stretch, 1030 cm*$^{-1}$ *(9.71 µm).* **G.** *Out-of-plane* C—H *bend, 800-740 cm*$^{-1}$ *(12.6-13.5 µm).* **H.** *Out-of-plane ring* C$\overset{...}{-}$C *bend, 680 cm*$^{-1}$ *(14.7 µm).*

Branching on the carbon atoms adjacent to the oxygen usually leads to splitting of the C—O—C band. Isopropyl ether shows a triplet structure in the 1170-1114 cm^{-1} (8.55-8.98 µm) region, the principal band occurring at 1114 cm^{-1} (8.98 µm).

Spectra of aryl alkyl ethers display an asymmetrical C—O—C stretching band at 1275-1200 cm^{-1} (7.84-8.33 µm) with symmetrical stretching near 1075-1020 cm^{-1} (9.30-9.80 µm). Strong absorption due to asymmetrical C—O—C stretching in vinyl ethers occurs in the 1225-1200 cm^{-1} (8.16-8.33 µm) region with a strong symmetrical band at 1075-1020 cm^{-1} (9.30-9.80 µm). Resonance, which results in strengthening of the C—O bond, is responsible for the shift in the asymmetric absorption band of aryl alkyl and vinyl ethers.

The C=C stretching band of vinyl ethers occurs in the 1660-1610 cm^{-1} (6.02-6.21 µm) region. This olefinic band is characterized by its higher intensity compared with the C=C stretching band in olefinic hydrocarbons. This band frequently appears as a doublet resulting from absorption of rotational isomers.

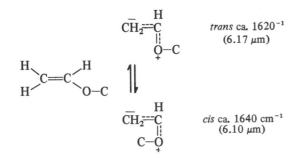

$H_2C=CH—O—C \rightleftharpoons$ $\overset{-}{C}H_2$---CH=$\overset{+}{O}$—C *trans* ca. 1620^{-1} (6.17 µm)

$\overset{-}{C}H_2$---CH=$\overset{+}{O}$—C *cis* ca. 1640 cm^{-1} (6.10 µm)

Coplanarity in the *trans* isomer allows maximum resonance, thus more effectively reducing the double-bond character of the olefinic linkage. Steric hindrance reduces resonance in the *cis* isomer.

The 2 bands arising from =C—H wagging in terminal olefins occur near 1000 cm^{-1} (10.00 µm) and near 909 cm^{-1} (11.00 µm). In the spectra of vinyl ethers, these bands are shifted to longer wavelengths because of resonance.

R—O—CH=CH$_2$

terminal CH$_2$ wag, 813 cm^{-1} (12.30 µm)
trans CH wag, 960 cm^{-1} (10.42 µm)

Alkyl and aryl peroxides display C—C—O absorption in the 1198-1176 cm^{-1} (8.35-8.50 µm) region. Acyl and aroyl peroxides display two carbonyl absorption bands in the 1818-1754 cm^{-1} (5.50-5.70 µm) region. Two bands are observed because of mechanical interaction between the stretching modes of the two carbonyl groups.

The symmetrical stretching, or ring breathing frequency, of the epoxy ring, all ring bonds stretching and contracting in phase, occurs near 1250 cm^{-1} (8.00 µm). Another band appears in the 950-810 cm^{-1} (10.53-12.35 µm) region attributed to asymmetrical ring stretching in which the C—C bond is stretching during contraction of the C—O bond. A third band, referred to as the "12 µ band," appears in the 840-750 cm^{-1} (11.90-13.33 µm) region. The C—H stretching vibrations of epoxy rings occur in the 3050-2990 cm^{-1} (3.28-3.34 µm) region of the spectrum.

KETONES

C=O Stretching Vibrations

Ketones, aldehydes, carboxylic acids, carboxylic esters, lactones, acid halides, anhydrides, amides, and lactams show a strong C=O stretching absorption band in the region of 1870-1540 cm^{-1} (5.35-6.50 μm). Its relatively constant position, high intensity, and relative freedom from interfering bands make this one of the easiest bands to recognize in infrared spectra.

Within its given range, the position of the C=O stretching band is determined by the following factors: (1) the physical state, (2) electronic and mass effects of neighboring substituents, (3) conjugation, (4) hydrogen bonding (intermolecular and intramolecular), and (5) ring strain. Consideration of these factors leads to a considerable amount of information about the environment of the C=O group.

In a discussion of these effects, it is customary to refer to the absorption frequency of a neat sample of a saturated aliphatic ketone, 1715 cm^{-1} (5.83 μm), as "normal." Changes in the environment of the carbonyl can either lower or raise the absorption frequency from this "normal" value. A typical ketone spectrum is displayed in Figure 20.

The absorption frequency observed for a neat sample is increased when absorption is observed in non-polar solvents. Polar solvents reduce the frequency of absorption. The overall range of solvent effects doesn't exceed 25 cm^{-1}.

Replacement of an alkyl group of a saturated aliphatic ketone by a heteroatom (X) shifts the carbonyl absorption. The direction of the shift depends on whether the inductive effect (a) or resonance effect (b) predominates:

$X(R)C{=}O$ (a) $\qquad X^{+}{=}C(R){-}\bar{O}^{-}$ (b)

The inductive effect reduces the length of the C=O bond and thus increases its force constant and the frequency of absorption. The resonance effect increases the C=O bond length and reduces the frequency of absorption.

The absorptions of several carbonyl compound classes are summarized in Table III.

Conjugation with a C=C bond results in delocalization of the π electrons of both unsaturated groups. Delocalization of the π electrons of the C=O group reduces the double bond character of the C to O bond, causing absorption at lower wavenumbers (longer wavelengths). Conjugation with an olefinic or phenyl group causes absorption in the 1685-1666 cm^{-1} (5.93-6.00 μm) region. Additional conjugation may cause a slight further reduction in frequency. This effect of conjugation is illustrated in Figure 21.

Steric effects which reduce the coplanarity of the conjugated system reduce the effect of conjugation. In the absence of steric hindrance, a conjugated system will tend toward a planar conformation. Thus, α,β-unsaturated ketones may exist in *s-cis* and *s-trans* conformations. When both forms are present, absorption for each of the forms is observed. The absorption of benzalacetone in carbon

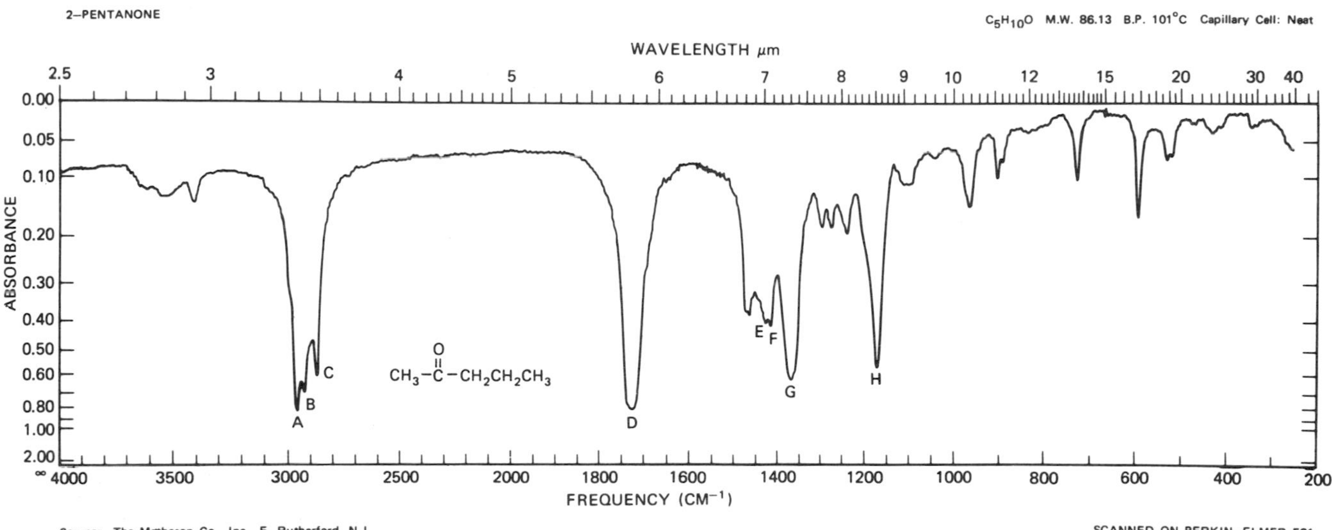

Figure 20. **A.** ν_{as}, *methyl, 2955* cm^{-1} *(3.39 μm).* **B.** ν_{as}, *methylene, 2930* cm^{-1} *(3.41 μm).* **C.** ν_s, *methyl, 2866* cm^{-1} *(3.49 μm).* **D.** *Normal** C=O *stretch, 1725* cm^{-1} *(5.80 μm).* **E.** δ_{as}, CH_3, *ca. 1430* cm^{-1} *(ca. 7.0 μm), see Table I, Appendix D.* **F.** δ_s, CH_2, *ca. 1430* cm^{-1} *(ca. 7.0 μm), see Table II, Appendix D.* **G.** δ_s, CH_3 *of* CH_3CO *unit, 1370* cm^{-1} *(7.30 μm), see Table I, Appendix D.* **H.** C–CO–C *stretch and bend, 1172* cm^{-1} *(8.53 μm).*

*See above for a definition of normal.

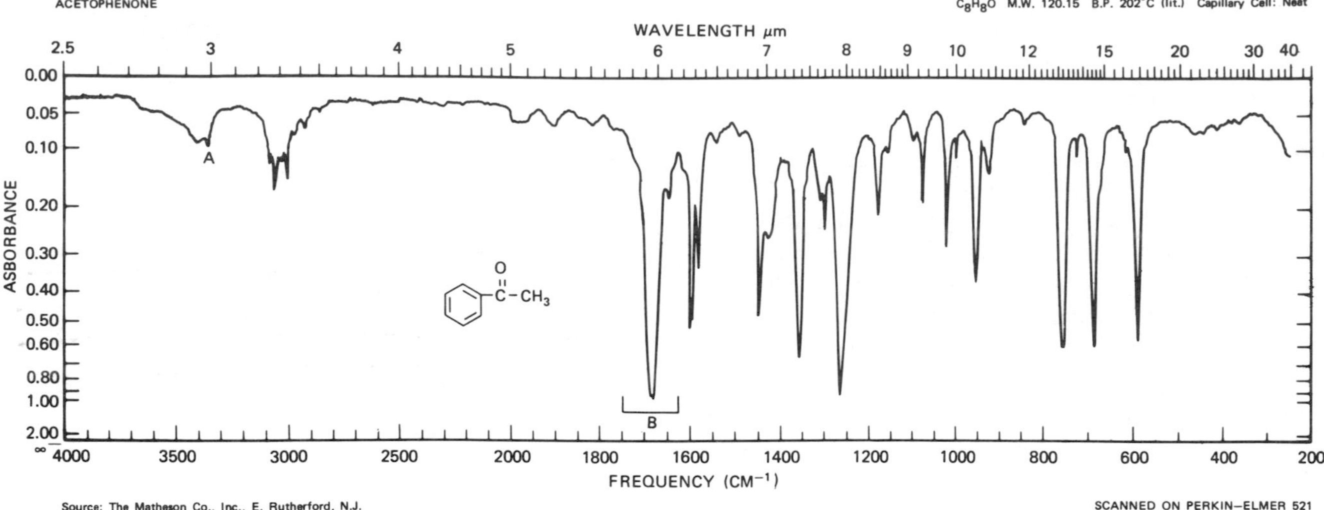

Figure 21. **A.** *Overtone of* C=O *stretch, 3350 cm*$^{-1}$ *(2.98 μm); frequency twice that of* C=O *stretch.* **B.** C=O stretch, *1683 cm*$^{-1}$ *(5.94 μm), lower frequency (longer wavelength) than observed in Figure 20 because of the conjugation with the phenyl group.*

disulfide serves as an example; both the *s-cis* and *s-trans* forms are present at room temperature.

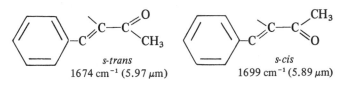

s-trans
1674 cm^{-1} (5.97 μm)

s-cis
1699 cm^{-1} (5.89 μm)

The absorption of the olefinic bond in conjugation with the carbonyl group occurs at a lower frequency than that of an isolated C=C bond; the intensity of the conjugated double bond absorption, when in an *s-cis* system, is greater than that of an isolated double bond.

Table III. The Carbonyl Absorption of Various RCX (with C=O) Compounds

X	νC=O
X Effect Predominantly Inductive	
Cl	1815-1785 cm^{-1} (5.51-5.60 μm)
F	ca. 1869 cm^{-1} (5.35 μm)
Br	1812 cm^{-1} (5.52 μm)
OH (monomer)	1760 cm^{-1} (5.68 μm)
OR	1750-1735 cm^{-1} (5.71-5.76 μm)
X Effect Predominantly Resonance	
X	νC=O
NH_2	1695-1650 cm^{-1} (5.90-6.06 μm)
SR	1720-1690 cm^{-1} (5.82-5.92 μm)

Intermolecular hydrogen bonding between a ketone and a hydroxylic solvent such as methyl alcohol causes a slight decrease in the absorption frequency of the carbonyl group. For example, a neat sample of methyl ethyl ketone absorbs at 1715 cm^{-1} (5.83 μm), whereas a 10% solution of the ketone in methanol absorbs at 1706 cm^{-1} (5.86 μm).

β-Diketones usually exist as mixtures of tautomeric keto and enol forms. The enolic form does not show the normal absorption of conjugated ketones. Instead, a broad band appears in the 1640-1580 cm^{-1} (6.10-6.33 μm) region, many times more intense than normal carbonyl absorption. The intense and displaced absorption results from intramolecular hydrogen bonding, the bonded structure being stabilized by resonance.

$$\mathrm{R{-}\overset{OH\cdots O}{C{=}CR'{-}C}{-}R''} \longleftrightarrow \mathrm{R{-}\overset{\overset{+}{O}H\cdots \bar{O}}{C{-}CR'{=}C}{-}R''}$$

Acetylacetone as a liquid at 40°C exists to the extent of 64% in the enolic form that absorbs at 1613 cm^{-1} (6.20 μm). The keto form and a small amount of unbonded enolic form may be responsible for 2 bands centering near 1725 cm^{-1} (5.80 μm). Interaction between the 2 carbonyl groups in the keto form has also been suggested as a cause for this doublet. The enolic O–H stretching absorption is seen as a broad shallow band at 3000-2700 cm^{-1} (3.33-3.70 μm).

α-Diketones, in which carbonyl groups exist in formal conjugation, show a single absorption band near the frequency observed for the corresponding monoketone. Biacetyl absorbs at 1718 cm^{-1} (5.82 μm), benzil at

1681 cm^{-1} (5.92 μm). Conjugation is ineffective for α-diketones and the C=O groups of these diketones do not couple as do, for example, the corresponding groups in acid anhydrides (see below).

Quinones, which have both carbonyl groups in the same ring, absorb in the 1690-1655 cm^{-1} (5.92-6.04 μm) region. With extended conjugation, in which the carbonyl groups appear in different rings, the absorption shifts to the 1655-1635 cm^{-1} (6.04-6.12 μm) region.

Acyclic α-chloro ketones absorb at 2 frequencies due to rotational isomerism. When the chlorine atom is near the oxygen, its negative field repels the nonbonding electrons of the oxygen atom, thus increasing the force constant of the C=O bond. This conformation absorbs at a higher frequency (1745 cm^{-1}, 5.73 μm) than that in which the carbonyl oxygen and chlorine atom are widely separated (1725 cm^{-1}, 5.80 μm). In rigid molecules such as the monoketo-steroids, α-halogenation results in equatorial or axial substitution. In the equatorial orientation, the halogen atom is near the carbonyl group and the "field effect" causes an increase in the C=O stretching frequency. In the isomer in which the halogen atom is axial to the ring, and distant from the C=O, no shift is observed.

In cyclic ketones, the bond angle of the C(C)C=O group influences the absorption frequency of the carbonyl group. The C=O stretching undoubtedly is affected by adjacent C—C stretching. In noncyclic ketones and in ketones with a 6-membered ring, the angle is near 120°. In strained rings in which the angle is less than 120°, interaction with C—C bond stretching increases the energy required to produce C=O stretching and thus increases the stretching frequency. Cyclohexanone absorbs at 1715 cm^{-1} (5.83 μm), cyclopentanone absorbs at 1751 cm^{-1} (5.71 μm), and cyclobutanone absorbs at 1775 cm^{-1} (5.63 μm).

C—C(=O)—C *Stretching and Bending Vibrations*

Ketones show moderate absorption in the 1300-1100 cm^{-1} (7.70-9.09 μm) region as a result of C—C—C stretching and bending in the C—C(=O)—C group. The absorption may consist of multiple bands. Aliphatic ketones absorb in the 1230-1100 cm^{-1} (8.13-9.09 μm) region; aromatic ketones absorb at the higher frequency end of the general absorption region.

The spectra of 2-butanone (methyl ethyl ketone-no. 18), acetone-no.17, and cyclohexanone-no.19 in Appendix B illustrate ketonic absorptions.

ALDEHYDES

The spectrum of 2-phenylpropionaldehyde, illustrating typical aldehydic absorption characteristics, is shown in Figure 22.

C=O Stretching Vibrations

The carbonyl groups of aldehydes absorb at slightly higher frequencies than that of the corresponding methyl ketones. Aliphatic aldehydes absorb near 1740-1720 cm^{-1} (5.75-5.82 μm). Aldehydic carbonyl absorption responds to structural changes in the same manner as ketones. Electro-

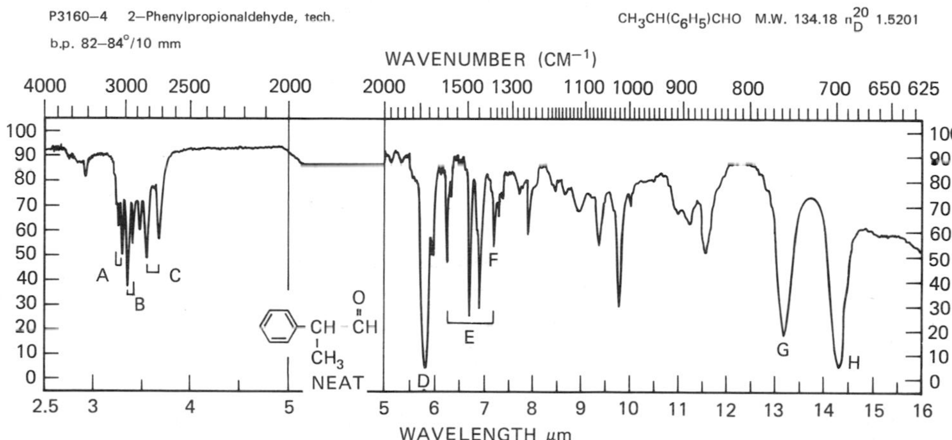

Figure 22. *2-Phenylpropionaldehyde.** **A.** *Aromatic, 3077, 3040* cm^{-1} *(3.25, 3.29 μm)(see Figure 13).* **B.** *Aliphatic, 2985, 2941, 2874* cm^{-1} *(3.35, 3.40, 3.48 μm)(see Figures 8 and 13).* **C.** *Aldehydic* C—H *stretch, 2825, 2717* cm^{-1} *(3.54, 3.68 μm). Doublet due to Fermi resonance with overtone of band at F.* **D.** *Normal aldehydic* C=O *stretch, 1730* cm^{-1} *(5.78 μm). Conjugated* C=O *stretch would be ca. 1700* cm^{-1}*, (5.89 μm), for example, in* C_6H_5CHO. **E.** *Ring* C⋯C *stretch, 1600, 1497, 1453* cm^{-1} *(6.25, 6.68, 6.88 μm).* **F.** *Aldehydic* C—H *bend, 1389* cm^{-1} *(7.20 μm).* **G.** *Out-of-plane* C—H *bend, 749* cm^{-1} *(13.35 μm).* **H.** *Out-of-plane* C—C *bend, 699* cm^{-1} *(14.30 μm).*

***A** through **C** are C—H stretch absorptions.
Courtesy of Aldrich Chemical Company.

negative substitution on the α-carbon increases the frequency of carbonyl absorption. Acetaldehyde absorbs at 1730 cm^{-1} (5.78 μm), trichloroacetaldehyde absorbs at 1768 cm^{-1} (5.65 μm). Conjugate unsaturation, as in α,β-unsaturated aldehydes and benzaldehydes, reduces the frequency of carbonyl absorption. α,β-Unsaturated aldehydes and benzaldehydes absorb in the region of 1710-1685 cm^{-1} (5.85-5.94 μm). Internal hydrogen bonding, such as occurs in salicylaldehyde, shifts the absorption (1666 cm^{-1}, 6.00 μm, for salicylaldehyde) to lower wavenumbers (longer wavelengths). Glyoxal, like the α-diketones, shows only one carbonyl absorption peak with no shift from the normal absorption position of mono-aldehydic absorption.

C–H Stretching Vibrations

The majority of aldehydes show aldehydic C–H stretching absorption in the 2830-2695 cm^{-1} (3.53-3.71 μm) region. Two moderately intense bands are frequently observed in this region. The appearance of two bands is attributed to Fermi resonance between the fundamental aldehydic C–H stretch and the first overtone of the aldehydic C–H bending vibration that usually appears near 1390 cm^{-1} (7.20 μm). Only one C–H stretching band is observed for aldehydes whose C–H bending band has been shifted appreciably from 1390 cm^{-1} (7.20 μm).

Some aromatic aldehydes with strongly electronegative groups in the ortho position may absorb as high as 2900 cm^{-1} (3.45 μm).

Medium intense absorption near 2720 cm^{-1} (3.68 μm), accompanied by a carbonyl absorption band is good evidence for the presence of an aldehyde group.

The absorption frequencies of methyl and methylene groups attached to a carbonyl are summarized in Tables I and II of Appendix D.

CARBOXYLIC ACIDS

O–H Stretching Vibrations

In the liquid or solid state, and in carbon tetrachloride solution at concentrations much over 0.01 *M*, carboxylic acids exist as dimers due to strong hydrogen bonding.

$$\text{R–C}\begin{matrix}\text{O–H}\cdots\text{O}\\ \text{O}\cdots\text{H–O}\end{matrix}\text{C–R} \longleftrightarrow \text{R–C}\begin{matrix}\bar{\text{O}}\cdots\text{H–}\overset{+}{\text{O}}\\ \underset{+}{\text{O}}\text{–H}\cdots\underline{\text{O}}\end{matrix}\text{C–R}$$

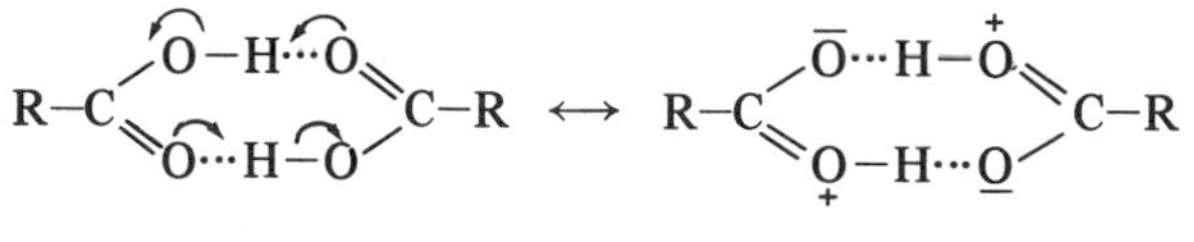

The exceptional strength of the hydrogen bonding is explained on the basis of the large contribution of the ionic resonance structure. Because of the strong bonding, a free hydroxyl stretching vibration (near 3520 cm^{-1}, 2.84 μm) is observed only in very dilute solution in nonpolar solvents or in the vapor phase. In each case, however, there is a mixture of monomer and dimer.

Carboxylic acid dimers display very broad, intense O–H stretching absorption in the region of 3300-2500 cm^{-1} 3.03-4.00 μm). The band usually centers near 3000 cm^{-1} (3.33 μm). The weaker C–H stretching bands are generally seen superimposed upon the broad O–H band. Fine structure observed on the long-wavelength side of the broad O–H band represents overtones and combination tones of fundamental bands occurring at longer wavelengths. The spectrum of a typical aliphatic carboxylic acid is displayed in Figure 23.

Other structures with strong hydrogen bonding, such as β-diketones, also absorb in the 3300-2500 cm^{-1} (3.03-4.00 μm) region, but the absorption is usually less intense. Also, the C=O stretching vibrations of structures such as β-diketones are shifted to lower frequencies than those observed for carboxylic acids.

Carboxylic acids can bond intermolecularly with ethers, such as dioxane and tetrahydrofuran, or with other solvents that can act as proton acceptors. Spectra determined in such solvents show bonded O–H absorption near 3100 cm^{-1} (3.23 μm).

C=O Stretching Vibrations

The C=O stretching bands of acids are considerably more intense than ketonic C=O stretching bands. The monomers of saturated aliphatic acids absorb near 1760 cm^{-1} (5.68 μm).

The carboxylic dimer has a center of symmetry; only the asymmetrical C=O stretching mode absorbs in the infrared. Hydrogen bonding and resonance weaken the C=O bond, resulting in absorption at a lower frequency than the monomer. The C=O group in dimerized saturated aliphatic acids absorbs in the region of 1720-1706 cm^{-1} (5.81-5.86 μm).

Internal hydrogen bonding reduces the frequency of the carbonyl stretching absorption to a greater degree than does intermolecular hydrogen bonding. For example, salicylic acid absorbs at 1665 cm^{-1} (6.01 μm), whereas *p*-hydroxybenzoic acid absorbs at 1680 cm^{-1} (5.95 μm).

Unsaturation in conjugation with the carboxylic carbonyl group decreases the frequency (increases the wavelength) of absorption of both the monomer and dimer forms only slightly. In general, α,β-unsaturated and aryl conjugated acids show absorption for the dimer in the 1710-1680 cm^{-1} (5.85-5.95 μm) region. Extension of conjugation beyond the α,β-position results in very little additional shifting of the C=O absorption.

Substitution in the α-position with electronegative groups, such as the halogens, brings about a slight increase in the C=O absorption frequency (10-20 cm^{-1}, 0.03-0.07 μm). The spectra of acids with halogens in the α-position, determined in the liquid state or in solution, show dual carbonyl bands due to rotational isomerism (field effect). The higher frequency (shorter wavelength) band corresponds to the conformation in which the halogen is in proximity to the carbonyl group.

C–O Stretching and O–H Bending Vibrations

Two bands arising from C–O stretching and O–H bending appear in the spectra of carboxylic acids near 1320-1210 cm^{-1} (7.58-8.26 μm) and near 1440-1395 cm^{-1} (6.94-7.17 μm), respectively. Both of these bands involve some interaction between C–O stretching and in-plane C–O–H bending. The more intense band, near 1315-1280 cm^{-1} (7.60-7.81 μm) for dimers, is generally referred to as the C–O stretching band and usually appears as a doublet in the spectra of long-chain fatty acids. The C–O–H bending band near 1440-1395 cm^{-1} (6.94-7.17 μm) is of moderate intensity and occurs in the same region as the CH_2 scissoring vibration of the CH_2 group adjacent to the carbonyl.

One of the characteristic bands in the spectra of dimeric carboxylic acids results from the out-of-plane bending of the bonded O–H. The band appears near 920 cm^{-1} (10.87 μm) and is characteristically broad with medium intensity.

CARBOXYLATE ANION

The carboxylate anion has two strongly coupled carbon to oxygen bonds with bond strengths intermediate between C=O and C–O.

$$-C\overset{\cdots O}{\underset{\cdots O}{}}\quad -$$

The carboxylate ion gives rise to 2 bands: a strong asymmetrical stretching band near 1650-1550 cm^{-1} (6.06-6.45 μm), and a weaker, symmetrical stretching band near 1400 cm^{-1} (7.15 μm).

The conversion of a carboxylic acid to a salt can serve as confirmation of the acid structure. This is conveniently done by the addition of a tertiary, aliphatic amine, such as triethylamine, to a solution of the carboxylic acid in chloroform (no reaction occurs in carbon tetrachloride). The carboxylate ion, thus formed, shows the two characteristic carbonyl absorption bands in addition to an "ammonium" band in the 2700-2200 cm^{-1} (3.70-4.55 μm) region. The O–H stretching band, of course, disappears. The spectrum of ammonium benzoate, Figure 24, demonstrates most of these features.

ESTERS AND LACTONES

Esters and lactones have two characteristically strong absorption bands arising from C=O and C–O stretching.

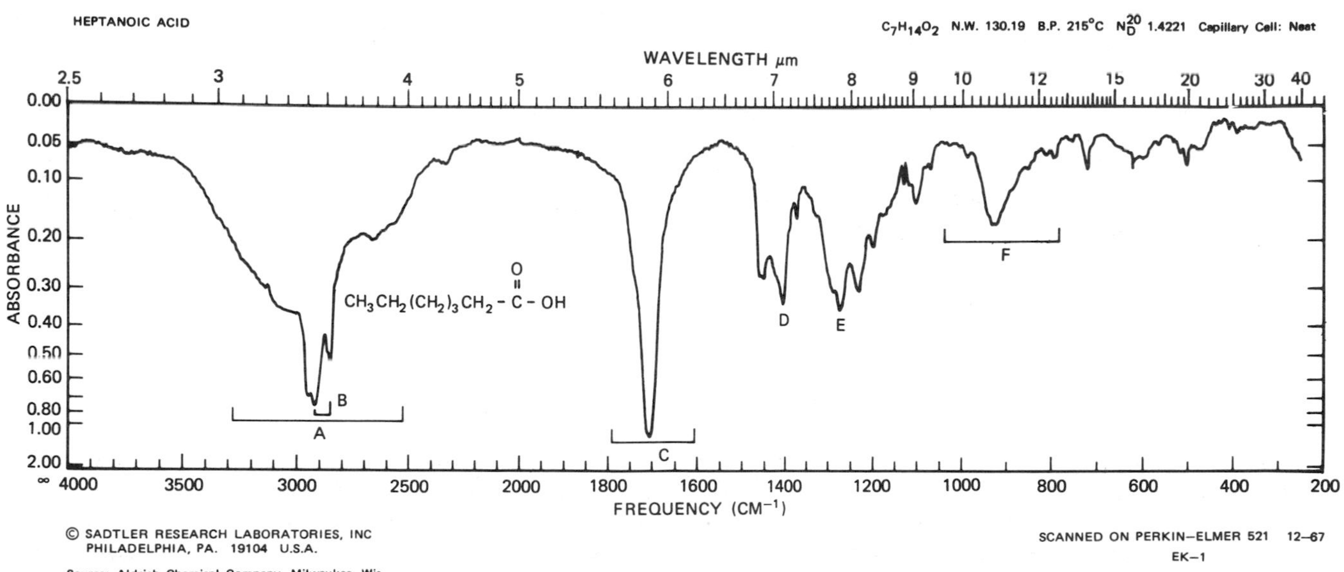

Figure 23. **A.** *Broad* O–H *stretch, 3300-2500* cm^{-1} *(3.03-4.00 μm).* **B.** C–H *stretch, (see Figure 8) 2950, 2920, 2850* cm^{-1} *(3.39, 3.43, 3.51 μm). (Superimposed upon* O–H *stretch.* **C.** *Normal, dimeric carboxylic* C=O *stretch, 1715* cm^{-1} *(5.83 μm).* **D.** C–O–H *in-plane bend,* 1408* cm^{-1} *(7.10 μm).* **E.** C–O *stretch,* dimer, 1280* cm^{-1} *(7.81 μm).* **F.** O–H *out-of-plane bend, 930* cm^{-1} *(10.75 μm).*

*Bands at **D** and **E** involve C–O–H interaction.

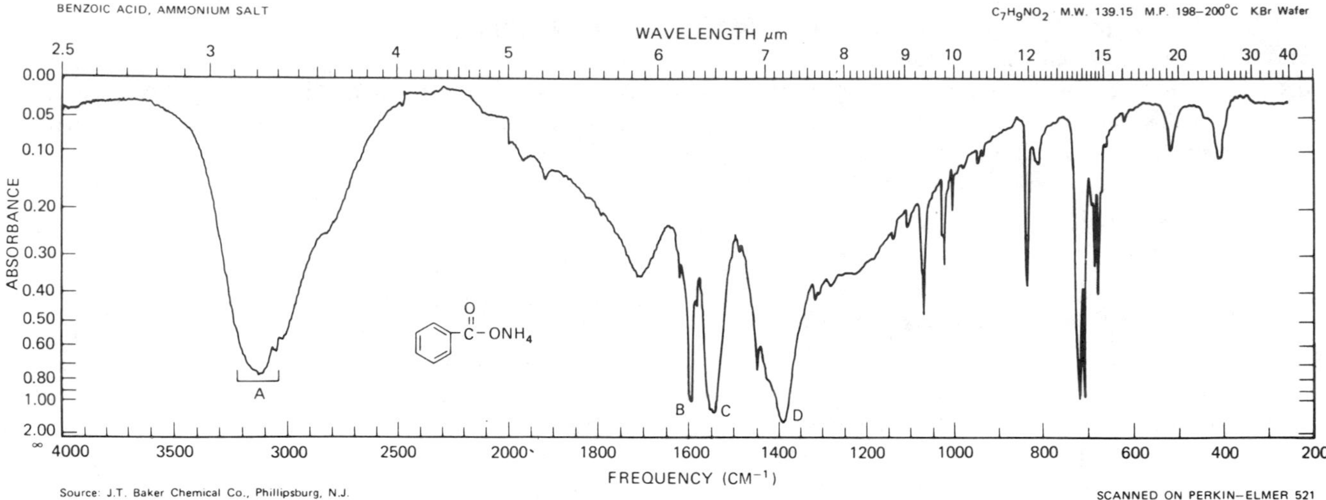

Figure 24. **A.** N–H *and* C–H *stretch, 3600-2500 cm*$^{-1}$ *(2.78-4.00 μm).* **B.** *Ring C∸C stretch 1600 cm*$^{-1}$ *(6.25 μm).* **C.** *Asymmetric carboxylate anion C (∸0)*$_2^-$ *stretch 1550 cm*$^{-1}$ (6.45 μm). **D.** *Symmetric carboxylate C (∸0)*$_2^-$ *stretch 1385 cm*$^{-1}$ *(7.22 μm).*

The intense C=O stretching vibration occurs at higher frequencies (shorter wavelength) than that of normal ketones. The force constant of the carbonyl bond is increased by the electron attracting nature of the adjacent oxygen atom (inductive effect). Overlapping occurs between esters in which the carbonyl frequency is lowered, and ketones in which the normal ketone frequency is raised. A distinguishing feature of esters and lactones, however, is the strong C–O stretching band in the region where a weaker band occurs for ketones. There is overlapping in the C=O frequency of esters or lactones and acids, but the OH stretching and bending vibrations and the possibility of salt formation distinguish the acids.

The frequency of the ester carbonyl responds to environmental changes in the vicinity of the carbonyl group in much the same manner as ketones. The spectrum of phenyl acetate, Figure 25, illustrates most of the important absorption characteristics for esters.

C=O Stretching Vibrations

The C=O absorption band of saturated aliphatic esters (except formates) is in the 1750-1735 cm^{-1} (5.71-5.76 μm) region. The C=O absorption bands of formates, α,β-unsaturated, and benzoate esters are in the region of 1730-1715 cm^{-1} (5.78-5.83 μm). Further conjugation has little or no additional effect upon the frequency of the carbonyl absorption.

In the spectra of vinyl or phenyl esters, with unsaturation adjacent to the C–O– group, a marked rise in the carbonyl frequency is observed along with a lowering of the C–O frequency. Vinyl acetate has a carbonyl band at 1776 cm^{-1} (5.63 μm); phenyl acetate absorbs at 1770 cm^{-1} (5.65 μm).

α-Halogen substitution results in a rise in the C=O stretching frequency. Ethyl trichloroacetate absorbs at 1770 cm^{-1} (5.65 μm).

In oxalates and α-keto esters, as in α-diketones, there appears to be little or no interaction between the two carbonyl groups so that normal absorption occurs in the region of 1755-1740 cm^{-1} (5.70-5.75 μm). In the spectra of β-keto esters, however, where enolization can occur, a band is observed near 1650 cm^{-1} (6.06 μm) that results from bonding between the ester C=O and the enolic hydroxyl group.

The carbonyl absorption of saturated δ-lactones (6-membered ring) occurs in the same region as straight-chain, unconjugated esters. Unsaturation α to the C=O reduces

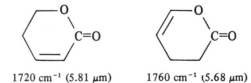

O C=O — 1720 cm^{-1} (5.81 μm)

O C=O — 1760 cm^{-1} (5.68 μm)

the C=O absorption frequency. Unsaturation α to the –O– group increases it.

α-Pyrones frequently display 2 carbonyl absorption bands in the 1775-1715 cm^{-1} (5.63-5.83 μm) region, probably due to Fermi resonance.

Saturated γ-lactones (5-membered ring) absorb at shorter wavelengths than esters or δ-lactones: 1795-1760 cm^{-1} (5.57-5.68 μm); γ-valerolactone absorbs at 1770 cm^{-1} (5.65 μm). Unsaturation in the γ-lactone molecule affects the carbonyl absorption in the same manner as unsaturation in δ-lactones.

ACETIC ACID, PHENYL ESTER

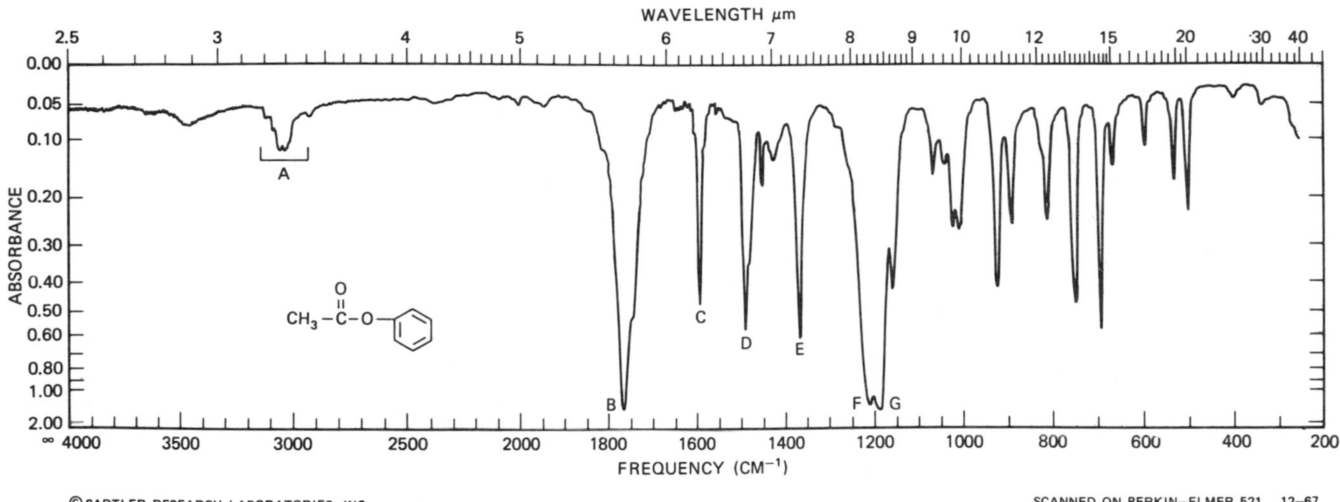

SCANNED ON PERKIN–ELMER 521 12–67

Source: The Matheson Company, Inc., East Rutherford, New Jersey

Figure 25. **A.** *Aromatic* C–H *stretch, 3070, 3040 cm^{-1} (3.26, 3.29 μm).* **B.** C=O *stretch, 1770 cm^{-1} (5.65 μm); this is higher frequency than that due to normal ester* C=O *stretch (ca. 1740 cm^{-1}, 5.75 μm, see Table III), due to phenyl conjugation with alcohol oxygen; conjugation of an aryl group or other unsaturation with the carbonyl group causes this* C=O *stretch to be at lower than normal frequency (e.g., benzoates absorb at ca. 1724 cm^{-1}, ca. 5.80 μm).* **C.** *Ring* C⩦C *stretch, 1593 cm^{-1} (6.28 μm).* **D.** δ_{as}, CH_3, *1493 cm^{-1} (6.70 μm).** **E.** δ_s, CH_3, *1360 cm^{-1} (7.35 μm).** **F.** *Acetate* CC(=O)–O *stretch, 1205 cm^{-1} (8.30 μm). G.* O–C⩦C *asym. stretch, 1183 cm^{-1} (8.45 μm).*

*See Table I, Appendix D.

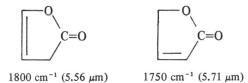

1800 cm^{-1} (5.56 μm) 1750 cm^{-1} (5.71 μm)

In unsaturated lactones, when the double bond is adjacent to the –O–, a strong C=C absorption is observed in the 1685-1660 cm^{-1} (5.94-6.02 μm) region.

C–O Stretching Vibrations

The "C–O stretching vibrations" of esters actually consists of 2 asymmetric coupled vibrations: C–C(=O)–O and O–C–C, the former being more important. These bands occur in the region of 1300-1000 cm^{-1} (7.70-10.00 μm). The corresponding symmetric vibrations are of little importance. The C–O stretch correlations are less reliable than the C=O stretch correlations.

The C–C(=O)–O band of saturated esters, except for acetates, shows strongly in the 1210-1163 cm^{-1} (8.26-8.60 μm) region. It is often broader and stronger than the C=O stretch absorption. Acetates of saturated alcohols display this band at 1240 cm^{-1} (8.07 μm). Vinyl and phenyl acetates absorb at a somewhat lower frequency, 1190-1140 cm^{-1} (8.40-8.77 μm); for example, see Figure 25. The C–C(=O)–O stretch of esters of α,β-unsaturated acids results in multiple bands in the 1300-1160 cm^{-1} (7.69-8.62 μm) region. Esters of aromatic acids absorb strongly in the 1310-1250 cm^{-1} (7.63-8.00 μm) region. The analogous type of stretch in lactones is observed in the 1250-1111 cm^{-1} (8.00-9.00 μm) region.

The O–C–C band of esters ("alcohol" carbon-oxygen stretch) of primary alcohols occurs at about 1064-1031 cm^{-1} (9.40-9.70 μm) and that of esters of secondary alcohols occurs at about 1100 cm^{-1} (9.09 μm). Aromatic esters of primary alcohols show this absorption near 1111 cm^{-1} (9.00 μm).

Methyl esters of long-chain fatty acids present a 3 band pattern with bands near 1250 cm^{-1} (8.00 μm), 1205 cm^{-1} (8.30 μm), and 1175 cm^{-1} (8.51 μm). The band near 1175 cm^{-1} (8.51 μm) is the strongest.

The influence of the –C(=O)–O group upon adjacent CH_2 and CH_3 groups is summarized in Appendix C, Tables I and II. Some spectra showing typical ester absorptions are shown in Appendix B: ethyl acetate-no.20 and diethyl phthalate-no.21.

ACID HALIDES

C=O Stretching Vibrations

Acid halides show strong absorption in the C=O stretching region. Unconjugated acid chlorides absorb in the 1815-1785 cm^{-1} (5.51-5.60 μm) region. Acetyl fluoride in the gas phase absorbs near 1869 cm^{-1} (5.35 μm). Conjugated acid halides absorb at a slightly lower frequency because resonance reduces the force constant of the C=O bond; aromatic acid chlorides absorb strongly at 1800-1770 cm^{-1} (5.56-5.65 μm). A weak band near 1750-1735 cm^{-1} (5.71-5.76 μm) appearing in the spectra of aroyl chlorides probably results from Fermi resonance between the C=O band and the overtone of a longer wavelength band near 875 cm^{-1} (11.43 μm). The annotated spectrum of benzoyl chloride is given in Figure 26.

CARBOXYLIC ACID ANHYDRIDES

C=O Stretching Vibrations

Anhydrides display 2 stretching bands in the carbonyl region. The 2 bands result from asymmetrical and symmetrical C=O stretching modes. Saturated noncyclic anhydrides absorb near 1818 cm^{-1} (5.50 μm) and near 1750 cm^{-1} (5.71 μm). Conjugated noncyclic anhydrides show absorption near 1775 cm^{-1} (5.63 μm) and near 1720 cm^{-1} (5.81 μm); the decrease in the frequency of absorption is due to resonance. The higher frequency band is the more intense.

Cyclic anhydrides with 5-membered rings show absorption at higher frequencies (lower wavelengths) than noncyclic anhydrides because of ring strain; succinic anhydride absorbs at 1865 cm^{-1} (5.37 μm) and at 1782 cm^{-1} (5.62 μm); The lower frequency (longer wavelength) C=O band is the stronger of the 2 carbonyl bands in 5-membered-ring cyclic anhydrides.

C–O Stretching Vibrations

Other strong bands appear in the spectra of anhydrides as a result of $\mathrm{C{-}\overset{O}{\overset{\|}{C}}{-}O{-}\overset{O}{\overset{\|}{C}}{-}C}$ stretching vibrations. Unconjugated straight chain anhydrides absorb near 1047 cm^{-1} (9.55 μm). Cyclic anhydrides display bands near 952-909 cm^{-1} (10.50-11.00 μm). and near 1299-1176 cm^{-1} (7.70-8.50 μm). The C–O stretching band for acetic anhydride is at 1125 cm^{-1} (8.89 μm).

The spectrum of Figure 27 is that of a typical aliphatic anhydride.

AMIDES

All amides show a carbonyl absorption band known as the Amide I band. Its position depends on the degree of hydrogen bonding and, thus, on the physical state of the compound.

Primary amides show 2 N–H stretching bands resulting from symmetrical and asymmetrical N–H stretching. Secondary amides and lactams show only 1 N–H stretching band. As in the case of O–H stretching, the frequency of the N–H stretching is reduced by hydrogen bonding,

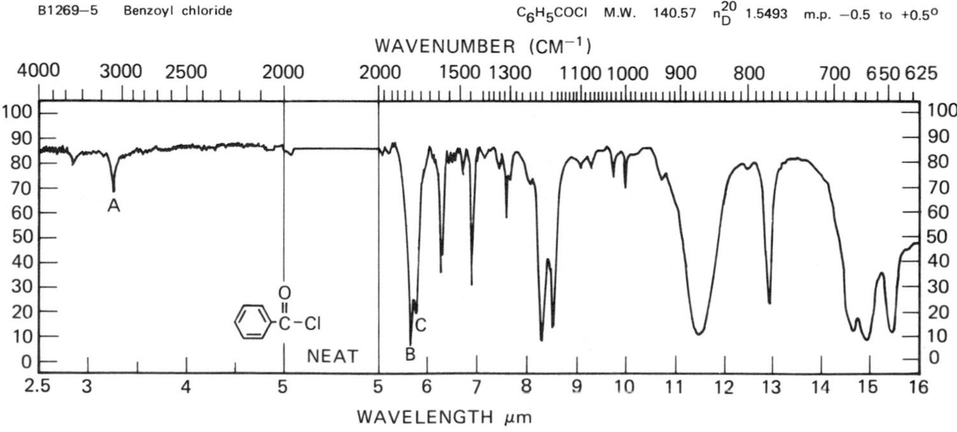

Figure 26. **A.** *Aromatic* C–H *stretch, 3080 cm^{-1} (3.25 μm).* **B.** C=O *stretch, 1790 cm^{-1} (5.58 μm see Table III p. 118). (Acid chloride* C=O *stretch position shows very small dependence upon conjugation; aroyl chlorides identified by band such as at* C.*)* **C.** *Fermi resonance band (of* C=O *stretch and overtone of 875 cm^{-1} band), 1745 cm^{-1} (5.73 μm).*

Source: Courtesy of Aldrich Chemical Co.

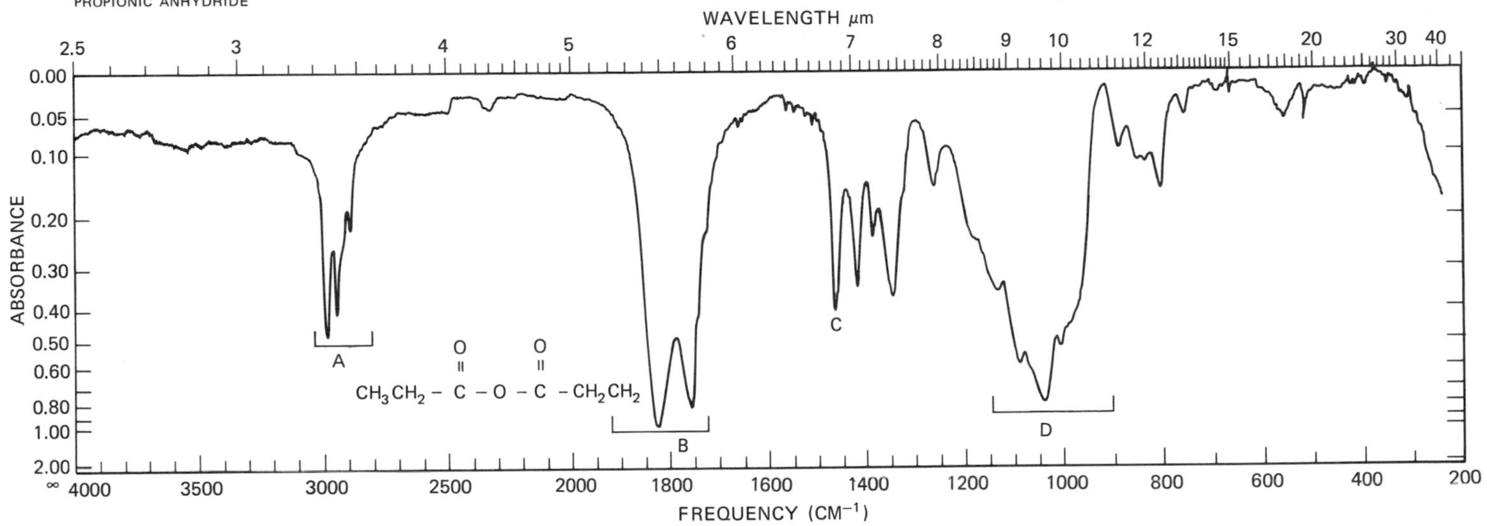

Figure 27. A. C–H *stretch, 2990, 2950, 2880 cm^{-1} (3.34, 3.39, 3.47 μm).* **B.** *Asymmetric and symmetric* C=O *coupled stretching, respectively: 1825, 1758 cm^{-1} (5.48, 5.69 μm). See Table III. p. 118.* **C.** δ_s CH_2 *(scissoring), 1465 cm^{-1} (6.83 μm), See Table II, Appendix D.***D.** C–CO–O–CO–C *stretch, 1040 cm^{-1} (9.62 μm).*

though to a lesser degree. Overlapping occurs in the observed position of N–H and O–H stretching frequencies so that an unequivocal differentiation in structure is sometimes impossible.

Primary amides and secondary amides, and a few lactams, display a band or bands in the region of 1650-1515 cm^{-1} (6.06-6.60 μm) due primarily to NH_2 or NH bending: the amide II band. This absorption involves coupling between N–H bending and other fundamental vibrations and requires a *trans* geometry.

Out-of-plane NH wagging is responsible for a broad band of medium intensity in the 800-666 cm^{-1} (12.5-15.0 μm) region.

The spectrum of Figure 28 is that of a typical primary amide of an aliphatic acid. The spectrum of DMF (*N,N*-dimethylformamide, Appendix B, no. 28) displays typical amide absorptions.

N–H Stretching Vibrations

In dilute solution in nonpolar solvents, primary amides show two moderately intense NH stretching frequencies corresponding to the asymmetrical and symmetrical NH stretching vibrations. These bands occur near 3520 cm^{-1} (2.84 μm) and 3400 cm^{-1} (2.94 μm), respectively. In the spectra of solid samples, these bands are observed near 3350 cm^{-1} (2.99 μm) and 3180 cm^{-1} (3.15 μm) because of hydrogen bonding.

In infrared spectra of secondary amides, which exist mainly in the *trans* conformation, the free NH stretching vibration observed in dilute solutions occurs near 3500-3400 cm^{-1} (2.86-2.94 μm). In more concentrated solutions and in solid samples, the free NH band is replaced by multiple bands in the 3330-3060 cm^{-1} (3.00-3.27 μm) region. Multiple bands are observed since the amide group can bond to produce dimers, with a *cis* conformation, and polymers, with a *trans* conformation.

```
R      O----H     R'
  \  //      \   /
    C          N
    |          |
    N          C
  /  \      //  \
R'     H----O     R
```

cis

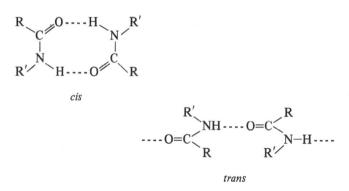

```
              R'              R
                \            /
                 NH----O=C
                           \
----O=C                     N-H----
       \                   /
        R                R'
```

trans

C=O Stretching Vibrations (Amide I Band)

The C=O absorption of amides occurs at longer wavelengths than "normal" carbonyl absorption due to the resonance effect (see p. 118). The position of absorption depends on the same environmental factors as the carbonyl absorption of other compounds.

Primary amides (except acetamide, whose C=O bond absorbs at 1694 cm^{-1}, 5.90 μm) have a strong amide I band in the region of 1650 cm^{-1} (6.06 μm) when examined in the solid phase. When the amide is examined in dilute solution, the absorption is observed at a higher frequency,

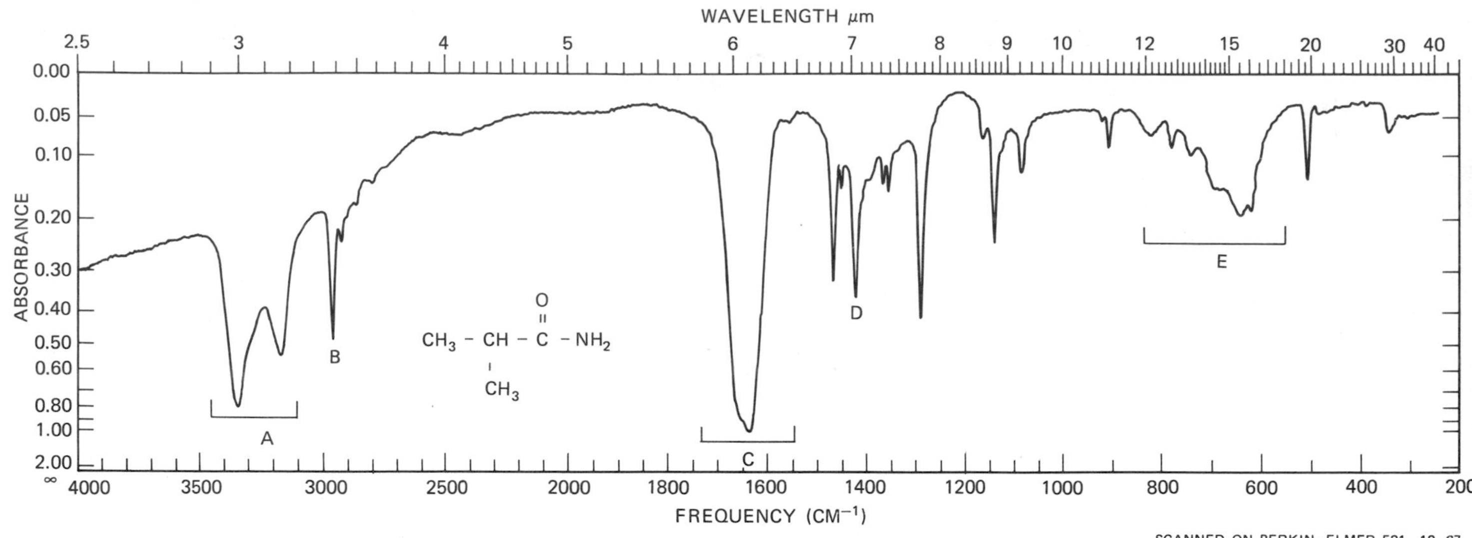

Source: Aldrich Chemical Company, Milwaukee, Wis.

Figure 28. **A**. N–H *stretch, coupled, primary amide, hydrogen bonded. Asymmetric, 3350* cm^{-1} *(2.99 μm). Symmetric, 3170* cm^{-1} *(3.15 μm).* **B**. *Aliphatic* C–H *stretch, 2960* cm^{-1} *(3.15 μm).* **C**. *Overlap* C=O *stretch, Amide I band, 1640* cm^{-1} *(6.10 μm), see Table III.* N–H *bend, 1640* cm^{-1} *(6.10 μm), Amide II band.* **D**. C–N *stretch, 1425* cm^{-1} *(7.02 μm).* **E**. *Broad* N–H *out-of-plane bend, 700-600* cm^{-1} *(14.28-16.67 μm).*

near 1690 cm^{-1} (5.92 μm). In more concentrated solutions, the C=O frequency is observed at some intermediate value, depending on the degree of hydrogen bonding.

Simple, open-chain, secondary amides absorb near 1640 cm^{-1} (6.10 μm) when examined in the solid state. In dilute solution, the frequency of the amide I band may be raised to 1680 cm^{-1} (5.95 μm) and even to 1700 cm^{-1} (5.88 μm) in the case of the anilides. In the anilide structure there is competition between the ring and the C=O for the nonbonded electron pair of the nitrogen.

The carbonyl frequency of tertiary amides is independent of the physical state since hydrogen bonding with another tertiary amide group is impossible. The C=O absorption occurs in the range of 1680-1630 cm^{-1} (5.95-6.13 μm). The absorption range of tertiary amides in solution is influenced by hydrogen bonding with the solvent. *N,N*-Diethylacetamide absorbs at 1647 cm^{-1} (6.07 μm) in dioxane and at 1615 cm^{-1} (6.20 μm) in methanol.

Electron-attracting groups attached to the nitrogen increase the frequency of absorption since they effectively compete with the carbonyl oxygen for the electrons of the nitrogen, thus increasing the force constant of the C=O bond.

N–H Bending Vibrations (Amide II Band)

All primary amides show a sharp absorption band in dilute solution (amide II band) resulting from NH_2 bending at a somewhat lower frequency than the C=O band. This band has an intensity of one-half to one-third of the C=O absorption band. In mulls and pellets the band occurs near 1655-1620 cm^{-1} (6.04-6.17 μm) and is usually under the envelope of the amide I band. In dilute solutions, the band appears at lower frequency, 1620-1590 cm^{-1} (6.17-6.29 μm), and normally is separated from the amide I band. Multiple bands may appear in the spectra of concentrated solutions, arising from the free and associated states. The nature of the R group ($R-\overset{\overset{O}{\|}}{C}-NH_2$) has little effect upon the amide II band.

Secondary acyclic amides in the solid state display an amide II band in the region of 1570-1515 cm^{-1} (6.37-6.60 μm). In dilute solution, the band occurs in the 1550-1510 cm^{-1} (6.45-6.62 μm) region. This band results from interaction between the N–H bending and the C–N stretching of the C–N–H group. A second, weaker band near 1250 cm^{-1} (8.00 μm) also results from interaction between the N–H bending and C–N stretching.

Other Vibration Bands

The C–N stretching band of primary amides occurs near 1400 cm^{-1} (7.14 μm). A broad, medium band in the 800-666 cm^{-1} (12.5-15.0 μm) region in the spectra of primary and secondary amides results from out-of-plane N–H wagging.

LACTAMS

In lactams of medium ring size, the amide group is forced into the *cis* conformation. Solid lactams absorb strongly near 3200 cm^{-1} (3.12 μm) because of the N—H stretching vibration. This band does not shift appreciably with dilution since the *cis* form remains associated at relatively low concentrations.

C=O Stretching Vibrations

The C=O absorption of lactams with 6-membered rings or larger is near 1650 cm^{-1} (6.06 μm). Five-membered ring (γ) lactams absorb in the 1750-1700 cm^{-1} (5.71-5.88 μm) region. Four-membered-ring (β) lactams, unfused, absorb at 1760-1730 cm^{-1} (5.68-5.78 μm). Fusion of the lactam ring to another ring generally increases the frequency by 20-50 cm^{-1} (0.07-0.17 μm).

Most lactams do not show a band near 1550 cm^{-1} (6.45 μm) that is characteristic of *trans* noncyclic secondary amides. The N—H out-of-plane wagging in lactams causes broad absorption in the 800-700 cm^{-1} (12.5-14.3 μm) region.

AMINES

The spectrum of a typical primary, aliphatic amine appears in Figure 29.

N—H Stretching Vibrations

Primary amines, examined in dilute solution, display two weak absorption bands: one near 3500 cm^{-1} (2.86 μm), the other near 3400 cm^{-1} (2.94 μm). These bands represent, respectively, the "free" asymmetrical and symmetrical N—H stretching modes. Secondary amines show a single weak band in the 3350-3310 cm^{-1} (2.98-3.02 μm) region. These bands are shifted to longer wavelengths by hydrogen bonding. The associated N—H bands are weaker and frequently sharper than the corresponding O—H bands. Aliphatic primary amines (neat) absorb at 3400-3330 cm^{-1} (2.94-3.00 μm) and at 3330-3250 cm^{-1} (3.00-3.08 μm). Aromatic primary amines absorb at slightly higher frequencies (shorter wavelengths). In the spectra of liquid primary and secondary amines, a shoulder usually appears on the low-frequency (long-wavelength) side of the N—H stretching band, arising from the overtone of the NH bending band intensified by Fermi resonance. Tertiary amines do not absorb in this region of the spectrum.

N—H Bending Vibrations

The N—H bending (scissoring) vibration of primary amines is observed in the 1650-1580 cm^{-1} (6.06-6.33 μm) region of the spectrum. The band is medium to strong in intensity and is moved to slightly higher frequencies when the compound is associated. The N—H bending band is seldom

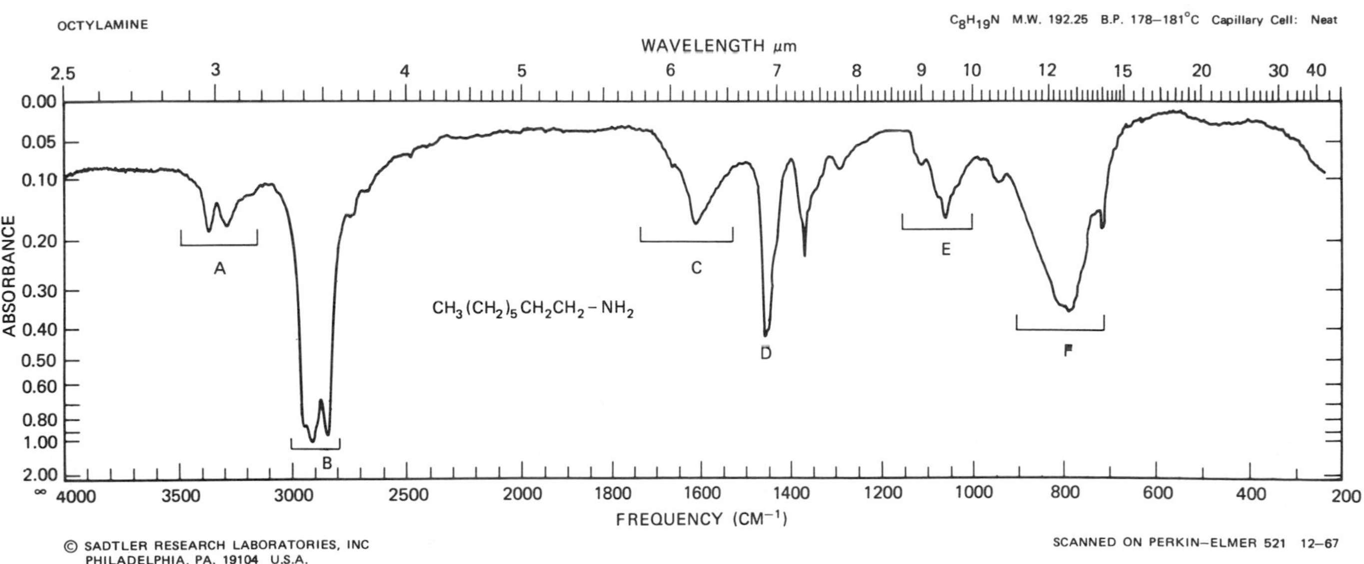

Source: The Matheson Company, Inc., East Rutherford, N.J.

Figure 29. **A.** N—H *stretch, hydrogen bonded, primary amine coupled doublet: Asymmetric, 3365 cm^{-1} (2.97 μm). Symmetric, 3290 cm^{-1} (3.04 μm). (Shoulder at ca. 3200 cm^{-1}, ca. 3.12 μm, Fermi resonance band with overtone of band at C.)* **B.** *Aliphatic* C—H *stretch, 2910, 2850 cm^{-1} (3.44, 3.51 μm);* ν_s, CH_2, *2817 cm^{-1} (3.55 μm), see Table II, Appendix C.* **C.** N—H *bend (scissoring), 1620 cm^{-1} (6.17 μm).* **D.** δ_s, CH_2 *(scissoring), 1458 cm^{-1} (6.86 μm), see Table II, Appendix D.* **E.** C—N *stretch, 1063 cm^{-1} (9.41 μm).* **F.** N—H *wag (neat sample), 909-666 cm^{-1} (11.00-15.00 μm).*

detectable in the spectra of aliphatic secondary amines, whereas secondary aromatic amines absorb near 1515 cm^{-1} (6.60 μm).

Liquid samples of primary and secondary amines display medium-to-strong broad absorption in the 909-666 cm^{-1} (11.00-15.00 μm) region of the spectrum arising from NH wagging. The position of this band depends on the degree of hydrogen bonding.

C–N Stretching Vibrations

Medium-to-weak absorption bands for the unconjugated C–N linkage in primary, secondary, and tertiary aliphatic amines appear in the region of 1250-1020 cm^{-1} (8.00-9.80 μm). The vibrations responsible for these bands involve C–N stretching coupled with the stretching of adjacent bonds in the molecule. The position of absorption in this region depends on the class of the amine and the pattern of substitution on the α-carbon.

Aromatic amines display strong C–N stretching absorption in the 1342-1266 cm^{-1} (7.45-7.90 μm) region. The absorption appears at higher frequencies (shorter wavelengths) than the corresponding absorption of aliphatic amines because the force constant of the C–N bond is increased by resonance with the ring.

Characteristic strong C–N stretching bands in the spectra of aromatic amines have been assigned as in Table IV.

The absorption frequencies of methyl and methylene groups attached to the nitrogen atom of an amine are summarized in Tables I and II of Appendix D.

AMINE SALTS

N–H Stretching Vibrations

The ammonium ion displays strong, broad absorption in the 3300-3030 cm^{-1} (3.03-3.30 μm) region because of N–H stretching vibrations (see Figure 24). There is also a combination band in the 2000-1709 cm^{-1} (5.00-5.85 μm) region.

Salts of primary amines show strong, broad absorption between 3000-2800 cm^{-1} (3.33-3.57 μm) arising from asymmetrical and symmetrical stretching in the NH_3^+ group. In addition, multiple combination bands of medium intensity occur in the 2800-2000 cm^{-1} (3.57-5.00 μm) region, the most prominent being the band near 2000 cm^{-1} (5.00 μm). Salts of secondary amines absorb strongly in the 3000-2700 cm^{-1} (3.33-3.70 μm) region with multiple bands extending to 2273 cm^{-1} (4.00 μm). A medium band near 2000 cm^{-1} (5.00 μm) may be observed. Tertiary amine salts absorb at longer wavelengths than the salts of primary and secondary amines; 2700-2250 cm^{-1} (3.70-4.44 μm). Quaternary ammonium salts can have no N–H stretching vibrations.

Table IV. C–N Stretch of Primary, Secondary, and Tertiary Aromatic Amines

Primary	1340-1250 cm^{-1} (7.46-8.00 μm)
Secondary	1350-1280 cm^{-1} (7.41-7.81 μm)
Tertiary	1360-1310 cm^{-1} (7.35-7.63 μm)

N–H Bending Vibrations

The ammonium ion displays a strong, broad NH_4^+ bending band near 1429 cm^{-1} (7.00 μm). The NH_3^+ group of the salt of a primary amine absorbs near 1600-1575 cm^{-1} (6.25-6.35 μm) and near 1550-1504 cm^{-1} (6.45-6.65 μm). These bands originate in asymmetrical and symmetrical NH_3^+ bending, analogous to the corresponding bands of the CH_3 group. Salts of secondary amines absorb near 1620-1560 cm^{-1} (6.17-6.41 μm). The N–H bending band of the salts of tertiary amines is weak and of no practical value.

AMINO ACIDS AND SALTS OF AMINO ACIDS

Amino acids are encountered in three forms: the free amino acid (zwitterion),

$$-\overset{|}{\underset{\underset{NH_3^+}{|}}{C}}-COO^-$$

the hydrochloride (or other salt),

$$-\overset{|}{\underset{\underset{NH_3^+\,Cl^-}{|}}{C}}-COOH$$

and the sodium (or other cation) salt,

$$-\overset{|}{\underset{\underset{NH_2}{|}}{C}}-COO^-\,Na^+$$

Free primary amino acids are characterized by the following absorptions (most of the work has been done with α-amino acids, but the relative positions of the amino and carboxyl groups seem to have little effect):

1. A broad, strong NH_3^+ stretching band in the 3100-2600 cm^{-1} (3.23-3.85 μm) region. Multiple combination and overtone bands extend the absorption to about 2000 cm^{-1} (5.00 μm). This overtone region usually contains a prominent band near 2222-2000 cm^{-1} (4.50-5.00 μm) assigned to a combination of the asymmetrical NH_3^+ bending vibration and the torsional oscillation of the NH_3^+ group. The torsional oscillation occurs near 500 cm^{-1} (20.00 μm). The 2000 cm^{-1} (5.00 μm) band is absent if the nitrogen atom of the amino acid is substituted.
2. A weak asymmetric NH_3^+ bending band near 1660-1610 cm^{-1} (6.03-6.21 μm); a fairly strong symmetrical bending band near 1550-1485 cm^{-1} (6.45-6.73 μm).
3. The carboxylate ion group $\left(-C\begin{smallmatrix}\nearrow O\\ \searrow O\end{smallmatrix}-\right)$ absorbs strongly near 1600-1590 cm^{-1} (6.25-2.29 μm) and more weakly near 1400 cm^{-1} (7.14 μm). These bands result, respectively, from asymmetrical and symmetrical $C(\dot{=}O)_2$ stretching.

The spectrum of the amino acid leucine, including assignments corresponding to the preceding three categories, is shown in Figure 30.

Hydrochlorides of amino acids present the following patterns:

1. Broad strong absorption in the 3333-2380 cm^{-1} (3.00-4.20 μm) region resulting from superimposed O–H and NH_3^+ stretching bands. Absorption in this region is characterized by multiple fine structure on the long wavelength side of the band.
2. A weak, asymmetrical $\overset{+}{N}H_3$ bending band near 1610-1590 cm^{-1} (6.21-6.29 μm); a relatively strong, symmetrical $\overset{+}{N}H_3$ bending band at 1550-1481 cm^{-1} (6.45-6.75 μm).
3. A strong band at 1220-1190 cm^{-1} (8.20-8.40 μm) arising from $C-\overset{O}{\overset{\|}{C}}-O$ stretching.
4. Strong carbonyl absorption at 1755-1730 cm^{-1} (5.70-5.78 μm) for α-amino acid hydrochlorides, and at 1730-1700 cm^{-1} (5.78-5.88 μm) for other amino acid hydrochlorides.

Sodium salts of amino acids show the normal N–H stretching vibrations at 3400-3200 cm^{-1} (2.94-3.13 μm) common to other amines. The characteristic carboxylate ion bands appear near 1600-1590 cm^{-1} (6.25-6.29 μm) and near 1400 cm^{-1} (7.14 μm).

NITRILES

The spectra of nitriles (R–C≡N) are characterized by weak to medium absorption in the triple-bond stretching region of the spectrum. Aliphatic nitriles absorb near 2260-2240 cm^{-1} (4.42-4.46 μm). Electron attracting atoms, such as

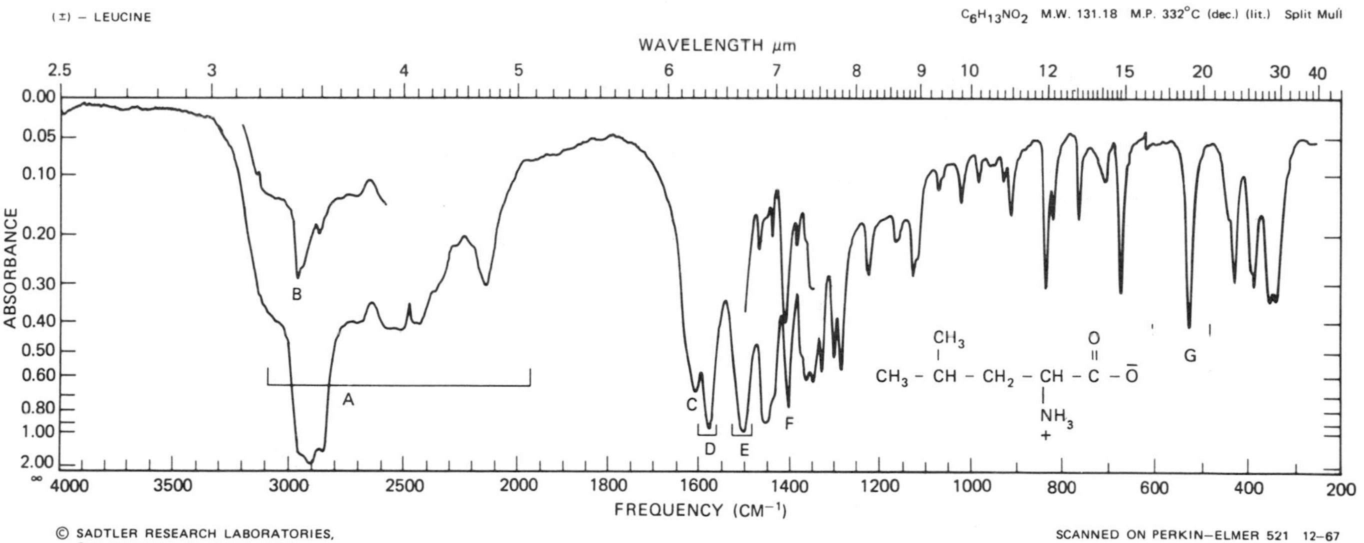

Source: The Matheson Company, Inc., E. Rutherford, N.J.

Figure 30. A. *Broad* $(-NH_3^+)$ N–H *stretch, 3100-2000* cm^{-1} *(3.25-5.00 μm), extended by combination band at 2140* cm^{-1} *(4.67 μm), and other combination-overtone bands.* **B.** *Aliphatic* C–H *stretch (superimposed on* N–H *stretch), 2967* cm^{-1} *(3.37 μm).* **C.** *Asymmetric* $(-NH_3^+)$ N–H *bend, 1610* cm^{-1} *(6.21 μm).* **D.** *Asymmetric carboxylate* $(C\dot{=}O)_2$ *stretch, 1580* cm^{-1} *(6.33 μm).* **E.** *Symmetric* $(-NH_3^+)$ N–H *bend, 1505* cm^{-1} *(6.65 μm).* **F.** *Symmetric carboxylate* $(C\dot{=}O)_2$ *stretch, 1405* cm^{-1} *(7.12 μm).* **G.** *Torsional* $(-NH_3^+)$ N–H *oscillation, 525* cm^{-1} *(19.0 μm).*

oxygen or chlorine, attached to the carbon atom alpha to to the C≡N group reduce the intensity of absorption. Conjugation, such as occurs in aromatic nitriles, reduces the frequency of absorption to 2240-2222 cm^{-1} (4.46 to 4.50 μm) and enhances the intensity. The spectrum of a typical nitrile, with an aryl group in conjugation with the cyano function, is shown in Figure 31.

COMPOUNDS CONTAINING C≡N, C=N, –N=C=O AND –N=C=S GROUPS

Isocyanides (isonitriles), isocyanates, thiocyanates, and isothiocyanates all show C≡N stretch or cumulated double bond (–Y=C=X; X, Y = N, S or O) stretch in the 2273-2000 cm^{-1} (4.40-5.00 μm) region (Appendix C). Schiff's bases (RCH=NR, imines), oximes, thiazoles, iminocarbonates, guanidines, etc., show the C=N stretch in the 1689-1471 cm^{-1} (5.92-6.80 μm) region. Although the intensity of the C=N stretch is variable, it is usually more intense than the C=C stretch.

COMPOUNDS CONTAINING –N=N– GROUP

The N=N stretching vibration of a symmetrical *trans* azo compound is forbidden in the infrared but absorbs in the 1576 cm^{-1} (6.35 μm) region of the Raman spectrum. Unsymmetrical para-substituted azobenzenes in which the substituent is an electron donating group absorb near 1429 cm^{-1} (7.00 μm). The bands are weak because of the nonpolar nature of the bond.

COVALENT COMPOUNDS CONTAINING NITROGEN-OXYGEN BONDS

Nitro compounds, nitrates, and nitramines contain an NO_2 group. Each of these classes shows absorption due to asymmetrical and symmetrical stretching of the NO_2 group. Asymmetrical absorption results in a strong band in the 1661-1499 cm^{-1} (6.02-6.67 μm) region; symmetrical absorption occurs in the region between 1389-1259 cm^{-1} (7.20-7.94 μm). The exact position of the bands is dependent on substitution and unsaturation in the vicinity of the NO_2 group.

NITRO COMPOUNDS. In the nitro-alkanes, the bands occur near 1550 cm^{-1} (6.45 μm) and near 1372 cm^{-1} (7.29 μm). Conjugation lowers the frequency of both bands, resulting in absorption near 1550-1500 cm^{-1} (6.45-6.67 μm) and near 1360-1290 cm^{-1} (7.36-7.75 μm). Attachment of electronegative groups to the α-carbon of a nitro compound causes an increase in the frequency of the asymmetrical NO_2 band and a reduction in the frequency of the symmetrical band; chloropicrin, Cl_3CNO_2, absorbs at 1610 cm^{-1} (6.21 μm) and at 1307 cm^{-1} (7.65 μm).

Aromatic nitro groups absorb near the same frequencies as observed for conjugated aliphatic nitro compounds. Interaction between the NO_2 out-of-plane bending and ring C–H out-of-plane bending frequencies destroys the reliability of the substitution pattern observed for nitro-aromatics in the long wavelength region of the spectrum. Nitro-aromatic compounds show a C–N stretching vibration near 870 cm^{-1} (11.49 μm). The spectrum of nitrobenzene, with assignments corresponding to the preceding discussion, is shown in Figure 32.

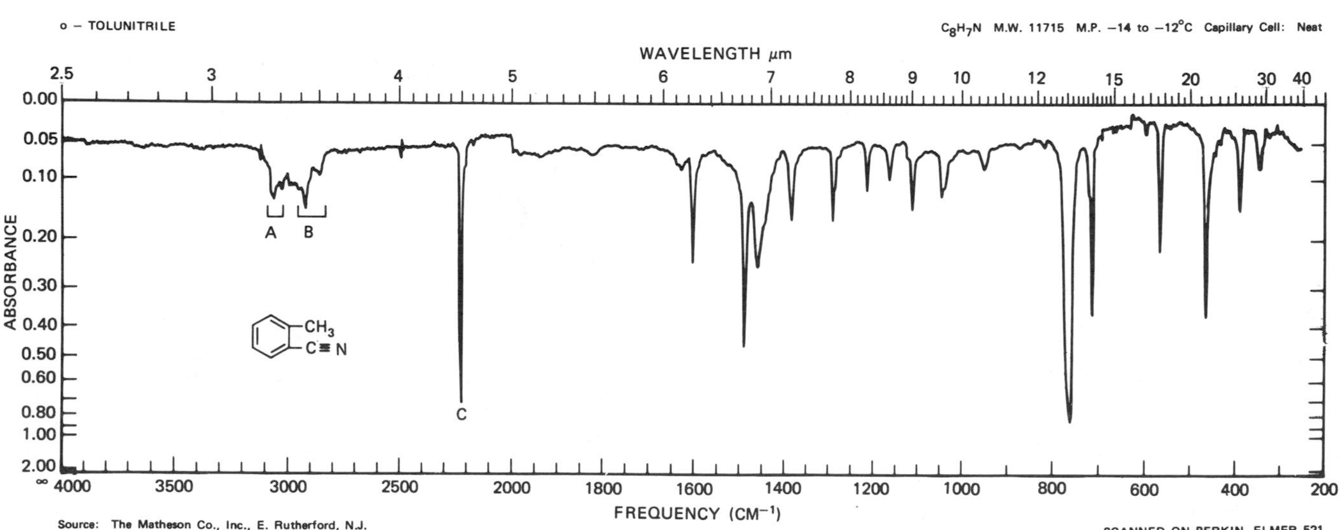

Source: The Matheson Co., Inc., E. Rutherford, N.J.

Figure 31. **A.** *Aromatic* C–H *stretch, 3070* cm^{-1}*, 3025* cm^{-1} *(3.26, 3.31 μm).* **B.** *Aliphatic* C–H *stretch, 2910, 2860* cm^{-1} *(3.44, 3.50 μm).* **C.** C≡N *stretch, 2210* cm^{-1} *(4.53 μm)(intensified by aryl conjugation); aliphatic nitriles absorb at higher frequency (shorter wavelength).*

Because of strong resonance in aromatic systems containing NO_2 groups and electron-donating groups such as the amino group, ortho or para to one another, the symmetrical NO_2 vibration is shifted to lower frequencies and increases in intensity. *p*-Nitroaniline absorbs at 1475 cm^{-1} (6.78 μm) and 1310 cm^{-1} (7.64 μm).

The positions of asymmetric and symmetric NO_2 stretching bands of nitramines $\left(\rangle N{-}NO_2\right)$ and the NO stretch of nitrosoamines are given in Appendix C.

NITRATES. Organic nitrates show absorption for N–O stretching vibrations of the NO_2 group and for the O–N linkage. Asymmetrical stretching in the NO_2 group results in strong absorption in the 1660-1625 cm^{-1} (6.02-6.15 μm) region; the symmetrical vibration absorbs strongly near 1300-1255 cm^{-1} (7.69-7.97 μm). Stretching of the π bonds of the N–O linkage produces absorption near 870-833 cm^{-1} (11.50-12.00 μm). Absorption observed at longer wavelengths, near 763-690 cm^{-1} (13.10-14.50 μm), likely results from NO_2 bending vibrations.

NITRITES. Nitrites display two strong N=O stretching bands. The band near 1680-1650 cm^{-1} (5.95-6.06 μm) is attributed to the *trans* isomer; the *cis* isomer absorbs in the 1625-1610 cm^{-1} (6.16-6.21 μm) region. The N–O stretching band appears in the region between 850-750 cm^{-1} (11.76-13.33 μm). The nitrite absorption bands are among the strongest observed in infrared spectra.

NITROSO COMPOUNDS. Primary and secondary aliphatic *C*-nitroso compounds are usually unstable and rearrange to oximes or dimerize. Tertiary and aromatic nitroso compounds are reasonably stable, existing as monomers in the gaseous phase or in dilute solution and as dimers in neat samples. Monomeric, tertiary, aliphatic nitroso compounds show N=O absorption in the 1585-1539 cm^{-1} (6.31-6.50 μm) region; aromatic monomers absorb between 1511-1495 cm^{-1} (6.62-6.69 μm).

The N→O stretching absorption of dimeric nitroso compounds are categorized in Appendix C as to *cis* vs. *trans* and aliphatic vs. aromatic. Nitrosoamine absorptions are given in Appendix C.

ORGANIC SULFUR COMPOUNDS

MERCAPTANS. Aliphatic mercaptans and thiophenols, as liquids or in solution, show S–H stretching absorption in the range of 2600-2550 cm^{-1} (3.85-3.92 μm). The S–H stretching band is characteristically weak and may go undetected in the spectra of dilute solutions or thin films. However, since few other groups show absorption in this region, it is useful in detecting S–H groups. The spectrum of benzyl mercaptan in Figure 33 is that of a mercaptan with a detectable S–H stretch band. The band may be obscured by strong carboxyl absorption in the same region. Hydrogen bonding is much weaker for S–H groups than for O–H and N–H groups.

The S–H group of thiol acids absorbs in the same region as mercaptans and thiophenols.

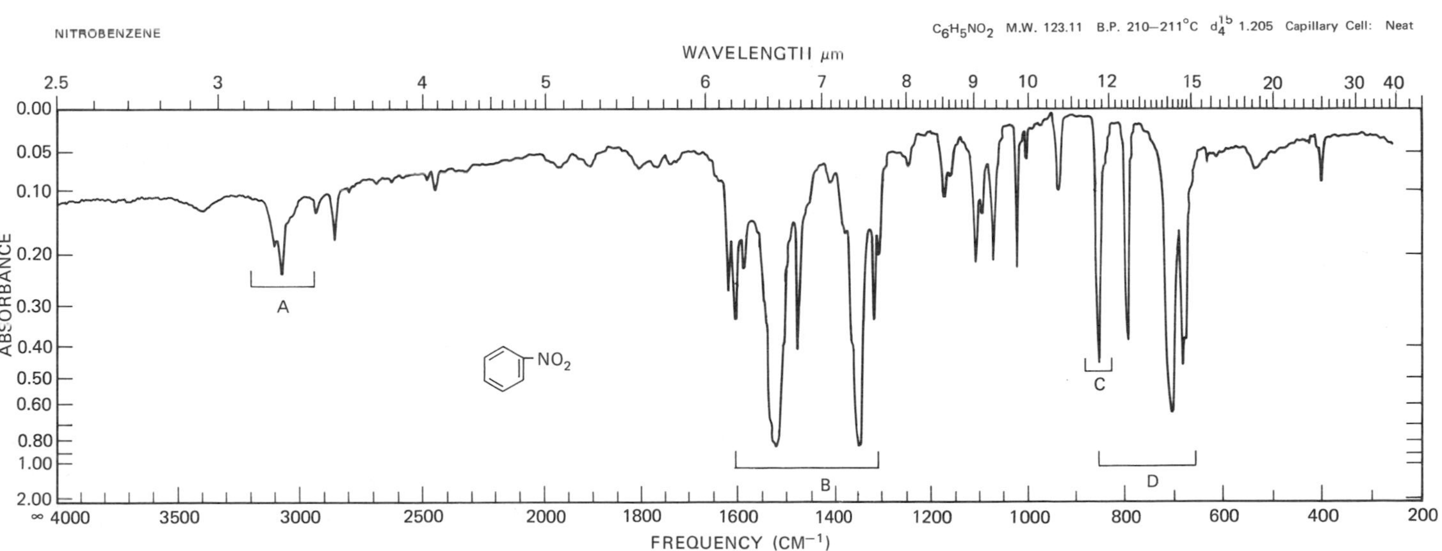

Source: J.T. Baker Chemical Co., Phillipsburg, N.J.

Figure 32. **A.** *Aromatic* C–H *stretch, 3100, 3080 cm^{-1} (3.23, 3.25 μm).* **B.** *Asymmetric* ($ArNO_2$) N⋯O *stretch, 1520 cm^{-1} (6.58 μm). Symmetric* ($ArNO_2$) N⋯O *stretch 1345 cm^{-1} (7.44 μm).* **C.** $ArNO_2$ C–N *stretch, 850 cm^{-1} (11.76 μm).* **D.** *Low frequency bands are of little use in determining the nature of ring substitution since these absorption patterns are due to interaction of* NO_2 *and* C–H *out-of-plane bending frequencies.*

SULFIDES. The stretching vibrations assigned to the C—S linkage occur in the region of 700-600 cm^{-1} (14.3-16.7 μm). The weakness of absorption and variability of position make this band of little value in structural determination.

DISULFIDES. The S—S stretching vibration is very weak and falls between 500-400 cm^{-1} (20-25 μm), outside the range of sodium chloride optics. The characteristic absorptions of methyl and methylene groups attached to sulfur are summarized in Tables I and II in Appendix D.

THIOCARBONYL COMPOUNDS. Aliphatic thials or thiones exist as trimeric, cyclic sulfides. Aralkyl thiones may exist either as monomers or trimers, whereas diaryl thiones, such as thiobenzophenone, exist only as monomers. The C=S group is less polar than the C=O group and has a considerably weaker bond. In consequence, the band is not intense, and it falls at lower frequencies, where it is much more susceptible to coupling effects. Identification is therefore difficult and uncertain.

Compounds that contain a thiocarbonyl group show absorption in the 1250-1020 cm^{-1} (8.00-9.70 μm) region. Thiobenzophenone and its derivatives absorb moderately in the 1224-1207 cm^{-1} (8.17-8.29 μm) region. Since the absorption occurs in the same general region as C—O and C—N stretching, considerable interaction can occur between these vibrations within a single molecule.

Spectra of compounds in which the C=S group is attached to a nitrogen atom show an absorption band in the general C=S stretching region. In addition, several other bands in the broad region of 1563-700 cm^{-1} (6.40-14.30 μm) can be attributed to vibrations involving interaction between C=S stretching and C—N stretching.

Thioketo compounds that can undergo enolization exist as thioketo-thioenol tautomeric systems; such systems show S—H stretching absorption. The thioenol tautomer of ethyl thiobenzoylacetate,

$$\begin{array}{c} \quad\quad\quad\quad H \\ \quad\quad\quad\quad | \\ C_6H_5-C=C-C-OC_2H_5 \\ \quad | \quad\quad\quad \| \\ \quad SH---O \end{array}$$

absorbs broadly at 2415 cm^{-1} (4.14 μm) due to bonded S—H stretching absorption.

COMPOUNDS CONTAINING SULFUR–OXYGEN BONDS

SULFOXIDES. Alkyl and aryl sulfoxides as liquids or in solution show strong absorption in the 1070-1030 cm^{-1} (9.35-9.71 μm) region. This absorption occurs at 1050 cm^{-1} (9.52 μm) for dimethyl sulfoxide (DMSO, methyl sulfoxide) as may be seen in Appendix B, spectrum 26. Conjugation brings about a small change (10-20 cm^{-1}) in the observed frequency in contrast to the marked reduction in frequency of the C=O bond accompanying conjugation. Diallyl sulfoxide absorbs at 1047 cm^{-1} (9.55 μm). Phenyl methyl sulfoxide and cyclohexyl methyl sulfoxide absorb at 1055 cm^{-1} (9.48 μm) in dilute solution in carbon tetrachloride. The sulfoxide group is susceptible to hydrogen bonding, the absorption shifting to slightly lower frequencies from dilute solution to the liquid phase. The frequency

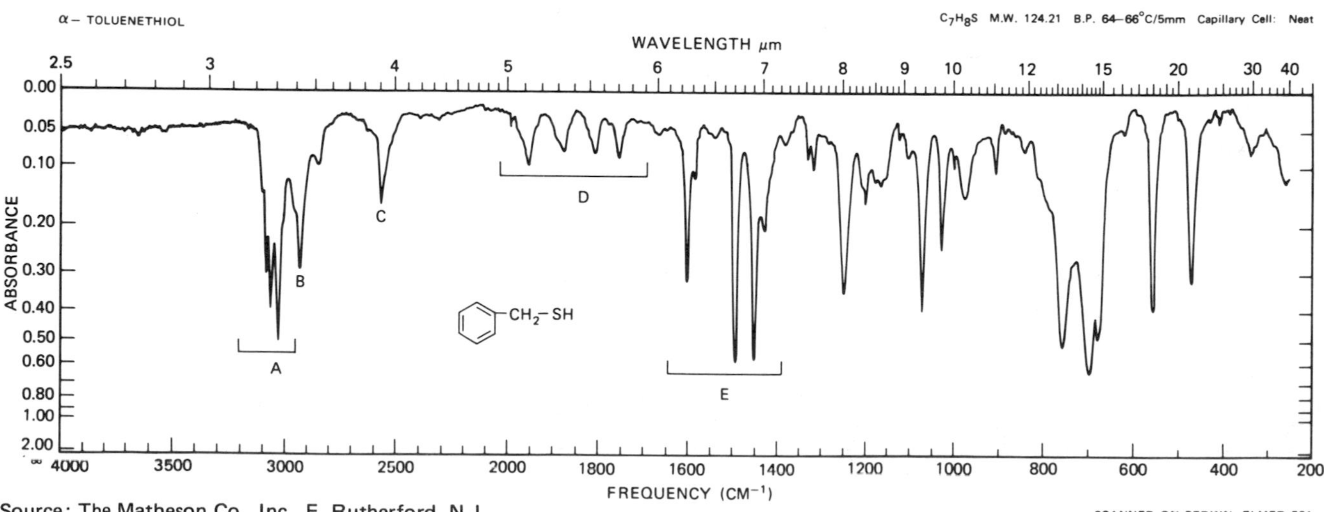

Figure 33. **A.** *Aromatic* C—H *stretch, 3085, 3060, 3030* cm^{-1} *(3.24, 3.27, 3.30 μm).* **B.** *Aliphatic* C—H *stretch, 2930* cm^{-1} *(3.41 μm).* **C.** *Moderately weak* S—H *stretch, 2565* cm^{-1} *(3.90 μm).* **D.** *Overtone or combination band pattern indicative of monosubstituted aromatic (see Figure 14), 2000-1667* cm^{-1} *(5.0-6.0 μm).* **E.** C⋯C *ring stretch, 1600, 1495, 1455* cm^{-1} *(6.25, 6.69, 6.87 μm).*

of S=O absorption is increased by electronegative substitution.

SULFONES. Spectra of sulfones show strong absorption bands at 1350-1300 cm^{-1} (7.41-7.70 μm) and at 1160-1120 cm^{-1} (8.62-8.93 μm). These bands arise from asymmetric and symmetric SO_2 stretching, respectively. Hydrogen bonding results in absorption near 1300 cm^{-1} (7.70 μm) and near 1125 cm^{-1} (8.90 μm). Splitting of the high frequency band often occurs in carbon tetrachloride solution or in the solid state.

SULFONYL CHLORIDES. Sulfonyl chlorides absorb strongly in the regions of 1410-1380 cm^{-1} (7.09-7.25 μm) and 1204-1177 cm^{-1} (8.31-8.49 μm). This increase in frequency, compared with the sulfones, results from the electronegativity of the chlorine atom.

SULFONAMIDES. Solutions of sulfonamides absorb strongly at 1370-1335 cm^{-1} (7.30-7.49 μm) and at 1170-1155 cm^{-1} (8.55-8.66 μm). In the solid phase, these frequencies are lowered by 10-20 cm^{-1}. In solid samples, the high frequency band is broadened and several submaxima usually appear.

Primary sulfonamides show strong N—H stretching bands at 3390-3330 cm^{-1} (2.95-3.00 μm) and 3300-3247 cm^{-1} (3.03-3.08 μm) in the solid state; secondary sulfonamides absorb near 3265 cm^{-1} (3.06 μm).

SULFONATES, SULFATES, AND SULFONIC ACIDS. The asymmetric (higher frequency, shorter wavelength) and symmetric S=O stretching frequency ranges for these compounds are as follows:

Class	cm^{-1}	μm
Sulfonates (covalent)	1372-1335, 1195-1168	7.29-7.49, 8.37-8.56
Sulfates (organic)	1415-1380, 1200-1185	7.06-7.24, 8.33-8.44
Sulfonic acids	1350-1342, 1165-1150	7.41-7.45, 8.58-8.69
Sulfonate salts	ca. 1175 ca. 1055	ca. 8.5 ca. 9.5

The spectrum of a typical alkyl arenesulfonate is given in Figure 34. In virtually all sulfonates, the asymmetric stretch occurs as a doublet. Alkyl and aryl sulfonates show negligible differences; electron donating groups in the para position of arenesulfonates cause higher frequency absorption.

Sulfonic acids are listed in narrow ranges above; these apply only to anhydrous forms. Such acids hydrate readily to give bands that are probably a result of the formation of hydronium sulfonate salts, in the 1230-1120 cm^{-1} (8.13-8.93 μm) range.

ORGANIC HALOGEN COMPOUNDS

The strong absorption of halogenated hydrocarbons arises from the stretching vibrations of the carbon to halogen bond.

Aliphatic C—Cl absorption is observed in the broad region between 850 and 550 cm^{-1} (11.76-18.18 μm). When several chlorine atoms are attached to one carbon atom, the band is usually more intense and at the high frequency end of the assigned limits. Carbon tetrachloride (see Appendix B, spectrum 10) shows an intense band at 797 cm^{-1} (12.55 μm). The first overtones of the intense fundamental bands are frequently observed. Spectra of typical chlorinated hydrocarbons are shown in Appendix B: nos. 10, 11, 12, and 13. Brominated compounds absorb in the 690-515 cm^{-1} (14.49-19.42 μm) region, iodo-compounds in the 600-500 cm^{-1} (16.67-20.00 μm) region. A strong CH_2 wagging band is observed for the CH_2X (X = Cl, Br, I) group in the 1300-1150 cm^{-1} (7.69-8.70 μm) region.

Fluorine-containing compounds absorb strongly over a wide range between 1400-730 cm^{-1} (7.14-13.70 μm) due to C—F stretching modes. A monofluoroalkane shows a strong band in the 1100-1000 cm^{-1} (9.09-10.00 μm) region. As the number of fluorine atoms in an aliphatic molecule increases, the band pattern becomes more complex, with multiple strong bands appearing over the broad region of C—F absorption. The CF_3 and CF_2 groups absorb strongly in the 1350-1120 cm^{-1} (7.41-8.93 μm) region. The spectrum of Fluorolube®, Appendix B, spectrum 14, illustrates many of the preceding absorption characteristics.

Chlorobenzenes absorb in the 1096-1089 cm^{-1} (9.12-9.18 μm) region. The position within this region depends on the substitution pattern. Aryl fluorides absorb in the 1250-110 cm^{-1} (8.00-9.10 μm) region of the spectrum. A monofluorinated benzene ring displays a strong, narrow absorption band near 1230 cm^{-1} (8.13 μm).

SILICON COMPOUNDS

Tables of characteristic absorptions of silicon compounds are found in Appendix E. Special features related to these tables are discussed below.

Si—H Vibrations

Vibrations for the Si—H bond are listed in Appendix E, Table I. The Si—H stretching frequencies are increased by the the attachment of an electronegative group to the silicon.

The absorption bands of methylene and methyl groups attached to silicon are summarized in Appendix D, Tables I and II.

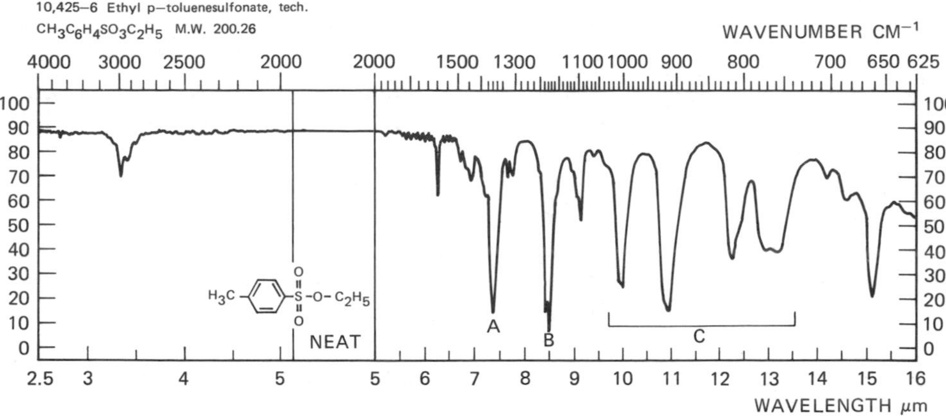

Courtesy of Aldrich Chemical Co.

Figure 34. **A.** *Asymmetric* $S(=O)_2$ *stretch, 1351 cm^{-1} (7.40 µm).* **B.** *Symmetric* $S(=O)_2$ *stretch, 1176 cm^{-1} (8.50 µm).* **C.** *Various strong* S–O–C *stretching, 1000-769 cm^{-1} (10.0-13.0 µm).*

Si–O Stretching Vibration

The Si–O stretching vibrations are summarized in Appendix E, Table II.

The OH stretching vibrations of the SiOH group absorb in the same region as the alcohols, 3700-3200 cm^{-1} (2.70-3.12 µm). As in alcohols, the absorption characteristics depend on the degree of hydrogen bonding.

Silicon-Halogen Stretching Vibrations

The absorption of groups containing Si–F bonds is summarized in Appendix E, Table III.

Bands resulting from Si–Cl stretching occur at frequencies below 666 cm^{-1} (wavelengths longer than 15 µm).

The spectrum of silicone lubricant, Appendix B-no.27, illustrates some of the preceding absorptions.

PHOSPHORUS COMPOUNDS

P–H Stretching

The P–H group absorbs in the regions described in Appendix E, Table IV.

P=O Stretching Vibrations

Such absorptions are listed in Appendix E, Table IV.

P–O Stretching Vibrations

Absorption characteristics of P–O stretching vibrations are summarized in Appendix E, Table V.

The absorptions of methylene and methyl groups attached to a phosphorus atom are summarized in Appendix D, Tables I and II.

HETEROAROMATIC COMPOUNDS

The spectra of heteroaromatic compounds result primarily from the same vibrational modes as observed for the aromatics.

C–H Stretching Vibrations

Heteroaromatics, such as pyridines, pyrazines, pyrroles, furans, and thiophenes, show C–H stretching bands in the 3077-3003 cm^{-1} (3.25-3.33 µm) region.

N–H Stretching Frequencies

Heteroaromatics containing an N–H group show N–H stretching absorption in the region of 3500-3220 cm^{-1} (2.86-3.11 µm). The position of absorption within this general region depends upon the degree of hydrogen bonding, and hence upon the physical state of the sample or the polarity of the solvent. Pyrrole and indole in dilute solution in nonpolar solvents show a sharp band near 3495 cm^{-1} (2.86 μm); concentrated solutions show a widened band near 3400 cm^{-1} (2.94 μm). Both bands may be seen at intermediate concentrations.

Ring-Stretching Vibrations (Skeletal Bands)

Ring stretching vibrations occur in the general region between 1600-1300 cm^{-1} (6.25-7.69 μm). The absorption involves stretching and contraction of all of the bonds in the ring and interaction between these stretching modes. The band pattern and the relative intensities depend on the substitution pattern and the nature of the substituents.

Pyridine (Figure 35) shows four bands in this region and, in this respect, closely resembles a monosubstituted benzene. Furans, pyrroles, and thiophenes display 2 to 4 bands in this region.

C–H Out-of-Plane Bending

The C–H out-of-plane bending (γCH) absorption pattern of the heteroaromatics is determined by the number of adjacent hydrogen atoms bending in phase. The C–H out-of-plane and ring bending (β ring) absorption of the alkylpyridines are summarized in Table VI, Appendix E.

Absorption data for the out-of-phase C–H bending (γCH) and ring bending (β ring) modes of 3 common 5-membered heteroaromatic rings are presented in Table VII, Appendix E. The ranges in Table VII (p. 177) include polar as well as nonpolar substituents on the ring.

REFERENCES

Introductory and Theoretical

1. Conley, R. T. *Infrared Spectroscopy.* 2nd ed. Boston: Allyn and Bacon, 1972.
2. Nakanishi, Koji, Solomon, P. H. *Infrared Absorption Spectroscopy–Practical.* 2nd ed. San Francisco: Holden-Day, 1977.
3. Colthup, N. B. Daly, L. H. Wiberley, S. E. *Introduction to Infrared and Raman Spectroscopy.* 2nd ed. New York and London: Academic Press, 1975.
4. Dyer, J. R. *Applications of Absorption Spectroscopy of Organic Compounds.* Englewood Cliffs, N. J.: Prentice-Hall, 1965.
5. Miller, K. G. J. and Stace, B. C. (eds.) *Laboratory Methods in Infrared Spectroscopy.* 2nd ed., New York: Heyden, 1972.
6. Herzberg, G. *Infrared and Raman Spectra of Polyatomic Molecules.* New York: D. Van Nostrand, 1945.

Compilations of Spectra or Data

7. Pouchert, C. J. (ed.) *The Aldrich Library of Infrared Spectra.* Milwaukee, Wisc.: Aldrich, 1975.
8. Bellamy, L. J. *The Infrared Spectra of Complex Molecules.* 3rd ed. London: Chapman and Hall Ltd., New York: Halsted-Wiley, 1975.
9. Bellamy, L. J. *Advances in Infrared Group Frequencies.* London: Methuen, 1968.
10. Flett, M. St. C. *Characteristic Frequencies of Chemi-*

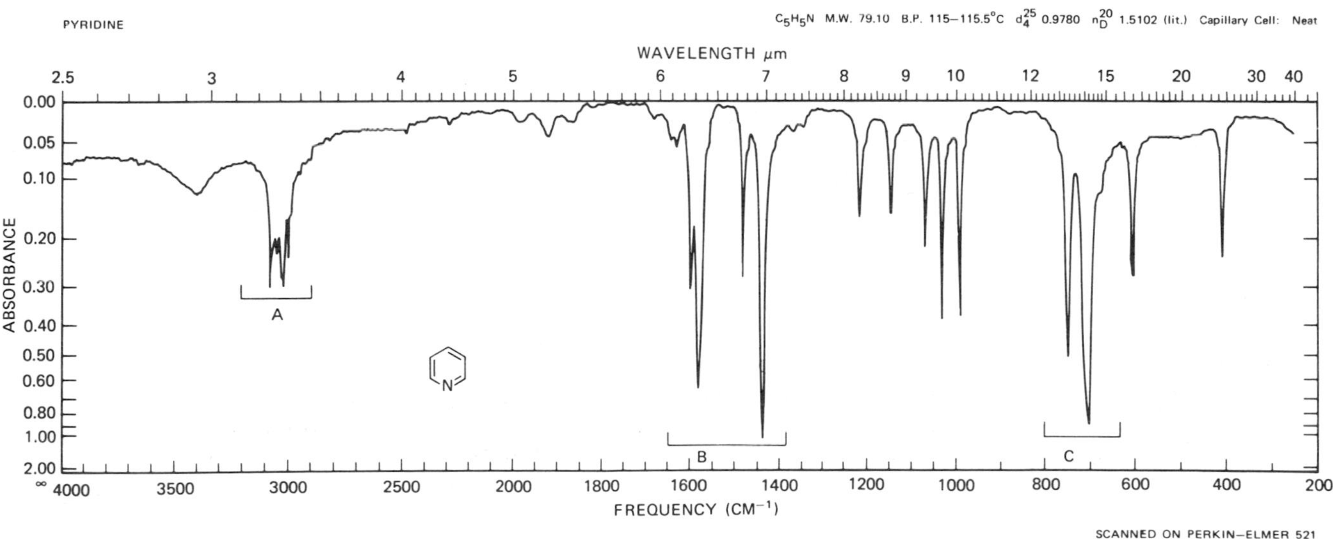

Source: The Matheson Co., Inc. E. Rutherford, N.J.

Figure 35. **A.** *Aromatic* C–H *stretch, 3080-3010 cm^{-1} (3.25-3.32 μm).* **B.** C⋯C, C⋯N *ring stretching (skeletal bands), 1600-1430 cm^{-1} (6.25-7.00 μm).* **C.** C–H *out-of-plane bending, 748, 703 cm^{-1} (13.37, 14.22 μm). See Appendix E, Table VII, for patterns in region* C *for substituted pyridines.*

cal Groups in the Infrared. New York: American Elsevier, 1969.

11. Dolphin, D. and Wick, A. E. *Tabulation of Infrared Spectral Data.* New York: Wiley, 1977.
12. *Catalog of Infrared Spectrograms,* American Petroleum Institute Research Project 44. Pittsburgh, Pa: Carnegie Institute of Techmology.
    Catalog of Infrared Spectral Data, Manufacturing Chemists Association Research Project, Chemical and Petroleum Research Laboratories. Pittsburgh, Pa.: Carnegie Institute of Technology, to June 30, 1960; College Station, Tex.: Chemical Thermodynamics Properties Center, Agriculture and Mechanical College of Texas, from July 1, 1960.
13. *ASTM-Wyandotte Index, Molecular Formula List of Compounds, Names and References to Published Infrared Spectra.* Philadelphia, Pa.: ASTM Special Technical Publications 131 (1962) and 131-A (1963). Lists about 57,000 compounds. Covers infrared, near infrared, and far infrared spectra.
14. Hershenson, H. H. *Infrared Absorption Spectra Index.* New York and London: Academic Press, 1959, 1964. Two volumes cover 1945-1962.
15. *Documentation of Molecular Spectroscopy (DMS),* London: Butterworths Scientific Publications, and Weinheim/Bergstrasse, West Germany: Verlag Chemie GMBH, in cooperation with the infrared Absorption Data Joint Committee, London, and the Institut für Spectrochemie und Angewandte Spectroskopie, Dortmund. Spectra are presented on coded cards. Coded cards containing abstracts of articles relating to infrared spectrometry are also issued.
16. IRDC Cards: *Infrared Data Committee of Japan* (S. Mizushima). Haruki-cho, Tokyo: Handled by Nankodo Co.
17. *The NRC-NBS (Creitz) File of Spectrograms,* issued by National Research Council-National Bureau of Standards Committee on Spectral Absorption Data, National Bureau of Standards, Washington 25, D. C. Spectra presented on edge-punched cards.
18. Bentley, F. F., L. D. Smithson, A. L. Rozek, *Infrared Spectra and Characteristic Frequencies ~ 700 to 300 cm^{-1}, A Collection of Spectra, Interpretation and Bibliography.* New York: Wiley Interscience, 1968.
19. Ministry of Aviation Technical Information and Library Services (ed.) *An Index of Published Infra-Red Spectra.* vol. 1 and 2. London: Her Majesty's Stationery Office, 1960.
20. Brown, C. R., Ayton, M. W., Goodwin, T. C., Derby, T. J. *Infrared–A Bibliography.* Washington, D. C.: Library of Congress, Technical Information Division, 1954.
21. Dobriner, K., Katzenellenbogen, E. R., Jones, R. N. *Infrared Absorption Spectra of Steroids–An Atlas.* vol. 1. New York: Wiley-Interscience, 1953.
22. Szymanski, H. A. *Correlation of Infrared and Raman Spectra of Organic Compounds.* Hertillon Press, 1969.
23. Szymanski, H. A. *IR Theory and Practice of Infrared Spectroscopy.* New York: Plenum Press, 1964.
    Szymanski, H. A. *Interpreted Infrared Spectra.* vol. I, II, III. New York: Plenum Press, 1964-1967.
24. Szymanski, H. A. *Infrared Band Handbook.* New York: Plenum, 1963.
    Infrared Band Handbook. Supplements 1 and 2. New York: Plenum Press, 1964, 259 pp.
    Infrared Band Handbook. Supplements 3 and 4. New York: Plenum Press, 1965.
    Infrared Band Handbook. 2nd rev. ed. New York: IFI/Plenum, 1970, 2 vol.
25. *Catalog of Infrared Spectrograms.* Philadelphia: Sadtler Research Laboratories, Pa. 19104. Spectra are indexed by name and by the Spec-Finder. The latter is an index which tabulates major bands by wavelength intervals. This allows quick identification of unknown compounds.
    Sadtler Standard Infrared Grating Spectra. Philadelphia: Sadtler Research Laboratories, Inc., 1972. 26,000 spectra in 26 volumes.
    Sadtler Standard Infrared Prism Spectra. Philadelphia: Sadtler Research Laboratories, Inc., 1972. 43,000 spectra in 43 volumes.
    Sadtler Reference Spectra–Commonly Abused Drugs IR & UV Spectra. Philadelphia: Sadtler Research Laboratories, Inc., 1972. 600 IR Spectra, 300 UV.
    Sadtler Reference Spectra–Gases & Vapors High Resolution Infrared. Philadelphia: Sadtler Research Laboratories, Inc., 1972. 150 Spectra.
    Sadtler Reference Spectra–Inorganics IR Grating. Philadelphia: Sadtler Research Laboratories, Inc., 1967. 1300 spectra.
    Sadtler Reference Spectra–Organometallics IR Grating. Philadelphia: Sadtler Research Laboratories, Inc., 1966. 400 spectra.

Special Applications

26. Maslowsky, E. *Vibrational Spectra of Organometallic Compounds.* New York: Wiley-Interscience, 1977.
27. Katritzky, A. R. (ed.) *Physical Methods in Heterocyclic Chemistry.* vol. II. New York and London: Academic Press, 1963.
28. Szymanski, H. A. (ed.) *Raman Spectroscopy: Theory and Practice.* vols. 1, 2. New York: Plenum, 1969, 1970.

29. Loader, E. J. *Basic Laser Raman Spectroscopy.* London, New York: Heyden-Sadtler, 1970.

30. Finch, A., Gates, P. N., Radcliffe, K., Dickson, F. N., Bentley, F. F. *Chemical Applications of Far Infrared Spectroscopy.* New York: Academic Press, 1970.

31. Stewart, J. E. "Far Infrared Spectroscopy" in *Interpretive Spectroscopy.* Freeman, S. K. (ed.) New York: Reinhold, 1965, p. 131.

32. Tichy, M. "The Determination of Intramolecular Hydrogen Bonding by Infrared Spectroscopy and Its Applications in Stereochemistry," in *Advances in Organic Chemistry: Methods and Results.* vol. 5. R. R. Raphael (ed.) New York: Wiley-Interscience, 1964.

PROBLEMS

For problems 3.1-3.5, match the name from each list to the proper infrared spectrum. Identify the diagnostic bands in each spectrum. These spectra are on pp. 138-148.

3.1 Spectra A-D
1,3-cyclohexadiene
diphenylacetylene
1-octene
2-pentene

3.2 Spectra E-I
n-butyl acetate
butyramide
isobutylamine
lauric acid
sodium propionate

3.3 Spectra J-M
allyl phenyl ether
benzaldehyde
o-cresol
m-toluic acid

3.4 Spectra N-R
aniline
azobenzene
benzophenone oxime
benzylamine
dimethylamine hydrochloride

3.5 Spectra S-U
1,5-dichloropentane
epichlorohydrin
4-hydroxy-4-methyl-2-pentanone

PROB. 3.1 SPECTRUM A

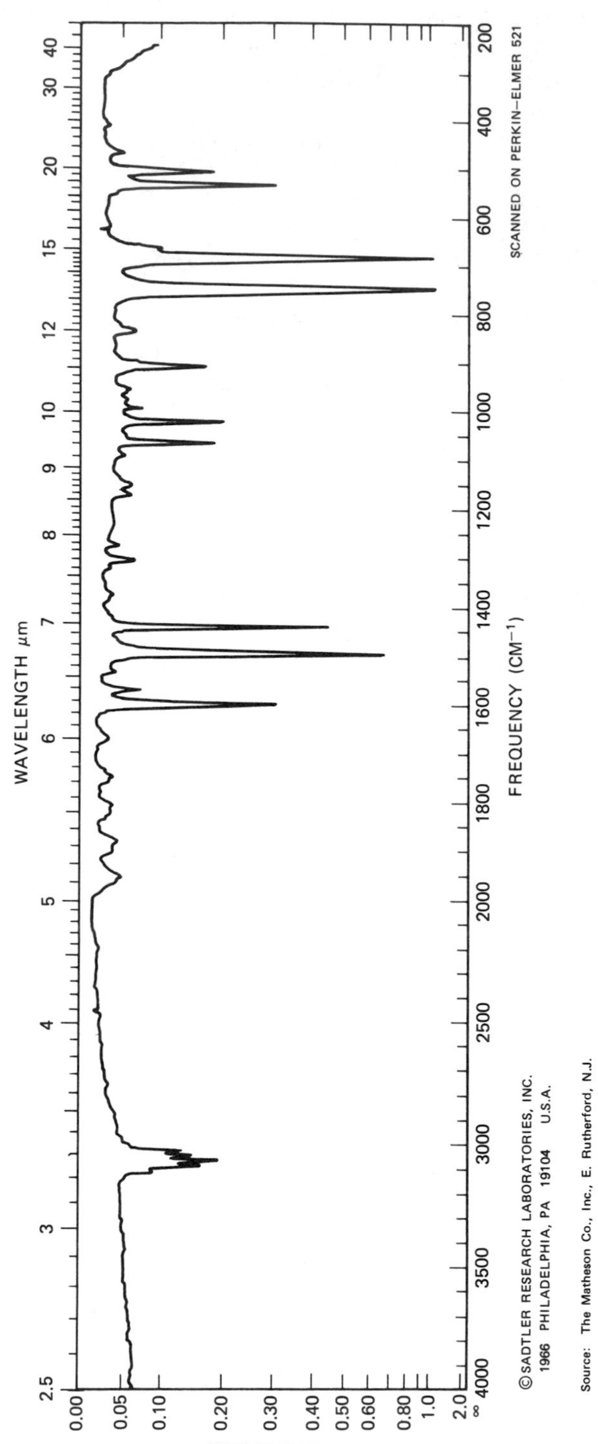
WAVELENGTH μm
FREQUENCY (CM^{-1})
ABSORBANCE
SCANNED ON PERKIN–ELMER 521
©SADTLER RESEARCH LABORATORIES, INC.
1966 PHILADELPHIA, PA 19104 U.S.A.
Source: The Matheson Co., Inc., E. Rutherford, N.J.

PROB. 3.1 SPECTRUM B

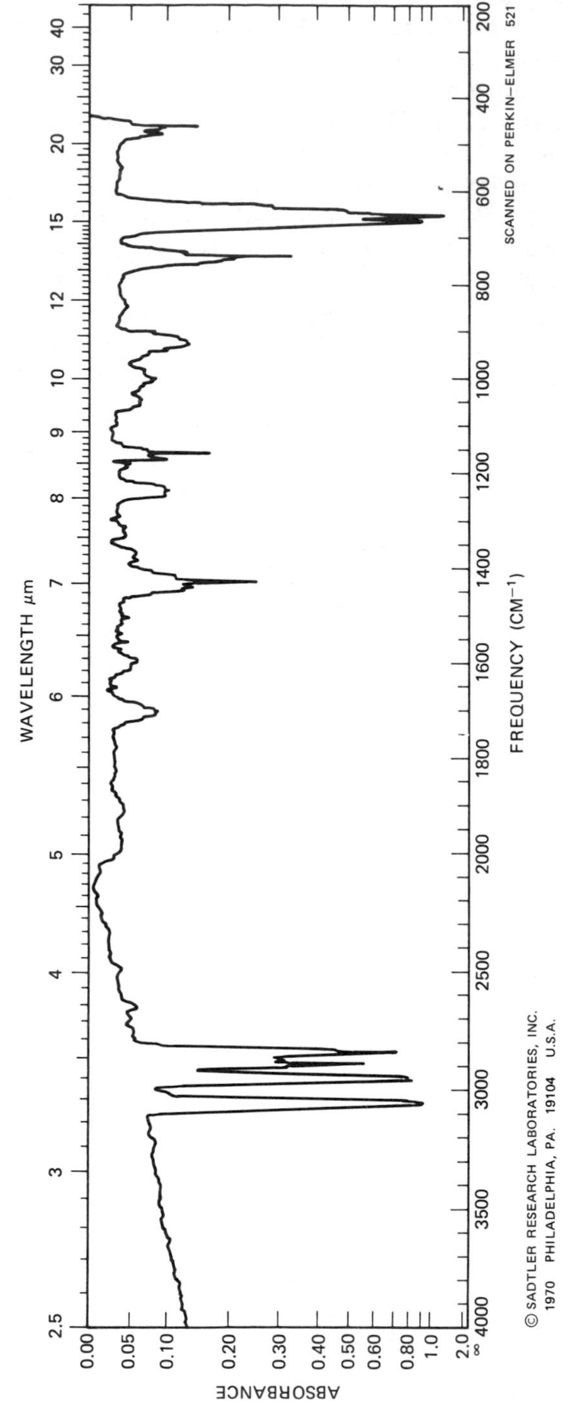
WAVELENGTH μm
FREQUENCY (CM^{-1})
ABSORBANCE
SCANNED ON PERKIN–ELMER 521
© SADTLER RESEARCH LABORATORIES, INC.
1970 PHILADELPHIA, PA. 19104 U.S.A.
Source: Chemical Samples Company, Columbus, Ohio

PROB. 3.1 SPECTRUM C

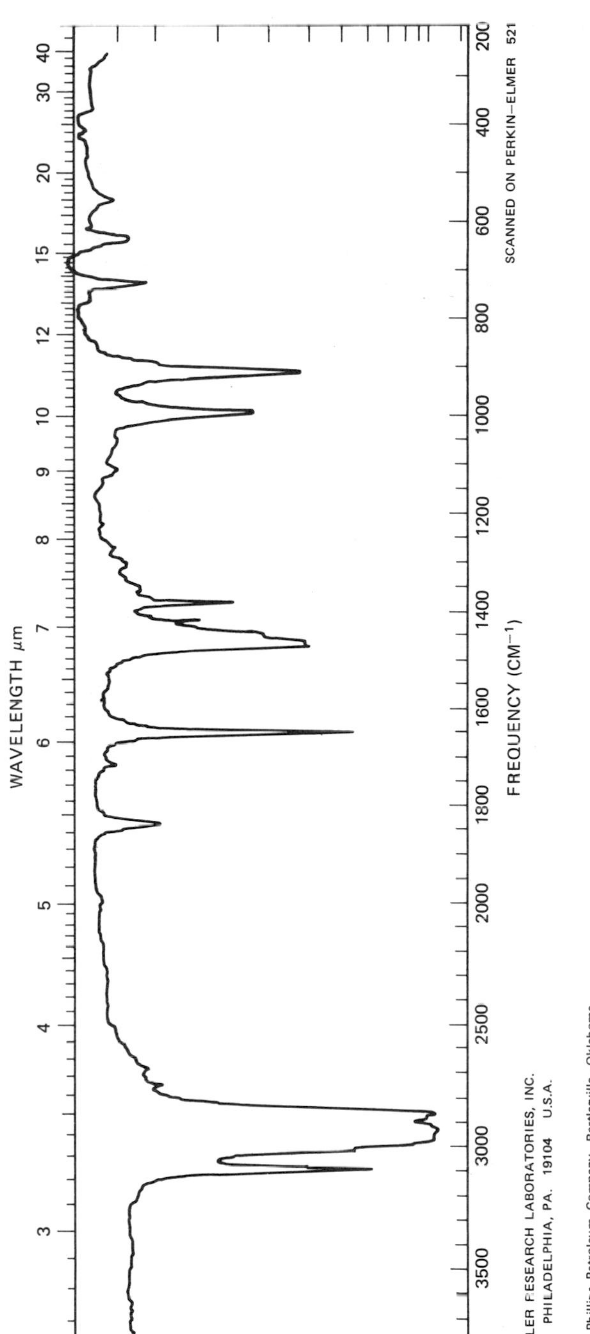

WAVELENGTH μm
FREQUENCY (CM⁻¹)
ABSORBANCE
SCANNED ON PERKIN-ELMER 521
© SADTLER RESEARCH LABORATORIES, INC.
1967 PHILADELPHIA, PA. 19104 U.S.A.
Source: Phillips Petroleum Company, Bartlesville, Oklahoma

PROB. 3.1 SPECTRUM D

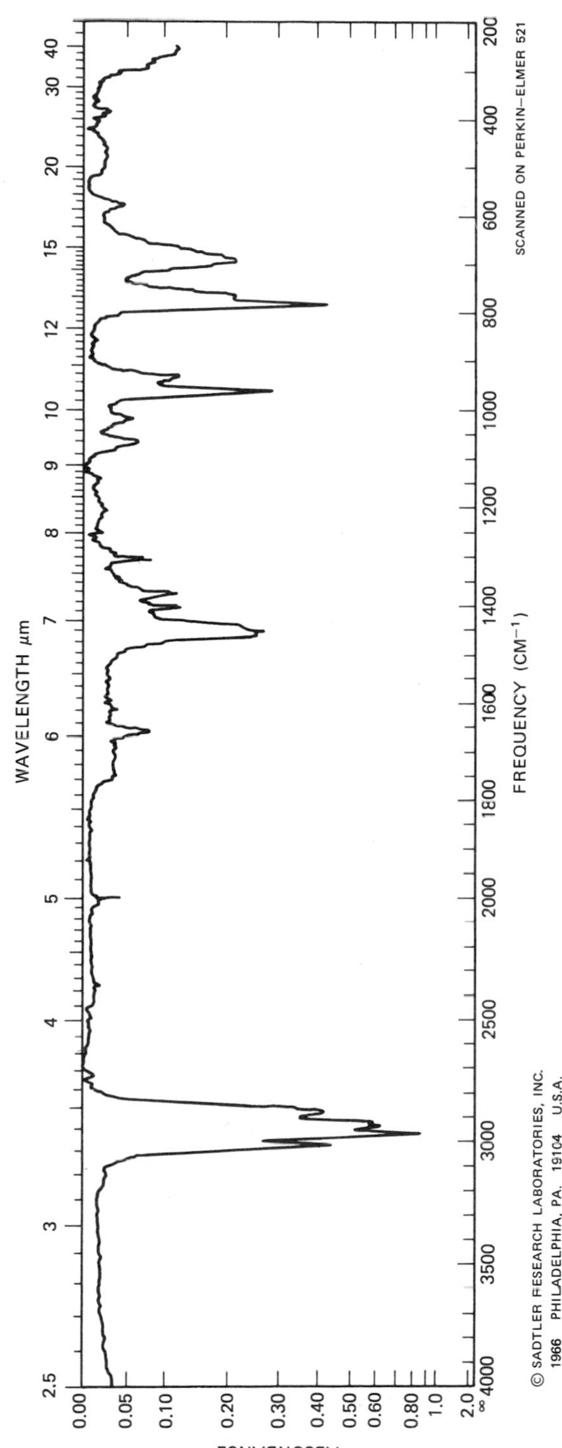

WAVELENGTH μm
FREQUENCY (CM⁻¹)
ABSORBANCE
SCANNED ON PERKIN-ELMER 521
© SADTLER RESEARCH LABORATORIES, INC.
1966 PHILADELPHIA, PA. 19104 U.S.A.
Source: The Matheson Co., Inc., E. Rutherford, N.J.

PROB. 3.2 SPECTRUM E

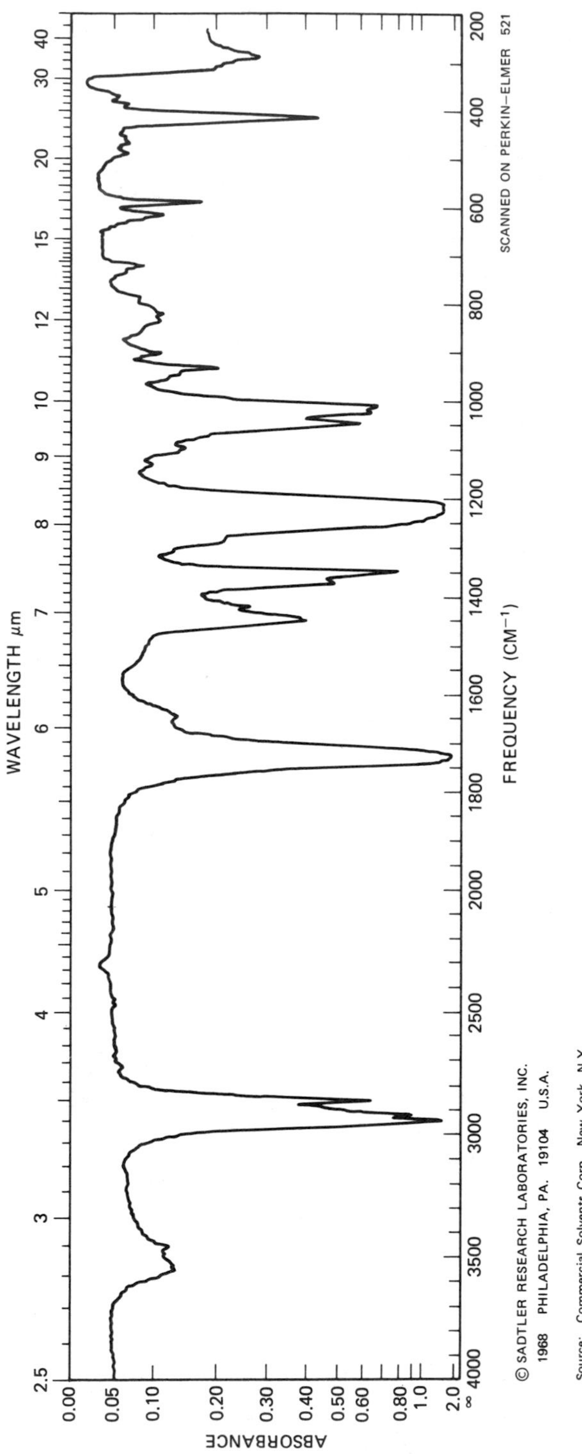

PROB. 3.2 SPECTRUM F

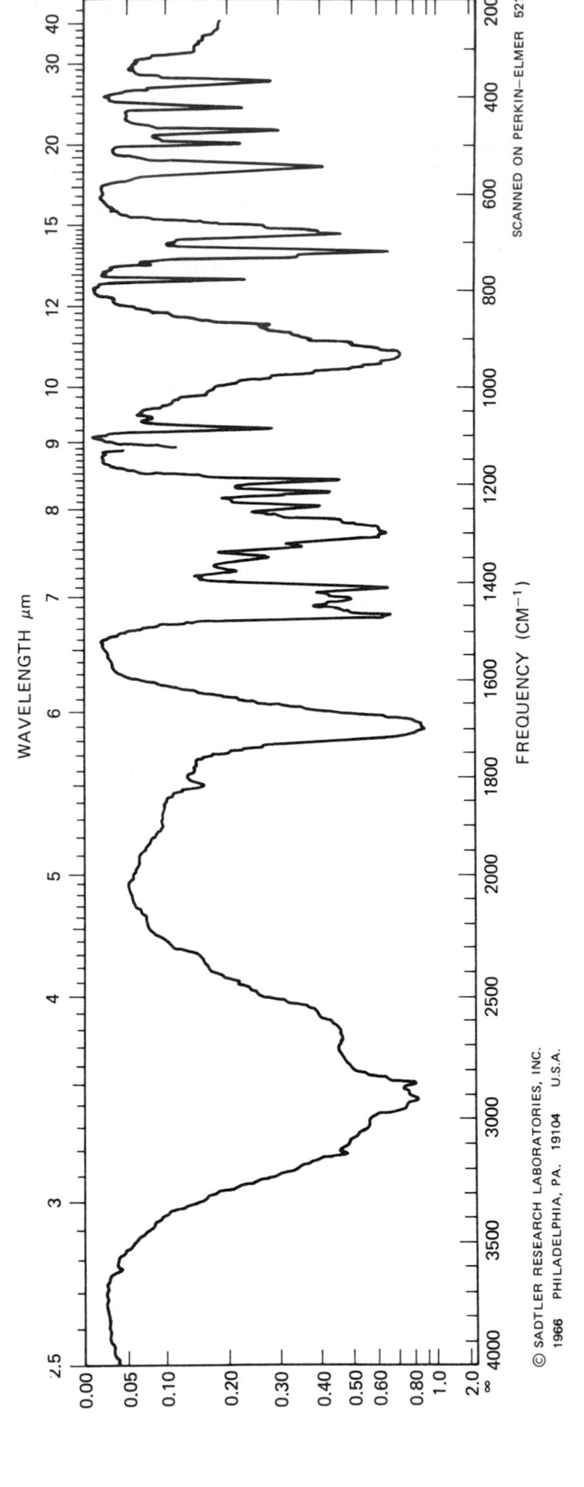

PROB. 3.2 SPECTRUM G

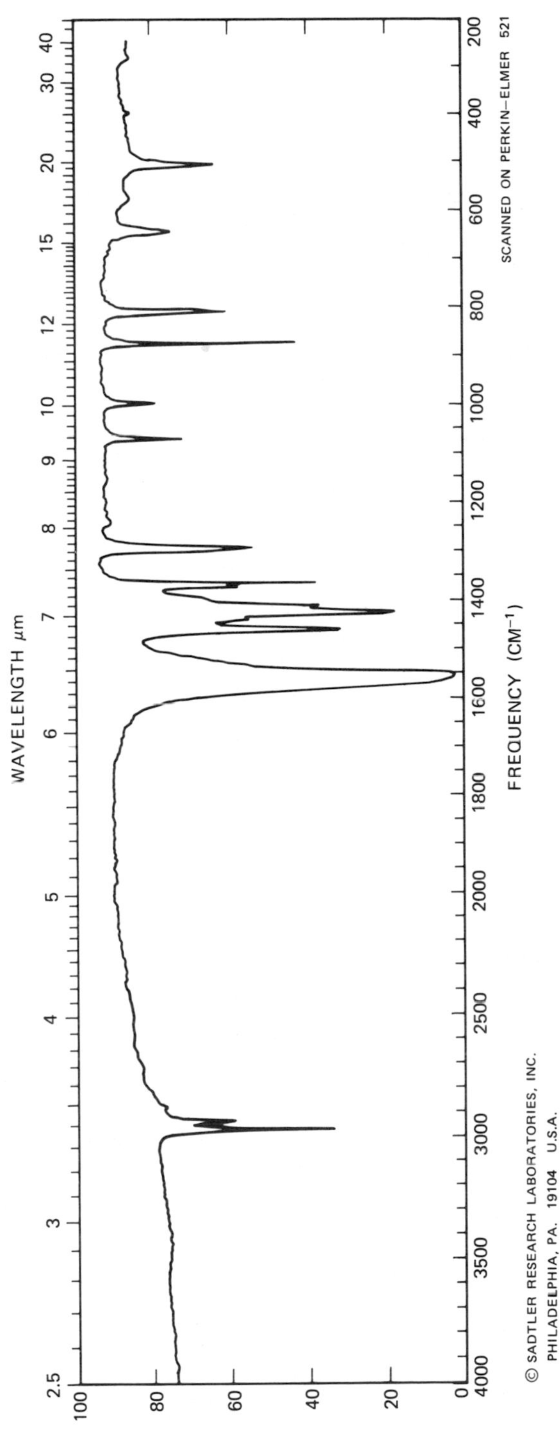

PROB. 3.3 SPECTRUM H

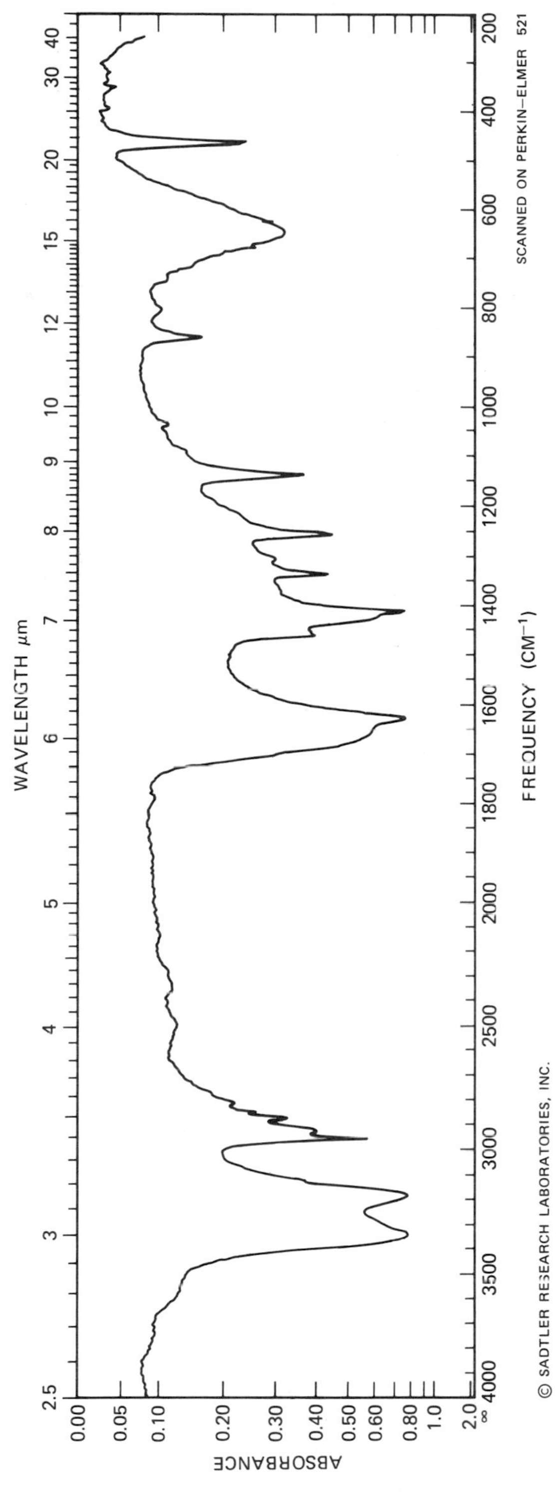

PROB. 3.2 SPECTRUM I

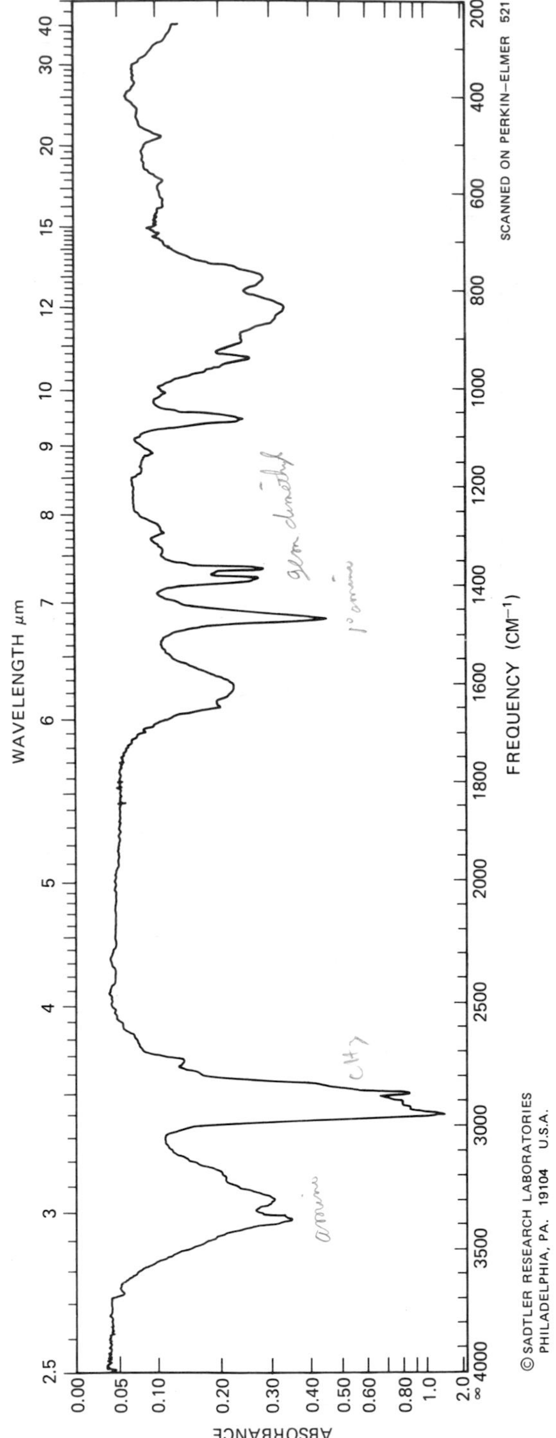

PROB. 3.3 SPECTRUM J

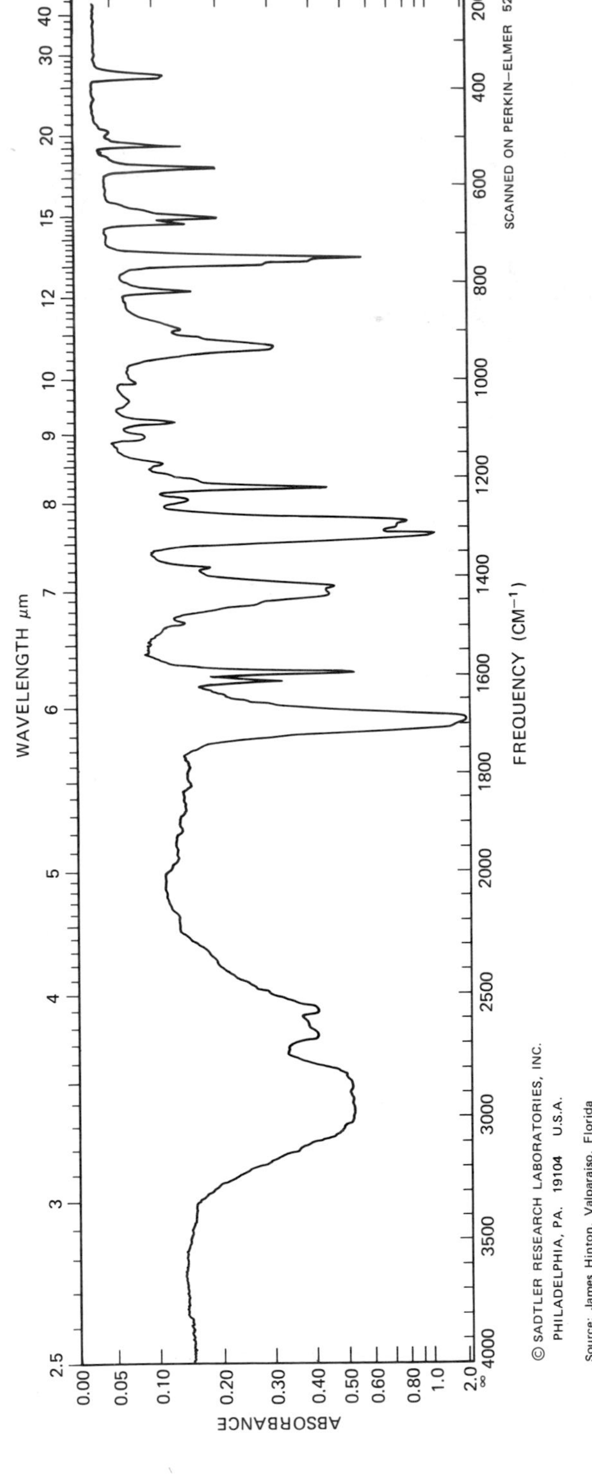

PROB. 3.3 SPECTRUM K

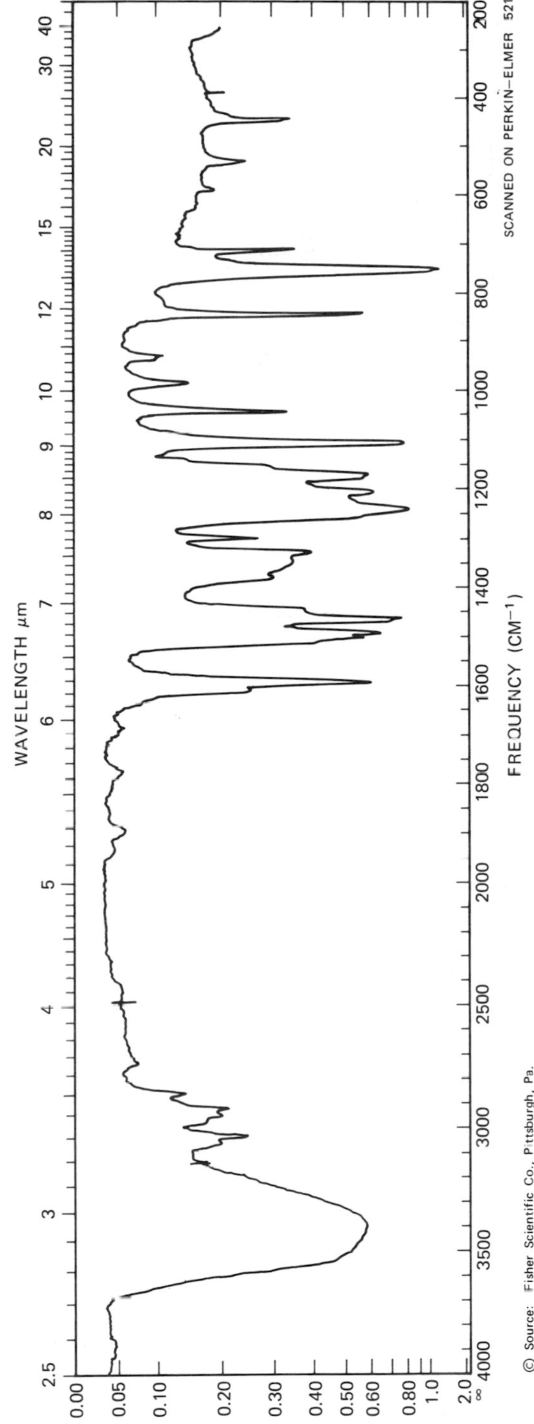

PROB. 3.3 SPECTRUM L

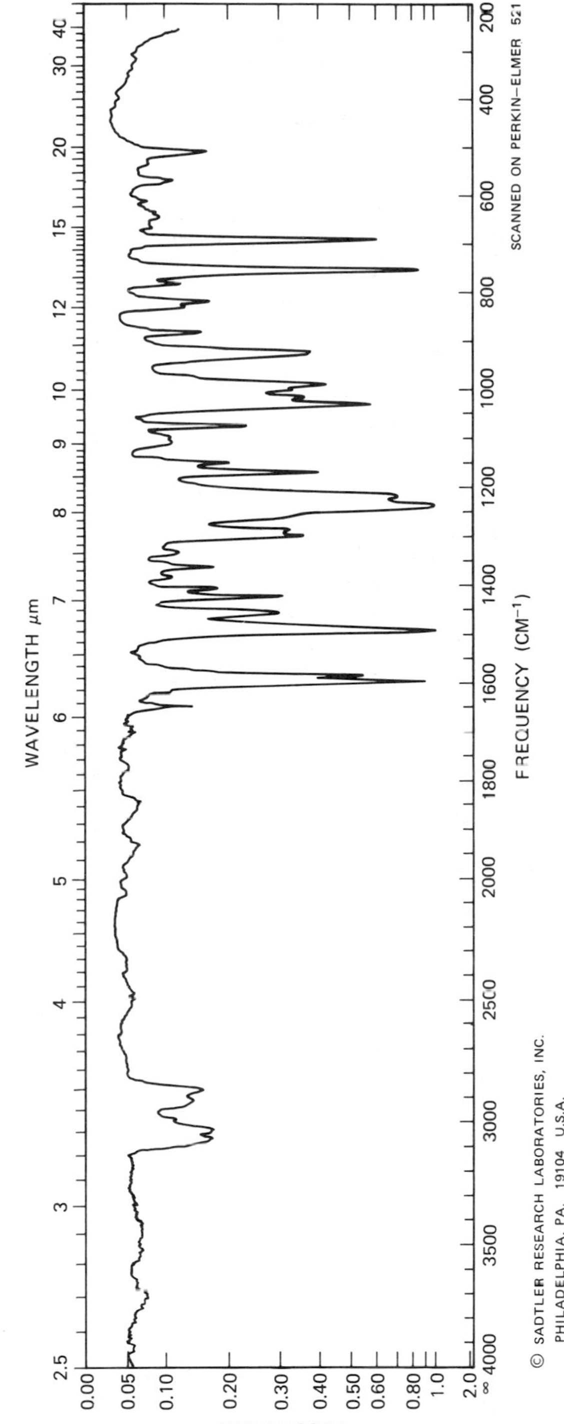

PROB. 3.3 SPECTRUM M

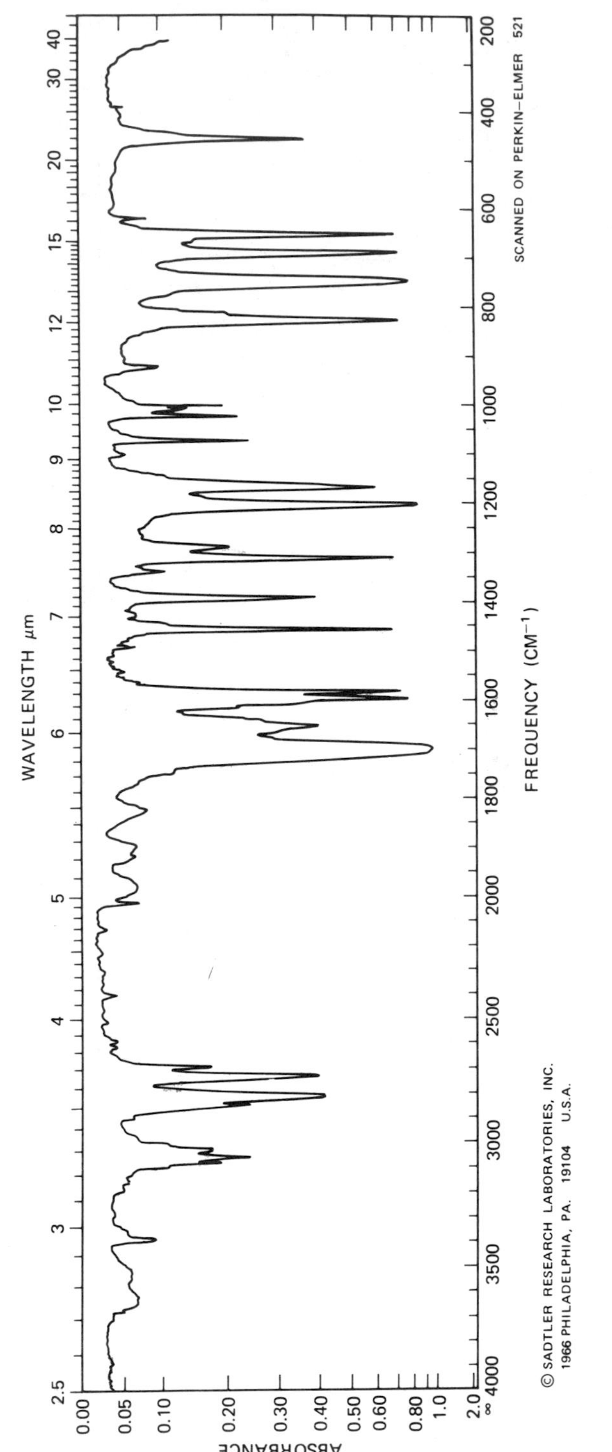

PROB. 3.4 SPECTRUM N

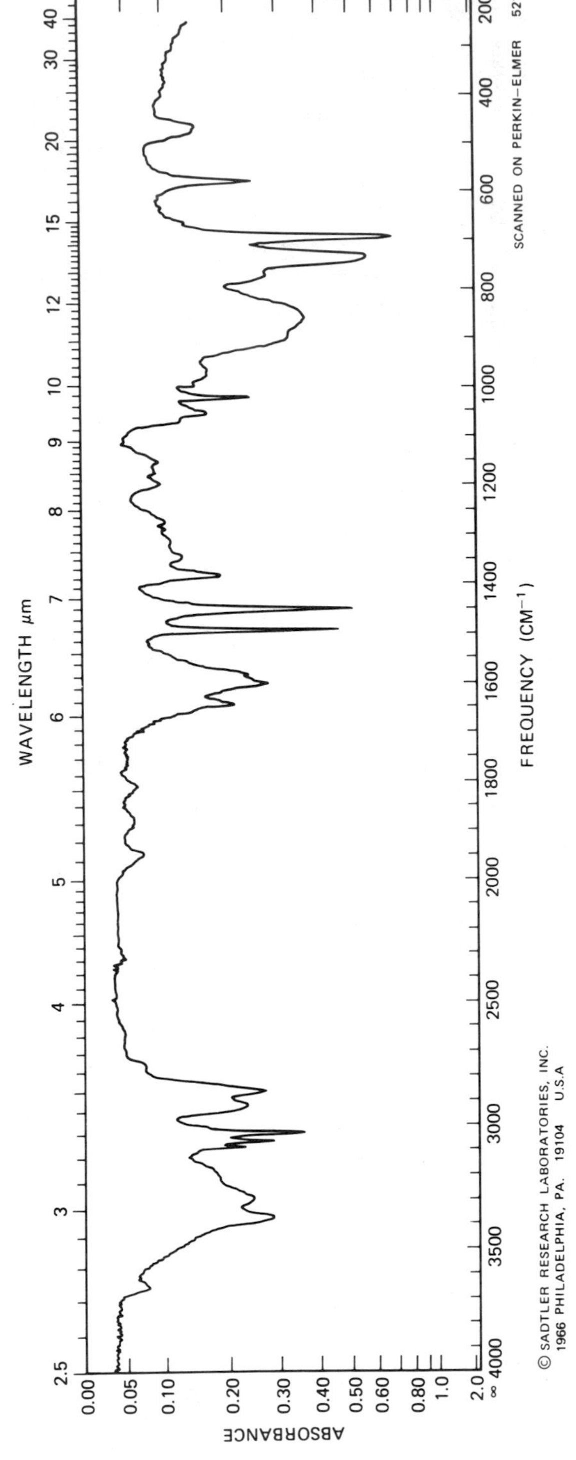

PROB. 3.4 SPECTRUM O

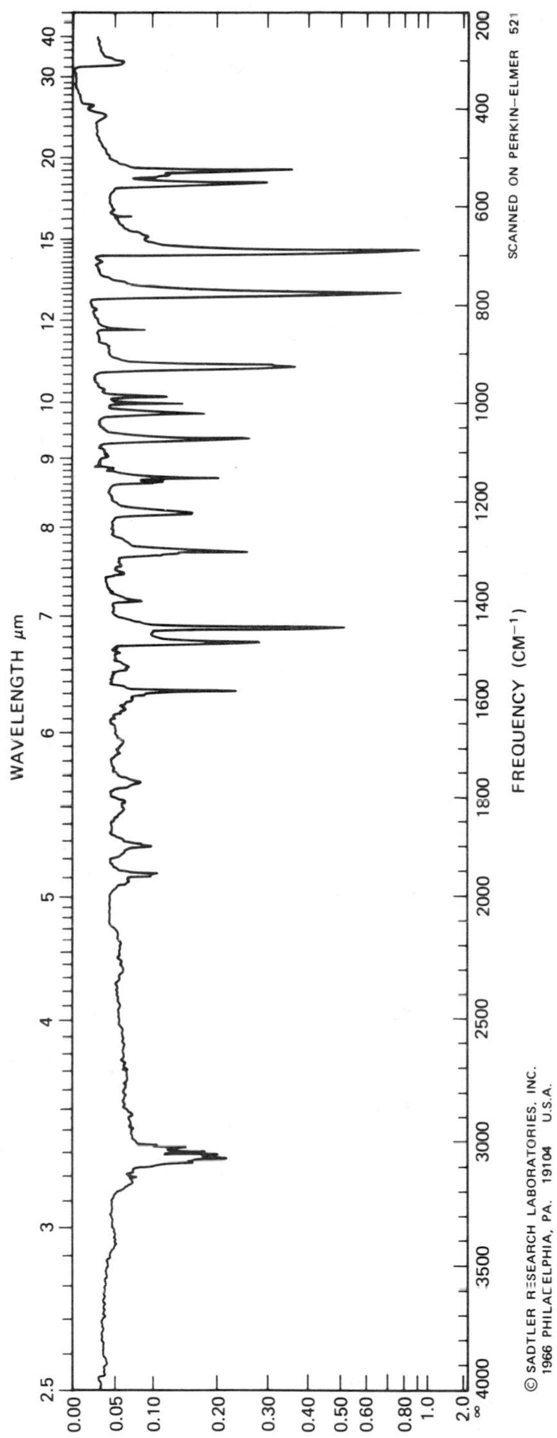

PROB. 3.4 SPECTRUM P

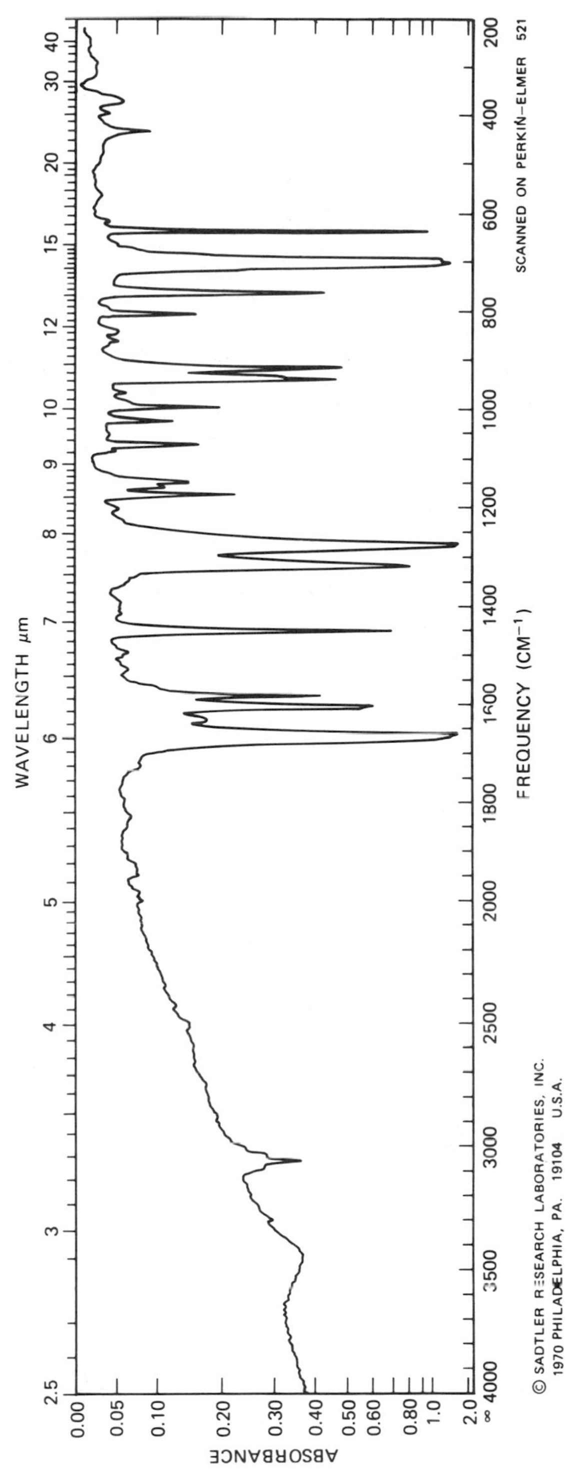

PROB. 3.4 SPECTRUM Q

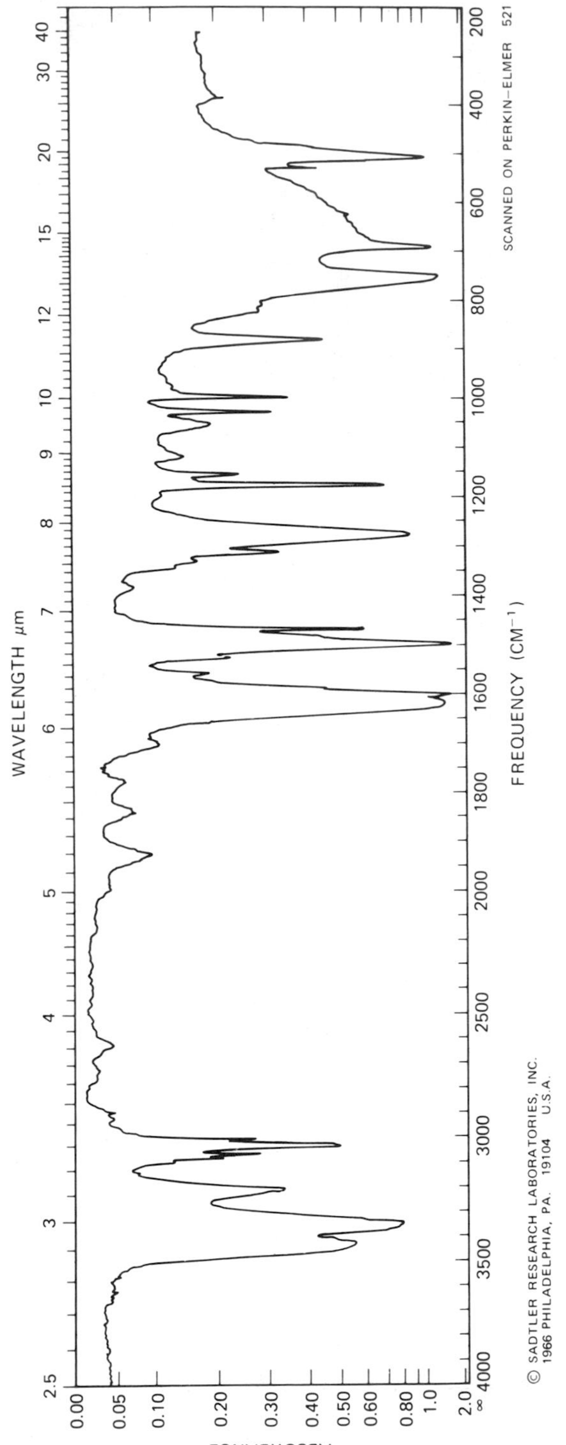

PROB. 3.4 SPECTRUM R

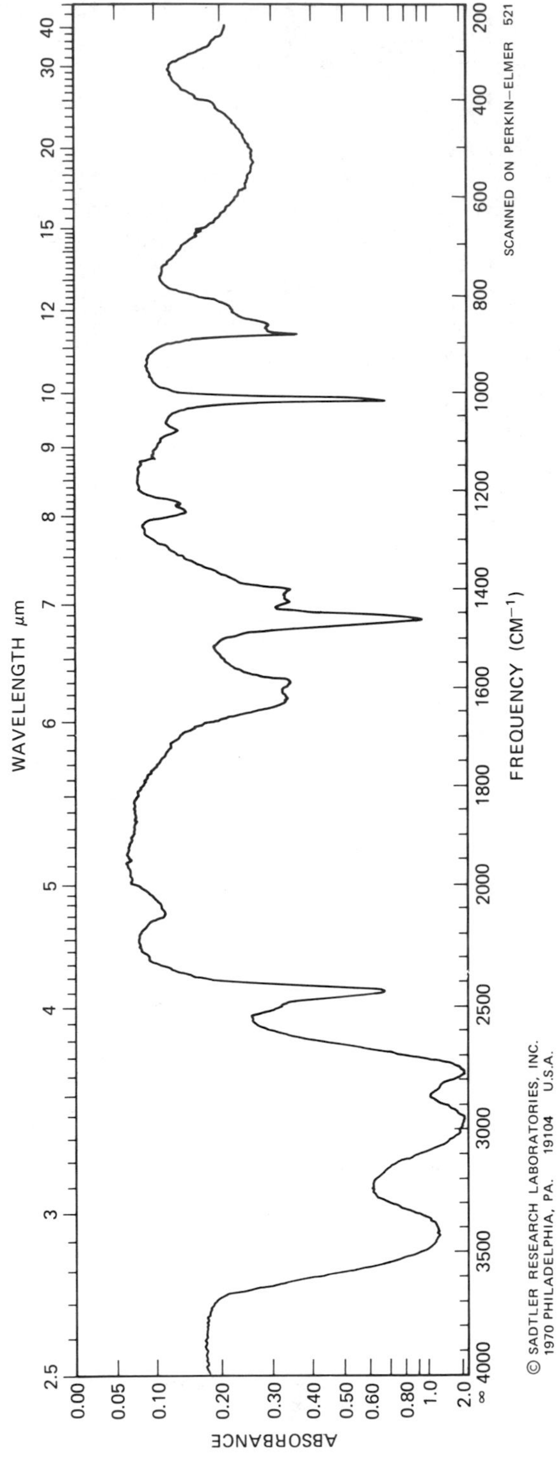

PROB. 3.5 SPECTRUM S

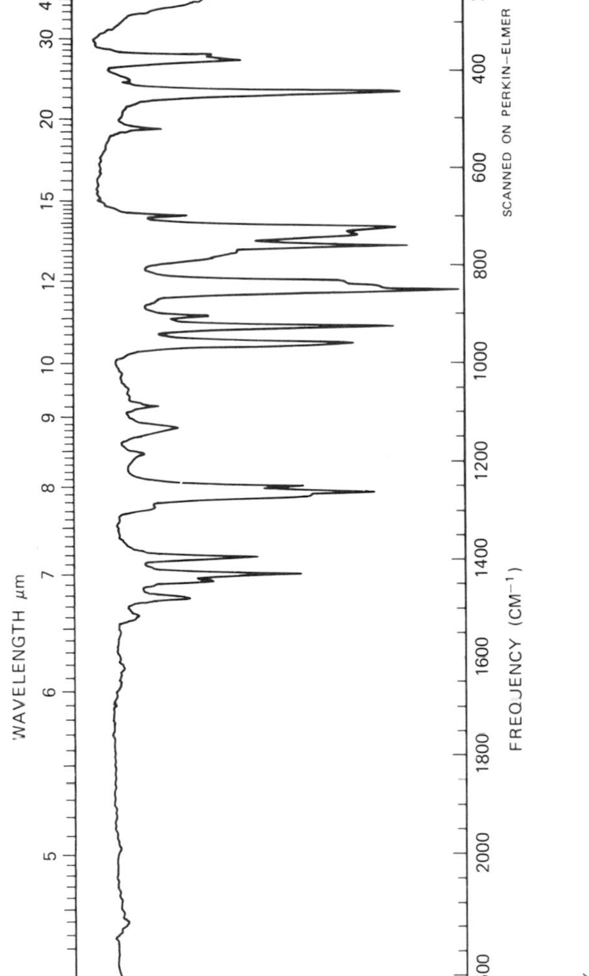

PROB. 3.5 SPECTRUM T

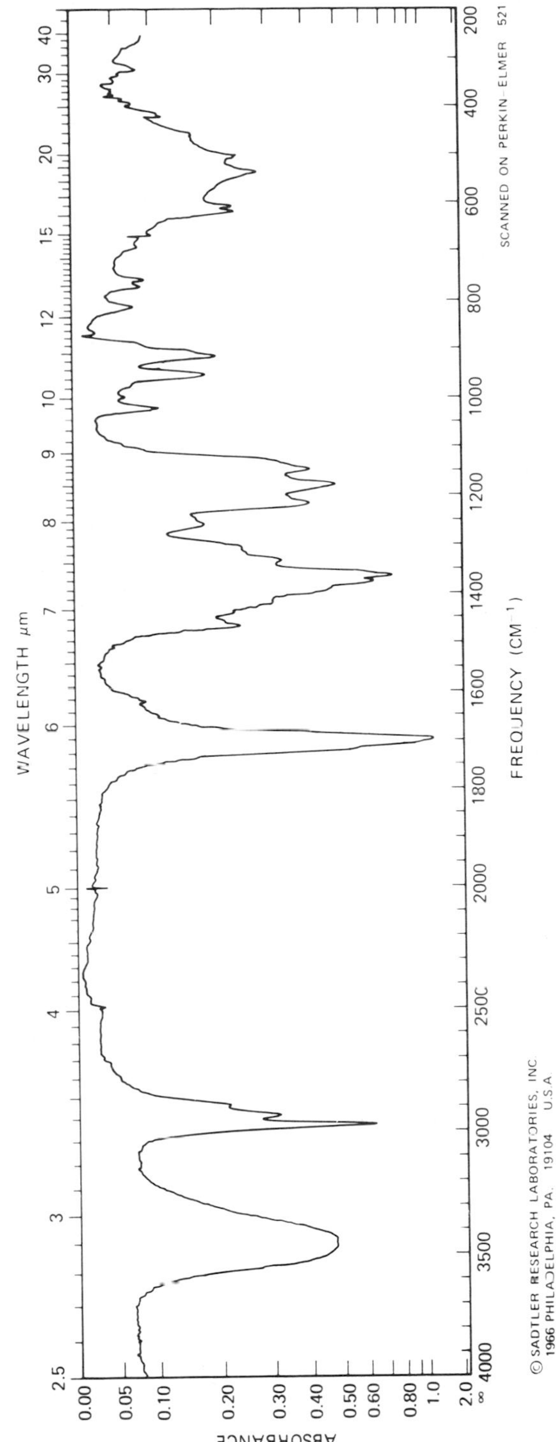

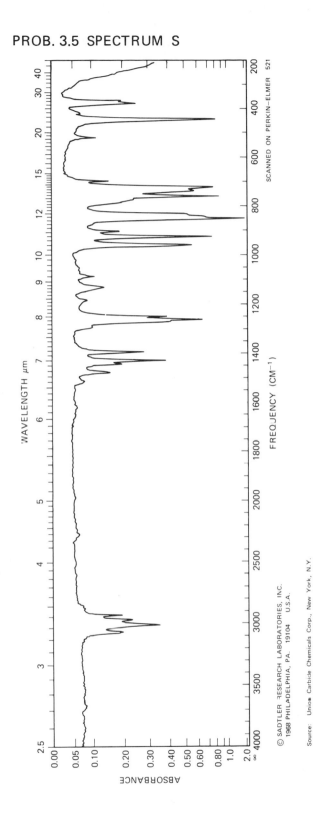

PROB. 3.5 SPECTRUM U

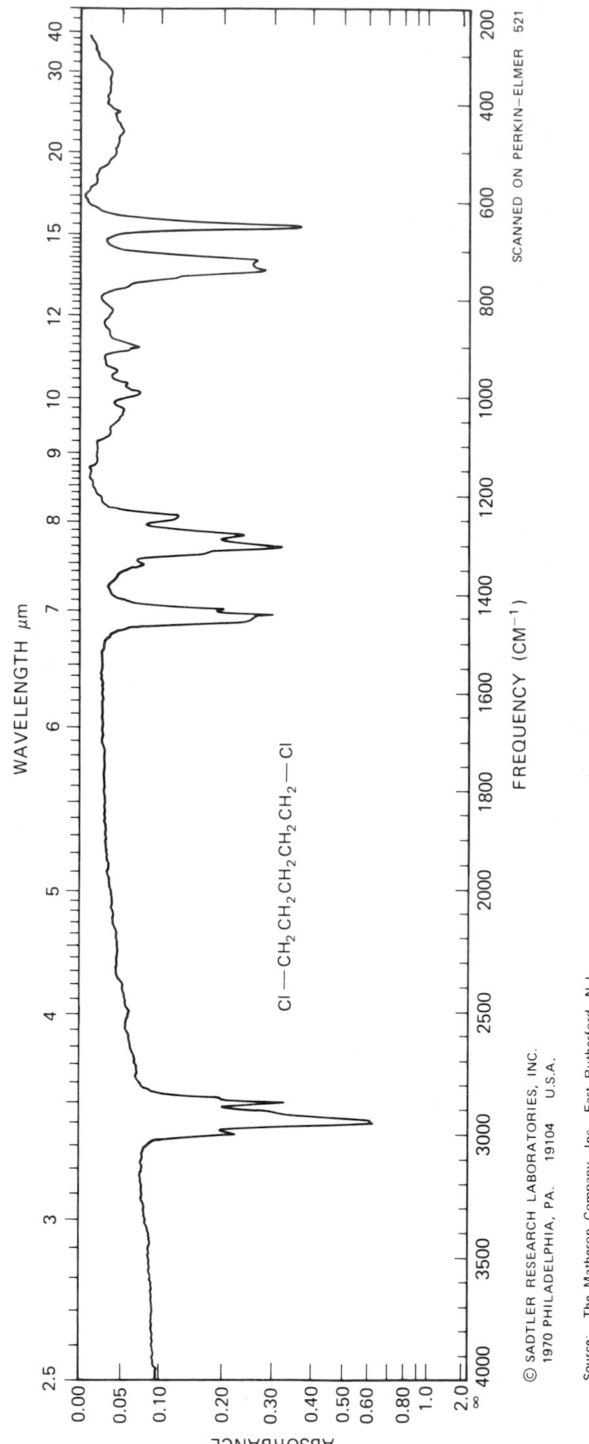

3.6 Deduce the structure of a compound whose formula is C_2H_3NS from the following spectrum.

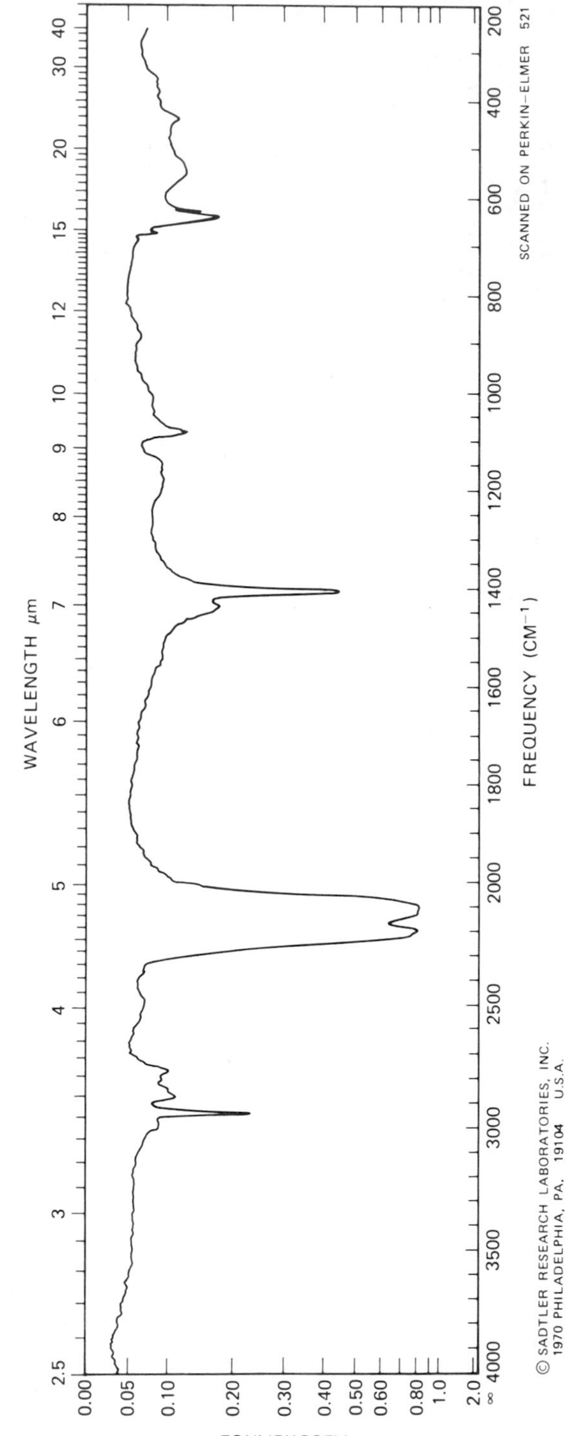

3.7 Point out evidence for enol formation in the infrared spectrum of 2,4-pentanedione (The band at 3400 cm^{-1} is due to water.

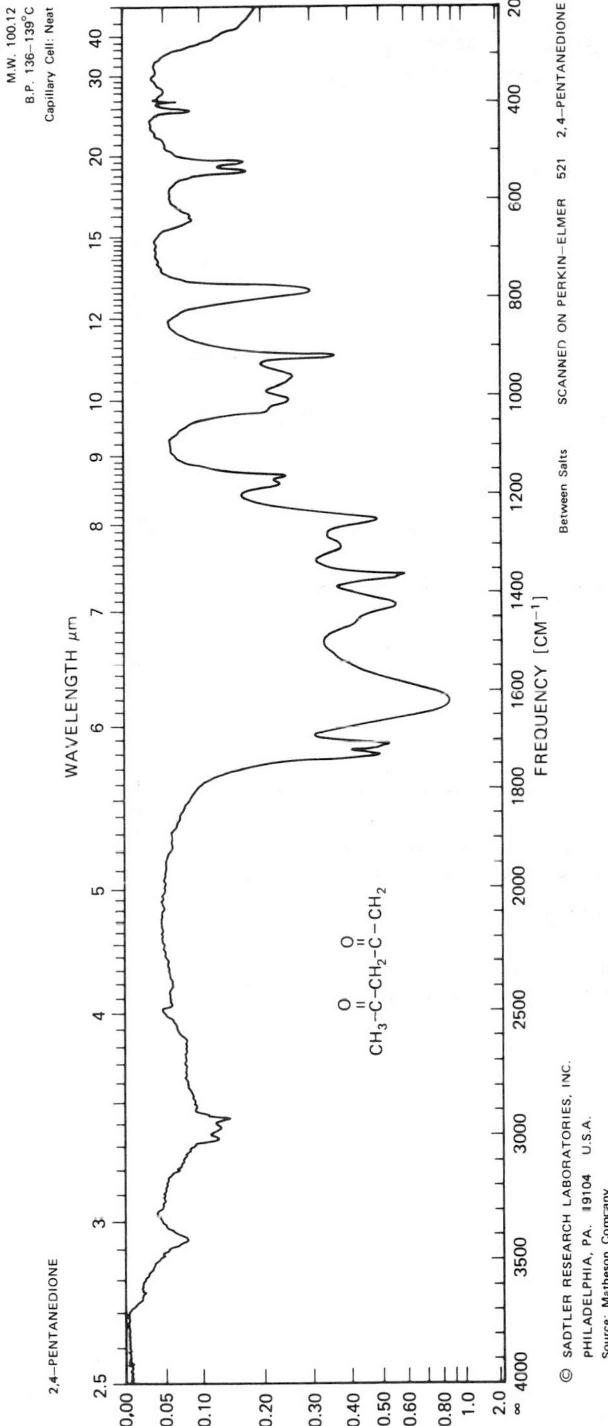

appendix a

CHART AND SPECTRAL PRESENTATIONS OF ORGANIC SOLVENTS, MULLING OILS, AND OTHER COMMON LABORATORY SUBSTANCES

Transparent Regions of Solvents and Mulling Oils.

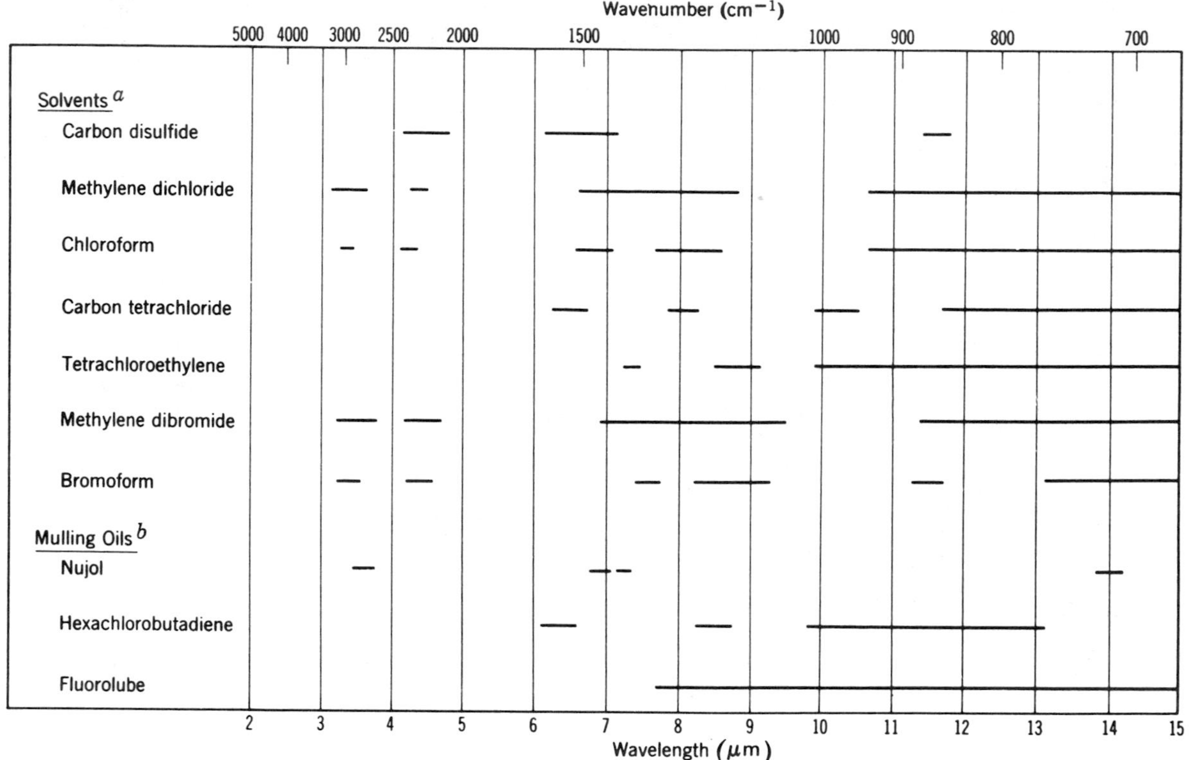

Transparent regions of solvents and mulling oils. [a] The open regions are those in which the solvent transmits more than 25% of the incident light at 1 mm thickness. *b* The open regions for mulling oils indicate transparency of thin films.

appendix b

SPECTRA OF COMMON LABORATORY SUBSTANCES*

Alphabetical Listing of Spectra Shown on Succeeding Pages (pp. 152-165).

	Spectrum No.		Spectrum No.
Acetone	17	Ethyl ether	22
Benzene	4	Fluorolube®	14
2-Butanone	18	Hexane	1
Carbon disulfide	25	Indene	8
Carbon tetrachloride	10	Methanol	15
Chloroform	9	Nujol®	2
Cyclohexane	3	Petroleum ether	24
Cyclohexanone	19	Phthalic acid, diethyl ester	21
1,2-Dichloroethane	11	Polystyrene	7
N,N-Dimethylformamide (DMF)	28	Silicone lubricant	27
Dimethyl sulfoxide (DMSO)	26	Tetrachloroethylene	13
p-Dioxane	23	Toluene	5
Ethyl acetate	20	Trichloroethylene	12
Ethyl alcohol	16	*m*-Xylene	6

Spectra courtesy of Sadtler Laboratories and Aldrich Chemical Co.

*Carbon dioxide (Sadtler Prism Spectrum No. 1924) has bands in the 3700-3550 cm^{-1} (2.70-2.82 μm) and 2380-2222 cm^{-1} (4.20-4.50 μm) regions and at *ca.* 720 cm^{-1} (13.9 μm).

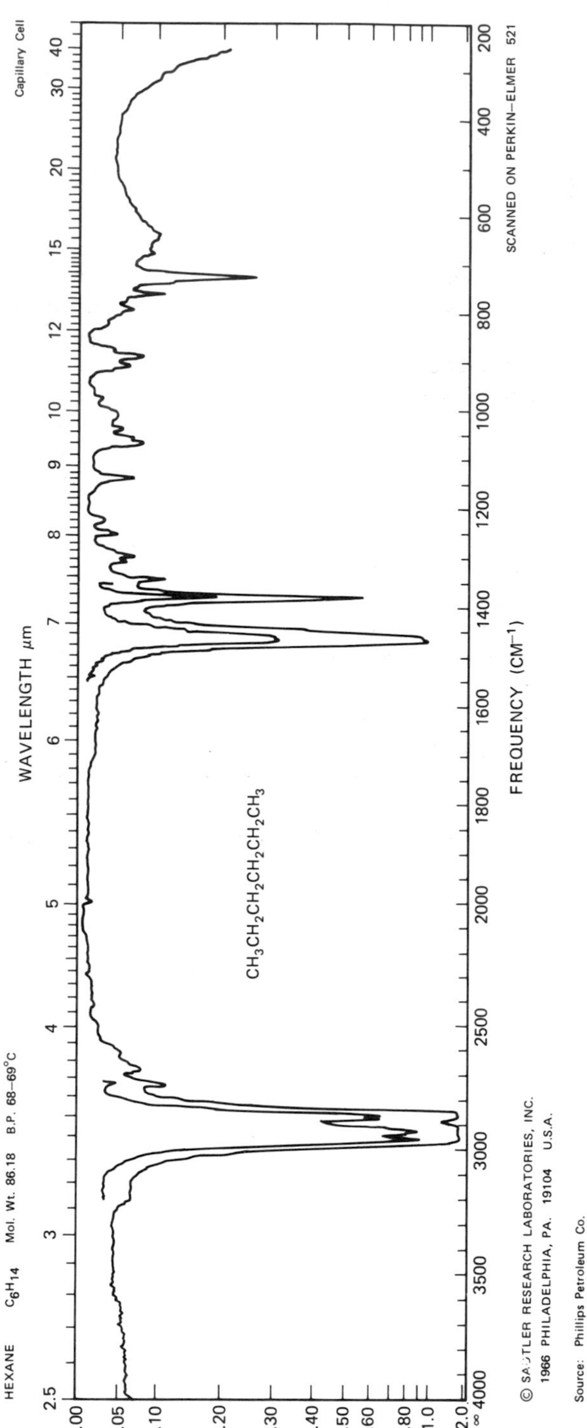
HEXANE
C_6H_{14}
Mol. Wt. 86.18
B.P. 68–69°C
Capillary Cell
WAVELENGTH μm
$CH_3CH_2CH_2CH_2CH_2CH_3$
FREQUENCY (CM^{-1})
SCANNED ON PERKIN-ELMER 521
ABSORBANCE
© SADTLER RESEARCH LABORATORIES, INC.
1966 PHILADELPHIA, PA. 19104 U.S.A.
Source: Phillips Petroleum Co.

No. 1

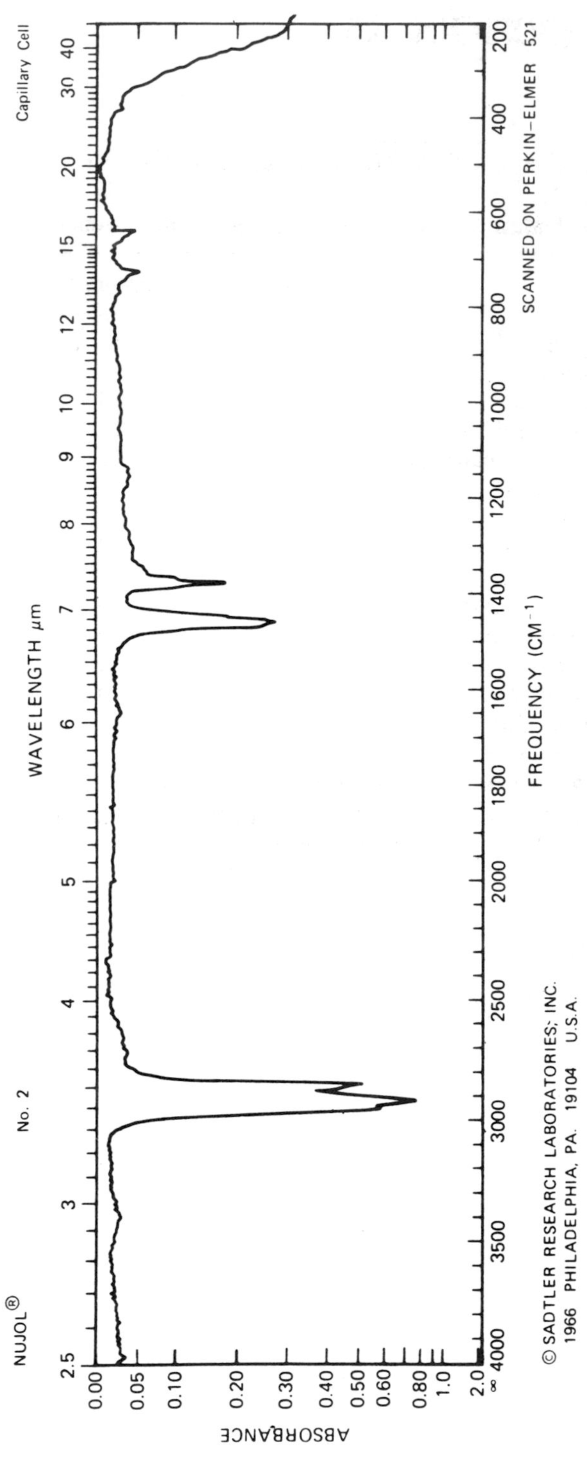
NUJOL®
No. 2
Capillary Cell
WAVELENGTH μm
FREQUENCY (CM^{-1})
SCANNED ON PERKIN-ELMER 521
ABSORBANCE
© SADTLER RESEARCH LABORATORIES, INC.
1966 PHILADELPHIA, PA. 19104 U.S.A.
Source: Plough, Inc.

No. 2

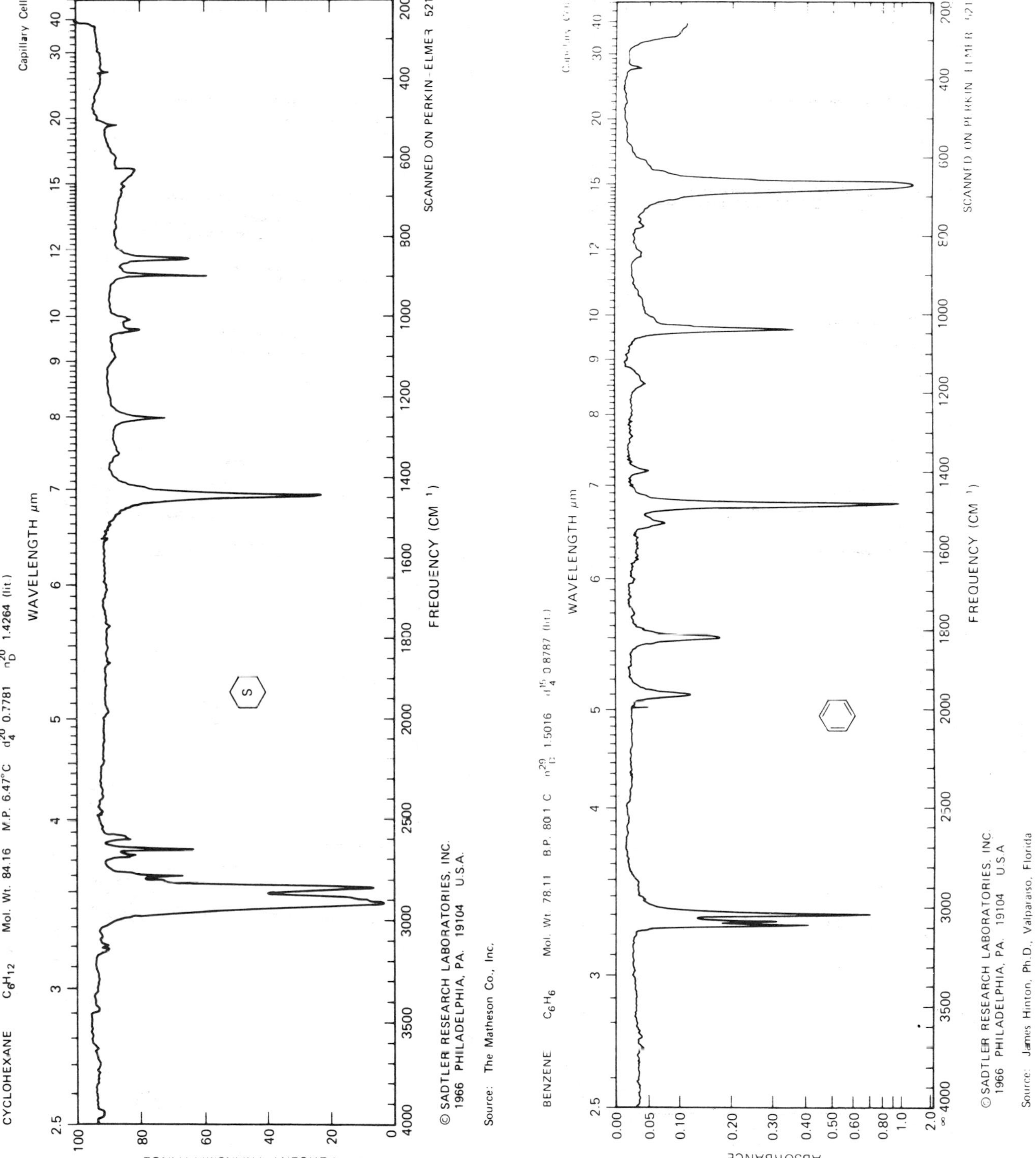

CYCLOHEXANE
C_6H_{12}
Mol. Wt. 84.16
M.P. 6.47°C
d_4^{20} 0.7781
n_D^{20} 1.4264 (lit.)
Capillary Cell
WAVELENGTH μm
PERCENT TRANSMITTANCE
FREQUENCY (CM⁻¹)
SCANNED ON PERKIN-ELMER 521
© SADTLER RESEARCH LABORATORIES, INC.
1966 PHILADELPHIA, PA. 19104 U.S.A.
Source: The Matheson Co., Inc.
BENZENE
C_6H_6
Mol. Wt. 78.11
B.P. 80.1 C
n_D^{29} 1.5016
d_4^{15} 0.8787 (lit.)
WAVELENGTH μm
ABSORBANCE
FREQUENCY (CM⁻¹)
SCANNED ON PERKIN-ELMER 521
© SADTLER RESEARCH LABORATORIES, INC.
1966 PHILADELPHIA, PA. 19104 U.S.A.
Source: James Hinton, Ph.D., Valparaiso, Florida

No. 3

No. 4

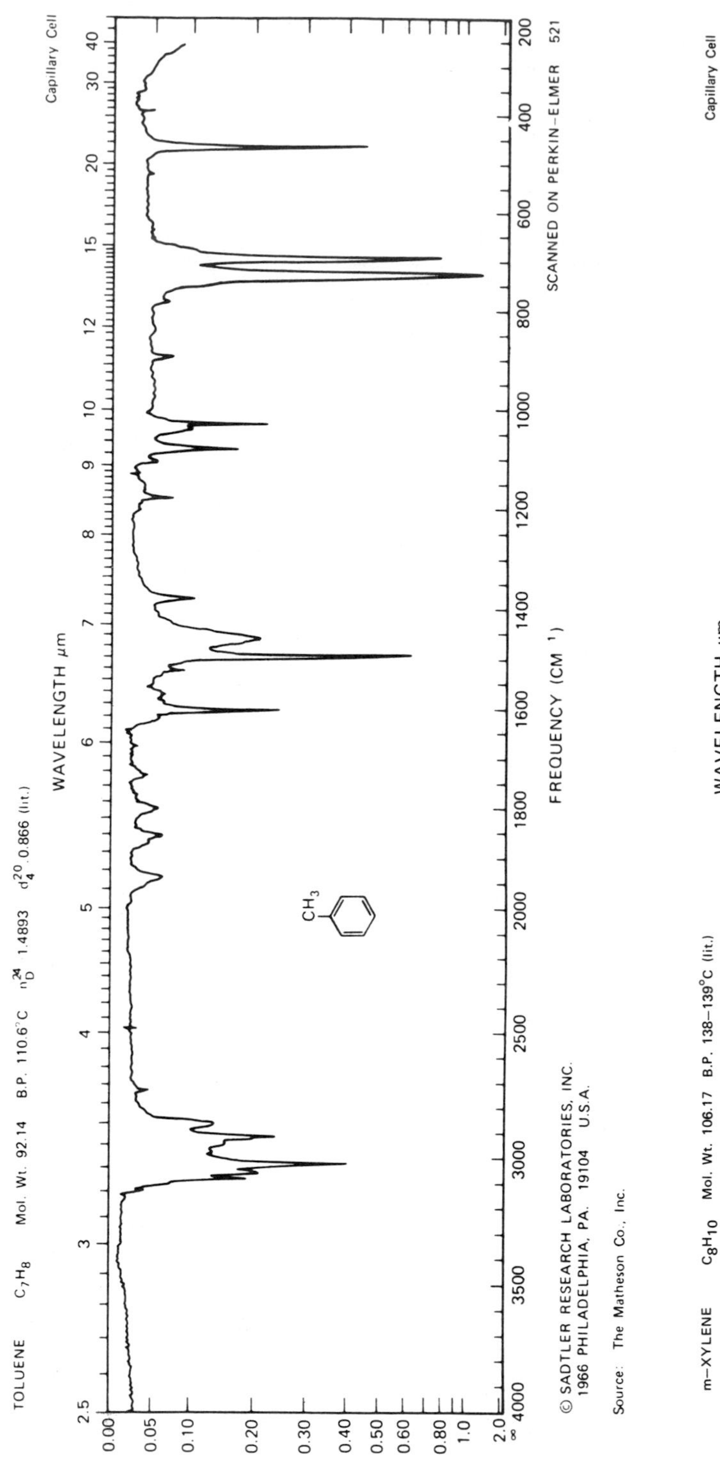

TOLUENE
C_7H_8
Mol. Wt. 92.14
B.P. 110.6°C
n_D^{24} 1.4893
d_4^{20} 0.866 (lit.)
Capillary Cell
WAVELENGTH μm
FREQUENCY (CM⁻¹)
ABSORBANCE
CH_3
SCANNED ON PERKIN–ELMER 521
© SADTLER RESEARCH LABORATORIES, INC.
1966 PHILADELPHIA, PA. 19104 U.S.A.
Source: The Matheson Co., Inc.

No. 5

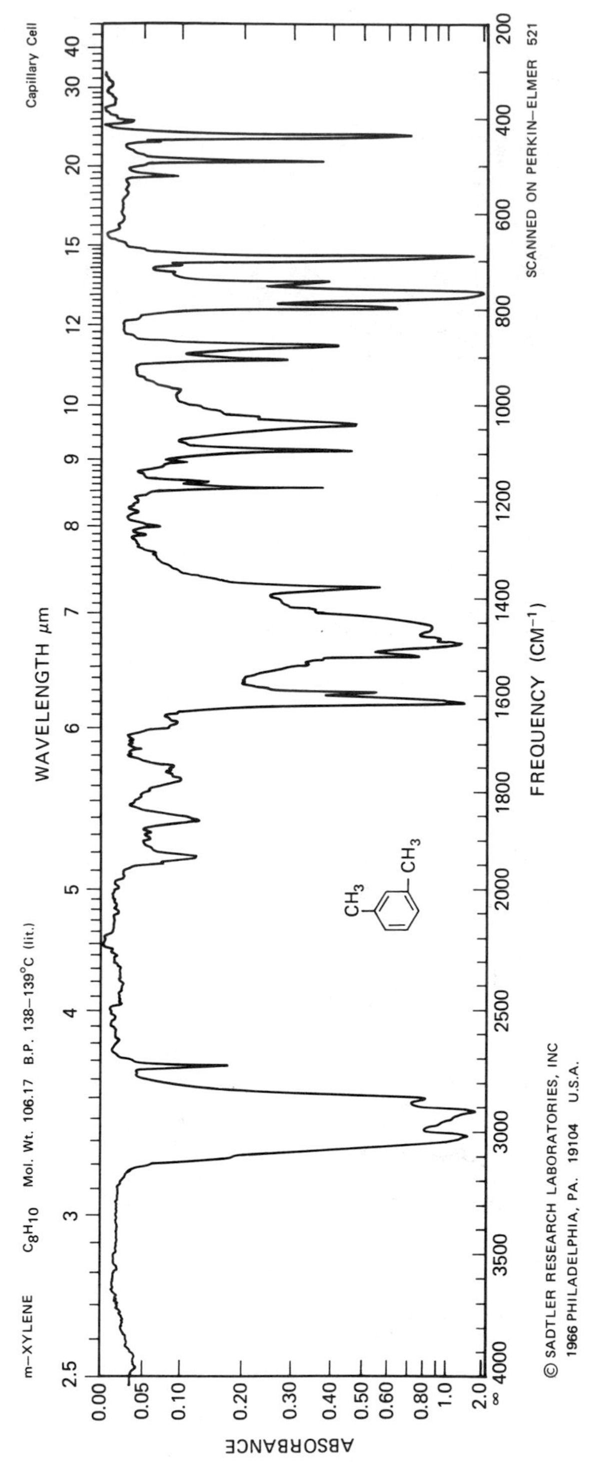

m–XYLENE
C_8H_{10}
Mol. Wt. 106.17
B.P. 138–139°C (lit.)
Capillary Cell
WAVELENGTH μm
FREQUENCY (CM^{-1})
ABSORBANCE
CH_3
CH_3
SCANNED ON PERKIN–ELMER 521
© SADTLER RESEARCH LABORATORIES, INC
1966 PHILADELPHIA, PA. 19104 U.S.A.
Source: Givaudan–Delawanna, Inc.

No. 6

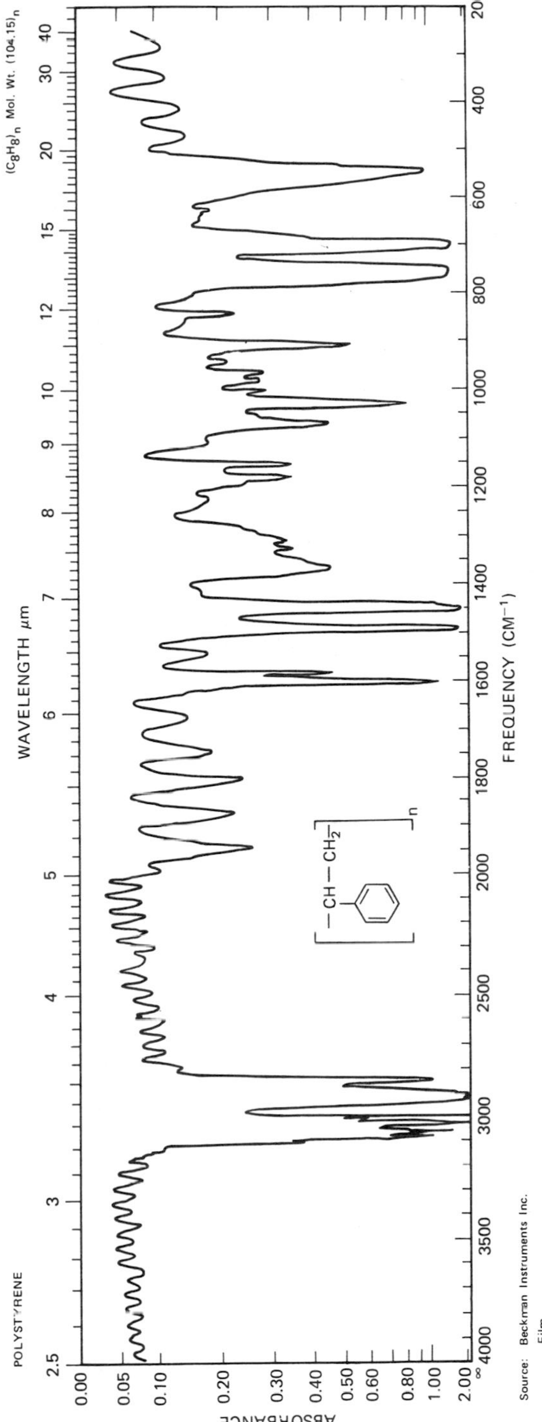
POLYSTYRENE
(C8H8)n Mol. Wt. (104.15)n
WAVELENGTH μm
FREQUENCY (CM⁻¹)
ABSORBANCE
Source: Beckman Instruments Inc.
Film

No. 7

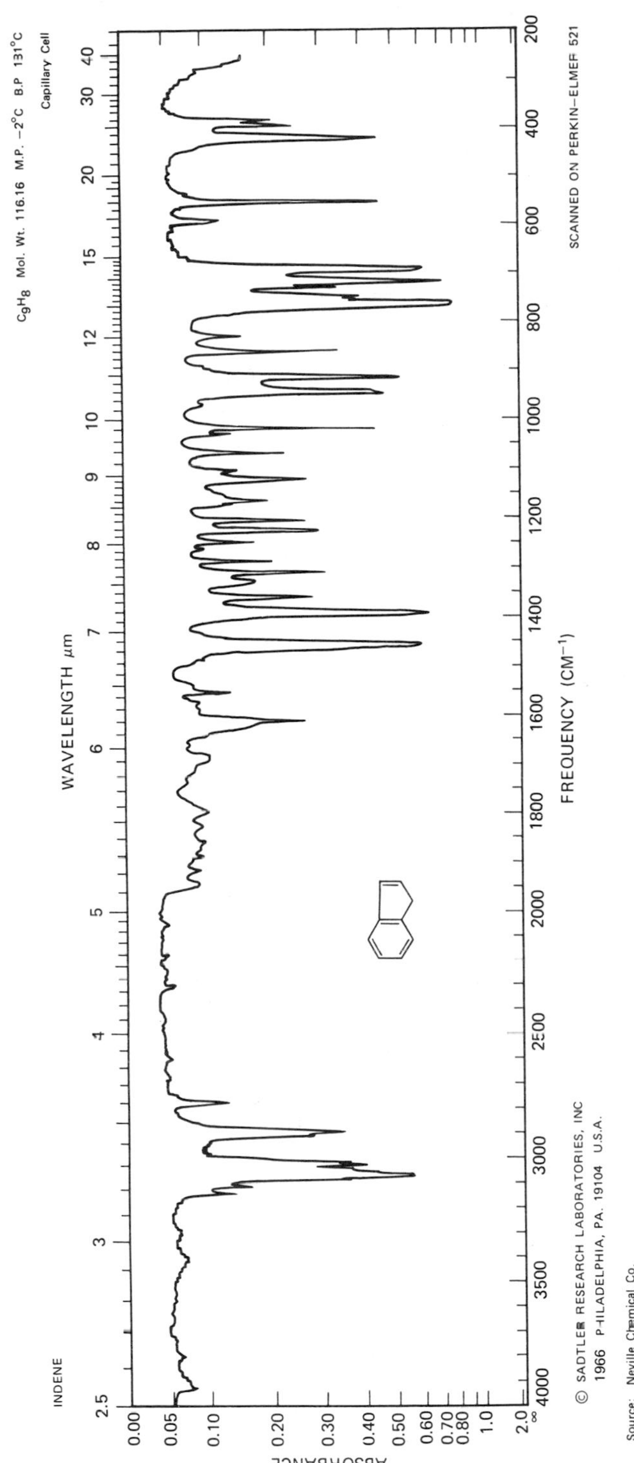
INDENE
C9H8 Mol. Wt. 116.16 M.P. −2°C B.P 131°C
Capillary Cell
WAVELENGTH μm
FREQUENCY (CM⁻¹)
ABSORBANCE
SCANNED ON PERKIN–ELMER 521
© SADTLER RESEARCH LABORATORIES, INC
1966 PHILADELPHIA, PA. 19104 U.S.A.
Source: Neville Chemical Co.

No. 8

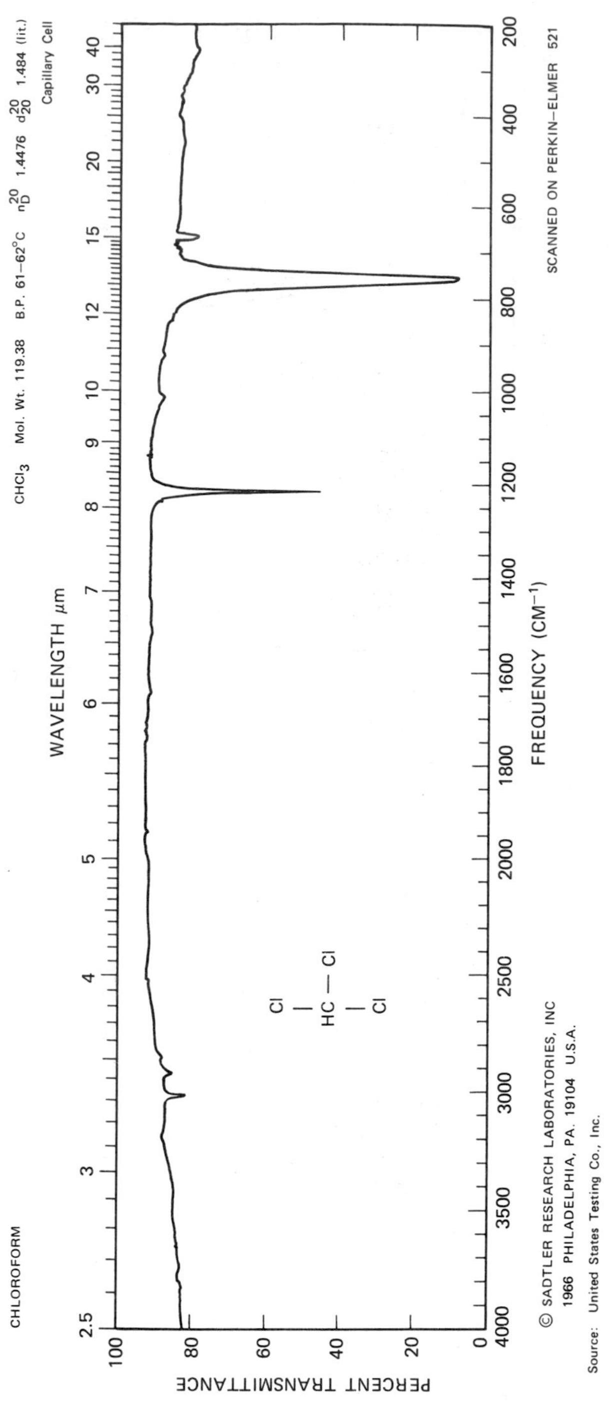
CHLOROFORM
$CHCl_3$ Mol. Wt. 119.38 B.P. 61–62°C n_D^{20} 1.4476 d_{20}^{20} 1.484 (lit.)
Capillary Cell
WAVELENGTH μm
FREQUENCY (CM^{-1})
PERCENT TRANSMITTANCE
SCANNED ON PERKIN–ELMER 521
© SADTLER RESEARCH LABORATORIES, INC
1966 PHILADELPHIA, PA. 19104 U.S.A.
Source: United States Testing Co., Inc.

No. 9

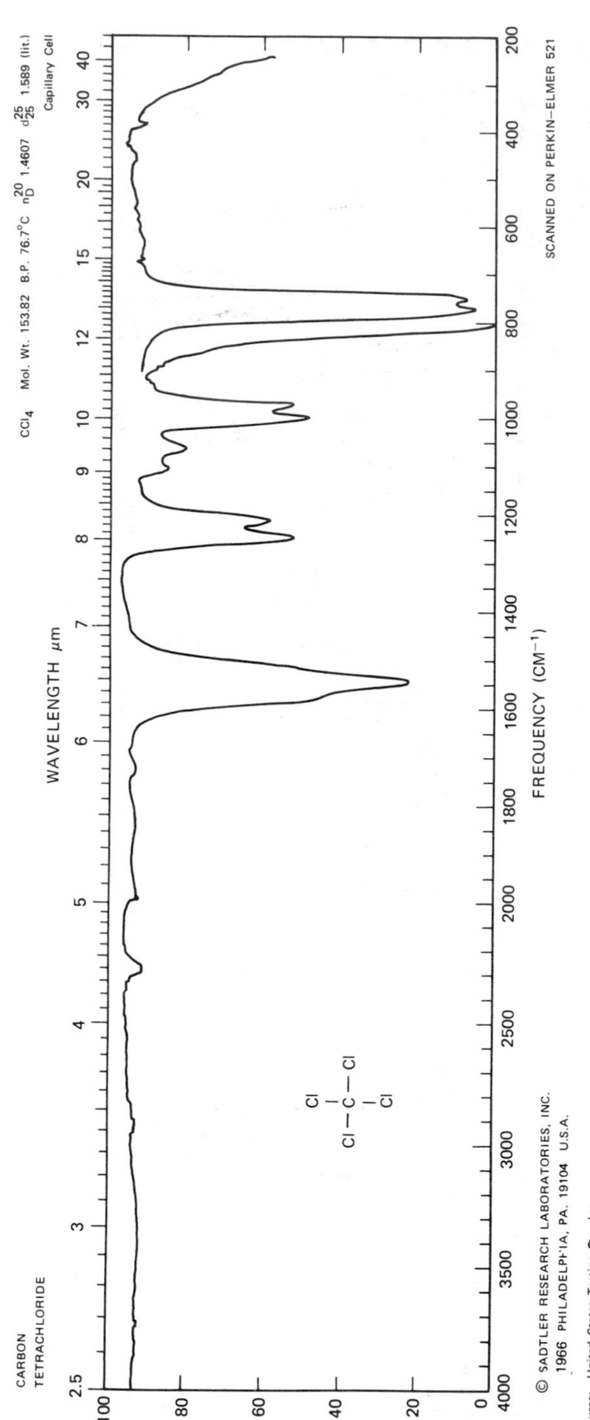
CARBON TETRACHLORIDE
CCl_4 Mol. Wt. 153.82 B.P. 76.7°C n_D^{20} 1.4607 d_{25}^{25} 1.589 (lit.)
Capillary Cell
WAVELENGTH μm
FREQUENCY (CM^{-1})
PERCENT TRANSMITTANCE
SCANNED ON PERKIN–ELMER 521
© SADTLER RESEARCH LABORATORIES, INC.
1966 PHILADELPHIA, PA. 19104 U.S.A.
Source: United States Testing Co. Inc.

No. 10

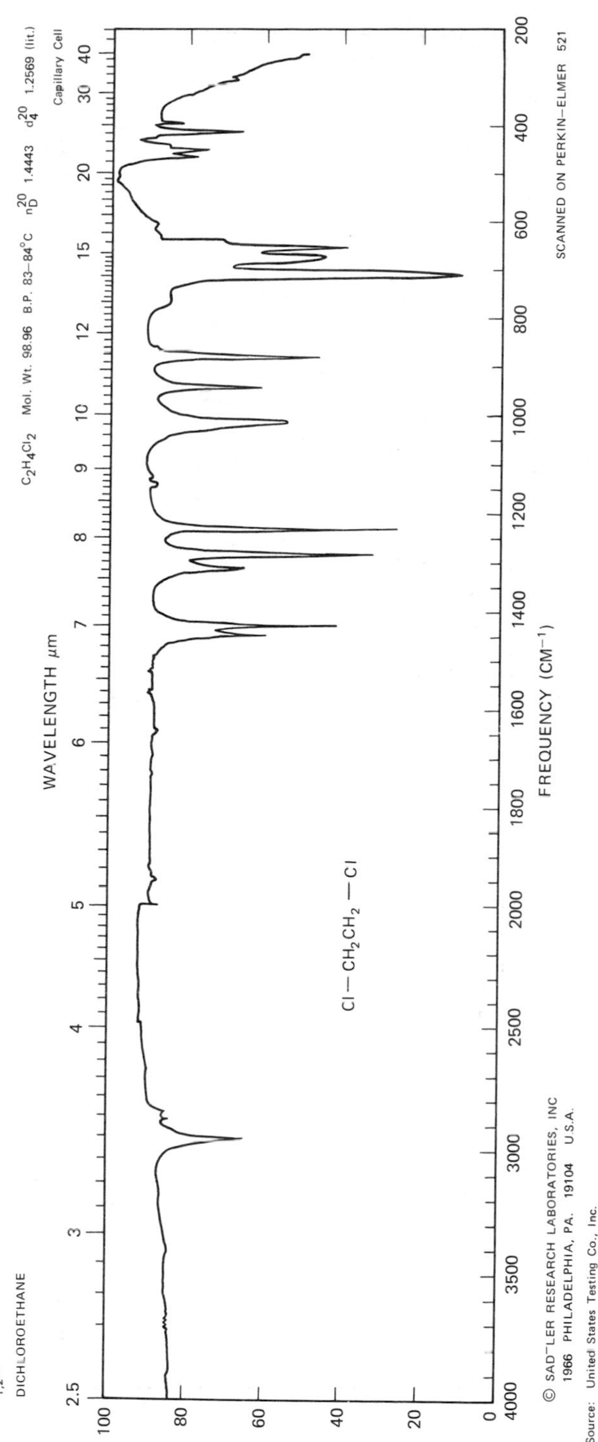
1,2-
DICHLOROETHANE
C2H4Cl2 Mol. Wt. 98.96 B.P. 83–84°C n20D 1.4443 d20 4 1.2569 (lit.)
Capillary Cell
WAVELENGTH μm
PERCENT TRANSMITTANCE
Cl — CH2CH2 — Cl
FREQUENCY (CM−1)
SCANNED ON PERKIN-ELMER 521
© SADTLER RESEARCH LABORATORIES, INC
1966 PHILADELPHIA, PA. 19104 U.S.A.
Source: United States Testing Co., Inc.

No. 11

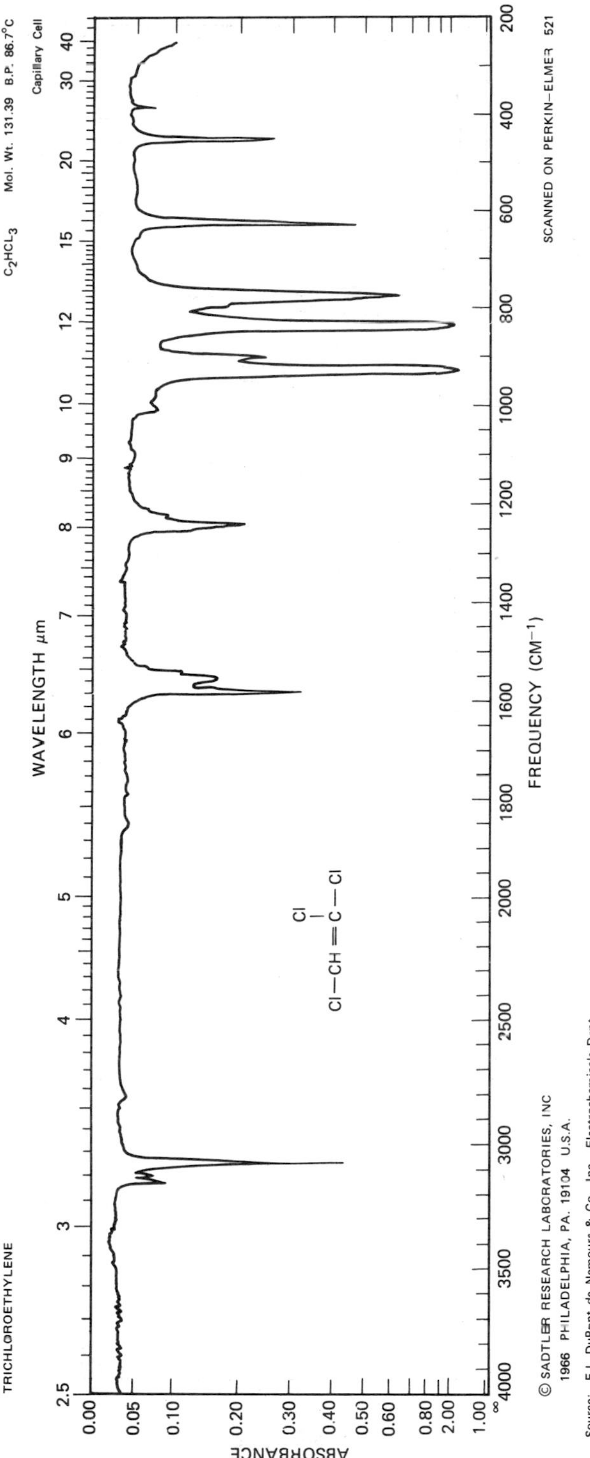
TRICHLOROETHYLENE
C2HCL3 Mol. Wt. 131.39 B.P. 86.7°C
Capillary Cell
WAVELENGTH μm
ABSORBANCE
Cl
Cl — CH = C — Cl
FREQUENCY (CM−1)
SCANNED ON PERKIN-ELMER 521
© SADTLER RESEARCH LABORATORIES, INC
1966 PHILADELPHIA, PA. 19104 U.S.A.
Source: E.I. DuPont de Nemours & Co., Inc., Electrochemicals Dept.

No. 12

No. 13

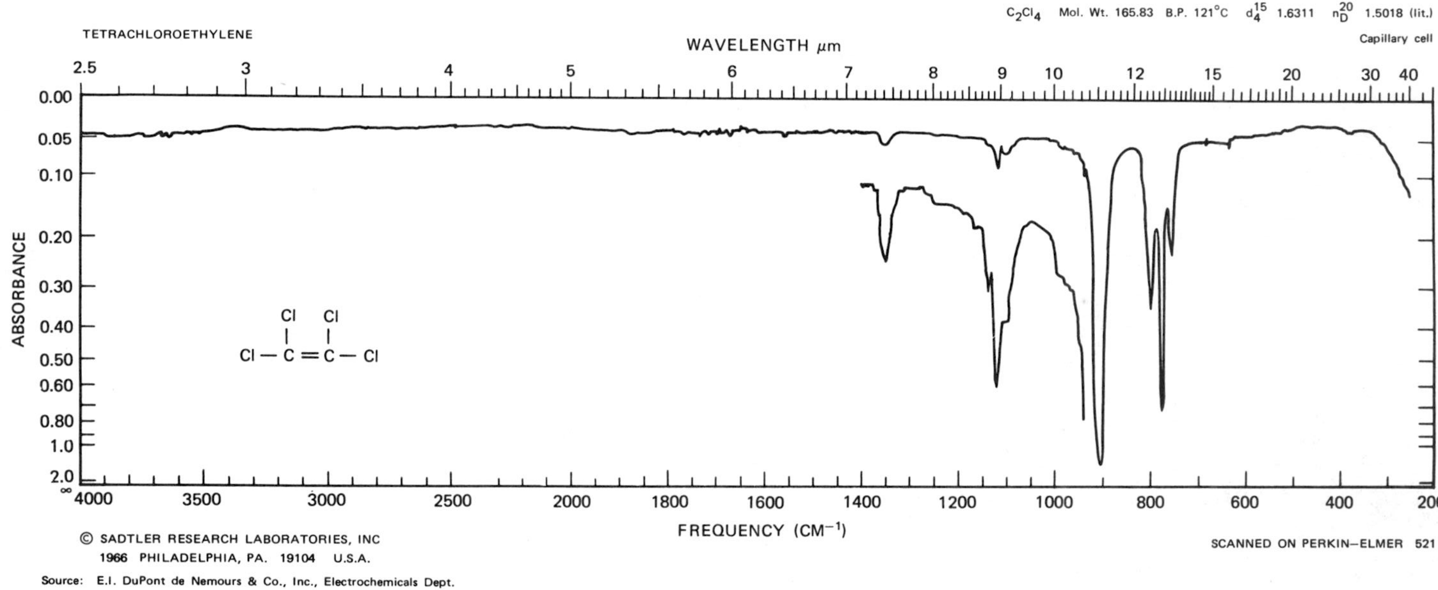

TETRACHLOROETHYLENE
C_2Cl_4 Mol. Wt. 165.83 B.P. 121°C d_4^{15} 1.6311 n_D^{20} 1.5018 (lit.)
Capillary cell
WAVELENGTH μm
ABSORBANCE
FREQUENCY (CM⁻¹)
© SADTLER RESEARCH LABORATORIES, INC
1966 PHILADELPHIA, PA. 19104 U.S.A.
Source: E.I. DuPont de Nemours & Co., Inc., Electrochemicals Dept.
SCANNED ON PERKIN–ELMER 521

No. 14

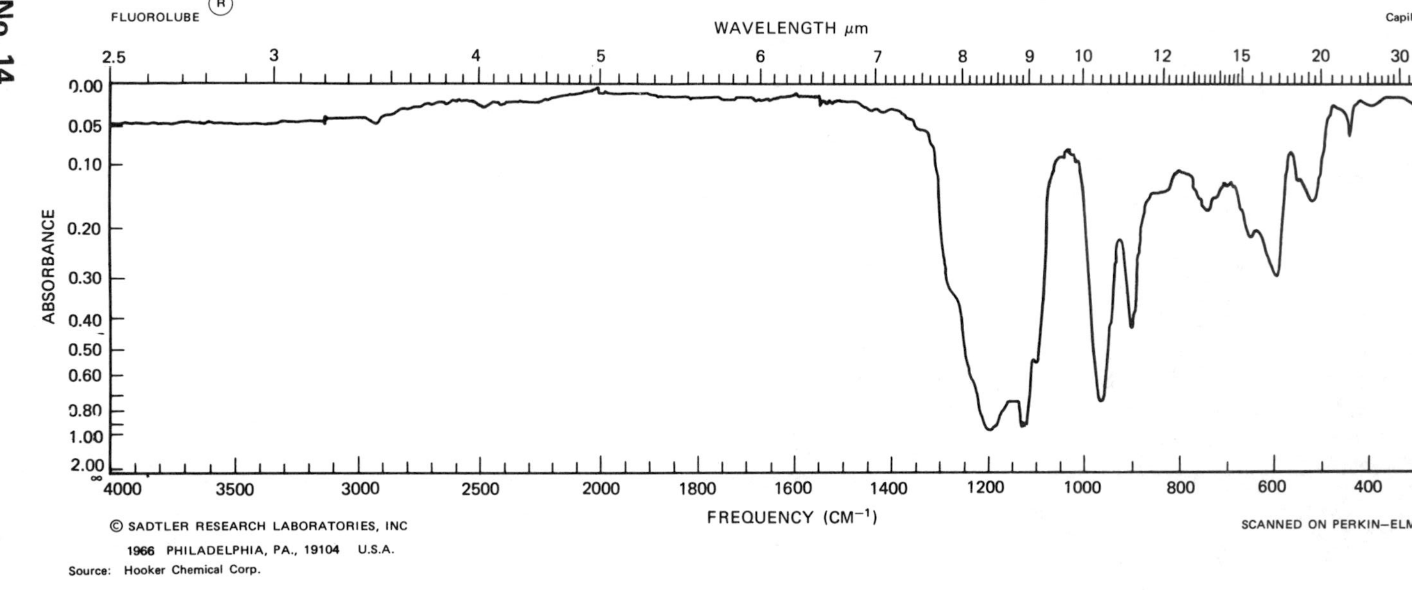

FLUOROLUBE (R)
Capillary Cell
WAVELENGTH μm
ABSORBANCE
FREQUENCY (CM⁻¹)
© SADTLER RESEARCH LABORATORIES, INC
1966 PHILADELPHIA, PA., 19104 U.S.A.
Source: Hooker Chemical Corp.
SCANNED ON PERKIN–ELMER 521

No. 15

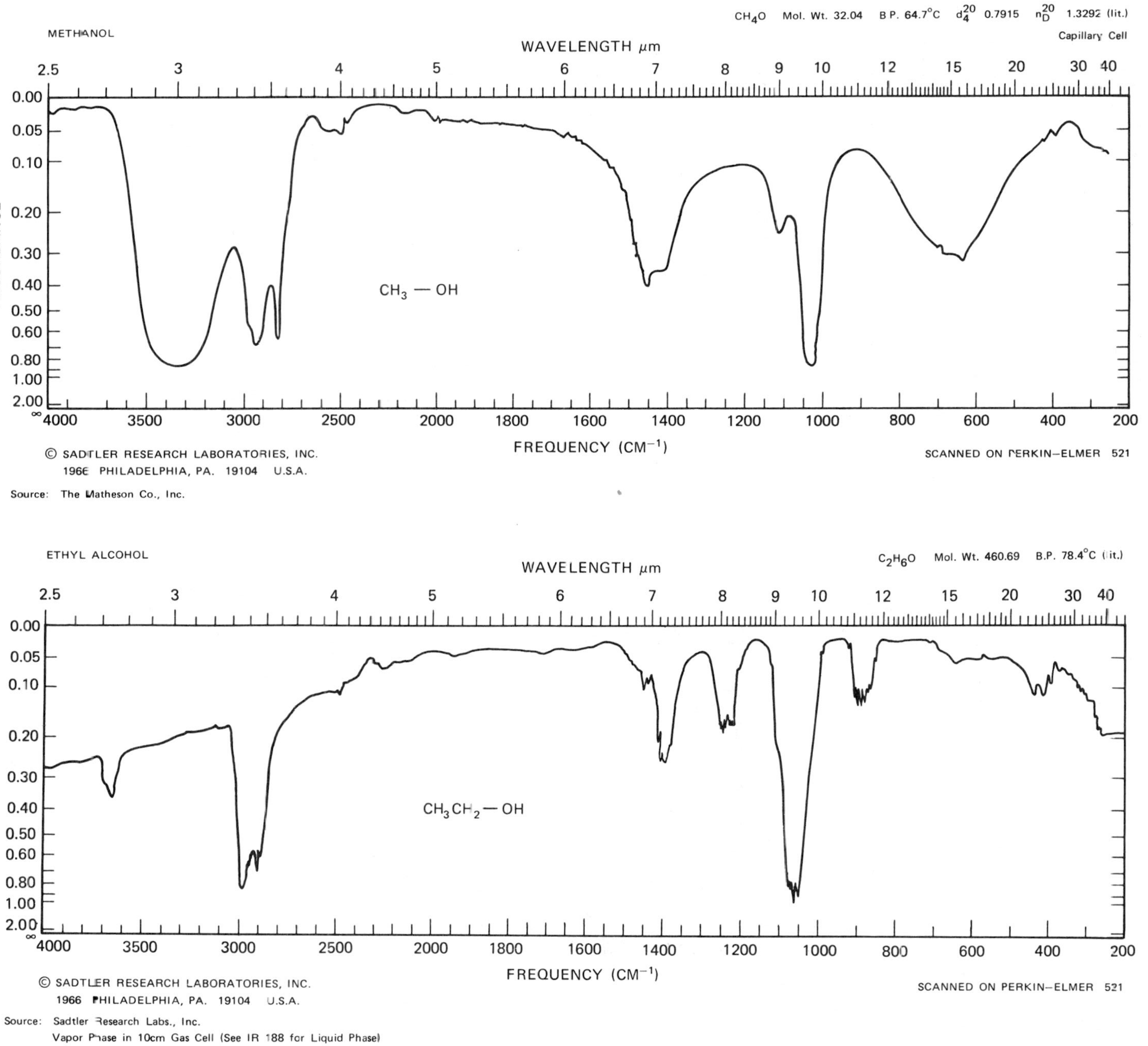

METHANOL
CH_4O Mol. Wt. 32.04 B.P. 64.7°C d_4^{20} 0.7915 n_D^{20} 1.3292 (lit.)
Capillary Cell
WAVELENGTH μm
ABSORBANCE
FREQUENCY (CM^{-1})
CH_3 — OH
© SADTLER RESEARCH LABORATORIES, INC.
1966 PHILADELPHIA, PA. 19104 U.S.A.
SCANNED ON PERKIN-ELMER 521
Source: The Matheson Co., Inc.
ETHYL ALCOHOL
C_2H_6O Mol. Wt. 460.69 B.P. 78.4°C (lit.)
WAVELENGTH μm
ABSORBANCE
FREQUENCY (CM^{-1})
CH_3CH_2 — OH
© SADTLER RESEARCH LABORATORIES, INC.
1966 PHILADELPHIA, PA. 19104 U.S.A.
SCANNED ON PERKIN-ELMER 521
Source: Sadtler Research Labs., Inc.
Vapor Phase in 10cm Gas Cell (See IR 188 for Liquid Phase)

No. 16

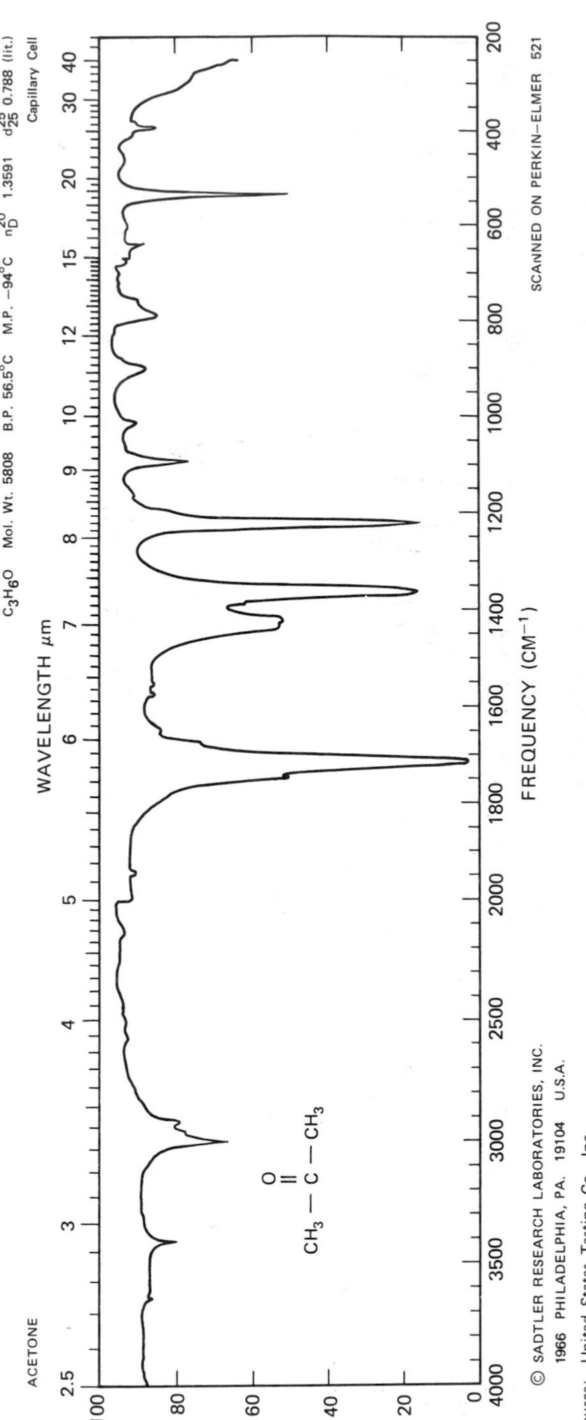

ACETONE
C_3H_6O Mol. Wt. 58.08 B.P. 56.5°C M.P. −94°C n_D^{20} 1.3591 d_{25}^{25} 0.788 (lit.) Capillary Cell
WAVELENGTH μm
PERCENT TRANSMITTANCE
FREQUENCY (CM^{-1})
$CH_3 - \overset{O}{\overset{\|}{C}} - CH_3$
© SADTLER RESEARCH LABORATORIES, INC.
1966 PHILADELPHIA, PA. 19104 U.S.A.
Source: United States Testing Co., Inc.
SCANNED ON PERKIN–ELMER 521

No. 17

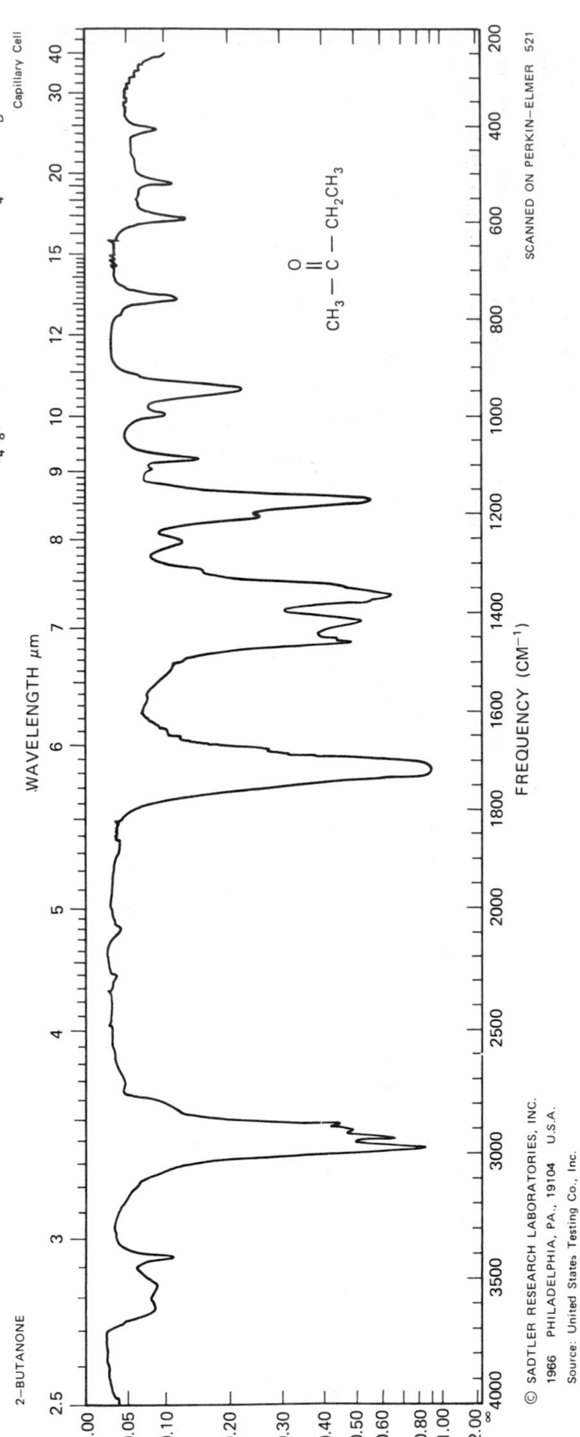

2–BUTANONE
C_4H_8O Mol. Wt. 72.11 B.P. 79.6°C d_4^{20} 0.805 n_D^{15} 1.3814 (lit.) Capillary Cell
WAVELENGTH μm
ABSORBANCE
FREQUENCY (CM^{-1})
$CH_3 - \overset{O}{\overset{\|}{C}} - CH_2CH_3$
© SADTLER RESEARCH LABORATORIES, INC.
1966 PHILADELPHIA, PA., 19104 U.S.A.
Source: United States Testing Co., Inc.
SCANNED ON PERKIN–ELMER 521

No. 18

No. 19

No. 20

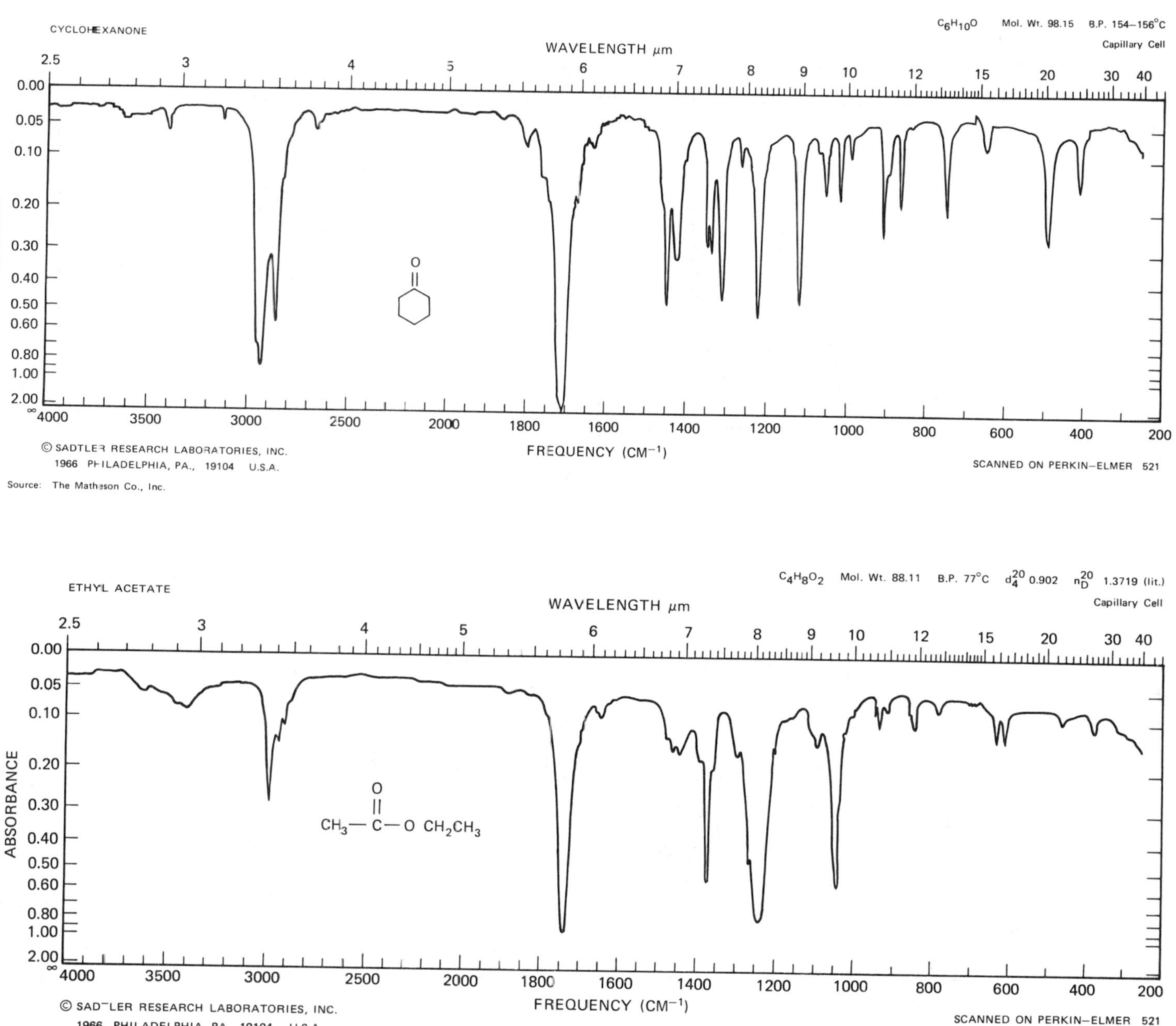

CYCLOHEXANONE
$C_6H_{10}O$ Mol. Wt. 98.15 B.P. 154–156°C
Capillary Cell
WAVELENGTH μm
ABSORBANCE
FREQUENCY (CM^{-1})
© SADTLER RESEARCH LABORATORIES, INC.
1966 PHILADELPHIA, PA., 19104 U.S.A.
Source: The Matheson Co., Inc.
SCANNED ON PERKIN-ELMER 521
ETHYL ACETATE
$C_4H_8O_2$ Mol. Wt. 88.11 B.P. 77°C d_4^{20} 0.902 n_D^{20} 1.3719 (lit.)
Capillary Cell
WAVELENGTH μm
ABSORBANCE
$CH_3—\overset{O}{\overset{\|}{C}}—O\ CH_2CH_3$
FREQUENCY (CM^{-1})
© SADTLER RESEARCH LABORATORIES, INC.
1966 PHILADELPHIA, PA., 19104 U.S.A.
Source: Dodge & Olcott, Inc.
SCANNED ON PERKIN-ELMER 521

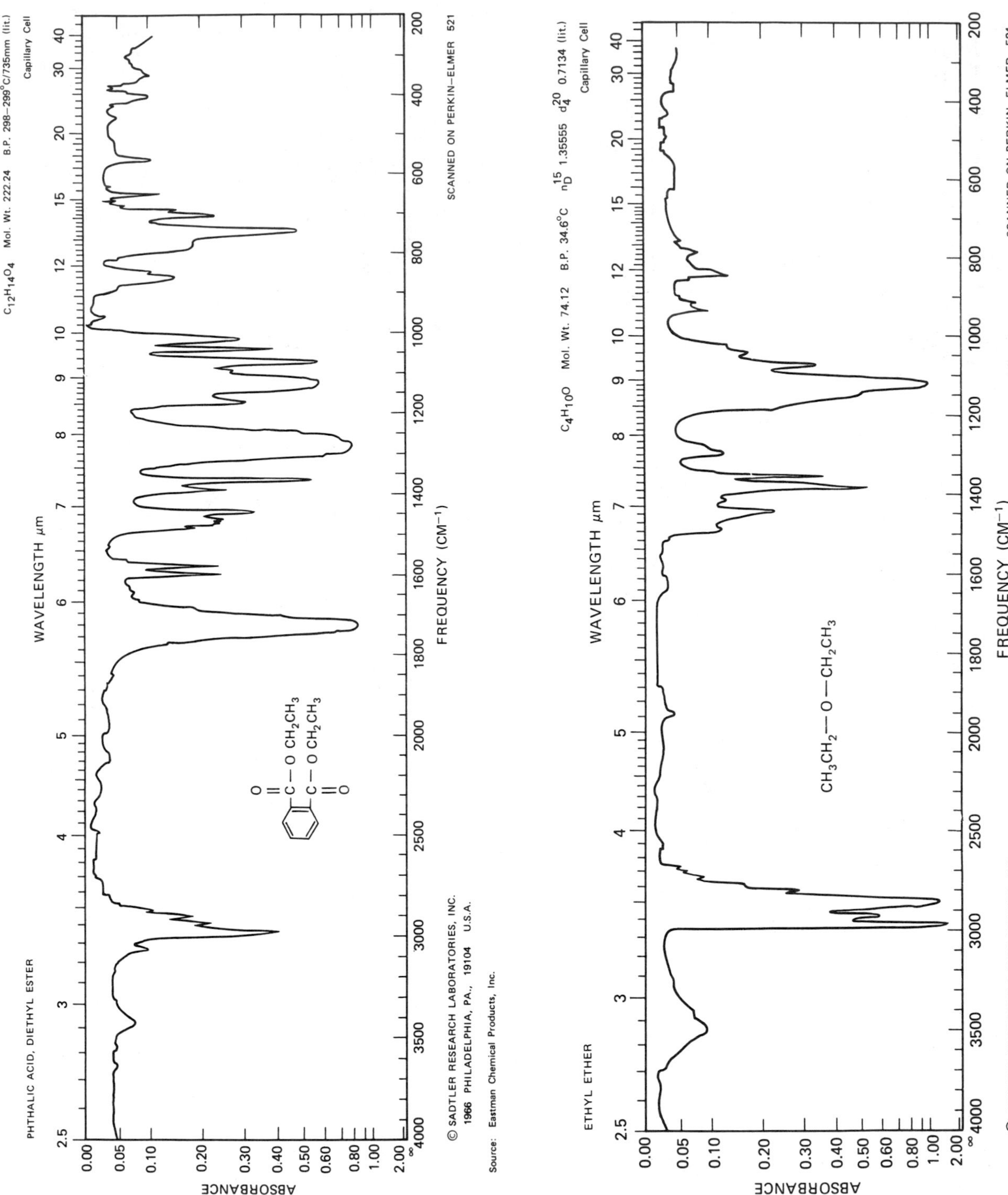

PHTHALIC ACID, DIETHYL ESTER
C12H14O4 Mol. Wt. 222.24 B.P. 298–299°C/735mm (lit.)
Capillary Cell
WAVELENGTH μm
FREQUENCY (CM−1)
ABSORBANCE
SCANNED ON PERKIN-ELMER 521
© SADTLER RESEARCH LABORATORIES, INC.
1966 PHILADELPHIA, PA., 19104 U.S.A.
Source: Eastman Chemical Products, Inc.
ETHYL ETHER
C4H10O Mol. Wt. 74.12 B.P. 34.6°C n15D 1.35555 d20 4 0.7134 (lit.)
Capillary Cell
CH3CH2—O—CH2CH3
WAVELENGTH μm
FREQUENCY (CM−1)
ABSORBANCE
SCANNED ON PERKIN-ELMER 521
© SADTLER RESEARCH LABORATORIES, INC.
1966 PHILADELPHIA, PA., 19104 U.S.A.
Source: United States Testing Co., Inc.

No. 21

No. 22

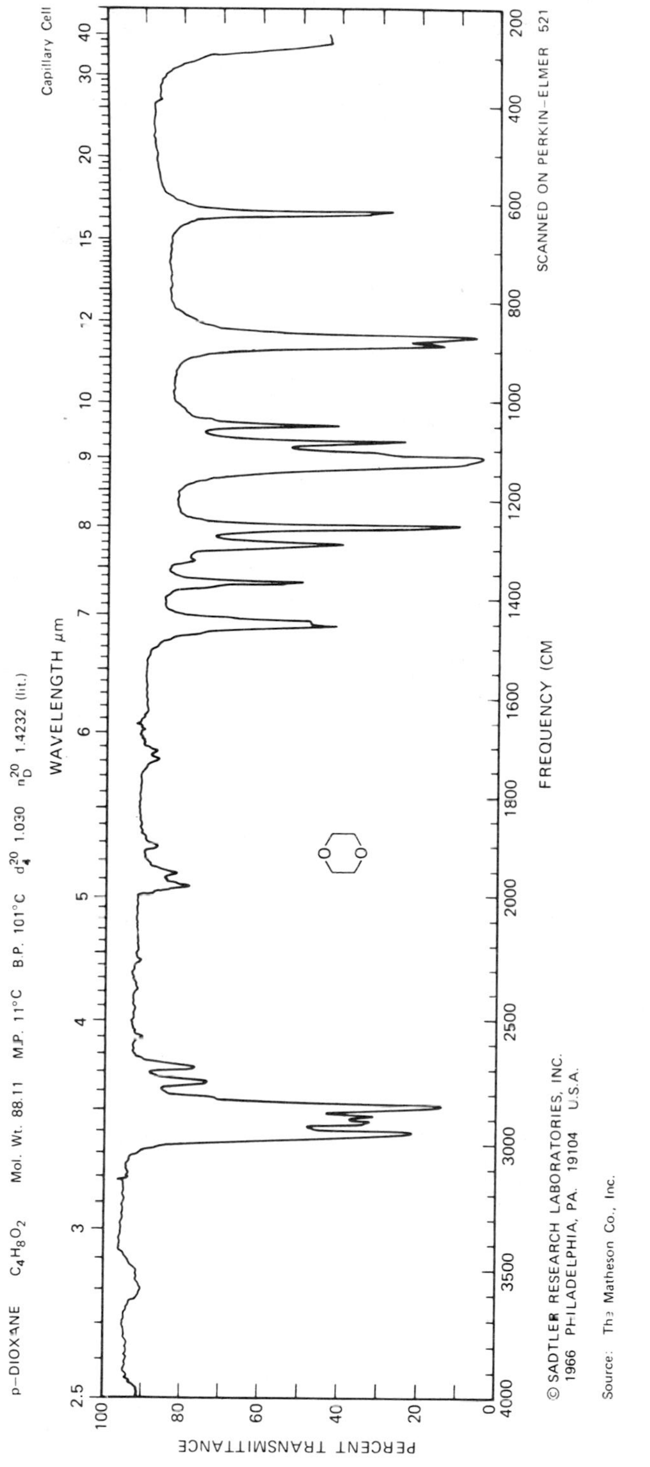
p-DIOXANE C4H8O2 Mol. Wt. 88.11 M.P. 11°C B.P. 101°C d20 4 1.030 n20 D 1.4232 (lit.)
Capillary Cell
WAVELENGTH μm
PERCENT TRANSMITTANCE
FREQUENCY (CM
SCANNED ON PERKIN-ELMER 521
© SADTLER RESEARCH LABORATORIES, INC. 1966 PHILADELPHIA, PA. 19104 U.S.A.
Source: The Matheson Co., Inc.

No. 23

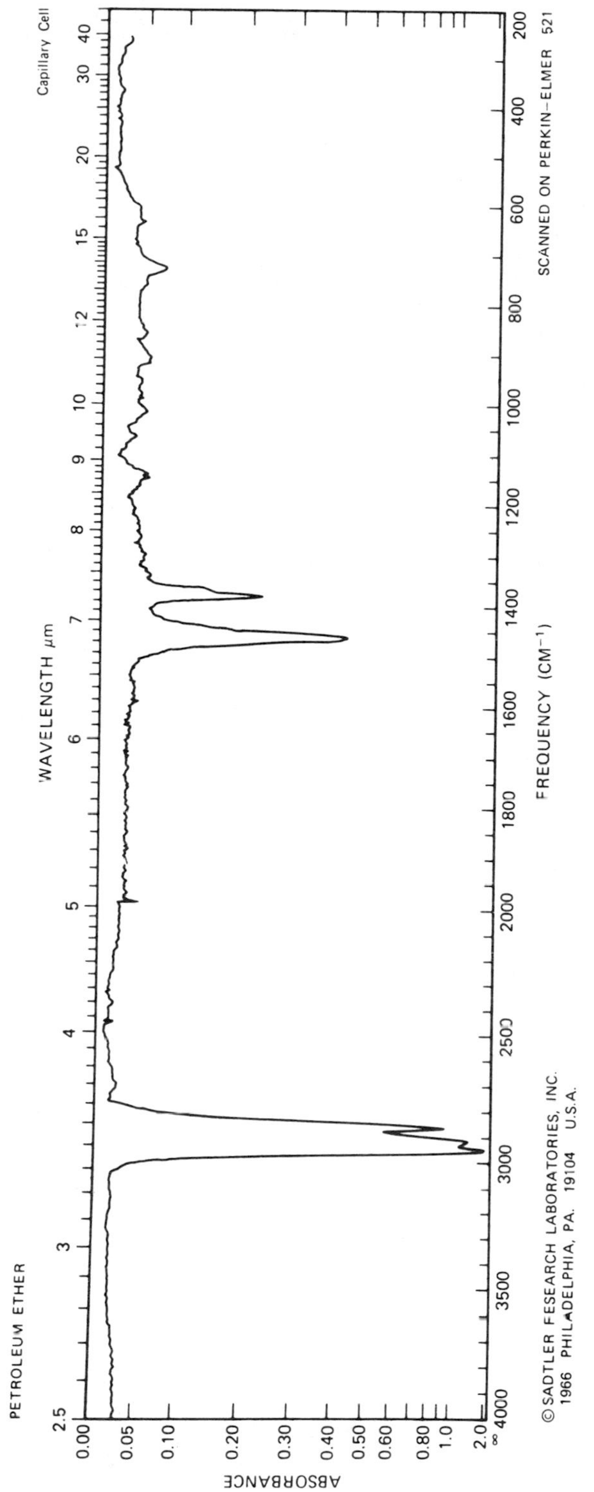
PETROLEUM ETHER
Capillary Cell
WAVELENGTH μm
ABSORBANCE
FREQUENCY (CM⁻¹)
SCANNED ON PERKIN-ELMER 521
© SADTLER RESEARCH LABORATORIES, INC. 1966 PHILADELPHIA, PA. 19104 U.S.A.
Source: Sadtler Research Labs., Inc.

No. 24

No. 25

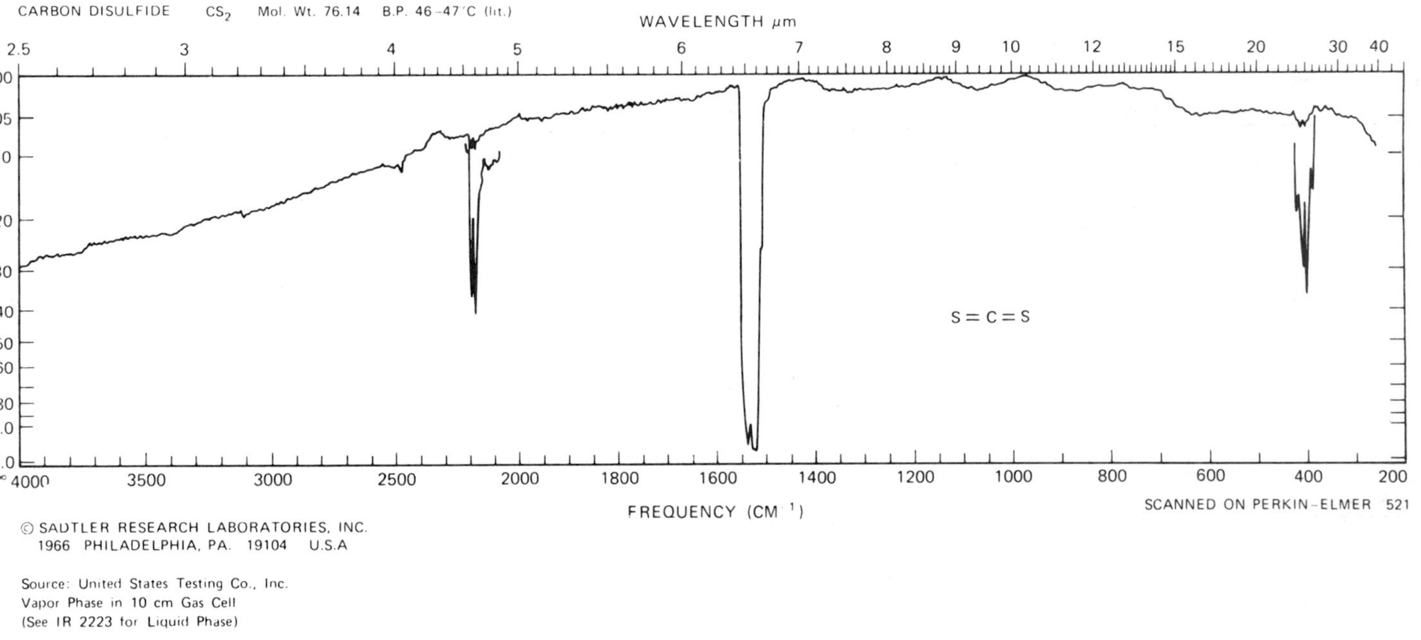
CARBON DISULFIDE CS2 Mol. Wt. 76.14 B.P. 46–47°C (lit.)
WAVELENGTH μm
S=C=S
ABSORBANCE
FREQUENCY (CM⁻¹)
SCANNED ON PERKIN-ELMER 521
© SADTLER RESEARCH LABORATORIES, INC. 1966 PHILADELPHIA, PA. 19104 U.S.A
Source: United States Testing Co., Inc.
Vapor Phase in 10 cm Gas Cell
(See IR 2223 for Liquid Phase)

No. 26

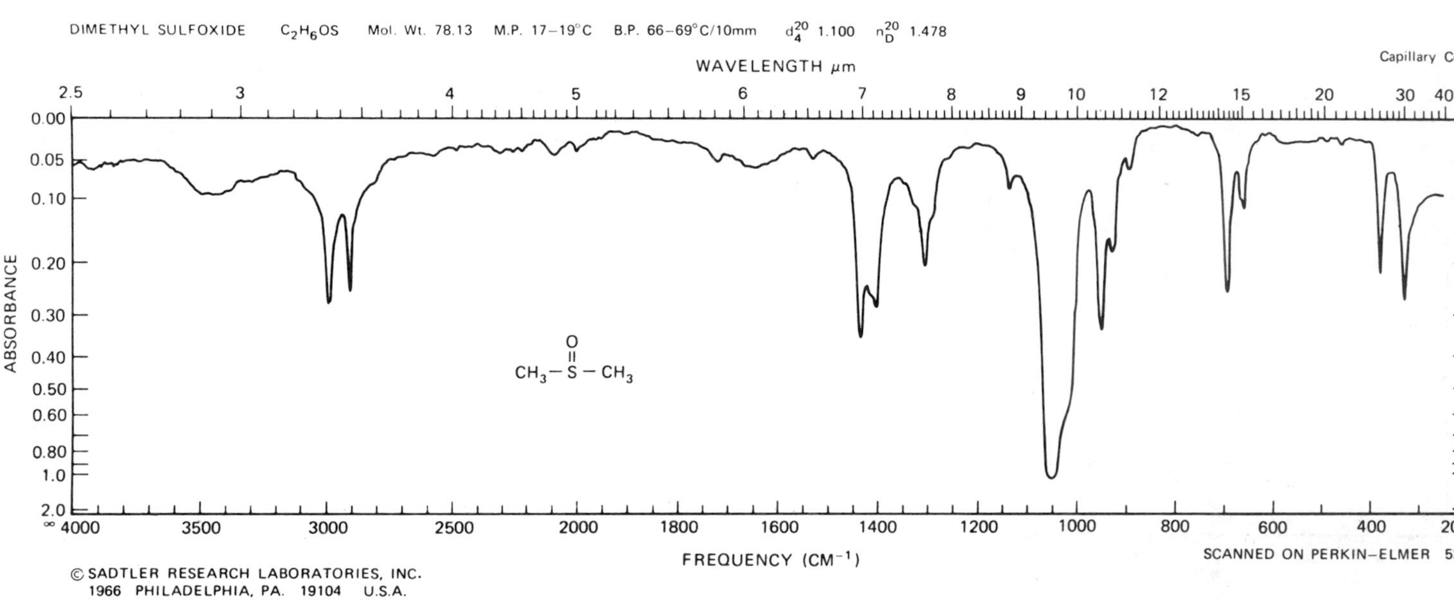
DIMETHYL SULFOXIDE C2H6OS Mol. Wt. 78.13 M.P. 17–19°C B.P. 66–69°C/10mm d420 1.100 nD20 1.478
Capillary Cell
WAVELENGTH μm
CH3–S(=O)–CH3
ABSORBANCE
FREQUENCY (CM⁻¹)
SCANNED ON PERKIN-ELMER 521
© SADTLER RESEARCH LABORATORIES, INC. 1966 PHILADELPHIA, PA. 19104 U.S.A.
Source: Fluka AG, Buchs, Switzerland

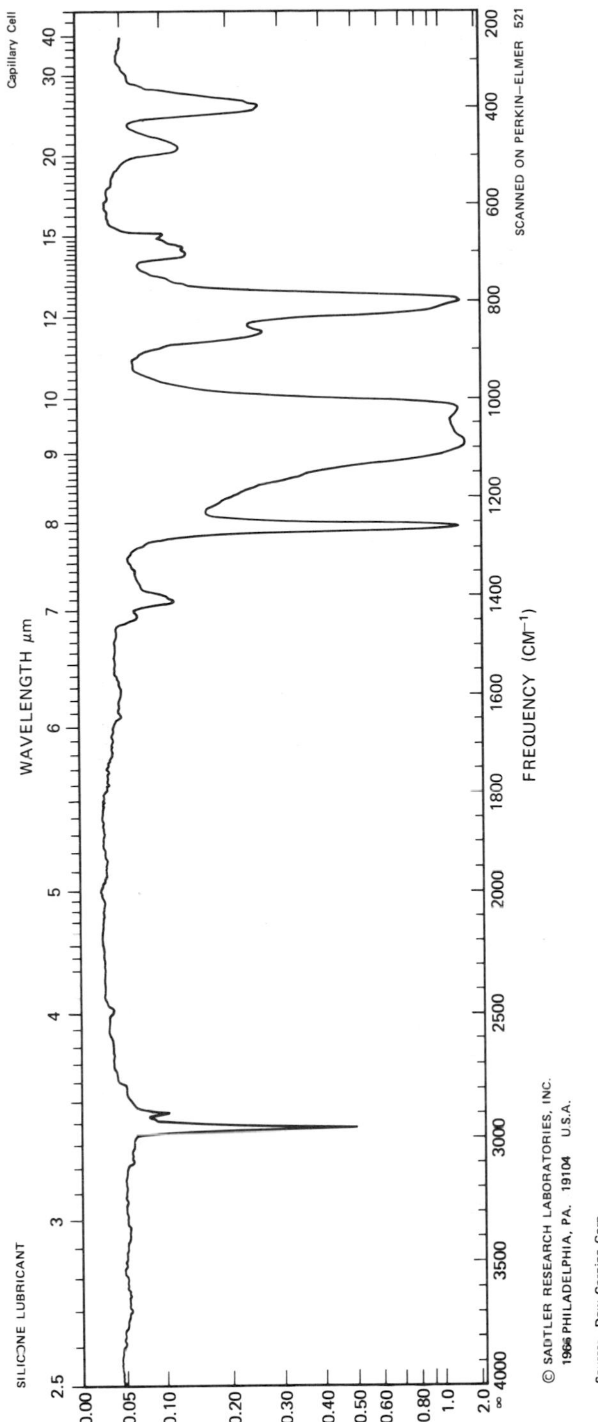

SILICONE LUBRICANT
Capillary Cell
WAVELENGTH μm
FREQUENCY (CM⁻¹)
ABSORBANCE
SCANNED ON PERKIN–ELMER 521
© SADTLER RESEARCH LABORATORIES, INC.
1966 PHILADELPHIA, PA. 19104 U.S.A.
Source: Dow Corning Corp.

No. 27

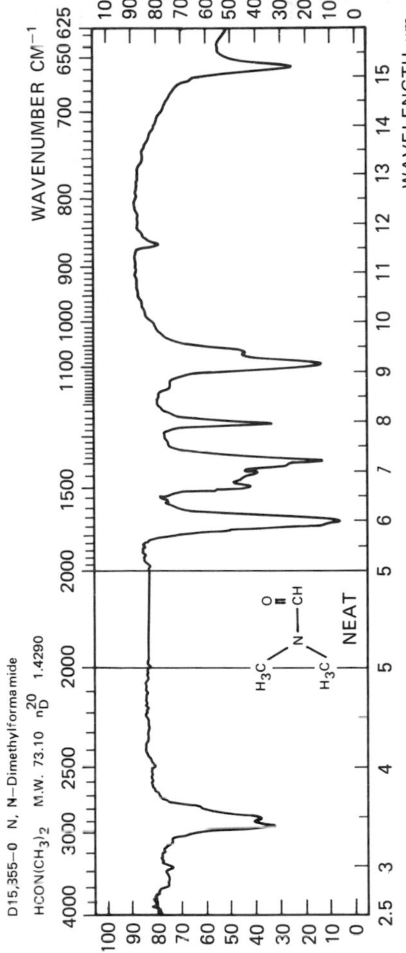

D15,355–0 N, N–Dimethylformamide
HCON(CH3)2 M.W. 73.10 n20D 1.4290
NEAT
WAVENUMBER CM⁻¹
WAVELENGTH μm

No. 28

appendix c CHARACTERISTIC GROUP ABSORPTIONS

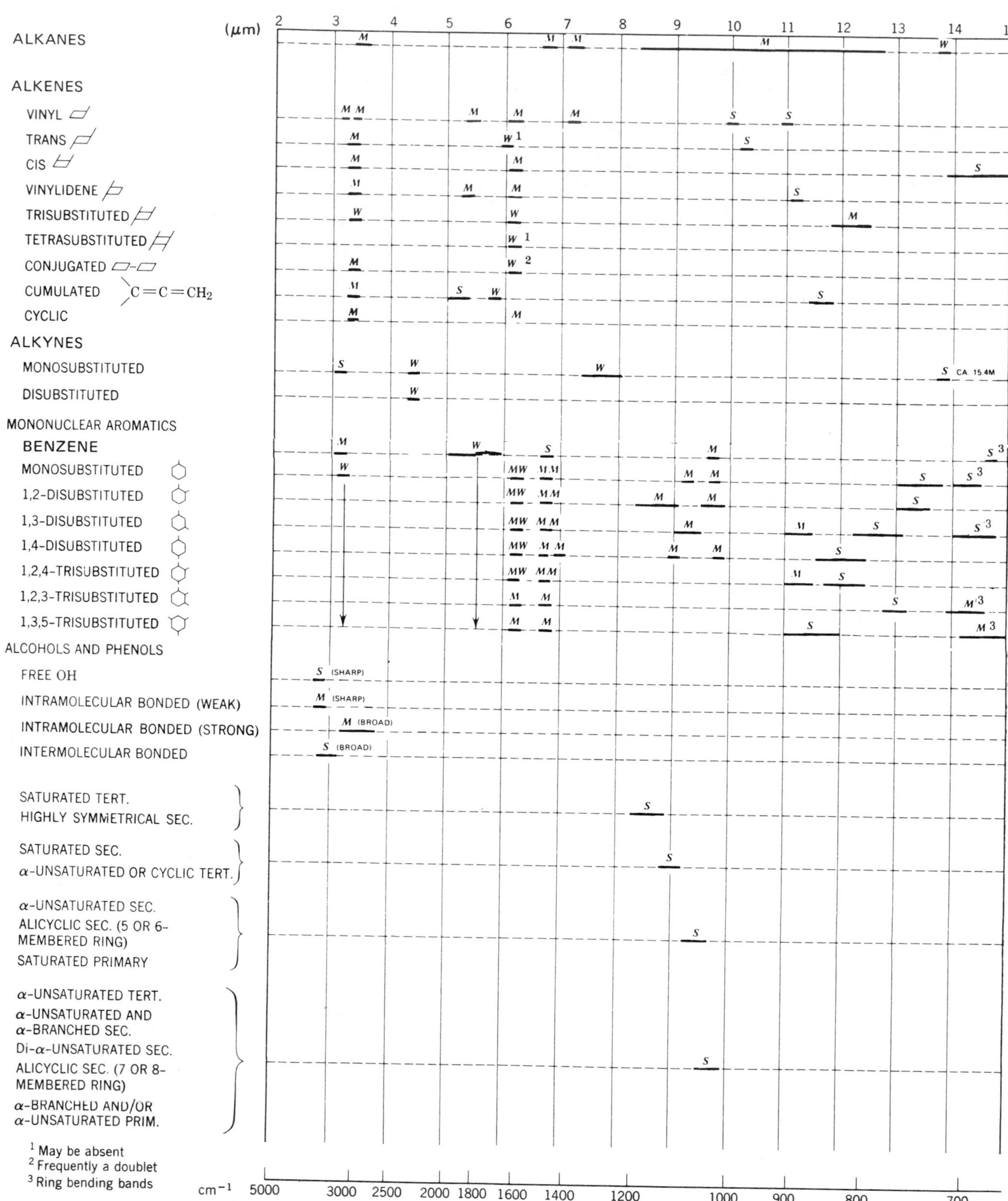

Characteristic group frequencies. The position of narrow absorption ranges is covered by a single letter indicating an average intensity. Broader absorption regions are indicated by a heavy bar. For example, a monosubstituted mononuclear aromatic may have four bands between 6 and 7 μm, three of medium intensity, and one weak.

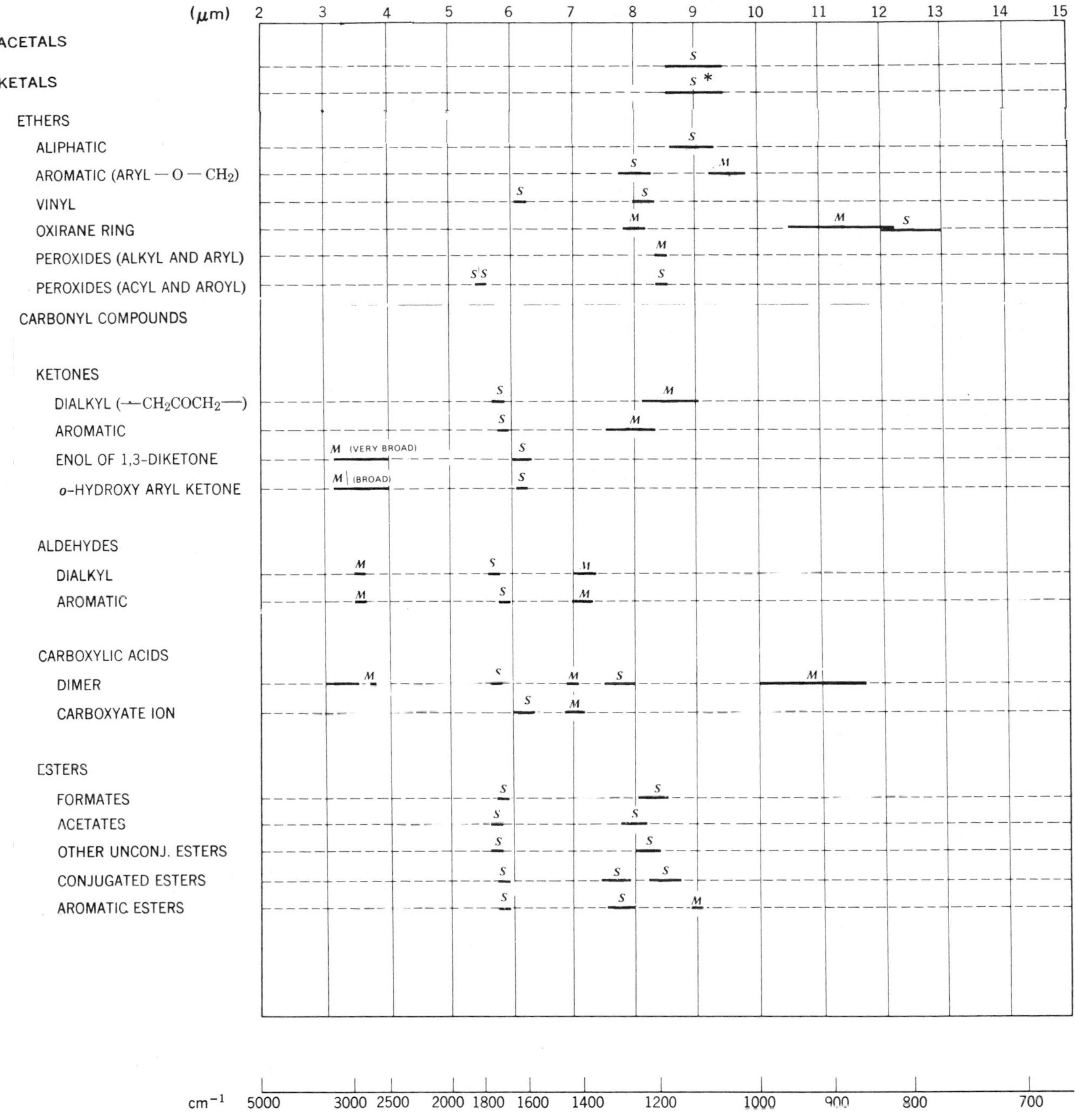

*Three bands, sometimes a fourth band for ketals and a fifth band for acetals.

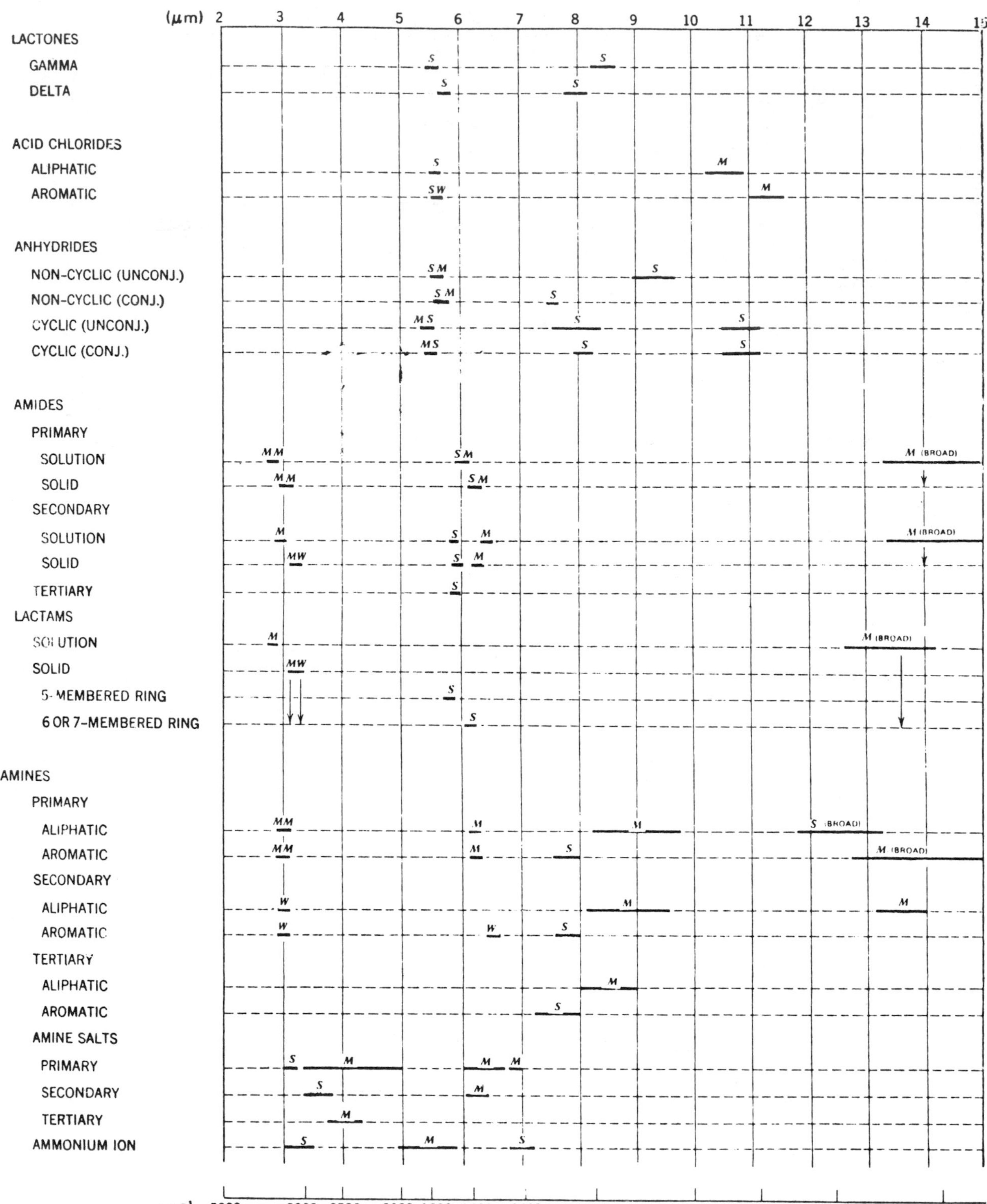

(μm) 2 3 4 5 6 7 8 9 10 11 12 13 14 15
LACTONES
GAMMA
DELTA
ACID CHLORIDES
ALIPHATIC
AROMATIC
ANHYDRIDES
NON-CYCLIC (UNCONJ.)
NON-CYCLIC (CONJ.)
CYCLIC (UNCONJ.)
CYCLIC (CONJ.)
AMIDES
PRIMARY
SOLUTION
SOLID
SECONDARY
SOLUTION
SOLID
TERTIARY
LACTAMS
SOLUTION
SOLID
5-MEMBERED RING
6 OR 7-MEMBERED RING
AMINES
PRIMARY
ALIPHATIC
AROMATIC
SECONDARY
ALIPHATIC
AROMATIC
TERTIARY
ALIPHATIC
AROMATIC
AMINE SALTS
PRIMARY
SECONDARY
TERTIARY
AMMONIUM ION
M (BROAD)
S (BROAD)
cm⁻¹ 5000 3000 2500 2000 1800 1600 1400 1200 1000 900 800 700

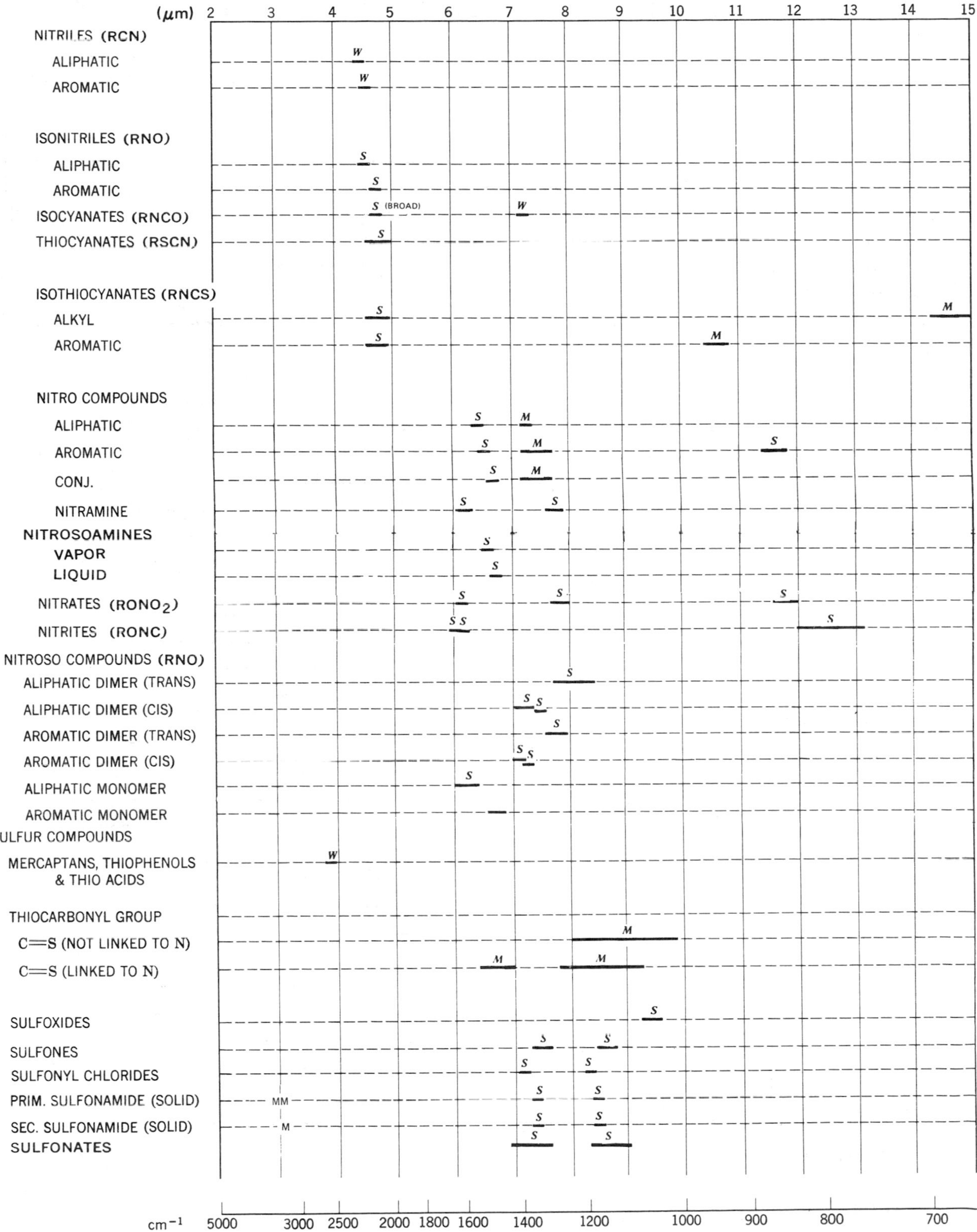

(μm) 2 3 4 5 6 7 8 9 10 11 12 13 14 15
NITRILES (RCN)
ALIPHATIC
AROMATIC
ISONITRILES (RNO)
ALIPHATIC
AROMATIC
ISOCYANATES (RNCO)
S (BROAD)
THIOCYANATES (RSCN)
ISOTHIOCYANATES (RNCS)
ALKYL
AROMATIC
NITRO COMPOUNDS
ALIPHATIC
AROMATIC
CONJ.
NITRAMINE
NITROSOAMINES
VAPOR
LIQUID
NITRATES (RONO2)
NITRITES (RONC)
NITROSO COMPOUNDS (RNO)
ALIPHATIC DIMER (TRANS)
ALIPHATIC DIMER (CIS)
AROMATIC DIMER (TRANS)
AROMATIC DIMER (CIS)
ALIPHATIC MONOMER
AROMATIC MONOMER
SULFUR COMPOUNDS
MERCAPTANS, THIOPHENOLS & THIO ACIDS
THIOCARBONYL GROUP
C=S (NOT LINKED TO N)
C=S (LINKED TO N)
SULFOXIDES
SULFONES
SULFONYL CHLORIDES
PRIM. SULFONAMIDE (SOLID)
MM
SEC. SULFONAMIDE (SOLID)
M
SULFONATES
cm⁻¹ 5000 3000 2500 2000 1800 1600 1400 1200 1000 900 800 700

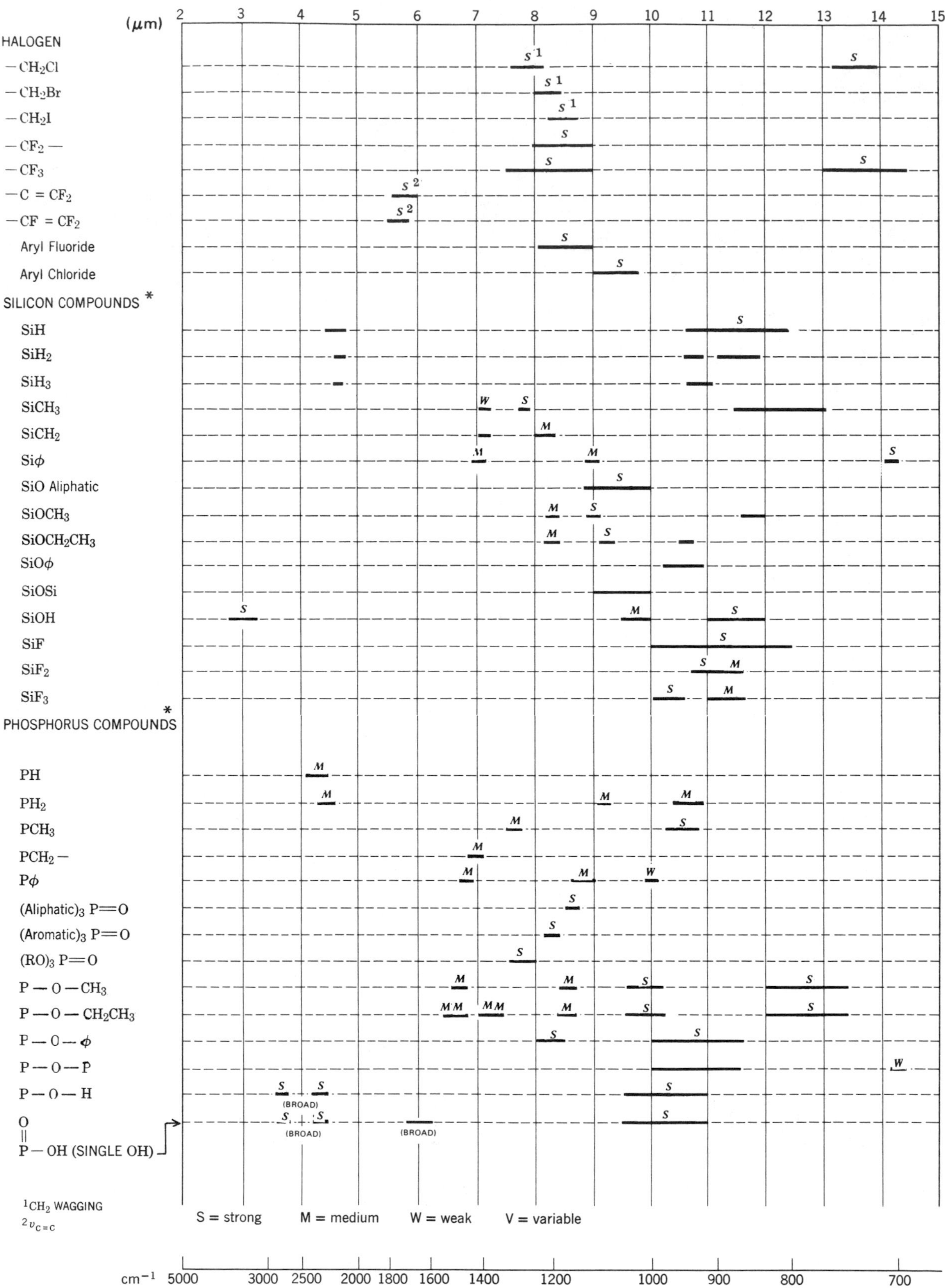

*Tables relating to these absorptions are found in Appendix E.

appendix d

TABLES AND CHARTS OF SPECIFIC GROUP FREQUENCIES[a]

Table I. CH_3 Group Absorption

Structure	$\nu_s CH_3$	$\delta_{as} CH_3$	$\delta_s CH_3$
	cm^{-1} (μm)	cm^{-1} (μm)	cm^{-1} (μm)
$CH_3-\overset{|}{\underset{|}{C}}$	2872 (3.48)	1450 (6.90)	1375 (7.28)
$CH_3-\overset{O}{\overset{\|}{C}}-$ (ketone)	3000-2900[a]w	1450-1400 (6.90-7.15)	1375-1350 (7.28-7.41)
CH_3-O-R	2832-2815[b] (3.53-3.55)	1470-1440 (6.80-6.95)	1470-1440 (6.80-6.95)
$CH_3-\overset{O}{\overset{\|}{C}}-O-$	–	1450-1400 (6.90-7.15)	1400-1340 (7.15-7.46)
$CH_3-N\langle$ (amines and imines)	2820-2760 (3.55-3.62)	–	1440-1390 (6.95-7.20)
$CH_3-\overset{|}{N}\underset{\underset{O}{\|}}{C}-$ (amides)	–	1500-1450 (6.67-6.90)	1420-1405 (7.04-7.11)
$CH_3-O-\overset{O}{\overset{\|}{C}}R$	–	ca. 1440 (ca. 6.95)	ca. 1360 (ca. 7.35)
CH_3-X (X = halogen)	–	–	1500-1250 (6.67-8.00)
CH_3-S-	–	1440-1415 (6.95-7.07)	1330-1290 (7.52-7.75)
$CH_3-Si\equiv$	–	1429-1389 (7.00-7.20)	1280-1255 (7.81-7.97)
$CH_3-P\langle$	–	–	1330-1290 (7.52-7.81)

[a]w = weak; m = medium; s = strong.
[b]Overtone of CH_3 bending band.

Table II. CH_2 Group Absorption

Structure	$\nu_s CH_2$ cm^{-1} (μm)	$\delta_s CH_2$ (scissoring) cm^{-1} (μm)	ωCH_2 (wagging) cm^{-1} (μm)
$-CH_2-R$	–	–	–
$-CH_2-C(=O)-$ (acyclic)	–	1435-1405 (6.97-7.11)	–
$-CH_2-C(=O)-$ (small ring)	–	1475-1425[a] (6.78-7.02)	–
$-CH_2-O-$ (acyclic)	–	1470-1435 (6.80-6.97)	–
$-CH_2-O-$ (small ring)	–	1500-1470[a] (6.67-6.80)	–
$-CH_2-N-$ (sec. and tert. amines)	2820-2760 (3.55-3.62)	1475-1445 (6.78-6.92)	–
$-CH_2-N(-)C(=O)-$ (amides)	–	1450-1405 (6.89-7.11)	–
$-CH_2-O-C(=O)-$	–	1475-1460 (6.78-6.85)	–
$-CH_2-S-$	–	1440-1415 (6.94-7.06)	–
$-CH_2-$halogen	–	1460-1430 (6.85-7.00)	1275-1170 (7.85-8.55)
$-CH_2NO_2$	–	1425-1415 (7.02-7.06)	–
$-CH_2-CN$	–	1425 (7.02)	–
$-CH_2-N=C=S$	–	–	1351-1316 (7.40-7.60)
$-Si(-)(-)-CH_2-$	–	near 1410 (7.09)	1250-1200 (8.00-8.33)
$-CH_2-P<$	–	1445-1405 (6.92-7.22)	–

[a]Multiple bands.

Table III. Olefinic Absorption

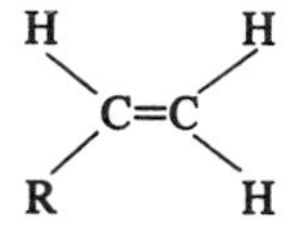

$$\underset{R}{\overset{H}{}}\!>C=C<\!\underset{H}{\overset{H}{}}$$

vinyl

1648–1638 cm^{-1} (6.07–6.10 μm)
995–985 cm^{-1} (10.05–10.15 μm) (*s*)[a]
915–905 cm^{-1} (10.93–11.05 μm) (*s*)

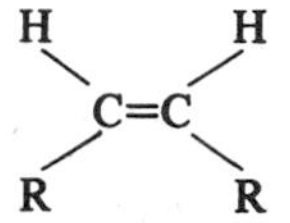

$$\underset{R}{\overset{H}{}}\!>C=C<\!\underset{R}{\overset{H}{}}$$

cis

1662–1626 cm^{-1} (6.02–6.15 μm) (*v*)
730–665 cm^{-1} (13.70–15.04 μm) (*s*)

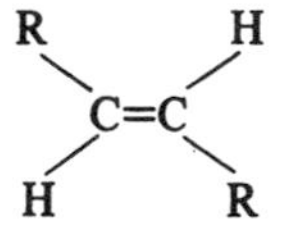

$$\underset{H}{\overset{R}{}}\!>C=C<\!\underset{R}{\overset{H}{}}$$

trans

1678–1668 cm^{-1} (5.96–6.00 μm) (v)
980–960 cm^{-1} (10.20–10.42 μm) (*s*)[b]

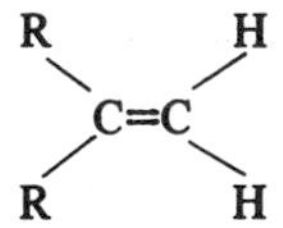

$$\underset{R}{\overset{R}{}}\!>C=C<\!\underset{H}{\overset{H}{}}$$

vinylidine

1658–1648 cm^{-1} (6.03–6.07 μm) (*m*)
895–885 cm^{-1} (11.17–11.30 μm) (*s*)

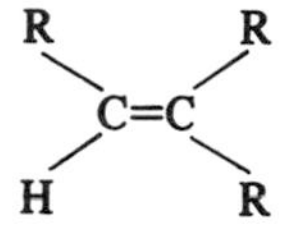

$$\underset{H}{\overset{R}{}}\!>C=C<\!\underset{R}{\overset{R}{}}$$

trisubstituted

1675–1665 cm^{-1} (5.97–6.01 μm) (*w*)
840–790 cm^{-1} (11.91–12.66 μm) (*m*)

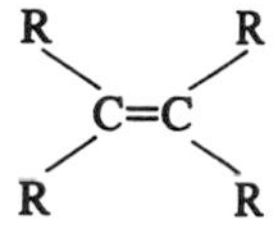

$$\underset{R}{\overset{R}{}}\!>C=C<\!\underset{R}{\overset{R}{}}$$

tetrasubstituted

1675–1665 cm^{-1} (5.97–6.01 μm very weak or absent.

[a]This band also shows a strong overtone band.
[b]This band occurs near 1000 cm^{-1} (10.00 μm) in conjugated *trans-trans* systems such as the esters of sorbic acid.
s = strong; m = medium; w = weak; v = variable.

Table IV. C=C Stretching Frequencies in Acyclic Systems

R_1	$CH_2=CHR_1$	$CH_2=C(R_1)_2$	$R_1-CH=CH-R_1$	$CHCH_3=C(R_1)_2$
H	1623[a] (6.16)	1623[a] (6.16)	1623[a] (6.16)	1648 (6.07)
D	–	–	–	–
F	1650 (6.06)	1728 (5.79)	–	–
Cl	1610 (6.21)	1620 (6.17)	*cis* 1590 (5.79) *trans* 1653 (6.05)	–
Br	1605 (6.23)	1593 (6.28)	–	–
I	1593 (6.28)	–	–	–
CH_3	1648 (6.07)	1661 (6.07)	*cis* 1661 (6.02) *trans* 1676 (5.97)	1681 (5.95)
CF_3	–	–	1700 (*cis*) (5.88)	–
OCH_3	1618 (6.18)	1639 (6.10)	–	1675 (5.97)
SCH_3	1586 (6.30)	–	–	1605 (6.23)

[a]Raman value.

Table V. C = C Stretching Frequencies in Cyclic and Acyclic Systems, cm^{-1} (μm)

Ring or Chain	(H)(C)C=C(H)(C)	(H)(C)C=C(CH_3)(C)	(CH_3)(C)C=C(CH_3)(C)	(C)(C)C=CH_2
Chain *cis*	1661 (6.02)	1681 (5.95)	1672 (5.98)	1661 (6.02)
Chain *trans*	1676 (5.97)			
Three-membered ring	1641 (6.10)		1890 (5.29)	1780 (5.62)
Four-membered ring	1566 (6.33)		1685 (5.94)	1678 (5.96)
Five-membered ring	1611 (6.21)	1658	1686 (5.93)	1657 (6.03)
Six-membered ring	1649 (6.06)	1678	1685 (5.93)	1651 (6.06)
Seven-membered ring	1651 (6.06	1673	–	
Eight-membered ring	1653 (6.05)	–	–	

appendix e ABSORPTIONS FOR Si AND P COMPOUNDS

Table I Si–H Absorptions

Mode	Position cm^{-1} (μm) intensity
Si–H stretch	2250-2100 (4.44-4.76) m-s
SiH_2 bend (scissoring)	942-923 (10.62-10.83)
SiH_3, two bands	
sym. and asym. bend	959-900 (10.43-11.11)
SiH_2 wag	900-840 (11.11-11.90)
R_3SiH bend	950-800 (10.53-12.50)

Table II Si–O Stretching Vibrations

Group	$\nu_{Si\text{-}O}$ Bands
Si–O–R (aliphatic)	1110-1000 cm^{-1} (9.01-10.00 μm)s
Si–O–Si	1110-1000 cm^{-1} (9.01-10.00 μm)s
Si–O–R (aromatic)	970-920 cm^{-1} (10.31-10.87 μm)s
Si–OH	910-830 cm^{-1} (11.00-12.05 μm)s

s = strong.

Table III Absorption of Si–F Bonds

Group	$\nu_{Si\text{-}F}$ Bands	
$\geq$SiF		1000-800 cm^{-1} (10.00-12.50 μm)
$>SiF_2$	943-910 cm^{-1} (10.60-11.00 μm)s	910-870 cm^{-1} (11.00-11.50 μm)m
$-SiF_3$	980-945 cm^{-1} (10.20-10.58 μm)s	910-860 cm^{-1} (11.00-11.63 μm)m

s = strong; m = medium.

Table IV PH, P = O Absorptions

Mode	Position cm^{-1} (μm) intensity
P-H stretch	2440-2275 (4.10-4.40)m
alkyl, aryl, P-H stretch	2326-2275 (4.30-4.40)
PH_2 bend	
scissoring	1090-1080 (9.19-9.26)m
wagging	940-909 (10.64-11.00)
P=O stretch	
phosphine oxides	
aliphatic	ca. 1150 (8.70)
aromatic	ca. 1190 (8.40)
phosphate esters[a]	1299-1250 (7.70-8.00)

[a]The increase in P = O stretching frequency of the ester, relative to the oxides, results from the electronegativity of the attached alkoxy groups.

Table V P–O Stretching Vibrations

Group	$\nu_{P\text{-}O}$ Bands	
P–OH	1040-910 cm^{-1} (9.62-11.00 μ)s	
P–O–P	1000-870 cm^{-1} (10.00-11.50 μm)s	ca. 700 cm^{-1} (14.30 μm)w
P–O–C (aliph.)	1050-970 cm^{-1} (9.52-10.30 μm)s[a]	830-740 cm^{-1} (12.05-13.51 μm)s[b]
P–O–C (arom.)	1260-1160 cm^{-1} (7.94-8.62 μm)s	994-855 cm^{-1} (10.06-11.70 μm)s

[a]May be a doublet.
bMay be absent.
s = strong; w = weak.

Table VI γCH and Ring Bending (β-Ring) Bands of Pyridines

Substitution	Number Adjacent H's	γCH	β Ring
2-	4	781-740 cm^{-1} (12.80-13.50 μm)	752-746 cm^{-1} (13.30-13.40 μm)
3-	3	810-789 cm^{-1} (12.35-12.67 μm)	715-712 cm^{-1} (13.99-14.04 μm)
4-	2	820-794 cm^{-1} (12.19-12.60 μm)	775-709 cm^{-1} (12.90-14.10 μm)

The γ and β notations are explained on p. 135 and in reference 27 listed on p. 136.

Table VII. Characteristic γCH or β Ring Bands of Furans, Thiophenes, and Pyrroles

Nucleus	Position of Substitution	Phase	γCH or β Ring Modes			
			cm^{-1} (μm)	cm^{-1} (μm)	cm^{-1} (μm)	cm^{-1} (μm)
Furan	2-	$CHCl_3$	ca. 925 (10.8)	ca. 884 (11.3)	835-780 (12.0-12.8)	
	2-	liq.	960-915 (10.4-10.9)	890-875 (11.2-11.4)		780-725 (12.8-13.8)
	2-	Solid	955-906 (10.5-11.0)	887-860 (11.3-11.6)	821-793 (12.2-12.6)	750-723 (13.3-13.8)
	3-	liq.		885-870 (11.3-11.5)	741 (13.5)	
Thiophene	2-	$CHCl_3$	ca. 925 (10.8)	ca. 853 (11.7)	843-803 (11.9-12.5)	
	3-	liq.				755 (13.2)
Pyrrole	2-Acyl	Solid			774-740 (12.9-13.5)	ca. 755 (13.2)

appendix f

WAVELENGTH–WAVE NUMBER CONVERSION TABLE

wavelength (μm)	Wave number (cm^{-1}) 0	1	2	3	4	5	6	7	8	9
2.0	5000	4975	4950	4926	4902	4878	4854	4831	4808	4785
2.1	4762	4739	4717	4695	4673	4651	4630	4608	4587	4566
2.2	4545	4525	4505	4484	4464	4444	4425	4405	4386	4367
2.3	4348	4329	4310	4292	4274	4255	4237	4219	4202	4184
2.4	4167	4149	4232	4115	4098	4082	4065	4049	4032	4016
2.5	4000	3984	3968	4953	3937	3922	3006	3891	3876	3861
2.6	3846	3831	3817	3802	3788	3774	3759	3745	3731	3717
2.7	3704	3690	3676	3663	3650	3636	3623	3610	3597	3584
2.8	3571	3559	3546	3534	3521	3509	3497	3484	3472	3460
2.9	3448	3436	3425	3413	3401	3390	3378	3367	3356	3344
3.0	3333	3322	3311	3300	3289	3279	3268	3257	3247	3236
3.1	3226	3215	3205	3195	3185	3175	3165	3155	3145	3135
3.2	3125	3115	3106	3096	3086	3077	3067	3058	3049	3040
3.3	3030	3021	3012	3003	2994	2985	2976	2967	2959	2950
3.4	2941	2933	2924	2915	2907	2899	2890	2882	2874	2865
3.5	2857	2849	2841	2833	2825	2817	2809	2801	2793	2786
3.6	2778	2770	2762	2755	2747	2740	2732	2725	2717	2710
3.7	2703	2695	2688	2681	2674	2667	2660	2653	2646	2639
3.8	2632	2625	2618	2611	2604	2597	2591	2584	2577	2571
3.9	2654	2558	2551	2545	2538	2532	2525	2519	2513	2506
4.0	2500	2494	2488	2481	2475	2469	2463	2457	2451	2445
4.1	2439	2433	2427	2421	2415	2410	2404	2398	2387	2387
4.2	2381	2375	2370	2364	2358	2353	2347	2342	2336	2331
4.3	2326	2320	2315	2309	2304	2299	2294	2288	2283	2278
4.4	2273	2268	2262	2257	2252	2247	2242	2237	2232	2227
4.5	2222	2217	2212	2208	2203	2198	2193	2188	2183	2179
4.6	2174	2169	2165	2160	2155	2151	2146	2141	2137	2132
4.7	2128	2123	2119	2114	2110	2105	2101	2096	2092	2088
4.8	2083	2079	2075	2070	2066	2062	2058	2053	2049	2045
4.9	2041	2037	2033	2028	2024	2020	2016	2012	2008	2004
5.0	2000	1996	1992	1988	1984	1980	1976	1972	1969	1965
5.1	1961	1957	1953	1949	1946	1942	1938	1934	1931	1927
5.2	1923	1919	1916	1912	1908	1905	1901	1898	1894	1890
5.3	1887	1883	1880	1876	1873	1869	1866	1862	1859	1855
	0	1	2	3	4	5	6	7	8	9

Wavelength (μm)	0	1	2	3	4	5	6	7	8	9
	Wave number (cm^{-1})									
5.4	1852	1848	1845	1842	1838	1835	1832	1828	1825	1821
5.5	1818	1815	1812	1808	1805	1802	1799	1795	1792	1788
5.6	1786	1783	1779	1776	1773	1770	1767	1764	1761	1757
5.7	1754	1751	1748	1745	1742	1739	1736	1733	1730	1727
5.8	1724	1721	1718	1715	1712	1709	1706	1704	1701	1698
5.9	1695	1692	1689	1686	1684	1681	1678	1675	1672	1669
6.0	1667	1664	1661	1658	1656	1653	1650	1647	1645	1642
6.1	1639	1637	1634	1631	1629	1626	1623	1621	1618	1616
6.2	1613	1610	1608	1605	1603	1600	1597	1595	1592	1590
6.3	1587	1585	1582	1580	1577	1575	1572	1570	1567	1565
6.4	1563	1560	1558	1555	1553	1550	1548	1546	1543	1541
6.5	1538	1536	1534	1531	1529	1527	1524	1522	1520	1517
6.6	1515	1513	1511	1508	1506	1504	1502	1499	1497	1495
6.7	1493	1490	1488	1486	1484	1481	1479	1477	1475	1473
6.8	1471	1468	1466	1464	1462	1460	1458	1456	1453	1451
6.9	1449	1447	1445	1443	1441	1439	1437	1435	1433	1431
7.0	1429	1427	1425	1422	1420	1418	1416	1414	1412	1410
7.1	1408	1406	1404	1403	1401	1399	1397	1395	1393	1391
7.2	1389	1387	1385	1383	1381	1379	1377	1376	1374	1372
7.3	1370	1368	1366	1364	1362	1361	1359	1357	1355	1353
7.4	1351	1350	1348	1346	1344	1342	1340	1339	1337	1335
7.5	1333	1332	1330	1328	1326	1325	1323	1321	1319	1318
7.6	1316	1314	1312	1311	1309	1307	1305	1304	1302	1300
7.7	1299	1297	1295	1294	1292	1290	1289	1287	1285	1284
7.8	1282	1280	1279	1277	1276	1274	1272	1271	1269	1267
7.9	1266	1264	1263	1261	1259	1258	1256	1255	1253	1252
8.0	1250	1248	1247	1245	1244	1242	1241	1239	1238	1236
8.1	1235	1233	1232	1230	1229	1227	1225	1224	1222	1221
8.2	1220	1218	1217	1215	1214	1212	1211	1209	1208	1206
8.3	1205	1203	1202	1200	1199	1198	1196	1195	1193	1192
8.4	1190	1189	1188	1186	1185	1183	1182	1181	1179	1178
8.5	1176	1175	1174	1172	1171	1170	1168	1167	1166	1164
8.6	1163	1161	1160	1159	1157	1156	1155	1153	1152	1151
8.7	1149	1148	1147	1145	1144	1143	1142	1140	1139	1138
8.8	1136	1135	1134	1133	1131	1130	1129	1127	1126	1125
8.9	1124	1122	1121	1120	1119	1117	1116	1115	1114	1112
9.0	1111	1110	1109	1107	1106	1105	1104	1103	1101	1100
9.1	1099	1098	1096	1095	1094	1093	1092	1091	1089	1088
9.2	1087	1086	1085	1083	1082	1081	1080	1079	1078	1076
9.3	1075	1074	1073	1072	1071	1070	1068	1067	1066	1065
9.4	1064	1063	1062	1060	1059	1058	1057	1056	1055	1054
9.5	1053	1052	1050	1049	1048	1047	1046	1045	1044	1043
9.6	1042	1041	1040	1038	1037	1036	1035	1034	1033	1032
9.7	1031	1030	1029	1028	1027	1026	1025	1024	1022	1021
9.8	1020	1019	1018	1017	1016	1015	1014	1013	1012	1011
9.9	1010	1009	1008	1007	1006	1005	1004	1003	1002	1001
10.0	1000	999	998	997	996	995	994	993	992	991
	0	1	2	3	4	5	6	7	8	9

Wavelength (μm)	Wave number (cm^{-1})									
	0	1	2	3	4	5	6	7	8	9
10.1	990	989	988	987	986	985	984	983	982	981
10.2	980	979	978	978	977	976	975	974	973	972
10.3	971	970	969	968	967	966	965	964	963	962
10.4	962	961	960	959	958	957	956	955	954	953
10.5	952	951	951	950	949	948	947	946	945	944
10.6	943	943	942	941	940	939	938	937	936	935
10.7	935	934	933	932	931	930	929	929	928	927
10.8	926	925	924	923	923	922	921	920	919	918
10.9	917	917	916	915	914	913	912	912	911	910
11.0	909	908	907	907	906	905	904	903	903	902
11.1	901	900	899	898	898	897	896	895	894	894
11.2	893	892	891	890	890	889	888	887	887	886
11.3	885	884	883	883	882	881	880	880	879	878
11.4	877	876	876	875	874	873	873	872	871	870
11.5	870	869	868	867	867	866	865	864	864	863
11.6	862	861	861	860	859	858	858	857	856	855
11.7	855	854	853	853	852	851	850	850	849	848
11.8	847	847	846	845	845	844	843	842	842	841
11.9	840	840	839	838	838	837	836	835	835	834
12.0	833	833	832	831	831	830	829	829	828	827
12.1	826	826	825	824	824	823	822	822	821	820
12.2	820	819	818	818	817	816	816	815	814	814
12.3	813	812	812	811	810	810	809	808	808	807
12.4	806	806	805	805	804	803	803	802	801	801
12.5	800	799	799	798	797	797	796	796	795	794
12.6	794	793	792	792	791	791	790	789	789	788
12.7	787	787	786	786	785	784	784	783	782	782
12.8	781	781	780	779	779	778	778	777	776	776
12.9	775	775	774	773	773	772	772	771	770	770
13.0	769	769	768	767	767	766	766	765	765	764
13.1	763	763	762	762	761	760	760	759	759	758
13.2	758	757	756	756	755	755	754	754	753	752
13.3	752	751	751	750	750	749	749	748	747	747
13.4	746	746	745	745	744	743	743	742	742	741
13.5	741	740	740	739	739	738	737	737	736	736
13.6	735	735	734	734	733	733	732	732	731	730
13.7	730	729	729	728	728	727	727	726	726	725
13.8	725	724	724	723	723	722	722	721	720	720
13.9	719	719	718	718	717	717	716	716	715	715
14.0	714	714	713	713	712	712	711	711	710	710
14.1	709	709	708	708	707	707	706	706	705	705
14.2	704	704	703	703	702	702	701	701	700	700
14.3	699	699	698	698	697	697	696	696	695	695
14.4	694	694	693	693	693	692	692	691	691	690
14.5	690	689	689	688	688	687	687	686	686	685
14.6	685	684	684	684	683	683	682	682	681	681
14.7	680	680	679	679	678	678	678	677	677	676
14.8	676	675	675	674	674	673	673	672	672	672
14.9	671	671	670	670	669	669	668	668	668	667
	0	1	2	3	4	5	6	7	8	9

four

proton magnetic resonance spectrometry

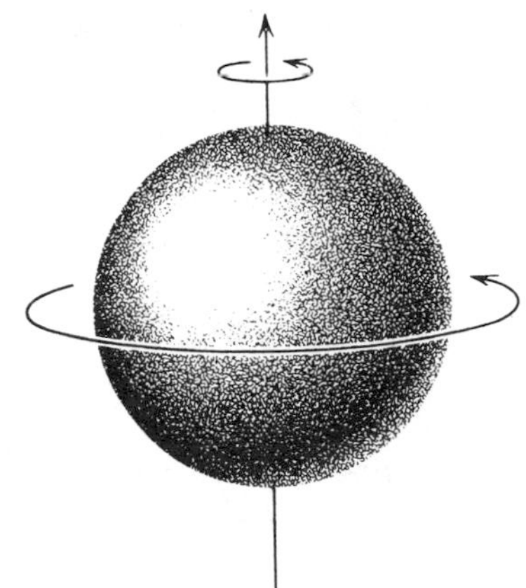

Figure 1. *Spinning charge in proton generates magnetic dipole.*

I. INTRODUCTION AND THEORY

Nuclear magnetic resonance (NMR) spectrometry is basically another form of absorption spectrometry, akin to infrared or ultraviolet spectrometry. Under appropriate conditions, a sample can absorb electromagnetic radiation in the radio-frequency region at frequencies governed by the characteristics of the sample. Absorption is a function of certain nuclei in the molecule. A plot of the frequencies of the absorption peaks versus peak intensities constitutes an NMR spectrum. This chapter covers proton magnetic resonance (^{1}H NMR or pmr) spectrometry.

With some mastery of basic theory, interpretation of NMR spectra merely by inspection is usually feasible in greater detail than is the case for infrared or ultraviolet spectra. To supplement the material presented here, several nonmathematical introductions to nuclear magnetic resonance are recommended.[1-12] For greater depth several treatises are available.[13-17] Collections of spectra[18-21] and spectra-structure correlations[22-26] are essential for practical use. The present account will suffice for the immediate limited objective: identification of organic compounds in conjunction with other spectrometric information.

We begin by describing some magnetic properties of nuclei. All nuclei carry a charge. In some nuclei this charge "spins" on the nuclear axis, and this circulation of nuclear charge generates a magnetic dipole along the axis (Figure 1). The angular momentum of the spinning charge can be described in terms of spin numbers I; these numbers have values of 0, 1/2, 1, 3/2, and so forth ($I = 0$ denotes no spin). The intrinsic magnitude of the generated dipole is expressed in terms of nuclear magnetic moment, μ.

Each proton and neutron has its own spin, and I is a resultant of these spins. If the sum of protons and neutrons is even, I is zero or integral (0, 1, 2, . . .); if the sum is odd, I is half-integral (1/2, 3/2, 5/2, . . .); if both protons and neutrons are even-numbered, I is zero. Both ^{12}C and ^{16}O fall in the last category and give no NMR signal

Several nuclei (^{1}H, ^{19}F, ^{13}C, and ^{31}P) have a spin number I of 1/2 and a uniform spherical charge distribution (Figure 1). Nuclei with a spin number I of 1 or higher have a nonspherical charge distribution. This asymmetry is described by an electrical quadrupole moment which, as we shall see later, affects the relaxation time and, consequently, the coupling with neighboring nuclei. ^{14}N and ^{2}H have a spin number I of 1. ^{11}B, ^{35}Cl, ^{79}Br, and ^{81}Br are examples of nuclei with $I = 3/2$.

In quantum mechanical terms, the spin number I determines the number of orientations a nucleus may assume in an external uniform magnetic field in accordance with the formulas $2I + 1$. We are concerned primarily with the proton whose spin number I is ½ (Figure 2); thus, there are 2 energy levels, and a slight excess population in the lower energy state.

Two energy levels for the proton having been established, it should be possible to introduce quanta of energy $h\nu$ (h is Planck's constant; ν is the frequency of electro-

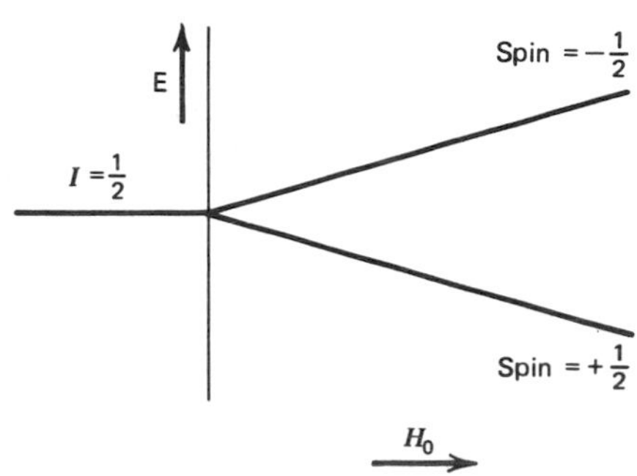

Figure 2. *Two energy levels for a proton in a magnetic field, H_0.*

magnetic radiation) to effect a transition between these energy levels in a magnetic field of given strength H_0. The fundamental NMR equation correlating electromagnetic frequency with magnetic field strength is

$$\nu = \frac{\gamma H_0}{2\pi}$$

A frequency of 60 megahertz (MHz) is needed at a magnetic field H_0 of 14,092 gauss for the proton (or any other desired combination in the same ratio). The constant γ is called the magnetogyric ratio and is a fundamental nuclear constant; it is the proportionality constant between the magnetic moment μ and the spin number I:

$$\gamma = \frac{2\pi\mu}{hI}$$

The problem is how to inject radiofrequency electromagnetic energy into protons aligned in a magnetic field, and how to measure the energy thus absorbed as the proton is raised to the higher spin state. This can best be explained in classical mechanical terms wherein we visualize the proton as spinning in an external magnetic field: the magnetic axis of the proton will precess about the axis of the external magnetic field in the same manner in which a spinning gyroscope precesses under the influence of gravity (Figure 3). The precessional angular velocity (Larmor frequency ω_0) is equal to the product of the magnetogyric ratio and the strength of the applied magnetic field H_0.

$$\omega_0 = \gamma H_0$$

We recall from the fundamental NMR equation that

$$\gamma H_0 = 2\pi\nu$$

Therefore,

$$\omega_0 = 2\pi\nu$$

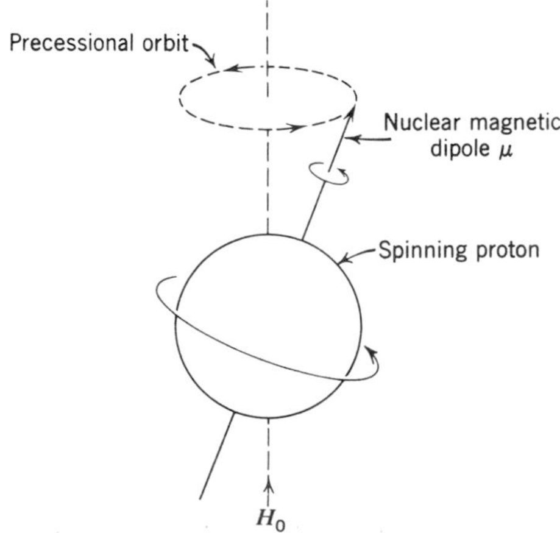

Figure 3. *Proton precessing in a magnetic field H_0.*

We may now arrange the geometry for an NMR experiment. We subject the protons to a powerful, uniform magnetic field and consider those protons aligned with the field and precessing about the axis of the applied magnetic field H_0.

Because of thermal disorder, actually only a small fraction of the total population of the protons is properly aligned, but this fraction is sufficient. The radiofrequency electromagnetic energy is applied in such a way that its magnetic component H_1 is at right angles to the main magnetic field H_0 and is rotating with the precessing proton. This is accomplished by a coil with its axis at right angles to the axis of the main magnetic field H_0. Such a coil will generate an oscillating magnetic field H_1 along the direction of the coil axis as shown in Figure 4. An oscillating magnetic field can be resolved into two components rotating in opposite directions. One of these components is rotating in the same direction as the precessional orbit of the nuclear magnetic dipole (the proton); the oppositely rotating component of H_1 is disregarded. If H_0 is held constant and the oscillator frequency is varied, the angular velocity of the component of rotating magnetic field H_1 will vary until it is equal to (in resonance with) the angular velocity ω_0 of the precessing proton. At this point, the absorbed energy is at a maximum, and the precessing nuclei are tilted away from alignment with H_0 toward the horizontal plane in Figure 4. The magnetic component thus generated in that plane can be detected. Alternatively, the oscillator frequency is held constant, and H_0 is swept over a narrow range.

Now that we have briefly discussed how a nucleus is excited to a higher energy state by absorption of energy, we need to describe a mechanism for the return of the nucleus to the ground state. In the absence of such a mechanism, all of the small excess population of nuclei in the lower energy state will be raised to the high energy state, and no more energy could be absorbed. Fortunately, there exists a mechanism whereby the nucleus in the higher energy state can lose energy to its environment and thus return to its lower energy state. The mechanism is called a spin-lattice or longitudinal relaxation process and involves transfer of energy from the nucleus in its high-energy state to the molecular lattice. Its efficiency is described as the half-life, T_1, for the transfer, an efficient relaxation process involving a short time T_1. The line width is inversely proportional to the lifetime of the excited state. In neat liquids, solutions, and gases, a peak of usable width is obtained. In solids, this mechanism is not effective; T_1 is therefore very long, and, in the absence of any other effects, a crystalline solid would show extremely narrow lines. There is another effect called spin-spin or transverse relaxation, described in terms of time T_2. This involves transfer of energy from one high-energy nucleus to another. There is no net loss of energy, but the spread of energy among the nuclei concerned results in line broadening. Solids show very broad bands.

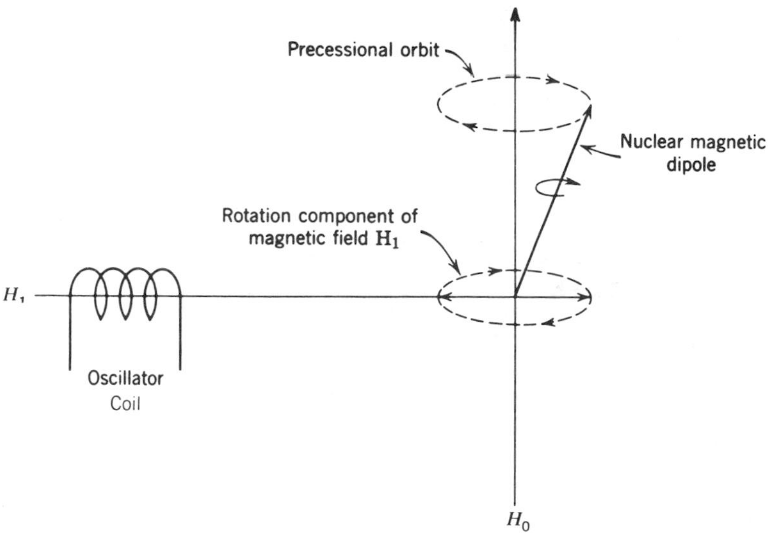

Figure 4. *Oscillator generates rotating component of magnetic field* H_1.

We can now consider very briefly the instrumentation necessary to obtain a nuclear magnetic resonance spectrum.

II. APPARATUS AND SAMPLE HANDLING

Proton NMR spectra are commonly run at 60, 80, 90, or 100 MHz* corresponding to magnetic fields of 14,092, 18,667, ~21,000 or ~23,500 gauss, respectively. Higher frequencies (up to 400 MHz) are used in conjunction with superconducting, helium-cooled magnets. In addition, ^{19}F, ^{11}B, ^{13}C, ^{2}H, ^{15}N, and ^{31}P and many other nuclei can be examined at appropriate combinations of frequencies and magnetic-field strength.

A schematic diagram of a simple NMR spectrometer is shown in Figure 5. The instrument can be described in terms of the following components:

1. A strong magnet with a homogeneous field that can be varied continuously and precisely over a relatively narrow range. This is accomplished by the sweep generator.
2. A radiofrequency transmitter.
3. A radiofrequency receiver.
4. A recorder, calibrator, and integrator.
5. A sample holder that positions the sample relative to the main magnetic field, the transmitter coil, and the receiver coil. The sample holder also spins the sample to increase the apparent homogeneity of the magnetic field. A variable-temperature sample holder is also available.

*One hertz, Hz, is 1 cycle per second. MHz is 1×10^6 Hz.

By measuring frequency shifts from a reference marker, an accuracy of <1 Hz can be achieved. The recording is presented as a series of peaks with the peak areas being proportional to the number of protons they represent. Peak areas are measured by an electronic integrator that traces a series of steps with heights proportional to the peak areas. The steps can be superimposed on the peaks. As will be obvious in Chapter 7, proton counting with the integrator is extremely useful. Peaks hidden under other peaks or in the baseline noise can be detected. Proton counting is often useful for determining sample purity and, of course, for quantitative analytical work.

A routine sample on a 60-MHz instrument would consist of about 5-50 mg of the compound in about 0.4 ml of solvent in a 5-ml O.D. glass tube. A microtube consisting of a thick-wall capillary leading to a 25-μl spherical cavity allows usable spectra to be obtained on less than 1 mg of sample. To overcome the lack of inherent sensitivity of NMR spectrometry, repetitive scans can be used. Signals can be accumulated and noise averaged out by a computer attachment to give usable spectra on samples of the order of 150 μg. On more sophisticated instruments, a capillary probe together with Fourier Transform (FT, see Chapter 5) makes it possible to obtain usable spectra on amounts as small as 5 μg.

The ideal solvent should contain no protons in its structure and be inexpensive, low-boiling, nonpolar, and inert. Carbon tetrachloride is ideal when the sample is sufficiently soluble in it. The most widely used solvent is deuterated

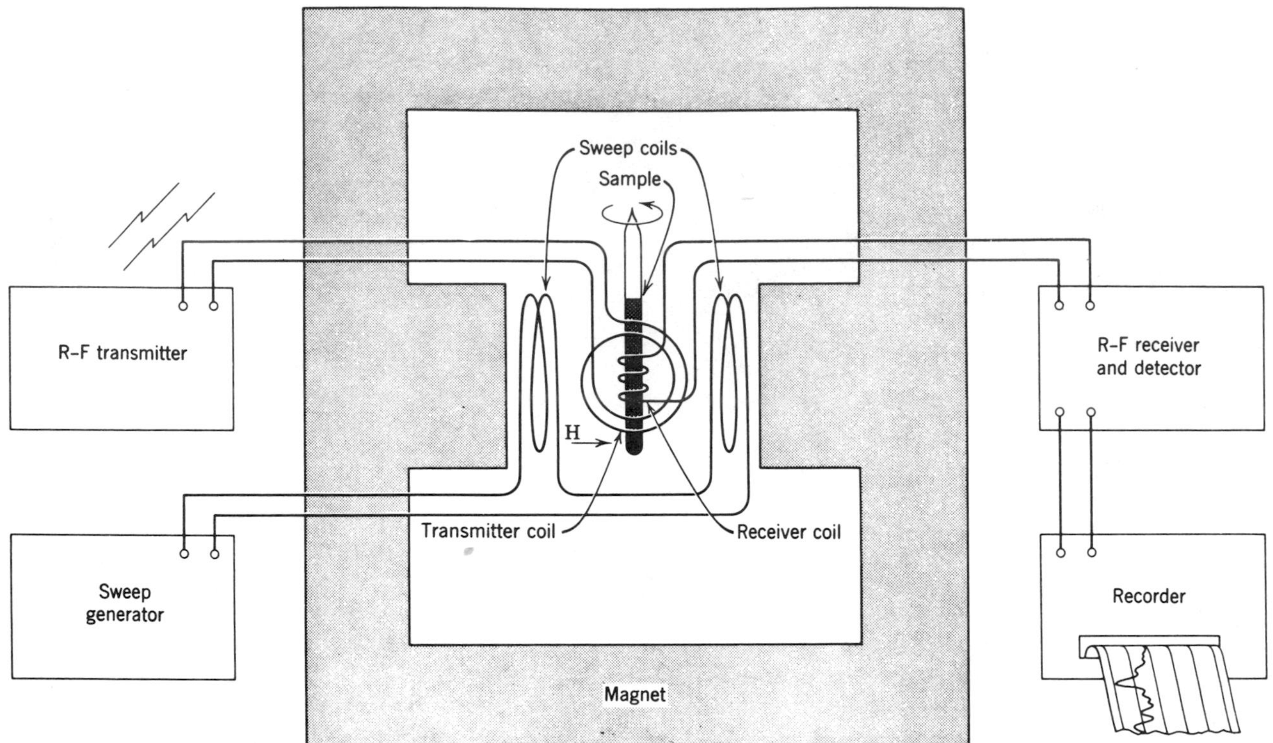

Figure 5. *Schematic diagram of an NMR spectrometer.*

Courtesy of Varian Associates, Palo Alto, Calif.

chloroform ($CDCl_3$). The small sharp proton peak from the $CHCl_3$ impurity present (see Appendix A) rarely interferes seriously. Carbon disulfide is also a useful solvent. Almost all of the common solvents are available in the deuterated form with an isotopic purity (atom %D) of 98-99.8%. A list of these solvents and the positions of their proton impurity peaks are given in Appendix A.

Small "spinning side bands" (see Figure 6) are often seen, symmetrically disposed on both sides of a strong absorption peak; these result from inhomogeneities in the magnetic field and in the spinning tube. They are readily recognized because of their symmetrical appearance and because their separation from the absorption peak is equal to the rate of spinning. The oscillations often seen at the high-field end of a strong sharp peak are called "ringing" (Figure 7). These are "beat" frequencies resulting from "fast" (normal operation) passage through the absorption peak.

Traces of ferromagnetic impurities cause severe broadening of absorption peaks. Common sources are tap water, steel wool, Raney nickel, and particles from metal spatulas or fittings (Figure 8). These impurities can be removed by dipping a thin bar magnet into the NMR tube, by filtration, or by centrifugation.

III. CHEMICAL SHIFT

Only a single peak should be obtainable from the interaction of radio frequency energy and a strong magnetic field on a proton in accordance with the basic NMR equation in which γ, the magnetogyric ratio, is an intrinsic property of the nucleus. The peak area (measured by the integrator) is proportional to the number of protons it represents. Fortunately, the situation is not quite so simple. The nucleus is shielded to a small extent by its electron cloud whose density varies with the environment. This variation gives rise to different absorption positions usually within the range of about 750 Hz in a magnetic field corresponding to 60 MHz or about 1250 Hz in a field corresponding to 100 MHz. The ability to discriminate among the individual absorptions describes high resolution NMR spectrometry.

Electrons under the influence of a magnetic field will circulate, and, in circulating, will generate their own magnetic field opposing the applied field, hence, the shielding effect (Figure 9). This effect accounts for the diamagnetism exhibited by all organic materials. In the case of materials with an unpaired electron, the paramagnetism associated with the net electron spin far overrides the diamagnetism of the circulating, paired electrons.

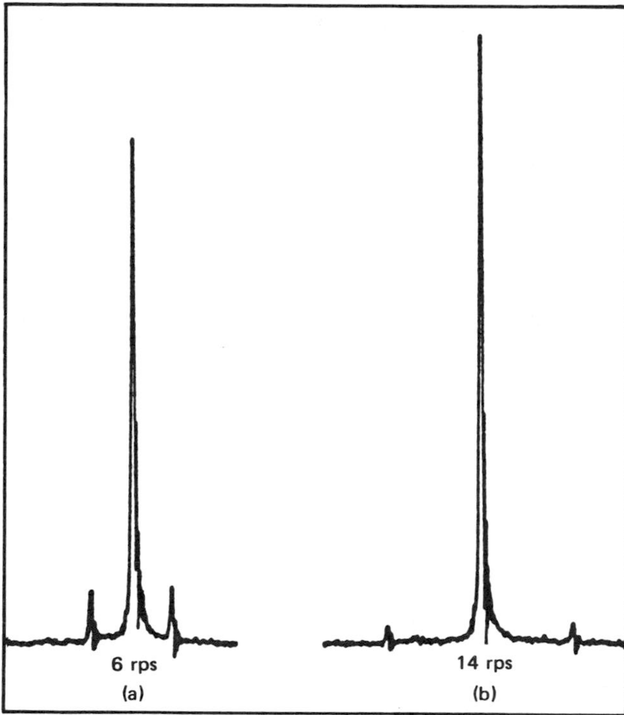

Figure 6. *Signal of neat chloroform with spinning side bands produced by spinning rate of (a) 6 rps and (b) 14 rps.*

From Reference 1, with permission.

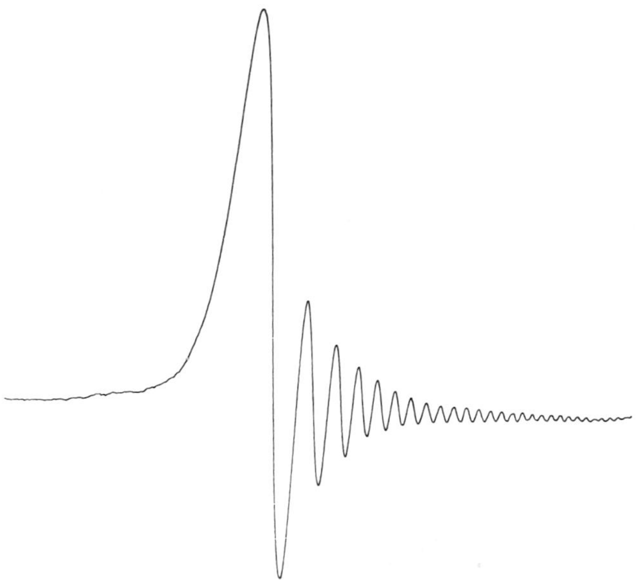

Figure 7. *Ringing (or "wiggles") seen after passage through resonance. Direction of scan is from left to right.*

(From Reference 7, with permission.)

The degree of shielding depends on the density of the circulating electrons, and, as a first, very rough approximation, the degree of shielding of a proton on a carbon atom will depend on the inductive effect of other groups attached to the carbon atom. The difference in the absorption position of a particular proton from the absorption position of a reference proton is called the *chemical shift* of the particular proton.

We now have the concept that protons in "different" chemical environments have different chemical shifts. Conversely, protons in the "same" chemical environment have the same chemical shift. But what do we mean by "different" and "same"? While it is intuitively obvious that the chemically different methylene groups of $ClCH_2CH_2OH$ have different chemical shifts and that the protons in either one of the methylene groups have the same chemical shift, it may not be so obvious, for example, that the protons of the methylene group of $C_6H_5CH_2CHBrCl$ do not have the same chemical shift. For the present, we shall deal with obvious cases and postpone a more rigorous treatment of chemical-shift equivalence to Section VII.

The most generally useful reference compound is tetramethylsilane (TMS).

$$\begin{array}{c} CH_3 \\ | \\ CH_3\text{—}Si\text{—}CH_3 \\ | \\ CH_3 \end{array}$$

It has several advantages: it is chemically inert, magnetically isotropic, volatile (b.p. 27°), and soluble in most organic solvents; it gives a single sharp absorption peak, and absorbs at higher field than almost all organic protons. When water or deuterium oxide is the solvent, TMS can be used as an "external reference," that is, sealed in a capillary immersed in the solution. The methyl protons of sodium 2,2-dimethyl-2-silapentane-5-sulfonate (DSS):

$$(CH_3)_3SiCH_2CH_2CH_2SO_3Na$$

are sometimes used as an internal reference in aqueous solution. Since only enough is used to give a small methyl peak, the diffuse pattern of the CH_2 peaks barely shows on the base line. Unless hydrogen-bonding effects are involved, a proton peak referenced to TMS in deuterochloroform will be within 0.01 to 0.03 ppm of the same peak referenced to DSS in water or deuterium oxide. Acetonitrile and dioxane are also used as references in aqueous solution.

Let us set up an NMR scale (Figure 10) and set the TMS peak at 0 Hz at the right-hand edge. The magnetic field increases toward the right. When chemical shifts are given in Hz (designated ν), the applied frequency must be specified. Chemical shifts can be expressed in dimensionless units, independent of the applied frequency, by dividing

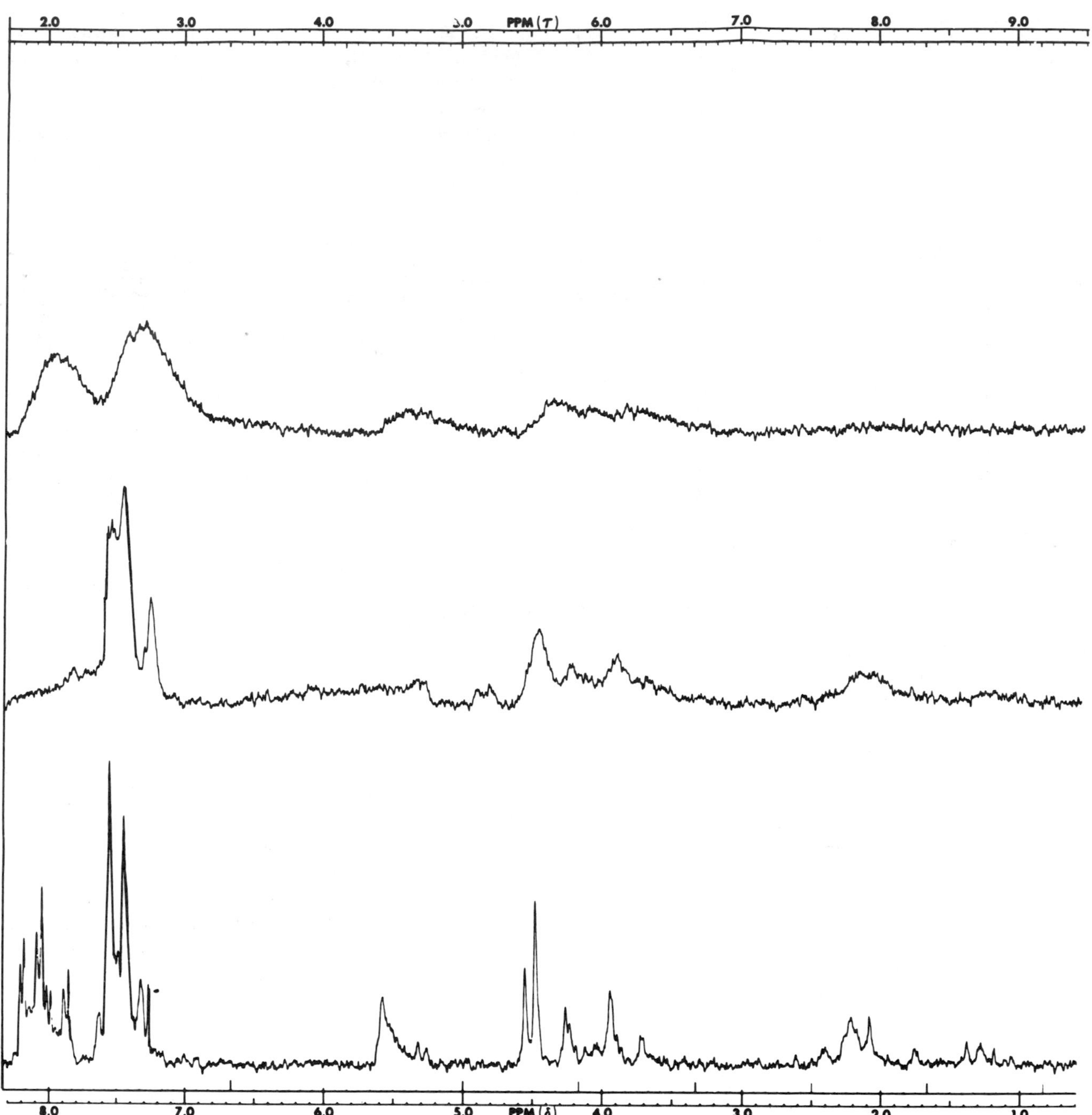

Figure 8. *The effect of a tiny ferromagnetic particle on the proton resonance spectrum of a benzoylated sugar. The top and middle curves are repeated runs with the particle present; the bottom curve is the spectrum with the particle removed. (From Reference 7, with permission)*

ν by the applied frequency and multiplying by 10^6. Thus, a peak at 60 Hz (ν 60) from TMS at an applied frequency of 60 MHz would be at $\delta 1.00$ (δ scale).

$$\frac{60}{60 \times 10^6} \times 10^6 = \delta 1.00$$

Since δ units are expressed in parts per million, the expression ppm is often used. The same peak at an applied frequency of 100 MHz would be at ν 100 but would still be at δ 1.00.

$$\frac{100}{100 \times 10^6} \times 10^6 = \delta\ 1.00$$

An alternative system assigns a value of 10.00 for tetra-

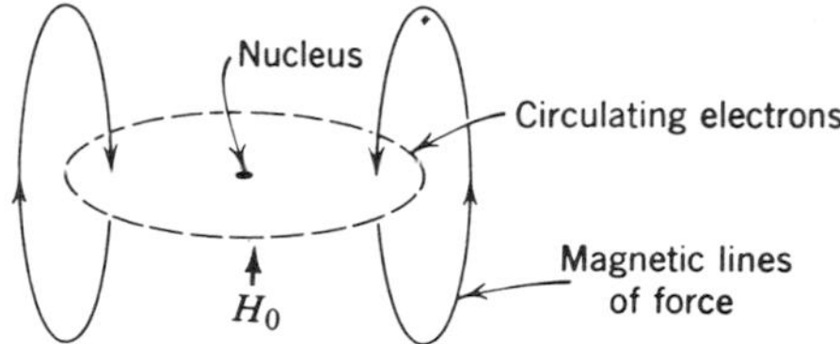

Figure 9. *Diamagnetic shielding of nucleus by circulating electrons.*

methylsilane, and describes chemical shifts in terms of τ values.

$$\tau = 10.00 - \delta$$

Shifts at higher field than TMS (δ 0.00, τ 10.00) will be encountered very rarely; δ values are then shown with a negative sign, and τ values merely increase numerically. The δ system is now used almost universally, and will be used here.

It is important to realize that the chemical shift in Hz is directly proportional to the strength of the applied field H_0 and therefore to the applied frequency. This is understandable because the chemical shift is dependent on the diamagnetic shielding induced by H_0. The strongest magnetic field consistent with field homogeneity should be used to spread out the chemical shifts. This is made clear in Figure 10, and in Figure 11 in which increased applied magnetic field in the NMR spectrum of acrylonitrile means increased separation of signals.

The concept of electronegativity is a dependable guide, up to a point, to chemical shifts. It tells us that the electron density around the protons of TMS is high (silicon is electropositive relative to carbon), and these protons will therefore be highly shielded and their peak will be found at high field. We could make a number of good estimates as to chemical shifts, using concepts of electronegativity and proton acidity. For example, the following values are reasonable on these grounds:

	δ
$(CH_3)_2O$	3.27
CH_3F	4.30
RCOOH	10.8 (approx.)

But finding the protons of acetylene at δ 2.35, that is, more shielded than ethylene protons (δ 4.60), is unsettling. Finding the aldehydic proton of acetaldehyde at δ 9.97 definitely calls for some augmentation of the electronegativity concept. We shall use diamagnetic anisotropy to explain these and other apparent anomalies, such as the unexpectedly large deshielding effect of the benzene ring (benzene protons δ 7.27).

Let us begin with acetylene. The molecule is linear, and the triple bond is symmetrical about the axis. If this axis is aligned with the applied magnetic field, the π-electrons of the bond can circulate at right angles to the applied field, thus inducing their own magnetic field opposing the applied field. Since the protons lie along the magnetic axis, the magnetic lines of force induced by the circulating electrons

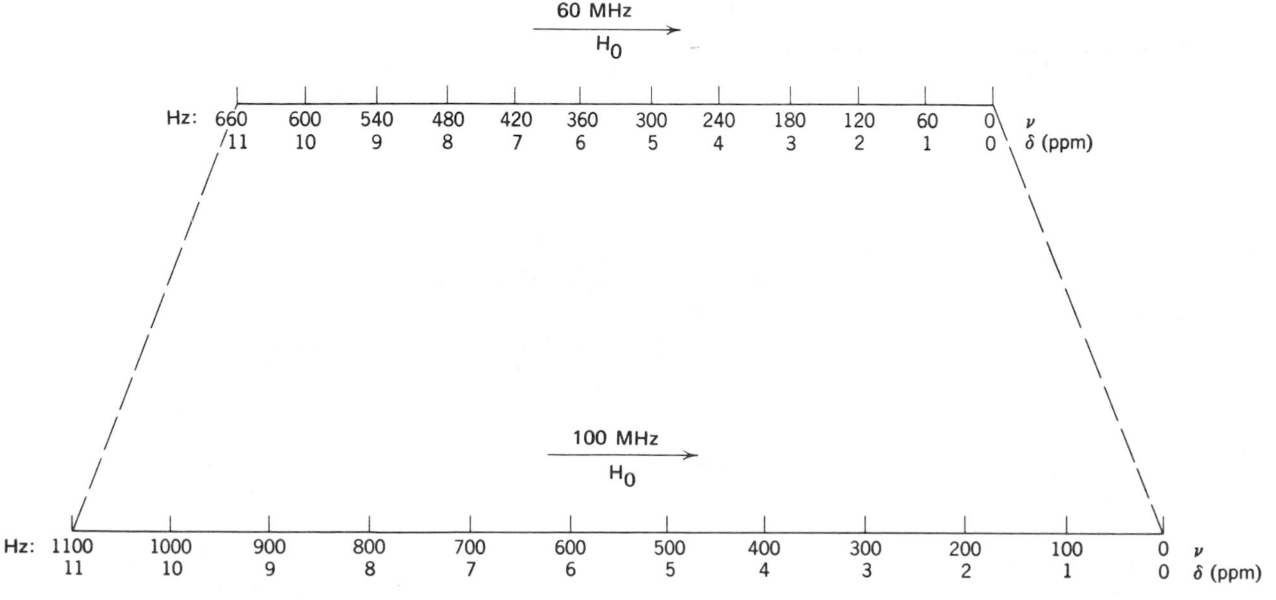

Figure 10. *NMR Scale at 60 MHz and 100 MHz.*

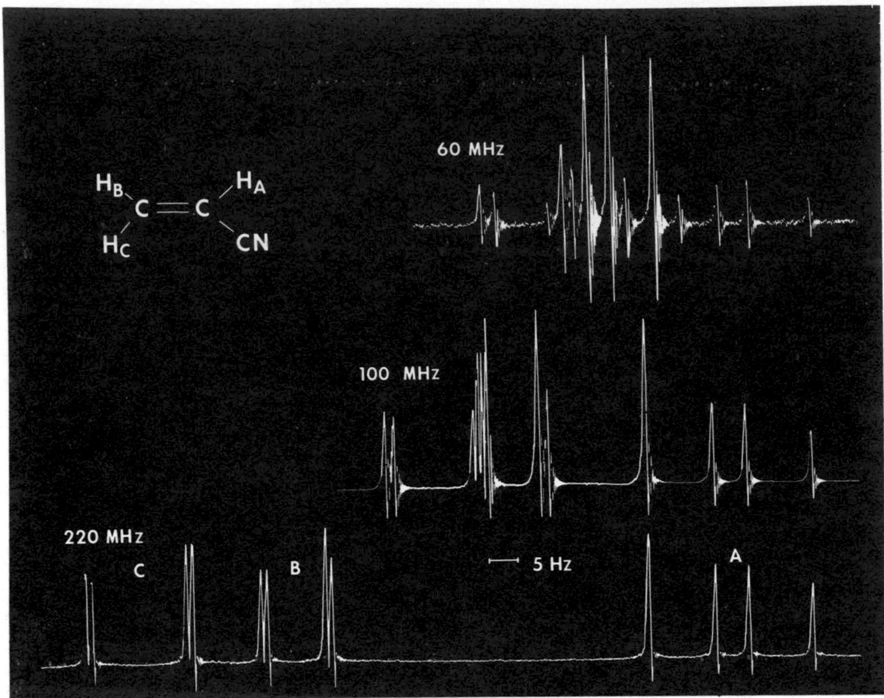

Figure 11. *60-, 100-, and 220-MHz spectra of acrylonitrile. Reprinted from* Anal. Chem. *Copyright 1971 by the American Chemical Society. Reprinted by permission of the copyright owner.*

act to shield the protons (Figure 12), and the NMR peak is found further upfield than electronegativity would predict. Of course, only a small number of the rapidly tumbling molecules are aligned with the magnetic field, but the overall average shift is affected by the aligned molecules.

This effect depends on diamagnetic anisotropy, which means that shielding and deshielding depend on the orientation of the molecule with respect to the applied magnetic field. Similar arguments can be adduced to rationalize the unexpected low-field position of the aldehydic proton. In this case, the effect of the applied magnetic field is greatest along the transverse axis of the C=O bond (i.e., in the plane of the page in Figure 13). The geometry is such that the aldehydic proton, which lies in front of the page, is in the deshielding portion of the induced magnetic field. The same argument can be used to account for at least part of the rather large amount of deshielding of olefinic protons.

The so-called "ring-current effect" is another example of diamagnetic anisotropy and accounts for the large deshielding of benzene ring protons. Figure 14 shows this effect. It also indicates that a proton held directly above or below the ring should be shielded. This has actually been found to be the case for some of the methylene protons in 1,4-polymethylenebenzenes.

All the ring protons of acetophenone are found down-

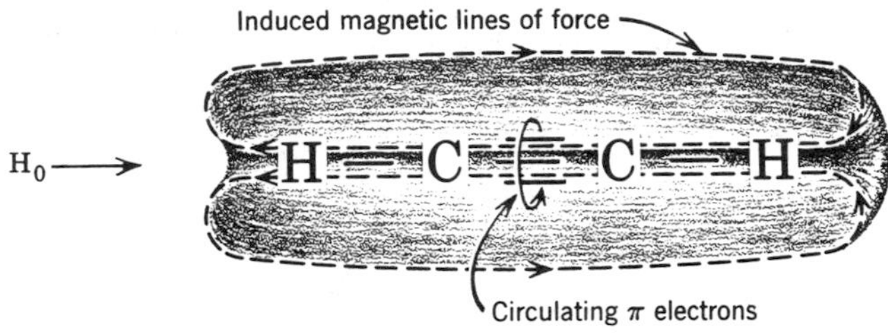

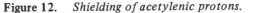

Figure 12. *Shielding of acetylenic protons.*

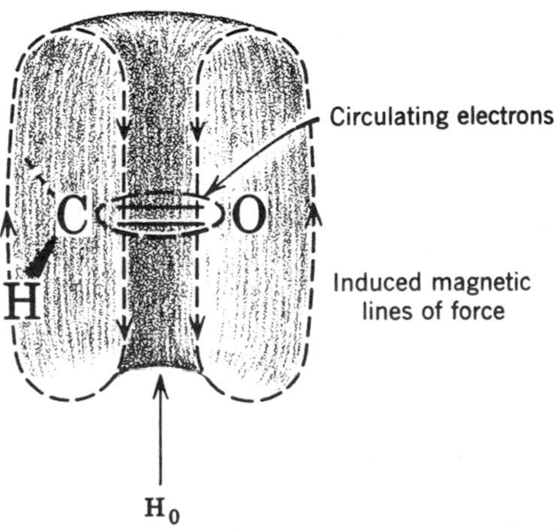

Figure 13. *Deshielding of aldehydic protons.*

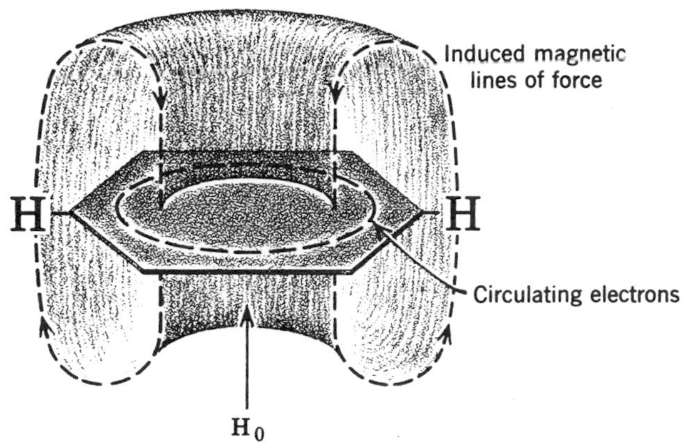

Figure 14. *Ring current effects in benzene.*

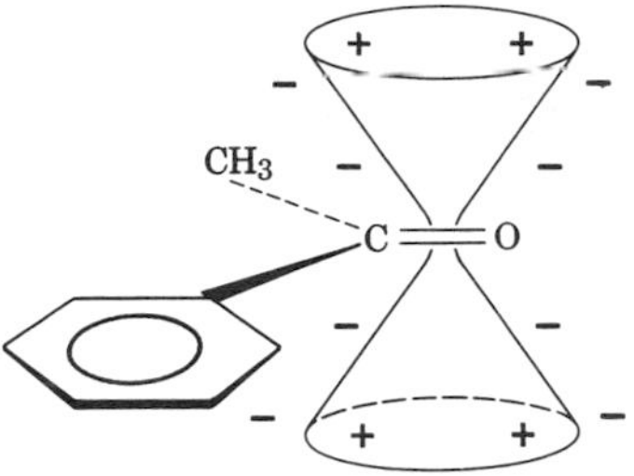

Figure 15. *Shielding (+) and deshielding (−) zones of acetophenone.*

field because of the ring current effect. Moreover, the ortho protons are shifted slightly further downfield (meta, para $\delta \sim 7.40$, ortho $\delta \sim 7.85$) because of the additional deshielding effect of the carbonyl group. In Figure 15 the carbonyl bond and the benzene ring are coplanar. If the molecule is oriented so that the applied magnetic field H_0 is perpendicular to the plane of the molecule, the circulating π electrons of the C=O bond shield the conical zones above and below them, and deshield the lateral zones in which the ortho proton is located. Both ortho protons are equally deshielded since another, equally populated, conformation can be written in which the "left hand" ortho proton is deshielded by the anisotropy cone. Nitrobenzene shows a similar effect.

A spectacular example of shielding and deshielding by ring currents is furnished by some of the annulenes.[17] The protons outside the ring of [18]annulene are strongly deshielded (δ 8.9) and those inside are strongly shielded (δ − 1.8).

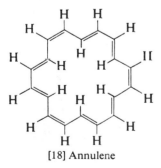

[18] Annulene

Demonstration of such a ring current is probably the best evidence available for aromaticity.

In contrast with the striking anisotropic effects of circulating π electrons, the σ electrons of a C−C bond produce a small effect. The axis of the C−C bond is the axis of the deshielding cone (Figure 16).

This figure accounts for the deshielding effect of successive alkyl substituents on a proton attached to a carbon atom. Thus, the protons are found progressively downfield in the sequence RCH_3, R_2CH_2, and R_3CH. The observation that an equatorial proton is consistently found further downfield by 0.1-0.7 ppm than the axial proton on the same carbon atom in a rigid six-membered ring can also be rationalized (Figure 17). The axial and equatorial protons on C_1 are oriented similarly with respect to C_1-C_2 and C_1-C_6, but the equatorial proton is within the deshielding cone of the C_2-C_3 bond (and C_5-C_6).

Extensive tables and charts of chemical shifts in the Appendices give the useful impression that chemical shifts of protons in organic compounds fall roughly into eight regions as shown in Figure 18.*

IV. SIMPLE SPIN-SPIN COUPLING

We have obtained a series of absorption peaks representing protons in different chemical environments, each absorp-

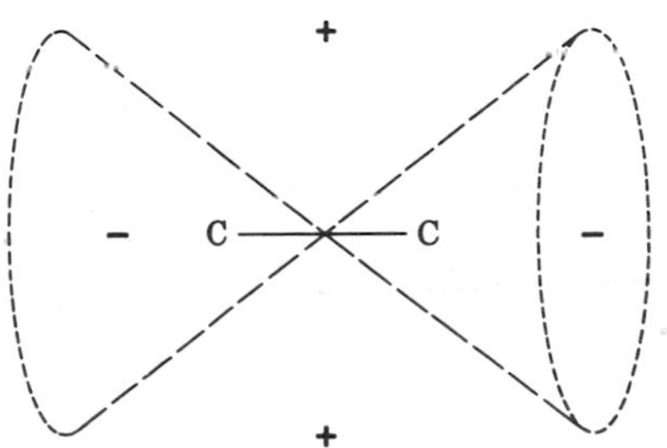

Figure 16. *Shielding (+) and deshielding (−) zones of C−C.*

*A rapid method for estimating proton shifts has been published. J. A. D. Jeffreys, *J. Chem. Ed.*, **56**, 806 (1979).

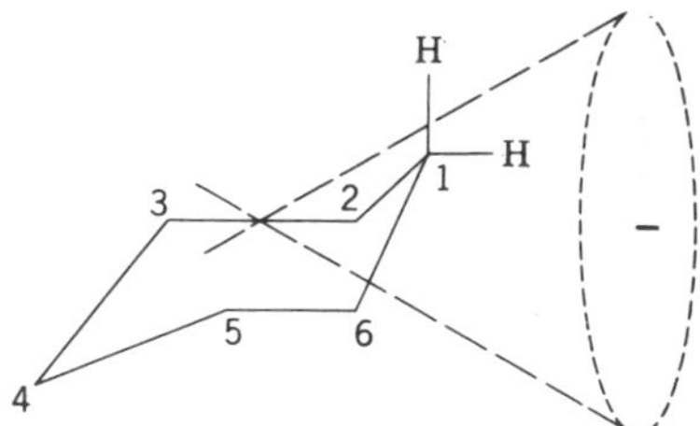

Figure 17. *Deshielding of equatorial proton of a rigid six-membered ring.*

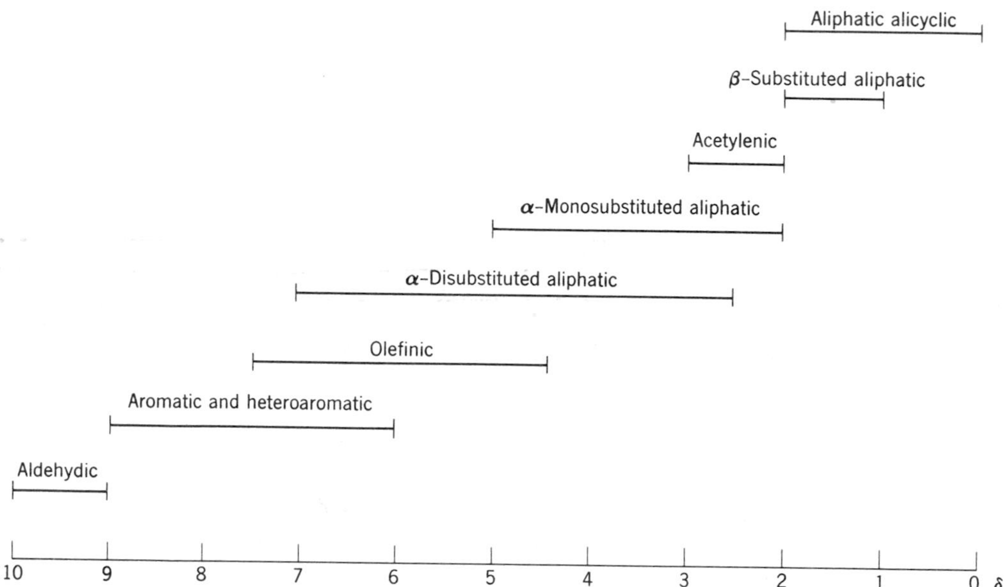

Figure 18. *General regions of chemical shifts.*

tion area being proportional to the number of protons it represents. We have now to consider one further phenomenon, spin-spin coupling. This can be described as the indirect coupling of proton spins through the intervening bonding electrons. Very briefly, it occurs because there is some tendency for a bonding electron to pair its spin with the spin of the nearest proton; the spin of a bonding electron having been thus influenced, the electron will affect the spin of the other bonding electron and so on through to the next proton. Coupling is ordinarily not important beyond 3 bonds unless there is ring strain as in small rings or bridged systems, or bond delocalization as in aromatic or unsaturated systems.

Suppose that two vicinal protons are in very different chemical environments from one another as in the compound

$$\mathrm{RO{-}\underset{|}{\overset{OR}{C}}H{-}\overset{CR_3}{C}H{-}CR_3}$$

RO–CH(OR)–CH(CR₃)–CR₃ ...

(Figure 19). The frequency differences betweeen the component peaks of a doublet is proportional to the effectiveness of the coupling and is denoted by a coupling constant, J, which is independent of the applied magnetic field H_0. Whereas chemical shifts usually range over about 1250 Hz at 100 MHz, coupling constants between protons rarely exceed 20 Hz (see Appendix F).

So long as the chemical shift difference in Hz ($\Delta\nu$) is much larger than the coupling constant ($\Delta\nu/J$ is greater than about 10), the simple pattern of 2 doublets appears. As $\Delta\nu/J$ becomes smaller, the doublets approach one another, the inner 2 peaks increase in intensity, and the outer two peaks decrease (Figure 20). The shift position of each proton is no longer midway between its 2 peaks as in Figure 19 but is at the "center of gravity" (Figure 21); it can be estimated with fair accuracy by inspection or determined precisely by the following formula in which the peak positions (1,2,3, and 4 from left to right) are given in Hz from TMS.

$$(1 - 3) = (2 - 4) = \sqrt{(\Delta\nu)^2 + J^2}$$

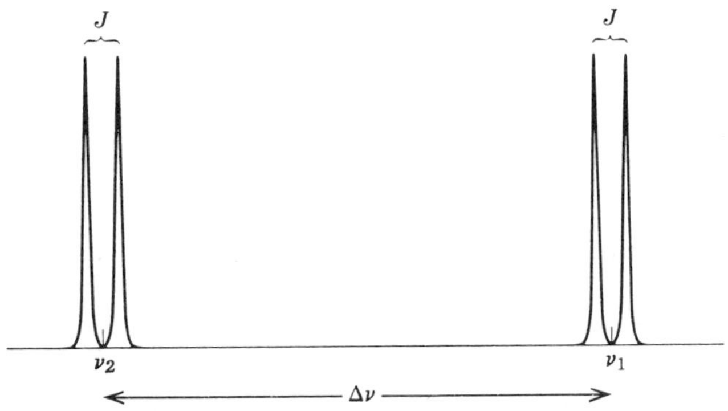

Figure 19. *Spin-spin coupling between two protons with very different chemical shifts.*

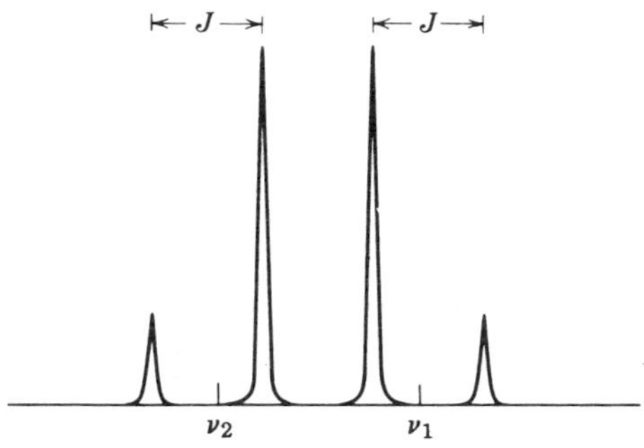

Figure 21. *"Center of gravity," instead of linear mid-points, for shift position location (due to "low" $\Delta\nu/J$ ratio).*

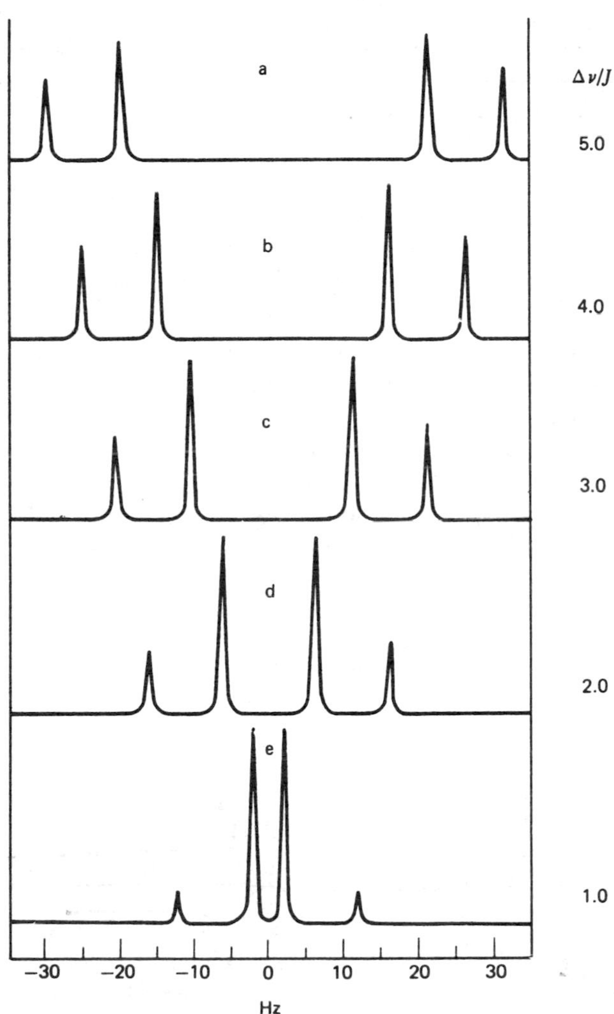

Figure 20. *A two-proton system spin coupling with a decreasing difference in chemical shifts and a large J value (10 Hz); the difference between AB and AX notation is explained in the text.*

The shift position of each proton is $\Delta\nu/2$ from the mid-point of the pattern. When $\Delta\nu = J\sqrt{3}$, the 2 pairs resemble a quartet that we shall see results from splitting by 3 equivalent vicinal protons (Figure 20d is almost at this stage). Failure to note the small outer peaks (i.e., 1 and 4) may lead to mistaking the 2 large inner peaks for a doublet (Figure 20e is almost at this stage). When the chemical shift difference becomes zero, the middle peaks coalesce to give a single peak, and the end peaks vanish; that is, the protons are equivalent. (Equivalent protons do spin-spin couple with one another, but splitting is not observed). A further point to be noted is the obvious one that the spacing between the peaks of 2 coupled multiplets is the same.

The dependence of chemical shift on the applied magnetic field and the independence of the spin-spin coupling afford a method of distinguishing between them. The spectrum is merely run on two different instruments, for example at 60 MHz and 100 MHz. Chemical shifts are also solvent dependent, but J values are usually only slightly affected by change of solvent, at least to a far lesser degree than are chemical shifts.

The chemical shifts of the methyl and acetylenic protons of methylacetylene are (fortuitously) coincident (δ 1.80) when the spectrum is obtained in a $CDCl_3$ solvent, whereas the spectrum of a neat sample of this alkyne shows the acetylenic proton at δ 1.80 and the methyl protons at δ 1.76. Figure 22 illustrates the chemical shift dependence of the protons of biacetyl on solvent. The change from a chlorinated solvent (e.g., $CDCl_3$) to an aromatic solvent (e.g., C_6D_6) often drastically influences the position and appearance of NMR signals.

Look at the next stage in complexity of spin-spin coupling (Figure 23). Consider the system $-HC-CH_2-$ in the compound $RO-CH(OR)-CH_2-CR_3$ in which the single methine proton is in a very different chemical environment from the 2 methylene protons. As before, we see 2 sets of absorp-

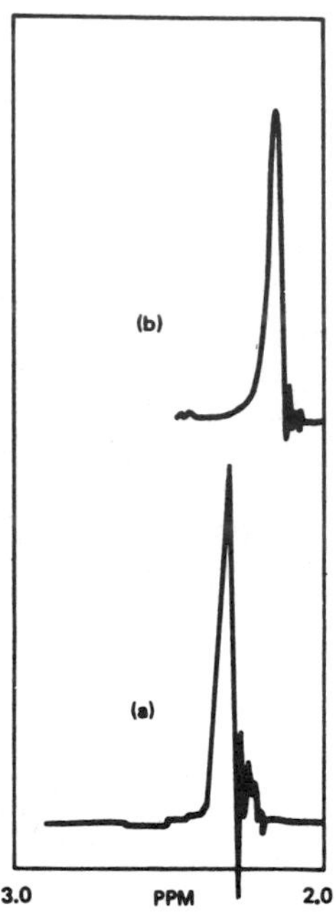

Figure 22. *The NMR spectrum of biacetyl (2,3-butanedione): a. in $CDCl_3$; b. in C_6D_6.*

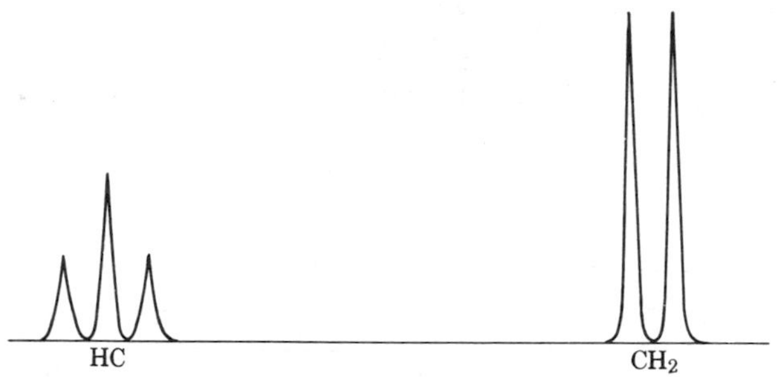

Figure 23. *Spin-spin coupling between CH and CH_2 with very different chemical shifts.*

tions widely separated, and now the absorption areas are in the ratio of 1:2. The methine proton couples with the methylene protons and splits the methylene proton absorption into a symmetrical doublet, as explained above. The 2 methylene protons split the methine proton absorption into a triplet because 3 combinations of proton spins exist in the two methylene protons (*a* and *b*) of Figure 24. Since there are 2 equivalent combinations of spin (pairs 2 and 3) that do not produce any net opposing or concerted field relative to the applied field, there is an absorption of relative intensity 2 at the center of the multiplet. Since there are single pairs (1 and 4) respectively opposed and in concert with the applied field, there are equally spaced (J) lines of relative intensity 1 upfield and 1 downfield from the center line. In summary, the intensities of the peaks in the triplet are in the ratio 1:2:1.

$\vec{H}_0$

	$\rightarrow b$ $\rightarrow a$ 4	
$\leftarrow b$ $\rightarrow a$ 2		$\rightarrow b$ $\leftarrow a$ 3
	$\leftarrow b$ $\leftarrow a$ 1	

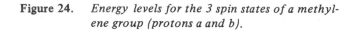

Figure 24. *Energy levels for the 3 spin states of a methylene group (protons a and b).*

When the methine and methylene protons in the system $-CH-CH_2-$ are in similar environments (i.e., $\Delta\nu/J$ is small), the simple doublet-triplet pattern degenerates to a complex pattern of from 7 to 9 lines as a result of second-order splitting; analysis by inspection is no longer possible, since the peak spacings may not correspond to the coupling constants.

Simple splitting patterns that are produced by the coupling of protons that have very different chemical shifts ($\Delta\nu/J$ is greater than about 10 or so) are called *first-order* splitting patterns. These can usually be interpreted by using two rules:

1. Splitting of a proton absorption is done by neighboring protons, and the multiplicity of the split is determined by the number of these protons. Thus, 1 proton causes a doublet, and 2 equally coupled neighboring protons cause a triplet. The multiplicity then is $n + 1$, n being the number of neighboring equally coupled protons. The general formula, which covers all nuclei, is $2nI + 1$, I being the spin number.
2. The relative intensities of the peaks of a multiplet also depend on n. We have seen that doublet ($n = 1$) peaks are in the ratio 1:1, and triplet peaks are in the ratio 1:2:1. Quartets are in the ratio 1:3:3:1. The general formula is $(a + b)^n$; when this is expanded to the desired value of n, the coefficients give the relative intensities. The multiplicity and relative intensities may be easily obtained from Pascal's triangle (Figure 25), in which n is the number of equivalently coupled protons.

Following Pople[14], we place protons that have the same chemical shift into sets (see Section VII), and we designate sets of protons separated by a small chemical shift with the letters A, B, and C, and sets separated by a large chemical shift ($\Delta\nu/J > \sim 10$) with the letters A, M, and X. The number of protons in each set is denoted by a subscript number. Thus, the first case we examined (Figure 19) is an AX system. The second case (Figure 21) is an AB system, and the third case (Figure 23) is an A_2X system. As $\Delta\nu/J$ decreases, the A_2X system approaches an A_2B system, and the simple first-order splitting of the A_2X system becomes more complex (see Section VII).

Thus far, we have dealt with 2 sets of protons; every proton in each set is equally coupled to every proton in the

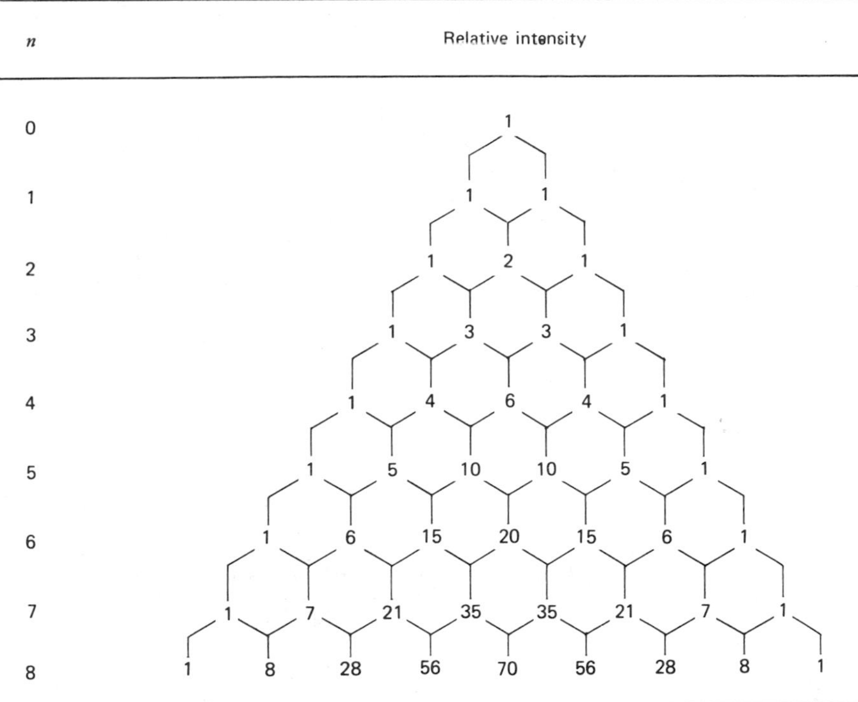

n
Relative intensity
0 1
1 1 1
2 1 2 1
3 1 3 3 1
4 1 4 6 4 1
5 1 5 10 10 5 1
6 1 6 15 20 15 6 1
7 1 7 21 35 35 21 7 1
8 1 8 28 56 70 56 28 8 1

Figure 25. *Pascal's triangle. Relative intensities of first-order multiplets; n = no. of equivalent coupling nuclei of spin 1/2 (e.g., protons).*

other set, that is, a single coupling constant is involved. Given these conditions and the conditions that $\Delta\nu/J$ be large (about 10), the two rules above apply, and we obtain a first-order pattern. In general, these are the A_aX_x systems (a and x are the number of protons in each set); the first-order rules apply only to these systems, but as we have seen, there is a gradual change in the appearances of spectra changing from an AX to an AB pattern. In a similar way, it is frequently possible to relate complex patterns back to first-order patterns. With practice, a fair amount of deviation from first-order may be tolerated. Wiberg's collection of calculated spectra[27] can be used to match fairly complex splitting patterns.

A system of 3 sets of protons, each set separated by a large chemical shift, can be designated $A_aM_mX_x$. If 2 sets are separated from each other by a small chemical shift, and the third set is widely separated from the other two, we use an $A_aB_bX_x$ designation. If all shift positions are close, the system is $A_aB_bC_c$. Both end sets may be coupled to the middle set with the same or different coupling constants, whereas the end sets may or may not be coupled to one another. AMX systems are first-order; ABX systems approximate first-order, but ABC systems cannot be analyzed by inspection. These more complex patterns are treated in Section VIII.

We can now appreciate the three main features of an NMR spectrum: chemical shifts, peak intensities, and spin-spin couplings that are first-order or that approximate first-order patterns. We can now analyze first-order NMR spectra.

The 60 MHz NMR spectrum of ethyl chloride is shown in Figure 26. The peak at δ 0.00 is the internal reference tetramethylsilane (TMS) and that at δ 7.25 is the $CHCl_3$ impurity in the $CDCl_3$ solvent. The methylene group (α)

$$\overset{\beta}{C}H_3-\overset{\alpha}{C}H_2-Cl$$

(δ 3.57) is more deshielded by the chlorine than is the methyl group (δ 1.48), and thus the methylene group is downfield (higher δ) from the methyl by 2.09 ppm or about 125 Hz. Since the coupling is about 9 Hz, $\Delta\nu/J$ is about 14, a large enough ratio for first order analysis. The system is A_3X_2, and the first order rules correctly predict a triplet and a quartet with relative intensities (see integration "steps") of 3 to 2 corresponding to the number of protons causing the absorptions. Note that even at $\Delta\nu/J$ = 14 there is a "leaning" of the two interacting signals toward each other; that is, the intensity of the upfield lines of the methylene signal and the downfield lines of the methyl signal are somewhat larger than they would be if these signals were perfectly symmetrical. This fact, together with the same spacing in both multiplets, is valuable in identifying interacting proton signals in more complex spectra.

We stated earlier that "the peak area (measured by the integrator) is proportional to the number of protons it represents." The "steps" in the integration curve in Figure

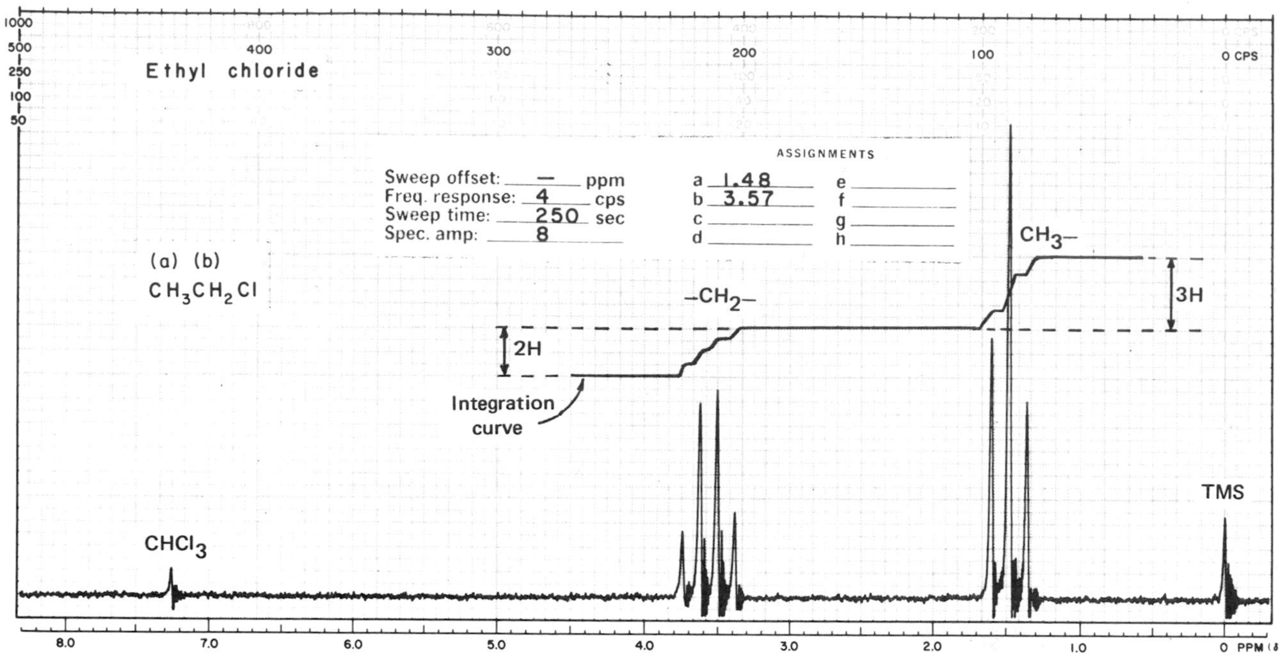

Figure 26. *Ethyl chloride in $CDCl_3$ at 60 MHz. (Courtesy of Varian Associates, Palo Alto, Calif.)*

26 thus provide the ratios of the different kinds of protons in the molecule. This count can be compared with the proton count obtained from the off-resonance decoupled ^{13}C spectrum (Chapter 5).

Consider the NMR spectrum of cumene in Figure 27. The five aromatic protons, δ 7.25, although actually not chemical shift equivalent (see Section VII), are fortuitously equivalent and occur as a single absorption downfield from the remaining absorptions (because of the benzene ring current, Figure 14). The side chain is treated as an A_6X system. The methyl signal occurs as a doublet at δ 1.25, the methine proton as a 1:6:15:20:15:6:1 septet at δ 2.90. Note that this signal is completely seen only when the sample is run at high gain (upper lines). Outer lines of complex multiplets may be overlooked, especially when these lines are part of a single proton absorption and when base line noise is substantial.

The vicinal coupling described (H—C—C—H) involves 3 bonds and can be designated by J_{HCCH} or ^3J. When geminal protons are not "equivalent" to one another (Section VII), we see coupling designated as J_{HCH} or ^2J.

V. PROTONS ON HETEROATOMS

Protons on a heteroatom differ from protons on a carbon atom in that: (1) they are exchangeable, (2) they are subject to hydrogen bonding, and (3) they are subject to partial or complete decoupling by electrical quadrupole effects of some heteroatoms. Shift ranges for protons on heteroatoms are given in Appendix E.

PROTONS ON OXYGEN

Alcohols

Unless special precautions are taken (see below) the spectrum of neat ethanol usually shows the hydroxylic proton as a slightly broadened peak at δ 5.35. At the commonly used concentration of about 5 to 20% in a nonpolar solvent, such as carbon tetrachloride or deuterochloroform, the hydroxylic peak is found between δ 2 and δ 4. On extrapolation to infinite dilution or in the vapor phase, the peak is near δ 0.5. A change in solvent or temperature will also shift the hydroxylic peak.

Hydrogen bonding explains why the shift position of the hydroxylic proton depends on concentration, temperature, and solvent. Hydrogen bonding decreases the electron density around the proton, and thus moves the proton absorption to lower field. The extent of intermolecular hydrogen bonding is decreased by dilution with a nonpolar solvent and with increased temperature. Polar solvents introduce the additional complication of hydrogen bonding between the hydroxylic proton and the solvent. Intramolecular hydrogen bonds are less affected by their environment than are intermolecular hydrogen bonds. In fact the enolic

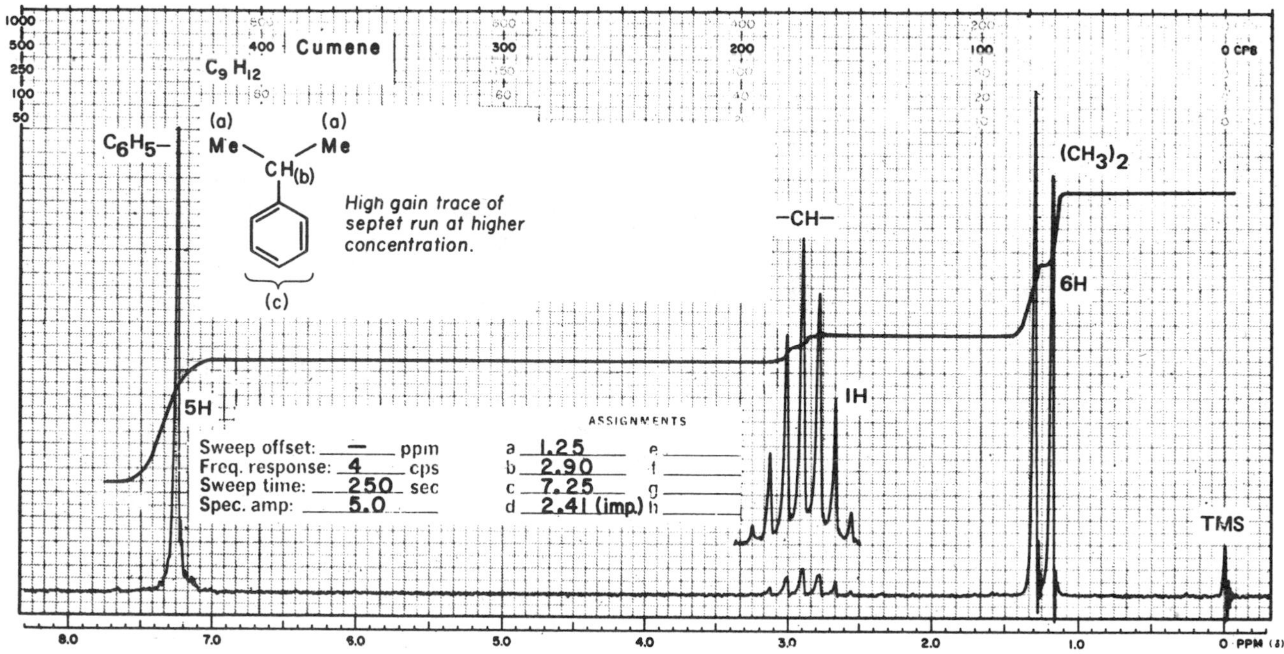

Figure 27. *Cumene in $CDCl_3$ at 60 MHz. (Courtesy of Varian Associates, Palo Alto, Calif.)*

hydroxylic absorption of β-diketones (O⋯H–O chelate: R–C(=O)–CH=C(–OH)–R)

for example, is hardly affected by change of concentration or solvent, though it can be shifted upfield somewhat by warming. NMR spectrometry is a powerful tool for studying hydrogen bonding.

Exchangeability explains why the hydroxylic peak of ethanol is usually seen as a singlet. Under ordinary conditions, enough acidic impurities are present in solution to catalyze rapid exchange of the hydroxylic proton. The proton is not on the oxygen atom long enough for it to be affected by the three states of the methylene protons, and there is no coupling. The rate of exchange can be decreased by treating the solution or the solvent with anhydrous sodium carbonate, alumina, or molecular sieves immediately before obtaining the spectrum. Purified deuterated dimethyl sulfoxide or acetone, in addition to allowing a lower rate of exchange, shifts the hydroxylic proton to lower field, even in dilute solution, by hydrogen bonding between solute and solvent.* Since the hydroxylic proton can now couple with the protons on the α-carbon, a primary alcohol will show a triplet, a secondary alcohol a doublet, and a tertiary alcohol a singlet. This is illustrated in Figure 28 and a list of successful applications is given in Table I. Exceptions have been reported, but these may be due to the concentration dependence of this phenome-

*O. L. Chapman and R. W. King, *J. Am. Chem. Soc.* **86**, 1256(1964). D. E. McGreer and M. M. Mocek, *J. Chem. Ed.* **40**, 358(1963).

Table I. Hydroxyl Proton Resonances in Dimethyl Sulfoxide*

Compound[a]	Chemical Shift, δ	Multiplicity
Methanol	4.08	q
Ethanol	4.35	t
Isopropyl alcohol	4.35	d
t-Butyl alcohol	4.16	s
t-Pentyl alcohol	3.99	s
Propylene glycol, 1-OH	4.45	t
2-OH	4.38	d
Cyclohexanol	4.38	d
cis-4-*t*-Butylcyclohexyl alcohol	4.11	d
trans-4-*t*-Butylcyclohexyl alcohol	4.45	d
Benzyl alcohol	5.16	t
Phenol	9.25	s
β-L-Arabinopyranose, O–C–OH	5.98	d
α-D-Glucopyranose, O–C–OH	6.16	d
α-D-Fructopyranose, O–C–OH	5.12	s

[a]All spectra were taken of dimethyl sulfoxide solutions with concentrations 10 mole % or less.

*O. L. Chapman and R.W. King, *J. Am. Chem. Soc.*, **86**, 1256 (1964).

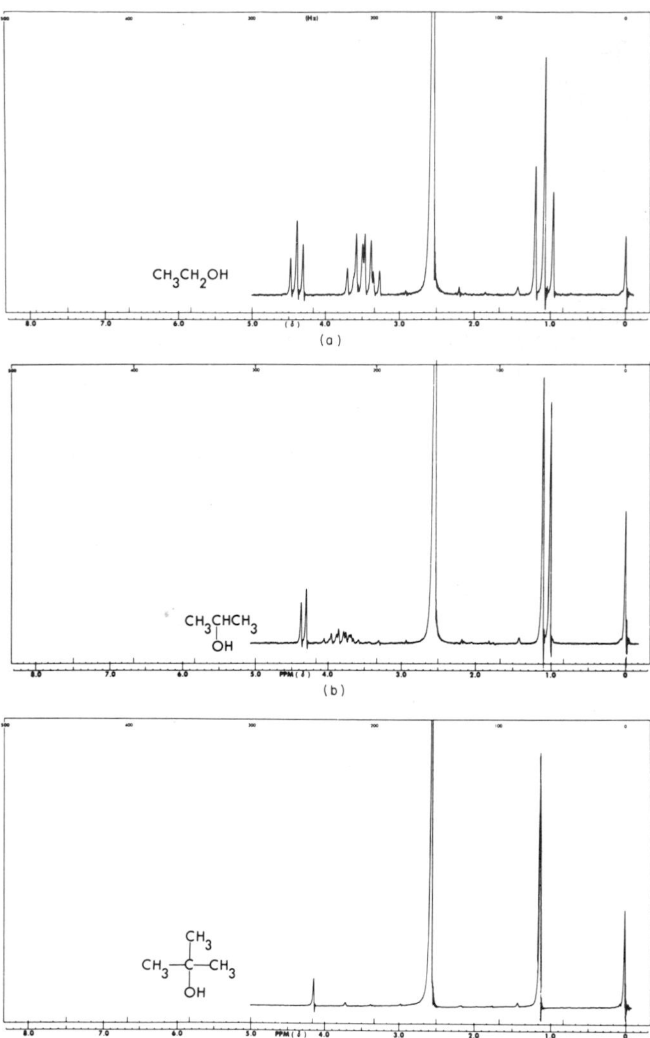

Figure 28. *NMR spectra of typical primary, secondary, and tertiary alcohols run in dimethyl sulfoxide. Absorption at δ 2.6 is due to dimethyl sulfoxide.*

Source: Daniel J. Pasto and Carl R. Johnson, *Organic Structure Determination,* © 1969, pp. 358-359. Reprinted by permission of Prentice-Hall, Inc., Englewood Cliffs, N.J.)

non. At intermediate rates of exchange, the multiplet merges into a broad absorption band; at this point, the exchange rate in Hz is equal to $\pi J/\sqrt{2}$.

If water is present in the alcohol solution, rapid interchange results in a single peak at a position intermediate between the actual shift positions of the HOH and ROH peaks. Of course, the presence of water is reflected in the integration for the single peak.

A dihydroxy alcohol may show separate absorption peaks for each hydroxylic proton; in this case, the rate of exchange in Hz is much less than the difference in Hz between the separate absorptions. As the rate increases (trace of acid catalyst), the two absorption peaks broaden, then merge to form a single broad peak; at this point, the exchange rate in Hz is equal to the original separation in Hz. As the rate increases, the single peak becomes sharper. The relative position of each peak depends on the extent of hydrogen bonding of each hydroxylic proton; steric hindrance to hydrogen bonding frequently accounts for a relative upfield absorption.

The spectrum of a compound containing exchangeable protons can be simplified, and the exchangeable proton absorption removed, simply by shaking the solution with excess deuterium oxide or by obtaining a spectrum in deuterium oxide solution if the compound is soluble. A peak due to HOD will appear, generally between δ 5 and δ 4.5 in nonpolar solvents, and near δ 3.3 in dimethyl sulfoxide (see Appendix E). A $CDCl_3$ or CCl_4 solution in the NMR tube may be shaken vigorously for several minutes with 1 or 2 drops of D_2O, and the mixture allowed to stand (or centrifuged) until the layers are clearly separated. The top aqueous layer does not interfere.

Acetylation or benzoylation of a hydroxyl group moves the carbinyl protons of a primary alcohol downfield about 0.5 ppm, and that of a secondary alcohol about 1.0-1.2 ppm. Such shifts provide a confirmation of the presence of a primary or secondary alcohol.

Phenols

The behavior of a phenolic proton resembles that of an alcoholic proton. The phenolic proton peak is usually a sharp singlet (rapid exchange, no coupling), and its range, depending on concentration, solvent, and temperature, is generally downfield (δ ~7.5 - δ ~4.0) compared with the alcoholic proton. This is illustrated in Figure 29. Note the concentration dependence of the OH peak. A carbonyl group in the ortho position shifts the phenolic proton absorption downfield to the range of about δ 12.0-δ 10.0 because of intramolecular hydrogen bonding. Thus, *o*-hydroxyacetophenone shows a peak at about δ 12.05 almost completely invariant with concentration. The much weaker intramolecular hydrogen bonding in *o*-chlorophenol explains its shift range (δ ~6.3 at 1 molar concentration to δ ~5.6 at infinite dilution), which is broad compared with that of *o*-hydroxyacetophenone but narrow compared with that of phenol.

Enols

Enols are usually stabilized by intramolecular hydrogen bonding, which varies from very strong in aliphatic β-diketones to weak in cyclic α-diketones. The enolic proton is downfield relative to alcohol protons and, in the case of the enolic form of some β-diketones, may be found as far downfield as δ 16.6 (the enolic proton of acetylacetone

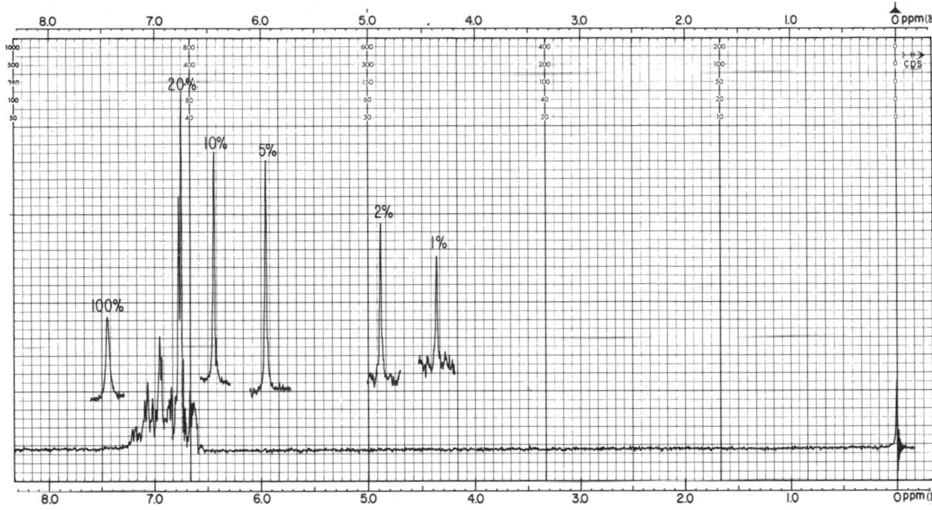

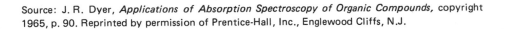

Figure 29. *Phenol, in CCl_4, at various w/v %, at 60 MHz. Complete sweep is at 20%; single absorptions represent the OH proton at the indicated w/v %.*

Source: J. R. Dyer, *Applications of Absorption Spectroscopy of Organic Compounds,* copyright 1965, p. 90. Reprinted by permission of Prentice-Hall, Inc., Englewood Cliffs, N.J.

absorbs at δ 15.0 and that of dibenzoylmethane at δ 16.6). The enolic proton peak is frequently broad at room temperature because of slow exchange. Furthermore, the keto-enol conversion is slow enough so that absorption peaks of both forms can be observed, and the equilibrium measured.

When strong intramolecular bonding is not involved, the enolic proton absorbs in about the same range as the phenolic proton.

Carboxylic Acids

Carboxylic acids exist as stable hydrogen-bonded dimers in nonpolar solvents even at high dilution. The carboxylic proton therefore absorbs in a characteristically narrow range, δ ~13.2 to δ ~10.0, and is affected only slightly by concentration. Polar solvents partially disrupt the dimer and shift the peak accordingly.

The peak width at room temperature ranges from sharp to broad, depending on the exchange rate of the particular acid. The carboxylic proton exchanges quite rapidly with protons of water and alcohols (or hydroxyl groups of hydroxy acids) to give a single peak whose position depends on concentration. Sulfhydryl or enolic protons do not exchange rapidly with carboxylic protons, and individual peaks are observed.

PROTONS ON NITROGEN

The ^{14}N nucleus has a spin number $I = 1$ and, in accordance with the formula $2I + 1$, should cause a proton attached to it and a proton on an adjacent carbon atom to show three equally intense peaks. There are two factors, however, that complicate the picture: the rate of exchange of the proton on the nitrogen atom, and the electrical quadrupole moment of the ^{14}N nucleus.

The proton on a nitrogen atom may undergo rapid, intermediate, or slow exchange. If the exchange is rapid, the NH proton(s) is decoupled from the N atom and from protons on adjacent carbon atoms. The NH peak is therefore a sharp singlet, and the adjacent CH protons are not split by NH. Such is the case for most aliphatic amines.* At an intermediate rate of exchange, the NH proton is partially decoupled, and a broad NH peak results. The adjacent CH protons are not split by the NH proton. Such is the case for *N*-methyl-*p*-nitroaniline. If the NH exchange rate is low, the NH peak is still broad because the electrical quadrupole moment of the nitrogen nucleus induces a moderately efficient spin relaxation and, thus, an intermediate lifetime for the spin states of the nitrogen nucleus. The proton thus sees three spin states of the nitrogen nucleus (spin number = 1) which are changing at a moderate rate, and the proton responds by giving a broad peak. In this case, coupling of the NH proton to the adjacent protons is observed. Such is the case for pyrroles, indoles, secondary and primary amides, and carbamates (Figure 30). Note that H—N—C—H coupling takes place through the C—H, C—N, and N—H bonds, but coupling between nitro-

*H—C—N—H coupling in several amines has been observed following rigorous removal (with Na-K alloy) of traces of water. This effectively stops proton exchange on the NMR time scale (K. L. Henold, *Chem. Comm.* 1340 (1970).

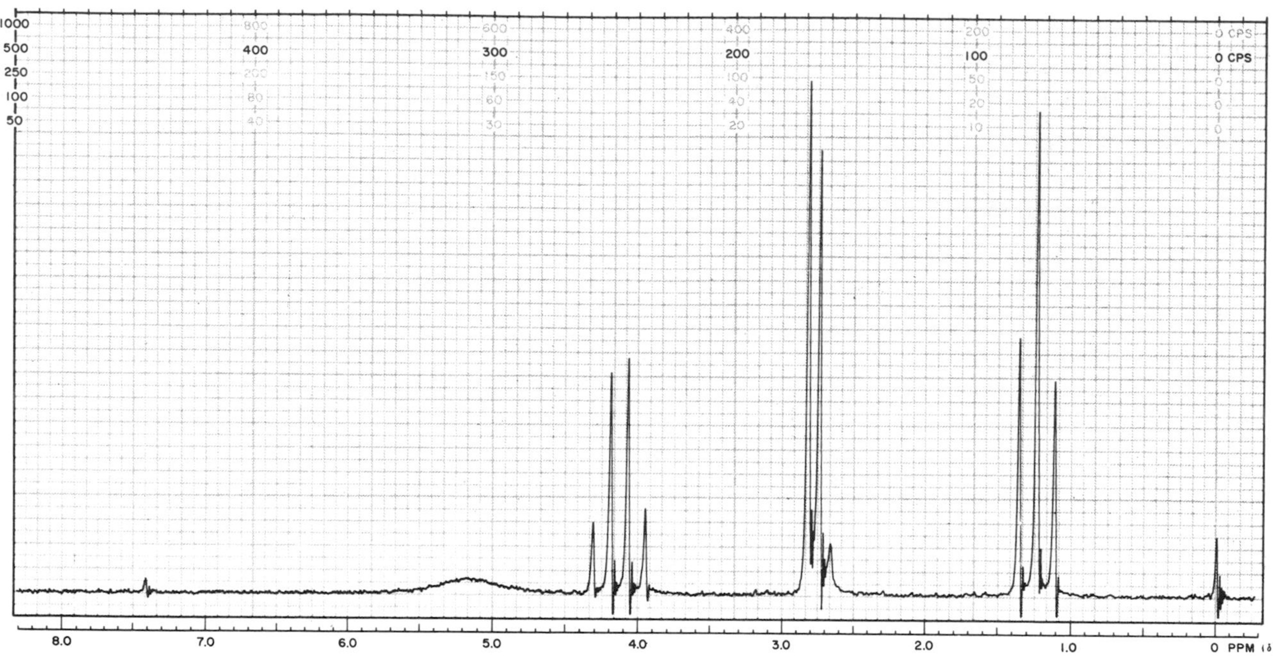

Figure 30. *Ethyl N-methylcarbamate,* $CH_3NHC(=O)OCH_2CH_3$*, 60 MHz.*

gen and protons on adjacent carbons is negligible. The coupling is observed in the signal due to hydrogen on carbon; the N—H proton signal is severely broadened by the quadrupolar interaction with nitrogen. In the spectrum of ethyl *N*-methylcarbamate (Figure 30) $CH_3NHC(=O)OCH_2CH_3$, the NH proton shows a broad absorption centered about δ 5.16, and the $N-CH_3$ absorption at δ 2.78 is split into a doublet ($J \sim 5$ Hz) by the NH proton. The ethoxy protons are represented by the triplet at δ 1.23 and the quartet at δ 4.14. The small peak at δ 2.67 is an impurity.

Aliphatic and cyclic amine NH protons absorb from $\delta \sim 3.0$-$\delta \sim 0.5$; aromatic amines absorb from $\delta \sim 5.0$-$\delta \sim 3.0$. Because amines are subject to hydrogen bonding, the shift position depends on concentration, solvent, and temperature. Amides, pyrroles, and indoles absorb from $\delta \sim 8.5$-$\delta \sim 5.0$; the effect on the absorption position of concentration, solvent, and temperature is generally smaller than in the case of amines. The nonequivalence of the protons on the nitrogen atom of a primary amide and of the methyl groups of *N,N*-dimethylamides is caused by hindered rotation around the $C(=O)-N$ bond because of the contribution of the resonance form $C(-O^-)=N^+$.

Protons on the nitrogen atom of an amine salt exchange at a low rate; they are seen as a broad peak downfield ($\delta \sim 8.5$-$\delta \sim 6.0$), and they are coupled to protons on adjacent carbon atoms ($J \sim 7$ Hz); the α-protons are recognized by their downfield position in the salt compared with that in the free amine. The use of trifluoroacetic acid as both a protonating agent and a solvent frequently allows classification of amines as primary, secondary, or tertiary. This is illustrated in Table II in which the number of protons on nitrogen determines the multiplicity of the methylene unit in the salt (Figure 31). Sometimes the broad ^+NH, $^+NH_2$, or $^+NH_3$ absorption can be seen to consist of three broad humps. These humps represent splitting by the nitrogen nucleus ($J \sim 50$ Hz). With good resolution, it is sometimes possible to observe splitting of each of the humps by the protons on adjacent carbon atoms ($J \sim 7$ Hz).

PROTONS ON SULFUR

Sulfhydryl protons usually exchange at a low rate so that at room temperature they are coupled to protons on adja-

Table II. Classification of Amines by NMR of Their Ammonium Salts in Trifluoroacetic Acid*

Amine Precursor Class	Ammonium Salt Structure	Multiplicity of Methylene Unit
Primary	$C_6H_5CH_2NH_3^+$	Quartet (Figure 31)
Secondary	$C_6H_5CH_2NH_2R^+$	Triplet
Tertiary	$C_6H_5CH_2NHR_2^+$	Doublet

*W.R. Anderson, Jr., and R.M. Silverstein, *Anal. Chem.*, **37** 1417 (1965).

Figure 31. *NMR spectrum of α methylene unit of a primary amine in CF_3CO_2H; corresponds to Table II, first line.*

cent carbons ($J \sim 8$ Hz). Nor do they exchange rapidly with hydroxyl, carboxylic, or enolic protons on the same or on other molecules; thus, separate peaks are seen. However, exchange is rapid enough that shaking for a few minutes with deuterium oxide replaces sulfhydryl protons with deuterium. The absorption range for aliphatic sulfhydryl protons is $\delta \sim 1.6$-$\delta \sim 1.2$; for aromatic sulfhydryl protons, $\delta \sim 3.6$-$\delta \sim 2.8$. Concentration, solvent, and temperature affect the position within these ranges.

PROTONS ON HALOGENS

Chlorine, bromine, and iodine nuclei are completely decoupled from protons directly attached, or on adjacent carbon atoms, because of strong electrical quadrupole moments.

The ^{19}F atom has a spin number of 1/2 and couples strongly with protons (see Appendix F). The rules for coupling of protons with fluorine are the same as for proton-proton coupling; in general, the proton-fluorine coupling constants are somewhat larger, and long-range effects are frequently found. The ^{19}F nucleus can be observed at 56.4 MHz at 14,092 gauss. Of course, its spin is split by proton and fluorine spins, and the multiplicity rules are the same as those observed in proton spectra.

VI. COUPLING OF PROTONS TO OTHER NUCLEI

The organic chemist may encounter proton coupling with such other nuclei (besides ^1H, ^{19}F, and ^{14}N) as ^{31}P, ^{13}C, ^2H, and ^{29}Si. Three factors must be considered: natural abundance, spin number, and electrical quadrupole moment; the nuclear magnetic moment and relative sensitivity are important when a spectrum of the particular nucleus is considered. These properties are listed for a number of nuclei in Appendix I.

The magnetogyric ratio, γ, describes the combined effect of the magnetic moment and spin number of a given type of nucleus. The ratio of the magnetogyric ratios of one type of nucleus to that of another type of nucleus is a measure of the relative coupling constants of those two nuclei to a given reference nucleus. Consider the relative magnitudes of coupling of hydrogen and of deuterium, D, to a particular nucleus, X. Since

$$\frac{J_{HX}}{J_{DX}} = \sim \frac{\gamma_H}{\gamma_D}$$

we calculate from the magnetogyric ratios in Appendix I that

$$\frac{J_{HX}}{J_{DX}} = \sim \frac{26.753}{4.107} = 6.51$$

Thus, it is anticipated that deuterium coupling is less than the corresponding hydrogen coupling by a factor of about 6.5.

^{31}P has a natural abundance of 100% and a spin number of 1/2 (therefore no electrical quadrupole moment). The multiplicity rules for proton-phosphorus splitting are the same as those for proton-proton splitting. The coupling constants are large ($J_{H-P} \sim 200$-700 Hz, and J_{HC-P} is 0.5-20 Hz) and are observable through at least four bonds. The ^{31}P nucleus can be observed at the appropriate frequency and magnetic field.

^{13}C has a natural abundance relative to ^{12}C of 1.1%, and a spin number of 1/2. Typically, absorption due to coupling of a proton with a ^{13}C nucleus is seen as weak "^{13}C satellite peaks" on both sides of a very strong proton peak. Coupling constants for H—^{13}C have been correlated with hybridization of the ^{13}C atom: $J_{sp^3} \sim 120$ Hz, $J_{sp^2} \sim 170$ Hz, and $J_{sp} \sim 250$ Hz. The proton attached to the ^{13}C is also split by protons on adjacent ^{12}C atoms with the usual H—C—C—H coupling constant of about 7 Hz. Thus, the configuration $^{13}CH_3-^{12}CH_2-$ shows a triplet on both sides of an amplified $^{12}CH_3$ absorption from the ^{13}C—H and the $^{13}CH_3-^{12}CH_2$ couplings. Other molecules in the magnetic field will have the configuration $^{12}CH_3-^{13}CH_2-$; these molecules also show a triplet on each side of the very large $^{12}CH_3$ absorption, but since $J_{^{12}CH_3-^{13}C}$ is only ~ 4.7 Hz, these peaks are buried on the sides of the $^{12}CH_3$ absorption The ^{13}C satellite peaks can be distinguished from spinning side bands by their invariance with the rate of spinning of the sample tube. Parenthetically, note that each spinning side band has the same

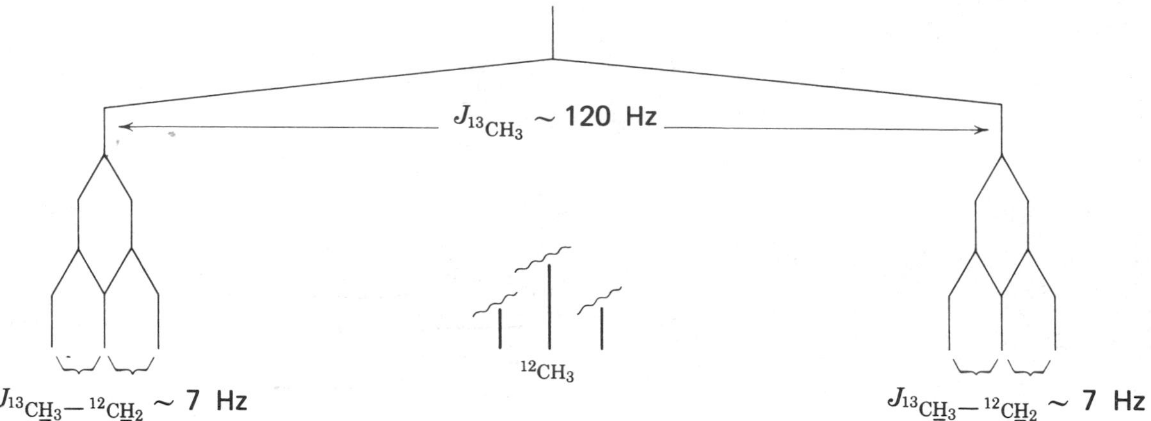

multiplicity as its "parent," but the satellite peaks do not necessarily have the same multiplicity as the parent peak (see page 141 of Jackman and Sternhell[4]).

Deuterium (^2H or D) usually is introduced into a molecule to detect a group or to simplify a spectrum. ^2H has a spin number of 1, a small coupling constant with protons, and a small electrical quadrupole moment. A proton-deuterium coupling constant is approximately 15% of the corresponding proton-proton constant (see the section above on ratio of the magnetogyric ratios). Suppose the protons on the α-carbon atom of a ketone

$$\overset{\gamma}{X{-}CH_2}{-}\overset{\beta}{CH_2}{-}\overset{\alpha}{CH_2}{-}\overset{O}{\overset{\|}{C}}{-}$$

were replaced by deuterium:

$$\overset{\gamma}{X{-}CH_2}{-}\overset{\beta}{CH_2}{-}\overset{\alpha}{CD_2}{-}\overset{O}{\overset{\|}{C}}{-}$$

The original spectrum would consist of a triplet for the α protons, a quintet (assuming equal coupling) for the β-protons, and a triplet for the γ protons. In the deuterated compound, the α-proton absorption would be absent, the β-protons would appear as a slightly broadened triplet, and the γ protons would be unaffected. Actually, each peak of the β-proton triplet is a very closely spaced quintet (J_{CH-CD} ~ 1 Hz or less), but the effect under ordinary resolution is peak broadening. Even this modest broadening due to the deuterium coupling can be removed by double resonance experiments (see below) involving irradiation at the deuterium resonance; this technique has been used to obtain a more exact measurement of the remaining proton-proton coupling. Most deuterated solvents have residual protonated impurities; the CHD_2 group in deuterated acetone or dimethyl sulfoxide, for example, is frequently encountered as a closely spaced quintet (J ~ 2 Hz, intensities 1:2:3:2:1).

^{29}Si has a natural abundance of 5.1% (based on ^{28}Si = 100%), and a spin number of 1/2. $J_{^{29}Si-CH}$ is about 6 Hz. The small doublet caused by the $^{29}Si-CH_3$ coupling can often be seen straddling (± 3 Hz) an amplified peak of tetramethylsilane; the $^{13}C-H_3$ doublet can also be seen at ± 59 Hz.

VII. CHEMICAL SHIFT EQUIVALENCE AND MAGNETIC EQUIVALENCE

CHEMICAL SHIFT EQUIVALENCE

The Pople notation (Section IV) is based on the concept of *sets* of nuclei within a *spin system.* A set of nuclei consists of *chemical shift equivalent* (defined below) nuclei. A spin system consists of sets of nuclei that "interact (spin couple) among each other but do not interact with any nuclei outside the spin system. It is not necessary for all nuclei within a spin system to be coupled with all the other nuclei" in the spin system.[4] Spin systems are "insulated" from one another; for example, the ethyl protons in ethyl isopropyl ether constitute one spin system, and the isopropyl protons another.

If nuclei are interchangeable by a symmetry operation or a rapid process, they are chemical shift equivalent (*isochronous*); that is, they have exactly the same chemical shift under all achiral conditions. Nuclei are interchangeable if the structures before and after the operation are indistinguishable. A rapid process means one that occurs faster than once in about 10^{-3} second. Pople proposed that the chemical shift equivalent nuclei be ordered into sets designated A_3, B, X_2 and so forth.

The symmetry operations are: rotation about a symmetry axis (C_n); inversion at a center of symmetry (i); reflection at a plane of symmetry (σ); or higher orders of rotation about an axis followed by reflection in a plane normal to this axis (S_n). The symmetry element (axis, center, or plane) must be a symmetry element for the entire molecule. The term "interchange" will be clarified by the examples below.

Protons a and b in *trans*-1,2-dichlorocyclopropane are chemical shift equivalent, as are the protons c and d (Figure 32). The molecule has an axis of symmetry passing through C_3 and bisecting the C_1-C_2 bond. Rotation of the molecule by 180° around the axis of symmetry interchanges proton H_a for H_b and H_c for H_d. If the protons were not labeled, it would not be possible to tell if the symmetry operation had been performed merely by inspecting the molecule before and after the operation; the rotated structure is superimposable on the original structure.

Protons that are interchangeable through an axis of symmetry are *homotopic*: that is, they are chemical shift equivalent in any environment (solvent or reagent) whether achiral or chiral. Protons that are interchangeable through any other symmetry operation are called *enantiotopic* (nonsuperimposable mirror images), and these are chemical shift equivalent only in an achiral environment. Non-interchangeable geminal protons are termed *diastereotopic*, and they are not chemical shift equivalent in any environment. Noninterchangeable protons on different carbon atoms are termed *heterotopic*, and they are not chemical shift equivalent in any environment. However, enantiotopic protons in a chiral environment, or diastereotopic or heterotopic protons, may fortuitously absorb at the same position for a given instrumental resolution.

The concept of interchange through a rapid mechanism can be illustrated by the rapidly interchanging protons on some heteroatoms and by protons in some groups in molecules that are rapidly changing conformations. If the interchange is rapid enough, a single peak will result from, say, the carboxylic acid proton and the hydroxylic proton of a hydroxy carboxylic acid. Chemical shift equivalence of protons on a CH_3 group results from rapid rotation around a carbon-carbon single bond even in the absence of a symmetry element. Figure 33 shows Newman projections of the three staggered rotamers of a molecule containing a methyl group attached to another sp^3 carbon atom having four different substituents, that is, a chiral center. In any single rotamer, none of the protons can be interchanged by a symmetry operation. However, the protons are rapidly changing position. The time spent in any one rotamer is short ($\sim 10^{-6}$ second), because the energy barrier for rotation around a C–C single bond is small. The chemical shift of the methyl group is an average of the shifts of each of the 3 protons. In other words, each proton can be interchanged with the others by a rapid rotational operation. Thus, without the a, b, c labels, the rotamers are indistinguishable.

The chemical shift of cyclohexane protons is an average of the shifts of the axial and the equatorial protons. The chemical shift of equivalence of the cyclohexane protons results from rapid interchange of axial and equatorial protons as the molecule flips between chair forms. The chemical equivalence of its protons is not a consequence of achieving a planar conformation because the planar conformation of cyclohexane is not a reasonable one.

At room temperature, *N,N*-dimethylformamide shows 2 CH_3 peaks, which merge at ~ 123° because of the rate of rotation around the hindered C⋯N bond.

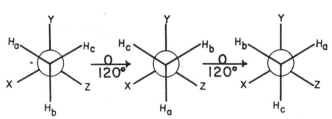

$$\delta^- O{=}C(H){=}N^{\delta+}(CH_3(a))(CH_3(b)) \rightleftharpoons \delta^- O{=}C(H){=}N^{\delta+}(CH_3(b))(CH_3(a))$$

MAGNETIC EQUIVALENCE

A further refinement involves the concept of *magnetic equivalent* nuclei. If nuclei in the same set (i.e, chemical shift equivalent nuclei) couple equally to each nucleus (probe nucleus) in every other set of the spin system, they are magnetic equivalent, and the designations A_2, X_2, and so on apply. However, if the nuclei are not magnetic equivalent, the designations AA′, XX′ are used.

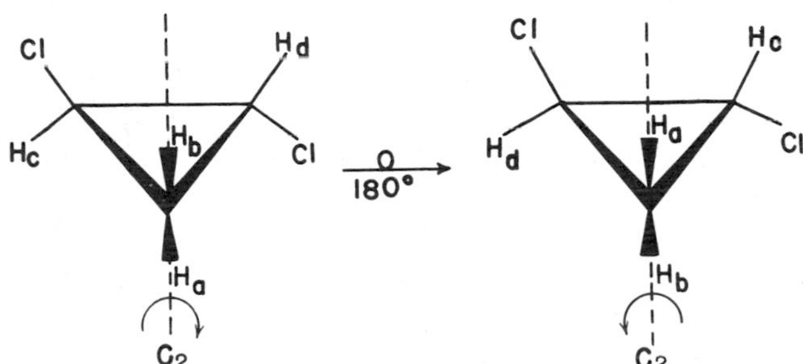

Figure 32. trans-*1,2-Dichlorocyclopropane showing axis of symmetry and effect of rotation around the axis.*

Figure 33. *Newman projection of the staggered rotamers of a molecule with a methyl group attached to a chiral sp^3 carbon atom.*

Magnetic equivalence presupposes chemical shift equivalence. To determine whether chemical shift equivalent nuclei are magnetic equivalent, it must be determined whether they are coupled equally to each nucleus (probe nucleus) in every other set in the spin system. This is done by examining geometrical relationships. If the bond distances and angles from each nucleus in relation to the probe nucleus are identical, the nuclei in question are magnetic equivalent. In other words, two chemical shift equivalent nuclei are magnetic equivalent if they are symmetrically disposed with respect to each nucleus (probe) in any other set in the spin system. This means that the two nuclei under consideration can be interchanged through a reflection plane passing through the probe nucleus and perpendicular to a line joining the chemical shift equivalent nuclei. Note that a test for magnetic equivalence is valid only when the *two nuclei are chemical shift equivalent.* The latter point bears repeating: only chemical shift equivalent nuclei are tested for magnetic equivalence.

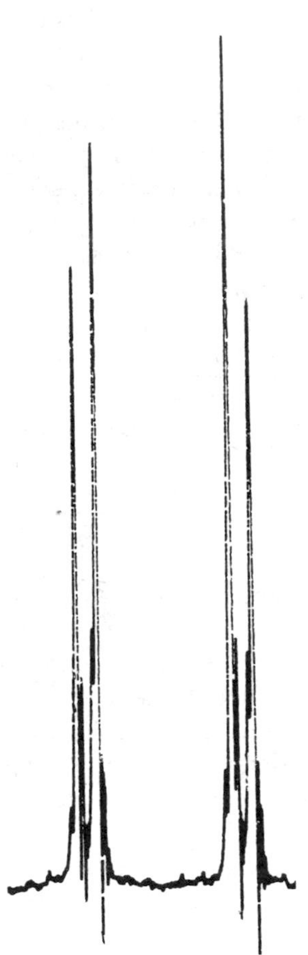

Figure 34. *p-Chloronitrobenzene, 100 MHz in CCl_4.*

These rules are applied readily to conformationally rigid structures. Thus, in *p*-chloronitrobenzene (Figure 34) the protons ortho to the nitro group (H_A and $H_{A'}$) are chemical shift equivalent to each other, and the protons ortho to the chlorine group (H_X and $H_{X'}$) are chemical shift equivalent to each other. In general for *p*-disubstituted benzene rings, J_{AX} and $J_{A'X'}$ are the same, approximately 7 to 10 Hz; $J_{A'X}$ and $J_{AX'}$ are also the same but much smaller, approximately 0 to 1 Hz. Since H_A and $H_{A'}$ couple differently to another specific proton, they are not magnetic equivalent, and first-order rules do not apply. (Similarly, H_X is not magnetic equivalent to $H_{X'}$). The system is described as $AA'XX'$, and the spectrum is very complex (Figure 34). Fortunately the pattern is readily recognized because of its symmetry and apparent simplicity; under ordinary resolution it resembles an AB pattern of two distorted doublets. Closer inspection reveals many additional splittings. As the para substituents become more similar to each other (in their shielding properties), the system tends toward $AA'BB'$; even these absorptions resemble AB patterns until they overlap.

The aromatic protons of symmetrically *ortho*-disubstituted benzenes also give $AA'BB'$ spectra. An example is *o*-dichlorobenzene (Figure 35).

The three isomeric difluoroethylenes furnish additional examples of chemical shift equivalent nuclei that are not magnetic equivalent.

F_b H_b / C=C / F_a H_a F_a F_b / C=C / H_a H_b F_a H_b / C=C / H_a F_b

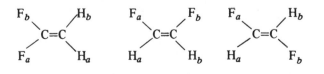

In each case, the protons H_a and H_b comprise a set, and fluorines F_a and F_b comprise a set (of chemical shift equivalent nuclei), but the nuclei in each set are not magnetic equivalent and the spectra are complex.

In conformationally mobile systems, the situation becomes involved. We have seen that the CH_3 protons are chemical shift equivalent, even when the molecule has no symmetry element, by rapid rotational interchange. If one of the substituents on the chiral center is a proton (Y=H in Figure 33), the CH_3 protons are also magnetic equivalent by rapid rotational averaging of the coupling constants among identical rotamers; the system is A_3B or A_3X. Consider, in contrast with the methyl group, a methylene group next to a chiral center, as in 1-bromo-1,2-dichloroethane (Figure 36). Protons H_a and H_b are not chemical shift equivalent, since they cannot be interchanged by a symmetry operation or by rapid rotation. The molecule has no axis, plane, center, or reflection axis of symmetry.

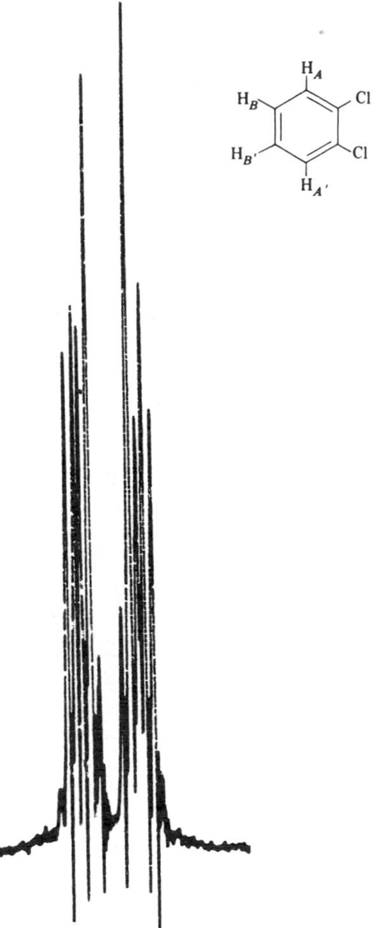

Figure 35. *o-Dichlorobenzene, 100 MHz, in CCl_4.*

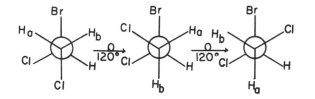

Figure 36. *Newman projection of the rotamers of 1-bromo-1,2-dichloroethane.*

Although there is a rapid rotation around the carbon-carbon single bond, the protons are not interchangeable by a rotational operation. An observer can detect the difference before and after rotating the methylene group; the rotamers are not superimposable.

The question of magnetic equivalence does not arise since there is no chemical shift equivalence; the system is *ABX*.

In ZCH_2CH_2Y molecules, we are dealing with an *anti* rotamer and 2 enantiomeric *gauche* rotamers (Figure 37). The analysis is quite complex*, but in essence, these are $AA'XX'$ systems, which usually give apparent A_2X_2 spectra. The spectrum of 2-dimethylaminoethyl acetate (Figure 38a) is a case in point. Figure 38 b-d shows the progressive distortions as $AA'XX' \rightarrow AA'BB'$ (i.e., $\Delta\nu/J$ decreases) in compounds of type ZCH_2CH_2Y. As the absorptions move closer together, the inner peaks increase in intensity, additional splitting occurs, and some of the outer peaks disappear in the baseline noise. The general appearance of symmetry throughout aids recognition of the type of spin system involved. At the extreme, the two methylene groups become chemical shift equivalent and a single A_4 peak results. Although 1,3-dichloropropane (Figure 39) is properly described as $XX'AA'XX'$, it presents an apparent A_2X_4 system (triplet and quintet).

A_2B_2 or A_2X_2 systems are quite rare (examples include difluoromethane, 1,1-difluoroallene and 1,1,3,3-tetrachloropropane); most systems described in the literature as A_2X_2 are really $AA'XX'$.

It turns out that the question of magnetic equivalence in rotamers is readily resolved by examining the individual rotamers in which the protons in question are chemical shift equivalent by a symmetry operation. If the protons are magnetic équivalent in these rotamers, they are so for the molecule.

An interesting situation occurs in a molecule such as 1,3-dibromo-1,3-diphenylpropane, which has a methylene group between two chiral centers (Figure 40). In (1*R*,3*R*)-1,3-dibromo-1,3-diphenylpropane (one of a racemic pair), H_a and H_b are chemical shift equivalent and so are H_c and H_d, because of an axis of symmetry (C_2) in the molecule. Rotation around the axis interchanges H_a with H_b, and H_c with H_d. On the other hand, in (1*S*,3*R*)-1,3-dibromo-1,3,-diphenylpropane (a meso compound), H_c and H_d are not chemical shift equivalent: they cannot be interchanged by a symmetry operation. (1*S*,3*R*)-1,3-Dibromo-1,3-diphenylpropane does have a plane of symmetry (σ), but both H_c and H_d are in that plane and cannot be inter-

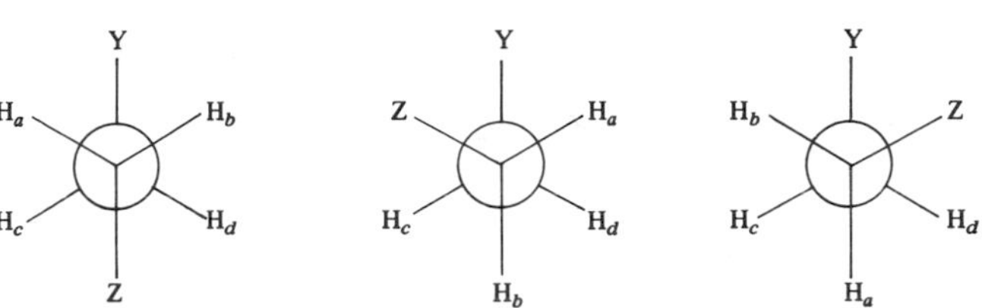

Figure 37. *Newman projection of the rotamers of the YCH_2CH_2Z system.*

*See R. M. Silverstein and R. T. LaLonde, *J. Chem. Ed.*, **57**, 343 (1980).

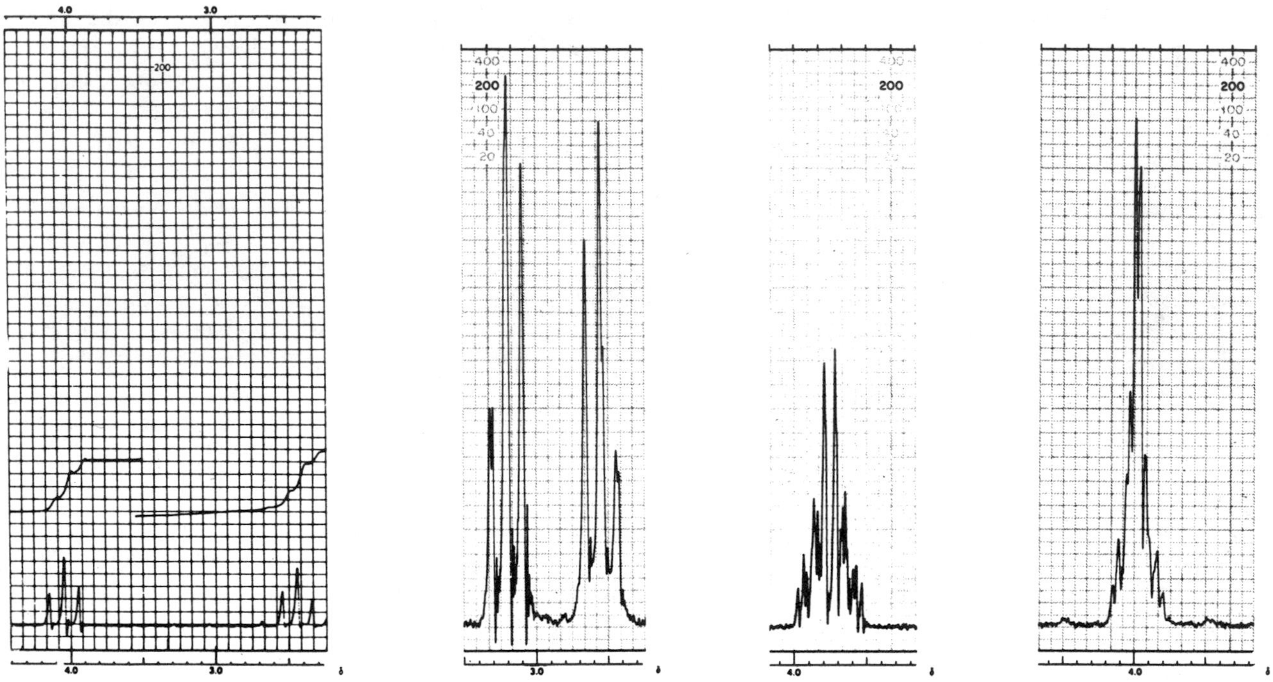

$(CH_3)_2NCH_2CH_2OCCH_3$ (with C=O)
2-dimethylaminoethyl acetate

(a)

$ND_2CH_2CH_2COOD$
β-alanine

(b)

$ClCH_2CH_2OH$
2-chloroethanol

(c)

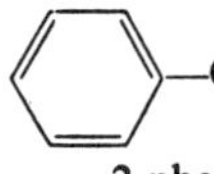

C_6H_5—OCH_2CH_2OH
2-phenoxyethanol

(d)

Figure 38. *Progressive distortions as $AA'XX' \rightarrow AA'BB'$ in $Z-CH_2-CH_2Y$. 60 MHz.*

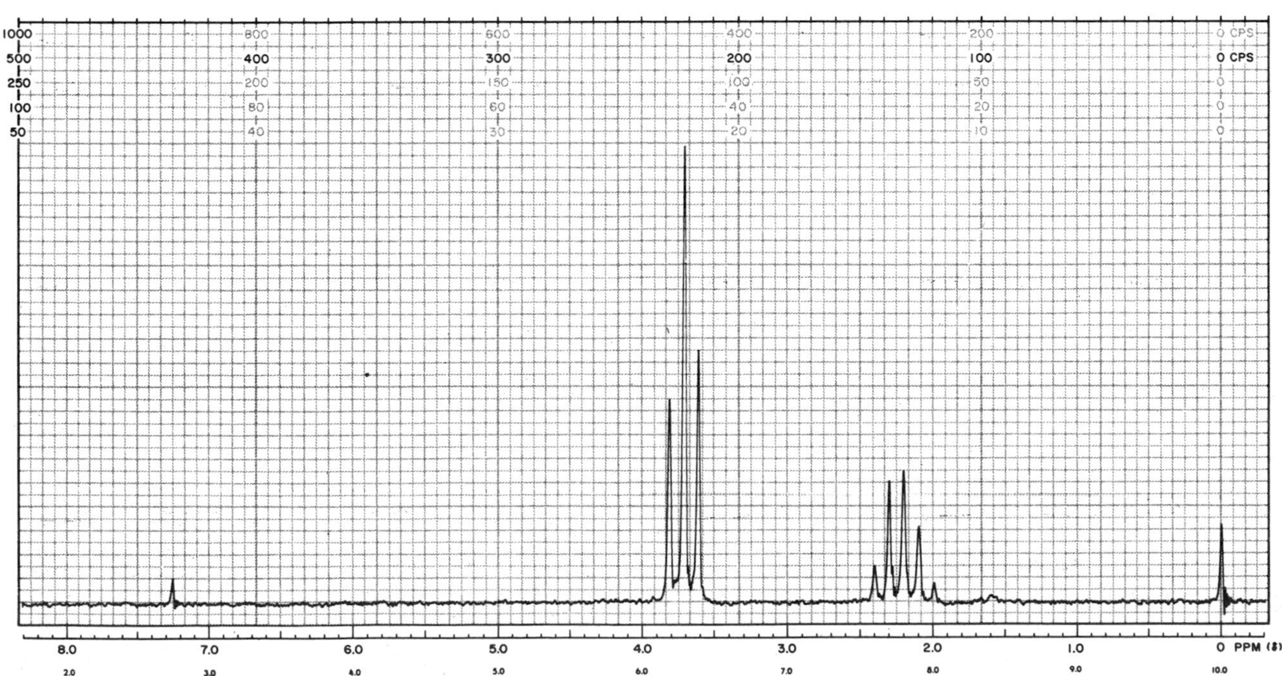

Figure 39. *1,3-Dichloropropane in $CDCl_3$. 60 MHZ. $XX'AA'XX'$ system*

changed. H_a and H_b are chemical shift equivalent because they can be interchanged by reflection through the plane of symmetry shown in Figure 40.

In the $1R,3R$-compound, H_a and H_b are not magnetic equivalent since they do not identically couple to H_c or to H_d; H_c and H_d also are not magnetic equivalent since they do not identically couple to H_a or H_b. Note that only those rotamers are examined in which the protons in question were chemical shift equivalent. The system is $AA'XX'$ or more descriptively A-XX'-A'. In the $1S,3R$ compound of Figure 40, $J_{ad} = J_{bd}$ and $J_{ac} = J_{bc}$; thus, in this molecule, H_aH_b are magnetic equivalent. The question of magnetic equivalence of H_c and H_d is not relevant since they are not chemical shift equivalent. The system is ABX_2 or more descriptively, X-AB-X.

VIII. AMX, ABX, AND ABC SYSTEMS WITH THREE COUPLING CONSTANTS

The spectrum of methyl 2-furoate (Chapter 7, compound 7-2) represents a nearly first-order *AMX* systm with three coupling constants. Each proton is represented by a pair of doublets.

Although *ABX* systems with three coupling constants are not first-order, the patterns are frequently recognized if the distortions are not too severe. The degree of distortion depends on the ratios of the separation of the *A* and *B* protons to their coupling constants. The vinylic structure gives an *ABX* or an *ABC* system depending on the nature of the substituent *Y*, which determines the shift positions of the protons.

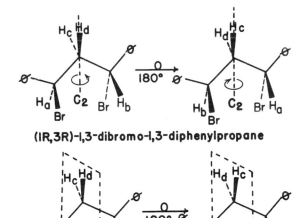

(1R,3R)-1,3-dibromo-1,3-diphenylpropane

(1S,3R)-1,3-dibromo-1,3-diphenylpropane

Figure 40. *Two isomers of 1,3-dibromo-1,3-diphenylpropane. In the (1R,3R) isomer, H_a and H_b are chemical shift equivalent, as are H_c and H_d. In the (1S,3R) isomer, H_a and H_b are chemical shift equivalent, but H_c and H_d are not.*

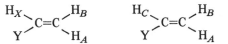

p-Chlorostyrene (Figure 41) shows an *ABX* spectrum that can readily be related to an *AMX* pattern. The following analysis, though not rigorous, is useful.

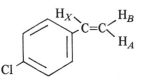

Protons *A* and *B* are not chemical shift equivalent. Proton *A* ($\delta \sim 5.70$) is deshielded about 25 Hz compared with proton *B*, because of its relative proximity to the ring. Proton *X* ($\delta \sim 6.70$) is strongly deshielded by the ring and is split by proton *A* ($J \sim 18$ Hz) and by proton *B* ($J \sim 11$ Hz). The *A* proton signal is split by the *X* proton ($J \sim 18$ Hz) and by the *B* proton ($J \sim 2$ Hz). The *B* proton signal is split by the *X* proton ($J \sim 11$ Hz) and by the *A* proton ($J \sim 2$ Hz).

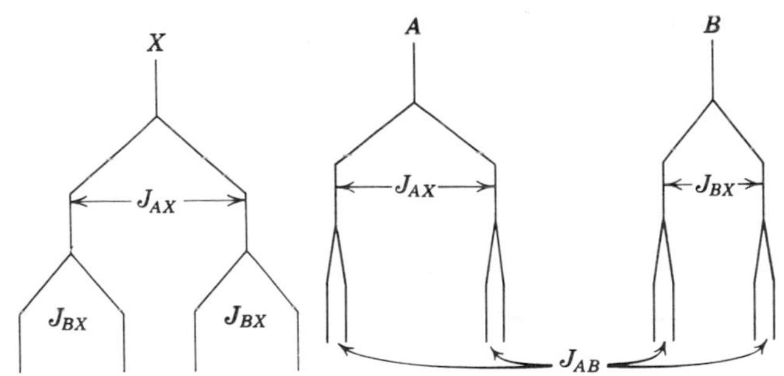

The coupling constants for a vinylic system are characteristic; the *trans* coupling is larger than the *cis*, and the geminal coupling is very small.

The preceding analysis lacks rigor in several details. The splittings of proton *X* do not correspond exactly to J_{BX} and J_{AX}, and although this is a good approximation when $\Delta\nu$ is greater than about 10 for protons *AB*, the only exact information obtainable from the *X* pattern is that the spread between the outside peaks is equal to $J_{AX} + J_{BX}$. The distortions in peak intensities from an *AMX* pattern are obvious. The pattern of two closely spaced pairs should not be confused with the quartet resulting from splitting by three equivalent protons.

As the shift positions of protons *A* and *B* approach each other so that $\Delta\nu$ becomes much smaller than 10, the

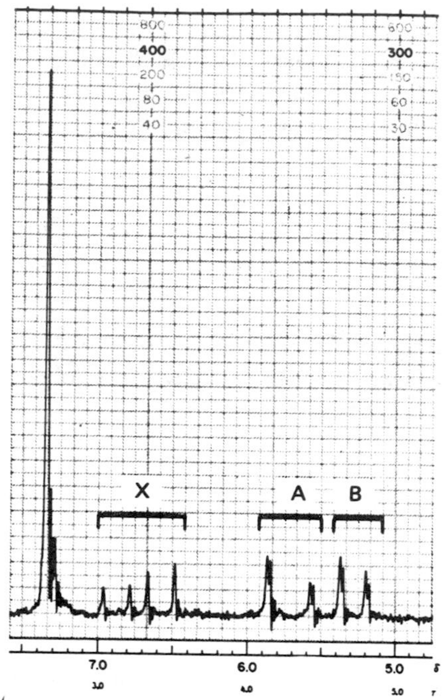

Figure 41. *p-Chlorostyrene, 60 MHz.*

deviations from a first-order spectrum become severe; the A and B patterns overlap, and the middle peaks of the X pattern merge. In the extreme case that $\nu_A = \nu_B$ and $J_{AX} = J_{BX}$, protons A and B are magnetic equivalent, the spectrum is simply A_2X, and the X proton absorption is a triplet. Note that a very similar spectrum would arise if the A pair were chemical, but not magnetic, equivalent: that is, an $AA'X$ system. As the shift position of the X proton approaches the A and B absorptions, the spectrum degenerates to a very complex ABC pattern.

An equatorial and an axial proton on the same carbon of a conformationally locked 6-membered ring may form the AB part of an ABX or an ABC pattern, depending on the nature of Y on an adjacent carbon. Typically, H_A (equatorial) is deshielded relative to H_B (axial) by about 0.1 to 0.7 ppm (Section III), J_{AB} is about 12 to 15 Hz, J_{BX} (ax.-ax.) is about 5–10 Hz, and J_{AX} (eq.-ax.) is about 2 to 3 Hz (Section XI).

H_B H_A Y H_X

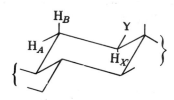

IX. STRONGLY AND WEAKLY COUPLED SPIN SYSTEMS

The 60 MHz spectrum of 1-nitropropane (Figure 42) is an $A_3MM'XX'$ system that resembles an $A_3M_2X_2$ system with J_{AM} very similar to J_{MX}. We interpret the upfield triplet as the methyl group split by the adjacent methylene group, the downfield triplet as the methylene group (adjacent to the nitro group) split by the adjacent

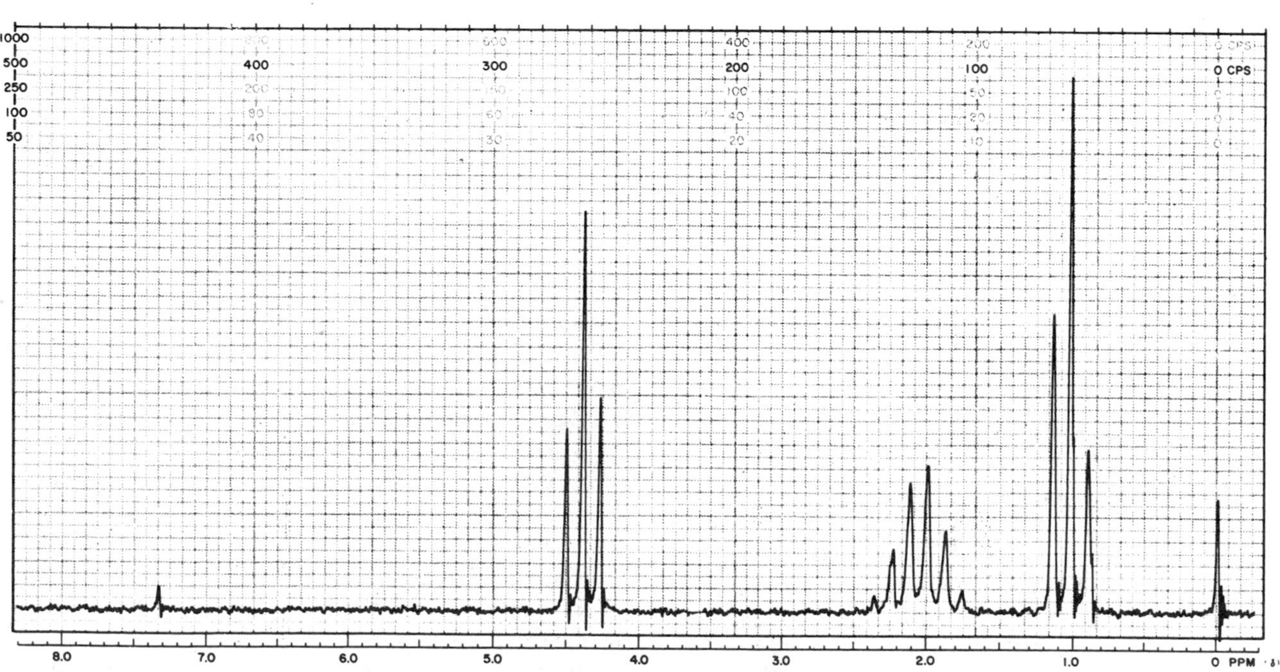

Figure 42. *1-Nitropropane in $CDCl_3$. 60 MHz. CH_3-CH_2-CH_2-NO_2, $A_3MM'XX'$*

methylene group that in turn is split by 5 neighboring protons to give a sextet. We note some broadening of the sextet peaks since J_{AM} is not exactly equal to J_{MX}.

What we have done, in effect, is to mentally cleave the spin system into segments, and are successful in rationalizing the spectrum. Why is it that we can use this kind of oversimplified analysis for 1-nitropropane, but we run into trouble when we predict a triplet for the methyl group (Figure 43a) of heptaldehyde (A_3B_2 portion of the system), or a doublet for the methyl group (Figure 43b) of β-methylglutaric acid, $HOOC{-}CH_2{-}\underset{\displaystyle CH_3}{\underset{|}{CH}}{-}CH_2COOH$ (A_2B portion of the system)?

The answer is that we have weakly coupled protons in 1-nitropropane, but strongly coupled protons in the latter two compounds. Or, to put it another way, the spins of the methyl protons in 1-nitropropane are affected only (or almost entirely) by the adjacent methylene protons. But in heptaldehyde, the methyl proton spins are affected by all of the adjoining methylene groups that have practically the same chemical shift; that is, $\Delta\nu/J$ for four of the methylene protons is almost zero, and they are considered to be strongly coupled.

A similar situation exists in β-methylglutaric acid. in which the methylene groups and the methine group are strongly coupled (the system is $A_3BCC'DD'$). To explain the difficulties arising from attempting to segregate portions of a strongly coupled system, the concept of virtual coupling is invoked. Even though the coupling constant of the methyl group to the methylene group once removed in heptaldehyde or in β-methylglutaric acid is almost zero, the methyl group is considered to be "virtually coupled" to those once removed methylene groups in each compound.

The concept of virtual coupling, then, is convenient but only necessary when one incorrectly considers a portion of a strongly coupled spin system as a separate entity and attempts to apply first-order rules to that protion.

The modus operandi of the organic chemist in interpreting an NMR spectrum is to look initially for first-order absorptions, and then for patterns that can be recognized as distortions of first-order absorptions. There is not usually very much interest in precise solutions of multispin systems. Frequently the emphasis is on portions of spin systems. The above considerations should be of help in setting limits for this convenient, although not rigorous, approach to the interpretation of NMR spectra.

A well-defined multiplet like any one in Figure 42 has a chemical shift that is at the midpoint of the signal (e.g., the triplet at δ 4.37 in Figure 42). In contrast, an ill-defined multiplet, such as in Figure 43b, should be reported as the range, in Hz, over which the signal extends; the instrument frequency must also be reported.*

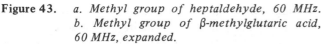

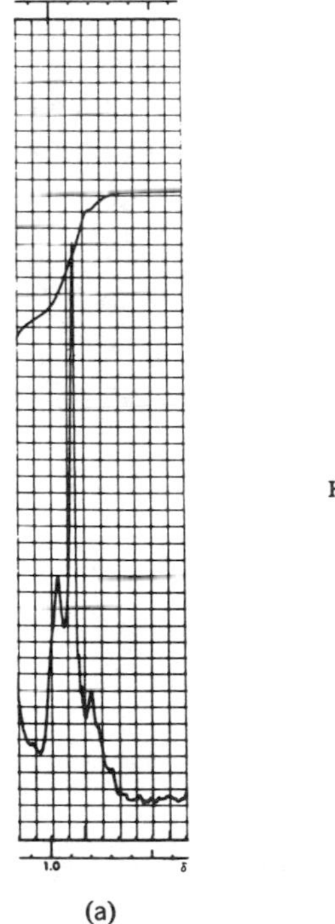

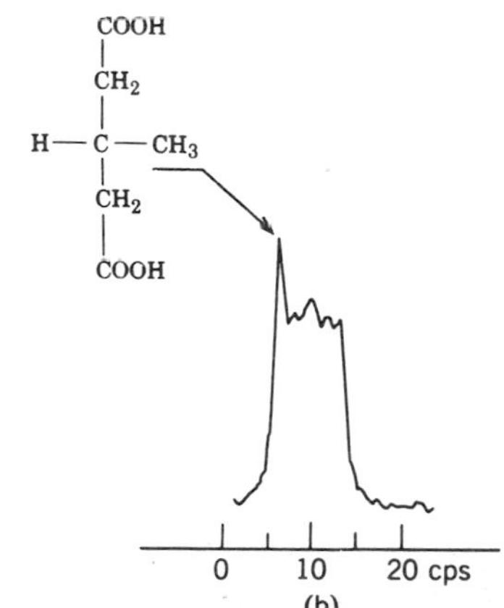

Figure 43. *a. Methyl group of heptaldehyde, 60 MHz. b. Methyl group of β-methylglutaric acid, 60 MHz, expanded.*

X. EFFECTS OF A CHIRAL CENTER

The protons of a methylene group near a chiral center are not chemical shift equivalent, as explained in Section VII. They couple with each other, and each may have a different coupling to a vicinal proton. Even assuming fast rotation around the C–C bond, the methylene protons of a compound

$$\begin{array}{c} H_a\ M \\ |\quad | \\ R{-}C{-}C{-}L \\ |\quad | \\ H_b\ S \end{array}$$

may show an AB pattern; that is, they are diastereotopic since they cannot be operationally interchanged (Section VII).

However, a large part of the nonequivalence may result from unequal populations of conformers even at fast rotation. The spectrum may be further complicated by slow rotation caused by low temperature or bulky substituents. Nonequivalence of the methylene protons persists in a compound such as

$$\begin{array}{c} H_a\ M\ H_a \\ |\quad |\quad | \\ R{-}C{-}C{-}C{-}R \\ |\quad |\quad | \\ H_b\ S\ H_b \end{array}$$

even though the middle carbon atom is not a chiral center.

*G. Slomp, *J. Am. Chem. Soc.*, **84**, 673 (1962).

The methylene group may display nonequivalence even though it is once removed from the chiral center. Examples of the type $R{-}\overset{H_a}{\underset{H_b}{C}}{-}O{-}\overset{M}{\underset{S}{C}}{-}L$ have been reported. The chiral center need not be a carbon atom. The phenomenon has been reported for the methylene protons in quaternary ammonium salts, sulfites, sulfoxides, diethyl sulfide-borane complexes, and thiophosphonates.

Chemical shift nonequivalence of the methyl groups of an isopropyl moiety near a chiral center is frequently observed; the effect has been measured through as many as 7 bonds between the chiral center and the methyl protons.

The protons of the methyl groups in the terpene alcohol 2-methyl-6-methylene-7-octen-4-ol are not chemical shift

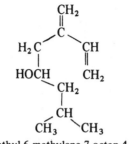

2-Methyl-6-methylene-7-octen-4-ol

equivalent (Figure 44), and even though they are 4 bonds removed from the chiral center, the nonequivalence is detectable at 100 MHz (δ 0.92 and δ 0.90); each absorption is split by the vicinal CH, giving rise to two overlapping doublets ($J = 7$ Hz). The nonequivalent protons of the CH_2 (*a* and *b*) between the $CH_2{=}C$ and CHOH groups

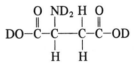

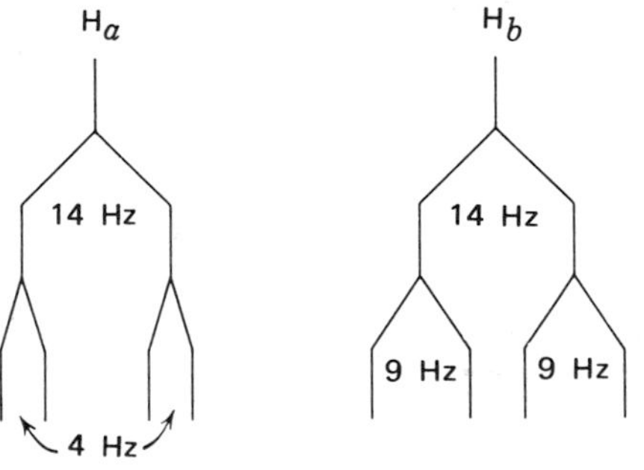

absorb, respectively, at ~δ2.47 and ~δ2.18. Each proton is split by the other (J_{gem} = ~14Hz) and unequally by the neighboring proton (J_{vic} = ~9Hz and 4 Hz). The protons of the other CH_2 group are also nonequivalent; additional splitting by the adjacent methine proton results in a partially resolved multiplet.

The methylene protons (upfield absorption) in aspartic acid (Figure 45) in D_2O are nonequivalent, and the shift difference between them is small compared with the geminal coupling constant. Thus, the *AB* pattern from the geminal

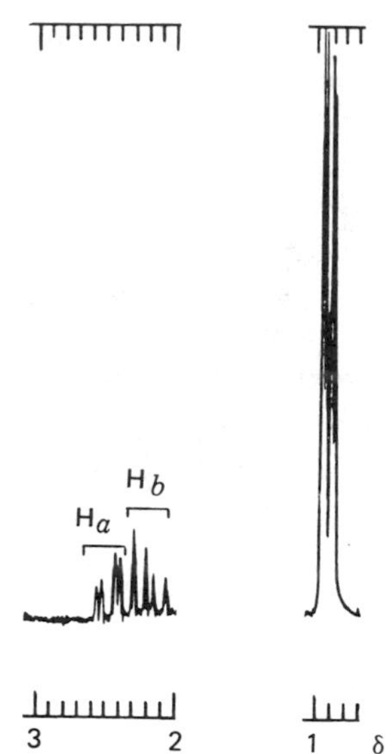

Figure 44. *5-Methylene and methyl protons of 2-methyl-6-methylene-7-octen-4-ol, 100 MHz.*

aspartic acid-*N*-d_2-*O*-*d*

coupling is quite distorted; the inner peaks are strong, the outer peaks weak. Each methylene proton is also split by the vicinal proton with different coupling constants. Two of the peaks coincide, the end peaks are lost in the baseline noise, and the net result is three peaks. The methine proton absorption consists of two pairs.

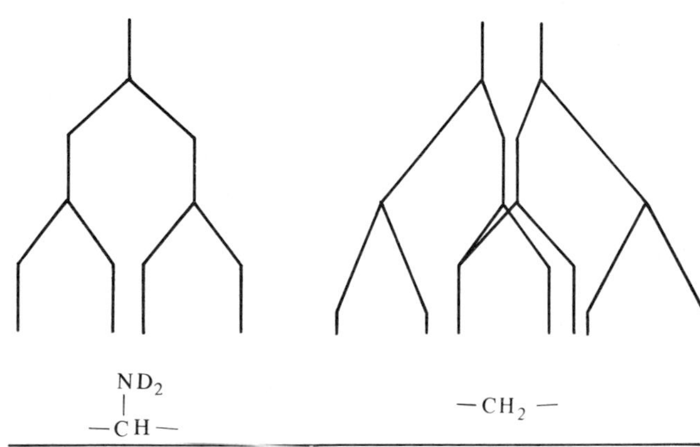

XI. VICINAL AND GEMINAL COUPLING IN RIGID SYSTEMS

Coupling between protons on vicinal carbon atoms in rigid systems depends primarily on the dihedral angle ϕ between the H—C—C′ and the C—C′—H′ planes. This angle can be visualized by an end-on view of the bond between the

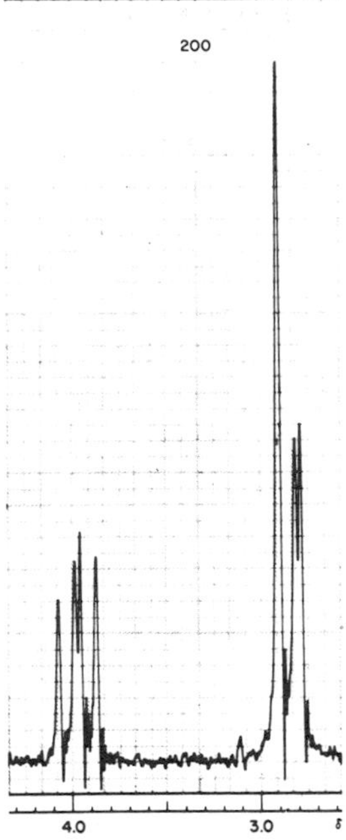

Figure 45. *Aspartic acid in D_2O, 60 MHz.*

H ϕ H′

vicinal carbon atoms and by the perspective in Figure 46 in which the calculated relationship between dihedral angle and vicinal coupling constant is graphed. Karplus emphasized that his calculations are approximations and do not take into account such factors as electronegative substituents, the bond angles θ ($\angle$H–C–C′ and $\angle$C–C′–H′), and bond lengths. Deductions of dihedral angles from measured coupling constants are safely made only by comparison with closely related compounds. The correlation has been very useful in cyclopentanes, cyclohexanes, carbohydrates, and bridged polycyclic systems. In cyclopentanes, the observed values of about 8 Hz for vicinal *cis* protons and about 0 Hz for vicinal *trans* protons are in accord with the corresponding angles of about 0° and about 90°, respectively. In substituted or fused cyclohexane rings, the following relations obtain:

	Calc. J	Observed J (Hz)
axial-axial	9	8-14 (usually 8-10)
axial-equatorial	1.8	1-7 (usually 2-3)
equatorial-equatorial	1.8	1-7 (usually 2-3)

A modified Karplus equation can be applied to vicinal coupling in olefins. The prediction of a larger *trans* coupling ($\varphi = 180°$) than *cis* coupling ($\varphi = 0°$) is borne out. The *cis* coupling in unsaturated rings decreases with decreasing ring size (increasing bond angle θ) as follows: cyclohexenes $J = 8.8$ to 10.5, cyclopentenes $J = 5.1$ to 7.0, cyclobutenes $J = 2.5$ to 4.0, and cyclopropenes $J = 0.5$ to 2.0.

The calculated relationship due to the H–C–H angle of geminal protons is shown in Figure 47. This relationship is quite susceptible to other influences and should be used with due caution. However, it is useful for characterizing methylene groups in a fused cyclohexane ring (approximately tetrahedral, $J \sim 12$ to 18), methylene groups of a cyclopropane ring ($J \sim 5$), or a terminal methylene group ($J \sim 0$ to 3). Geminal coupling constants are actually negative numbers, but this can be ignored except for calculations.

In view of the many factors other than angle dependence that influence coupling constants, it is not surprising that there have been abuses of the Karplus correlation. Direct "reading off" of the angle from the magnitude of the J value is risky. The safest application of the relationships is to structure determinations in which molecular geometries have provided the extrema of the high and low expected J values and for which the 0° and 90° and 180° structures are known for a given system. The limitations of the Karplus correlations are discussed in Jackman and Sternhell.[4]

XII. LONG-RANGE COUPLING

Proton-proton coupling beyond 3 bonds may occur in olefins, acetylenes, aromatics, and heteroaromatics, and in strained ring systems (small or bridged rings). Allylic (H–C–C=C–H) coupling constants are about 0 to 3 Hz. Homoallylic (H–C–C=C–CH) couplings are usually negligible but may be as much as 1.6 Hz. Coupling through conjugated polyacetylenic chains may occur through as many as 9 bonds. Meta coupling in a benzene ring is 1 to 3 Hz, and para, 0 to 1 Hz. In 5-membered heteroaromatic rings, coupling between the 2,4 protons is 0 to 2 Hz.

J_{AB} in the bicyclo[2.1.1]hexane system is about 7 Hz.

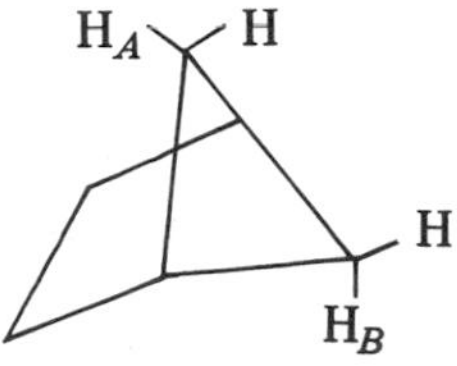

This unusually high long-range coupling constant is attributed to the "W-conformation" of the 4 sigma bonds between H_A and H_B:

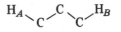

H_A–C–C–C–H_B

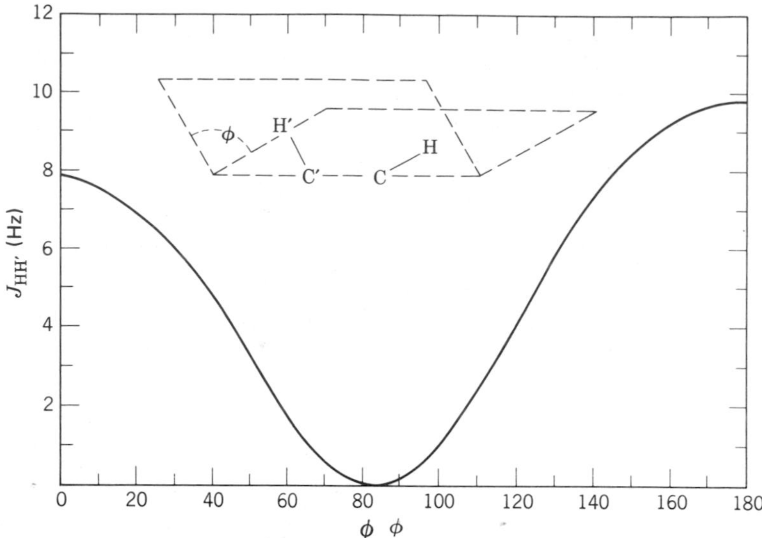

Figure 46. *The vicinal Karplus correlation. Relationship between dihedral angle and coupling constant for vicinal protons.*

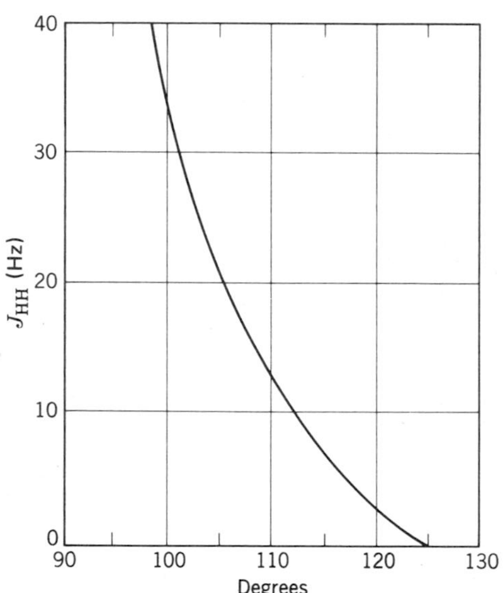

Figure 47. *The geminal Karplus correlation. J_{HH} for CH_2 groups as function of ∠ H–C–H.*

XIII. SPIN-SPIN DECOUPLING

Spin-spin decoupling (double irradiation or double resonance) is a powerful tool for simplifying a spectrum, for determining the relative positions of protons in a molecule, or for locating a buried absorption.

Spin-spin decoupling is simply a technique for irradiating a nucleus with a strong radiofrequency signal at its resonance frequency, while scanning other nuclei to detect which ones are affected by decoupling from the irradiated nucleus. The early experiments in spin-spin decoupling were limited to irradiation of other nuclei such as ^{14}N, ^{31}P, or ^{19}F because of the very large frequency difference from the proton resonance. For example, decoupling of the N atom of a pyrrole or an amide has been used to sharpen the proton absorption, which can be so broad as to be merely a slight bulge on the baseline. As previously pointed out (Section V), this broadening is a result of partial decoupling by the nitrogen electrical quadrupole moment. Irradiation of the nitrogen atom of an amine salt changes the broad ^{+}NH absorption to a sharp absorption whose multiplicity depends on the number of α protons.

Protons can be readily decoupled provided they are more than about 20 Hz apart at 100 MHz. The utility of proton-proton decoupling is shown in the 100 MHz partial spectrum of methyl 2,3,4-tri-*O*-benzoyl-β-L-lyxopyranoside (Figure 48). The integration (not shown) gives the following ratios in Figure 48a from high field to low: 3:1:1:1:1:2. The sharp peak at δ 3.53 is the OCH_3 group. Decoupling the 2-proton multiplet at δ 5.75 causes the multiplet at δ 5.45 to collapse to four peaks, and the doublet at δ 5.00 to a sharp singlet (Figure 48b). Decoupling the multiplet at δ 5.45 partially collapses the multiplet at δ 5.75 and collapses the 2 upfield pairs of doublets (at δ 4.45 and δ 3.77) to 2 doublets (Figure 48c). The H_5 absorption should be upfield since these 2 protons are deshielded by 2 ether oxygens, and H_2, H_3, and H_4 are deshielded by benzoyl groups. The 2 H_5 protons are the *AM* portion of an *AMX* pattern; the H_4 proton is the *X* portion (with additional splitting). The pair of doublets at δ 4.45 is 1 (H_5') proton at C_5 strongly deshielded by the benzoyl group on C_4, and the pair of doublets at δ 3.77 is

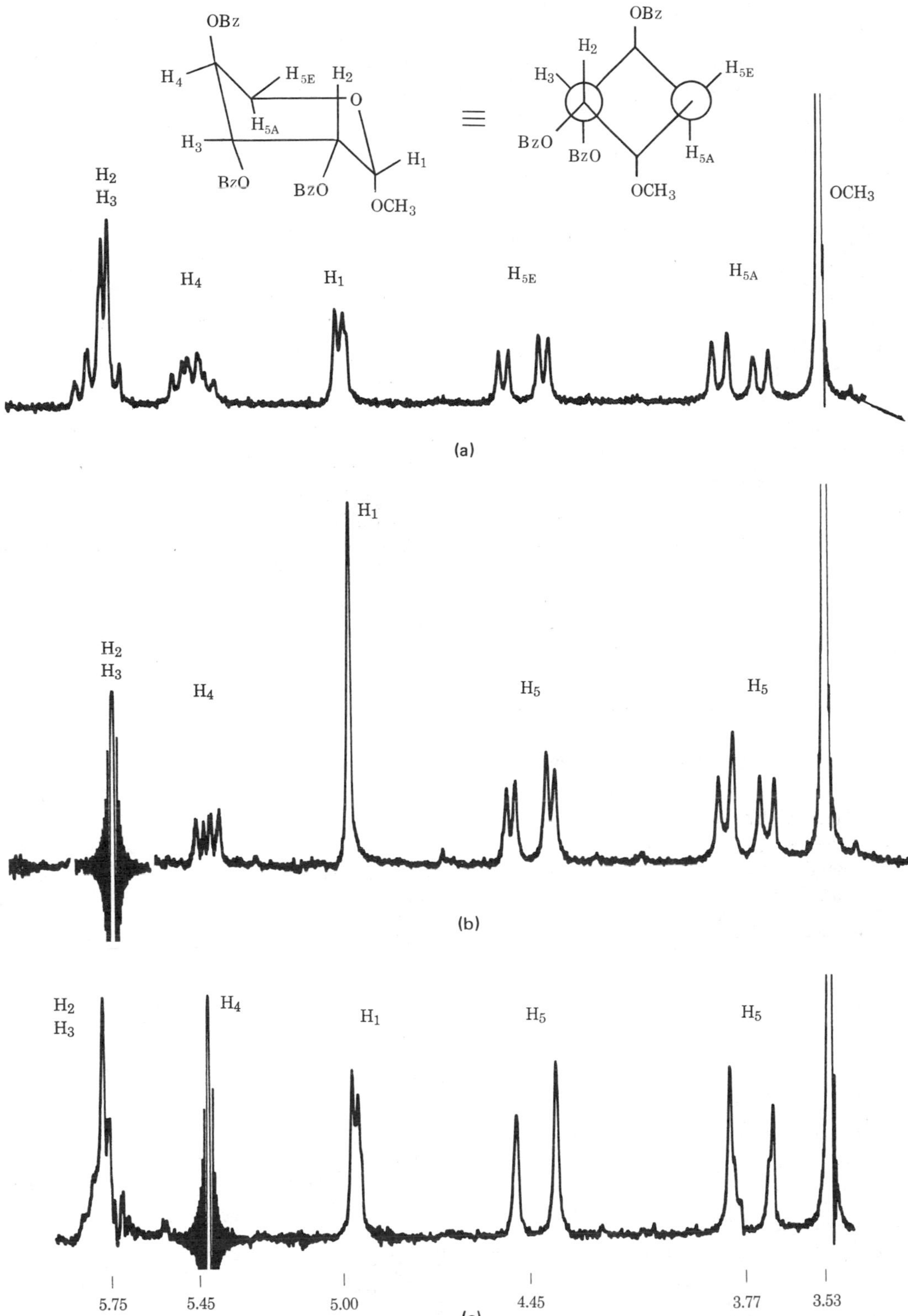

Figure 48. *(a) Partial spectrum of methyl-2,3,4-tri-O-benzoyl-β-L-lyxopyranoside, 100 MHz, $CDCl_3$. (b) H_2 and H_3 decoupled. (c) H_4 decoupled. Note that there are two H_5 protons. Bz = C_6H_5CO.*

the other (H_5) absorption. Further confirmation is provided by the collapse of each pair of doublets to doublets with the characteristic large geminal coupling ($J = 12.5$ Hz) on irradiation of the multiplet at δ 5.45, which must therefore be the H_4 absorption (i.e., the X proton that was further split). The multiplet at δ 4.75 must therefore represent H_2 and H_3 since irradiation of this multiplet collapsed the multiplet that we identified as H_4 (this now appears as the X portion of the AMX) pattern), and also the doublet at δ 5.00, which must be the H_1 absorption.

A multiplet caused by coupling to 2 nonequivalent protons can be completely collapsed to a singlet by irradiating both coupling protons simultaneously.

XIV. SHIFT REAGENTS

Shift reagents, introduced in 1969, provide a method for spreading out NMR absorption patterns without increasing the strength of the applied magnetic field. Addition of shift reagents to appropriately functionalized samples results in substantial magnification of the chemical shift differences of nonequivalent protons. The shift reagents are ions in the rare earth (lanthanide) series coordinated to organic ligands. Although it had been long known that some metal ions caused shifts, it was not until the application of the more recent shift reagents, $Eu(dpm)_3$ and $Eu(fod)_3$ that these shifts could be effected without substantial line broadening.

The notation for the more commonly used shift reagents, $Eu(dpm)_3$ and $Eu(fod)_3$, come from their names *tris*-(*di*pivalo*m*ethanato)europium and *tris*-1,1,1,2,2,3,3-hepta*f*luoro-7,7-dimethyl-3,5-*o*ctane*d*ionatoeuropium. A more systematic name for the former is *tris*-2,2,6,6-*t*etramethyl-3,5-*h*eptane*d*ionatoeuropium and thus the (less common) substitute $Eu(thd)_3$ for $Eu(dpm)_3$.

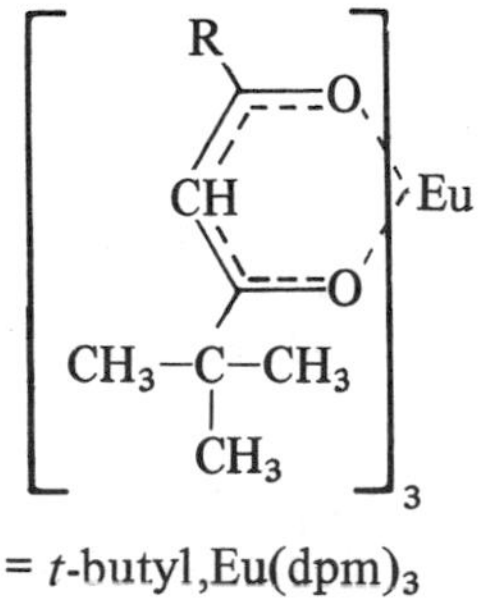

R

O

CH Eu

O

CH_3-C-CH_3

CH_3]$_3$

R = *t*-butyl, $Eu(dpm)_3$
R = $CF_2CF_2CF_3$, $Eu(fod)_3$

The use of such shift reagents is illustrated in Figure 49 in which the NMR spectrum of 1-heptanol is simplified by the addition of $Eu(dpm)_3$. As is usual for such an alcohol in the absence of shift reagents, the only interpretable signal (Figure 49a) is that of the methylene adjacent to OH (proton set A, δ 3.8, triplet) and the terminal methyl (distorted set G, triplet, δ 0.9). Upon addition of the shift reagent $Eu(dpm)_3$, the signals of the methylene groups closer to the OH group are moved downfield so that a separate signal is available for each of the methylene units (Figure 49c):

$$\underset{G}{CH_3}-\underset{F}{CH_2}-\underset{E}{CH_2}-\underset{D}{CH_2}-\underset{C}{CH_2}-\underset{B}{CH_2}-\underset{A}{CH_2}-OH$$

The closer the group to the functional group, the greater is the shift per increment of shift reagent (Figure 50). Not all plots analogous to Figure 50 show linearity.

There are two major applications of the shift reagent to structure determination: (1) to simplify the spectrum and (2) to assign the protons from data on the response curves as in Figure 50. The former is straightforward and is subject only to the limit of the dependence of the coupling constant on shift reagent concentration. The latter is less certain because of the many parameters on which these slopes depend.

It is necessary to report the solvent and the concentration of the substrate and shift reagent for a given shift reagent experiment (see Figure 49 legend). Another way to report results is to give the slope of such graphs as Figure 49 as ppm per mole of shift reagent per mole of substrate.

Table III shows the general correlation between basicity

Table III. Variation of the Magnitude of Induced Shift with Functionality[28]

Functional Group	Ppm per Mol of $Eu(dpm)_3$ per Mol of Substrate in CCl_4
$RCH_2N\underline{H}_2$	~150
$RCH_2O\underline{H}$	~100
$R_2C{=}NO\underline{H}$	~40
$RC\underline{H}_2NH_2$	30-40
$RC\underline{H}_2OH$	20-25
$R\underline{H}C{=}NOH$	14-30
$RC\underline{H}_2RC{=}NOH$	14-10
$RC\underline{H}_2COR$	10-17
$RC\underline{H}_2CHO$	11-19
$RC\underline{H}_2SOR$	9-11
$RC\underline{H}_2-O-C\underline{H}_2R$	10 (17-28 in $CDCl_3$)
$RC\underline{H}_2CO_2Me$	7
$RCH_2CO_2C\underline{H}_3$	6-5
$RC\underline{H}_2CN$	3-7
$RC\underline{H}_2NO_2$	~0
halides, indoles, alkenes	0
RCO_2H and phenols*	decompose reagent

*Phenols can be studied using $Eu(fod)_3$.

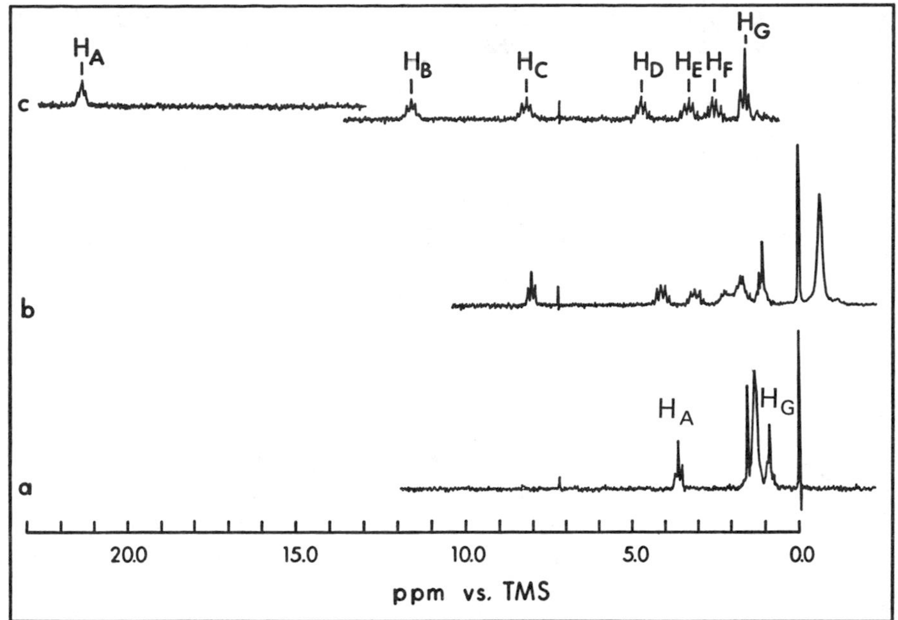

Figure 49. *60 MHz proton NMR spectra of 0.40 ml of $CDCl_3$ solution containing 0.300M 1-heptanol at various mole ratios [moles of $Eu(dpm)_3$ per mole of 1-heptanol]: (a) 0.00, (b) 0.19, (c) 0.78. Temperature, 30°C. From* Anal. Chem., *Vol. 43, p. 1599, Copyright 1971 by the American Chemical Society. Reprinted by permission of the copyright owner.*

of the functional group of the substrate and shift magnitude as induced by $Eu(dpm)_3$. $Eu(fod)_3$ is normally a "stronger" shift reagent, since it is a stronger Lewis acid; it has the additional advantage of much greater solubility. However quantitative assessments, such as Table II, should be treated with caution, since such data were assembled before it was fully appreciated that shift reagent performance is susceptible to many parameters including the purity of the shift reagent, the method of data handling, and the presence of water. Because of the hygroscopicity of the reagents and the adverse effect of small amounts of water, it is advisable to work in a dry-box for quantitative results.

Experimental limitations are set by the solvents in which the reagents can be used and by the absorption of the protons of the organic ligands on the reagent in the NMR spectrum. Shift reagents are most effective in "noncompetitive" solvents, that is, those that do not coordinate strongly to the shift reagent. The most effective solvent is carbon tetrachloride; benzene and deuterochloroform are also useful, giving shifts that are approximately 90% and 50% as great as in CCl_4. One limitation of such solvents is the solubility of the shift reagent alone in the NMR solvent; $Eu(dpm)_3$ dissolves to the extent of about 100 mg/ml in benzene and about 200-300 mg/ml in the chlorinated solvents. Effectiveness of the shift reagent analysis may be increased by the use of low temperature NMR. The typical resonances of the ligands in the shift reagents in the presence of substrate are: $Eu(fod)_3$, δ 0.4 to 2.0; $Eu(dpm)_3$, δ −1.0 to −2.0; $Pr(dpm)_3$, δ 3.0 to 5.0. Deuterated reagents are available to alleviate this interference. The substrates under study may be recovered by the use of tlc or glc.

Most explanations for the effect of shift reagents are based on a pseudocontact mechanism. The dependence of the magnitude of the shift on the distance of the protons of interest from the functional group is only an approximation in the pseudocontact shift equation; this dependence breaks down when the shift reagent coordination with the substrate is not a simple, linear situation, that is, when unique molecular geometry causes an acute proton-functional group-rare earth atom angle. The equilibrium for the shift reagent with the substrate is rapidly established and thus only a single set of time-averaged peaks is observed. Upfield shifts can be effected by the use of diamagnetic shift reagents, for example complexes of Pr. Optical purity (enantiomeric composition) of chiral substrates may be determined by using shift reagents with chiral ligands. Excellent reviews have been published.[28-32]

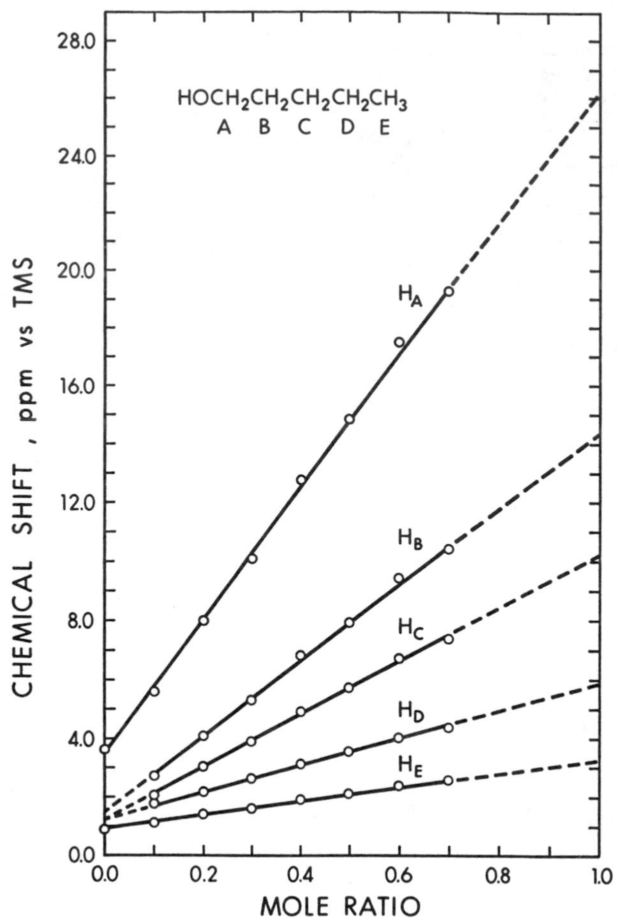

Figure 50. *Variation in chemical shift for carbon-bonded protons of 1-pentanol in $CDCl_3$ (0.40 ml of 0.300M solution) with increasing concentration of $Eu(dpm)_3$. Temperature, 30° C. From* Anal. Chem., *Vol. 43, p. 1599. Copyright by the American Chemical Society (1971) Printed by permission of the copyright owner.*

REFERENCES

1. Bovey, F. A. *NMR Spectrometry.* New York: Academic Press, 1969.
2. Bible, R. H., Jr. *Interpretation of NMR Spectra.* New York: Plenum Press, 1965.
3. Bhacca, N. S. and Williams, D. H. *Applications of NMR Spectrosopy in Organic Chemistry.* San Francisco: Holden-Day, 1964.
4. Jackman, L. M., and Sternhell, S. *Applications of NMR Spectroscopy in Organic Chemistry.* 2nd ed. New York: Pergamon, 1969.
5. Bible, R. H., Jr. *Guide to the Empirical Method: A Workbook.* New York: Plenum Press, 1967.
6. Mathieson, D. W. (ed.). *Nuclear Magnetic Resonance for Organic Chemists,* New York: Academic Press, 1967.
7. Becker, E. D. *High Resolution NMR.* New York: Academic Press, 1969.
8. Ault, A. and Dudek, G. O. *NMR, An Introduction to Proton Nuclear Magnetic Resonance Spectrometry.* San Francisco: Holden-Day, 1976.
9. Leyden, D. E. and Cox, R. H. *Analytical Applications of NMR.* New York: Wiley, 1977.
10. Shoolery, J. N. *A Basic Guide to NMR.* Palo Alto: Varian Assoc., 1972.
11. McFarlane, W. *Proton Magnetic Resonance,* in *Elucidation of Organic Structure by Physical and Chemical Methods.* 2nd ed. Part 1. K. W. Bentley and G. W. Kirby (eds). New York: Wiley, 1972.
12. Abraham, R. J. and Loftus, P. *Proton and Carbon –13 NMR Spectroscopy.* London: Heyden, 1978.
13. Jackman, L. M. and Cotton, F. A. (eds). *Dynamic NMR Spectroscopy.* New York: Academic Press, 1975.
14. Pople, J. A., Schneider, W. G., Bernstein, H. J. *High-Resolution Nuclear Magnetic Resonance.* New York: McGraw-Hill, 1959.
15. Carrington, A. and McLachlan, A. D. *Introduction to Magnetic Resonance.* New York: Harper and Row, 1967.
16. Ionin, B. I., and Ershov, B. A. *NMR Spectroscopy in Organic Chemistry.* New York: Plenum Press, 1970.
17. Emsley, J. W., Feeney, J., Sutcliffe, L. H. *High Resolution Nuclear Magnetic Resonance Spectroscopy.* New York: Pergamon, vol. 1, 1975; vol 2, 1966.
18. Varian Associates. *High Resolution NMR Spectra Catalogue,* Vol. 1, 1962; Vol. 2, 1963.
19. *The Sadtler Collection of High Resolution Spectra.* Philadelphia: Sadtler Research Laboratories.
20. Pouchert, C. J. and Campbell, J. R. *Aldrich Library of NMR Spectra,* Vols I-XI. Milwaukee, Wisc: Aldrich Chem. Co., 1975.
21. Batterman, T. J. *NMR Spectra of Simple Heterocycles.* New York: Wiley, 1973.
22. Brugel, W. *Handbook of NMR Spectral Parameters.* Philadelphia: Heyden, 1979.
23. Bovey, F. A. *NMR Data Tables for Organic Compounds.* vol. I. New York: Wiley-Interscience, 1967.
24. Szymanski, H. A. and Yelin, R. E. *NMR Band Handbook.* New York: IFI/Plenum, 1968.
25. Chamberlain, N. F. *The Practice of NMR Spectroscopy with Spectra-Structure Correlation for* ^{1}H. New York: Plenum, 1974.

26. Zwolinski, B. J., et al. *Selected Nuclear Magnetic Resonance Spectral Data.* American Petroleum Institute Project 44, vols. I-III. College Station, Texas: Thermodynamics Research Center, Texas A & M.
27. Wiberg, K. B., and Nist, B. J. *The Interpretation of NMR Spectra.* New York: W. A. Benjamin, 1962.
28. Campbell, J. R. *Aldrichimica Acta.* Aldrich Chemical Co., 4:55 (1971).
29. Sievers, R. E. *NMR Shift Reagents.* New York: Academic Press, 1973.
30. Hofer, O. *Topics in Stereochemistry*, vol. 9, N. Allinger and E. L. Eliel (eds). New York: Wiley-Interscience, p. 111 ff.
31. Willcott, M. R., III and Davis, R. E. *Science* **190**, 850 (1975).
32. Cockerill, A., *et al.*, *Chem. Rev.* **73**, 553 (1973).
33. Kime, K.A. and Sievers, R. E. *Aldrichimica Acta.* Aldrich Chemical Co., 10:54(1977).

PROBLEMS

4.1 In each of the following pairs, predict which of the designated protons occur at lower field. Explain your choice.

a. $CH_3CH_2CH_2C{\equiv}CH$ and $CH_3CH_2CH_2CH{=}CH_2$

b. 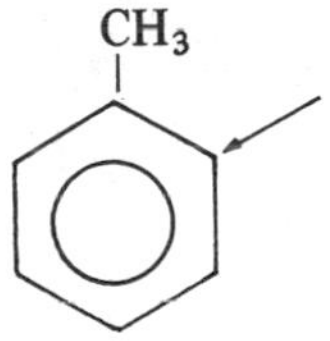 (toluene, ring proton ortho to CH_3) and (acetophenone, ring proton ortho to $COCH_3$)

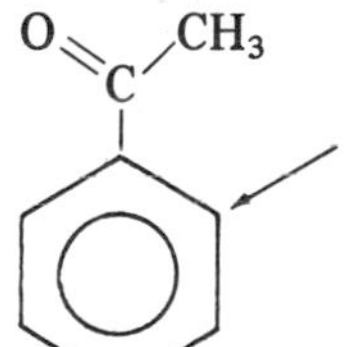

c. 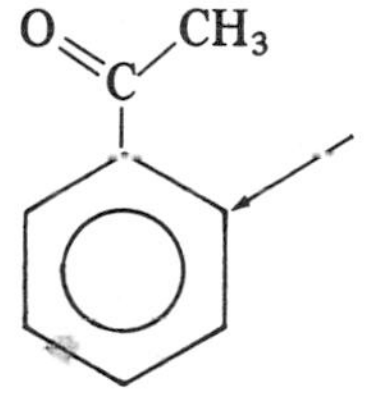 (acetophenone, ring proton ortho) and (benzophenone, $C_6H_5COC_6H_5$, ring proton ortho)

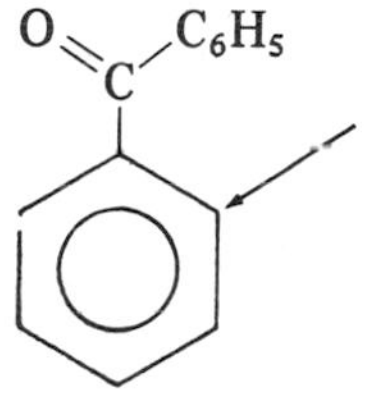

d. 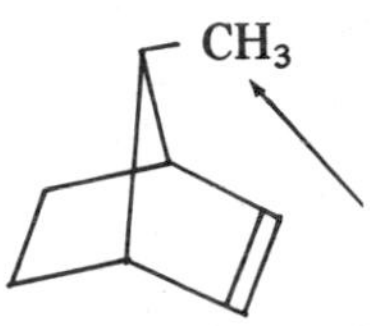 (7-methylnorbornene, CH_3 anti to the double bond) and (7-methylnorbornene, CH_3 syn to the double bond)

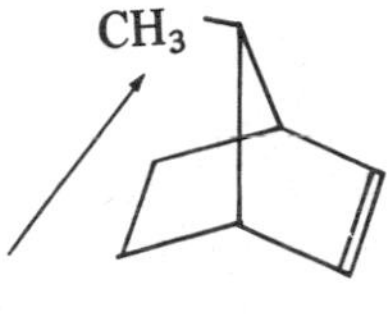

4.2 Draw an NMR spectrum for each of the following compounds. Assume a first-order spectrum in all cases.

a. CH_3CH_2Cl

b. $(CH_3)_2CHBr$

c. $CH_3C(CH_3)_2CH_2Br$

d. cycloheptane

e. $CH_3COCH_2CH_3$ (with C=O)

f. 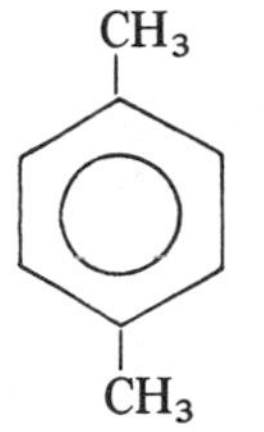 p-xylene ($CH_3C_6H_4CH_3$)

g. $BrCH_2CHO$

4.3 The pattern in Figure 20d at first glance may represent the 2 protons of an *AB* system or the *X* proton of an A_3X system. How is it possible to distinguish between these possibilities?

4.4 a. For ethylbenzene, designate the spin systems and the sets of protons in each system. Designate each set in terms of magnetic equivalence, and write the Pople notation for the spin systems.

b. The pair of protons at C-3 of *cis*-1,2-dichlorocyclopropane are diastereotopic. Explain.

c. Give the Pople notation for 1,1-dideuteroethylene:

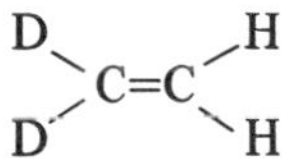

$D_2C{=}CH_2$

4.5 Consider the following two representations of the same compound:

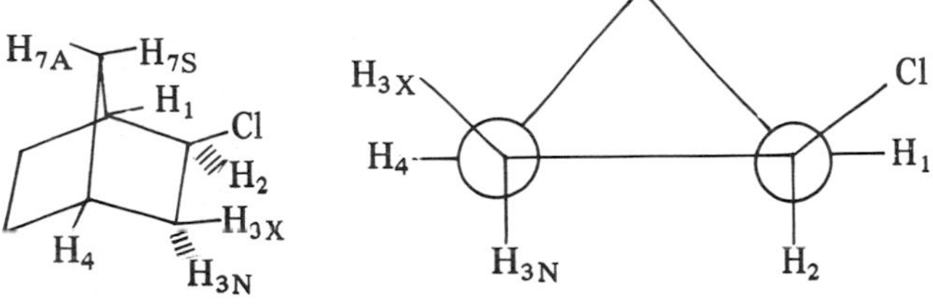

Using the discussion in Sections XI and XII of this chapter, explain why one expects ranges of $J_{1,2}$ = ~0-2 Hz, $J_{2,3N}$ = 8-10 Hz, $J_{3X,2}$ = ~ 3-5 Hz and $J_{7A,7S}$ = ~9-11 Hz. Which would be the larger long-range coupling: $J_{2,7A}$ or $J_{2,7S}$? Why?

4.6 Show the structure of each compound represented by the following spectra (80 MHz). The molecular formula for each compound is given. The signals at δ2.0-4.0 for compounds C, E, and F are impurities.

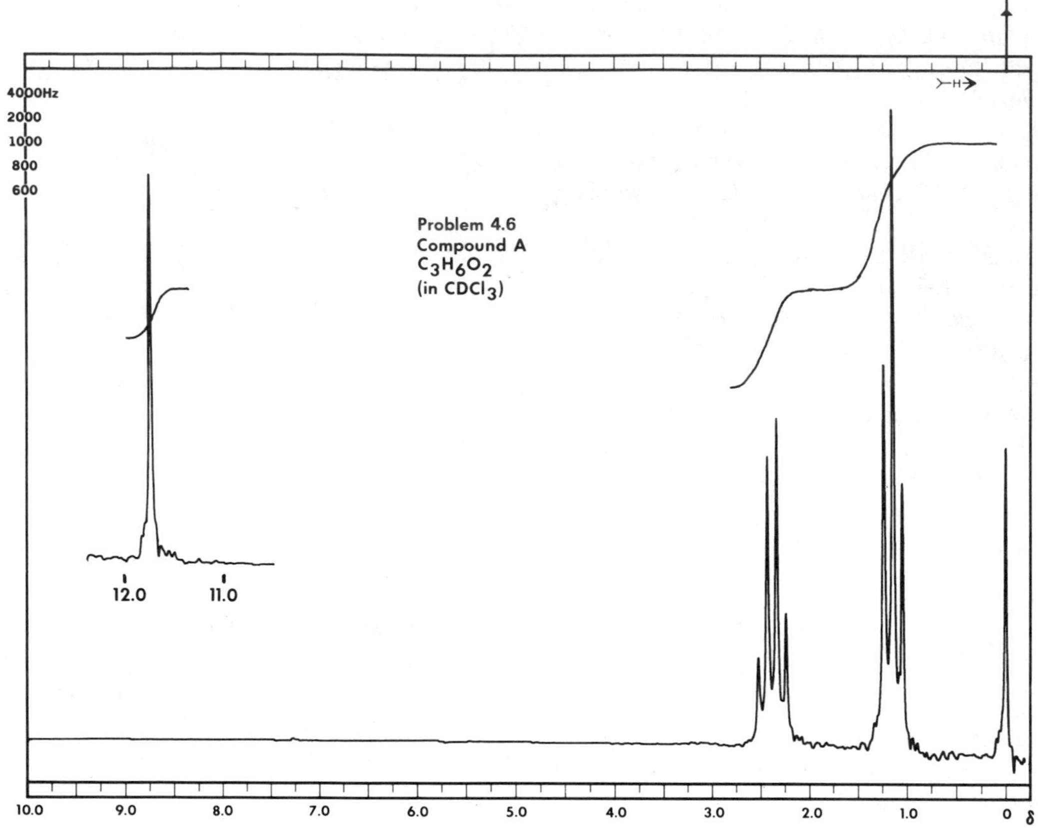
4000Hz
2000
1000
800
600
Problem 4.6
Compound A
$C_3H_6O_2$
(in $CDCl_3$)
12.0
11.0
10.0
9.0
8.0
7.0
6.0
5.0
4.0
3.0
2.0
1.0
0 δ

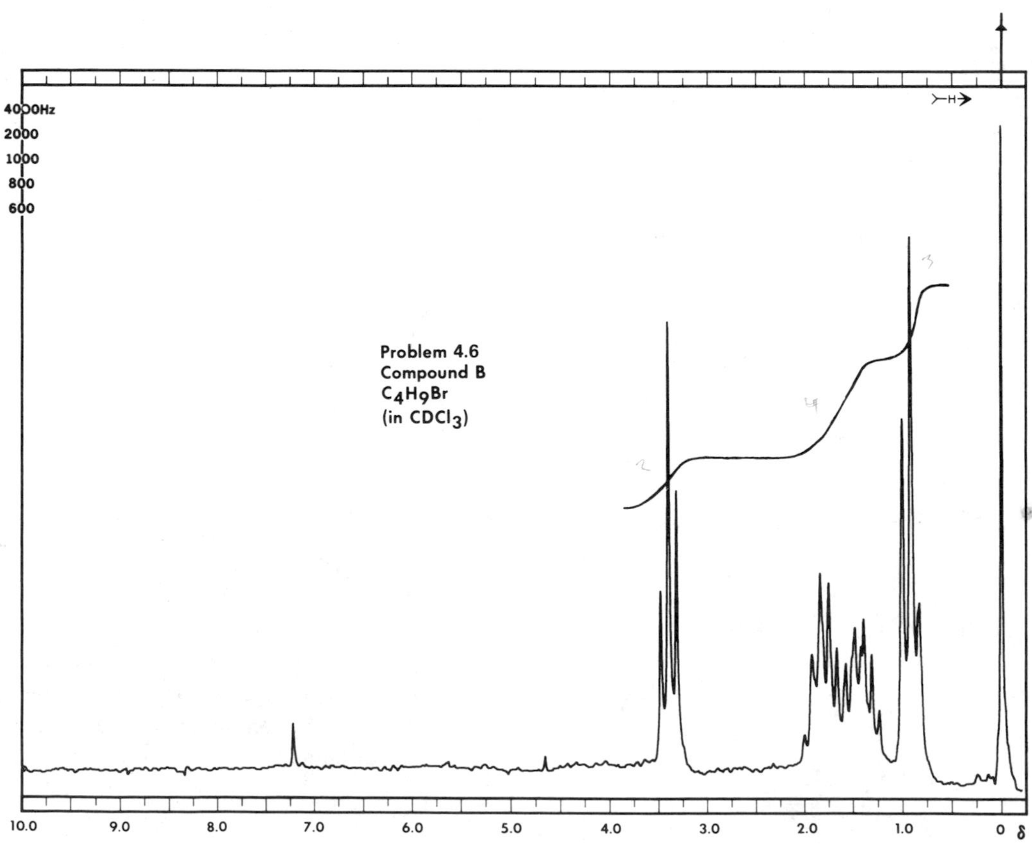
4000Hz
2000
1000
800
600
Problem 4.6
Compound B
C_4H_9Br
(in $CDCl_3$)
10.0
9.0
8.0
7.0
6.0
5.0
4.0
3.0
2.0
1.0
0 δ

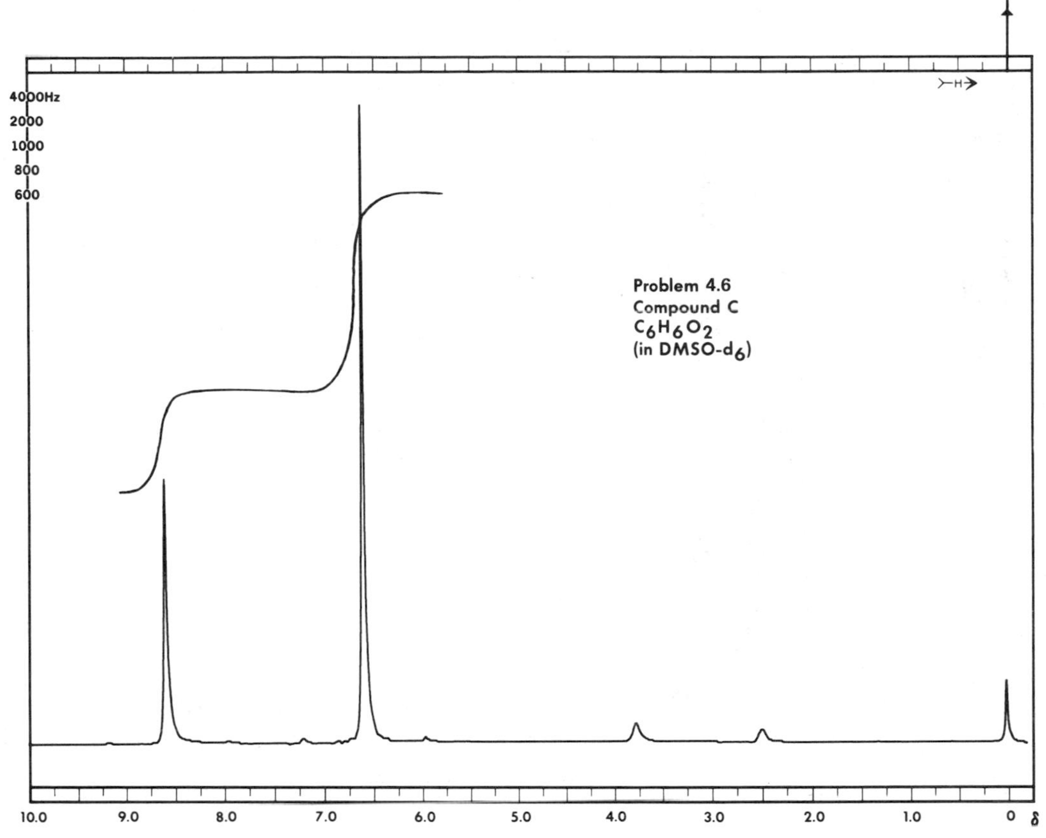
4000Hz
2000
1000
800
600
Problem 4.6
Compound C
$C_6H_6O_2$
(in DMSO-d_6)
10.0
9.0
8.0
7.0
6.0
5.0
4.0
3.0
2.0
1.0
0 δ

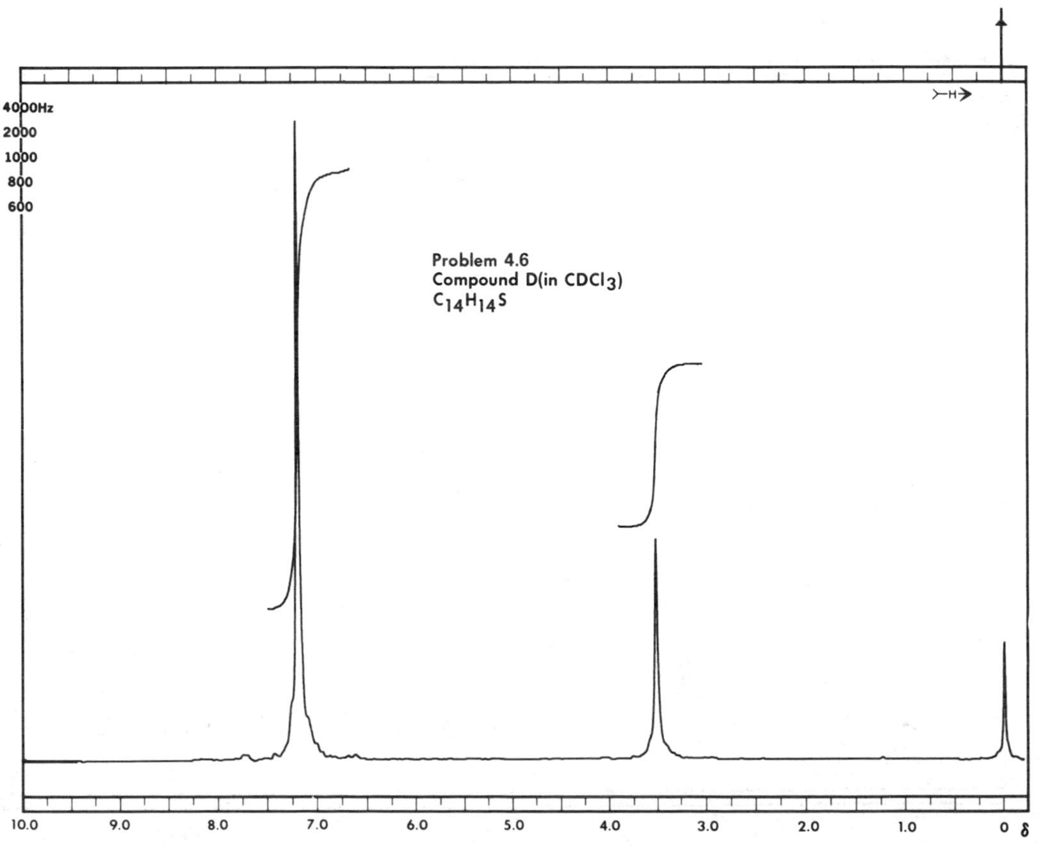
4000Hz
2000
1000
800
600
Problem 4.6
Compound D(in $CDCl_3$)
$C_{14}H_{14}S$
10.0
9.0
8.0
7.0
6.0
5.0
4.0
3.0
2.0
1.0
0 δ

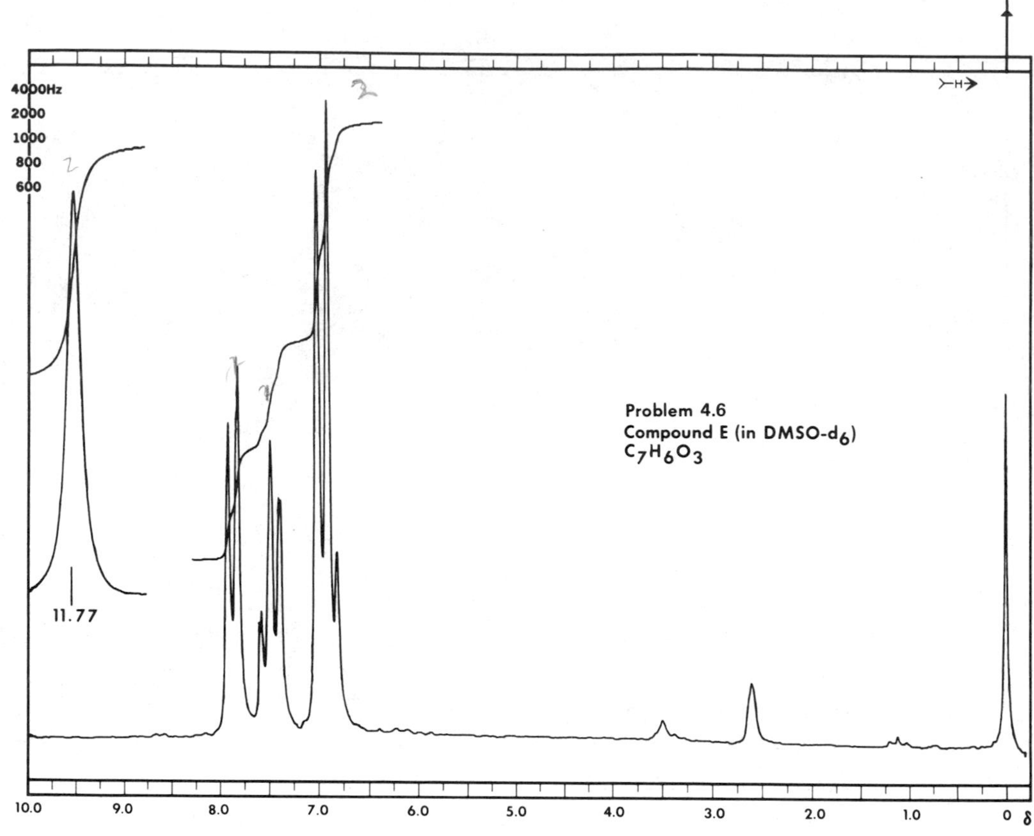
4000Hz
2000
1000
800
600
11.77
Problem 4.6
Compound E (in DMSO-d6)
C7H6O3
10.0
9.0
8.0
7.0
6.0
5.0
4.0
3.0
2.0
1.0
0 δ

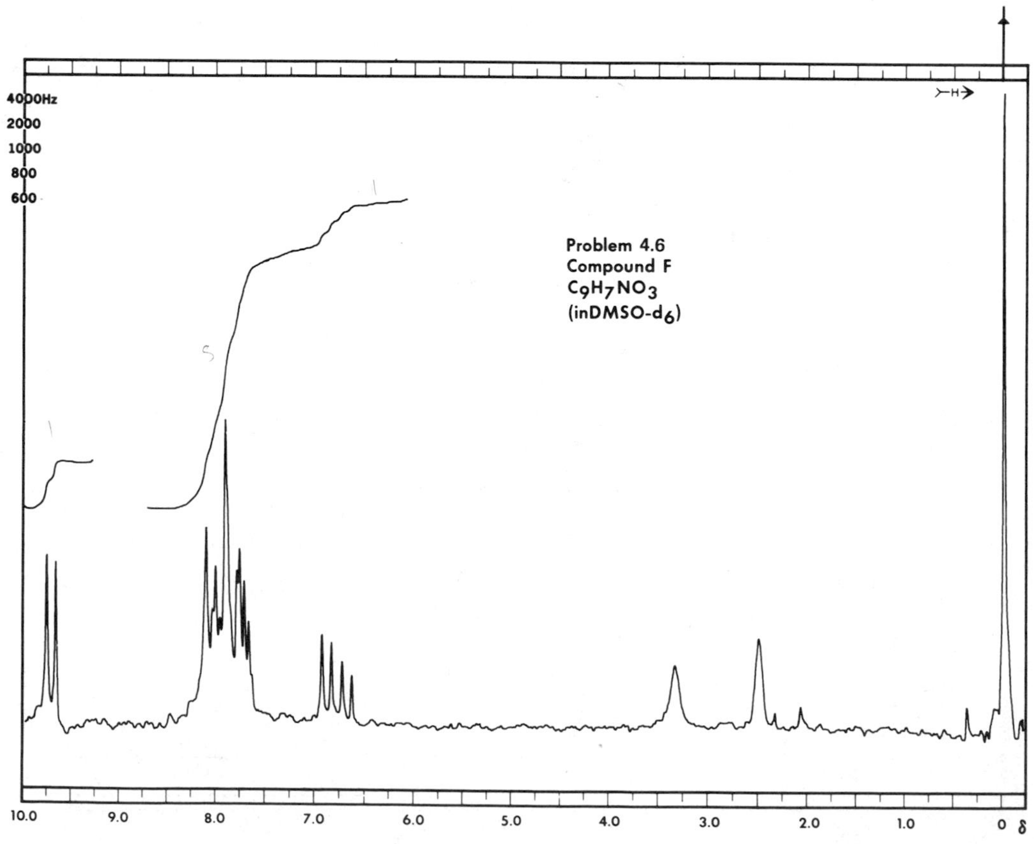
4000Hz
2000
1000
800
600
Problem 4.6
Compound F
C9H7NO3
(inDMSO-d6)
10.0
9.0
8.0
7.0
6.0
5.0
4.0
3.0
2.0
1.0
0 δ

appendix a. effect on shift positions of a single functional group

Charts 1 and 2 give the chemical shifts of CH_3, CH_2, and CH groups that are adjacent to, or once removed from, a functional group in aliphatic compounds. In a hydrocarbon, the methyl protons are at about δ 0.90, the methylene protons at about δ 1.25, and the methine protons at about δ 1.50. The first line of the first chart, for example, shows that the protons of CH_3Br absorb at about δ 2.70, the methylene protons of RCH_2Br at about δ 3.40, and the methine protons of R_2CHBr at about δ 4.10.

The shift positions vary with changes in solvent or concentration, but in general these variations are slight in the absence of a very polar solvent or of hydrogen bonding. The positions shown are average values for ranges that are not greater than 0.5 ppm wide if the common solvents CCl_4 or $CDCl_3$ are used. These values were gathered from several sources: Jackman and Sternhell[4], Chamberlain[25], Bovey[26], K. Nukada, et al. *Anal. Chem.* **35**, 1892 (1963), and G. V. D. Tiers (*NMR Summary;* 3M Company, St. Paul, Minnesota). The functional groups are arranged in the order in which they are covered in organic chemistry textbooks.

CHART 1. CHEMICAL SHIFTS OF PROTONS ON A CARBON ATOM ADJACENT (α - POSITION) TO A FUNCTIONAL GROUP IN ALIPHATIC COMPOUNDS (M—Y).

- ❙ M = methyl
- ○ M = methylene
- ● M = methine

δ

.4 .2 5 .8 .6 .4 .2 4 .8 .6 .4 .2 3 .8 .6 .4 .2 2 .8 .6 .4 .2 1 .8 .6 .4 .2 0

- $M-CH_2R$
- M—C=C
- M—C≡C
- M—φ
- M—F
- M—Cl
- M—Br
- M—I
- M—OH
- M—OR
- M—Oφ
- M—OC(=O)R
- M—OC(=O)φ
- $M-OC(=O)CF_3$
- M—OTs*
- M—C(=O)H
- M—C(=O)R
- M—C(=O)φ
- M—C(=O)OH
- M—C(=O)OR
- $M-C(=O)NR_2$
- M—C≡N
- $M-NH_2$
- $M-NR_2$
- M—NφR
- $M-N^+R_3$
- M—NHC(=O)R
- $M-NO_2$

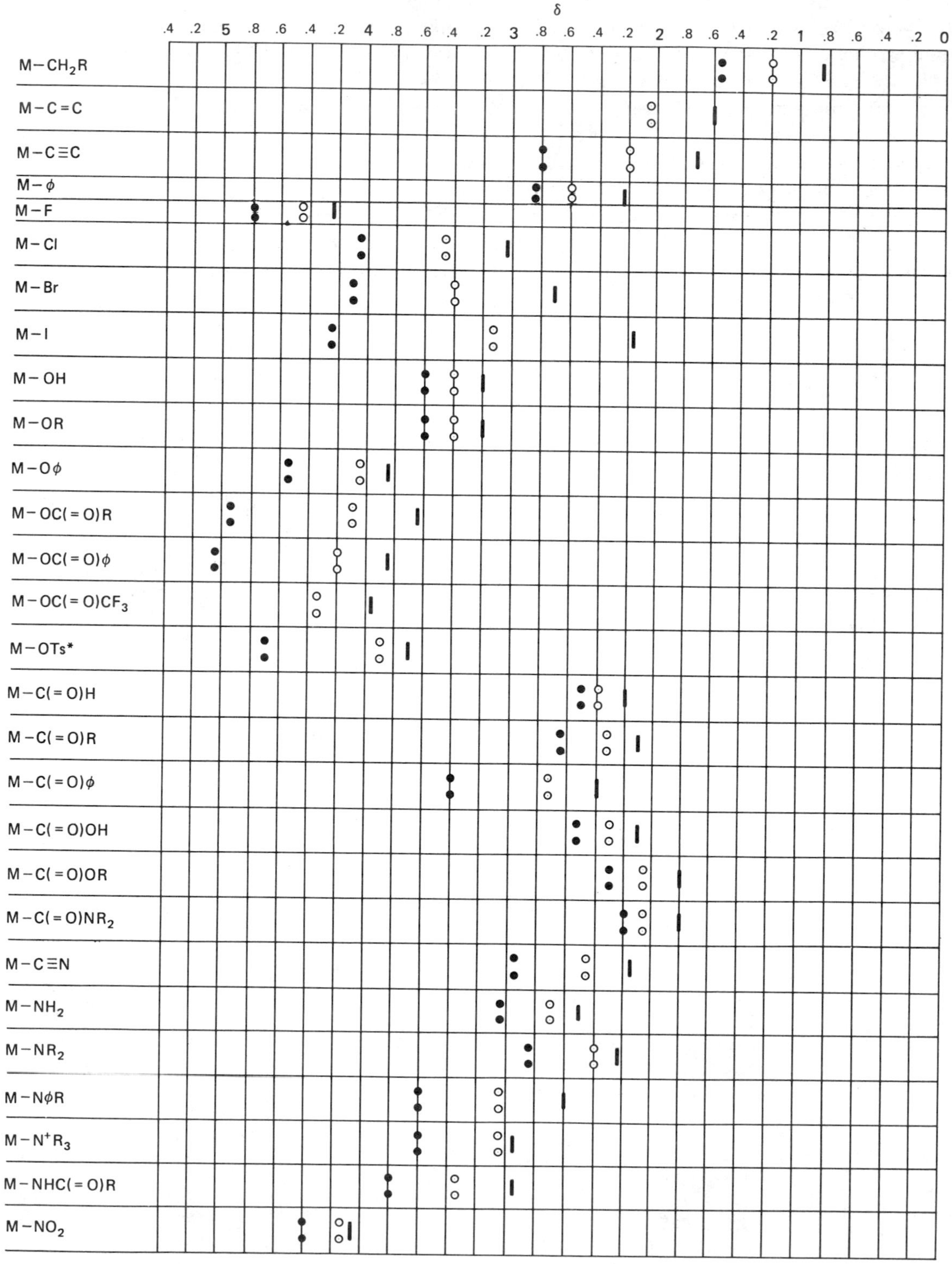

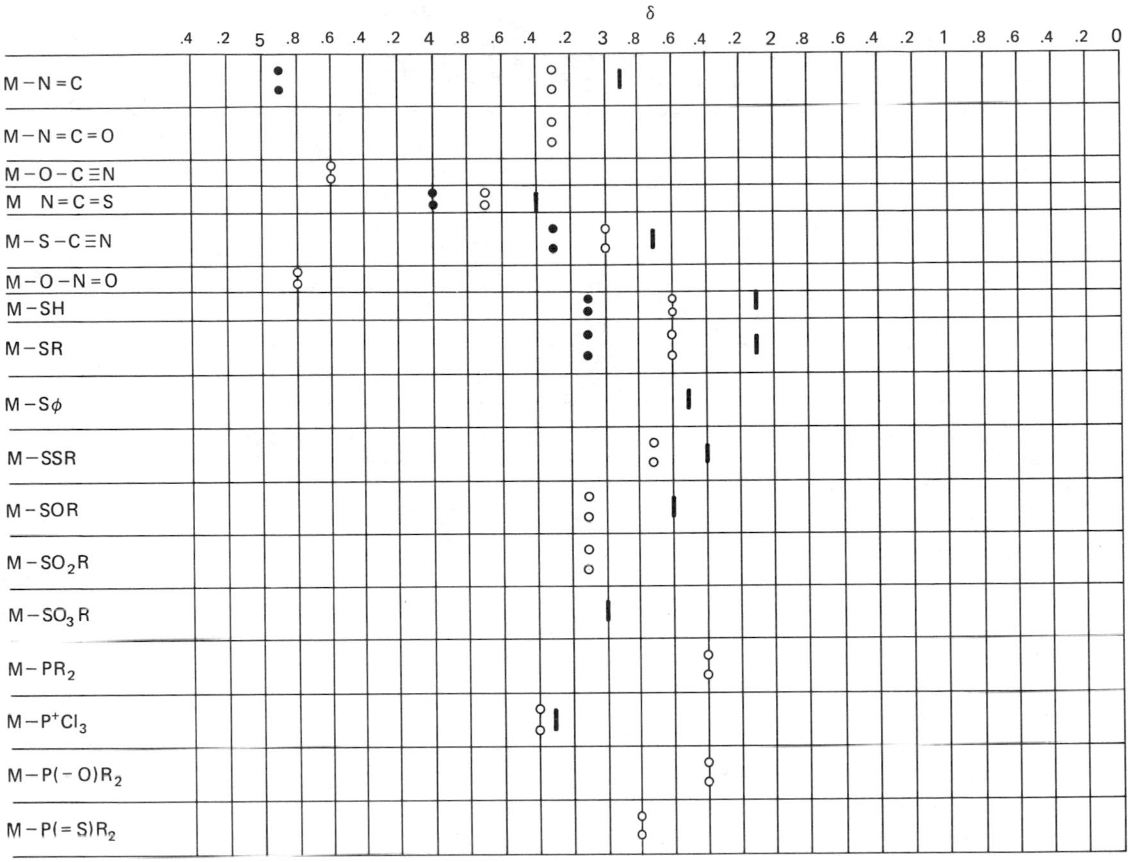

δ

.4 .2 5 .8 .6 .4 .2 4 .8 .6 .4 .2 3 .8 .6 .4 .2 2 .8 .6 .4 .2 1 .8 .6 .4 .2 0

M−N=C

M−N=C=O

M−O−C≡N

M N=C=S

M−S−C≡N

M−O−N=O

M−SH

M−SR

M−Sϕ

M−SSR

M−SOR

M−SO_2R

M−SO_3R

M−PR_2

M−P^+Cl_3

M−P(−O)R_2

M−P(=S)R_2

*OTs is $-O_2S-C_6H_4-CH_3$

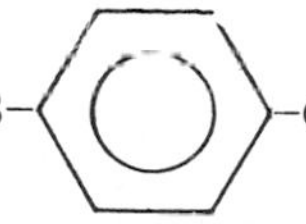

CHART 2. CHEMICAL SHIFTS OF PROTONS ON A CARBON ATOM ONCE REMOVED (β- POSITION) FROM A FUNCTIONAL GROUP IN ALIPHATIC COMPOUNDS (M—C—Y).

| M is methyl
o M is methylene
• M is methine

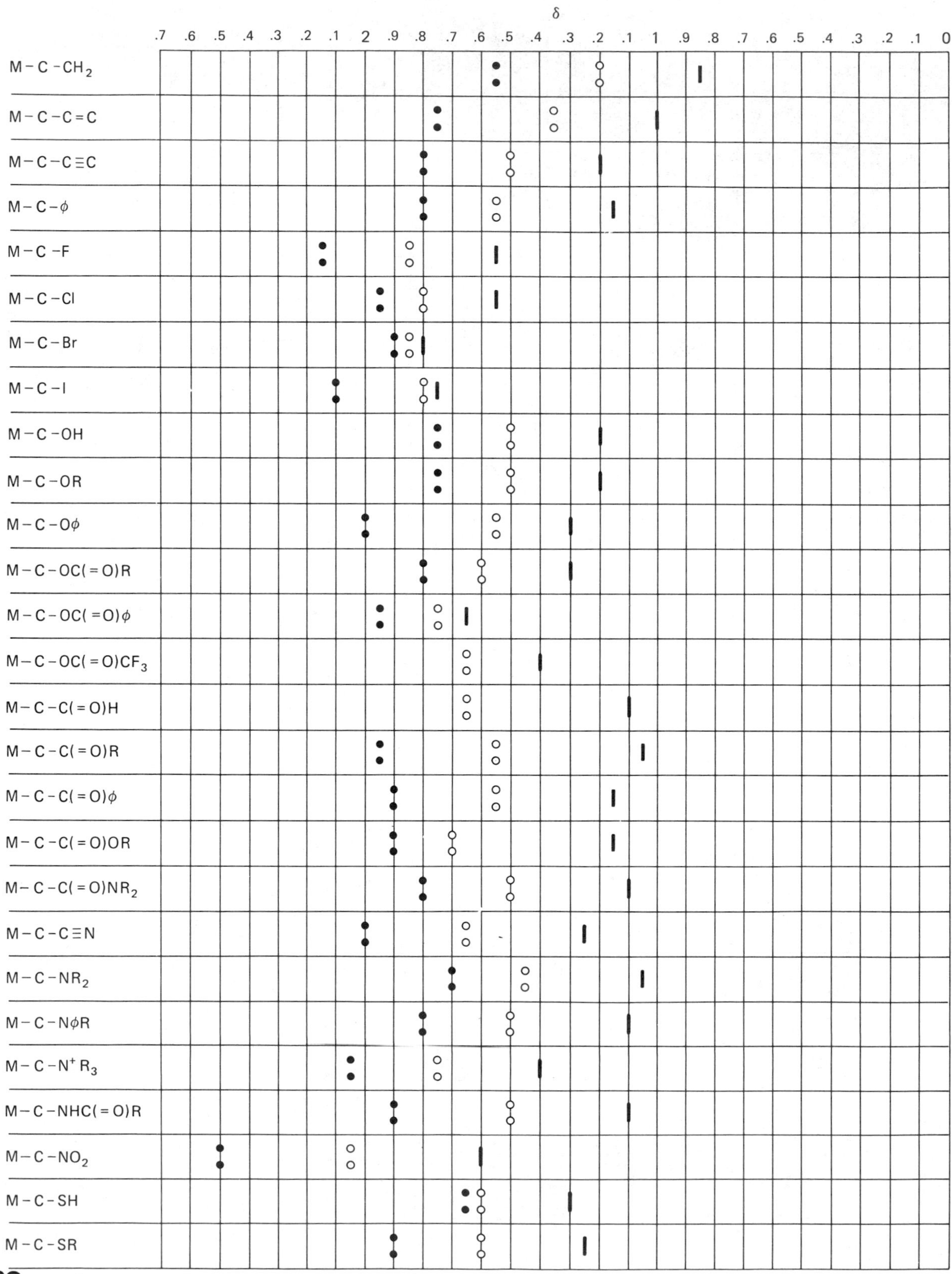

appendix b. effect on chemical shifts of two or three functional groups (Y-CH_2-Z, and Y-CH(W)-Z)

Shoolery's rules (B. P. Dailey and J. W. Shoolery, *J. Am. Chem. Soc.*, 77, 3977 (1955) permit calculation of a shift position of a methylene group attached to two functional groups by the additive effect of the shielding constants in Table 1, below. The sum of the constants is added to δ 0.23, the position for CH_4.

Thus, to calculate the shift for the $-CH_2-$ protons of $C_6H_5CH_2Br$:

C_6H_5 = 1.85	0.23	
Br = 2.33	4.18	
4.18	4.41	= δ value for the $-CH_2-$ group.

Table I. Shielding Constants

Y or *Z*	Shielding Constants	*Y* or *Z*	Shielding Constants
$-CH_3$	0.47	−C(=O)NR_2	1.59
−C=C	1.32	−C≡N	1.70
−C≡C	1.44	$-NR_2$	1.57
−ϕ	1.85	−NHC(=O)R	2.27
$-CF_2$	1.21	$-N_3$	1.97
$-CF_3$	1.14	−SR	1.64
−Cl	2.53	$-OSO_2R$	3.13
−Br	2.33		
−I	1.82		
−OH	2.56		
−OR	2.36		
−Oϕ	3.23		
−OC(=O)R	3.13		
−C(=O)R	1.70		
−C(=O)ϕ	1.84		
−C(=O)OR	1.55		

The shielding constants have been used to prepare the chart on page 224. Several values have been added to the original set of constants.

Alternatively, Chart 1 can be used to find the shift position of a methylene group attached to two functional groups from the δ values in the box at the intersection of the horizontal and diagonal groups ("mileage chart"). The upper number in each box is an experimental value; the lower number is calculated from Shoolery's constants.

CHART 1. CHEMICAL SHIFTS FOR METHYLENE GROUPS ATTACHED TO TWO FUNCTIONAL GROUPS ($Y-CH_2-Z$).

GROUP	$-CH_3$	–C=C	–C≡C	–φ	$-CF_2$	$-CF_3$	–Cl	–Br	–I	–OH	–OR	–Oφ	–OC(=O)R	–C(=O)R	–C(=O)φ	–C(=O)OR	$-C(=O)NR_2$	–C≡N	$-NR_2$	–NHC(=O)R	$-N_3$	–SR
$-CH_3$	1.17	1.90 2.02	2.14	2.55	1.91	1.84	3.57 3.23	3.43 3.03	3.20 2.52	3.70 3.26	3.40 3.06	3.93	4.25 3.83	2.47 2.40	2.54	2.25 2.25	2.23 2.29	2.40	2.63 2.27	2.97	2.67	2.53 2.34
–C=C		2.60 2.87	3.39 2.99	3.30 3.40	2.76	2.69	4.08	3.93 3.88	3.87 3.37	4.13	3.95 3.91	4.78	4.68	3.25	3.39	3.00 3.10	3.14	3.15 3.25	3.30 3.12	3.82	3.52	3.08 3.19
–C≡C			3.11	3.52	2.88	2.81	4.09 4.20	3.90 4.00	3.49	4.28	4.03	4.90	4.71 4.80	3.37	3.51	3.22	3.26	3.37	3.50 3.24	3.97	3.67	3.31
–φ				3.97 3.93	3.29	3.50 3.22	4.50 4.61	4.35 4.41	3.90	4.58	4.70 4.44	4.91 5.31	5.08 5.21	3.55 3.78	3.92	3.40 3.63	3.66	3.65 3.78	3.48 3.65	4.34	4.04	3.68 3.72
$-CF_2$					2.63	2.58	3.97	3.76	3.26	4.01 4.01	3.80	4.67	4.57	3.12	3.28	2.99	3.03	3.12	3.01	3.71	3.41	3.08
$-CF_3$						2.51	3.90	3.70	3.56 3.19	3.93	4.73	4.60	4.54	3.07	3.21	2.92	2.96	3.07	2.94	3.64	3.34	3.01
–Cl							4.99 5.29	5.16 5.09	4.99 4.58	5.32	5.40 5.12	5.99	5.89	4.46	4.60	4.05 4.31	4.17 4.35	4.07 4.46	4.37	5.13	4.73	4.40
–Br								4.94 4.89	4.38	5.12	4.92	6.36 5.79	5.69	4.26	4.40 3.40	3.70 4.11	3.92 4.15	4.26	4.13	4.83	4.53	4.20
–I									3.90 3.87	4.61	4.41	5.16	5.06	3.75	3.89	3.60	3.65 3.64	3.65 3.75	3.62	4.32	4.02	3.69
–OH										4.55 4.35	5.15	6.02	5.92	4.49	4.63	4.34	4.38	4.49	4.35	5.06	4.76	4.43
–OR											5.55 4.95	5.82	5.72	4.29	4.43	4.13 4.22	4.18	4.20 4.29	4.15	4.94	4.64	4.23
–Oφ												6.69	6.59	5.16	5.30	5.09	4.45 4.05	5.16	5.03	5.86	5.56	5.10
–OC(=O)R													6.46	4.42 5.10	5.20	4.91	4.95	5.10	4.92	5.63	5.37	5.00
–C(=O)R														3.60 3.63	3.77	3.37 3.48	3.52	3.63	3.50	4.10 4.20	3.90	3.57
–C(=O)φ															3.91	3.62	3.66	3.77	3.64	4.34	4.02	3.71
–C(=O)OR																3.35 3.33	3.37	3.48	3.17 3.35	4.05	3.75	3.42
$-C(=O)NR_2$																	3.30 3.41	3.52	3.39	4.09	3.79	3.46
–C≡N																		3.63	3.50	4.20	3.90	3.57
$-NR_2$																			3.10 3.37	4.07	3.77	3.44
–NHC(=O)R																				4.75	4.45	4.14
$-N_3$																					4.15	3.84
–SR																						3.51

Table II can be used to calculate chemical shifts of Y–CH_3, Y–CH_2–Z, or Y–CH(W)–Z groups. Thus, the CH_2 chemical shifts in $BrCH_2\overset{2}{C}H_2O\overset{1}{C}H_2CH_2Br$ can be calculated.

1–CH_2		2–CH_2	
CH_2 standard (footnote b)	1.20	CH_2 standard	1.20
α–OR	2.35	α–Br	2.18
β–Br	0.60	β–OR	0.15
	4.15		3.53

Determined: 1–CH_2 at ~δ 3.80; 2–CH_2 at ~δ 3.40.

Table II. Substituent Effects on Chemical Shift[a]

C–C–H (β, α)

Substituent	Type of Hydrogen[b]	Alpha Shift	Beta Shift
–C=C–	CH_3	0.78	–
	CH_2	0.75	0.10
	CH	–	–
–C=C–C(=X)–R (X=C or O)	CH_3	1.08	–
Aryl	CH_3	1.40	0.35
	CH_2	1.45	0.53
	CH	1.33	–
–Cl	CH_3	2.43	0.63
	CH_2	2.30	0.53
	CH	2.55	0.03
–Br	CH_3	1.80	0.83
	CH_2	2.18	0.60
	CH	2.68	0.25
–I	CH_3	1.28	1.23
	CH_2	1.95	0.58
	CH	2.75	0.00
–OH	CH_3	2.50	0.33
	CH_2	2.30	0.13
	CH	2.20	–
–OR (R is saturated)	CH_3	2.43	0.33
	CH_2	2.35	0.15
	CH	2.00	
–OC(=O)–R, –OC(=O)–OR, –OAr	CH_3	2.88	0.38
	CH_2	2.98	0.43
	CH	3.43 (esters only)	–
–C(=O)R, where R is alkyl, aryl, OH, OR′, H, CO, or N	CH_3	1.23	0.18
	CH_2	1.05	0.31
	CH	1.05	–
–NRR′	CH_3	1.30	0.13
	CH_2	1.33	0.13
	CH	1.33	–

[a]From the Ph.D. dissertation of T.J. Curphey, Harvard University, by permission.
[b]Standard positions are CH_3, δ0.87; CH_2, δ1.20; CH, δ1.55.

appendix c. chemical shifts in alicyclic and heterocyclic rings

Table I. Chemical Shifts in Alicyclic Rings

0.22
1.96
1.51
1.44
1.54
1.78
O
1.65
O
1.96
3.03
O
2.06
2.02
O
2.22
~1.8
~1.8
O
2.38
O
2.30
~1.94
~1.52
~1.52

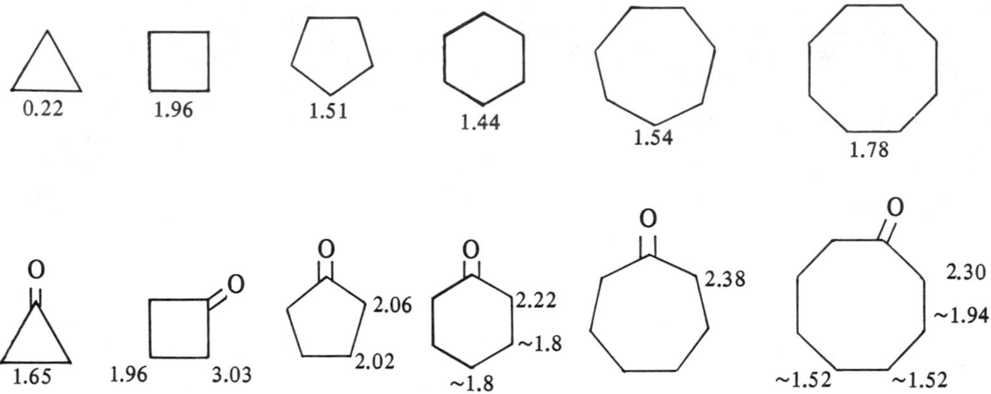

Table II. Chemical Shifts in Heterocyclic Rings

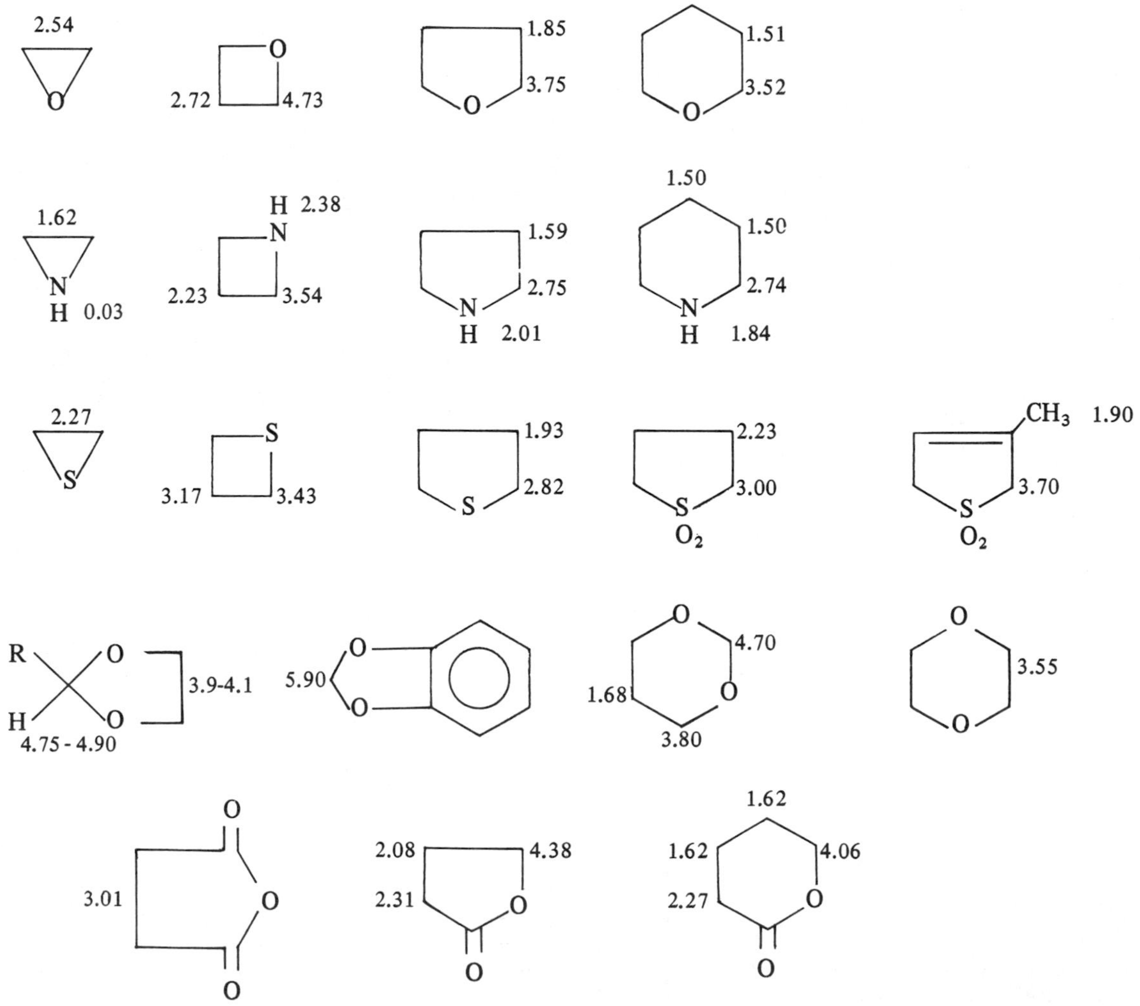

2.54
O
O
2.72
4.73
1.85
3.75
O
1.51
3.52
O
1.62
N
H 0.03
H 2.38
N
2.23
3.54
1.59
2.75
N
H 2.01
1.50
1.50
2.74
N
H 1.84
2.27
S
S
3.17
3.43
1.93
2.82
S
2.23
3.00
S
O2
CH3 1.90
3.70
S
O2
R
O
H
O
3.9-4.1
4.75 - 4.90
5.90
O
O
O
4.70
1.68
O
3.80
O
3.55
O
O
3.01
O
O
2.08
4.38
2.31
O
O
1.62
1.62
4.06
2.27
O
O

appendix d. chemical shifts in unsaturated systems

(See Table on next page)

$$\begin{matrix} R_{cis} & & H \\ & \rangle C{=}C\langle & \\ R_{trans} & & R_{gem} \end{matrix} \qquad \delta_H = 5.28 + Z_{gem} + Z_{cis} + Z_{trans}$$

For example, the chemical shifts of the olefinic protons in

$$\begin{matrix} C_6H_5 & & OC_2H_5 \\ & \rangle C{=}C\langle & \\ H_a & & H_b \end{matrix}$$

are calculated:

H_a:	$C_6H_{5\,gem}$	1.35	5.28
	OR_{trans}	−1.28	0.07
		0.07	δ 5.35
H_b:	OR_{gem}	1.18	5.28
	$C_6H_{5\,trans}$	−0.10	1.08
		1.08	δ 6.36

Table I. Substituent Constants (Z) for Chemical Shifts of Substituted Ethylenes (in CCl_4)

Substituent R	Z gem	Z cis	Z trans	Substituent R	Z gem	Z cis	Z trans
–H	0	0	0	–C(H)=O	1.03	0.97	1.21
–Alkyl	0.44	–0.26	–0.29	–C(N)=O	1.37	0.93	0.35
–Alkyl–Ring[a]	0.71	–0.33	–0.30	–C(Cl)=O	1.10	1.41	0.99
$-CH_2O$, $-CH_2I$	0.67	–0.02	–0.07	–OR, R:aliph.	1.18	–1.06	–1.28
$-CH_2S$	0.53	–0.15	–0.15	–OR, R:conj.[b]	1.14	–0.65	–1.05
$-CH_2Cl$, $-CH_2Br$	0.72	0.12	0.07	–OCOR	2.09	–0.40	–0.67
$-CH_2N$	0.66	–0.05	–0.23	–Aromatic	1.35	0.37	–0.10
–C≡C	0.50	0.35	0.10	–Cl	1.00	0.19	0.03
–C≡N	0.23	0.78	0.58	–Br	1.04	0.40	0.55
–C=C	0.98	–0.04	–0.21	–N(R)(R) R:aliph.	0.69	–1.19	–1.31
–C=C conj.[b]	1.26	0.08	–0.01	–N(R)(R) R:conj.[b]	2.30	–0.73	–0.81
–C=O	1.10	1.13	0.81	–SR	1.00	–0.24	–0.04
–C=O conj.[b]	1.06	1.01	0.95	$-SO_2$	1.58	1.15	0.95
–COOH	1.00	1.35	0.74				
–COOH conj.[b]	0.69	0.97	0.39				
–COOR	0.84	1.15	0.56				
–COOR conj.[b]	0.68	1.02	0.33				

[a]Alkyl-ring indicates that the double bond is part of the ring: R(C=C)

[b]The Z factor for the conjugated substituent is used when either the substituent or the double bond is further conjugated with other groups.

Source: C. Pascual, J. Meier, and W. Simon, *Helv. Chim. Acta*, 49, 164 (1966).

Table II. Chemical Shifts of Miscellaneous Olefins

5.70 6.50 5.02
RCH=CR–CH=CH_2

6.01 6.08 7.16 5.62
CH_3–CH=CH–CH=CH–C(=O)OC_2H_5

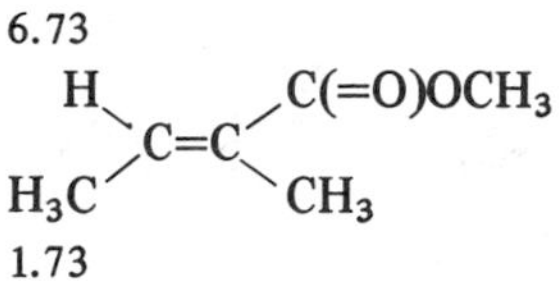

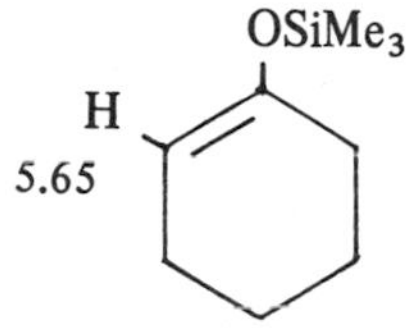

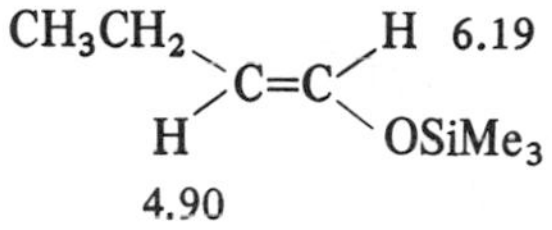

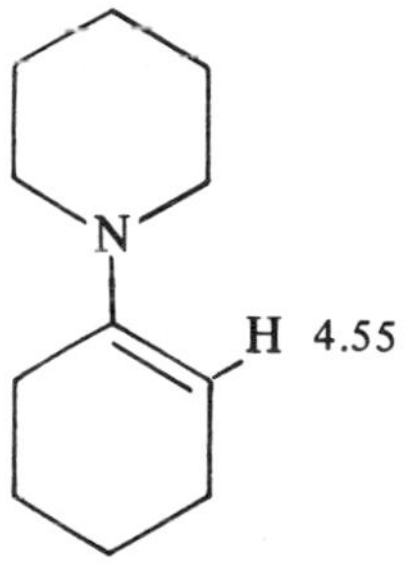

6.73
H, C(=O)OCH_3
C=C
H_3C, CH_3
1.73

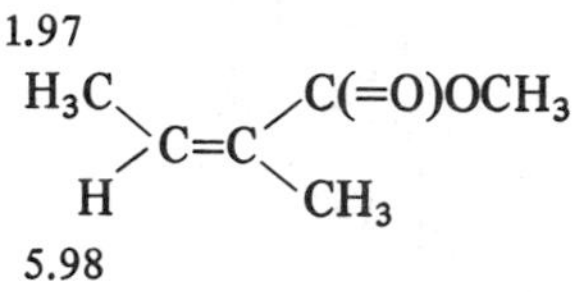

1.97
H_3C, C(=O)OCH_3
C=C
H, CH_3
5.98

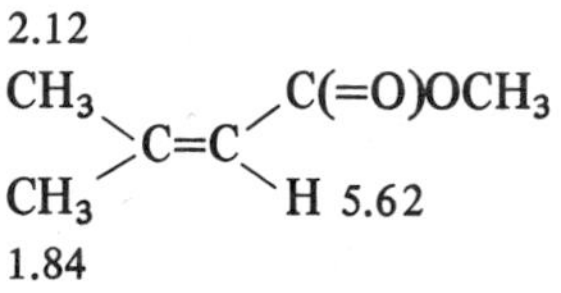

2.12
CH_3, C(=O)OCH_3
C=C
CH_3, H 5.62
1.84

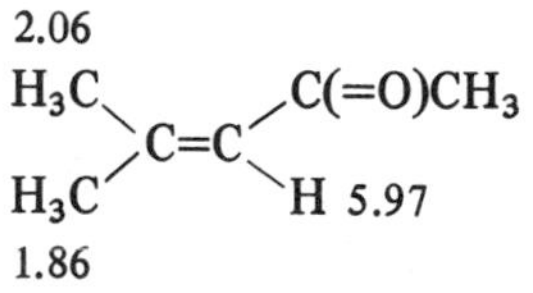

2.06
H_3C, C(=O)CH_3
C=C
H_3C, H 5.97
1.86

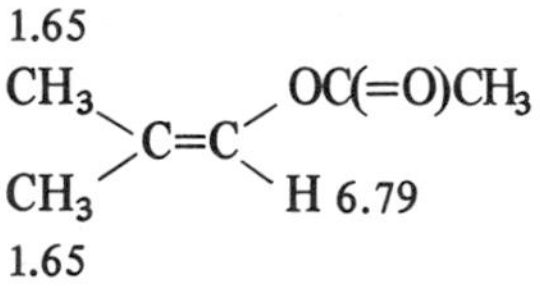

1.65
CH_3, OC(=O)CH_3
C=C
CH_3, H 6.79
1.65

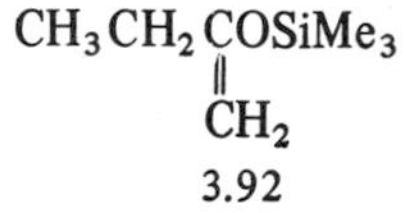

CH_3CH_2C(=CH_2)$OSiMe_3$
3.92

$OSiMe_3$
H 5.65 (1-trimethylsilyloxycyclohexene)

CH_3CH_2, H 6.19
C=C
H, $OSiMe_3$
4.90

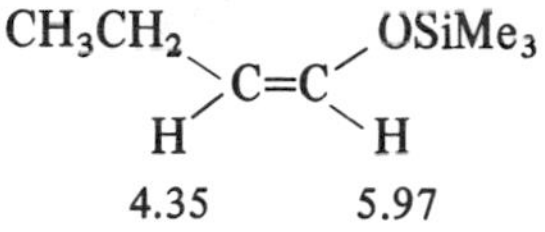

CH_3CH_2, $OSiMe_3$
C=C
H, H
4.35 5.97

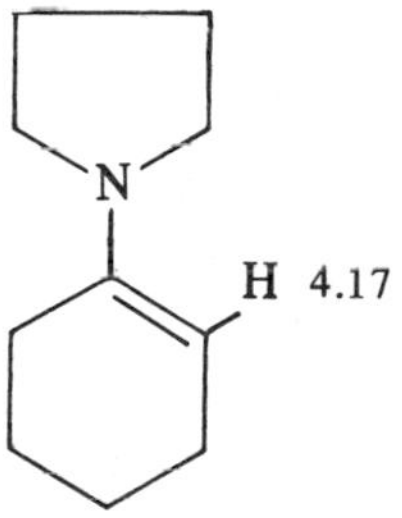

1-pyrrolidinocyclohexene: H 4.17

1-piperidinocyclohexene: H 4.55

Continued on next page

Table II, continued

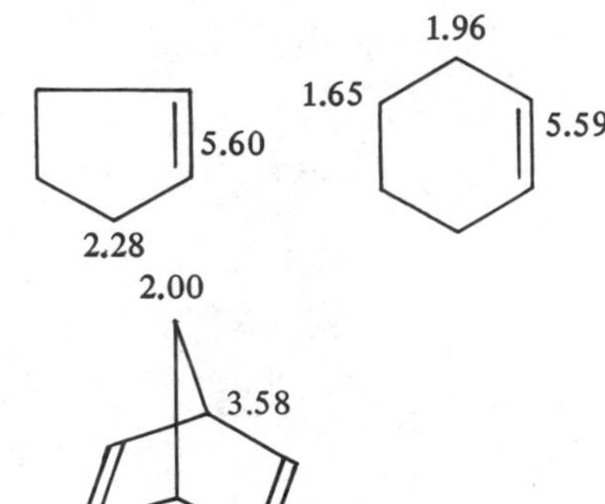

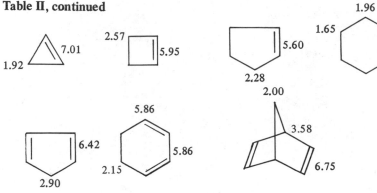

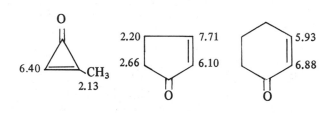

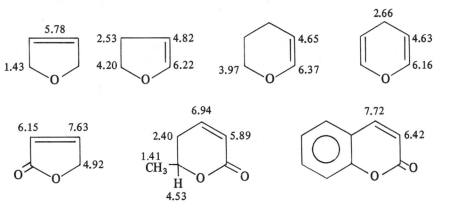

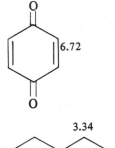

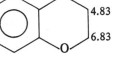

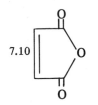

Table III. Chemical Shifts of Alkyne Protons

HC≡CR	1.73-1.88
HC≡C−COH	2.23
HC≡C−C≡CR	1.95
HC≡CH	1.80
HC≡CAr	2.71-3.37
HC≡C−C≡CR	2.60-3.10

CHART 1. CHEMICAL SHIFTS OF PROTONS ON MONOSUBSTITUTED* BENZENE RINGS

δ	9–.8	.8–.6	.6–.4	.4–.2	.2–8	8–.8	.8–.6	.6–.4	.4–.2	.2–7	7–.8	.8–.6	.6–.4	.4–.2	.2–6
Benzene									:						
CH_3 (omp)										:					
CH_3CH_2 (omp)										:					
$(CH_3)_2CH$ (omp)										:					
$(CH_3)_3C$ o,m,p									:	: :					
$C{=}CH_2$ (omp)									:						
$C{\equiv}CH$ o, (mp)								:	:						
Phenyl o, m, p								:	:	:					
CF_3 (omp)								:							
CH_2Cl (omp)									:						
$CHCl_2$ (omp)								.	:						
CCl_3 o, (mp)					:			:							
CH_2OH (omp)										:					
CH_2OR (omp)									:						
$CH_2OC({=}O)CH_3$ (omp)									:						
CH_2NH_2 (omp)									:						
F m,p.o									:	:	:				
Cl (omp)									:						
Br o, (pm)								:	:						
I o,p,m							:		:	:					
OH m,p,o										:	:	:			
OR m, (op)									:			:			
$OC({=}O)CH_3$ (mp), o									:	:					
OTs**(mp), o									:	:					
$CH({=}O)$o,p,m						:	:	:							
$C({=}O)CH_3$ o, (mp)						:		:							
$C({=}O)OH$ o, p, m						:		:	:						
$C({=}O)OR$ o, p, m					:			: :							
$C({=}O)Cl$ o, p, m					:			: :							
$C{\equiv}N$								:							
NH_2 m,p,o										:	:		:		
$N(CH_3)_2$ m(op)										:		:			
$NHC({=}O)R$ o								:							
NH_3^+ o							:								
NO_2 o,p,m				:			:	:							
SR (omp)									:						
$N{=}C{=}O$ (omp)										:					

*The principle of additivity can be used to calculate approximate chemical shifts for polysubstituted rings.
**OTs = *p*-toluenesulfonyloxy group

Table IV. Chemical Shifts of Protons on Fused Aromatic Rings.

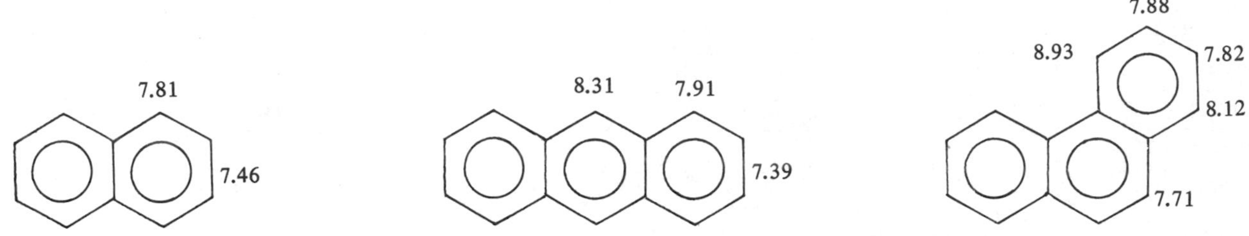

Table V. Chemical Shifts of Protons on Heteroaromatic Rings

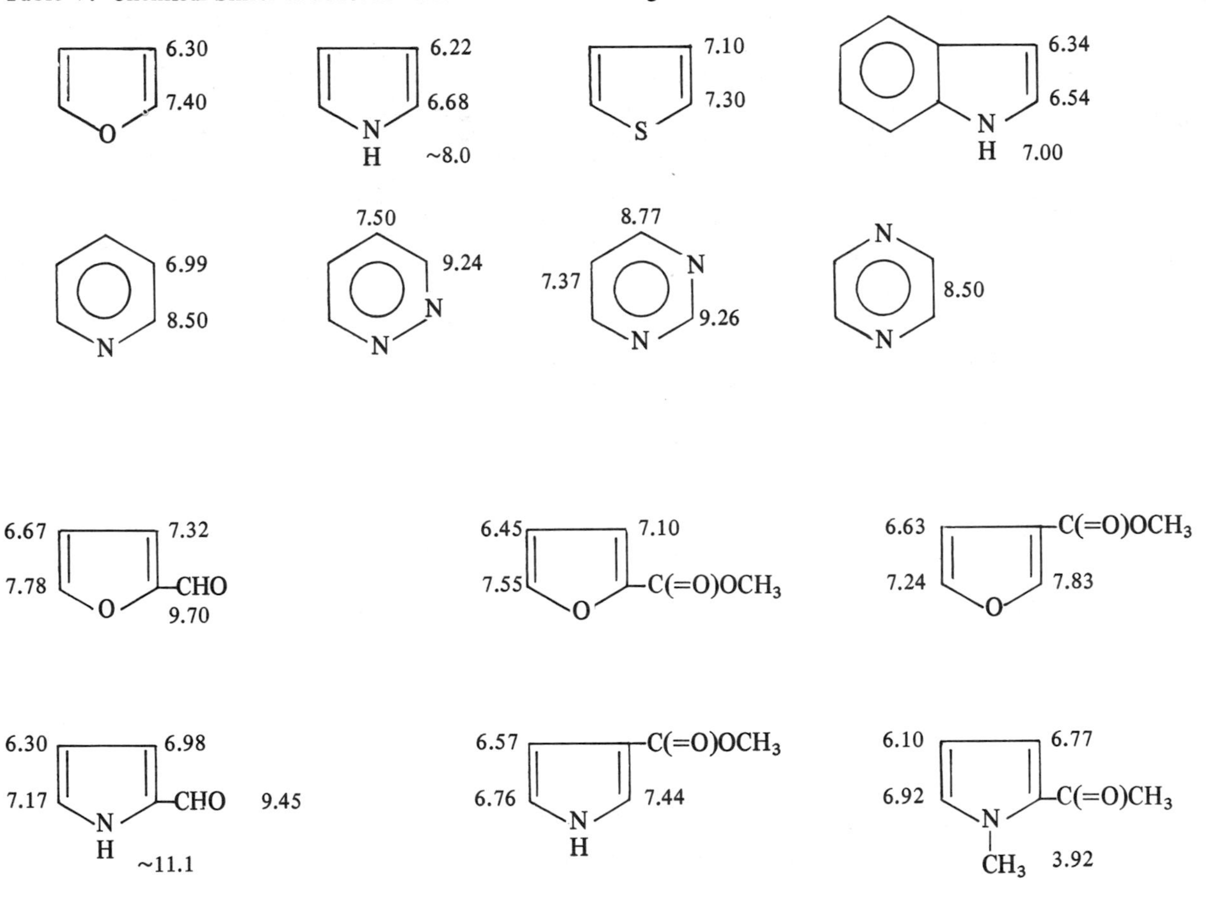

7.22 7.78

7.78 CHO

S

9.92

Table VI. Chemical Shifts of HC=O and HC=N Protons (also $HC(O)_3$)

RCH=O	9.70	HC(=O)OR	8.05	RC$\underline{H}$=NOH *cis*	7.25
ϕCH=O	9.98	HC(=O)NR_2	8.05	RC$\underline{H}$=NOH *trans*	6.65
RCH=CHCH=O	9.78	$HC(OR)_3$	5.00		
				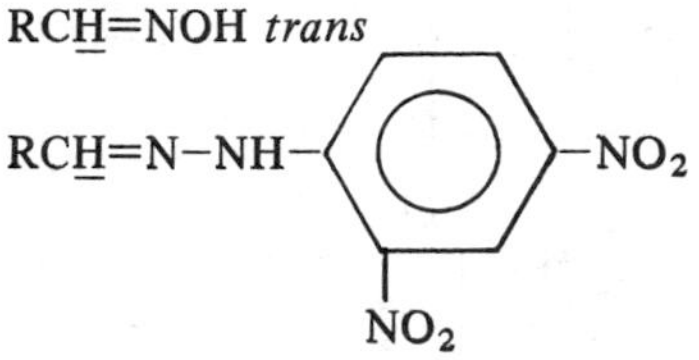 RC$\underline{H}$=N–NH–(2,4-$(NO_2)_2C_6H_3$)	6.05

appendix e

PROTONS SUBJECT TO HYDROGEN-BONDING EFFECTS (Protons on Heteroatoms)*

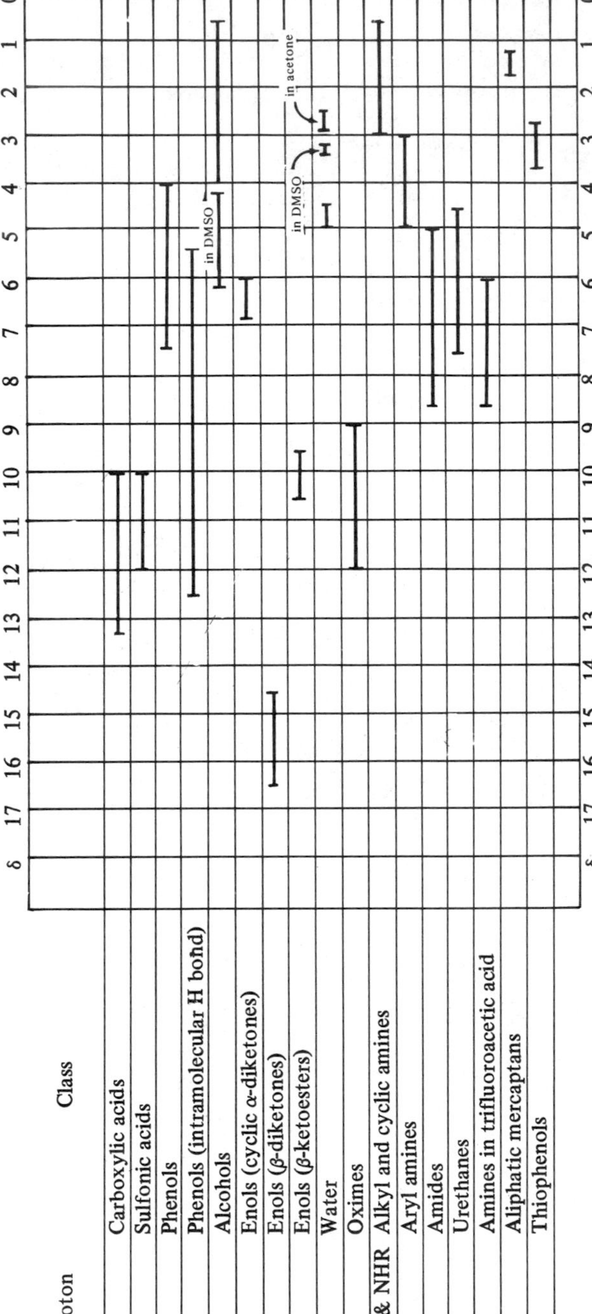

*Solvent $CDCl_3$.

appendix f

PROTON SPIN-SPIN-COUPLING CONSTANTS* (Absolute Values)†

Type		J_{ab} (Hz)	J_{ab} Typical
$>C<$ (H_a, H_b)		0 to 30	12 to 15
CH_a-CH_b (free rotation)		6-8	7
$CH_a-\overset{\mid}{\underset{\mid}{C}}-CH_b$		0-1	0
ax.-ax.		6-14	8-10
ax.-eq.		0-5	2-3
eq.-eq.		0-5	2-3
H_a, H_b (*cis* or *trans*)		0-7	4-5
H_a, H_b (*cis* or *trans*)		6-10	8
H_a, H_b (*cis* or *trans*)			3-5
H_a, H_b, O			6
H_a, H_b, O			4
H_a, H_b, O			2.5
H_a C=C H_b (ring)	3 mem.	0.5-2.0	
	4 mem.	2.5-4.0	
	5 mem.	5.1-7.0	
	6 mem.	8.8-11.0	
	7 mem.	9-13	
	8 mem.	10-13	

Type		J_{ab} (Hz)	J_{ab} Typical
$>C=C<$ (CH_a, H_b)		4-10	7
H_a $>C=C<$ CH_b		0-3	1.5
H_a $>C=C<$ CH_b		0-3	2
$C=CH_a-CH_b=C$		9-13	10
$CH_a-C\equiv CH_b$		2-3	
$-CH_a-C\equiv C-CH_b-$		2-3	
CH_a-OH_b (no exchange)		4-10	5
$>CH_a-\overset{O}{\overset{\|}{C}}H_b$		1-3	2-3
$C=CH_a-\overset{O}{\overset{\|}{C}}H_b$		5-8	6
H_a $>C=C<$ H_b		12-18	17
$>C=C<$ (H_a, H_b)		0-3	0-2
H_a $>C=C<$ H_b		6-12	10
CH_a $>C=C<$ CH_b		0-3	1-2
4, 3, 5, 2, N–H	J (1-2)	2-3	
	J (1-3)	2-3	
	J (2-3)	2-3	
	J (3-4)	3-4	
	J (2-4)	1-2	
	J (2-5)	1.5-2.5	
4, 5, N, 6, 2, N	J (4-5)	4-6	
	J (2-5)	1-2	
	J (2-4)	0-1	
	J (4-6)	?	
4, N, 5, 2, S	J (4-5)	3-4	
	J (2-5)	1-2	
	J (2-4)	~0	

Structure		Range	Typical
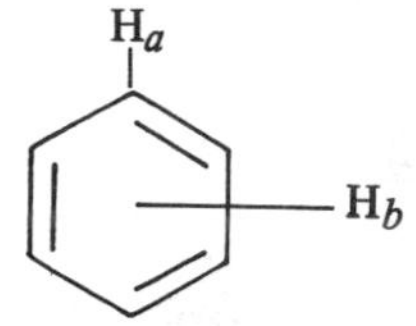 Benzene (H_a, H_b)	J (ortho)	6-10	9
	J (meta)	1-3	3
	J (para)	0-1	~0
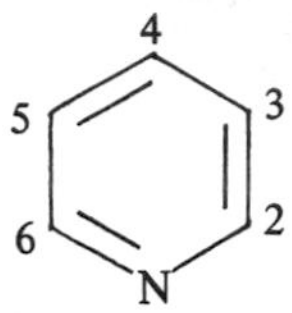 Pyridine	J (2-3)	5-6	5
	J (3-4)	7-9	8
	J (2-4)	1-2	1.5
	J (3-5)	1-2	1.5
	J (2-5)	0-1	1
	J (2-6)	0-1	~0
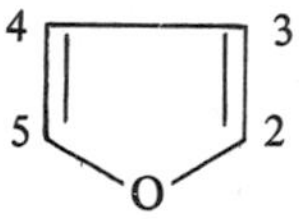 Furan	J (2-3)	1.3-2.0	1.8
	J (3-4)	3.1-3.8	3.6
	J (2-4)	0-1	~0
	J (2-5)	1-2	1.5
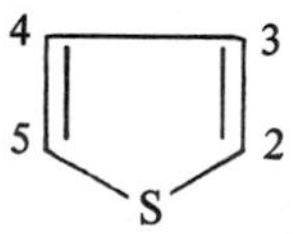 Thiophene	J (2-3)	4.9-6.2	5.4
	J (3-4)	3.4-5.0	4.0
	J (2-4)	1.2-1.7	1.5
	J (2-5)	3.2-3.7	3.4

Proton-Fluorine

Structure	J
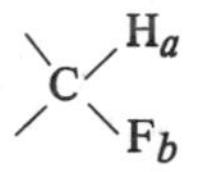 $>C(H_a)(F_b)$	44-81
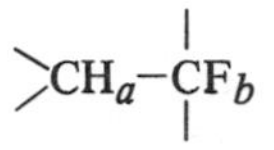 $>CH_a-CF_b$	3-25
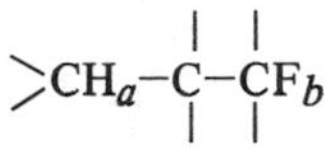 $>CH_a-C-CF_b$	0
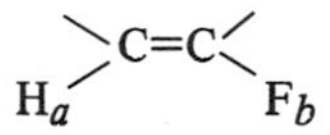 H_a and F_b cis on C=C	1-8
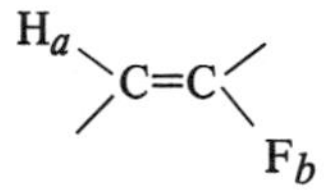 H_a and F_b trans on C=C	12-40
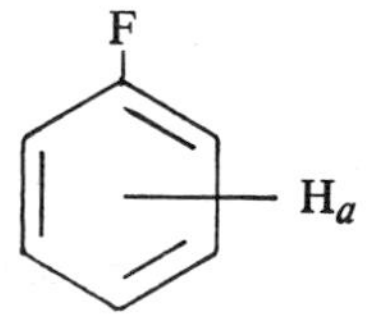 Fluorobenzene (H_a)	o 6-10 m 5-6 p 2

Proton-Phosphorus

Compound		
$>P(=O)H$	630-707	
$(CH_3)_3P$	2.7	
$(CH_3)_3P{=}O$	13.4	
$(CH_3CH_2)_3P$	13.7 (HCCP)	0.5 (HCP)
$(CH_3CH_2)_3P{=}O$	16.3 (HCCP)	11.9 (HCP)
$CH_3P(=O)(OR)_2$	10-13	
$CH_3C-P(=O)(OR)_2$	15-20	
$CH_3OP(OR)_2$	10.5-12	
$P[N(CH_3)_2]_3$	8.8	
$O{=}P[N(CH_3)_2]_3$	9.5	

*Compiled by Varian Associates.
†Signs of typical coupling constants can be obtained from the Appendixes of Bovey's book.[1]

appendix g. chemical shifts of residual protons in commercially available deuterated solvents. (source: merck sharp and dohme of canada, ltd.)

Solvent	Isotopic Purity Atom % D	Positions of Residual Protons					
		Group	δ	Group	δ	Group	δ
Acetic Acid-d_4	99.5	methyl	2.05	hydroxyl	11.53[a]		
Acetone-d_6	99.5	methyl	2.05				
Acetonitrile-d_3	98	methyl	1.95				
Benzene-d_6	99.5	methine	7.20				
Chloroform-d	99.8	methine	7.25				
Cyclohexane-d_{12}	99	methylene	1.40				
Deuterium Oxide	99.8	hydroxyl	4.75[a]				
1,2-Dichloroethane-d_4	99	methylene	3.69				
Diethyl-d_{10} Ether	98	methyl	1.16	methylene	3.36		
Dimethylformamide-d_7	98	methyl	2.76	methyl	2.94	formyl	8.05
Dimethyl-d_6 Sulfoxide	99.5	methyl	2.50				
p-Dioxane-d_8	98	methylene	3.55				
Ethyl Alcohol-d_6 (anh.)	98	methyl	1.17	methylene	3.59	hydroxyl	2.60[a]
Hexafluoroacetone Deuterate	99.5	hydroxyl	9.0[a]				
Methyl Alcohol-d_4	99	methyl	3.35	hydroxyl	4.84[a]		
Methylcyclohexane-d_{14}	99	methyl	0.92	methylene	1.54	methine	1.65
Methylene-d_2 Chloride	99	methylene	5.35				
Pyridine-d_5	99	alpha	8.70	beta	7.20	gamma	7.58
Silanar[c]-C ($CDCl_3$ + 1% TMS)	99.8	methyl	0.00	methine	7.25		
Tetrahydrofuran-d_8	98	α-methylene	3.60	β-methylene	0.75		
Tetramethylene-d_8 Sulfone	98	α-methylene	2.92	β-methylene	2.16		

[a]This value may vary considerably, depending upon the solute.
[b]By definition.
[c]Trademark.

Note: Several solvents are available in "100%" isotopic purity.

appendix h

NMR Spectra of Common Solvents and Common Impurities (Source: Sadtler Research Laboratories, Inc.) (See Following Pages)

	Spectrum Number		Spectrum Number
Acetone	1	Ethylene glycol	14
Antifoam A®	2	Isopropyl Alcohol	15
Apiezon® L Grease	3	Lubriseal® (Stopcock Grease)	16
Benzene	4	Methanol (+ Acid)	17
Chloroform	5	Methanol	18
Cyclohexane	6	Methyl ethyl ketone	19
Difluoroacetic acid	7	Phthalic acid, bis(2-ethylhexyl) ester	20
N,N-Dimethylformamide	8	Pyridine-d_5	21
Dimethyl Sulfoxide-d_6	9	Toluene	22
p-Dioxane	10	3-(Trimethylsilyl) propanesulfonic Acid, sodium salt (DSS)	23
Dow Corning Stopcock Grease	11	Vertex® flakes (soap)	24
Ethanol	12		
Ethyl Acetate	13		

No. 1

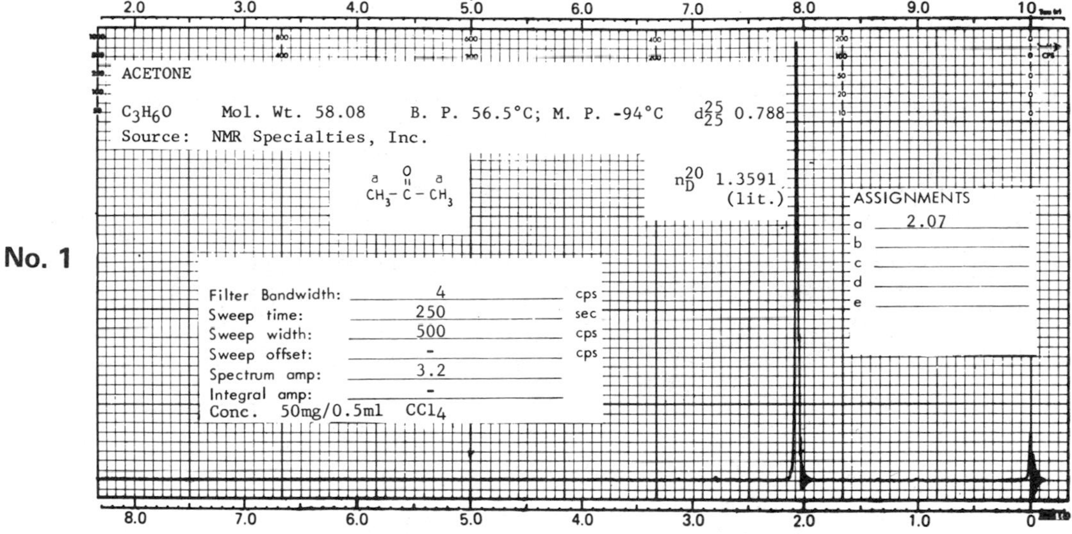

ACETONE
C3H6O Mol. Wt. 58.08 B. P. 56.5°C; M. P. -94°C d25/25 0.788
Source: NMR Specialties, Inc.
a O a
CH3- C - CH3
n20/D 1.3591 (lit.)
ASSIGNMENTS
a 2.07
b
c
d
e
Filter Bandwidth: 4 cps
Sweep time: 250 sec
Sweep width: 500 cps
Sweep offset: - cps
Spectrum amp: 3.2
Integral amp: -
Conc. 50mg/0.5ml CCl4
2.0 3.0 4.0 5.0 6.0 7.0 8.0 9.0 10
8.0 7.0 6.0 5.0 4.0 3.0 2.0 1.0 0

No. 2

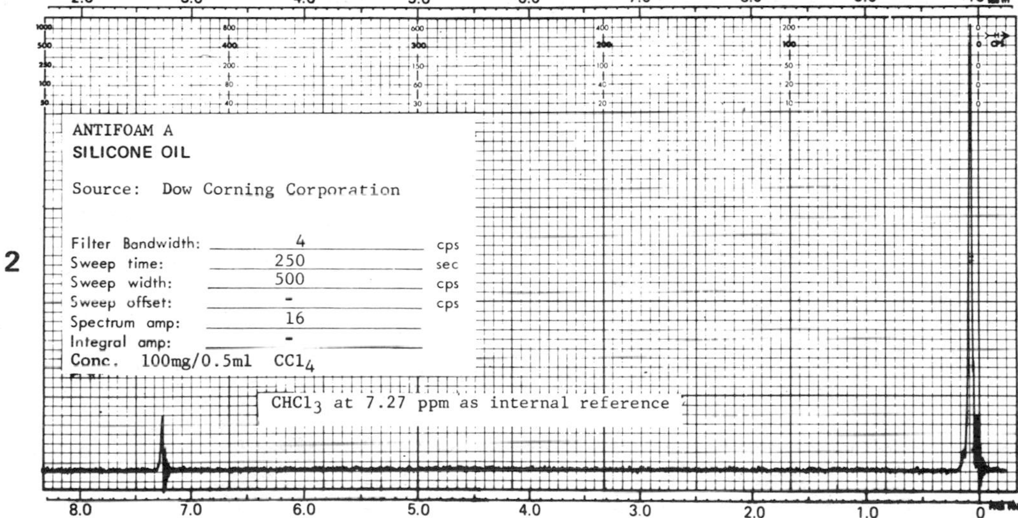

ANTIFOAM A
SILICONE OIL
Source: Dow Corning Corporation
Filter Bandwidth: 4 cps
Sweep time: 250 sec
Sweep width: 500 cps
Sweep offset: - cps
Spectrum amp: 16
Integral amp: -
Conc. 100mg/0.5ml CCl4
CHCl3 at 7.27 ppm as internal reference
2.0 3.0 4.0 5.0 6.0 7.0 8.0 9.0 10
8.0 7.0 6.0 5.0 4.0 3.0 2.0 1.0 0

No. 3

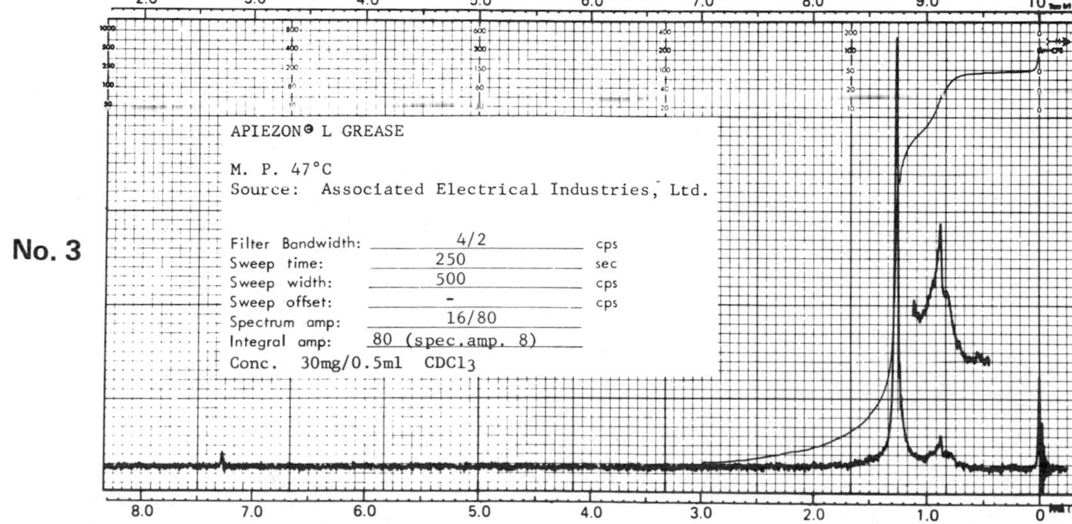

APIEZON® L GREASE
M. P. 47°C
Source: Associated Electrical Industries, Ltd.
Filter Bandwidth: 4/2 cps
Sweep time: 250 sec
Sweep width: 500 cps
Sweep offset: - cps
Spectrum amp: 16/80
Integral amp: 80 (spec.amp. 8)
Conc. 30mg/0.5ml CDCl3
2.0 3.0 4.0 5.0 6.0 7.0 8.0 9.0 10
8.0 7.0 6.0 5.0 4.0 3.0 2.0 1.0 0

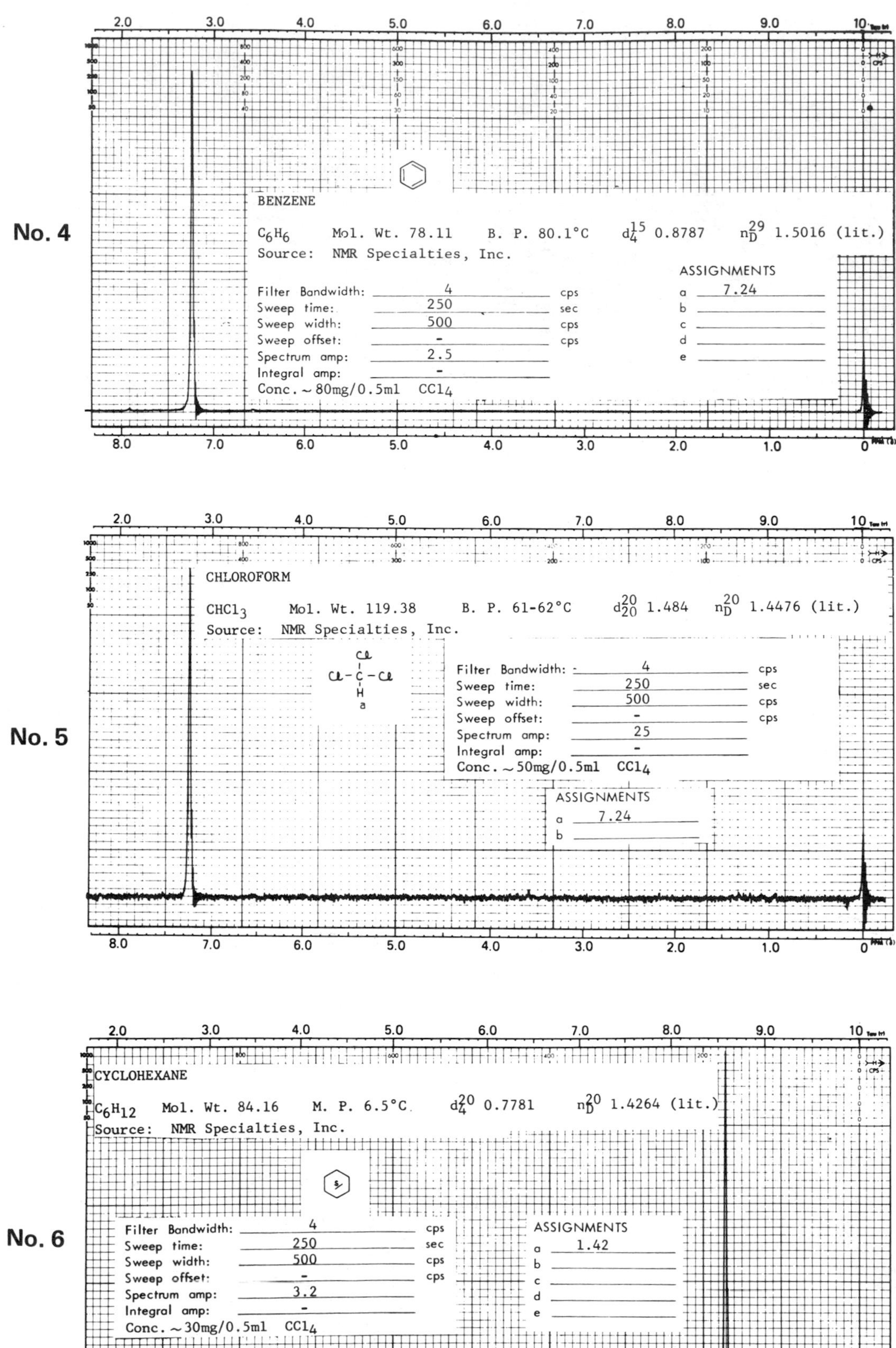
No. 4
BENZENE
C6H6 Mol. Wt. 78.11 B. P. 80.1°C d15/4 0.8787 n29/D 1.5016 (lit.)
Source: NMR Specialties, Inc.
Filter Bandwidth: 4 cps
Sweep time: 250 sec
Sweep width: 500 cps
Sweep offset: - cps
Spectrum amp: 2.5
Integral amp: -
Conc. ~80mg/0.5ml CCl4
ASSIGNMENTS
a 7.24
b
c
d
e
No. 5
CHLOROFORM
CHCl3 Mol. Wt. 119.38 B. P. 61-62°C d20/20 1.484 n20/D 1.4476 (lit.)
Source: NMR Specialties, Inc.
Filter Bandwidth: 4 cps
Sweep time: 250 sec
Sweep width: 500 cps
Sweep offset: - cps
Spectrum amp: 25
Integral amp: -
Conc. ~50mg/0.5ml CCl4
ASSIGNMENTS
a 7.24
b
No. 6
CYCLOHEXANE
C6H12 Mol. Wt. 84.16 M. P. 6.5°C d20/4 0.7781 n20/D 1.4264 (lit.)
Source: NMR Specialties, Inc.
Filter Bandwidth: 4 cps
Sweep time: 250 sec
Sweep width: 500 cps
Sweep offset: - cps
Spectrum amp: 3.2
Integral amp: -
Conc. ~30mg/0.5ml CCl4
ASSIGNMENTS
a 1.42
b
c
d
e

No. 7

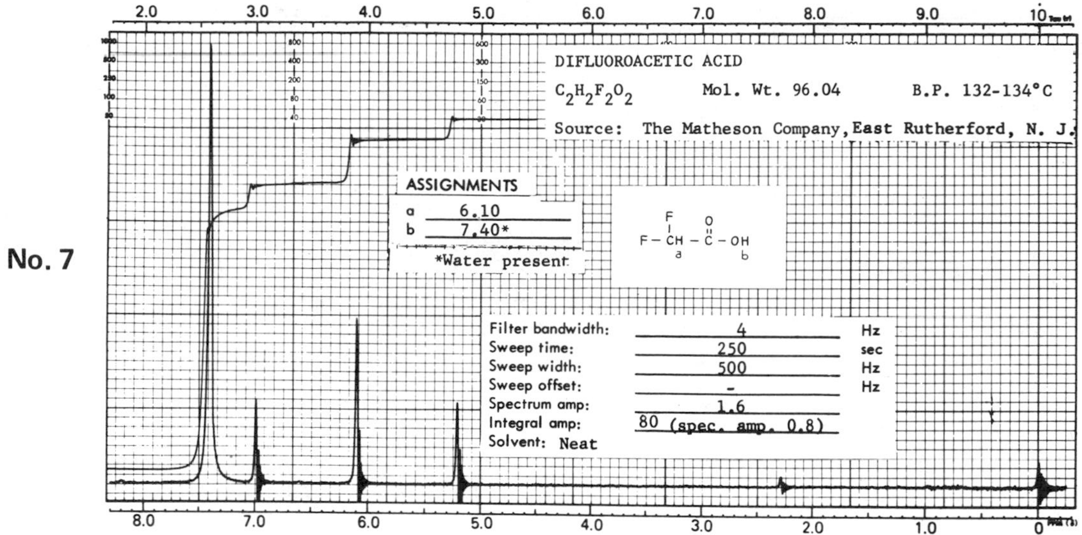
DIFLUOROACETIC ACID
$C_2H_2F_2O_2$ Mol. Wt. 96.04 B.P. 132-134°C
Source: The Matheson Company, East Rutherford, N. J.
ASSIGNMENTS
a 6.10
b 7.40*
*Water present
Filter bandwidth: 4 Hz
Sweep time: 250 sec
Sweep width: 500 Hz
Sweep offset: - Hz
Spectrum amp: 1.6
Integral amp: 80 (spec. amp. 0.8)
Solvent: Neat

No. 8

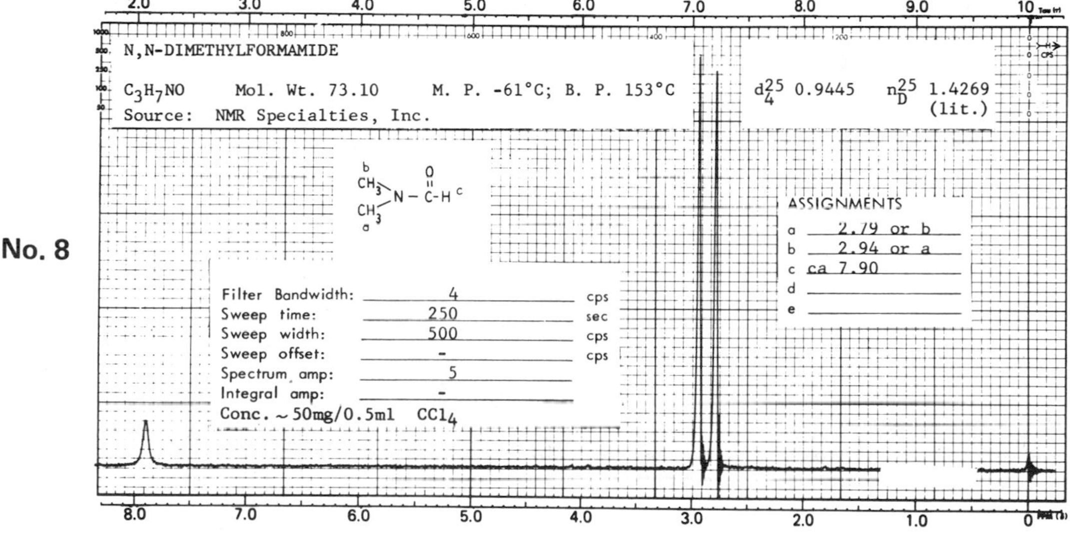
N,N-DIMETHYLFORMAMIDE
C_3H_7NO Mol. Wt. 73.10 M. P. -61°C; B. P. 153°C
d_4^{25} 0.9445 n_D^{25} 1.4269 (lit.)
Source: NMR Specialties, Inc.
ASSIGNMENTS
a 2.79 or b
b 2.94 or a
c ca 7.90
d
e
Filter Bandwidth: 4 cps
Sweep time: 250 sec
Sweep width: 500 cps
Sweep offset: - cps
Spectrum amp: 5
Integral amp: -
Conc. ~50mg/0.5ml CCl_4

No. 9

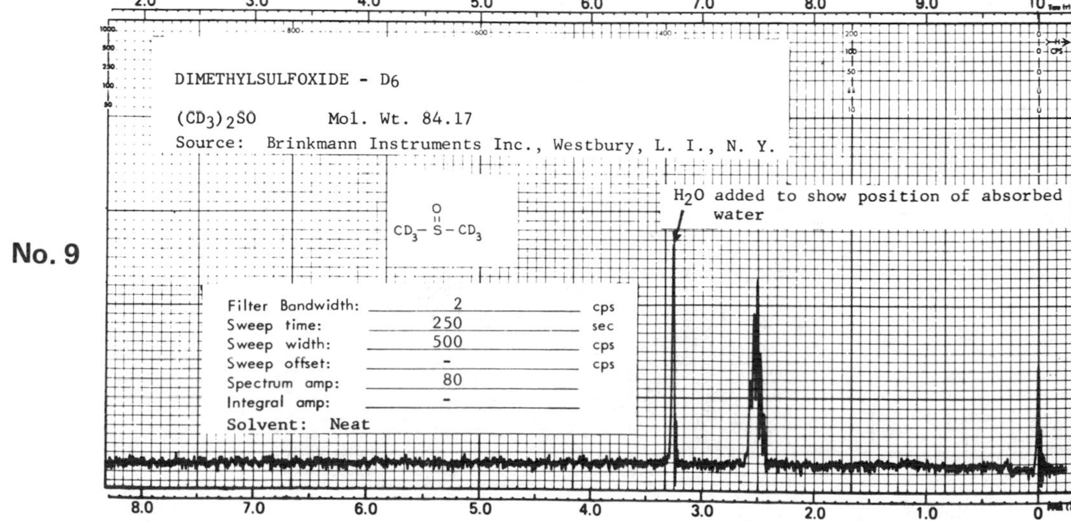
DIMETHYLSULFOXIDE - D6
$(CD_3)_2SO$ Mol. Wt. 84.17
Source: Brinkmann Instruments Inc., Westbury, L. I., N. Y.
H_2O added to show position of absorbed water
Filter Bandwidth: 2 cps
Sweep time: 250 sec
Sweep width: 500 cps
Sweep offset: - cps
Spectrum amp: 80
Integral amp: -
Solvent: Neat

No. 10

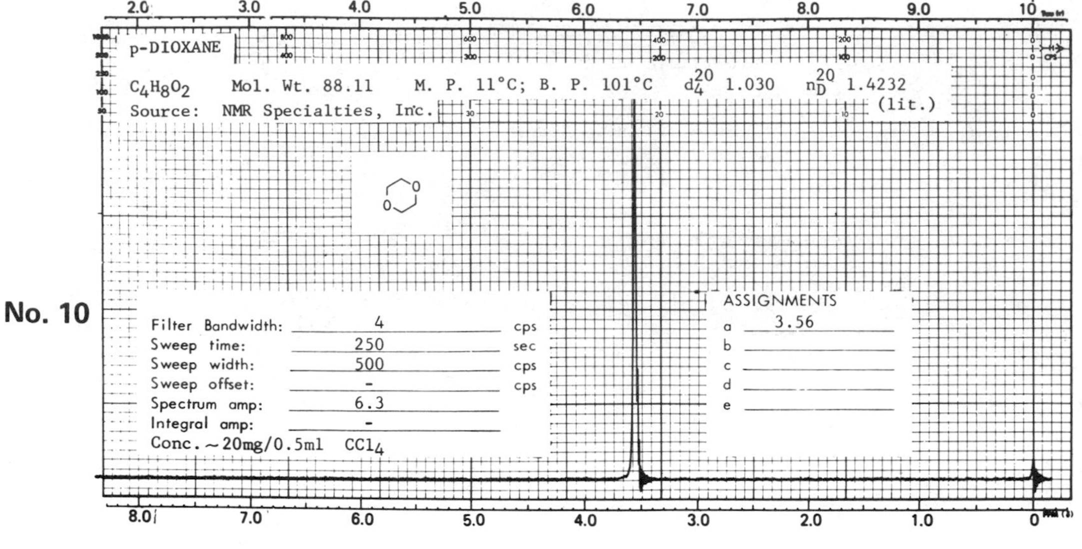

No. 11

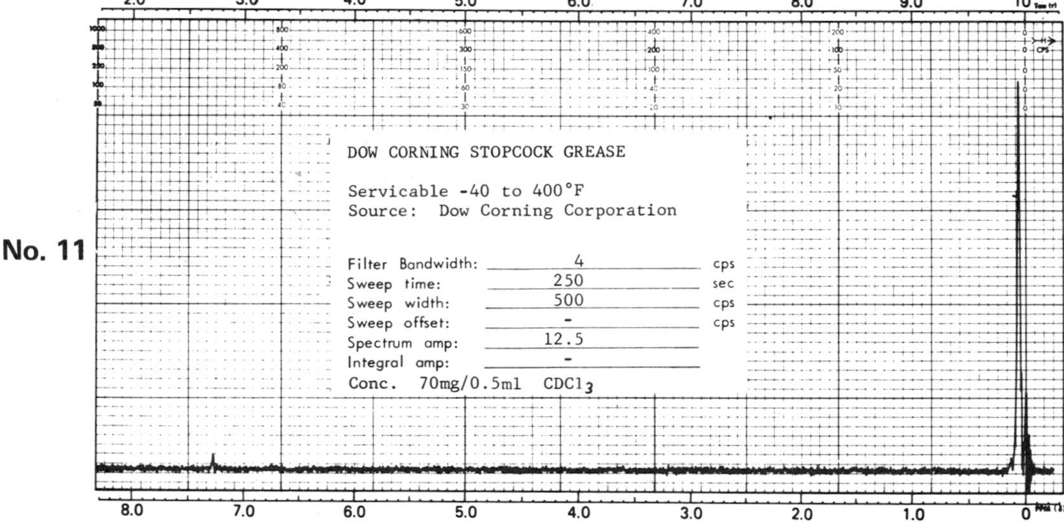

DOW CORNING STOPCOCK GREASE
Servicable -40 to 400°F
Source: Dow Corning Corporation
Filter Bandwidth: 4 cps
Sweep time: 250 sec
Sweep width: 500 cps
Sweep offset: - cps
Spectrum amp: 12.5
Integral amp: -
Conc. 70mg/0.5ml $CDCl_3$

No. 12

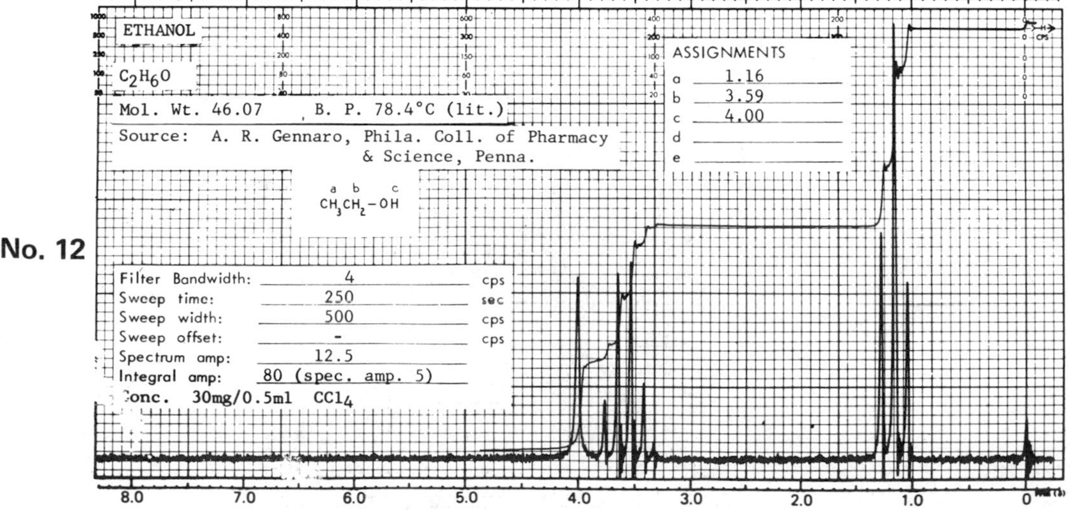

ETHANOL
C_2H_6O
Mol. Wt. 46.07 B. P. 78.4°C (lit.)
Source: A. R. Gennaro, Phila. Coll. of Pharmacy & Science, Penna.
$\overset{a}{CH_3}\overset{b}{CH_2}-\overset{c}{OH}$
ASSIGNMENTS
a 1.16
b 3.59
c 4.00
d
e
Filter Bandwidth: 4 cps
Sweep time: 250 sec
Sweep width: 500 cps
Sweep offset: - cps
Spectrum amp: 12.5
Integral amp: 80 (spec. amp. 5)
Conc. 30mg/0.5ml CCl_4

No. 13

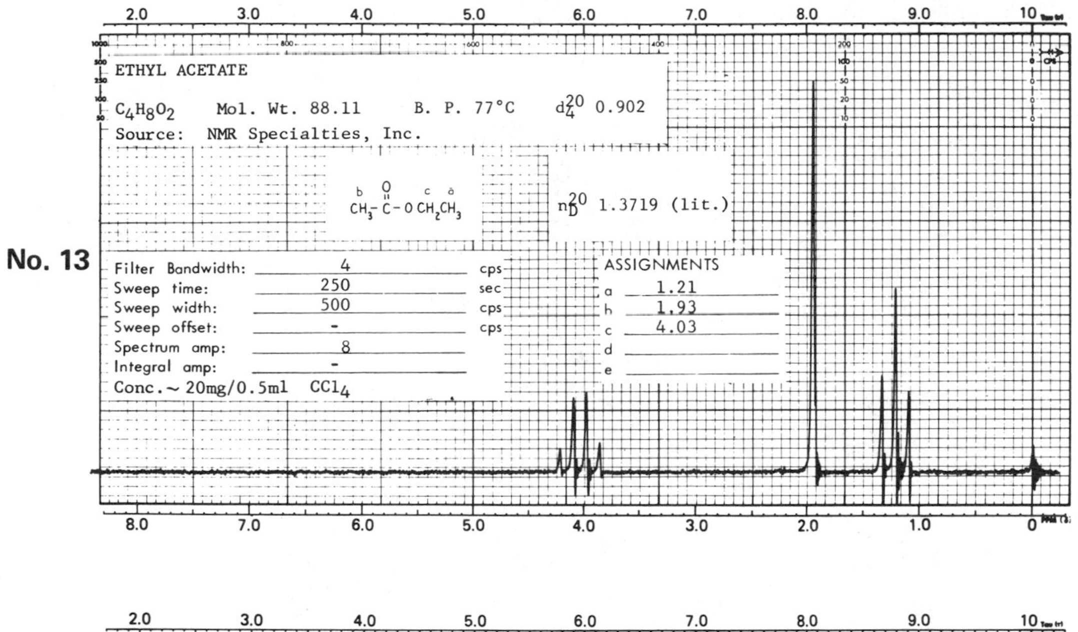

ETHYL ACETATE
$C_4H_8O_2$ Mol. Wt. 88.11 B. P. 77°C d_4^{20} 0.902
Source: NMR Specialties, Inc.
n_D^{20} 1.3719 (lit.)
Filter Bandwidth: 4 cps
Sweep time: 250 sec
Sweep width: 500 cps
Sweep offset: - cps
Spectrum amp: 8
Integral amp: -
Conc. ~ 20mg/0.5ml CCl_4
ASSIGNMENTS
a 1.21
b 1.93
c 4.03
d
e

No. 14

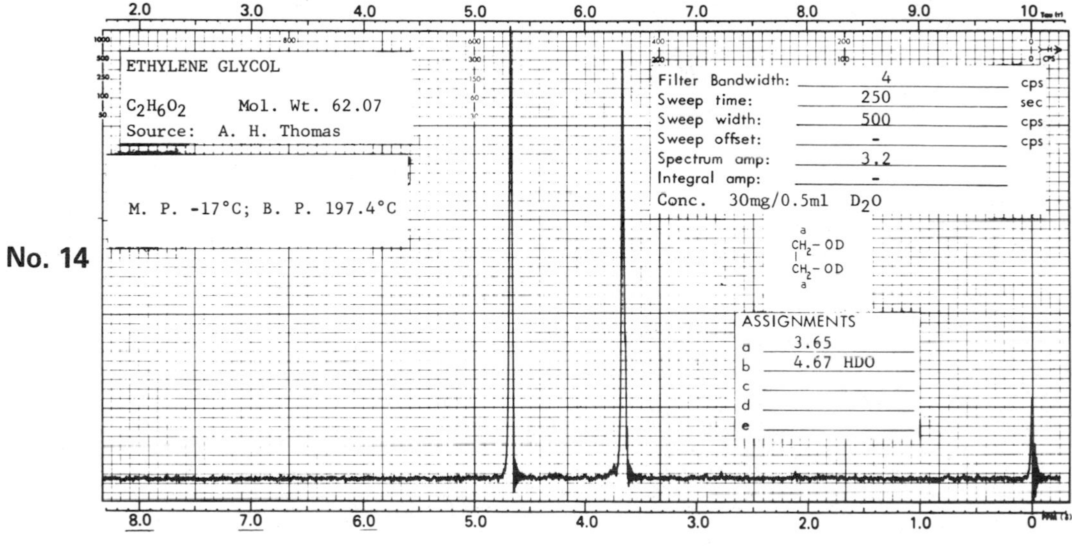

ETHYLENE GLYCOL
$C_2H_6O_2$ Mol. Wt. 62.07
Source: A. H. Thomas
M. P. -17°C; B. P. 197.4°C
Filter Bandwidth: 4 cps
Sweep time: 250 sec
Sweep width: 500 cps
Sweep offset: - cps
Spectrum amp: 3.2
Integral amp: -
Conc. 30mg/0.5ml D_2O
ASSIGNMENTS
a 3.65
b 4.67 HDO
c
d
e

No. 15

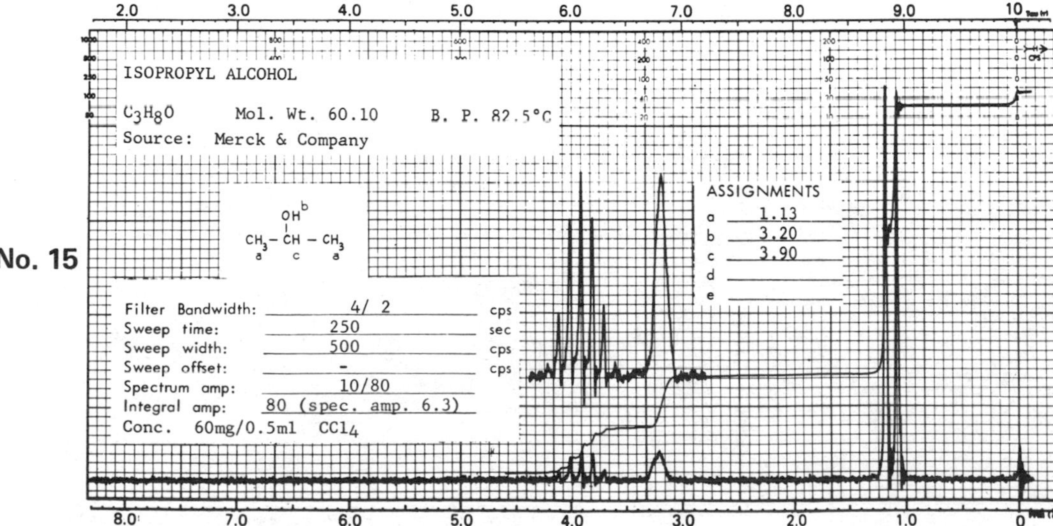

ISOPROPYL ALCOHOL
C_3H_8O Mol. Wt. 60.10 B. P. 82.5°C
Source: Merck & Company
ASSIGNMENTS
a 1.13
b 3.20
c 3.90
d
e
Filter Bandwidth: 4/ 2 cps
Sweep time: 250 sec
Sweep width: 500 cps
Sweep offset: - cps
Spectrum amp: 10/80
Integral amp: 80 (spec. amp. 6.3)
Conc. 60mg/0.5ml CCl4

No. 16

No. 17

No. 18

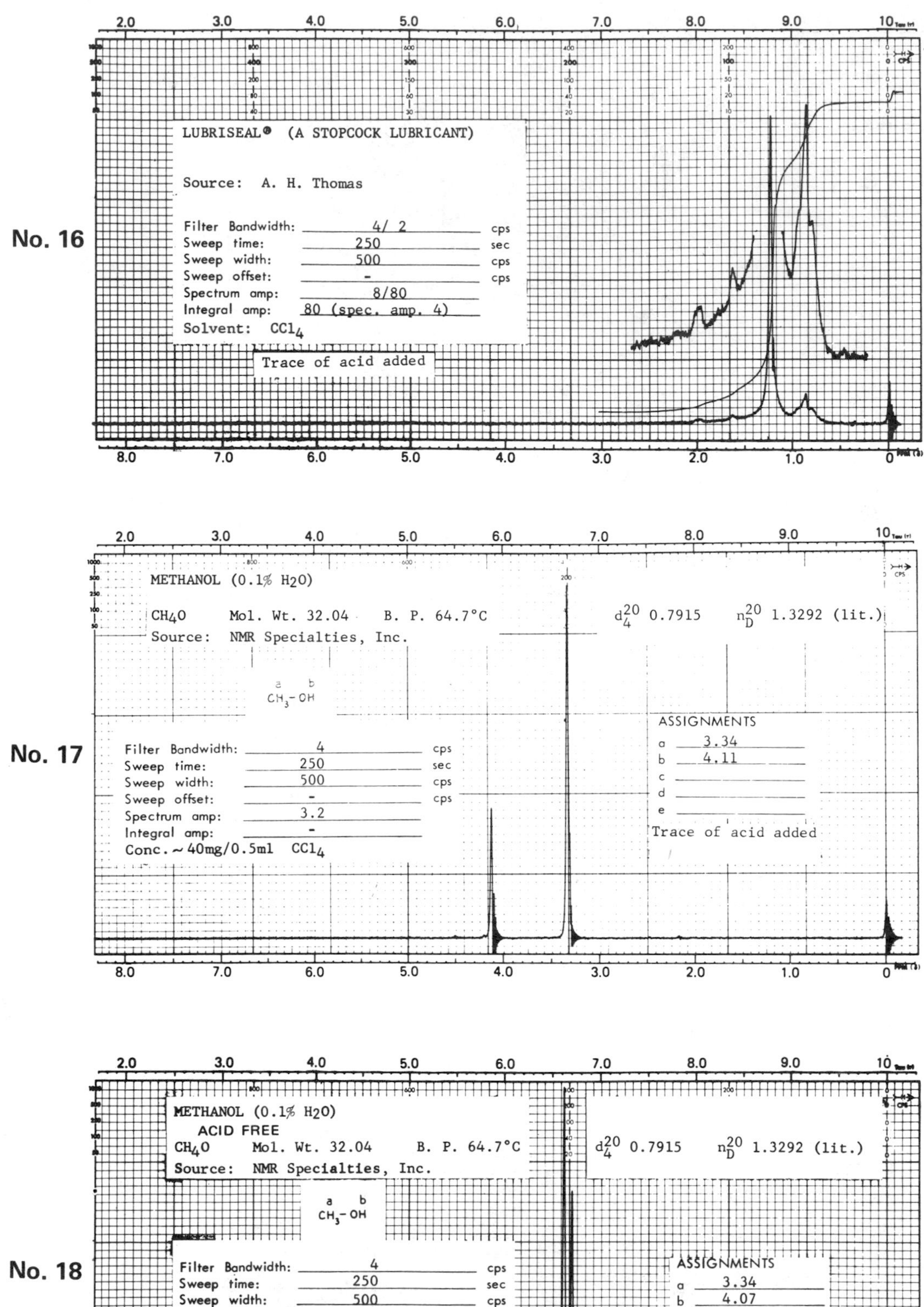

No. 19

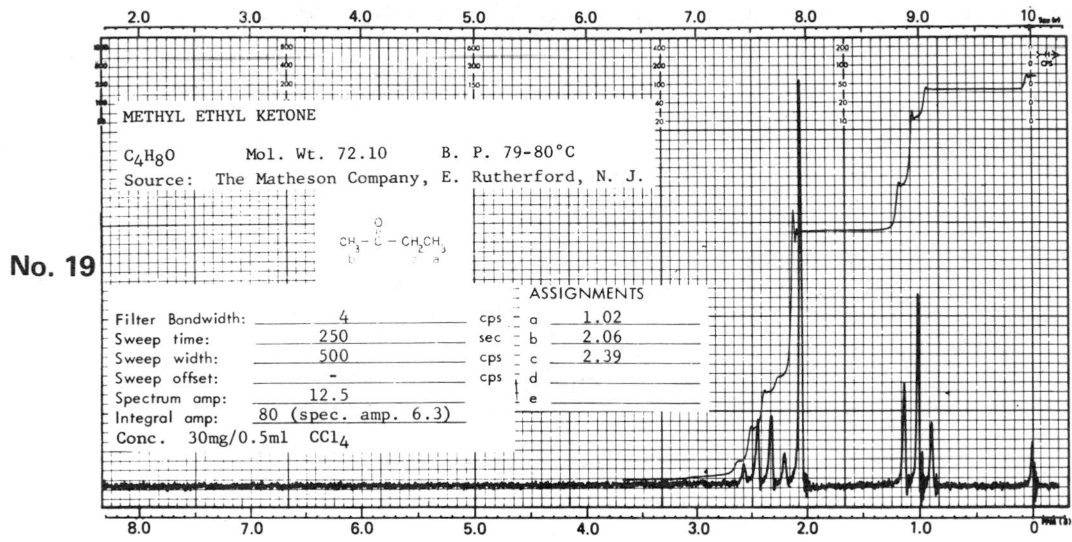

METHYL ETHYL KETONE
C_4H_8O Mol. Wt. 72.10 B. P. 79-80°C
Source: The Matheson Company, E. Rutherford, N. J.
ASSIGNMENTS
a 1.02
b 2.06
c 2.39
Filter Bandwidth: 4 cps
Sweep time: 250 sec
Sweep width: 500 cps
Sweep offset: - cps
Spectrum amp: 12.5
Integral amp: 80 (spec. amp. 6.3)
Conc. 30mg/0.5ml CCl_4

No. 20

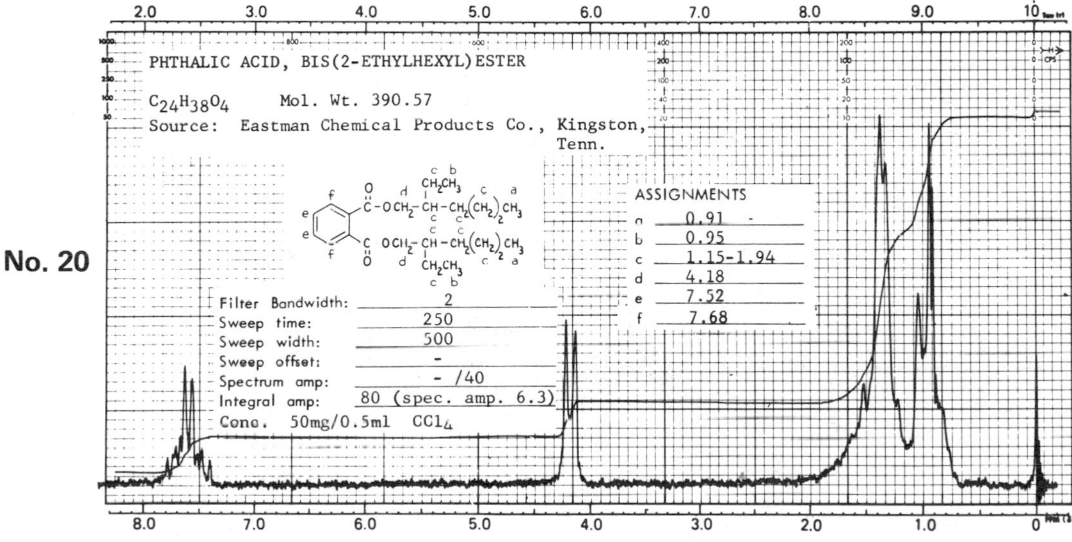

No. 21

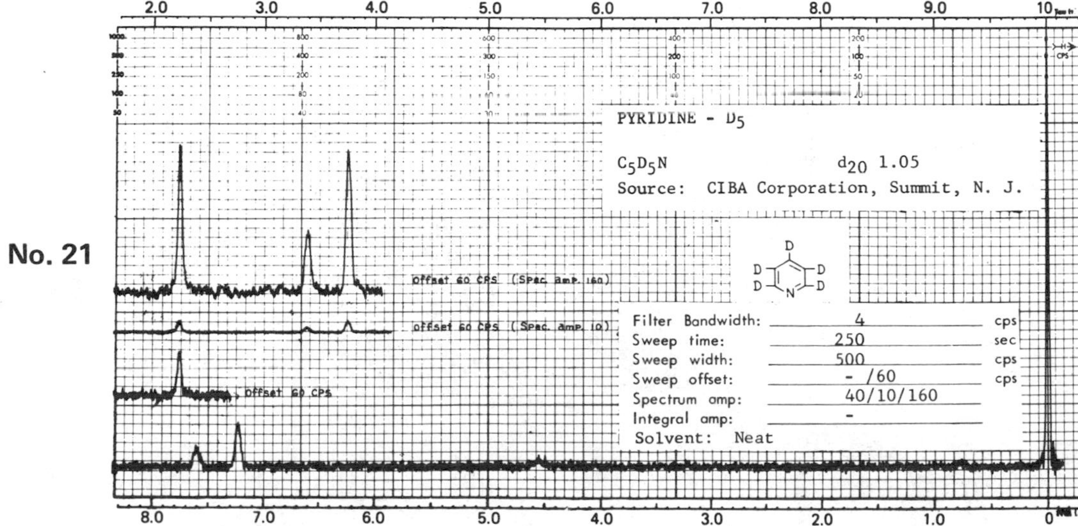

No. 22

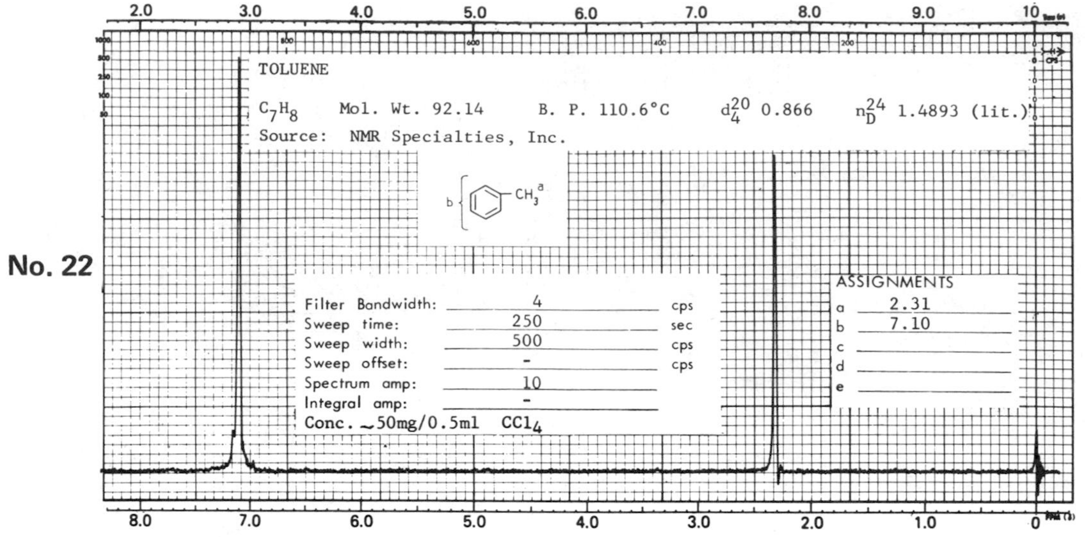

TOLUENE
C_7H_8 Mol. Wt. 92.14 B. P. 110.6°C d_4^{20} 0.866 n_D^{24} 1.4893 (lit.)
Source: NMR Specialties, Inc.
Filter Bandwidth: 4 cps
Sweep time: 250 sec
Sweep width: 500 cps
Sweep offset: - cps
Spectrum amp: 10
Integral amp: -
Conc. ~50mg/0.5ml CCl_4
ASSIGNMENTS
a 2.31
b 7.10
c
d
e

No. 23

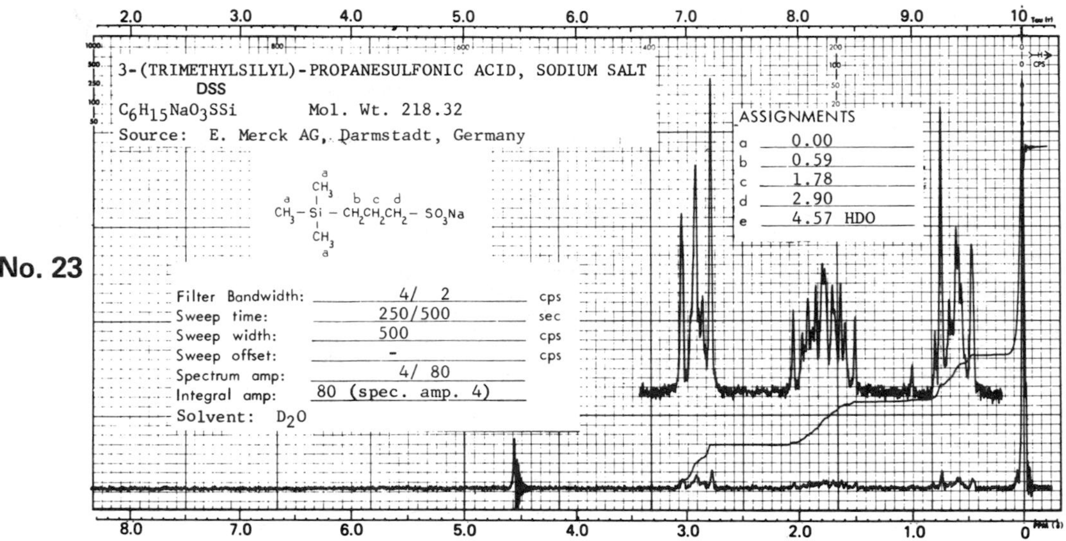

No. 24

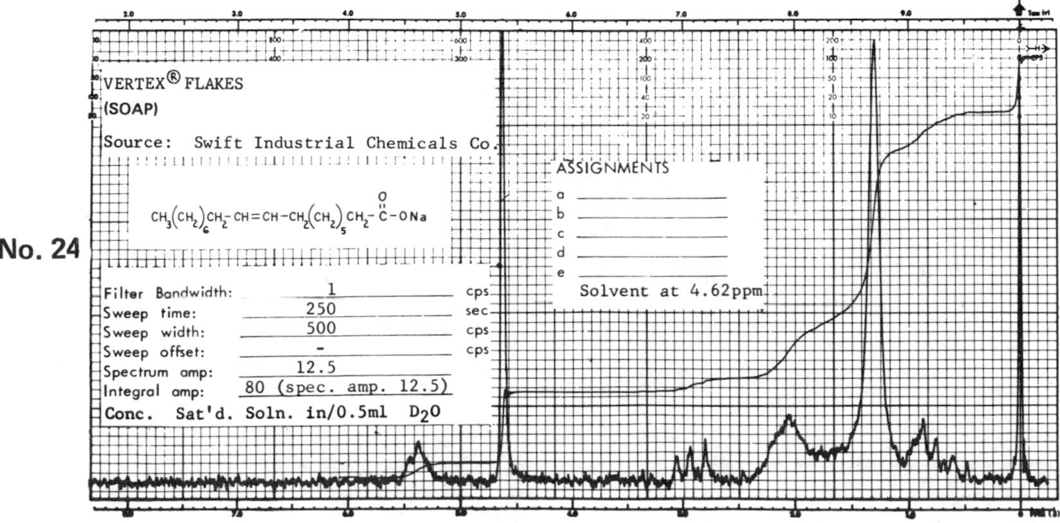

VERTEX® FLAKES
(SOAP)
Source: Swift Industrial Chemicals Co.
Filter Bandwidth: 1 cps
Sweep time: 250 sec
Sweep width: 500 cps
Sweep offset: - cps
Spectrum amp: 12.5
Integral amp: 80 (spec. amp. 12.5)
Conc. Sat'd. Soln. in/0.5ml D_2O
ASSIGNMENTS
a
b
c
d
e
Solvent at 4.62ppm

appendix i

PROPERTIES OF SEVERAL NUCLEI[a]

Isotope	NMR Frequency MHz for a 10-Kilogauss Field	Natural Abundance %	Relative Sensitivity at Constant Field	Magnetic Moment μ	Spin Number I	Electrical Quadrupole Moment $e \times 10^{-24}$ cm^2	Magnetogyric Ratio γ_N radians gauss^{-1}
^{1}H	42.576	99.9844	1.000	2.79268	1/2	–	26,753
^{2}H	6.5357	1.56×10^{-2}	9.64×10^{-2}	0.85739	1	2.77×10^{-3}	4,107
^{3}H	45.414	–	1.21	2.9788	1/2	–	–
^{10}B	4.575	18.83	1.99×10^{-2}	1.8005	3	7.4×10^{-2}	–
^{11}B	13.660	81.17	0.165	2.6880	3/2	3.55×10^{-2}	–
^{12}C	–	98.9	–	–	0	–	–
^{13}C	10.705	1.108	1.59×10^{-2}	0.70220	1/2	–	6,728
^{14}N	3.076	99.635	1.01×10^{-3}	0.40358	1	7.1×10^{-2}	–
^{15}N	4.315	0.365	1.04×10^{-3}	−0.28304	1/2	–	−2,712
^{16}O	–	99.76	–	–	0	–	–
^{17}O	5.772	3.7×10^{-2}	2.91×10^{-2}	−1.8930	5/2	-4.0×10^{-3}	−3,628
^{19}F	40.055	100	0.834	2.6273	1/2	–	25,179
^{28}Si	–	92.28	–	–	0	–	–
^{29}Si	8.458	4.70	7.85×10^{-2}	−0.55548	1/2	–	−5,319
^{30}Si	–	3.02	–	–	0	–	–
^{31}P	17.236	100	6.64×10^{-2}	1.1305	1/2	–	10,840
^{32}S	–	95.06	–	–	0	–	–
^{33}S	3.266	0.74	2.26×10^{-3}	0.64274	3/2	−0.053	2,054
^{34}S		4.2	–	–	0	–	–
^{35}Cl	4.172	75.4	4.71×10^{-3}	0.82091	3/2	-7.9×10^{-2}	2,624
^{37}Cl	3.472	24.6	2.72×10^{-3}	0.68330	3/2	-6.21×10^{-2}	2,184
^{79}Br	10.667	50.57	7.86×10^{-2}	2.0991	3/2	0.34	–
^{81}Br	11.499	49.43	9.84×10^{-2}	2.2626	3/2	0.28	–
^{127}I	8.519	100	9.35×10^{-2}	2.7937	5/2	−0.75	–

[a]Varian Associates NMR Table, 4th ed., 1964.

five

^{13}C nmr spectrometry

I. INTRODUCTION*

Direct observation of the carbon skeleton as well as the carbon atoms in carbon-containing functional groups has been available on a practical basis only since the early 1970s. The ^{12}C nucleus is not magnetically "active" (spin number, I, is zero), but the ^{13}C nucleus, like the ^{1}H nucleus, has a spin number of ½. However, since the natural abundance of ^{13}C is only 1.1% that of ^{12}C, and its sensitivity is only about 1.6% that of ^{1}H, the overall sensitivity of ^{13}C compared with ^{1}H is about 1/5700.

The conventional, continuous-wave, slow-scan procedure requires a large sample and a prohibitively long time to obtain a ^{13}C spectrum, but the availability of Fourier transform instrumentation, which permits simultaneous irradiation of all ^{13}C nuclei, has resulted in an explosion of ^{13}C spectrometry, beginning in the early 1970s, comparable to the burst of activity in ^{1}H spectrometry that began in the late 1950s.

An important development was the use of wide-band heteronuclear (noise) decoupling. Because of the large J values for ^{13}C–H (~ 110-320 Hz) and appreciable values for $^{13}C–C–H$ and $^{13}C–C–C–H$, nondecoupled (proton coupled) ^{13}C spectra usually show complex overlapping multiplets that are difficult to interpret, but some nondecoupled spectra such as that of diethyl phthalate (Figure 1a) are quite simple. Noise decoupling (Figure 1b) of the protons by means of a wideband (noise) generator removes these couplings. The result, in the absence of other coupling nuclei, such as ^{31}P or ^{19}F, is a *single sharp peak*** *for each chemically nonequivalent* ^{13}*C atom*, except for the infrequent fortuitous coincidence of ^{13}C chemical shifts. Furthermore, an increase in signal (up to a factor of ~ 3) accrues from the nuclear Overhauser effect (NOE). This enhancement results from an increase in population of the lower energy level of the ^{13}C nuclei concomitant with the increase in population of the high energy level of the ^{1}H nuclei on irradiation (decoupling) of the ^{1}H nuclei. The net effect is a very large reduction in the time needed to obtain a noise-decoupled spectrum (Figure 1b) as compared with a nondecoupled spectrum (Figure 1a).

In the fourier transform mode, a short powerful rf pulse (on the order of a few microseconds) excites all the ^{13}C nuclei simultaneously. Since the frequencies in the pulse are slightly off-resonance (pulse center at approximately the frequency of ^{13}C in TMS) for all of the nuclei, each nucleus shows a free induction decay (FID), which is an exponentially decaying sine wave with a frequency equal to the difference between the applied frequency and the resonance frequency for that nucleus (Figure 2a).

The free induction decay display for a compound containing more than one ^{13}C nucleus consists of superimposed sine waves, each with its characteristic frequency, and an interference ("beat") pattern results (Figure 2b). These data are automatically digitized and stored in a computer, and a series of repetitive pulses, with signal acquisition and accumulation between pulses, builds up the signal. Fourier transformation by the computer converts this information to the conventional presentation of a ^{13}C NMR spectrum. Figure 2a represents the FID and the conventional ^{13}C spectra for CH_3OH. Figure 2b shows the same spectra for C_2H_5OH. The free induction decay is a time domain spectrum, whereas the transformed, conventional presentation is a frequency domain spectrum. The sequence (Fig. 2c) is: pulse, acquisition, and pulse delay if required (see below).

The ^{13}C spectra in this text were run on a Varian XL-100 (25.2 MHz for ^{13}C) or a Varian CFT 20 (20.0 MHz for ^{13}C) spectrometer. Since these instruments lock on deuterium, the common deuterated solvents (usually $CDCl_3$) are used to furnish the internal lock.* (A list of common deuterated solvents is given in Appendix A.) As it is with ^{1}H spectrometry, the common internal reference is tetramethylsilane (TMS), and the scale is given in δ units (ppm) downfield (deshielding) from TMS in positive numbers, and upfield (shielding) from TMS in negative numbers. The shifts encountered in routine ^{13}C spectra range about 240 ppm downfield from TMS; this is a range of about 20 times that of routine ^{1}H spectra (~ 12 ppm). Several cations have been recorded at ~ 335 ppm downfield, and CI_4 absorbs at ~ −290 ppm (upfield from TMS).

On the Varian XL-100 instrument, the ^{13}C frequency is 25.2 MHz, and the routine sweep width is 5000 Hz (198.4 ppm). Each of the major divisions on the bottom scale of

*Familiarity with Chapter 4 is assumed.

**Because of the low natural abundance of ^{13}C, the occurrence of adjacent ^{13}C atoms has a low probability; thus, we are free of the complication of ^{13}C–^{13}C coupling.

*A field-frequency internal lock provides corresponding changes in the irradiating frequency for minor variations in field strength to furnish a constant field/frequency ratio.

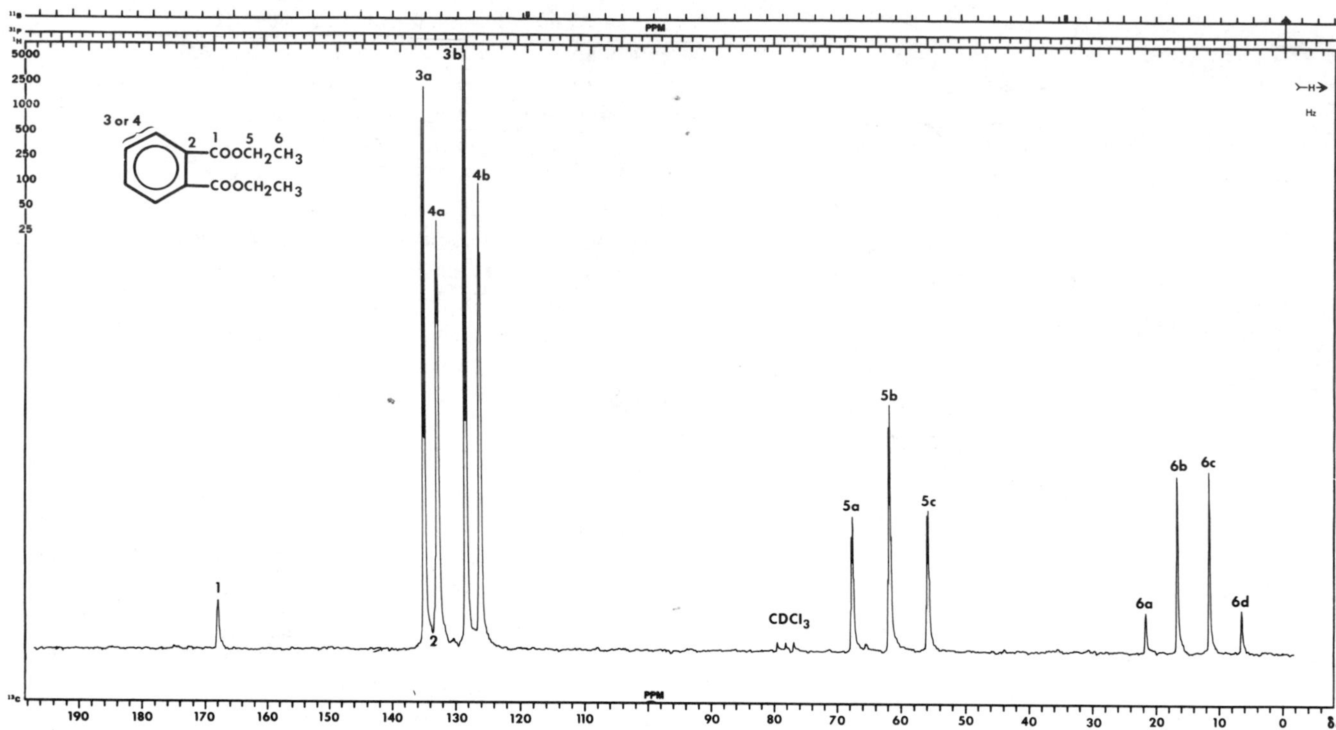

Figure 1a. ^{13}C *NMR spectrum of diethyl phthalate with the protons completely coupled.* $CDCl_3$ *solvent, 25.2 MHz.*

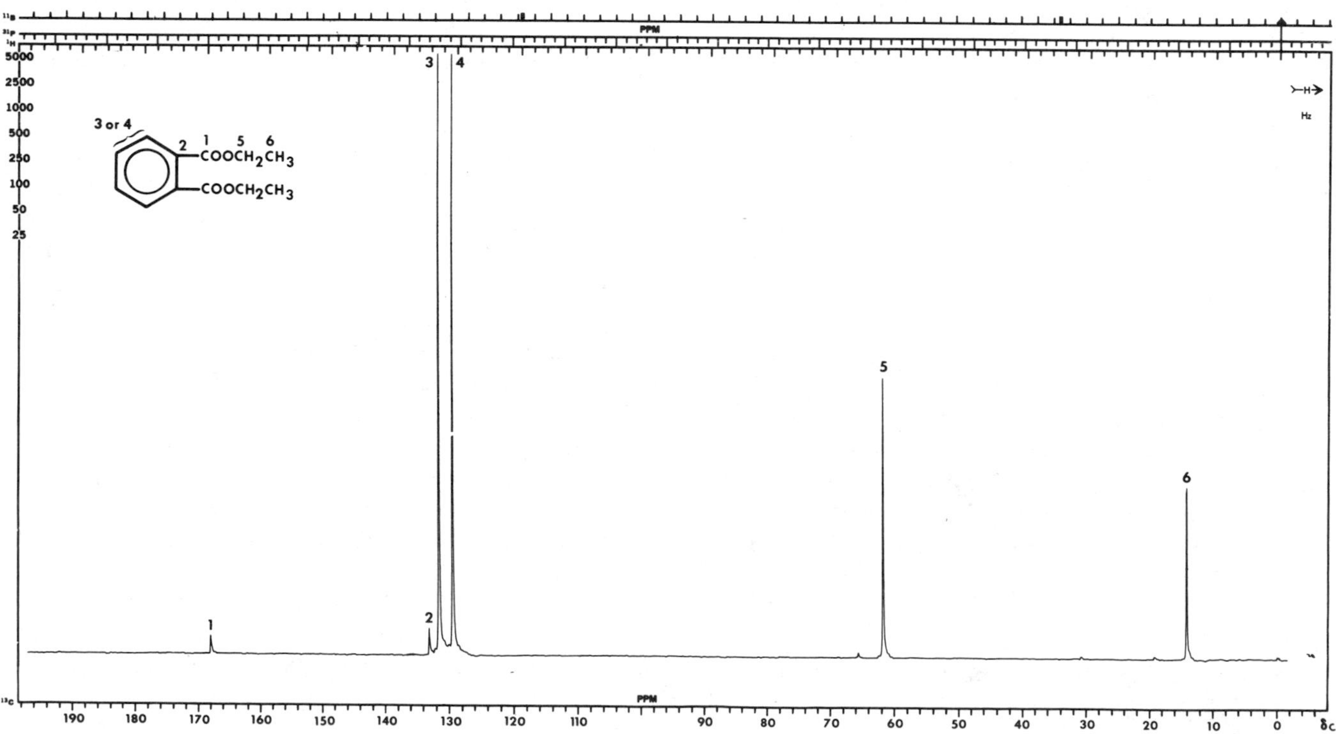

Figure 1b. ^{13}C *NMR spectrum of diethyl phthalate with the protons completely decoupled by broad-band noise.* $CDCl_3$ *solvent, 25.2 MHz.*

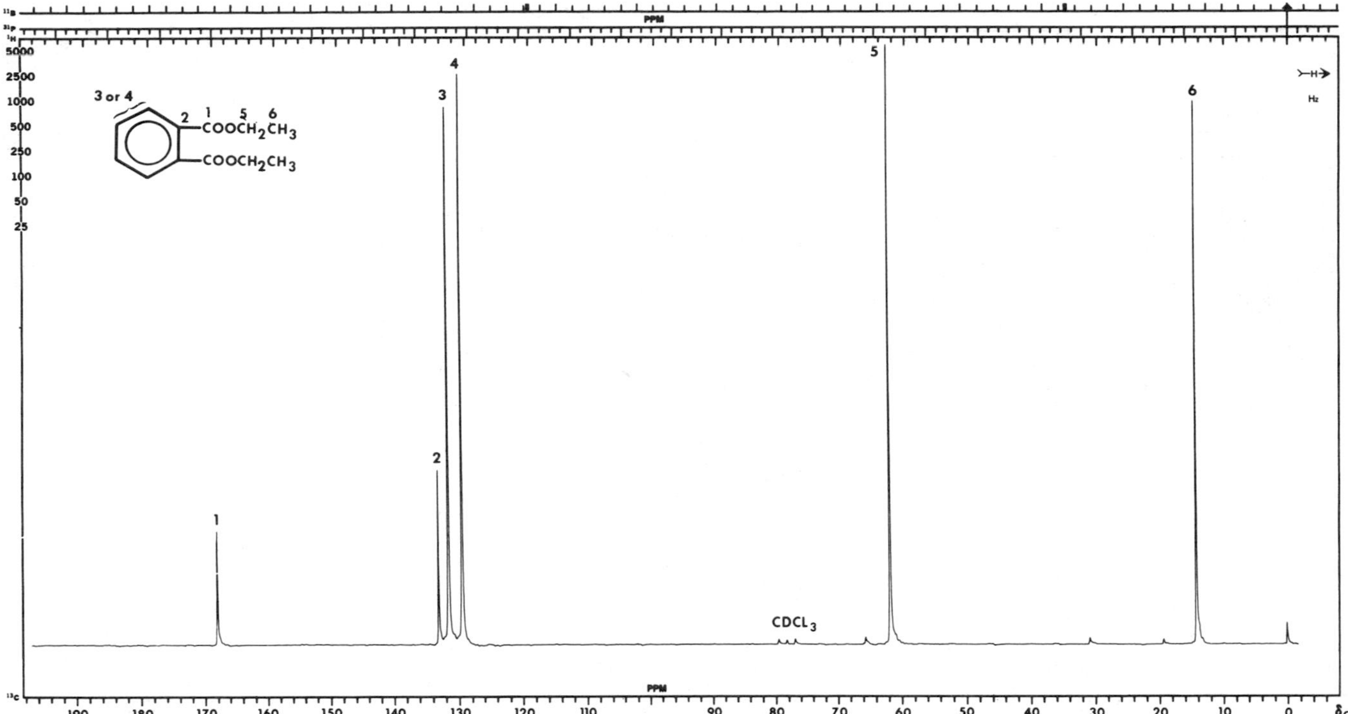

Figure 1c. ^{13}C *NMR spectrum of diethyl phthalate with the protons completely decoupled and a 10 sec. delay between pulses.* $CDCl_3$ *solvent, 25.2 MHz.*

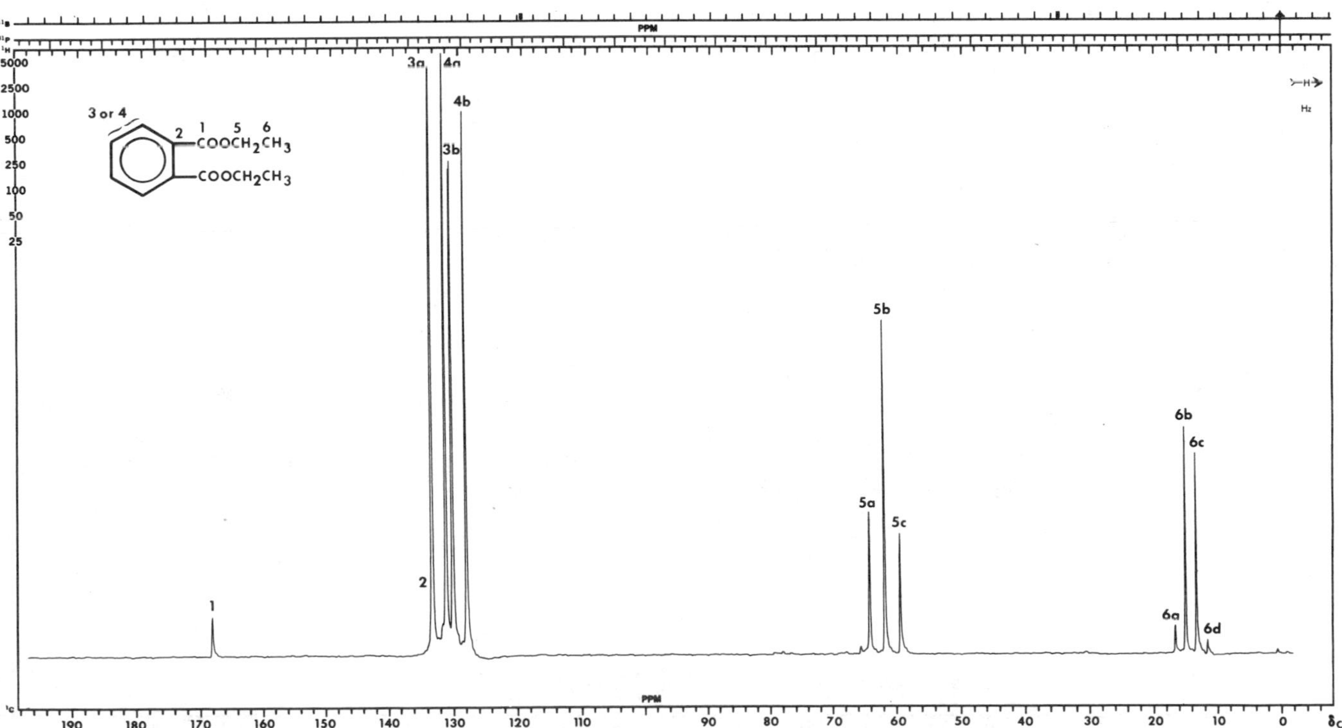

Figure 1d. ^{13}C *NMR spectrum of diethyl phthalate with off-resonance decoupled protons.* $CDCl_3$ *solvent, 25.2 MHz.*

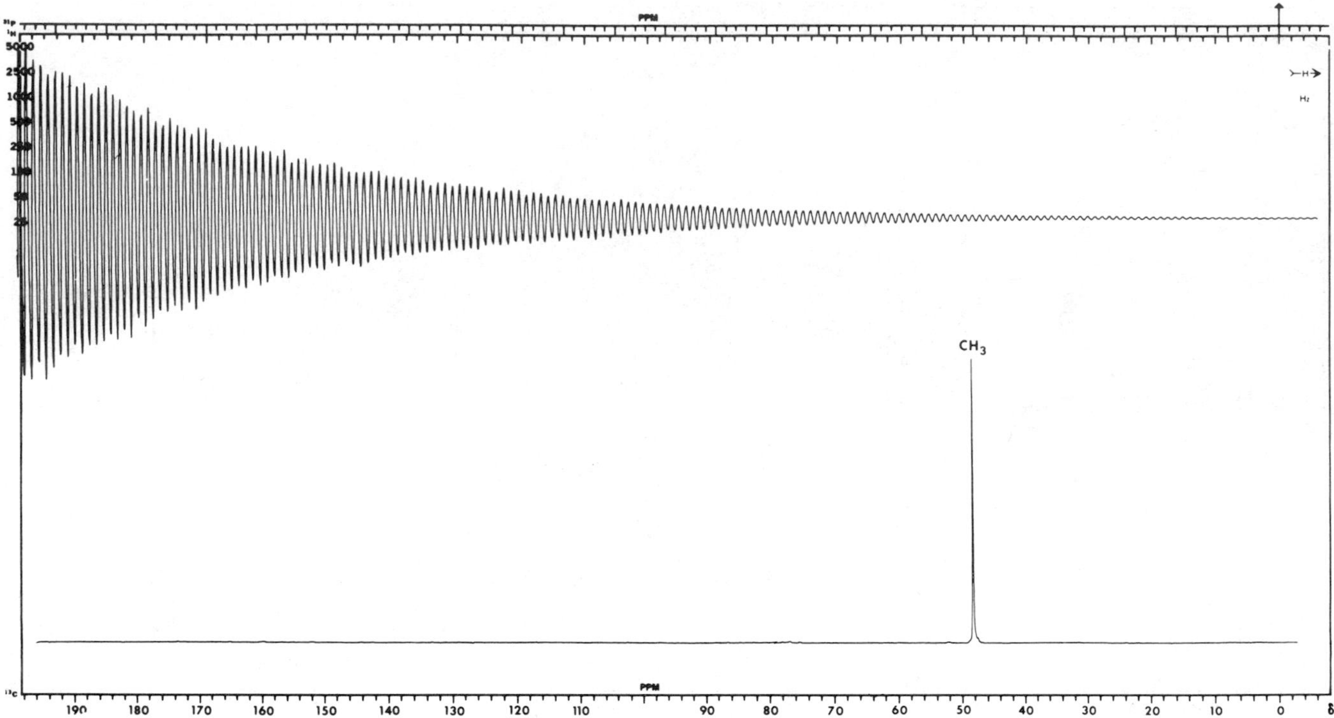

Figure 2a. *Free induction decay (above) and transformed ^{13}C spectrum of methanol (60% in $CDCl_3$, 25.2 MHz).*

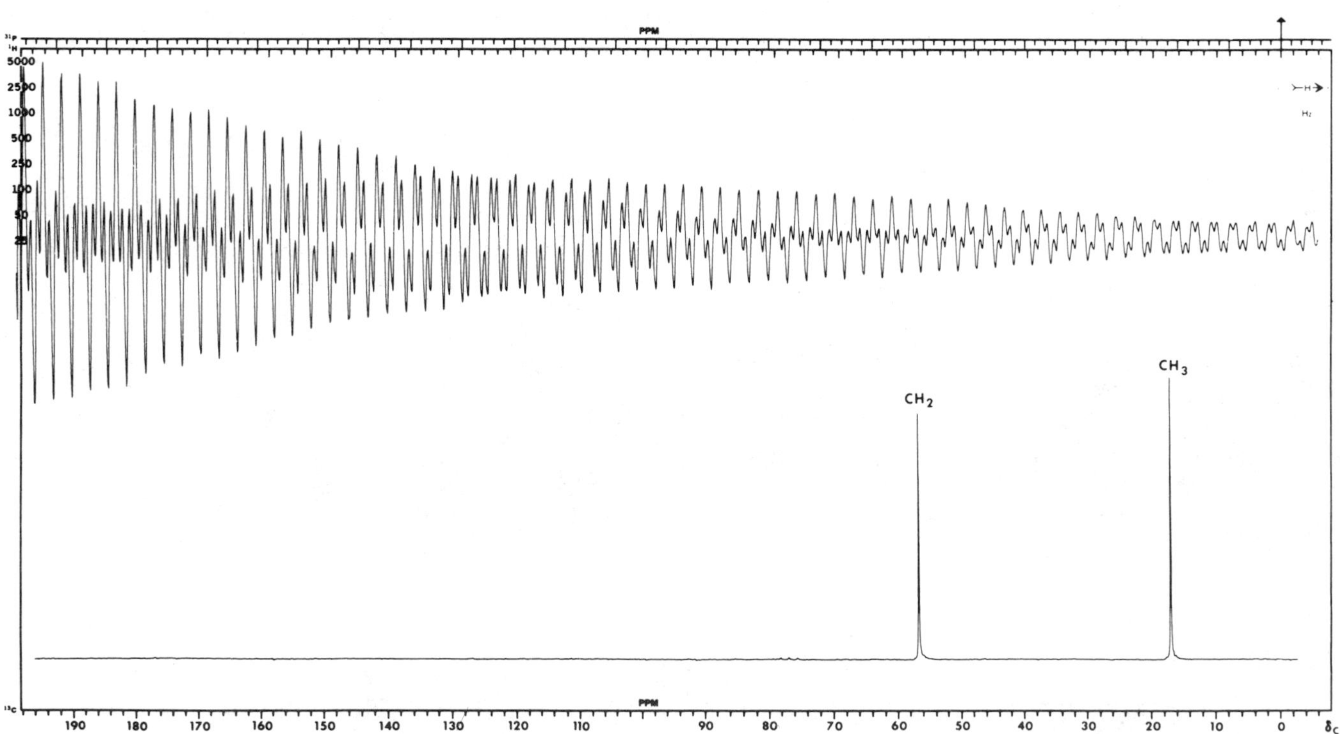

Figure 2b. *Free induction decay (above) and transformed ^{13}C spectrum of ethanol (60% in $CDCl_3$, 25.2 MHz).*

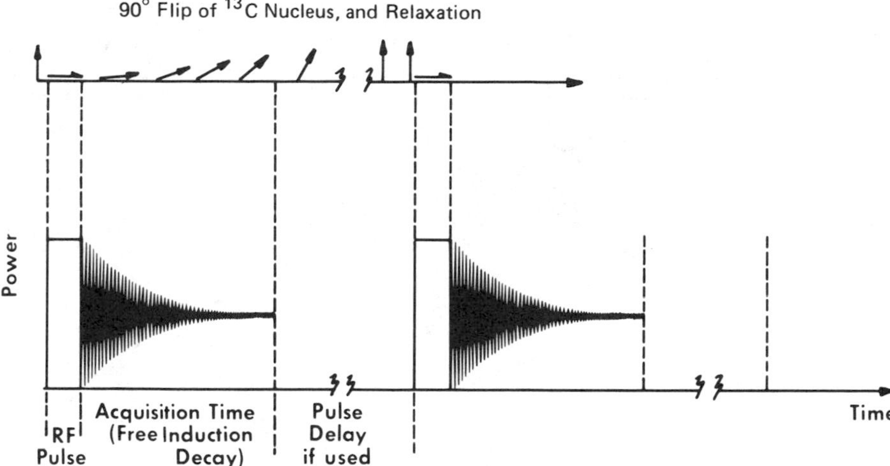

Figure 2c. *Schematic representation of the radiofrequency pulse followed by the free induction decay (acquisition time) and pulse delay.*

the spectrum represents 10 ppm (Figure 1). If the presence of carbon atoms that absorb further downfield is suspected, the sweep width is increased accordingly. For a sweep width of 5000 Hz (250 ppm) on the Varian CFT 20 instrument, each division of the scale represents 6.25 ppm (see Figure 3).

As a result of the large sweep width and the sharpness of the peaks, impurities (or mixtures) are readily detected. Even stereoisomers that are difficult to separate, or detect by other procedures, show up as discrete peaks.

Routine sample sizes for ^{13}C NMR are 0.5-2.0 g in 10-, 12-, or 15-mm OD tubes, and these samples require only a few (~ 5-10) pulses. Samples of 100-200 mg can be handled in the familiar 5-mm OD tubes if solubility is no problem, but the number of pulses required may take several hours. Samples of 10-100 mg may be handled in the conventional microcells with a volume of about 100 μl. By using a 1.7-mm capillary cell, samples of 1 mg or less can be handled if solubility is no problem.

Several texts on ^{13}C NMR spectrometry for the organic chemist are available.[1-5] A one-volume collection of ^{13}C spectra with peak assignments is very useful.[6] The Sadtler collection now includes ^{13}C spectra.[7] The principles and practice of Fourier Transform NMR are discussed[8-10], and a set of spectral problems in ^{13}C spectrometry is available.[11] Breitmaier et al. have recently assembled a three-volume set, *Atlas of Carbon-13 NMR Data.*[12]

II. INTERPRETATION OF ^{13}C SPECTRA (PEAK ASSIGNMENTS)

How are specific peaks assigned in a noise decoupled ^{13}C spectrum, such as Figure 1b? First, we do have reference material and empirical relationships to correlate chemical shift with structure. (See discussion under Chemical Shifts in Section III for the major chemical classes, and Appendices B and C.) However, we lack coupling information because of noise decoupling of protons, and we cannot depend on peak integrations for the following reasons.

Repetitive scan (continuous wave) ^{1}H spectrometry usually results in a satisfactory relationship between integrated peak areas and the number of nuclei under those areas because there is sufficient time between irradiations of an individual proton for relaxation to occur. Satisfactory integrations would also result for most ^{13}C spectra under these conditions, but the time needed for the number of scans required is prohibitive for routine work.

In routine Fourier Transform runs, the repetitive pulses are spaced at intervals of 0.1-1 second (acquisition time) during which the signal is averaged and stored. Under these conditions, ^{13}C nuclei, whose relaxation times (T_1) vary over a wide range, are not equally relaxed between pulses, and the resulting peak areas do not integrate to give the correct number of carbon atoms. Very long pulse delays (following the acquisition period) can be used, but the time required limits this technique to special situations. Furthermore, there is another complication: the Nuclear Overhauser Enhancement (see above) is not the same for all nuclei, and this results in further loss of quantitation.

However, one advantage does result. It is usually possible by "eyeballing" a ^{13}C spectrum to recognize the nuclei that do not bear protons by their low intensity (peaks 1 and 2 in Figure 1b). The common spin lattice relaxation mechanism for ^{13}C results from dipole-dipole interaction with directly attached protons. Thus, nonprotonated carbon atoms often have longer relaxation times and give small peaks. It is therefore often possible readily to detect carbonyl groups (except formyl), nitriles, nonprotonated

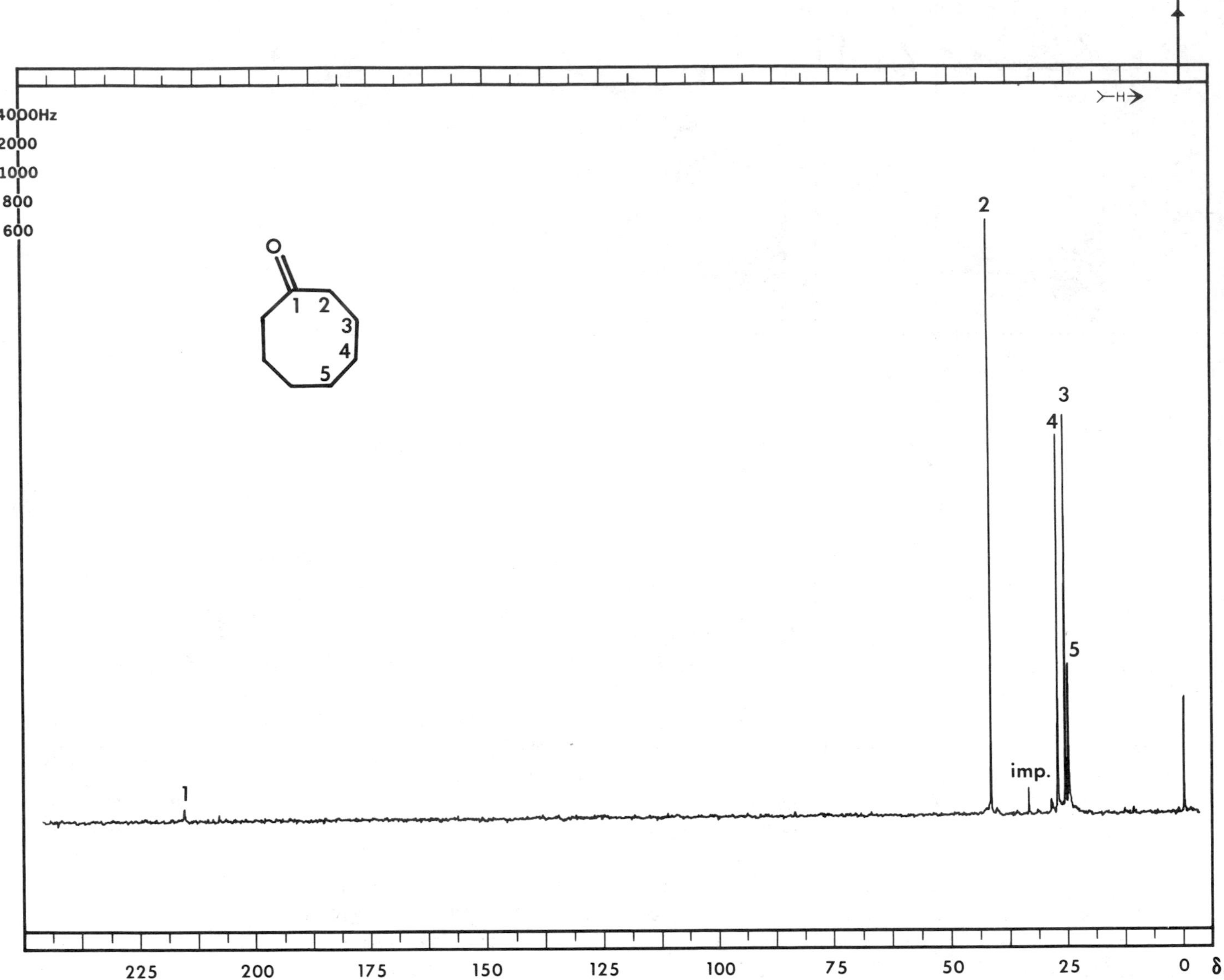

Figure 3a. ^{13}C *NMR spectrum of cyclooctanone with proton coupling completely removed by noise decoupling.* $CDCl_3$ *solvent, 20 MHz, sweepwidth 5000 Hz.*

olefinic and acetylenic carbon atoms, and other quaternary carbon atoms. However, care must be taken to allow a sufficient number of pulses or a long enough interval between pulses (to compensate for the long T_1) so that these weak signals are not completely lost in the baseline noise. In Figure 1c, a 10-second interval (pulse delay) was used to increase the relative intensities of peaks 1 and 2).

The most useful techniques for assigning peaks, in addition to structure-shift correlations, are off-resonance decoupling, selective proton decoupling, the use of chemical shift reagents, and deuterium substitution. The concept of chemical shift equivalence (Chapter 4) should always be kept in mind. The use of spin relaxation times is beyond the scope of this chapter.

In the noise decoupled spectrum of diethyl phthalate (Figure 1b),

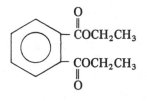

$$C_6H_4(\overset{O}{\overset{\|}{C}}OCH_2CH_3)_2$$

we can assign the very small peak at 167.75 ppm to the 2 equivalent C=O groups, the very small peak at 132.85 ppm to the equivalent substituted aromatic carbons, the large peaks at 131.33 ppm and 129.19 ppm to the remaining aromatic carbons, the medium peaks at 61.63 ppm to the 2 equivalent CH_2 groups, and the medium peak at 14.15 ppm to the 2 equivalent CH_3 groups. These assignments can be made on the basis of Appendices B and C and on the assumption that the quaternary carbon atoms are responsible for

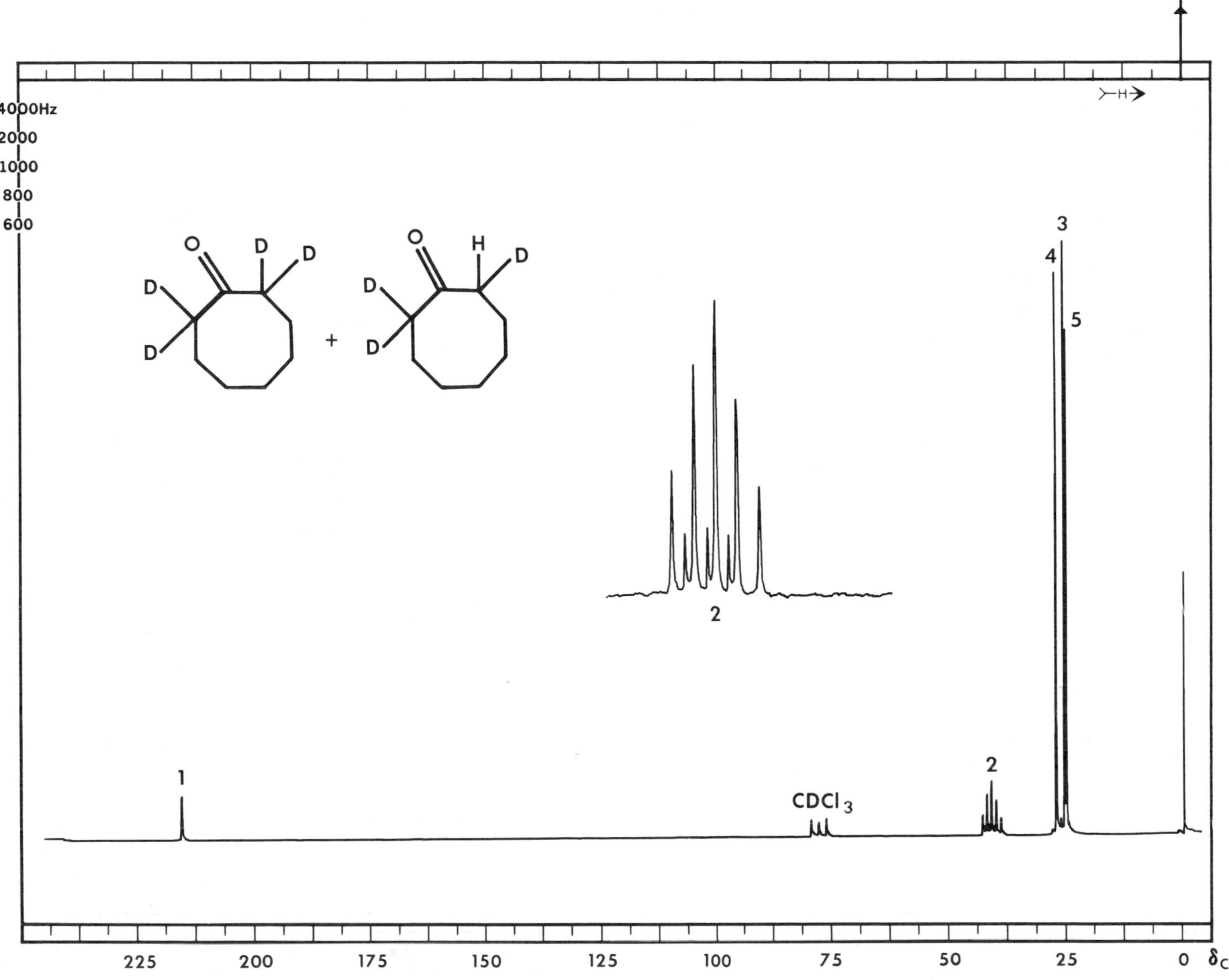

Figure 3b. *Noise decoupled ^{13}C spectrum of tetradeuterated (major component) and trideuterated (minor component) cyclooctanone. $CDCl_3$ solvent, 20 MHz, sweepwidth 5000 Hz. For insert, sweepwidth 1000 Hz.*

the weak peaks. Their relative intensity can be increased by inserting a pulse delay (an interval between the acquisition period and the next pulse) as in Figure 1c. Note that the 10-second pulse delay nearly equalizes the intensities of all the peaks except for those representing the quaternary carbons.

OFF-RESONANCE DECOUPLING

We have seen that proton noise decoupling simplifies the spectrum and increases peak heights, but with the loss of coupling information. Off-resonance decoupling (Figure 1d) gives a simplified spectrum yet retains "residual" $^{13}C{-}H$ coupling (J^r) information. Off-resonance decoupling is achieved by offsetting the high-power proton decoupler by about 1000-2000 Hz upfield or about 2000-3000 Hz downfield from the *proton* frequency of TMS without using the noise generator; that is, we irradiate upfield or downfield of the usual (1000 Hz sweepwidth) *proton* spectrum. This results in residual coupling from protons directly bonded to the ^{13}C atoms, while long-range coupling is usually lost. The observed residual coupling (J^r), which is determined by the amount of offset and the power of the decoupler, is smaller than the true coupling J.

$$J^r \simeq \frac{2\pi J \Delta\nu}{\gamma H_2}$$

where γ is the magnetogyric ratio for protons, $\Delta\nu$ is the difference between the resonance frequency of the proton of interest and the decoupler frequency, and H_2 is the

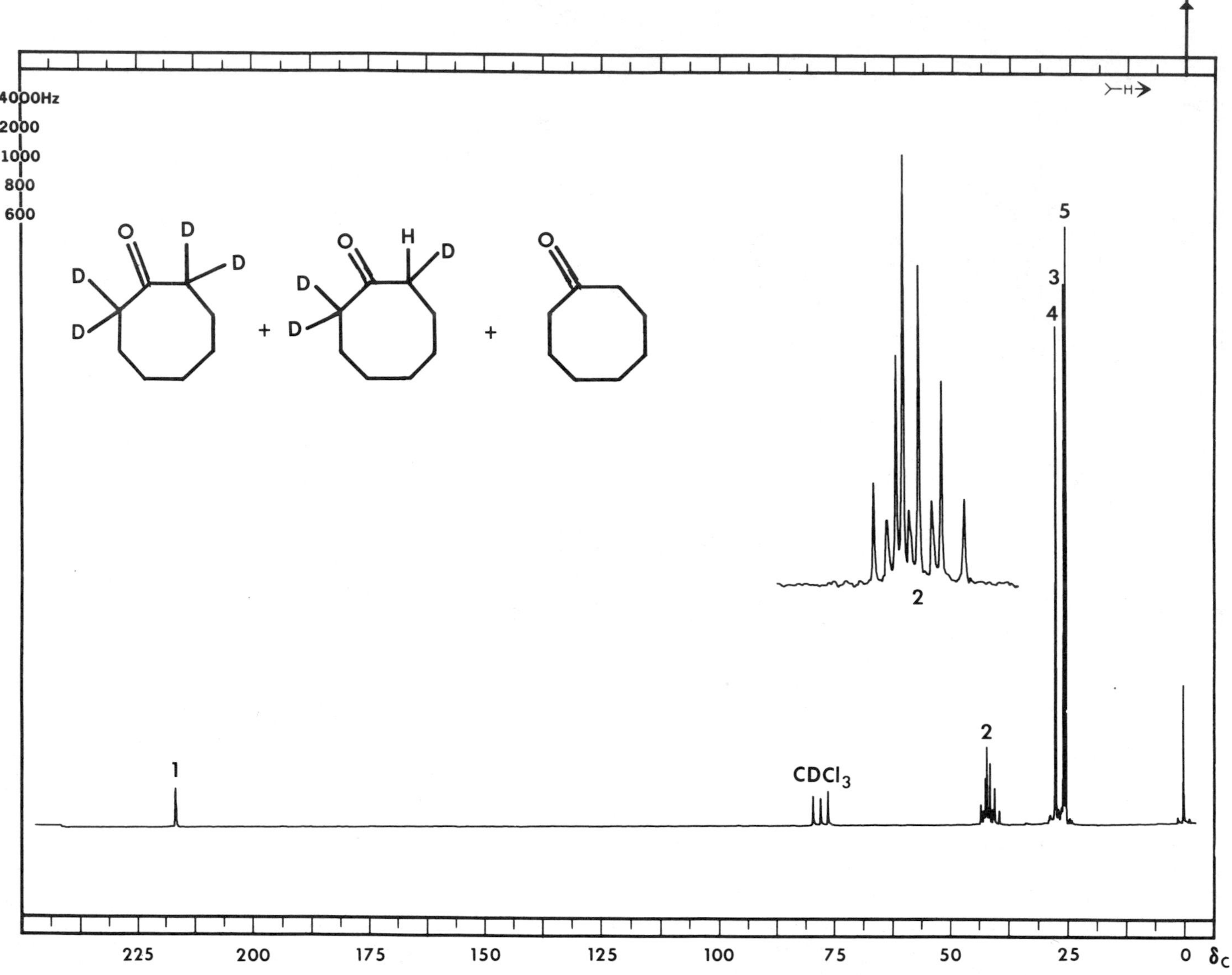

Figure 3c. *Noise-decoupled ^{13}C spectrum of deuterated sample undeuterated cyclooctanone. $CDCl_3$ solvent, 20 MHz, sweep width 5000 Hz. For insert, sweepwidth 1000 Hz.*

strength of the rotating magnetic field generated by the decoupler frequency. Thus, the multiplicities of the ^{13}C bands can be readily observed, usually without overlap with other ^{13}C bands. Methyl carbon atoms appear as quartets, methylenes as triplets (or as pair of doublets if the protons are not equivalent and their coupling constants are sufficiently different), methines as doublets, and quaternary carbon atoms as singlets.* The procedure, of course, also gives a proton count to corroborate the proton spectrum. A discrepancy between the number of protons obtained from the off-resonance decoupled ^{13}C spectrum and the molecular formula obtained from the mass spectrum results from elements of symmetry whose presence can be correlated with molecular structure.

In Figure 1d, only the doublets of peaks 3 and 4 overlap. The residual couplings become slightly larger as the frequency increases; in this case, the irradiation is upfield of the proton frequency of TMS. It is convenient to superimpose the off-resonance decoupled spectrum on the noise-decoupled spectrum on a light-table. Thus, we can see that the quaternary carbons (singlets) and the center peak of the CH_2 triplets are superimposed on the corresponding noise-decoupled singlets, whereas the CH doublet and CH_3

*In practice, the outer peaks are usually weaker than expected from the theoretical 1:2:1 or 1:3:3:1 peak height ratios because of decoupler field inhomogeneity. R. Freeman, et al., *J. Am. Chem. Soc.* **100**, 5637 (1979).

quartets "straddle" the corresponding noise-decoupled peaks.

A second-order effect is discussed in Section VI and the use of off-resonance decoupling to correlate a particular ^{13}C peak with the shift position of its attached proton is discussed in Section V.

The off-resonance decoupled spectrum of diethyl phthalate (Figure 1d) allows us to confirm the assignments made on the basis of chemical shifts and peak heights. The multiplicity of peak 1 is unchanged. Peak 2 is buried under peak 3a. Peaks 3 and 4 are overlapping doublets. Peak 5 is a triplet, and peak 6 a quartet. We count 6 carbon atoms and 7 protons, which, given an element of symmetry, corroborates the molecular formula $C_{12}H_{14}O_4$ and the *ortho* substitution. For this simple molecule, the nondecoupled spectrum (Figure 1a) presents no problem, but in more complex molecules, overlapping multiplets are often difficult to interpret.

SELECTIVE PROTON DECOUPLING

When a specific proton is irradiated at its exact frequency at a lower power level than is used for off-resonance decoupling, the absorption of the directly bonded ^{13}C becomes a singlet, while the other ^{13}C absorptions show residual coupling. This technique has been used for peak assignment, but satisfactory results depend on finding the precise frequency of the proton and the appropriate power level for decoupling it, which can be difficult.

CHEMICAL SHIFT REAGENTS

The lanthanide chemical shift reagents (see Chapter 4) can be used to "spread out" a ^{13}C spectrum in the same way as they are used in ^1H studies. Two useful results may be obtained: coincident peaks may separate and the proton shifts are also spread out; selective proton decoupling is thus facilitated. However, facile analogies between ^{13}C and ^1H effects are risky.

DEUTERIUM SUBSTITUTION

Deuteration of a carbon results in a dramatic diminution of its ^{13}C signal in a noise-decoupled spectrum for the following reasons. Since deuterium has a spin number of 1 and a magnetic moment 15% that of ^1H, it will split the ^{13}C absorption into 3 lines (ratio 1:1:1) with a J value equal to $0.15 \times J_{CH}$. Furthermore, T_1 for $^{13}C{-}D$ is longer than that for $^{13}C{-}H$ because of decreased dipole-dipole relaxation. Finally, the Nuclear Overhauser Effect is lost, since there is no irradiation of deuterium.* A separate peak may also be seen for any residual $^{13}C{-}H$, since the isotope effect usually results in a slight upfield shift of the $^{13}C{-}D$ absorption (~ 0.2 ppm per D atom). The isotope effect may also slightly shift the absorption of the carbon atoms once removed from the deuterated carbon.

The net result depends on the extent of deuteration. A noise-decoupled spectrum of cyclooctanone is shown in Figure 3a, and we note the following assignments.

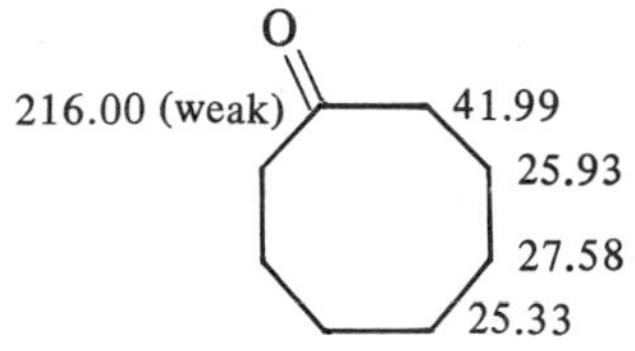

O
216.00 (weak)
41.99
25.93
27.58
25.33

On deuteration of cyclooctanone, the α-^{13}C peak at δ 41.99 in the undeuterated compound is replaced by a very weak cluster of peaks (Figure 3b). In the expanded insert, we can pick out the five larger peaks of a 1:2:3:2:1 quintet (representing the CD_2 absorption) and a smaller 1:1:1 triplet (representing the CHD absorption) of the two compounds.

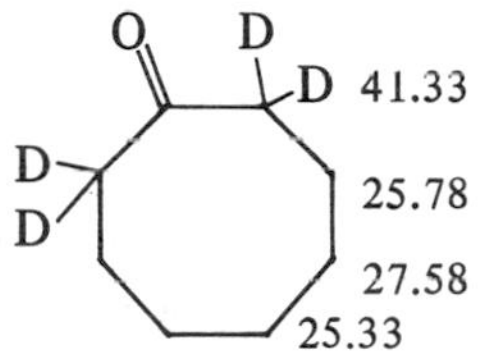

O D
D 41.33
D
D
25.78
27.58
25.33

Major component

and

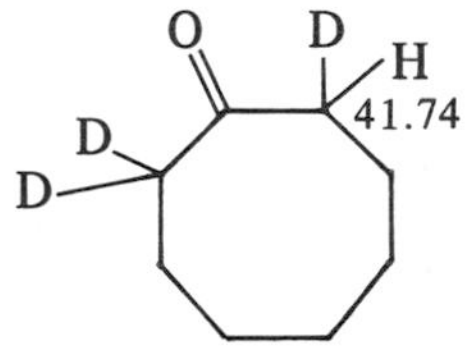

O D
H
41.74
D
D

Minor component

Note the slight upfield shift of the α- and β-carbon atoms in the deuterated compounds.

*The same explanation also accounts for the relatively weak signal shown by deuterated solvents. Deuterated chloroform, $CDCl_3$, shows a 1:1:1 *triplet*, deuterated *p*-dioxane a 1:2:3:2:1 *quintet*, and deuterated DMSO $(CD_3)_2SO$, a 1:3:6:7:6:3:1 *septet* in accordance with the $n2I + 1$ rule (Chapter 4).

Figure 3c is the spectrum of a deuterated cyclooctanone sample that contains approximately 15% of residual non-deuterated cyclooctanone in addition to the *tri-* and *tetra-*deuterated compounds. The CH_2 singlet is clearly seen in the insert in addition to the quintet and the triplet.

CHEMICAL SHIFT EQUIVALENCE

The definition of chemical shift equivalence given for protons also applies to carbon atoms: interchangeability by a symmetry operation or by a rapid mechanism. The presence of equivalent carbon atoms (or fortuitous coincidence of shift) in a molecule results in a discrepancy between the apparent number of peaks and the actual number of carbon atoms in the molecule.

Thus, ^{13}C atoms of the methyl groups in *t*-butyl alcohol (Figure 4) are equivalent by rapid rotation in the same sense in which the protons of a methyl group are equivalent. The ^{13}C spectrum of *t*-butyl alcohol shows 2 peaks, one much larger than the other, but not necessarily exactly 3 times as large; the carbinyl carbon peak (quaternary) is much less than 1/3 the intensity of the peak representing the carbons of the methyl groups.

In the chiral molecule 2,2,4-trimethyl-1,3-pentanediol (Figure 5), we note that CH_3a and CH_3b are not equivalent, and 2 peaks are seen. Even though the 2 methyl groups labelled *c* are not equivalent, they fortuitously show only 1 peak. Two peaks may be seen at higher resolution, or with the help of a shift reagent.

In Chapter 4, Section VII, we noted that the CH_3 protons of $(CH_3)_2NCH{=}O$ gave separate peaks at room temperature, but became chemical shift equivalent at ~ 123°. Of course, the ^{13}C peaks show the same behavior.

III. CHEMICAL SHIFTS

In this section, chemical shifts will be discussed under the headings of the common chemical classes of organic compounds. As noted earlier, the range of shifts generally encountered in routine ^{13}C studies is about 240 ppm.

As a first reassuring statement, we can say that trends in chemical shifts of ^{13}C are somewhat parallel to those of ^1H, so that some of the "feeling" for ^1H spectra may carry over to ^{13}C spectra. Furthermore, the concept of *additivity of substituent effects* is useful for both spectra. Shifts are related mainly to hybridization and substituent electronegativity; solvent effects are important in both spectra. Chemical shifts for ^{13}C are affected by substitutions as far removed as the δ position; in the benzene ring, pronounced shifts for ^{13}C are caused by substituents at the ortho, meta, and para positions. ^{13}C chemical shifts are also moved significantly upfield by steric compression. (See the γ-gauche effect, below.) Upfield shifts, as much as several ppm, may occur on dilution. Hydrogen-bonding effects with polar solvents may cause downfield shifts.

Appendix B gives credence to the statement that ^{13}C chemical shifts somewhat parallel those of ^1H, but we note some divergences that are not readily explainable and require development of another set of interpretive skills. In general, in comparison with ^1H spectra, it seems more difficult to correlate ^{13}C shifts with substituent electronegativity.

As in other types of spectrometry, peak assignments are made on the basis of reference compounds. Reference material for many classes of compounds is rapidly accumulating in the literature. The starting point is a general correlation chart for chemical shift regions of ^{13}C atoms in the major chemical classes (See Appendices B and C);

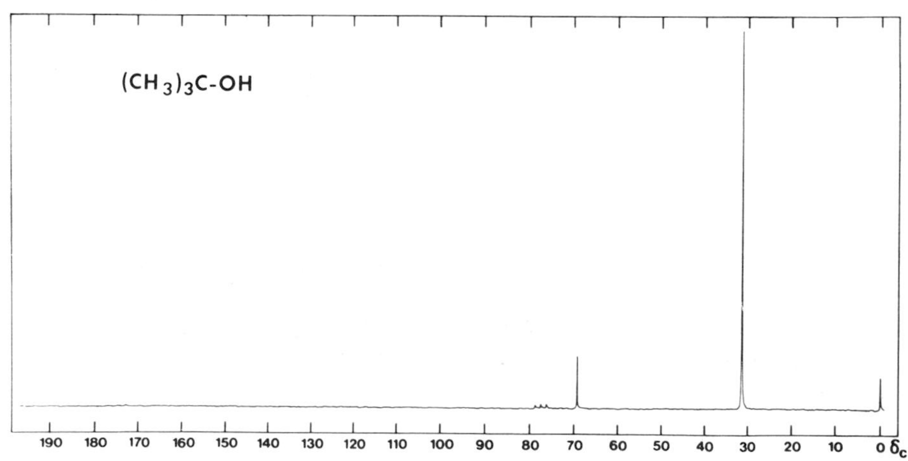

Figure 4. *Noise-decoupled ^{13}C spectrum of* t-butyl alcohol. *Solvent $CDCl_3$. 25.2 MHz, 5000 Hz sweep width. From reference 6, spectrum No. 88, with permission.*

then, minor changes within these regions are correlated with structure variations in the particular chemical class. The chemical shift values in the following Tables must not be taken too literally because of the use of various solvents and concentration. (Furthermore, much of the early work used various reference compounds, and the values were corrected to give ppm from TMS.) For example, the C=O absorption of acetophenone in $CDCl_3$ appears at 2.4 ppm further downfield than in CCl_4; the effect on the other carbon atoms of acetophenone ranges from 0.0-1.1 ppm.

ALKANES

Linear and Branched Alkanes

We know from the general correlation chart (Appendix C) that alkane groups unsubstituted by heteroatoms absorb downfield from TMS to about 60 ppm. (Methane absorbs at 2.5 ppm upfield from TMS.) Within this range, we can predict the chemical shifts of individual ^{13}C atoms in a straight chain or branched chain hydrocarbon from the data in Table I and the formula given below.

This table shows the additive shift parameters (A) in hydrocarbons: the α effect of +9.1 (downfield), the β effect of +9.4 ppm, the γ effect of −2.5 (upfield), the δ effect of +0.3, the ϵ effect of +0.1, and the corrections for branching effects. The calculated (and observed) shifts for the carbon atoms of 3-methylpentane are:

$$\begin{array}{ccccc} & & \overset{6}{CH_3}\ \ +19.3\ (18.6) & & \\ & & | & & \\ \overset{1}{CH_3} - & \overset{2}{CH_2} - & \overset{3}{CH} - & \overset{4}{CH_2} - & \overset{5}{CH_3} \\ +11.3 & +29.5 & +36.2 & & \\ (+11.3) & (+29.3) & (+36.7) & & \end{array}$$

Calculations of shift are made from the formula: $\delta = -2.5 + \Sigma nA$, where δ is the predicted shift for a carbon atom; A is the additive shift parameter; and n is the number of carbon atoms for each shift parameter (−2.5 is the shift of the ^{13}C of methane). Thus, for carbon atom 1, we have 1 α-carbon, 1 β-carbon, 2 γ-carbons, and 1 δ carbon:

$$\delta_1 = -2.5 + (9.1 \times 1) + (9.4 \times 1) + (-2.5 \times 2) + (0.3 \times 1) = +11.3$$

Carbon atom 2 has 2 α-carbons, 2 β-carbons and 1 γ-carbon. Carbon atom 2 is a 2° carbon with a 3° carbon attached [2°(3°) = −2.5].

$$\delta_2 = -2.5 + (9.1 \times 2) + (9.4 \times 2) + (-2.5 \times 1) + (-2.5 \times 1) = 29.5$$

Carbon atom 3 has 3 α-carbons and 2 β-carbons, and it is a 3° atom with two 2° atoms attached [3°(2°) = −3.7]. Thus,

$$\delta_3 = -2.5 + (9.1 \times 3) + (9.4 \times 2) + (-3.7 \times 2) = +36.2$$

Carbon atom 6 has 1 α-carbon, 2 β-carbons, and 2 γ-carbons, and it is a 1° atom with a 3° atom attached [1°(3°) = −1.1]. Thus,

$$\delta_6 = -2.5 + (9.1 \times 1) + (9.4 \times 2) + (-2.5 \times 2) + (-1.1 \times 1) = +19.3$$

The agreement for such calculations is very good. It is essential that the reference compounds used for such additivity calculations be structurally similar to the compound of interest. Another useful calculation has been given.* The ^{13}C upfield γ-shift due to the γ-carbon has been attributed to the steric compression of a gauche interaction but has no counterpart in ^1H spectra. It accounts, for example, for the upfield position of an axial methyl substituent on a conformationally rigid cyclohexane ring, relative to an equatorial methyl, and for the upfield shift of the γ carbon atoms of the ring.

Table I. ^{13}C Shift Parameters in Some Linear and Branched Hydrocarbons

^{13}C Atoms	Shift (ppm) (A)
α	+9.1
β	+9.4
γ	−2.5
δ	+0.3
ϵ	+0.1
1° (3°)[a]	−1.1
1° (4°)[a]	−3.4
2° (3°)[a]	−2.5
2° (4°)	−7.2
3° (2°)	−3.7
3° (3°)	−9.5
4° (1°)	−1.5
4° (2°)	−8.4

[a]The notations 1° (3°) and 1° (4°) denote a CH_3 group bound to a R_2CH group and to a R_3C group, respectively. The notation 2° (3°) denotes a RCH_2 group bound to a R_2CH group—and so forth.

*L. P. Lindeman and J. Q. Adams, *Anal. Chem.* **43**, 1245 (1971).

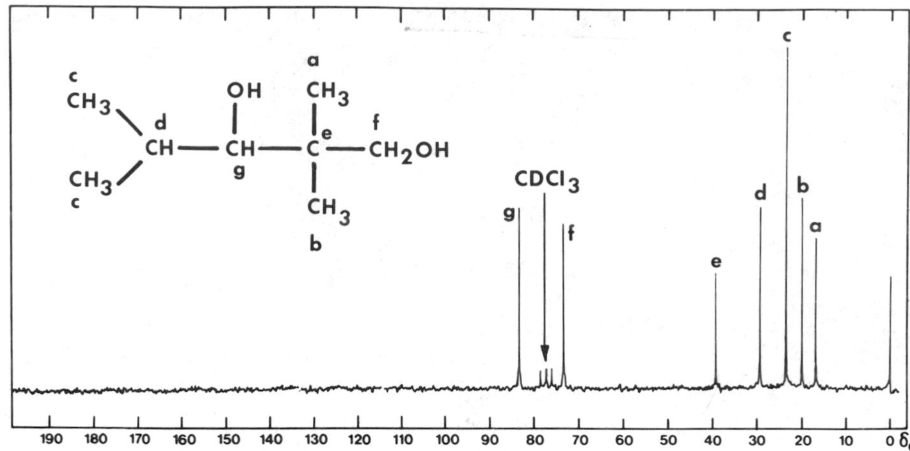

Figure 5. *Noise decoupled ^{13}C spectrum of 2,2,4-trimethyl-1-3-pentanediol. Solvent $CDCl_3$. 25.2 MHz, 5000 Hz sweepwidth. From reference 6, spectrum no. 324, with permission.*

Table II lists the shifts in some linear and branched alkanes.

EFFECT OF SUBSTITUENTS ON ALKANES

Table III shows the effects of a substituent on linear and branched alkanes. The effect on the α-carbon parallels electronegativity of the substituent except for bromine and iodine. The effect at the β-carbon seems fairly constant for all the substituents except for the carbonyl, cyano, and nitro groups. The upfield shift at the γ-carbon results (as above) from steric compression of a *gauche* interaction. For Y = N, O, and F, there is also an upfield shift with Y in the *anti* conformation, attributed to hyperconjugation.

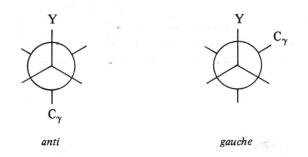

Y Y

C_γ C_γ

anti *gauche*

Table II. ^{13}**C Shifts for Some Linear and Branched-Chain Alkanes (ppm from TMS)**

Compound	C-1	C-2	C-3	C-4	C-5
Methane	−2.3				
Ethane	5.7				
Propane	15.8	16.3	15.8		
Butane	13.4	25.2	25.2		
Pentane	13.9	22.8	34.7	22.8	13.9
Hexane	14.1	23.1	32.2	32.2	23.1
Heptane	14.1	23.2	32.6	29.7	32.6
Octane	14.2	23.2	32.6	29.9	29.9
Nonane	14.2	23.3	32.6	30.0	30.3
Decane	14.2	23.2	32.6	31.1	30.5
Isobutane	24.5	25.4			
Isopentane	22.2	31.1	32.0	11.7	
Isohexane	22.7	28.0	42.0	20.9	14.3
Neopentane	31.7	28.1			
2,2-Dimethylbutane	29.1	30.6	36.9	8.9	
3-Methylpentane	11.5	29.5	36.9 (18.8, 3-CH_3)		
2,3-Dimethylbutane	19.5	34.3			
2,2,3-Trimethylbutane	27.4	33.1	38.3	16.1	
2,3-Dimethylpentane	7.0	25.3	36.3 (14.6, 3−CH_3)		

From Table III, the approximate shifts for the carbon atoms of, for example, 3-pentanol, may be calculated from the values for *n*-pentane in Table II; that is, the

$$\overset{\gamma}{CH_3}-\overset{\beta}{CH_2}-\underset{OH}{\overset{\alpha}{CH}}-\overset{\beta}{CH_2}-\overset{\gamma}{CH_3}$$

increment for the functional group in Table III is added to the appropriate value in Table II as follows.

	Calculated	Found (See Table IX)
C_α	34.7 + 41 = 75.8	73.8
C_β	22.8 + 8 = 30.8	30.0
C_γ	13.9 − 5 = 8.9	10.1

Table III. Incremental Substituent Effects (ppm) on Replacement of H by Y in Alkanes. Y is Terminal or Internal*

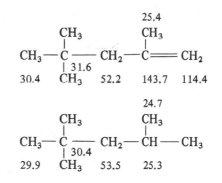

$\cdots\overset{\gamma}{}\;\underset{\beta}{}\;\overset{\alpha}{}\;Y$ (terminal) $\qquad \cdots\overset{\gamma}{}\;\underset{\beta}{}\;\overset{Y}{\underset{\alpha}{}}\;\underset{\beta}{}\;\overset{\gamma}{}\cdots$ (internal)

Y	α terminal	α internal	β terminal	β internal	γ
CH_3	+ 9	+ 6	+10	+ 8	−2
$CH{=}CH_2$	+20		+ 6		−0.5
$C{\equiv}CH$	+ 4.5		+ 5.5		−3.5
COOH	+21	+16	+ 3	+ 2	−2
COO^-	+25	+20	+ 5	+ 3	−2
COOR	+20	+17	+ 3	+ 2	−2
COCl	+33	+28		+ 2	
$CONH_2$	+22		+ 2.5		−0.5
COR	+30	+24	+ 1	+ 1	−2
CHO	+31		0		−2
Phenyl	+23	+17	+ 9	+ 7	−2
OH	+48	+41	+10	+ 8	−5
OR	+58	+51	+ 8	+ 5	−4
OCOR	+51	+45	+ 6	+ 5	−3
NH_2	+29	+24	+11	+10	−5
NH_3^+	+26	+24	+ 8	+ 6	−5
NHR	+37	+31	+ 8	+ 6	−4
NR_2	+42		+ 6		−3
NR_3^+	+31		+ 5		−7
NO_2	+63	+57	+ 4	+ 4	
CN	+ 4	+ 1	+ 3	+ 3	−3
SH	+11	+11	+12	+11	−4
SR	+20		+ 7		−3
F	+68	+63	+ 9	+ 6	−4
Cl	+31	+32	+11	+10	−4
Br	+20	+25	+11	+10	−3
I	− 6	+ 4	+11	+12	−1

From F.W. Wehrli and T. Wirthlin, *Interpretation of Carbon-13 NMR Spectra.*[3]

*Add these increments to the shift value of the appropriate carbon atom in Table II or to the shift value calculated from Table I.

Table IV. Chemical Shifts of Cycloalkanes (ppm from TMS)

C_3H_6	−3.5	C_7H_{14}	28.4
C_4H_8	22.4	C_8H_{16}	26.9
C_5H_{10}	25.6	C_9H_{18}	26.1
C_6H_{12}	26.9	$C_{10}H_{20}$	25.3

CYCLOALKANES AND SATURATED HETEROCYCLICS

The chemical shifts of the CH_2 groups in monocyclic alkanes are given in Table IV.

The striking feature here is the strong upfield shift of cyclopropane, analogous to the upfield shift of its proton absorptions.

Each ring skeleton has its own set of shift parameters, but a detailed listing of these is beyond the scope of this text. Rough estimates for substituted rings can be made with the substitution increments in Table III.

As mentioned above, one of the striking effects in rigid cyclohexane rings is the upfield shift caused by the *γ-gauche* steric compression. Thus an axial methyl group at C-1 causes an upfield shift of several ppm at C-3 and C-5.

Table IVa presents chemical shifts for several saturated heterocyclics.

ALKENES

The sp^2 carbon atoms of alkenes substituted only by alkyl groups absorb in the range of ~110-150 ppm downfield from TMS. The double bond has a rather small effect on the shifts of the sp^3 carbons in the molecule as the following comparisons demonstrate.

CH_3 (30.4)—$C(CH_3)_2$ (31.6)—CH_2 (52.2)—$C(CH_3)$ (143.7; CH_3 25.4)$=CH_2$ (114.4)

CH_3 (29.9)—$C(CH_3)_2$ (30.4)—CH_2 (53.5)—$CH(CH_3)$ (25.3; CH_3 24.7)—CH_3

The methyl signal of propene is at 18.7 ppm, and of propane at 15.8 ppm. In *cis*-2-butene, the methyl signals are at 12.1 ppm, compared with 17.6 ppm in *trans*-2-butene, because of the γ effect. (For comparison, the methyl signals of butane are at 13.4 ppm.) Note the γ-effect on one of the geminal methyl groups in 2-methyl-2-butene (Table V).

In general, the terminal $=CH_2$ group (usually a triplet in an off-resonance decoupled spectrum) absorbs upfield from an internal $=CH-$ group, and *cis* $-CH{=}CH-$ signals are upfield from those of corresponding *trans* groups. Calculations of approximate shifts can be made from the following parameters where α, β, and γ represent substituents on the same end of the double bond as the olefinic

Table IVa. Chemical Shifts for Saturated Heterocyclies (ppm from TMS, neat)

Unsubstituted

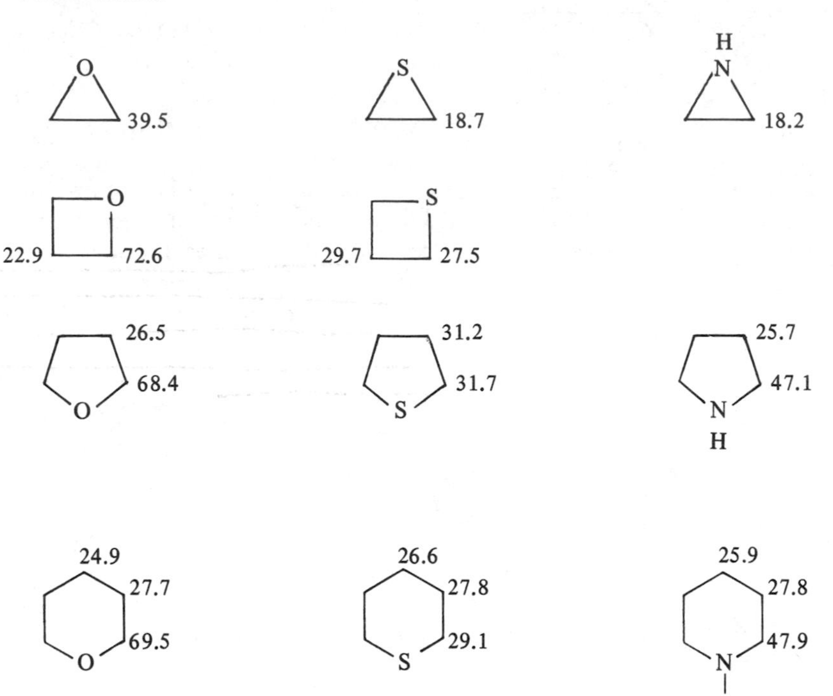

Substituted

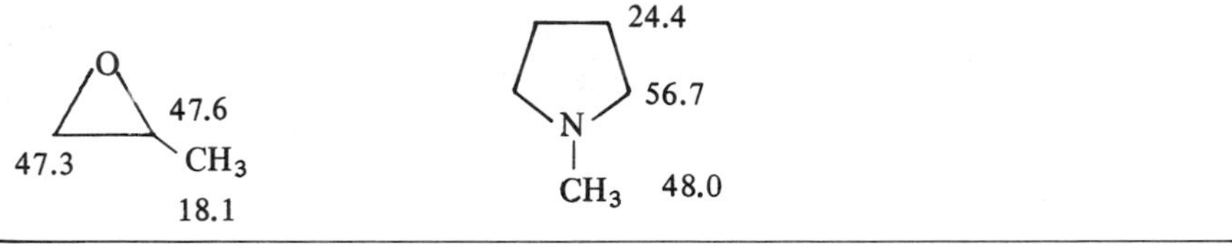

carbon of interest, and α', β', and γ' represent substituents on the far side.

α	+10.6
β	+ 7.2
γ	− 1.5
α'	− 7.9
β'	− 1.8
γ'	− 1.5
cis correction	− 1.1

These parameters are added to 123.3 ppm, the shift for ethylene. We can calculate the values for *cis* - 3-methyl-2-pentene as follows.

$$\overset{\beta}{H_3C}-\overset{\alpha}{CH_2}-\underset{3}{\overset{\overset{\alpha}{CH_3}}{C}}=\underset{2}{\overset{H}{C}}-\underset{1}{\overset{\alpha'}{CH_3}} \qquad \overset{\beta'}{H_3C}-\overset{\alpha'}{CH_2}-\underset{3}{\overset{\overset{\alpha'}{CH_3}}{C}}=\underset{2}{\overset{H}{C}}-\underset{1}{\overset{\alpha}{CH_3}}$$

(carbons numbered 5 4 3 2 1)

$$\delta_{C-3} = 123.3 + (2 \times 10.6) + (1 \times 7.2) + (1 \times -7.9) - 1.1 = 142.7 \text{ ppm}$$

$$\delta_{C-2} = 123.3 + (1 \times 10.6) + (2 \times -7.9) + (1 \times -1.8) - 1.1 = 115.2 \text{ ppm}$$

The measured values are C-3 = 137.2, C-2 = 116.8. The agreement is fair.

Olefinic carbons in polyenes are treated as though they were alkane carbon substituents on one of the double bonds. Thus, in calculating the shift of C-2 in 1,4-pentadiene, C-4 is treated like a β-sp^3 carbon atom.

Representative olefins are presented in Table V.

Table V. Alkene Chemical Shifts (ppm from TMS)

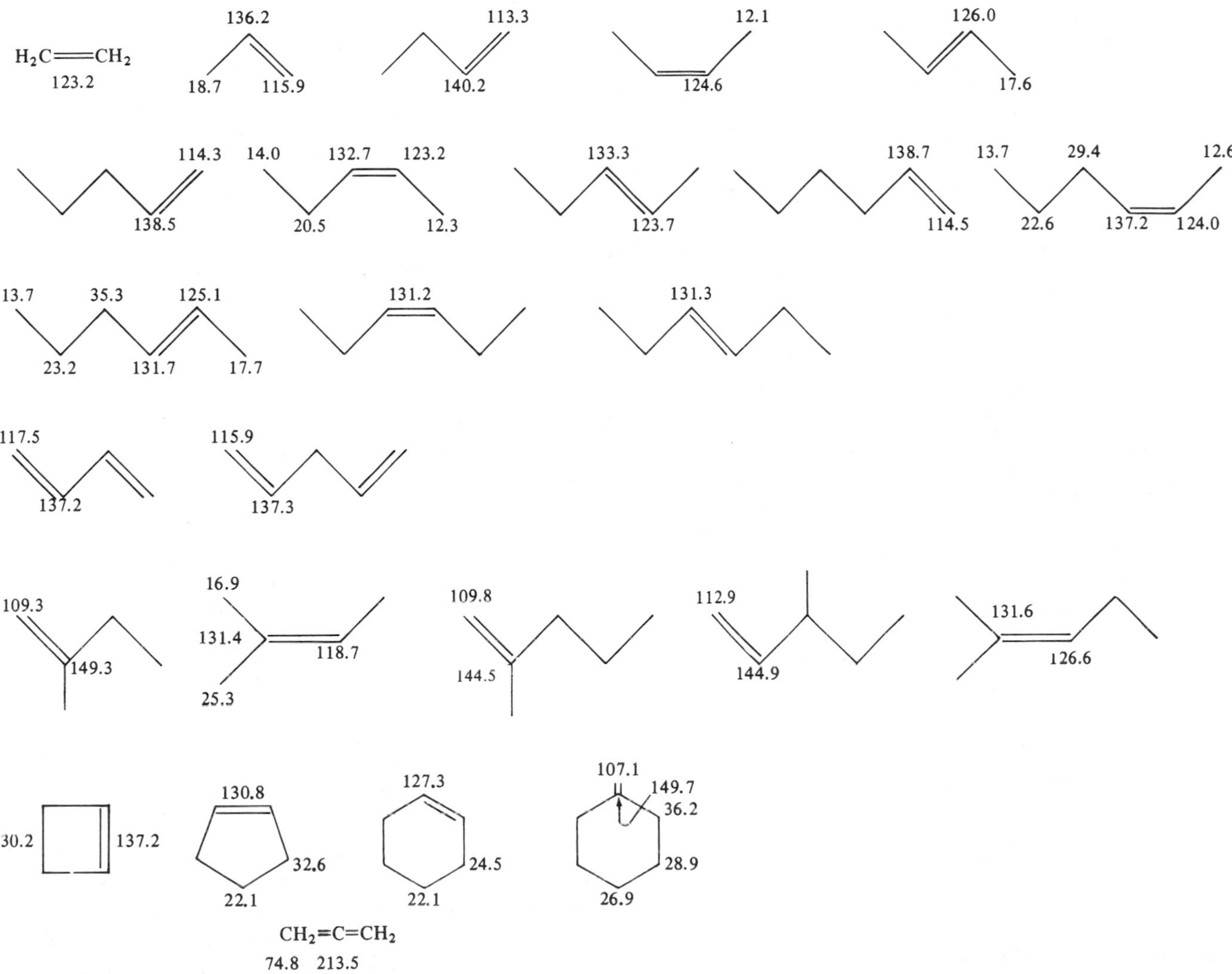

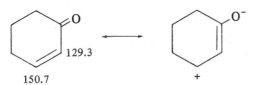

There are no simple rules to handle polar substituents on an olefinic carbon. The shifts for vinyl ethers can be rationalized on the basis of electron density of the contributory structures

$$\underset{84.2\ \ 153.2}{CH_2{=}CH{-}O{-}CH_3} \longleftrightarrow \overset{-}{C}H_2{-}CH{=}O^{+}{-}CH_3$$

as can the shifts for α,β-unsaturated ketones.

O

129.3

150.7

$\longleftrightarrow$

O^-

+

The same rationalization applies to the proton shifts in these compounds.

Shifts for several substituted alkenes are presented in Table Va.

The central carbon atom (=C=) of alkyl-substituted allenes absorbs far downfield in the range of ~ 200-215 ppm, whereas the terminal atoms (C=C=C) absorb upfield in the range of ~ 75-97 ppm.

ALKYNES

The sp carbons of alkynes substituted only by alkyl groups absorb in the range ~ 65-90 ppm (Table VI). The triple bond shifts the sp^3 carbons directly attached about 5-15 ppm upfield relative to the corresponding alkane. The terminal ≡CH absorbs upfield from the internal ≡CR. Off-resonance decoupling gives a doublet for the terminal ≡CH. Alkyne carbon atoms with a polar group directly attached absorb from ~ 20-95 ppm.

Table Va. Chemical Shifts of Substituted Alkenes (ppm from TMS)

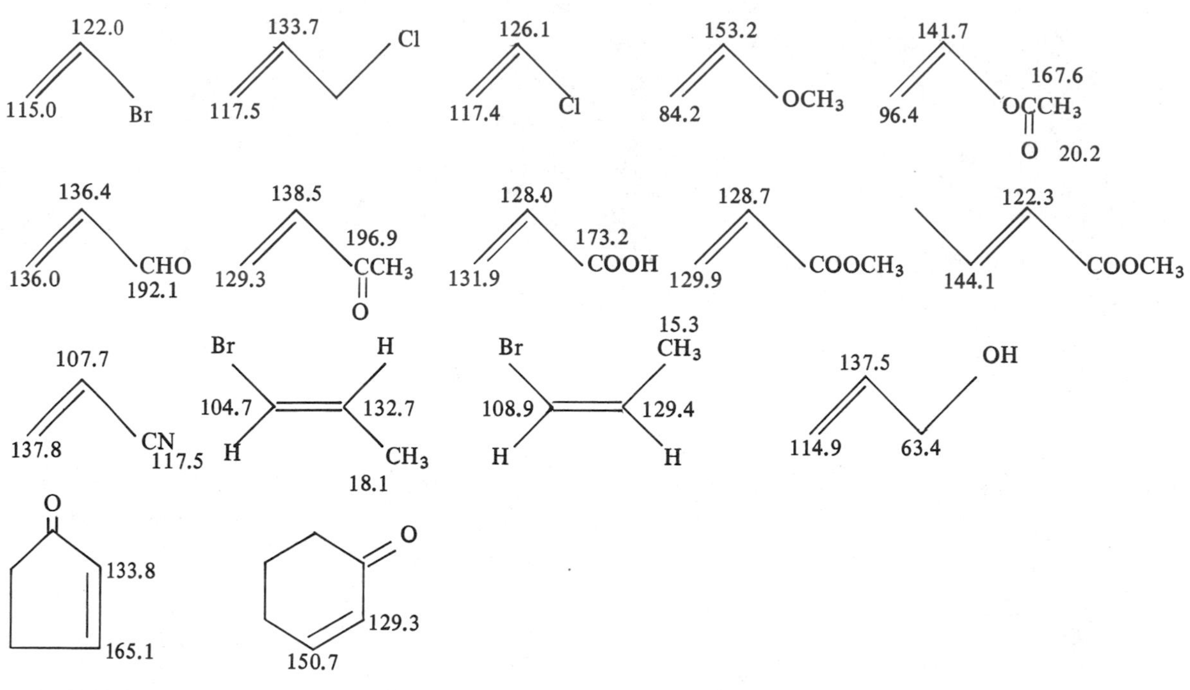

23.2 89.4
$HC{\equiv}C{-}OCH_2CH_3$

28.0 88.4
$CH_3{-}C{\equiv}C{-}O{-}CH_3$

Polar resonance structures explain these shifts as shown above for vinyl ethers.

AROMATIC COMPOUNDS

Benzene carbon atoms absorb at 128.5 ppm, neat or as a solution in $CDCl_3$ or CCl_4. Substituents shift the attached aromatic carbon atom as much as ±35 ppm. Fused-ring absorptions are as follows:

Naphthalene: C–1, 128.1; C–2, 125.9; C–4a, 133.7.
Anthracene: C–1, 130.1; C–2, 125.4; C–4a, 132.2; C–9, 132.6.

Table VI. Alkyne Chemical Shifts (ppm)

Compound	C-1	C-2	C-3	C-4	C-5	C-6
1-Butyne	67.0	84.7				
2-Butyne		73.6				
1-Hexyne	67.4	82.8	17.4	29.9	21.2	12.9
2-Hexyne	1.7	73.7	76.9	19.6	21.6	12.1
3-Hexyne	14.4	12.0	79.9			

Phenanthrene: C–1, 128.3; C–2, 126.3; C–3, 126.3; C–4, 122.2; C–4a, 131.9*; C–9, 126.6; C–10a, 130.1*.

Shifts of the aromatic carbon atom directly attached to the substituent have been correlated with substituent electronegativity after correcting for magnetic anisotropy effects; shifts at the *para* aromatic carbon have been correlated with the Hammett σ constant. *Ortho* shifts are not readily predictable and range over ~ 15 ppm. *Meta* shifts are generally small—up to several ppm for a single substituent.

The substituted aromatic carbon atoms can be distinguished from the unsubstituted aromatic carbon atom by its decreased peak height; that is, it lacks a proton and thus suffers from a longer T_1 and a diminished NOE.

Incremental shifts from benzene for the aromatic carbons of representative monosubstituted benzene rings (and shifts from TMS of carbon-containing substituents) are given in Table VII. Shifts from benzene for polysubstituted benzene ring carbons can be approximated by applying the principle of substituent additivity. For example, the shift from benzene for C–2 of the disubstituted compound 4-chlorobenzonitrile is calculated by adding the effect for an *ortho* CN group (+3.6) to that for a *meta* Cl group (+1.3):

*Not distinguishable.

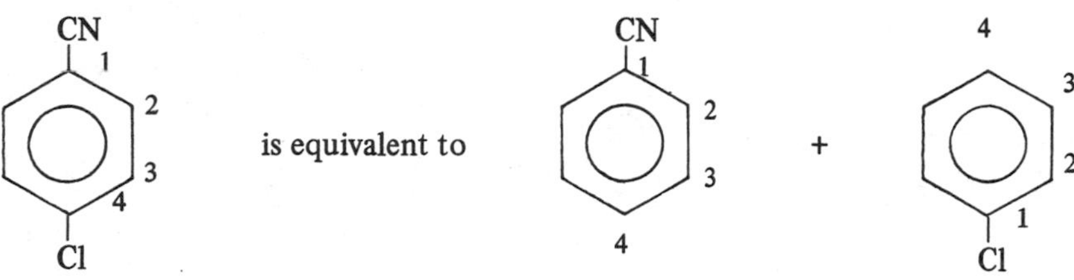

	Calcd.	Observed		Observed		Observed
1.	−17.3	−16.6	1.	−15.4	4.	−1.9
2.	+ 4.9	+ 5.1	2.	+ 3.6	3.	+1.3
3.	+ 1.0	+ 1.3	3.	+ 0.6	2.	+0.4
4.	+10.1	+10.8	4.	+ 3.9	1.	+6.2

Table VII. Incremental Shifts of the Aromatic Carbon Atoms of Monosubstituted Benzenes (ppm from Benzene at 128.5 ppm, + downfield, — upfield). C Atom of Substituents in ppm from TMS

Substituent	C-1 (Attachment)	C-2	C-3	C-4	C of Substituent (ppm from TMS)
H[b]	0.0	0.0	0.0	0.0	
$\mathrm{CH_3}$[b]	+8.9	+0.7	−0.1	−2.9	21.3
$\mathrm{CH_2CH_3}$[a]	+15.6	−0.5	0.0	−2.6	29.2 ($\mathrm{CH_2}$), 15.8 ($\mathrm{CH_3}$)
$\mathrm{CH(CH_3)_2}$[a]	+20.1	−2.0	0.0	−2.5	34.4 (CH), 24.1 ($\mathrm{CH_3}$)
$\mathrm{C(CH_3)_3}$[a]	+22.2	−3.4	−0.4	−3.1	34.5 (C), 31.4 ($\mathrm{CH_3}$)
$\mathrm{CH{=}CH_2}$[a]	+9.5	−2.0	+0.2	−0.5	135.5 (CH), 112.0 ($\mathrm{CH_2}$)
$\mathrm{C{\equiv}CH}$[b]	−6.1	+3.8	+0.4	−0.2	
$\mathrm{C_6H_5}$[b]	+13.1	−1.1	+0.4	−1.2	
$\mathrm{CH_2OH}$[a]	+12.3	−1.4	−1.4	−1.4	64.5
$\mathrm{CH_2O\underset{\underset{O}{\|}}{C}CH_3}$[c]	+7.7	~0.0	~0.0	~0.0	20.7 ($\mathrm{CH_3}$), 66.1 ($\mathrm{CH_2}$), 170.5 (C=O)
OH[b]	+26.9	−12.7	+1.4	−7.3	
$\mathrm{OCH_3}$[b]	+31.4	−14.4	+1.0	−7.7	54.1
$\mathrm{OC_6H_5}$[a]	+29.2	−9.4	+1.6	−5.1	
$\mathrm{O\overset{\overset{O}{\|}}{C}CH_3}$[a]	+23.0	−6.4	+1.3	−2.3	
$\mathrm{\overset{\overset{O}{\|}}{C}H}$[a]	+8.6	+1.3	+0.6	+5.5	192.0
$\mathrm{\overset{\overset{O}{\|}}{C}CH_3}$[b]	+9.1	+0.1	0.0	+4.2	25.0 ($\mathrm{CH_3}$), 195.7 (C=O)
$\mathrm{\overset{\overset{O}{\|}}{C}C_6H_5}$[b]	+9.4	+1.7	−0.2	+3.6	
$\mathrm{\overset{\overset{O}{\|}}{C}CF_3}$[b]	−5.6	+1.8	+0.7	+6.7	
$\mathrm{\overset{\overset{O}{\|}}{C}OH}$[b]	+2.1	+1.5	0.0	+5.1	172.6
$\mathrm{\overset{\overset{O}{\|}}{C}OCH_3}$[a]	+1.3	−0.5	−0.5	+3.5	51.0 ($\mathrm{CH_3}$)
$\mathrm{\overset{\overset{O}{\|}}{C}Cl}$[b]	+4.6	+2.4	0.0	+6.2	

Table VII (continued)

Substituent	C-1 (Attachment)	C-2	C-3	C-4	C of Substituent (ppm from TMS)
$C(=O)CF_3$[b]	−5.6	+1.8	−0.7	+6.7	
$C{\equiv}N$[b]	−15.4	+3.6	+0.6	+3.9	118.7
NH_2[b]	+18.0	−13.3	+0.9	−9.8	
$N(CH_3)_2$[a]	+22.4	−15.7	+0.8	−15.7	
$NHC(=O)CH_3$[a]	+11.1	−9.9	+0.2	−5.6	
NO_2[b]	+20.0	−4.8	+0.9	+5.8	
$N{=}C{=}O$[b]	+5.7	−3.6	+1.2	−2.8	129.5
F[b]	+34.8	−12.9	+1.4	−4.5	
Cl[b]	+6.2	+0.4	+1.3	−1.9	
Br[b]	−5.5	+3.4	+1.7	−1.6	
I[a]	−32.2	+9.9	+2.6	−7.4	
CF_3[b]	−9.0	−2.2	+0.3	+3.2	
SH[c]	+2.3	+1.1	+1.1	−3.1	
SCH_3[c]	+10.2	−1.8	+0.4	−3.6	
SO_2NH_2[c]	+15.3	−2.9	+0.4	+3.3	
$Si(CH_3)_3$[a]	+13.4	+4.4	−1.1	−1.1	

[a]Neat
[b]In CCl_4
[c]In $CDCl_3$

HETEROAROMATIC COMPOUNDS

Complex rationalizations have been offered for the shifts of carbon atoms in heteroaromatic compounds. As a general rule, C−2 of oxygen- and nitrogen-containing rings is further downfield than C−3. Large solvent and pH effects have been recorded. Table VIII gives values for neat samples of several 5- and 6-membered heterocyclic compounds.

ALCOHOLS

Substitution of H in an alkane by an OH group causes downfield shifts of 35-52 ppm for C−1, 5-12 ppm for C−2, and upfield shifts of about 0-6 ppm for C−3. Shifts for several acyclic and alicyclic alcohols are given in Table IX. Acetylation provides a useful diagnostic test for an alcohol, the C−1 absorption moving downfield by ~ 2.5-4.5 ppm, and the C−2 absorption moving upfield by a similar amount; a 1,3-diaxial interaction may cause a slight (~ 1 ppm) downfield shift of C−3. Table III may be used to calculate shifts for alcohols as described earlier.

ETHERS

An alkoxy substituent causes a somewhat larger downfield shift at C−1 (~ 10 ppm larger) than that of a hydroxy substituent. This is attributed to the C−1′ of the alkoxy group having the same effect as a β−C relative to C−1. The O atom is regarded here as an "α−C" to C−1.

$$\underset{17.6}{CH_3}-\underset{57.0}{CH_2}-OH \qquad \underset{14.7}{\overset{2}{CH_3}}-\underset{67.9}{\overset{1}{CH_2}}-O-\underset{57.6}{\overset{1'}{CH_3}}$$

Note also that the "γ-effect" (upfield shift) on C−2 is explainable by similar reasoning. Conversely, the ethoxy group affects the OCH_3 group (compare CH_3OH).

Table X gives shifts of several ethers.

The dioxygenated carbon of acetals and ketals absorbs in the range of ~ 88-112 ppm.

The alkyl carbons of aralkyl ethers have shifts similar to those of dialkyl ethers. Note the strong upfield shift of the ring *ortho* carbon resulting from electron delocalization as in the vinyl ethers.

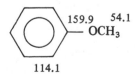

Anisole: C_6H_5−OCH_3 — C-1 159.9, OCH_3 54.1, ortho C 114.1

Table VIII. Shifts for C Atoms of Heteroaromatics (neat, ppm from TMS)

Compound	C-2	C-3	C-4	C-5	C-6	Substituent
Furan	142.7	109.6				
2-Methylfuran	152.2	106.2	110.9	141.2		13.4
Furan-2-carboxaldehyde	153.3	121.7	112.9	148.5		178.2
Pyrrole	118.4	108.0				
2-Methylpyrrole	127.2	105.9	108.1	116.7		12.4
Pyrrole-2-carboxaldehyde	134.0	123.0	112.0	129.0		
Thiophene	124.4	126.2				
2-Methylthiophene	139.0	124.7	126.4	122.6		14.8
Thiophene-2-carbox-aldehyde	143.3	136.4	128.1	134.6		182.8
Thiazole	152.2		142.4	118.5		
Imidazole	136.2		122.3	122.3		
Pyridine	150.2	123.9	135.9			
Pyrimidine	159.5		157.4	122.1	157.4	
Pyrazine	145.6					
2-Methylpyrazine	154.0	141.8*	143.8*	144.7*		21.6

*Assignment not certain.

HALIDES

The effect of halide substitution is complex. A single fluorine atom (in CH_3F) causes a large downfield shift from CH_4 as electronegativity considerations would suggest. Successive geminal substitution by Cl (CH_3Cl, CH_2Cl_2, $CHCl_3$, CCl_4) results in increasing downfield effects—again expected on the basis of electronegativity. But with Br and I, the "heavy atom effect" supervenes. The carbon shifts of CH_3Br and CH_2Br_2 move progressively downfield, but those of $CHBr_3$ and CBr_4 move progressively upfield. A strong upfield progression for I commences with CH_3I, which is upfield from CH_4. There is a progressive downfield effect at C—2 in the order I > Br > F. Cl and Br show γ-gauche shielding at C—3, but I does not, presumably because of the low population of the hindered gauche rotamer. Table XI shows these trends.

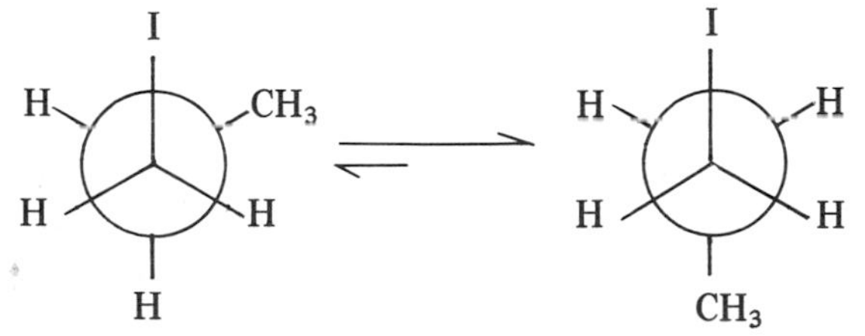

Halides may show large solvent effects; for example, C—1 for iodoethane is at −6.6 in cyclohexane, and at −0.4 in dimethylformamide.

AMINES

An NH_2 group attached to an alkyl chain causes a downfield shift of ~30 ppm at C—1, a down field shift of ~11 ppm at C—2, and an upfield shift of ~4.0 ppm at C—3. The NH_3^+ group shows a somewhat smaller effect. N-alkylation increases the downfield effect of the NH_2 group at C—1. Shift positions for selected acyclic and alicyclic amines are given in Table XII (See Table IVa for heterocyclic amines).

FUNCTIONAL GROUPS CONTAINING CARBON

^{13}C NMR spectrometry permits direct observation of carbon-containing functional groups; the shift ranges for these are given in the Appendix. With the exception of CH=O, the presence of these groups could not be directly ascertained by ^1H NMR.

Ketones and Aldehydes

The R_2C=O and the RCH=O carbons absorb in a characteristic region, downfield. Acetone absorbs at 203.8 ppm and acetaldehyde at 199.3 ppm. Alkyl substitution on the α—C causes a downfield shift of the C=O absorption of 2-3 ppm until steric effects supervene. Replacement of the CH_3 of acetone or acetaldehyde by a phenyl group causes an upfield shift of the C=O absorption (acetophenone 195.7 ppm, benzaldehyde 190.7 ppm); similarly α,β-unsaturation causes upfield shifts (acrolein 192.1 ppm, compared with propionaldehyde 201.5 ppm). Presumably, charge delocalization by the benzene ring or the double bond makes the carbonyl carbon less electron deficient. Of the cycloalkanones, cyclopentanone has an anomalously low shift position. Table XIII presents chemical shifts of the C=O group of some ketones and aldehydes. Because of

Table IX. Chemical Shifts of Alcohols (neat, ppm from TMS)

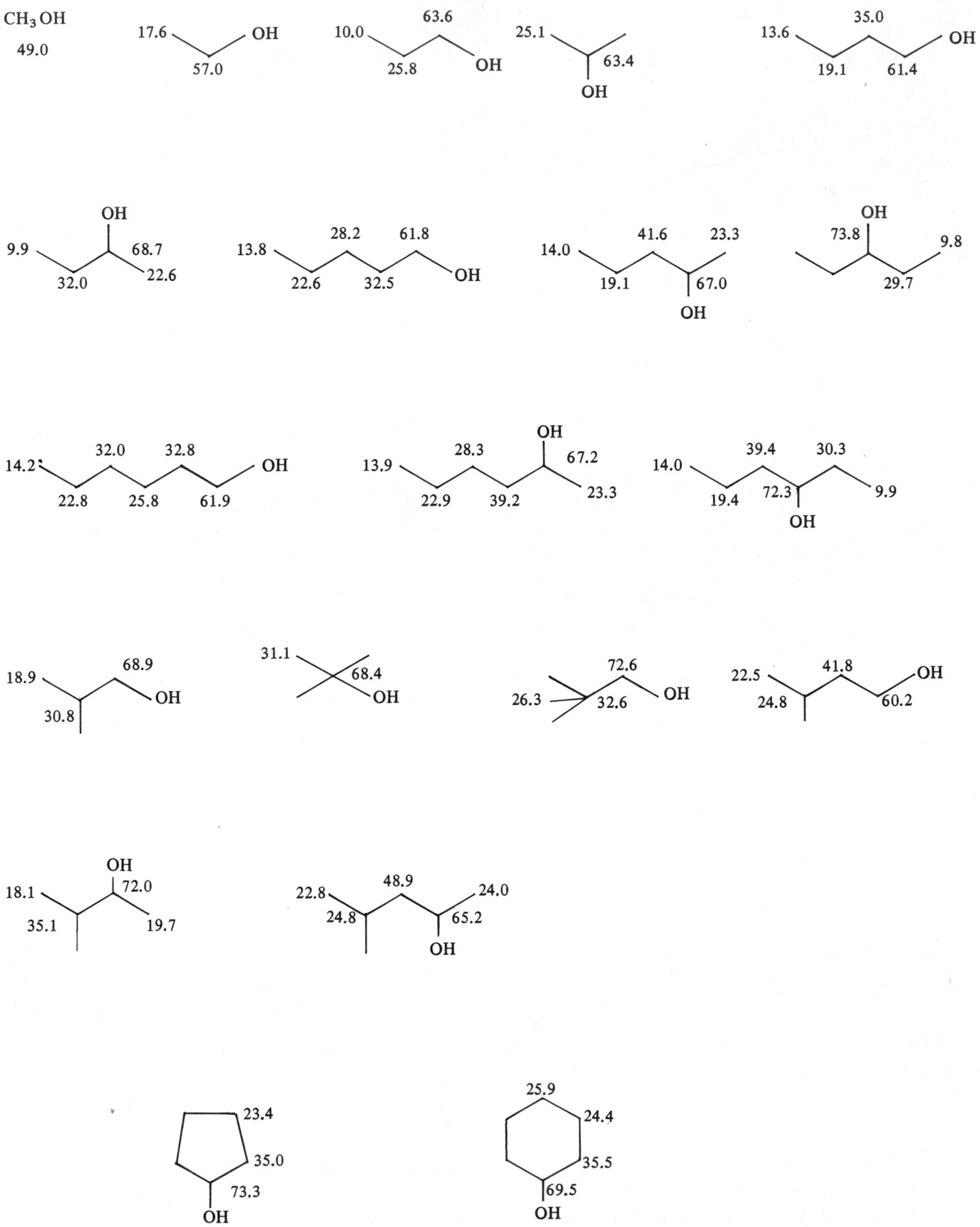

CH_3OH 49.0

17.6 OH 57.0

10.0 63.6 25.8 OH

25.1 63.4 OH

13.6 35.0 OH 19.1 61.4

OH 9.9 68.7 32.0 22.6

13.8 28.2 61.8 22.6 32.5 OH

14.0 41.6 23.3 19.1 67.0 OH

OH 73.8 9.8 29.7

14.2 32.0 32.8 OH 22.8 25.8 61.9

OH 13.9 28.3 67.2 22.9 39.2 23.3

14.0 39.4 30.3 19.4 72.3 9.9 OH

18.9 68.9 OH 30.8

31.1 68.4 OH

72.6 26.3 32.6 OH

22.5 41.8 OH 24.8 60.2

OH 18.1 72.0 35.1 19.7

22.8 48.9 24.0 24.8 65.2 OH

23.4 35.0 73.3 OH

25.9 24.4 35.5 69.5 OH

Table X. **Chemical Shifts of Ethers (neat, ppm from TMS)**

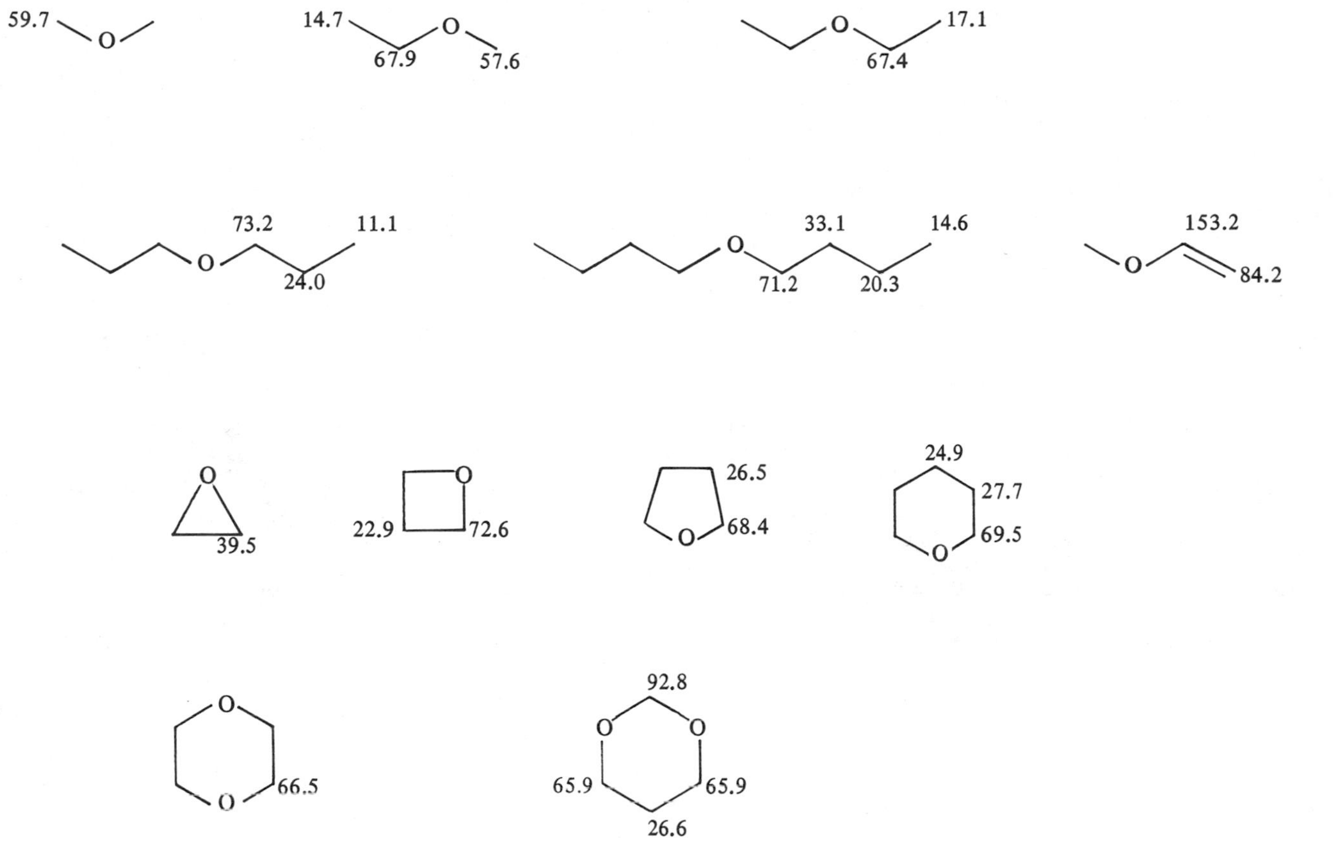

59.7 O 14.7 O 67.9 57.6 O 17.1 67.4

73.2 11.1 O 24.0 33.1 14.6 O 71.2 20.3 153.2 O 84.2

O 39.5 O 22.9 72.6 26.5 O 68.4 24.9 27.7 O 69.5

O O 66.5 92.8 O O 65.9 65.9 26.6

Table XI. **Shift Positions for Alkyl Halides (neat, ppm from TMS)**

Compound	C-1	C-2	C-3
CH_4	−2.3		
CH_3F	75.4		
CH_3Cl	24.9		
CH_2Cl_2	54.0		
$CHCl_3$	77.5		
CCl_4	96.5		
CH_3Br	10.0		
CH_2Br_2	21.4		
$CHBr_3$	12.1		
CBr_4	−28.5		
CH_3I	−20.7		
CH_2I_2	−54.0		
CHI_3	−139.9		
CI_4	−292.5		
CH_3CH_2F	79.3	14.6	
CH_3CH_2Cl	39.9	18.7	
CH_3CH_2Br	28.3	20.3	
CH_3CH_2I	−0.2	21.6	
$CH_3CH_2CH_2Cl$	46.7	26.5	11.5
$CH_3CH_2CH_2Br$	35.7	26.8	13.2
$CH_3CH_2CH_2I$	10.0	27.6	16.2

rather large solvent effects, there are differences of several ppm from different literature sources. Replacement of CH_2 of alkanes by C=O causes a downfield shift at the α—C (~ 10-14 ppm) and an upfield shift at the β—C (several ppm in acyclic compounds).

Carboxylic Acids, Esters, Chlorides, Anhydrides, Amides, and Nitriles

The C=O groups of carboxylic acids and derivatives are in the range of 150-185 ppm. Dilution and solvent effects are marked for carboxylic acids; anions appear at lower field. The effects of substituents and electron delocalization are generally similar to those for ketones. Nitriles absorb in the range of 115-125 ppm. Alkyl substituents on the nitrogen of amides cause a small (up to several ppm) upfield shift of the C=O group. (See Table XIV.)

Oximes

The simple oximes absorb in the range of 145-165 ppm. It is possible to distinguish between *syn* and *anti* isomers since the C=N shift is upfield in the sterically more com-

Table XII. Shift Positions of Acyclic and Alicyclic Amines (neat, ppm from TMS)

Compound	C-1	C-2	C-3	C-4
CH_3NH_2	26.9			
$CH_3CH_2NH_2$	35.9	17.7		
$CH_3CH_2CH_2NH_2$	44.9	27.3	11.2	
$CH_3CH_2CH_2CH_2NH_2$	42.3	36.7	20.4	14.0
$(CH_3)_3N$	47.5			
$CH_3CH_2N(CH_3)_2$	58.2	13.8		
Cyclohexylamine	50.4	36.7	25.7	25.1
N-Methylcyclohexylamine	58.6	33.3	25.1	26.3 ($N-CH_3$ 33.5)

Table XIII. Shift Positions of the C=O Group and Other Carbon Atoms of Ketones and Aldehydes (ppm from TMS).

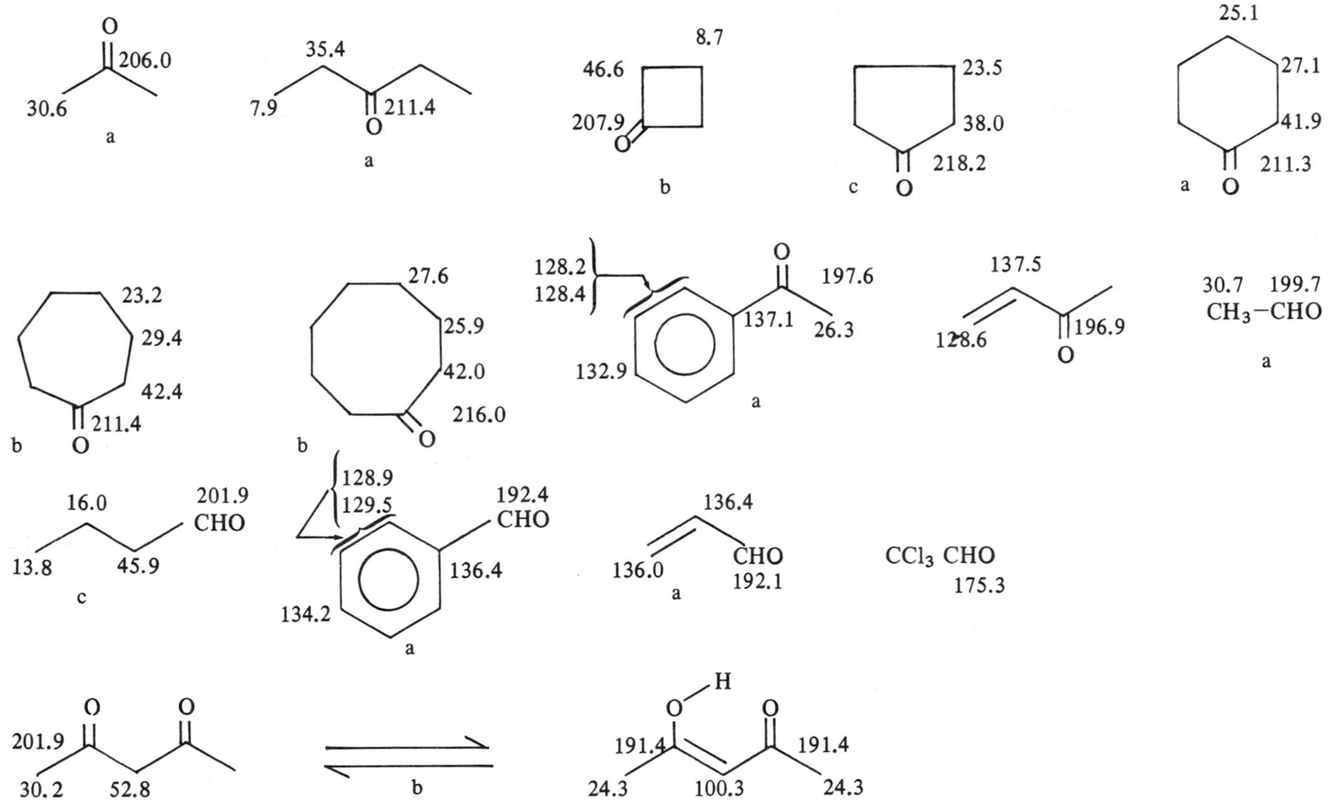

O 206.0 30.6 a

35.4 7.9 211.4 O a

8.7 46.6 207.9 O b

23.5 38.0 218.2 c O

25.1 27.1 41.9 211.3 a O

23.2 29.4 42.4 211.4 b O

27.6 25.9 42.0 216.0 b O

128.2 128.4 O 197.6 137.1 26.3 132.9 a

137.5 128.6 196.9 O

30.7 199.7 CH_3-CHO a

16.0 201.9 CHO 13.8 45.9 c

128.9 129.5 192.4 CHO 136.4 134.2 a

136.4 CHO 136.0 192.1 a

CCl_3 CHO 175.3

O O 201.9 30.2 52.8 ⇌ b H O O 191.4 191.4 24.3 100.3 24.3

[a]~50% in $CDCl_3$.

[b]Contains ~10% dioxane or methylene chloride.

[c]~50% in dioxane.

Table XIV. Shift Positions for the C=O group and other Carbon Atoms of Carboxylic Acids, Esters, Lactones, Chlorides Anhydrides, Amides, Carbamates, and Nitriles (ppm from TMS)

CH_3-COOH
20.6 178.1
a

184.8
—COOH

128.0
131.9 COOH
173.2

89.1 168.0
CCl_3-COOH

115.0 163.0
$F_3C-COOH$
a

172.6
COOH
129.4
133.7 130.2
128.4
a

181.5
$CH_3-CO\bar{O}Na^+$
b

NH_2
51.5
17.2 COOH
176.5
d

O
60.0
170.3 O
20.0 13.8
a

27.2
9.2 O
173.3 O 50.8
a

23.4 25.5 O
172.1 51.1
O
14.3 32.2 33.9
a

O
170.0 66.8
O
21.4 20.4
a

128.7
O
129.9 164.5
O
c

O
167.7 141.7
O
96.4
c

O
115.3 158.1 64.7
F_3C C O 13.8
a

167.0 51.0
$COOCH_3$

111.9 117.9
159.1 51.8
$COOCH_3$
146.4 O 144.8

O
CH_3CCl
169.5
c

O
168.0
Cl
133.1
135.3 131.3
128.9
a

$(CH_3C(O))_2O$
167.3
c

O
28.6 172.9
O
O
a

O
165.9
O
O

28.7
$(CH_3-CH_2-C(O))_2O$
8.4 170.3
a

O
27.7 177.9
O
22.2
68.6
a

O
$HCNH_2$
165.5
c

O
CH_3CNH_2
172.7
c

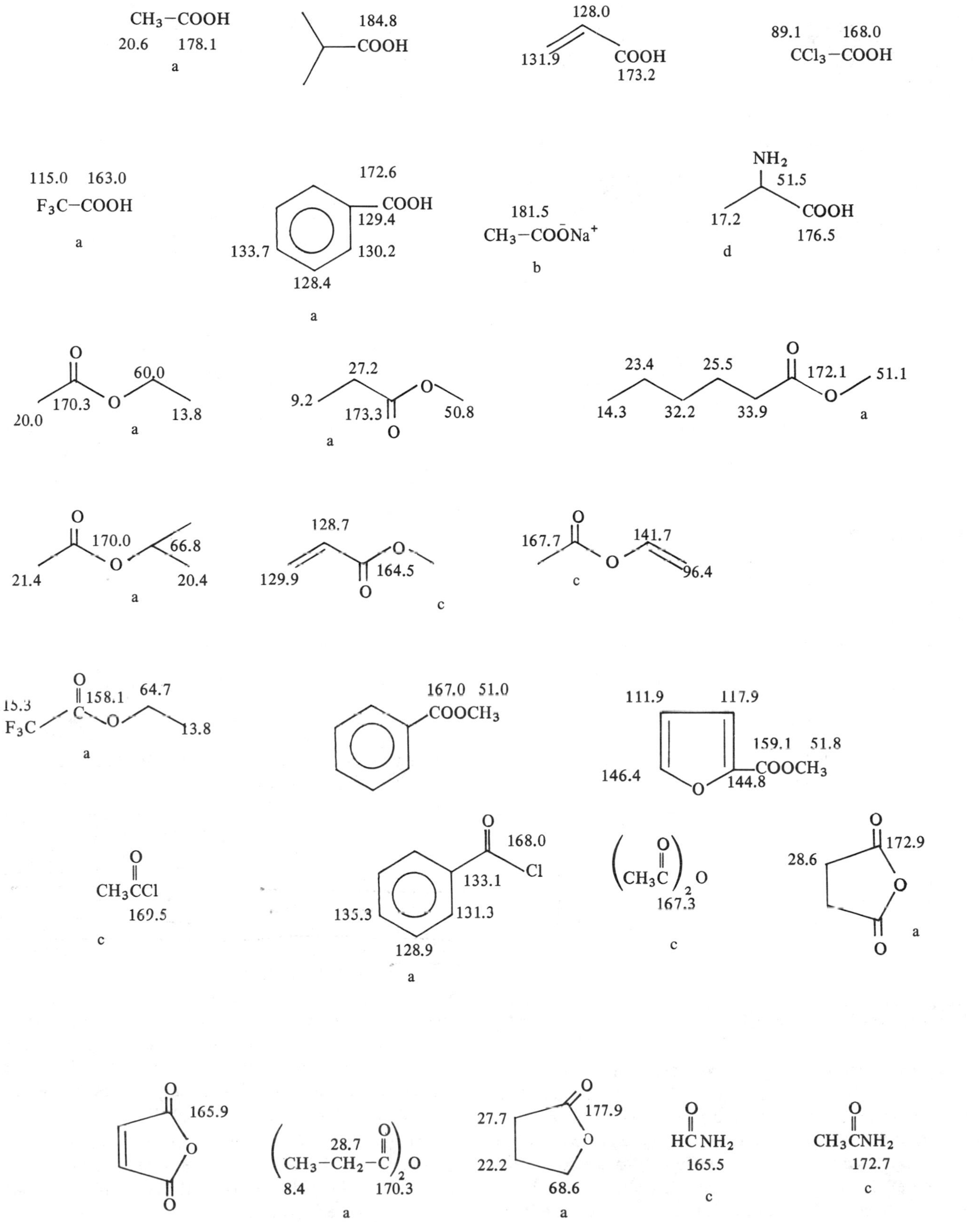

Table XIV (continued)

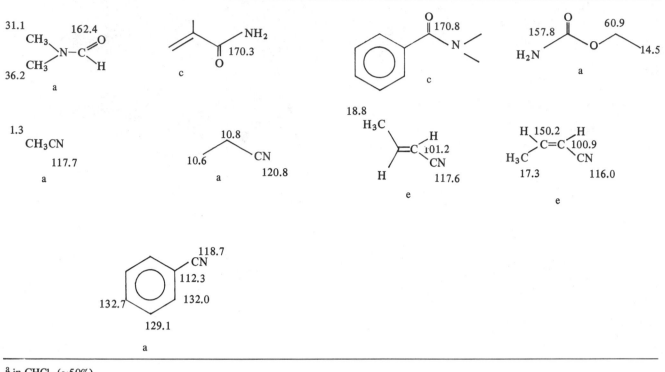

$(CH_3)_2N{-}CHO$: CH_3 31.1, CH_3 36.2, C=O 162.4 (a)	$CH_2{=}C(CH_3){-}CONH_2$: C=O 170.3 (c)	$C_6H_5CON(CH_3)_2$: C=O 170.8 (c)	$H_2N{-}CO{-}OCH_2CH_3$: C=O 157.8, OCH_2 60.9, CH_3 14.5 (a)
CH_3CN: CH_3 1.3, CN 117.7 (a)	CH_3CH_2CN: CH_3 10.6, CH_2 10.8, CN 120.8 (a)	$H_3C(H)C{=}C(H)CN$: CH_3 18.8, =C(H)CN 101.2, CN 117.6 (e)	$H_3C(H)C{=}C(H)CN$: CH_3 17.3, =CH 150.2, =C(H)CN 100.9, CN 116.0 (e)
C_6H_5CN: CN 118.7, C-1 112.3, C-2 132.0, C-3 129.1, C-4 132.7 (a)			

[a] in $CHCl_3$ (~50%).
[b] saturated aqueous solution of CH_3COONa
[c] neat or saturated solution
[d] in H_2O
[e] in DMSO
[f] in dioxane (~50%).

pressed form, and the more hindered substituent (*syn* to the OH) is upfield of the less hindered:

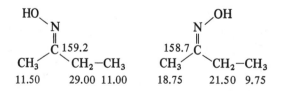

HO–N=C(CH$_3$)(CH$_2$–CH$_3$): C 159.2; CH$_3$ 11.50; CH$_2$ 29.00; CH$_3$ 11.00	OH–N=C(CH$_3$)(CH$_2$–CH$_3$): C 158.7; CH$_3$ 18.75; CH$_2$ 21.50; CH$_3$ 9.75

IV. SPIN-SPIN COUPLING

Coupling—at least as an initial consideration—is less important in ^{13}C NMR than in ^{1}H NMR, since routine ^{13}C spectra are usually noise-decoupled. Thus coupling information is discarded in the interest of obtaining a spectrum in a shorter time or on a smaller sample—a spectrum, furthermore, free of complex overlapping absorptions. However, as mentioned earlier, information from residual coupling can be regained through off-resonance decoupling.

^{13}C-^{13}C coupling is usually not observed, except in compounds that have been deliberately enriched with ^{13}C, because of the low probability of two adjacent ^{13}C atoms in a molecule. One-bond ^{13}C–H coupling ($^1J_{CH}$) ranges from about 110 to 320 Hz, increasing with increased *s* character of the ^{13}C–H bond, with substitution on the carbon atom of electron-withdrawing groups, and with angular distortion. Appreciable ^{13}C–H coupling also extends over two or more (n) bonds ($^nJ_{CH}$). Table XV gives some representative $^1J_{CH}$ values. Table XVI gives some representative $^2J_{CH}$ values, which range from about –5 to 60 Hz.

$^3J_{CH}$ values are roughly comparable to $^2J_{CH}$ values for sp^3 carbons. In aromatic rings, however, the $^3J_{CH}$ values are characteristically larger than $^2J_{CH}$ values. In benzene itself, $^3J_{CH}$ = 7.4 Hz, and $^2J_{CH}$ = 1.0 Hz.

Coupling of ^{13}C to several other nuclei, the most important of which are ^{31}P, ^{19}F, and D, may be observed in proton-decoupled spectra. Representative coupling constants are given in Table XVII.

Table XV. Some ${}^1J_{CH}$ Values

Compound	J (Hz)
sp^3	
CH_3CH_3	124.9
$CH_3\underline{C}H_2CH_3$	119.2
$(CH_3)_3CH$	114.2
CH_3NH_2	133.0
CH_3OH	141.0
CH_3Cl	150.0
CH_2Cl_2	178.0
$CHCl_3$	209.0
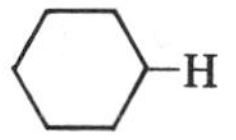 (cyclohexane)–H	123.0
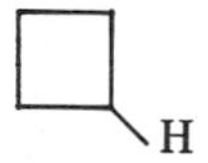 (cyclobutane)–H	134.0
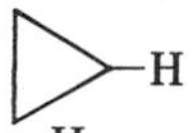 (cyclopropane)–H	161.0
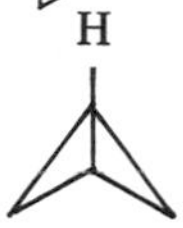 (bicyclobutane) H	205.0
sp^2	
$CH_2{=}CH_2$	156.2
$CH_3\underline{C}H{=}C(CH_3)_2$	148.4
$C_2H_5C\underline{C}H{=}O$	225.6
$CH_3\underline{C}H{=}O$	172.4
$NH_2\underline{C}H{=}O$	188.3
C_6H_6	159.0
sp	
$CH{\equiv}CH$	249.0
$C_6H_5C{\equiv}CH$	251.0
$HC{\equiv}N$	269.0

Table XVI. Some ${}^2J_{CH}$ Values

Compound	J (Hz)
sp^3	
$C\underline{H}_3\underline{C}H_3$	−4.5
$C\underline{H}_3\underline{C}Cl_3$	5.9
$\underline{C}H_3C\underline{H}{=}O$	26.7
sp^2	
$C\underline{H}_2{=}\underline{C}H_2$	−2.4
$(C\underline{H}_3)_2\underline{C}{=}O$	5.5
$CH_2{=}\underline{C}HC\underline{H}{=}O$	26.9
C_6H_6	1.0
sp	
$C\underline{H}{\equiv}\underline{C}H$	49.3
$C_6H_5O\underline{C}{\equiv}C\underline{H}$	61.0

V. PEAK ASSIGNMENT PROBLEM

We may review some of the above material by assigning the peaks of *p*-methoxybenzaldehyde from Figure 6a and 6b by using the chemical shift correlation tables, peak heights, off-resonance decoupling, and the concept of symmetry.

We immediately assign the downfield absorption in Figure 6a at δ 191.0 (peak 1) to the CHO group (Table XIII gives δ 190.7 for benzaldehyde), and the upfield absorption at δ 55.6 (peak 6) to the OCH_3 group (Table VII gives δ 54.1 for anisole). The small peaks 2 and 4 must represent the quaternary ring carbon atoms, and the 2 large peaks 3 and 5 must each represent the 2 pairs of equivalent ring carbons. At this point the following assignments are evident.

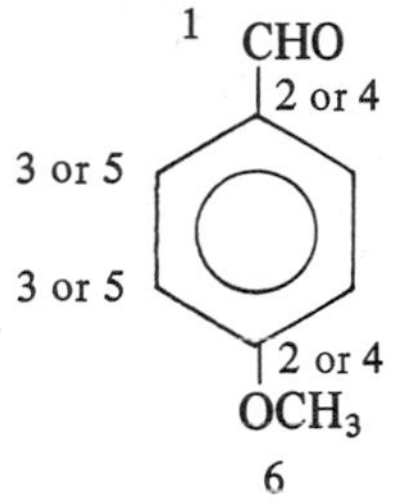

1 CHO
2 or 4
3 or 5
3 or 5
2 or 4
OCH_3
6

These assignments are confirmed by the off-resonance decoupled spectrum (Figure 6b): doublet for CHO, quartet for OCH_3, singlets for the quaternary carbons. Peak 4 is buried under the doublet of peak 3.

The ambiguous assignments can be resolved by applying the principle of substituent additivity and the data of Table VII as follows:

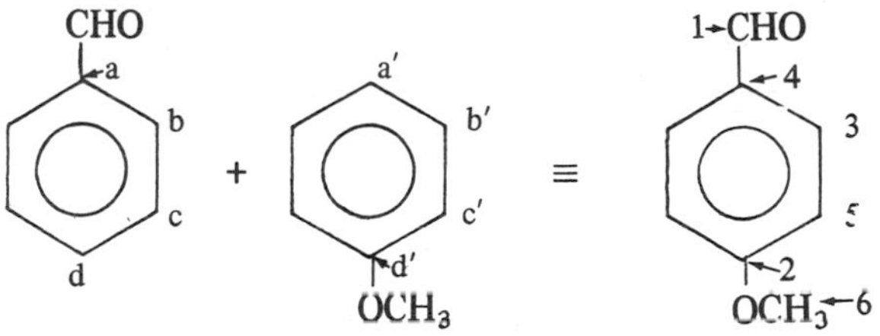

CHO (a, b, c, d) + OCH_3 (a′, b′, c′, d′) ≡ 1–CHO (4, 3, 5, 2), OCH_3–6

ppm from benzene (128.5 ppm) (see Table VII)		ppm from TMS (δ)	
for CHO	for OCH_3	Calcd.	Observed
a. +8.6	a.′ −7.7	4. 129.4	4. 130.2
b. +1.3	b.′ +1.0	3. 130.8	3. 132.1
c. +0.6	c.′ −14.4	5. 114.7	5. 114.5
d. +5.5	d.′ +31.4	2. 165.4	2. 164.9

Table XVII. Coupling constants for ^{31}P, ^{19}F, and D coupled to ^{13}C

Compound	^1J (Hz)	^2J (Hz)	^3J (Hz)
CH_3CF_3	271		
CF_2H_2	235		
CF_3COOH	284	43.7	
C_6H_5F	245	21.0	
$(CH_3CH_2)_3P$	5.4	10.0	
$(CH_3CH_2)_4P^+ Br^-$	49	4.3	
$(C_6H_5)_3P^+CH_3 I^-$	88 (CH_3 52)	10.9	
$CH_3CH_2P(=O)(OCH_2CH_3)_2$	143	7.1 (J_{COP} 6.9)	J_{CCOP} 6.2
$CDCl_3$	31.5		
$CD_3C(=O)CD_3$	19.5		
$(CD_3)_2SO$	22.0		
C_6D_6	25.5		

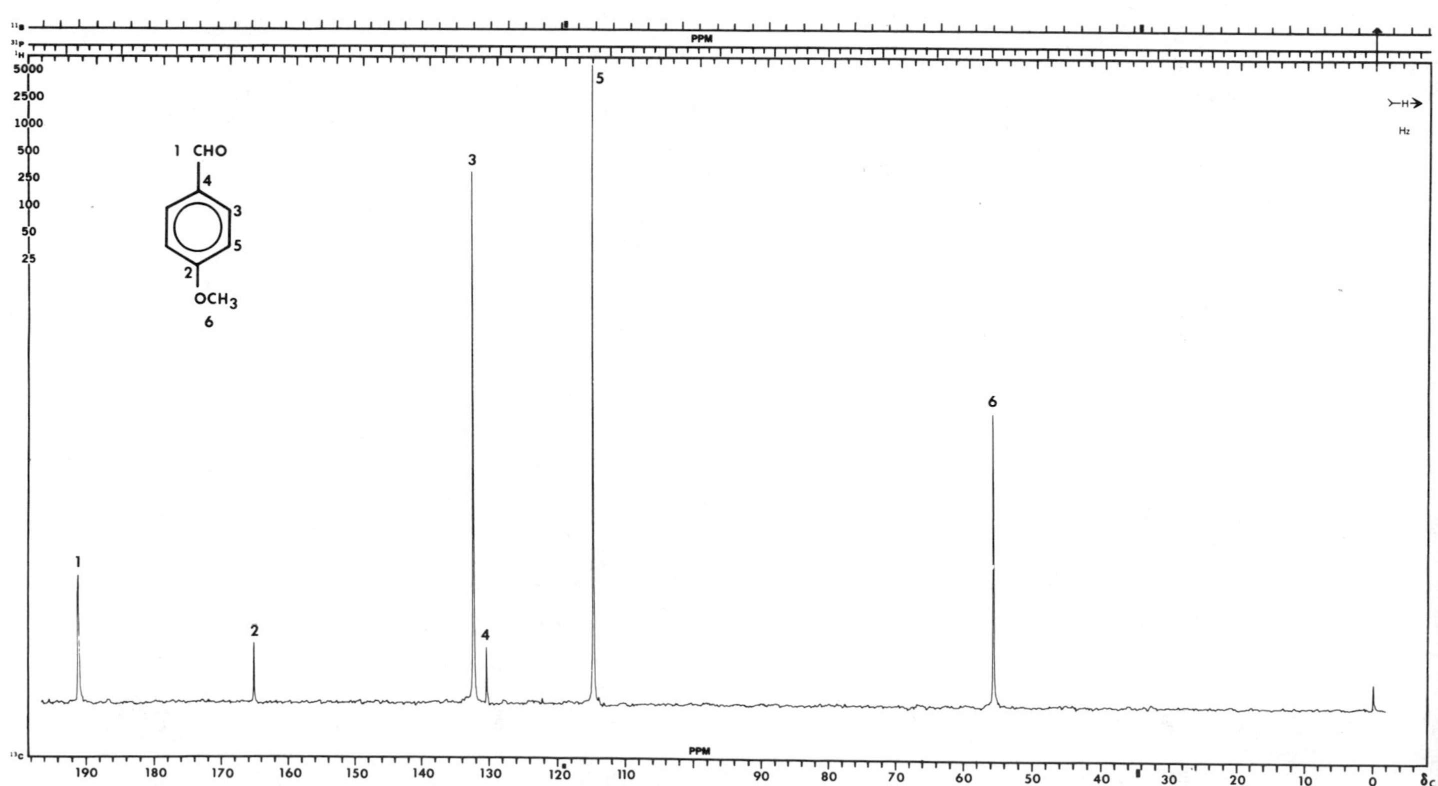

Figure 6a. *(a). Noise-decoupled ^{13}C spectrum of* p-methoxybenzaldehyde. *Solvent $CDCl_3$, 25.2 MHz, 5000 Hz sweep-width.*

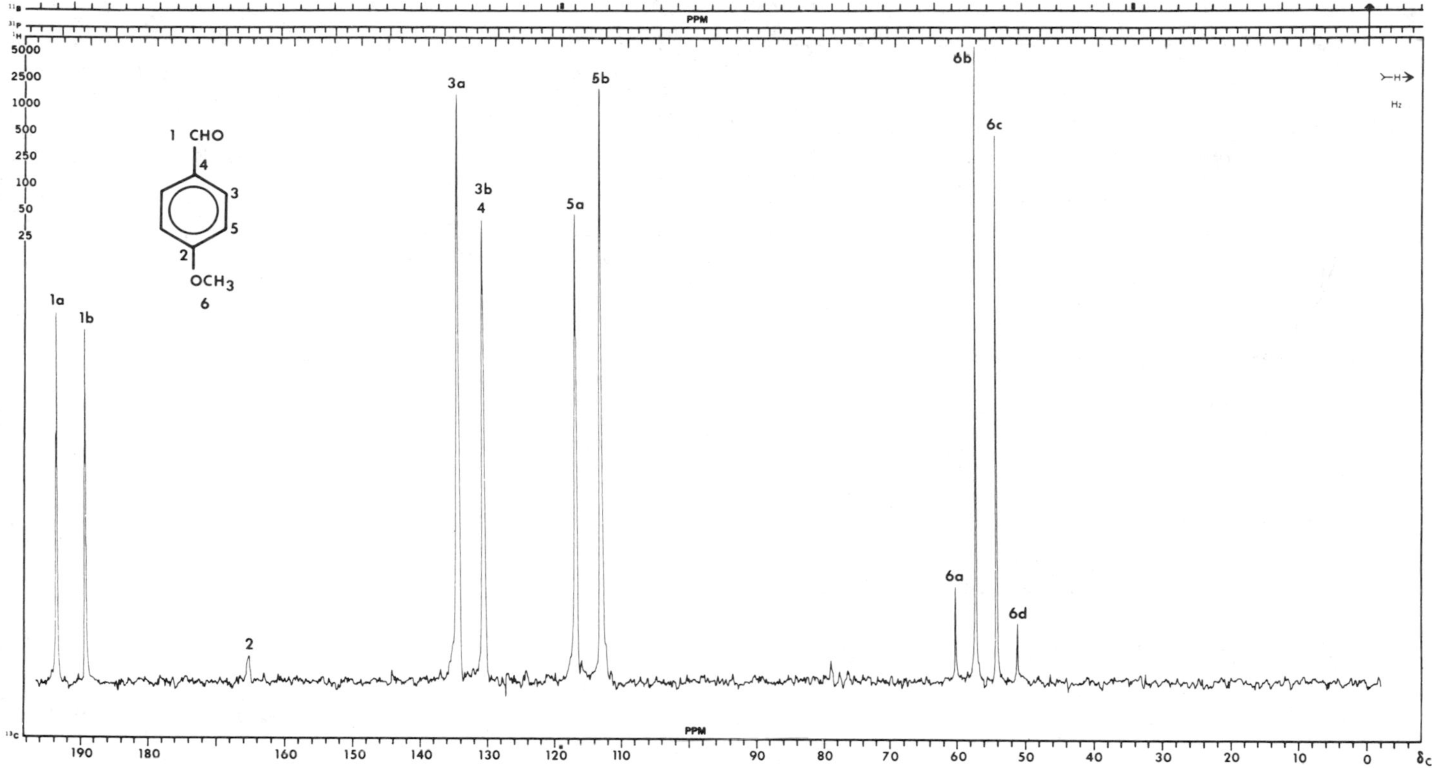

Figure 6b. *Off-resonance decoupled* ^{13}C *spectrum of* p-methoxybenzaldehyde. *Solvent* $CDCl_3$, *25.2 MHz, 5000 Hz sweepwidth.*

Note also that we count 6 carbon atoms in the noise-decoupled spectrum and 6 protons in the off-resonance decoupled spectrum. The discrepancy between this count and the molecular formula $C_8H_8O_2$ confirms the *para* substitution in the benzene ring.

VI. THE "CROSSOVER" EXPERIMENT

It is possible to correlate a particular ^{13}C peak with the shift position of its attached proton(s) by carrying out a "crossover" experiment (Figure 7a). This is done by obtaining ^{13}C off-resonance decoupled spectra at 4 different frequencies, 2 upfield of the proton range of frequencies and 2 downfield. Thus, for example, the decoupler might be set (a) at the proton frequency of TMS, (b) at 1000 Hz upfield from TMS, (c) at 1000 Hz downfield from TMS, and at 2000 Hz downfield.

(d) — 1000 Hz — (c) — 1000 Hz — (a) — 1000 Hz — (b)

Proton sweepwidth TMS

The residual couplings (J^r) in the spectrum obtained with the decoupler set at (a) are less than those obtained with the decoupler set at (b). Similarly, J^r with the decoupler at (c) is less than J^r with the decoupler at (d). The J^r values can be obtained accurately from the computer readout. The ratio J^r_b/J^r_a for a particular ^{13}C is a measure of how far upfield (in the proton spectrum) the attached protons(s) is—the larger the ratio, the further upfield. Likewise, the larger the ratio J^r_d/J^r_c, the further downfield is the attached proton. Either ratio gives the relative shift positions for the attached protons, but the additional assurance provided is worth the effort of obtaining 2 downfield as well as 2 upfield decouplings.

This procedure is shown applied to the problem of assigning the peaks in the ^{13}C spectrum of furfural (Figure 7b), given the proton assignments; the ^1H spectrum of the same solution is shown in Figure 7c. The proton assignments for furfural are as follows (see Chapter 4, Appendix D, Table 5).

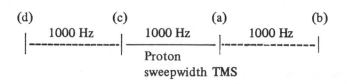

H 4 6.67; H 3 7.32; H 5 7.78; O; 2 CHO 9.70

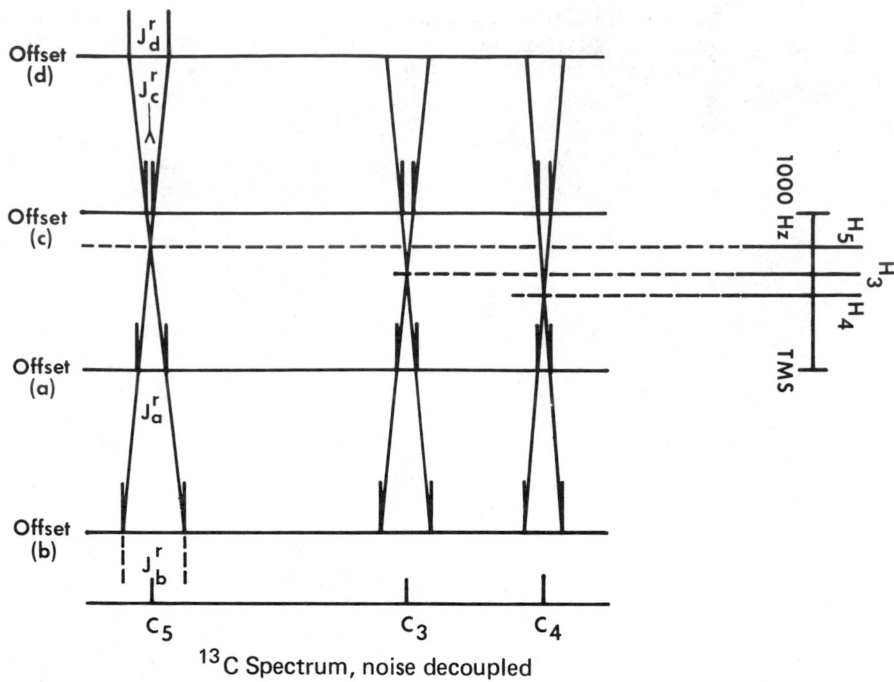

Figure 7a. *"Crossover" schematic for furfural. Scale on bottom locates the three carbons bearing protons, and the scale on the right locates these protons.*

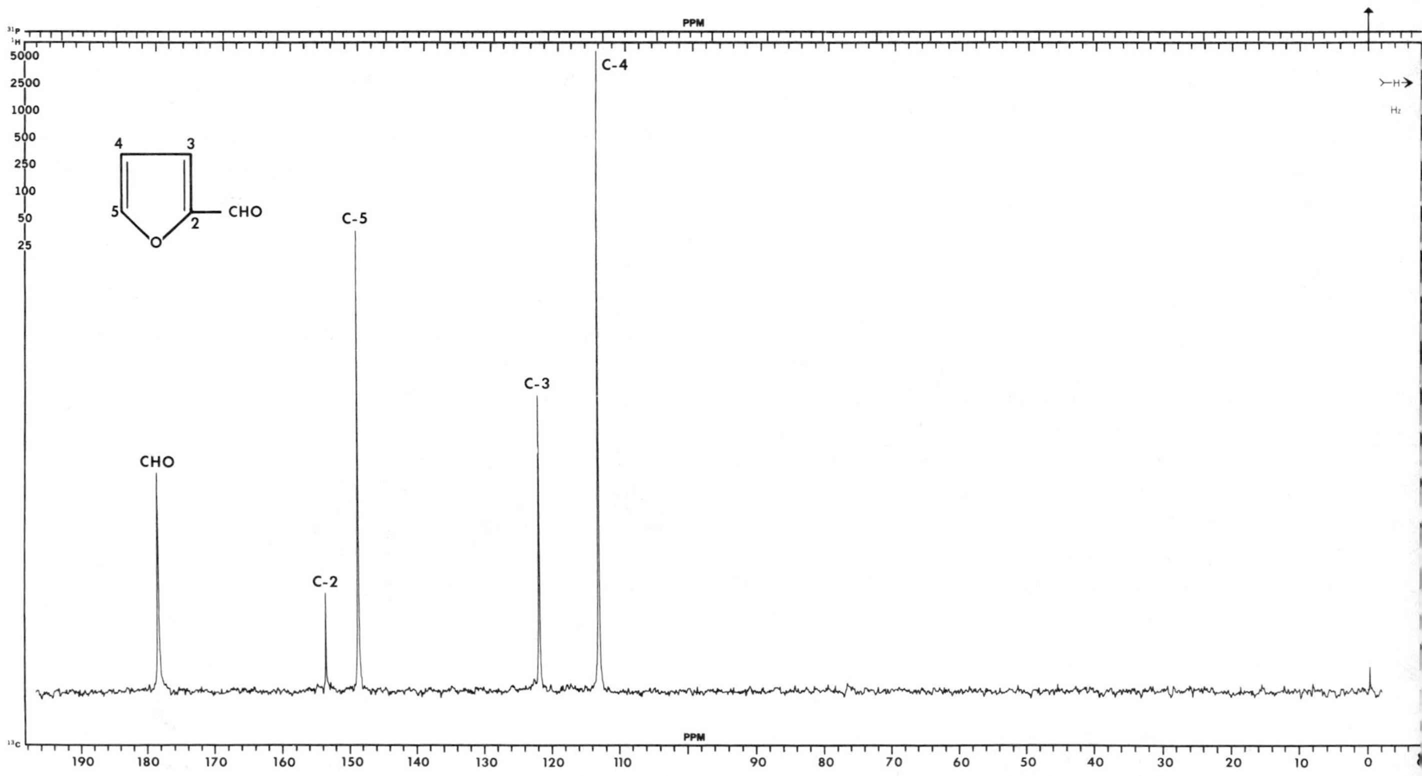

Figure 7b. *Noise-decoupled ^{13}C spectrum for furfural (50% solution in $CDCl_3$). 25.2 MHz, 5000 Hz sweepwidth.*

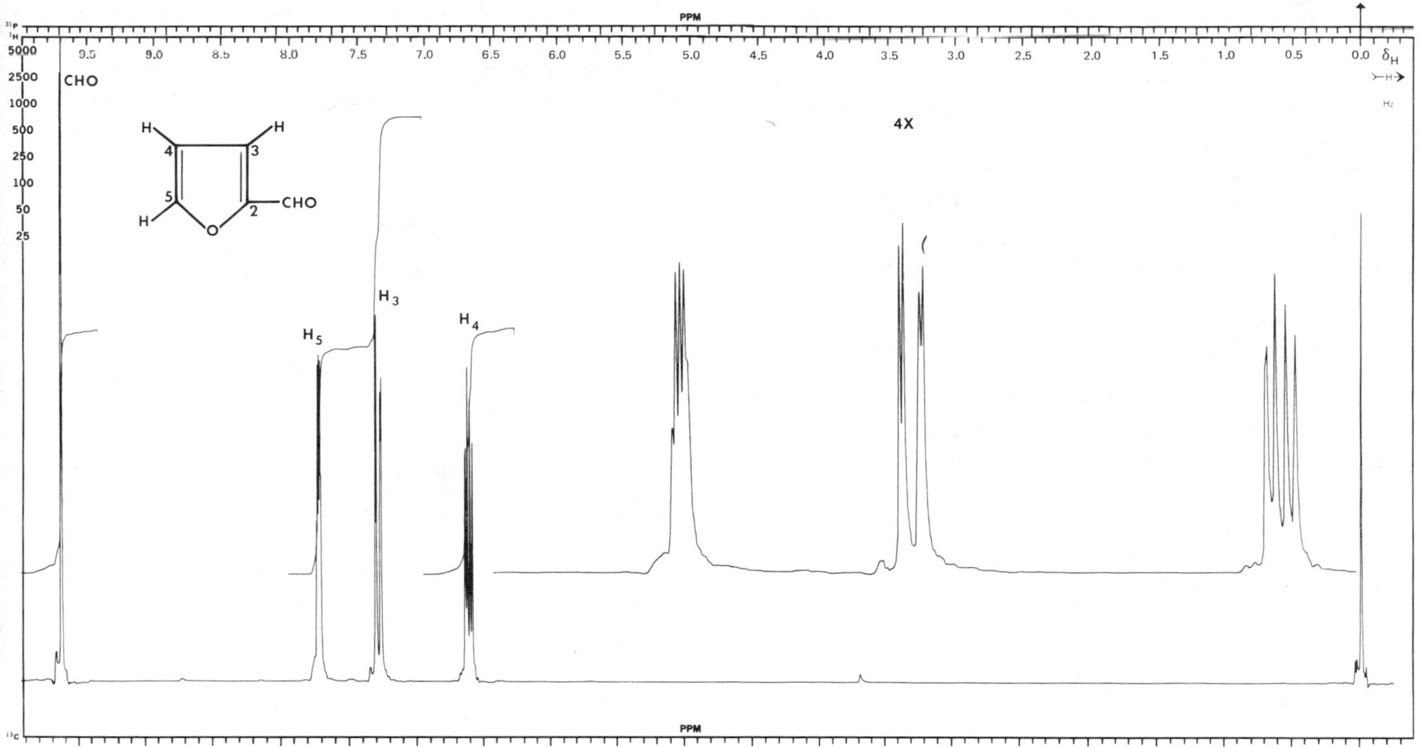

Figure 7c. [1]*H NMR spectrum of the same furfural solution used for Fig. 7b. $CDCl_3$ solvent, 100 MHz, 1000 Hz sweep-width.*

The ring ^{13}C assignments of C_3, C_4, and C_5 can be made from the following J^r values and ratios as described above; the downfield position of the ^{13}CHO (178.15 ppm) is obvious, as is the assignment of C_2 (153.28 ppm) by virtue of low intensity and its singlet in the off-resonance decoupled spectra.

^{13}C(ppm)	J_b^r	J_a^r	J_b^r/J_a^r	J_d^r	J_c^r	J_d^r/J_c^r
148.51	99.31	47.55	2.092	76.78	15.93	4.831
121.66	85.00	39.54	2.151	67.86	17.11	3.968
112.90	83.01	36.99	2.310	71.08	21.02	3.356

We see that the ratios J_b^r/J_a^r and J_d^r/J_c^r both indicate that the downfield proton (on C_5) is connected to the 148.51 ppm ^{13}C peak; the next upfield proton (on C_3) is connected to the 121.66 ppm ^{13}C peak; and the furthest upfield proton (on C_4) is connected to the furthest upfield (112.90 ppm) ^{13}C peak. Although the ^{13}C peak at 148.51 ppm might reasonably be assigned to C_5 because of its proximity to the oxygen atom, the distinction between C_3 and C_4 is not obvious. The "crossover" experiment gives an unambiguous result despite the fact that the proton shifts are quite close. Here the progression of shift positions for the ^{13}C atoms and for their attached protons is the same, but this is not always true (see Appendix B). The ^{13}C assignments in furfural are as follows:

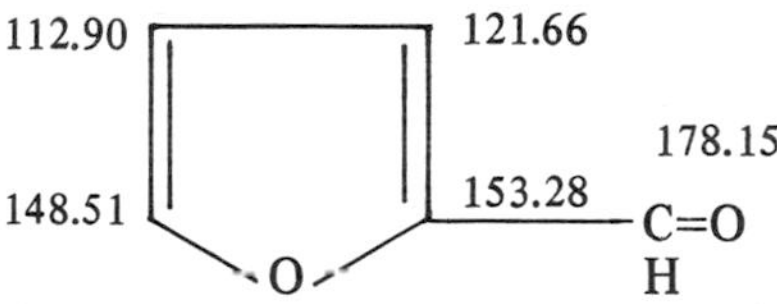

The term "crossover" derives from the appearance of a graphical presentation in which lines connect the peaks of ^{13}C absorptions at offset (b) with the corresponding peaks at offset (a), and similarly connect the peaks at offset (d) and offset (c). The intersection point is a function of J_b^r/J_a^r and J_d^r/J_c^r as shown in the schematic diagram in Figure 7a, and is related to the chemical shift of the attached proton. In practice, the graphical method is not satisfactory (because of the very small slopes of the converging lines) unless the protons are widely separated. The slopes in Figure 7a are greatly exaggerated.

VII. SECOND ORDER EFFECTS

Second-order effects may complicate off-resonance spectra. (Several examples are given on pages 72-75 of Reference 3.) The problem may occur in a $Y-CH_2CH_2-X$ system in which first order character requires that $^1J_{CH} >> {}^2J_{CCH}$ and $^3J_{HH}$. Under the influence of the high power decoupler, $^1J_{CH}$ is reduced to $^1J^r_{CH}$. Furthermore, the decoupler reduces the effective shift difference between the 2 sets of protons ($\Delta\nu$), which results in more strongly coupled protons$\left(\frac{\Delta\nu}{J} \text{ is small}\right)$. These effects result in a distorted "triplet" (with additional peaks) for each CH_2 rather than a clean first-order triplet. If the situation is understood, the deviation from first order is usually no problem. (Similar situations in proton spectra are described as "virtual coupling" in Chapter 4.) An off-resonance decoupled spectrum of thiamine hydrochloride that shows both first order triplets and distorted "triplets" is presented in Figure 8.

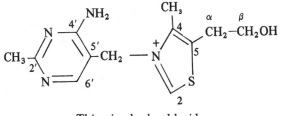

Thiamine hydrochloride

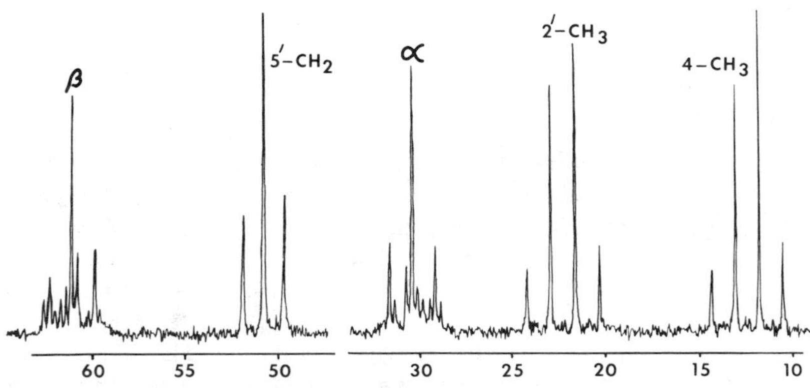

Figure 8. *High-field region of the 20 MHz ^{13}C spectrum of thiamine hydrochloride in D_2O; off-resonance decoupled. From ref. 3, page 73, with permission.*

Figure 9a is a noise-decoupled spectrum of 6-methyl-5-hepten-2-one, and the peak assignments are made on the basis of chemical shifts, peak intensity of the quaternary carbons, and the off-resonance decoupled spectrum (Figure 9b).

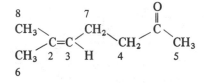

In an attempt to reduce the amount of overlapping in the off-resonance decoupled spectrum, the decoupler was moved to a smaller offset. The spectrum in Figure 9b was obtained with the decoupler at 1500 Hz upfield from TMS and in Figure 9c, the decoupler was at 750 Hz upfield from TMS. The smaller residual coupling in Figure 9c had the intended effect of reducing the overlap, but the triplets of peaks 4 and 7 in Figure 9b are distorted in Figure 9c by the second-order effects described above.

VIII. QUANTITATIVE ANALYSIS

Quantitative ^{13}C–NMR is desirable in two situations. First, in structural determinations, it is clearly useful to know if a signal is due to more than one shift-equivalent carbon.

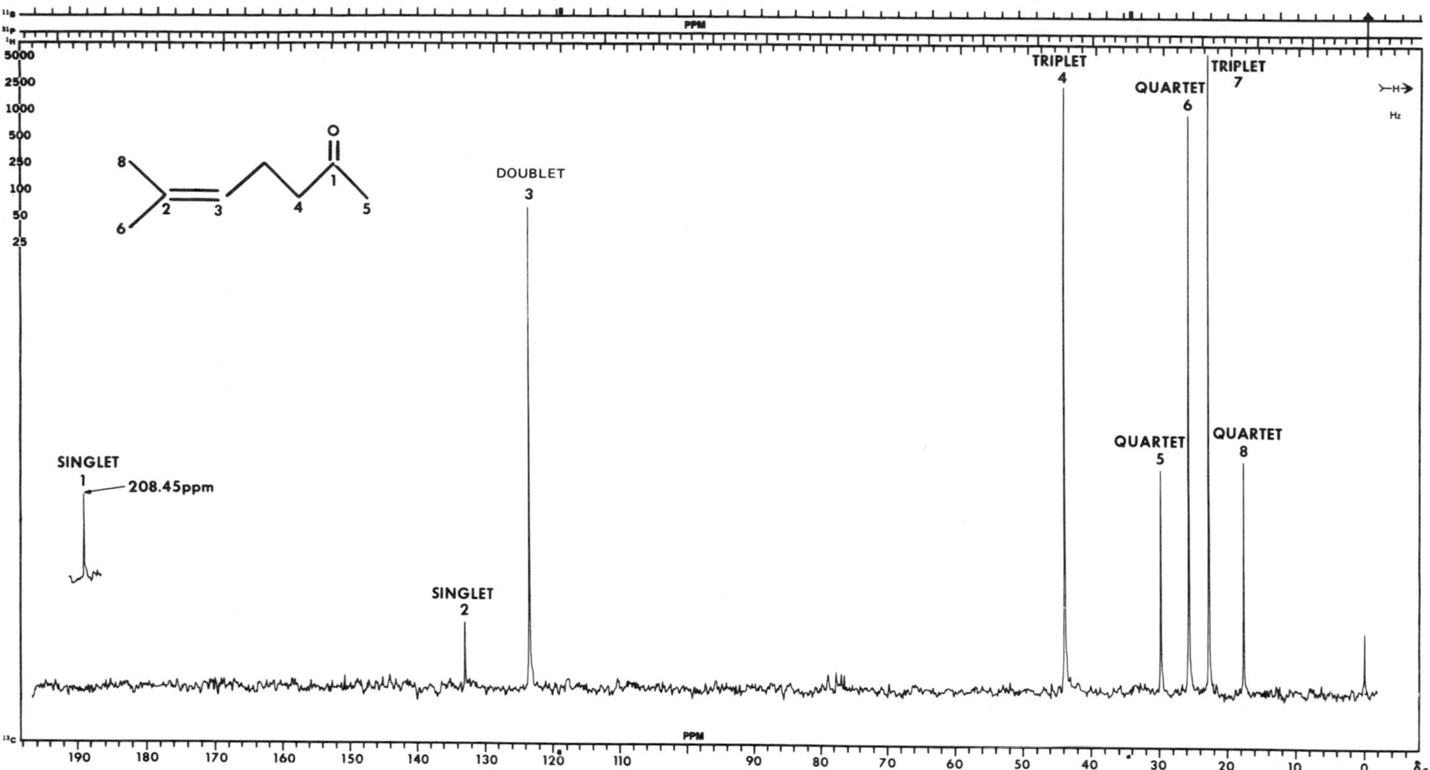

Figure 9a. *Noise-decoupled* ^{13}C *spectrum of 6-methyl-5-hepten-2-one;* $CDCl_3$ *solvent, 25.2 MHz, 5000 Hz sweepwidth.*

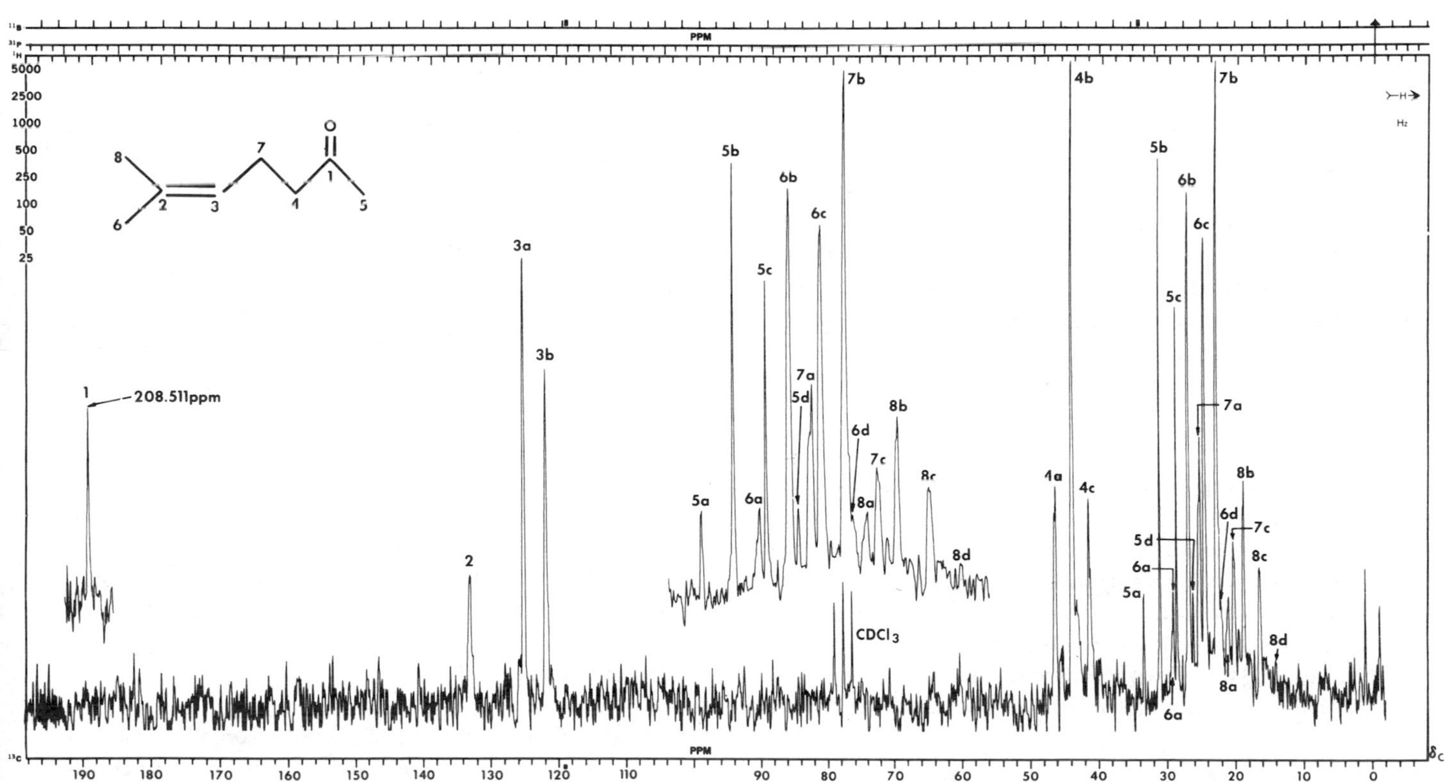

Figure 9b. ^{13}C*-spectrum of 6-methyl-5-hepten-2-one with off-resonance decoupling 1500 Hz upfield of the TMS proton frequency. Solvent* $CDCl_3$*, 25.2 MHz, 5000 Hz sweepwidth.*

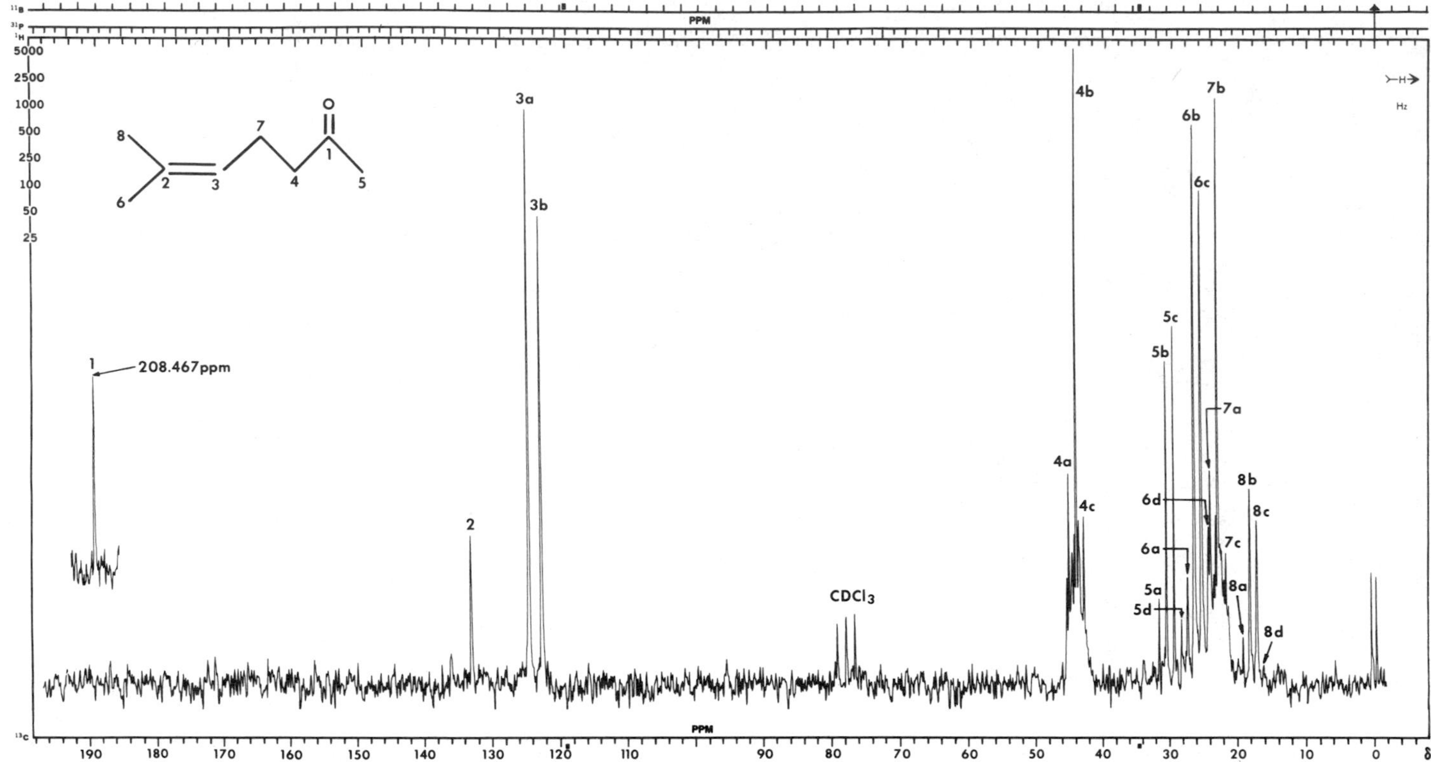

Figure 9c. ^{13}C *spectrum of 6-methyl-5-hepten-2-one with off-resonance decoupling 750 Hz upfield of the TMS proton frequency.*

Second, quantitative analysis of a mixture of two or more components requires that the area of a signal be proportional to the number of carbon atoms causing that signal.

There are two reasons why noise-decoupled spectra are usually not quantitative. First, carbons with long spin-lattice relaxation times (T_1) may not completely return to a Boltzmann's distribution between pulses, and the resulting signals will be considerably weaker than expected from the number of carbons causing those signals (see Figure 4 and the corresponding text). Second, some carbons give rise to exceptionally small signals due to weak NOE enhancement.

Three methods may be employed in an effort to obtain quantitative noise-decoupled spectra: long pulse delays, gated decoupling, and the addition of paramagnetic relaxation reagents.

Delays of tens to hundreds of seconds (to allow return of the nuclei to a Boltzmann's distribution) will remove signal area discrepancies caused by long relaxation times; such delays, unfortunately, will frequently require prohibitively long times to obtain the spectrum.

If the noise decoupler is gated on during the pulse and the early part of the free induction decay and then is gated off during the pulse delay, NOE enhancement will be minimized for all carbons. This is true because the free induction signal decays quickly in an exponential fashion, while the NOE factor slowly builds up in an exponential fashion. A weakness of the gated decoupling method is the removal of the NOE factor, whose loss requires extended signal acquisition.

Added paramagnetic relaxation reagents lead to more quantitative spectra by reducing all of the spin-lattice relaxation times (by means of an electronic relaxation mechanism) and by levelling all of the noE factors. These reagents suffer from the fact that the effects are not complete and from the requirement that the sample must be separated (e.g., by chromatography) if it is to be recovered. Successful quantitative analysis has been carried out by combining two or all three of these procedures.[13]

REFERENCES

1. Levy, G. C., Lichter, R. L., and Nelson, G. L. *Carbon-13 Nuclear Magnetic Resonance for Organic Chemists.* 2nd ed. New York: Wiley, 1980.
2. Stothers, J. B. *Carbon-13 NMR Spectroscopy.* New York: Academic, 1972.

3. Wehrli, F. W. and Wirthlin, T. *Interpretation of Carbon-13 NMR Spectra.* London: Heyden, 1976.
4. Breitmaier, E. and Voelter, W. *^{13}C NMR Spectroscopy,* 2nd ed. New York: Verlag Chemie, 1978.
5. Abraham, R. J. and Loftus, P. *Proton and ^{13}C NMR Spectroscopy.* London: Heyden, 1978.
6. Johnson, L. F., and Jankowski, W. C. *Carbon-13 NMR Spectra, a Collection of Assigned, Coded, and Indexed Spectra.* New York: Wiley, 1972.
7. *Nuclear Magnetic Resonance Spectra.* Philadelphia, PA.: Sadtler Research Laboratories.
8. Farrar, T. C. and Becker, E. D. *Pulse and Fourier Transform NMR* New York: Academic, 1971.
9. Shaw, D. *Fourier Transform NMR Spectroscopy.* North-Holland, N.Y.: Elsevier, 1976.
10. Muller, K. and Pregosin, P. S. *Fourier Transform NMR: A Practical Approach.* New York: Academic, 1976.
11. Fuchs, P. L. and Bunnell, C. A. *Carbon-13 NMR Based Organic Spectral Problems.* New York: Wiley, 1979.
12. Breitmaier, E., Haas, G., Voelter, W. *Atlas of Carbon-13 NMR Data.* vol. 1-3. Philadelphia, PA.: Heyden, 1978.
13. Shoolery, J. N., *Progress in NMR Spectroscopy,* **11**, 79(1977).

PROBLEMS

5.1 How many ^{13}C peaks should be seen in the noise-decoupled spectrum of each of the following compounds? Assign the multiplicity expected for each signal in the off-resonance decoupled spectrum.

a. benzene
b. toluene
c. naphthalene
d. dodecane

e.

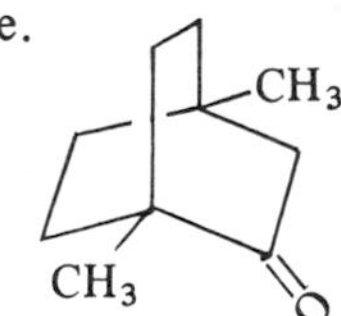

f.

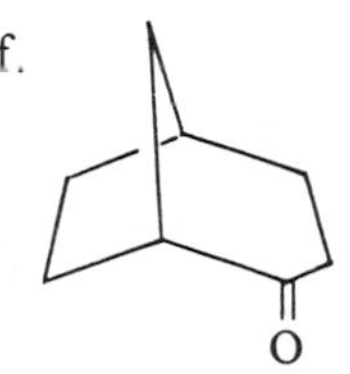

5.2 Because a compound of molecular formula C_6H_8 is highly symmetrical, it shows just two singlets in the noise-decoupled spectrum. The off-resonance decoupled spectrum shows only a triplet and a doublet. Draw the structure.

5.3 Predict the number of lines for $-^{13}CD$, $-^{13}CD_2$, and $-^{13}CD_3$. In each case, assume that only 1-bond C–D coupling is important, and check your results against the footnote on page 257. Recall the $CDCl_3$ signal that appears in many ^{13}C spectra and adapt Pascal's triangle (Figure 25, Chapter 4), to ^{13}C–D coupling.

5.4 Deduce the structure of compounds A-I from their spectra and assign as many ^{13}C signals as possible. The multiplicities of off-resonance decoupled multiplets are abbreviated: s = singlet, d = doublet, t = triplet, q = quartet. The solvent is $CDCl_3$ in all cases except E which is dissolved in D_2O.

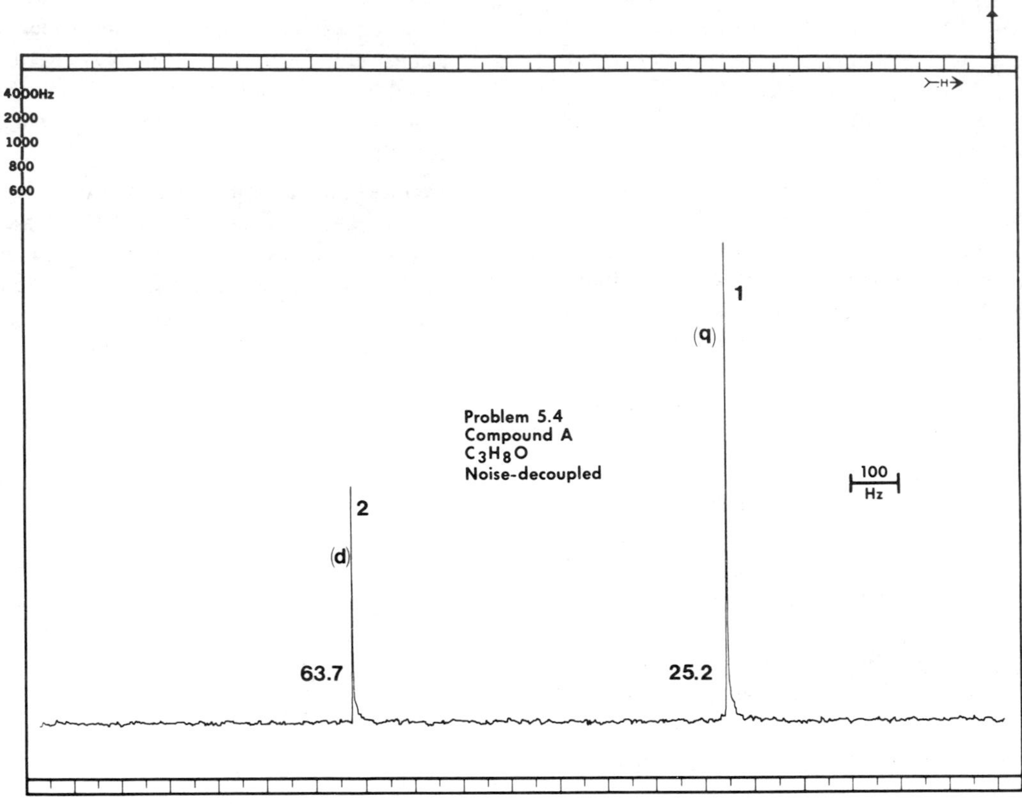
4000Hz
2000
1000
800
600
1
(q)
Problem 5.4
Compound A
C_3H_8O
Noise-decoupled
100
Hz
2
(d)
63.7
25.2

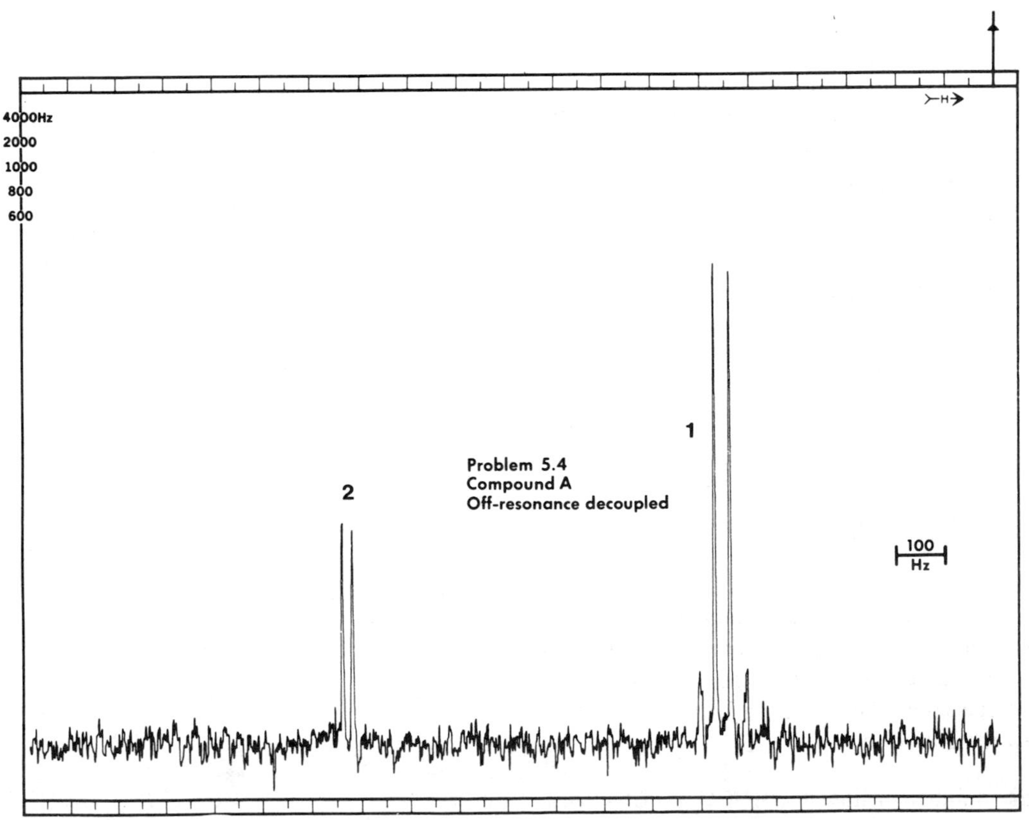
4000Hz
2000
1000
800
600
1
Problem 5.4
Compound A
Off-resonance decoupled
2
100
Hz

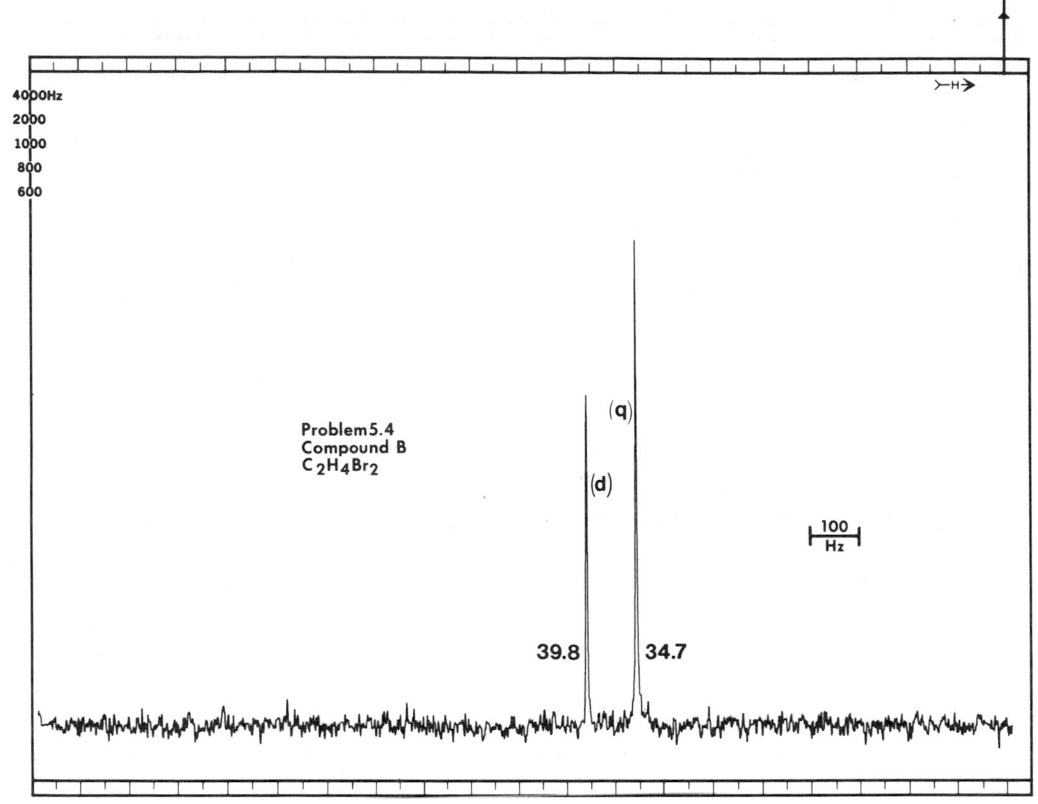
4000Hz
2000
1000
800
600
Problem 5.4
Compound B
C2H4Br2
(q)
(d)
100
Hz
39.8
34.7

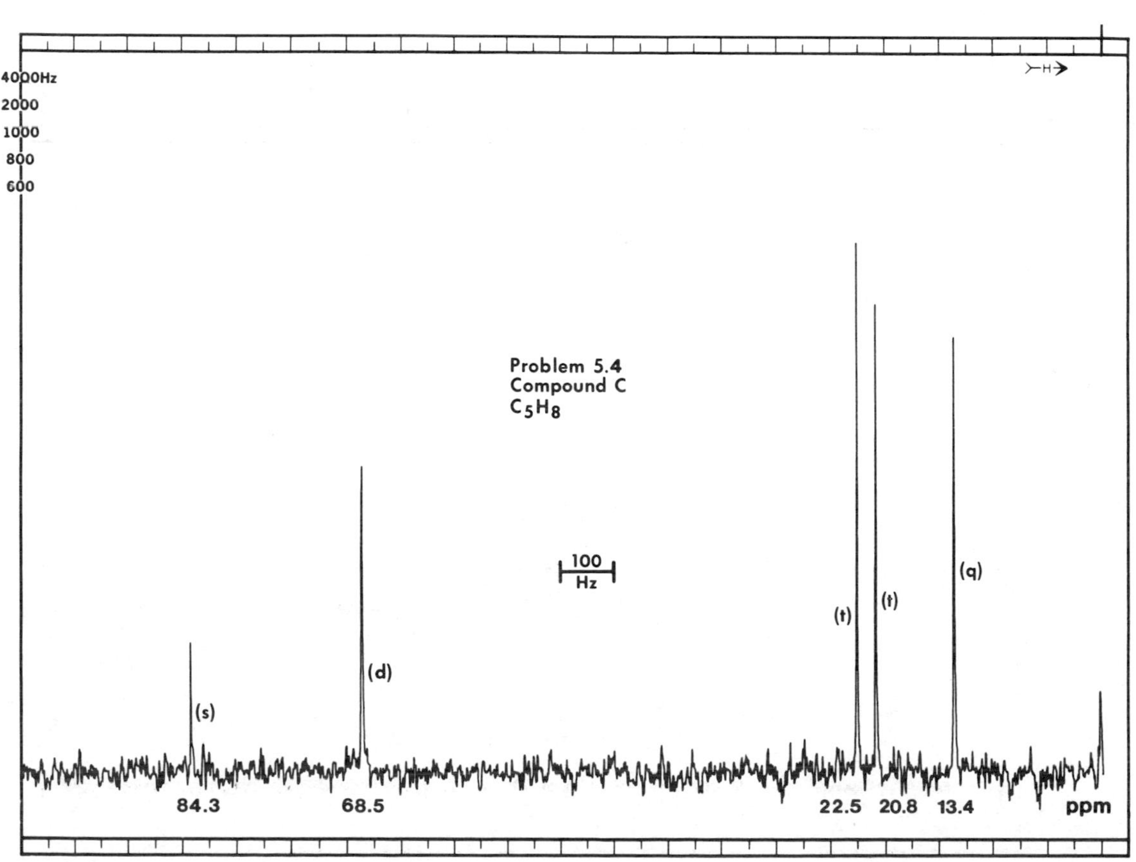
4000Hz
2000
1000
800
600
Problem 5.4
Compound C
C5H8
100
Hz
(t)
(t)
(q)
(d)
(s)
84.3
68.5
22.5
20.8
13.4
ppm

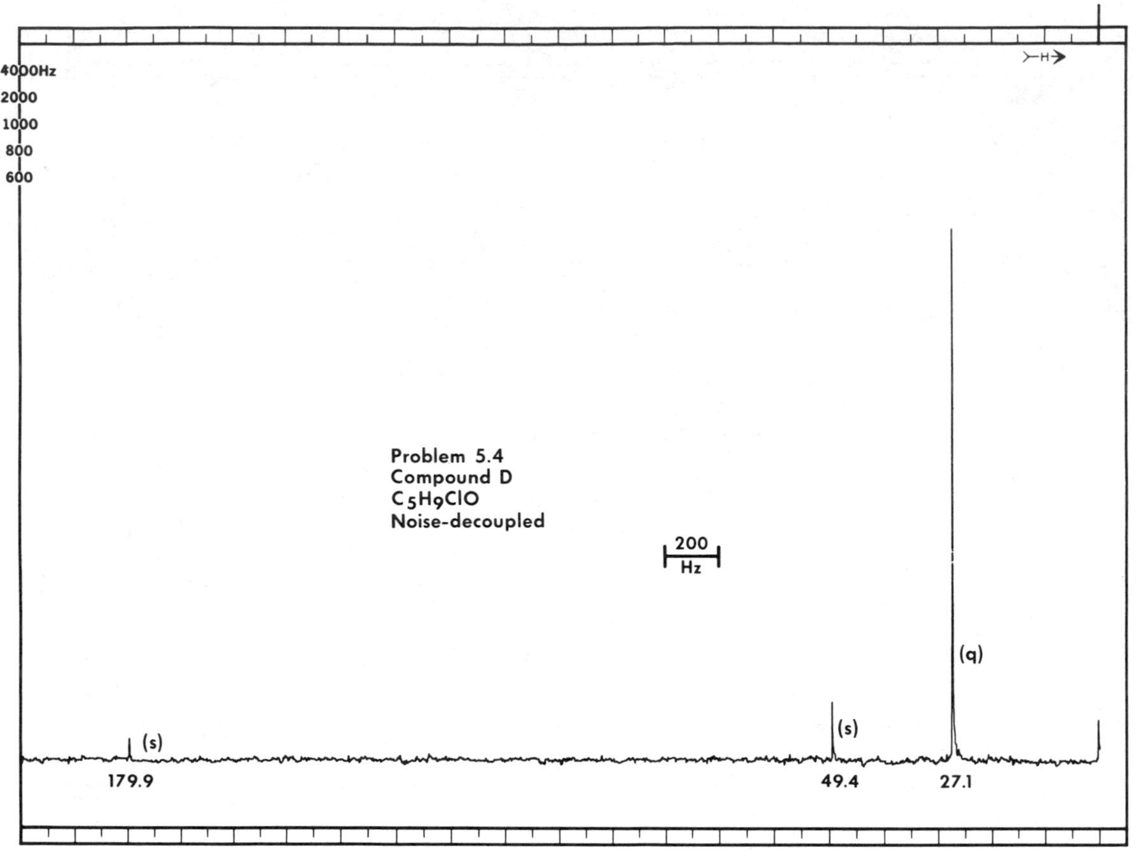
4000Hz
2000
1000
800
600
Problem 5.4
Compound D
C5H9ClO
Noise-decoupled
200
Hz
(q)
(s)
(s)
179.9
49.4
27.1

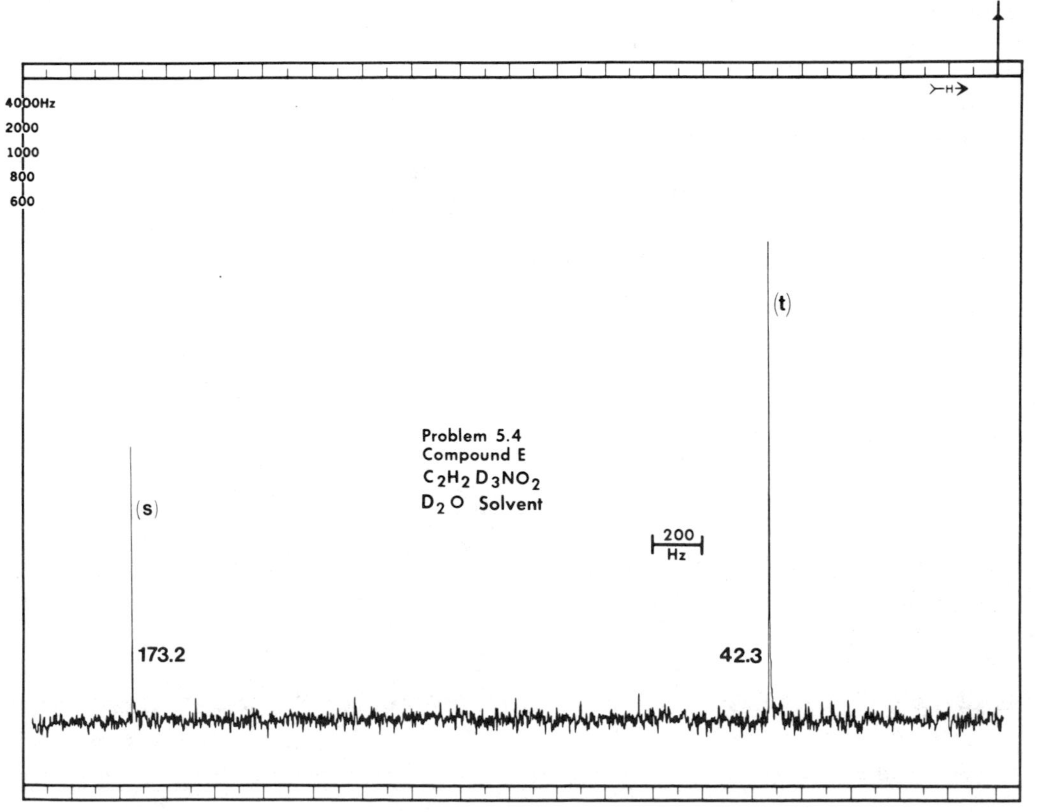
4000Hz
2000
1000
800
600
(t)
Problem 5.4
Compound E
C2H2D3NO2
D2O Solvent
(s)
200
Hz
173.2
42.3

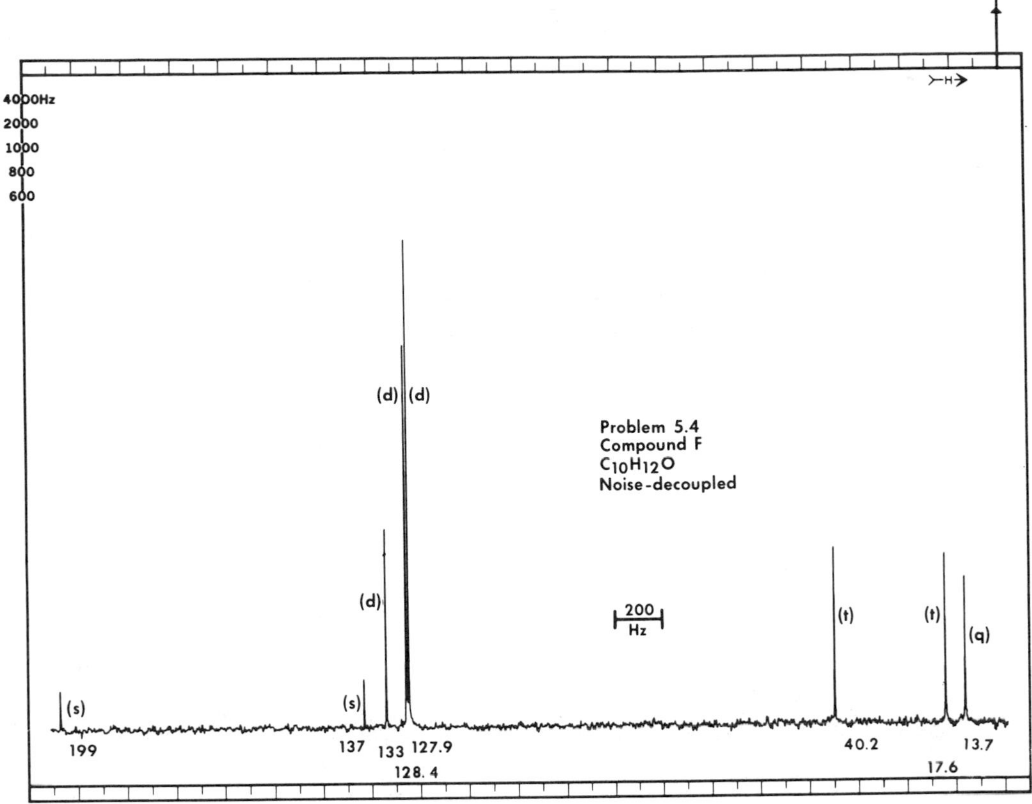
4000Hz
2000
1000
800
600
Problem 5.4
Compound F
$C_{10}H_{12}O$
Noise-decoupled
(s)
(s)
(d)
(d)
(d)
200 Hz
(t)
(t)
(q)
199
137
133
127.9
128.4
40.2
17.6
13.7

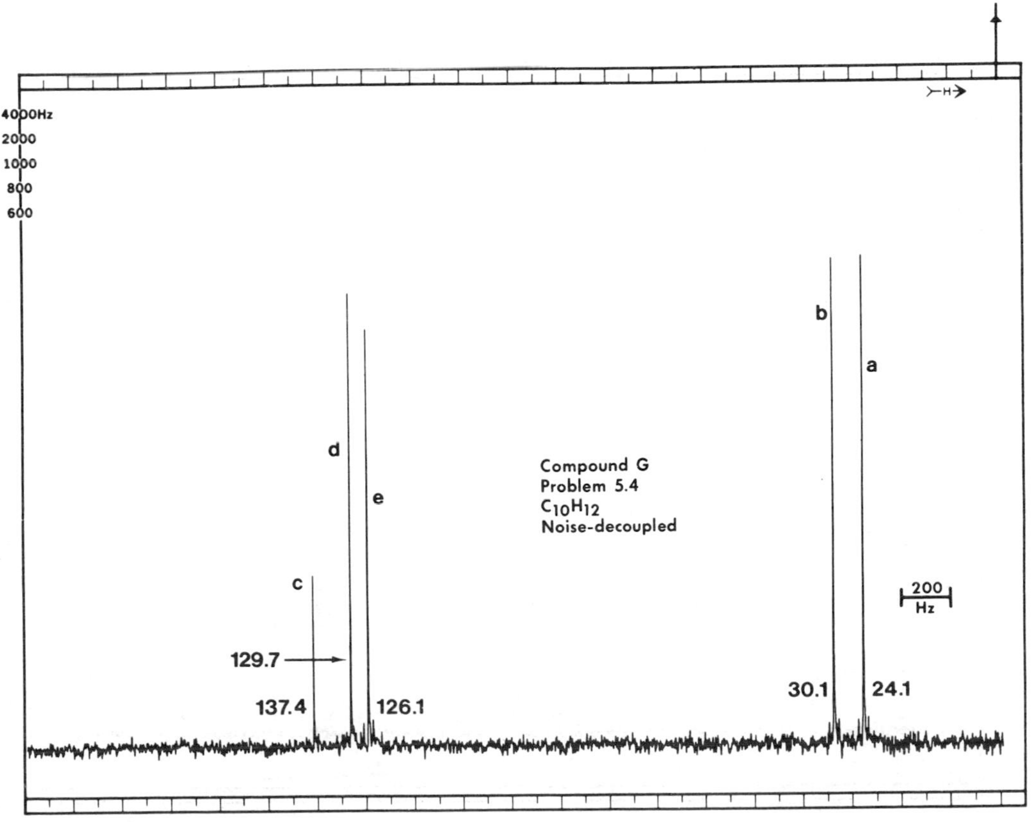
4000Hz
2000
1000
800
600
Compound G
Problem 5.4
$C_{10}H_{12}$
Noise-decoupled
c
d
e
b
a
200 Hz
129.7
137.4
126.1
30.1
24.1

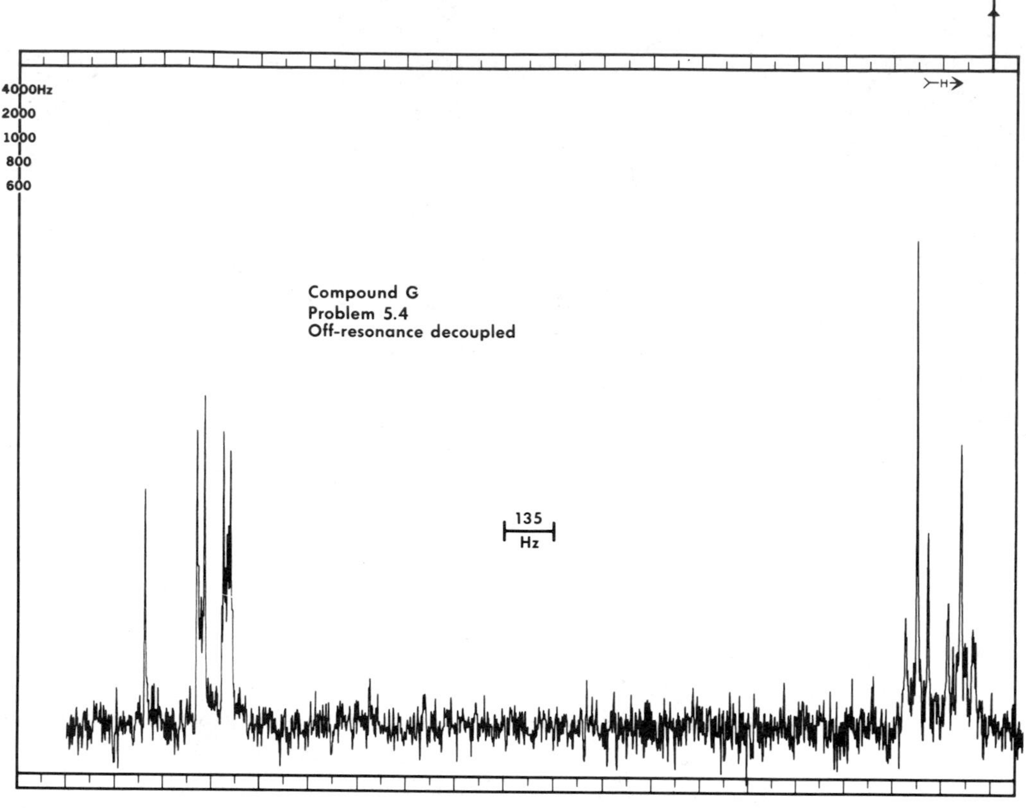
4000Hz
2000
1000
800
600
Compound G
Problem 5.4
Off-resonance decoupled
135
Hz

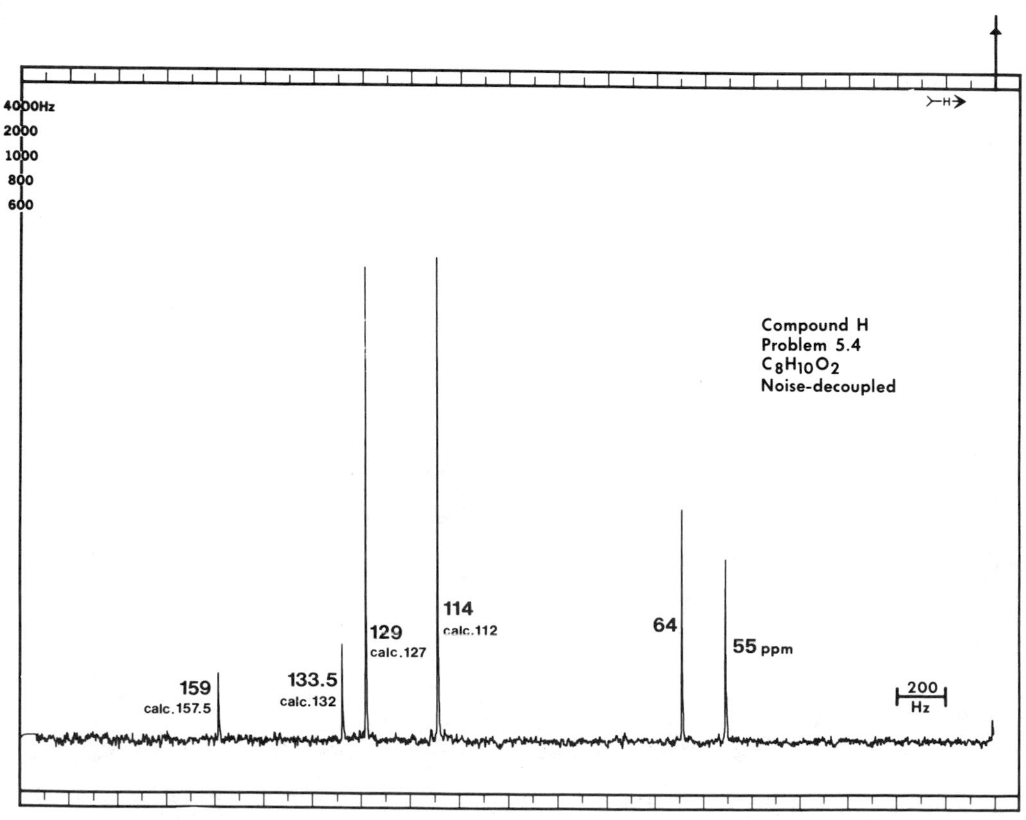
4000Hz
2000
1000
800
600
Compound H
Problem 5.4
C8H10O2
Noise-decoupled
159
calc.157.5
133.5
calc.132
129
calc.127
114
calc.112
64
55 ppm
200
Hz

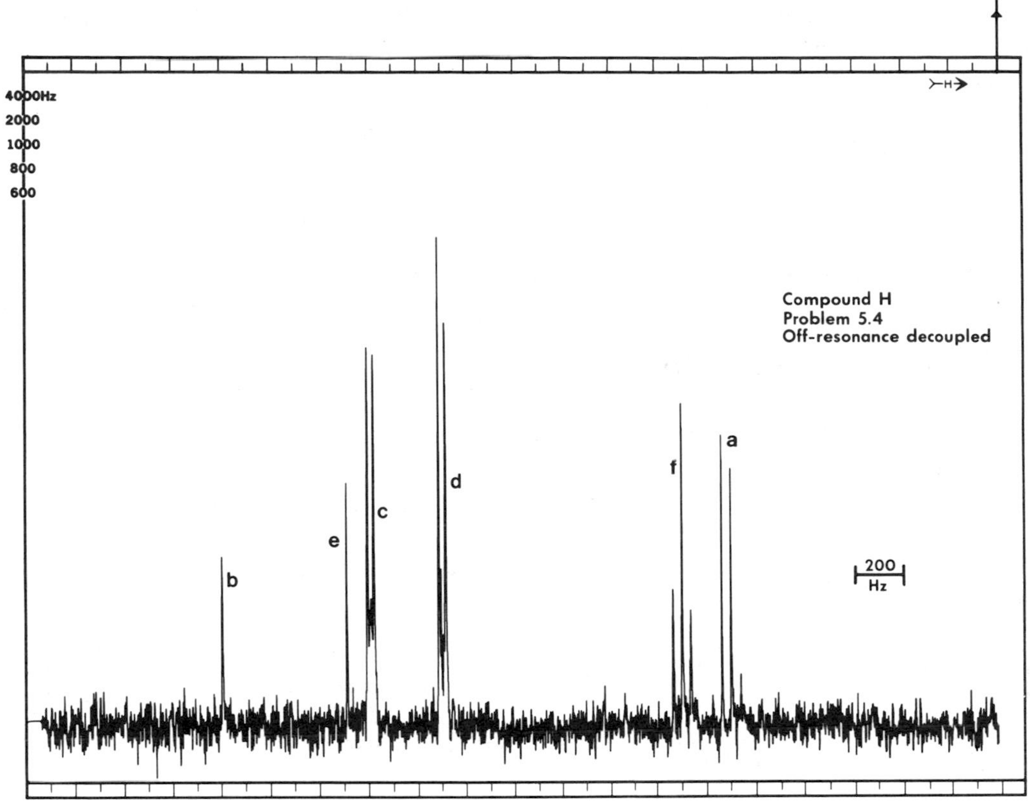
4000Hz
2000
1000
800
600
Compound H
Problem 5.4
Off-resonance decoupled
a
b
c
d
e
f
200
Hz

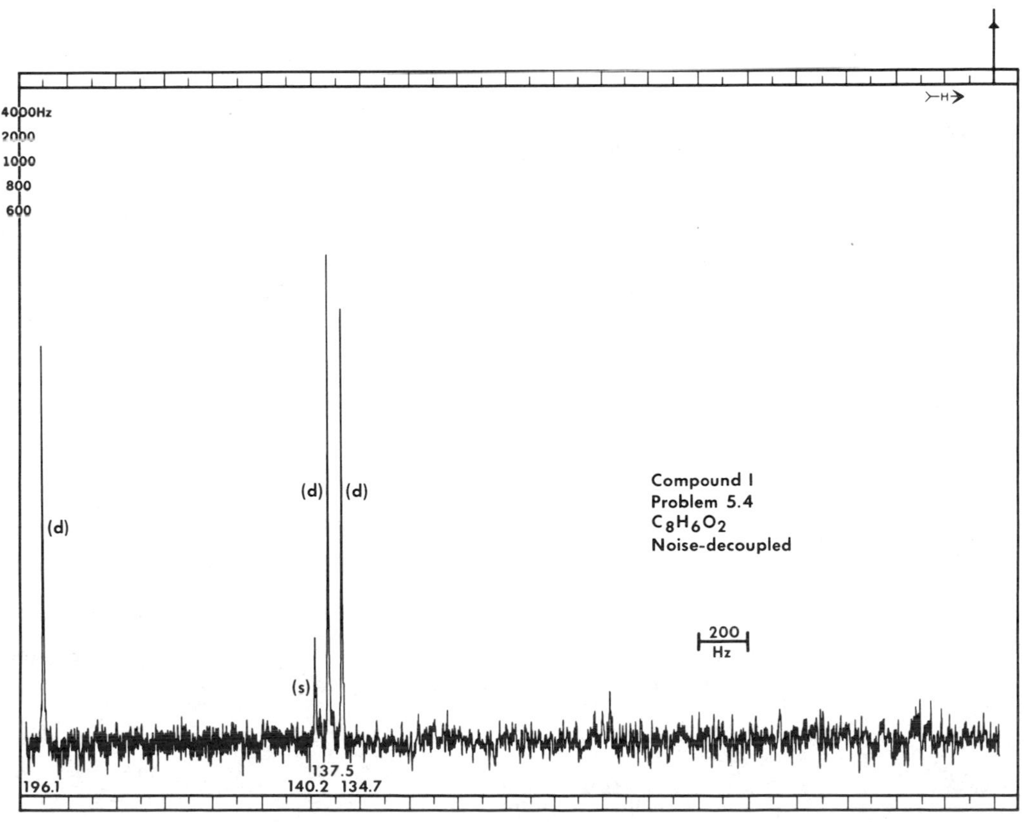
4000Hz
2000
1000
800
600
Compound I
Problem 5.4
$C_8H_6O_2$
Noise-decoupled
(d)
(d)
(d)
(s)
200
Hz
196.1
137.5
140.2
134.7

5.5 For each structure represented by the molecular formula, the ^{13}C peak shifts in ppm and the relative intensities in the noise-decoupled spectrum are given. The multiplicity of each peak in the off-resonance decoupled spectrum is shown as s, d, t, or q. Determine the structure.

A. $C_8H_{14}O_3$. 13.5 ppm (98, q), 18.1 ppm (95, t), 37.3 ppm (91, t), 169.6 ppm (34, s).
B. $C_7H_{12}O_4$. 14.2 ppm (80, q), 41.7 ppm (50, t), 61.4 ppm (100, t), 166.8 ppm (52, s).
C. $C_{13}H_{12}O$. 76.9 ppm (24, d), 128.3 ppm (99, d), 127.4 ppm (57, d), 129.3 ppm (87, d), 144.7 (12, s).
D. $C_5H_6O_4$, 38.2 ppm (100, t), 128.3 ppm (84, t), 136.1 ppm (99, s), 168.2 ppm (77, s), 172.7 (78, s).
E. $C_9H_{10}O$. 63.1 ppm (49, t), 126.4 ppm (98, ?), 127.5 ppm (44, ?), 128.5 ppm (87, ?), 128.7 ppm (60, ?), 130.7 ppm (42, d), 136.8 ppm (11, s).

5.6 Compare diethyl phthalate to the *meta* and *para* isomers with respect to the "carbon count" from the noise-decoupled spectrum, and the "proton count" from the off-resonance decoupled spectrum; that is, show how molecular symmetry leads to unique spectral results for each compound.

appendix a

^{13}C Chemical Shifts for Common Deuterated Solvents (neat, ppm from TMS)

	Shift	Multiplicity
Acetic Acid CD_3COOD	178.4, 20.0	1, 7
Acetone $(CD_3)_2CO$	206.0, 29.8	1, 7
Acetonitrile CD_3CN	117.7, 1.3	1, 7
Benzene C_6D_6	128.5	1
(Carbon tetrachloride CCl_4)	96.0	1
Chloroform $CDCl_3$	77.0	3
Dimethyl sulfoxide $(CD_3)_2SO$	39.5	7
Dioxane $C_4D_8O_2$	67.4	1
Methanol CD_3OD	49.0	7
Methylene chloride CD_2Cl_2	53.8	5
Nitromethane CD_3NO_2	57.3	7
Pyridine C_5D_5N	149.9, 135.5, 123.5	3, 3, 3
(Carbon disulfide CS_2)	192.8	1

appendix b. comparison of ^1H and ^{13}C chemical shifts.

From Application of Fourier Transform NMR to Carbon-13,
Varian Associates Lecture Booklet, 1974. Reprinted with permission.

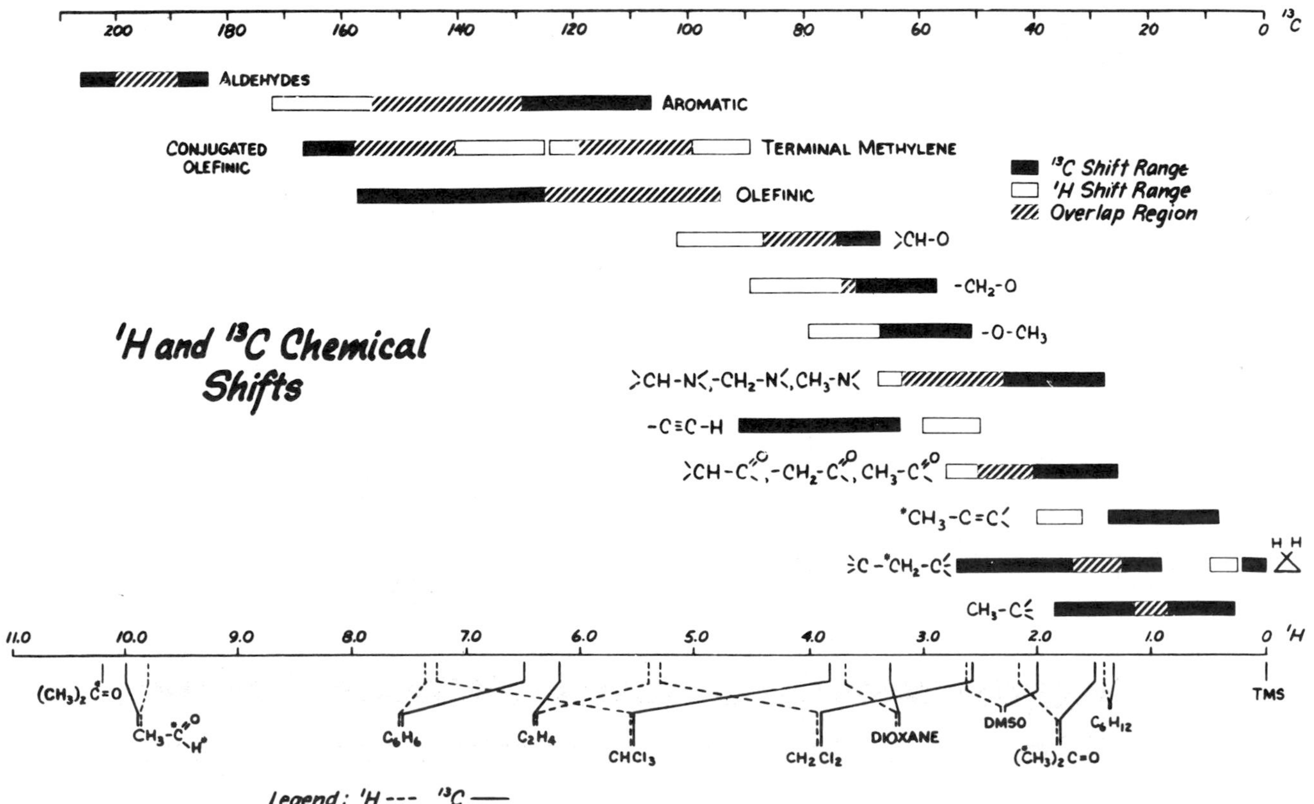

appendix c. ^{13}C correlation chart for chemical classes.

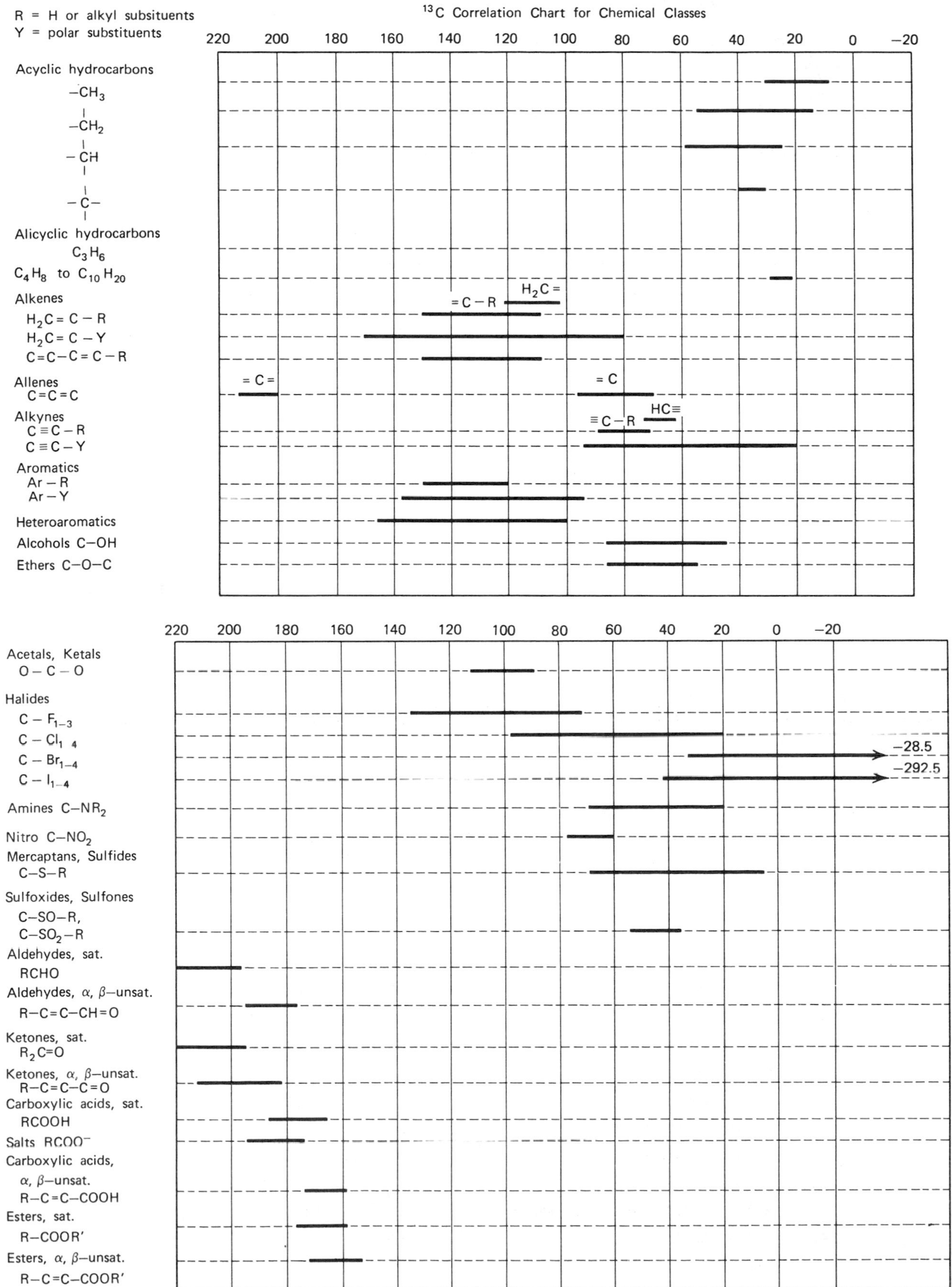

Continued on next page

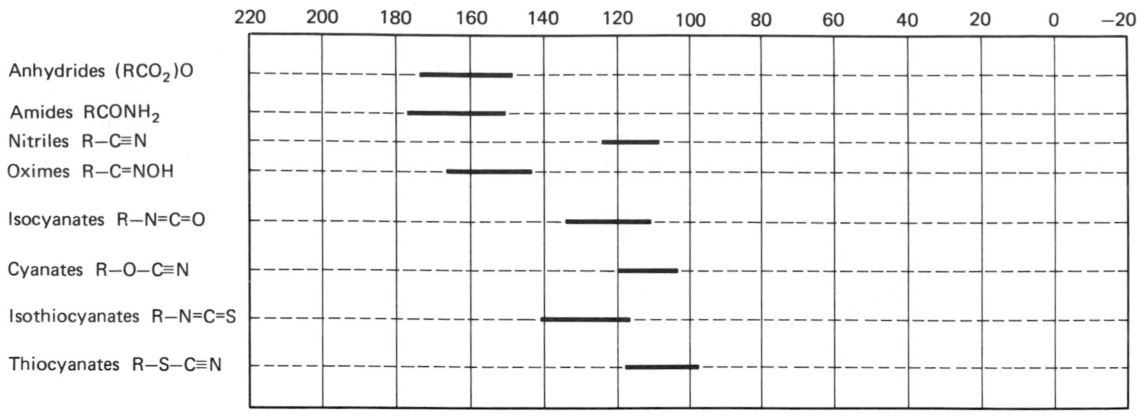

appendix d. ^{13}CNMR spectra of common organic compounds.

One the next several pages are ^{13}C—NMR spectra of common organic compounds. All signals are assigned. Permission for the publication herein of Sadtler Spectra® has been granted, and all rights are reserved by Sadtler Laboratories, Inc.

No.	Compound	No.	Compound
1	Acetone	14	Hexane
2	Acetophenone	15	1-Hexanol
3	Acrylonitrile	16	Iodoethane
4	Benzene	17	Methyl phenyl sulfide
5	Benzofuran	18	α-Methylstyrene
6	Bromocyclohexane	19	Myristic Acid, methyl ester
7	Cyclohexanone	20	Oxalic acid, diallyl ester
8	*p*-Dichlorobenzene	21	Phenol
9	Diethylene glycol	22	3-Picoline
10	Diphenylamine	23	Sodium 2,2-dimethyl-2-silapentane-5-sulfonate (DSS)
11	*N,N*-Dimethylformamide (DMF)	24	Tetrahydrofuran (THF)
12	Dipropylamine	25	Toluene
13	Furan		

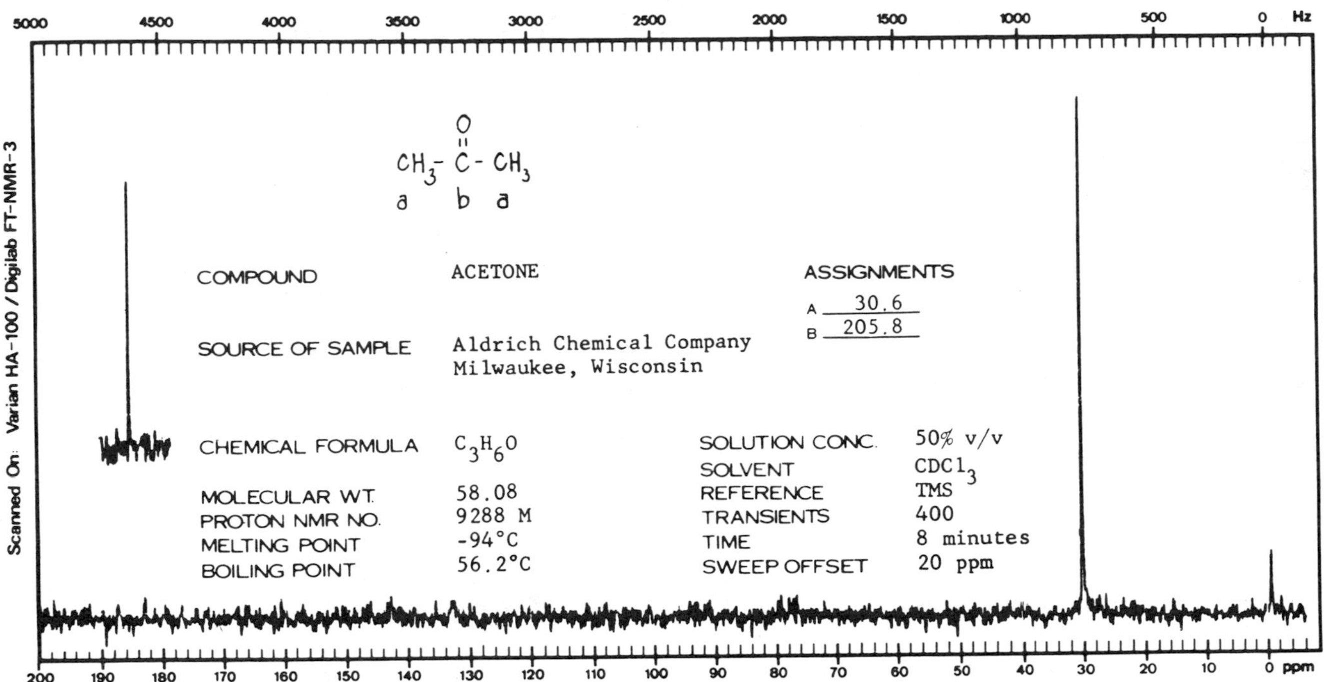
5000 4500 4000 3500 3000 2500 2000 1500 1000 500 0 Hz
Scanned On: Varian HA-100 / Digilab FT-NMR-3
$CH_3-\overset{O}{\overset{\|}{C}}-CH_3$
a b a
COMPOUND ACETONE
ASSIGNMENTS
A 30.6
B 205.8
SOURCE OF SAMPLE Aldrich Chemical Company
Milwaukee, Wisconsin
CHEMICAL FORMULA C_3H_6O
SOLUTION CONC. 50% v/v
SOLVENT $CDCl_3$
MOLECULAR WT. 58.08
REFERENCE TMS
PROTON NMR NO. 9288 M
TRANSIENTS 400
MELTING POINT -94°C
TIME 8 minutes
BOILING POINT 56.2°C
SWEEP OFFSET 20 ppm
200 190 180 170 160 150 140 130 120 110 100 90 80 70 60 50 40 30 20 10 0 ppm

No. 1

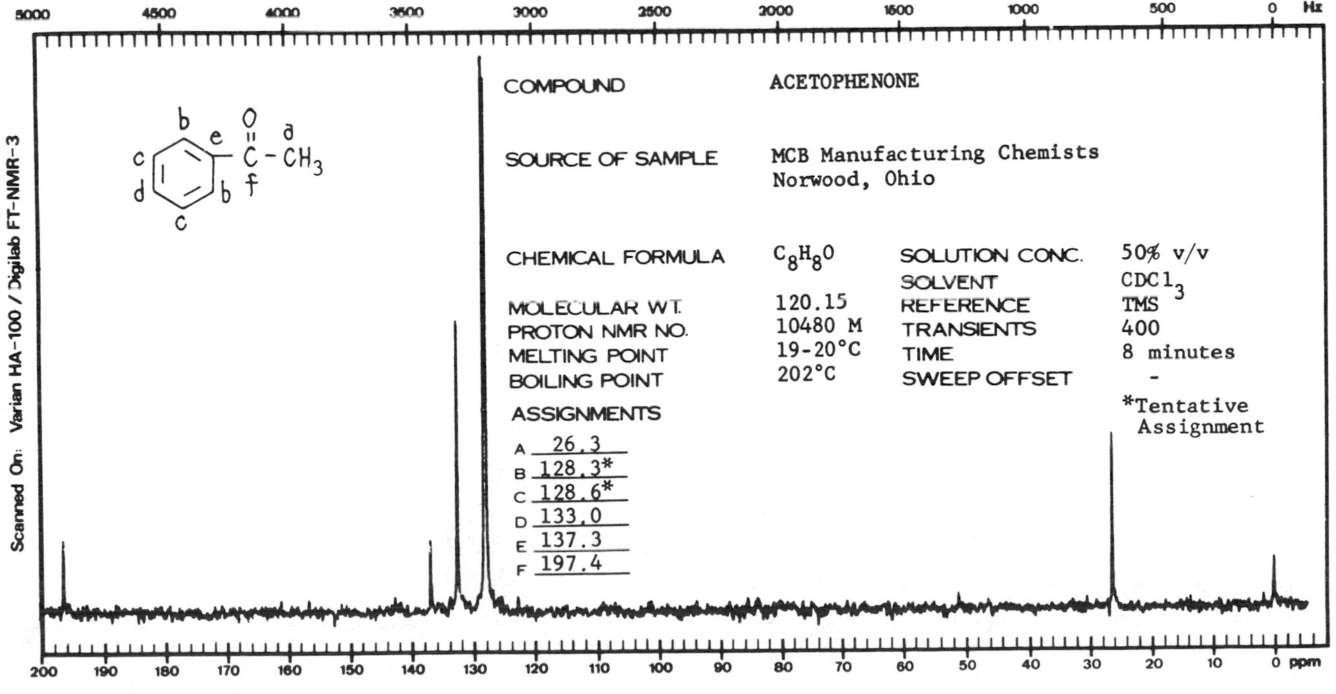
5000 4500 4000 3500 3000 2500 2000 1500 1000 500 0 Hz
Scanned On: Varian HA-100 / Digilab FT-NMR-3
b e O a
c C - CH_3
d b f
c
COMPOUND ACETOPHENONE
SOURCE OF SAMPLE MCB Manufacturing Chemists
Norwood, Ohio
CHEMICAL FORMULA C_8H_8O
SOLUTION CONC. 50% v/v
SOLVENT $CDCl_3$
MOLECULAR WT. 120.15
REFERENCE TMS
PROTON NMR NO. 10480 M
TRANSIENTS 400
MELTING POINT 19-20°C
TIME 8 minutes
BOILING POINT 202°C
SWEEP OFFSET -
ASSIGNMENTS
*Tentative Assignment
A 26.3
B 128.3*
C 128.6*
D 133.0
E 137.3
F 197.4
200 190 180 170 160 150 140 130 120 110 100 90 80 70 60 50 40 30 20 10 0 ppm

No. 2

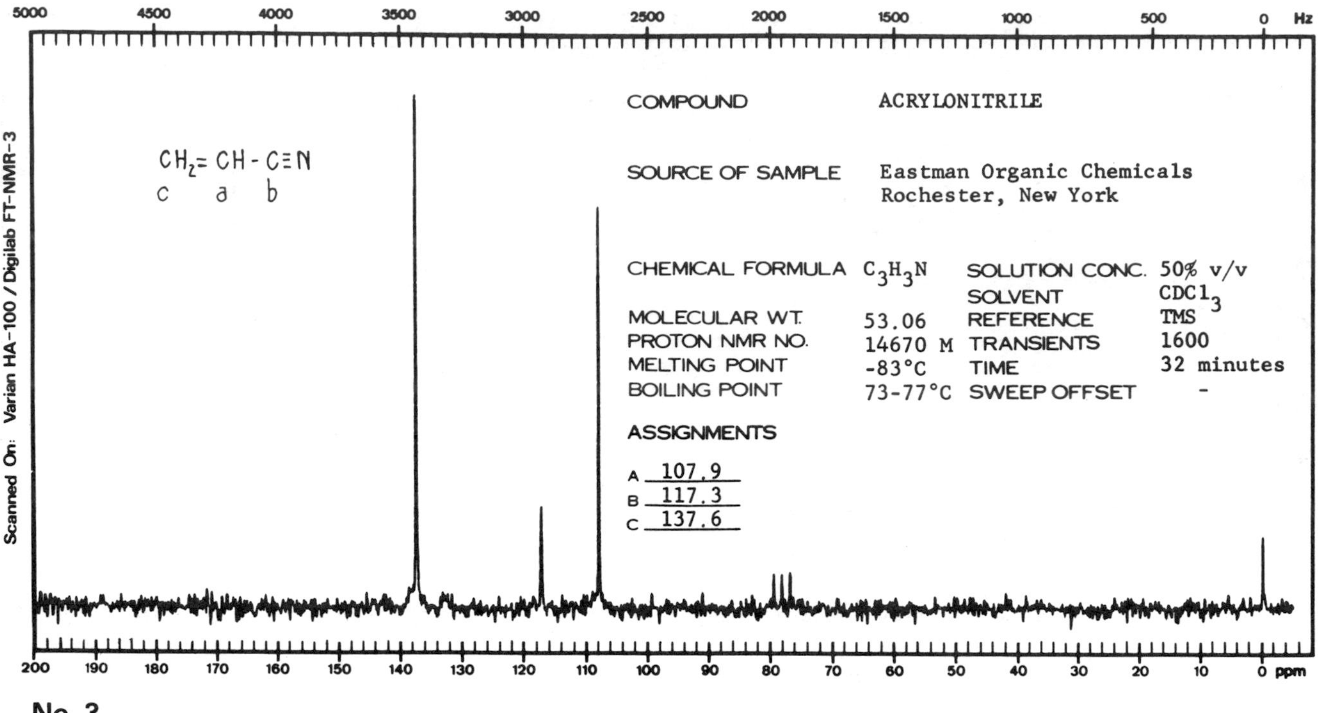
5000
4500
4000
3500
3000
2500
2000
1500
1000
500
0 Hz
Scanned On: Varian HA-100 / Digilab FT-NMR-3
CH₂= CH - C≡N
c a b
COMPOUND ACRYLONITRILE
SOURCE OF SAMPLE Eastman Organic Chemicals
Rochester, New York
CHEMICAL FORMULA C3H3N
SOLUTION CONC. 50% v/v
SOLVENT CDCl3
MOLECULAR WT. 53.06
REFERENCE TMS
PROTON NMR NO. 14670 M
TRANSIENTS 1600
MELTING POINT -83°C
TIME 32 minutes
BOILING POINT 73-77°C
SWEEP OFFSET -
ASSIGNMENTS
A 107.9
B 117.3
C 137.6
200
190
180
170
160
150
140
130
120
110
100
90
80
70
60
50
40
30
20
10
0 ppm

No. 3

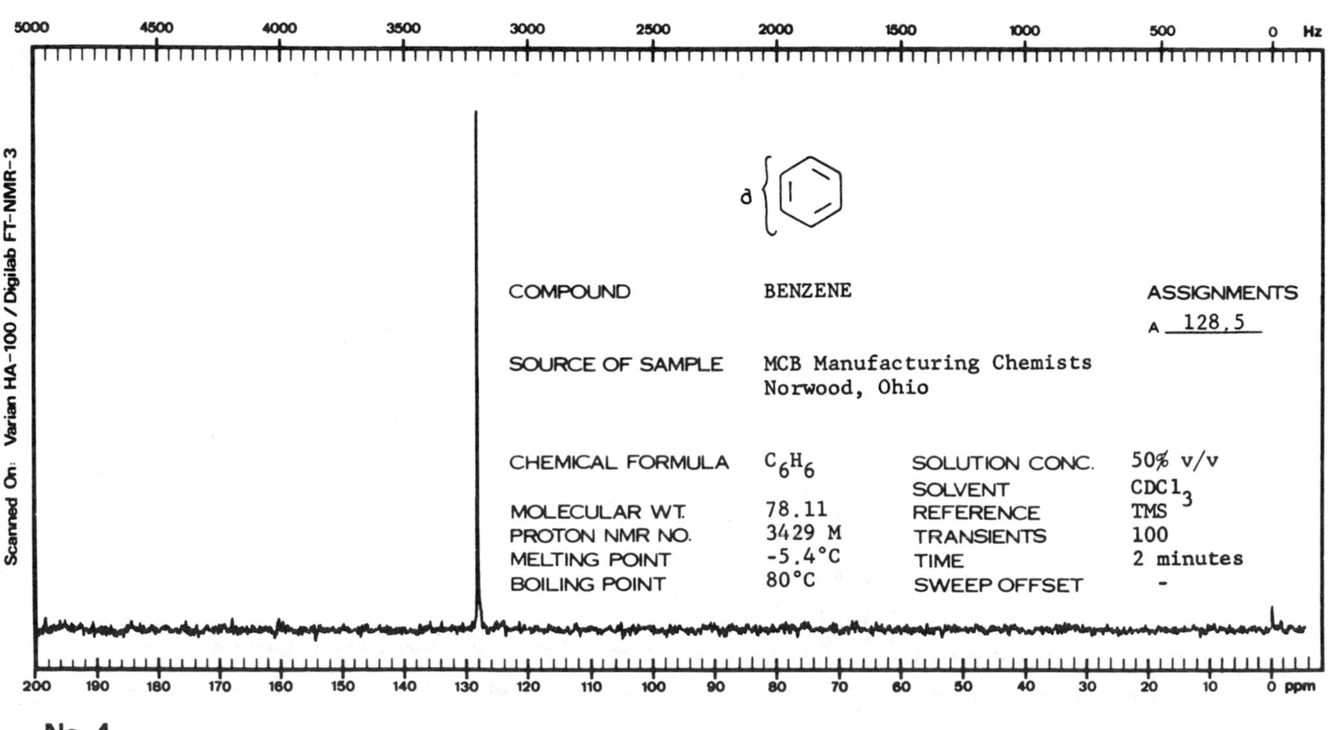
5000
4500
4000
3500
3000
2500
2000
1500
1000
500
0 Hz
Scanned On: Varian HA-100 / Digilab FT-NMR-3
a
COMPOUND BENZENE
ASSIGNMENTS
A 128.5
SOURCE OF SAMPLE MCB Manufacturing Chemists
Norwood, Ohio
CHEMICAL FORMULA C6H6
SOLUTION CONC. 50% v/v
SOLVENT CDCl3
MOLECULAR WT. 78.11
REFERENCE TMS
PROTON NMR NO. 3429 M
TRANSIENTS 100
MELTING POINT -5.4°C
TIME 2 minutes
BOILING POINT 80°C
SWEEP OFFSET -
200
190
180
170
160
150
140
130
120
110
100
90
80
70
60
50
40
30
20
10
0 ppm

No. 4

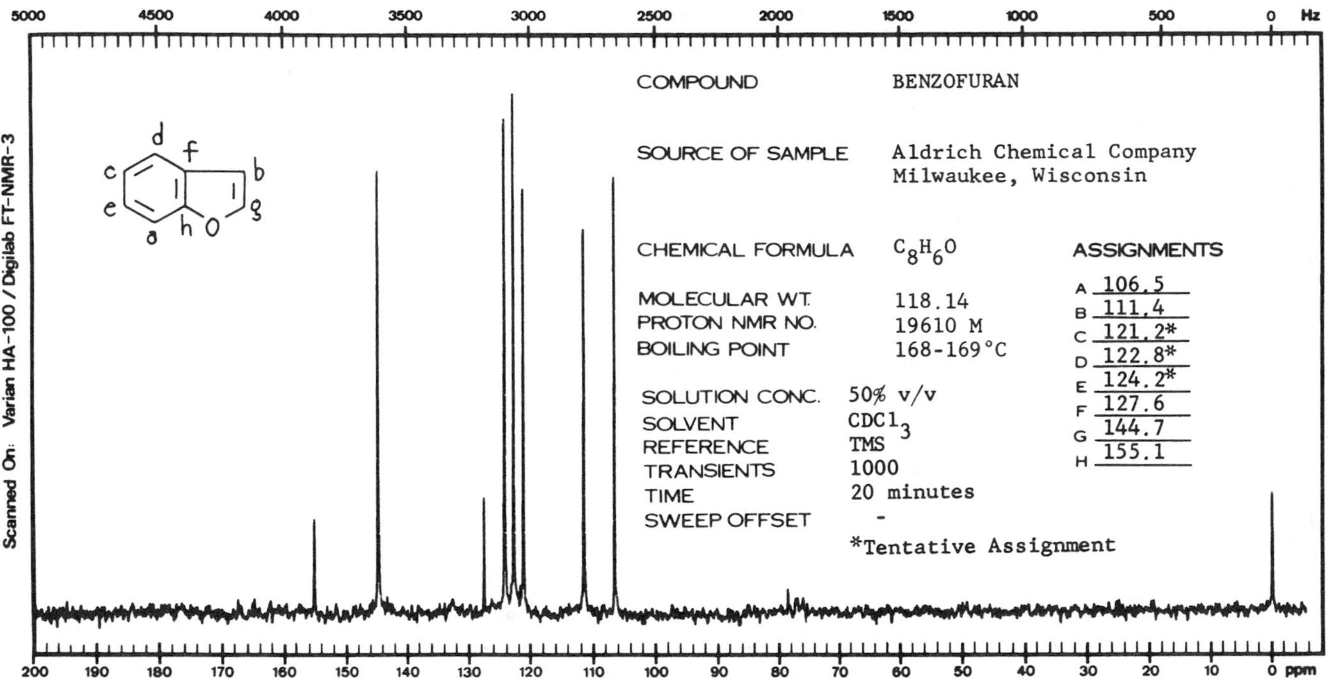

Scanned On: Varian HA-100 / Digilab FT-NMR-3
COMPOUND BENZOFURAN
SOURCE OF SAMPLE Aldrich Chemical Company
Milwaukee, Wisconsin
CHEMICAL FORMULA C8H6O
MOLECULAR WT. 118.14
PROTON NMR NO. 19610 M
BOILING POINT 168-169°C
SOLUTION CONC. 50% v/v
SOLVENT CDCl3
REFERENCE TMS
TRANSIENTS 1000
TIME 20 minutes
SWEEP OFFSET -
*Tentative Assignment
ASSIGNMENTS
A 106.5
B 111.4
C 121.2*
D 122.8*
E 124.2*
F 127.6
G 144.7
H 155.1

No. 5

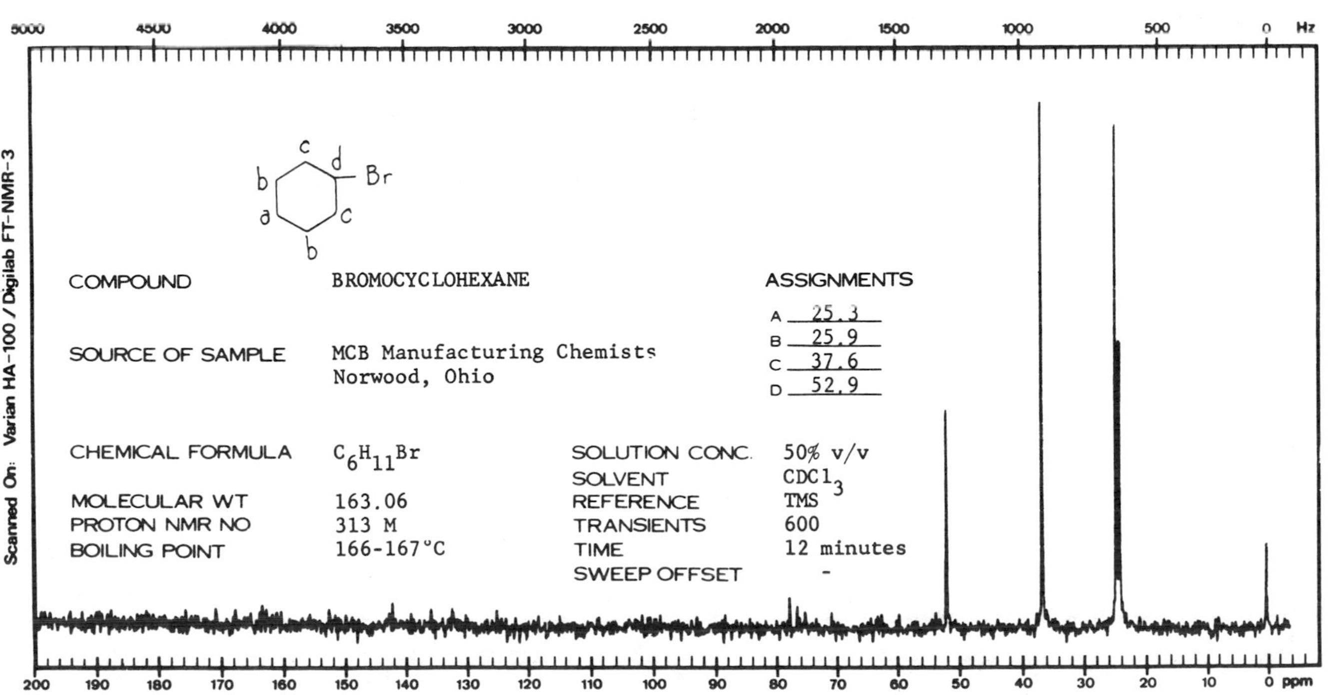

Scanned On: Varian HA-100 / Digilab FT-NMR-3
COMPOUND BROMOCYCLOHEXANE
SOURCE OF SAMPLE MCB Manufacturing Chemists
Norwood, Ohio
CHEMICAL FORMULA C6H11Br
MOLECULAR WT 163.06
PROTON NMR NO 313 M
BOILING POINT 166-167°C
SOLUTION CONC. 50% v/v
SOLVENT CDCl3
REFERENCE TMS
TRANSIENTS 600
TIME 12 minutes
SWEEP OFFSET -
ASSIGNMENTS
A 25.3
B 25.9
C 37.6
D 52.9

No. 6

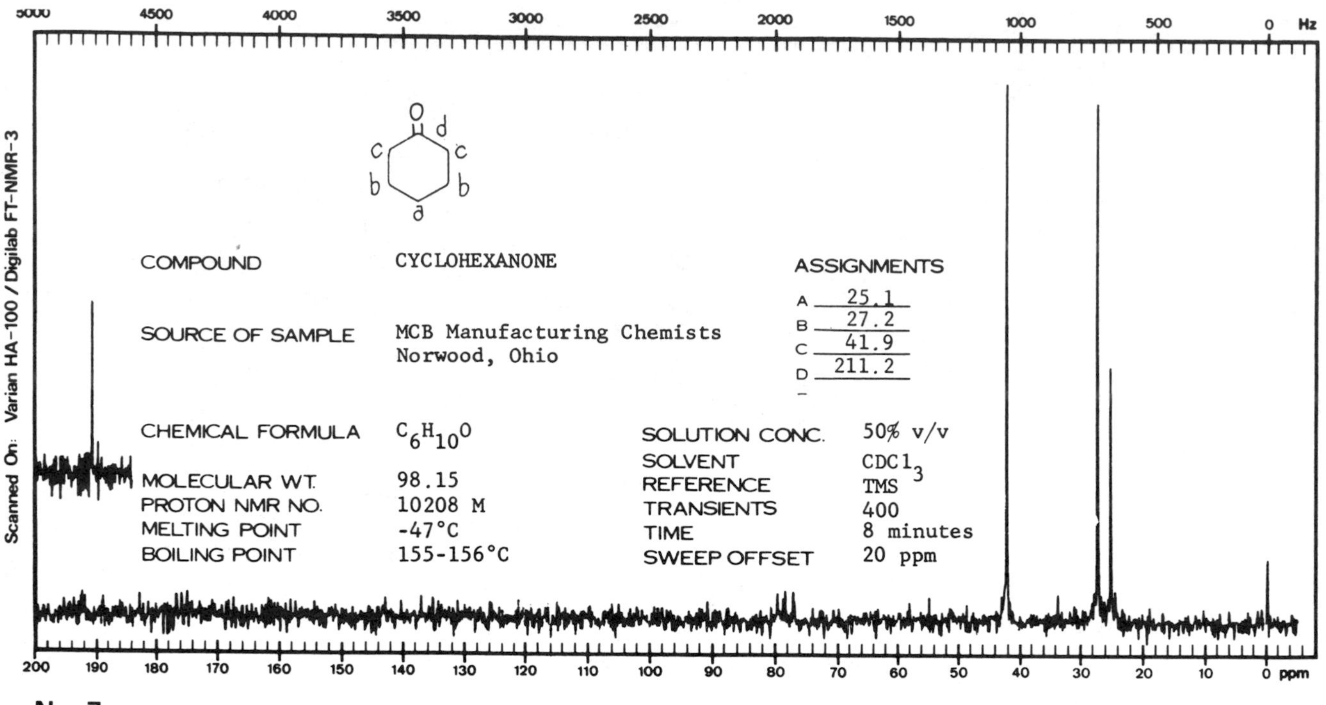
Scanned On: Varian HA-100 / Digilab FT-NMR-3
COMPOUND CYCLOHEXANONE
SOURCE OF SAMPLE MCB Manufacturing Chemists Norwood, Ohio
CHEMICAL FORMULA C6H10O
MOLECULAR WT. 98.15
PROTON NMR NO. 10208 M
MELTING POINT -47°C
BOILING POINT 155-156°C
SOLUTION CONC. 50% v/v
SOLVENT CDCl3
REFERENCE TMS
TRANSIENTS 400
TIME 8 minutes
SWEEP OFFSET 20 ppm
ASSIGNMENTS
A 25.1
B 27.2
C 41.9
D 211.2

No. 7

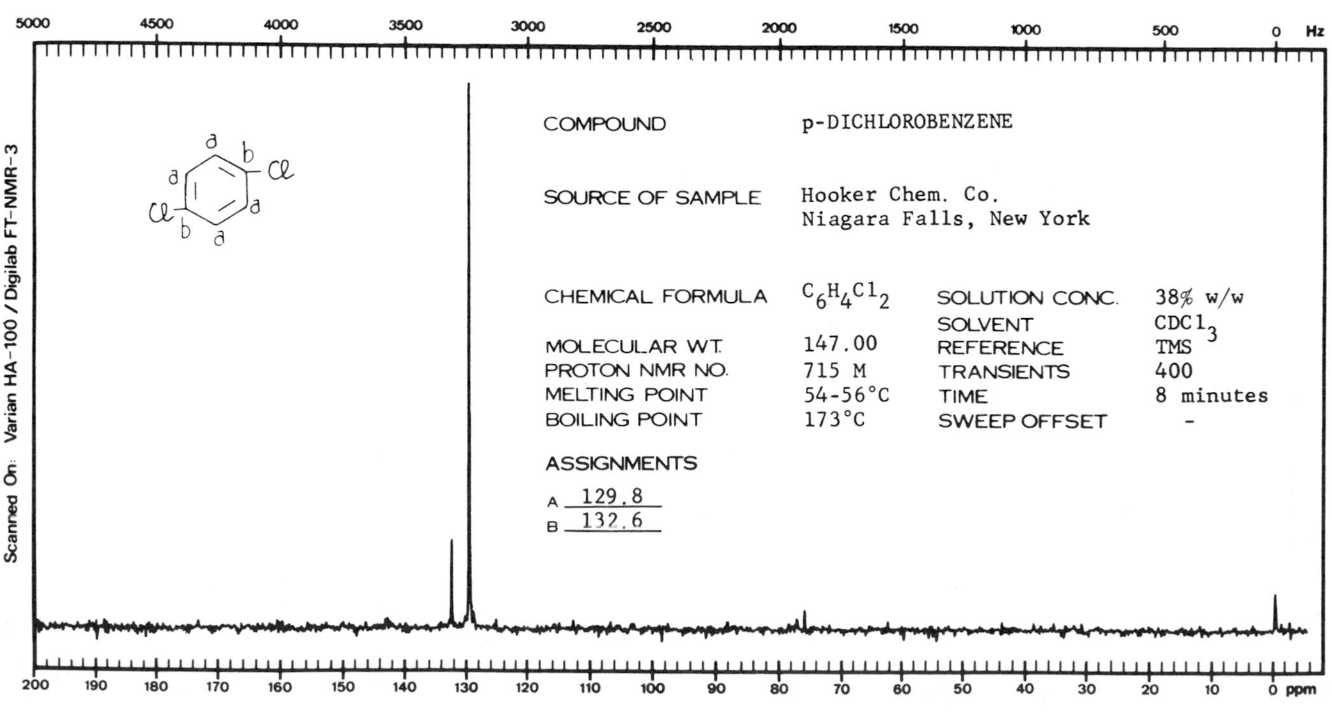
Scanned On: Varian HA-100 / Digilab FT-NMR-3
COMPOUND p-DICHLOROBENZENE
SOURCE OF SAMPLE Hooker Chem. Co. Niagara Falls, New York
CHEMICAL FORMULA C6H4Cl2
MOLECULAR WT. 147.00
PROTON NMR NO. 715 M
MELTING POINT 54-56°C
BOILING POINT 173°C
SOLUTION CONC. 38% w/w
SOLVENT CDCl3
REFERENCE TMS
TRANSIENTS 400
TIME 8 minutes
SWEEP OFFSET -
ASSIGNMENTS
A 129.8
B 132.6

No. 8

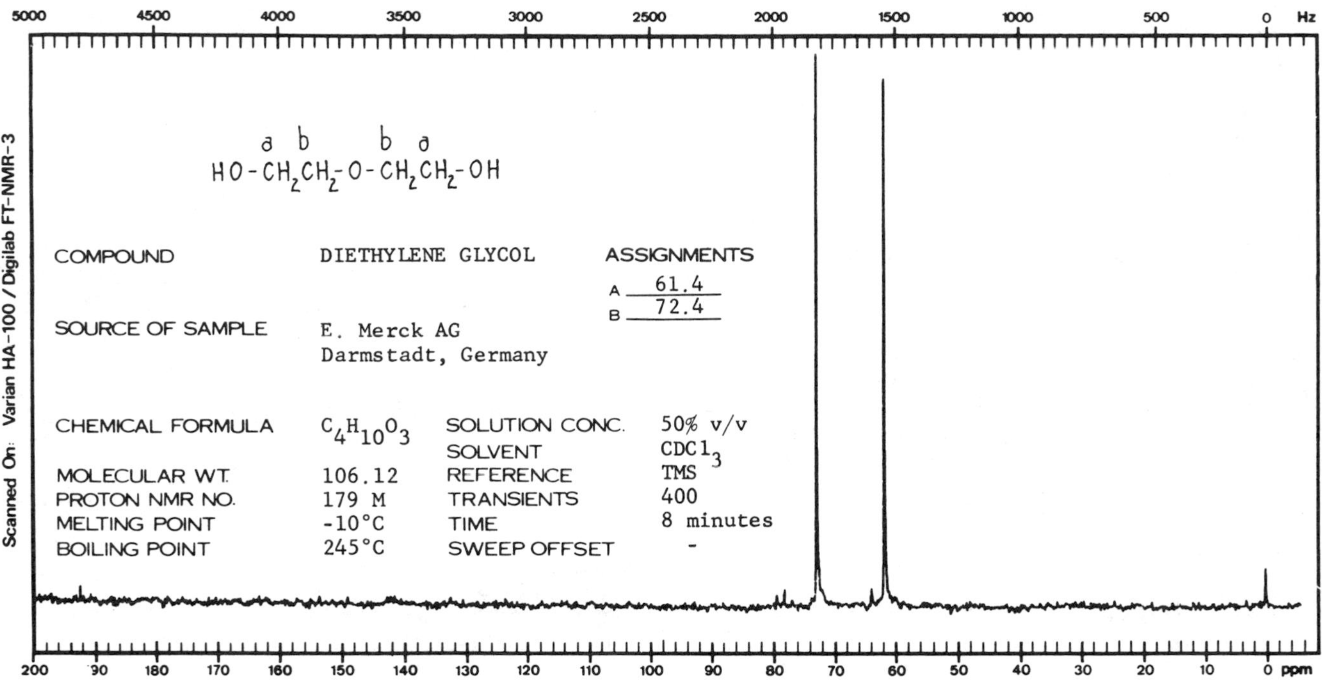

Scanned On: Varian HA-100 / Digilab FT-NMR-3
a b b a
HO-CH2CH2-O-CH2CH2-OH
COMPOUND DIETHYLENE GLYCOL
ASSIGNMENTS
A 61.4
B 72.4
SOURCE OF SAMPLE E. Merck AG
Darmstadt, Germany
CHEMICAL FORMULA C4H10O3
SOLUTION CONC. 50% v/v
SOLVENT CDCl3
MOLECULAR WT. 106.12
REFERENCE TMS
PROTON NMR NO. 179 M
TRANSIENTS 400
MELTING POINT -10°C
TIME 8 minutes
BOILING POINT 245°C
SWEEP OFFSET -
Hz
ppm

No. 9

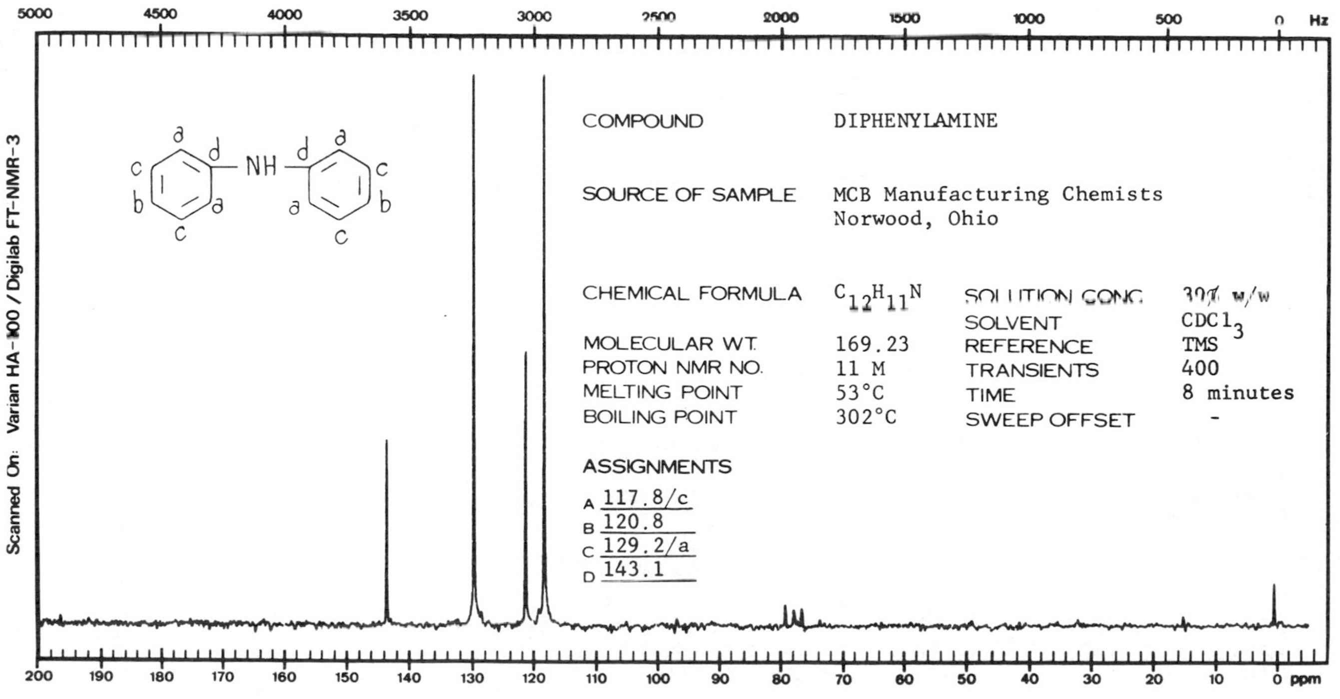

Scanned On: Varian HA-100 / Digilab FT-NMR-3
COMPOUND DIPHENYLAMINE
SOURCE OF SAMPLE MCB Manufacturing Chemists
Norwood, Ohio
CHEMICAL FORMULA C12H11N
SOLUTION CONC. 30% w/w
SOLVENT CDCl3
MOLECULAR WT. 169.23
REFERENCE TMS
PROTON NMR NO. 11 M
TRANSIENTS 400
MELTING POINT 53°C
TIME 8 minutes
BOILING POINT 302°C
SWEEP OFFSET -
ASSIGNMENTS
A 117.8/c
B 120.8
C 129.2/a
D 143.1
Hz
ppm

No. 10

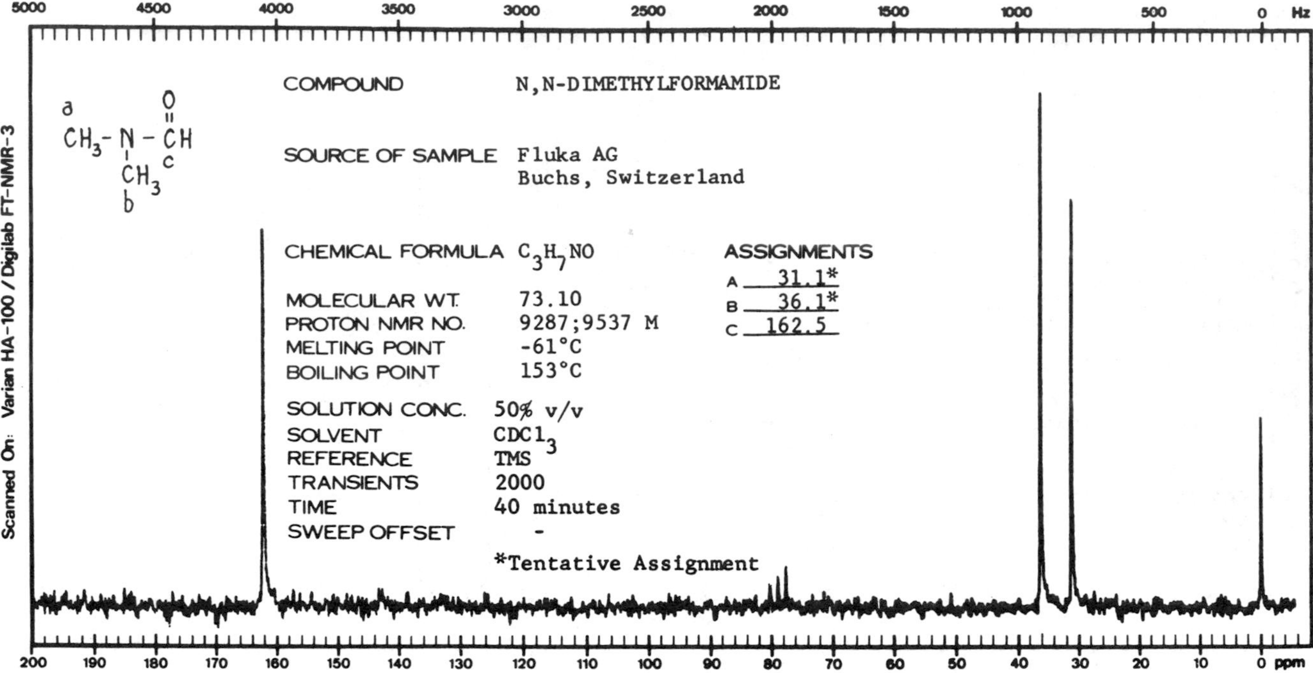

No. 11

5000 4500 4000 3500 3000 2500 2000 1500 1000 500 0 Hz

Scanned On: Varian HA-100 / Digilab FT-NMR-3

a b c c b a
$CH_3CH_2CH_2-NH-CH_2CH_2CH_3$

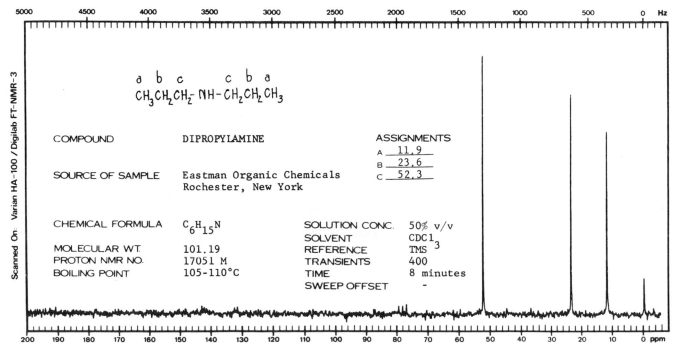

COMPOUND	DIPROPYLAMINE		
SOURCE OF SAMPLE	Eastman Organic Chemicals Rochester, New York		
CHEMICAL FORMULA	$C_6H_{15}N$	SOLUTION CONC.	50% v/v
		SOLVENT	$CDCl_3$
MOLECULAR WT.	101.19	REFERENCE	TMS
PROTON NMR NO.	17051 M	TRANSIENTS	400
BOILING POINT	105-110°C	TIME	8 minutes
		SWEEP OFFSET	-

ASSIGNMENTS
A 11.9
B 23.6
C 52.3

200 190 180 170 160 150 140 130 120 110 100 90 80 70 60 50 40 30 20 10 0 ppm

No. 12

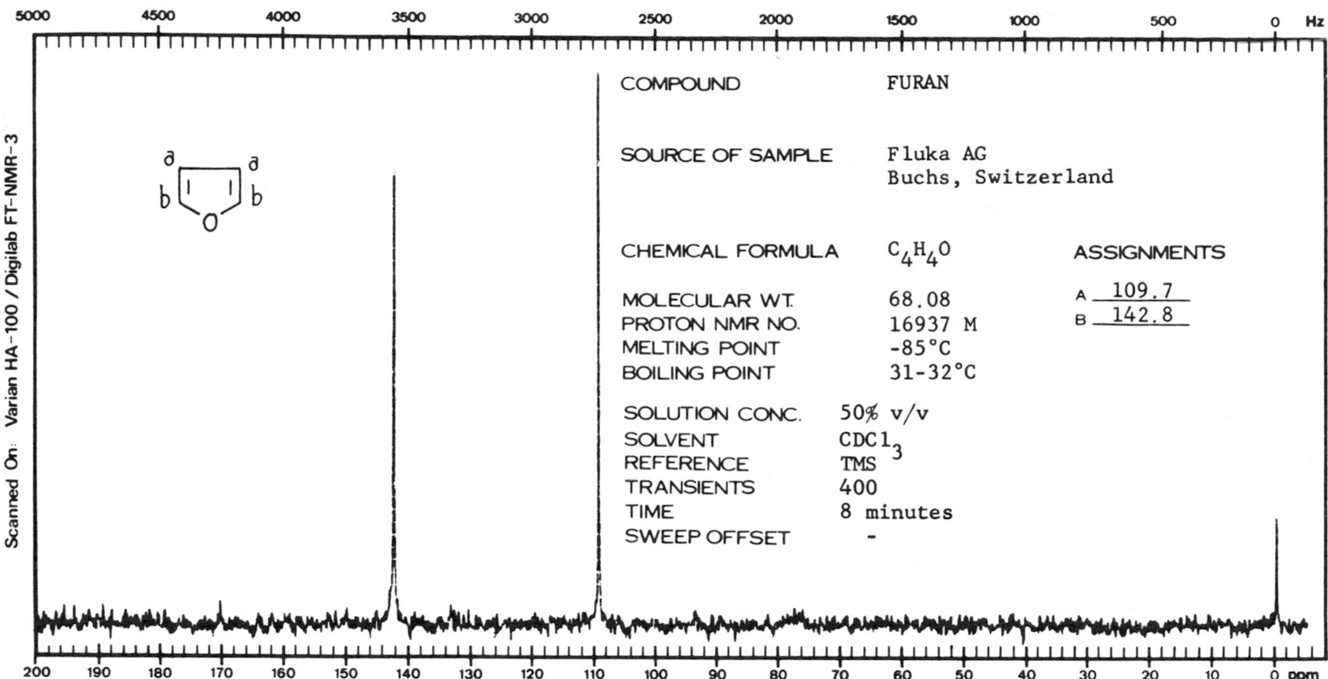
Scanned On: Varian HA-100 / Digilab FT-NMR-3
COMPOUND FURAN
SOURCE OF SAMPLE Fluka AG Buchs, Switzerland
CHEMICAL FORMULA C4H4O
MOLECULAR WT. 68.08
PROTON NMR NO. 16937 M
MELTING POINT -85°C
BOILING POINT 31-32°C
ASSIGNMENTS
A 109.7
B 142.8
SOLUTION CONC. 50% v/v
SOLVENT CDCl3
REFERENCE TMS
TRANSIENTS 400
TIME 8 minutes
SWEEP OFFSET -

No. 13

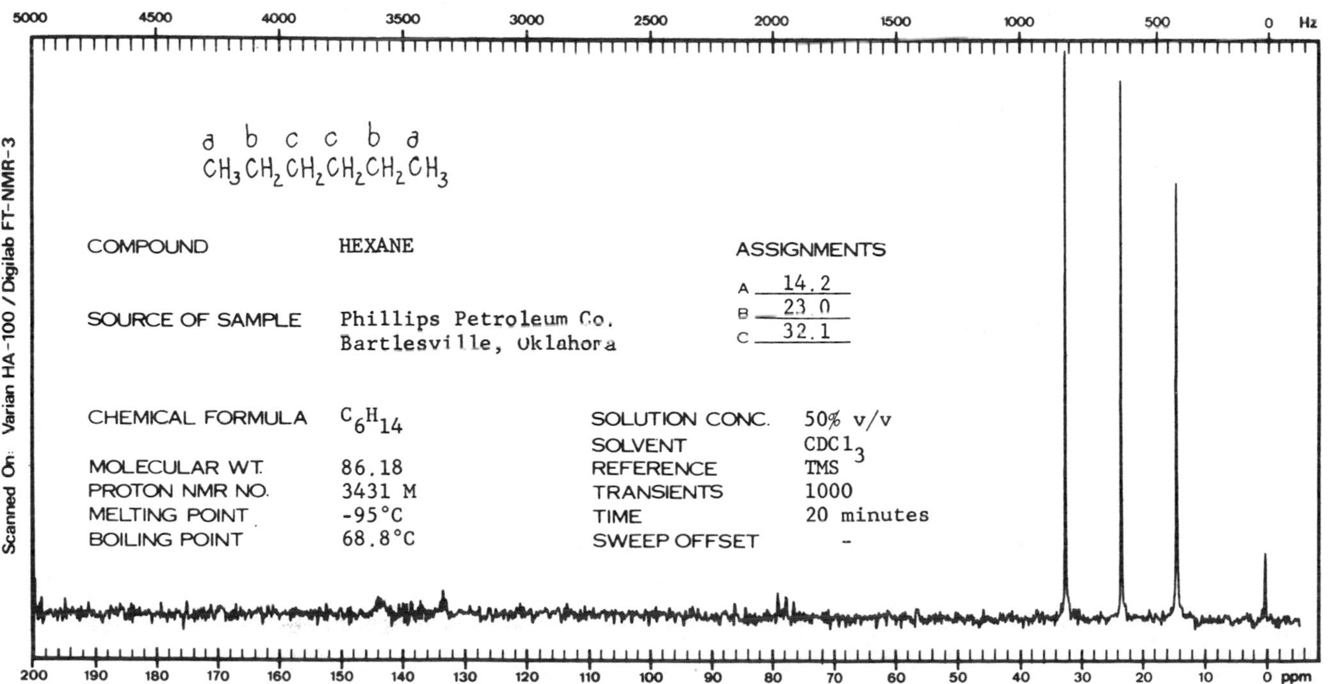
Scanned On: Varian HA-100 / Digilab FT-NMR-3
a b c c b a
CH3CH2CH2CH2CH2CH3
COMPOUND HEXANE
SOURCE OF SAMPLE Phillips Petroleum Co. Bartlesville, Oklahoma
ASSIGNMENTS
A 14.2
B 23.0
C 32.1
CHEMICAL FORMULA C6H14
MOLECULAR WT. 86.18
PROTON NMR NO. 3431 M
MELTING POINT -95°C
BOILING POINT 68.8°C
SOLUTION CONC. 50% v/v
SOLVENT CDCl3
REFERENCE TMS
TRANSIENTS 1000
TIME 20 minutes
SWEEP OFFSET -

No. 14

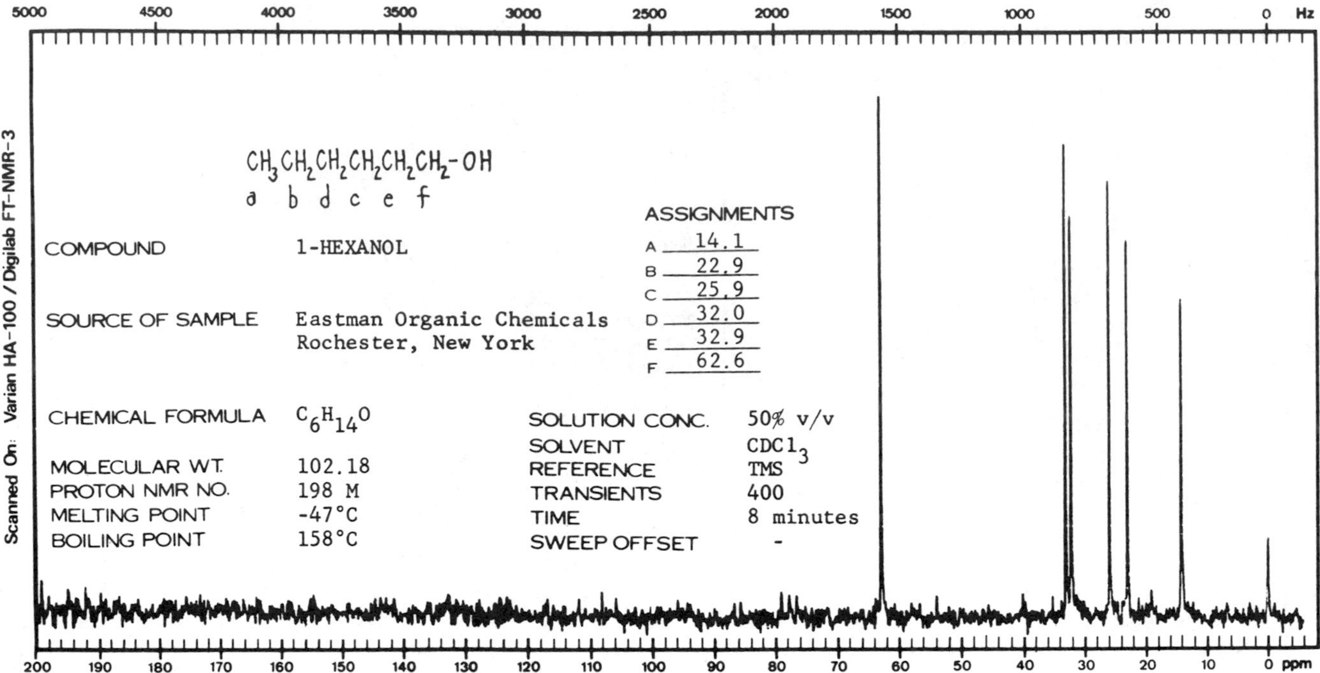
5000
4500
4000
3500
3000
2500
2000
1500
1000
500
0 Hz
Scanned On: Varian HA-100 / Digilab FT-NMR-3
CH3CH2CH2CH2CH2CH2-OH
a b d c e f
COMPOUND
1-HEXANOL
ASSIGNMENTS
A 14.1
B 22.9
C 25.9
D 32.0
E 32.9
F 62.6
SOURCE OF SAMPLE
Eastman Organic Chemicals
Rochester, New York
CHEMICAL FORMULA
C6H14O
SOLUTION CONC.
50% v/v
SOLVENT
CDCl3
MOLECULAR WT.
102.18
REFERENCE
TMS
PROTON NMR NO.
198 M
TRANSIENTS
400
MELTING POINT
-47°C
TIME
8 minutes
BOILING POINT
158°C
SWEEP OFFSET
-
200
190
180
170
160
150
140
130
120
110
100
90
80
70
60
50
40
30
20
10
0 ppm

No. 15

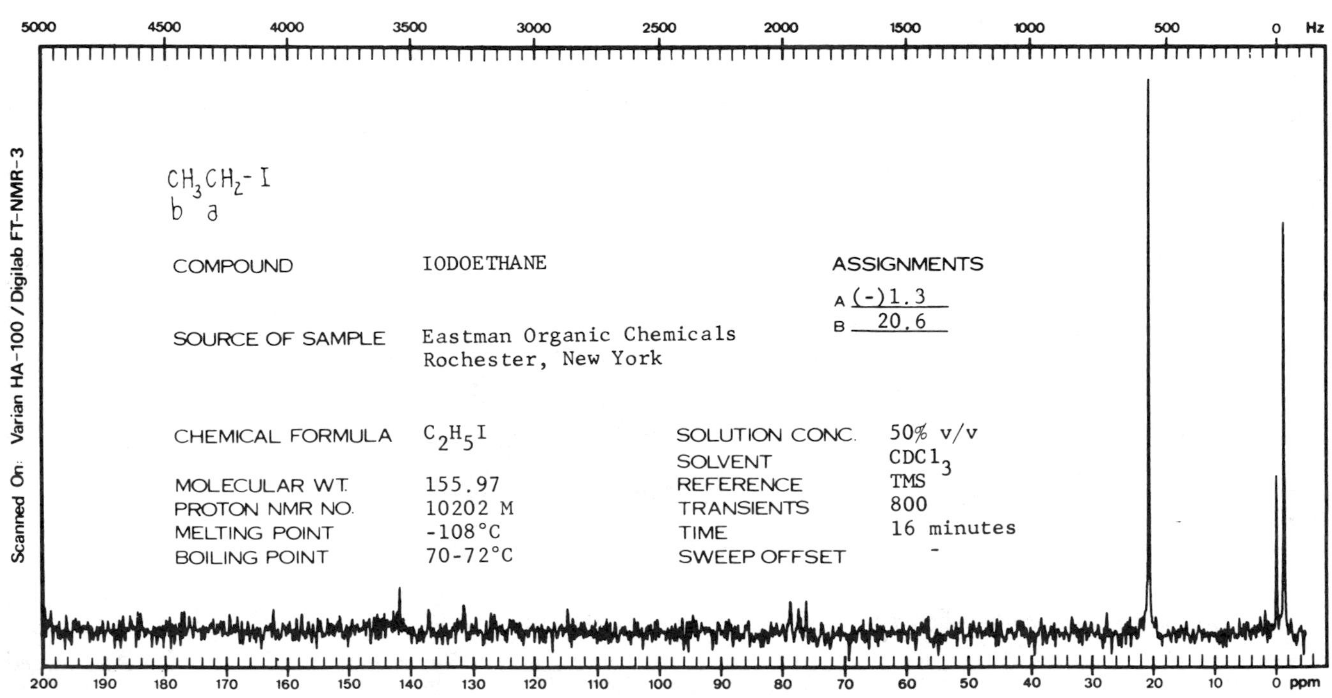
5000
4500
4000
3500
3000
2500
2000
1500
1000
500
0 Hz
Scanned On: Varian HA-100 / Digilab FT-NMR-3
CH3CH2-I
b a
COMPOUND
IODOETHANE
ASSIGNMENTS
A (-)1.3
B 20.6
SOURCE OF SAMPLE
Eastman Organic Chemicals
Rochester, New York
CHEMICAL FORMULA
C2H5I
SOLUTION CONC.
50% v/v
SOLVENT
CDCl3
MOLECULAR WT.
155.97
REFERENCE
TMS
PROTON NMR NO.
10202 M
TRANSIENTS
800
MELTING POINT
-108°C
TIME
16 minutes
BOILING POINT
70-72°C
SWEEP OFFSET
-
200
190
180
170
160
150
140
130
120
110
100
90
80
70
60
50
40
30
20
10
0 ppm

No. 16

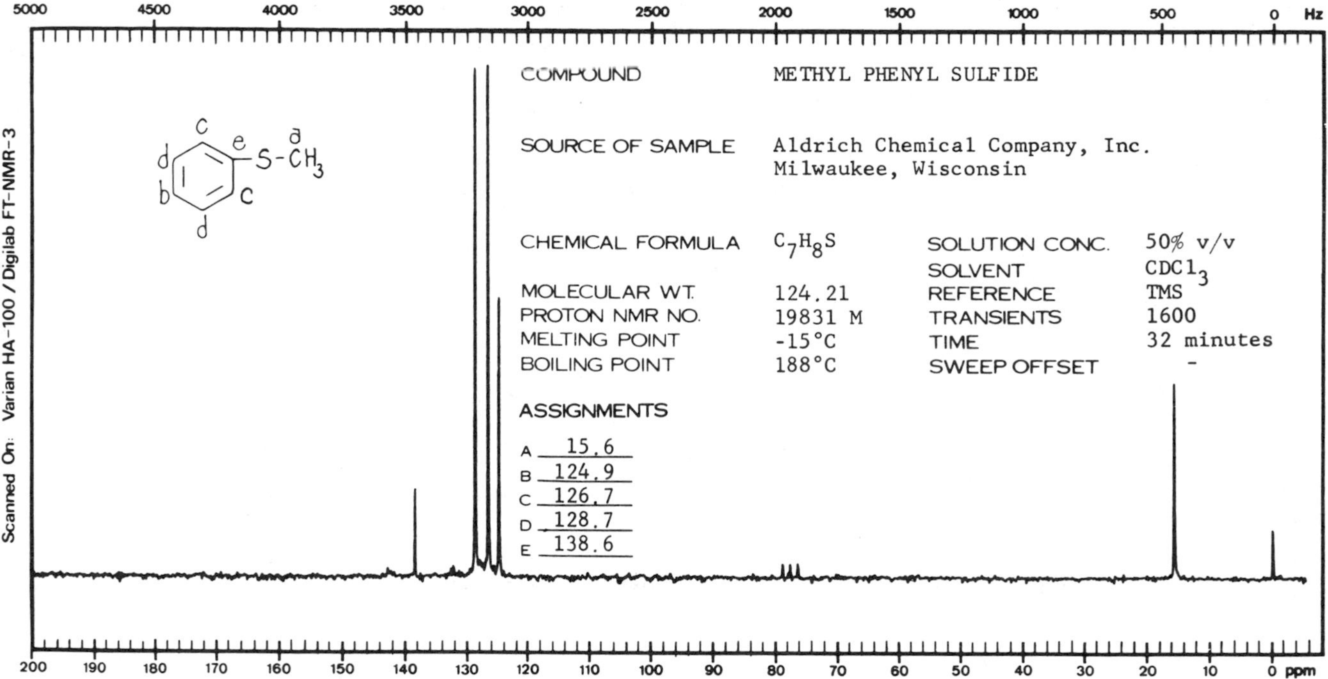
COMPOUND METHYL PHENYL SULFIDE
SOURCE OF SAMPLE Aldrich Chemical Company, Inc. Milwaukee, Wisconsin
CHEMICAL FORMULA C7H8S
MOLECULAR WT. 124.21
PROTON NMR NO. 19831 M
MELTING POINT -15°C
BOILING POINT 188°C
SOLUTION CONC. 50% v/v
SOLVENT CDCl3
REFERENCE TMS
TRANSIENTS 1600
TIME 32 minutes
SWEEP OFFSET -
ASSIGNMENTS
A 15.6
B 124.9
C 126.7
D 128.7
E 138.6
Scanned On: Varian HA-100 / Digilab FT-NMR-3

No. 17

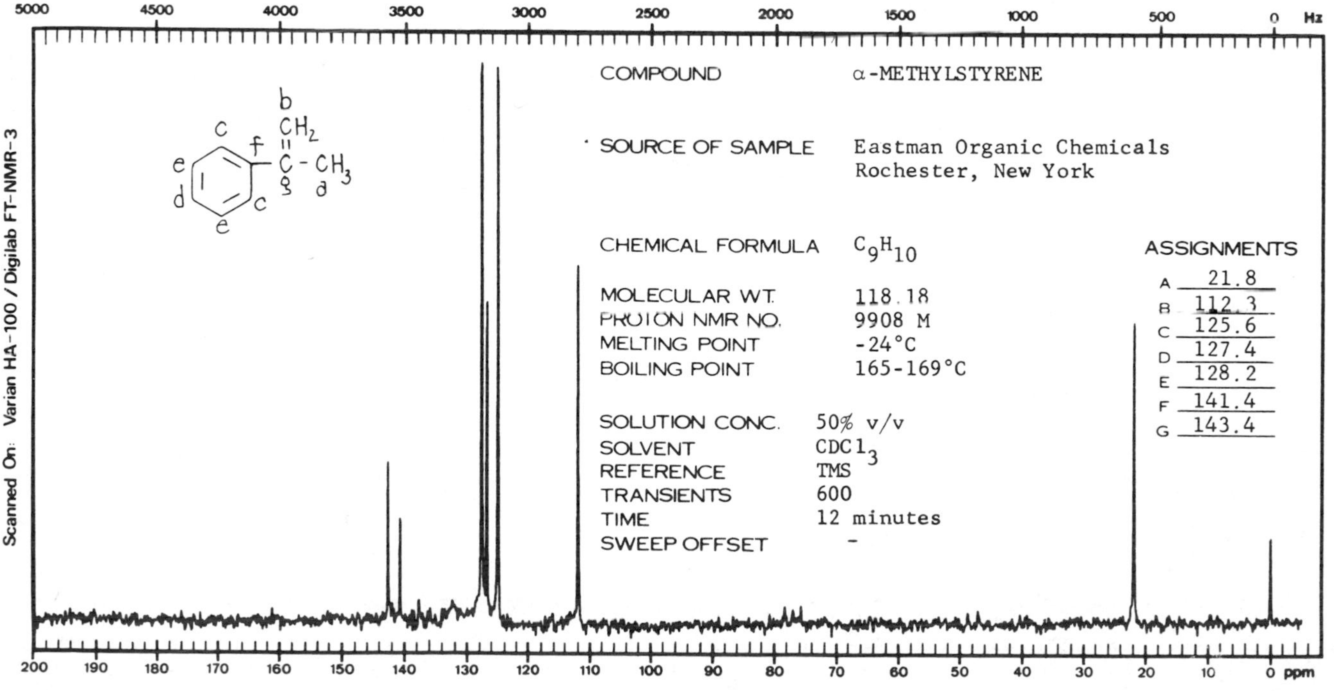
COMPOUND α-METHYLSTYRENE
SOURCE OF SAMPLE Eastman Organic Chemicals Rochester, New York
CHEMICAL FORMULA C9H10
MOLECULAR WT. 118.18
PROTON NMR NO. 9908 M
MELTING POINT -24°C
BOILING POINT 165-169°C
SOLUTION CONC. 50% v/v
SOLVENT CDCl3
REFERENCE TMS
TRANSIENTS 600
TIME 12 minutes
SWEEP OFFSET -
ASSIGNMENTS
A 21.8
B 112.3
C 125.6
D 127.4
E 128.2
F 141.4
G 143.4
Scanned On: Varian HA-100 / Digilab FT-NMR-3

No. 18

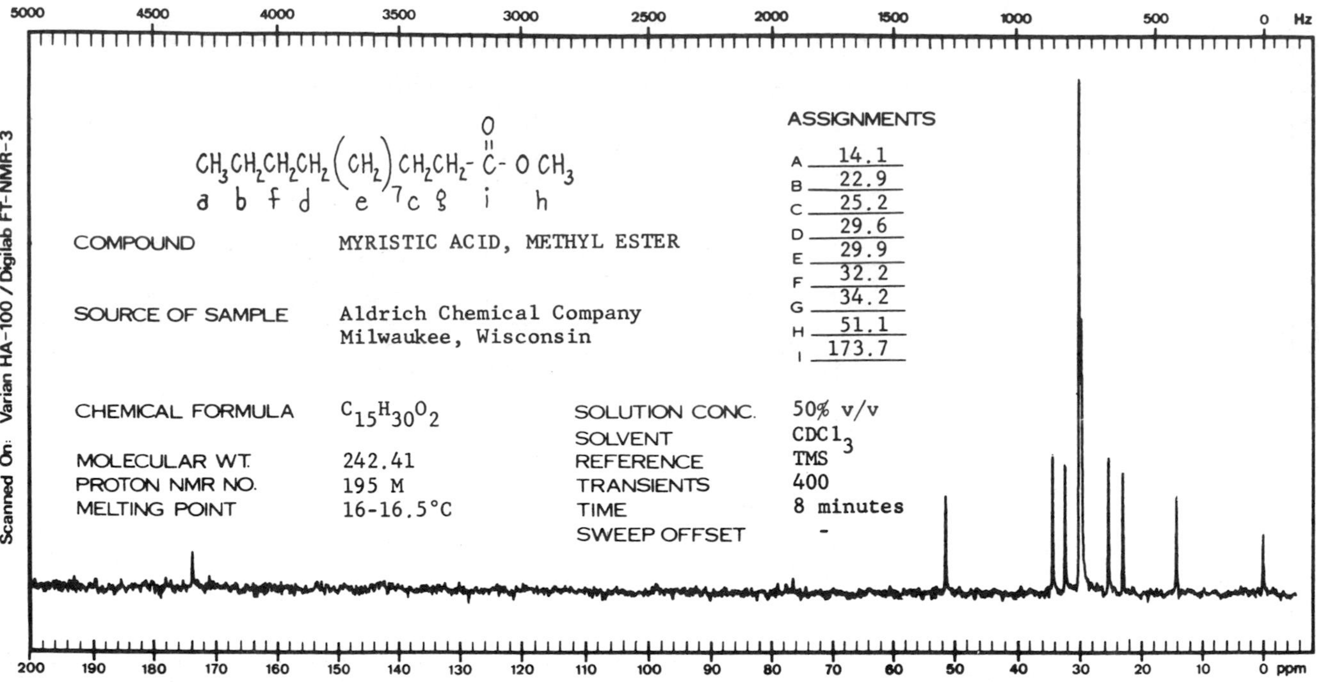

Scanned On: Varian HA-100 / Digilab FT-NMR-3
$CH_3CH_2CH_2CH_2(CH_2)_7CH_2CH_2-\overset{O}{\overset{\|}{C}}-OCH_3$
a b f d e c g i h
ASSIGNMENTS
A 14.1
B 22.9
C 25.2
D 29.6
E 29.9
F 32.2
G 34.2
H 51.1
I 173.7
COMPOUND MYRISTIC ACID, METHYL ESTER
SOURCE OF SAMPLE Aldrich Chemical Company Milwaukee, Wisconsin
CHEMICAL FORMULA $C_{15}H_{30}O_2$
MOLECULAR WT. 242.41
PROTON NMR NO. 195 M
MELTING POINT 16-16.5°C
SOLUTION CONC. 50% v/v
SOLVENT $CDCl_3$
REFERENCE TMS
TRANSIENTS 400
TIME 8 minutes
SWEEP OFFSET -

No. 19

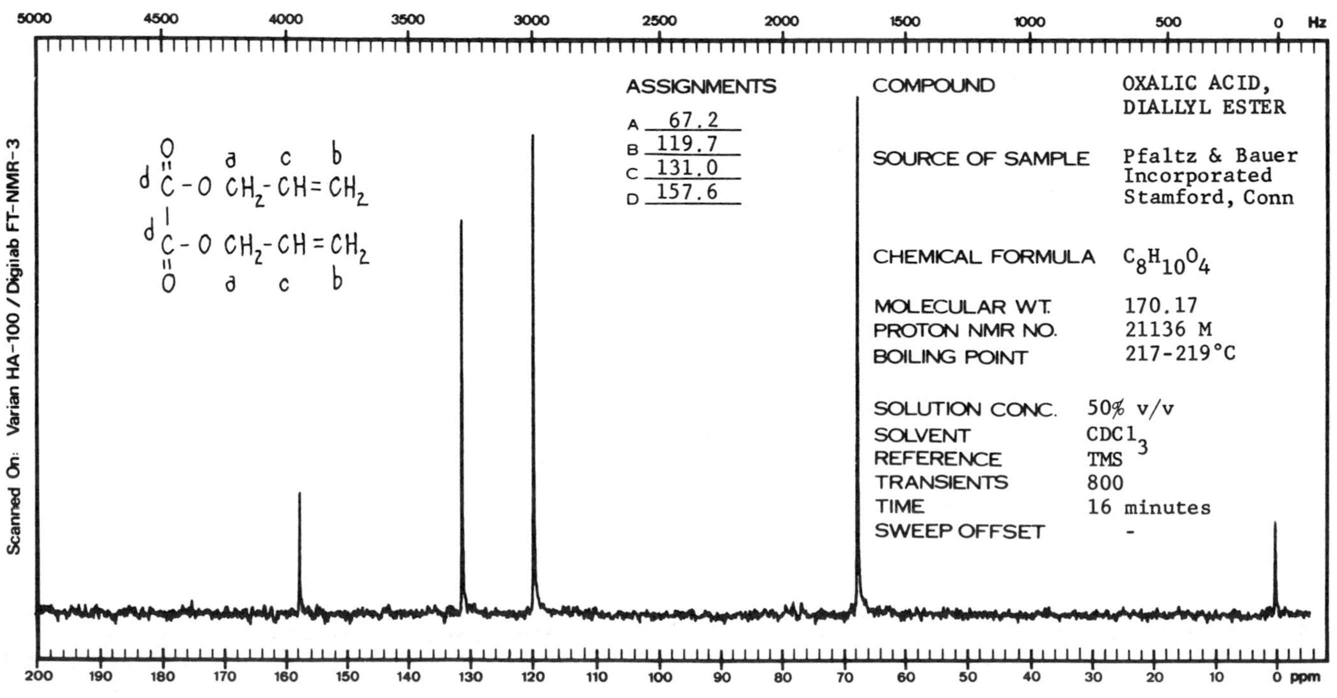

Scanned On: Varian HA-100 / Digilab FT-NMR-3
d $\overset{O}{\overset{\|}{C}}$-O CH_2-CH=CH_2 (a c b)
d $\underset{O}{\underset{\|}{C}}$-O CH_2-CH=CH_2 (a c b)
ASSIGNMENTS
A 67.2
B 119.7
C 131.0
D 157.6
COMPOUND OXALIC ACID, DIALLYL ESTER
SOURCE OF SAMPLE Pfaltz & Bauer Incorporated Stamford, Conn
CHEMICAL FORMULA $C_8H_{10}O_4$
MOLECULAR WT. 170.17
PROTON NMR NO. 21136 M
BOILING POINT 217-219°C
SOLUTION CONC. 50% v/v
SOLVENT $CDCl_3$
REFERENCE TMS
TRANSIENTS 800
TIME 16 minutes
SWEEP OFFSET -

No. 20

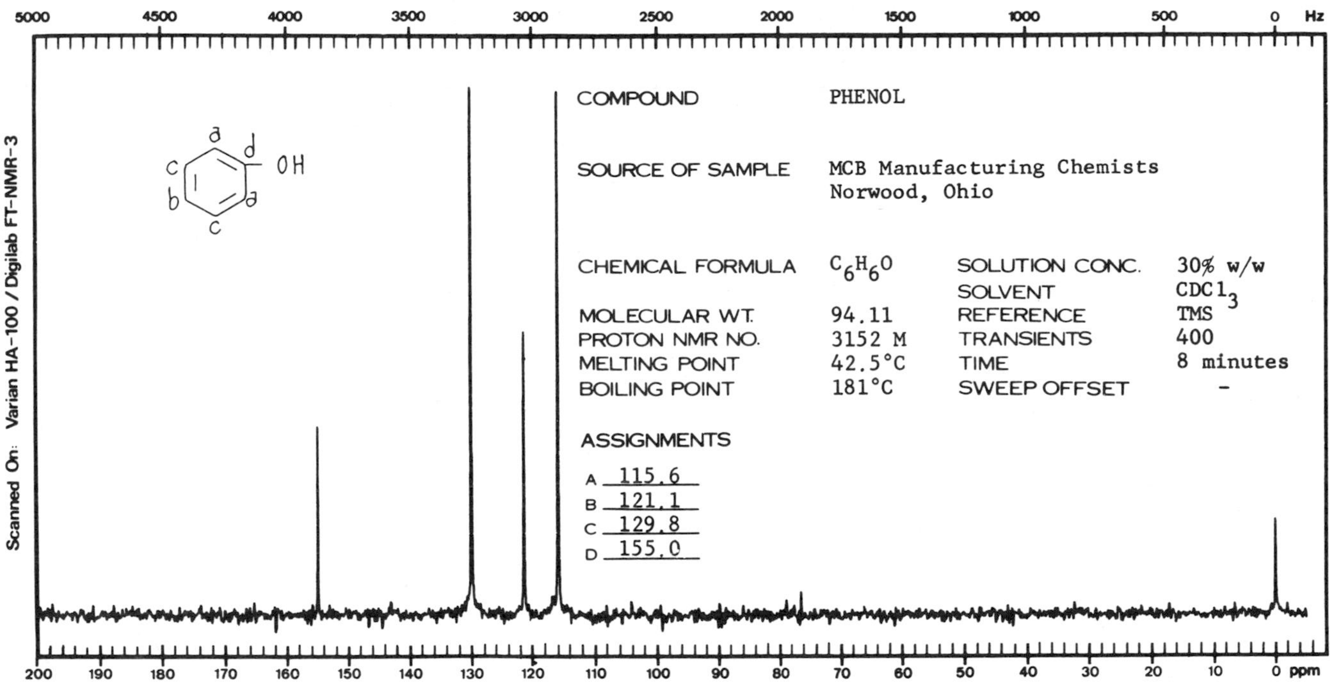
5000
4500
4000
3500
3000
2500
2000
1500
1000
500
0 Hz
Scanned On: Varian HA-100 / Digilab FT-NMR-3
COMPOUND PHENOL
SOURCE OF SAMPLE MCB Manufacturing Chemists
Norwood, Ohio
CHEMICAL FORMULA C6H6O
MOLECULAR WT. 94.11
PROTON NMR NO. 3152 M
MELTING POINT 42.5°C
BOILING POINT 181°C
SOLUTION CONC. 30% w/w
SOLVENT CDCl3
REFERENCE TMS
TRANSIENTS 400
TIME 8 minutes
SWEEP OFFSET -
ASSIGNMENTS
A 115.6
B 121.1
C 129.8
D 155.0
200
190
180
170
160
150
140
130
120
110
100
90
80
70
60
50
40
30
20
10
0 ppm

No. 21

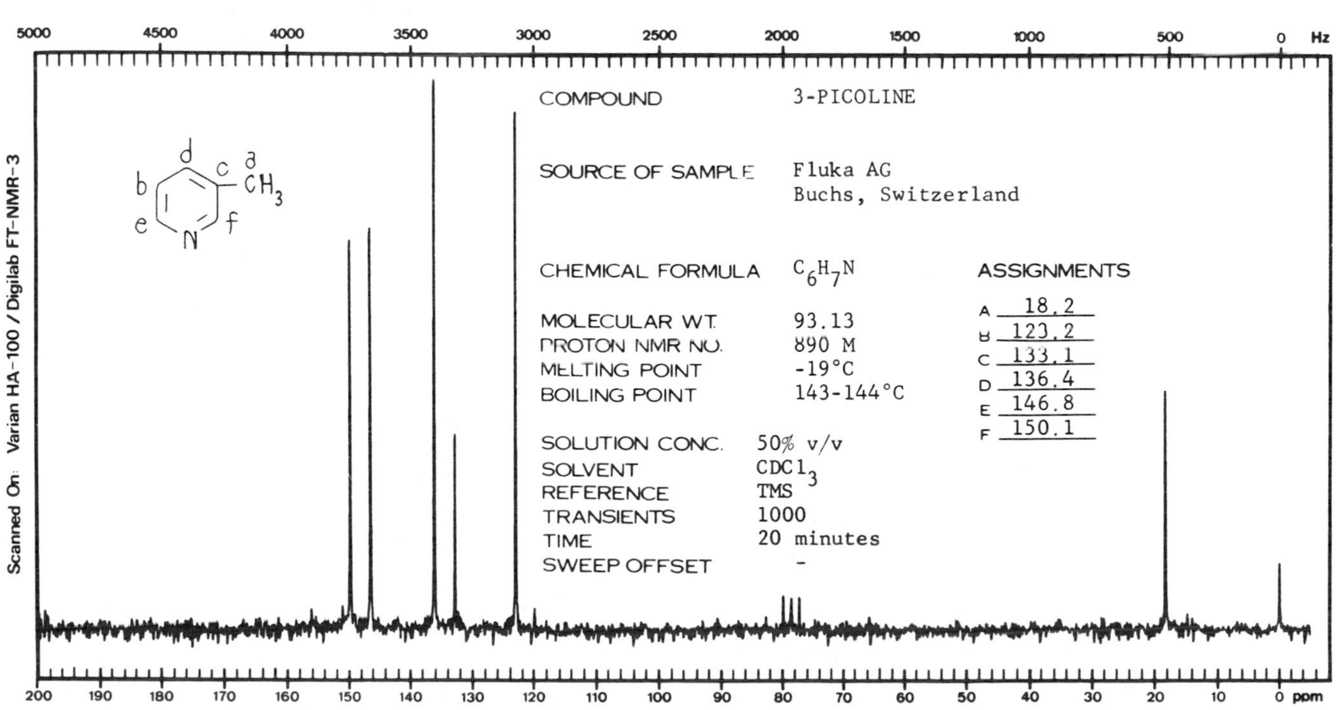
5000
4500
4000
3500
3000
2500
2000
1500
1000
500
0 Hz
Scanned On: Varian HA-100 / Digilab FT-NMR-3
COMPOUND 3-PICOLINE
SOURCE OF SAMPLE Fluka AG
Buchs, Switzerland
CHEMICAL FORMULA C6H7N
MOLECULAR WT. 93.13
PROTON NMR NO. 890 M
MELTING POINT -19°C
BOILING POINT 143-144°C
SOLUTION CONC. 50% v/v
SOLVENT CDCl3
REFERENCE TMS
TRANSIENTS 1000
TIME 20 minutes
SWEEP OFFSET -
ASSIGNMENTS
A 18.2
B 123.2
C 133.1
D 136.4
E 146.8
F 150.1
200
190
180
170
160
150
140
130
120
110
100
90
80
70
60
50
40
30
20
10
0 ppm

No. 22

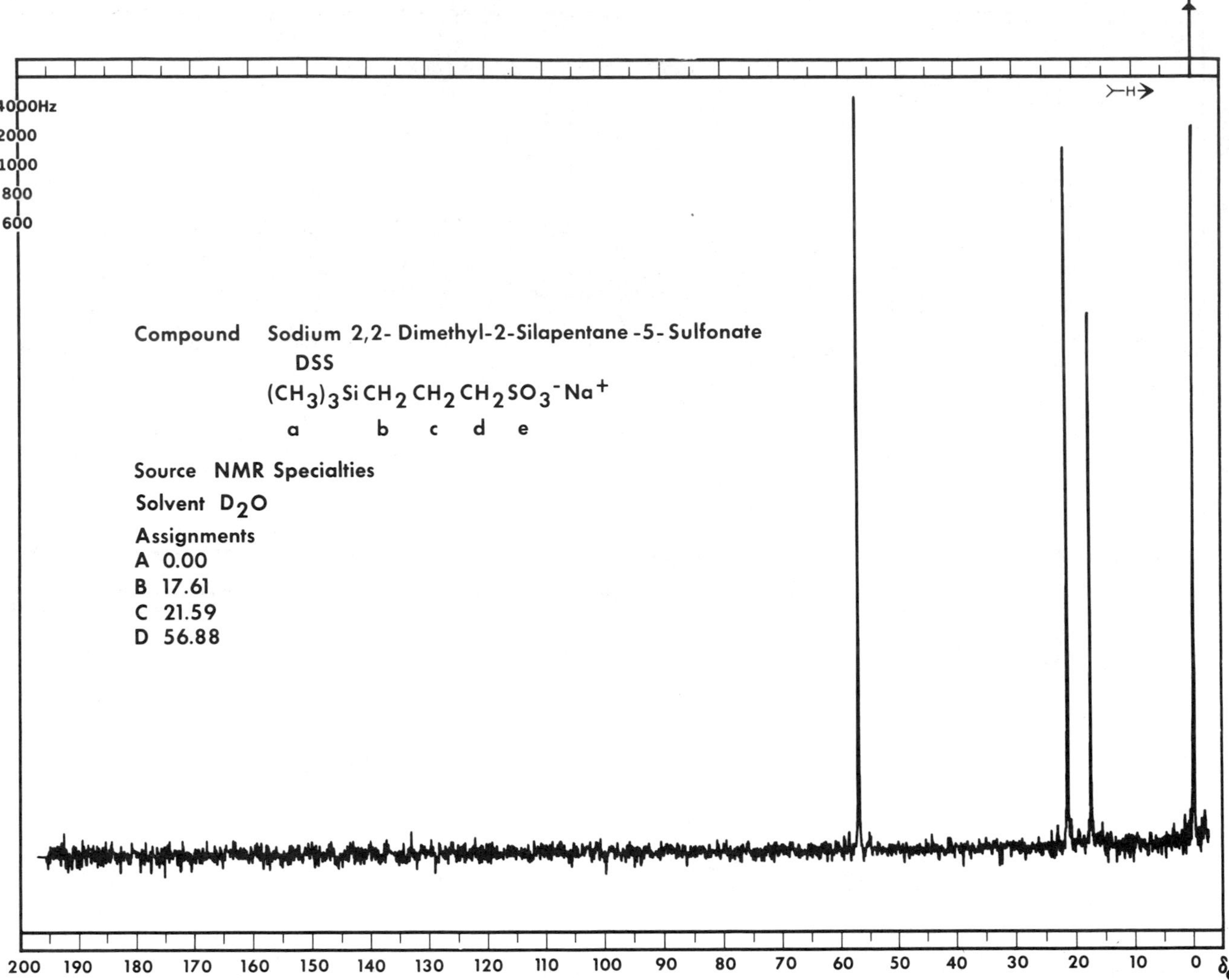

4000Hz
2000
1000
800
600
Compound Sodium 2,2- Dimethyl-2-Silapentane -5- Sulfonate
DSS
(CH3)3 Si CH2 CH2 CH2 SO3⁻ Na+
a b c d e
Source NMR Specialties
Solvent D2O
Assignments
A 0.00
B 17.61
C 21.59
D 56.88
200 190 180 170 160 150 140 130 120 110 100 90 80 70 60 50 40 30 20 10 0 δc

No. 23

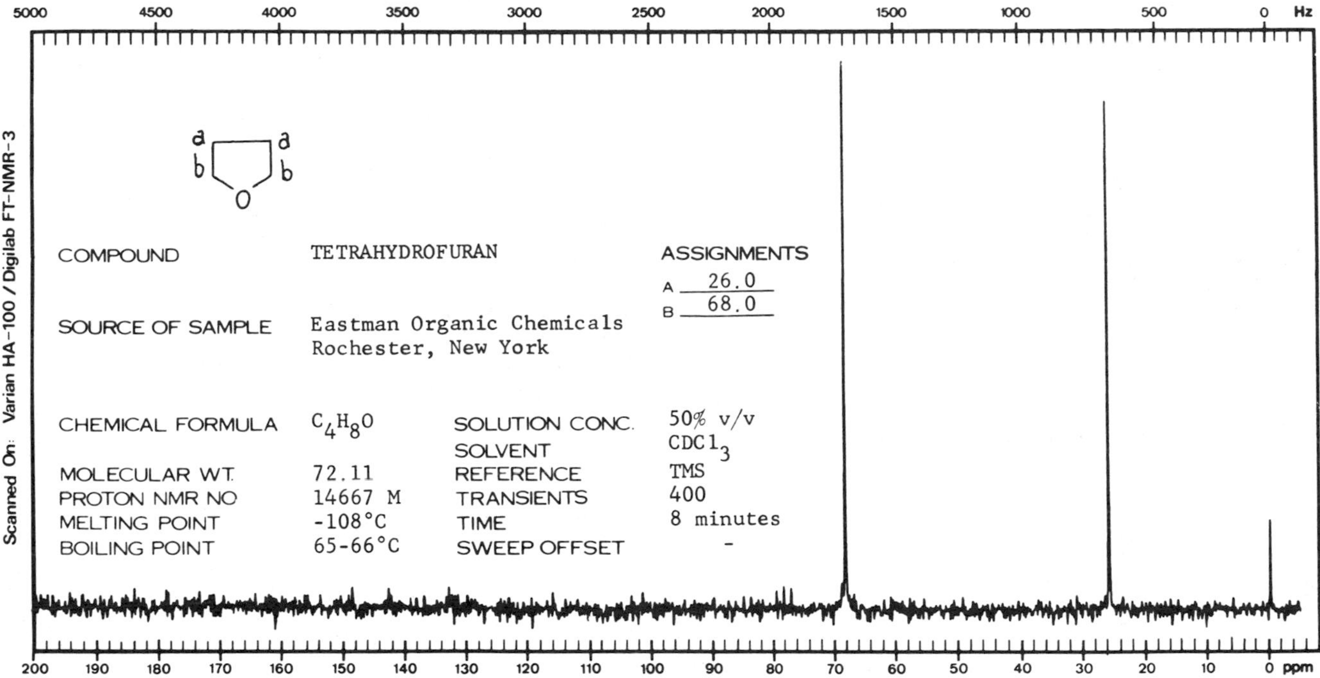
5000
4500
4000
3500
3000
2500
2000
1500
1000
500
0 Hz
Scanned On: Varian HA-100 / Digilab FT-NMR-3
COMPOUND
TETRAHYDROFURAN
ASSIGNMENTS
A 26.0
B 68.0
SOURCE OF SAMPLE
Eastman Organic Chemicals
Rochester, New York
CHEMICAL FORMULA
C_4H_8O
SOLUTION CONC.
50% v/v
SOLVENT
$CDCl_3$
MOLECULAR WT.
72.11
REFERENCE
TMS
PROTON NMR NO
14667 M
TRANSIENTS
400
MELTING POINT
-108°C
TIME
8 minutes
BOILING POINT
65-66°C
SWEEP OFFSET
-
200
190
180
170
160
150
140
130
120
110
100
90
80
70
60
50
40
30
20
10
0 ppm

No. 24

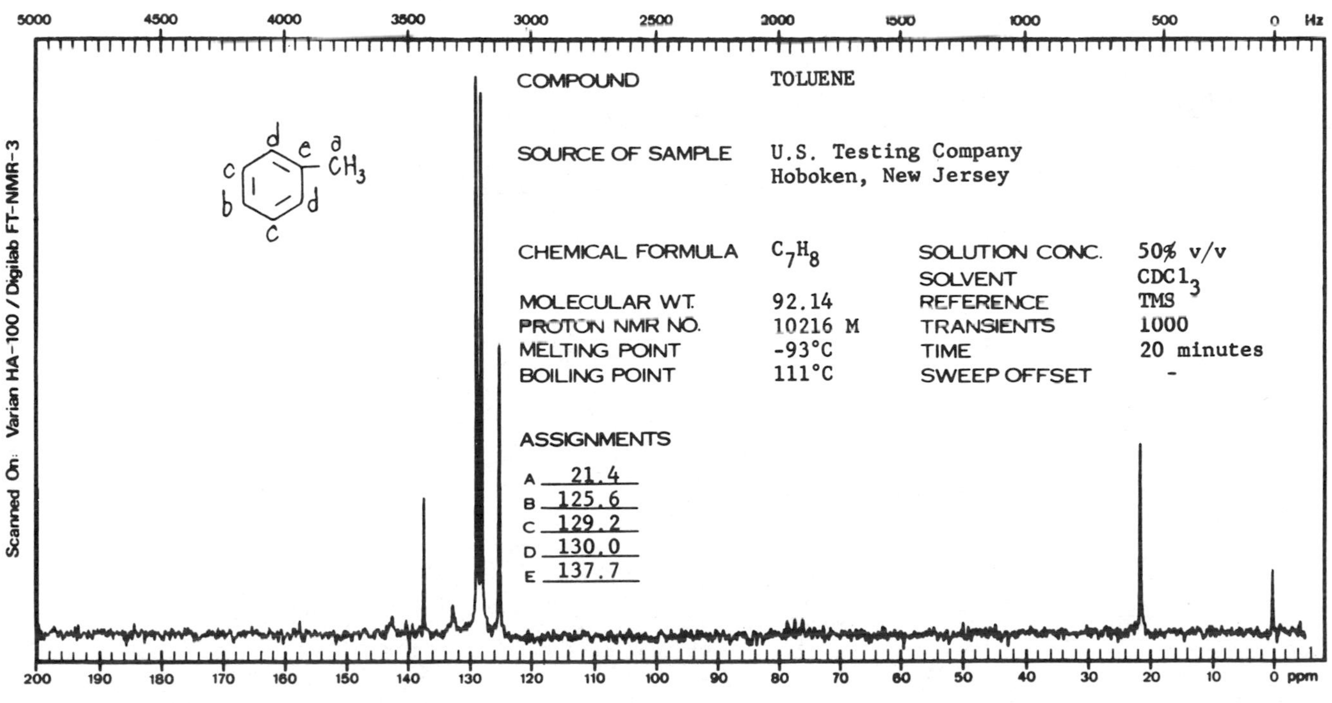
5000
4500
4000
3500
3000
2500
2000
1500
1000
500
0 Hz
Scanned On: Varian HA-100 / Digilab FT-NMR-3
CH_3
COMPOUND
TOLUENE
SOURCE OF SAMPLE
U.S. Testing Company
Hoboken, New Jersey
CHEMICAL FORMULA
C_7H_8
SOLUTION CONC.
50% v/v
SOLVENT
$CDCl_3$
MOLECULAR WT.
92.14
REFERENCE
TMS
PROTON NMR NO.
10216 M
TRANSIENTS
1000
MELTING POINT
-93°C
TIME
20 minutes
BOILING POINT
111°C
SWEEP OFFSET
-
ASSIGNMENTS
A 21.4
B 125.6
C 129.2
D 130.0
E 137.7
200
190
180
170
160
150
140
130
120
110
100
90
80
70
60
50
40
30
20
10
0 ppm

No. 25

six

ultraviolet spectrometry

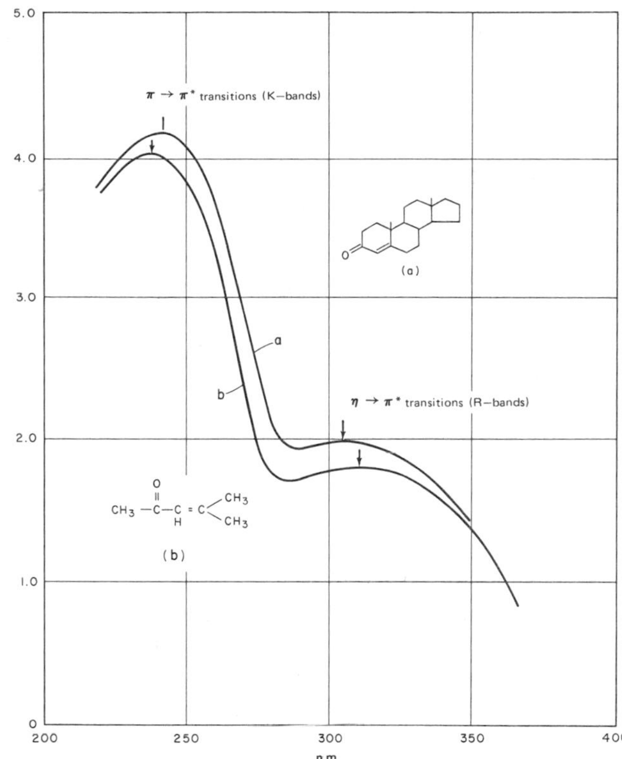

Figure 1. *Ultraviolet spectra of (a) cholest-4-en-3-one and (b) mesityl oxide.*

I. INTRODUCTION

Molecular absorption in the ultraviolet and visible region of the spectrum is dependent on the electronic structure of the molecule. Absorption of energy is quantized, resulting in the elevation of electrons from orbitals in the ground state to higher-energy orbitals in an excited state. For many electronic structures, the absorption does not occur in the readily accessible portion of the ultraviolet region. In practice, ultraviolet spectrometry is limited to conjugated systems for the most part.

There is, however, an advantage to the selectivity of ultraviolet absorption: characteristic groups may be recognized in molecules of widely varying complexities. A large portion of a relatively complex molecule may be transparent in the ultraviolet so that we may obtain a spectrum similar to that of a much simpler molecule. Thus, the spectrum of cholest-4-en-3-one closely resembles the spectrum of mesityl oxide. The absorption results from the conjugated enone portion of the 2 compounds (Figure 1).

A detailed mathematical treatment of the theoretical basis for ultraviolet or electronic spectra is beyond the scope of this chapter. Rather, it is our objective to point out correlations between spectra and structure to be used by the organic chemist. However, enough theory to rationalize these correlations will be presented.

An ultraviolet spectrum obtained directly from an instrument is simply a plot of wavelength (or frequency) of absorption vs. the absorption intensity (absorbance or transmittance). The data are frequently converted to a graphical plot of wavelength vs. the molar absorptivity (ϵ_{max} or log ϵ_{max} as in Figure 1). The use of molar absorptivity as the unit of absorption intensity has the advantage that all intensity values refer to the same number of absorbing species. A tabular presentation is used in Chapters 7 and 8 of this text.

An abundance of reference material relating to the theory and interpretation of ultraviolet spectra is available.[1 27] Two of the most useful references for the organic chemist are the texts by Stern and Timmons[3] and by A. I. Scott.[2] The latter is particularly recommended to the natural-products chemist. The text by Jaffé and Orchin is an excellent source of theory for both the spectroscopist and the organic chemist.[1] Several compilations of ultraviolet spectra and absorption data are available.[16-27] The Sadtler spectra[27] and the volumes of "Electronic Spectral Data"[19] are particularly useful to the organic chemist.

II. THEORY

The ultraviolet portion of the electromagnetic spectrum is indicated in Figure 2. Wavelengths in the ultraviolet region are usually expressed in nanometers (1 nm = 10^{-9} m) or angstroms, Å (1Å = 10^{-8} m). Occasionally, absorption is reported in wavenumbers ($\bar{\nu}$, units = cm^{-1}). We are primarily interested in the near ultraviolet (quartz) region extending from 200-380 nm. The atmosphere is transparent in this region and quartz optics may be used to scan from 200 to 380 nm. Atmospheric absorption starts near 200 nm and extends into the shorter-wavelength region, which is accessible through vacuum ultraviolet spectrometry.

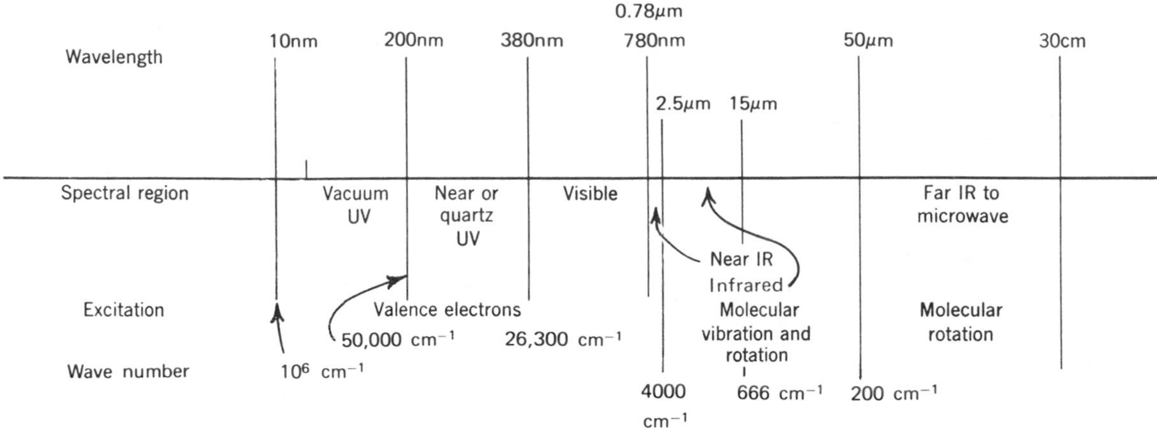

The total energy of a molecule is the sum of its electronic energy, its vibrational energy, and its rotational energy. The magnitude of these energies decreases in the following order: E_{elec}, E_{vib}, and E_{rot}. Energy absorbed in the ultraviolet region produces changes in the electronic energy of the molecule resulting from transitions of valence electrons in the molecule. These transitions consist of the excitation of an electron from an occupied molecular orbital (usually a nonbonding p or bonding π-orbital) to the next higher energy orbital (an antibonding, π^* or σ^*, orbital). The antibonding orbital is designated by an asterisk. Thus, the promotion of an electron from a π-bonding orbital to an antibonding (π^*) orbital is indicated $\pi \to \pi^*$.

The concept of an antibonding orbital can be explained simply by considering the ultraviolet absorption of ethylene. The ethylenic double bond, in the ground state, consists of a pair of bonding σ-electrons and a pair of bonding π-electrons. On absorption of ultraviolet radiation near 165 nm, one of the bonding π-electrons is raised to the next higher energy orbital, an antibonding π-orbital.

The orbitals occupied by the π-electron in the ground state and in the excited state are diagrammed in Figure 3. The shaded volumes indicate regions of maximum electron density. It can be seen that the antibonding π-electron no longer contributes appreciably to the overlap of the C-to-C bond. In fact, it negates the bonding power of the remaining unexcited π-electron; the olefinic bond has considerable single-bond character in the excited state.

The relationship between the energy absorbed in an electronic transition and the frequency (ν), wavelength (λ), and wavenumber ($\bar{\nu}$) of radiation producing the transition is

$$\Delta E = h\nu = \frac{hc}{\lambda} = h\bar{\nu}c$$

where h is Planck's constant and c is the velocity of light. ΔE is the energy absorbed in an electronic transition in a molecule from a low-energy state (ground state) to a high-energy state (excited state). The energy absorbed is dependent on the energy difference between the ground state and the excited state; the smaller the difference in energy, the longer the wavelength of absorption. The excess energy in the excited state may result in dissociation or ionization of the molecule, or it may be reemitted as heat

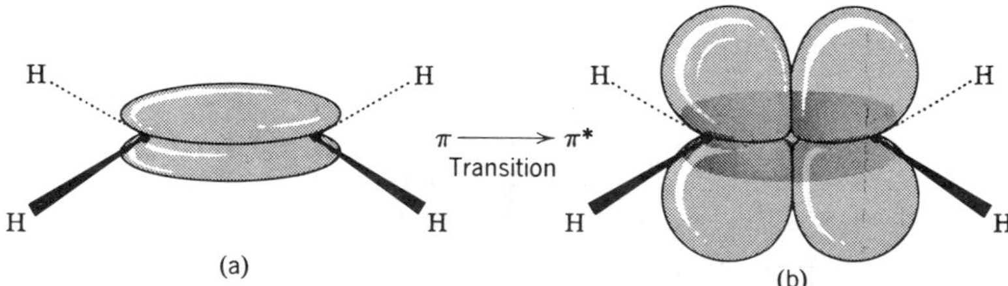

Figure 3. *π and π* orbitals of ethylene. (a.) Bonding orbital, π. Both π electrons occupying bonding orbital. A bonding π orbital has only 1 nodal plane in the plane of the molecular skeleton. (b.) Antibonding orbital, π*. One π electron in bonding orbital, one in antibonding orbital. An (antibonding) π* orbital has an additional nodal plane perpendicular to the plane of the molecule and bisecting the carbon-carbon bond.*

or light. The release of energy as light results in fluorescence or phosphorescence.

Since ultraviolet energy is quantized, the absorption spectrum arising from a single electronic transition should consist of a single, discrete line. A discrete line is not obtained since electronic absorption is superimposed on rotational and vibrational sublevels. The spectra of simple molecules in the gaseous state consist of narrow absorption peaks, each representing a transition from a particular combination of vibrational and rotational levels in the electronic ground state to a corresponding combination in the excited state. This is shown schematically in Figure 4, in which the vibrational levels are designated ν_0, ν_1, ν_2, and so forth. At ordinary temperatures, most of the molecules in the electronic ground state will be in the zero vibrational level ($G\nu_0$); consequently, there are many electronic transitions from that level. In molecules containing more atoms, the multiplicity of vibrational sublevels and the closeness of their spacing cause the discrete bands to coalesce, and broad absorption bands or "band envelopes" are obtained.

The principal characteristics of an absorption band are its position and intensity. The position of absorption corresponds to the wavelength of radiation whose energy is equal to that required for an electronic transition. The intensity of absorption is largely dependent on two factors: the probability of interaction between the radiation energy and the electronic system and the difference between the ground and the excited state. The probability of transition is proportional to the square of the transition moment. The transition moment, or dipole moment of transition, is proportional to the change in the electronic charge distribution occurring during excitation. Intense absorption occurs when a transition is accompanied by a large change in the transition moment. Absorption with ϵ_{max} values $> 10^4$ is high-intensity absorption; low-intensity absorption corresponds to ϵ_{max} values $< 10^3$. Transitions of low probability are "forbidden" transitions. The intensity of absorption may be expressed as transmittance (T), defined by

$$T = I/I_0$$

where I_0 is the intensity of the radiant energy striking the sample, and I is the intensity of the radiation emerging from the sample. A more convenient expression of absorption intensity is that derived from the Lambert-Beer law which establishes a relationship between the transmittance, the sample thickness, and the concentration of the absorbing species. This relationship is expressed as

$$\log_{10}(I_0/I) = kcb = A$$

where k = a constant characteristic of the solute,
c = concentration of solute,
b = path length through the sample,
A = absorbance (optical density in the older literature).

When c is expressed in moles per liter, and the path length (b) through the sample is expressed in centimeters, the preceding expression becomes:

$$A = \epsilon cb$$

The term ϵ is known as the molar absorptivity, formerly called the molar extinction coefficient.

If the concentration (c) of the solute is now defined as g/liter, the equation becomes:

$$A = abc$$

where a is the absorptivity and is thus related to the molar absorptivity by

$$\epsilon = aM$$

where M is the molecular weight of the solute.

The intensity of an absorption band in the ultraviolet spectrum is usually expressed as the molar absorptivity at maximum absorption, ϵ_{max} or $\log \epsilon_{max}$. When the molecular weight of an absorbing material is unknown, the intensity of absorption may be expressed as

$$E^{1\%}_{1\text{cm}} = \frac{A}{cb}$$

where c = concentration in grams per 100 ml,
b = path length through the sample in centimeters.

At this point it is desirable to define certain terms that are frequently used in the discussion of electronic spectra.

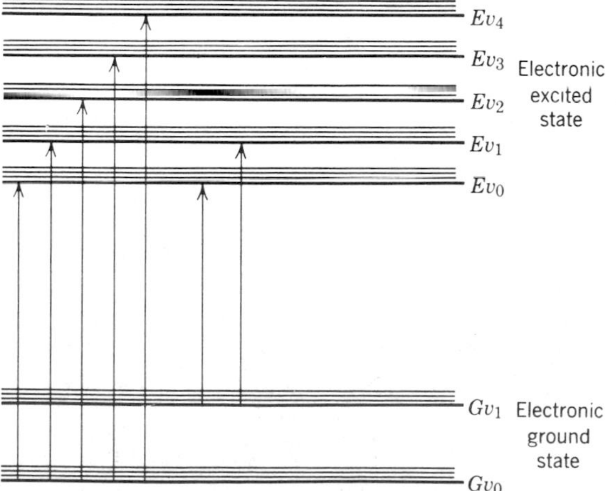

Table I. Summary of Electronic Transitions

Example	Electronic Transition	λ_{max} (*nm*)	ϵ_{max}	Band[a]
Ethane	$\sigma \rightarrow \sigma^*$	135	–	–
Water	$n \rightarrow \sigma^*$	167	7,000	–
Methanol	$n \rightarrow \sigma^*$	183	500	–
1-Hexanethiol	$n \rightarrow \sigma^*$	224	126	–
n-Butyl iodide	$n \rightarrow \sigma^*$	257	486	–
Ethylene	$\pi \rightarrow \pi^*$	165	10,000	–
Acetylene	$\pi \rightarrow \pi^*$	173	6,000	–
Acetone	$\pi \rightarrow \pi^*$	about 150	–	–
	$n \rightarrow \sigma^*$	188	1,860	–
	$n \rightarrow \pi^*$	279	15	*R*
1,3-Butadiene	$\pi \rightarrow \pi^*$	217	21,000	*K*
1,3,5-Hexatriene	$\pi \rightarrow \pi^*$	258	35,000	*K*
Acrolein	$\pi \rightarrow \pi^*$	210	11,500	*K*
	$n \rightarrow \pi^*$	315	14	*R*
Benzene	Aromatic $\pi \rightarrow \pi^*$	about 180	60,000	E_1
	Aromatic $\pi \rightarrow \pi^*$	about 200	8,000	E_2
	Aromatic $\pi \rightarrow \pi^*$	255	215	*B*
Styrene	Aromatic $\pi \rightarrow \pi^*$	244	12,000	*K*
	Aromatic $\pi \rightarrow \pi^*$	282	450	*B*
Toluene	Aromatic $\pi \rightarrow \pi^*$	208	2,460	E_2
	Aromatic $\pi \rightarrow \pi^*$	262	174	*B*
Acetophenone	Aromatic $\pi \rightarrow \pi^*$	240	13,000	*K*
	Aromatic $\pi \rightarrow \pi^*$	278	1,110	*B*
	$n \rightarrow \pi^*$	319	50	*R*
Phenol	Aromatic $\pi \rightarrow \pi^*$	210	6,200	E_2
	Aromatic $\pi \rightarrow \pi^*$	270	1,450	*B*

[a]R Band, German *radikalartig;* K-Band, German, *konjugierte;* B-Band, benzenoid; E-Band ethylenic; see: A. Burawoy, *Ber.*, **63**, 3155 (1930); *J. Chem. Soc.*, 1177 (1939); also see E.A. Braude.[5]

CHROMOPHORE. A covalently unsaturated group responsible for electronic absorption (for example, C=C, C=O, and NO_2).

AUXOCHROME. A saturated group with nonbonded electrons which, when attached to a chromophore, alters both the wavelength and the intensity of the absorption (e.g., $\ddot{\underset{..}{O}}H$, $\ddot{N}H_2$, and $:\ddot{\underset{..}{Cl}}$).

BATHOCHROMIC SHIFT. The shift of absorption to a longer wavelength due to substitution or solvent effect (a red shift).

HYPSOCHROMIC SHIFT. The shift of absorption to a shorter wavelength due to substitution or solvent effect (a blue shift).

HYPERCHROMIC EFFECT. An increase in absorption intensity.

HYPOCHROMIC EFFECT. A decrease in absorption intensity.

The absorption characteristics of organic molecules in the ultraviolet region depend on the electronic transitions that can occur and the effect of the atomic environment on the transitions. A summary of electronic structures and transitions that are involved in ultraviolet absorption is presented in Table I.

Representative UV spectra (plots of log ϵ vs. λ) are shown in Figure 1. Note that the spectrum of the relatively simple model compound mesityl oxide very closely approximates the spectrum of the more complex steroid. Increased molecular structure complexity normally results in increased spectral complexity in NMR, infrared, and mass spectra; Figure 1 shows that this is not necessarily true for UV spectra.

The relative ease with which the various transitions can occur is summarized in Figure 5. Although the energy changes are not shown in scale, it is readily seen, for example, that an $n \rightarrow \pi^*$ transition requires less energy than a $\pi \rightarrow \pi^*$ or a $\sigma \rightarrow \sigma^*$ transition. Several notation systems are used to designate UV absorption bands (Appendix 1 of Jaffe and Orchin[1]). It seems simplest to use electronic transitions or the letter designation assigned by Burawoy as shown in Table I and described below.

The $n \rightarrow \pi^*$ transitions (also called *R*-bands) of single

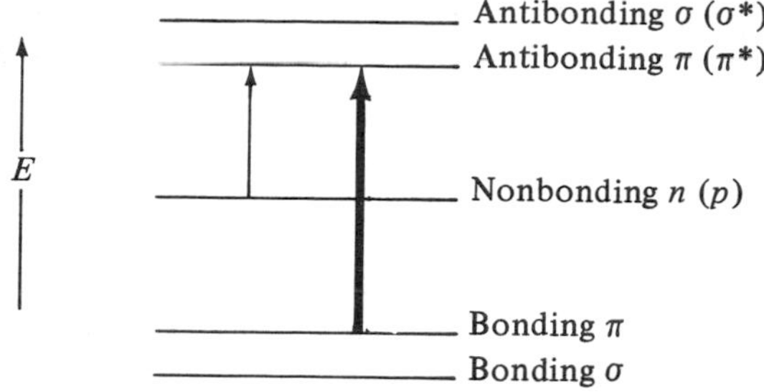

Figure 5. *Summary of electronic energy levels. Both $n \rightarrow \pi^*$ and $\pi \rightarrow \pi^*$ (heavy arrow) transitions are represented.*

chromophoric groups, such as the carbonyl or nitro group are forbidden and the corresponding bands are characterized by low molar absorptivities, ϵ_{max} generally less than 100. They are further characterized by the hypsochromic or blue shift observed with an increase in solvent polarity. They frequently remain in the spectrum when modifications in molecular structure introduce additional bands at shorter wavelengths. When additional bands make their appearance, the $n \rightarrow \pi^*$ transition is shifted to a longer wavelength but may be submerged by more intense bands.

Bands attributed to $\pi \rightarrow \pi^*$ transitions (*K*-Bands) appear in the spectra of molecules that have conjugated π-systems such as butadiene or mesityl oxide. Such absorptions also appear in the spectra of aromatic molecules possessing chromophoric substitution–styrene, benzaldehyde, or acetophenone. These $\pi \rightarrow \pi^*$ transitions are usually characterized by high molar absorptivity, $\epsilon_{max} > 10{,}000$.

The $\pi \rightarrow \pi^*$ transitions (*K*-bands) of conjugated di- or poly-ene systems can be distinguished from those of enone systems by observing the effect of changing solvent polarity. The $\pi \rightarrow \pi^*$ transitions of diene or polyene systems are essentially unresponsive to solvent polarity; the hydrocarbon double bonds are nonpolar. The corresponding absorptions of enones, however, undergo a bathochromic shift, frequently accompanied by increasing intensity, as the polarity of the solvent is increased. The red shift presumably results from a reduction in the energy level of the excited state accompanying dipole-dipole interaction and hydrogen bonding.

The effect of solvent has been measured for the $n \rightarrow \pi^*$ transition of acetone. This maximum is at 279 nm in hexane, and decreases to 272 and 264.5 nm for the solvents ethanol and water, respectively.

B-Bands (benzenoid bands) are characteristic of the spectra of aromatic or heteroaromatic molecules. Benzene shows a broad absorption band, containing multiple peaks or fine structure, in the near ultraviolet region between 230 and 270 nm (ϵ of most intense peak *ca.* 255 nm). The fine structure arises from vibrational sublevels affecting the electronic transitions. When a chromophoric group is attached to an aromatic ring, the *B*-bands are observed at longer wavelengths than the more intense $\pi \rightarrow \pi^*$ transitions. For example, styrene has a $\pi \rightarrow \pi^*$ transition at λ_{max} 244 nm (ϵ_{max} 12,000), and a *B*-band at λ_{max} 282 nm (ϵ_{max} 450). When an $n \rightarrow \pi^*$ transition appears in the spectrum of an aromatic compound that contains $\pi \rightarrow \pi^*$ transitions (including *B*-bands), the $n \rightarrow \pi^*$ transition is shifted to longer wavelengths. The characteristic fine structure of the *B*-bands may be absent in spectra of substituted aromatics. The fine structure is often destroyed by the use of polar solvents.

E-Bands (ethylenic bands), like the *B*-bands, are characteristic of aromatic structures. The E_1- and E_2-bands of benzene are observed near 180 nm and 200 nm, respectively. Auxochromic substitution brings the E_2 band into the near ultraviolet region, although in many cases it may not appear at wavelengths much over 210 nm. In auxochromic substitution, the heteroatom with the lone pair of electrons shares these electrons with the π-electron system of the ring, facilitating the $\pi \rightarrow \pi^*$ transition and thus causing a red shift of the *E*-bands. The molar absorptivity of *E*-bands generally varies between 2000 and 14,000.

A bathochromically displaced E_2-band is probably responsible for the intense, fine-structured bands of polynuclear aromatics. With the appearance of the *E*-bands as a results of auxochromic substitution, the *B*-band shifts to longer wavelengths and frequently increases in intensity. Molecules such as benzylideneacetone in which more complex conjugated chromophoric substitution occurs, produce spectra with both *E*- and *K*-bands; the *B*-bands are obscured by the displaced *K*-bands.

III. SAMPLE HANDLING

Ultraviolet spectra of compounds are usually determined either in the vapor phase or in solution.

A variety of quartz cells is available for the determination of spectra in the gas phase. These cells are equipped with gas inlets and outlets and have path lengths from 1.0 mm to 100 mm. Path lengths of 0.5-120 meters can be reached by using cells containing mirrors. Cell jackets are available through which liquids may be circulated for temperature control.

Cells used for the determination of spectra in solution vary in path length from 1 cm to 10 cm. Quartz cells, 1-cm square, are commonly used. These require about 3 ml of solution. Filler plugs are available to reduce the volume and the path length of the 1-cm square cell. Small-volume cells with 1-cm path lengths are also available. Microcells may be used when only a small amount of solution is available: the use of a beam condenser to minimize the loss of energy is advisable when the microcells are used. Microcells with an internal width of 2 mm and a path length as short as 5 mm are available.

In preparing a solution, a sample is accurately weighed and made up to volume in a volumetric flask. Aliquots are then removed, and additional dilutions made until the desired concentration has been acquired. Clean cells are of utmost importance. The cells should be rinsed several times with solvent and checked for absorption between successive determinations. It may be necessary to clean the cells with a detergent or hot nitric acid to remove traces of previous samples.

Many solvents are available for use in the ultraviolet region. Three common solvents are cyclohexane, 95% ethanol, and 1,4-dioxane. Cyclohexane may be freed of aromatic and olefinic impurities by percolation through activated silica gel and is transparent down to about 210 nm. Aromatic compounds, particularly the polynuclear aromatics, are usually soluble and their spectra generally retain their fine-line structure when determined in cyclohexane. The fine structure is often lost in more polar solvents.

Ninety-five percent ethanol is generally a good choice when a more polar solvent is required. This solvent generally can be used as purchased, but absolute ethanol must be freed of benzene used in its preparation. The last traces of benzene are removed by careful fractional distillation, or by preparative gas chromatography. The lower limit of transparency for ethanol is near 210 nm.

1,4-Dioxane can be purified by distillation from sodium. Benzene contamination can be removed by the addition of methanol followed by distillation to remove the benzene-methanol azeotrope. 1,4-Dioxane is transparent down to about 220 nm.

Many types of "spectral grade" solvents for ultraviolet analysis are now commercially available. These are usually highly purified and are free of interfering absorptions in the shaded regions indicated in Figure 6. Care should be exercised to choose a solvent that will be inert to the solute. For example, the spectra of aldehydes should not be determined in alcohols. Photochemical reactions may be detected by checking for changes in absorbance with time after exposure to the ultraviolet beam in the instrument.

IV. CHARACTERISTIC ABSORPTION OF ORGANIC COMPOUNDS

In our discussion of the theory of electronic or ultraviolet spectra, it was shown that the ability of an organic compound to absorb ultraviolet radiation is dependent on its electronic structure. In the following sections we shall discuss the characteristic absorption of basic electronic structures and the effects of molecular geometry and substitution on the absorption.

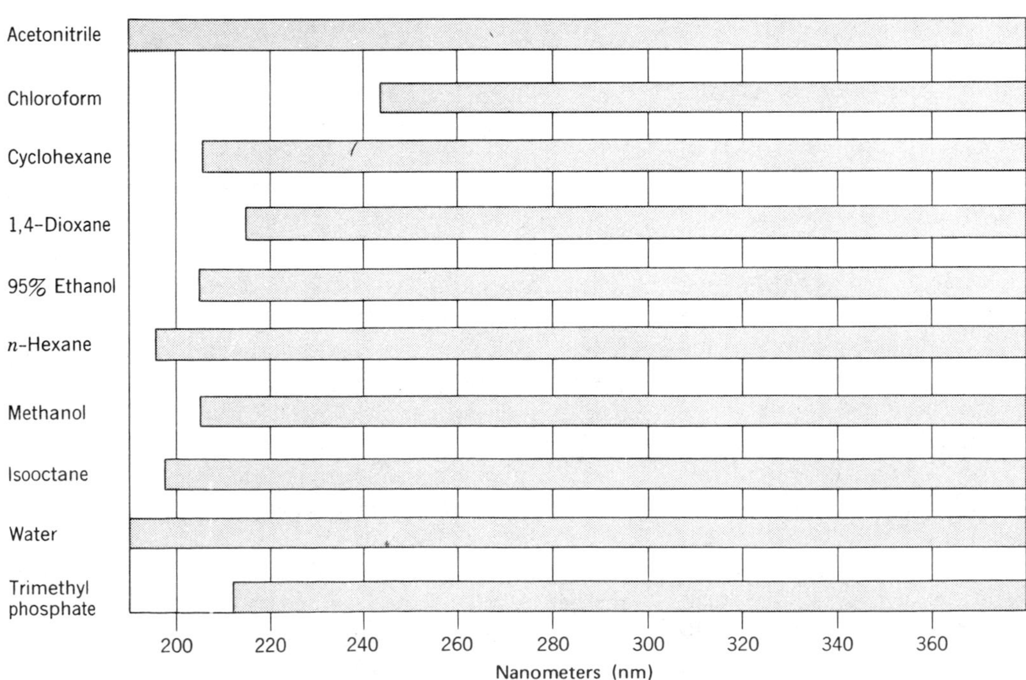

Figure 6. *Useful transparency ranges of solvents in near ultraviolet region.*

COMPOUNDS CONTAINING ONLY σ-ELECTRONS

Saturated hydrocarbons contain σ-electrons exclusively. Since the energy required to bring about a $\sigma \rightarrow \sigma^*$ transition is of the order of 185 kcal per mole and is available only in the far (vacuum) ultraviolet region, saturated hydrocarbons are transparent in the near ultraviolet region and thus can be used as solvents.

SATURATED COMPOUNDS CONTAINING n-ELECTRONS

Saturated compounds containing heteroatoms, such as oxygen, nitrogen, sulfur, or the halogens, possess nonbonding electrons (n- or p-electrons) in addition to σ-electrons. The $n \rightarrow \sigma^*$ transition requires less energy than the $\sigma \rightarrow \sigma^*$ transition (see Figure 4*a*), but the majority of compounds in this class still show no absorption in the near ultraviolet.

Alcohols and ethers absorb at wavelengths shorter than 185 nm (Table II) and therefore are commonly used for work in the near ultraviolet region. When these compounds are used as solvents, the intense absorption extends into the near ultraviolet, producing end-absorption or cut-off in the 200-220 nm region (Figure 6).

Sulfides, disulfides, thiols, amines, bromides, and iodides may show weak absorption in the near ultraviolet. Frequently, the absorption appears only as a shoulder or an inflection so that its diagnostic value is questionable.

Absorption data for several saturated compounds bearing heteroatoms are presented in Table II.

Table II. Absorption Characteristics of Saturated Compounds Containing Heteroatoms ($n \rightarrow \sigma^*$)

Compound	λ_{max} (*nm*)	ϵ_{max}	Solvent
Methanol	177	200	Hexane
Di-*n*-butyl sulfide	210	1200	Ethanol
	229 (*s*)		
Di-*n*-butyl disulfide	204	2089	Ethanol
	251	398	
1-Hexanethiol	224 (*s*)	126	Cyclohexane
Trimethylamine	199	3950	Hexane
N-Methylpiperidine	213	1600	Ether
Methyl chloride	173	200	Hexane
n-Propyl bromide	208	300	Hexane
Methyl iodide	259	400	Hexane
Diethyl ether	188	1995	Gas Phase
	171	3981	

(*s*) Shoulder inflection

COMPOUNDS CONTAINING π-ELECTRONS (CHROMOPHORES)

The absorption characteristics of a list of compounds containing single, isolated chromophoric groups are presented in Table III. All the compounds contain π-electrons, and many also contain nonbonding electron pairs. An examination of the absorption data shows that many of these single chromophoric groups absorb strongly in the far ultraviolet region with no absorption in the near ultraviolet. Those groups containing both π- and n-electrons can undergo 3 transitions: $n \rightarrow \sigma^*$, $\pi \rightarrow \pi^*$, and $n \rightarrow \pi^*$. Weak absorption of single chromophores in the near ultraviolet region results from the low-energy, forbidden $n \rightarrow \pi^*$ transition.

Ethylenic Chromophore

The isolated ethylenic chromophore is responsible for intense absorption that almost always occurs in the far ultraviolet region. Absorption is due to a $\pi \rightarrow \pi^*$ transition. Ethylene in the vapor phase absorbs at 165 nm (ϵ_{max} 10,000). A second band near 200 nm has been attributed to the elevation of two π electrons to π^* orbitals. The intensity of olefinic absorption is essentially independent of solvent because of the nonpolar nature of the olefinic bond.

Alkyl substitution of the parent compound moves the absorption to longer wavelengths. The bathochromic effect is progressive as the number of alkyl groups increases. A double bond exocyclic to 2 rings absorbs near 204 nm. The bathochromic shift accompanying alkyl substitution results from hyperconjugation in which the σ-electrons of the alkyl group are mobile enough to interact with the chromophoric group.

Attachment of a heteroatom bearing a nonbonded electron pair to the ethylenic linkage brings about a bathochromic shift. Nitrogen and sulfur atoms are most effective, bringing the absorption well into the near ultraviolet region. Methyl vinyl sulfide, for example, absorbs at 228 nm (ϵ_{max} 8000).

The absorption characteristics of cyclic monoolefins resemble those of the acyclic compounds, the absorption bearing no apparent relationship to ring size.

When two or more ethylenic linkages appear in a single molecule, isolated from one another by at least 1 methylene group, the molecule absorbs at the same position as the single ethylenic chromophore. The intensity of absorption is proportional to the number of isolated chromophoric groups in the molecule.

Allenes, which possess the C=C=C unit, show a strong absorption band in the far ultraviolet region near 170 nm, with a shoulder on the long wavelength side sometimes extending into the near ultraviolet region.

Table III. Absorption Data for Isolated Chromophores

Chromophoric Group	System	Example	λ_{max} (*nm*)	ϵ_{max}	Transition	Solvent
Ethylenic	RCH=CHR	Ethylene	165	15,000	$\pi \rightarrow \pi^*$	Vapor
			193	10,000	$\pi \rightarrow \pi^*$	
Acetylenic	R−C≡C−R	Acetylene	173	6,000	$\pi \rightarrow \pi^*$	Vapor
Carbonyl	$RR_1C=O$	Acetone	188	900	$\pi \rightarrow \pi^*$	*n*-Hexane
			279	15	$n \rightarrow \pi^*$	
Carbonyl	RHC=O	Acetaldehyde	290	16	$n \rightarrow \pi^*$	Heptane
Carboxyl	RCOOH	Acetic acid	204	60	$n \rightarrow \pi^*$	Water
Amido	$RCONH_2$	Acetamide	<208	–	$n \rightarrow \pi^*$	–
Azomethine	>C=N−	Acetoxime	190	5,000	$\pi \rightarrow \pi^*$	Water
Nitrile	−C≡N	Acetonitrile	<160	–	$\pi \rightarrow \pi^*$	–
Azo	−N=N−	Azomethane	347	4.5	$n \rightarrow \pi^*$	Dioxane
Nitroso	−N=O	Nitrosobutane	300	100		Ether
			665	20		
Nitrate	$-ONO_2$	Ethyl nitrate	270	12	$n \rightarrow \pi^*$	Dioxane
Nitro	$-N(=O)(\rightarrow O)$	Nitromethane	271	18.6	$n \rightarrow \pi^*$	Alcohol
Nitrite	−ONO	Amyl nitrite	218.5	1,120	$\pi \rightarrow \pi^*$	Petroleum ether
			346.5[a]		$n \rightarrow \pi^*$	
Sulfoxide	S=O	Cyclohexyl methyl sulfoxide	210	1,500		Alcohol
Sulfone	>S(=O)=O	Dimethyl sulfone	<180	–		–

[a]Most intense peak of fine structure group.

Source: A.E. Gillam and E.S. Stern, *An Introduction to Electronic Absorption Spectroscopy in Organic Chemistry*, 2nd ed., London: Edward Arnold, 1957.

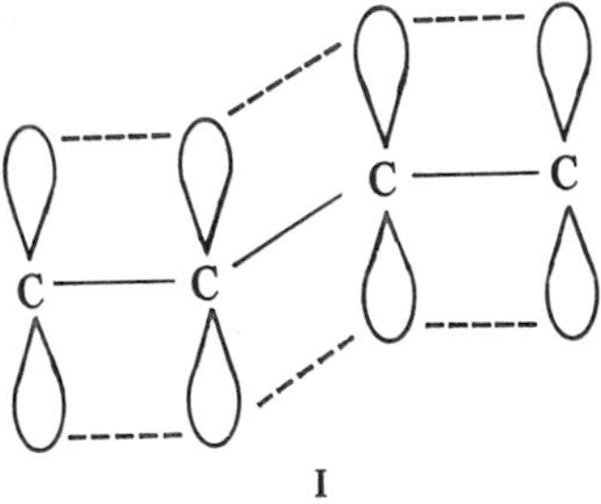

I

The olefinic bond is described by two π orbitals of different energy, bonding and antibonding. In conjugated diene molecules such as 1,3-butadiene, when coplanarity permits, there is an effective overlap of π orbitals resulting in a π–π conjugated system (I).

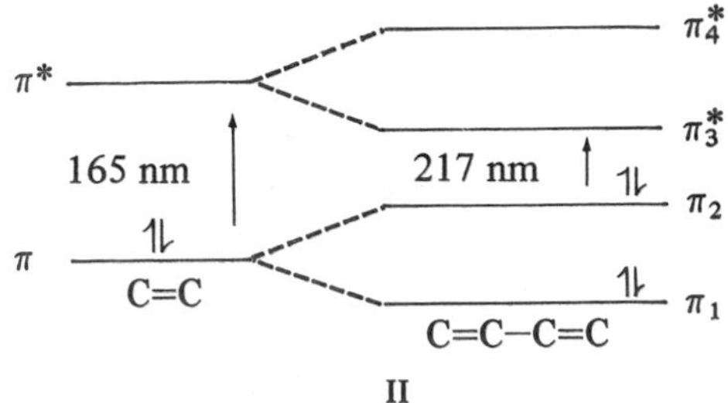

II

This overlap or interaction results in the creation of two new energy levels in butadiene (II). Thus, the $\pi_2 \rightarrow \pi_3^*$ transition of butadiene is bathochromically shifted relative to the $\pi \rightarrow \pi^*$ transition of ethylene. There are other $\pi \rightarrow \pi^*$ transitions possible in the conjugated system. Their intensities depend on the "allowedness" of the transitions.

Acyclic conjugated dienes show intense $\pi \rightarrow \pi^*$ transition bands (*K*-bands) in the 215-230 nm region. 1,3-Butadiene absorbs at 217 nm (ϵ_{max} 21,000). Further conjugation, in open-chain trienes and polyenes, results in additional bathochromic shifts accompanied by increases in absorption intensity. The spectra of polyenes are characterized by fine structure, particularly when the spectra are determined in the vapor phase or in nonpolar solvents. Absorption data for several conjugated olefins are presented in Table IV.

The bathochromic effect of alkyl substitution in 1,3-butadiene is apparent from the data for 2,3-dimethyl-1,3-butadiene.

In cases where *cis* and *trans* isomers are possible, the *trans* isomer absorbs at the longer wavelength with the greater intensity (see Table XXII). This difference becomes more pronounced as the length of the conjugated system

Table IV. Absorption Data for Conjugated Olefins

Compound	$\pi \rightarrow \pi^*$ Transition (*K*-band) λ_{max} (*nm*)	ϵ_{max}	Solvent
1,3-Butadiene	217	21,000	Hexane
2,3-Dimethyl-1,3-butadiene	226	21,400	Cyclohexane
1,3,5-Hexatriene	253	~50,000	Isooctane
	263	52,500	
	274	~50,000	
1,3-Cyclohexadiene	256	8,000	Hexane
1,3-Cyclopentadiene	239	3,400	Hexane

increases. Coplanarity is required for the most effective overlap of the π-orbitals and lower energy of the $\pi \rightarrow \pi^*$ transition. Of the two isomers, the *cis* isomer is more likely to be forced into a nonplanar conformation by steric effects (see discussion of stilbene below). The greater absorption intensity of the *trans* isomer results from the greater overall magnitude of the transition moment of the excited molecule.

An empirical method for predicting the bathochromic effect of alkyl substitution in 1,3-butadiene has been formulated by Woodward. These rules can be summarized as follows: (1) Each alkyl group, or ring residue, attached to the parent diene (1,3-butadiene) shifts the absorption 5 nm toward the long wavelength region; and (2) the creation of an exocyclic double bond causes an additional bathochromic shift of 5 nm, the shift being 10 nm if the double bond is exocyclic to two rings. For example, 217 + (2 X 5) is the predicted λ_{max} value for 2,3-dimethyl-1,3-butadiene; the observed λ_{max} is 226.

Examination of the data in Table IV shows a marked bathochromic shift and a decrease in absorption intensity for the conjugated, monocyclic diene system compared with 1,3-butadiene. Butadiene exists in the preferred *s-trans* (transoid) conformation, whereas the cyclic dienes are forced into an *s-cis* (cisoid) conformation. The reason for the bathochromic shift in the *s-cis* structure brought about by cyclization is not clear. The decrease in intensity is more easily explained, since the transition moment of the cyclic or homoannular system is less than that of the acyclic or the heteroannular systems.

Homoannular Heteroannular

Heteroannular and acyclic dienes usually display molar absorptivities in the 8000-20,000 range, whereas homoannular dienes usually display molar absorptivities in the 5000-8000 range.

Table V. Rules of Diene Absorption

Base value for heteroannular diene	214
Base value for homoannular diene	253
Increments for	
Double bond extending conjugation	+30
Alkyl substituent or ring residue	+5
Exocyclic double bond	+5
Polar groupings: OAc	+0
OAlk	+6
SAlk	+30
Cl, Br	+5
$N(Alk)_2$	+60
Solvent correction*	+0
	λ_{calc} = Total

See L.M. Fieser and M. Fieser, *Steroids.* New York: Reinhold, 1959, pp. 15-24; R.B. Woodward, *J. Am. Chem. Soc.*, **63**, 1123 (1941); **64**, 72, 76 (1942); A.I. Scott, *Interpretation of the Ultraviolet Spectra of Natural Products.* New York: Pergamon Press (MacMillan), 1964.

*Solvents have negligible effects upon the λ_{max} of these π-π^* transitions.

Rules for predicting the position of absorption of homo- and heteroannular systems are due largely to the work of Fieser. These rules are summarized in Table V.

Poor correlations are obtained when the data of Table V are applied to cross-conjugated polyene systems such as

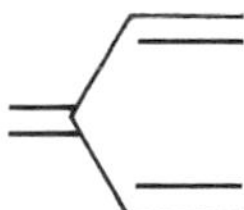

The value of these rules in structural studies of natural products will be obvious from two examples:

1. Cholesta-3,5-diene

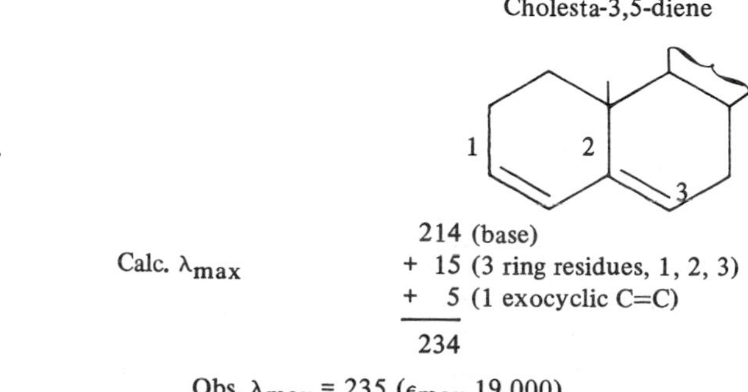

Calc. λ_{max}
214 (base)
+ 15 (3 ring residues, 1, 2, 3)
+ 5 (1 exocyclic C=C)
234

Obs. λ_{max} = 235 (ϵ_{max} 19,000)

2. Cholesta-2,4-diene

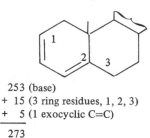

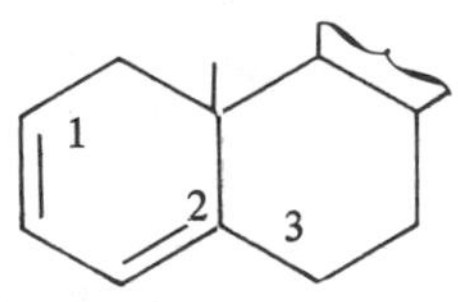

Calc. λ_{max} 253 (base)
+ 15 (3 ring residues, 1, 2, 3)
+ 5 (1 exocyclic C=C)
273

Obs. λ_{max} = 275 (ϵ_{max} 10,000)

The examples above illustrate the typically much higher molar absorptivity of the heteroannular (*transoid*) diene compared to the homoannular (*cisoid*) diene.

The rules outlined in Table V work reasonably well for conjugated systems of 4 double bonds or less. If a conjugated polyene contains more than 4 double bonds, the Fieser-Kuhn rules are used. In this approach both the λ_{max} and ϵ_{max} are related to the number of conjugated double bonds, as well as other structural units, by the following equations:

$$\lambda_{max} = 114 + 5M + n(48.0 - 117n) - 16.5\ R_{endo} - 10R_{exo}$$

$$\epsilon_{max} = (1.74 \times 10^4)n$$

where n = no. of conjugated double bonds
M = no. of alkyl or alkyl-like substituents on the conjugated system
R_{endo} = no. of rings with endocyclic double bonds in the conjugated system
R_{exo} = no. of rings with exocyclic double bonds

The equations may be applied to lycopene as follows:

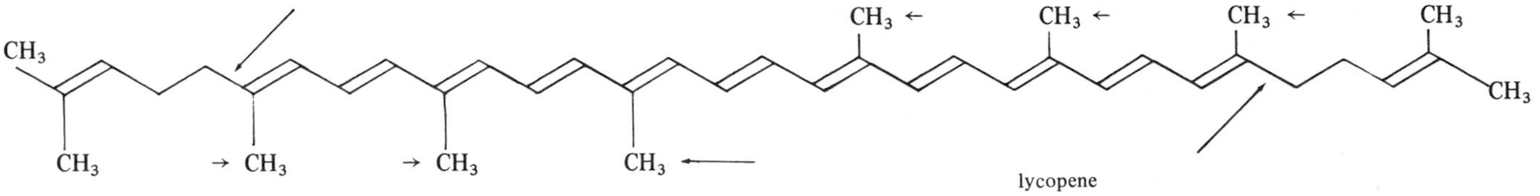

lycopene

Since only the 11 internal double bonds of the 13 double bonds of lycopene are conjugated, $n = 11$. In addition, since there are 8 substituents (arrows, methyl groups and chain residues) on the polyene system, then M = 8. Finally, since there are no ring systems, there are neither exocyclic nor endocyclic double bonds in this conjugated system and $R_{exo} = R_{endo} = 0$. Thus,

$$\lambda_{max}^{calc} = 114 + 5(8) + 11[48.0 - 1.7(11)] - 0 - 0$$

$$= 476 \text{ nm}$$

$$\lambda_{max}^{obs} = 474 \text{ nm (hexane)}$$

$$\epsilon_{max}^{cal.} = 1.74 \times 10^4(11) = 19.1 \times 10^4$$

$$\epsilon_{max}^{obs} = 18.6 \times 10^4 \text{ (hexane)}$$

A similar calculation can be carried out for β-carotene:

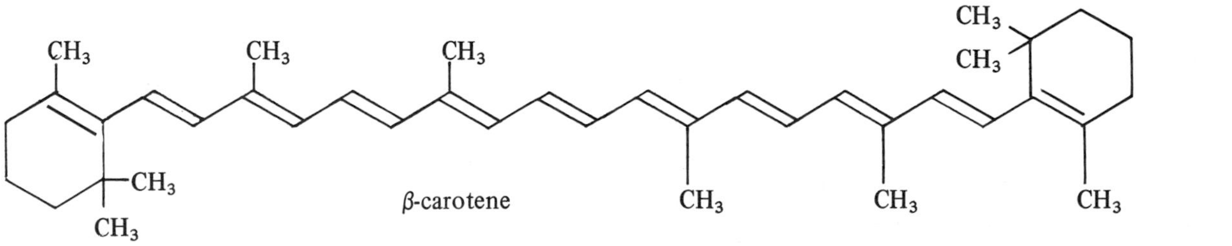

β-carotene

$\lambda_{max}^{calc.} = 453.3$ nm $\qquad \epsilon_{max}^{calc.} = 19.1 \times 10^4$

$\lambda_{max}^{obs} = 452$ nm (hexane) $\qquad \epsilon_{max}^{obs} = 15.2 \times 10^4$

Acetylenic Chromophore

The absorption characteristics of the acetylenic chromophore are more complex than those of the ethylenic chromophore. Acetylene shows a relatively weak band at 173 nm resulting from a $\pi \rightarrow \pi^*$ transition. Conjugated polyynes show 2 principal bands in the near ultraviolet which are characterized by fine structure. The short-wavelength band is extremely intense, and all of these bands arise from $\pi \rightarrow \pi^*$ transitions. Typical absorption data for conjugated polyynes are shown in Table VI.

Table VI. Absorption Data for Conjugated Polyynes

	$\lambda_{max}(nm)$	ϵ_{max}	$\lambda_{max}(nm)$	ϵ_{max}
2,4-Hexadiyne	–	–	227	360
2,4,6-Octatriyne	207	135,000	268	200
2,4,6,8-Decatetrayne	234	281,000	306	180

Carbonyl Chromophore

The carbonyl group contains, in addition to a pair of σ-electrons, a pair of π-electrons and 2 pairs of nonbonding (n or p) electrons. Saturated ketones and aldehydes display 3 absorption bands, 2 of which are observed in the far ultraviolet region. A $\pi \rightarrow \pi^*$ transition absorbs strongly near 150 nm; an $n \rightarrow \sigma^*$ transition absorbs near 190 nm. The third band (R-band) appears in the near ultraviolet in the 270-300 nm region. The R-band is weak ($\epsilon_{max} < 30$) and results from the forbidden transition of a loosely held n-electron to the π^* orbital, the lowest unoccupied orbital of the carbonyl group. R-bands undergo a blue shift as the polarity of the solvent is increased. Acetone absorbs at 279 nm in n-hexane; in water the λ_{max} is 264.5. The blue shift results from hydrogen bonding which lowers the energy of the n orbital. The blue shift can be used as a measure of the strength of the hydrogen bond.

SATURATED KETONES AND ALDEHYDES. Absorption data for several saturated ketones and aldehydes are presented in Table VII.

The bathochromic effect accompanying the introduction of larger and more highly branched alkyl groups in the aliphatic ketones can be seen from the data in Table VII.

The $n \rightarrow \pi^*$ absorption of ketones and aldehydes is weak. Spectra of derivatives, such as the semicarbazones or 2,4-dinitrophenylhydrazones, are often used for identification work.

Table VII. Absorption Data for Saturated Aldehydes and Ketones

Compound	$n \rightarrow \pi^*$ Transition (R-band) λ_{max} (nm)	ϵ_{max}	Solvent
Acetone	279	13	Isooctane
Methyl ethyl ketone	279	16	Isooctane
Diisobutyl ketone	288	24	Isooctane
Hexamethylacetone	295	20	Alcohol
Cyclopentanone	299	20	Hexane
Cyclohexanone	285	14	Hexane
Acetaldehyde	290	17	Isooctane
Propionaldehyde	292	21	Isooctane
Isobutyraldehyde	290	16	Hexane

Table VIII. Shifts in Absorption Maxima for Substituted Cyclohexanones

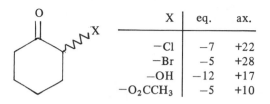

X	eq.	ax.
–Cl	–7	+22
–Br	–5	+28
–OH	–12	+17
$-O_2CCH_3$	–5	+10

The introduction of an α-halogen atom in an aliphatic ketone has little effect on the $n \rightarrow \pi^*$ transition. However, α-substitution of halogen atoms in saturated cyclic ketones has a marked effect on the absorption characteristics (Table VIII). The λ_{max} of the parent compound is reduced by 5-10 nm when the substituent is equatorial and a bathochromic shift of 10-30 nm occurs when the substituent is axial. The bathochromic shift is usually accompanied by a strong hyperchromic effect. These effects are valuable in the structure determination of halogenated steroids and terpenes. Some sort of conformational rigidity must be present in the cyclohexanone system so that the substituent, X, is clearly either in an equatorial or an axial position.

The attachment of groups containing lone electron pairs to carbonyl groups has a marked effect upon the $n \rightarrow \pi^*$ transition. The R-band is shifted to shorter wavelengths with little effect upon intensity. The shift in absorption results from a combination of inductive and resonance effects. Substitution may change the energy levels of both the ground state and the excited state, but the important factor is the relative energies of the two levels. Absorption values for the $n \rightarrow \pi^*$ transitions of several simple carbonyl compounds are presented in Table IX.

α-DIKETONES AND α-KETO ALDEHYDES. Acyclic α-diketones, such as biacetyl, exist in the *s-trans* conformation (with a dihedral angle, ϕ, of 180°). The spectrum of biacetyl shows the normal weak R-band at 275 nm and a

Table IX. $n \rightarrow \pi^*$ Transitions (R-Bands) of Simple Carbonyl-Containing Compounds

Compound	λ_{max} (nm)	ϵ_{max}	Solvent
Acetaldehyde	293	11.8	Hexane
Acetic acid	204	41	Ethanol
Ethyl acetate	207	69	Pet. ether
Acetamide	220 (*s*)		Water
Acetyl chloride	235	53	Hexane
Acetic anhydride	225	47	Isooctane
Acetone	279	15	Hexane

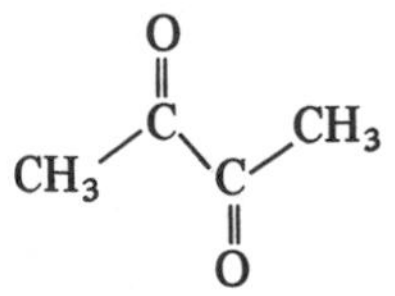

$\phi = \sim 180°$

$\lambda_{max} = 450$ nm

$\epsilon = 10$

weak band near 450 nm resulting from interaction between the carbonyl groups. The position of the long-wavelength band of α-diketones incapable of enolization reflects the effect of coplanarity upon resonance, and hence depends on the dihedral angle ϕ between the carbonyl groups (I, II, III):

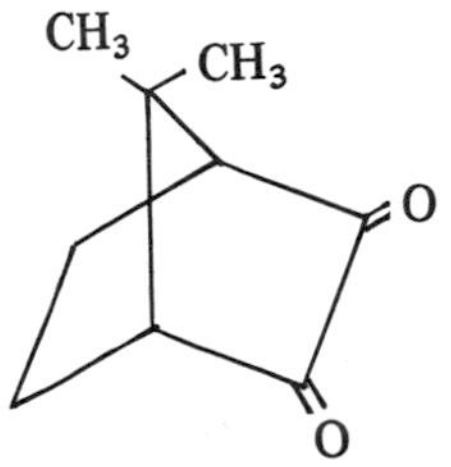

I Camphorquinone

$\phi = 0\text{-}10°$ λ_{max} 488 *nm*

$\epsilon = 17$

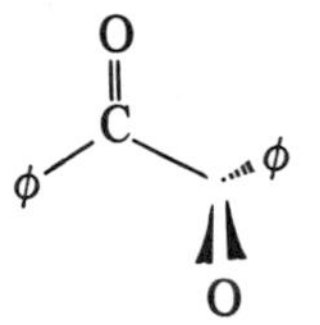

II Benzil

$\phi = 90°$ λ_{max} 370 *nm*

$\epsilon = 40$

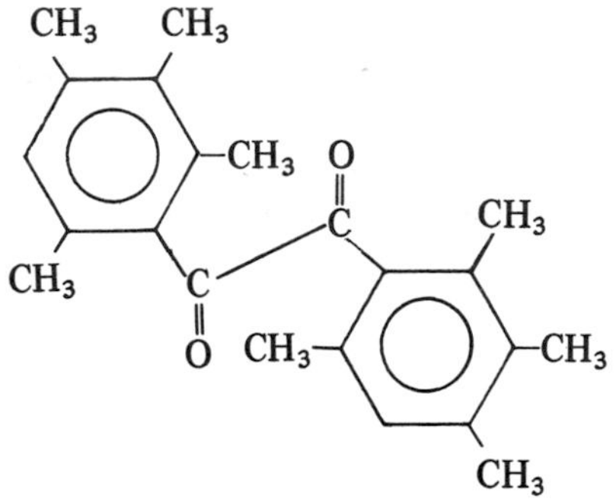

III Isoduril

$\phi = 180°$ λ_{max} 490 *nm*

Cyclic α-diketones with α-hydrogen atoms exist in the enolic form, for example, diosphenol. The absorption characteristics of diosphenol appear later in this chapter.

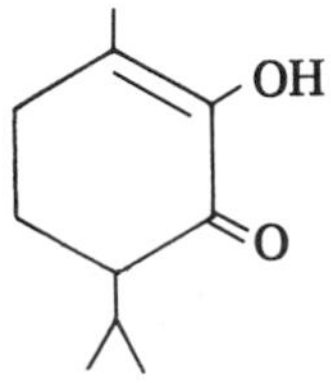

Diosphenol

β-DIKETONES. The ultraviolet spectra of β-diketones depend on the degree of enolization. The enolic form is stabilized when steric considerations permit intramolecular hydrogen bonding. Acetylacetone is a classic example. The enolic species exists to the extent of about 15% in aqueous solution and 91-92% in the vapor phase or in solution in nonpolar solvents. The absorption is directly dependent on the concentration of the enol tautomer.

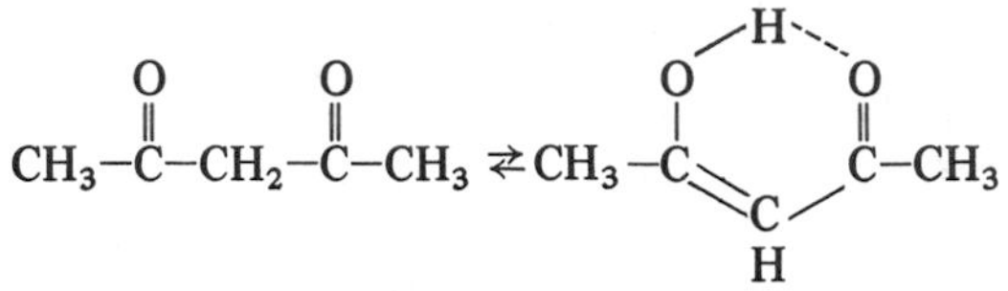

$\lambda_{max}^{H_2O}$ 274, ϵ_{max} 2050

$\lambda_{max}^{Isooctane}$ 272, ϵ_{max} 12,000

Cyclic β-diketones, such as 1,3-cyclohexanedione, exist almost exclusively in the enolic form even in polar solvents. The enolic structures show strong absorption in the 230-260 nm region due to the $\pi \rightarrow \pi^*$ transition in the *s-trans* enone system. 1,3-Cyclohexanedione, in ethanol, absorbs at 253 nm (ϵ_{max} 22,000). The formation of the enolate ion, in alkaline solution, shifts the strong absorption band into the 270 to 300 nm region.

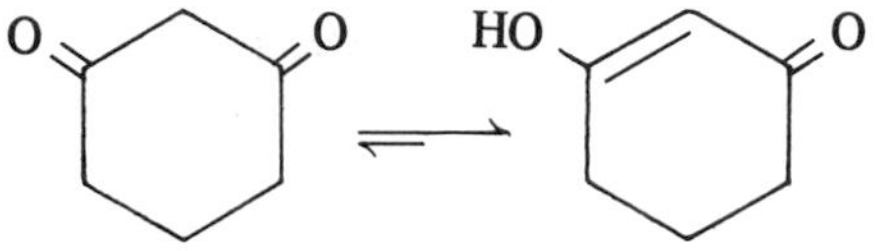

enone

α,β-UNSATURATED KETONES AND ALDEHYDES. Compounds containing a carbonyl group in conjugation with an ethylenic group are called *enones*. Spectra of enones are characterized by an intense absorption band (*K*-band) in the 215-250 nm region (ϵ_{max} usually 10,000-20,000), and a weak $n \rightarrow \pi^*$ band (*R*-band) at 310-330 nm. The weak *R*-band is frequently poorly defined. Absorption data for several conjugated ketones and aldehydes are presented in Table X.

Since carbonyl compounds are polar, the positions of the *K*- and *R*-bands of enones are both dependent on the solvent. The hypsochromic effect on the *R*-band with increasing solvent polarity has already been discussed above.

The *K*-bands of enones undergo a bathochromic shift with increasing solvent polarity. The solvent effect on the spectrum of mesityl oxide is summarized in Table XI.

Table X. Absorption Data for Conjugated Ketones and Aldehydes

Compound	*K*-band λ_{max} (*nm*)	*K*-band log ϵ_{max}	*R*-band λ_{max} (*nm*)	*R*-band log ϵ_{max}	Solvent
Methyl vinyl ketone	212.5	3.85	320	1.32	Ethanol
Methyl isopropenyl ketone	218	3.90	315	1.4	Ethanol
Acrolein	210	4.06	315	1.41	Water
Crotonaldehyde	220	4.17	322	1.45	Ethanol
Crotonaldehyde	214	4.20	329	1.39	Isooctane
			341	1.38	
			352 (*s*)	1.25	

(*s*) shoulder.

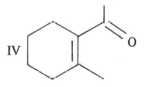

$(CH_3)_2C{=}CH{-}C({=}O){-}CH_3$ mesityl oxide

Table XI. Effect of Solvent Polarity on the Spectrum of Mesityl Oxide

Solvent	Transition $\pi \rightarrow \pi^*$ (λ_{max} *nm*)	Transition $n \rightarrow \pi^*$ (λ_{max} *nm*)
Isooctane	230.6	321
Chloroform	237.6	314
Water	242.6	Submerged under *K*-band

The intensity of the *K*-band may be reduced to less than 10^4 where steric hindrance prevents coplanarity. This frequently occurs in cyclic systems, such as IV.

IV

λ_{max} 243 *nm*, ϵ_{max} 1400

Table XII. Maxima of α,β-Unsaturated Ketones

Substitution		Probable λ_{max} (*nm*)
Unsubstituted		215
α or β	No exocyclic C=C	225
αβ or ββ	No exocyclic C=C	235
αβ or ββ	One exocyclic C=C	240
αββ	No exocyclic C=C	247
αββ	One exocyclic C=C	252

Woodward has derived empirical generalizations for the effect of substitution on the position of the $\pi \rightarrow \pi^*$ transition (*K*-band) in the spectra of α,β-unsaturated ketones. The positions of the *K*-bands that result from substitution in the basic formula

$$R{-}\overset{O}{\overset{\|}{C}}{-}\overset{\alpha}{\overset{|}{C}}{=}C\begin{matrix}\beta\\ \beta\end{matrix}$$

are summarized in Table XII.

The spectra of α,β-unsaturated aldehydes (Table XIII) are similar to those of the α,β-unsaturated ketones. The *R*-bands occur in the 350-370 nm region and exhibit some fine structure when the spectra are determined in nonpolar solvents.

Table XIII. Maxima of Unsaturated Aldehydes

Substitution	Probable λ_{max} (*nm*)[a]
Unsubstituted	208
α or β	220
αβ or ββ	230
αββ	242

[a]exocyclic double bond +5

Similar rules apply to the cyclopentenone system:

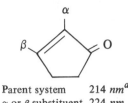

Parent system	214 *nm*[a]
α or β substituent	224 *nm*
α, β substituents	236 *nm*

[a]This value can be calculated from data in Table XIV
202 (Base value to form 5-membered cyclic enone)
12 (β-substituent, bond to form ring)
214

More extensive rules of enone absorption have been summarized by Fieser. These rules appear as Table XIV.

A few examples will serve to illustrate the usefulness of these correlations.

1. 1-Acetylcyclohexene

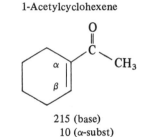

Calc. λ_{max}^{EtOH} 215 (base)
10 (α-subst)
12 (β-subst)
237

Obs. λ_{max}^{EtOH} = 232

2. Cholesta-1,4-dien-3-one

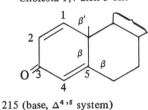

Calc. λ_{max}^{EtOH} 215 (base, $\Delta^{4,5}$ system)
24 (2 β-subst)
5 (1 exo C=C)
244

Obs. λ_{max}^{EtOH} = 245

Cross conjugation has little effect on the λ_{max} of cholesta-1,4-diene-3-one. The calculations used were for an enone (β,β-disubstituted) with no correction required for the 1,2 double bond or for the β′ group. The corresponding calculation using the $\Delta^{1,2}$ system yields 227 nm as the predicted λ_{max}; this is an indication that, when such

Table XIV. Rules of Enone and Dienone Absorption (α,β-Unsaturated Carbonyls of Ketones)

β–C=C–C=O (Enone) and δ–C=C–C=C–C=O (Dienone)

Base values:			
Acyclic α,β-unsaturated ketones			215
Six-membered cyclic α,β-unsaturated ketones			215
Five-membered cyclic α,β-unsaturated ketones			202
α,β-Unsaturated aldehydes			210
α,β-Unsaturated carboxylic acids and esters			195
Increments for			
Double bond extending conjugation			+30
Alkyl group, ring residue		α	+10
		β	+12
		γ and higher	+18
Polar groupings:	–OH	α	+35
		β	+30
		δ	+50
	–OAc	α,β,δ	+6
	–OMe	α	+35
		β	+30
		γ	+17
		δ	+31
	–SAlk	β	+85
	–Cl	α	+15
		β	+12
	–Br	α	+25
		β	+30
	$-NR_2$	β	+95
Exocyclic double bond			+5
Homodiene component[a]			+39
Solvent correction (see table below)			variable
			λ_{calc} = Total[b]

[a]Two conjugated double bonds, both in the same ring.
[b]The calculated values usually fall within ±3 nm of the observed values. The molar absorptivities of *cisoid* enones are usually less than 10,000, whereas the molar absorptivities of *transoid* enones are greater than 10,000.

Solvent Corrections*

Solvent	Correction (*nm*)
Ethanol	0
Methanol	0
Dioxane	+5
Chloroform	+1
Ether	+7
Water	–8
Hexane	+11
Cyclohexane	+11

**E.g.*, if the absorption was determined in cyclohexane, 11 nm would be added to the observed $\lambda_{max}^{cyclohexane}$ to compare it to a calculated $\lambda_{max}^{ethanol}$.

a choice is presented, the more reliable prediction arises from the system that is more highly substituted (longer wavelength $\pi \rightarrow \pi^*$ transition).

3. Enol of 1,2-cyclopentanedione

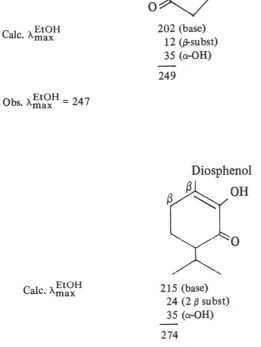

HO, β, O

Calc. λ_{max}^{EtOH}

202 (base)
12 (β-subst)
35 (α-OH)
249

Obs. λ_{max}^{EtOH} = 247

4. Diosphenol

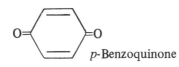

β, β, OH, O

Calc. λ_{max}^{EtOH}

215 (base)
24 (2 β subst)
35 (α-OH)
274

Obs. λ_{max}^{EtOH} = 270

The spectrum of *p*-benzoquinone is similar to

p-Benzoquinone

that of a typical α,β-unsaturated ketone, the strong *K*-band appearing at 245 nm with a weak *R*-band near 435 nm.

There are cases where the C=O and the C=C are non-conjugated in the formal sense but where interaction does occur to produce an absorption band. In structures where this occurs, the C=O and C=C groups must be oriented so that there can be effective overlap of the π orbitals. For example, the structure CH_2= ◇ =O shows a moderately strong band near 214 nm with a normal weak *R*-band at 284 nm. Similar effects are observed when there can be effective overlap of the π orbital for the C=O group and the *p*(*n*-) orbitals of a heteroatom. For example,

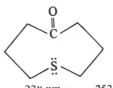

λ_{max} 238 *nm*, ϵ_{max} 2535

The interaction in these apparently nonconjugated systems is known as transannular conjugation.

CARBOXYLIC ACIDS. Saturated carboxylic acids show a weak absorption band near 200 nm resulting from the forbidden $n \rightarrow \pi^*$ transition. The position of the band undergoes a small bathochromic shift with an increase in chain length. This band is of little diagnostic value.

α,β-Unsaturated acids display a strong *K*-band characteristic of the conjugated system. In the first member of the series, acrylic acid, the absorption occurs near 200 nm, ϵ_{max} 10,000. Alkyl substitution in this basic structure results in a bathochromic shift of the *K*-band in much the same manner as observed for α,β-unsaturated ketones. The extension of conjugation produces further bathochromic shifts accompanied by an increase in the band intensity and the appearance of fine structure. The attachment of an electronegative group on the α-carbon also produces a bathochromic shift. We can calculate expected maxima for the compounds described in Table XV, utilizing the data of Table XIV.

The absorption characteristics of several α,β-unsaturated acids are summarized in Table XVI.

ESTERS AND LACTONES. Esters and sodium salts of carboxylic acids show absorption at wavelengths and intensities comparable to the parent acid. Conjugated, unsaturated lactones display spectra similar to unsaturated esters. The spectra of simple unsaturated lactones show end absorption in the 200-240 nm region. Extended conjugation produces a bathochromic shift of the $\pi \rightarrow \pi^*$ transi-

Table XV. Absorption Maxima of Unsaturated Carboxylic Acids and Esters

	λ_{max}^{EtOH} (±5 *nm*)
α or β-monosubstituted	208
α,β- or β,β-disubstituted	217
α,β,β-trisubstituted	225
exocyclic α,β double bond	+5
endocyclic α,β double bond in 5 or 7 membered ring	+5

(Data of Table XIV may be applied here.)
A.T. Nielson, *J. Org. Chem.*, **22**, 1539 (1957).

tion (*K*-band). Table XIV may be used to predict the maxima of such compounds.

AMIDES AND LACTAMS. α,β-Unsaturated amides and lactams show absorption in the near ultraviolet at λ_{max} 200-220, $\epsilon_{max} < 10{,}000$. α,β-Unsaturated lactams show a second band near 250 nm ($\epsilon_{max} \sim 1000$).

Azomethines (Imines) and Oximes

These structures show only weak absorption in the near ultraviolet unless the >C=N− group is involved in conjugation. In the spectra of conjugated azomethines and oximes the $\pi \rightarrow \pi^*$ transition (*K*-band) appears in the 220-230 nm region, $\epsilon_{max} > 10{,}000$. Acidification of the azomethines, producing a positive charge on the nitrogen, shifts the absorption to the 270-290 nm region. Simple imines show a weak $n \rightarrow \pi^*$ transition. For example, for $CH_3CH_2CH_2N{=}C(CH_3)_2$
λ_{max} = 246 nm (ϵ = 140, cyclohexane)
λ_{max} = 232 nm (ϵ = 200, ethanol)

Nitriles and Azo Compounds

α,β-Unsaturated nitriles absorb just inside the near ultraviolet region, near 213 nm, $\epsilon_{max} \sim 10{,}000$.

The azo group is analogous to the ethylenic linkage with two σ bonds being replaced by two lone pairs of electrons ($-\ddot{N}{=}\ddot{N}-$). The $\pi \rightarrow \pi^*$ transition occurs in the far (vacuum) ultraviolet. The $n \rightarrow \pi^*$ band in aliphatic azo compounds appears near 350 nm with the expected low intensity, $\epsilon_{max} < 30$. *trans*-Azobenzene absorbs at 320 nm (ϵ_{max} 21,000). Comparable absorption for *trans*-stilbene occurs at 295 nm (ϵ_{max} 28,000).

Compounds with N to O Bonds

Four groups contain multiple nitrogen-to-oxygen linkages: nitro, nitroso, nitrates, and nitrites. All of these structures show weak absorption in the near ultraviolet region resulting from an $n \rightarrow \pi^*$ transition.

The absorption of several typical compounds containing nitrogen-to-oxygen linkages are presented in Table XVII.

The effect of conjugation upon the absorption characteristics of the nitro group is apparent from the data for 1-nitro-1-propene. The strong $\pi \rightarrow \pi^*$ transition (*K*-band) submerges the weak $n \rightarrow \pi^*$ transition (*R*-band).

Multiply Bonded Sulfur Groups

Aliphatic sulfones are transparent in the near ultraviolet region. The sulfur atom in sulfones has no lone-pair electrons, and the lone pairs of electrons associated with the oxygen atoms appear to be tightly bound. In an α,β-unsaturated sulfone, such as ethyl vinyl sulfone, a band appears in the 210 nm region resulting from resonance between the S to O linkage and the ethylenic linkage.

Saturated sulfoxides absorb near 220 nm with intensities of the order of 1500. This absorption involves an $n \rightarrow \pi^*$

Table XVI. Absorption of α,β-Unsaturated Acids

Compound	λ_{max}^{EtOH}	ϵ_{max}
$CH_2{=}CH{-}COOH$	200 nm	10,000
$CH_3CH{=}CH{-}COOH$ (*trans*)	205	14,000
$CH_3CH{=}CH{-}COOH$ (*cis*)	205.5	13,500
$CH_2{=}C(CH_3){-}COOH$	210	–
(tetrahydrothiopyranylidene)=C(H)−COOH	220	14,000
(tetrahydrothiopyranylidene)=C(CN)−COOH	235	12,500
$CH_3{-}(CH{=}CH)_2{-}COOH$	254	25,000
$CH_3{-}(CH{=}CH)_3{-}COOH$	294	37,000
$CH_3{-}(CH{=}CH)_4{-}COOH$	332	49,000

Table XVII. Absorption of Compounds Containing Nitrogen-Oxygen Linkages

Compound	$n \rightarrow \pi^*$ Transition (*R*-band) λ_{max}	ϵ_{max}	Solvent
Nitromethane	275 nm	15	Heptane
2-Methyl-2-nitropropane	280.5	23	Heptane
1-Nitro-1-propene	229[a]	9400	Ethanol
	235[a]	9800	
Nitrosobutane	300	100	Ether
	665	20	
Octyl nitrate	270[b]	15	Pentane
Cyclohexyl nitrate	270[b]	22	–
n-Butyl nitrite	218	1050	Ethanol
	313-384[c]	17-45	

[a]These are $\pi \rightarrow \pi^*$ transitions.
[b]This is typically a point of inflection in the spectra of nitrates.
[c]This region is one of fine structure with bands roughly 10 nm apart. The band with maximum absorption occurs at 357 nm.

transition in the S=O group and thus undergoes a hypsochromic shift as solvent polarity is increased. Aromatic sulfoxides show an intense *K*-band in addition to the displaced *B*-band.

In compounds containing the $-\overset{\overset{\displaystyle O}{\|}}{S}-X$ grouping, the position of the $n \rightarrow \pi^*$ band depends on the electronegativity of X; the greater the electronegativity, the shorter will be the wavelength of absorption.

Simple thioketones of the dialkyl or alkylaryl types are unstable and generally exist as trimers. Thiobenzophenone is monomeric as are compounds in which the C=S group is attached to an electron-donating group, such as $-NR_2$. The $n \rightarrow \pi^*$ transition of the C=S group in thioketones occurs at a longer wavelength than the analogous C=O transition because the energy of the nonbonding electron pair of the sulfur atom lies at a higher level than the corresponding electrons of the oxygen atom. Compounds containing the C=S group also display intense bands in the 250-320 nm region which presumably arise from $\pi \rightarrow \pi^*$ and $n \rightarrow \sigma^*$ transitions in the C=S group.

The $n \rightarrow \pi^*$ bands of some thiocarbonyl compounds are summarized in Table XVIII.

Table XVIII. $n \rightarrow \pi^*$ Transitions in Thiocarbonyl Compounds

Compound	$n \rightarrow \pi^*$ Transition (*R*-band)	
	λ_{max} (*nm*)	log ϵ_{max}
Thiobenzophenone	599	2.81
Thioacetamide	358	1.25
Thiourea	291 (*s*)	1.85

(*s*) shoulder.

Benzene Chromophore

Benzene displays 3 absorption bands: 184 nm (ϵ_{max} 60,000), 204 nm (ϵ_{max} 7900), and 256 nm (ϵ_{max} 200). These bands originate from $\pi \rightarrow \pi^*$ transitions. The intense band near 180 nm results from an allowed transition, whereas the weaker bands near 200 and 260 nm result from forbidden transitions in the highly symmetrical benzene molecule. Different notations have been used to designate the absorption bands of banzene; these are summarized in Table XIX. We shall discuss these bands using Braude's *E* and *B* notation.

The *B*-band of benzene and many of its homologs is characterized by considerable fine structure. This is particularly true of spectra determined in the vapor phase or in nonpolar solvents. The fine structure originates from sublevels of vibrational absorption upon which the electronic absorption is superimposed (Figure 7, p. 327). In polar solvents, interactions between solute and solvent tend to reduce the fine structure.

Table XIX. Benzene Bands

184 *nm*	204 *nm*	256 *nm*	Ref.
E_1	E_2	*B*	a
–	*K*	*B*	b

[a]See Reference 5 at the end of the Chapter.
[b]A. Burawoy, *Ber.*, **63**, 3155 (1930); *J. Chem. Soc.*, 1177 (1939).

Substitution of alkyl groups on the benzene ring produces a bathochromic shift of the *B*-band, but the effect of alkyl substitution upon the *E*-bands is not clearly defined. The absorption characteristics of the *B*-bands of several alkylbenzenes are presented in Table XX.

The bathochromic shift is attributed to hyperconjugation in which the σ-electrons of an alkyl C–H bond participate in resonance with the ring. The methyl group is more effective in hyperconjugation than other alkyl groups.

H⁺
H–C=⟨ring :⟩–
H

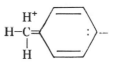

The addition of a second alkyl group to the molecule is most effective in producing a red shift if it is in the *para* position. The *para* isomer absorbs at the longest wavelength with the largest ϵ_{max}. The *ortho* isomer generally absorbs at the shortest wavelength with reduced ϵ_{max}. This effect is attributed to steric interactions between the *ortho* substituents, which effectively reduce hyperconjugation.

Substitution on the benzene ring of auxochromic groups (OH, NH_2, etc.) shifts the *E*- and *B*-bands to longer wavelengths, frequently with intensification of the *B*-band and loss of its fine structure, because of *n*-π conjugation (Table XXI).

Table XX. Absorption Data for Alkylbenzenes (*B*-Bands)

Compound	λ_{max} (*nm*)*	ϵ_{max}
Benzene	256	200
Toluene	261	300
m-Xylene	262.5	300
1,3,5-Trimethylbenzene	266	305
Hexamethylbenzene	272	300

*λ_{max} of most intense peak in band with fine structure.

Table XXI. Effect of Auxochromic Substitution on the Spectrum of Benzene

Compound	E_2-band		B-band		Solvent
	λ_{max} (*nm*)	ϵ_{max}	λ_{max} (*nm*)	ϵ_{max}	
Benzene	204	7,900	256	200	Hexane
Chlorobenzene	210	7,600	265	240	Ethanol
Thiophenol	236	10,000	269	700	Hexane
Anisole	217	6,400	269	1,480	2% Methanol
Phenol	210.5	6,200	270	1,450	Water
Phenolate anion	235	9,400	287	2,600	Aq. alkali
o-Catechol	214	6,300	276	2,300	Water (pH 3)
o-Catecholate anion	236.5	6,800	292	3,500	Water (pH 11)
Aniline	230	8,600	280	1,430	Water
Anilinium cation	203	7,500	254	160	Aq. acid
Diphenyl ether	255	11,000	272	2,000	Cyclohexane
			278	1,800	

Conversion of a phenol to the corresponding anion results in a bathochromic shift of the E_2- and B-bands and an increase in ϵ_{max} because the nonbonding electrons in the anion are available for interaction with the π-electron system of the ring. When aniline is converted to the anilinium cation, the pair of nonbonding electrons of aniline is no longer available for interaction with the π-electrons of the ring, and a spectrum almost identical to that of benzene results.

Confirmation of a suspected phenolic structure may be obtained by comparison of the ultraviolet spectra obtained for the compound in neutral and in alkaline solution (pH 13). Similar confirmatory information for a suspected aniline derivative may be obtained by a comparison of spectra determined in neutral and in acid solution (pH 1).

When the spectra of pure acid and of pure conjugate base cross at some point, the spectra of all solutions containing various ratios of these two species (and no other

Table XXII. Absorption Characteristics of Benzenes Substituted with Chromophores

Compound	$\pi \rightarrow \pi^*$ Transition K-band		B-band		$n \rightarrow \pi^*$ Transition R-band		Solvent
	λ_{max} (nm)	ϵ_{max}	λ_{max} (nm)	ϵ_{max}	λ_{max} (nm)	ϵ_{max}	
Benzene	–	–	255	215	—	–	Alcohol
Styrene	244	12,000	282	450	—	–	Alcohol
Phenylacetylene	236	12,500	278	650	—	–	Hexane
Benzaldehyde	244	15,000	280	1,500	328	20	Alcohol
Acetophenone	240	13,000	278	1,100	319	50	Alcohol
Nitrobenzene	252	10,000	280	1,000	330	125	Hexane
Benzoic acid	230	10,000	270	800	–	–	Water
Benzonitrile	224	13,000	271	1,000	–	–	Water
Diphenyl sulfoxide	232	14,000	262	2,400	–	–	Alcohol
Phenyl methyl sulfone	217	6,700	264	977	–	–	–
Benzophenone	252	20,000	–	–	325	180	Alcohol
Biphenyl	246	20,000	submerged		–	–	Alcohol
Stilbene (*cis*)	283	12,300*	submerged		–	–	Alcohol
Stilbene (*trans*)	295†	25,000*	submerged		–	–	Alcohol
1-Phenyl-1,3-butadiene							
cis-	268	18,500	–	–	–	–	Isooctane
trans-	280	27,000	–	–	–	–	Isooctane
1,3-Pentadiene							
cis-	223	22,600	–	–	–	–	Alcohol
trans-	223.5	23,000	–	–	–	–	Alcohol

*Intense bands also occur in the 200-230 nm region.

†Most intense band of fine structure.

absorbing species) must also go through this point, the isosbestic point, provided that the sum of the concentrations of both species is constant, and that the spectral characteristics (i.e., the absorption coefficients) of both species are insensitive to the effect of pH and of the buffer used. An isosbestic point's presence can be used to verify that one is dealing with a simple acid-base reaction that is not complicated by further equilibria or other phenomena.

Interaction between the nonbonding electron pair(s) of a heteroatom attached to the ring and the π-electrons of the ring is most effective when the p orbital of the nonbonding electrons is parallel to the π orbitals of the ring. Thus, bulky substitution in the ortho position of molecules such as *N,N*-dimethylaniline causes a hypsochromic shift in the E_2-band, accompanied by a marked reduction in ϵ_{max}.

N,N-Dimethylaniline	λ_{max} 251	ϵ_{max} 15,500
2-Methyl-*N,N*-dimethylaniline	λ_{max} 248	ϵ_{max} 6360

Direct attachment of an unsaturated group (chromophore) to the benzene ring produces a strong bathochromic shift of the *B*-band, and the appearance of a *K*-band (ϵ_{max} > 10,000) in the 200-250 nm region (Table XXII). The overlap of absorption positions of the *K*-band, and the displaced *E*-bands of auxochromically substituted benzenes, may lead to confusion in the interpretation of ultraviolet spectra. Generally *E*-bands are less intense. The *B*-bands are sometimes buried under the *K*-bands.

The data in Table XXII show that in certain structures, such as acetophenone and benzaldehyde, displaced *B*- and *R*-bands can still be recognized.

The data of Table XXIII allow calculation of an expected maximum for aromatic aldehydes, ketones, carboxylic acids and esters. Application of Table XXIII to structure confirmation is illustrated with the following examples:

Table XXIII. Calculation of the Principal Band ($\pi \rightarrow \pi^*$ Transition) of Substituted Benzene Derivatives, Ar—COG (in EtOH)[2,*]

ArCOR/ArCHO/$ArCO_2H$/$ArCO_2R$		λ_{max}^{EtOH} (nm)
Parent chromophore: Ar = C_6H_5		
G = Alkyl or ring residue, (e.g., ArCOR)		246
G = H, (ArCHO)		250
G = OH, OAlk, ($ArCO_2H$, $ArCO_2R$)		230
Increment for each substituent on Ar:		
—Alkyl or ring residue	*o*-, *m*-	+3
	p-	+10
—OH, —OCH_3, —OAlk	*o*-, *m*-	+7
	p-	+25
—O^- (oxyanion)	*o*-	+11
	m-	+20
	p-	+78[a]
—Cl	*o*-, *m*-	+0
	p-	+10
—Br	*o*-, *m*-	+2
	p-	+15
—NH_2	*o*-, *m*-	+13
	p-	+58
—$NHCOCH_3$	*o*-, *m*-	+20
	p-	+45
—$NHCH_3$	*p*-	+73
—$N(CH_3)_2$	*o*-, *m*-	+20
	p-	+85

[a]This value may be decreased markedly by steric hindrance to coplanarity.

*Use of the table is illustrated by the examples on pp. 323-324 and in A.I. Scott's book.[2]

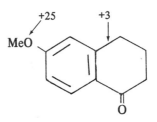

Example 1. 6-Methoxytetralone

Calc: λ_{max}^{EtOH} = 246 (parent chromophore) + 3(*o*-ring residue) + 25 (*p*-OMe) = 274 nm

Obs: λ_{max}^{EtOH} = 276 nm

Example 2. 3-Carbethoxy-4-methyl-5-chloro-8-hydroxytetralone

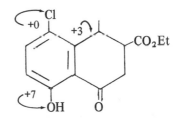

Calc.: λ_{max}^{EtOH} = 246 + 3 (*o*-ring residue) + 7 (*o*-OH) = 256 nm

Obs.: λ_{max}^{EtOH} = 257 nm

Example 3. 3,4-Dimethoxy-4b,5,6,7,8,8a,9,10-octahydrophenanthren-10-one

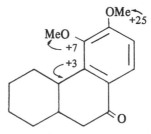

Calc.: λ_{max}^{EtOH} = 246 + 25 + 7 + 3 = 281 nm

Obs.: λ_{max}^{EtOH} = 278 nm

Application of Table XXIII does not give accurate predicted maxima in the 2,6-disubstituted phenyl carbonyl compounds; that this is due to disruption of the planarity necessary for conjugation of the carbonyl and phenyl groups is supported by the large decrease in molar absorptivity accompanying such substitution.

When auxochromic groups appear on the same ring as the chromophore, both groups influence the absorption. The influence is most pronounced when an electron donating group and electron attracting group are para to one another (complementary substitution, Table XXIV).

The red shift and increase in intensity of the *K*-band are related to contributions of the following polar resonance forms:

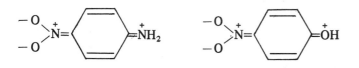

$^{-}O_2\overset{+}{N}=C_6H_4=\overset{+}{N}H_2$ $^{-}O_2\overset{+}{N}=C_6H_4=\overset{+}{O}H$

Biphenyl is the parent molecule of a series of compounds in which two aromatic rings are in conjugation. Resonance energy is at a maximum when the rings are coplanar and essentially zero when the rings are at 90° to one another.

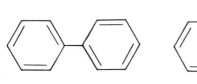

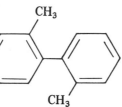

K-Band, λ_{max} 252, ϵ_{max} 19,000
Biphenyl

B-Band, λ_{max} 270, ϵ_{max} 800
2, 2′-Dimethylbiphenyl

The effect of forcing the rings out of coplanarity is readily seen from a comparison of the absorption characteristics of biphenyl and its 2,2′-dimethyl homolog whose absorption characteristics are similar to those of *o*-xylene.

Table XXIV. Absorption Characteristics of Disubstituted Benzenes

Compound	$\pi \rightarrow \pi^*$ Transition *K*-band		*B*-band	
	λ_{max}	ϵ_{max}	λ_{max}	ϵ_{max}
o-NO_2 Phenol	279	6,600	351	3,200
m-NO_2 Phenol	274	6,000	333	1,960
p-NO_2 Phenol	318	10,000	submerged	
o-NO_2 Aniline	283	5,400	412	4,500
m-NO_2 Aniline	280	4,800	358	1,450
p-NO_2 Aniline	381	13,500	submerged	

Introduction of a methylene group between two chromophores is generally considered capable of destroying conjugation. Compare the data of diphenylmethane

$C_6H_5-CH_2-C_6H_5$ λ_{max} = 262 *nm* ϵ_{max} = 5000

with the ϵ_{max} for biphenyl above and the substituted diphenylmethanes below. In some substituted diphenylmethanes, there is an effective overlap of π orbitals of the two rings resulting in homoconjugation. The ϵ_{max} of 4-nitro-4′-methoxydiphenylmethane is not merely the sum of the ϵ_{max} of *p*-nitrotoluene and *p*-methoxytoluene:

$O_2N-C_6H_4-CH_3$

λ_{max}	ϵ_{max}
274	9,490

$CH_3-C_6H_4-OCH_3$

λ_{max}	ϵ_{max}
277	2,190
285.5	1,786

$O_2N-C_6H_4-CH_2-C_6H_4-OCH_3$

λ_{max}	ϵ_{max}
280	24,400
287	16,800

Another approach to predicting the λ_{max} of the primary band of substituted benzenes involves the use of Table XXV. This table has been successfully used with disubstituted compounds when the following rules are used:

1. Para substitution. a. Both groups are either electron donating or electron withdrawing: Only the effect of the group causing the larger shift is used. For example, the λ_{max} of p-nitrobenzoic acid would be expected to be the same as that of nitrobenzene, ~(203.5 + 65.0) = ~268.5 (in alcohol solvent). b. One group is electron donating and the other, electron withdrawing: The calculated λ_{max} here would usually be much lower than the observed λ_{max} for the reasons that were discussed above for *p*-nitrophenol and Table XXIV. Table XXIII may be used for some of the above compounds for which Table XXV cannot be used.
2. Ortho, meta substitution. The shift effects are additive.

Benzenoid compounds containing three or more substituents have not been thoroughly tested by the method of Table XXV.

The types of compounds for which there are empirical methods for calculating the λ_{max} of important bands are listed in Table XXVI.

Table XXV. Calculation of the Primary Band ($\pi \rightarrow \pi^*$ transition) of Substituted Benzenes (CH_3OH Solvent). Base value: 203.5 nm

Substituent	Shift	Substituent	Shift	Substituent	Shift
$-CH_3$	3.0	$-Br$	6.5	$-OH$	7.0
$-CN$	20.5	$-Cl$	6.0	$-O^-$	31.5
$-CHO$	46.0	$-NH_2$	26.5	$-OCH_3$	13.5
$-COCH_3$	42.0	$-NHCOCH_3$	38.5		
$-CO_2H$	25.5	$-NO_2$	65.0		

C.N.R. Rao. *J. Sci. Res. (India)*, **17B**, 56 (1958); *Current Sci. (India)*, **26**, 276 (1957).

Table XXVI. Empirical Correlations Useful for Predicting the λ_{max} of Organic Compounds

Molecular Class	General Structure	Table No.	Comment
Small Polyenes	$(>C=C<)_x$ $x = 1 - 4$	V	Woodward-Fieser Rules
Larger Polyenes	as above, x > 4	VI	Fieser-Kuhn Rules
Substituted Cyclohexanones	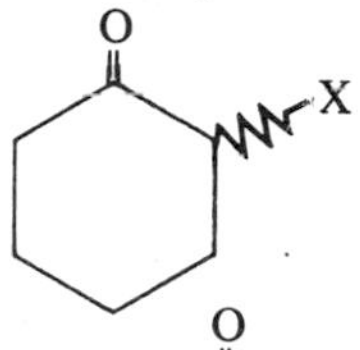 [cyclohexanone with X substituent]	VIII	includes halo ketone effect
α,β-Unsaturated Ketones, etc.	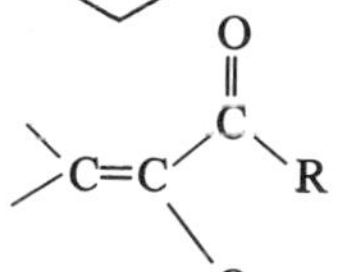 $>C=C<C(=O)R$	XII, XIV	Woodward-Fieser Rules
α,β-Unsaturated Aldehydes	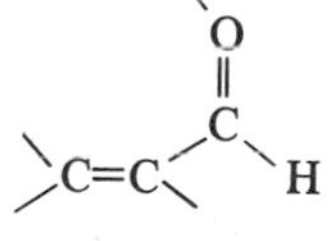 $>C=C<C(=O)H$	XIII, XIV	
α,β-Unsaturated Esters	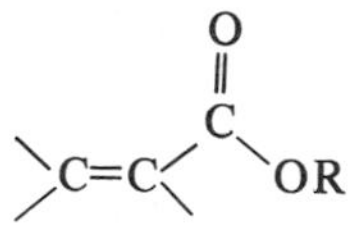 $>C=C<C(=O)OR$	XV	
Aryl Carbonyl Compounds	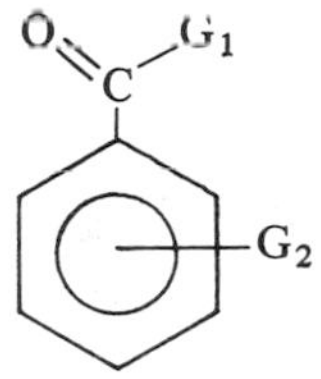 [benzene ring with $C(=O)G_1$ and G_2] G_1 = H, R, OH, OR	XXIII	
Disubstituted Benzenes	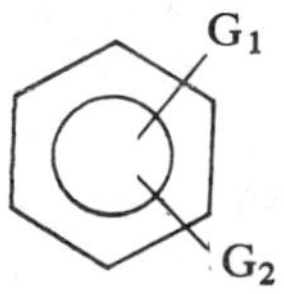 [benzene ring with G_1 and G_2]	XXV	limited for certain *para* compounds

The first homolog in the diphenylpolyene series (ϕ-(C=C)$_n$-ϕ) is stilbene. Stilbene offers an interesting example of steric effects in electronic spectra.

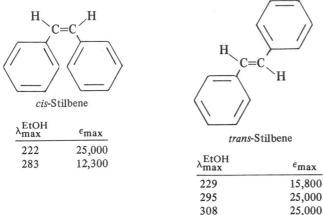

cis-Stilbene

λ_{max}^{EtOH}	ϵ_{max}
222	25,000
283	12,300

trans-Stilbene

λ_{max}^{EtOH}	ϵ_{max}
229	15,800
295	25,000
308	25,000
320 (shoulder)	15,800

The destruction of coplanarity by steric interference, in the *cis*-structure, is reflected by the lower intensity of the 283 nm band compared with the corresponding band (295 nm) in the *trans* isomer. The *B*-band appears to be swamped by this intense absorption.

The absorption bands of the parent stilbene molecule move to longer wavelengths and increase in intensity as *n*, in ϕ-(CH=CH)$_n$-ϕ, increases. When *n* equals 7, the molecule absorbs in the 400-465 region with ϵ_{max} 135,000.

Two common series of aromatic compounds are the linear series, such as anthracene, and the angular series such as phenanthrene. Although the polynuclear aromatics might well be treated as individual chromophores, a correlation between the bands of benzene and the acenes, as well as naphthalene, can be made. These correlations appear in Table XXVII.

As the number of condensed rings increases in the acene series, the absorption moves to progressively longer wavelengths until it occurs in the visible region.

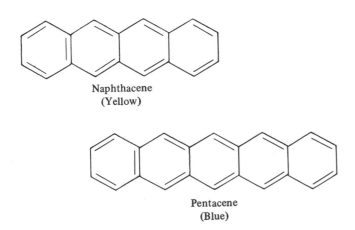

Naphthacene
(Yellow)

Pentacene
(Blue)

Table XXVII. Correlation of Aromatic Absorption

	E_1 band	E_2-band	B-band	
Compound	λ_{max} (ϵ_{max})	λ_{max} (ϵ_{max})	λ_{max} (ϵ_{max})	λ_{max} (ϵ_{max})
Benzene	184 (60,000)	204 (7900)	256 (200)	
Naphthalene	221 (133,000)	286 (9300)	312 (289)	
Anthracene	256 (180,000)	375 (9000)	Submerged	221 (14,500)

The angular polycyclic compounds, the aphenes, also show a bathochromic shift of the 3-band system with an increase in the number of rings. However, the increase in λ_{max}, per ring added, is less than for the acenes. The 3-band system is still distinct for phenanthrene but in the spectrum of anthracene the E_2-band has already swamped the *B*-band.

The spectra of polynuclear aromatics are characterized by vibrational fine-structure as observed in the spectrum of benzene. Spectra of some polynuclear aromatics are shown in Figure 7.

Heteroaromatic Compounds

Saturated 5- and 6-membered heterocyclic compounds are transparent at wavelengths longer than 200 nm. Only the unsaturated heterocyclic compounds (heteroaromatics) show absorption in the near ultraviolet region.

FIVE-MEMBERED RINGS. The theoretical interpretation of the spectra of 5-membered-ring heteroaromatic compounds is not simple. The absorption of these compounds has been compared to that of cyclopentadiene, the *cis*-diene analog, which shows strong diene absorption near 200 nm, and moderately intense absorption near 238 nm. The aromatic properties increase in the order cyclopentadiene, furan, pyrrole, and thiophene. The absorption of some 5-membered heteroaromatics is compared with cyclopentadiene in Table XXVIII. No attempt has been made to classify the bands, although the band near 200 nm has been likened to the E_2-band of benzene, and the long-wavelength band frequently has fine structure analogous to that of the *B*-band of benzene.

Auxochromic or chromophoric substitution of the five-membered unsaturated heterocyclics causes a bathochromic shift and an increase in the intensity of the bands of the parent molecule (Table XXIX).

SIX-MEMBERED RINGS. The spectrum of pyridine is similar to that of benzene. The *B*-band of pyridine, however, is somewhat more intense and has less distinct fine structure than that of benzene (Figure 8). This transition is allowed

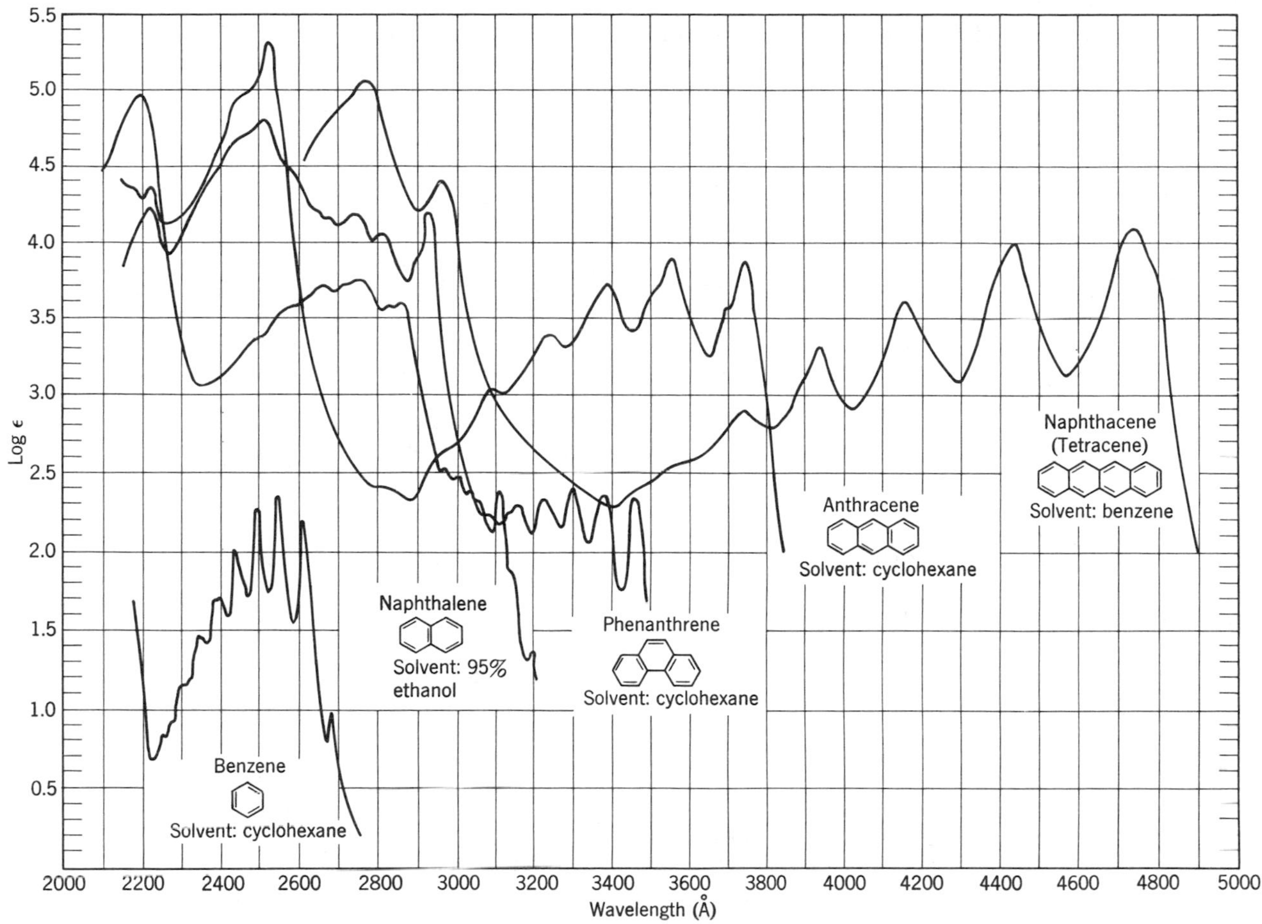

Figure 7. *Electronic absorption spectra of benzene, naphthalene, phenanthrene, anthracene, and naphthacene.*

for pyridine, but forbidden for the more symmetrical benzene molecule. The weak *R*-band expected for an $n \rightarrow \pi^*$ transition in pyridine has been observed in vapor phase spectra. This band is generally swamped by the more intense *B*-band when the spectrum is determined in solution.

An increase in solvent polarity has little or no effect on the position or intensity of the *B*-band of benzene, but produces a marked hyperchromic effect on the *B*-band of pyridine and its homologs. The hyperchromic effect undoubtedly results from hydrogen bonding through the lone pair of electrons of the nitrogen atom. The extreme case is the absorption of a pyridinium salt. The absorption character-

Table XXVIII. Absorption Data for Some 5-Membered Heteroaromatics

Compound	Band I λ_{max}	Band I ϵ_{max}	Band II λ_{max}	Band II ϵ_{max}	Solvent
[Cyclopentadiene]	200	10,000	238.5	3,400	Hexane
Furan	200	10,000	252	1[a]	Cyclohexane
Pyrrole[b]	209	6,730	240	300[a]	Hexane
Thiophene	231	7,100	269.5	1.5[a]	Hexane
Pyrazole	214	3,160	–	–	Ethanol

[a]These weak bands may be due to impurities rather than a forbidden transition ($n \rightarrow \pi^*$) of a heteroaromatic molecule.

[b]See Table XXIX for different assignment of bands in the spectrum of pyrrole.

Table XXIX. Absorption Characteristics of 5-Membered Heteroaromatics

Parent	Substituent	Band I λ_{max} (*nm*)	Band I ϵ_{max}	Band II λ_{max} (*nm*)	Band II ϵ_{max}
Furan		200	10,000	252	1
Furan	2-CHO	227	2,200	272	13,000
Furan	$2\text{-}\overset{\text{O}}{\overset{\|}{\text{C}}}\text{-CH}_3$	225	2,300	270	12,900
Furan	2-COOH	214	3,800	243	10,700
Furan	2-NO_2	225	3,400	315	8,100
Furan	2-Br, 5-NO_2	–	–	315	9,600
Pyrrole		183	–	209	6,730
Pyrrole	2-CHO	252	5,000	290	16,600
Pyrrole	$2\text{-}\overset{\text{O}}{\overset{\|}{\text{C}}}\text{-CH}_3$	250	4,400	287	16,000
Pyrrole	2-COOH	228	4,500	258	12,600
Pyrrole	$1\text{-}\overset{\text{O}}{\overset{\|}{\text{C}}}\text{-CH}_3$	234	10,800	288	760
Thiophene		231	7,100	–	–
Thiophene	2-CHO	265	10,500	279	6,500
Thiophene	$2\text{-}\overset{\text{O}}{\overset{\|}{\text{C}}}\text{-CH}_3$	252	10,500	273	7,200
Thiophene	2-COOH	249	11,500	269	8,200
Thiophene	2-NO_2	268-272	6,300	294-298	6,000
Thiophene	2-Br	236	9,100	–	–

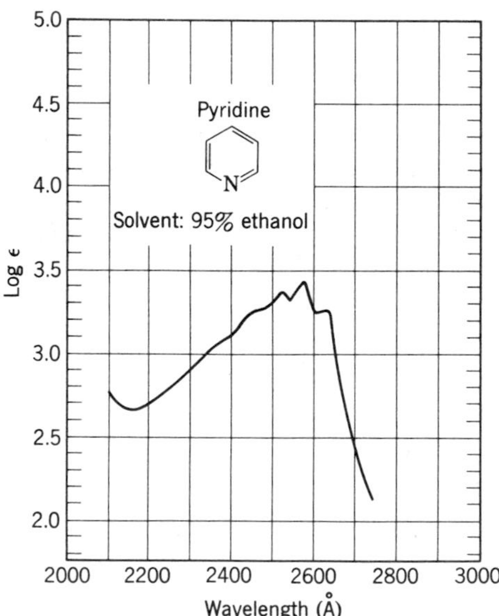

Figure 8. *Ultraviolet spectrum of pyridine.*

istics of 2-methylpyridine (α-picoline), in several solvents, are shown in Table XXX.

The effect of substitution on the 257 nm band (*B*-band) of pyridine is illustrated in the data presented in Table XXXI.

The absorption of the 2-OH and 4-OH pyridines is attributed to the pyridone structures. Tautomerism in hydroxy- and aminopyridines is discussed thoroughly in the book by A. I. Scott.[2]

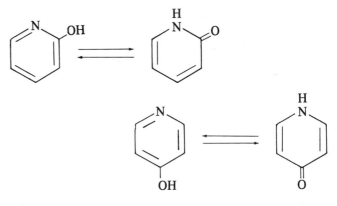

Table XXX. Absorption Characteristics of 2-Methylpyridine

Solvent	λ_{max} (*nm*)	ϵ_{max}
Hexane	260	2000
Chloroform	263	4500
Ethanol	260	4000
Water	260	4000
Ethanol-HCl(1:1)	262	5200

Table XXXI. Absorption Characteristics of Pyridine Derivatives

Derivative	$\lambda_{max}^{(pH > 7)}$	ϵ_{max}
Pyridine	257	2,750
	270	450
2-CH_3	262	3,560
3-CH_3	263	3,110
4-CH_3	255	2,100
2-F	257	3,350
2-Cl	263	3,650
2-Br	265	3,750
2-I	272	400
2-OH	230	10,000
	295	6,300
4-OH	239	14,100
3-OH	260	2,200

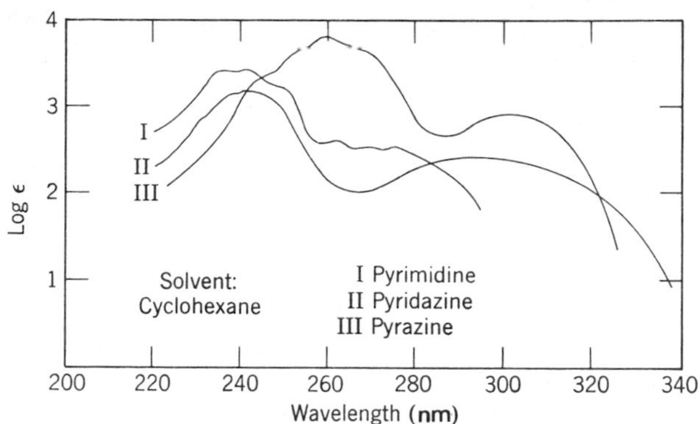

Figure 9.

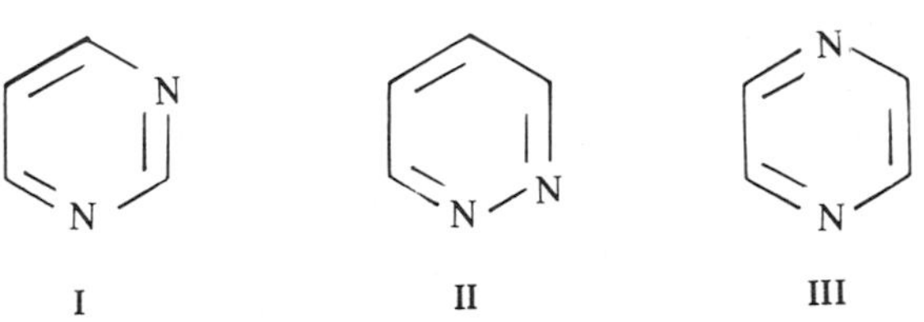

The spectra of heteroaromatics appear to be related to their isocyclic analogs (Table XXXII).

The spectra of the diazines are similar to those of pyridine (Figure 9). In addition to the enhanced *B*-band, still retaining some fine structure, the enhanced $n \rightarrow \pi^*$ bands are quite prominent.

The absorption of the diazines, for example the pyrazines, respond to solvent polarity in a manner similar to the pyridines. The *B*-band undergoes a hyperchromic effect with an increase in solvent polarity with little or no effect upon λ_{max}. The *R*-band, of course, disappears in acid solution, since the nonbonding electrons of the free base are now involved in salt formation.

REFERENCES

General Monographs

1. Jaffé, H. H. and Orchin, Milton. *Theory and Application of Ultraviolet Spectroscopy.* New York: Wiley, 1962.
2. Scott, A. I. *Interpretation of the Ultraviolet Spectra of Natural Products.* New York: Pergamon Press (The Macmillan Co.), 1964.
3. Stern, E. S. and Timmons, T. C. J. *Electronic Absorption Spectroscopy in Organic Chemistry.* New York: St. Martin's Press, 1971. (effectively the third edition of Gillam and Stern).

Selections from Broader References

4. Ayling, G. M. *Appl. Spec. Rev.,* 8, 1 (1974).
5. Braude, E. A., "Ultraviolet and Visible Light Absorption," in *Determination of Organic Structures by Physical Methods.* New York: Academic Press, 1955, chap. 4, pp. 131-94.
6. Braude, E. A. "Ultra-Violet Light Absorption and the Structure of Organic Compounds," *Ann. Repts. on Progress Chem.* Chemical Society of London, XLII 105-130 (1945).

Table XXXII. Absorption Characteristics of Some Heteroaromatics Containing Nitrogen and Their Isocyclic Analogs

	E_1-band[a]		E_2-band[a]		B-band[a]		
Compound	λ_{max} (nm)	ϵ_{max}	λ_{max} (nm)	ϵ_{max}	λ_{max} (nm)	ϵ_{max}	Solvent
Benzene	184	60,000	204	7,900	256	200	–
Naphthalene	221	100,000	286	9,300	312	280	–
Quinoline	228	40,000	270	3,162	315	2,500	Cyclohexane
Isoquinoline	218	63,000	265	4,170	313	1,800	Cyclohexane
Anthracene	256	180,000	375	9,000	–	–	–
Acridine	250	200,000	358	10,000	–	–	Ethanol

[a]All of the bands contain fine structure.

7. Ferguson, L. N. and Nnadi, J. C. "Electronic Interactions between Nonconjugated Groups," *J. Chem. Educ.*, **42**: 529(1965).
8. Lambert, J. B., Shurvell, H. F., Verbit, L., Cooks, R. G., Stout, G. H. *Organic Structural Analysis.* New York: MacMillan, 1976, pp. 317-401.
9. May, L. "IUPAC Definitions in Ultraviolet Spectroscopy," *Appl. Spectroscopy*, **27**: 419 (1973).
10. Scheinman, F. *An Introduction to Spectroscopic Methods for Identification of Organic Compounds.* New York: Pergamon Press, 1973, pp. 93-152.
11. Schönig, A. G. "A Computer-based UV Storage/ Retrieval System, *Anal. Chim. Acta*, **71**: 17 (1974).
12. "Spectrometry Nomenclature," *Anal. Chem.* annual December issue.
13. Ultraviolet Spectrometry, *Anal. Chem.* Periodically this journal publishes literature reviews that include the title subject.

Monographs–Narrow Topics

14. Bolshakow, G. F., Vatago, V. S., Agrest, F. B. *Ultraviolet Spectra of Hetero-Organic Compounds.* Leningrad, U.S.S.R.: Khimia Publ. House, 1969.
15. Katrizky, A. R. (ed), *Physical Methods in Heterocyclic Chemistry.* Vol. II, New York and London: Academic Press, 1963.

Compilations–Data, Spectra, Indices

16. American Petroleum Institute Research Project 44, *Selected Ultraviolet Spectral Data.* vol I-IV. College Station, Texas: Thermodynamics Research Center, Texas A&M University, 1945-1977, 1178 compounds, 1279 pages.
17. *ASTM Index to Ultraviolet and Visible Spectra.* ASTM Technical Publication No. 357, American Society for Testing Materials, 1916 Race St., Philadelphia Pa., 1963.
18. *DMS UV Atlas of Organic Compounds.* Vol. I-V. New York: Plenum Press, 1966-1971.
19. *Electronic Spectral Data; Organic Electronic Spectral Data.* vol. I-XIII. New York: Wiley-Interscience, 1946-1968.
20. Friedel, R. A. and Orchin, Milton. *Ultraviolet Spectra of Aromatic Compounds.* New York: Wiley, 1951.
21. Graselli, J. G. and Ritchey, W. M. (eds.) *Atlas of Spectral Data and Physical Constants.* Cleveland: CRC Publishing, 1975, pp. 399-408. This includes a listing, in groups by molar absorptivity ($\epsilon <$ 100,000; = 10,000-99,999; 1000 = 9,999; 0-1,000; and no listed ϵ); of compounds in order of the λ_{max} of their principal absorption, along with all remaining UV maxima for each compound.
22. Hershenson, H. M. *Ultraviolet Absorption Spectra. Index for 1954-1957.* New York: Academic Press, 1959.
23. Hirayama, K. *Handbook of Ultraviolet and Visible Absorption Spectra of Organic Compounds.* New York: Plenum Press Data Div., 1967.
24. Holubek, J. and Strouf, O. (eds.) *Spectral Data and Physical Constants of Alkaloids.* New York: Heyden and Sons, 1962.
25. Lang, L. *Absorption Spectra in the Ultraviolet and Visible Region.* New York: Academic Press.
26. *UV-Atlas of Organic Compounds.* Photoelectric Spectrometry Group, London, and Institut für Spektrochemie und Angewandte Spektroskopie, Dortmund, London: Butterworths.
27. *Ultraviolet Reference Spectra.* Philadelphia: Sadtler Research Laboratories, 3314-20 Spring Garden St.
28. *Ultraviolet Spectral Data.* Pittsburgh, Pa.: Manufacturing Chemists Association Research Project, Carnegie Institute of Technology.

PROBLEMS

6.1 Match the 3 steroid structures shown below to the following values for λ_{max}^{hexane}: compound A, 257 nm; B, 304 nm, and C, 356 nm.

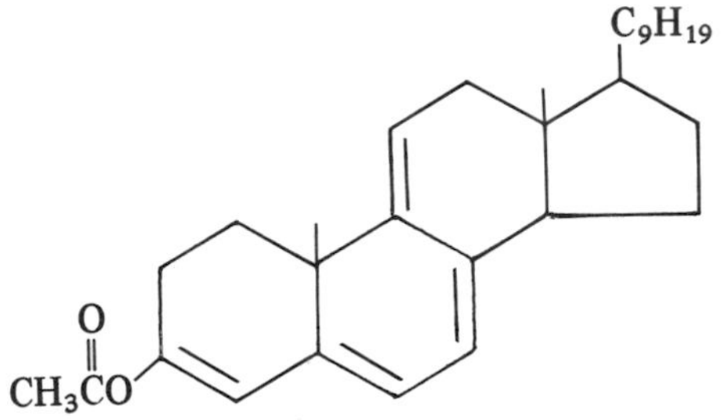

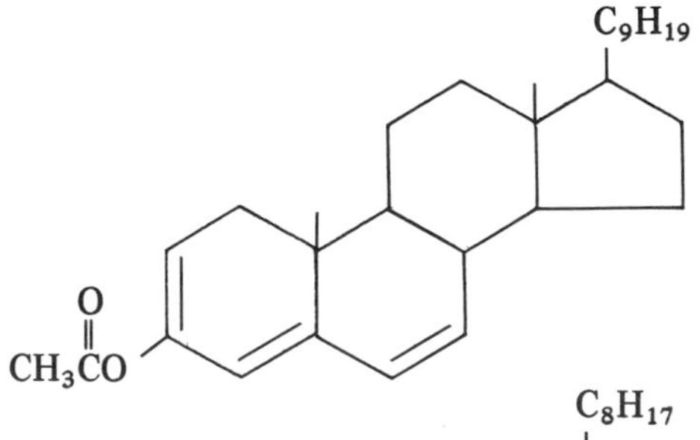

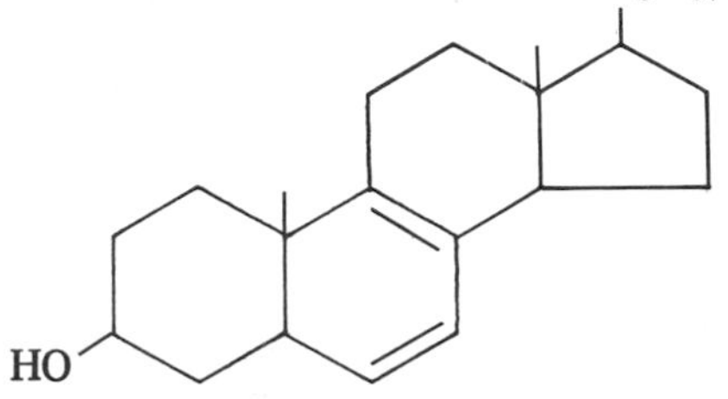

C_9H_{19}

$CH_3\overset{O}{\overset{\|}{C}}O$

C_9H_{19}

$CH_3\overset{O}{\overset{\|}{C}}O$

C_8H_{17}

HO

6.2 Partial hydrogenation of the triene shown below results in 2 compounds, D and E, both of molecular formula $C_{10}H_{14}$. Compound D shows a λ_{max}^{hexane} = 235 nm and E, 275 nm. Assign the structures.

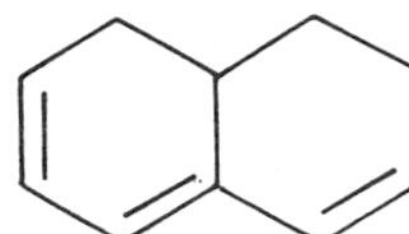

6.3 Compound F ($C_7H_{10}O$) gives a positive test with 2,4-dinitrophenylhydrazine, with sodium hydroxide/iodine and with bromine in carbon tetrachloride. The UV spectrum of F shows a $\lambda_{max}^{C_2H_5OH}$ at 257 nm. Deduce the structure for this compound. Is only one structure possible?

6.4 An acetone solution of compound G ($C_7H_7N_2O_2$) shows ^1H NMR signals at δ 6.63-7.90, 4H, and at δ 6.55, 3H. The intensity of the δ 6.55 signal is greatly reduced on treatment with heavy water. The UV spectrum of G shows a $\lambda_{max}^{C_2H_5OH}$ at 288 nm. Deduce the structure of compound G. Compound G can be prepared by reducing a nitrobenzoic acid with a mineral acid-metal mixture.

6.5 A Nujol mull of compound H ($C_7H_5BrO_2$) shows, among others, the following IR bands: a broad band, with a number of maxima, from 3100 to 2500 cm^{-1}, and a strong band at 1686 cm^{-1}. The ^1H NMR spectrum (DMSO-d_6/$CDCl_3$) shows signals at δ 7.5 to 8.0 (4H) and at δ 12.15 (1H). The latter signal is greatly diminished upon heavy water treatment. The UV spectrum shows a $\lambda_{max}^{C_2H_5OH}$ at 245 nm. Deduce the structure of compound H.

introduction to seven and eight

In practice, identification of an organic compound begins with a history, and large areas are quickly excluded from further consideration. Since, with a few exceptions, we dispense with a history, the limits will be set as follows: The compounds are quite pure; they may contain carbon, hydrogen, oxygen, nitrogen, sulfur, and the halogens in any combination; since the table of isotope contributions runs only to a molecular weight of 250, the examples will be limited to this range. Within these restrictions, we can still cover a vast expanse of organic chemistry.

Additional information, usually a history, would probably be necessary to identify a compound that contained boron, silicon, or phosphorus in addition to the above elements, but it is hardly likely that a chemist would encounter such compounds without prior knowledge that these elements are present.

The interpreter and correlator of data has always lived on the edge of uncertainty. How pure is his compound, how reliable his data, how relevant his reference material? There are few unequivocal data, either spectrometric or chemical; thus, there is no substitute for experience and a broad knowledge of chemistry. Nor is there a prescribed procedure. In general, we attempt as a first step to establish a tentative molecular formula from the molecular ion peak and the isotope contributions. In a number of cases, the molecular ion peak is so small that the isotope contributions cannot be accurately measured, and we settle for the molecular weight. In some cases, the molecular ion peak may be missing. We then try to establish the molecular weight from other evidence. Often, this can be done from the fragmentation pattern and from the other spectra at hand. The minimum number of carbon atoms can be obtained from the ^{13}C NMR spectrum. A difference between the number of peaks in the ^{13}C spectrum and the number of C atoms in the molecular formula indicates the presence of symmetry elements. (Coincidence of ^{13}C peaks, though rare, must nonetheless be considered.) The off-resonance-decoupled ^{13}C spectrum gives the minimum number of protons, which can be compared with the integrated ^{1}H spectrum and with the molecular formula. In other cases, we may resort to preparation of appropriate derivatives, to other methods of obtaining molecular weights, or to other methods of determining elemental composition. Recent improvements in combustion analysis permit C, H, and N determinations on less than a milligram of sample. If a high-resolution mass spectrometer is available, we may obtain the molecular forumula and the formula of each of the fragments directly.

We use the obvious features of one spectrum to bring out the more subtle aspects of another. The power of this methodology lies in the complementary features of the spectra. For example, we can calculate the index of hydrogen deficiency from the molecular formula and determine the presence or absence of certain unsaturated functional groups from the IR spectrum; this information can be confirmed by the ^{1}H and ^{13}C spectra. The UV spectrum quickly establishes the presence or absence of conjugated or aromatic structures. The ^{1}H spectrum (with decoupling if necessary) is a powerful tool for distinguishing between isomeric possibilities. When enough information is accumulated, a structure is postulated and spectral features predicted on the basis of the postulated structure are compared with the spectra actually observed. Appropriate structural modifications are made to accommodate the discrepancies. Confirmation is obtained by comparison with the spectra of an authentic sample. Let us work through the samples comprising Chapter 7 as preparation for solving the problems in Chapter 8.

We may not always be able to select an unequivocal structure; nor can we in any system of organic analysis. However, we should at least be able to narrow down the possibilities to several structures (often isomers) and to indicate the steps required to complete the identification.

Mass spectra were obtained on a Consolidated Electrodynamics Corporation Model 21-103C and a Perkin Elmer Model RMU-6E. The lower limit of the fragmentation patterns is mass 20. Peaks of less than 3% relative intensity are not reported except for the molecular ion and isotope peaks. Infrared spectra were run on the Perkin-Elmer Corporation Models 221 and 137. Proton NMR spectra were run on Varian Associates Models A-60, HR-60, HA-100 and a Perkin Elmer Model R-20; samples were dissolved in carbon tetrachloride or deuterated chloroform, and 1% tetramethylsilane in the solution was used as a reference. ^{13}C NMR spectra were run on Varian XL-100 and CFT-20 instruments with tetramethylsilane as an internal reference. Unless stated otherwise, the samples were dissolved in deuterated chloroform solvent, and the instrument was locked on the deuterium in the solvent. Ultraviolet spectra were obtained from the literature, or were run on an Applied Physics Corporation's Cary Model 14M at pH 7, pH 1, and pH 13. Only the pH 7 spectrum is recorded unless a change occurred at pH 1 or pH 13. In all ^{13}C data tables in Chapter 8, the intensities are for the completely decoupled spectra.

seven

sets of spectra translated into compounds

NUMBERS	NAME, PAGE	COMPOUND CLASS
7-1	Benzyl Acetate, 337	aromatic ester
7-2	Methyl 2-furoate, 341	heteroaromatic ester
7-3	β, β'-Dibromodiethyl ether, 343	halo ether
7-4	2-Mercaptoethanol, 346	hydroxy mercaptan
7-5	Isoeugenol, 350	aromatic, olefinic, phenol ether
7-6	Levulinic Acid, 355	keto carboxylic acid
7-7	Dibutyl(difluoromethyl)phosphine oxide, 357	halo-aliphatic, organophosphorus
7-8	Ethyl sorbate, 359	olefinic ester

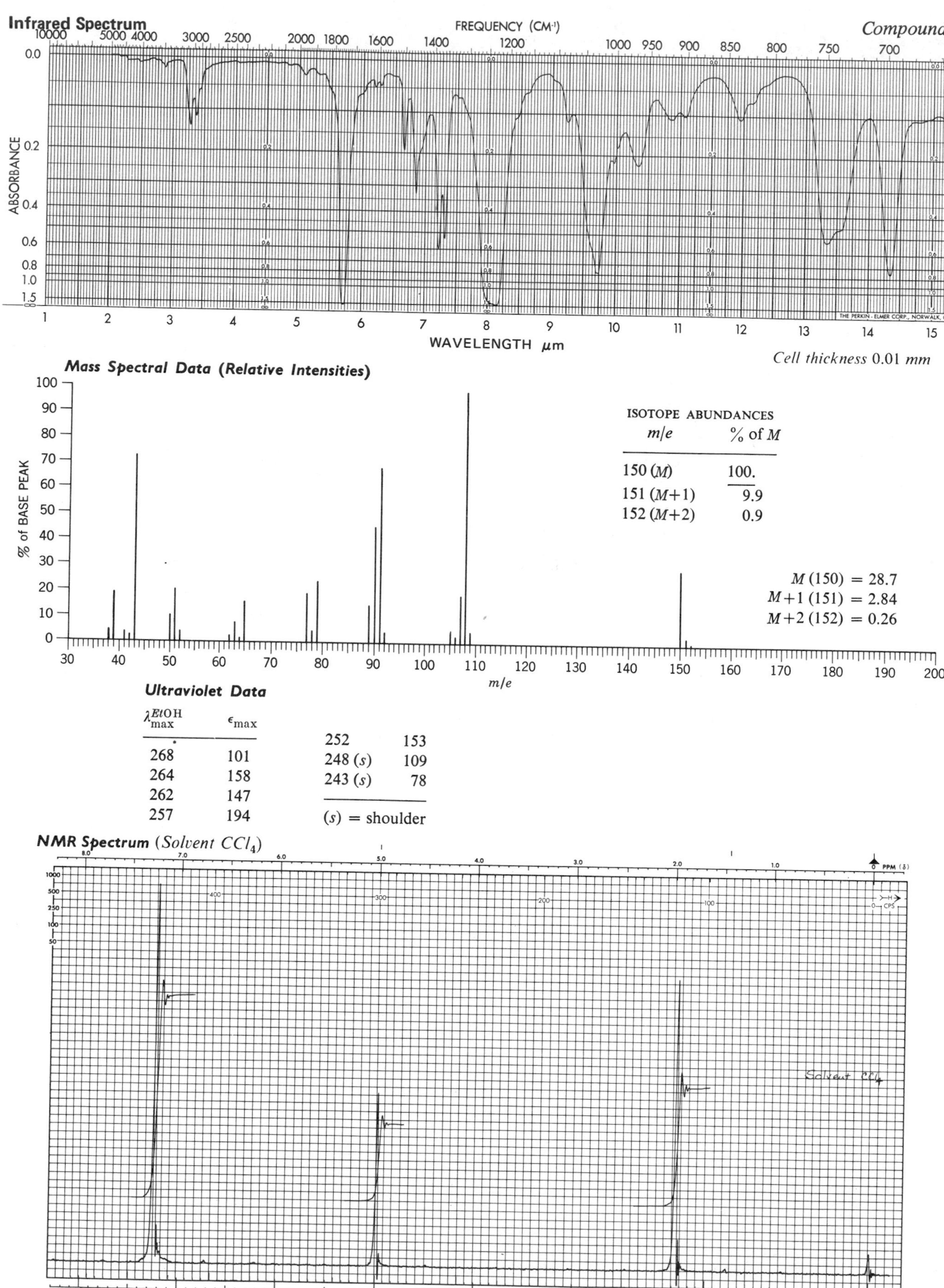

ISOTOPE ABUNDANCES

m/e	% of M
150 (M)	100.
151 ($M+1$)	9.9
152 ($M+2$)	0.9

Ultraviolet Data

λ_{max}^{EtOH}	ϵ_{max}
268	101
264	158
262	147
257	194
252	153
248 (*s*)	109
243 (*s*)	78

(*s*) = shoulder

compound number 7-1

The first step in translating these four spectra into a molecular structure is to establish a molecular formula. The molecular ion peak is 150; this is the molecular weight. The molecular ion peak is an even number. We are, therefore, permitted either no nitrogen atoms, or an even number of them. The $M + 2$ peak obviously does not allow for the presence of sulfur or halogen atoms.

We now look in Appendix A of Chapter 2 under molecular weight 150, and find 29 formulas of molecular weight 150 containing only CHN and O. Our $M + 1$ peak is 9.9% of the parent peak. We list the formulas with calculated isotopic contributions to the $M + 1$ peak falling—to be arbitrary—between 9.0 and 11.0; we also list their $M + 2$ values:

Formula	$M + 1$	$M + 2$
$C_7H_{10}N_4$	9.25	0.38
$C_8H_8NO_2$	9.23	0.78
$C_8H_{10}N_2O$	9.61	0.61
$C_8H_{12}N_3$	9.98	0.45
$C_9H_{10}O_2$	9.96	0.84
$C_9H_{12}NO$	10.34	0.68
$C_9H_{14}N_2$	10.71	0.52

Three of these formulas can be eliminated because they contain an odd number of nitrogen atoms. The $M + 2$ peak is 0.9% of the parent; this best fits $C_9H_{10}O_2$, which we shall tentatively designate as our molecular formula. We make a mental note that both the intensity of the molecular ion peak and the C to H ratio of the formula indicate aromaticity.

The infrared spectrum shows a C=O band at about 1745 cm^{-1} (5.73 μm). This, together with the presence of 2 O atoms in the formula, suggests an ester. We look for confirmation in the C—O—C stretching region and note the large broad band at about 1225 cm^{-1} (8.15 μm) characteristic of an acetate. Two large bands at about 749 cm^{-1} (13.35 μm) and 697 cm^{-1} (14.35 μm) suggest a singly substituted benzene ring.

The presence of a benzene ring and an acetate group is established. Furthermore, we note from the position of the carbonyl band that the C=O moiety is not conjugated with the ring. This is confirmed by the wavelengths and intensities of the ultraviolet absorption peaks which also eliminate a ketone from consideration. Subtraction of a singly substituted benzene ring and an acetate group from the molecular formula gives the following:

Molecular formula $C_9H_{10}O_2$

$$C_6H_5 + CH_3\overset{\overset{\displaystyle O}{\|}}{C}{-}O = C_8H_8O_2$$

Remaining: CH_2

It takes no great imagination to insert the CH_2 between the ring and the acetate group, and write

$$C_6H_5{-}CH_2O\overset{\overset{\displaystyle O}{\|}}{C}CH_3$$

The ^{1}H NMR spectrum provides almost conclusive confirmation for the above structure. We see 3 sharp unsplit peaks in the following positions and with the following integrated intensities

δ	Intensity
7.22	5
5.00	2
1.96	3

The 5 protons at δ 7.22 of course are the 5 benzene-ring protons. The singlet of 2 protons at δ 5.00 represents the methylene group substituted by a phenyl and an ester group. And, of course, the singlet of 3 protons at δ 1.96 represents the methyl group.

We can obtain additional confirmation by returning to the mass spectrum and considering the fragmentation pattern (see Appendix B, Chapter 2) in view of the information at hand. The base peak at 108 is a rearrangement peak representing cleavage of an acetyl group (43) and rearrangement of a single hydrogen atom (Chapter 2, page 29). The large peak at mass 91 is the benzyl (or tropylium) ion formed by cleavage beta to the ring. And the large peak at mass 43, of course, represents the acetyl fragment. The peaks at 77, 78, and 79 are additional evidence for the benzene ring.

We can state with a high degree of confidence that the compound represented by these spectra is:

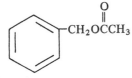

$$C_6H_5{-}CH_2O\overset{\overset{\displaystyle O}{\|}}{C}CH_3$$

Benzyl acetate

There are, of course a number of other sequences to the identity of this compound. Having established the molecular formula, we could note at once the characteristic benzene ring peak at δ 7.22 in the NMR spectrum. We could confirm this by the typical "benzenoid" fine structure absorption in the ultraviolet spectrum. The base peak in the mass spectrum is a common rearrangement peak, and the mass 91 peak immediately calls to mind the benzyl (or tropylium) structure. The large mass 43 peak strongly suggests the CH_3CO group in view of the C=O peak in the infrared. Subtraction of a benzyl and an acetyl group from the formula leaves a mass of 16; consideration of the infrared spectrum leaves very little question as to how to handle this oxygen atom.

The student will find it instructive to write the possible isomeric structures and to eliminate them on spectrometric grounds.

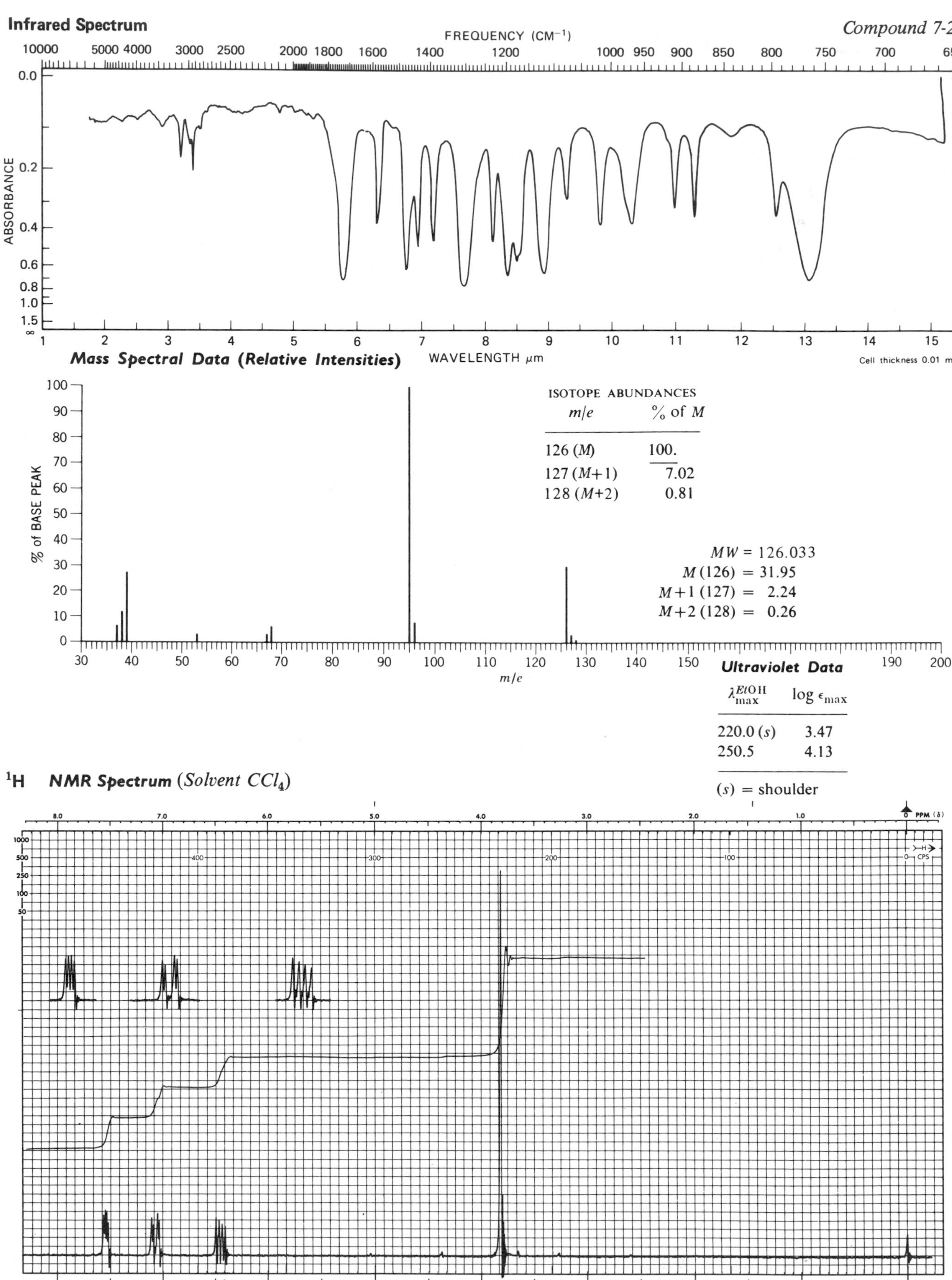

ISOTOPE ABUNDANCES

m/e	% of M
126 (M)	100.
127 (M+1)	7.02
128 (M+2)	0.81

λ_{max}^{EtOH}	$\log \epsilon_{max}$
220.0 (s)	3.47
250.5	4.13

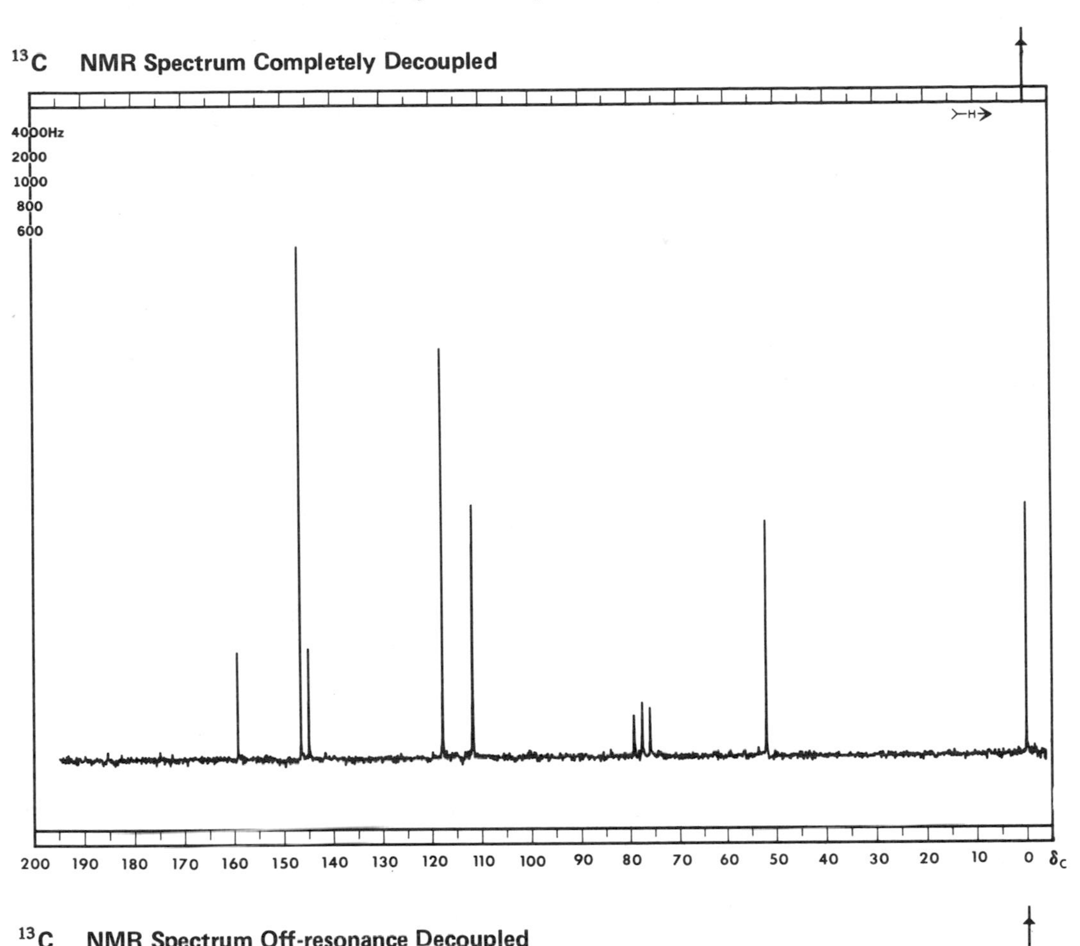
13C NMR Spectrum Completely Decoupled
4000Hz
2000
1000
800
600
200 190 180 170 160 150 140 130 120 110 100 90 80 70 60 50 40 30 20 10 0 δC

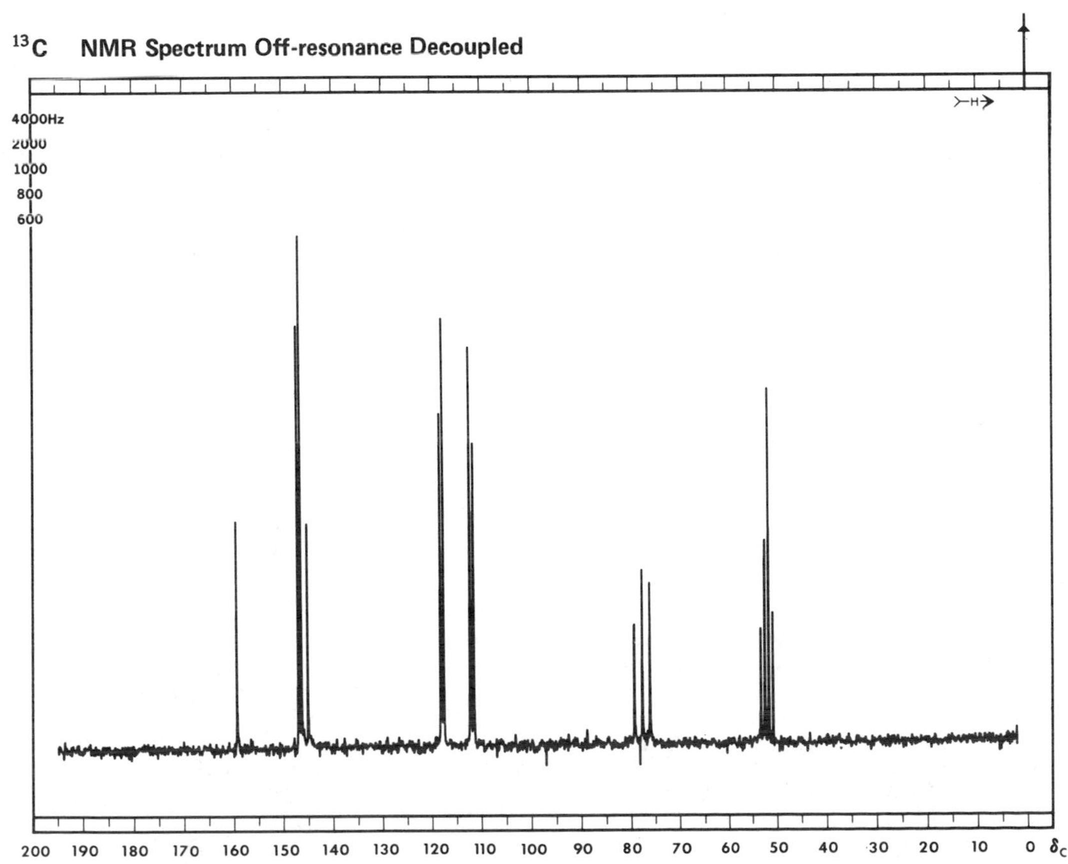
13C NMR Spectrum Off-resonance Decoupled
4000Hz
2000
1000
800
600
200 190 180 170 160 150 140 130 120 110 100 90 80 70 60 50 40 30 20 10 0 δC

compound number 7-2

The following molecular formulas fit our data for the molecular ion peak and the $M + 1$ peak:

Formula	$M + 1$	$M + 2$
$C_5H_6N_2O_2$	6.34	0.57
$C_5H_{10}N_4$	7.09	0.22
$C_6H_6O_3$	6.70	0.79
$C_6H_{10}N_2O$	7.45	0.44

The formula that best fits our $M + 2$ peak is $C_6H_6O_3$. We should bear in mind, however, that the $M + 2$ peak may be higher than the calculated figure, and we accept $C_6H_6O_3$ only as a tentative formula.

The high-resolution molecular weights for each of the four formulas are significantly different:

Formula	Molecular Wt.
$C_5H_6N_2O_2$	126.0430
$C_5H_{10}N_4$	126.0907
$C_6H_6O_3$	126.0317
$C_6H_{10}N_2O$	126.0794

The observed formula mass fits $C_6H_6O_3$.

A quick check of the ^1H and ^{13}C NMR spectra provides support for the tentative molecular formula. The integration of ^1H signals, from low to high field, is 1:1:1:3 which supports the total (6) of hydrogens from the mass spectrum above. The proton-decoupled ^{13}C spectrum shows 6 singlets (aside from the $CDCl_3$ signals at 75-80 ppm) downfield from TMS. Thus, there are 6 nonequivalent carbons in our compound, again consistent with the formula deduced from the mass spectrum. Finally, the ^{13}C off-resonance decoupled NMR spectrum supports the total of 6 protons since, from low to high field, the signals include a singlet, doublet, singlet, doublet, doublet and quartet, indicating that the corresponding carbons bear 0, 1, 0, 1, 1, and 3 protons.

The formula and the general appearance of the infrared spectrum suggest aromaticity. We note strong peaks beyond 800 cm^{-1} (12.5 μm), 2 strong peaks at 1587 cm^{-1} (6.30 μm) and 1479 cm^{-1} (6.76 μm), and a medium peak at 3106 cm^{-1} (3.22 μm). The intense band in the ultraviolet spectrum at 250 nm is suggestive of a chromophore conjugated with an aromatic ring. The striking pattern at the low-field end of the ^1H NMR spectrum demands an aromatic ring of some sort, as do the ^{13}C NMR signals at ca. δ 110-150.

A conspicuous feature of the infrared spectrum is the C=O peak at 1730 cm^{-1} (5.78 μm). Bearing in mind that we are probably dealing with a conjugated chromophore, we can make a choice between a ketone and an ester. We lean toward an ester because the conjugated ketone C=O bands are usually at lower frequency. We also look in vain for the long-wavelength *R*-band of conjugated ketones in the ultraviolet spectrum. (We should recall, however, that heteroaromatic ketones do not show a detectable R band.) There are a number of strong bands between 1420 and 1110 cm^{-1} (7.05 and 9.0 μm) some of which may be associated with an ester C–O absorption. And, of course, the empirical formula permits an ester group plus another oxygen.

Now we can profitably consider the base peak, mass 95, in the mass spectrum. The base peak arises from a loss of mass 31, and this loss is practically diagnostic for a methyl ester. We can write,

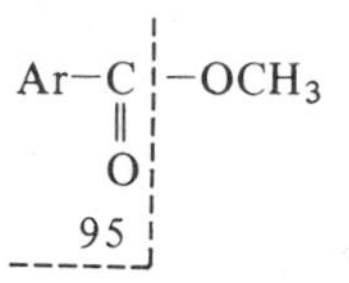

The mass 95 component then must be $C_4H_3O{-}C{=}O$, and we write the structural formula

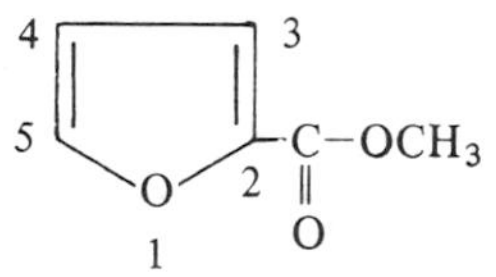

Methyl 2-furoate

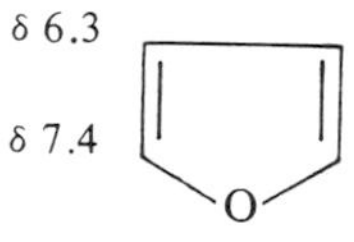

furan

^1H NMR

The ^1H NMR spectrum is nicely in accord with this structure. We see 3 separate ring-protons at low field, and the 3 protons of the methyl group as a sharp singlet at δ 3.81. From low to high field the 3 low-field peaks (multiplets) represent the 5-proton, the 3-proton, and the 4-proton, respectively. Each is shifted downfield, by the carboxylate substituent, from its position in an unsubstituted furan ring, the 3-proton being most strongly affected.

The ^{13}C NMR spectrum supports the furoate structure. The signal at δ 51.81 (a quartet in the off-resonance decoupled spectrum) corresponds to the methyl carbon of a $COOCH_3$ group (Table XIV, Chapter 5). Since the ^{13}C spectrum also shows a singlet at δ 159.10 which is still a singlet in the off-resonance decoupled spectrum, we have more support for a carbonyl group (see Chapter 5, Table XIV). Since the remaining ^{13}C NMR signals total only 4, we have support for the furan aromatic ring with all ring positions being nonequivalent. (Tables VII and XIV, Chapter 5.)

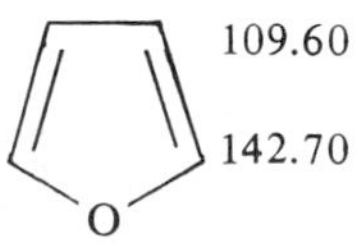

Thus, the lower field signals (δ 146.41 and 144.82 ppm) are assigned to the ring carbons attached to oxygen, and the signal at δ 144.82, since it is a singlet in the off-resonance decoupled spectrum, is assigned to C–2. Finally the 2 singlets between δ 110-120 ppm are assigned to C–3 and C–4, with C–3 being more deshielded. All of the ^{13}C signals due to ring carbons (except the δ 144.82 signal) are doublets in the off-resonance decoupled spectrum since these carbons each bear 1 proton.

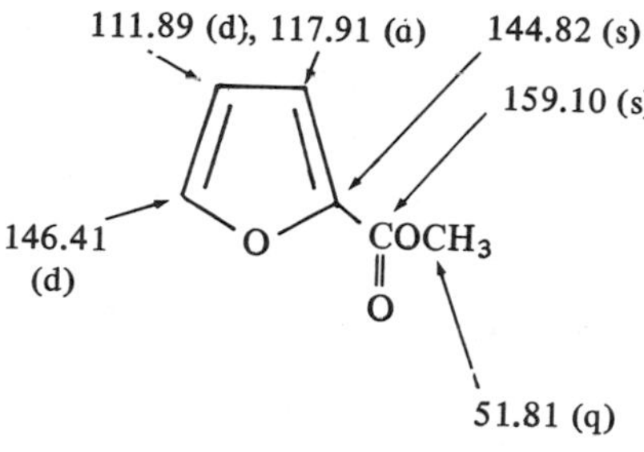

The ^1H NMR multiplets afford a tidy demonstration of spin-spin coupling. The system is *AMX* with 3 coupling constants. The 5-proton is coupled with the four-proton (J_{54} = 1 Hz) and with the 3-proton (J_{53} = 1 Hz).

The 5-proton, therefore, shows 2 pairs of peaks.

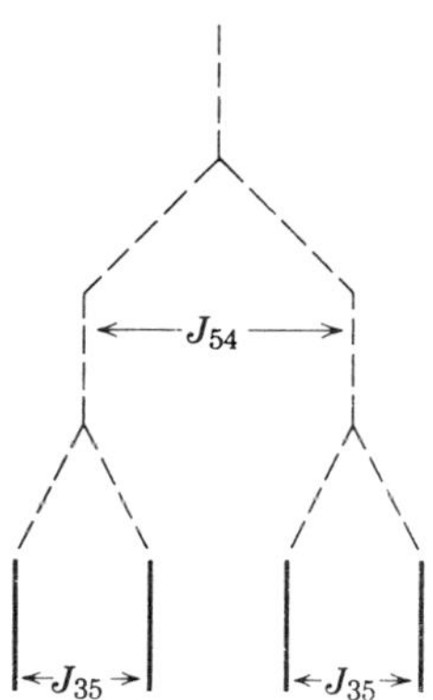

The 3-proton is coupled with the 4-proton (J_{34} = 3.5 Hz) and with the 5-proton (J_{35} = 1 Hz). Thus, the 3-proton shows a doublet of doublets.

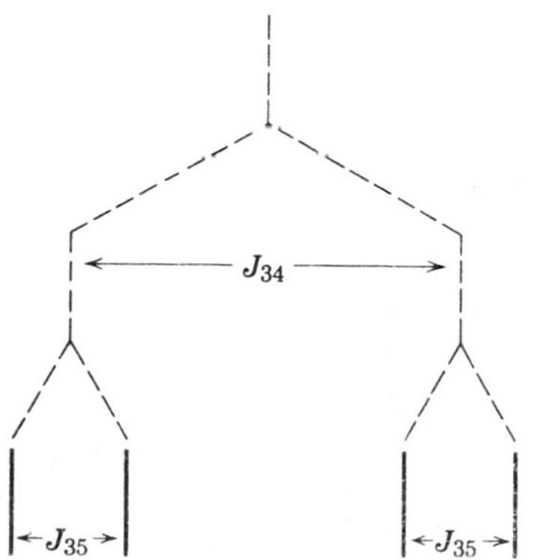

The 4-proton is coupled with the 3-proton (J_{43} = 3.5 Hz) and with the 5-proton (J_{45} = 2 Hz). Again we see a doublet of doublets.

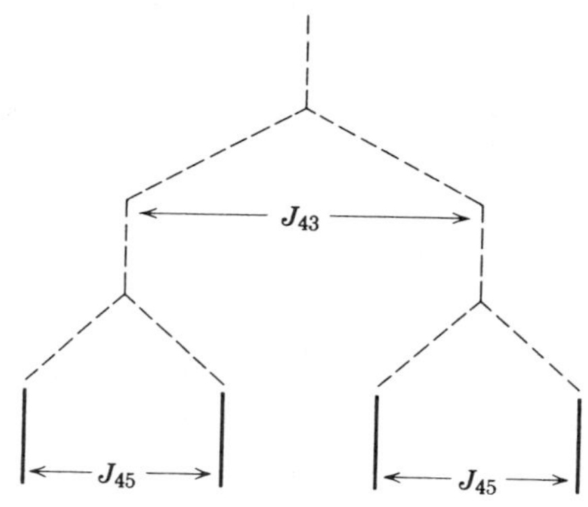

The couplings can be summed up as follows:

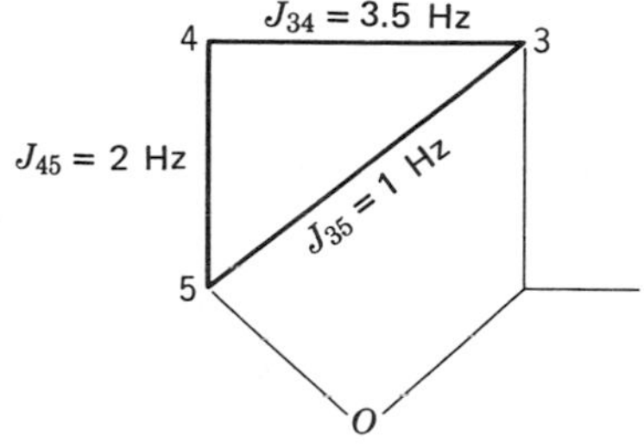

Note that methyl 3-furoate would give an entirely different ^1H NMR pattern.

Infrared Spectrum

Compound 7-3

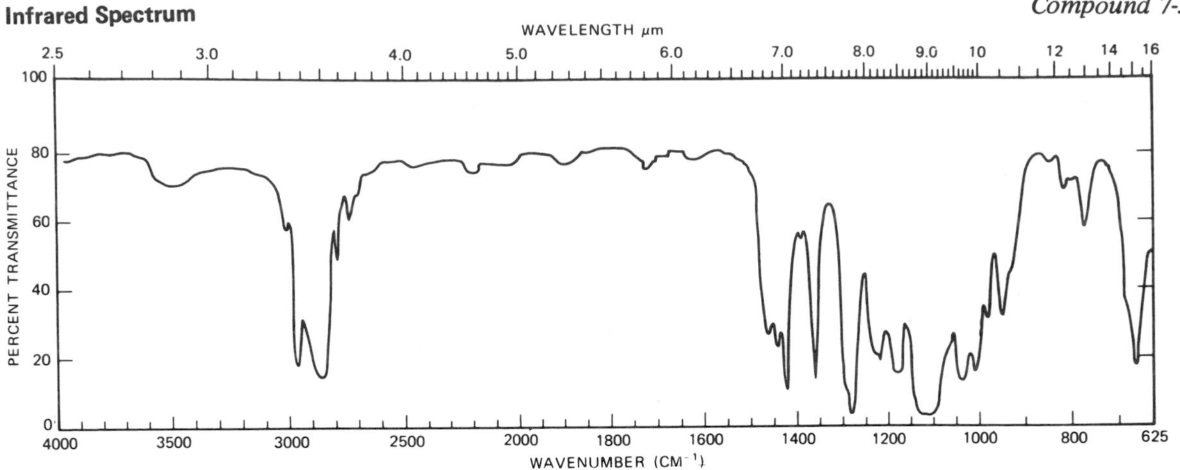

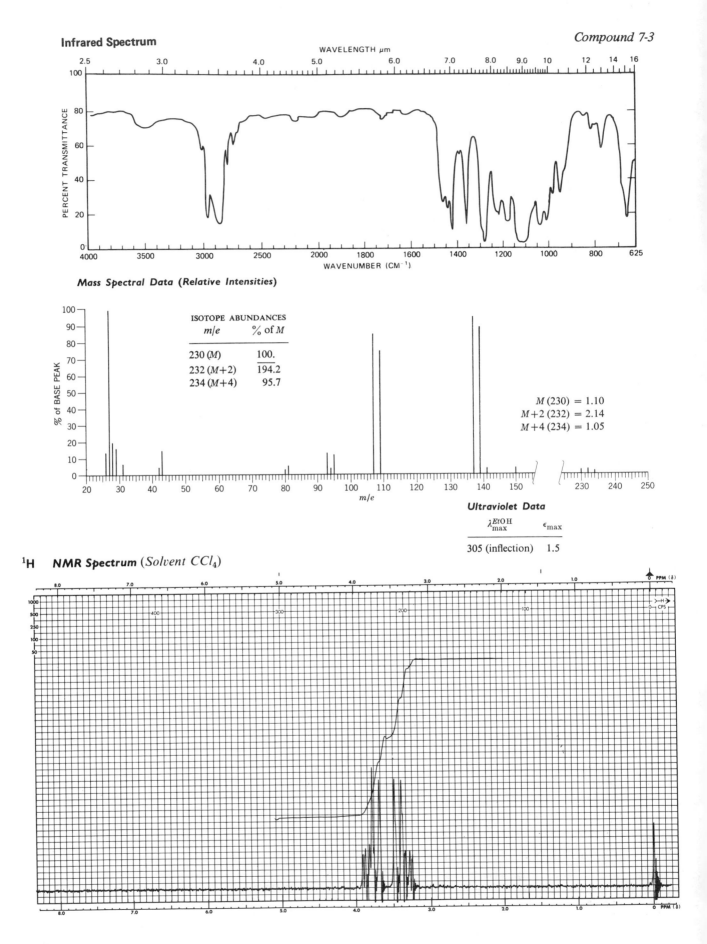

Mass Spectral Data (Relative Intensities)

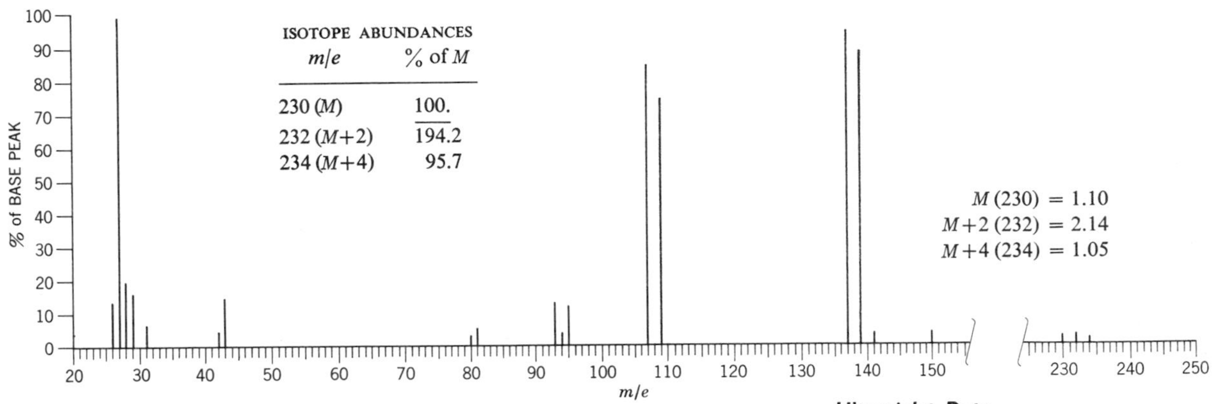

ISOTOPE ABUNDANCES

m/e	% of *M*
230 (*M*)	100.
232 (*M*+2)	194.2
234 (*M*+4)	95.7

M (230) = 1.10
M+2 (232) = 2.14
M+4 (234) = 1.05

Ultraviolet Data

λ_{max}^{EtOH}	ϵ_{max}
305 (inflection)	1.5

^{1}H NMR Spectrum (*Solvent* CCl_4)

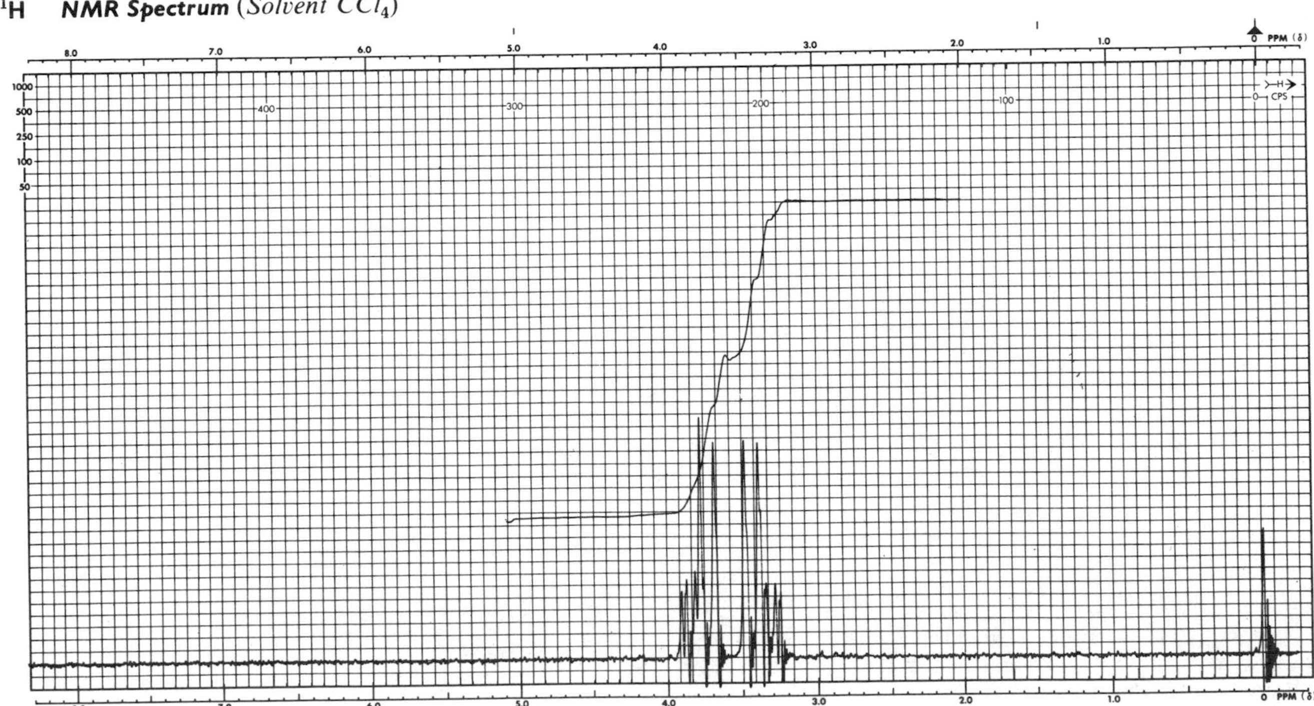

compound number 7-3

The characteristic $M+2$ and $M+4$ pattern indicates the presence of two bromine atoms. Subtracting 2×79 from the molecular weight leaves a mass of 72 for the rest of the molecule.

The strongest peak in the infrared spectrum at 1117 cm^{-1} (8.95 μm) strongly suggests an aliphatic ether. Since no nitrogen- or other oxygen-containing group appears to be present, we can assume, tentatively, that the remaining fragment is C_4H_8O. The weak inflection in the ultraviolet spectrum may simply result from the accumulation of atoms with nonbonding electrons.

The striking symmetry of the NMR spectrum suggests a symmetrical molecule. If we dispose our fragments, Br_2, $-O-$, and C_4H_8, in a symmetrical pattern, there are two possibilities

$$\underset{\text{Br}}{\text{CH}_3\text{CH}}\text{O}\underset{\text{Br}}{\text{CHCH}_3}$$

or

$$BrCH_2CH_2OCH_2CH_2Br$$

The first compound would give an A_3X NMR spectrum consisting of a 1-proton quartet quite far downfield ($\delta \sim 5.0$ and a doublet rather upfield ($\delta \sim 2.0$). The second compound would give an $AA'BB'$ pattern in accord with that observed.

The fragmentation pattern, though complex, affords ample confirmation. Characteristic bromine-containing pairs are discernible at m/e 137 and 139 (loss of CH_2Br), 107 and 109 (loss of OCH_2CH_2Br), and 93 and 95 (CH_2Br^+).

The compound is

$$BrCH_2CH_2OCH_2CH_2Br$$

β,β'-dibromodiethyl ether

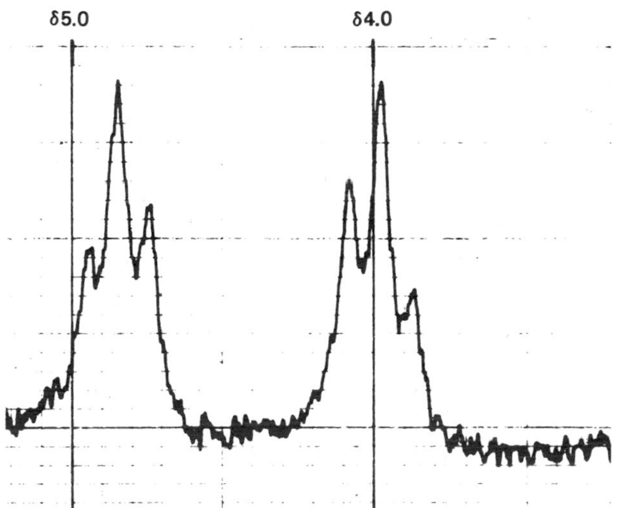

Sample 4 (see table, this page) of Shift Reagent study; 600 Hz Sweep Width, CCl_4 Solvent.

The strong peak in the infrared spectrum at 1279 cm^{-1} (7.82 μm) is attributed to a CH_2Br wag.

A CH_2 adjacent to O absorbs at δ 3.4 (Chart 1, p. 220), but the CH_2O in this compound is further deshielded by the Br atom once removed (Chart 2, p. 222). A CH_2 adjacent to Br also absorbs at δ 3.4 (Chart 1), but the CH_2Br in this compound is also further deshielded by the O atom once removed (Chart 2). Since the Br once removed is more effective than the O once removed, (Chart 2), we assign the downfield absorption to CH_2O.

	Mole Ratio	*Chemical Shifts (δ)*		*Change in Chemical Shifts*	
Sample	$Eu(fod)_3$ / ether	Upfield Signal	Downfield Signal	Upfield Signal	Downfield Signal
1	0.00	3.35	3.75	–	–
2	0.12	3.60	4.16	+0.25	+0.41
3	0.25	3.80	4.53	+0.45	+0.78
4[b]	0.38	3.98	4.87	+0.63	+1.12

[a]In all experiments, there was 30 mmoles of the bromo ether in about 0.5 ml of CCl_4.

[b]This spectrum is displayed at the lower left.

These NMR assignments are in accord with a shift reagent [$Eu(fod)_3$] study. The above table describes a downfield progression of both NMR signals as a function of increased amounts of shift reagent. Such results are consistent with the chemical shift changes being inversely dependent on the distance from the oxygen atom. The table shows that the separation has resulted from the greater downfield shift of the lower field signal per increment of shift reagent and was not the result of a cross-over. Since the shift reagent is expected to coordinate more strongly to the nucleophilic oxygen atom, the signal at lower field (δ 3.75) in the original spectrum is due to the methylene protons alpha to oxygen. The spectrum of Sample 4 shows that the shift reagent has begun to convert a higher-order system (sample 1, $\Delta\nu/J$ = ca. 24/6 = ca. 4) to a first-order system (sample 4, $\Delta\nu/J$ = ca. 53/6 = ca. 9). The benefits clearly outweigh the slight line broadening effect of the shift reagent.

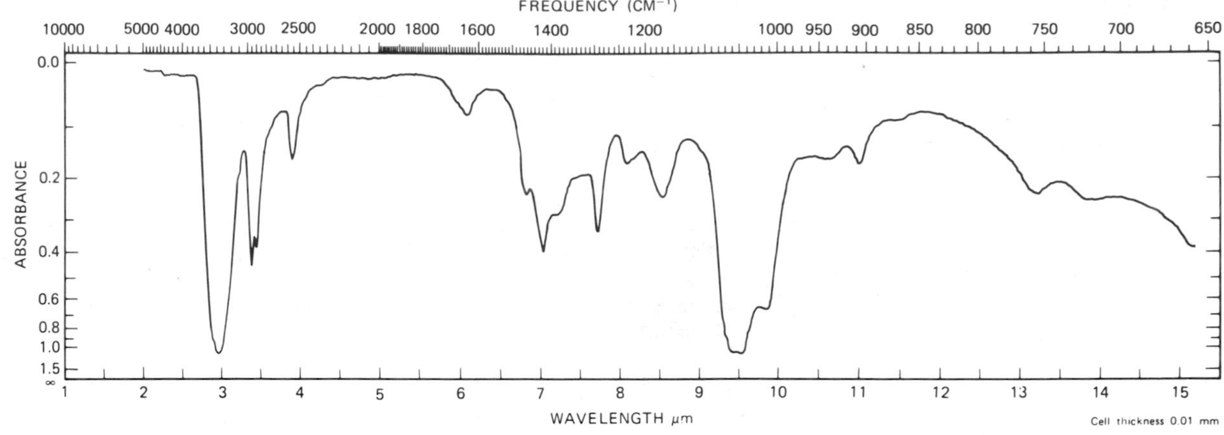

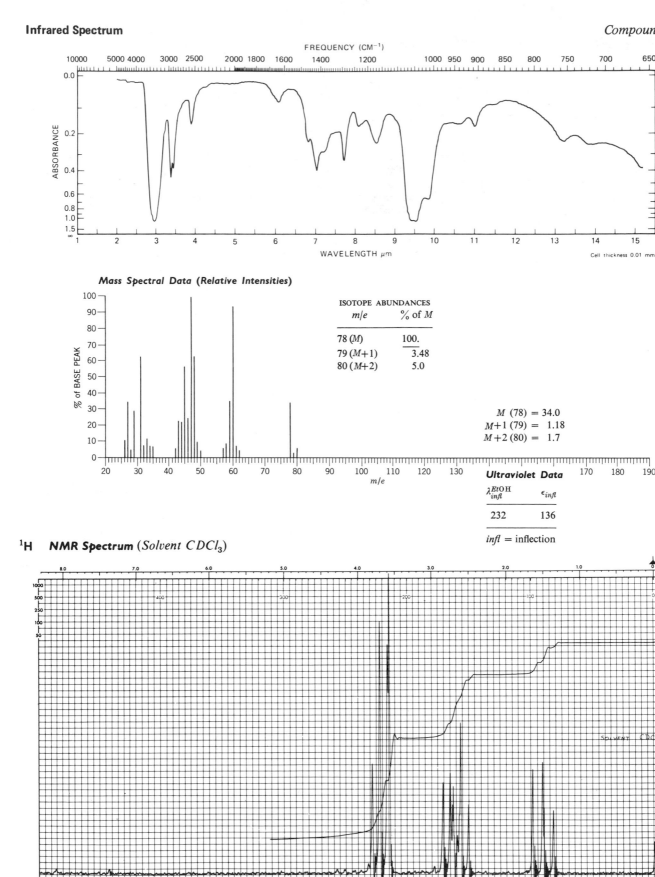

Mass Spectral Data (Relative Intensities)

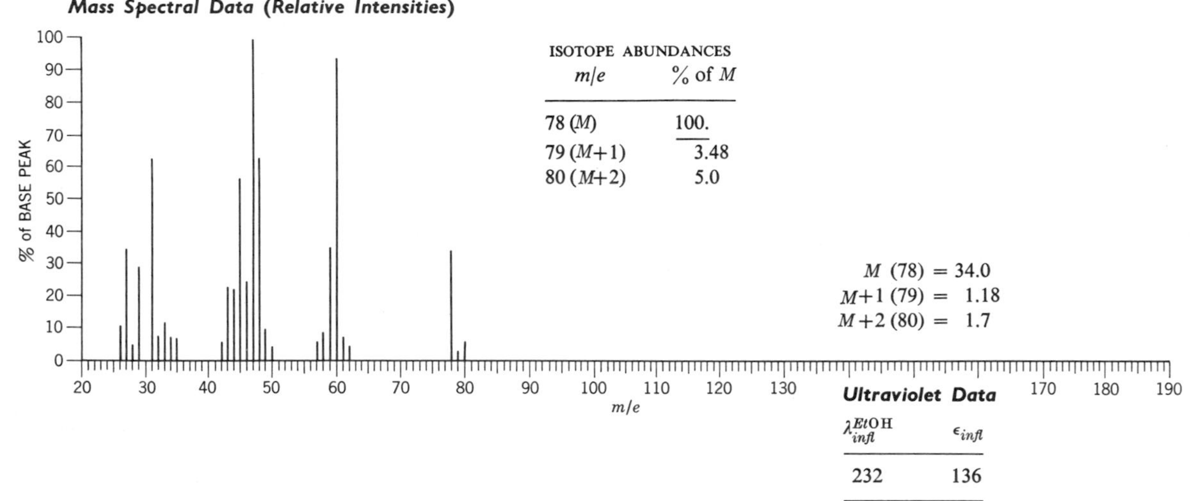

ISOTOPE ABUNDANCES

m/e	% of M
78 (M)	100.
79 ($M+1$)	3.48
80 ($M+2$)	5.0

M (78) = 34.0
$M+1$ (79) = 1.18
$M+2$ (80) = 1.7

Ultraviolet Data

λ_{infl}^{EtOH}	ϵ_{infl}
232	136

infl = inflection

^{1}H NMR Spectrum (*Solvent CDCl$_3$*)

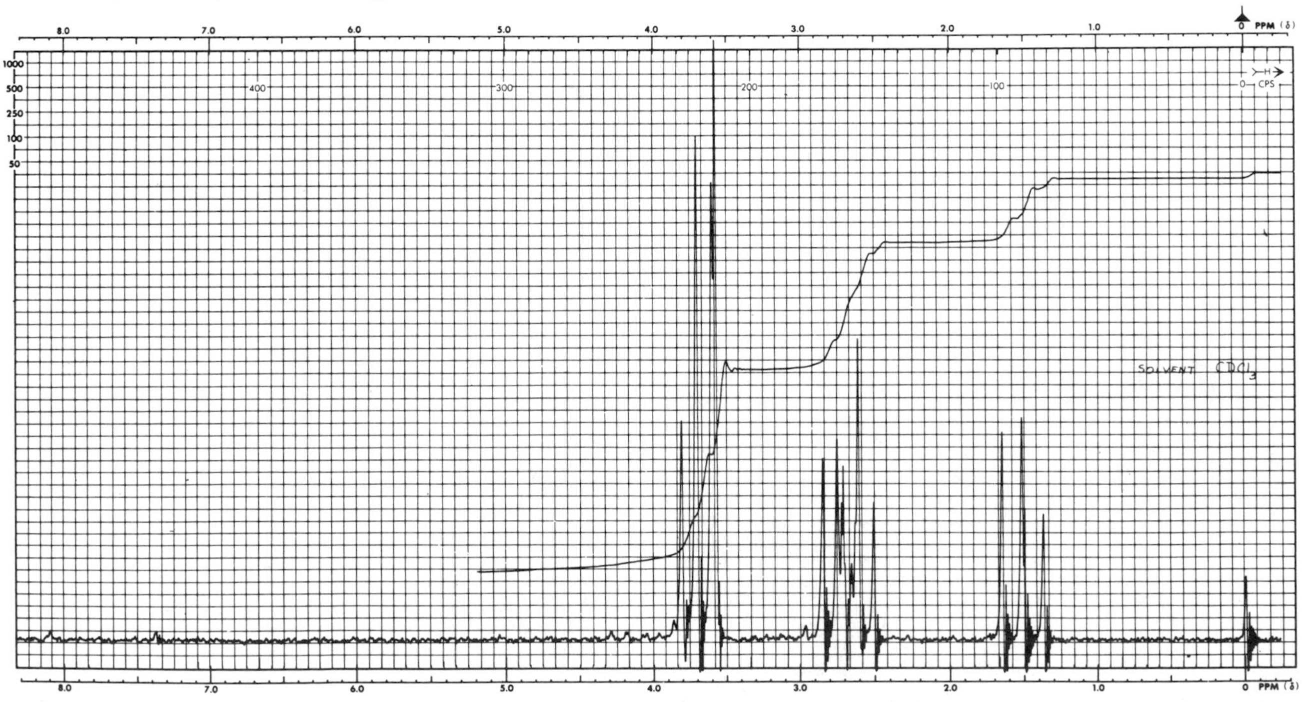

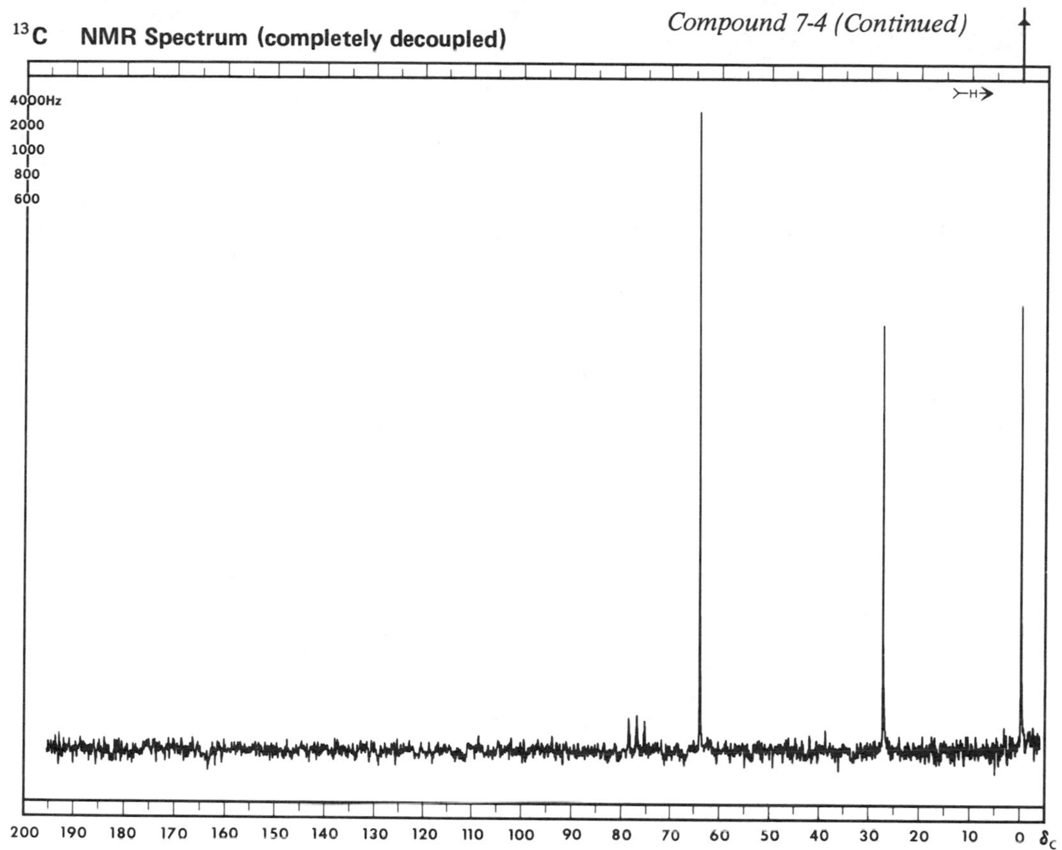
13C NMR Spectrum (completely decoupled)
4000Hz
2000
1000
800
600
200 190 180 170 160 150 140 130 120 110 100 90 80 70 60 50 40 30 20 10 0 δC

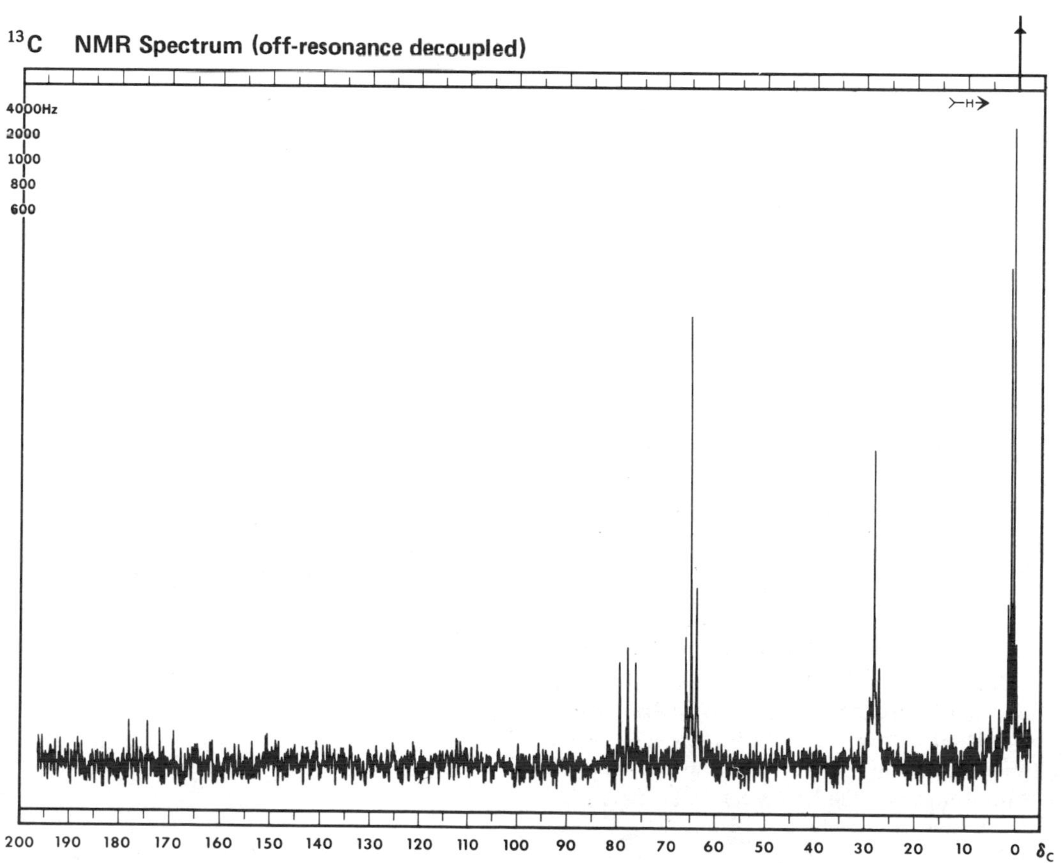
13C NMR Spectrum (off-resonance decoupled)
4000Hz
2000
1000
800
600
200 190 180 170 160 150 140 130 120 110 100 90 80 70 60 50 40 30 20 10 0 δC

compound number 7-4

We immediately note the large $M + 2$ peak which suggests that 1 sulfur atom is present. We then list the possible formulas under mass 46 (78 less 32). We should list all except the trivial formulas and those containing an odd number of N atoms; we also subtract 0.78 from the $M + 1$ peak, which leaves 2.70. But this turns out to leave us with only a single choice, C_2H_6O. The molecular formula, therefore, is C_2H_6OS.

^{13}C NMR supports a molecular formula with 2 carbon atoms since the noise-decoupled spectrum shows just 2 singlets (at δ 27.50 and δ 64.06). Since both ^{13}C signals are triplets in the off-resonance decoupled spectrum, we have shown that 4 of the 6 protons are in methylene groups.

The infrared spectrum shows a strong, rather broad band at 3367 cm^{-1} (2.97 μm). Our impression is that we are dealing with an alcohol. The very broad band at about 1050 cm^{-1} (3.91 μm) suggests a primary alcohol. Our attention is then caught by a rather weak band at 2558 cm^{-1} (3.91 μm), which practically spells out a mercaptan group. In this case, the infrared spectrometer is at some disadvantage with respect to the nose. Had this been a thin film spectrum, we might have missed the S–H stretching band.

We now have the fragments: CH_2OH and SH. This only leaves a CH_2 group to fit in, and we write

$HOCH_2CH_2SH$

2-Mercaptoethanol

Some of the major fragmentation peaks can be assigned as follows:

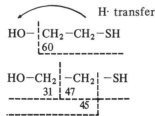

H· transfer

$HO-CH_2-CH_2-SH$ (cleavage: 60)

$HO-CH_2-CH_2-SH$ (cleavages: 31, 47, 45)

The ^1H NMR spectrum provides exhaustive confirmation for the structure written. It also shows a number of interesting features. The starting point is the distribution of protons as shown by the integration curve. If we assume that the triplet at the high-field position contains 1 proton, then the next cluster of peaks contains 2 protons, and the low-field peaks account for 3 protons. At first glance, this does not seem reasonable. But we must bear several things in mind. First, the positions of the OH peak and the SH peak depend on concentration. Since the OH-group forms stronger hydrogen bonds than the SH group, it is likely to be further downfield at a given concentration. Second, the OH proton will undergo rapid exchange under normal condition and will usually appear as a single peak; the SH proton, under the same conditions will not exchange rapidly (at least in a nonaqueous solvent), and the peak will be split by the adjacent methylene group.

The ^1H NMR spectrum represents an $AA'MM'X$ system (if we neglect the rapidly exchanging OH proton), but we can treat it as A_2M_2X and examine the A_2M_2 and M_2X couplings:

$$HO-CH_2-CH_2-SH$$
$$AA'MM'X = \sim A_2M_2X$$

We see then that the upfield triplet (with slight second-order splitting of the middle peak) represents the SH proton coupled with (split by) the adjacent methylene group; the coupling constant is 8 Hz. The adjacent methylene signal is split into a doublet by the SH proton (coupling constant, of course is 8 Hz), and again into a triplet by the other methylene group with a coupling constant of 6 Hz. This is a somewhat distorted A_2M_2X system with 2 coupling constants. An idealized diagram is as follows:

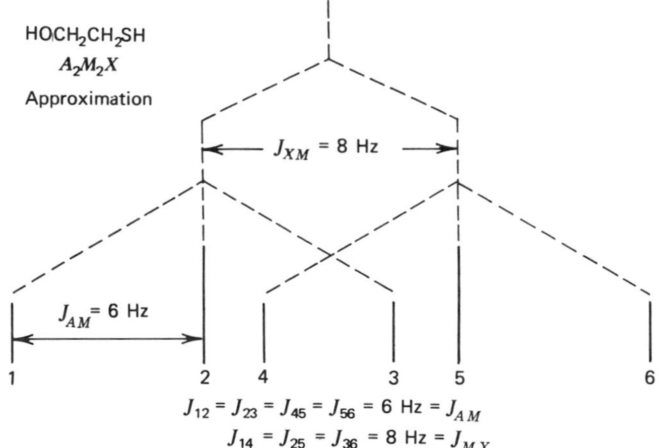

The low-field peaks must then consist of a triplet with a coupling constant of 6 Hz, and it must also contain the OH proton. We do indeed see a triplet with the upfield peak distorted by the peak of the OH proton with which it almost coincides. The hydroxyl proton peak can be shifted by change of concentration, solvent, or temperature. Both the OH and SH peaks can be removed by shaking with deuterium oxide; this, of course, would collapse the CH_2S multiplet to a symmetrical "apparent triplet" signal.

There are distortions in the spectrum because of marginal $\Delta\nu/J$ values.

notes

Infrared Spectrum *Compound 7-5*

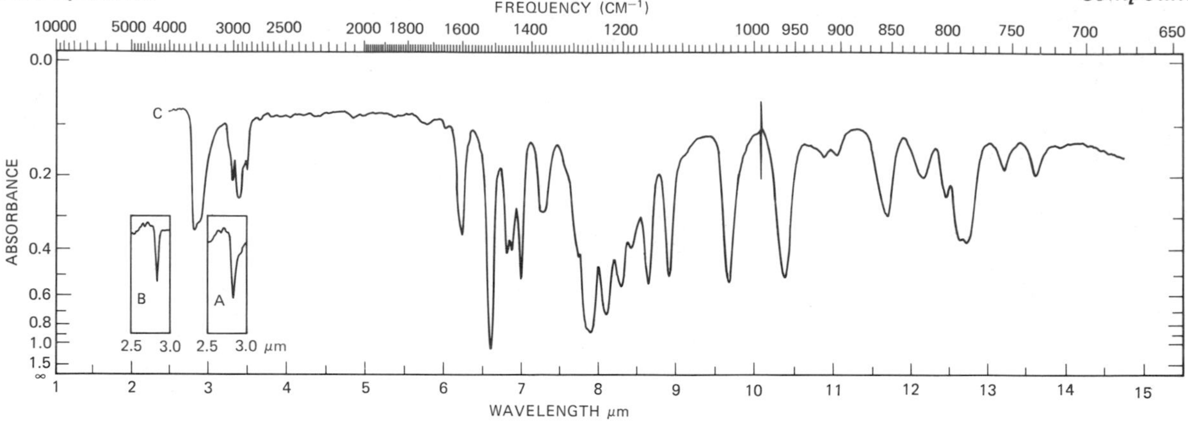

A. 0.03 M IN CCl_4, PATH LENGTH 0.406 mm
B. 1.0 M IN CCl_4, PATH LENGTH 0.01 mm
C. THIN FILM

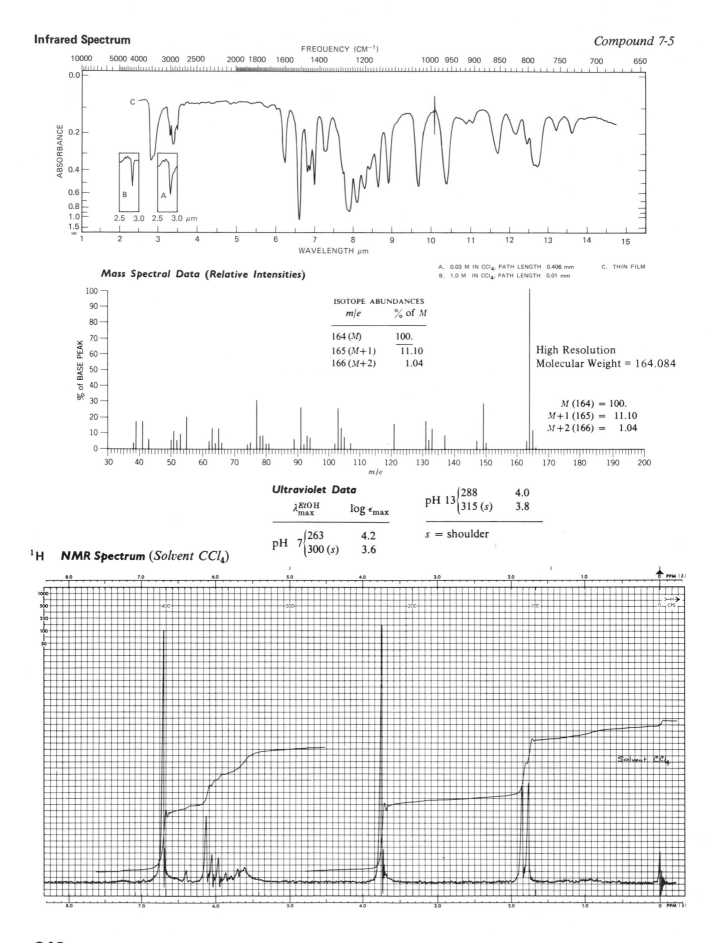

Mass Spectral Data (Relative Intensities)

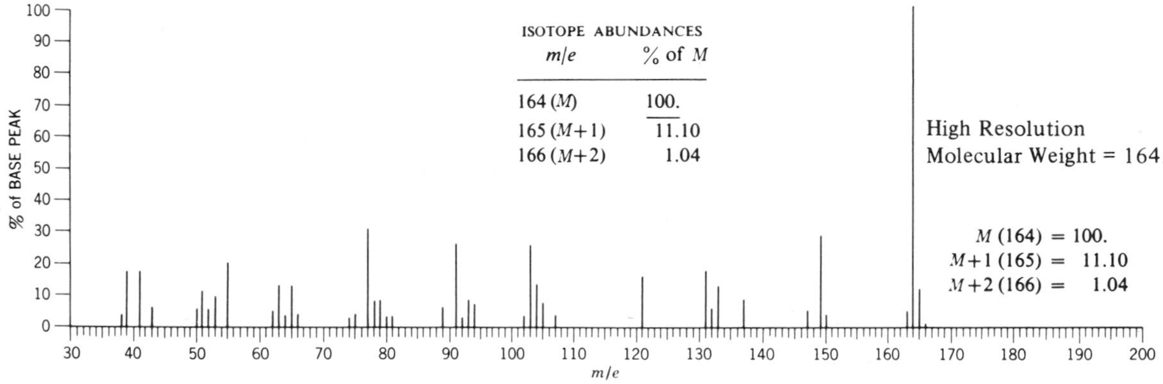

ISOTOPE ABUNDANCES

m/e	% of M
164 (M)	100.
165 ($M+1$)	11.10
166 ($M+2$)	1.04

High Resolution
Molecular Weight = 164.084

M (164) = 100.
$M+1$ (165) = 11.10
$M+2$ (166) = 1.04

Ultraviolet Data

	λ_{max}^{EtOH}	$\log \epsilon_{max}$
pH 7	263	4.2
	300 (*s*)	3.6
pH 13	288	4.0
	315 (*s*)	3.8

s = shoulder

^1H **NMR Spectrum** (*Solvent* CCl_4)

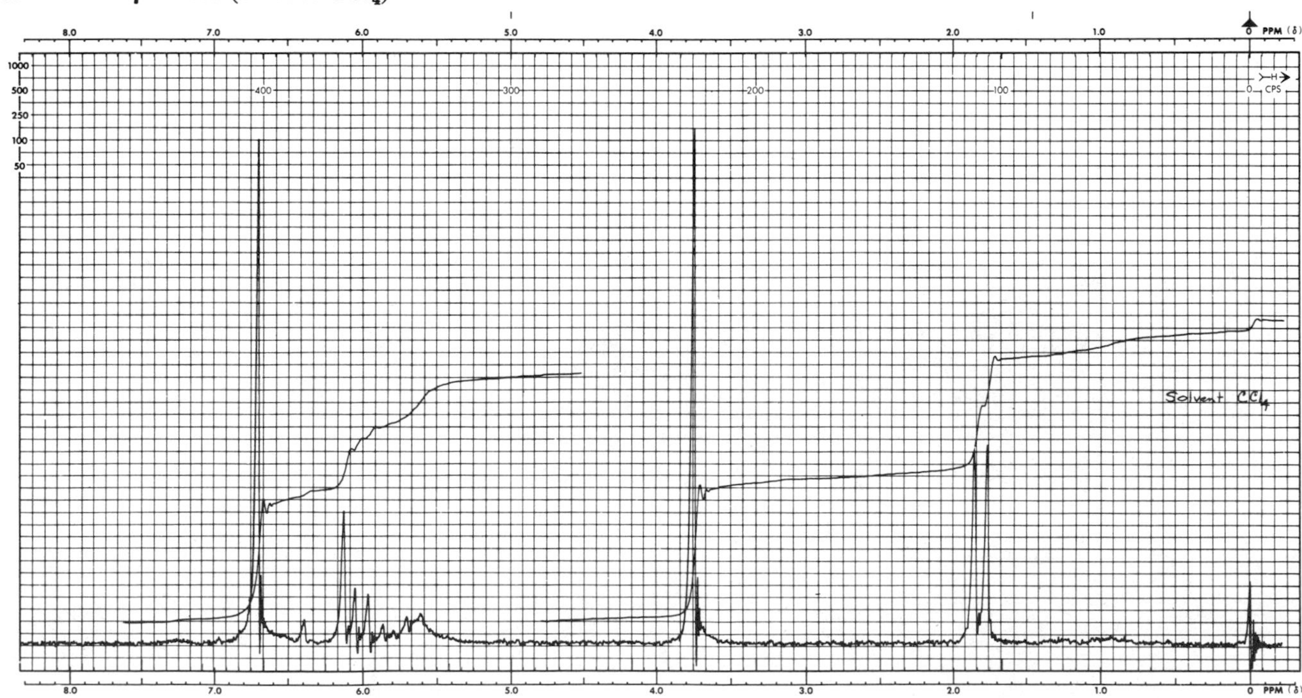

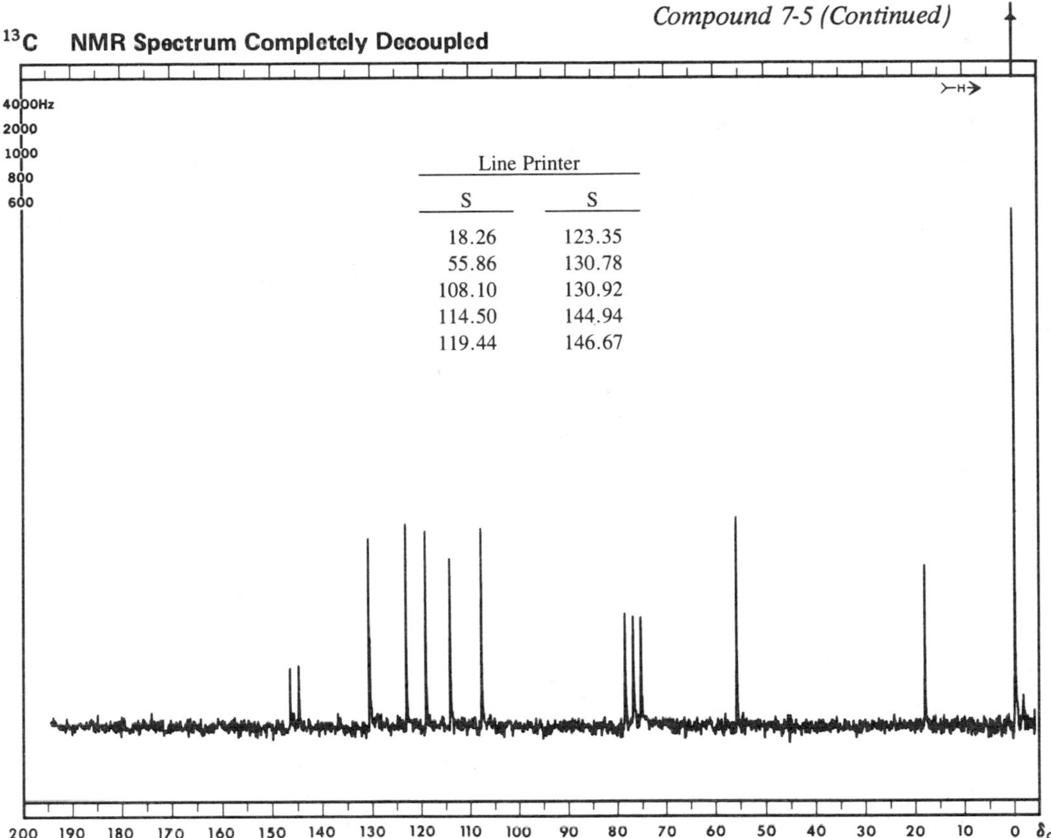

^{13}C NMR Spectrum Completely Decoupled
4000Hz
2000
1000
800
600
Line Printer
S
S
18.26
55.86
108.10
114.50
119.44
123.35
130.78
130.92
144.94
146.67
200 190 180 170 160 150 140 130 120 110 100 90 80 70 60 50 40 30 20 10 0 δC

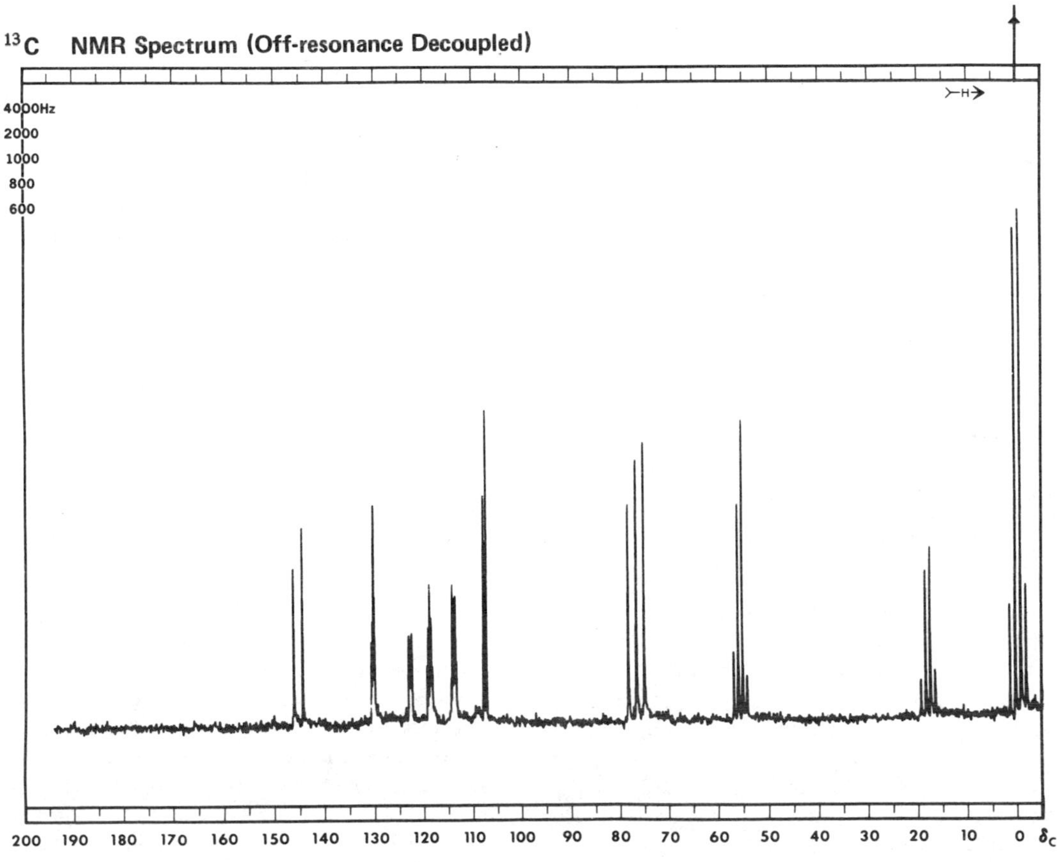

^{13}C NMR Spectrum (Off-resonance Decoupled)
4000Hz
2000
1000
800
600
200 190 180 170 160 150 140 130 120 110 100 90 80 70 60 50 40 30 20 10 0 δC

compound number 7-5

The following molecular formulas fit the data for the molecular ion peak and for the $M+1$ peak. We have eliminated trivial formulas and those containing an odd number of nitrogen atoms.

Formula	$M+1$	$M+2$	*FM*
$C_8H_8N_2O_2$	9.61	0.81	164.0584
$C_8H_{12}N_4$	10.36	0.49	164.1064
$C_9H_8O_3$	9.97	1.05	164.0473
$C_9H_{12}N_2O$	10.72	0.72	164.0951
$C_{10}H_{12}O_2$	11.08	0.96	164.0838
$C_{10}H_{16}N_2$	11.83	0.64	164.1315

The high-resolution formula mass determined to be 164.084, pinpoints $C_{10}H_{12}O_2$ as the formula of choice.

The ^{13}C NMR spectrum clearly shows 9 lines and thus we have at least 9 carbons. The computer line printer indicates 10 signals. The presence of more than 1 oxygen functional group is supported below. We have tabulated the ^{13}C multiplets (observed upon application of off-resonance decoupling) here:

δ	Mult.	δ	Mult.
18.26	q	123.35	d
55.86	q	130.78	s?
108.10	d	130.92	m?
114.50	m	144.94	s
119.44	m	146.67	s

The two signals near δ 130 overlap in both ^{13}C spectra making their interpretation difficult. Assuming the multiplets are due to carbons each bearing at least 1 proton, we can state that the compound must have at least 3 + 3 + 1 + 1 + 1 + 1 + 1 = 11 hydrogens.

The presence of an aromatic ring is indicated by the ^1H NMR peak at δ 6.70 (distinctly upfield from the position of an unsubstituted benzene ring) and the general appearance of the infrared spectrum. We note a small peak at 3030 cm^{-1} (3.30 μm), strong peaks between 1600 and 1430 cm^{-1} (6.2 and 7.0 μm), and several moderately strong peaks in the low-frequency (long wavelength) region. As additional confirmation of aromaticity, we note that the molecular ion peak in the mass spectrum is also the base peak.

A conspicuous feature of the infrared spectrum is the rather strong absorption at 3510 cm^{-1} (2.85 μm). This is either an OH or an NH stretching band. We look to the region between 1230 and 1010 cm^{-1} (8.13 and 9.90 μm) for confirmation of OH absorption, but we are frustrated by the large number of bands in this region.

The ultraviolet spectrum gives us a good deal of information. The shift in wavelength at pH 13 is diagnostic for a phenol. Furthermore, the intense *K*-band at 263 nm is indicative of a chromophore conjugated with the ring. In the absence of a carbonyl band in the infrared spectrum, we would suspect a C=C group. We quickly confirm this possibility by the olefinic proton absorption in the ^1H NMR spectrum at about δ 6.0. A strong peak in the infrared spectrum at 965 cm^{-1} (10.36 μm) is evidence for *trans* olefinic hydrogens. The peak at 1605 cm^{-1} (6.25 μm) can partially be ascribed to the conjugated C=C stretching vibration.

At this point we have the following information:

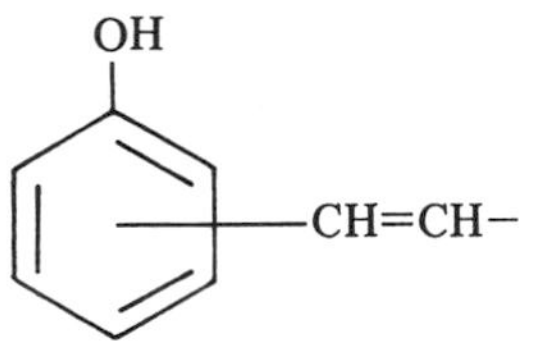

OH

—CH=CH—

We now turn our attention to the doublet centered at δ 1.81. The shift position is right for an allylic methyl group, and its existence as a doublet can be justified. We write

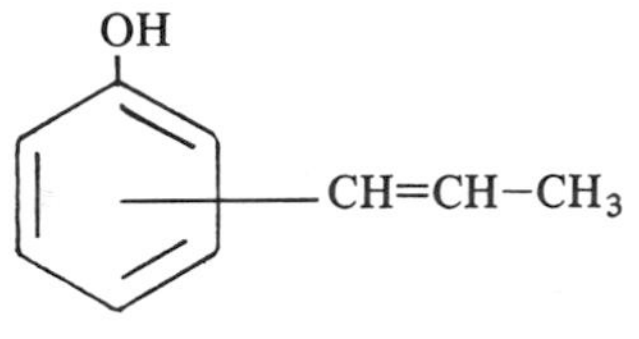

OH

—CH=CH—CH_3

The ^{13}C NMR data fit in very nicely at this point. There are 2 quartets in the off-resonance decoupled spectrum. The one at the higher field (δ 18.26) corresponds to the allylic methyl, and that at a lower field (δ 55.86) is consistent with a methoxyl methyl. The methoxyl protons occur at δ 3.75 in the ^1H NMR spectrum.

Thus, the tri-substituted benzene shown below accounts for all of the atoms in the molecular formula $C_{10}H_{12}O_2$.

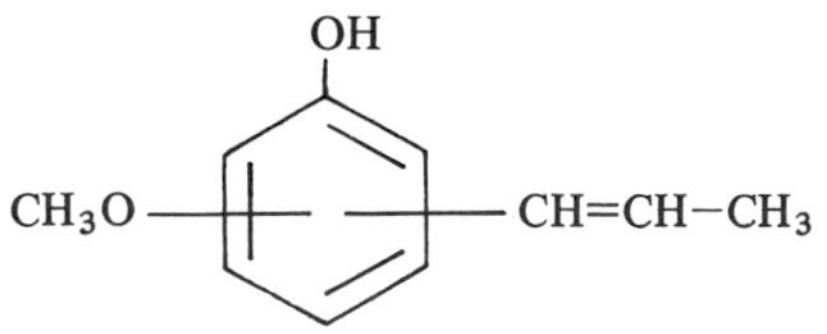

OH

CH_3O— —CH=CH—CH_3

In the ^1H NMR spectrum, the benzene peak at δ 6.70 contains 3 protons; the complex olefinic multiplet also accounts for 3 protons. Obviously, the phenolic proton is hidden under the olefinic multiplet; it could either be moved by change of temperature, solvent, or concentration, or eliminated by shaking with D_2O. It is probably the broadened peak at ~ δ 5.60. The olefinic proton near the ring is shown by the badly distorted doublet ($J \sim 16$) whose "center of gravity" is $\sim \delta$ 6.2.

Assignment of the positions of the substituents is not easy. We cannot rely upon the C–H out-of-plane deformations in the long-wavelength region of the infrared spectrum because of the polar nature of the substituents. The ^{1}H NMR spectrum, which is usually very helpful for distinguishing among isomers, is of little use because the ring protons are all at the same shift value. We might note that the phenolic OH band in the infrared spectrum is at an unusually high frequency (short wavelength) for a neat sample, and is not shifted on dilution; this suggests intramolecular hydrogen bonding to the ether oxygen. The methoxy group being ortho to the hydroxy group satisfies the geometric requirements for such intramolecular hydrogen bonding. The fragmentation in the mass spectrum is complex.

The ^{13}C NMR spectrum allows us to narrow down the choices. Since the oxygen-containing groups must be *ortho*, we are concerned with the 4 possible structures implied by this formula:

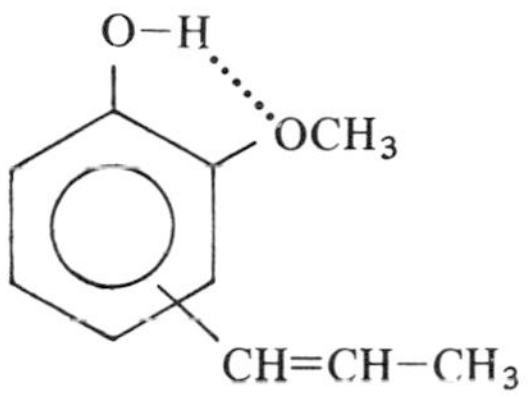

O–H ··· OCH_3

CH=CH–CH_3

Any structure containing a vicinal (1,2,3) arrangement of the 3 groups will require an unacceptably high δ value for 1 of the ring carbons. For example, for the structure

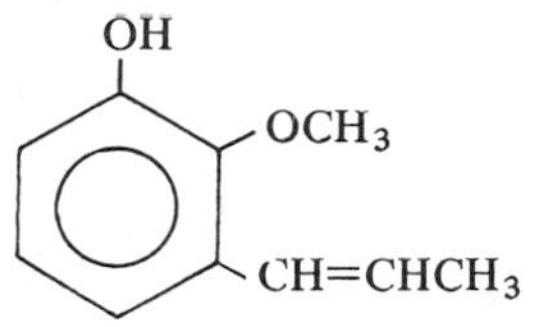

OH

OCH_3

CH=$CHCH_3$

we calculate (Table VII, Chapter 5) that the ring carbon bearing the methoxyl group should be at $\delta = 128.5 + 31.4 - 12.7 + 9.5 = 156.7$, and our spectrum provides no signals of shift greater than δ 150. We are thus limited to structures in which the C_3 unit is *para* either to the hydroxyl or to the methoxyl group. The ^{13}C spectrum, however, does not differentiate between these 2 isomers.

Familiarity with the structures of natural products might suggest isoeugenol, and comparison of our spectra with those of an authentic sample would provide confirmation.

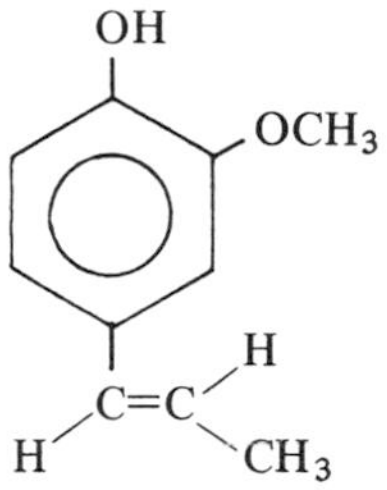

OH

OCH_3

H C=C H

CH_3

The ^{13}C NMR spectrum is consistent with the *trans*-isomer.

In the structure below, calculated values are in parentheses (Table VII, Chapter 5), The allyl methyl of the *cis*-isomer is at considerably higher field (δ 14.5).

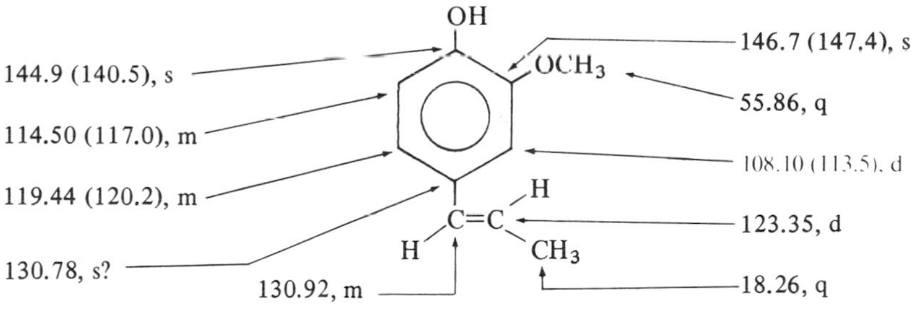

OH

OCH_3

144.9 (140.5), s

114.50 (117.0), m

119.44 (120.2), m

130.78, s?

130.92, m

146.7 (147.4), s

55.86, q

108.10 (113.5), d

123.35, d

18.26, q

H C=C H

CH_3

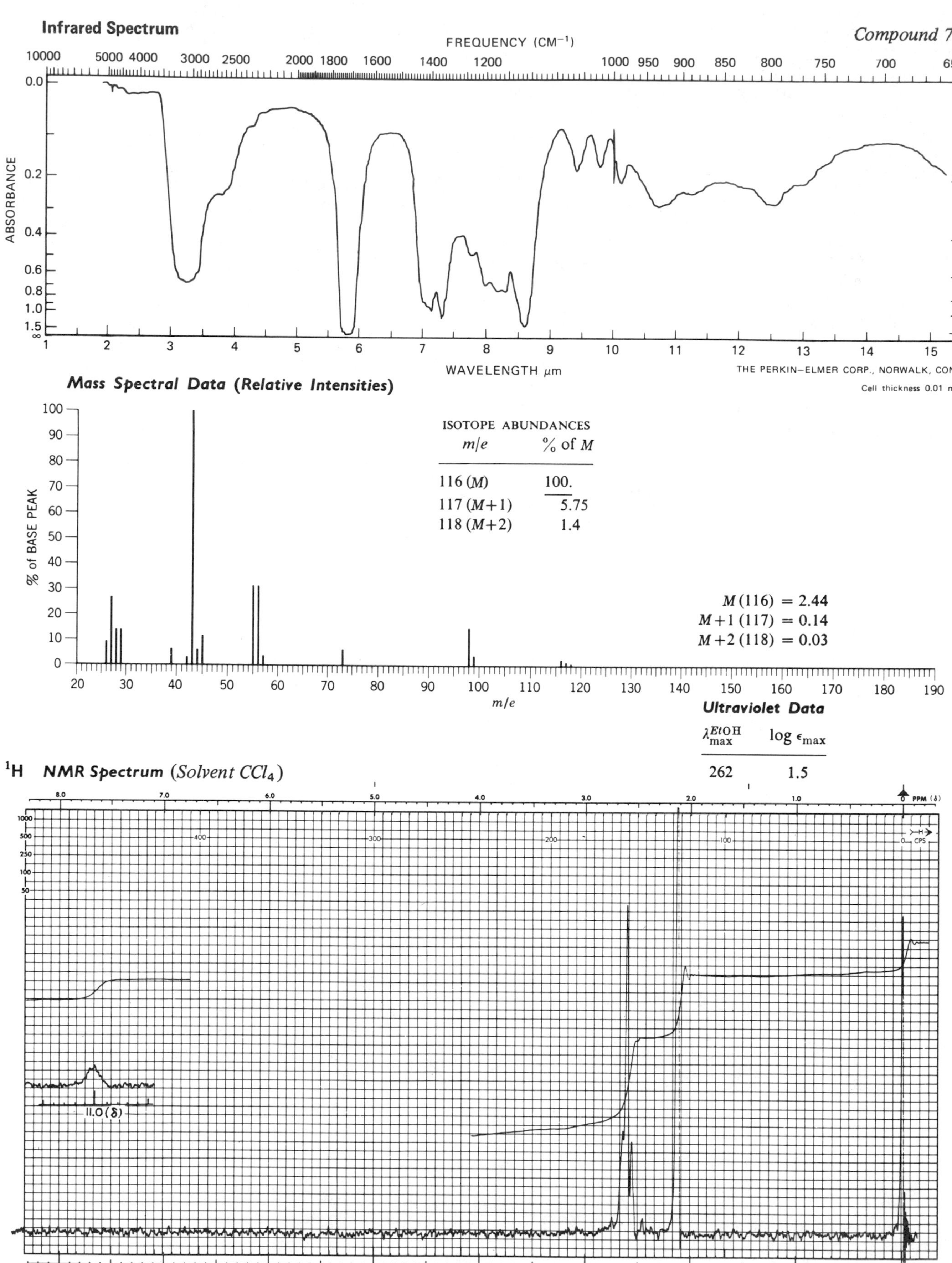

Infrared Spectrum
Compound 7-6
FREQUENCY (CM⁻¹)
ABSORBANCE
WAVELENGTH μm
THE PERKIN–ELMER CORP., NORWALK, CONN.
Cell thickness 0.01 mm
Mass Spectral Data (Relative Intensities)
% of BASE PEAK
m/e
ISOTOPE ABUNDANCES
m/e % of M
116 (M) 100.
117 (M+1) 5.75
118 (M+2) 1.4
M (116) = 2.44
M+1 (117) = 0.14
M+2 (118) = 0.03
Ultraviolet Data
λ max EtOH log ε max
262 1.5
¹H NMR Spectrum (Solvent CCl4)
PPM (δ)
CPS
11.0 (δ)

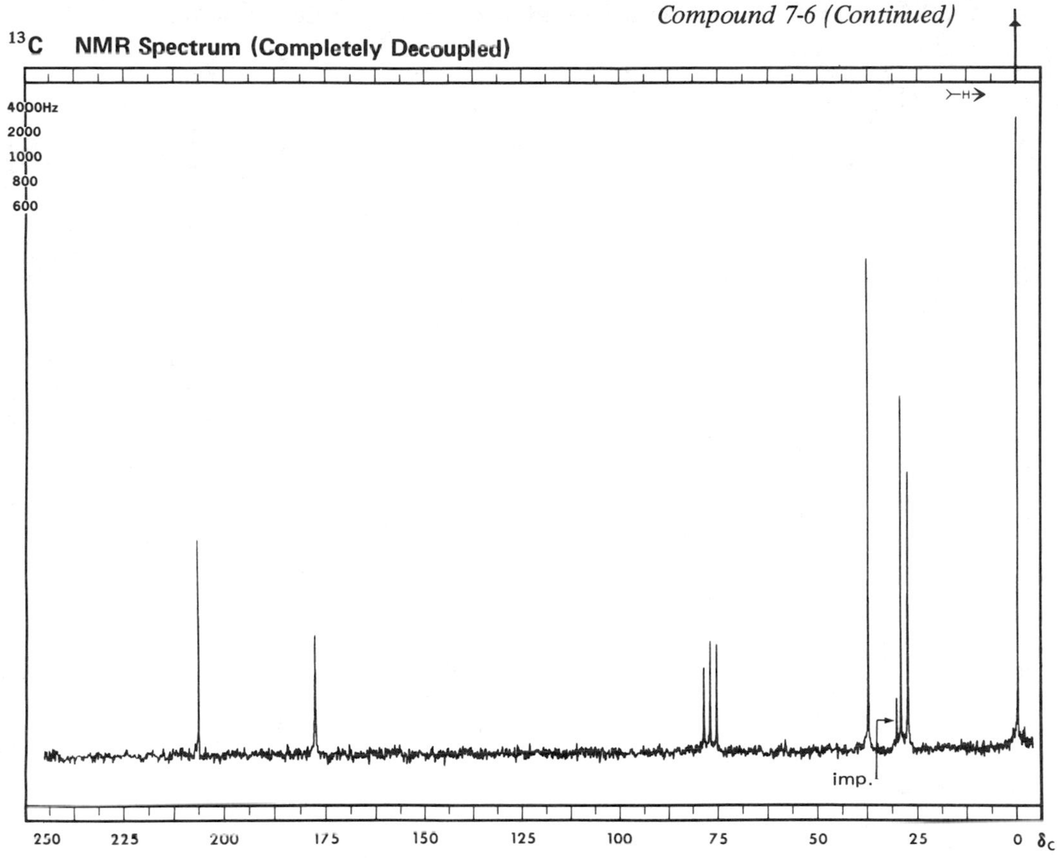

^{13}C NMR Spectrum (Completely Decoupled)
4000Hz
2000
1000
800
600
imp.
250
225
200
175
150
125
100
75
50
25
0
δ_C

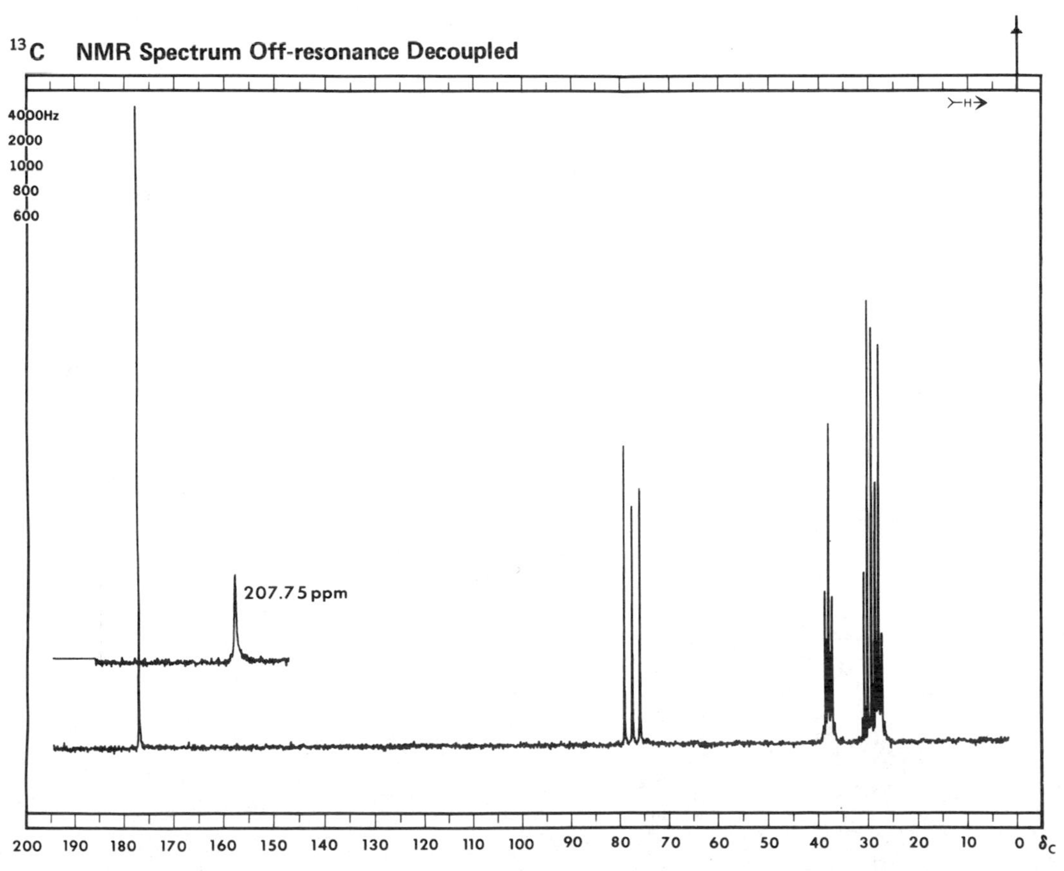

^{13}C NMR Spectrum Off-resonance Decoupled
4000Hz
2000
1000
800
600
207.75 ppm
200
190
180
170
160
150
140
130
120
110
100
90
80
70
60
50
40
30
20
10
0
δ_C

^{13}C **NMR Spectrum (Off-resonance Decoupled, Expanded)** *Compound 7-6 (Continued)*

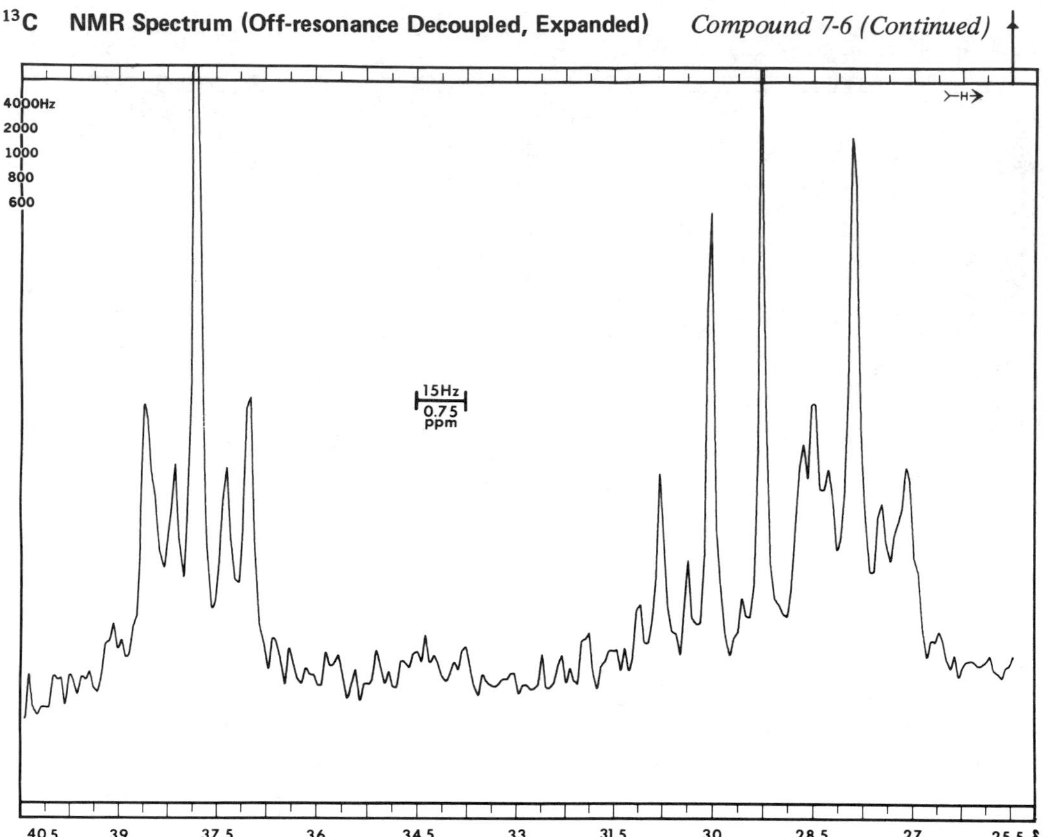

compound number 7-6

We list the molecular formulas under mass 116 (molecular ion) that give a reasonable fit for the $M + 1$ value of 5.75%.

Formula	$M + 1$	$M + 2$
$C_3H_8N_4O$	4.94	0.30
$C_4H_4O_4$	4.54	0.88
$C_4H_8N_2O_2$	5.29	0.52
$C_4H_{12}N_4$	6.04	0.16
$C_5H_8O_3$	5.65	0.73
$C_5H_{12}N_2O$	6.40	0.37
$C_6H_{12}O_2$	6.75	0.59

The formulas that best fit our $M + 1$ value and our $M + 2$ value (1.4%) are $C_4H_4O_4$, $C_4H_8N_2O_2$, $C_5H_8O_3$, and $C_6H_{12}O_2$.

The infrared spectrum points to an aliphatic carboxylic acid. We note the very broad, bonded OH-stretching absorption extending between about 3333 and 2300 cm^{-1} (3.0 and 4.3 μm) with its characteristic pattern on its low-frequency (long-wavelength) side. The strong C=O band at 1715 cm^{-1} (5.83 μm) satisfies the requirement for an aliphatic carboxylic acid. We find ready confirmation by noting the carboxylic acid proton at δ 11.0 in the ^{1}H NMR spectrum.

The 2 downfield ^{13}C-peaks are singlets in the off-resonance decoupled spectrum. They are clearly due to carbonyls, and we may tentatively assign the δ 178.23 peak to a carboxyl group and the δ 207.02 peak to a keto group (see Tables XIII and XIV, Chapter 5).

The UV spectrum provides a low-intensity absorption at 262 nm, suggesting an aliphatic unconjugated ketone. There is only one C=O band in the infrared spectrum, but it is rather broad; it must accommodate both the ketone and the acid C=O band. The fragmentation pattern does not resemble that of an ordinary aliphatic carboxylic acid; for example, there is no pronounced peak at mass 60. Since we are dealing with a keto group in the molecule, we assume that the base peak of 43 in the mass spectrum arises from $CH_3C{=}O$. We now have the fragments $CH_3C{=}O$ and COOH which add up to mass 88. These leave us with an unknown mass of 28 for which we can write either C=O or $CH_2{-}CH_2$ or N_2.

The NMR spectra allow us to make an unequivocal choice. In the ^{1}H NMR spectrum the large singlet at δ 2.12 (3H) must be the methyl protons of the CH_3CO group. If this peak contains 3 protons, then the absorption at δ 2.60 contains 4 protons. These must result from adjacent methylene groups which have almost the same chemical shift ($AA'BB'$ system). We now write the structure

```
5  43   2   1
CH3CCH2CH2COOH
    ||
    O
```

Levulinic acid

The multiplicity of the ^{13}C signals in the off-resonance decoupled spectrum would not be an easy starting point for determining the levulinic acid structure. However, now that the structure has been proposed, the ^{13}C spectra can be used for corroboration. The expanded off-resonance decoupled spectrum shows the δ 27.96 and δ 37.83 signals to be triplets, with additional complexity. The δ 29.71 signal is a quartet and it is thus assigned to the methyl carbon. Table III in Chapter 5 allows us to predict the chemical shifts of levulinic acid by calculating the shift due to the attachment of a carboxyl group to C–4 of 2-butanone.

```
   O
   ||
CH3CCH2CH3
```

2-Butanone

^{13}C Shifts for 2-butanone
C-1 = 28.8 ppm
C-2 = 206.3
C-3 = 36.4
C-4 = 7.6

A carboxyl group is expected to shift the α carbon +21 ppm, the β, +3 ppm and γ, −2 ppm. Thus, for levulinic acid we calculate:

$$\delta_2 = 7.6 + 21 = 28.6 \text{ ppm}$$
$$\delta_3 = 36.4 + 3 = 39.4 \text{ ppm}$$
$$\delta_4 = 206.3 - 2 = 204.3 \text{ ppm}$$
$$\delta_5 = 28.8 \text{ ppm}$$

These agree well with following assignments for the observed ^{13}C shifts:

```
    O
    ||
CH3CCH2CH2CO2H
```

29.71 (q) — CH_3; 207.02 (s) — C=O; 37.83 (t) — CH_2 (C-3); 27.96 (t) — CH_2 (C-2); 178.23 (s) — CO_2H

The signal at δ 30.82 is that of an impurity.

Infrared Spectrum

Compound 7-7

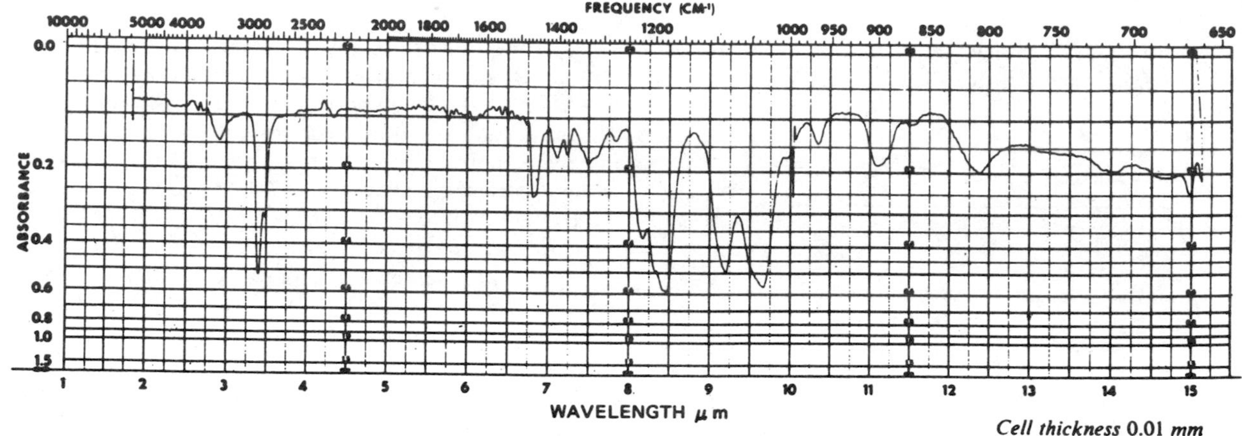

Cell thickness 0.01 *mm*

Mass Spectral Data (Relative Intensities)

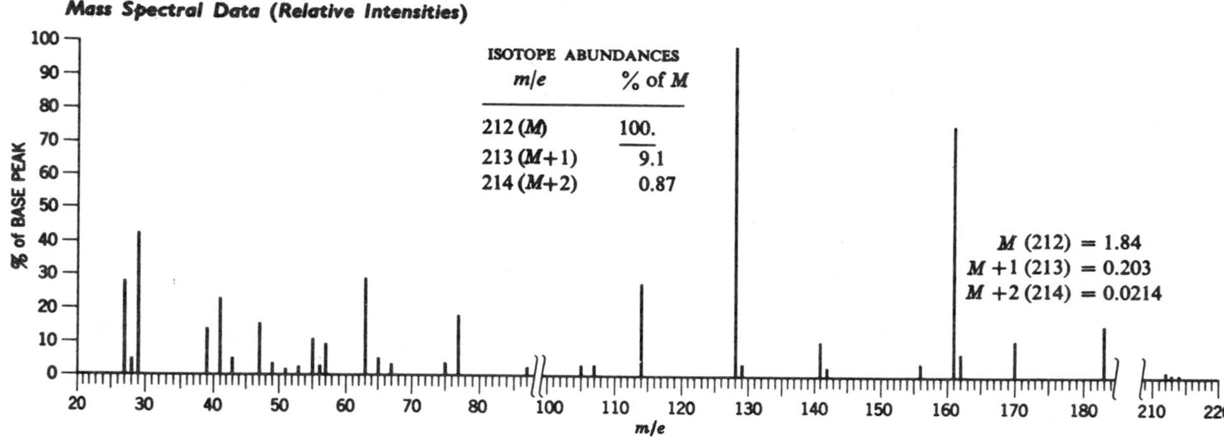

ISOTOPE ABUNDANCES

m/e	% of *M*
212 (*M*)	100.
213 (*M*+1)	9.1
214 (*M*+2)	0.87

M (212) = 1.84
M +1 (213) = 0.203
M +2 (214) = 0.0214

Ultraviolet Data

Featureless beyond 210 nm

^{1}H ***NMR Spectrum*** (*Solvent* CCl_4)

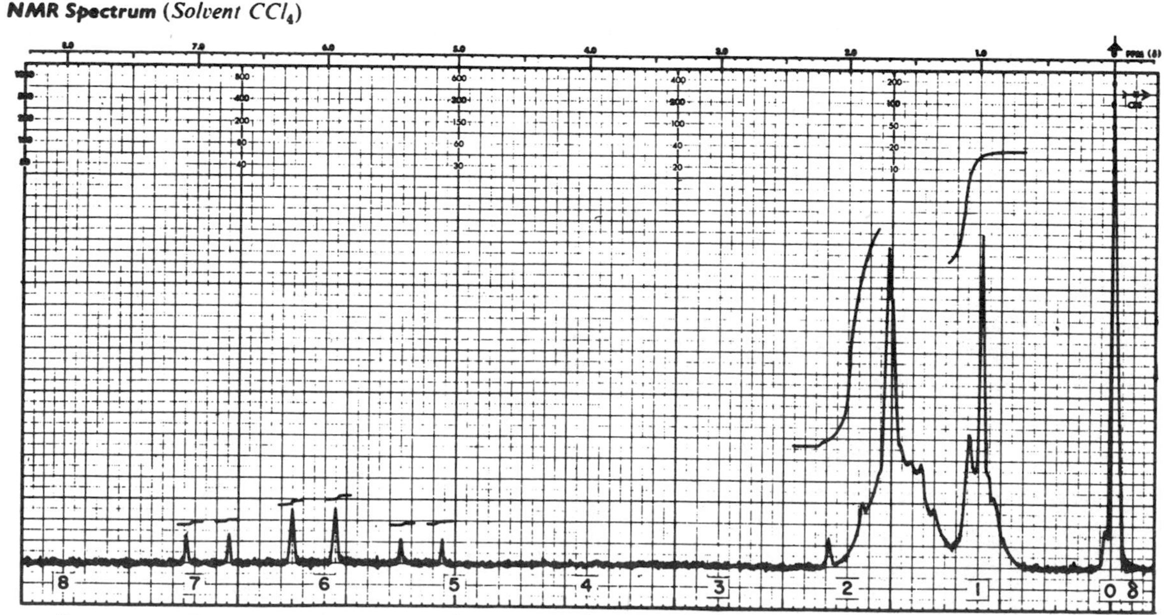

compound number 7-7

This example is presented to show some of the limitations of the methodology.

It is not likely that the four spectra presented would furnish sufficient evidence to identify this compound. However, given its history, we make the identification quite readily.

The compound was isolated as an unexpected, but rational, by-product from the following synthesis (S. A. Fuqua, W. G. Duncan, and R. M. Silverstein, *J. Org. Chem.*, **30**, 2543 (1965)):

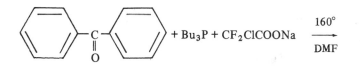

$$(C_6H_5)_2C{=}O + Bu_3P + CF_2ClCOONa \xrightarrow[\text{DMF}]{160^\circ}$$

$$\text{Compound 7-7} + Bu_3P \rightarrow O + (C_6H_5)_2C{=}CF_2$$

$Bu = n\text{-}C_4H_9$

The reaction should proceed by formation of a Wittig reagent from the phosphine and the CF_2 carbene formed by pyrolysis of sodium chlorodifluoroacetate. The Wittig reagent should react with the ketone.

$$Bu_3P{=}CF_2 + C_6H_5\overset{O}{\overset{\|}{C}}C_6H_5 \rightarrow C_6H_5\overset{O^-\ \ {}^+PBu_3}{C(C_6H_5)}{-}CF_2 \rightarrow$$

$$\rightarrow C_6H_5\underset{C_6H_5}{C}{=}CF_2 + O{\leftarrow}PBu_3$$

The molecular weight of the by-product is 212. It is obvious from the spectra (e.g., lack of aromatic character) that the material is not a benzophenone derivative. Rather, aliphatic absorption predominates in the spectra. The compound may contain phosphorus or fluorine or both. Obviously no chlorine is present. The strong absorption at 1182 cm^{-1} (8.46 μm) may be aliphatic P→O absorption, and that at 1087 cm^{-1} (9.20 μm) and 1033 cm^{-1} (9.68 μm) may result from C–F stretching. The puzzling thing is that if we have the P→O group present, we cannot possibly have sufficient remaining mass for 3 butyl groups; the compound is not the usual product $Bu_3P{\rightarrow}O$.

Suppose we assume displacement of one of the butyl group; this leaves mass 51 short of the molecular weight. It does not take long to propose CHF_2 as a possible fragment, and to confirm it with the striking NMR pattern. If the upfield protons represent the 18 protons of the butyl groups, the entire downfield pattern represents a single proton, centered at δ 6.07. The proton is split into a triplet by the geminal F atoms (J_{HF} 49.4 Hz), and each peak is again split into a doublet by the P atom (J_{HP} 19.4 Hz). The compound is therefore

$$Bu{-}\overset{O\uparrow}{P}{-}Bu \quad (P{-}CHF_2)$$

Dibutyl(difluoromethyl)phosphine oxide

Some of the major fragmentation peaks can be accounted for: m/e 183 (M minus CH_3CH_2), m/e 161 (M minus CHF_2), m/e 47 (P→O). The base peak, m/e 128 must result from complex rearrangements. The $M + 1$ and $M + 2$ peaks are in agreement with the calculated values for the proposed structure.

The rationale for this unexpected product depends on rearrangement of the initial Wittig reagent.

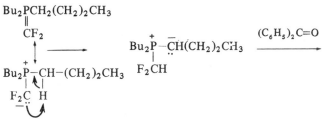

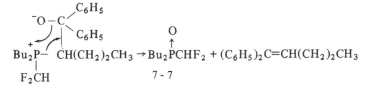

$$Bu_2\overset{}{P}CH_2(CH_2)_2CH_3 (=CF_2) \leftrightarrow Bu_2\overset{+}{P}{-}CH{-}(CH_2)_2CH_3 \,(F_2\ddot{C}^-,\ H) \longrightarrow Bu_2\overset{+}{P}{-}\overset{-}{\ddot{C}}H(CH_2)_2CH_3\,(F_2CH) \xrightarrow{(C_6H_5)_2C=O}$$

$$^-O{-}C(C_6H_5)_2 \cdots Bu_2\overset{+}{P}{-}CH(CH_2)_2CH_3\,(F_2CH) \rightarrow Bu_2\overset{O\uparrow}{P}CHF_2 + (C_6H_5)_2C{=}CH(CH_2)_2CH_3$$

7-7

Note: The P → O bond has also been described as a P=O bond.

Infrared Spectrum

Compound 7-8

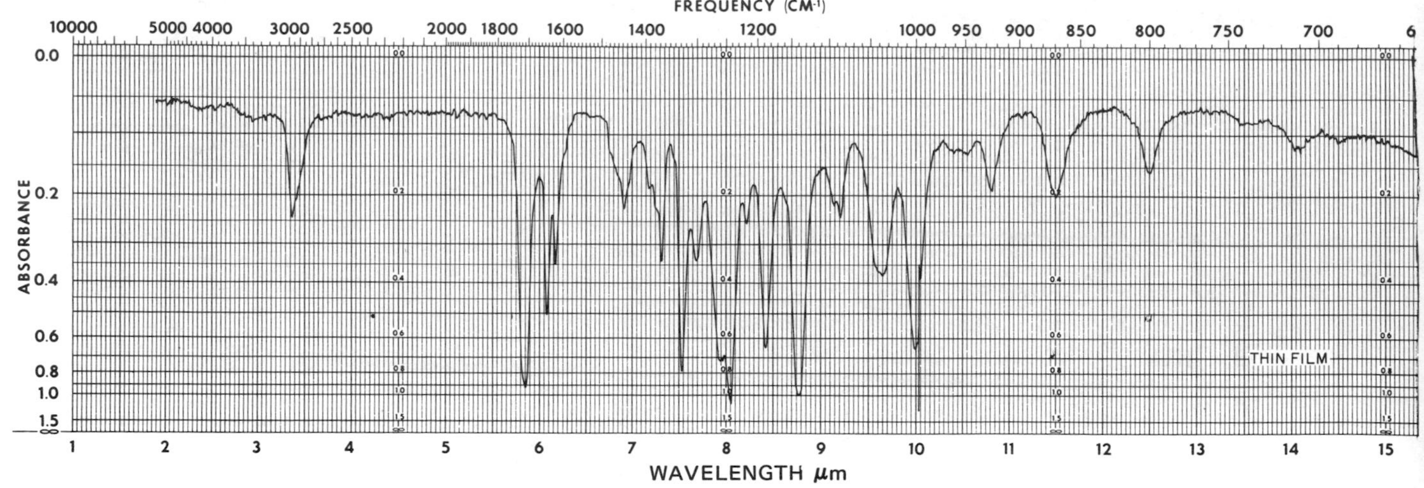

Mass Spectral Data (Relative Intensities)

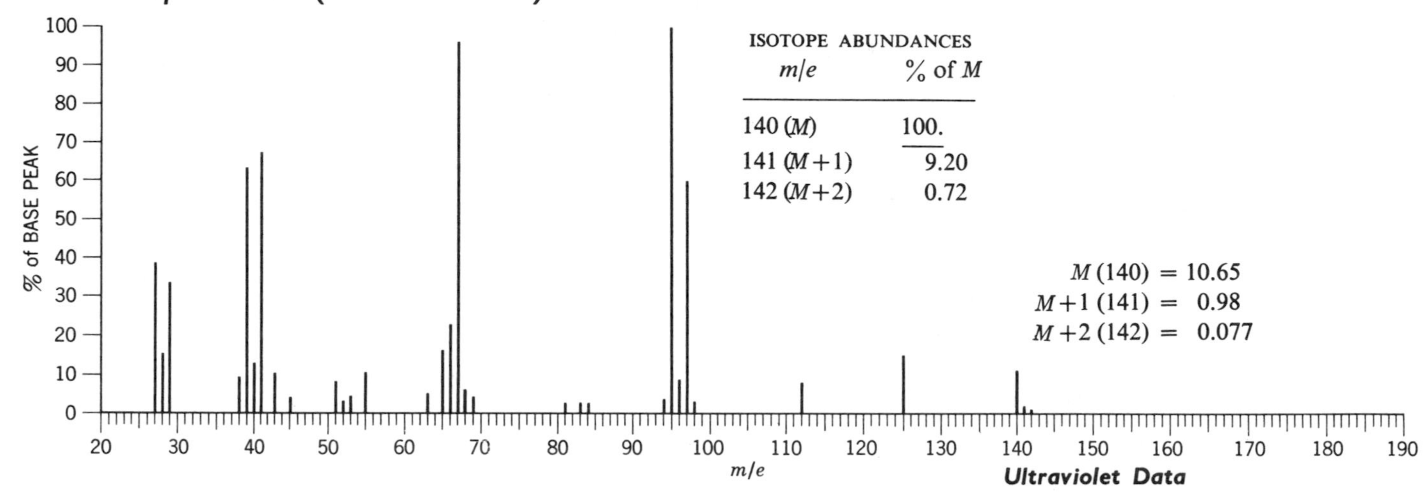

ISOTOPE ABUNDANCES

m/e	% of M
140 (M)	100.
141 ($M+1$)	9.20
142 ($M+2$)	0.72

M (140) = 10.65
$M+1$ (141) = 0.98
$M+2$ (142) = 0.077

Ultraviolet Data

λ_{max}^{EtOH}	$\log \epsilon_{max}$
259	4.4

^{1}H *NMR Spectrum*

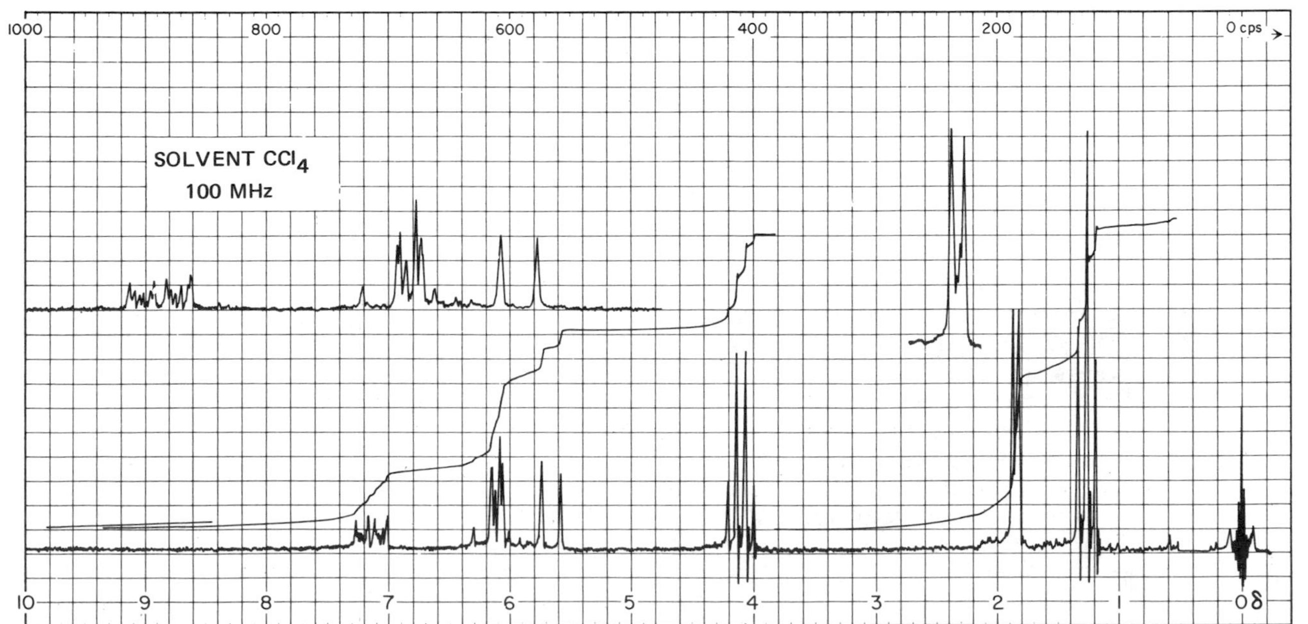

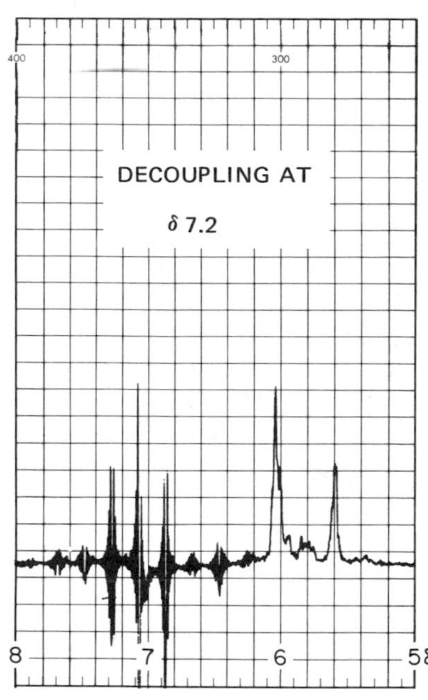

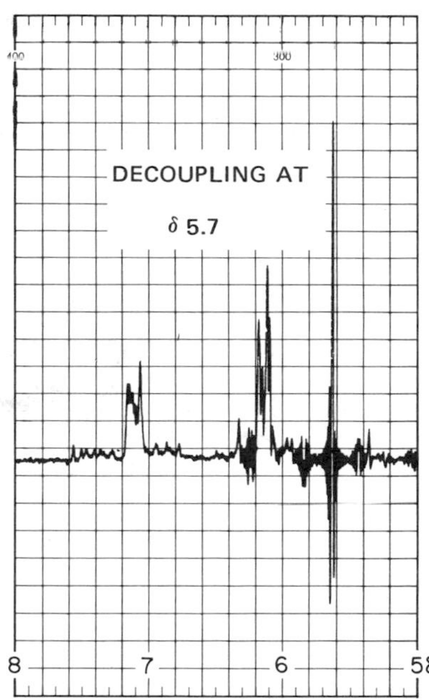

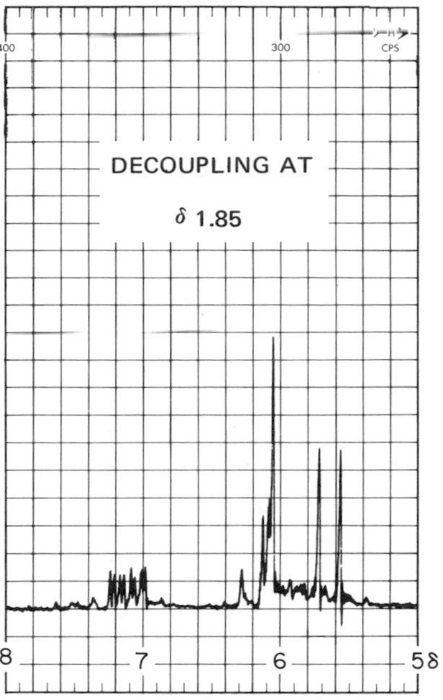

compound number 7-8

The molecular ion (m/e 140) and $M+1$ and $M+2$ data dictate, after elimination of formulas containing an odd number of N atoms, consideration of the following formulas:

Formula	$M+1$	$M+2$
$C_7H_{12}N_2O$	8.56	0.52
$C_8H_{12}O_2$	8.92	0.75
$C_8H_{16}N_2$	9.66	0.42

The formula containing 2 oxygen atoms is chosen on the basis of the infrared spectrum: the strong absorption at 1709 cm^{-1} (5.85 μm) and the multiple bands in the 1300-1160 cm^{-1} (7.69-8.62 μm) region are indicative of C=O and C–C(=O)–O stretching, respectively. We therefore anticipate an ester of molecular formula $C_8H_{12}O_2$. An ethyl ester is suggested by the ^{1}H NMR triplet at δ 1.25 (3 protons) and the quartet at δ 4.10 (2 protons); the chemical shift of the latter signal is consistent with a methylene group of $RC(=O)OCH_2CH_3$. The R group, by difference from the molecular formula, is C_5H_7- and contains two sites of unsaturation, at least one of which must be due to a double bond in view of the olefinic protons, δ 5.0-6.5, in the NMR spectrum. It is also clear that a double bond is in conjugation with the carbonyl group since the C=O stretch cited above is at a lower wavenumber (higher wavelength) than for normal, unconjugated, aliphatic esters (1730-1715 cm^{-1}, 5.71-5.76 μm). The UV spectrum also supports a highly conjugated structure. A C=C–C=C system is delineated by the coupled (respectively, symmetric and asymmetric) C=C stretching vibrations at 1645 and 1634 cm^{-1} (6.08 and 6.22 μm). Since there is no other reason for low-field protons in our structure (e.g., no indication of aromaticity), the signal at δ 7.10 is also an olefinic proton of unusually low field position. Integration shows a total of four olefinic protons. The remaining NMR absorption (δ 1.85) is quite consistent with a methyl group (integration intensity) attached to a double bond (see chemical shift tables) bearing a single proton (hence, the doublet). These data can be satisfied only by placing the methyl group at the end of the conjugated sequence (remote to the ester grouping), thus:

$$CH_3-CH=CH-CH=CH-CO_2CH_2CH_3$$

Ethyl sorbate

This structure is supported by data from Tables XIV-XVI in Chapter 6:

	nm
$R-C=CCO_2R$ basic unit	208
One C=C extending conjugation	+30
One δ substituent	+18
(λ_{max}^{EtOH}) calc =	256±5

The calculated value agrees very well with the observed 259 nm maximum.

The mass spectral fragmentation pattern is easily rationalized by the following cleavage patterns:

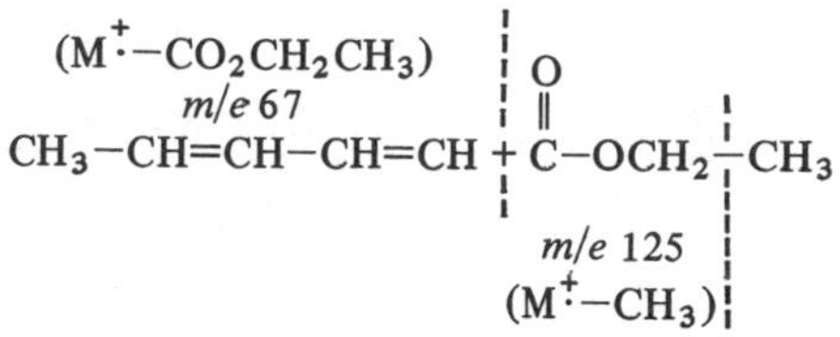

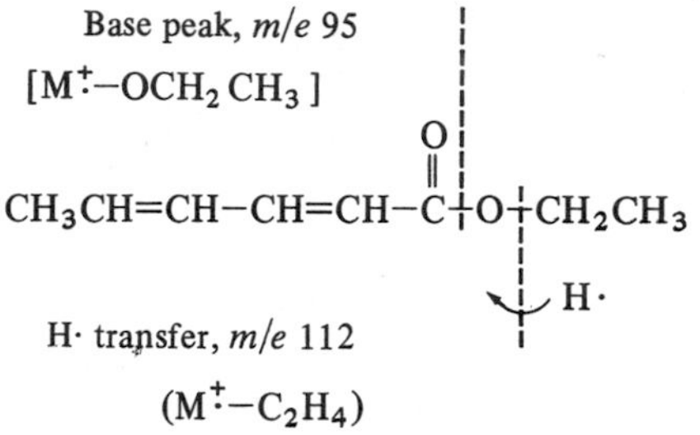

($M^{+\cdot}-CO_2CH_2CH_3$) *m/e* 67

$CH_3-CH{=}CH-CH{=}CH \,|\, \overset{O}{\overset{\|}{C}}-OCH_2 \,|\, CH_3$

m/e 125 ($M^{+\cdot}-CH_3$)

Base peak, *m/e* 95 [$M^{+\cdot}-OCH_2CH_3$]

$CH_3CH{=}CH-CH{=}CH-\overset{O}{\overset{\|}{C}} \,|\, O \,|\, CH_2CH_3$

H· transfer, *m/e* 112 ($M^{+\cdot}-C_2H_4$)

The infrared absorption at 1000 cm^{-1} (10.00 μm), (C—H bend) is evidence that the configuration about the C=C double bonds is *trans-trans* (see footnote on Table III, Appendix D, Chapter 3).

The NMR spectrum possesses some unusual features that can be interpreted only after spin decoupling experiments:

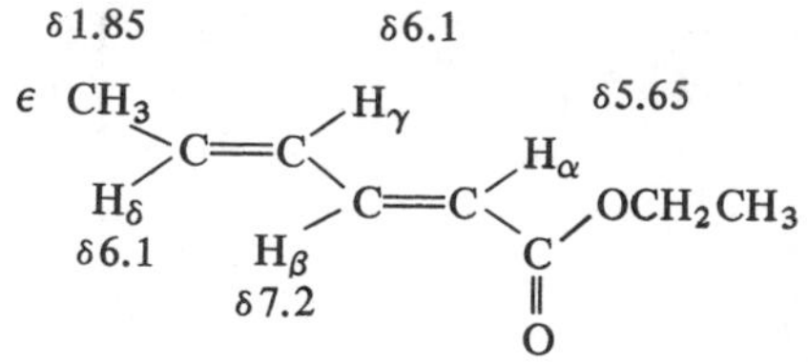

ϵ CH_3 (δ 1.85), H_δ (δ 6.1), H_γ (δ 6.1), H_β (δ 7.2), H_α (δ 5.65), $C(=O)OCH_2CH_3$

Irradiation at δ 7.2 (H_β) removes $J_{\beta\delta}$ and $J_{\beta\alpha}$, simplifying the signal for H_α at about δ 5.65. Note that the *trans* assignment for the α,β double bond is reinforced by the vicinal *J* value of about 18 Hz observed for the H_α absorption. The absorptions for H_δ and H_γ at about δ 6.1 are also partially collapsed. The unusually low field of H_β, compared to usual olefinic protons, can be rationalized by the resonance form that shows a decreased electron density at the β-carbon:

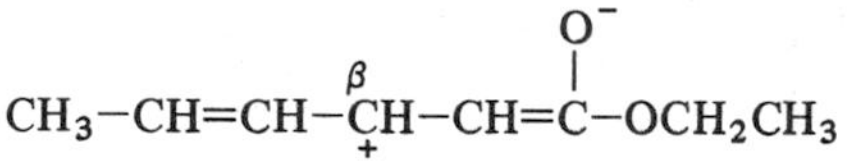

$$CH_3-CH{=}CH-\overset{\beta}{\underset{+}{C}}H-CH{=}\overset{O^-}{\overset{|}{C}}-OCH_2CH_3$$

Decoupling at δ 5.65 (H_α) results in partial collapse of the signal at δ 7.2, as does decoupling at δ 1.85 (H_ϵ).

eight

problems

Problems Involving Mass Spectra, IR, ^{1}H NMR, and UV Only: 8-1 to 8-5, 8-21 to 8-30, 8-36, 8-38 to 8-44, 8-47, and 8-49.

Problems Involving the Spectra Listed Above and Chemical Information: 8-6 to 8-10.

Problems Involving the Spectra Listed Above and ^{13}C-NMR: 8-11 to 8-20, 8-31 to 8-35, 8-37, 8-45, 8-46, 8-48 (Prob. 8-18, 8-19 and 8-46 have chemical information).

Beilstein references have been provided for the following twenty-seven problems: 8-1 to 8-4, 8-9, 8-13 to 8-18, 8-20 to 8-22, 8-24, 8-25, 8-27, 8-30 to 8-35, 8-39 to 8-41, and 8-49.

Infrared Spectrum *Compound 8-1* Beilstein Ref. 7,304

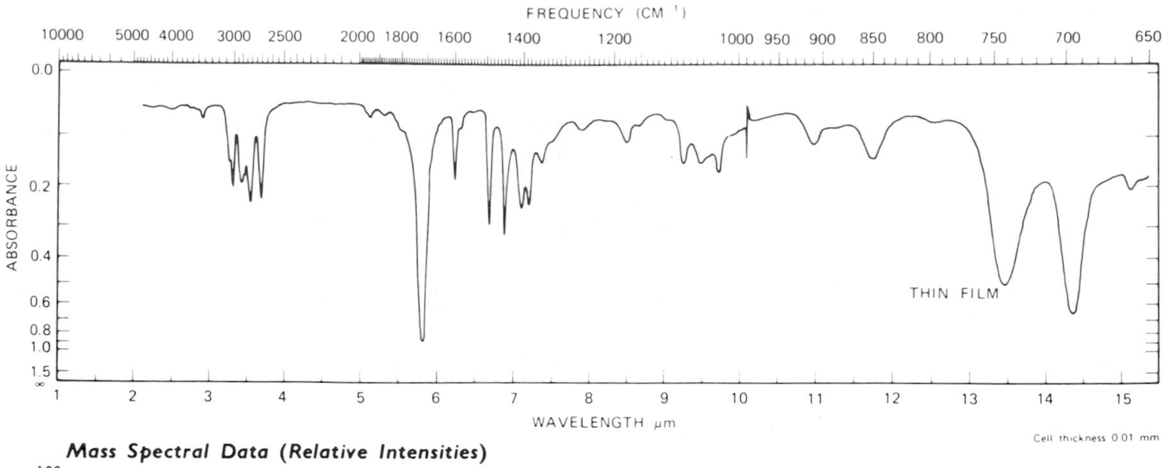

FREQUENCY (CM⁻¹)
ABSORBANCE
THIN FILM
WAVELENGTH μm
Cell thickness 0.01 mm

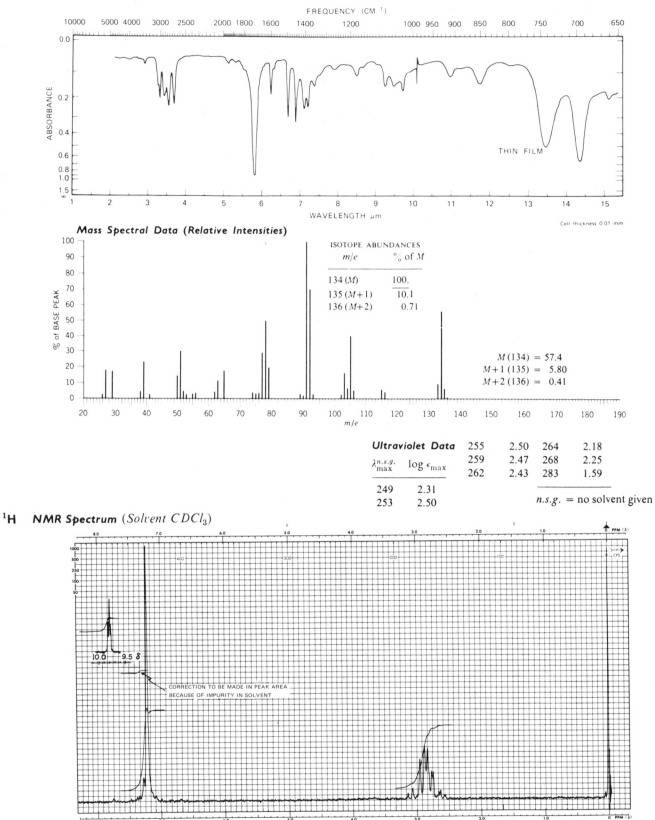

Mass Spectral Data (Relative Intensities)

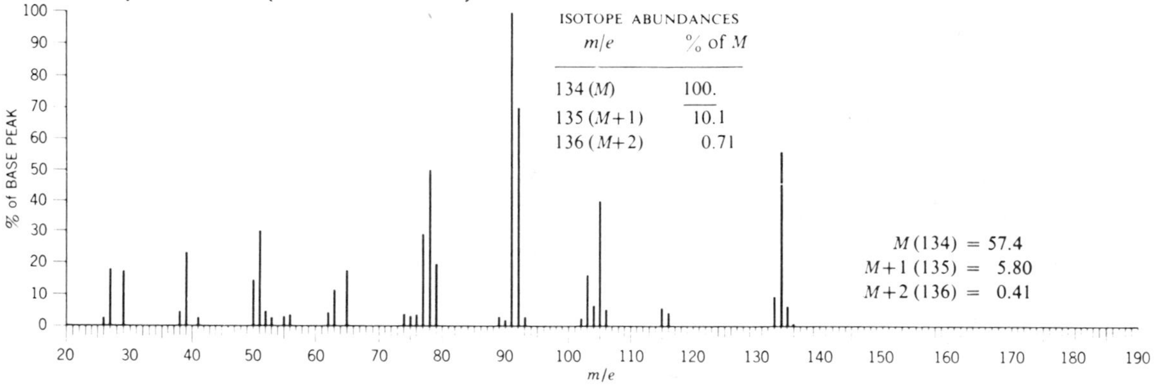

% of BASE PEAK
m/e

ISOTOPE ABUNDANCES

m/e	% of M
134 (M)	100.
135 (M+1)	10.1
136 (M+2)	0.71

M (134) = 57.4
M+1 (135) = 5.80
M+2 (136) = 0.41

Ultraviolet Data

$\lambda_{max}^{n.s.g.}$	log ϵ_{max}				
249	2.31	255	2.50	264	2.18
253	2.50	259	2.47	268	2.25
		262	2.43	283	1.59

n.s.g. = no solvent given

^{1}H ***NMR Spectrum*** (*Solvent* $CDCl_3$)

PPM (δ)
CPS
10.0
9.5 δ
CORRECTION TO BE MADE IN PEAK AREA
BECAUSE OF IMPURITY IN SOLVENT

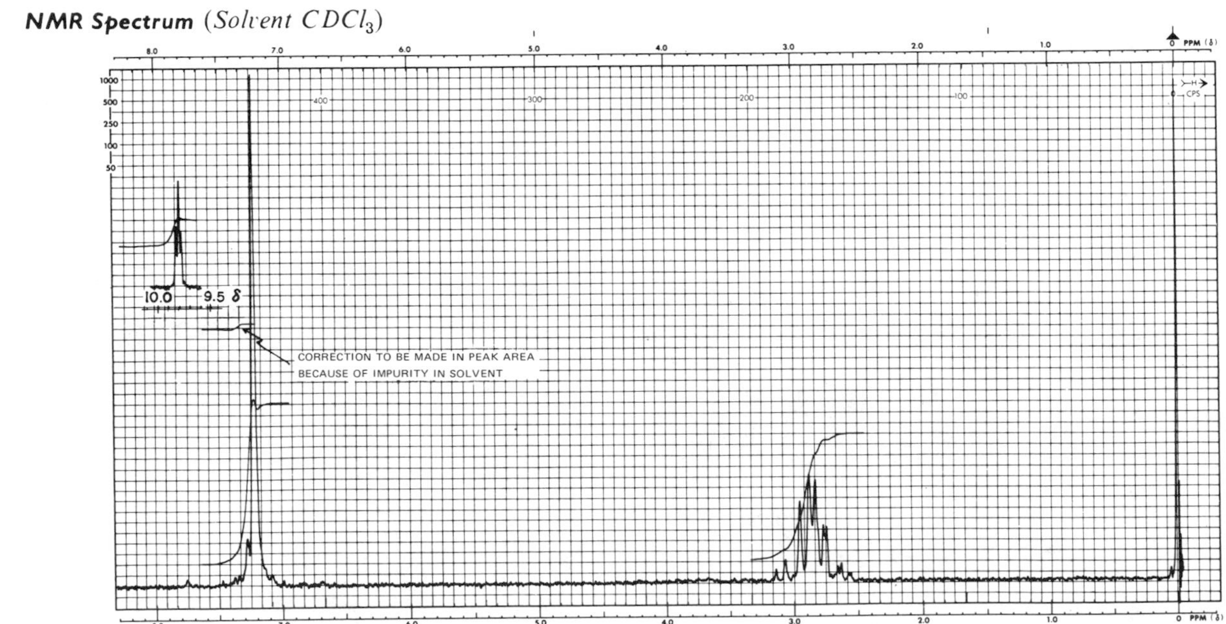

Infrared Spectrum *Compound 8-2*

Beilstein Ref. 5, 436
Capillary Cell: Neat

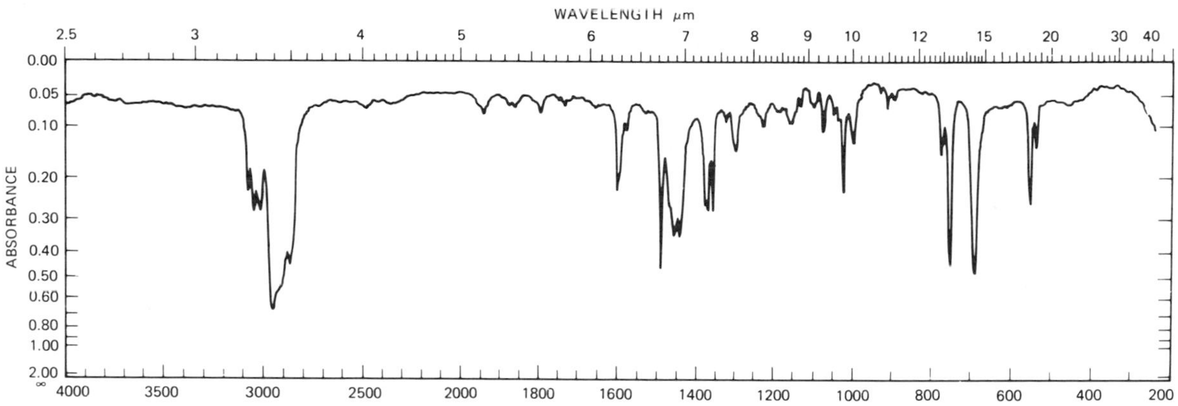

WAVELENGTH μm

ABSORBANCE

FREQUENCY (CM^{-1})

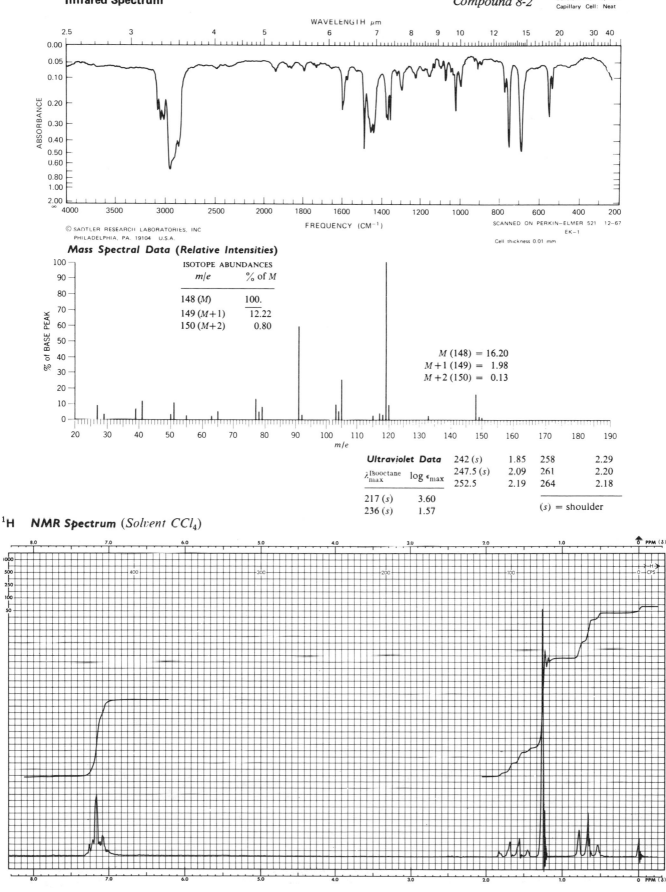

SCANNED ON PERKIN-ELMER 521 12-67
EK-1

Cell thickness 0.01 mm

Mass Spectral Data (Relative Intensities)

ISOTOPE ABUNDANCES

m/e	% of *M*
148 (*M*)	100.
149 (*M*+1)	12.22
150 (*M*+2)	0.80

M (148) = 16.20
M+1 (149) = 1.98
M+2 (150) = 0.13

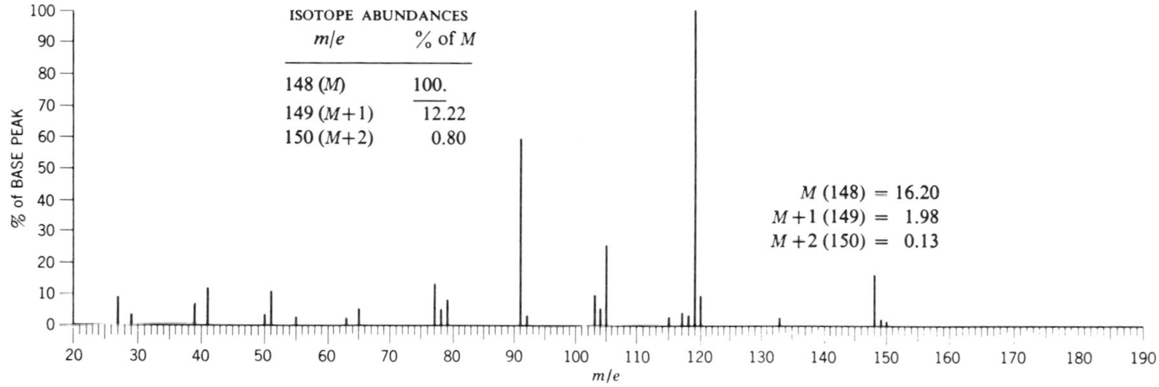

% of BASE PEAK

m/e

Ultraviolet Data

$\lambda_{max}^{isooctane}$	$\log \epsilon_{max}$
217 (*s*)	3.60
236 (*s*)	1.57
242 (*s*)	1.85
247.5 (*s*)	2.09
252.5	2.19
258	2.29
261	2.20
264	2.18

(*s*) = shoulder

^{1}H NMR Spectrum (*Solvent* CCl_4)

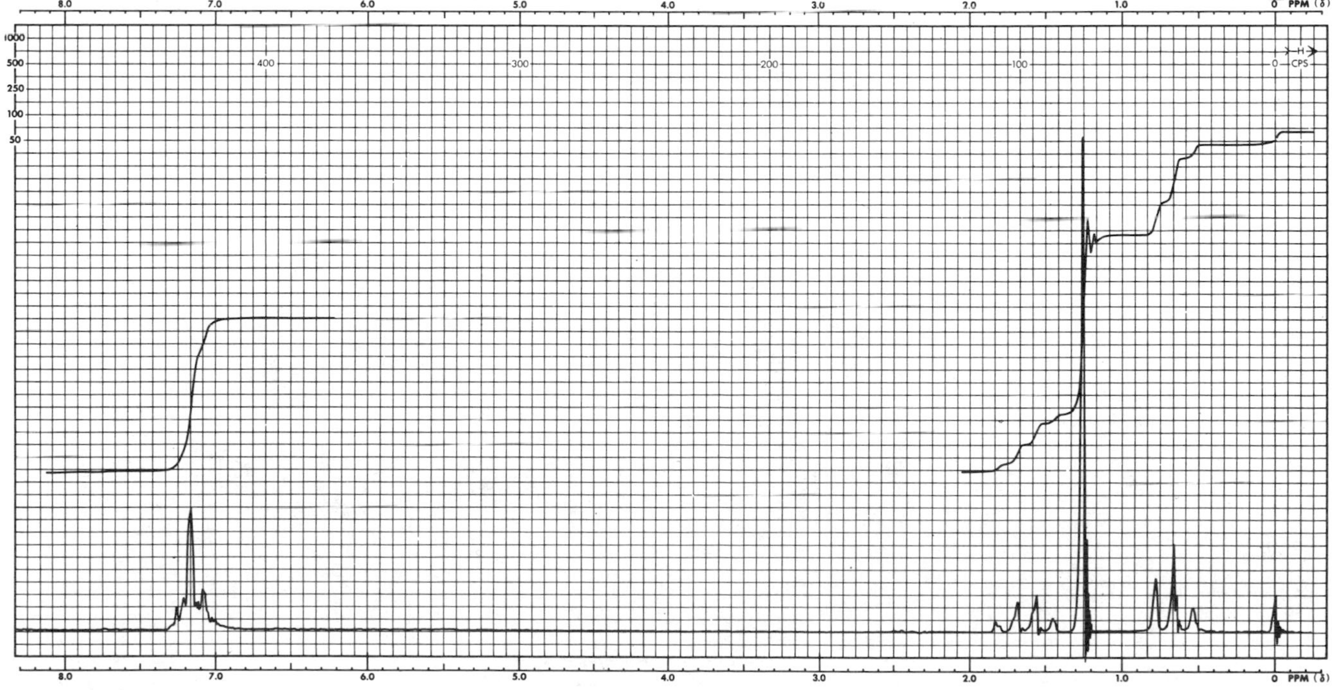

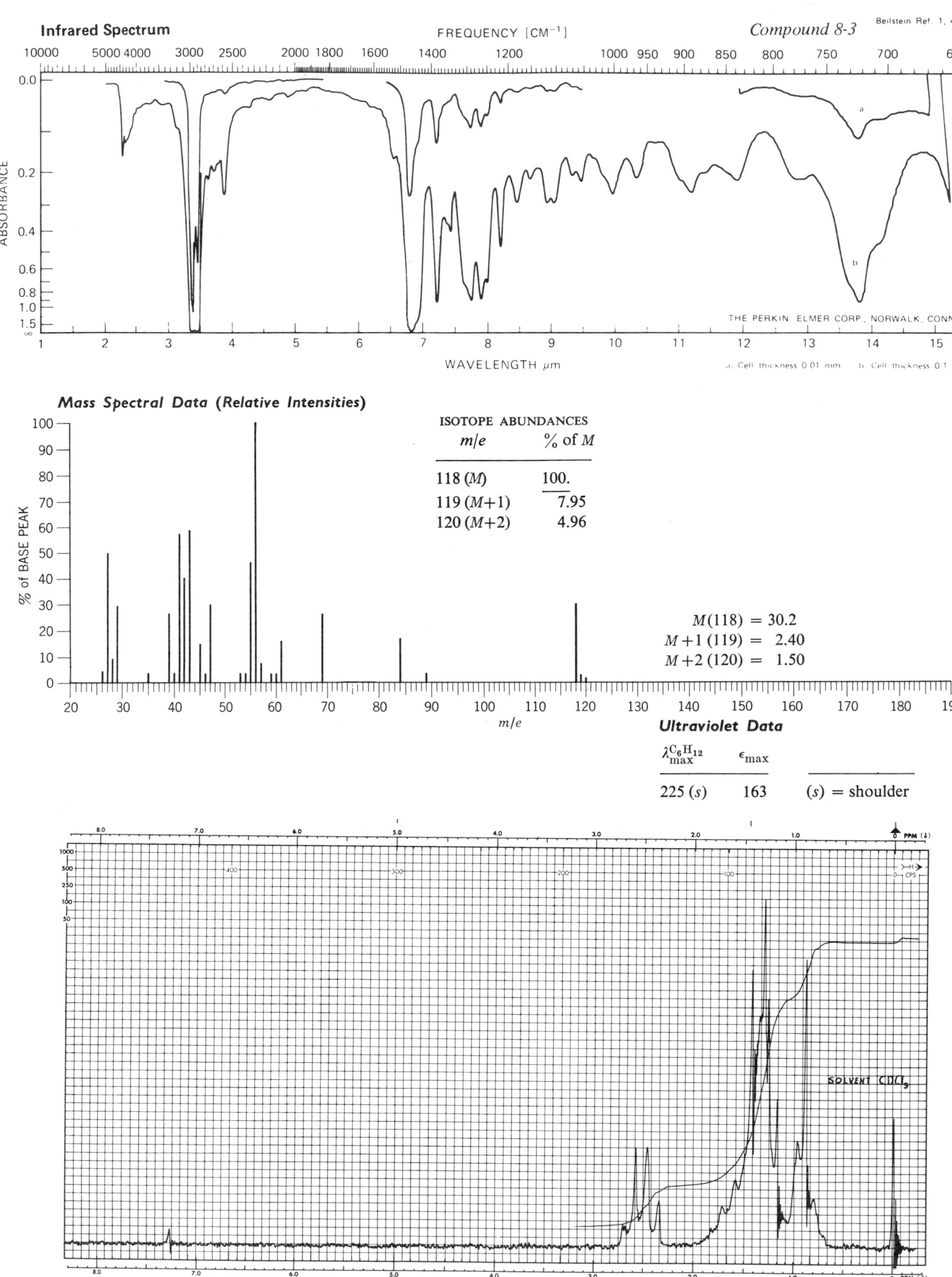

ISOTOPE ABUNDANCES

m/e	% of M
118 (M)	100.
119 (M+1)	7.95
120 (M+2)	4.96

$\lambda_{max}^{C_6H_{12}}$	ϵ_{max}
225 (s)	163

(s) = shoulder

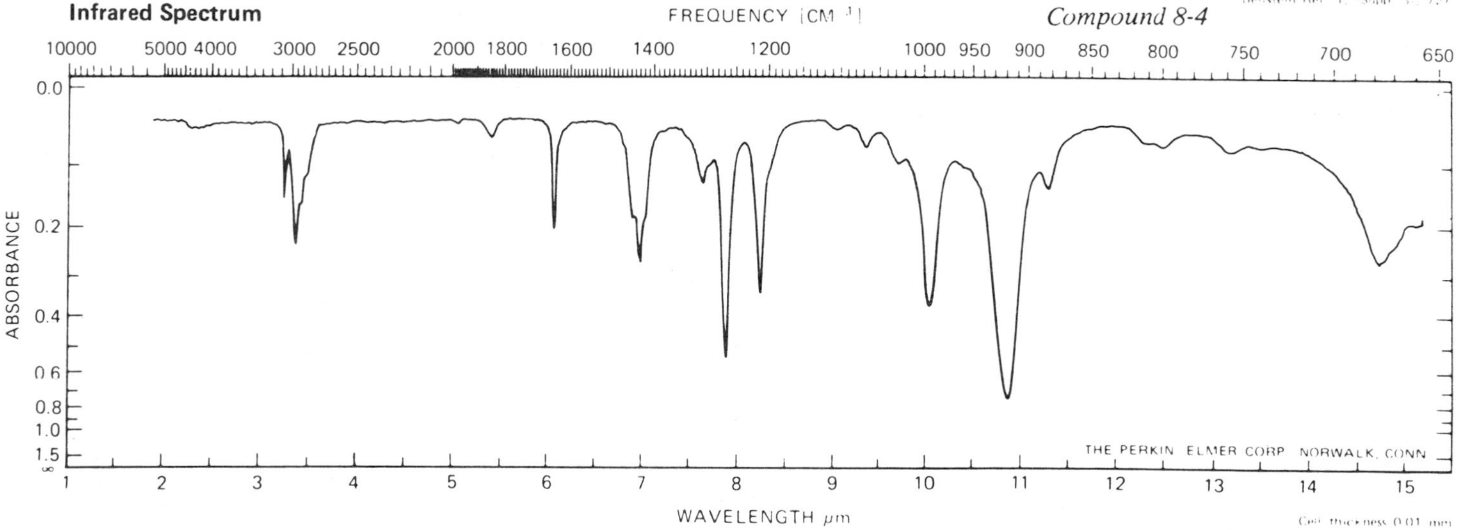

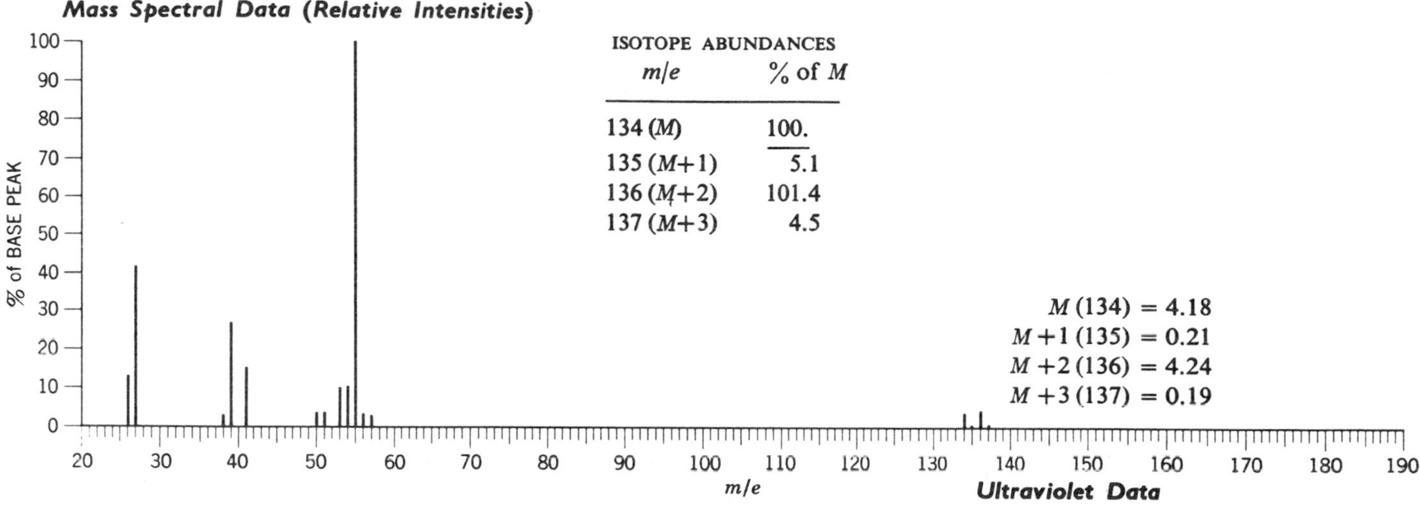

ISOTOPE ABUNDANCES

m/e	% of M
134 (M)	100.
135 (M+1)	5.1
136 (M+2)	101.4
137 (M+3)	4.5

M (134) = 4.18
M+1 (135) = 0.21
M+2 (136) = 4.24
M+3 (137) = 0.19

Ultraviolet Data

Featureless above 210 nm

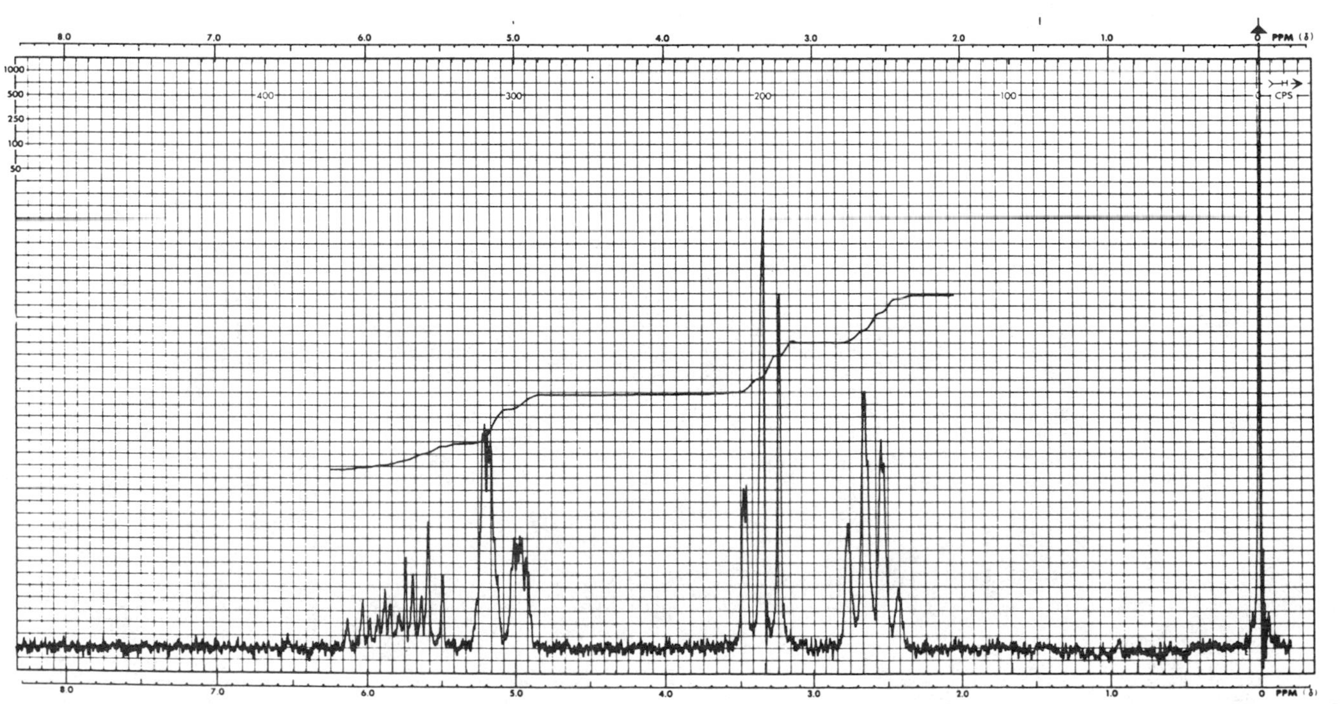

Compound 8-5

Infrared Spectrum

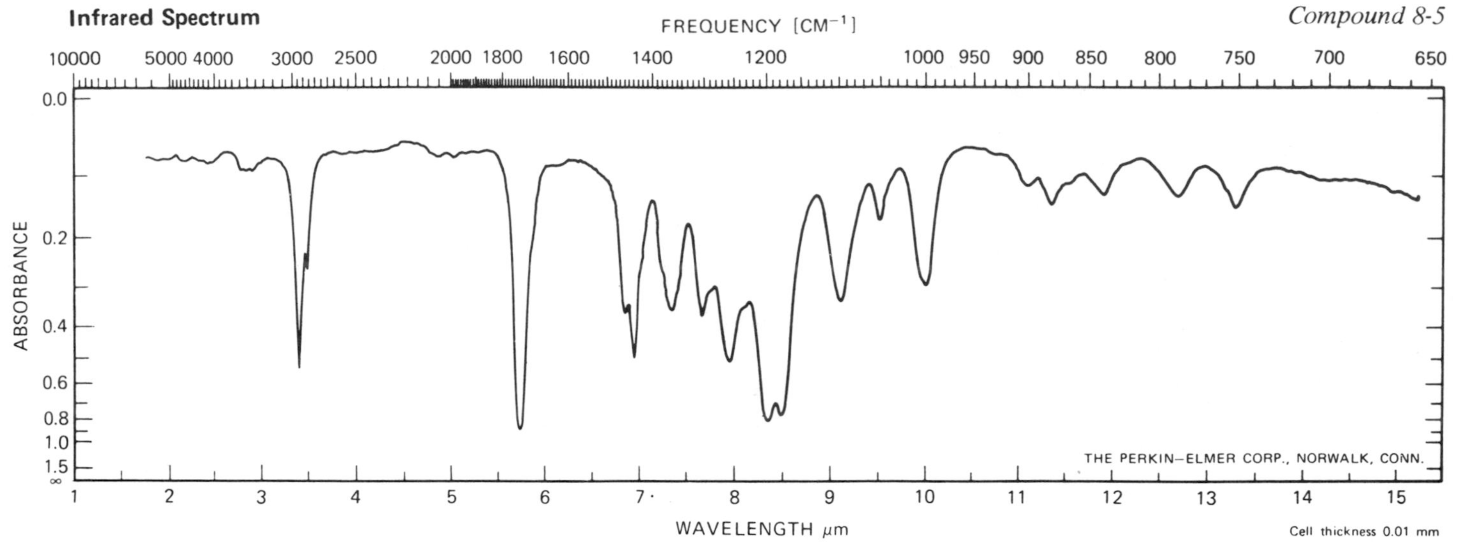

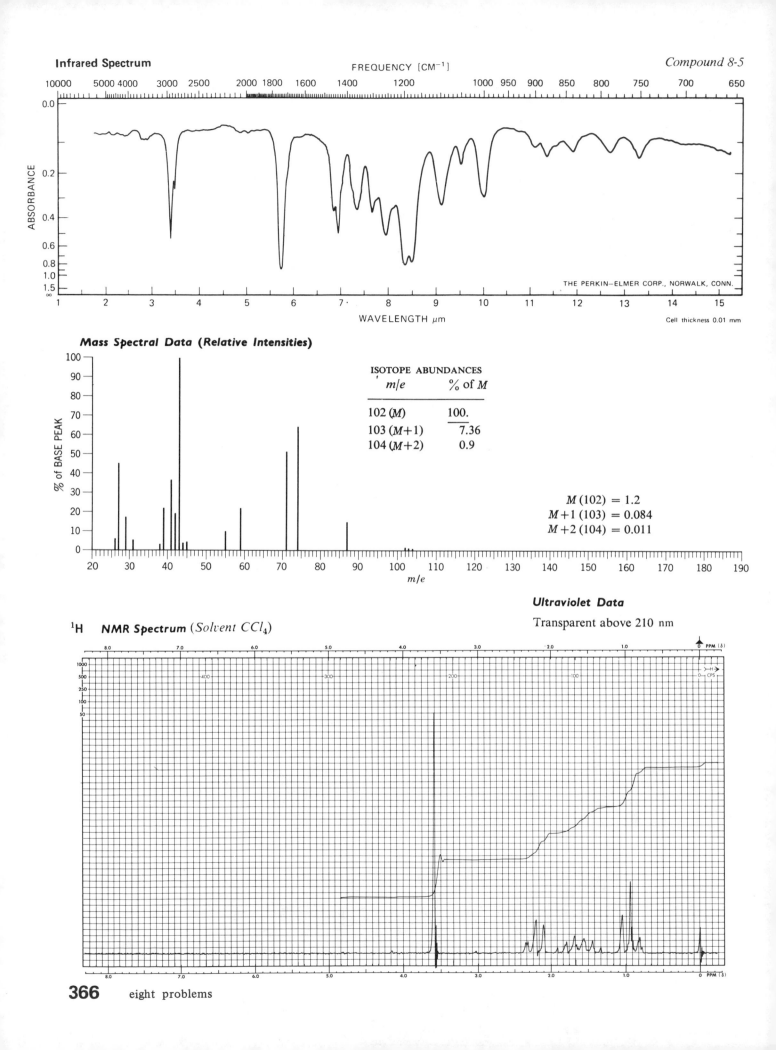

Mass Spectral Data (Relative Intensities)

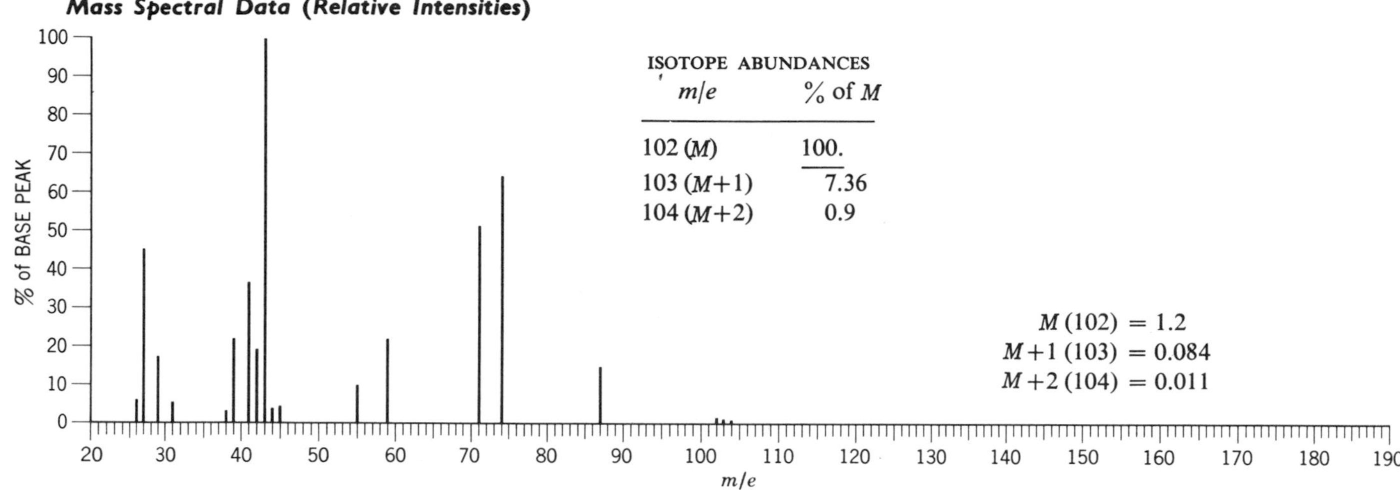

ISOTOPE ABUNDANCES

m/e	% of *M*
102 (*M*)	100.
103 (*M*+1)	7.36
104 (*M*+2)	0.9

M (102) = 1.2
$M+1$ (103) = 0.084
$M+2$ (104) = 0.011

Ultraviolet Data

Transparent above 210 nm

^{1}H **NMR Spectrum** (*Solvent* CCl_4)

ROAD MAP PROBLEM (COMPOUNDS 8-6 TO 8-10)

Determine the structure of compound 8-6 from the spectra provided. Treatment of isopropyl alcohol with NaH followed by addition of carbon disulfide results in the formation of an isolable sodium salt; treatment of compound 8-6 with this salt, in acetone solution, results in compound 8-7. Identify 8-7 from the spectra provided. Compound 8-7 was pyrolyzed at 160° for 7 hours to give compound 8-8; identify 8-8 from the spectra provided. Treatment of 8-8 with dicyanodichloroquinone (DDQ) in benzene for 20 minutes at room temperature resulted in compound 8-9, and treatment of compound 8-8 with refluxing glacial acetic acid containing hydrogen peroxide resulted in compound 8-10. Identify compounds 8-9 and 8-10 from the spectra.

Reaction Scheme

$$(CH_3)_2CHOH \xrightarrow[\text{2) } CS_2 \atop \text{3) 8-6/acetone}]{\text{1) NaH}} \text{8-7} \xrightarrow[\text{7 hours}]{160^\circ} \text{8-8}$$

$$\text{8-8} \xrightarrow[\text{benzene}]{\text{DDQ}} \text{8-9}$$

$$\text{8-8} \xrightarrow[\Delta/CH_3CO_2H]{CH_3CO_3H} \text{8-10}$$

Infrared Spectrum

Compound 8-6

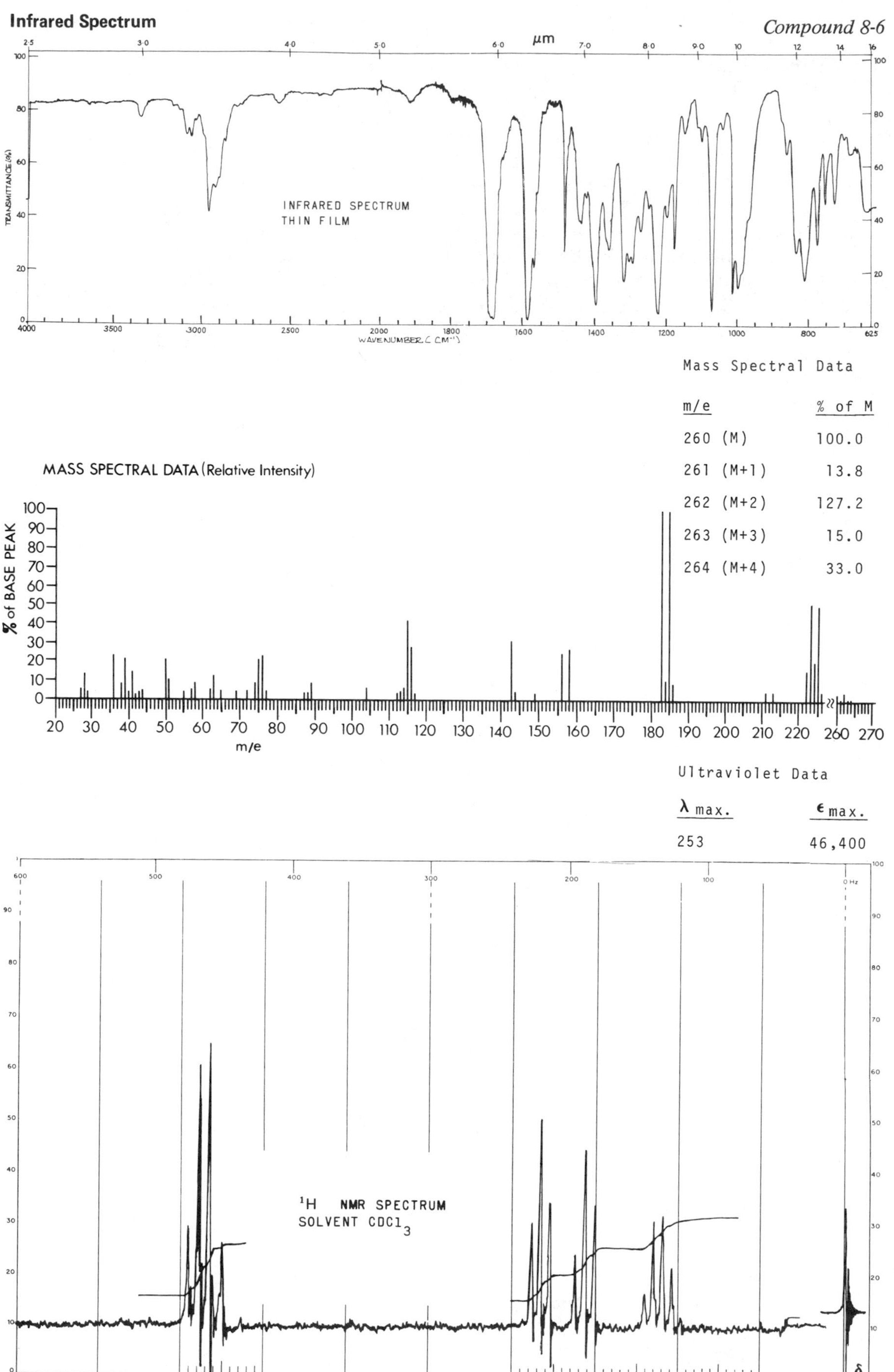

Mass Spectral Data

m/e	% of M
260 (M)	100.0
261 (M+1)	13.8
262 (M+2)	127.2
263 (M+3)	15.0
264 (M+4)	33.0

Ultraviolet Data

λ max.	ε max.
253	46,400

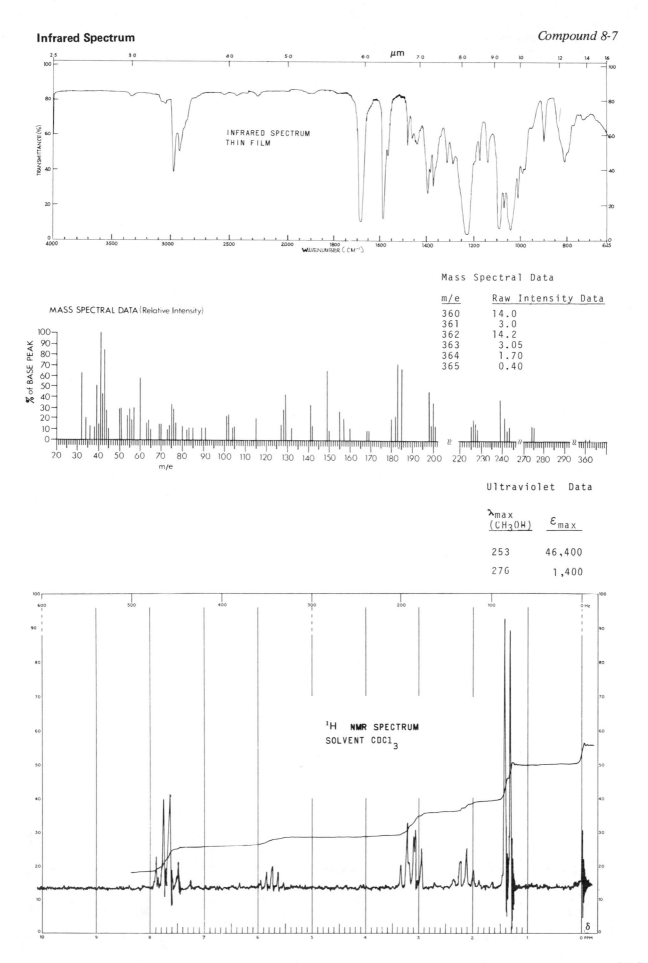

Infrared Spectrum

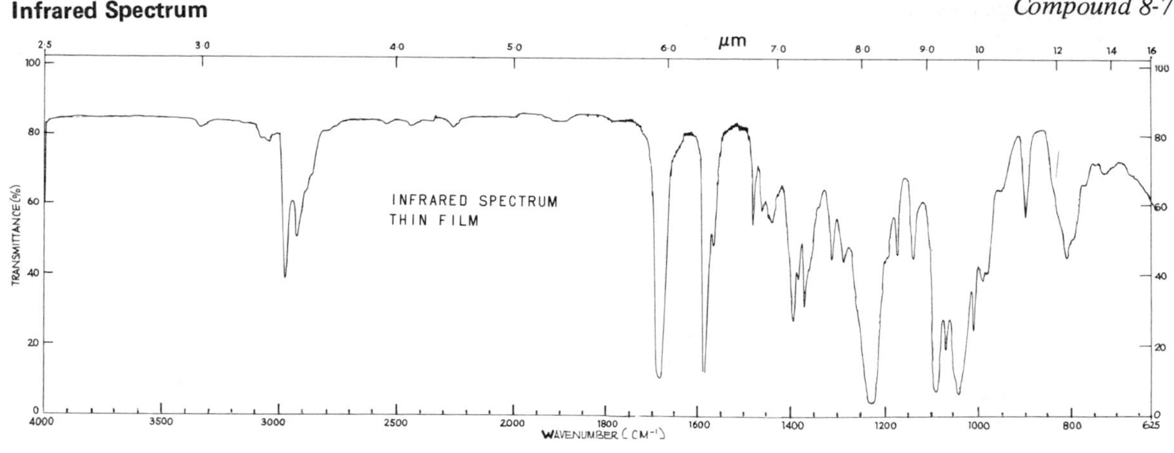

MASS SPECTRAL DATA (Relative Intensity)

% of BASE PEAK

m/e

Mass Spectral Data

m/e	Raw Intensity Data
360	14.0
361	3.0
362	14.2
363	3.05
364	1.70
365	0.40

Ultraviolet Data

λ_{max} (CH_3OH)	ϵ_{max}
253	46,400
276	1,400

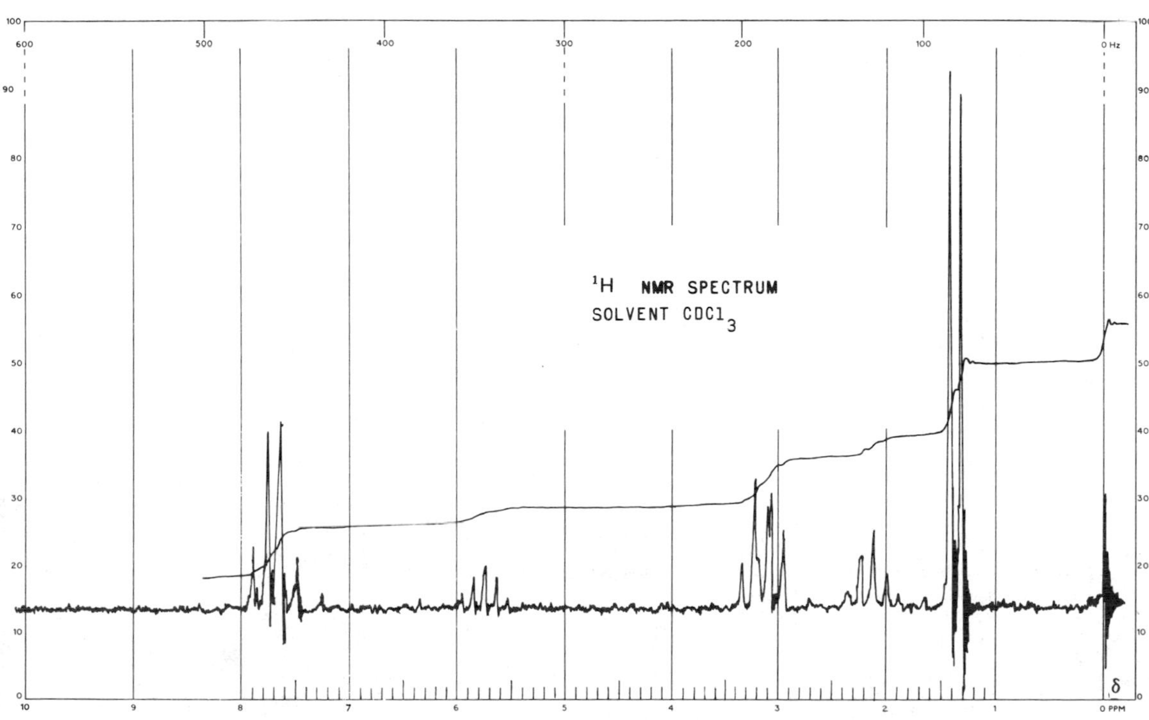

Infrared Spectrum

Compound 8-8

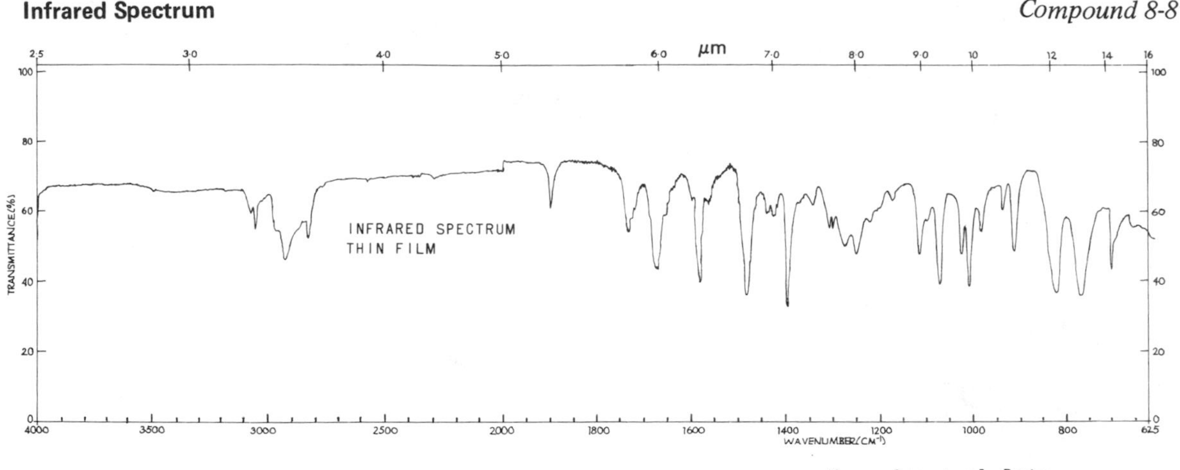

Mass Spectral Data

m/e	Raw Intensity Data
239	68.0
240	153.0
241	88.0
242	152.0
243	24.0
244	9.0

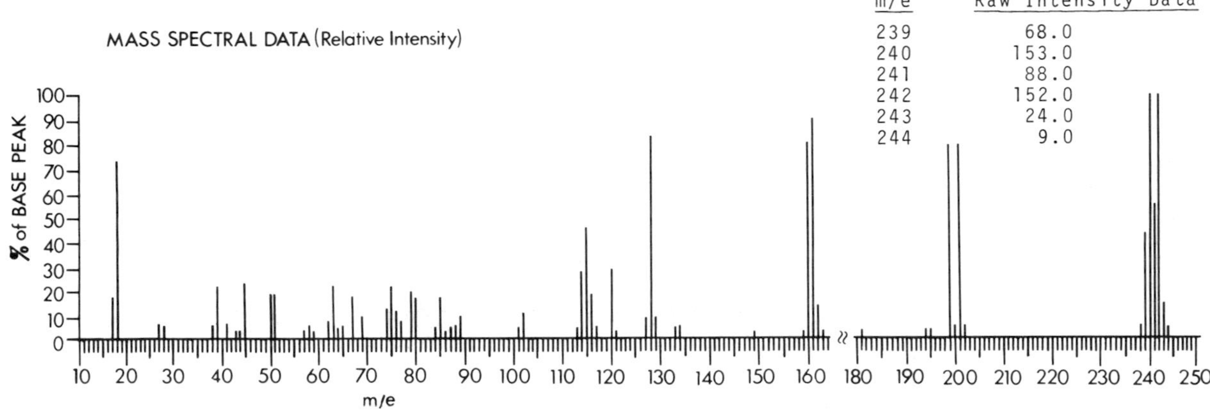

Ultraviolet Data

λ_{max} (CH_3OH)	ϵ_{max}
237	18,600
301	5,540

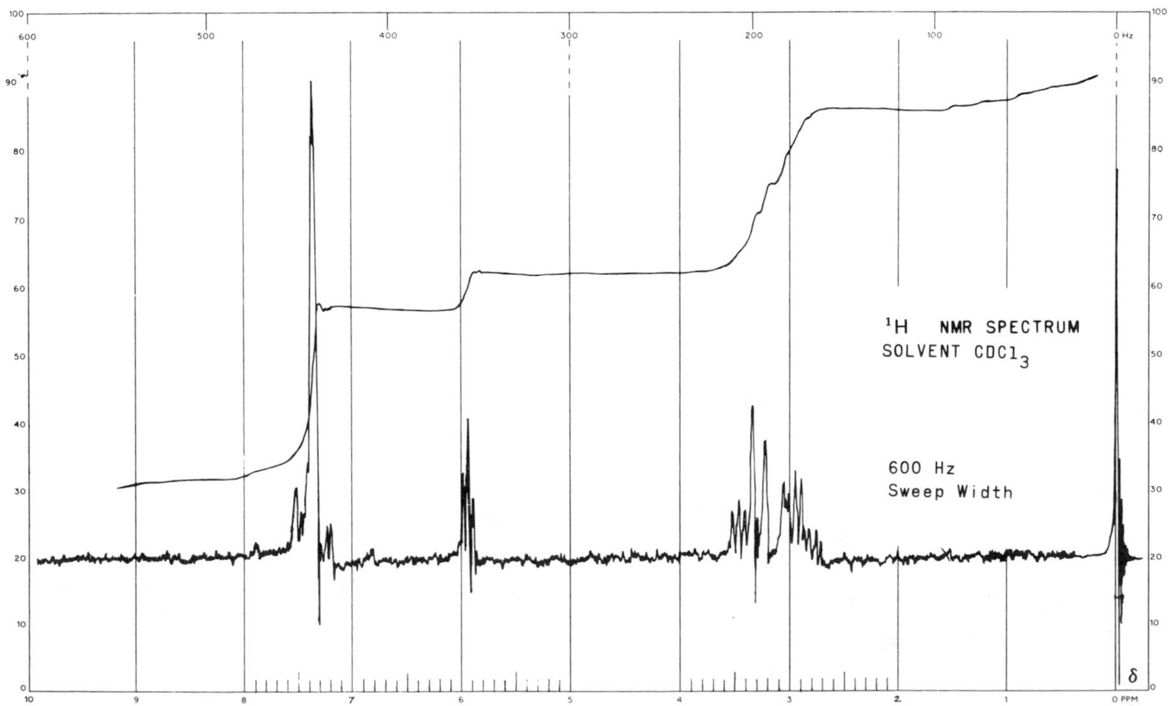

Infrared Spectrum *Compound 8-9* Beilstein Ref. 2nd Supplement, **17,** 64

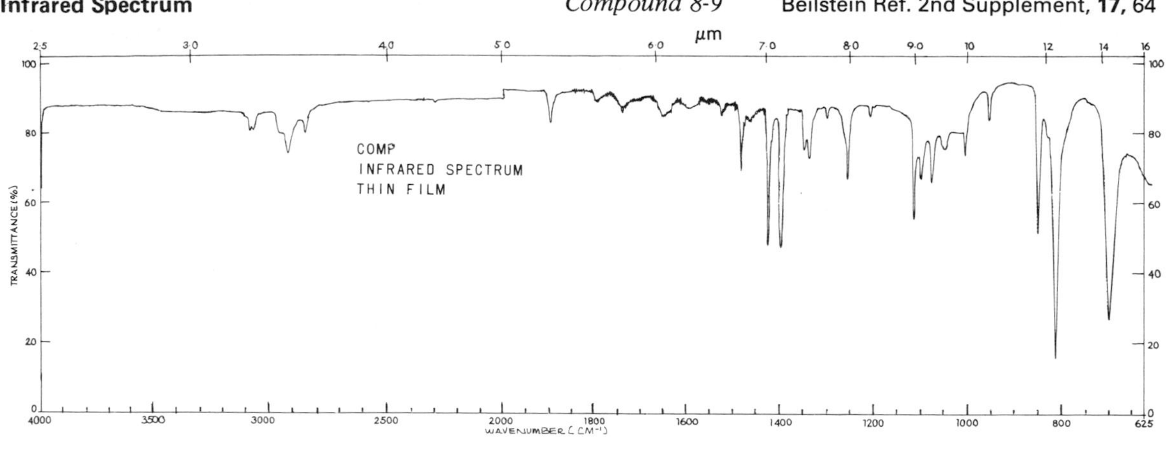

Mass Spectral Data

m/e	Raw Intensity Data
238	144
239	21
240	143
241	20
242	8

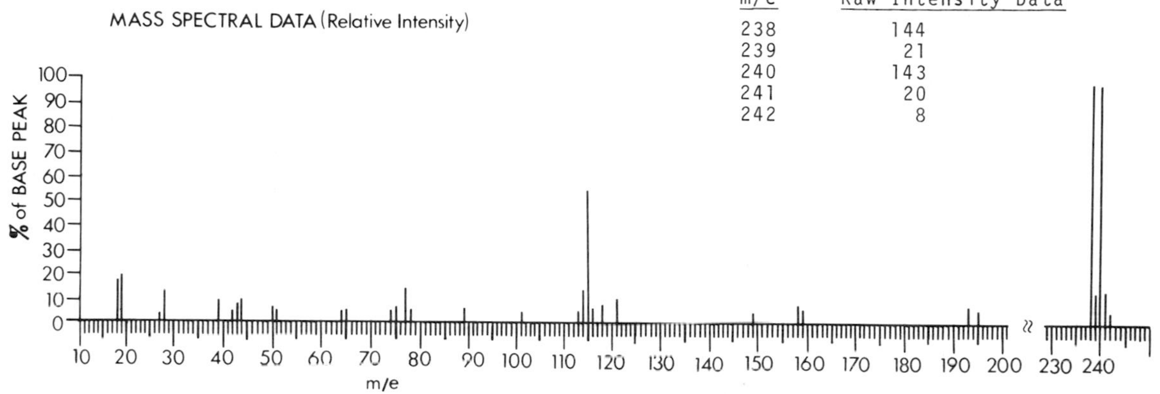

Ultraviolet Data

λ_{max} (CH_3OH)	ε_{max}
285	23,400

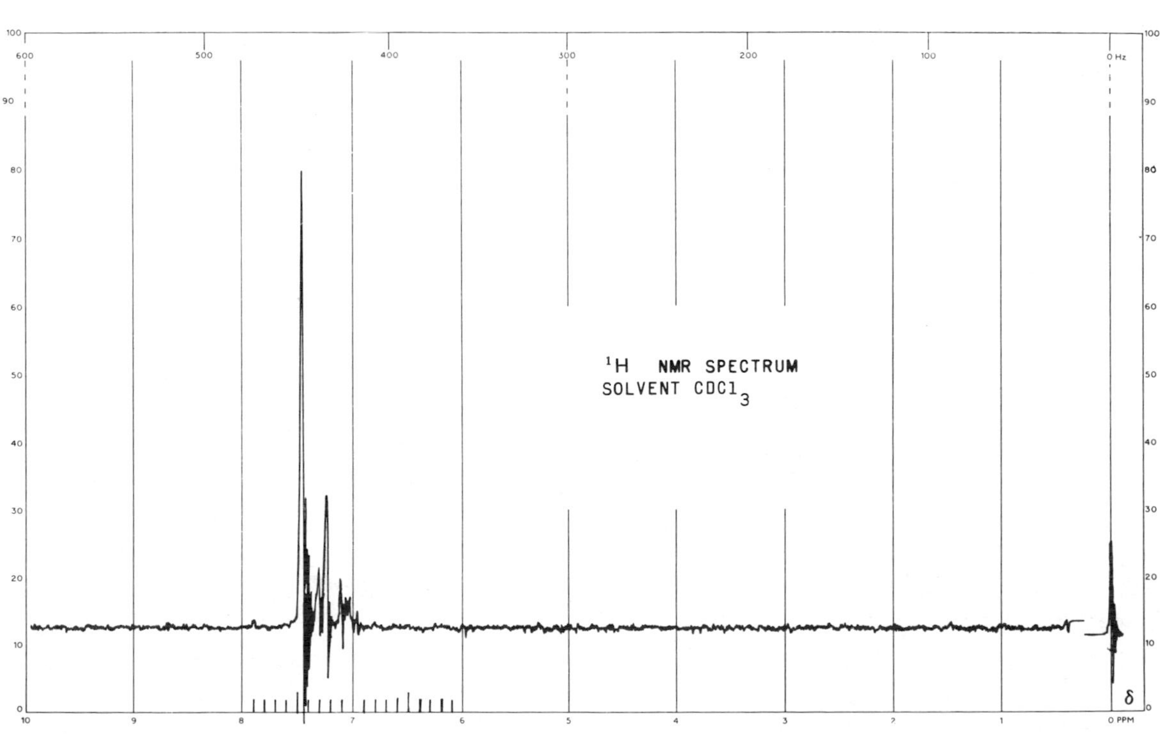

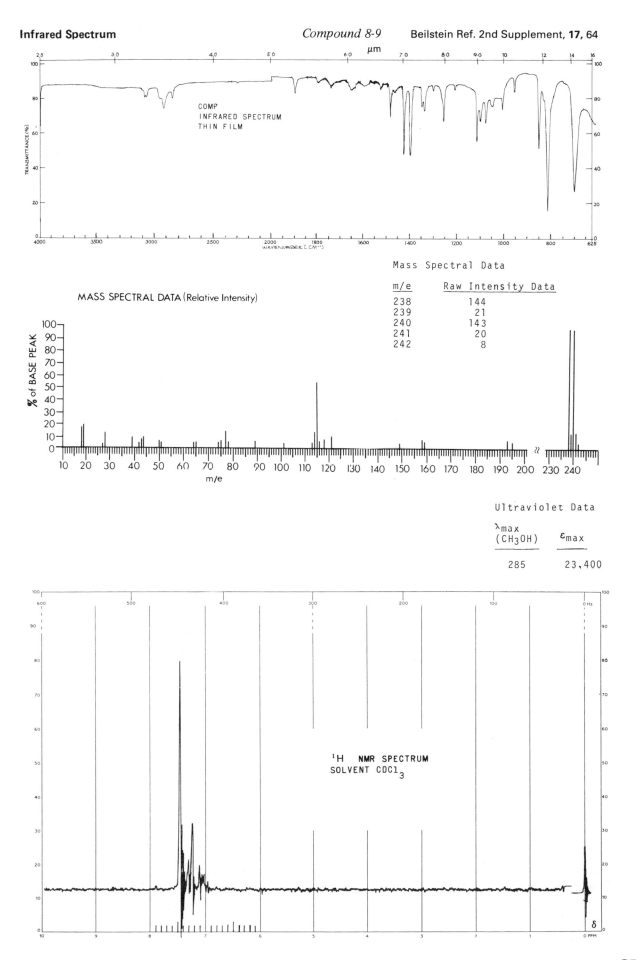

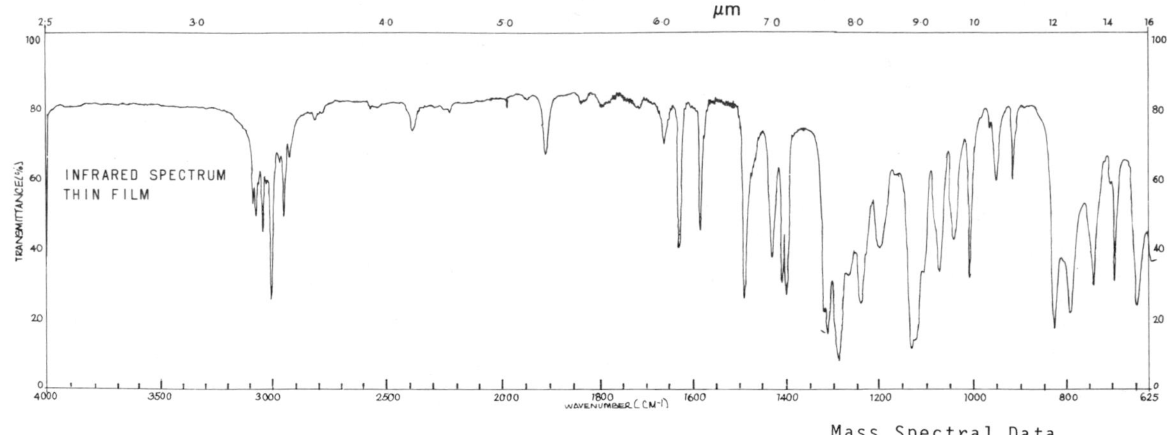

Mass Spectral Data

m/e	Raw Intensity Data
272	36.94
273	5.40
274	36.03
275	4.50
276	2.70

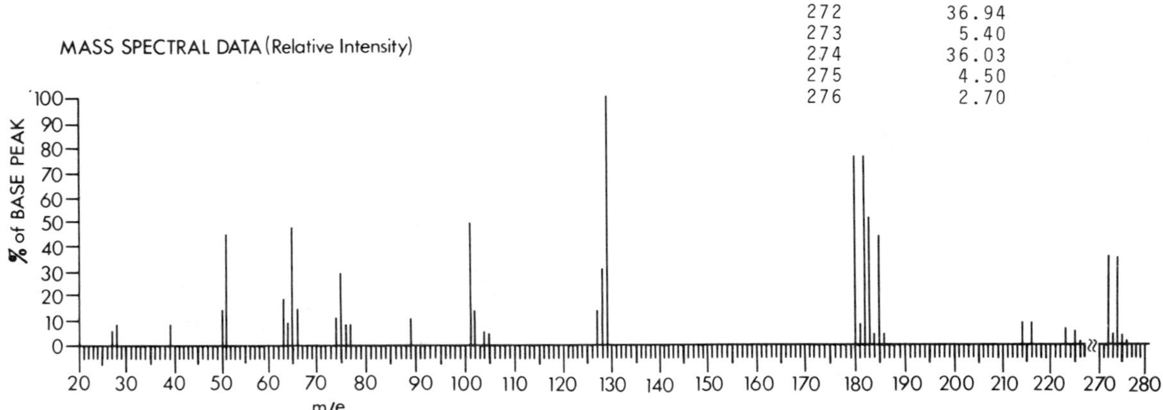

Ultraviolet Data

λ_{max} (CH_3OH)	ε_{max}
255	21,800

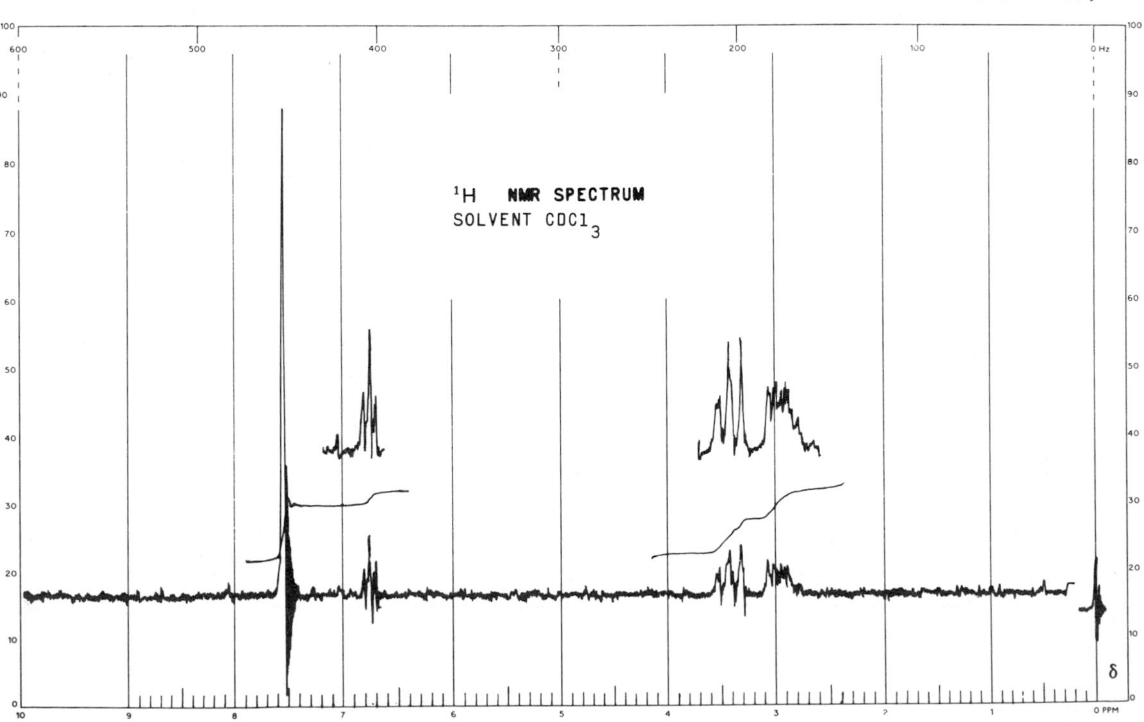

notes

Infrared Spectrum

Compound 8-11

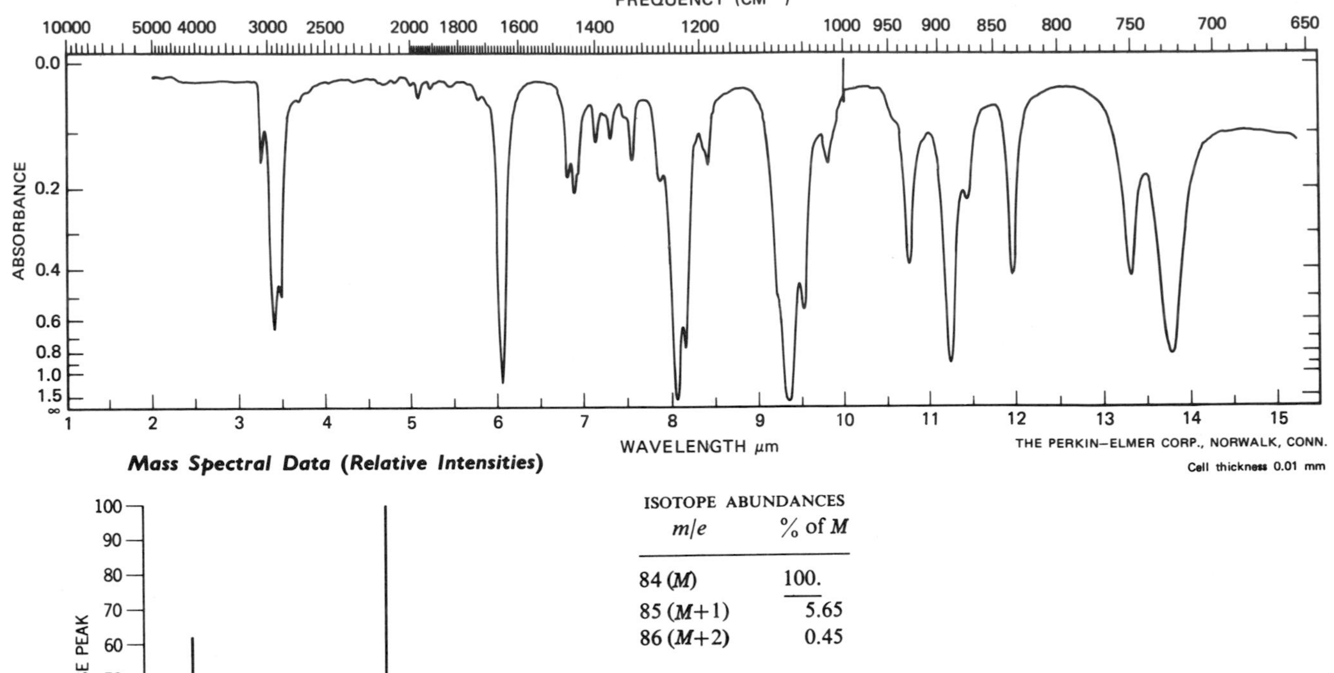

Cell thickness 0.01 mm

Mass Spectral Data (Relative Intensities)

% of BASE PEAK (0–100) vs. m/e (20–190)

ISOTOPE ABUNDANCES

m/e	% of M
84 (M)	100.
85 (M+1)	5.65
86 (M+2)	0.45

M (84) = 50.0
M+1 (85) = 2.83
M+2 (86) = 0.23

Ultraviolet Data

Transparent in near ultraviolet. (200–380 nm)

¹H NMR Spectrum (*Solvent CCl₄*)

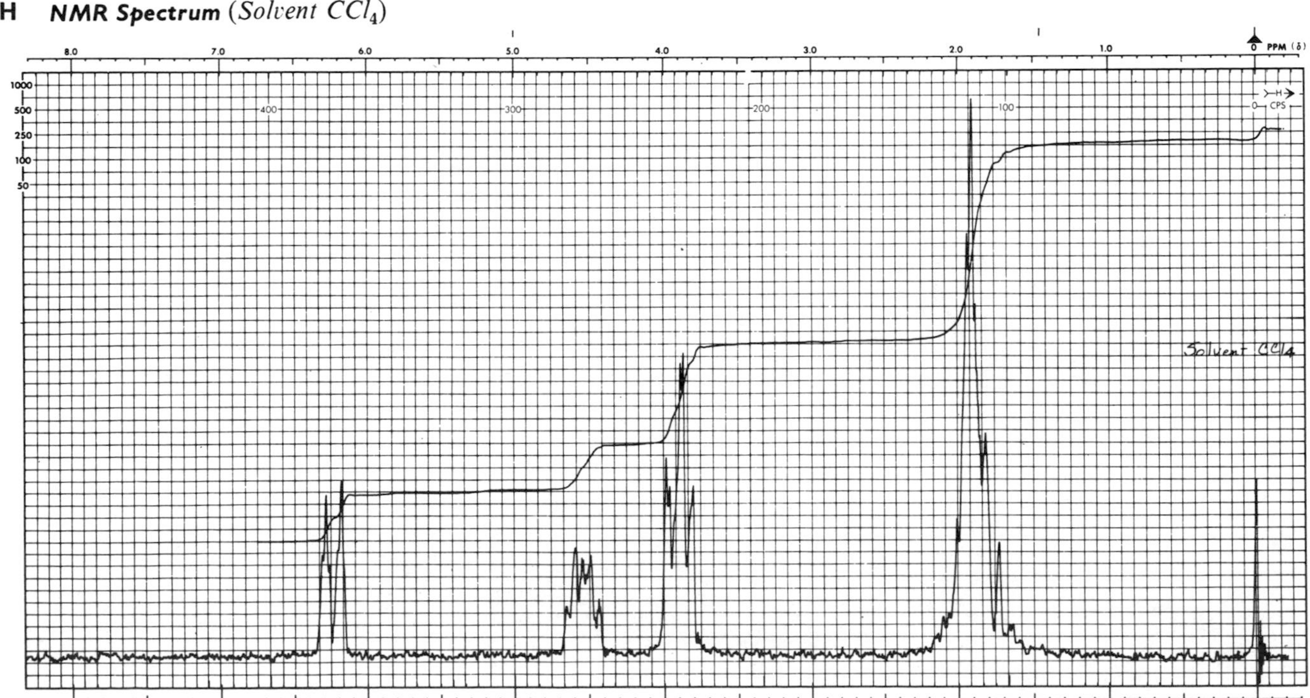

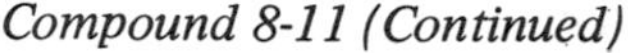

C–13 Spectrum Completely Decoupled

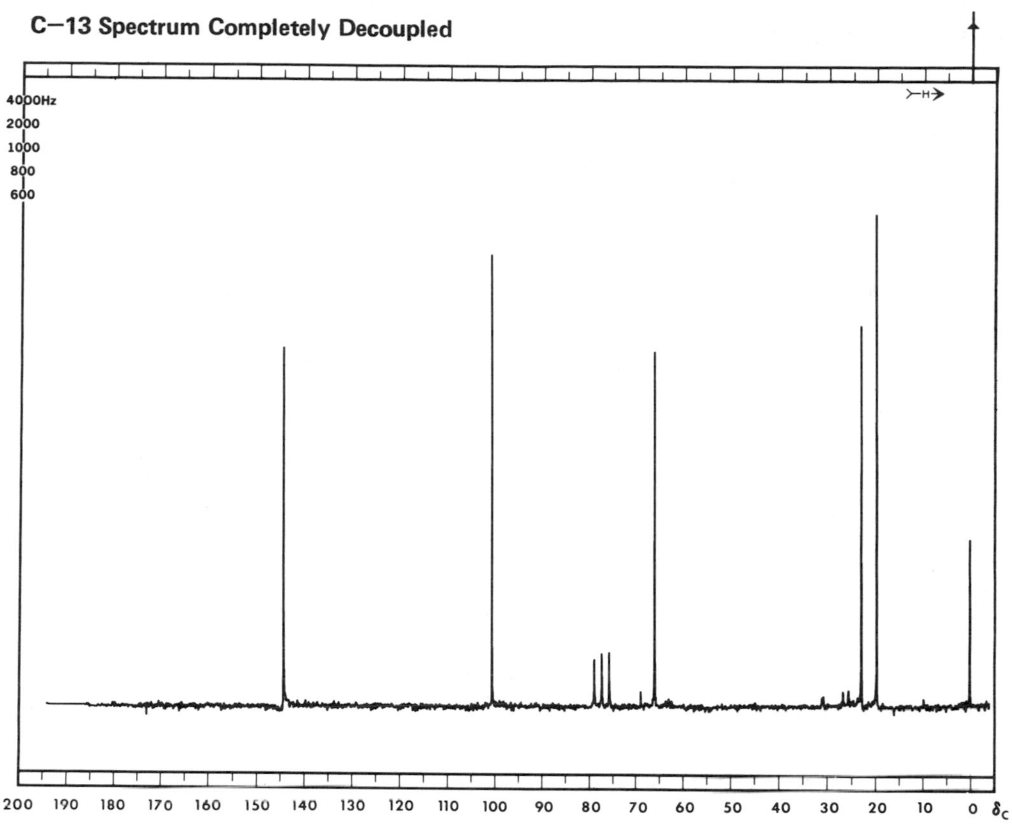

^{13}C NMR Spectrum Off-resonance Decoupled

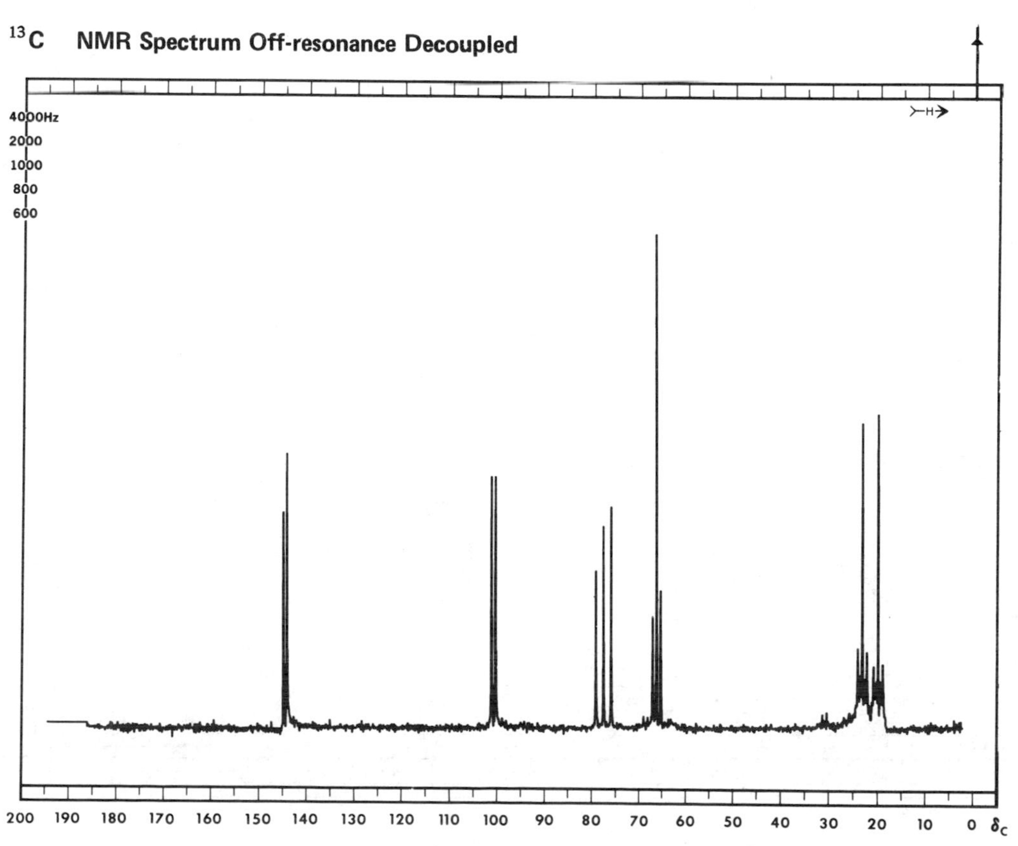

Infrared Spectrum

Compound 8-12

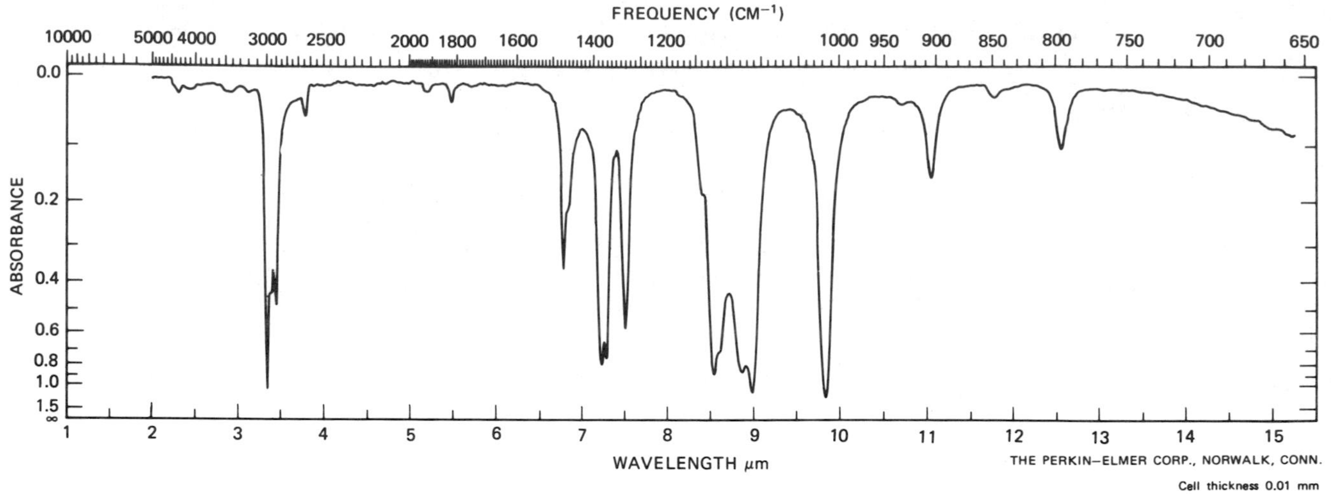

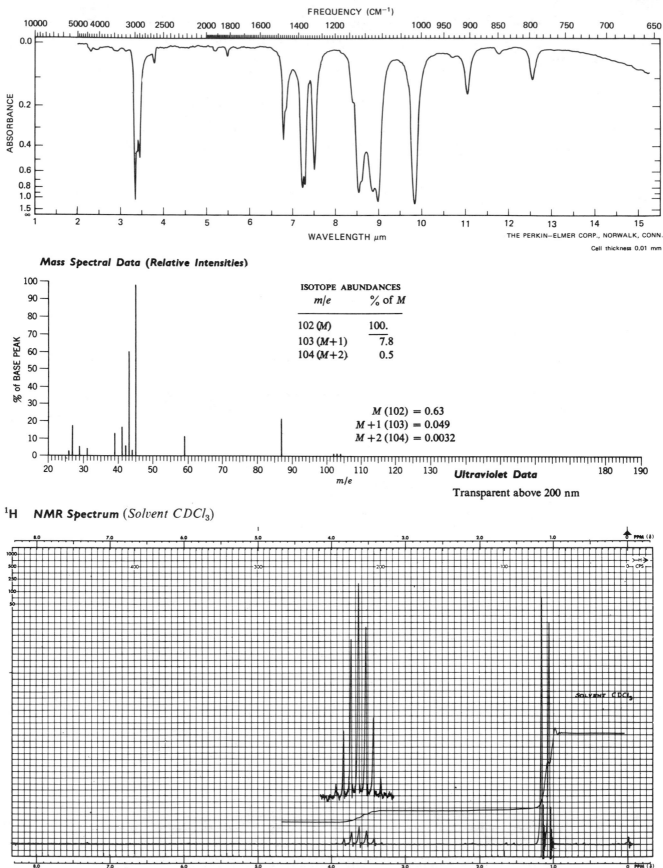

Mass Spectral Data (Relative Intensities)

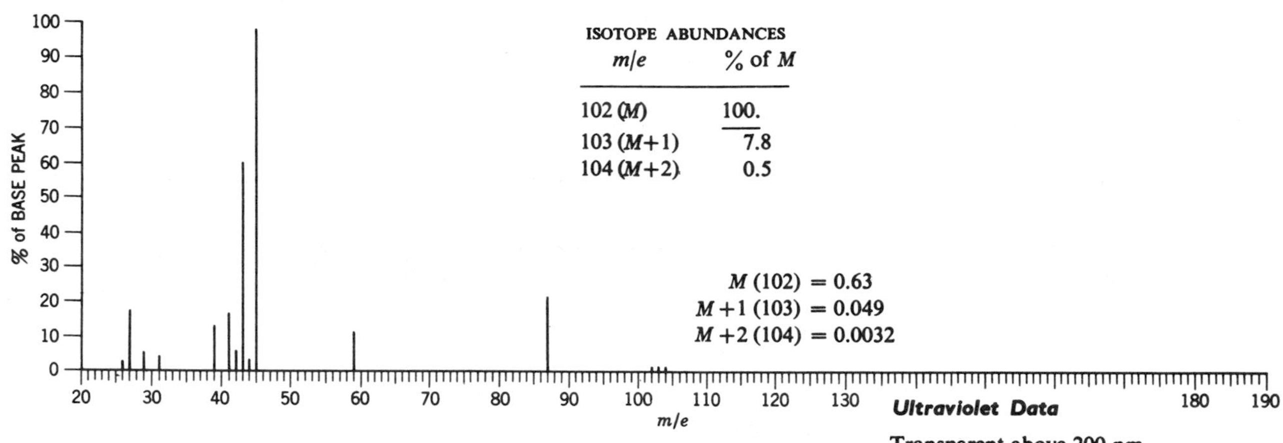

ISOTOPE ABUNDANCES	
m/e	% of *M*
102 (*M*)	100.
103 (*M*+1)	7.8
104 (*M*+2)	0.5

M (102) = 0.63
M +1 (103) = 0.049
M +2 (104) = 0.0032

Ultraviolet Data

Transparent above 200 nm

^{1}H NMR Spectrum (*Solvent CDCl$_3$*)

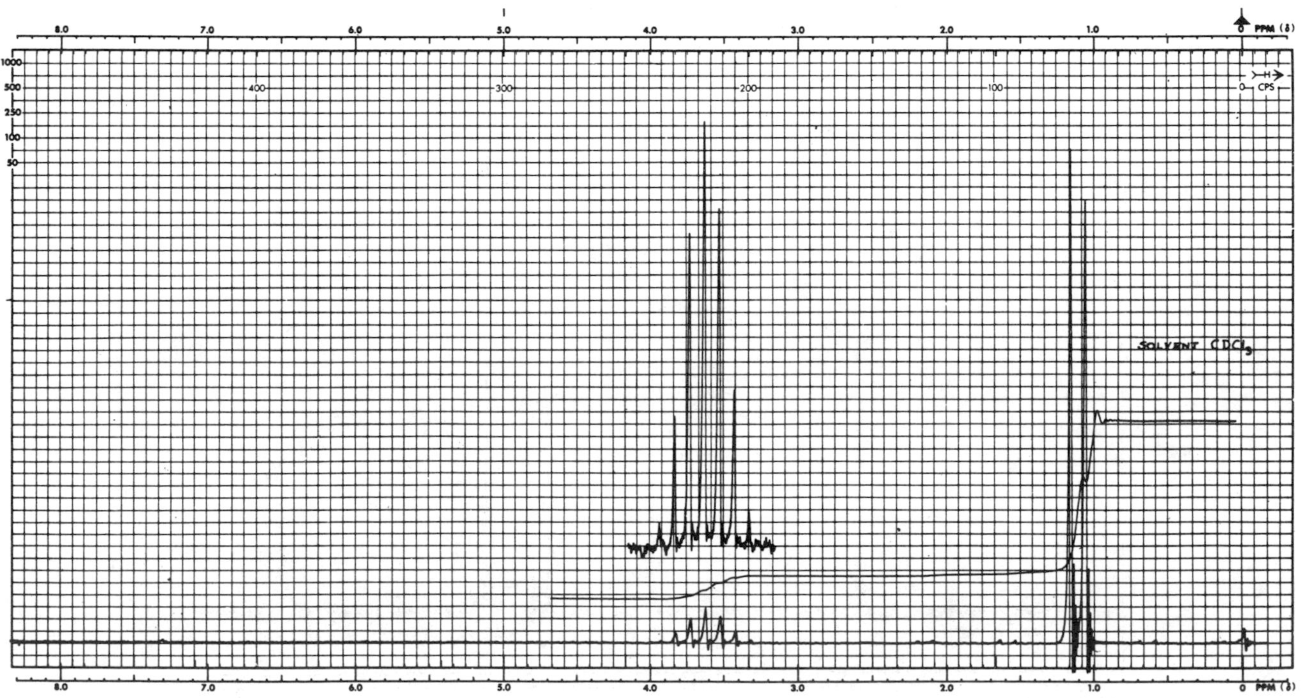

^{13}C **NMR Spectrum Completely Decoupled**

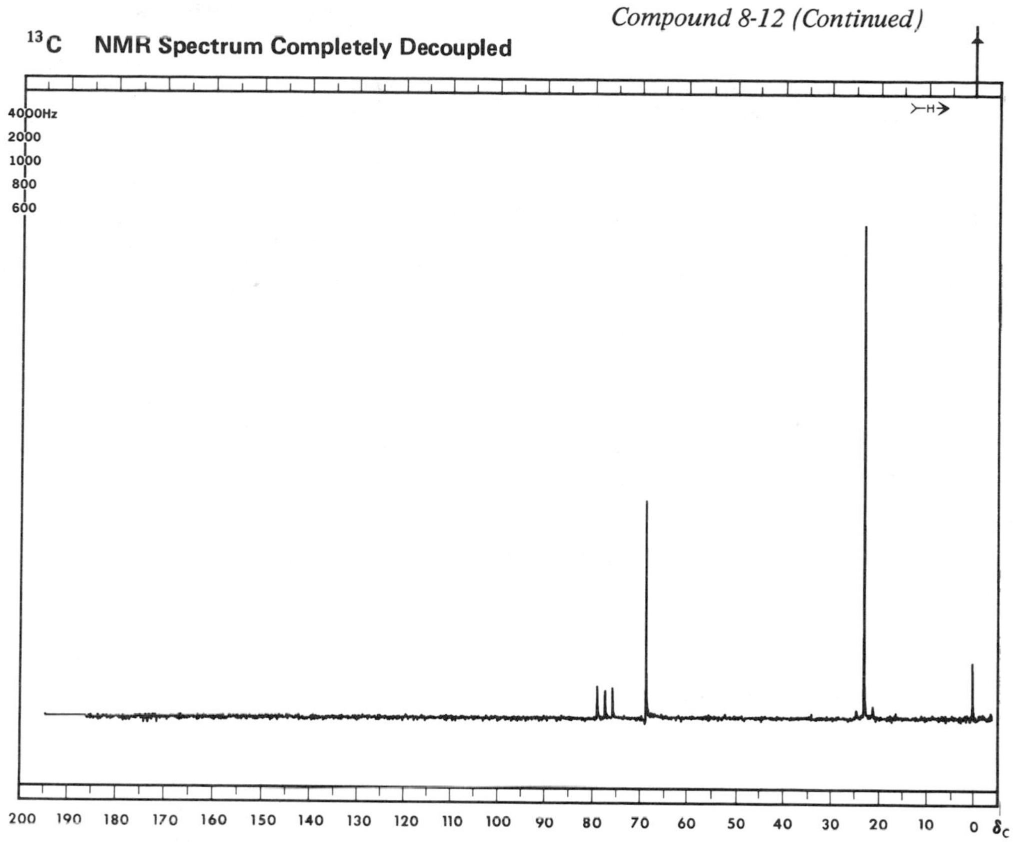

^{13}C **NMR Spectrum Off-resonance Decoupled**

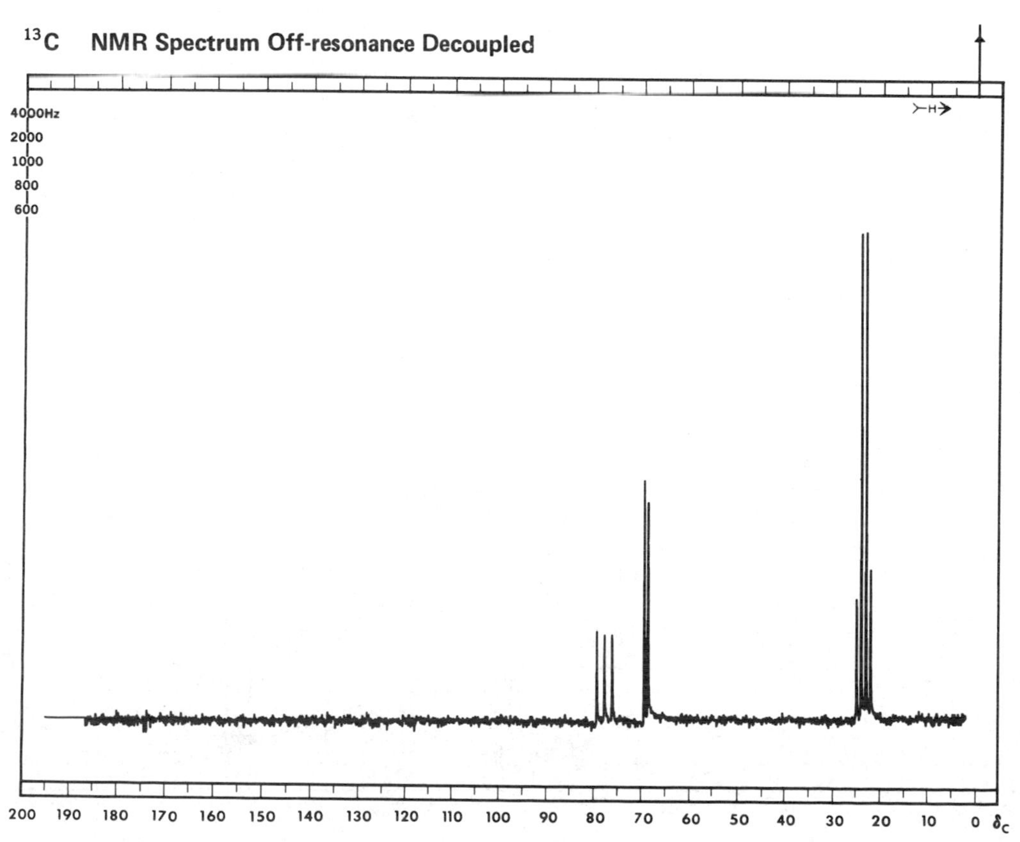

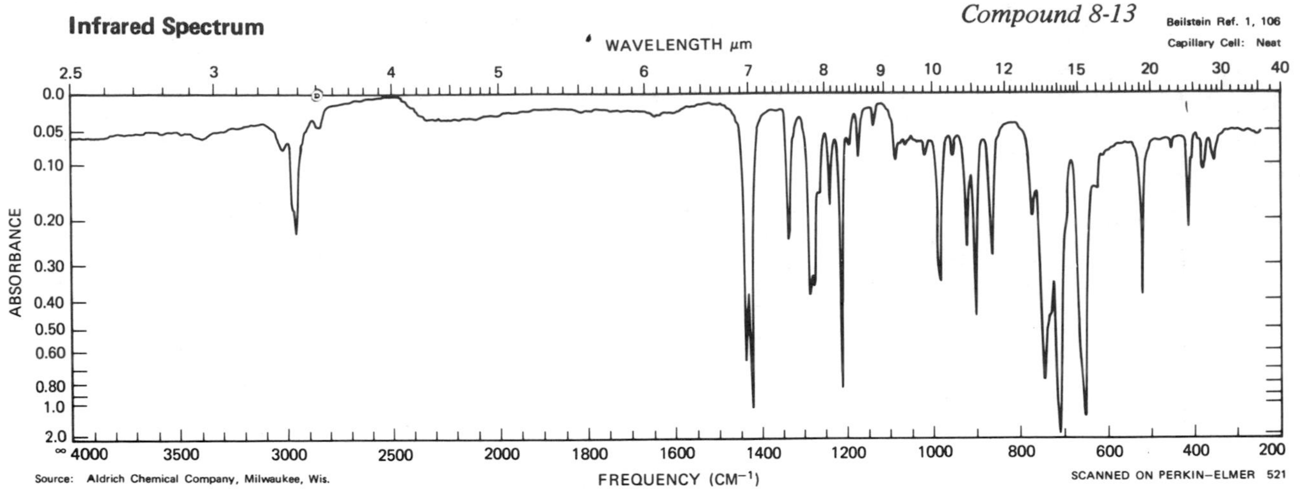

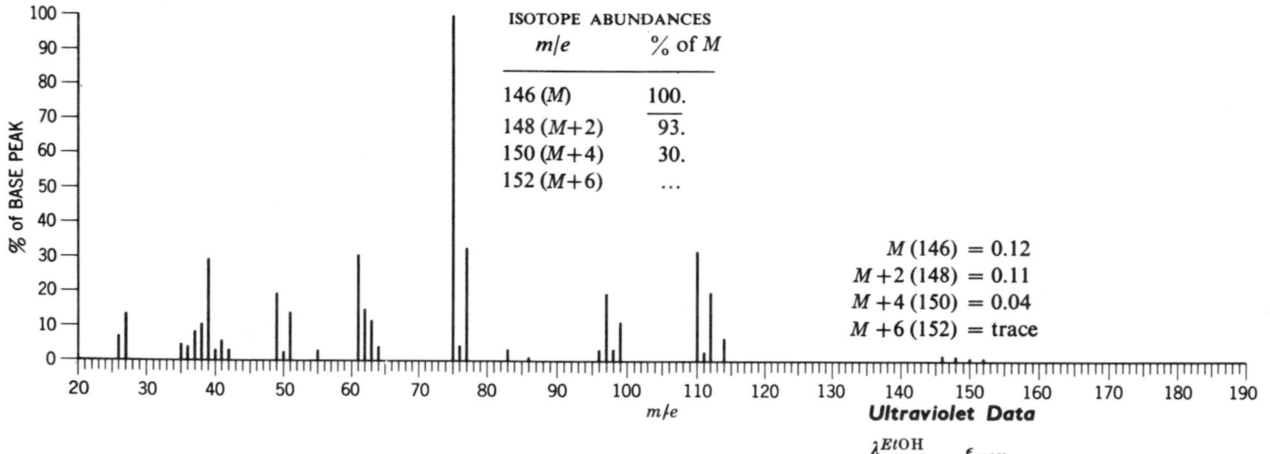

ISOTOPE ABUNDANCES

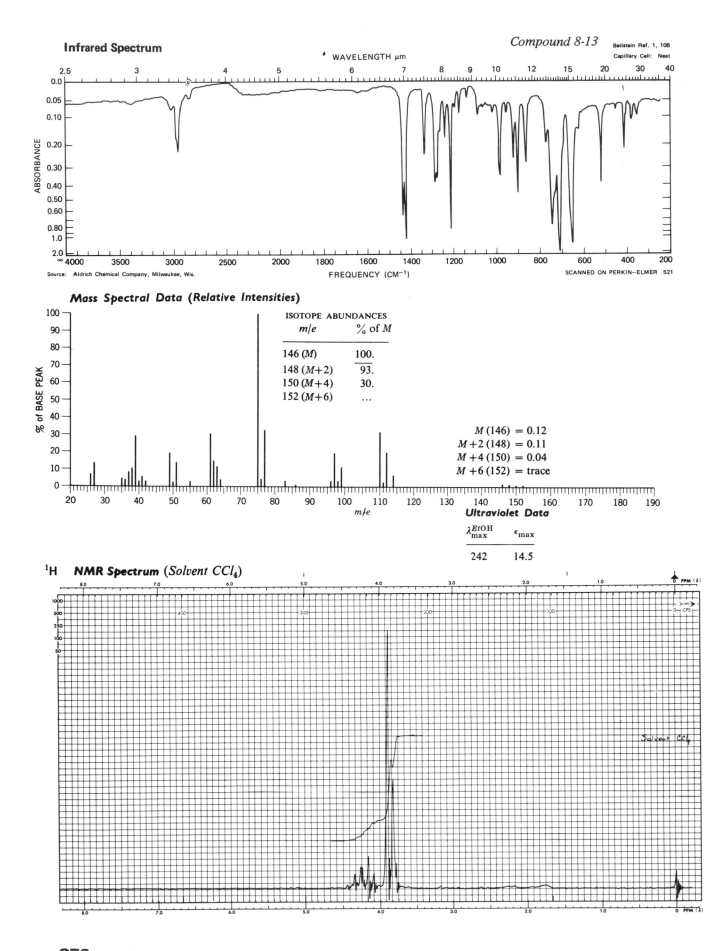

m/e	% of M
146 (M)	100.
148 (M+2)	93.
150 (M+4)	30.
152 (M+6)	...

M (146) = 0.12
M + 2 (148) = 0.11
M + 4 (150) = 0.04
M + 6 (152) = trace

Ultraviolet Data

λ_{max}^{EtOH}	ϵ_{max}
242	14.5

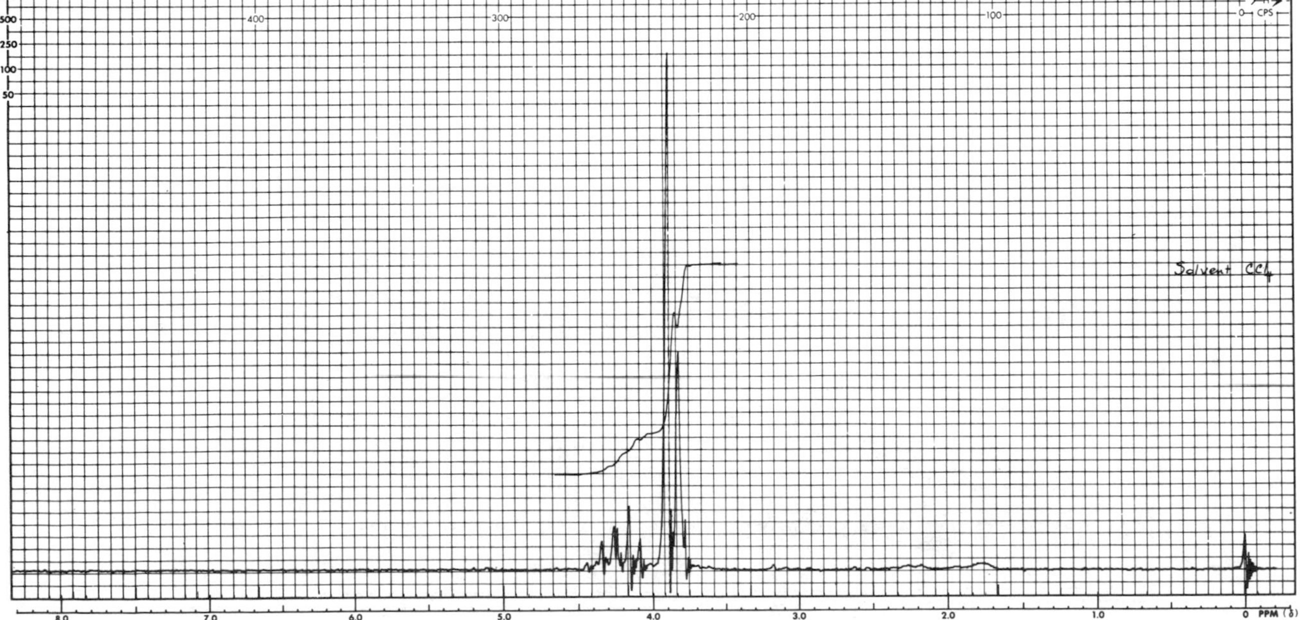

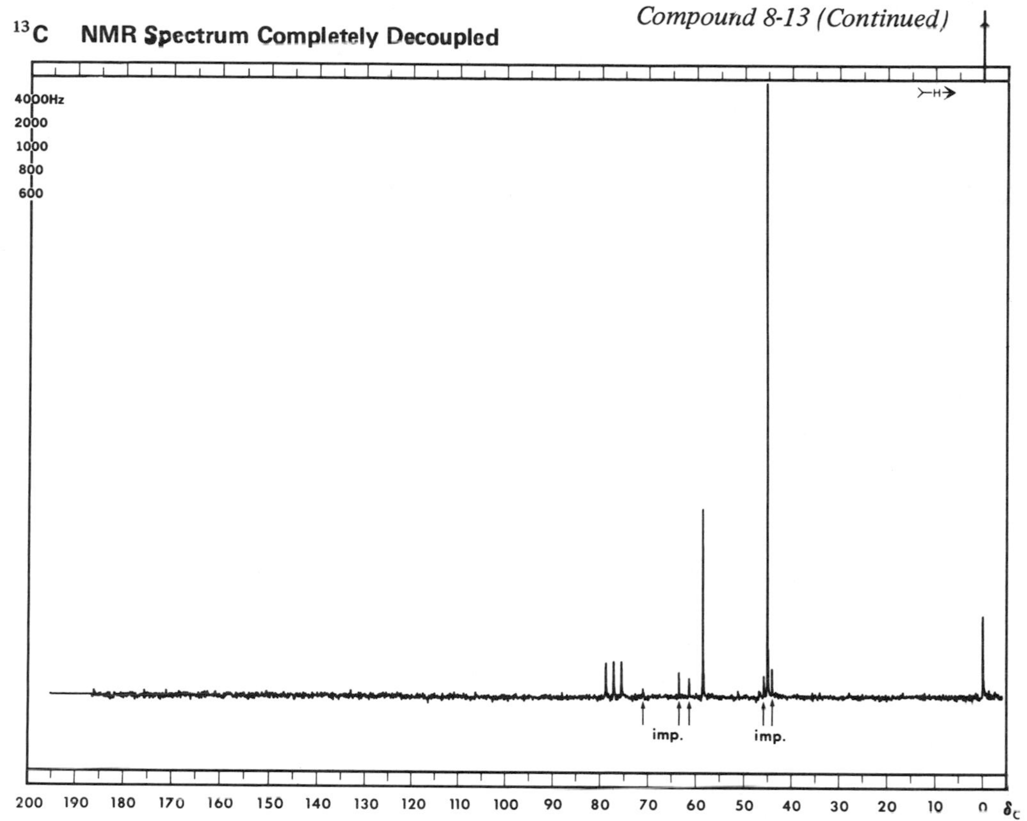
^{13}C NMR Spectrum Completely Decoupled
4000Hz
2000
1000
800
600
imp.
imp.
200 190 180 170 160 150 140 130 120 110 100 90 80 70 60 50 40 30 20 10 0 δ_C

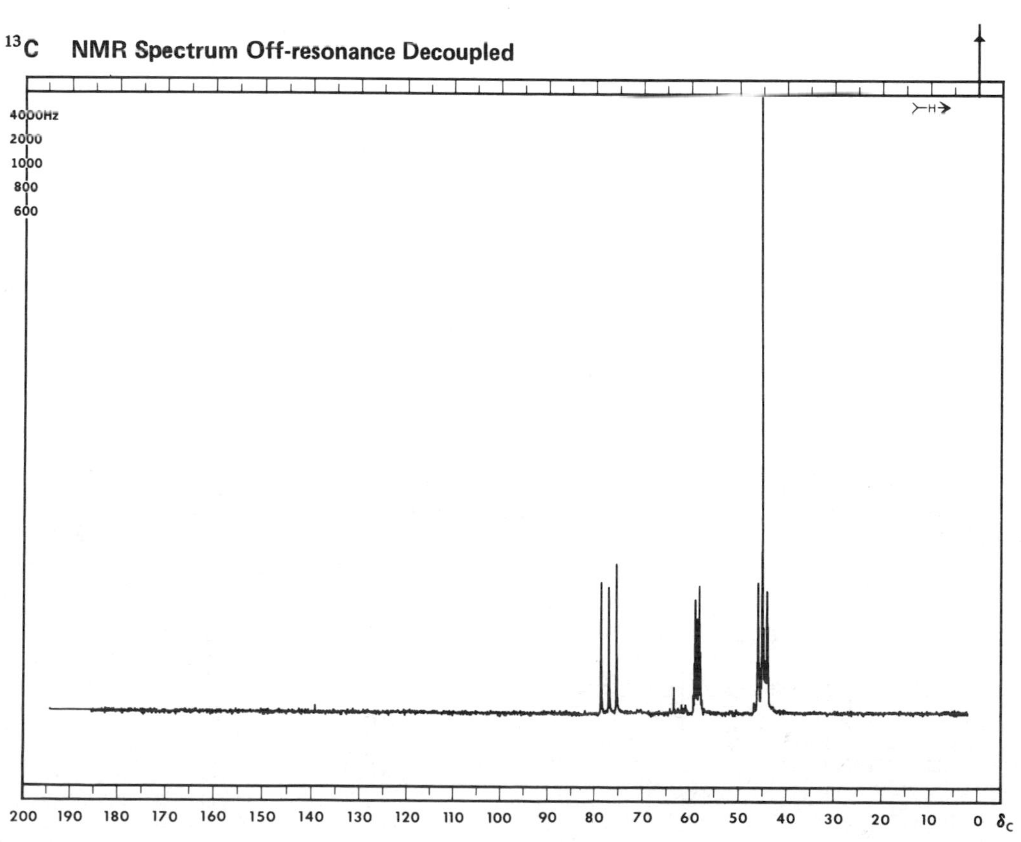
^{13}C NMR Spectrum Off-resonance Decoupled
4000Hz
2000
1000
800
600
200 190 180 170 160 150 140 130 120 110 100 90 80 70 60 50 40 30 20 10 0 δ_C

Infrared Spectrum

Compound 8-14

Beilstein Ref. 4, 156

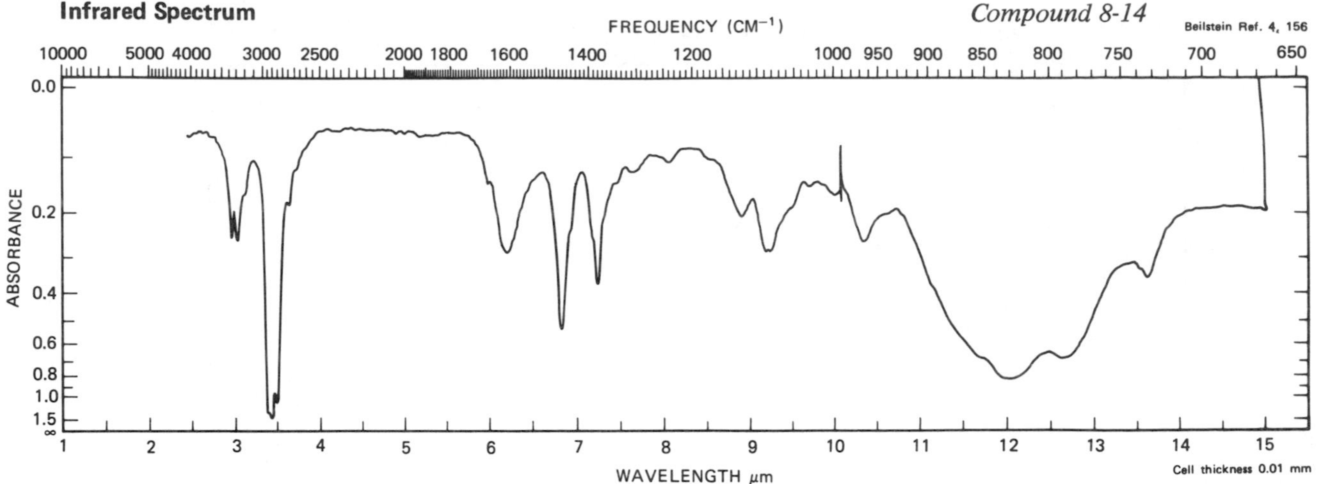

Mass Spectral Data (Relative Intensities)

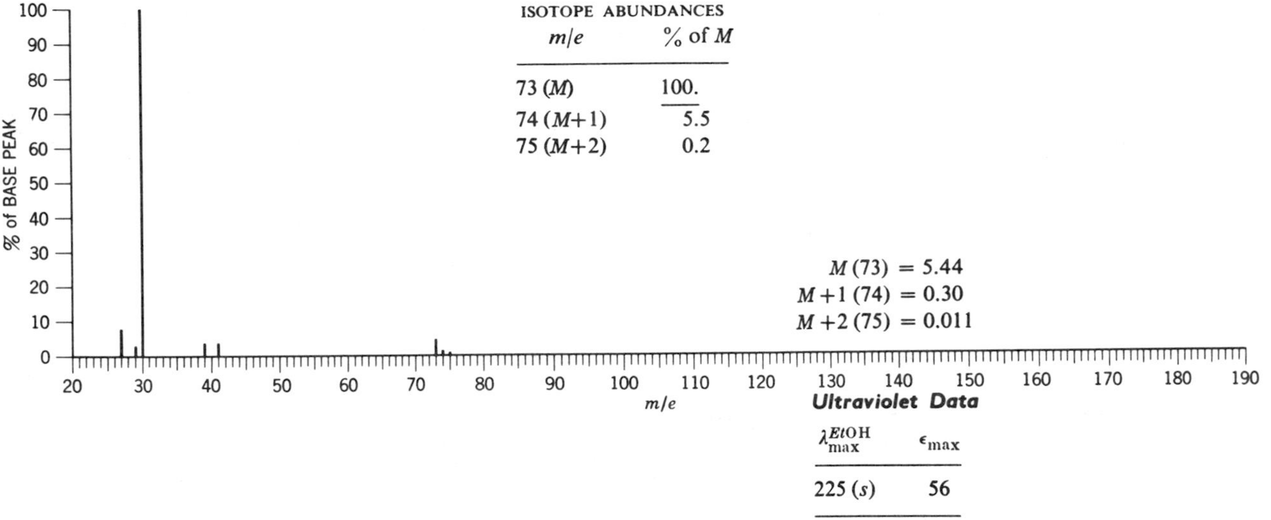

ISOTOPE ABUNDANCES

m/e	% of M
73 (M)	100.
74 (M+1)	5.5
75 (M+2)	0.2

$M(73) = 5.44$
$M+1\,(74) = 0.30$
$M+2\,(75) = 0.011$

Ultraviolet Data

λ_{max}^{EtOH}	ϵ_{max}
225 (s)	56

(s) = shoulder

^{1}H NMR Spectrum (Solvent CCl_4)

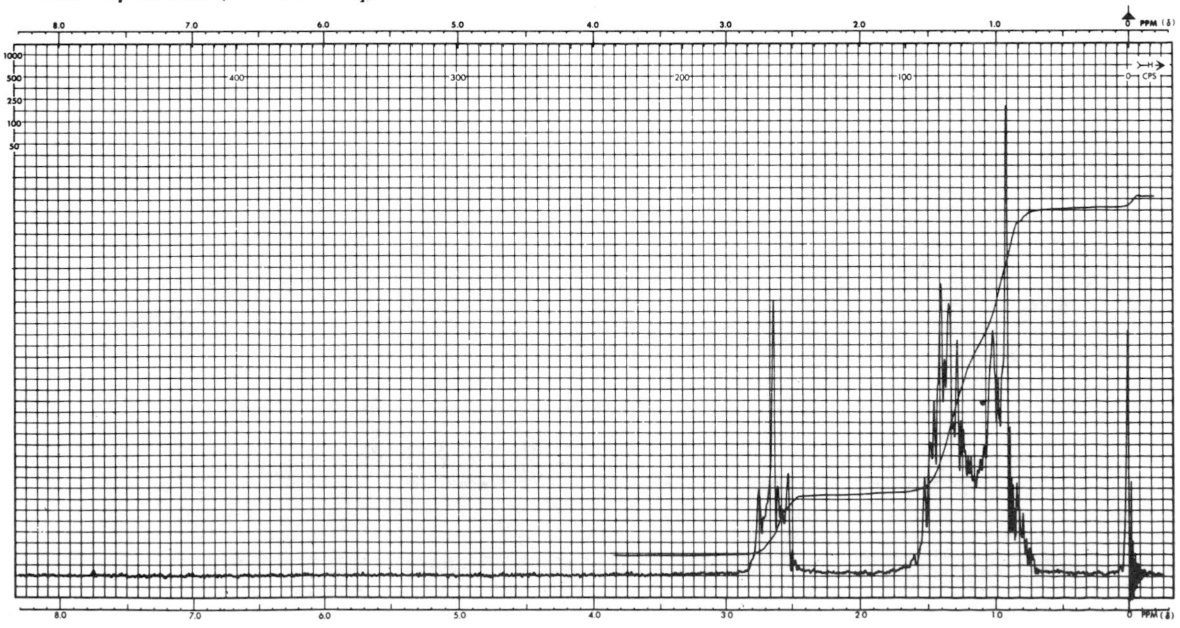

Compound 8-14 ^{13}C NMR Data ($CDCl_3$ solvent)

δ	Intensity	Off-resonance Decoupled Multiplet
13.9	95.3	*q*
20.2	100.0	*t*
36.3	85.3	*t*
42.0	93.0	*t*

Infrared Spectrum

Compound 8-15

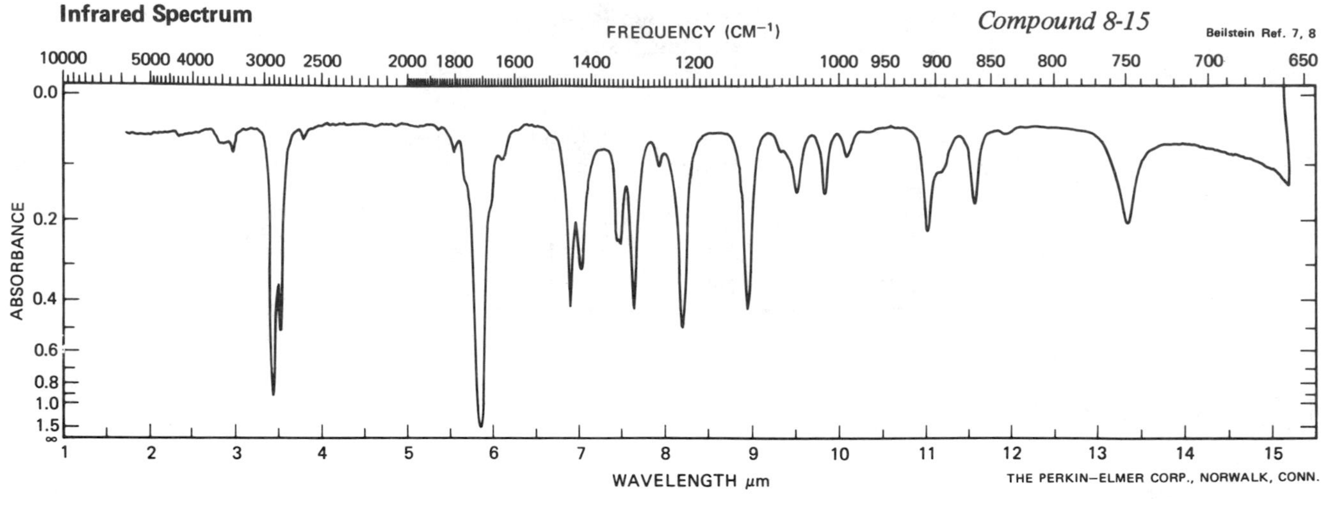

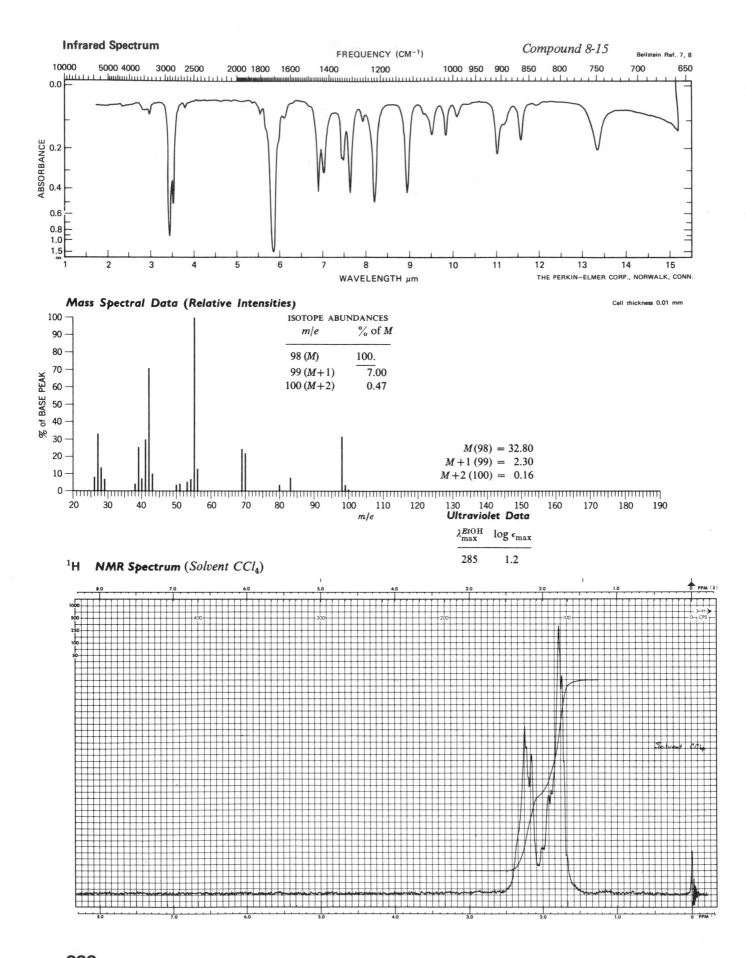

Mass Spectral Data (Relative Intensities)

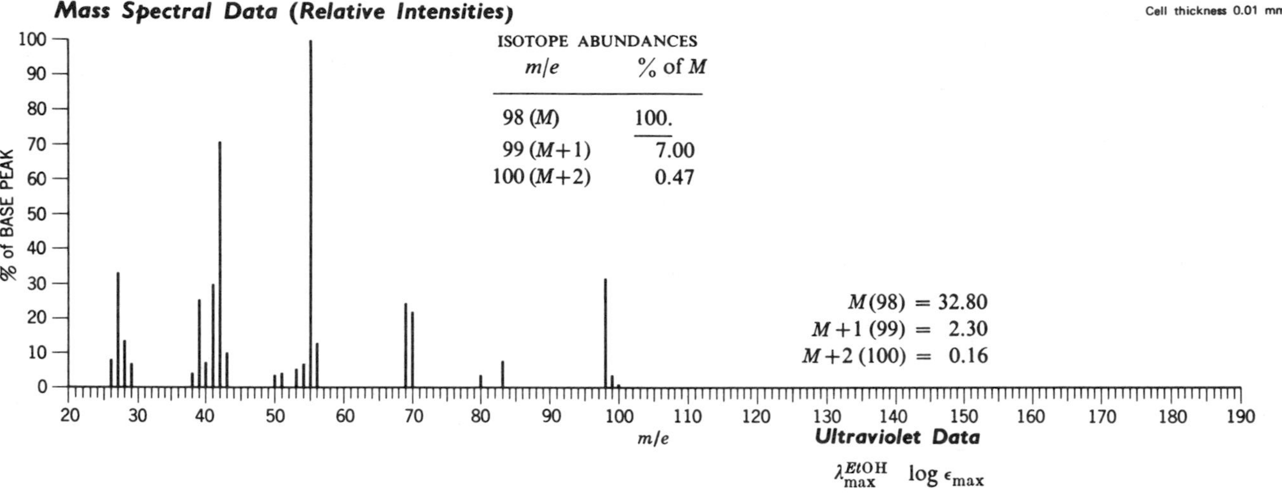

ISOTOPE ABUNDANCES

m/e	% of *M*
98 (*M*)	100.
99 (*M*+1)	7.00
100 (*M*+2)	0.47

$M(98) = 32.80$
$M+1\,(99) = 2.30$
$M+2\,(100) = 0.16$

Ultraviolet Data

λ_{max}^{EtOH}	$\log \epsilon_{max}$
285	1.2

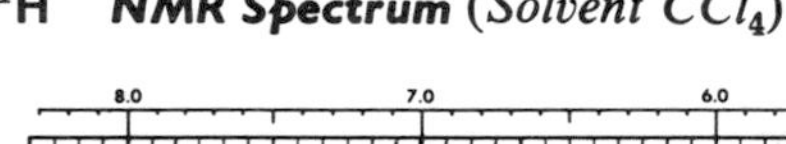

^{1}H NMR Spectrum (*Solvent* CCl_4)

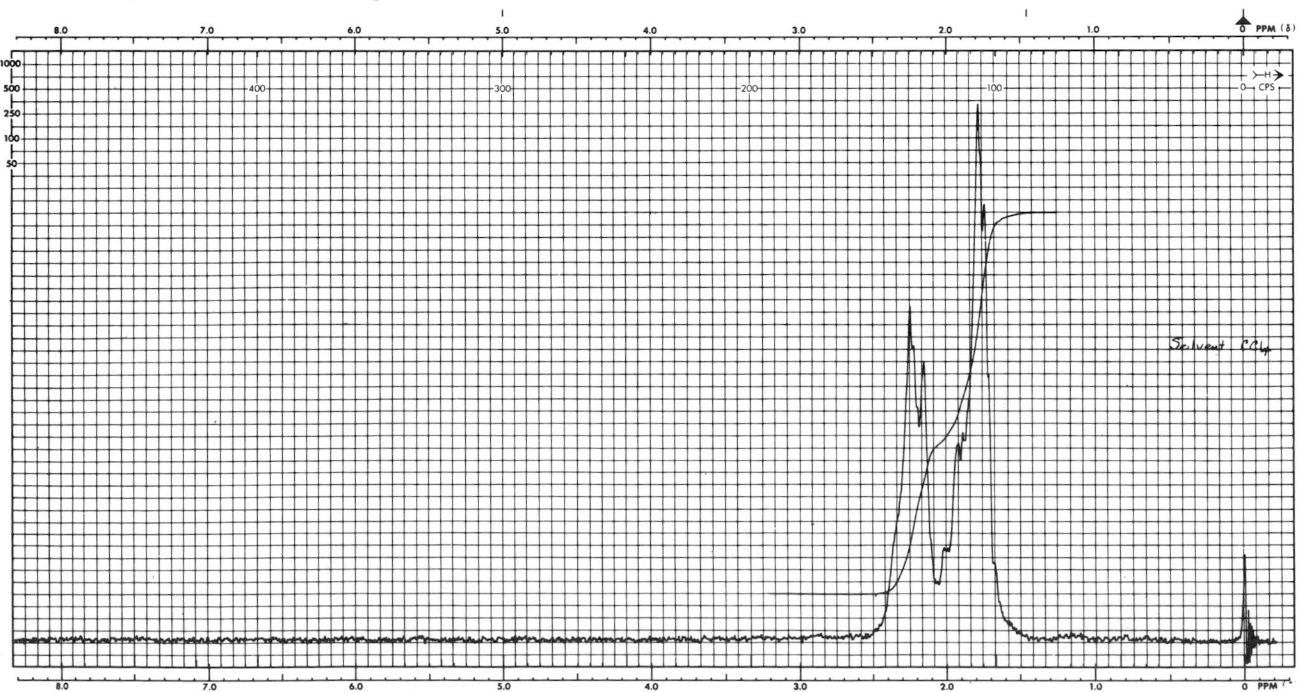

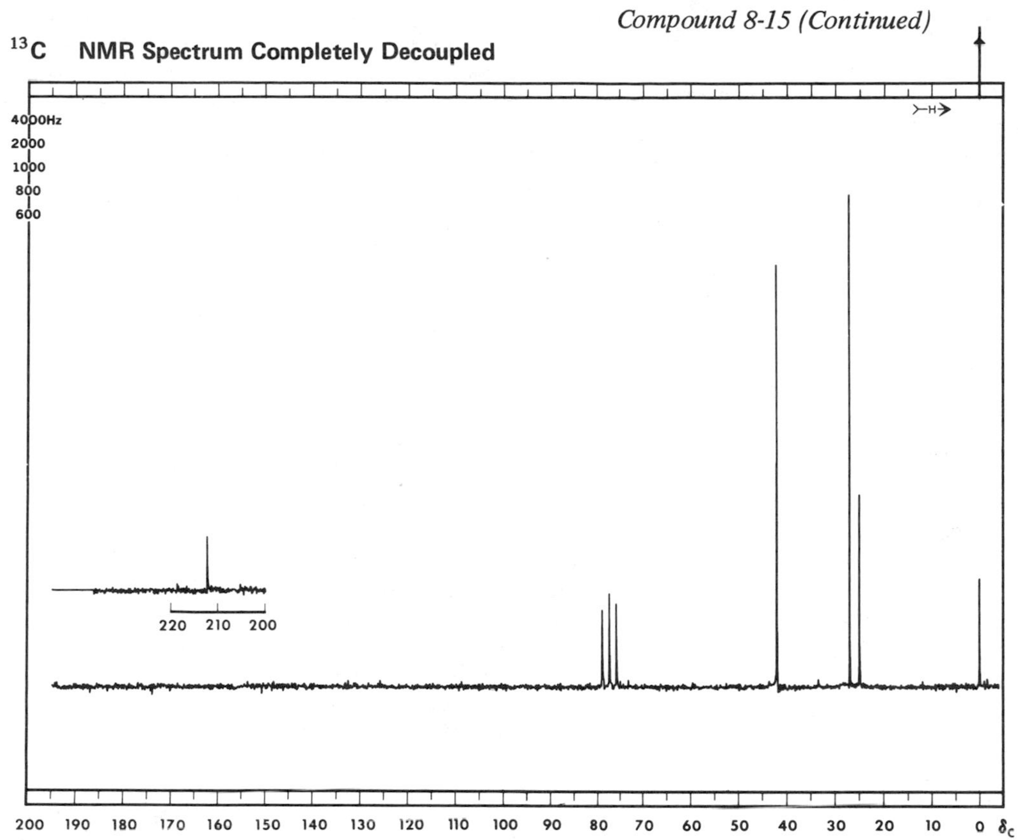

^{13}C NMR Spectrum Completely Decoupled
4000Hz
2000
1000
800
600
220 210 200
200 190 180 170 160 150 140 130 120 110 100 90 80 70 60 50 40 30 20 10 0 δ_C

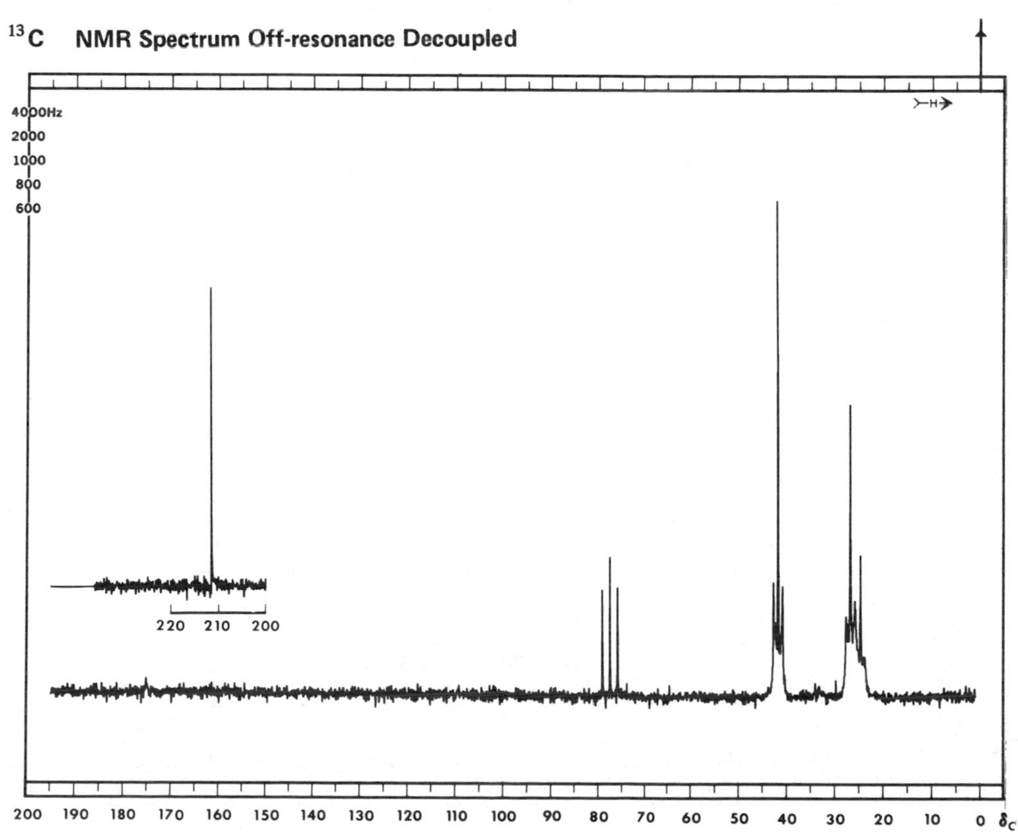

^{13}C NMR Spectrum Off-resonance Decoupled
4000Hz
2000
1000
800
600
220 210 200
200 190 180 170 160 150 140 130 120 110 100 90 80 70 60 50 40 30 20 10 0 δ_C

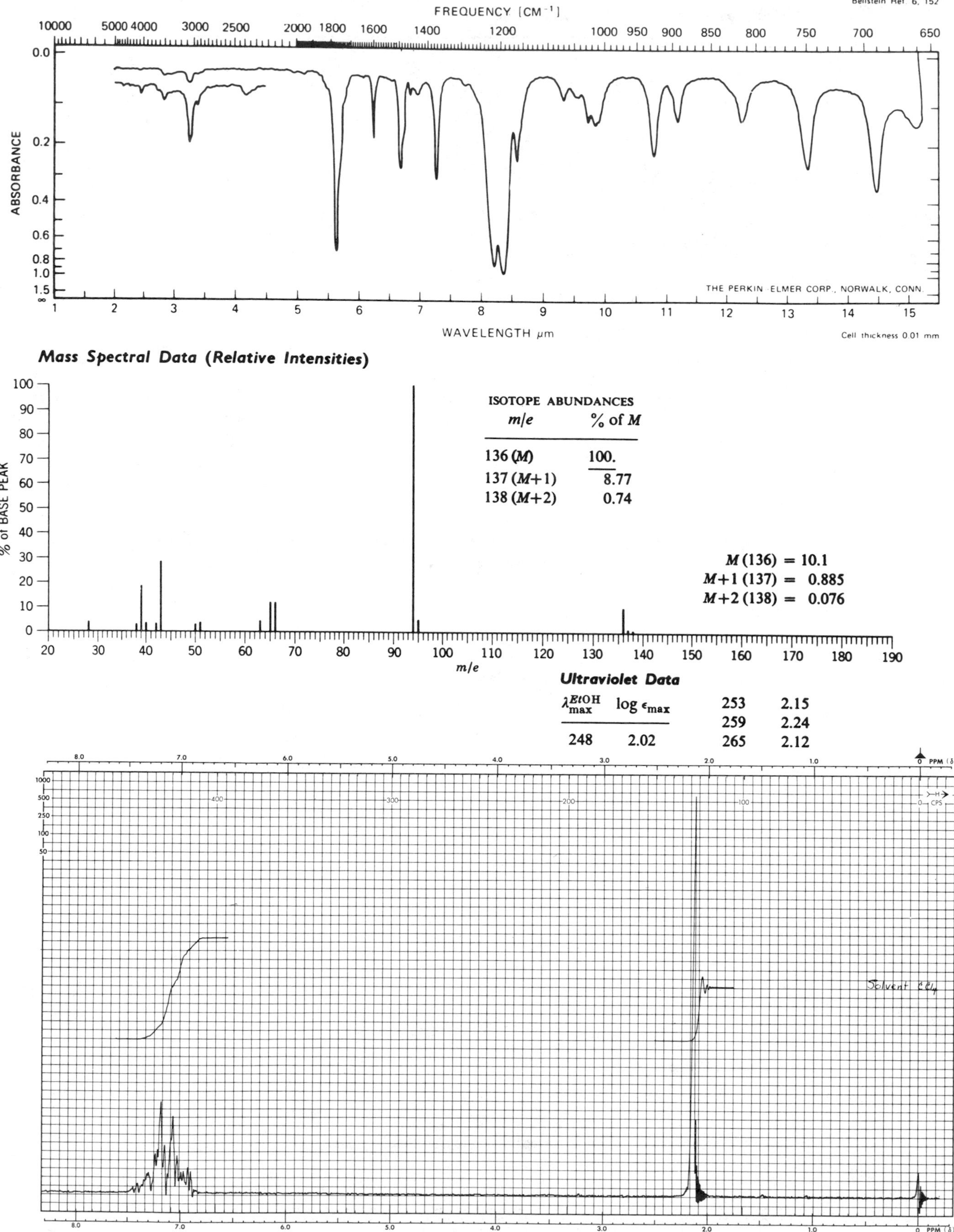
Infrared Spectrum
Compound 8-16
Beilstein Ref. 6, 152
FREQUENCY [CM⁻¹]
ABSORBANCE
WAVELENGTH μm
THE PERKIN-ELMER CORP., NORWALK, CONN.
Cell thickness 0.01 mm
Mass Spectral Data (Relative Intensities)
% of BASE PEAK
m/e
ISOTOPE ABUNDANCES
m/e | % of M
136 (M) 100.
137 (M+1) 8.77
138 (M+2) 0.74
M (136) = 10.1
M+1 (137) = 0.885
M+2 (138) = 0.076
Ultraviolet Data
λ max EtOH | log ε max
248 2.02
253 2.15
259 2.24
265 2.12
PPM (δ)
CPS
Solvent CCl4

Compound 8-16 ^{13}C NMR Data ($CDCl_3$ solvent)

δ	Intensity	Off-resonance Decoupled Multiplet
20.8	20	*q*
121.7	90	*d*
125.6	65	*d*
129.4	100	*d*
151.1	15	*s*
169.2	17	*s*

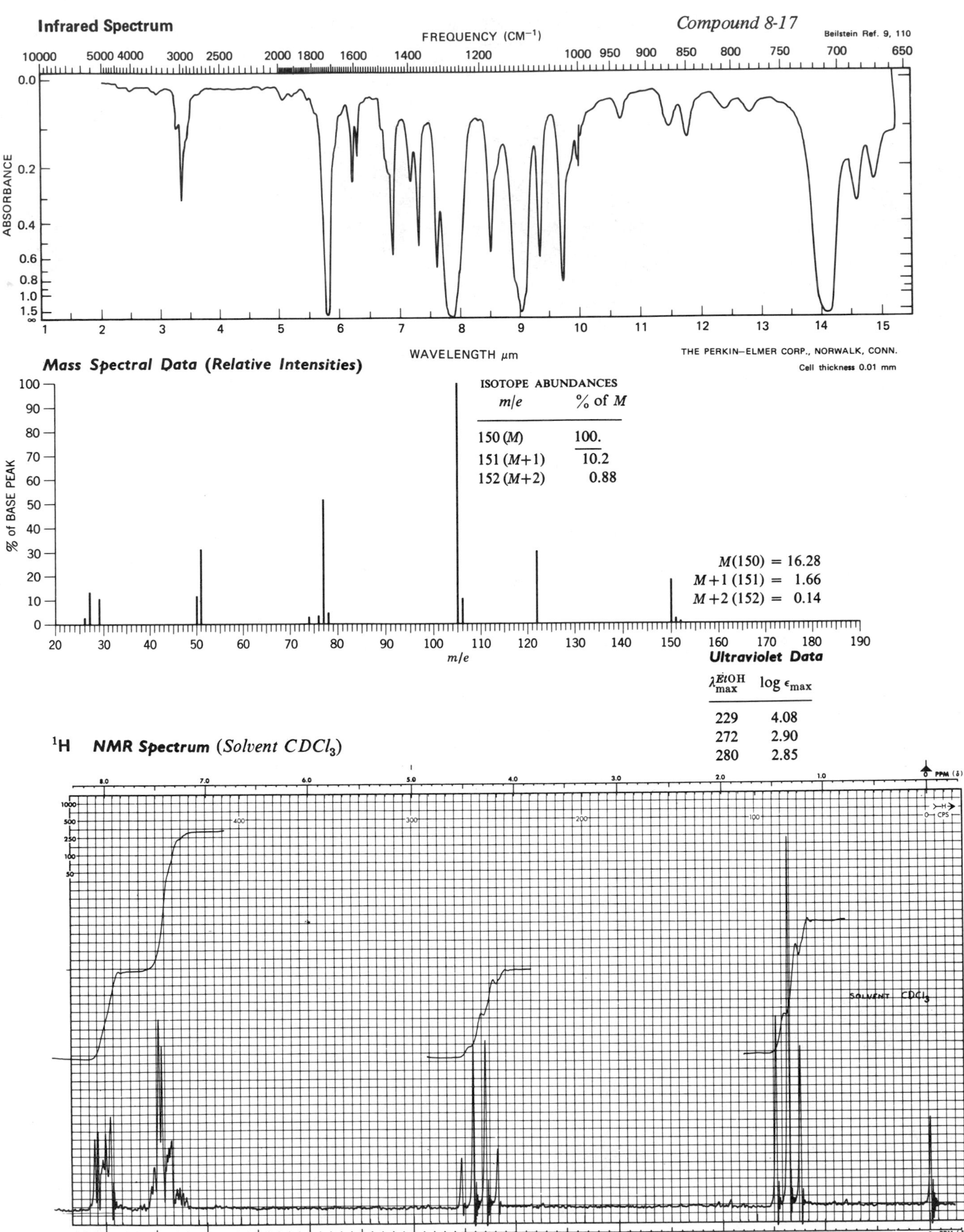
Infrared Spectrum
Compound 8-17
Beilstein Ref. 9, 110
FREQUENCY (CM⁻¹)
ABSORBANCE
WAVELENGTH μm
THE PERKIN–ELMER CORP., NORWALK, CONN.
Cell thickness 0.01 mm
Mass Spectral Data (Relative Intensities)
% of BASE PEAK
m/e
ISOTOPE ABUNDANCES
m/e % of M
150 (M) 100.
151 (M+1) 10.2
152 (M+2) 0.88
M(150) = 16.28
M+1 (151) = 1.66
M+2 (152) = 0.14
Ultraviolet Data
λ_{max}^{EtOH} log ϵ_{max}
229 4.08
272 2.90
280 2.85
¹H NMR Spectrum (Solvent $CDCl_3$)
PPM (δ)
CPS
SOLVENT $CDCl_3$

Compound 8-17 ^{13}C NMR Data ($CDCl_3$ solvent)

δ	Intensity	Off-resonance Decoupled Multiplet
14.4	47.0	*q*
60.8	39.7	*t*
128.4	87.0	*d*
129.7	100.0	*d*
130.9	8.7	*s*
132.8	50.2	*d*
166.3	9.2	*s*

Compounds 8-18 and 8-19

Determine the structure of compound 8-18 from the spectra provided. Compound 8-19 is a derivative of 8-18; determine the structure from the spectra provided. Compound 8-19 shows no molecular ion peak in the mass spectrum; however, the mass spectrum shows all the peaks that 8-18 shows (in roughly the same relative intensities). In addition, compound 8-19 shows two peaks, of essentially equal intensities, at m/e 136 and 138. Both compounds are transparent in the UV above 200 nm.

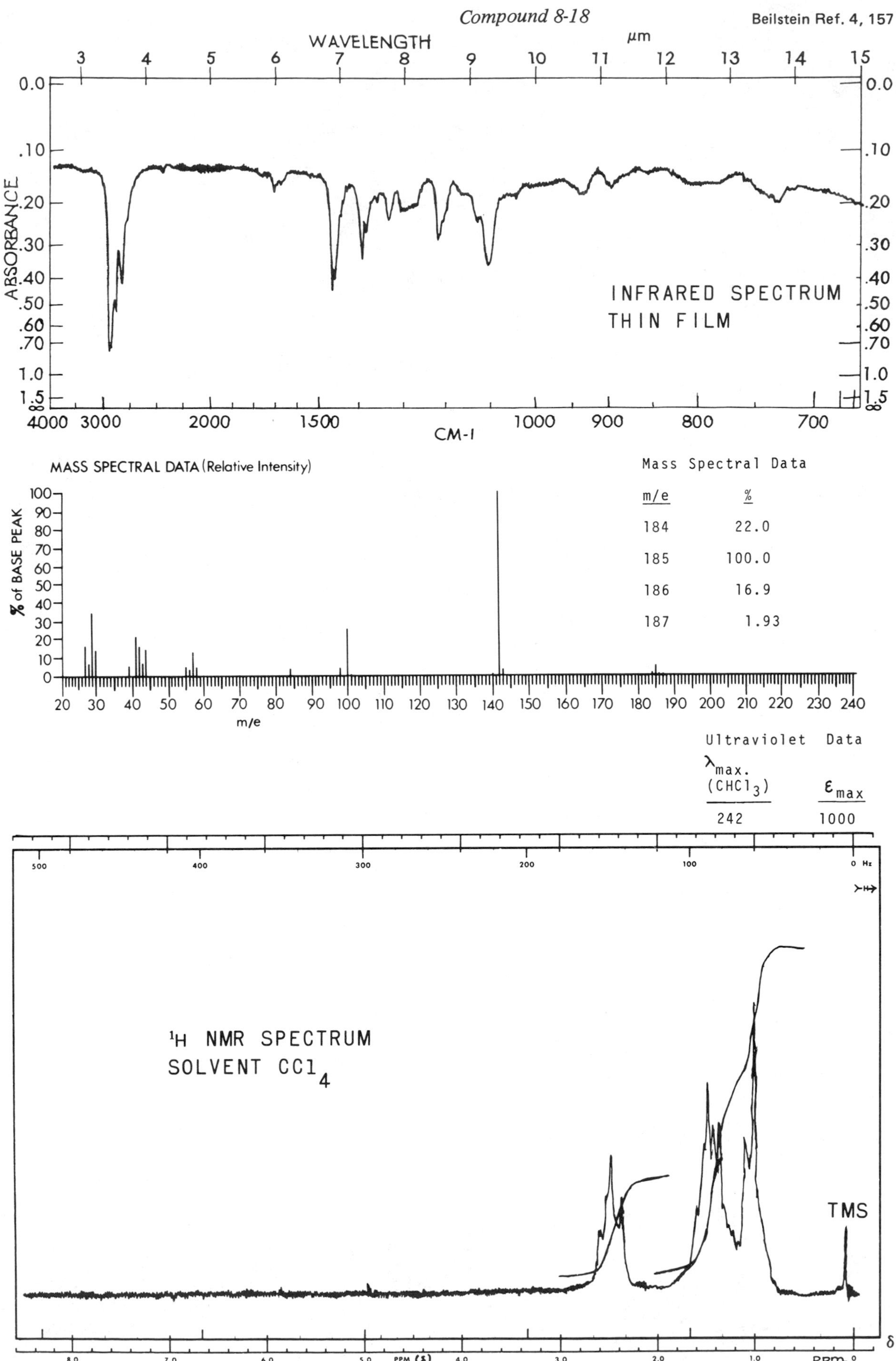

Compound 8-18
Beilstein Ref. 4, 157
WAVELENGTH
μm
ABSORBANCE
INFRARED SPECTRUM
THIN FILM
CM-1
MASS SPECTRAL DATA (Relative Intensity)
% of BASE PEAK
m/e
Mass Spectral Data
m/e %
184 22.0
185 100.0
186 16.9
187 1.93
Ultraviolet Data
λmax. (CHCl3) εmax
242 1000
¹H NMR SPECTRUM
SOLVENT CCl4
TMS
PPM (δ)
ppm
δ

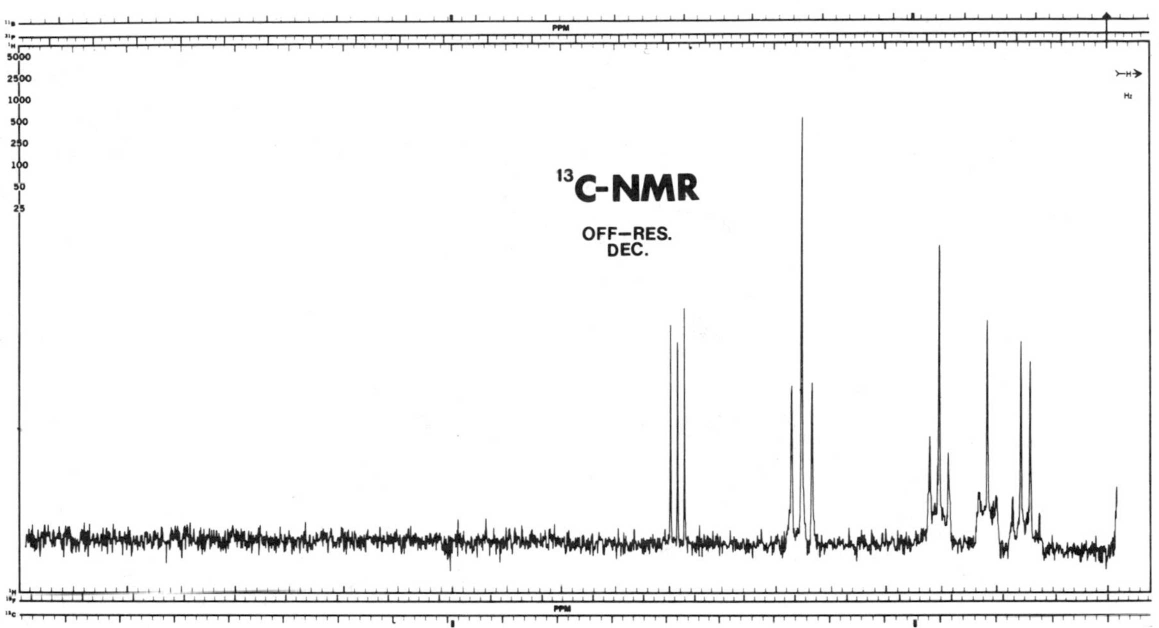

^{13}C-NMR
OFF—RES.
DEC.

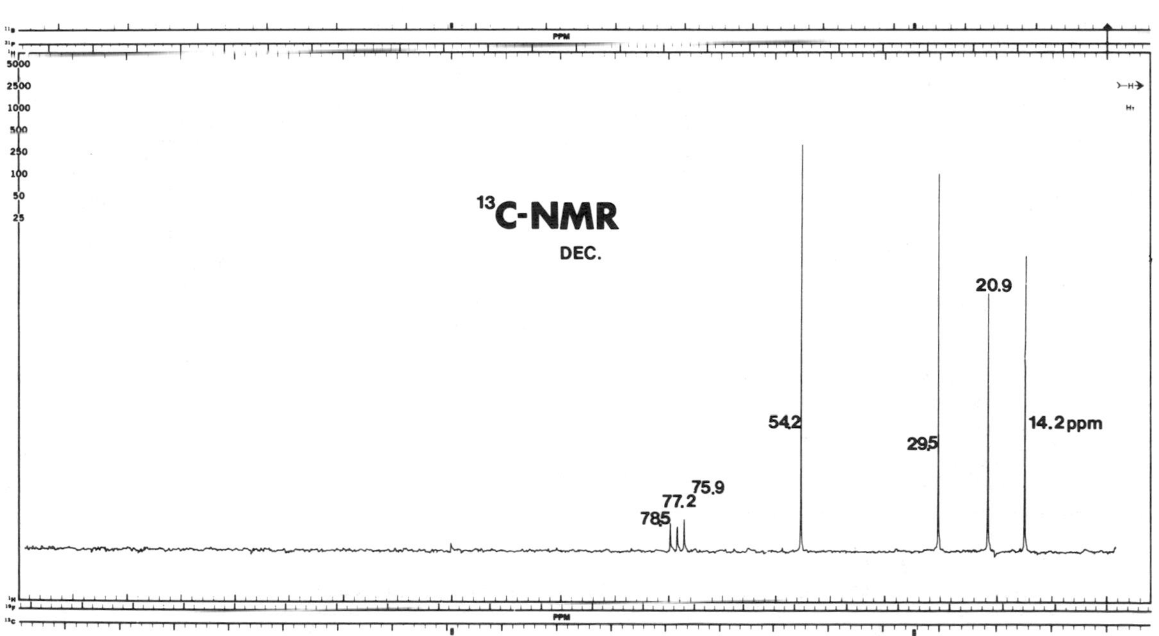

^{13}C-NMR
DEC.
78.5
77.2
75.9
54.2
29.5
20.9
14.2 ppm

Compound 8-19

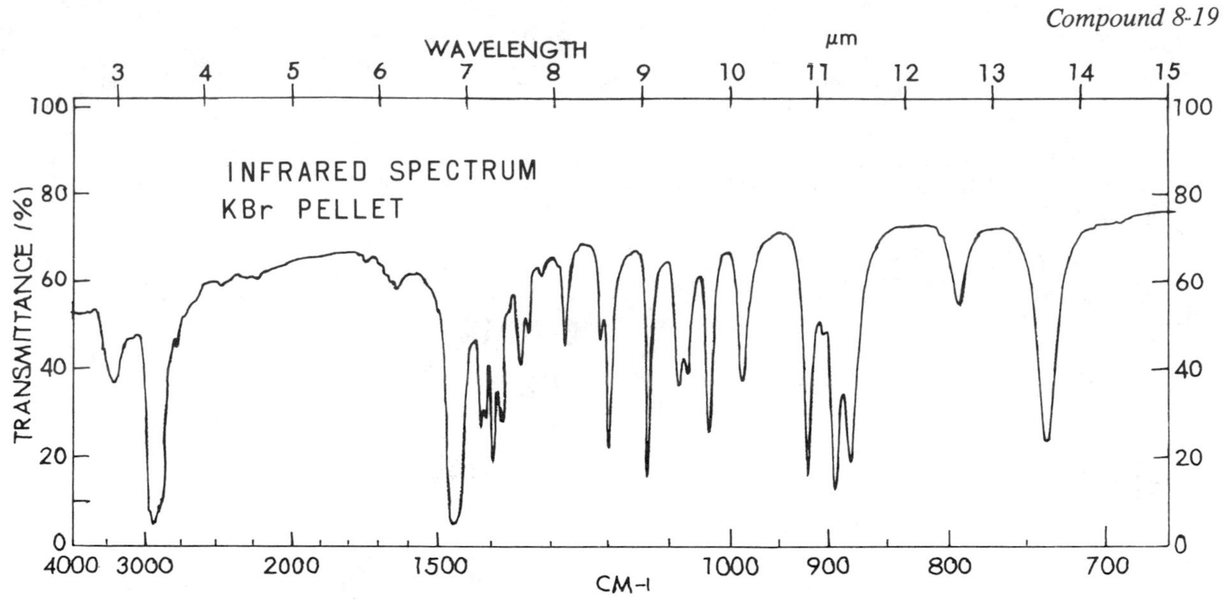

Mass Spectral Data
see p. 387

Ultraviolet Data
No λ_{max} above 200 nm

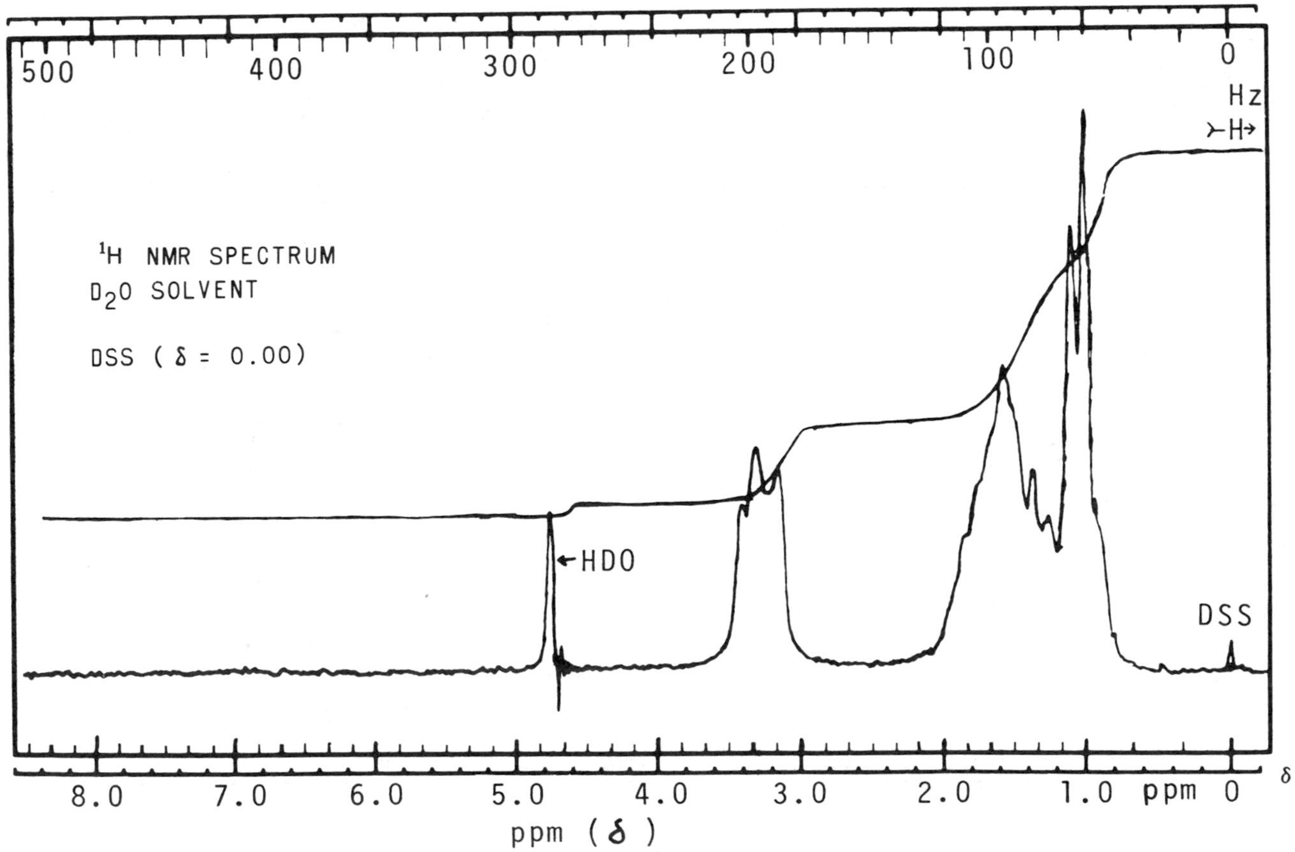

^{1}H NMR SPECTRUM
D_2O SOLVENT
DSS (δ = 0.00)

HDO

DSS

ppm (δ)

Compound 8-19 (Continued)

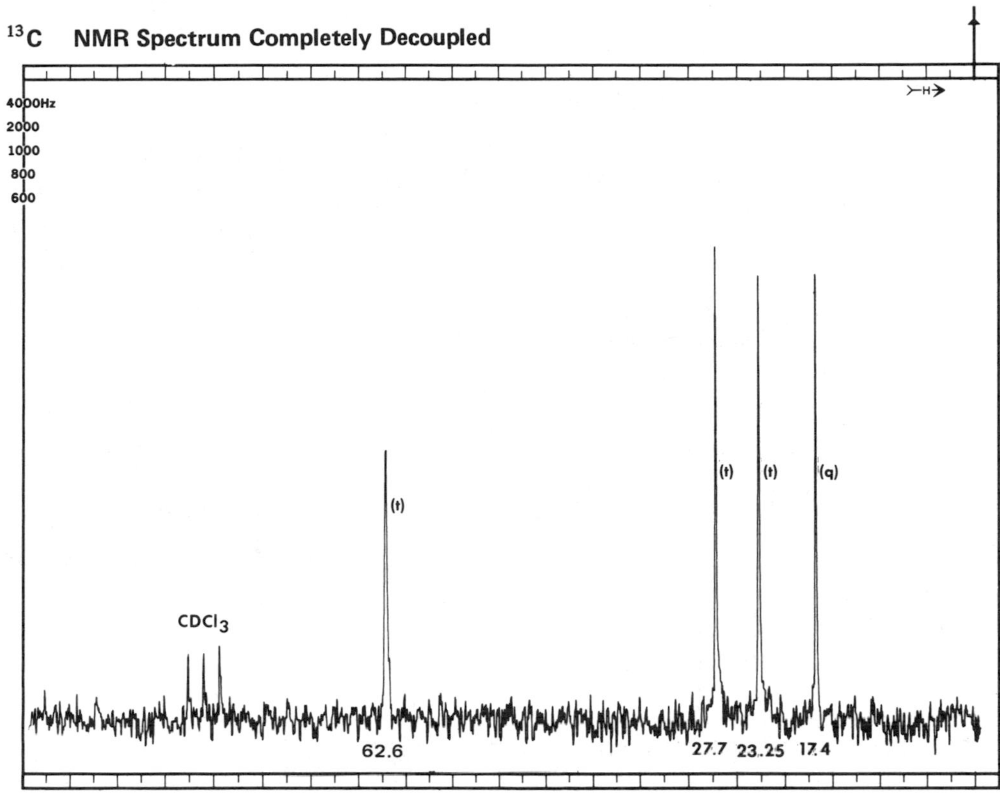

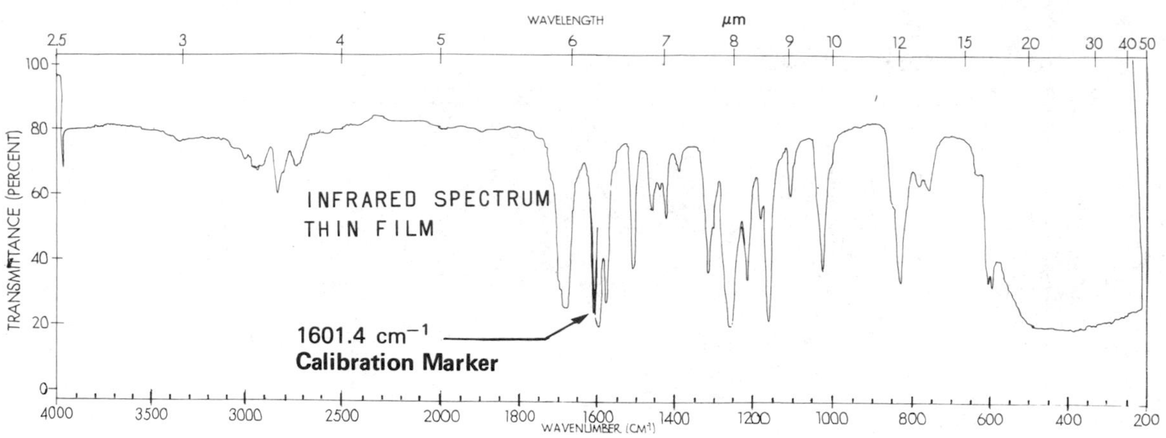

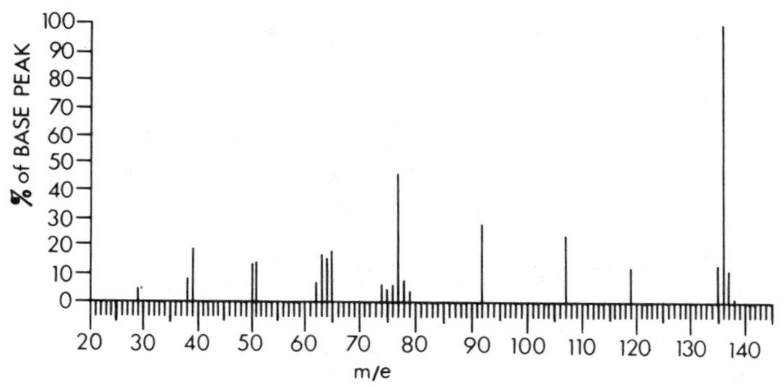

Mass Spectral Data

m/e	Relative Intensity
135	14.0
136	100.0
137	11.1
138	1.1

Ultraviolet Data

λ_{max} (cyclohexane)	ε_{max}
215	21,500
221	20,000
266	25,500
288	8,500
312	90
324	83
360	25

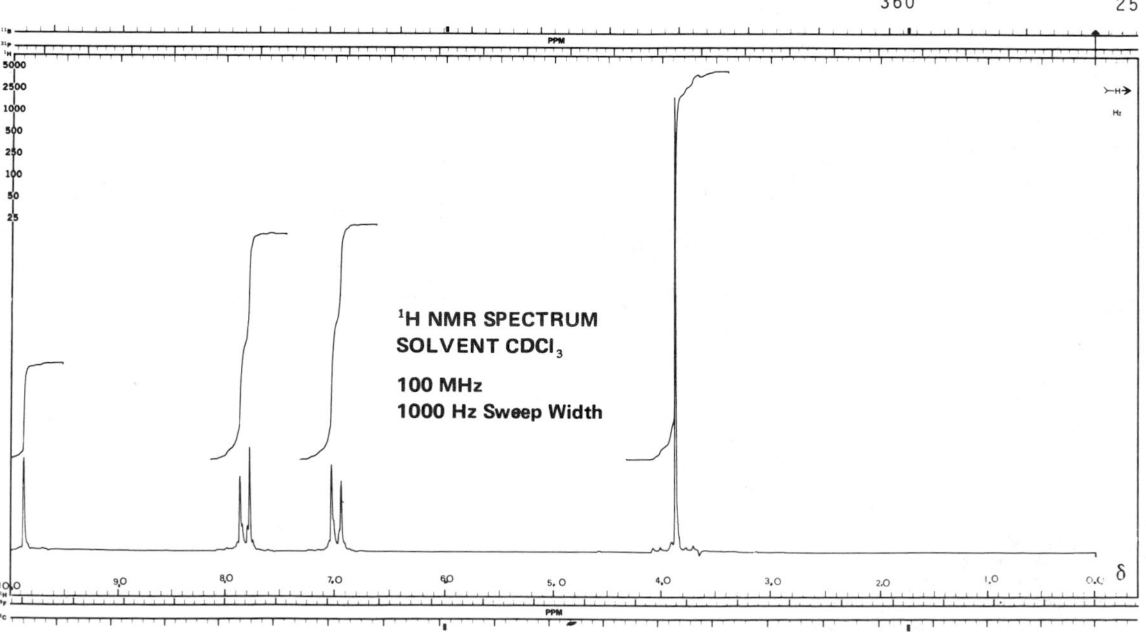

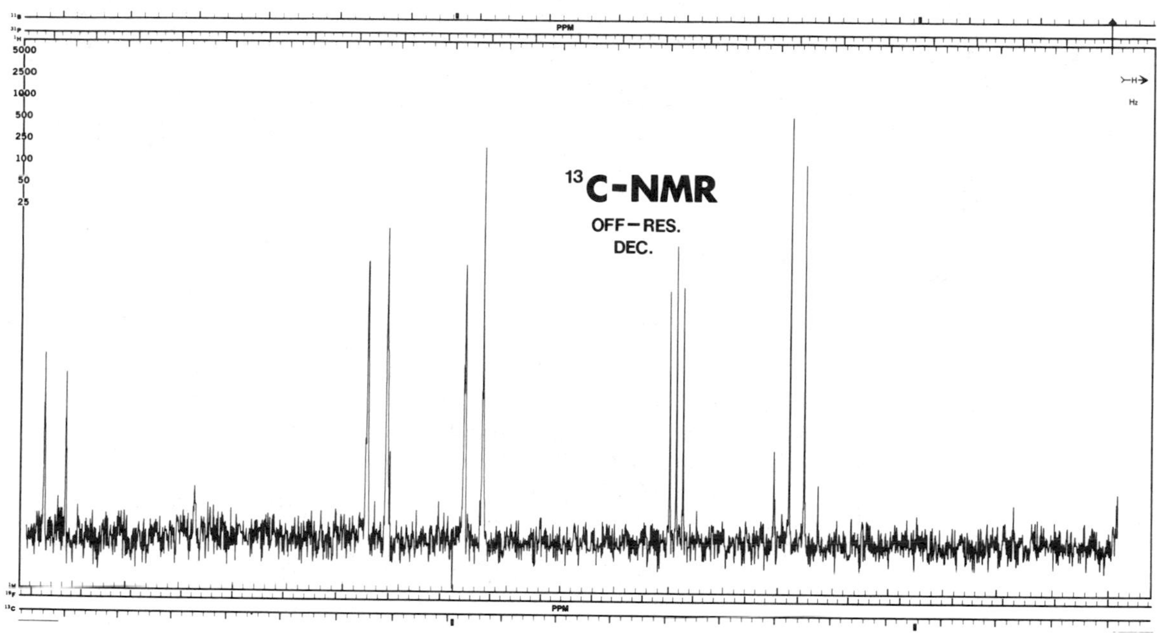

^{13}C-NMR
OFF-RES.
DEC.

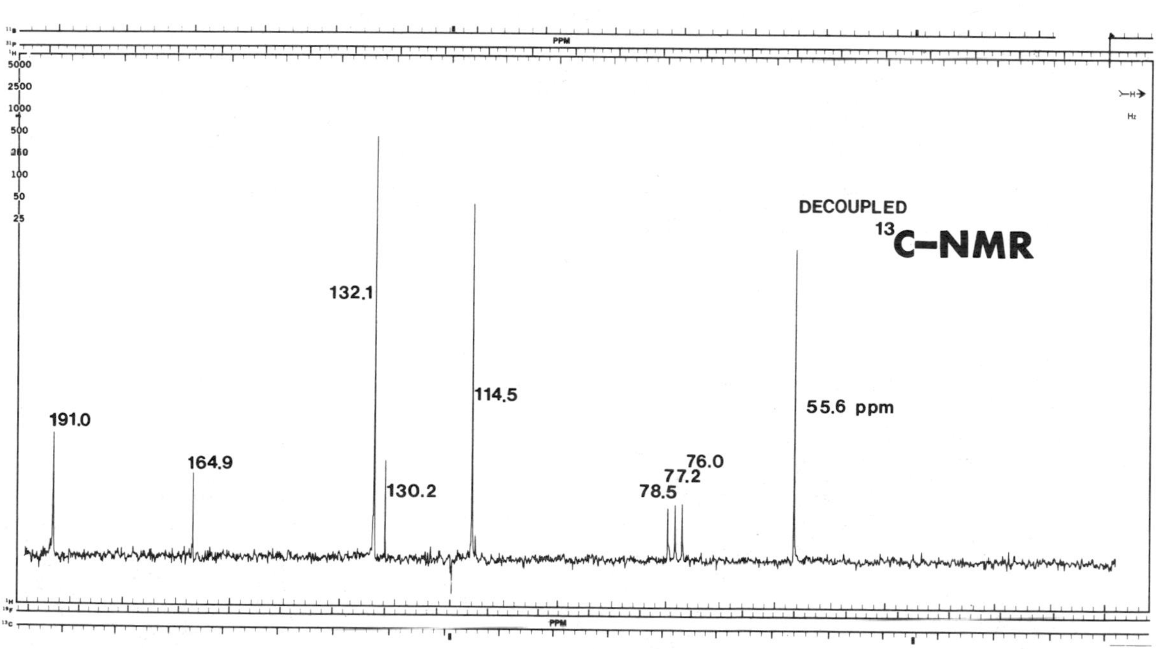

DECOUPLED
^{13}C-NMR
191.0
164.9
132.1
130.2
114.5
78.5
77.2
76.0
55.6 ppm

Compound 8-21

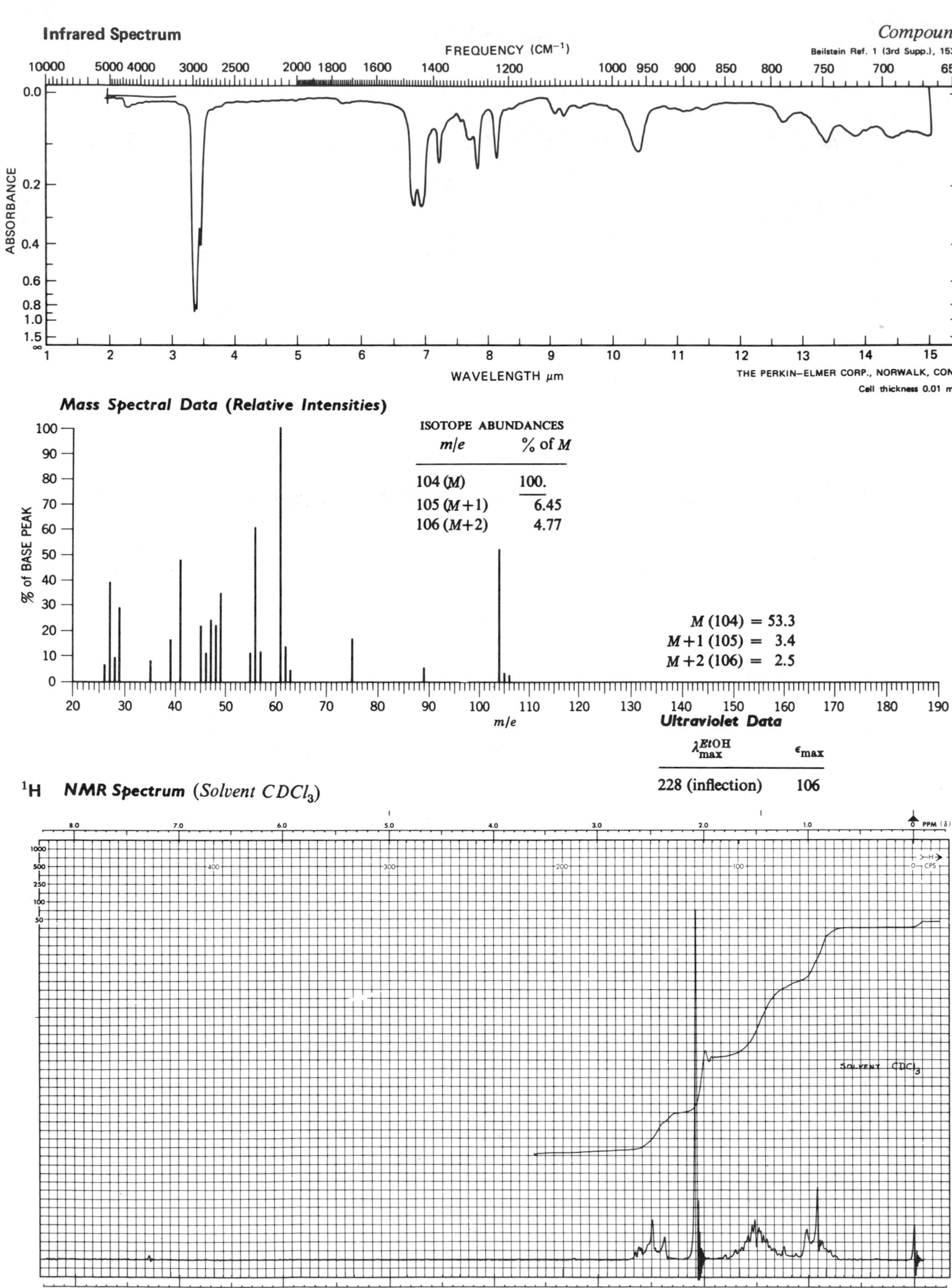

ISOTOPE ABUNDANCES	
m/e	% of M
104 (M)	100.
105 ($M+1$)	6.45
106 ($M+2$)	4.77

Ultraviolet Data

λ_{max}^{EtOH}	ϵ_{max}
228 (inflection)	106

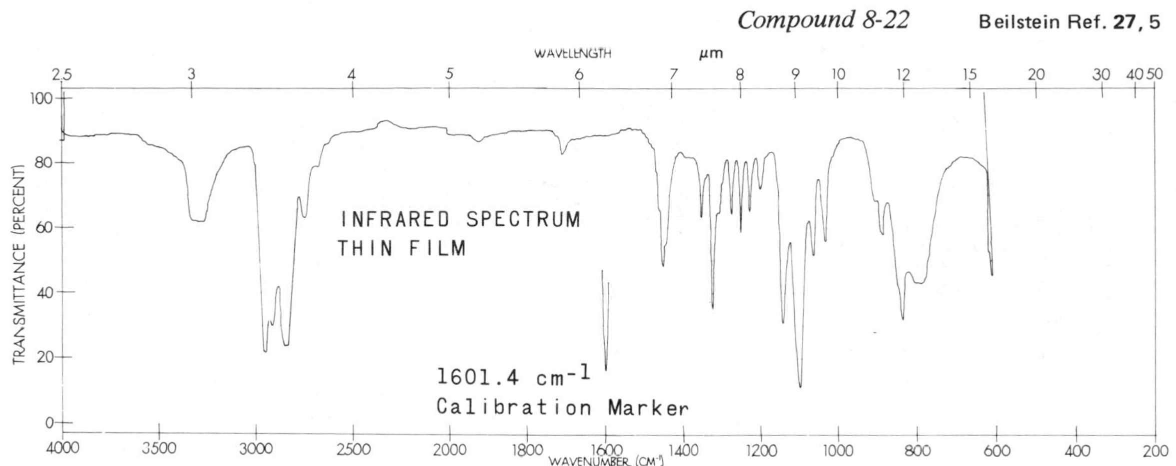

MASS SPECTRAL DATA (Relative Intensity)

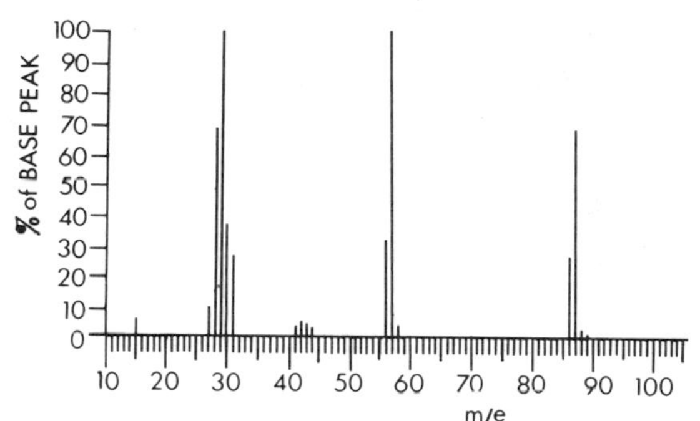

Mass Spectral Data

m/e	Relative Intensity
86	40.70
87	100.00
88	4.48
89	0.29

Ultraviolet Data

Transparent above 200nm

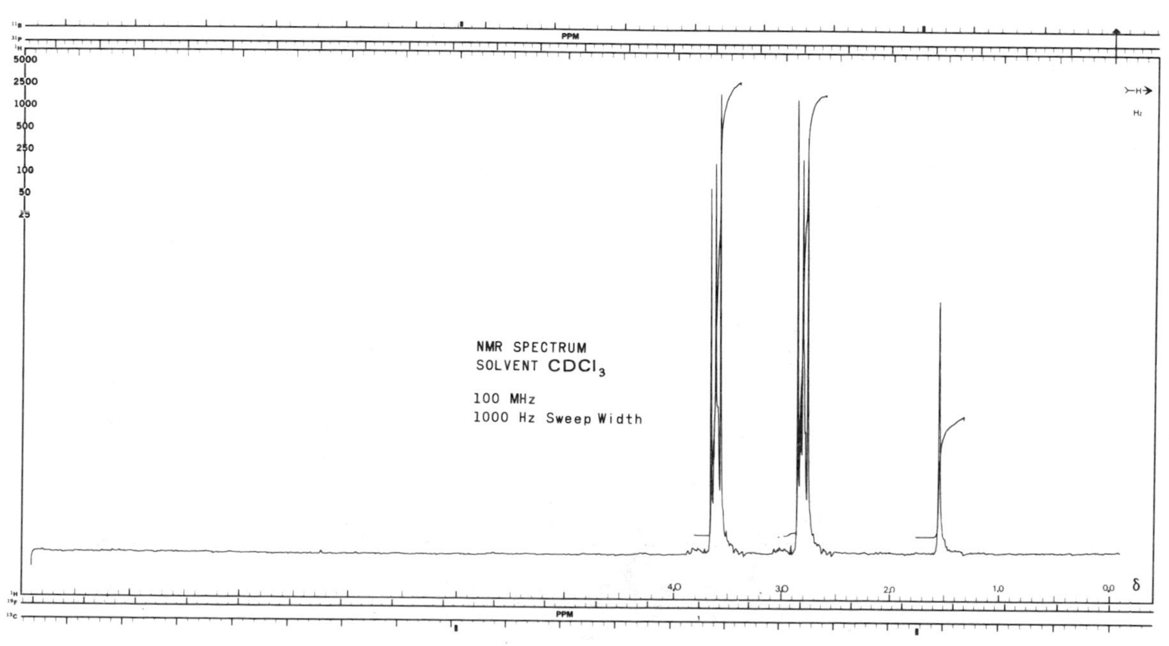

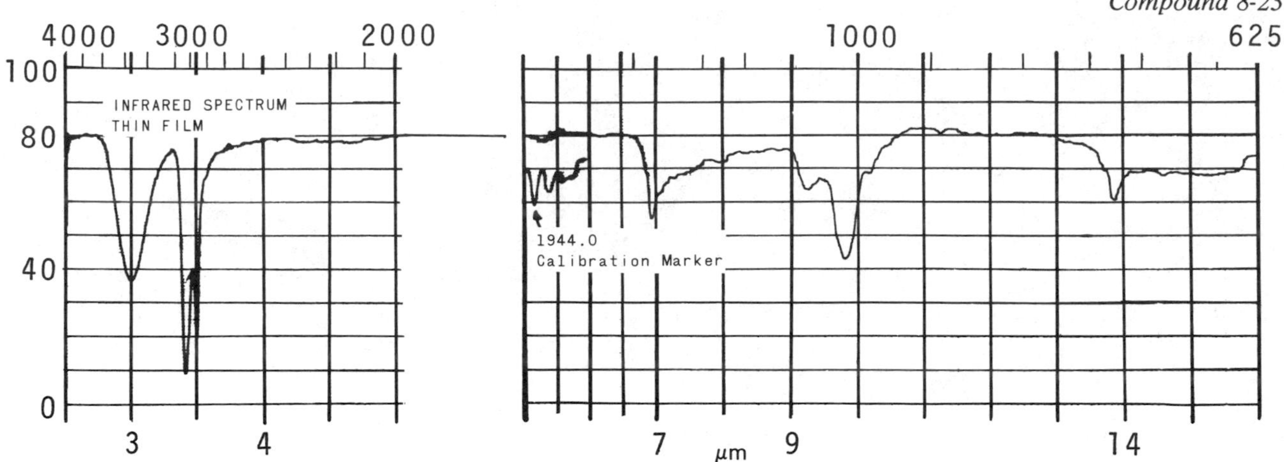

MASS SPECTRAL DATA (Relative Intensity)

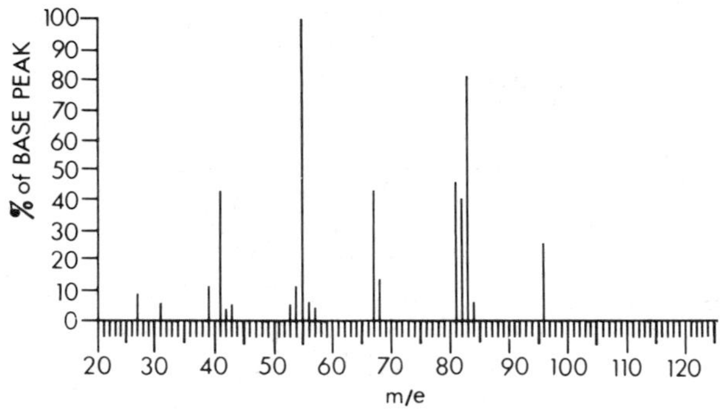

m/e 114, Trace

m/e 115, 116, too small to measure

Ultraviolet Data

Transparent above 200 nm

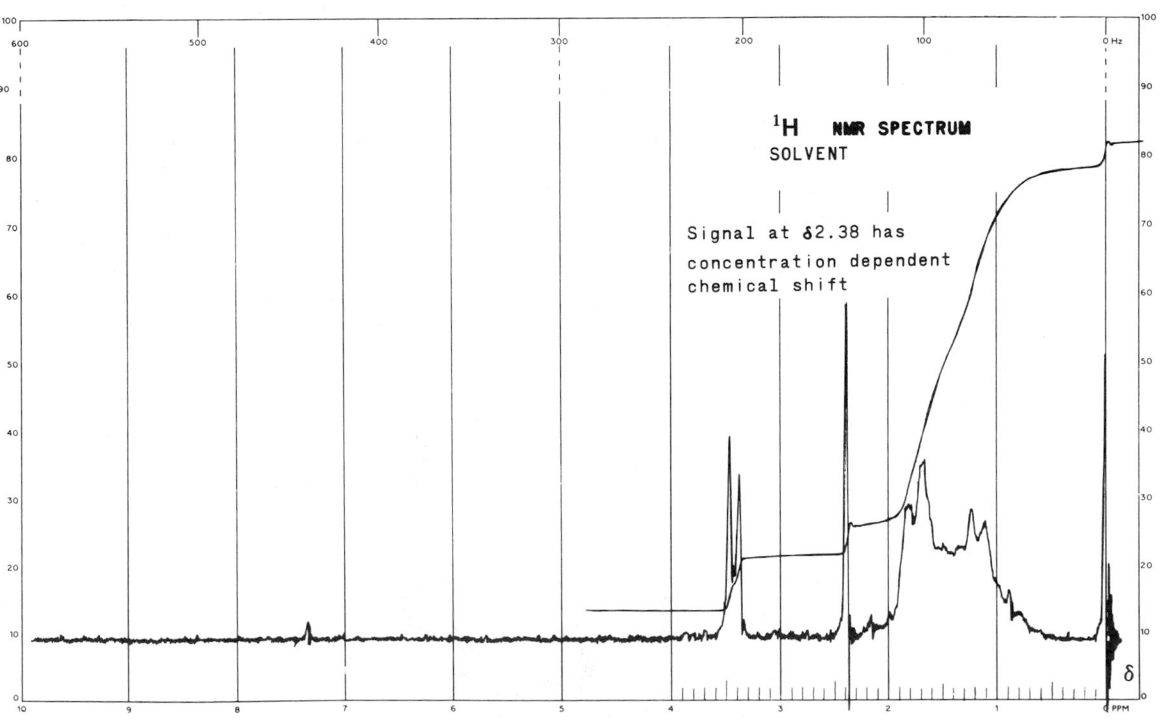

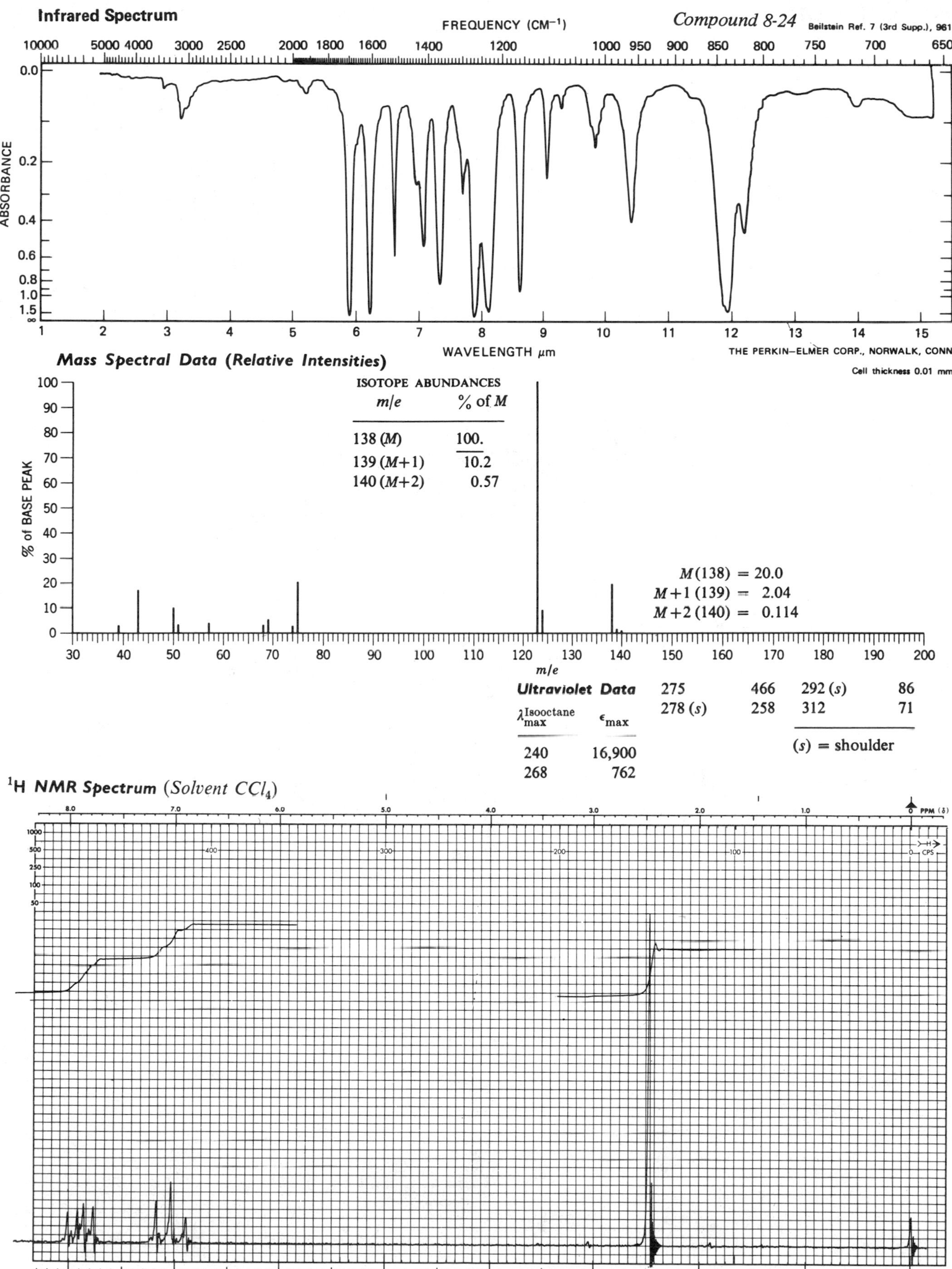

ISOTOPE ABUNDANCES

m/e	% of M
138 (M)	100.
139 (M+1)	10.2
140 (M+2)	0.57

Ultraviolet Data

$\lambda_{max}^{Isooctane}$	ϵ_{max}
240	16,900
268	762
275	466
278 (s)	258
292 (s)	86
312	71

(s) = shoulder

Infrared Spectrum

Compound 8-25

Beilstein Ref. 6, 140
Capillary Cell: Neat

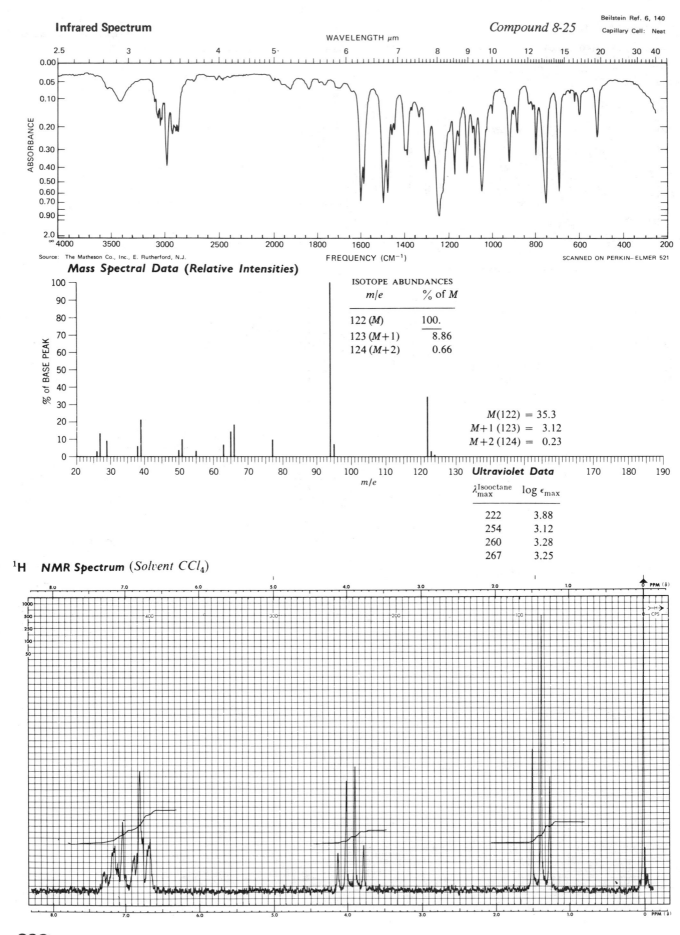

WAVELENGTH μm

2.5 3 4 5 6 7 8 9 10 12 15 20 30 40

ABSORBANCE

0.00 0.05 0.10 0.20 0.30 0.40 0.50 0.60 0.70 0.90 2.0 ∞

4000 3500 3000 2500 2000 1800 1600 1400 1200 1000 800 600 400 200

FREQUENCY (CM^{-1})

Source: The Matheson Co., Inc., E. Rutherford, N.J.

SCANNED ON PERKIN–ELMER 521

Mass Spectral Data (Relative Intensities)

% of BASE PEAK

100 90 80 70 60 50 40 30 20 10 0

20 30 40 50 60 70 80 90 100 110 120 130 170 180 190

m/e

ISOTOPE ABUNDANCES

m/e	% of *M*
122 (*M*)	100.
123 (*M*+1)	8.86
124 (*M*+2)	0.66

M(122) = 35.3
M+1 (123) = 3.12
M+2 (124) = 0.23

Ultraviolet Data

$\lambda_{max}^{Isooctane}$	$\log \epsilon_{max}$
222	3.88
254	3.12
260	3.28
267	3.25

^{1}H NMR Spectrum (*Solvent CCl_4*)

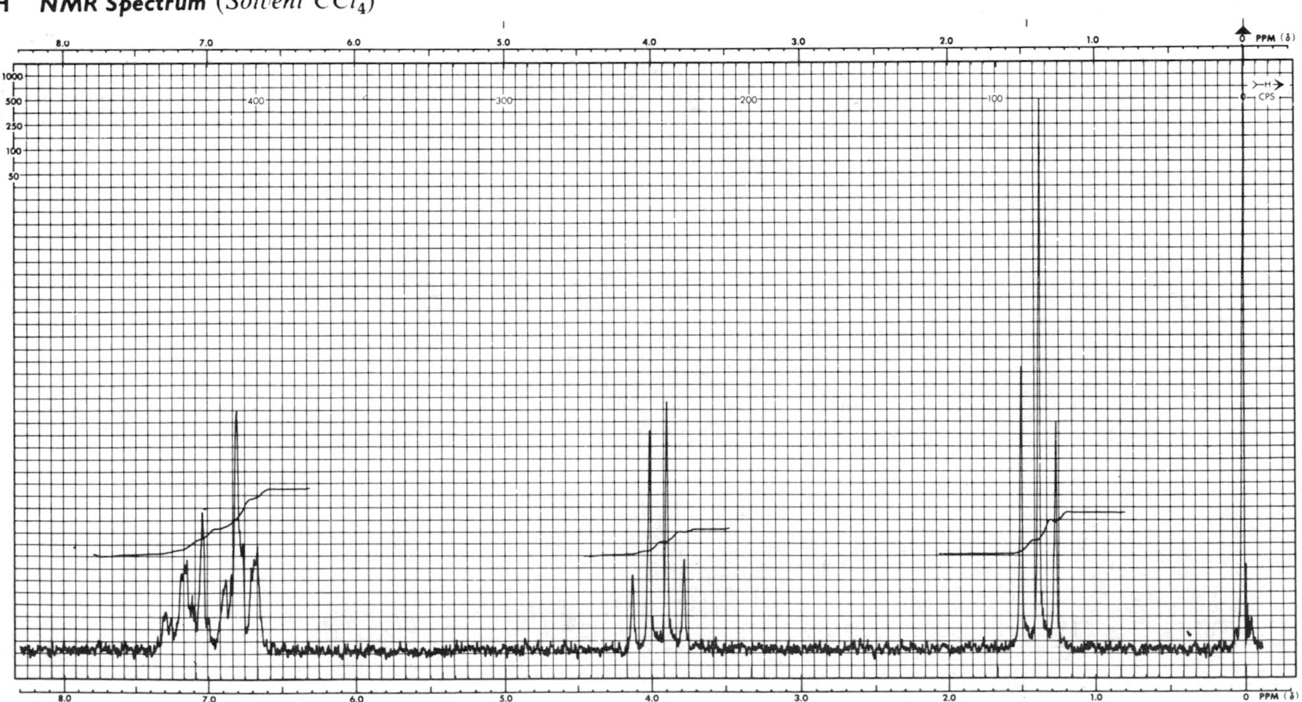

Compound 8-26

Infrared Spectrum

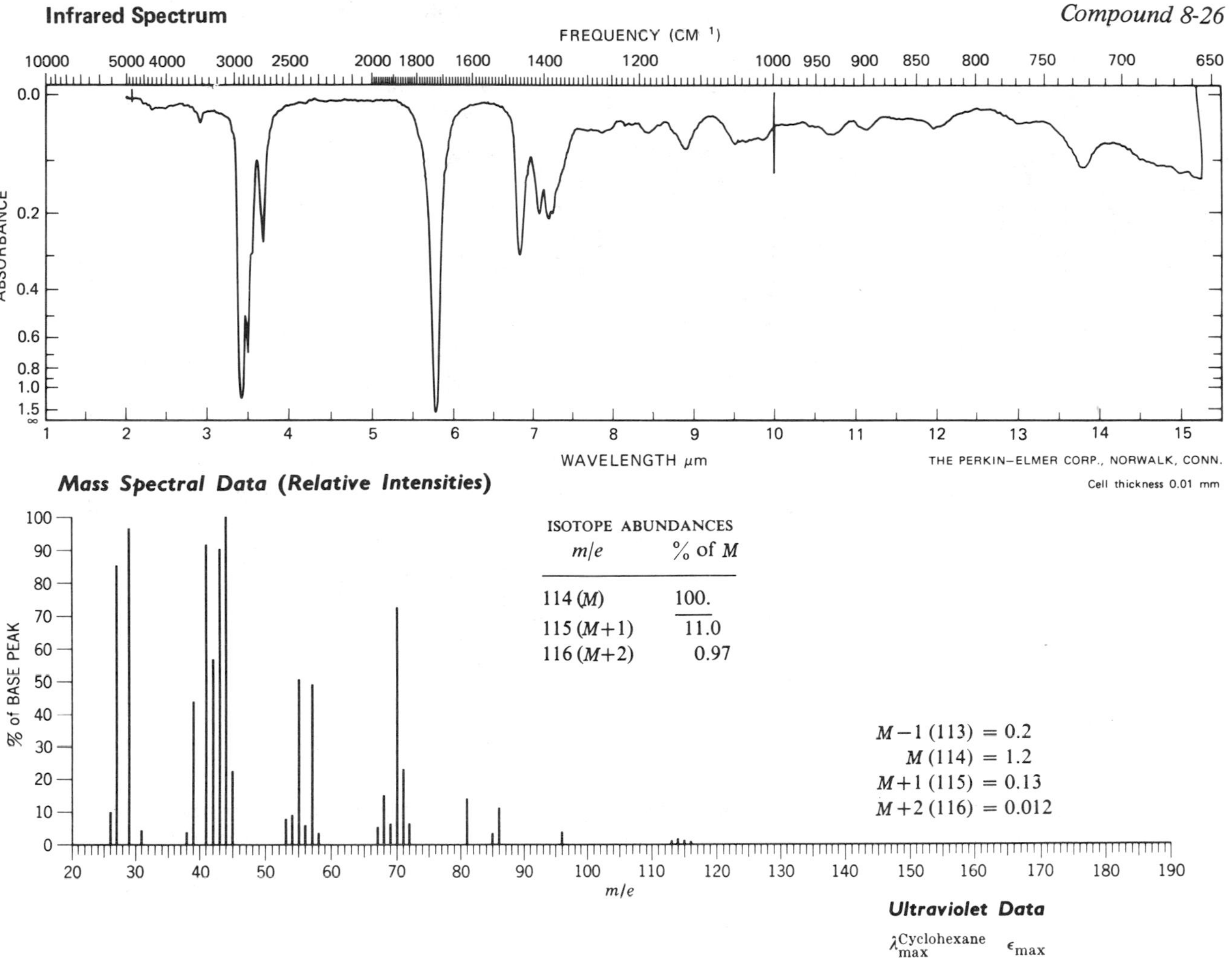

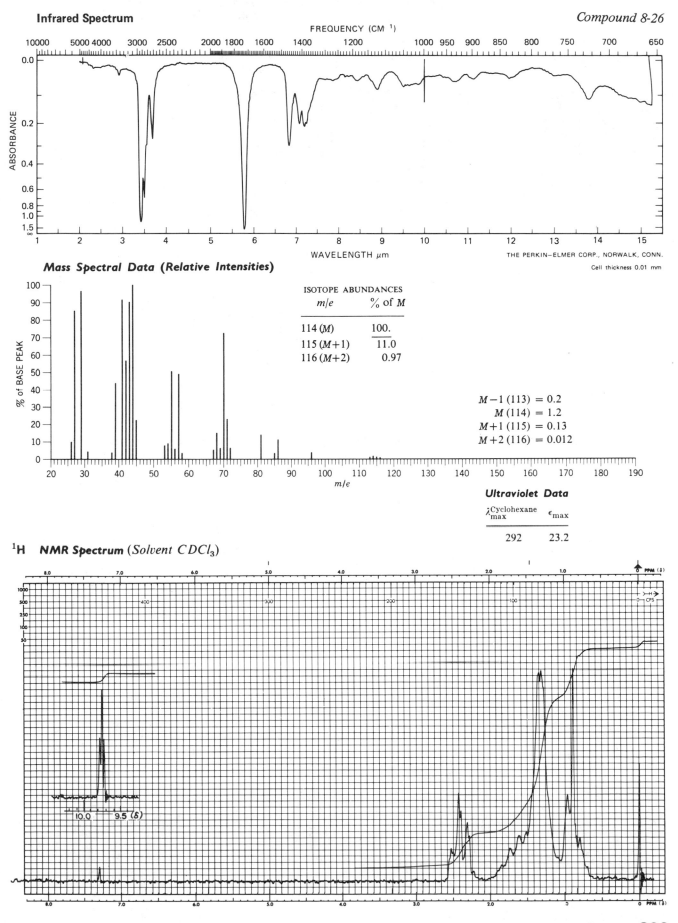

Mass Spectral Data (Relative Intensities)

ISOTOPE ABUNDANCES

m/e	% of M
114 (M)	100.
115 (M+1)	11.0
116 (M+2)	0.97

$M-1$ (113) = 0.2
M (114) = 1.2
$M+1$ (115) = 0.13
$M+2$ (116) = 0.012

Ultraviolet Data

$\lambda_{max}^{Cyclohexane}$	ϵ_{max}
292	23.2

^{1}H NMR Spectrum (Solvent $CDCl_3$)

Compound 8-27 Beilstein Ref. **II, 91,** 586

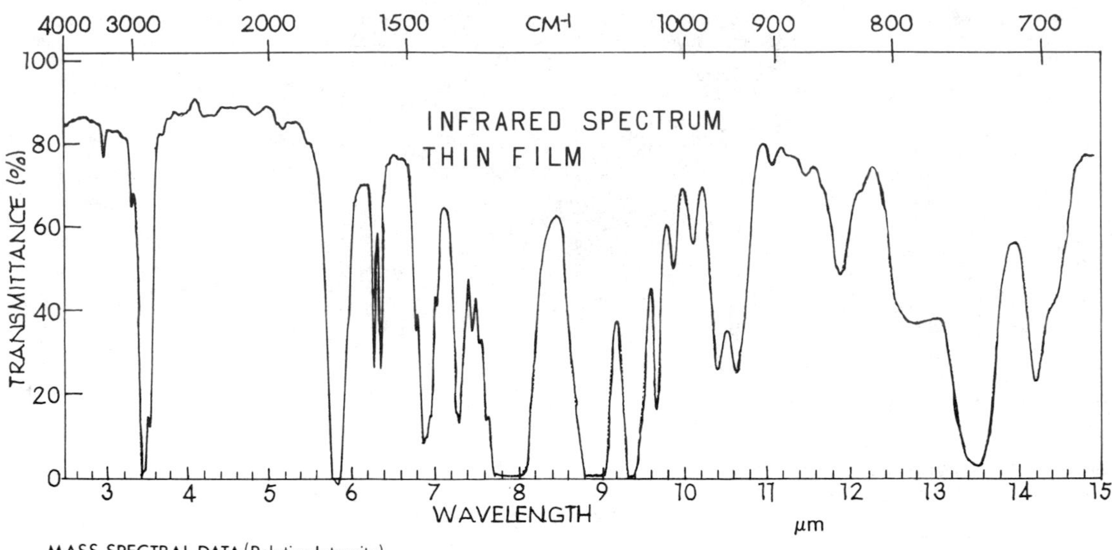

MASS SPECTRAL DATA (Relative Intensity)

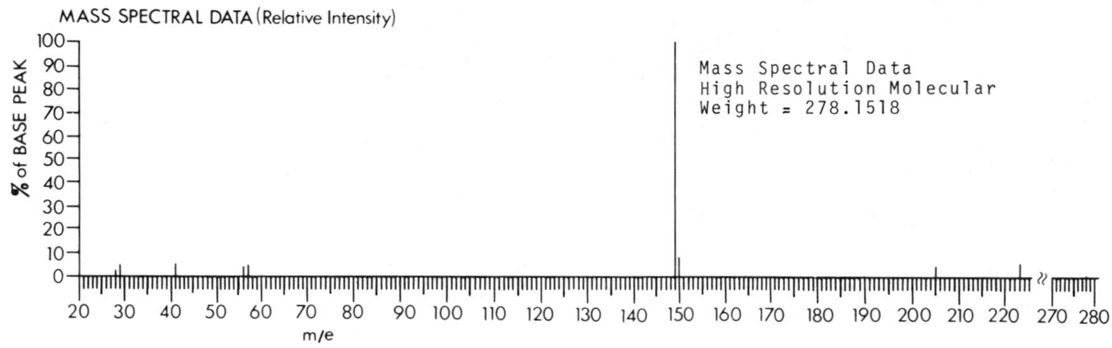

Ultraviolet Data

λ_{max} (CH_3OH)	ε_{max}
274	1230
225	7960

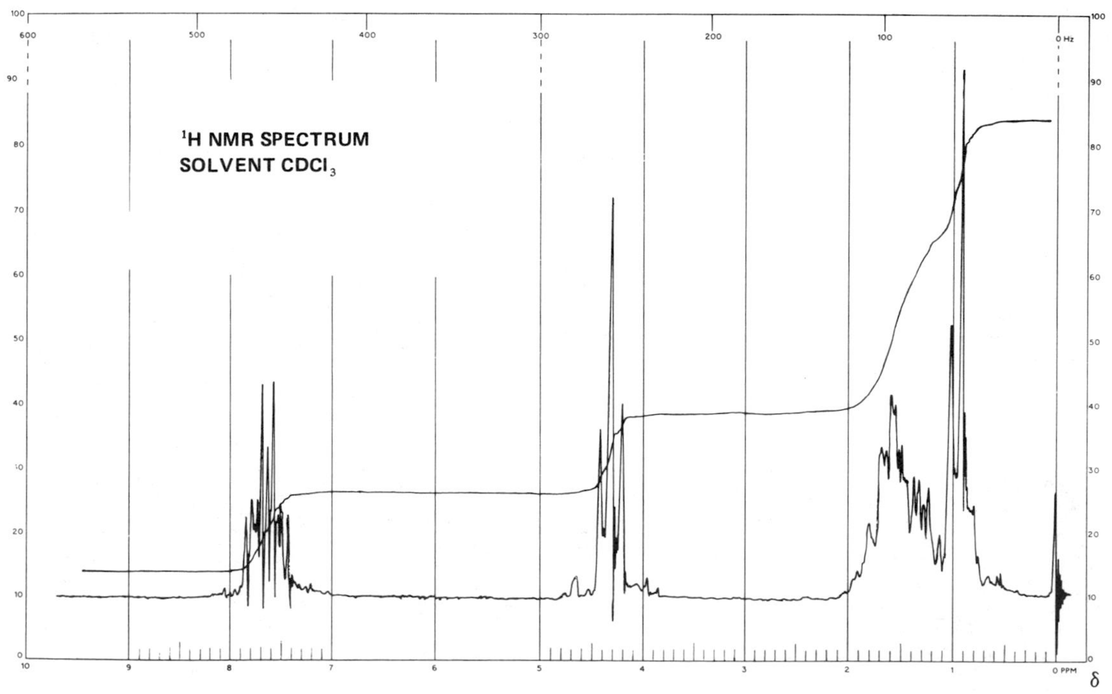

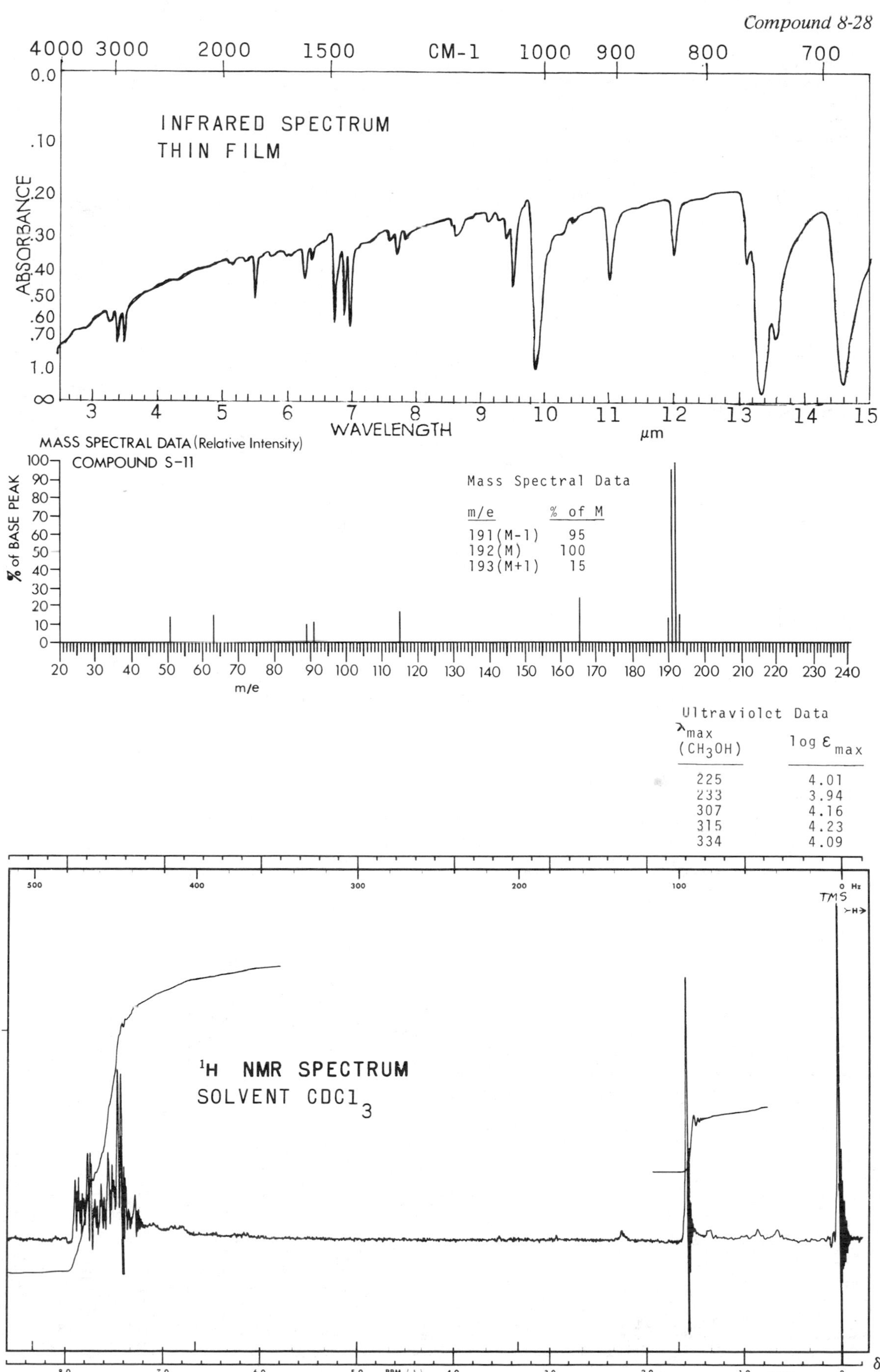

Ultraviolet Data

λ_{max} (CH_3OH)	log ε_{max}
225	4.01
233	3.94
307	4.16
315	4.23
334	4.09

Infrared Spectrum *Compound 8-29*

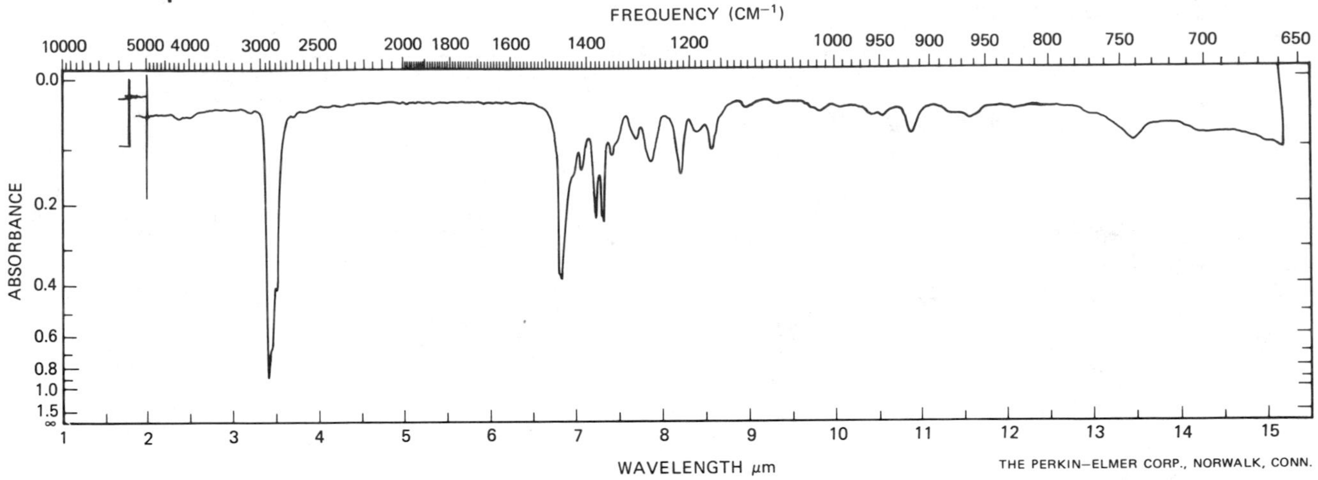

FREQUENCY (CM⁻¹)
ABSORBANCE
WAVELENGTH μm
THE PERKIN–ELMER CORP., NORWALK, CONN.
Cell thickness 0.01 mm

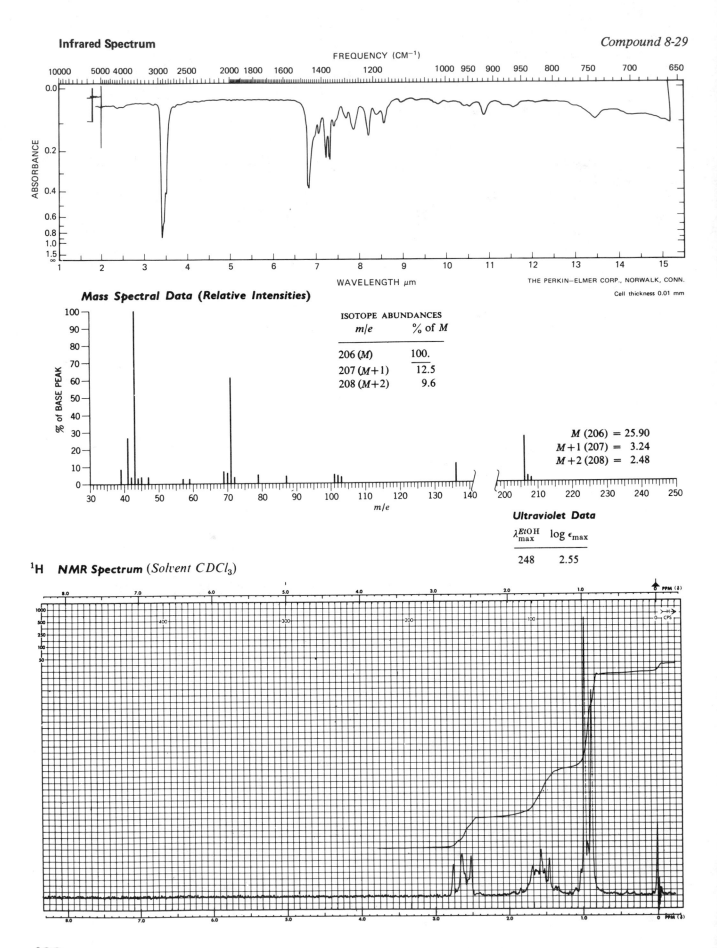

Mass Spectral Data (Relative Intensities)

% of BASE PEAK
m/e

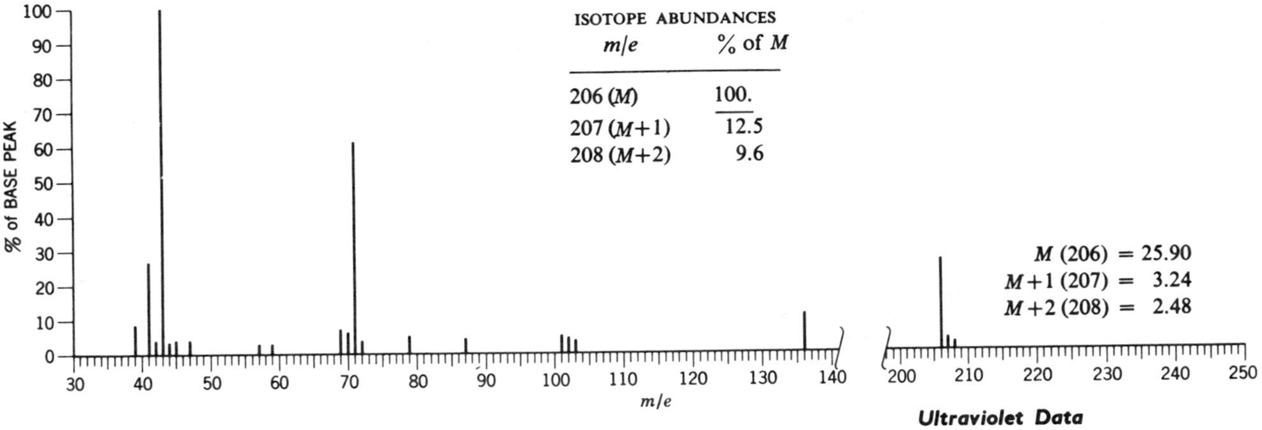

ISOTOPE ABUNDANCES

m/e	% of *M*
206 (*M*)	100.
207 (*M*+1)	12.5
208 (*M*+2)	9.6

M (206) = 25.90
M+1 (207) = 3.24
M+2 (208) = 2.48

Ultraviolet Data

λ_{max}^{EtOH}	log ϵ_{max}
248	2.55

^{1}H **NMR Spectrum** (*Solvent CDCl$_3$*)

PPM (δ)
CPS

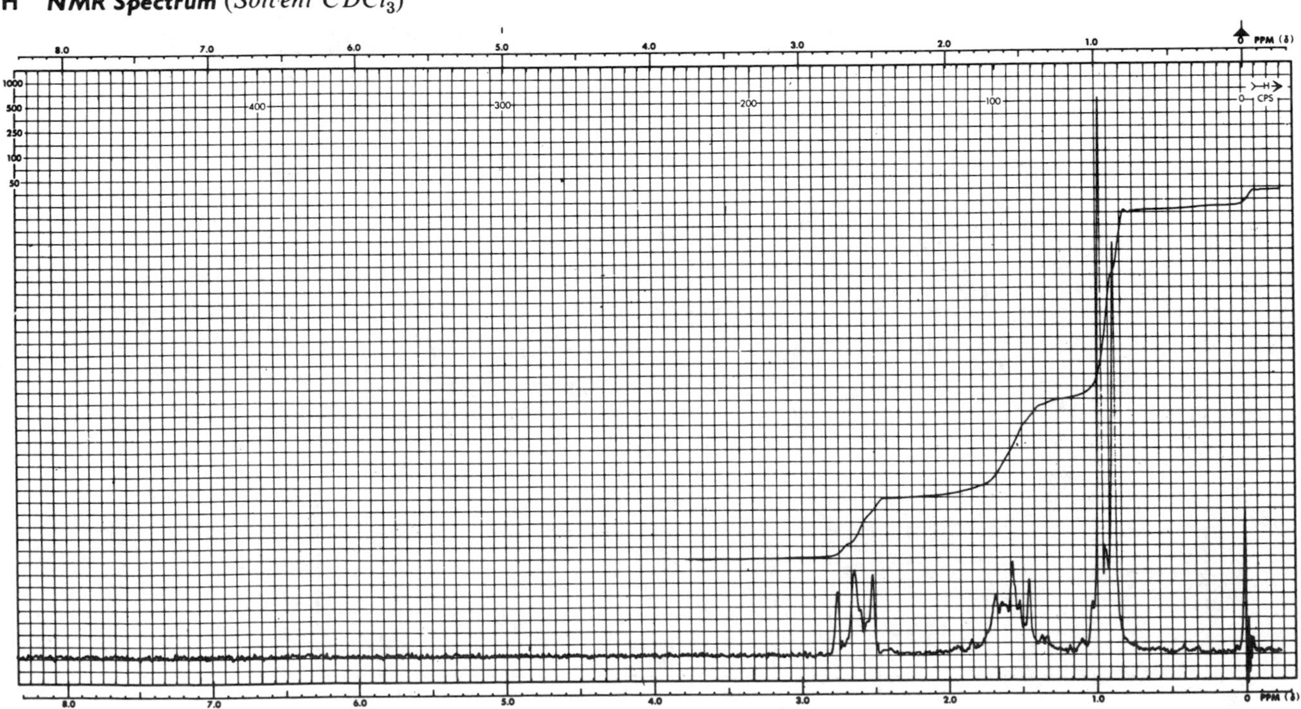

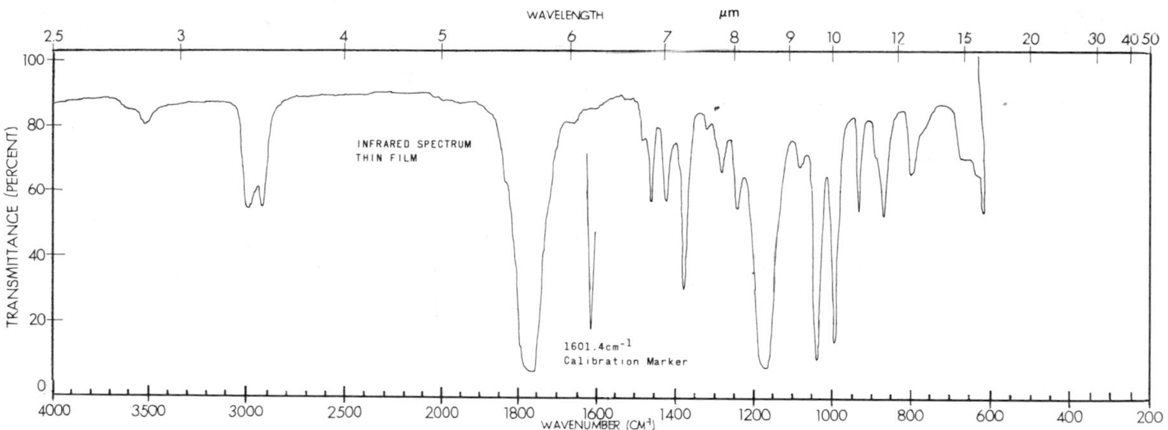

MASS SPECTRAL DATA (Relative Intensity)

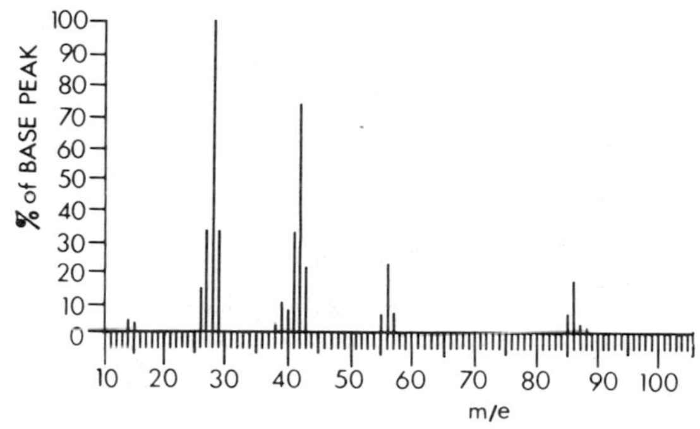

Mass Spectral Data

m/e	Relative Intensity
85	34.5
86	100.0
87	10.7
88	0.6

Ultraviolet Data

Transparent above 200nm

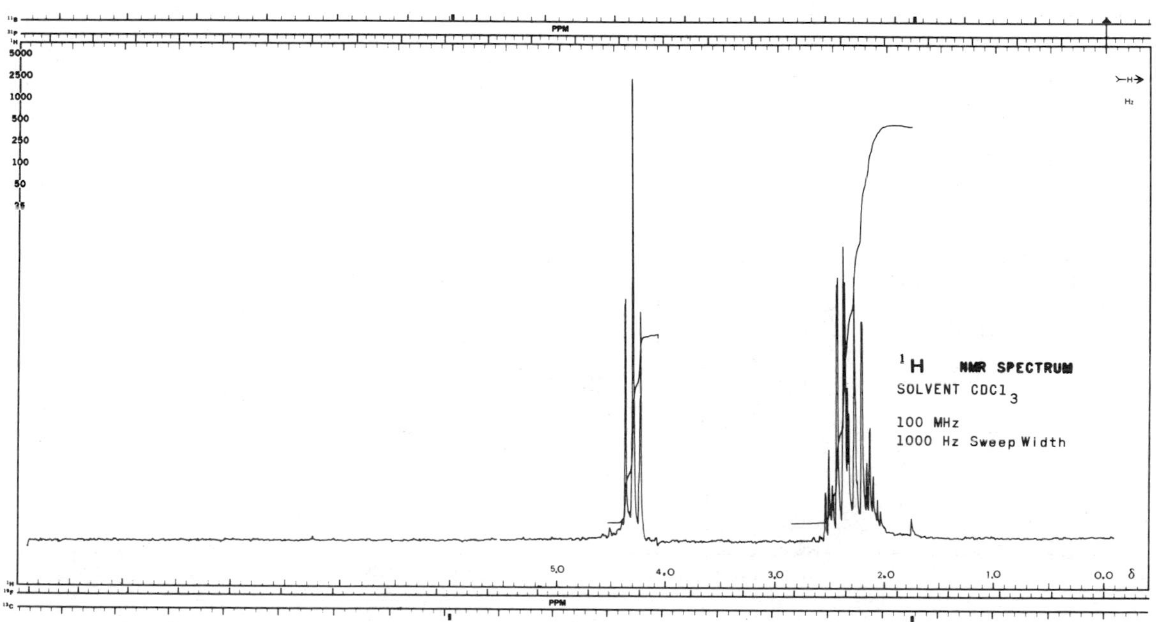

Infrared Spectrum *Compound 8-31* Beilstein Ref. 12, 437

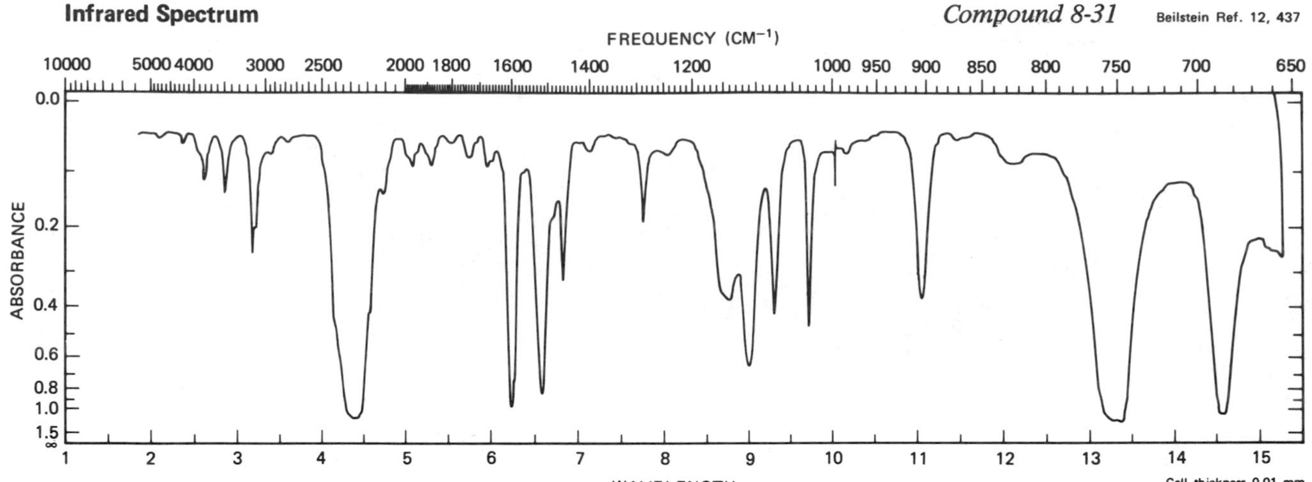

Mass Spectral Data (Relative Intensities)

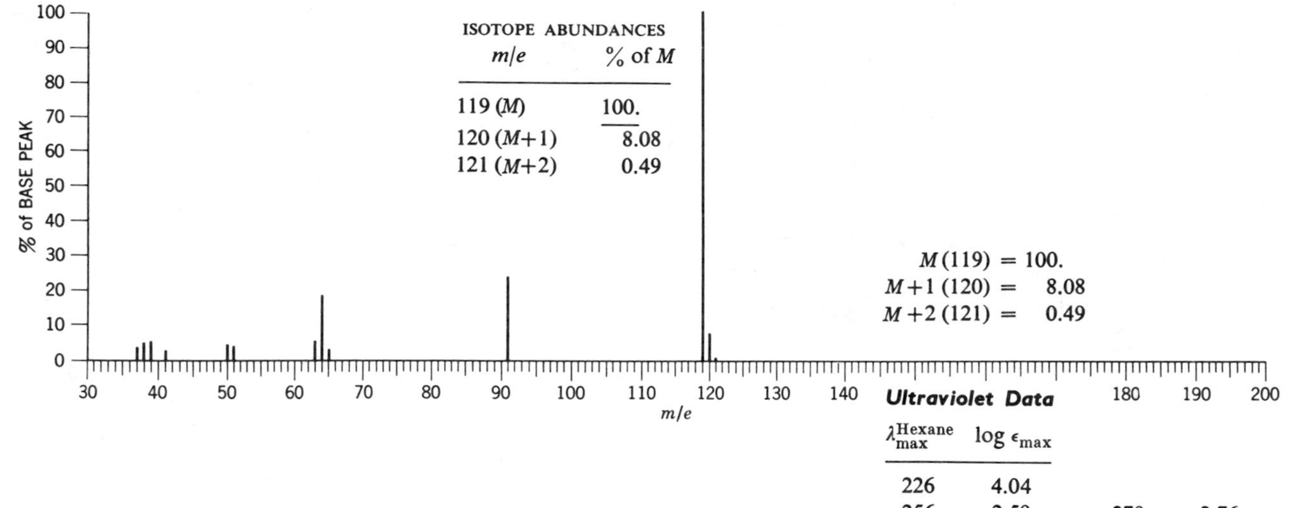

ISOTOPE ABUNDANCES

m/e	% of M
119 (M)	100.
120 (M+1)	8.08
121 (M+2)	0.49

Ultraviolet Data

λ_{max}^{Hexane}	log ϵ_{max}		
226	4.04		
256	2.59	270	2.76
263	2.66	277	2.67

^{1}H ***NMR Spectrum***

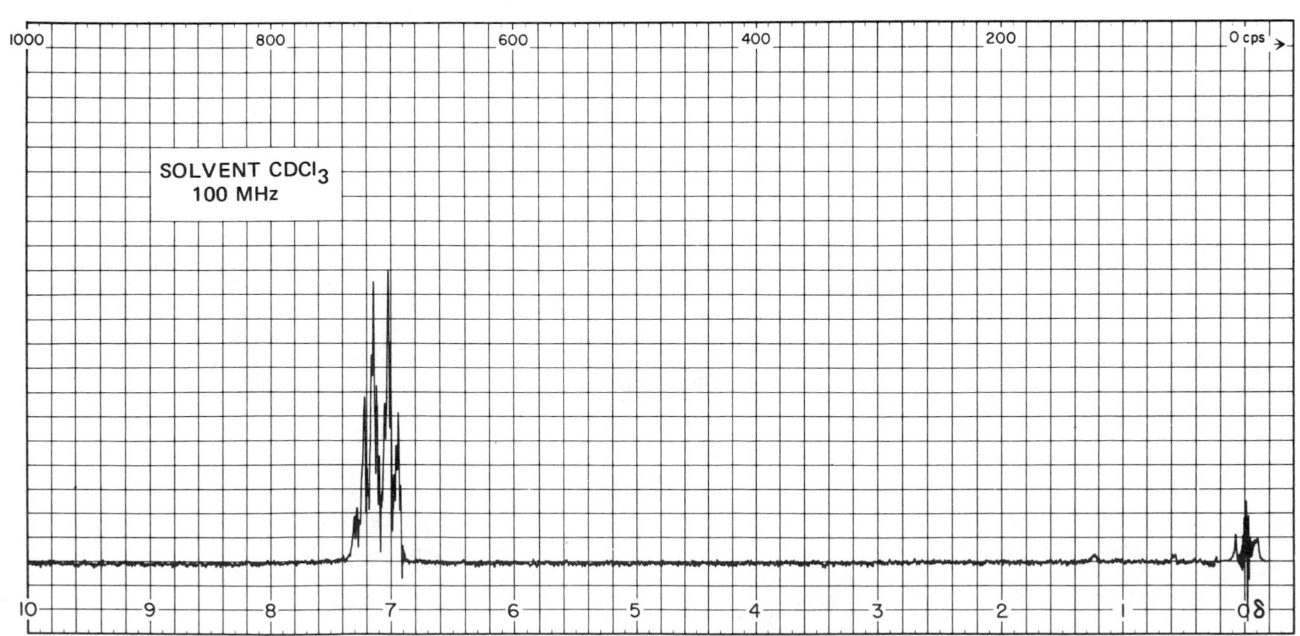

Compound 8-31 ^{13}C NMR Data ($CDCl_3$ solvent)

δ	Intensity	Off-resonance Decoupled Multiplet
124.7	90	*d*
125.7	60	*d*
129.5	100	*m*
133.6	10	*s*

Compound 8-32

Beilstein Ref., 2nd Supp., **4**, 938

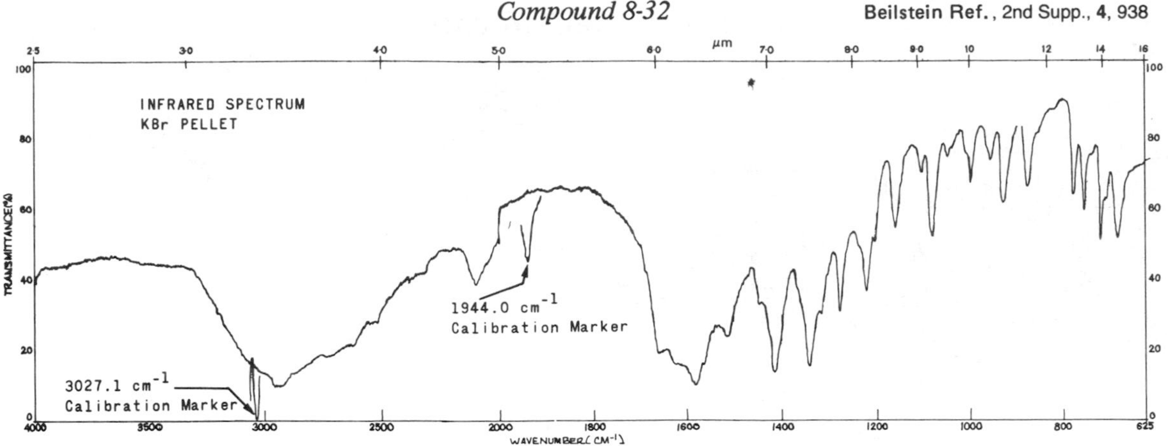

MASS SPECTRAL DATA (Relative Intensity)

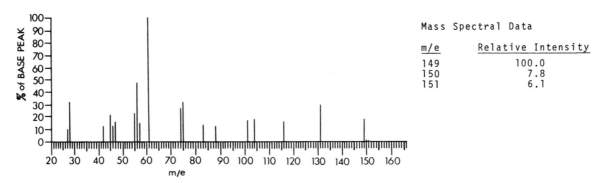

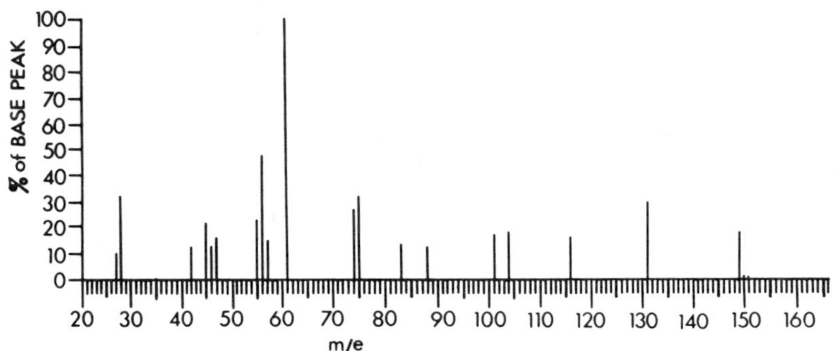

Mass Spectral Data

m/e	Relative Intensity
149	100.0
150	7.8
151	6.1

Ultraviolet Data

Transparent above 200nm

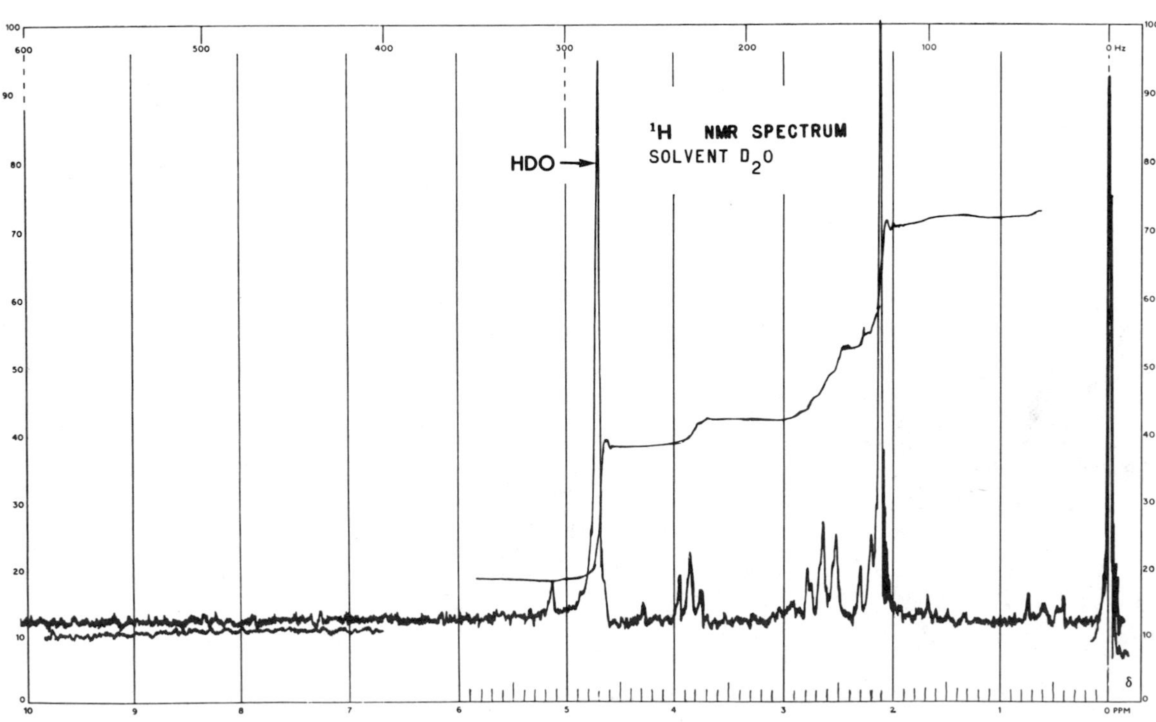

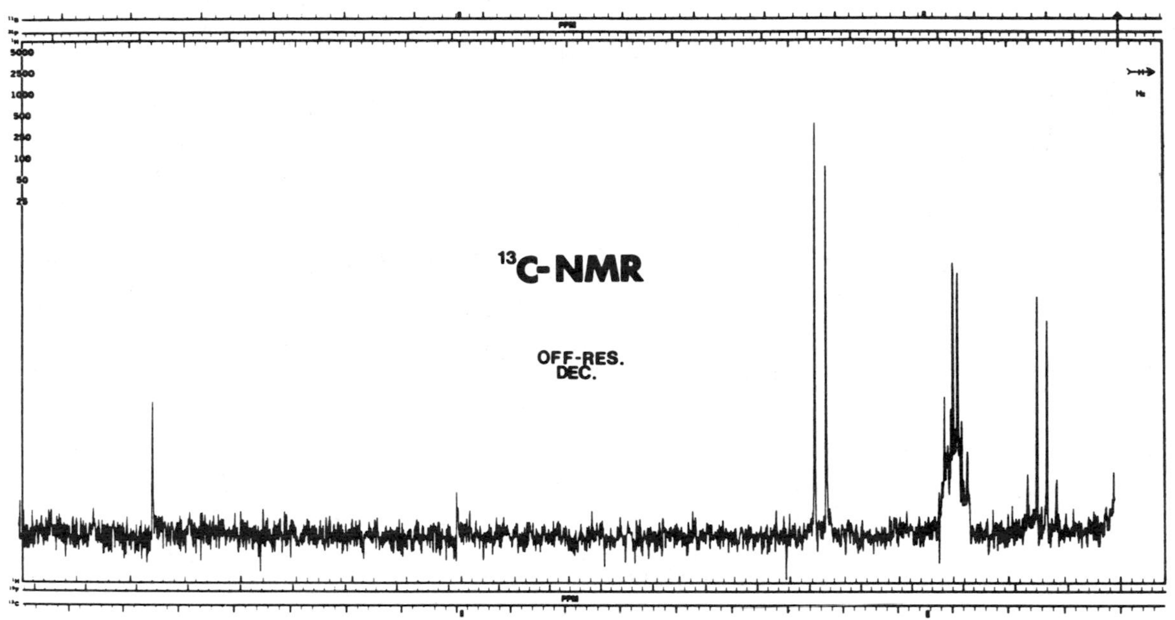

^{13}C-NMR
OFF-RES. DEC.

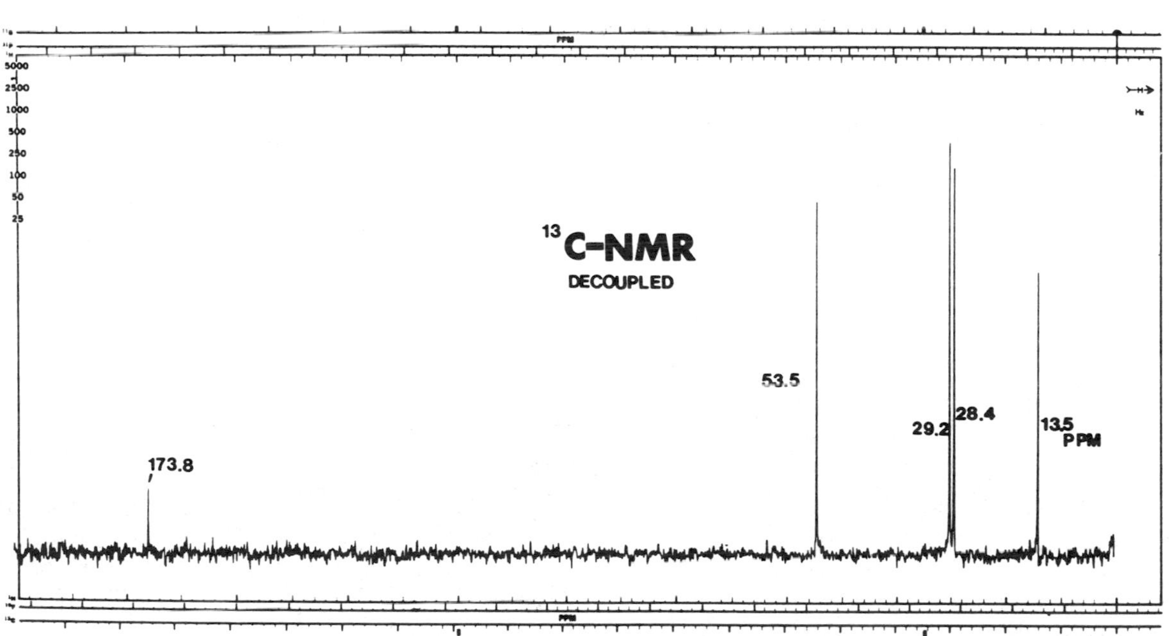

^{13}C-NMR
DECOUPLED
173.8
53.5
29.2
28.4
13.5
PPM

Infrared Spectrum

Compound 8-33

Beilstein Ref. 6, 146
Capillar Cell Neat

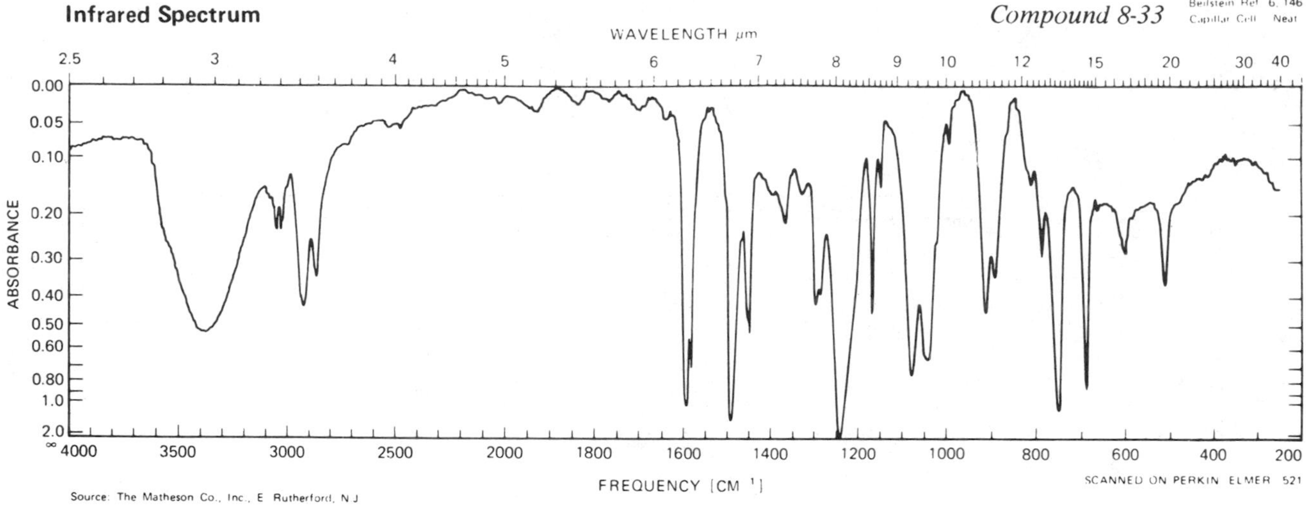

Source: The Matheson Co., Inc., E. Rutherford, N.J.

Mass Spectral Data (Relative Intensities)

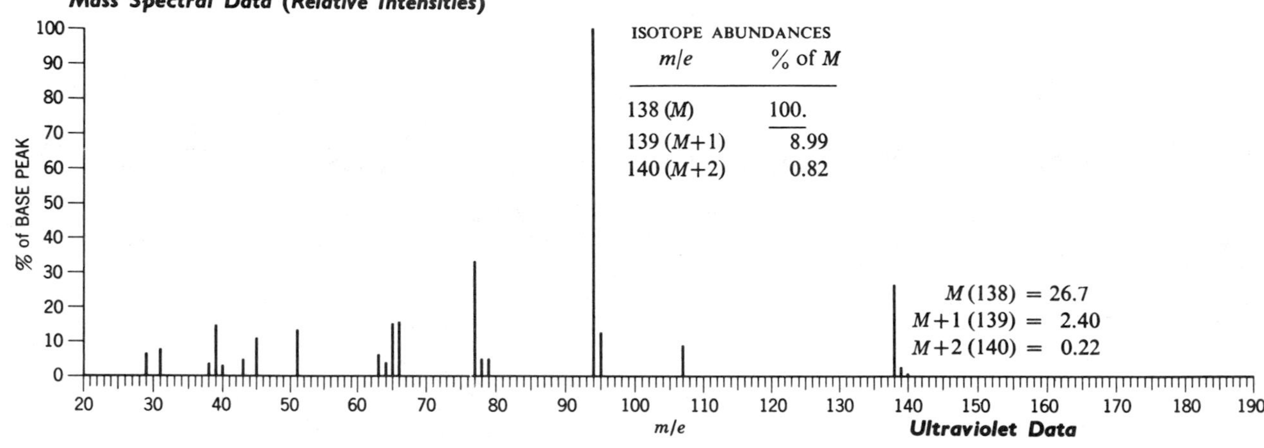

ISOTOPE ABUNDANCES

m/e	% of M
138 (M)	100.
139 ($M+1$)	8.99
140 ($M+2$)	0.82

$M(138) = 26.7$
$M+1\ (139) = 2.40$
$M+2\ (140) = 0.22$

Ultraviolet Data

$\lambda_{max}^{Isooctane}$	$\log \epsilon_{max}$
219	3.96
253	3.22
260	3.36
267	3.35

¹H NMR Spectrum (*Solvent CCl_4*)

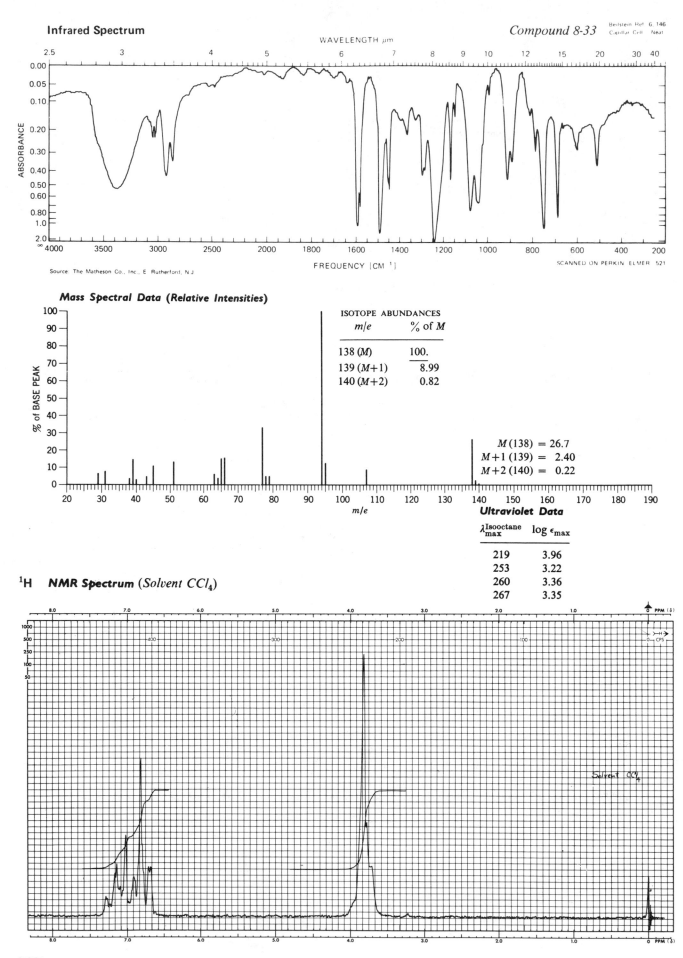

8.0 7.0 6.0 5.0 4.0 3.0 2.0 1.0 0 PPM (δ)

400 300 200 100 0 CPS

Solvent CCl_4

Compound 8-33 ^{13}C NMR Data ($CDCl_3$ solvent)

δ	Intensity	Off-resonance Decoupled Multiplet
60.9	40.0	*t*
69.1	47.1	*t*
114.6	88.4	*d*
120.9	38.8	*d*
129.5	100.0	*d*
158.8	18.1	*s*

Compound 8-34

Beilstein Ref., **17**, 287

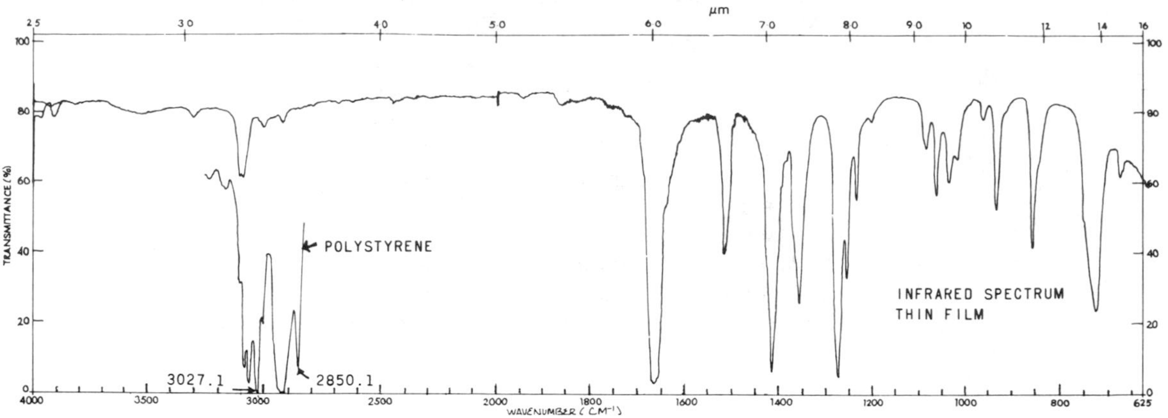

MASS SPECTRAL DATA (Relative Intensity)

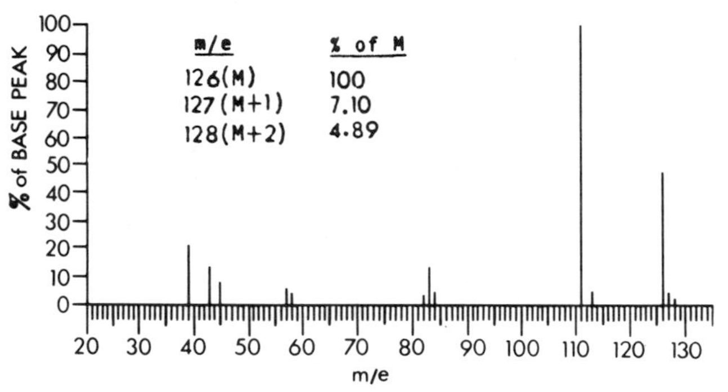

Mass Spectral Data

Molecular Weight 126.0135

(High Resolution Analysis)

Ultraviolet Data

λ_{max} (CH_3OH)	ε_{max}
257	10,500
280	7,200

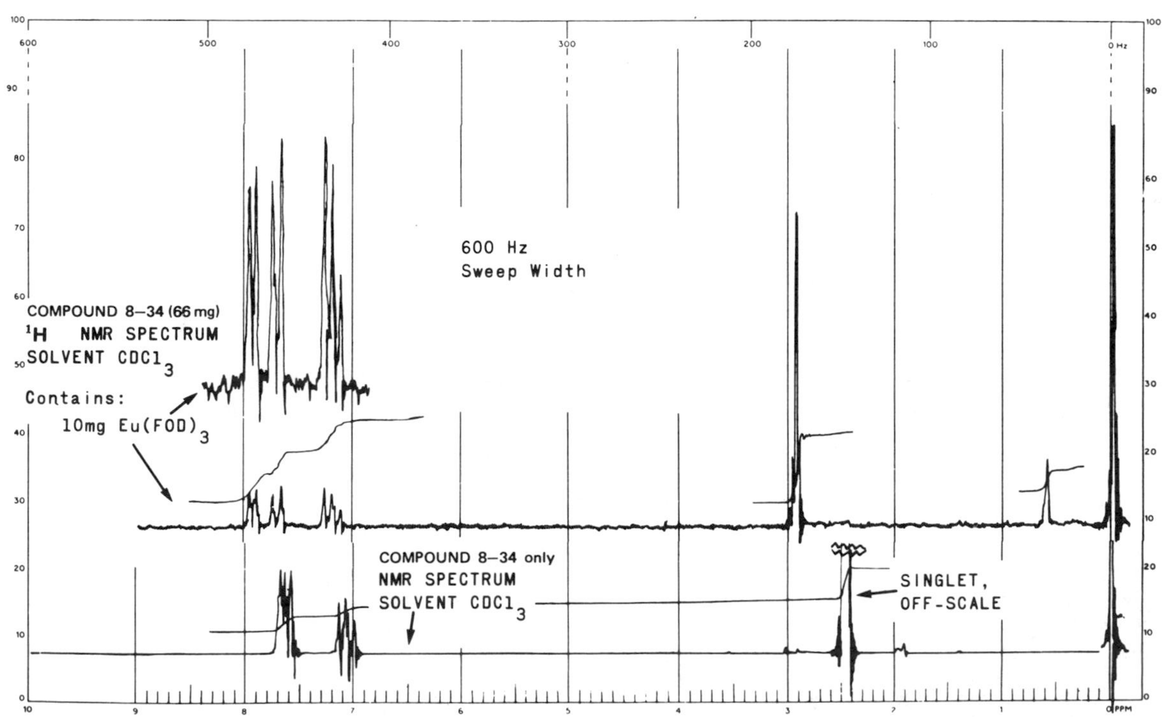

Compound 8-34 ^{13}C NMR Data ($CDCl_3$ solvent)

δ	Intensity	Off-resonance Decoupled Multiplet
26.6	76.0	*q*
128.3	89.8	*d*
132.8	100.0	*d*
133.8	98.4	*d*
144.6	25.0	*s*
190.4	34.9	*s*

Infrared Spectrum *Compound 8-35* Beilstein Ref. **6**, 5

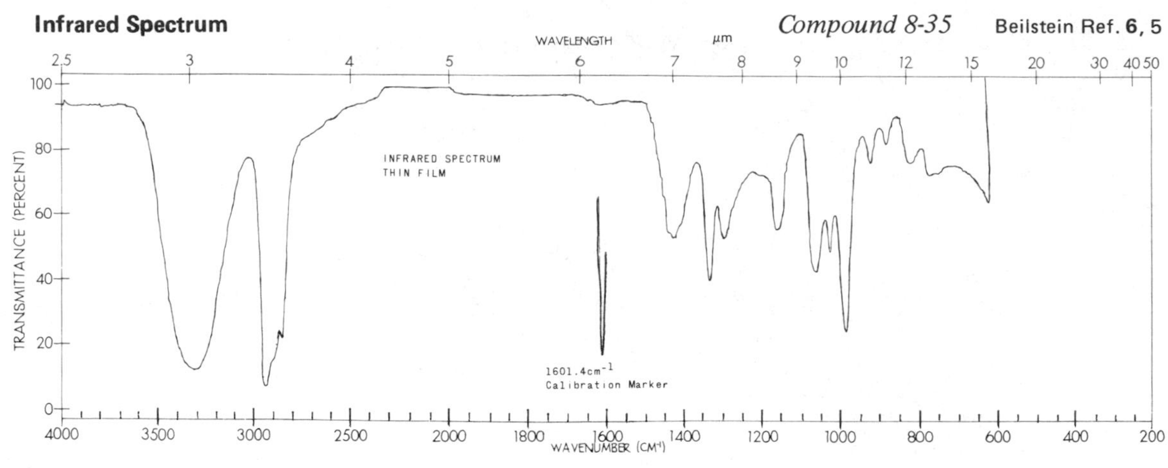

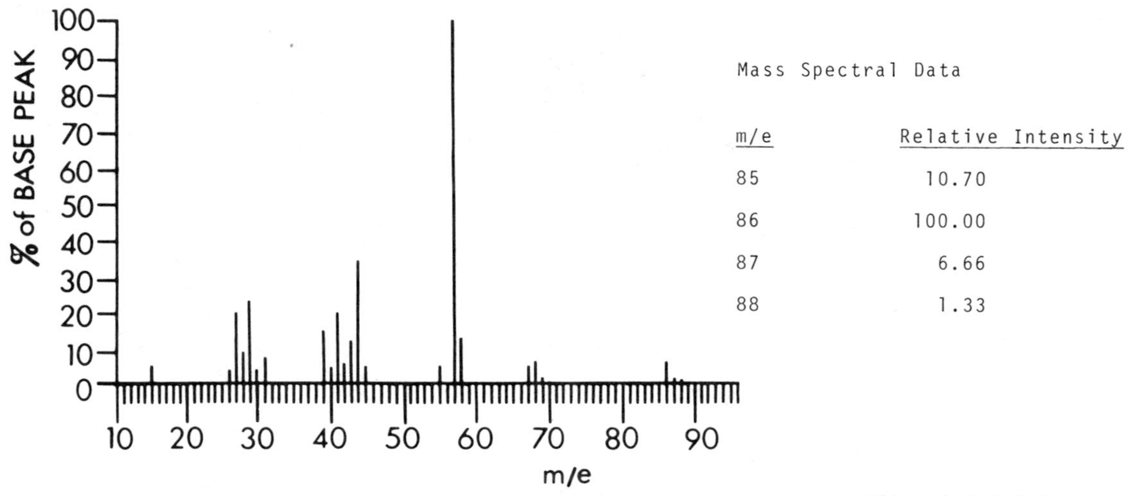

Mass Spectral Data

m/e	Relative Intensity
85	10.70
86	100.00
87	6.66
88	1.33

Ultraviolet Data

Transparent above 200nm

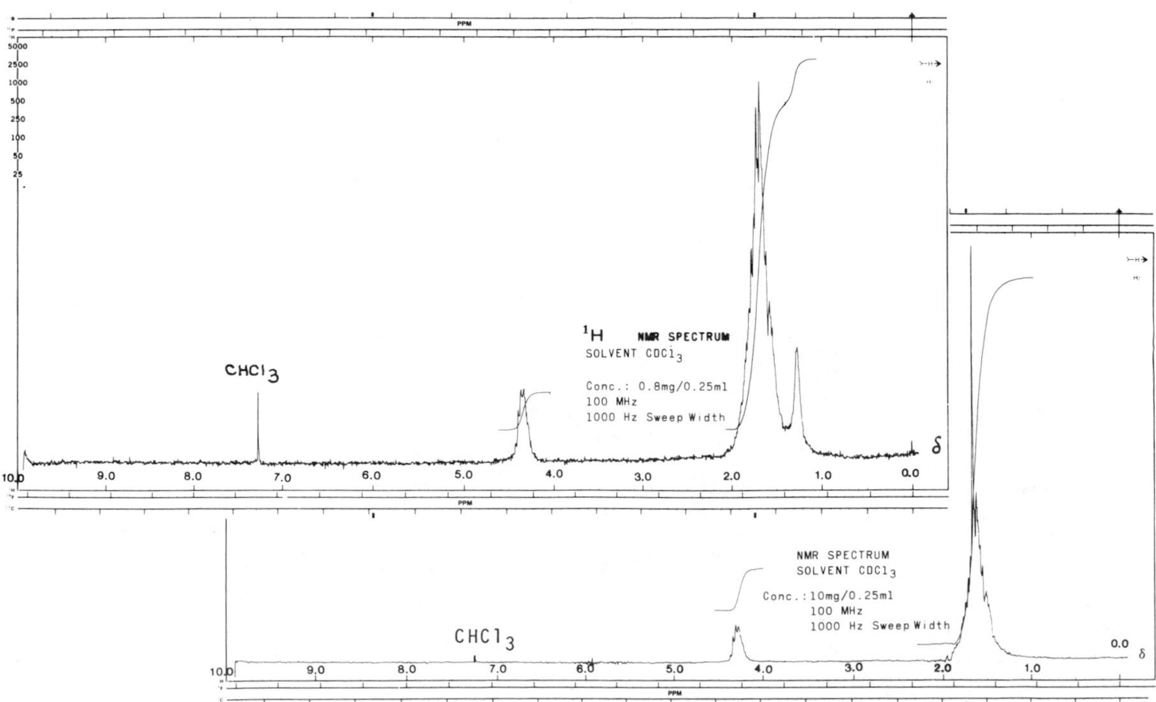

Compound 8-35 ^{13}C NMR Data (Dioxane solvent)

δ	Off-resonance Decoupled Multiplet
23.40	*t*
35.00	*t*
73.30	*d*

Infrared Spectrum *Compound 8-36*

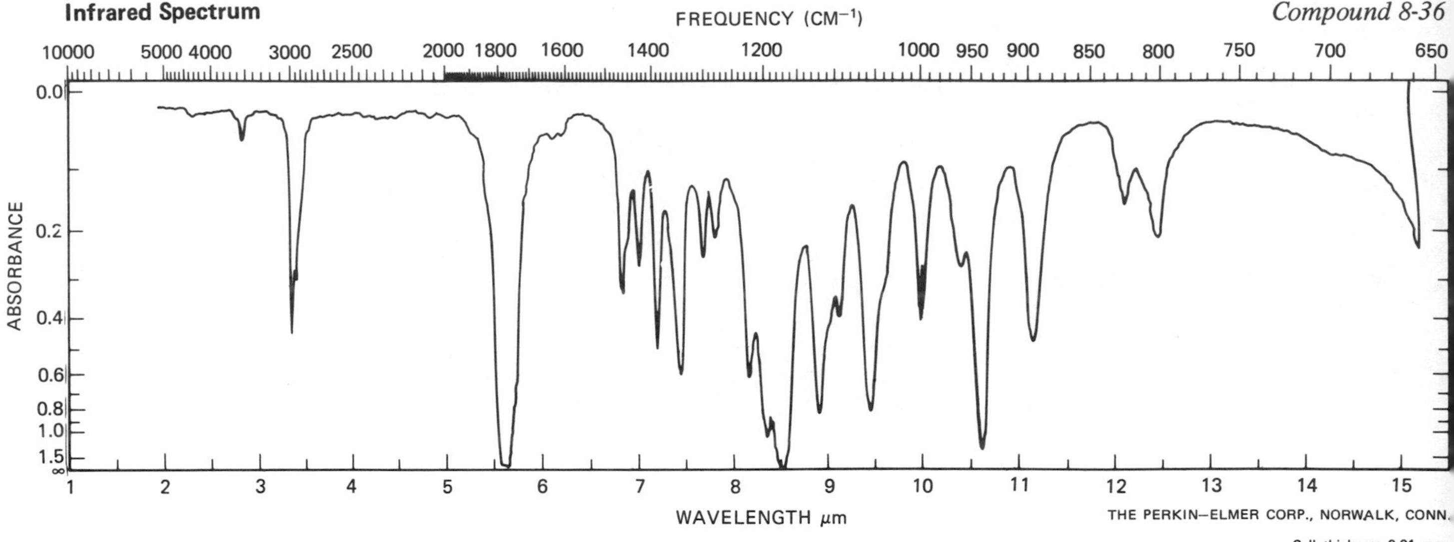

Mass Spectral Data (Relative Intensities)

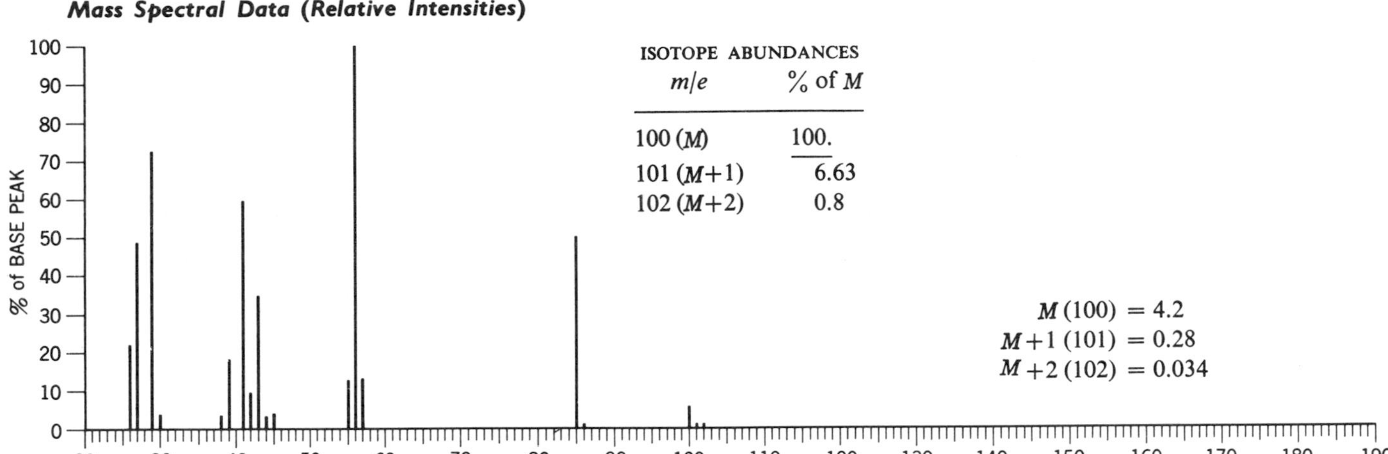

ISOTOPE ABUNDANCES

m/e	% of M
100 (M)	100.
101 ($M+1$)	6.63
102 ($M+2$)	0.8

M (100) = 4.2
$M+1$ (101) = 0.28
$M+2$ (102) = 0.034

Ultraviolet Data

Transparent above 200 nm

^{1}H **NMR Spectrum** *(Solvent CCl_4)*

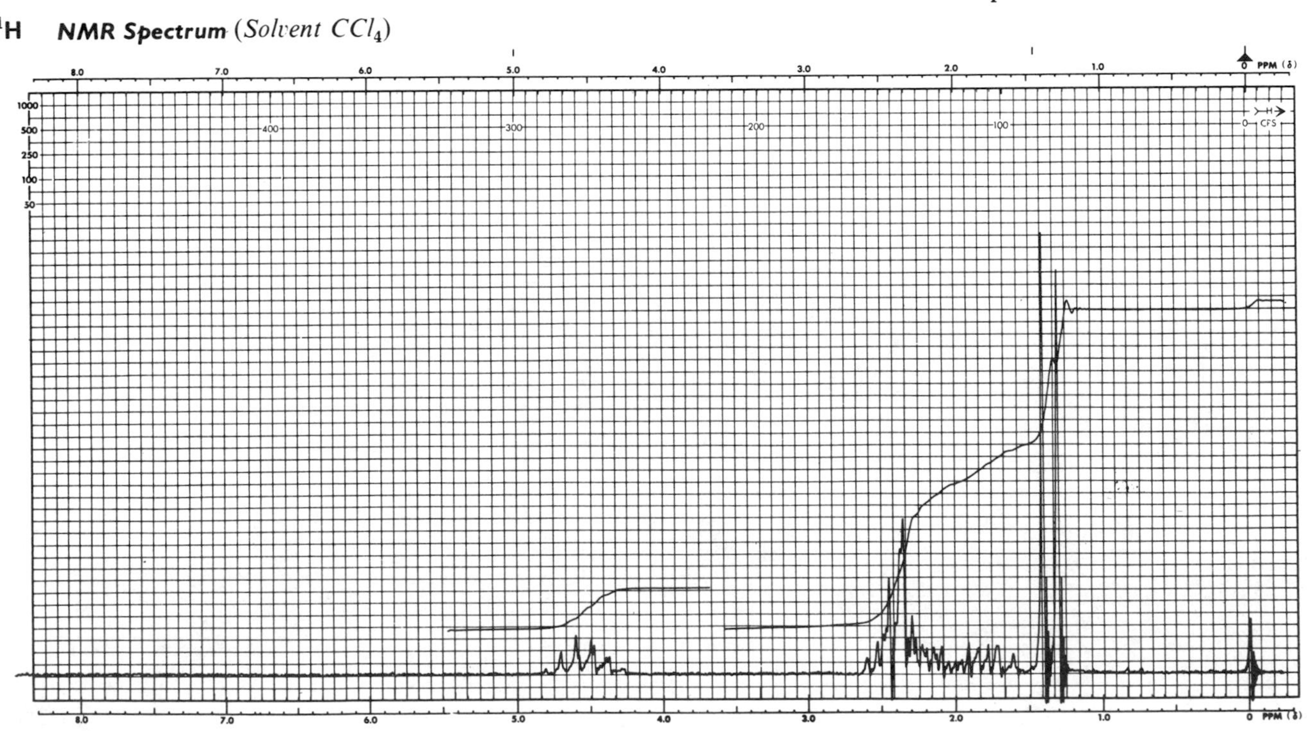

notes

Compound 8-37

Infrared Spectrum

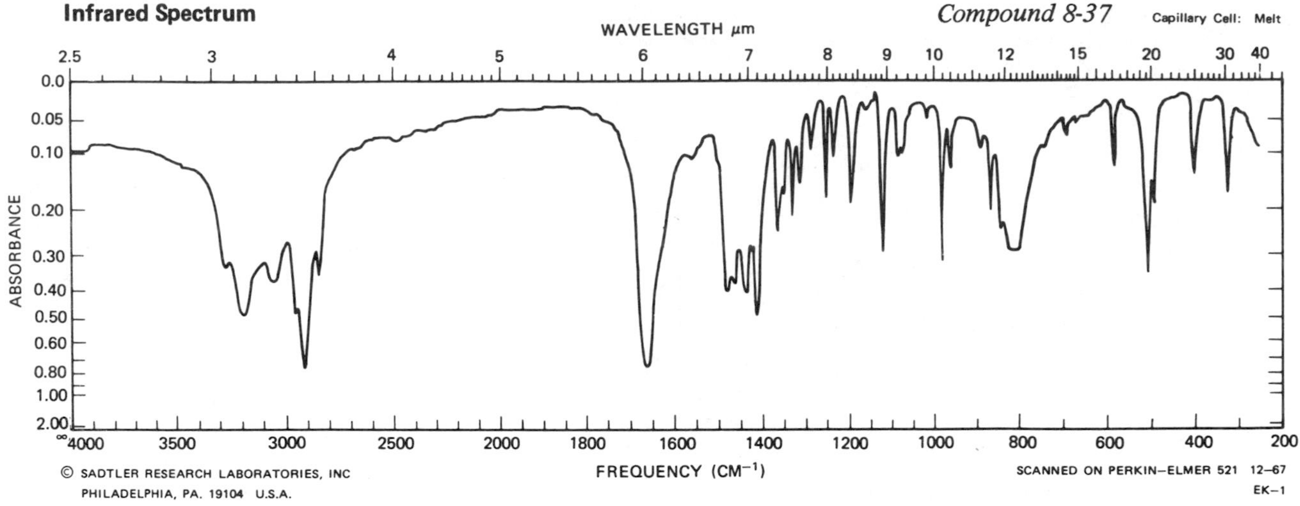

Mass Spectral Data (Relative Intensities)

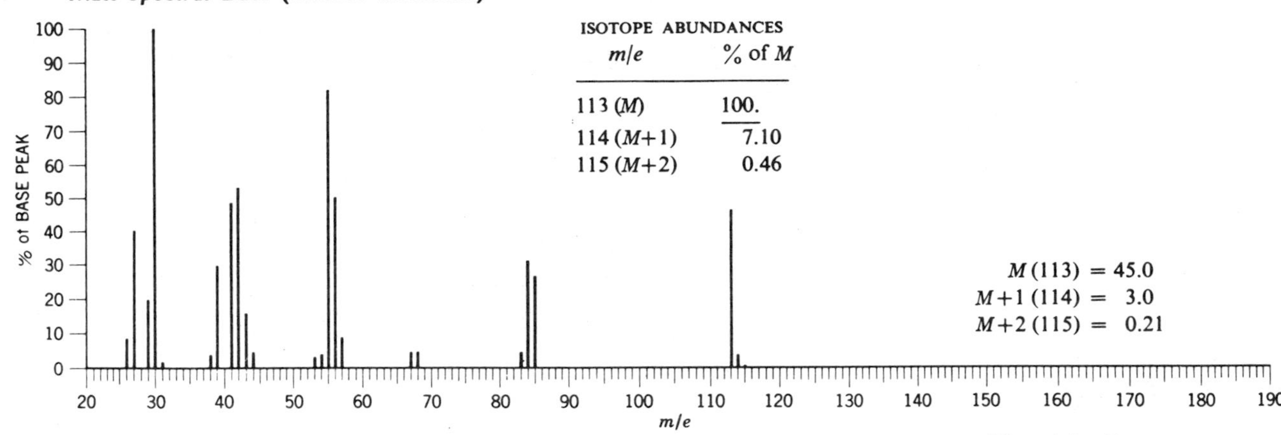

ISOTOPE ABUNDANCES

m/e	% of M
113 (M)	100.
114 (M+1)	7.10
115 (M+2)	0.46

M (113) = 45.0
M+1 (114) = 3.0
M+2 (115) = 0.21

Ultraviolet Data

EtOH–Featureless
above 210 nm

^{1}H NMR Spectrum (Solvent CCl_4)

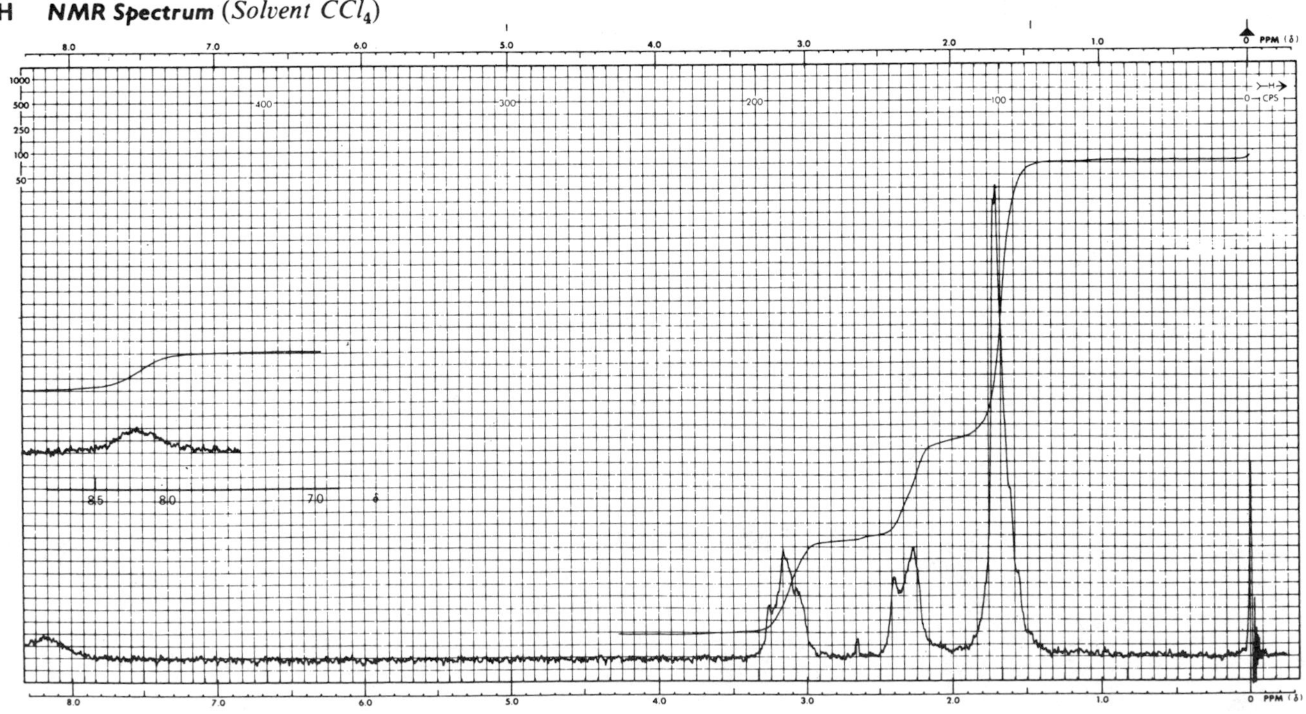

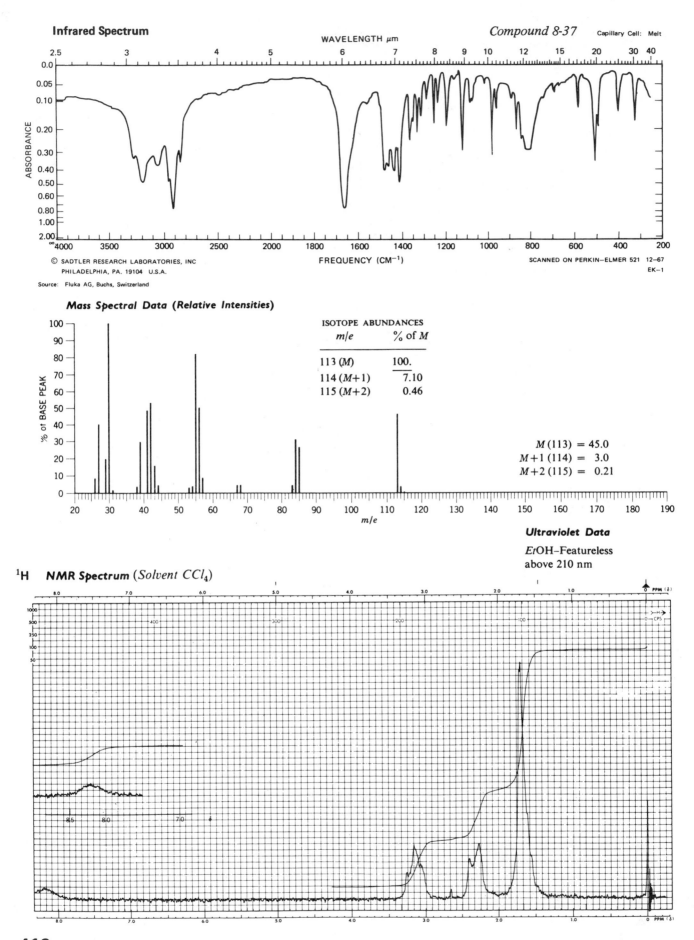

^{13}C NMR Spectrum Completely Decoupled

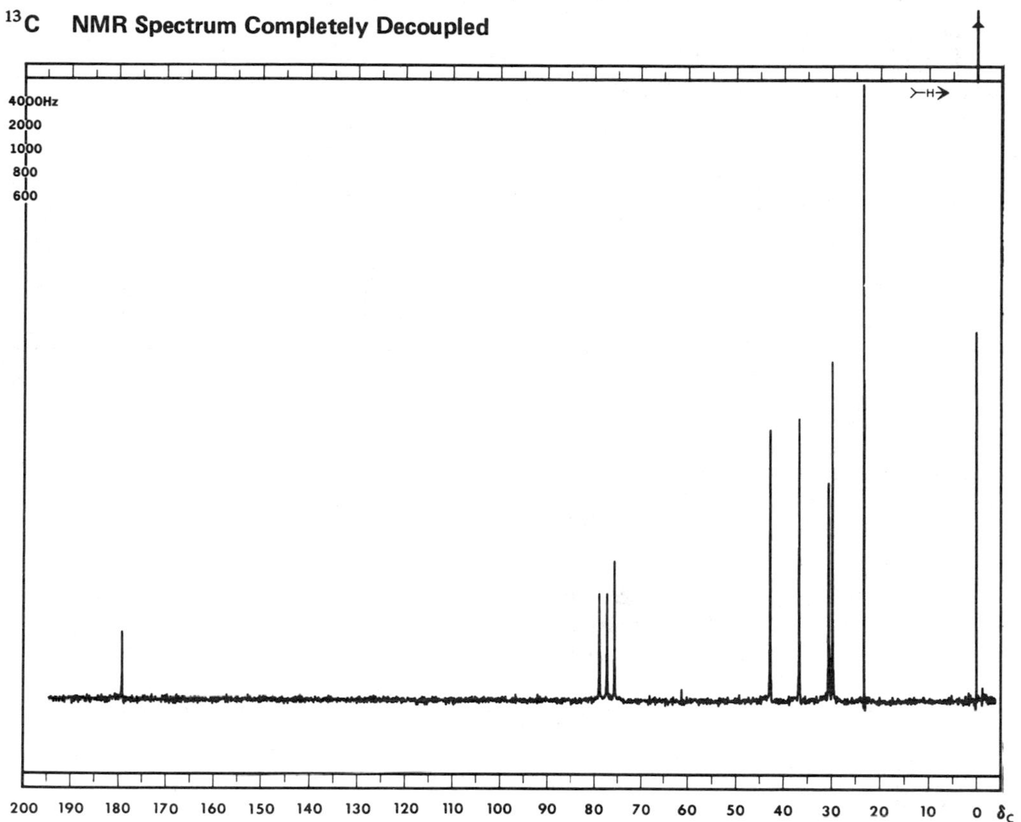

Infrared Spectrum *Compound 8-38*

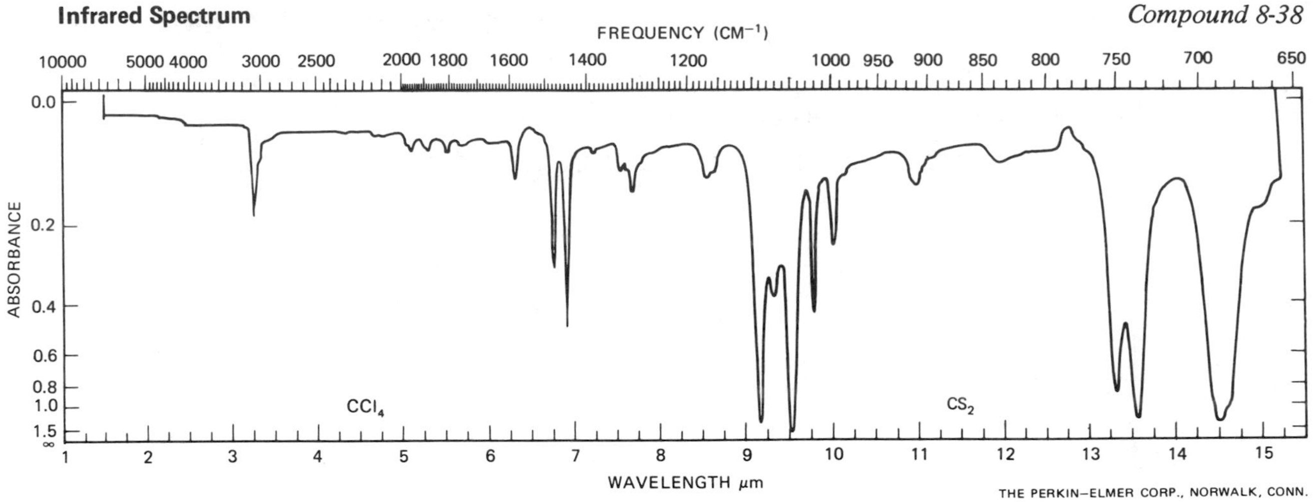

FREQUENCY (CM^{-1})

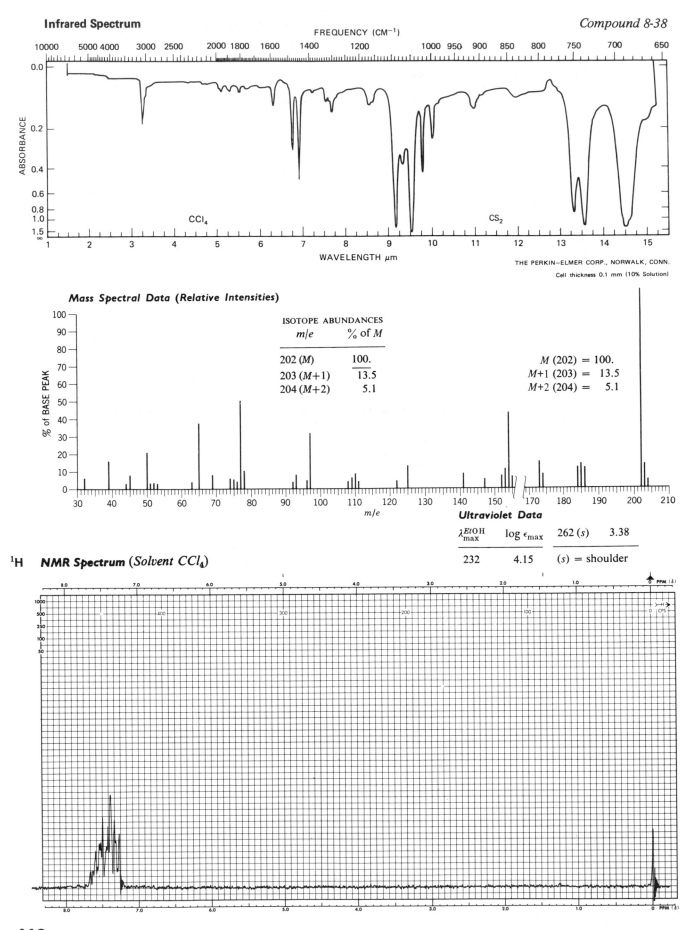

10000 5000 4000 3000 2500 2000 1800 1600 1400 1200 1000 950 900 850 800 750 700 650

ABSORBANCE 0.0 0.2 0.4 0.6 0.8 1.0 1.5 ∞

CCl_4 CS_2

1 2 3 4 5 6 7 8 9 10 11 12 13 14 15

WAVELENGTH μm

THE PERKIN–ELMER CORP., NORWALK, CONN.

Cell thickness 0.1 mm (10% Solution)

Mass Spectral Data (Relative Intensities)

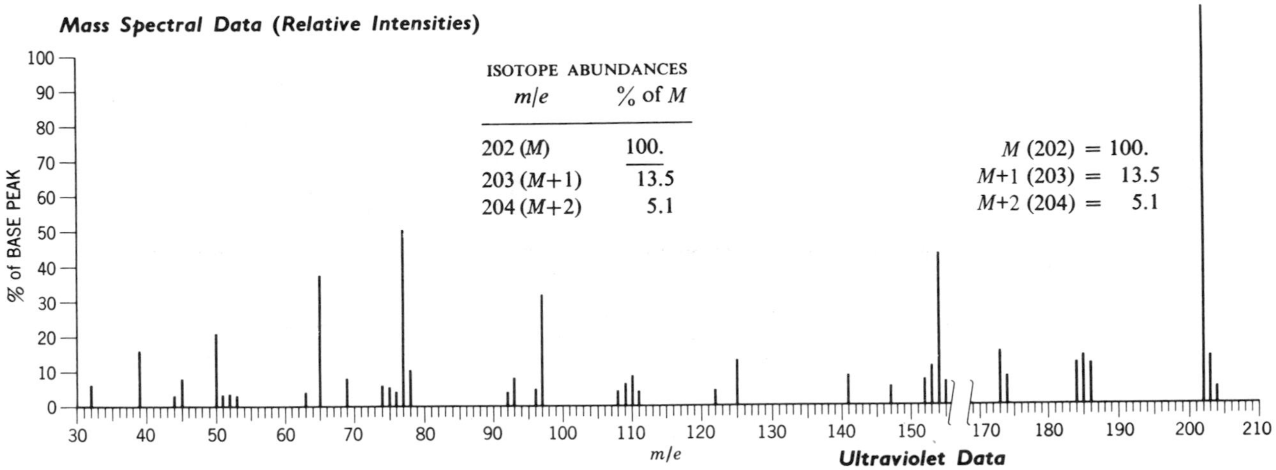

% of BASE PEAK 0 10 20 30 40 50 60 70 80 90 100

ISOTOPE ABUNDANCES

m/e	% of *M*
202 (*M*)	100.
203 (*M*+1)	13.5
204 (*M*+2)	5.1

M (202) = 100.
M+1 (203) = 13.5
M+2 (204) = 5.1

30 40 50 60 70 80 90 100 110 120 130 140 150 170 180 190 200 210

m/e

Ultraviolet Data

λ_{max}^{EtOH}	log ϵ_{max}	262 (*s*) 3.38
232	4.15	(*s*) = shoulder

^{1}H **NMR Spectrum** (*Solvent* CCl_4)

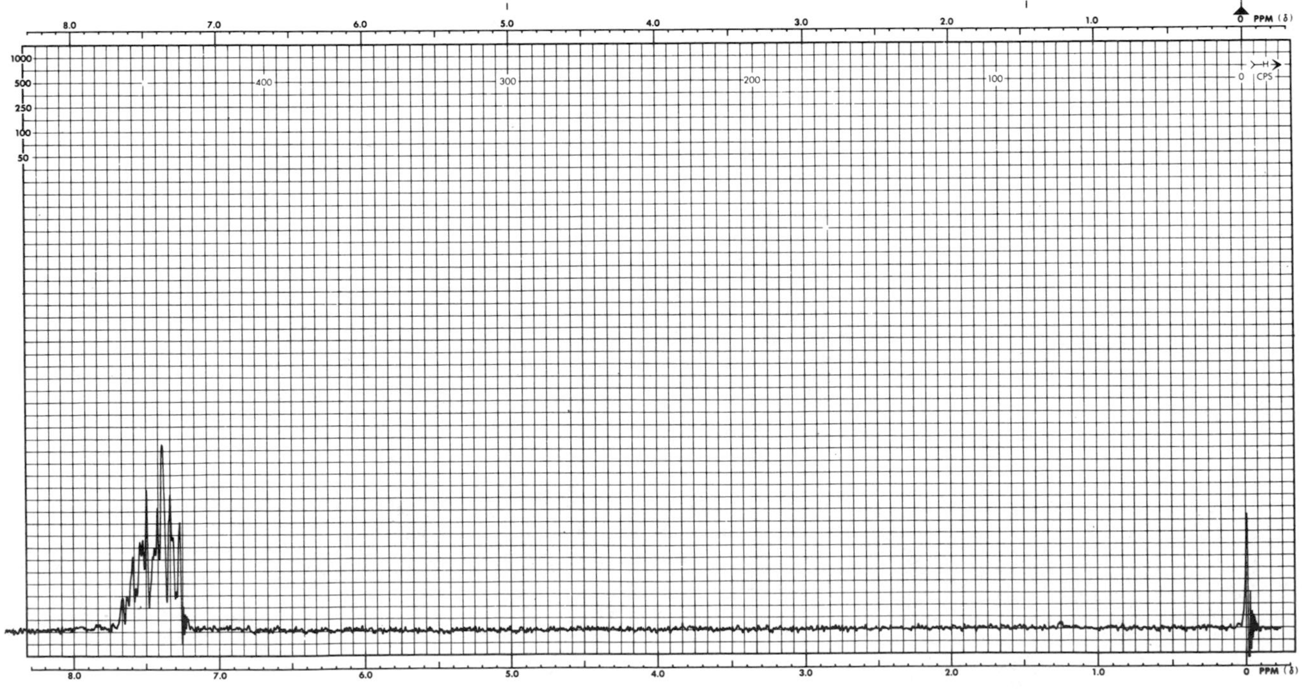

Infrared Spectrum

Compound 8-39 Beilstein Ref. 23, 94

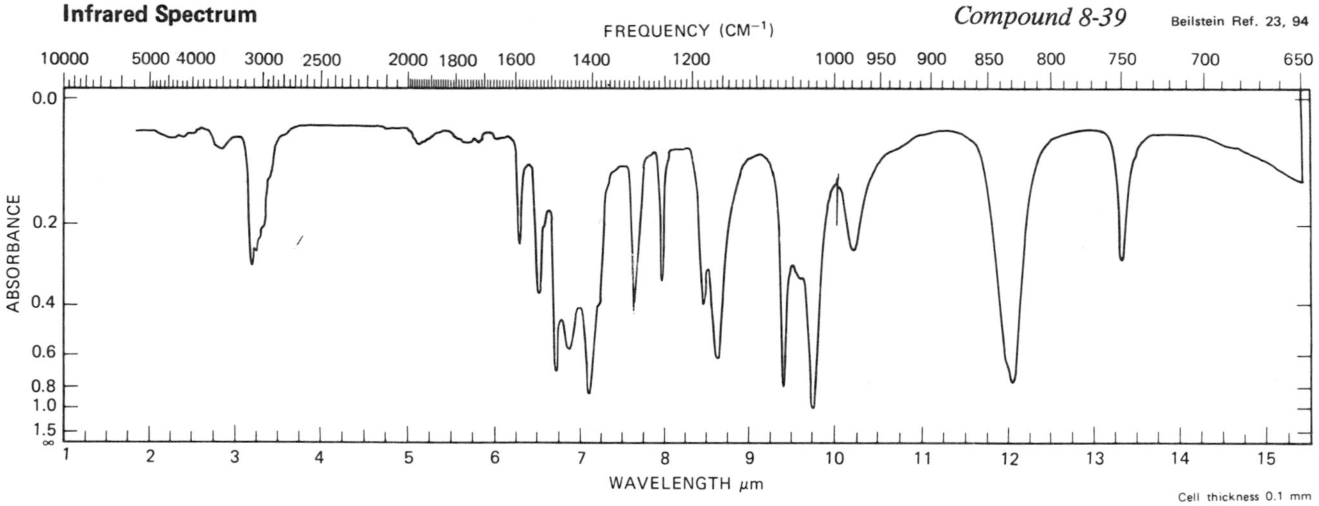

FREQUENCY (CM⁻¹)
ABSORBANCE
WAVELENGTH μm
Cell thickness 0.1 mm

Mass Spectral Data (Relative Intensities)

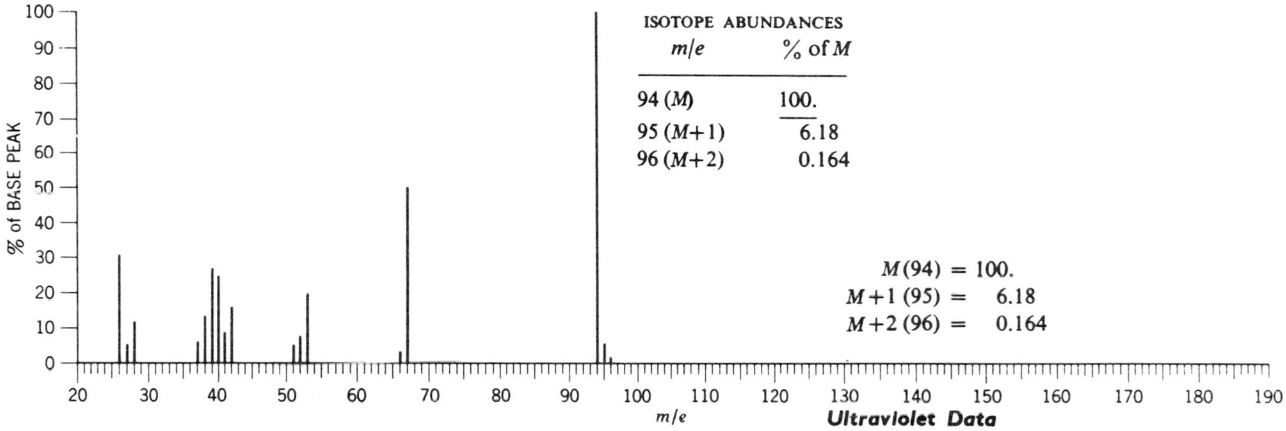

% of BASE PEAK
m/e

ISOTOPE ABUNDANCES

m/e	% of *M*
94 (*M*)	100.
95 (*M*+1)	6.18
96 (*M*+2)	0.164

$M(94) = 100.$
$M+1\ (95) = 6.18$
$M+2\ (96) = 0.164$

Ultraviolet Data

	λ_{max}^{EtOH}	ϵ_{max}			
pH 7	266	6,600	pH 1	266	10,200
	272.5	6,000		273	9,700
	305	900			

SOLVENT $CDCl_3$
100 MHz
0 cps
δ

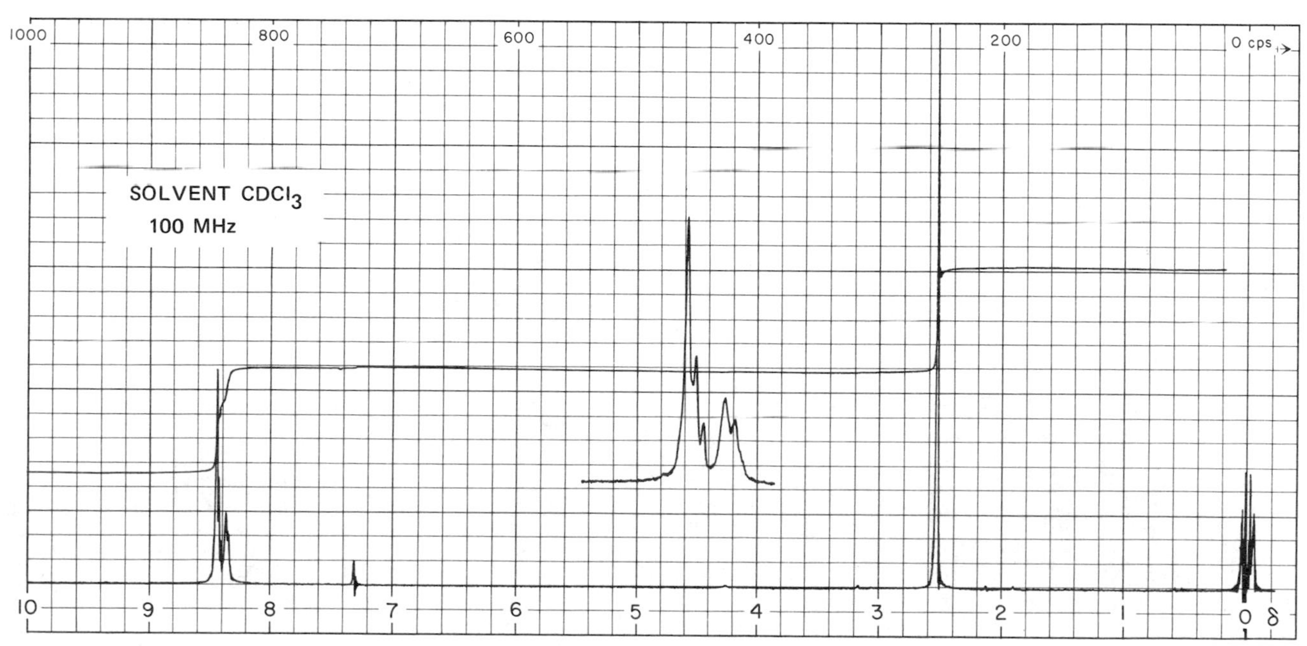

Infrared Spectrum

Compound 8-40

Beilstein Ref. 1 (3rd Supp.), 1994

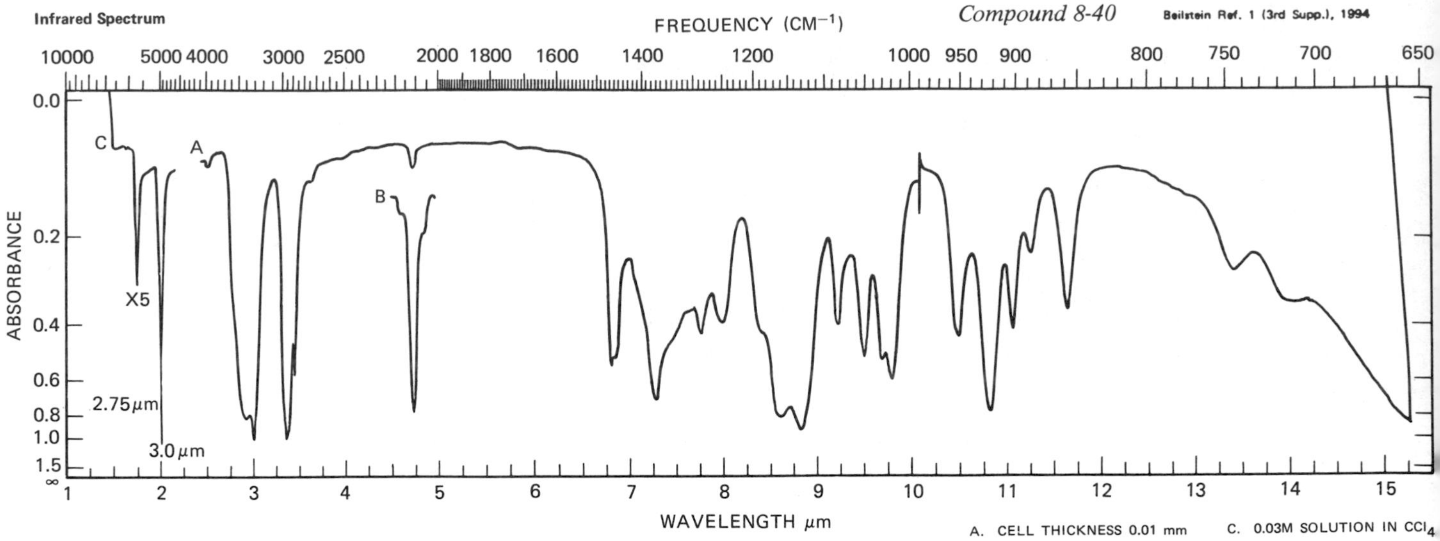

A. CELL THICKNESS 0.01 mm
B. CELL THICKNESS 0.40 mm
C. 0.03M SOLUTION IN CCl_4

Mass Spectral Data (Relative Intensities)

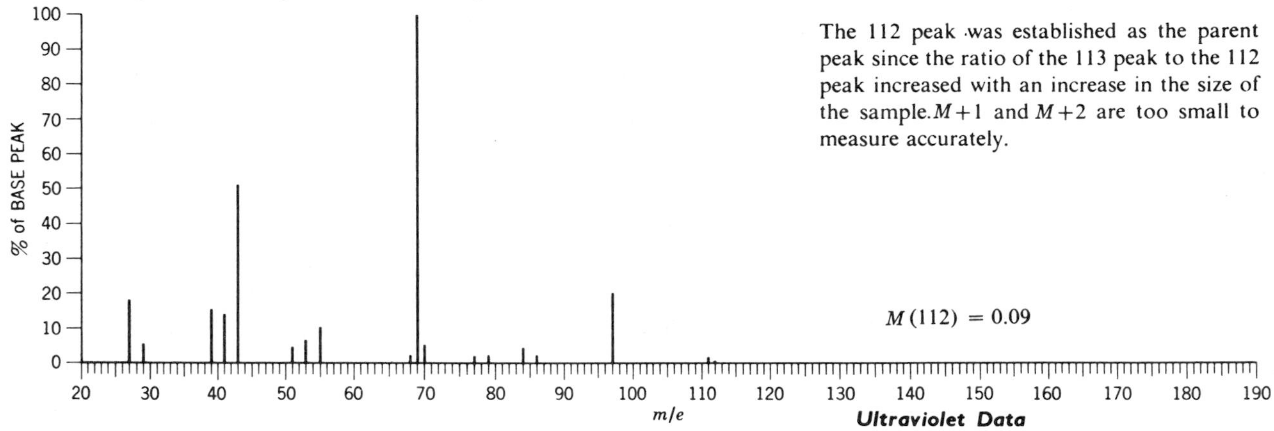

The 112 peak was established as the parent peak since the ratio of the 113 peak to the 112 peak increased with an increase in the size of the sample. $M+1$ and $M+2$ are too small to measure accurately.

$M(112) = 0.09$

Ultraviolet Data

Transparent beyond 200 nm

^{1}H **NMR Spectrum**

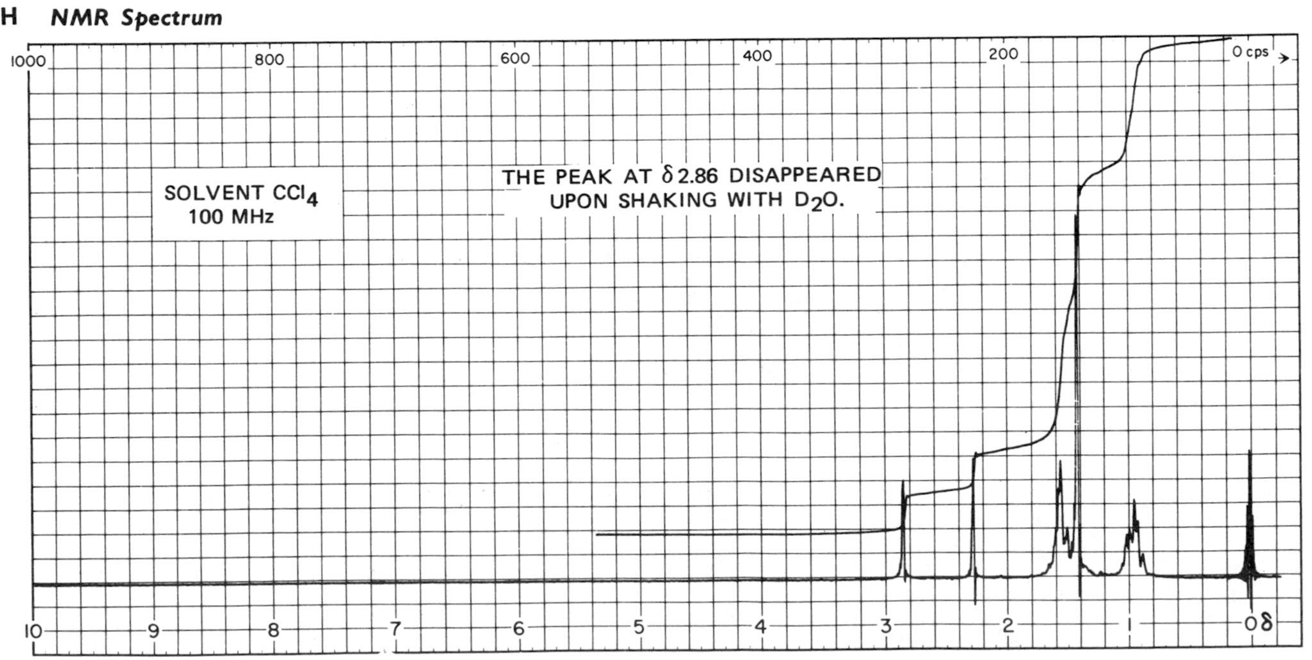

Infrared Spectrum

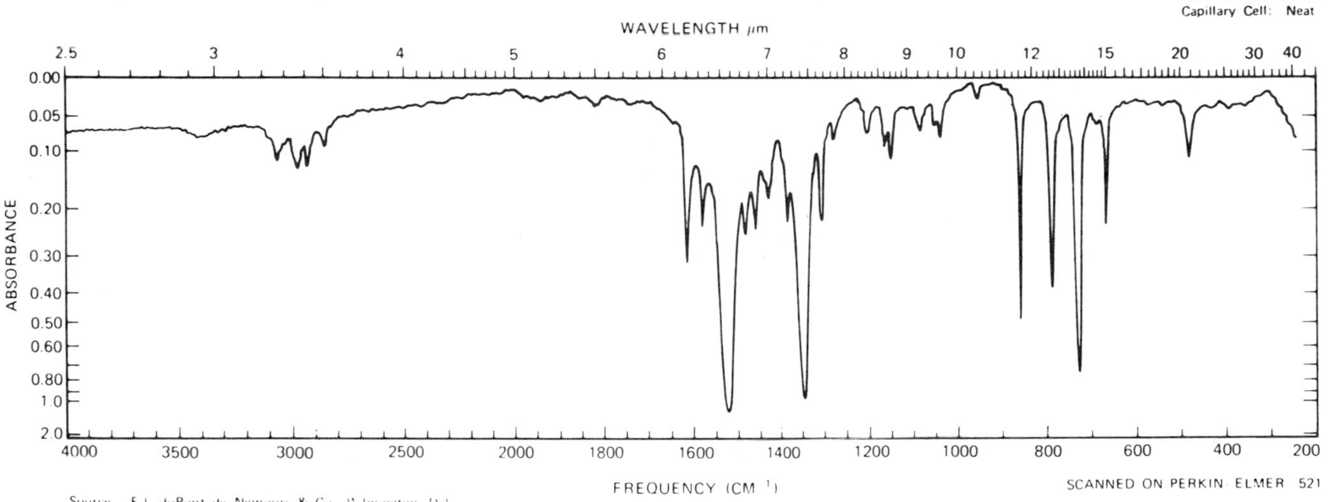

Mass Spectral Data (Relative Intensities)

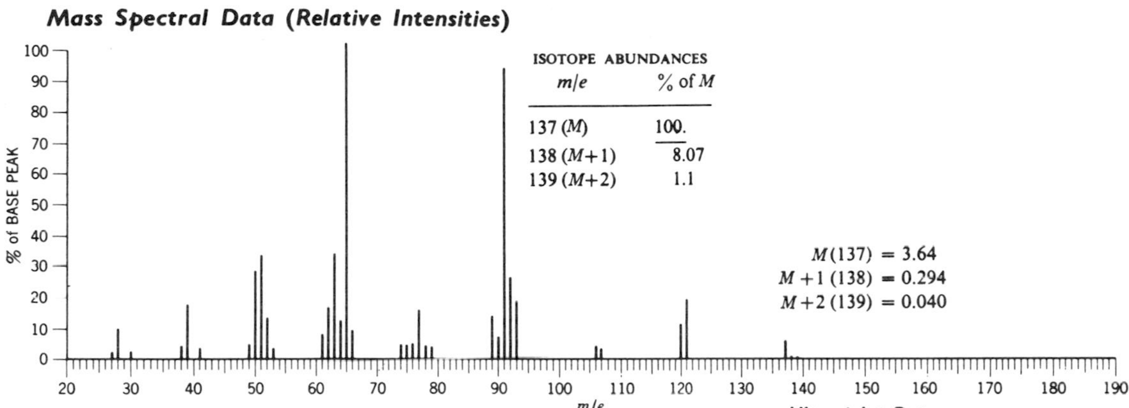

ISOTOPE ABUNDANCES

m/e	% of *M*
137 (*M*)	100.
138 (*M*+1)	8.07
139 (*M*+2)	1.1

M(137) = 3.64
M +1 (138) = 0.294
M+2 (139) = 0.040

Ultraviolet Data

$\lambda_{max}^{Isooctane}$	$\log \epsilon_{max}$
251	3.78
285 (*s*)	3.21
330	2.44

(*s*) = shoulder

¹H *NMR Spectrum*

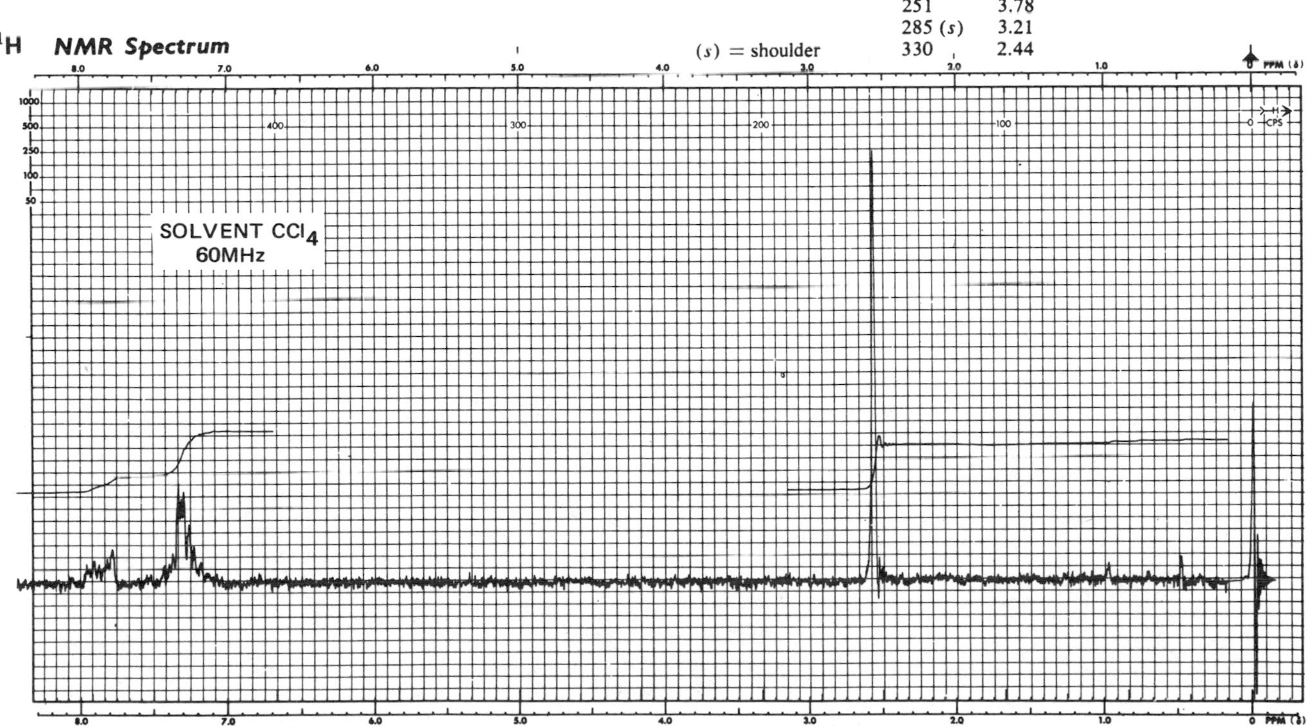

Treatment of an ethyl alcohol solution of α,α'-dimercapto-*p*-xylene with sodium hydroxide, followed by α,α'-dibromo-*o*-xylene treatment resulted in compound 8-42:

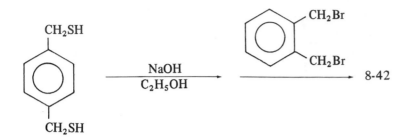

$$p\text{-}C_6H_4(CH_2SH)_2 \xrightarrow[C_2H_5OH]{NaOH} \xrightarrow{o\text{-}C_6H_4(CH_2Br)_2} \text{8-42}$$

Deduce the structure of 8-42 from the spectra provided.

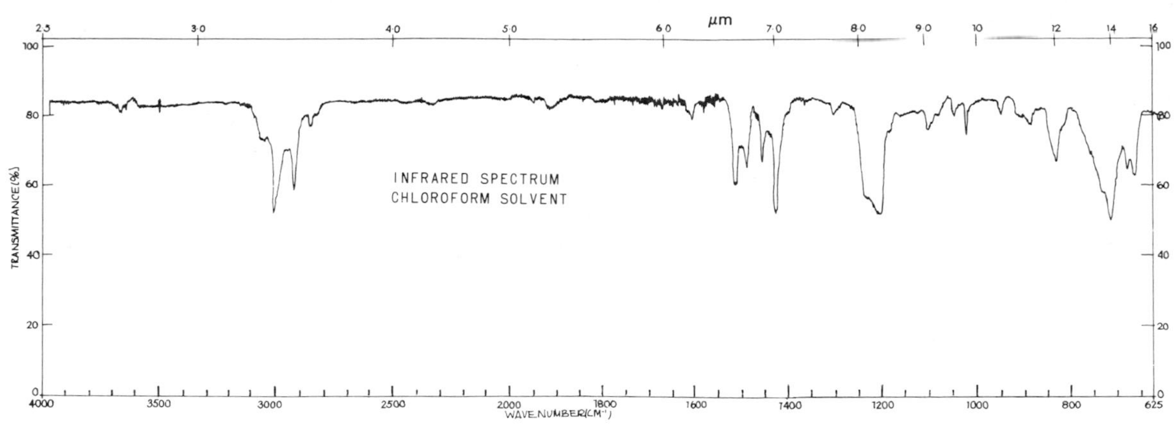

Mass Spectral Data

m/e	Raw Intensity Data
544	22.78
545	8.86
546	6.33
547	2.53

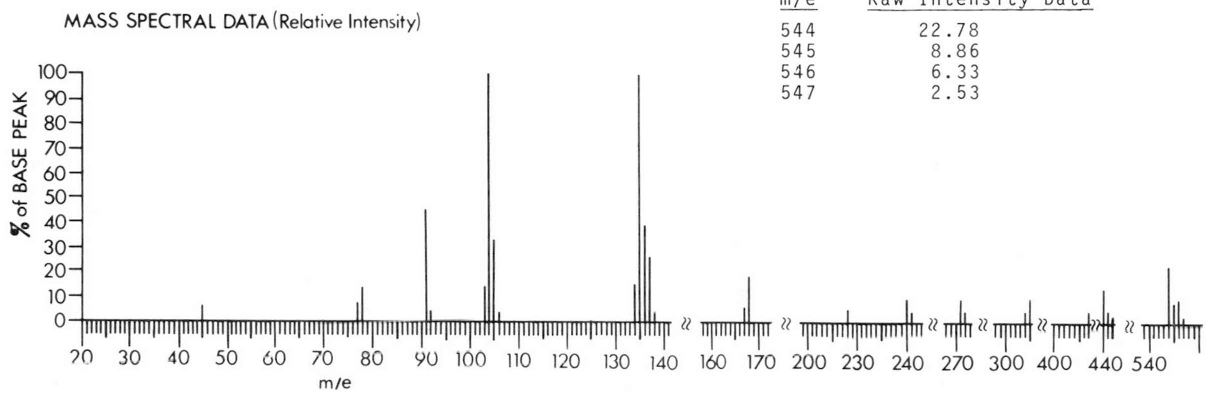

Ultraviolet Data

λ_{max} (CH_3CH_2OH)	ε_{max}
267	10,600
215(s)	11,780

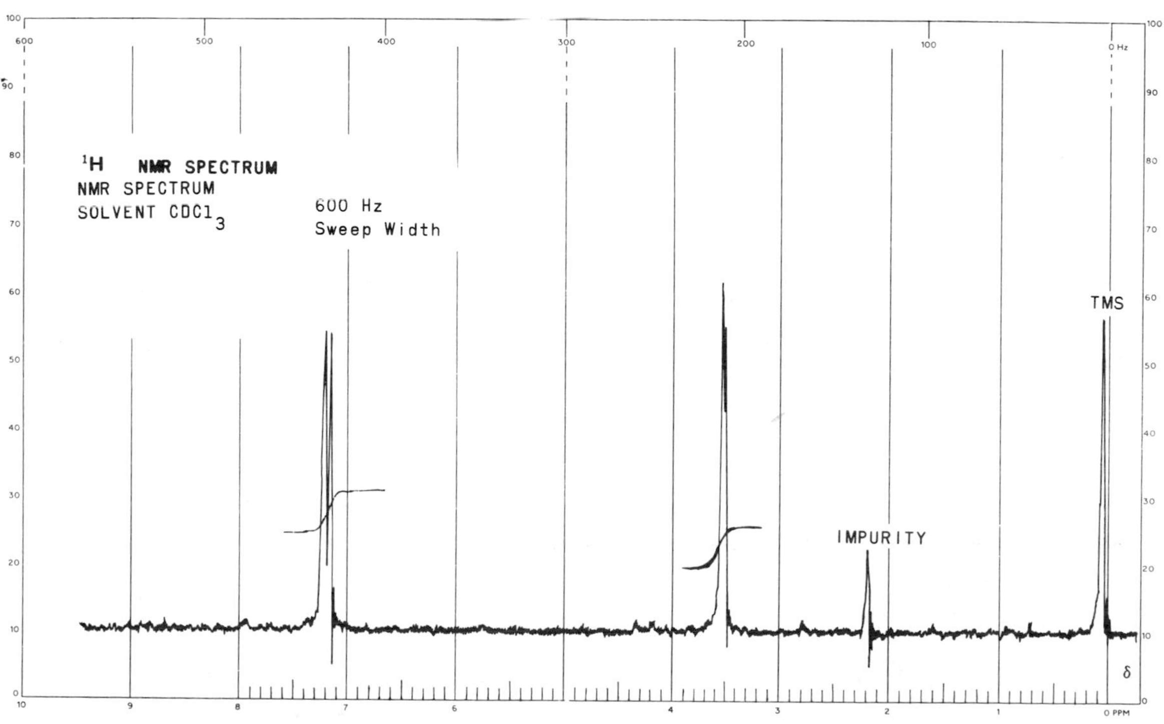

Compound 8-43

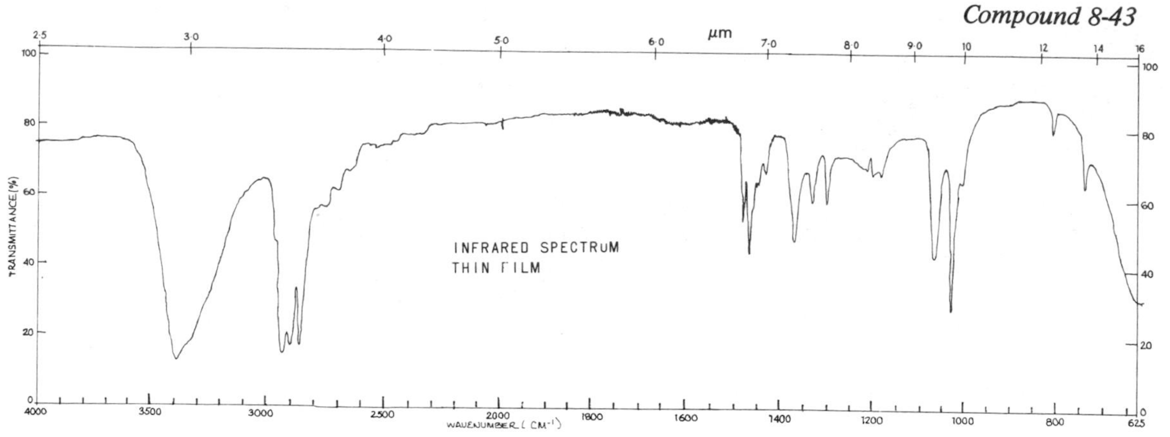

MASS SPECTRAL DATA (Relative Intensity)

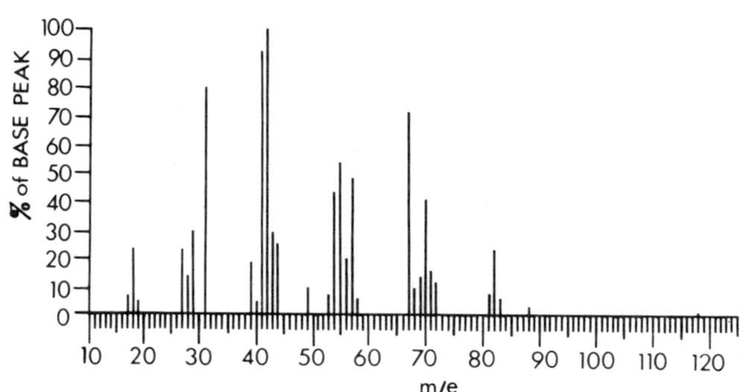

Mass Spectral Data

Molecular Ion at m/e 118;
M+1, M+2 too weak to be measured

Ultraviolet Data

Featureless above 200nm

^{1}H *NMR Spectrum*

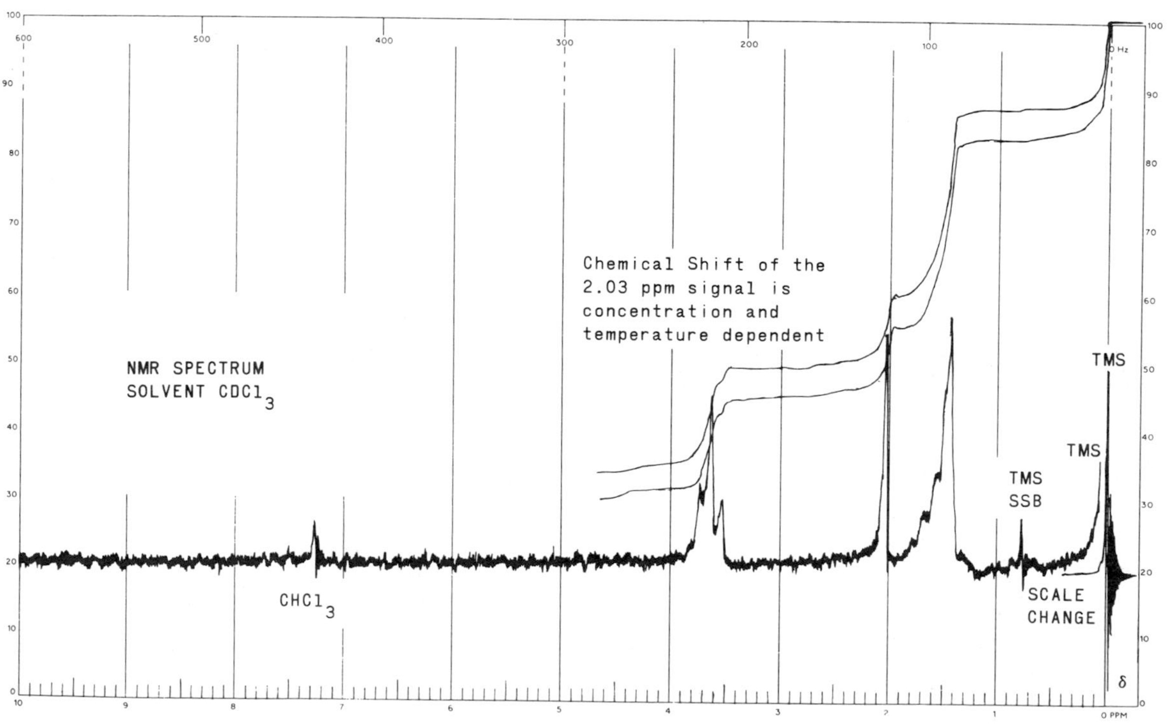

Infrared Spectrum

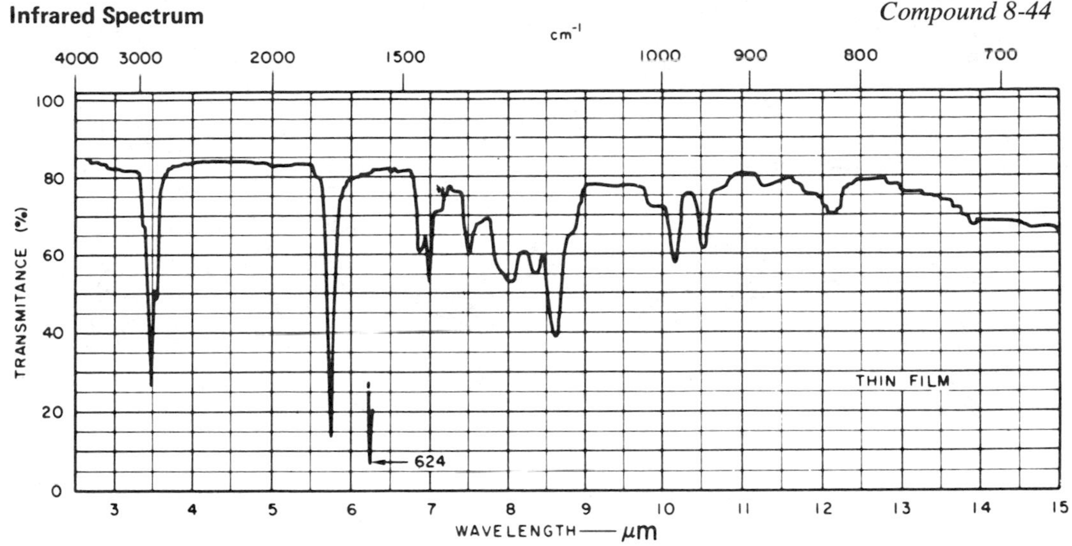

Mass Spectrum

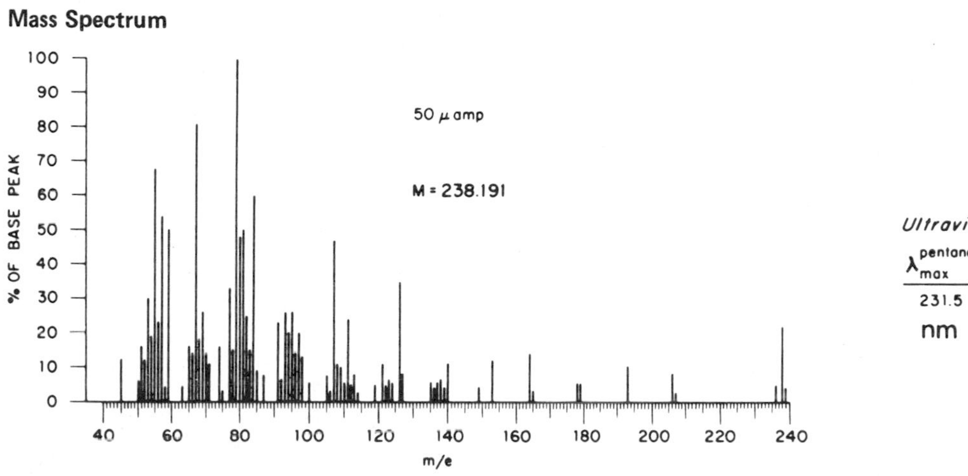

Ultraviolet Data

$\lambda_{max}^{pentane}$	ϵ_{max}
231.5	~30,000
nm	

^{1}H NMR Spectrum

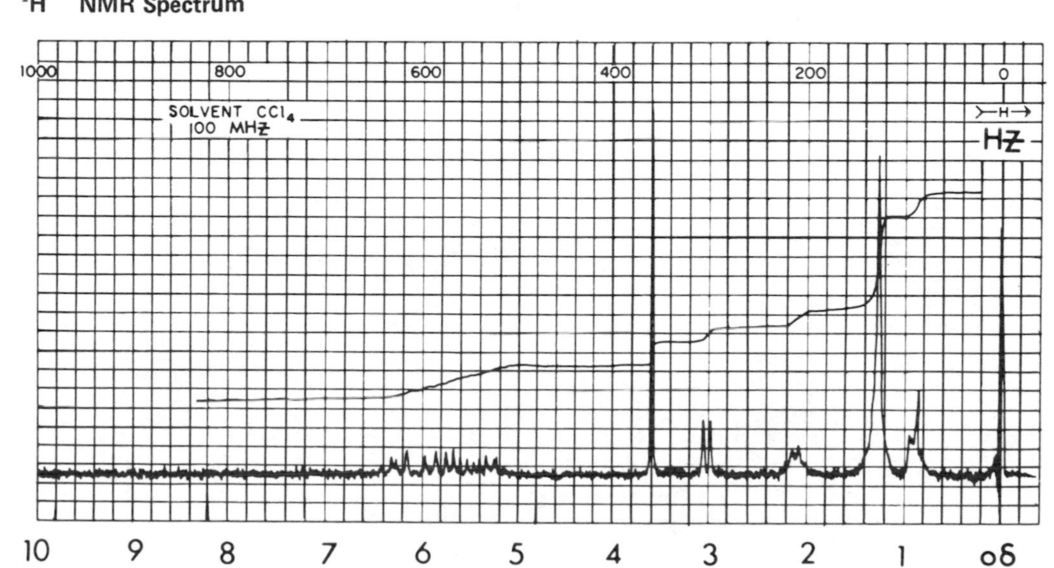

Infrared Spectrum *Compound 8-45*

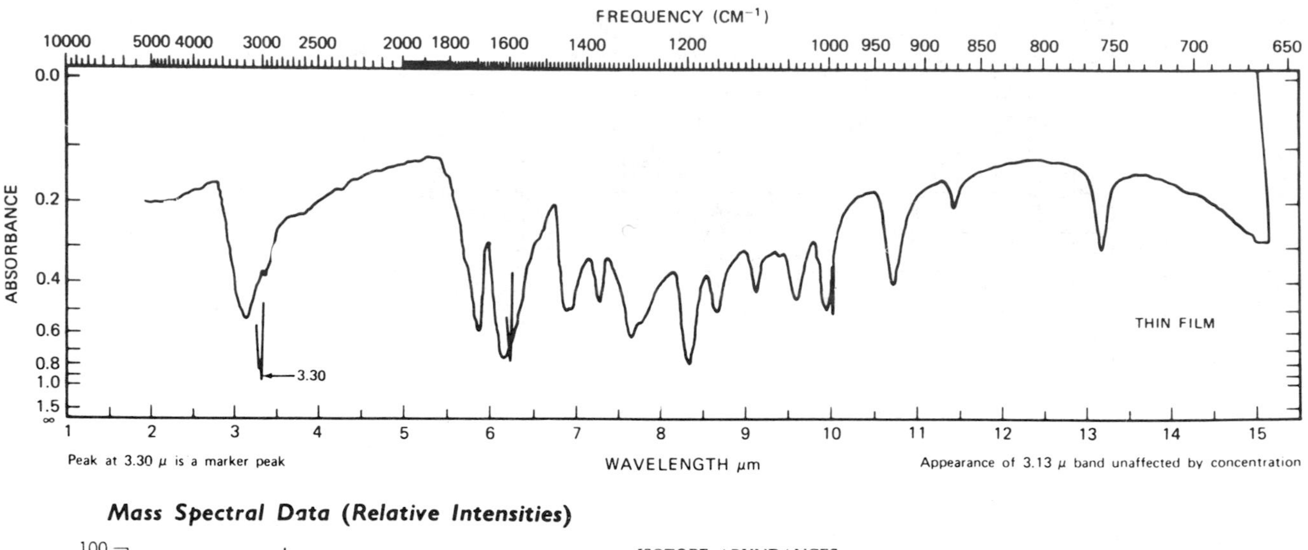

Mass Spectral Data (Relative Intensities)

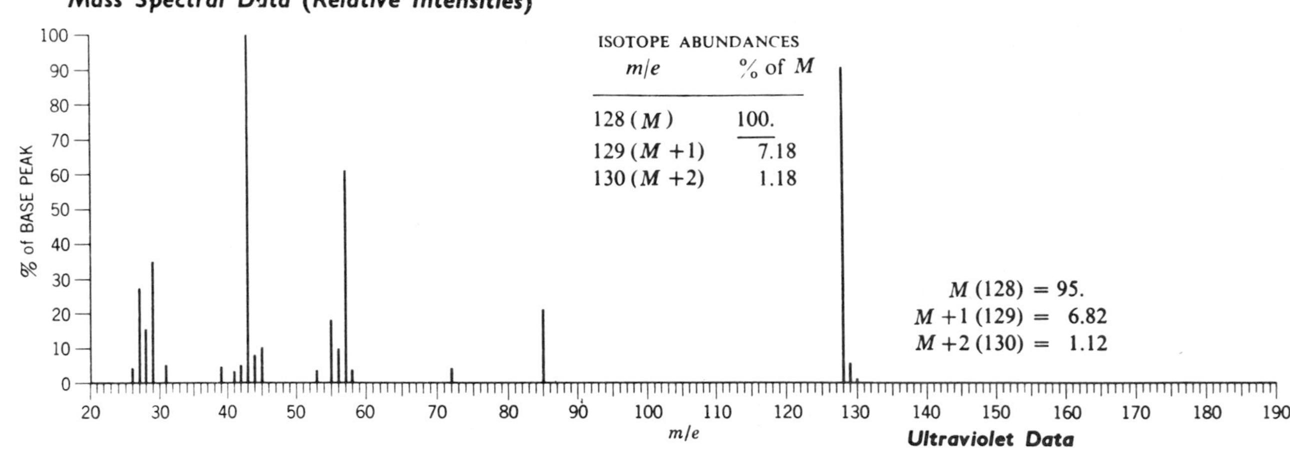

ISOTOPE ABUNDANCES

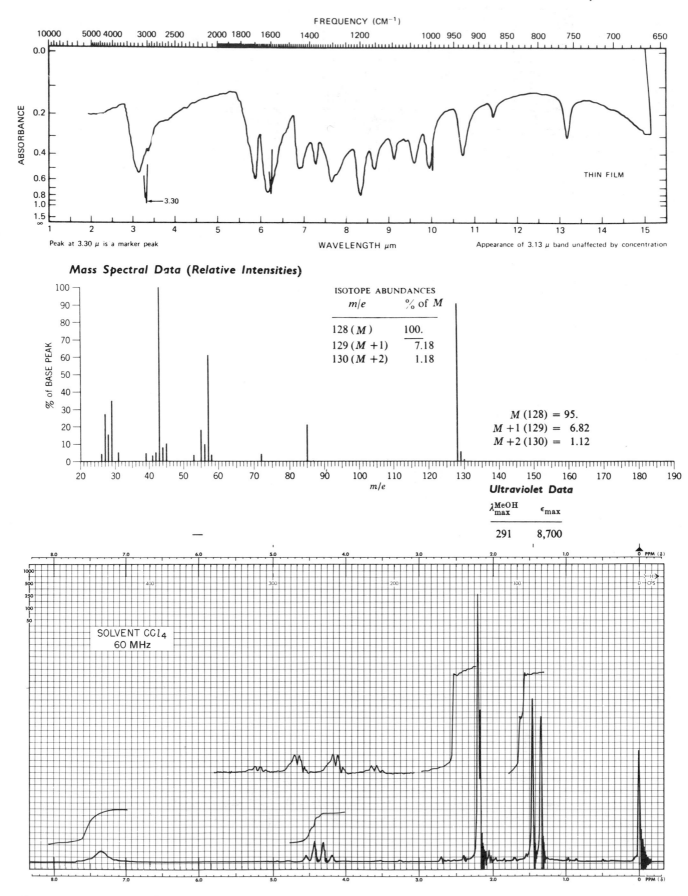

m/e	% of M
128 (M)	100.
129 (M +1)	7.18
130 (M +2)	1.18

M (128) = 95.
M +1 (129) = 6.82
M +2 (130) = 1.12

Ultraviolet Data

λ_{max}^{MeOH}	ϵ_{max}
291	8,700

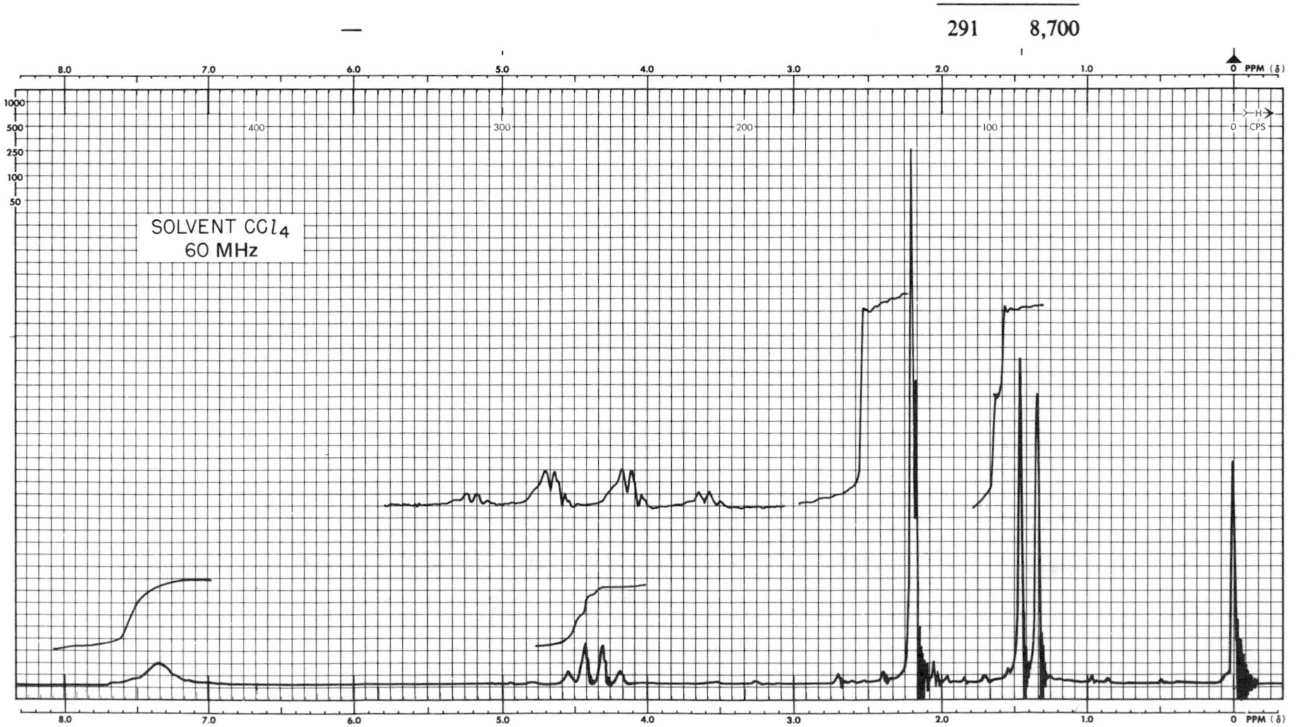

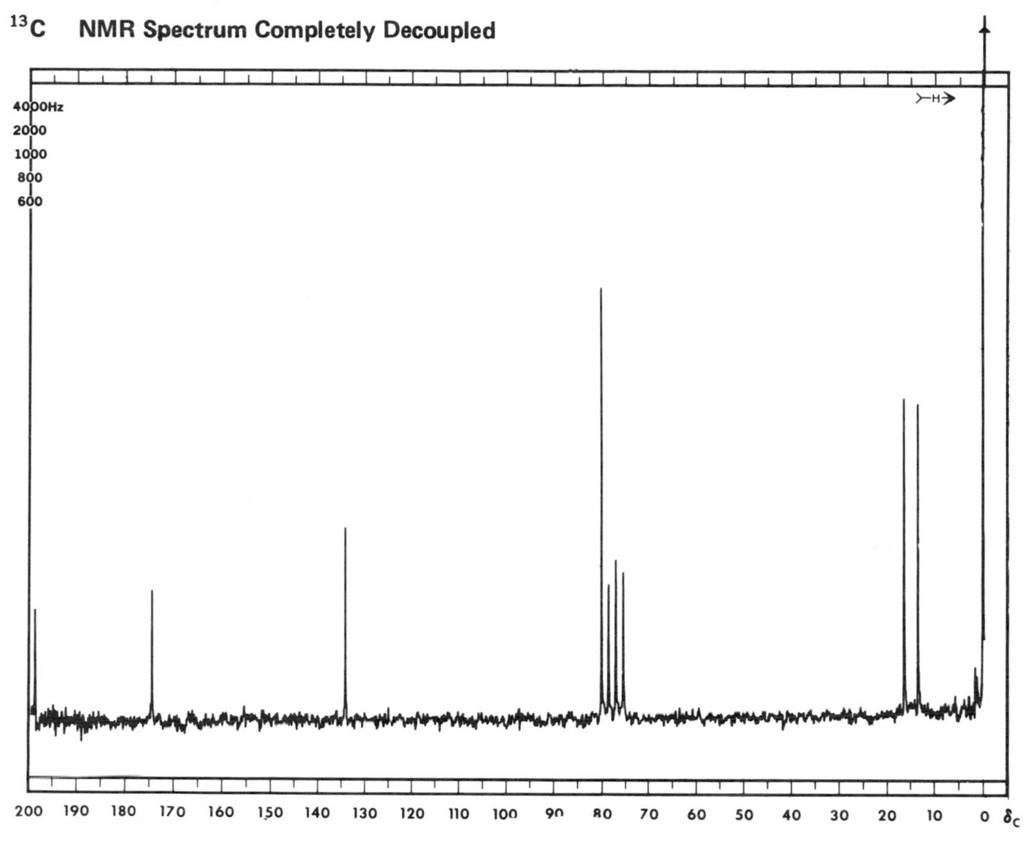

^{13}C NMR Spectrum Completely Decoupled
4000Hz
2000
1000
800
600
200 190 180 170 160 150 140 130 120 110 100 90 80 70 60 50 40 30 20 10 0 δ_C

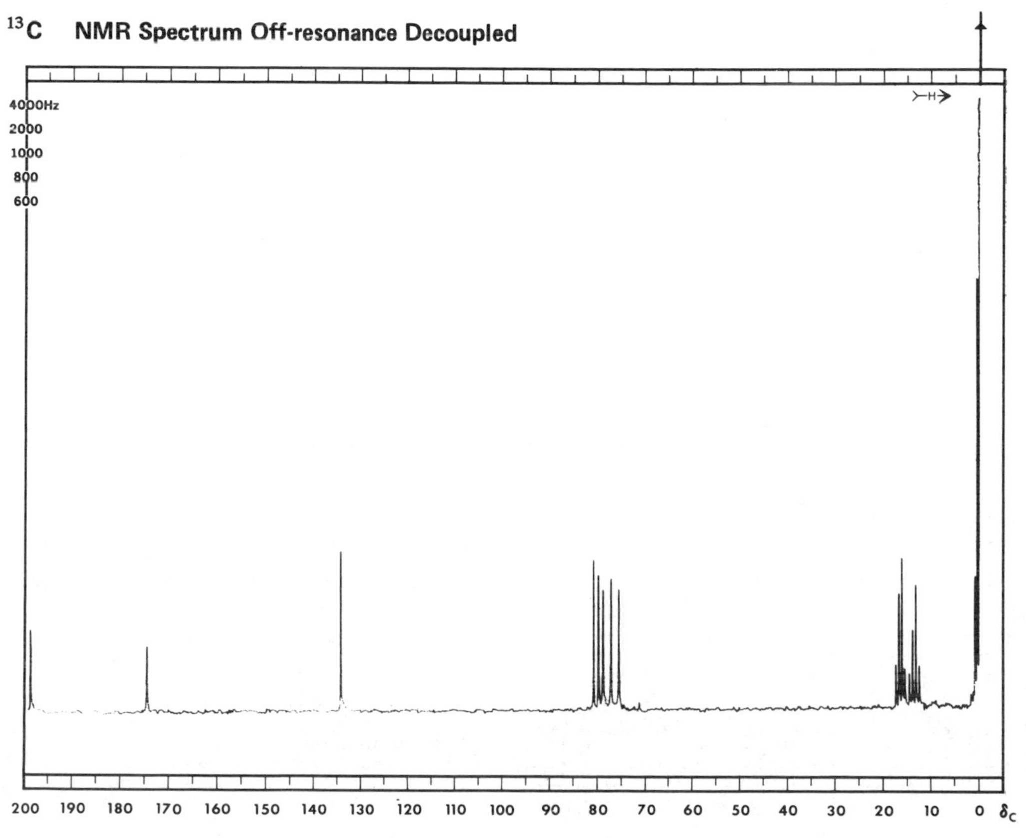

^{13}C NMR Spectrum Off-resonance Decoupled
4000Hz
2000
1000
800
600
200 190 180 170 160 150 140 130 120 110 100 90 80 70 60 50 40 30 20 10 0 δ_C

Infrared Spectrum *Compound 8-46*

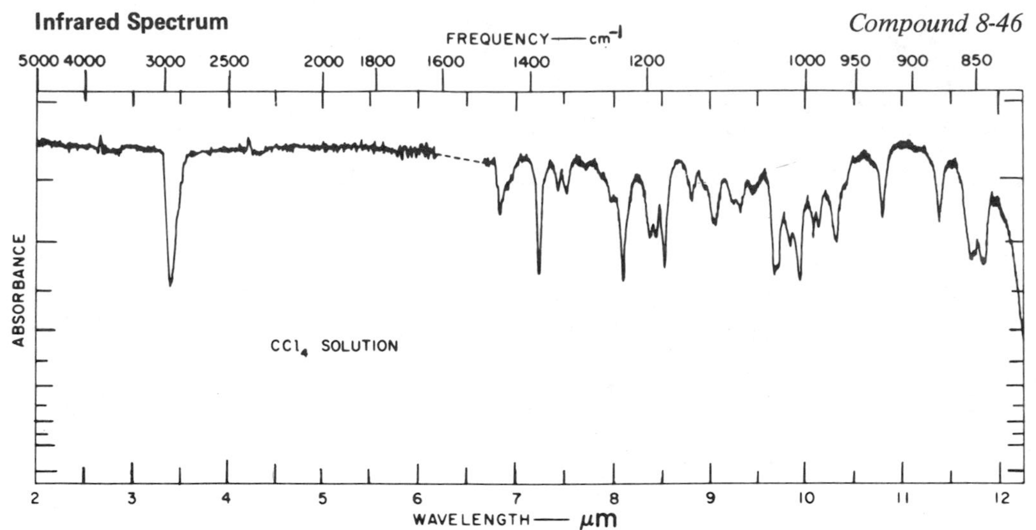

Mass Spectrum

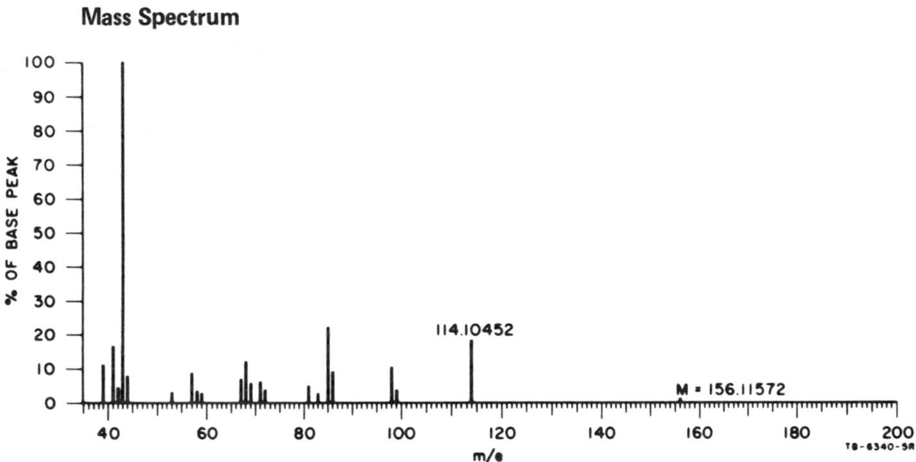

^{1}H NMR Spectrum

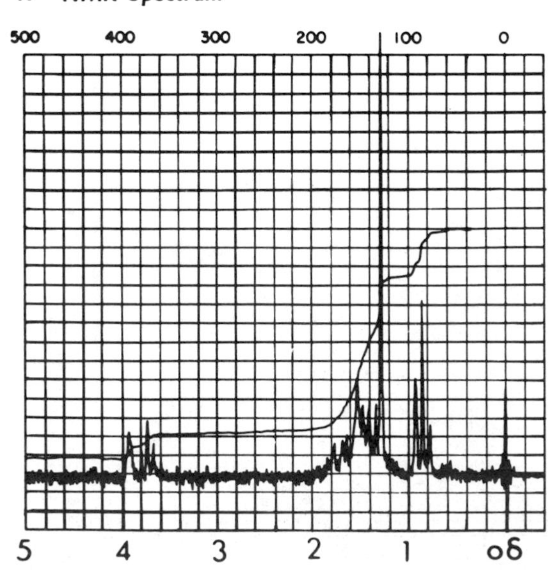

UV: No absorption above 200nm

Related Chemical Information

1. Recovered unchanged from treatment with $LiAlH_4$.
2. Catalytic hydrogenolysis (250°) gave nonane
3. The compound is part of the aggregation pheromone of the western pine beetle, Dendroctonus brevicomis.

Compound 8-46 ^{13}C **NMR Data ($CDCl_3$ solvent)**

δ	Intensity	Off-resonance Decoupled Multiplet
9.7	82.0	*q*
17.3	97.0	*t*
25.0	80.0	*q*
28.0	95.0	*t*
28.6	100.0	*t*
35.0	97.0	*t*
78.3	80.0	*d*
81.1	85.0	*d*
107.6	35.0	*s*

Infrared Spectrum

Compound 8-47

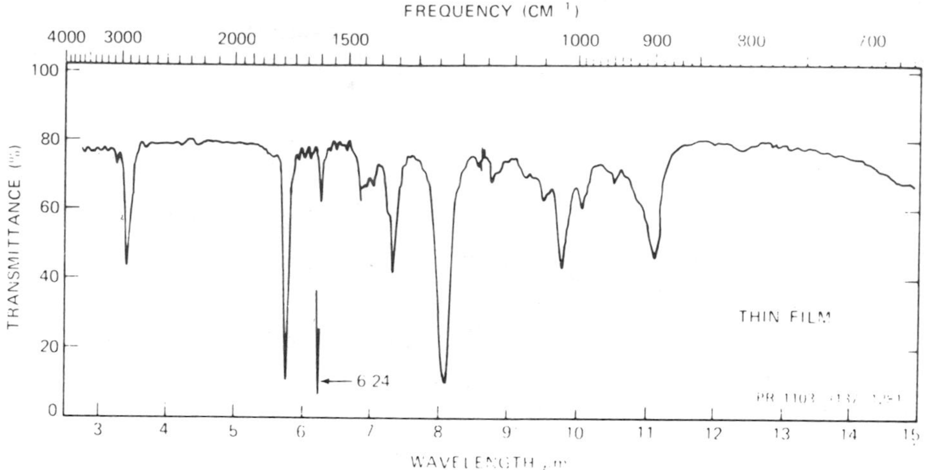

Mass Spectral Data (Relative Intensities)

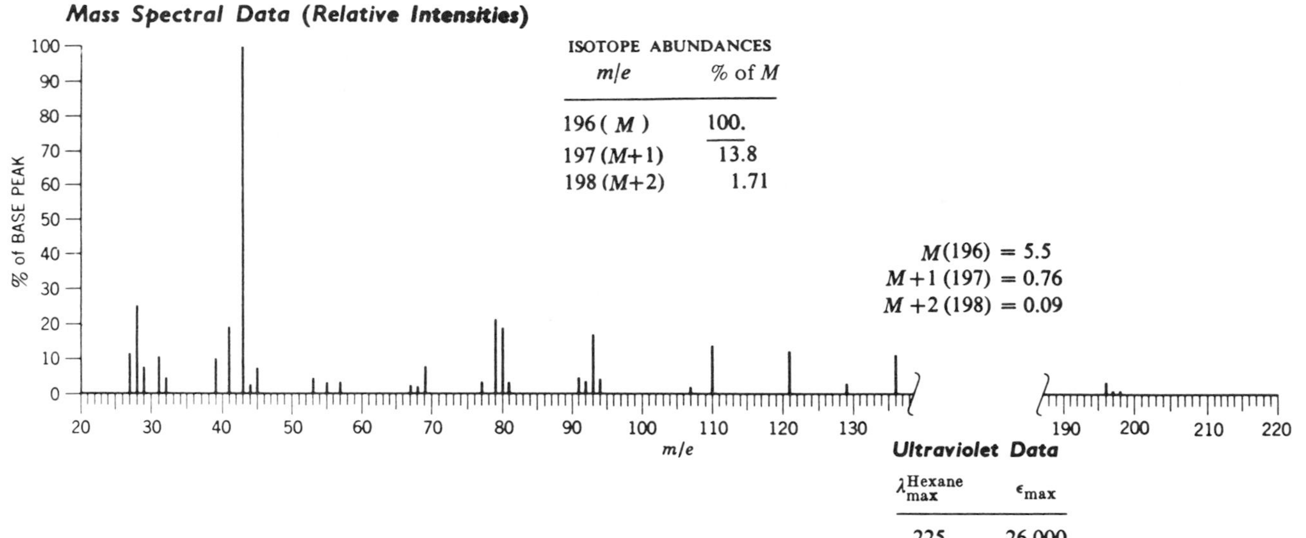

ISOTOPE ABUNDANCES

m/e	% of *M*
196 (*M*)	100.
197 (*M*+1)	13.8
198 (*M*+2)	1.71

M(196) = 5.5
M+1 (197) = 0.76
M+2 (198) = 0.09

Ultraviolet Data

λ_{max}^{Hexane}	ϵ_{max}
225	26,000

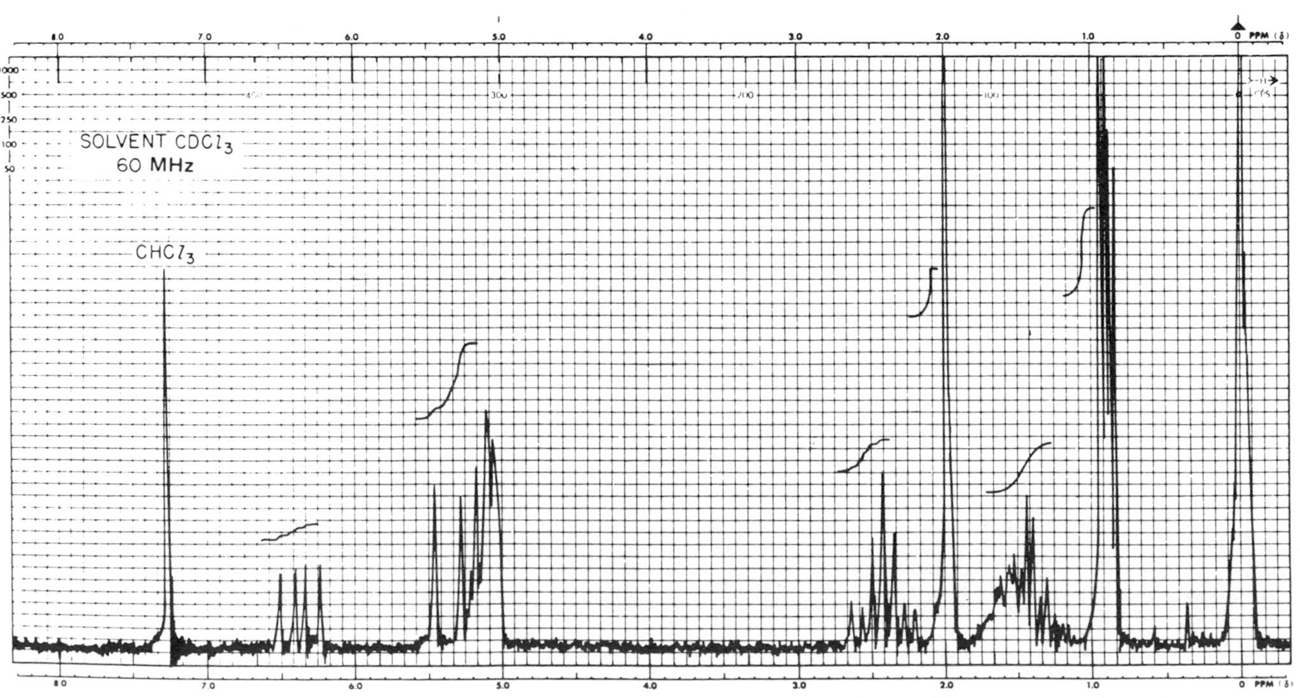

Infrared Spectrum *Compound 8-48*

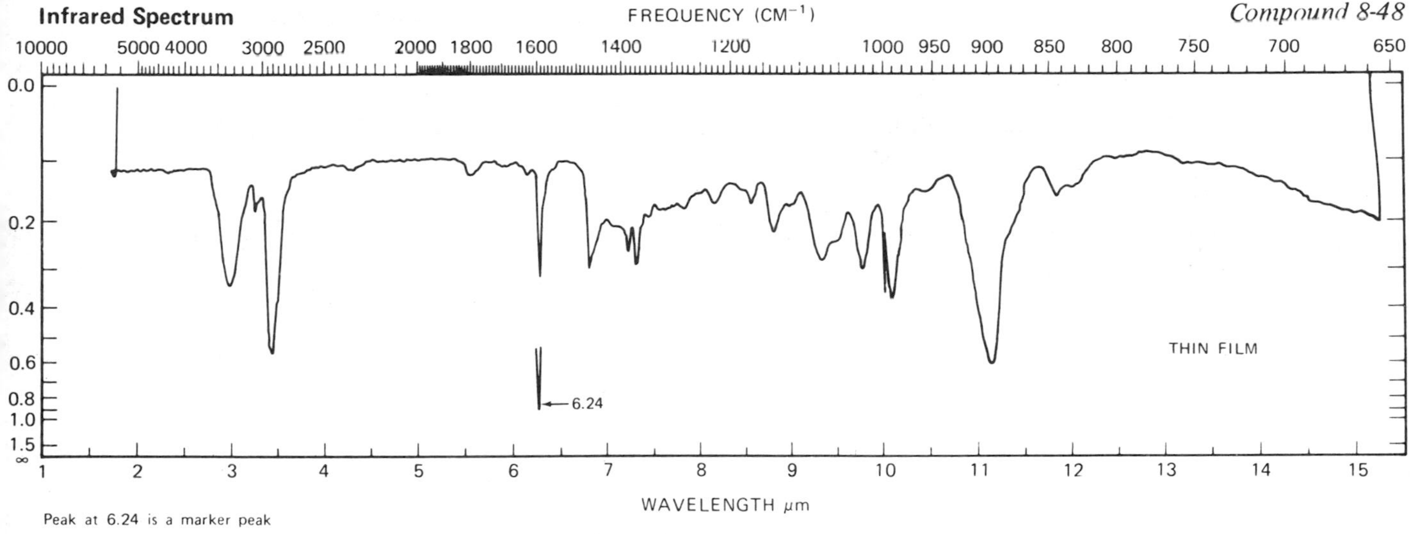

Peak at 6.24 is a marker peak

Mass Spectral Data (Relative Intensities)

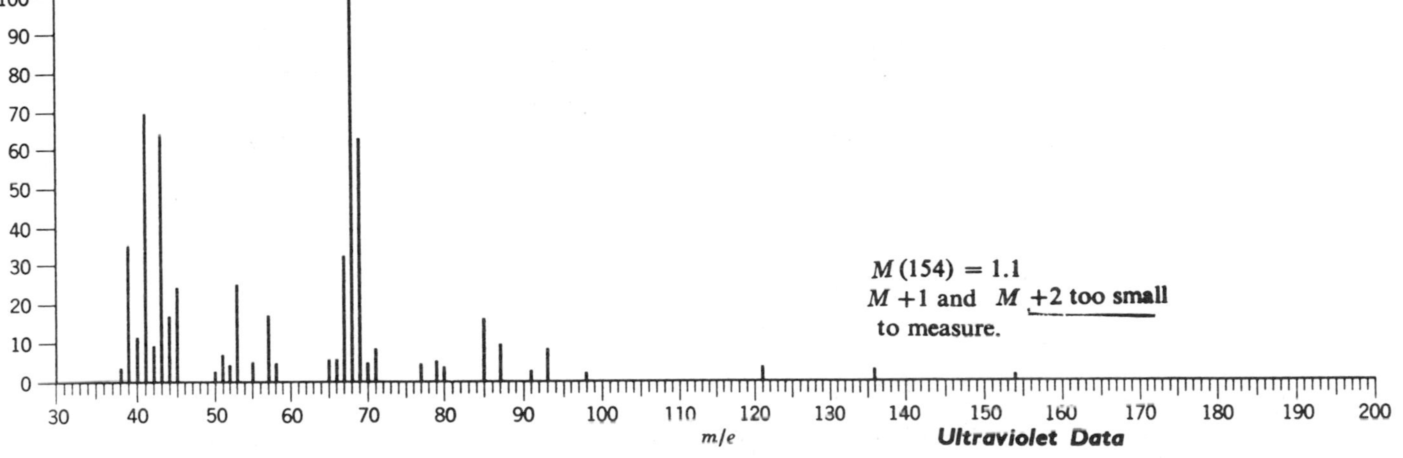

Ultraviolet Data

$\lambda_{max}^{C_6H_{12}}$	ϵ_{max}
226	20,000

NMR Spectrum

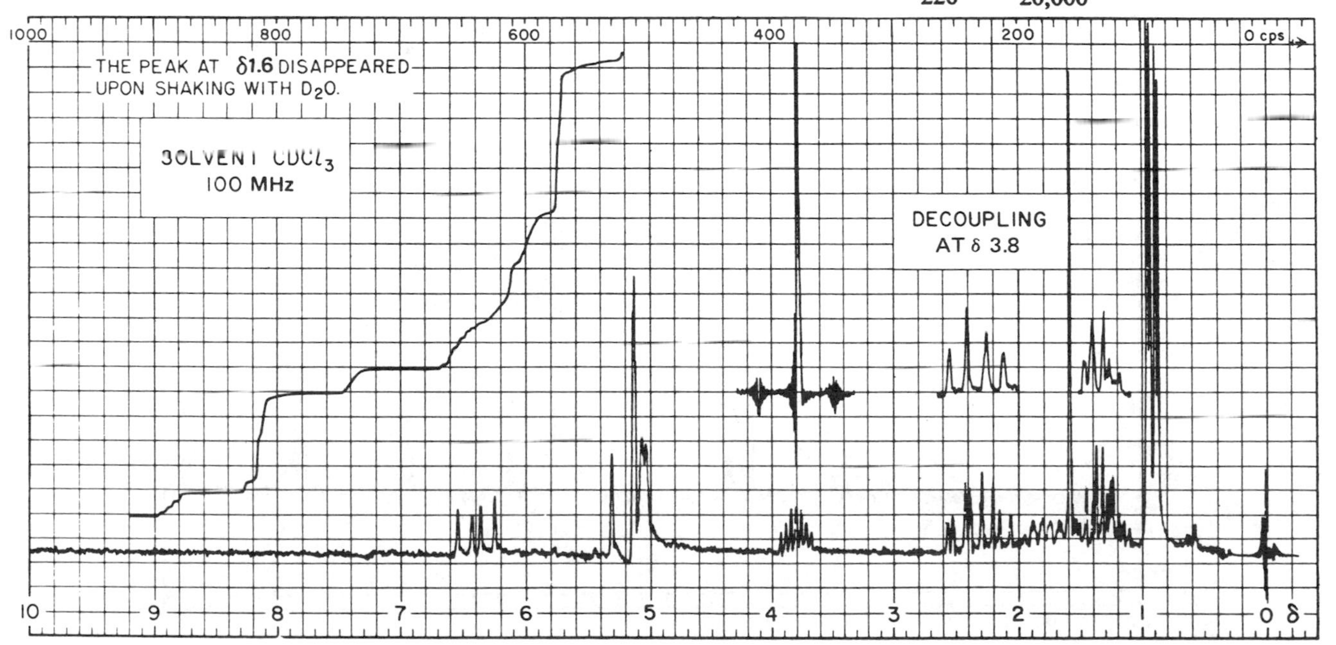

Continued on next page

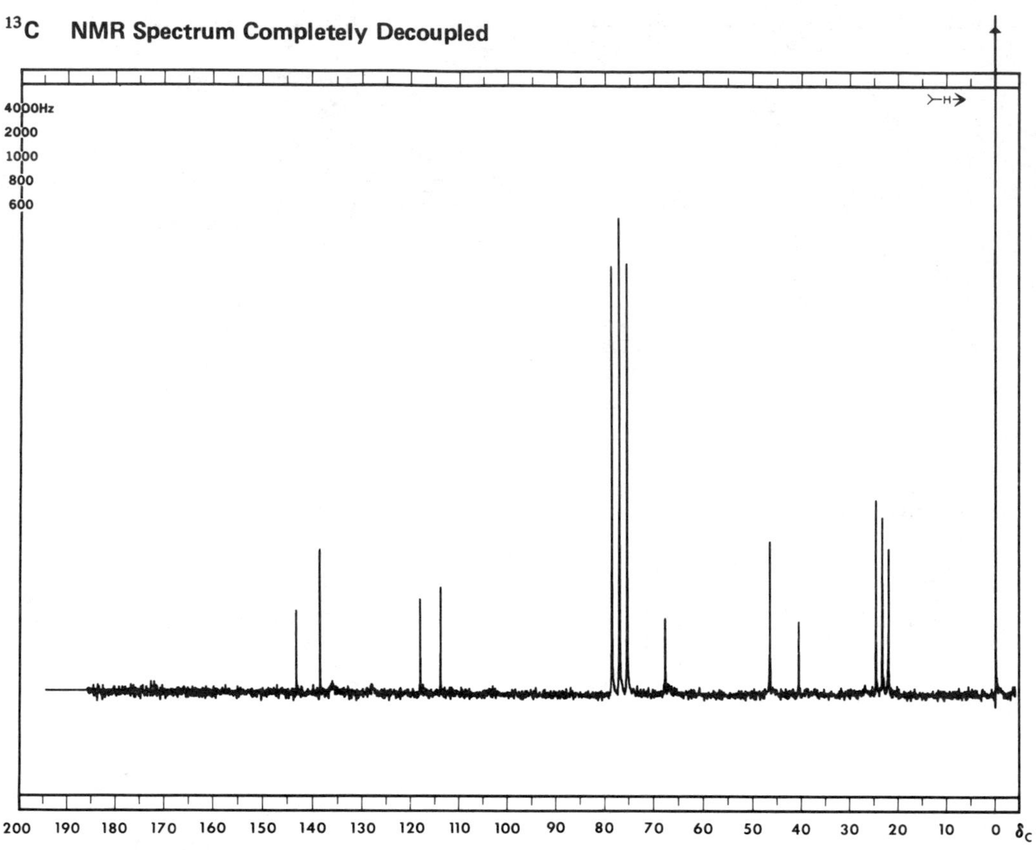
^{13}C NMR Spectrum Completely Decoupled
4000Hz
2000
1000
800
600
200 190 180 170 160 150 140 130 120 110 100 90 80 70 60 50 40 30 20 10 0 δ_C

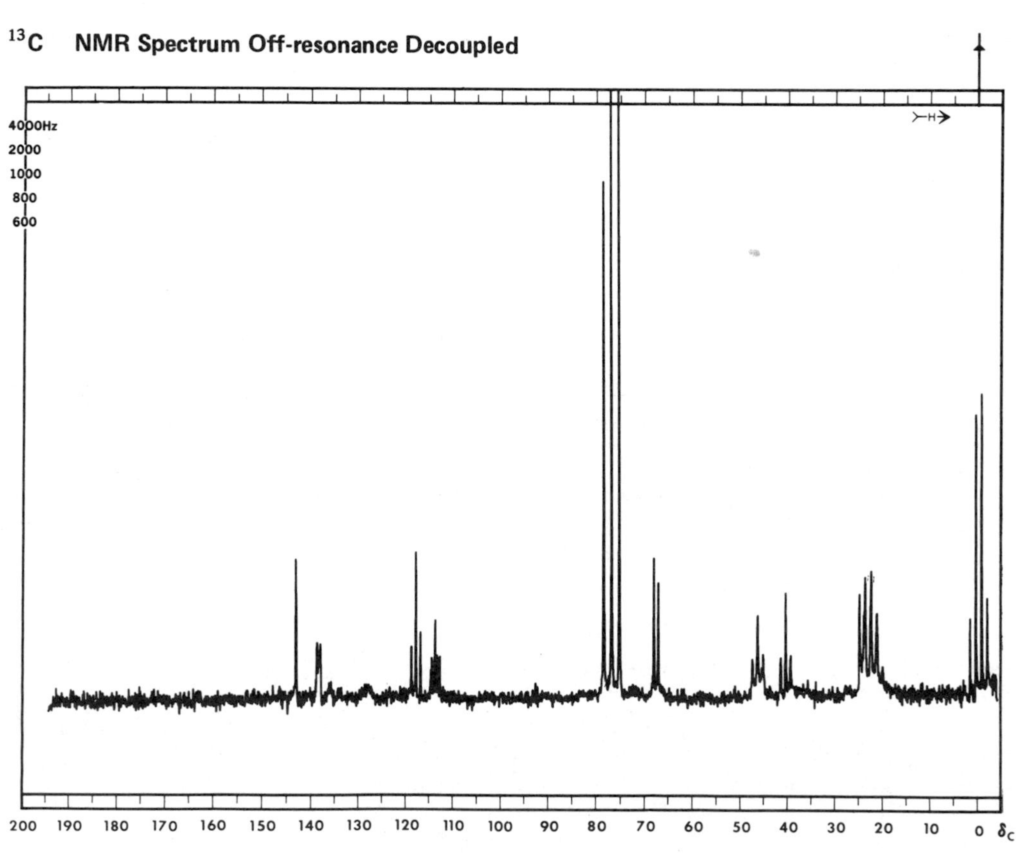
^{13}C NMR Spectrum Off-resonance Decoupled
4000Hz
2000
1000
800
600
200 190 180 170 160 150 140 130 120 110 100 90 80 70 60 50 40 30 20 10 0 δ_C

Infrared Spectrum

Compound 8-49 Beilstein Ref. 10, 70

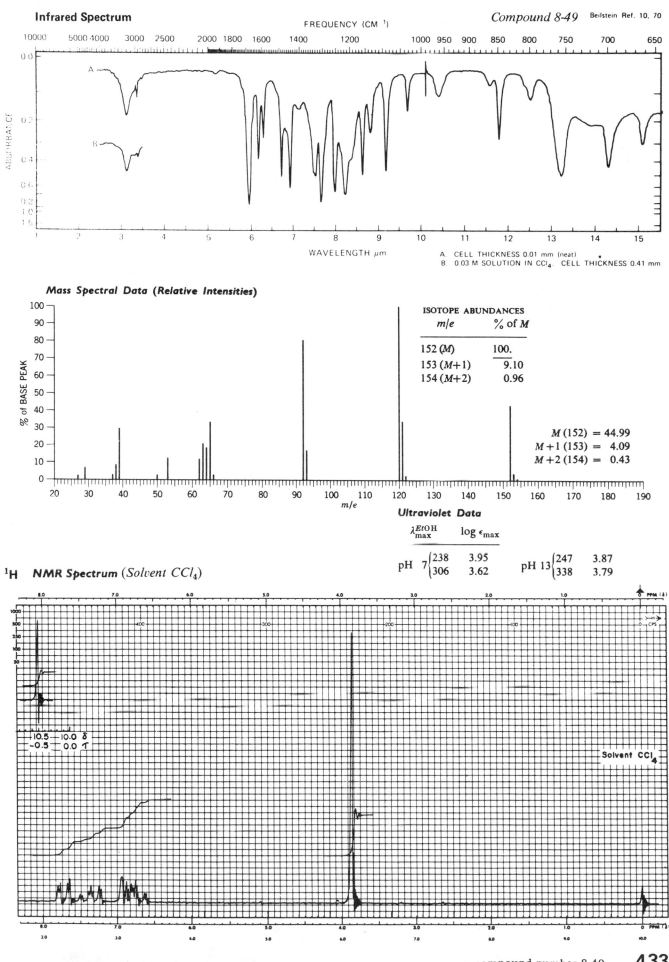

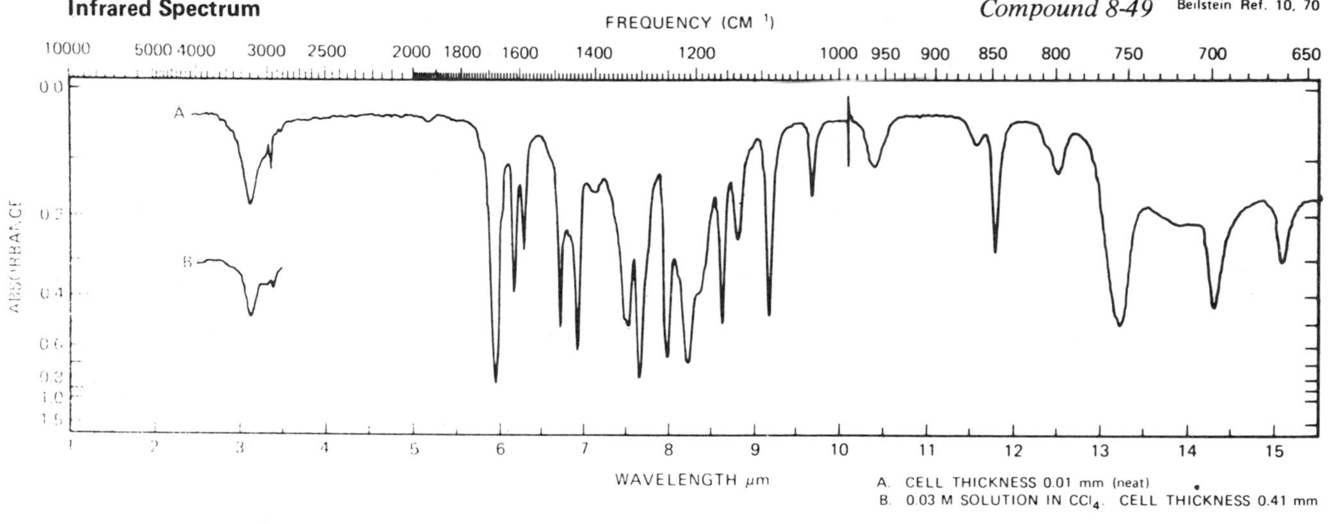

Mass Spectral Data (Relative Intensities)

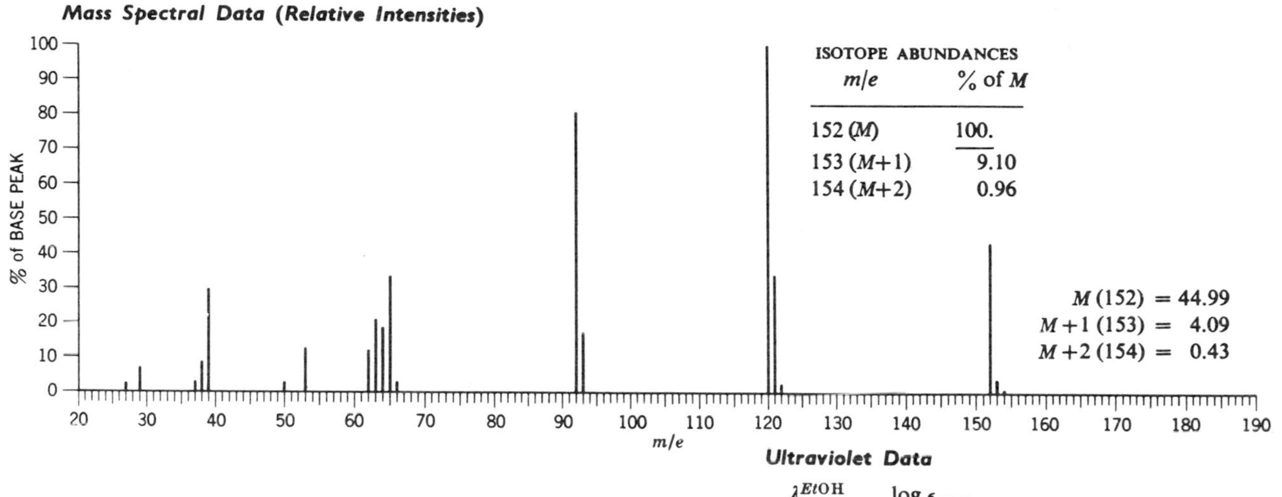

ISOTOPE ABUNDANCES

m/e	% of M
152 (M)	100.
153 (M+1)	9.10
154 (M+2)	0.96

M (152) = 44.99
$M+1$ (153) = 4.09
$M+2$ (154) = 0.43

Ultraviolet Data

	λ_{max}^{EtOH}	$\log \epsilon_{max}$
pH 7	238	3.95
	306	3.62
pH 13	247	3.87
	338	3.79

^{1}H NMR Spectrum (*Solvent CCl$_4$*)

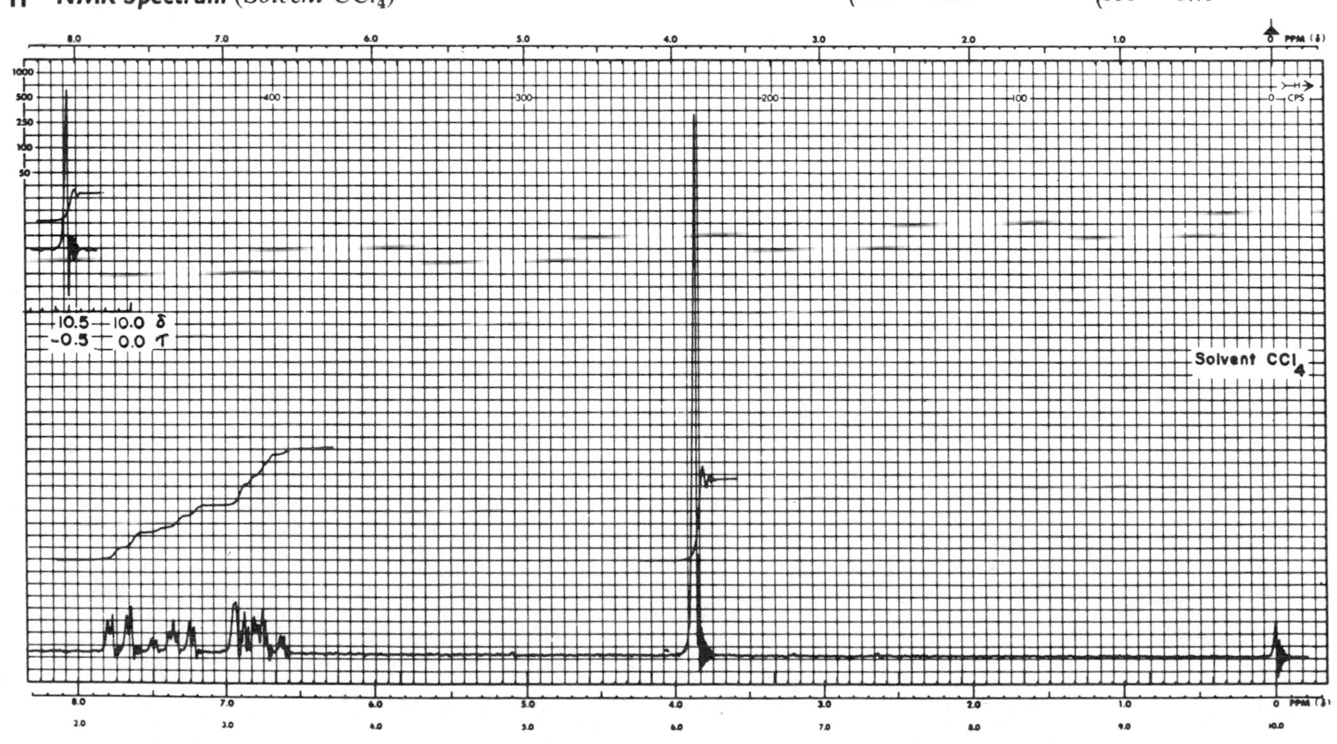

SUBJECT INDEX

Key:
^{13}C NMR = Carbon-13 nuclear magnetic resonance spectrometry (Chapter 5)
^{1}H NMR = Proton nuclear magnetic resonance spectrometry (Chapter 4)
IR = Infrared spectrometry (Chapter 3)
MS = Mass spectrometry (Chapter 2)
UV = Ultraviolet spectrometry (Chapter 6)
Specific compounds, for which complete spectra or spectral data are referenced, are listed in the Compound Index.

COMPOUND INDEX

Note: For many of the compounds below, complete spectra (of the type denoted in boldface) are referenced. Exceptions are **UV** and **^{13}C NMR** entries. For most of the **UV** entries the λ_{max} and ϵ_{max} for the major bands are referenced. For many of the **^{13}C** entries, shifts for most or all of the carbons in the molecule are referenced.